Lecture Notes in Computer Science 12077

Advanced Research in Computing and Software Science

Subline of Lecture Notes in Computer Science

More information about this series at http://www.springer.com/series/7407

Jean Goubault-Larrecq · Barbara König (Eds.)

Foundations of Software Science and Computation Structures

23rd International Conference, FOSSACS 2020
Held as Part of the European Joint Conferences on Theory and Practice of Software, ETAPS 2020
Dublin, Ireland, April 25–30, 2020
Proceedings

Springer Open

Editors
Jean Goubault-Larrecq
Université Paris-Saclay,
ENS Paris-Saclay, CNRS
Cachan, France

Barbara König
University of Duisburg-Essen
Duisburg, Germany

ISSN 0302-9743 ISSN 1611-3349 (electronic)
Lecture Notes in Computer Science
ISBN 978-3-030-45230-8 ISBN 978-3-030-45231-5 (eBook)
https://doi.org/10.1007/978-3-030-45231-5

LNCS Sublibrary: SL1 – Theoretical Computer Science and General Issues

This Springer imprint is published by the registered company Springer Nature Switzerland AG
The registered company address is: Gewerbestrasse 11, 6330 Cham, Switzerland

ETAPS Foreword

Welcome to the 23rd ETAPS! ETAPS 2020 was originally planned to take place in Ireland in its beautiful capital Dublin. Because of the Covid-19 pandemic, this was changed to an online event on July 2, 2020.

ETAPS 2020 is the 23rd instance of the European Joint Conferences on Theory and Practice of Software.

ETAPS is an annual federated conference established in 1998, and consists of four conferences: ESOP, FASE, FoSSaCS, and TACAS.

Each conference has its own Program Committee (PC) and its own Steering Committee (SC).

The conferences cover various aspects of software systems, ranging from theoretical computer science to foundations of programming language developments, analysis tools, and formal approaches to software engineering.

Organizing these conferences in a coherent, highly synchronized conference programme, enables researchers to participate in an exciting event, having the possibility to meet many colleagues working in different directions in the field, and to easily attend talks of different conferences.

On the weekend before the main conference, numerous satellite workshops take place that attract many researchers from all over the globe. Also, for the second time, an ETAPS Mentoring Workshop is organized.

This workshop is intended to help students early in the program with advice on research, career, and life in the fields of computing that are covered by the ETAPS conference.

ETAPS 2020 received 424 submissions in total, 129 of which were accepted, yielding an overall acceptance rate of 30.4%.

I thank all the authors for their interest in ETAPS, all the reviewers for their reviewing efforts, the PC members for their contributions, and in particular the PC (co-) chairs for their hard work in running this entire intensive process.

Last but not least, my congratulations to all authors of the accepted papers!

Because of the change to an online event, most of the original ETAPS program had to be cancelled. The ETAPS afternoon featured presentations of the three best paper awards, the Test-of-Time award and the ETAPS PhD award. The invited and tutorial speakers of ETAPS 2020 will be invited for ETAPS 2021, and all authors of accepted ETAPS 2020 papers will have the opportunity to present their work at ETAPS 2021.

ETAPS 2020 originally was supposed to place in Dublin, Ireland, organized by the University of Limerick and Lero. The local organization team consisted of Tiziana Margaria (UL and Lero, general chair), Vasileios Koutavas (Lero@UCD), Anila Mjeda (Lero@UL), Anthony Ventresque (Lero@UCD), and Petros Stratis (Easy Conferences). I would like to thank Tiziana and her team for all the preparations, and we hope there will be a next opportunity to host ETAPS in Dublin.

ETAPS 2020 is further supported by the following associations and societies: ETAPS e.V., EATCS (European Association for Theoretical Computer Science),

EAPLS (European Association for Programming Languages and Systems), and EASST (European Association of Software Science and Technology).

The ETAPS Steering Committee consists of an Executive Board, and representatives of the individual ETAPS conferences, as well as representatives of EATCS, EAPLS, and EASST.

The Executive Board consists of Holger Hermanns (Saarbrücken), Marieke Huisman (Twente, chair), Joost-Pieter Katoen (Aachen and Twente), Jan Kofron (Prague), Gerald Lüttgen (Bamberg), Tarmo Uustalu (Reykjavik and Tallinn), Caterina Urban (INRIA), and Lenore Zuck (Chicago).

Other members of the steering committee are:
Armin Biere (Linz)
Jordi Cabot (Barcelona)
Jean Goubault-Larrecq (Cachan)
Jan-Friso Groote (Eindhoven)
Esther Guerra (Madrid)
Jurriaan Hage (Utrecht)
Reiko Heckel (Leicester)
Panagiotis Katsaros (Thessaloniki)
Stefan Kiefer (Oxford)
Barbara König (Duisburg)
Fabrice Kordon (Paris)
Jan Kretinsky (Munich)
Kim G. Larsen (Aalborg)
Tiziana Margaria (Limerick)
Peter Müller (Zurich)
Catuscia Palamidessi (Palaiseau)
Dave Parker (Birmingham)
Andrew M. Pitts (Cambridge)
Peter Ryan (Luxembourg)
Don Sannella (Edinburgh)
Bernhard Steffen (Dortmund)
Mariëlle Stoelinga (Twente)
Gabriele Taentzer (Marburg)
Christine Tasson (Paris)
Peter Thiemann (Freiburg)
Jan Vitek (Prague)
Heike Wehrheim (Paderborn)
Anton Wijs (Eindhoven), and
Nobuko Yoshida (London)

I'd like to take this opportunity to thank all authors, attendants, organizers of the satellite workshops, and Springer-Verlag GmbH for their support.

I hope you all enjoyed the ETAPS 2020 afternoon.

July 2020

Marieke Huisman
ETAPS SC Chair
ETAPS e.V. President

Preface

This volume contains the papers accepted for the 23rd International Conference on Foundations of Software Science and Computation Structures (FoSSaCS). The conference series is dedicated to foundational research with a clear significance for software science. It brings together research on theories and methods to support the analysis, integration, synthesis, transformation, and verification of programs and software systems.

This volume contains 31 contributed papers selected from 98 full paper submissions, and also a paper accompanying an invited talk by Scott Smolka (Stony Brook University, USA). Each submission was reviewed by at least three Program Committee members, with the help of external reviewers, and the final decisions took into account the feedback from a rebuttal phase. The conference submissions were managed using the EasyChair conference system, which was also used to assist with the compilation of these proceedings.

We wish to thank all the authors who submitted papers to FoSSaCS 2020, the Program Committee members, the Steering Committee members, the external reviewers and the ETAPS 2020 organizers. To our great regret ETAPS 2020 in Dublin had to be cancelled due to the Covid-19 pandemic, but we hope to be able to listen to the authors' talks in the near future.

July 2020

Jean Goubault-Larrecq
Barbara König

Organization

Program Committee

Parosh Aziz Abdulla	Uppsala University, Sweden
Thorsten Altenkirch	University of Nottingham, UK
Paolo Baldan	Università di Padova, Italy
Nick Benton	Facebook, UK
Frédéric Blanqui	Inria and LSV, France
Michele Boreale	Università di Firenze, Italy
Corina Cirstea	University of Southampton, UK
Pedro R. D'Argenio	Universidad Nacional de Córdoba, CONICET, Argentina
Josée Desharnais	Université Laval, Canada
Jean Goubault-Larrecq	Université Paris-Saclay, ENS Paris-Saclay, CNRS, LSV, Cachan, France
Ichiro Hasuo	National Institute of Informatics, Japan
Delia Kesner	IRIF, Université de Paris, France
Shankara Narayanan Krishna	IIT Bombay, India
Barbara König	Universität Duisburg-Essen, Germany
Sławomir Lasota	University of Warsaw, Poland
Xavier Leroy	Collège de France and Inria, France
Leonid Libkin	University of Edinburgh, UK, and ENS Paris, France
Jean-Yves Marion	LORIA, Université de Lorraine, France
Dominique Méry	LORIA, Université de Lorraine, France
Matteo Mio	LIP, CNRS, ENS Lyon, France
Andrzej Murawski	University of Oxford, UK
Prakash Panangaden	McGill University, Canada
Amr Sabry	Indiana University Bloomington, USA
Lutz Schröder	Friedrich-Alexander-Universität Erlangen-Nürnberg, Germany
Sebastian Siebertz	Universität Bremen, Germany
Benoît Valiron	LRI, CentraleSupélec, Université Paris-Saclay, France

Steering Committee

Andrew Pitts (Chair)	University of Cambridge, UK
Christel Baier	Technische Universität Dresden, Germany
Lars Birkedal	Aarhus University, Denmark
Ugo Dal Lago	Università degli Studi di Bologna, Italy

Javier Esparza	Technische Universität München, Germany
Anca Muscholl	LaBRI, Université de Bordeaux, France
Frank Pfenning	Carnegie Mellon University, USA

Additional Reviewers

Accattoli, Beniamino
Alvim, Mario S.
André, Étienne
Argyros, George
Arun-Kumar, S.
Ayala-Rincon, Mauricio
Bacci, Giorgio
Bacci, Giovanni
Balabonski, Thibaut
Basile, Davide
Berger, Martin
Bernardi, Giovanni
Bisping, Benjamin
Bodeveix, Jean-Paul
Bollig, Benedikt
Bonchi, Filippo
Bonelli, Eduardo
Boulmé, Sylvain
Bourke, Timothy
Bradfield, Julian
Breuvart, Flavien
Bruni, Roberto
Bruse, Florian
Capriotti, Paolo
Carette, Jacques
Carette, Titouan
Carton, Olivier
Cassano, Valentin
Chadha, Rohit
Charguéraud, Arthur
Cho, Kenta
Choudhury, Vikraman
Ciancia, Vincenzo
Clemente, Lorenzo
Colacito, Almudena
Corradini, Andrea
Czerwiński, Wojciech
de Haan, Ronald
de Visme, Marc
Dell'Erba, Daniele
Deng, Yuxin
Eickmeyer, Kord
Exibard, Leo
Faggian, Claudia
Fijalkow, Nathanaël
Filali-Amine, Mamoun
Francalanza, Adrian
Frutos Escrig, David
Galletta, Letterio
Ganian, Robert
Garrigue, Jacques
Gastin, Paul
Genaim, Samir
Genest, Blaise
Ghica, Dan
Goncharov, Sergey
Gorla, Daniele
Guerrini, Stefano
Hirschowitz, Tom
Hofman, Piotr
Hoshino, Naohiko
Howar, Falk
Inverso, Omar
Iván, Szabolcs
Jaax, Stefan
Jeandel, Emmanuel
Johnson, Michael
Kahrs, Stefan
Kamburjan, Eduard
Katsumata, Shin-Ya
Kerjean, Marie
Kiefer, Stefan
Komorida, Yuichi
Kop, Cynthia
Kremer, Steve
Kuperberg, Denis
Křetínský, Jan
Laarman, Alfons

Laurent, Fribourg
Levy, Paul Blain
Li, Yong
Licata, Daniel R.
Liquori, Luigi
Lluch Lafuente, Alberto
Lopez, Aliaume
Malherbe, Octavio
Manuel, Amaldev
Manzonetto, Giulio
Matache, Christina
Matthes, Ralph
Mayr, Richard
Melliès, Paul-André
Merz, Stephan
Miculan, Marino
Mikulski, Łukasz
Moser, Georg
Moss, Larry
Munch-Maccagnoni, Guillaume
Muskalla, Sebastian
Nantes-Sobrinho, Daniele
Nestra, Härmel
Neumann, Eike
Neves, Renato
Niehren, Joachim
Padovani, Luca
Pagani, Michele
Paquet, Hugo
Patterson, Daniel
Pedersen, Mathias Ruggaard
Peressotti, Marco
Pitts, Andrew
Potapov, Igor
Power, John
Praveen, M.
Puppis, Gabriele
Péchoux, Romain
Pérez, Guillermo
Quatmann, Tim
Rabinovich, Roman
Radanne, Gabriel
Rand, Robert
Ravara, António
Remy, Didier
Reutter, Juan L.
Rossman, Benjamin
Rot, Jurriaan
Rowe, Reuben
Ruemmer, Philipp
Sammartino, Matteo
Sankaran, Abhisekh
Sankur, Ocan
Sattler, Christian
Schmitz, Sylvain
Serre, Olivier
Shirmohammadi, Mahsa
Siles, Vincent
Simon, Bertrand
Simpson, Alex
Singh, Neeraj
Sprunger, David
Srivathsan, B.
Staton, Sam
Stolze, Claude
Straßburger, Lutz
Streicher, Thomas
Tan, Tony
Tawbi, Nadia
Toruńczyk, Szymon
Tzevelekos, Nikos
Urbat, Henning
van Bakel, Steffen
van Breugel, Franck
van de Pol, Jaco
van Doorn, Floris
Van Raamsdonk, Femke
Vaux Auclair, Lionel
Verma, Rakesh M.
Vial, Pierre
Vignudelli, Valeria
Vrgoc, Domagoj
Waga, Masaki
Wang, Meng
Witkowski, Piotr
Zamdzhiev, Vladimir
Zemmari, Akka
Zhang, Zhenya
Zorzi, Margherita

Contents

Neural Flocking: MPC-based Supervised Learning of Flocking Controllers

(✉)Usama Mehmood[1], Shouvik Roy[1], Radu Grosu[2], Scott A. Smolka[1], Scott D. Stoller[1], and Ashish Tiwari[3]

[1] Stony Brook University, Stony Brook NY, USA
umehmood@cs.stonybrook.edu
[2] Technische Universitat Wien, Wien, Austria
[3] Microsoft Research, San Francisco CA, USA

Abstract. We show how a symmetric and fully distributed flocking controller can be synthesized using Deep Learning from a centralized flocking controller. Our approach is based on *Supervised Learning*, with the centralized controller providing the training data, in the form of trajectories of state-action pairs. We use Model Predictive Control (MPC) for the centralized controller, an approach that we have successfully demonstrated on flocking problems. MPC-based flocking controllers are high-performing but also computationally expensive. By learning a symmetric and distributed neural flocking controller from a centralized MPC-based one, we achieve the best of both worlds: the neural controllers have high performance (on par with the MPC controllers) and high efficiency. Our experimental results demonstrate the sophisticated nature of the distributed controllers we learn. In particular, the neural controllers are capable of achieving myriad flocking-oriented control objectives, including flocking formation, collision avoidance, obstacle avoidance, predator avoidance, and target seeking. Moreover, they generalize the behavior seen in the training data to achieve these objectives in a significantly broader range of scenarios. In terms of verification of our neural flocking controller, we use a form of statistical model checking to compute confidence intervals for its convergence rate and time to convergence.

Keywords: Flocking · Model Predictive Control · Distributed Neural Controller · Deep Neural Network · Supervised Learning

1 Introduction

With the introduction of Reynolds rule-based model [16, 17], it is now possible to understand the flocking problem as one of distributed control. Specifically, in this model, at each time-step, each agent executes a control law given in terms of the weighted sum of three competing forces to determine its next acceleration. Each of these forces has its own rule: *separation* (keep a safe distance away from your neighbors), *cohesion* (move towards the centroid of your neighbors), and *alignment* (steer toward the average heading of your neighbors). Reynolds

J. Goubault-Larrecq and B. König (Eds.): FOSSACS 2020, LNCS 12077, pp. 1–16, 2020.
https://doi.org/10.1007/978-3-030-45231-5_1

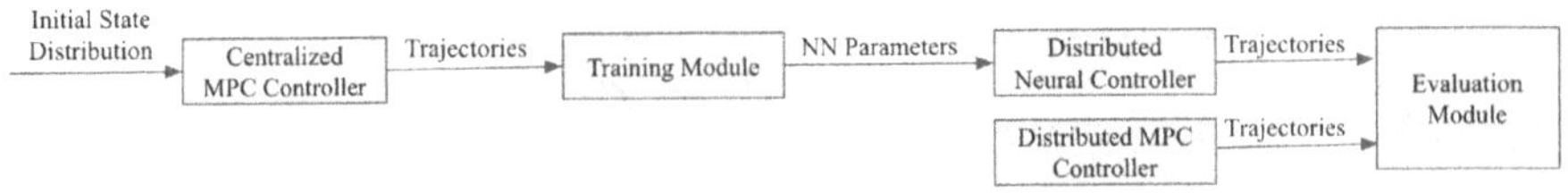

Fig. 1: Neural Flocking Architecture

controller is *distributed*; i.e., it is executed separately by each agent, using information about only itself and nearby agents, and without communication. Furthermore, it is *symmetric*; i.e., every agent runs the same controller (same code).

We subsequently showed that a simpler, more declarative approach to the flocking problem is possible [11]. In this setting, flocking is achieved when the agents combine to minimize a system-wide *cost function*. We presented centralized and distributed solutions for achieving this form of "declarative flocking" (DF), both of which were formulated in terms of Model-Predictive Control (MPC) [2].

Another advantage of DF over the ruled-based approach exemplified by Reynolds model is that it allows one to consider additional control objectives (e.g., obstacle and predator avoidance) simply by extending the cost function with additional terms for these objectives. Moreover, these additional terms are typically quite straightforward in nature. In contrast, deriving behavioral rules that achieve the new control objectives can be a much more challenging task.

An issue with MPC is that computing the next control action can be computationally expensive, as MPC searches for an action sequence that minimizes the cost function over a given prediction horizon. This renders MPC unsuitable for real-time applications with short control periods, for which flocking is a prime example. Another potential problem with MPC-based approaches to flocking is its performance (in terms of achieving the desired flight formation), which may suffer in a fully distributed setting.

In this paper, we present *Neural Flocking* (NF), a new approach to the flocking problem that uses Supervised Learning to learn a symmetric and fully distributed flocking controller from a centralized MPC-based controller. By doing so, we achieve the best of both worlds: high performance (on par with the MPC controllers) in terms of meeting flocking flight-formation objectives, and high efficiency leading to real-time flight controllers. Moreover, our NF controllers can easily be parallelized on hardware accelerators such as GPUs and TPUs.

Figure 1 gives an overview of the NF approach. A high-performing centralized MPC controller provides the labeled training data to the learning agent: a symmetric and distributed neural controller in the form of a deep neural network (DNN). The training data consists of trajectories of state-action pairs, where a state contains the information known to an agent at a time step (e.g., its own position and velocity, and the position and velocity of its neighbors), and the action (the label) is the acceleration assigned to that agent at that time step by the centralized MPC controller.

We formulate and evaluate NF in a number of essential flocking scenarios: basic flocking with inter-agent collision avoidance, as in [11], and more advanced scenarios with additional objectives, including obstacle avoidance, predator avoidance, and target seeking by the flock. We conduct an extensive performance evaluation of NF. Our experimental results demonstrate the sophisticated nature of NF controllers. In particular, they are capable of achieving all of the stated control objectives. Moreover, they generalize the behavior seen in the training data in order to achieve these objectives in a significantly broader range of scenarios. In terms of verification of our neural controller, we use a form of statistical model checking [5,10] to compute confidence intervals for its rate of convergence to a flock and for its time to convergence.

2 Background

We consider a set of n dynamic agents $\mathcal{A} = \{1, \ldots, n\}$ that move according to the following discrete-time equations of motion:

$$\begin{aligned} p_i(k+1) &= p_i(k) + dt \cdot v_i(k), \quad |v_i(k)| < \bar{v} \\ v_i(k+1) &= v_i(k) + dt \cdot a_i(k), \quad |a_i(k)| < \bar{a} \end{aligned} \tag{1}$$

where $p_i(k) \in \mathbb{R}^2$, $v_i(k) \in \mathbb{R}^2$, $a_i(k) \in \mathbb{R}^2$ are the position, velocity and acceleration of agent $i \in \mathcal{A}$ respectively at time step k, and $dt \in \mathbb{R}^+$ is the time step. The magnitudes of velocities and accelerations are bounded by $\bar{v}$ and $\bar{a}$, respectively. Acceleration $a_i(k)$ is the control input for agent i at time step k. The acceleration is updated after every η time steps i.e., $\eta \cdot dt$ is the control period. The flock *configuration* at time step k is thus given by the following vectors (in boldface):

$$\mathbf{p}(k) = [p_1^T(k) \cdots p_n^T(k)]^T \tag{2}$$

$$\mathbf{v}(k) = [v_1^T(k) \cdots v_n^T(k)]^T \tag{3}$$

$$\mathbf{a}(k) = [a_1^T(k) \cdots a_n^T(k)]^T \tag{4}$$

The configuration vectors are referred to without the time indexing as $\mathbf{p}$, $\mathbf{v}$, and $\mathbf{a}$. The *neighborhood* of agent i at time step k, denoted by $\mathcal{N}_i(k) \subseteq \mathcal{A}$, contains its $\mathcal{N}$-nearest neighbors, i.e., the $\mathcal{N}$ other agents closest to it. We use this definition (in Section 2.2 to define a distributed-flocking cost function) for simplicity, and expect that a radius-based definition of neighborhood would lead to similar results for our distributed flocking controllers.

2.1 Model-Predictive Control

Model-Predictive control (MPC) [2] is a well-known control technique that has recently been applied to the flocking problem [11,19,20]. At each control step, an optimization problem is solved to find the optimal sequence of control actions (agent accelerations in our case) that minimizes a given cost function with respect to a predictive model of the system. The first control action of the optimal control sequence is then applied to the system; the rest is discarded. In the computation of the cost function, the predictive model is evaluated for a finite prediction horizon of T control steps.

MPC-based flocking models can be categorized as *centralized* or *distributed.* A *centralized* model assumes that complete information about the flock is available to a single "global" controller, which uses the states of all agents to compute their next optimal accelerations. The following optimization problem is solved by a centralized MPC controller at each control step k:

$$\min_{\mathbf{a}(k|k),\ldots,\mathbf{a}(k+T-1|k) < \bar{a}} \quad J(k) + \lambda \cdot \sum_{t=0}^{T-1} \|\mathbf{a}(k+t \mid k)\|^2 \tag{5}$$

The first term $J(k)$ is the centralized model-specific cost, evaluated for T control steps (this embodies the predictive aspect of MPC), starting at time step k. It encodes the control objective of minimizing the cost function $J(k)$. The second term, scaled by a weight $\lambda > 0$, penalizes large control inputs: $\mathbf{a}(k+t \mid k)$ are the predictions made at time step k for the accelerations at time step $k+t$.

In *distributed MPC*, each agent computes its acceleration based only on its own state and its local knowledge, e.g., information about its neighbors:

$$\min_{a_i(k|k),\ldots,a_i(k+T-1|k) < \bar{a}} \quad J_i(k) + \lambda \cdot \sum_{t=0}^{T-1} \|a_i(k+t \mid k)\|^2 \tag{6}$$

$J_i(k)$ is the distributed, model-specific cost function for agent i, analogous to $J(k)$. In a distributed setting where an agent's knowledge of its neighbors' behavior is limited, an agent cannot calculate the exact future behavior of its neighbors. Hence, the predictive aspect of $J_i(k)$ must rely on some assumption about that behavior during the prediction horizon. Our distributed cost functions are based on the assumption that the neighbors have zero accelerations during the prediction horizon. While this simple design is clearly not completely accurate, our experiments show that it still achieves good results.

2.2 Declarative Flocking

Declarative flocking (DF) is a high-level approach to designing flocking algorithms based on defining a suitable cost function for MPC [11]. This is in contrast to the operational approach, where a set of rules are used to capture flocking behavior, as in Reynolds model. For basic flocking, the DF cost function contains two terms: (1) a *cohesion* term based on the squared distance between each pair of agents in the flock; and (2) a *separation* term based on the inverse of the squared distance between each pair of agents. The flock evolves toward a configuration in which these two opposing forces are balanced. The cost function J^C for centralized DF, i.e., centralized MPC (CMPC), is as follows:

$$J^C(\mathbf{p}) = \frac{2}{|\mathcal{A}| \cdot (|\mathcal{A}|-1)} \cdot \sum_{i \in \mathcal{A}} \sum_{j \in \mathcal{A}, i<j} \|p_{ij}\|^2 + \omega_s \cdot \frac{1}{\|p_{ij}\|^2} \tag{7}$$

where ω_s is the weight of the separation term and controls the density of the flock. The cost function is normalized by the number of pairs of agents, $\frac{|\mathcal{A}|\cdot(|\mathcal{A}-1|)}{2}$; as such, the cost does not depend on the size of the flock. The control law for CMPC is given by Eq. (5), with $J(k) = \sum_{t=1}^{T} J^{\mathrm{C}}(\mathbf{p}(k+t \mid k))$.

The basic flocking cost function for distributed DF is similar to that for CMPC, except that the cost function J_i^D for agent i is computed over its set of neighbors $\mathcal{N}_i(k)$ at time k:

$$J_i^{\mathrm{D}}(\mathbf{p}(k)) = \frac{1}{|\mathcal{N}_i(k)|} \cdot \sum_{j \in \mathcal{N}_i(k)} \|p_{ij}\|^2 + \omega_s \cdot \sum_{j \in \mathcal{N}_i(k)} \frac{1}{\|p_{ij}\|^2} \tag{8}$$

The control law for agent i is given by Eq. (6), with $J_i(k) = \sum_{t=1}^{T} J_i^{\mathrm{D}}(\mathbf{p}(k+t \mid k))$.

3 Additional Control Objectives

The cost functions for basic flocking given in Eqs. (7) and (8) are designed to ensure that in the steady state, the agents are well-separated. Additional goals such as obstacle avoidance, predator avoidance, and target seeking are added to the MPC formulation as weighted cost-function terms. Different objectives can be combined by including the corresponding terms in the cost function as a weighted sum.

Cost-Function Term for Obstacle Avoidance. We consider multiple rectangular obstacles which are distributed randomly in the field. For a set of m rectangular obstacles $\mathcal{O} = \{\mathcal{O}_1, \mathcal{O}_2, ..., \mathcal{O}_m\}$, we define the cost function term for obstacle avoidance as:

$$J_{OA}(\mathbf{p}, \mathbf{o}) = \frac{1}{|\mathcal{A}||\mathcal{O}|} \sum_{i \epsilon \mathcal{A}} \sum_{j \epsilon \mathcal{O}} \frac{1}{\left\| p_i - o_j^{(i)} \right\|^2} \tag{9}$$

where $\mathbf{o}$ is the set of points on the obstacle boundaries and $o_j^{(i)}$ is the point on the obstacle boundary of the jth obstacle $\mathcal{O}_j$ that is closest to the ith agent.

Cost-Function Term for Target Seeking. This term is the average of the squared distance between the agents and the target. Let g denote the position of the fixed target. Then the target-seeking term is as defined as

$$J_{TS}(\mathbf{p}) = \frac{1}{|\mathcal{A}|} \sum_{i \in \mathcal{A}} \|p_i - g\|^2 \tag{10}$$

Cost-Function Term for Predator Avoidance. We introduce a single predator, which is more agile than the flocking agents: its maximum speed and acceleration are a factor of f_p greater than $\bar{v}$ and $\bar{a}$, respectively, with $f_p > 1$. Apart from being more agile, the predator has the same dynamics as the agents, given by

Eq. (1). The control law for the predator consists of a single term that causes it to move toward the centroid of the flock with maximum acceleration.

For a flock of n agents and one predator, the cost-function term for predator avoidance is the average of the inverse of the cube of the distances between the predator and the agents. It is given by:

$$J_{PA}\left(\mathbf{p}, p_{pred}\right) = \frac{1}{|\mathcal{A}|}\sum_{i\epsilon\mathcal{A}} \frac{1}{\left\|p_i - p_{pred}\right\|^3} \tag{11}$$

where p_{pred} is the position of the predator. In contrast to the separation term in Eqs. (5)-(6), which we designed to ensure inter-agent collision avoidance, the predator-avoidance term has a cube instead of a square in the denominator. This is to reduce the influence of the predator on the flock when the predator is far away from the flock.

NF Cost-Function Terms. The MPC cost functions used in our examination of Neural Flocking are weighted sums of the cost function terms introduced above. We refer to the first term of our centralized DF cost function $J^C(\mathbf{p})$ (see Eq. (7)) as $J_{cohes}(\mathbf{p})$ and the second as $J_{sep}(\mathbf{p})$. We use the following cost functions J_1, J_2, and J_3 for basic flocking with collision avoidance, obstacle avoidance with target seeking, and predator avoidance, respectively.

$$J_1(\mathbf{p}) = J_{cohes}(\mathbf{p}) + \omega_s \cdot J_{sep}(\mathbf{p}) \tag{12a}$$

$$J_2(\mathbf{p}, \mathbf{o}) = J_{cohes}(\mathbf{p}) + \omega_s \cdot J_{sep}(\mathbf{p}) + \omega_o \cdot J_{OA}(\mathbf{p}, \mathbf{o}) + \omega_t \cdot J_{TS}(\mathbf{p}) \tag{12b}$$

$$J_3(\mathbf{p}, p_{pred}) = J_{cohes}(\mathbf{p}) + \omega_s \cdot J_{sep}(\mathbf{p}) + \omega_p \cdot J_{PA}(\mathbf{p}, p_{pred}) \tag{12c}$$

where ω_s is the weight of the separation term, ω_o is the weight of the obstacle avoidance term, ω_t is the weight of the target-seeking term, and ω_p is the weight of the predator-avoidance term. Note that J_1 is equivalent to J^C (Eq. (7)). The weight ω_s of the separation term is experimentally chosen to ensure that the distance between agents, throughout the simulation, is at least d_{min}, the minimum inter-agent distance representing collision avoidance. Similar considerations were given to the choice of values for ω_o and ω_p. The specific values we used for the weights are: $\omega_s = 2000$, $\omega_o = 1500$, $\omega_t = 10$, and $\omega_p = 500$.

We experimented with an alternative strategy for introducing inter-agent collision avoidance, obstacle avoidance, and predator avoidance into the MPC problem, namely, as *constraints* of the form $d_{min} - p_{ij} < 0$, $d_{min} - ||p_i - o_j^{(i)}|| < 0$, and $d_{min} - ||p_i - p_{pred}|| < 0$, respectively. Using the theory of exact penalty functions [12], we recast the constrained MPC problem as an equivalent unconstrained MPC problem by converting the constraints into a weighted *penalty term*, which is then added to the MPC cost function. This approach rendered the optimization problem difficult to solve due to the non-smoothness of the penalty term. As a result, constraint violations in the form of collisions were observed during simulation.

4 Neural Flocking

We learn a *distributed neural controller* (DNC) for the flocking problem using training data in the form of trajectories of state-action pairs produced by a CMPC controller. In addition to basic flocking with inter-agent collision avoidance, the DNC exhibits a number of other flocking-related behaviors, including obstacle avoidance, target seeking, and predator avoidance. We also show how the learned behavior exhibited by the DNC generalizes over a larger number of agents than what was used during training to achieve successful collision-free flocking in significantly larger flocks.

We use *Supervised Learning* to train the DNC. Supervised Learning learns a function that maps an input to an output based on example sequences of input-output pairs. In our case, the trajectory data obtained from CMPC contains both the training inputs and corresponding labels (outputs): the state of an agent in the flock (and that of its nearest neighbors) at a particular time step is the input, and that agent's acceleration at the same time step is the label.

4.1 Training Distributed Flocking Controllers

We use Deep Learning to synthesize a distributed and symmetric neural controller from the training data provided by the CMPC controller. Our objective is to learn basic flocking, obstacle avoidance with target seeking, and predator avoidance. Their respective CMPC-based cost functions are given in Sections 2.2 and 3. All of these control objectives implicitly also include inter-agent collision avoidance by virtue of the separation term in Eq. 7.

For each of these control objectives, DNC training data is obtained from CMPC trajectory data generated for $n = 15$ agents, starting from initial configurations in which agent positions and velocities are uniformly sampled from $[-15, 15]^2$ and $[0, 1]^2$, respectively. All training trajectories are 1,000 time steps in duration.

We further ensure that the initial configurations are *recoverable*; i.e., no two agents are so close to each other that they cannot avoid a collision by resorting to maximal accelerations. We learn a single DNC from the state-action pairs of all n agents. This yields a symmetric distributed controller, which we use for each agent in the flock during evaluation.

Basic Flocking. Trajectory data for basic flocking is generated using the cost function given in Eq. (7). We generate 200 trajectories, each of which (as noted above) is 1,000 time steps long. The input to the NN is the position and velocity of each agent along with the positions and velocities of its $\mathcal{N}$-nearest neighbors. This yields $200 \cdot 1,000 \cdot 15 = 3M$ total training samples.

Let us refer to the agent (the DNC) being learned as $\mathcal{A}_0$. Since we use neighborhood size $\mathcal{N} = 14$, the input to the NN is of the form $[p_0^x\ p_0^y\ v_0^x\ v_0^y\ p_1^x\ p_1^y\ v_1^x\ v_1^y\ \dots\ p_{14}^x\ p_{14}^y\ v_{14}^x\ v_{14}^y]$, where p_0^x, p_0^y are the position coordinates and v_0^x, v_0^y velocity coordinates for agent $\mathcal{A}_0$, and $p_{1\dots14}^x$, $p_{1\dots14}^y$ and $v_{1\dots14}^x$, $v_{1\dots14}^y$ are the position and velocity vectors of its neighbors. Since this input vector has 60 components, the input to the NN consists of 60 features.

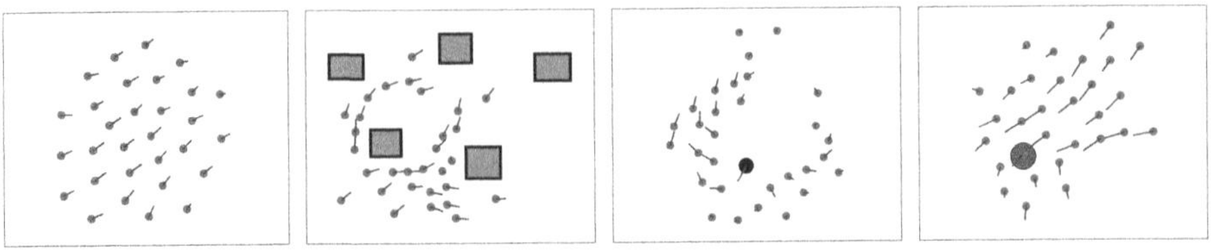

(a) Basic flocking (b) Obstacle avoid. (c) Predator avoid. (d) Target seeking

Fig. 2: Snapshots of DNC flocking behaviors for 30 agents

Obstacle Avoidance with Target Seeking. For obstacle avoidance with target seeking, we use CMPC with the cost function given in Eq. (12b). The target is located beyond the obstacles, forcing the agents to move through the obstacle field. For the training data, we generate 100 trajectories over 4 different obstacle fields (25 trajectories per obstacle field). The input to the NN consists of the 92 features $[p_0^x\ p_0^y\ v_0^x\ v_0^y\ o_0^x\ o_0^y \ldots p_{14}^x\ p_{14}^y\ v_{14}^x\ v_{14}^y\ o_{14}^x\ o_{14}^y\ g^x\ g^y]$, where o_0^x, o_0^y is the closest point on any obstacle to agent $\mathcal{A}_0$; $o_{1\ldots14}^x$, $o_{1\ldots14}^y$ give the closest point on any obstacle for the 14 neighboring agents, and g^x, g^y is the target location.

Predator Avoidance. The CMPC cost function for predator avoidance is given in Eq. (12c). The position, velocity, and the acceleration of the predator are denoted by p_{pred}, v_{pred}, a_{pred}, respectively. We take $f_p = 1.40$; hence $\bar{v}_{pred} = 1.40\,\bar{v}$ and $\bar{a}_{pred} = 1.40\,\bar{a}$. The input features to the NN are the positions and velocities of agent $\mathcal{A}_0$ and its $\mathcal{N}$-nearest neighbors, and the position and velocity of the predator. The input with 64 features thus has the form $[p_0^x\ p_0^y\ v_0^x\ v_0^y \ldots p_{14}^x\ p_{14}^y\ v_{14}^x\ v_{14}^y\ p_{pred}^x\ p_{pred}^y\ v_{pred}^x\ v_{pred}^y]$.

5 Experimental Evaluation

This section contains the results of our extensive performance analysis of the distributed neural flocking controller (DNC), taking into account various control objectives: basic flocking with collision avoidance, obstacle avoidance with target seeking, and predator avoidance. As illustrated in Fig. 1, this involves running CMPC to generate the training data for the DNCs, whose performance we then compare to that of the DMPC and CMPC controllers. We also show that the DNC flocking controllers generalize the behavior seen in the training data to achieve successful collision-free flocking in flocks significantly larger in size than those used during training. Finally, we use Statistical Model Checking to obtain confidence intervals for DNC's correctness/performance.

5.1 Preliminaries

The CMPC and DMPC control problems defined in Section 2.1 are solved using MATLAB fmincon optimizer. In the training phase, the size of the flock is

$n = 15$. For obstacle-avoidance with target-seeking, we use 5 obstacles with the target located at [60,50]. The simulation time is 100, $dt = 0.1$ time units, and $\eta = 3$, where (recall) $\eta \cdot dt$ is the control period. Further, the agent velocity and acceleration bounds are $\bar{v} = 2.0$ and $\bar{a} = 1.5$.

We use $d_{min} = 1.5$ as the minimum inter-agent distance for collision avoidance, $d_{min}^{obs} = 1$ as the minimum agent-obstacle distance for obstacle avoidance, and $d_{min}^{pred} = 1.5$ as the minimum agent-predator distance for predator avoidance. For initial configurations, recall that agent positions and velocities are uniformly sampled from $[-15, 15]^2$ and $[0, 1]^2$, respectively, and we ensure that they are *recoverable*; i.e., no two agents are so close to each other that they cannot avoid a collision when resorting to maximal accelerations. The predator starts at rest from a fixed location at a distance of 40 from the flock center.

For training, we considered 15 agents and 200 trajectories per agent, each trajectory 1,000 time steps in length. This yielded a total of 3,000,000 training samples. Our neural controller is a fully connected feed-forward Deep Neural Network (DNN), with 5 hidden layers, 84 neurons per hidden layer, and with a ReLU activation function. We use an iterative approach for choosing the DNN hyperparameters and architecture where we continuously improve our NN, until we observe satisfactory performance by the DNC.

For training the DNNs, we use Keras [3], which is a high-level neural network API written in Python and capable of running on top of TensorFlow. To generate the NN model, Keras uses the Adam optimizer [8] with the following settings: $lr = 10^{-2}$, $\beta_1 = 0.9$, $\beta_2 = 0.999$, $\epsilon = 10^{-8}$. The batch size (number of samples processed before the model is updated) is 2,000, and the number of epochs (number of complete passes through the training dataset) used for training is 1,000. For measuring training loss, we use the mean-squared error metric.

For basic flocking, DNN input vectors have 60 features and the number of trainable DNN parameters is 33,854. For flocking with obstacle-avoidance and target-seeking, input vectors have 92 features and the number of trainable parameters is 36,542. Finally, for flocking with predator-avoidance, input vectors have 64 features and the resulting number of trainable DNN parameters is 34,190.

To test the trained DNC, we generated 100 simulations (runs) for each of the desired control objectives: basic flocking with collision avoidance, flocking with obstacle avoidance and target seeking, and flocking with predator avoidance. The results presented in Tables 1, were obtained using the same number of agents and obstacles and the same predator as in the training phase. We also ran tests that show DNC controllers can achieve collision-free flocking with obstacle avoidance where the numbers of agents and obstacles are greater than those used during training.

5.2 Results for Basic Flocking

We use flock diameter, inter-agent collision count and velocity convergence [20] as performance metrics for flocking behavior. At any time step, the *flock diameter* $D(\mathbf{p}) = \max_{(i,j) \in \mathcal{A}} \|p_{ij}\|$ is the largest distance between any two agents in the flock. We calculate the average converged diameter by averaging the flock diameter

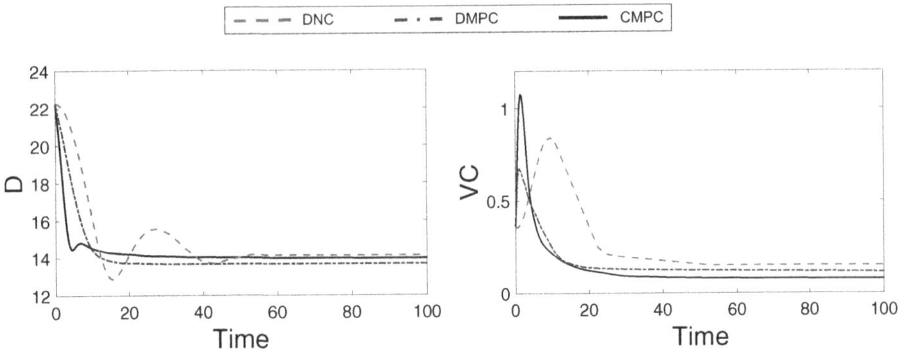

(a) Flock diameter (b) Velocity convergence

Fig. 3: Performance comparison for basic flocking with collision avoidance, averaged over 100 test runs.

in the final time step of the simulation over the 100 runs. An inter-agent collision (IC) occurs when the distance between two agents at any point in time is less than d_{min}. The IC rate (ICR) is the average number of ICs per test-trajectory time-step. The velocity convergence $VC(\mathbf{v}) = (1/n)\left(\sum_{i \in \mathcal{A}} \|v_i - (\sum_{j=1}^{n} v_j)/n\|^2\right)$ is the average of the squared magnitude of the discrepancy between the velocities of agents and the flock's average velocity. For all the metrics, lower values are better, indicating a denser and more coherent flock with fewer collisions. A successful flocking controller should also ensure that values of $D(\mathbf{p})$ and $VC(\mathbf{v})$ eventually stabilize.

Fig. 3 and Table 1 compare the performance of the DNC on the basic-flocking problem for 15 agents to that of the MPC controllers. Although the DMPC and CMPC outperform the DNC, the difference is marginal. An important advantage of the DNC over DMPC is that they are much faster. Executing a DNC controller requires a modest number of arithmetic operations, whereas executing an MPC controller requires simulation of a model and controller over the prediction horizon. In our experiments, on average, the CMPC takes 1209 msec of CPU time for the entire flock and DMPC takes 58 msec of CPU time per agent, whereas the DNC takes only 1.6 msec.

Table 1: Performance comparison for BF with 15 agents on 100 test runs

	Avg. Conv. Diameter	ICR	Velocity Convergence
DNC	14.13	0	0.15
DMPC	13.67	0	0.11
CMPC	13.84	0	0.10

Table 2: DNC Performance Generalization for BF

Agents	Avg. Conv. Diameter	Conv. Rate (%)	Avg. Conv. Time	ICR
15	14.13	100	52.15	0
20	16.45	97	58.76	0
25	19.81	94	64.11	0
30	23.24	92	72.08	0
35	30.57	86	83.84	0.008
40	38.66	81	95.32	0.019

5.3 Results for Obstacle and Predator Avoidance

For obstacle and predator avoidance, collision rates are used as a performance metric. An obstacle-agent collision (OC) occurs when the distance between an agent and the closest point on any obstacle is less than d^{obs}_{min}. A predator-agent collision (PC) occurs when the distance between an agent and the predator is less than d^{pred}_{min}. The OC rate (OCR) is the average number of OCs per test-trajectory time-step, and the PC rate (PCR) is defined similarly. Our test results show that the DNC, along with the DMPC and CMPC, is collision-free (i.e., each of ICR, OCR, and PCR is zero) for 15 agents, with the exception of DMPC for predator avoidance where PCR = 0.013. We also observed that the flock successfully reaches the target location in all 100 test runs.

5.4 DNC Generalization Results

Tables 2–3 present DNC generalization results for basic flocking (BF), obstacle avoidance (OA), and predator avoidance (PA), with the number of agents ranging from 15 (the flock size during training) to 40. In all of these experiments, we use a neighborhood size of $\mathcal{N} = 14$, the same as during training. Each controller was evaluated with 100 test runs. The performance metrics in Table 2 are the average converged diameter, convergence rate, average convergence time, and ICR.

The convergence rate is the fraction of successful flocks over 100 runs. The collection of agents is said to have converged to a flock (with collision avoidance) if the value of the global cost function is less than the convergence threshold. We use a convergence threshold of $J_1(\mathbf{p}) \leq 150$, which was chosen based on its proximity to the value achieved by CMPC. We use the cost function from Eq. 12a to calculate our success rate because we are showing convergence rate for basic flocking. The average convergence time is the time when the global cost function first drops below the success threshold and remains below it for the rest of the run, averaged over all 100 runs. Even with a local neighborhood of size 14, the results demonstrate that the DNC can successfully generalize to a large number of agents for all of our control objectives.

Table 3: DNC Generalization Performance for OA and PA

	OA		PA	
Agents	ICR	OCR	ICR	PCR
15	0	0	0	0
20	0	0	0	0
25	0	0	0	0
30	0	0	0	0
35	0.011	0.009	0.013	0.010
40	0.021	0.018	0.029	0.023

5.5 Statistical Model Checking Results

We use Monte Carlo (MC) approximation as a form of Statistical Model Checking [5, 10] to compute confidence intervals for the DNC's convergence rate to a flock with collision avoidance and for the (normalized) convergence time. The convergence rate is the fraction of successful flocks over N runs. The collection of agent is said to have converged to a successful flock with collision avoidance if the global cost function $J_1(\mathbf{p}) \leq 150$, where $J_1(\mathbf{p})$ is cost function for basic flocking defined in Eq. 12a.

The main idea of MC is to use N random variables, $Z_1, \ldots, Z_N$, also called samples, IID distributed according to a random variable Z with mean μ_Z, and to take the sum $\tilde{\mu}_Z = (Z_1 + \ldots + Z_N)/N$ as the value approximating the mean μ_Z. Since an exact computation of μ_Z is almost always intractable, an MC approach is used to compute an (ϵ, δ)-approximation of this quantity.

Additive Approximation [6] is an (ϵ, δ)-approximation scheme where the mean μ_Z of an RV Z is approximated with absolute error ϵ and probability $1 - \delta$:

$$Pr[\mu_Z - \epsilon \leq \tilde{\mu}_Z \leq \mu_Z + \epsilon] \geq 1 - \delta \tag{13}$$

where $\tilde{\mu}_Z$ is an approximation of μ_Z. An important issue is to determine the number of samples N needed to ensure that $\tilde{\mu}_Z$ is an (ϵ, δ)-approximation of μ_Z. If Z is a Bernoulli variable expected to be large, one can use the Chernoff-Hoeffding instantiation of the Bernstein inequality and take N to be $N = 4\ln(2/\delta)/\epsilon^2$, as in [6]. This results in the *additive approximation algorithm* [5], defined in Algorithm 1.

We use this algorithm to obtain a joint (ϵ, δ)-approximation of the mean convergence rate and mean normalized convergence time for the DNC. Each sample Z_i is based on the result of an execution obtained by simulating the system starting from a random initial state, and we take $Z = (B, R)$, where B is a Boolean variable indicating whether the agents converged to a flock during the execution, and R is a real value denoting the normalized convergence time. The normalized convergence time is the time when the global cost function first drops below the convergence threshold and remains below it for the rest of the run, measured as a fraction of the total duration of the run. The assumptions

Algorithm 1: Additive Approximation Algorithm

Input: (ϵ, δ) with $0 < \epsilon < 1$ and $0 < \delta < 1$
Input: Random variables Z_i, IID
Output: $\tilde{\mu}_Z$ approximation of μ_Z
$N = 4\ln(2/\delta)/\epsilon^2$;
for *(i=0; $i \leq N$; i++)* **do**
  $S = S + Z_i$;
$\tilde{\mu_Z} = S/N$; **return** $\tilde{\mu}_Z$;

Table 4: SMC results for DNC convergence rate and normalized convergence time; $\epsilon = 0.01$, $\delta = 0.0001$

Agents	$\tilde{\mu}_{CR}$	$\tilde{\mu}_{CT}$
15	0.99	0.53
20	0.97	0.58
25	0.94	0.65
30	0.91	0.71
35	0.86	0.84
40	0.80	0.95

about Z required for validity of the additive approximation hold, because RV B is a Bernoulli variable, the convergence rate is expected to be large (i.e., closer to 1 than to 0), and the proportionality constraint of the Bernstein inequality is also satisfied for RV R.

In these experiments, the initial configurations are sampled from the same distributions as in Section 5.1, and we set $\epsilon = 0.01$ and $\delta = 0.0001$, to obtain $N =$ 396,140. We perform the required set of N simulations for 15, 20, 25, 30, 35 and 40 agents. Table 4 presents the results, specifically, the (ϵ, δ)-approximations $\tilde{\mu}_{CR}$ and $\tilde{\mu}_{CT}$ of the mean convergence rate and the mean normalized convergence time, respectively. While the results for the convergence rate are (as expected) numerically similar to the results in Table 2, the results in Table 4 are much stronger, because they come with the guarantee that they are (ϵ, δ)-approximations of the actual mean values.

6 Related Work

In [18], a flocking controller is synthesized using multi-agent reinforcement learning (MARL) and natural evolution strategies (NES). The target model from which the system learns is Reynolds flocking model [16]. For training purposes, a list of metrics called *entropy* are chosen, which provide a measure of the collective behavior displayed by the target model. As the authors of [18] observe, this technique does not quite work: although it consistently leads to agents forming recognizable patterns during simulation, agents self-organized into a cluster instead of flowing like a flock.

In [9], reinforcement learning and flocking control are combined for the purpose of predator avoidance, where the learning module determines safe spaces in which the flock can navigate to avoid predators. Their approach to predator avoidance, however, isn't distributed as it requires a majority consensus by the flock to determine its action to avoid predators. They also impose an α-lattice structure [13] on the flock. In contrast, our approach is geometry-agnostic and achieves predator avoidance in a distributed manner.

In [7], an uncertainty-aware reinforcement learning algorithm is developed to estimate the probability of a mobile robot colliding with an obstacle in an unknown environment. Their approach is based on bootstrap neural networks using dropouts, allowing it to process raw sensory inputs. Similarly, a learning-based approach to robot navigation and obstacle avoidance is presented in [14]. They train a model that maps sensor inputs and the target position to motion commands generated by the ROS [15] navigation package. Our work in contrast considers obstacle avoidance (and other control objectives) in a multi-agent flocking scenario under the simplifying assumption of full state observation.

In [4], an approach based on Bayesian inference is proposed that allows an agent in a heterogeneous multi-agent environment to estimate the navigation model and goal of each of its neighbors. It then uses this information to compute a plan that minimizes inter-agent collisions while allowing the agent to reach its goal. Flocking formation is not considered.

7 Conclusions

With the introduction of Neural Flocking (NF), we have shown how machine learning in the form of Supervised Learning can bring many benefits to the flocking problem. As our experimental evaluation confirms, the symmetric and fully distributed neural controllers we derive in this manner are capable of achieving a multitude of flocking-oriented objectives, including flocking formation, inter-agent collision avoidance, obstacle avoidance, predator avoidance, and target seeking. Moreover, NF controllers exhibit real-time performance and generalize the behavior seen in the training data to achieve these objectives in a significantly broader range of scenarios.

Ongoing work aims to determine whether a DNC can perform as well as the centralized MPC controller for agent models that are significantly more realistic than our current point-based model. For this purpose, we are using transfer learning to train a DNC that can achieve acceptable performance on realistic quadrotor dynamics [1], starting from our current point-model-based DNC. This effort also involves extending our current DNC from 2-dimensional to 3-dimensional spatial coordinates. If successful, and preliminary results are encouraging, this line of research will demonstrate that DNCs are capable of achieving flocking with complex realistic dynamics.

For future work, we plan to investigate a distance-based notion of agent neighborhood as opposed to our current nearest-neighbors formulation. Furthermore, motivated by the quadrotor study of [21], we will seek to combine MPC with reinforcement learning in the framework of guided policy search as an alternative solution technique for the NF problem.

References

1. Bouabdallah, S.: Design and control of quadrotors with application to autonomous flying (2007)
2. Camacho, E.F., Bordons Alba, C.: Model Predictive Control. Springer (2007)
3. Chollet, F., et al.: Keras (2015), `https://github.com/keras-team/keras.git`
4. Godoy, J., Karamouzas, I., Guy, S.J., Gini, M.: Moving in a crowd: Safe and efficient navigation among heterogeneous agents. In: Proceedings of the Twenty-Fifth International Joint Conference on Artificial Intelligence. pp. 294–300. IJCAI'16, AAAI Press (2016)
5. Grosu, R., Peled, D., Ramakrishnan, C.R., Smolka, S.A., Stoller, S.D., Yang, J.: Using statistical model checking for measuring systems. In: 6th International Symposium, ISoLA 2014. Corfu, Greece (Oct 2014)
6. Hérault, T., Lassaigne, R., Magniette, F., Peyronnet, S.: Approximate probabilistic model checking. In: Steffen, B., Levi, G. (eds.) Verification, Model Checking, and Abstract Interpretation. pp. 73–84. Springer Berlin Heidelberg, Berlin, Heidelberg (2004)
7. Kahn, G., Villaflor, A., Pong, V., Abbeel, P., Levine, S.: Uncertainty-aware reinforcement learning for collision avoidance. arXiv preprint arXiv:1702.01182. pp. 1–12 (2017)
8. Kingma, D.P., Ba, J.: Adam: A method for stochastic optimization. In: 3rd International Conference on Learning Representations, ICLR 2015, San Diego, CA, USA, May 7-9, 2015, Conference Track Proceedings (2015)
9. La, H.M., Lim, R., Sheng, W.: Multirobot cooperative learning for predator avoidance. IEEE Transactions on Control Systems Technology **23**(1), 52–63 (2015)
10. Larsen, K.G., Legay, A.: Statistical model checking: Past, present, and future. In: 6th International Symposium, ISoLA 2014. Corfu, Greece (Oct 2014)
11. Mehmood, U., Paoletti, N., Phan, D., Grosu, R., Lin, S., Stoller, S.D., Tiwari, A., Yang, J., Smolka, S.A.: Declarative vs rule-based control for flocking dynamics. In: Proceedings of SAC 2018, 33rd Annual ACM Symposium on Applied Computing. pp. 816–823 (2018)
12. Nocedal, J., Wright, S.J.: Numerical Optimization. Springer, New York, NY, USA, second edn. (2006)
13. Olfati-Saber, R.: Flocking for multi-agent dynamic systems: Algorithms and theory. IEEE Transactions on automatic control **51**(3), 401–420 (2006)
14. Pfeiffer, M., Schaeuble, M., Nieto, J.I., Siegwart, R., Cadena, C.: From perception to decision: A data-driven approach to end-to-end motion planning for autonomous ground robots. In: 2017 IEEE International Conference on Robotics and Automation, ICRA 2017, Singapore, Singapore, May 29 - June 3, 2017. pp. 1527–1533 (2017)
15. Quigley, M., Conley, K., Gerkey, B.P., Faust, J., Foote, T., Leibs, J., Wheeler, R., Ng, A.Y.: ROS: an open-source robot operating system. In: ICRA Workshop on Open Source Software (2009)
16. Reynolds, C.W.: Flocks, herds and schools: A distributed behavioral model. SIGGRAPH Comput. Graph. **21**(4) (Aug 1987)
17. Reynolds, C.W.: Steering behaviors for autonomous characters. In: Proceedings of Game Developers Conference 1999. pp. 763–782 (1999)

18. Shimada, K., Bentley, P.: Learning how to flock: Deriving individual behaviour from collective behaviour with multi-agent reinforcement learning and natural evolution strategies. In: Proceedings of the Genetic and Evolutionary Computation Conference Companion. pp. 169–170. ACM (2018)
19. Zhan, J., Li, X.: Flocking of multi-agent systems via model predictive control based on position-only measurements. IEEE Transactions on Industrial Informatics **9**(1), 377–385 (2013)
20. Zhang, H.T., Cheng, Z., Chen, G., Li, C.: Model predictive flocking control for second-order multi-agent systems with input constraints. IEEE Transactions on Circuits and Systems I: Regular Papers **62**(6), 1599–1606 (2015)
21. Zhang, T., Kahn, G., Levine, S., Abbeel, P.: Learning deep control policies for autonomous aerial vehicles with MPC-guided policy search. In: 2016 IEEE International Conference on Robotics and Automation, ICRA 2016, Stockholm, Sweden, May 16-21, 2016. pp. 528–535 (2016)

On Well-Founded and Recursive Coalgebras*

Jiří Adámek[1,**], Stefan Milius[2,***,(✉)], and Lawrence S. Moss[3,†]

[1] Czech Technical University, Prague, Czech Republic
j.adamek@tu-braunschweig.de
[2] Friedrich-Alexander-Universität Erlangen-Nürnberg, Germany
mail@stefan-milius.eu
[3] Indiana University, Bloomington, IN, USA
lmoss@indiana.edu

Abstract This paper studies fundamental questions concerning category-theoretic models of induction and recursion. We are concerned with the relationship between well-founded and recursive coalgebras for an endofunctor. For monomorphism preserving endofunctors on complete and well-powered categories every coalgebra has a well-founded part, and we provide a new, shorter proof that this is the coreflection in the category of all well-founded coalgebras. We present a new more general proof of Taylor's General Recursion Theorem that every well-founded coalgebra is recursive, and we study conditions which imply the converse. In addition, we present a new equivalent characterization of well-foundedness: a coalgebra is well-founded iff it admits a coalgebra-to-algebra morphism to the initial algebra.

Keywords: Well-founded · Recursive · Coalgebra · Initial Algebra · General Recursion Theorem

1 Introduction

What is induction? What is recursion? In areas of theoretical computer science, the most common answers are related to *initial algebras*. Indeed, the dominant trend in abstract data types is initial algebra semantics (see e.g. [19]), and this approach has spread to other semantically-inclined areas of the subject. The approach in broad slogans is that, for an endofunctor F describing the type of algebraic operations of interest, the initial algebra μF has the property that for every F-algebra A, there is a unique homomorphism $\mu F \to A$, and this *is* recursion. Perhaps the primary example is *recursion on* $\mathbb{N}$, *the natural numbers*. Recall that $\mathbb{N}$ is the initial algebra for the set functor $FX = X + 1$. If A is any set, and $a \in A$ and $\alpha\colon A \to A + 1$ are given, then initiality tells us that there is a unique $f\colon \mathbb{N} \to A$ such that for all $n \in \mathbb{N}$,

$$f(0) = a \qquad f(n+1) = \alpha(f(n)). \tag{1.1}$$

* A full version of this paper including full proof details is available on arXiv [5].
** Supported by the Grant Agency of the Czech Republic under grant 19-00902S.
*** Supported by Deutsche Forschungsgemeinschaft (DFG) under project MI 717/5-2.
† Supported by grant #586136 from the Simons Foundation.

J. Goubault-Larrecq and B. König (Eds.): FOSSACS 2020, LNCS 12077, pp. 17–36, 2020.
https://doi.org/10.1007/978-3-030-45231-5_2

Then the first additional problem coming with this approach is that of how to "recognize" initial algebras: Given an algebra, how do we really know if it is initial? The answer – again in slogans – is that initial algebras are the ones with "no junk and no confusion."

Although initiality captures some important aspects of recursion, it cannot be a fully satisfactory approach. One big missing piece concerns recursive definitions based on well-founded relations. For example, the whole study of termination of rewriting systems depends on well-orders, the primary example of *recursion on a well-founded order*. Let (X, R) be a well-founded relation, i.e. one with no infinite sequences $\cdots x_2 \, R \, x_1 \, R \, x_0$. Let A be any set, and let $\alpha\colon \mathscr{P}A \to A$. (Here and below, $\mathscr{P}$ is the power set functor, taking a set to the set of its subsets.) Then there is a unique $f\colon X \to A$ such that for all $x \in X$,

$$f(x) = \alpha(\{f(y) : y \; R \; x\}). \tag{1.2}$$

The main goal of this paper is the study of concepts that allow one to extend the algebraic spirit behind initiality in (1.1) to the setting of recursion arising from well-foundedness as we find it in (1.2). The corresponding concepts are those of well-founded and recursive coalgebras for an endofunctor, which first appear in work by Osius [22] and Taylor [23, 24], respectively. In his work on categorical set theory, Osius [22] first studied the notions of well-founded and recursive coalgebras (for the power-set functor on sets and, more generally, the power-object functor on an elementary topos). He defined recursive coalgebras as those coalgebras $\alpha\colon A \to \mathscr{P}A$ which have a unique coalgebra-to-algebra homomorphism into every algebra (see Definition 3.2).

Taylor [23, 24] took Osius' ideas much further. He introduced well-founded coalgebras for a general endofunctor, capturing the notion of a well-founded relation categorically, and considered recursive coalgebras under the name 'coalgebras obeying the recursion scheme'. He then proved the General Recursion Theorem that all well-founded coalgebras are recursive, for every endofunctor on sets (and on more general categories) preserving inverse images. Recursive coalgebras were also investigated by Eppendahl [12], who called them algebra-initial coalgebras. Capretta, Uustalu, and Vene [10] further studied recursive coalgebras, and they showed how to construct new ones from given ones by using comonads. They also explained nicely how recursive coalgebras allow for the semantic treatment of (functional) divide-and-conquer programs. More recently, Jeannin et al. [15] proved the General Recursion Theorem for polynomial functors on the category of many-sorted sets; they also provide many interesting examples of recursive coalgebras arising in programming.

Our contributions in this paper are as follows. We start by recalling some preliminaries in Section 2 and the definition of (parametrically) recursive coalgebras in Section 3 and of well-founded coalgebras in Section 4 (using a formulation based on Jacobs' next time operator [14], which we extend from Kripke polynomial set functors to arbitrary functors). We show that every coalgebra for a monomorphism preserving functor on a complete and well-powered category has a well-founded part, and provide a new proof that this is the coreflection in the category of well-founded coalgebras (Proposition 4.19), shortening our previous proof [6]. Next we provide a new proof of Taylor's General Recursion Theorem (Theorem 5.1), generalizing this to endofunctors preserving monomorphisms on a complete and well-powered category having smooth monomorphisms (see Definition 2.8). For the category of sets, this implies that "well-founded $\Rightarrow$ recursive" holds for all endofunctors, strengthening Taylor's result. We then discuss the converse: is every recursive coalgebra well-founded? Here the assumption that F preserves inverse images cannot be lifted, and one needs additional assumptions. In fact, we present two results: one assumes universally smooth monomorphisms and that the functor has a pre-fixed point (see Theorem 5.5). Under these assumptions we also give a new equivalent characterization of recursiveness and well-foundedness: a coalgebra is recursive if it has a coalgebra-to-algebra morphism into the initial algebra (which exists under our assumptions), see Corollary 5.6. This characterization was previously established for finitary functors on sets [3]. The other converse of the above implication is due to Taylor using the concept of a subobject classifier (Theorem 5.8). It implies that 'recursive' and 'well-founded' are equivalent concepts for all set functors preserving inverse images. We also prove that a similar result holds for the category of vector spaces over a fixed field (Theorem 5.12).

Finally, we show in Section 6 that well-founded coalgebras are closed under coproducts, quotients and, assuming mild assumptions, under subcoalgebras.

2 Preliminaries

We start by recalling some background material. Except for the definitions of *algebra* and *coalgebra* in Subsection 2.1, the subsections below may be read as needed. We assume that readers are familiar with notions of basic category theory; see e.g. [2] for everything which we do not detail. We indicate monomorphisms by writing $\rightarrowtail$ and strong epimorphisms by $\twoheadrightarrow$.

2.1 Algebras and Coalgebras. We are concerned throughout this paper with *algebras* and *coalgebras* for an endofunctor. This means that we have an underlying category, usually written $\mathscr{A}$; frequently it is the category of sets or of vector spaces over a fixed field, and that a functor $F\colon \mathscr{A} \to \mathscr{A}$ is given. An F*-algebra* is a pair (A, α), where $\alpha\colon FA \to A$. An F*-coalgebra* is a pair (A, α), where $\alpha\colon A \to FA$. We usually drop the functor F. Given two algebras (A, α) and (B, β), an *algebra homomorphism* from the first to the second is $h\colon A \to B$ in $\mathscr{A}$ such that $h \cdot \alpha = \beta \cdot Fh$. Similarly, a *coalgebra homomorphism* satisfies $\beta \cdot h = Fh \cdot \alpha$. We denote by $\mathsf{Coalg}\, F$ the category of all coalgebras for F.

Example 2.1. (1) The power set functor $\mathscr{P}\colon \mathsf{Set} \to \mathsf{Set}$ takes a set X to the set $\mathscr{P}X$ of all subsets of it; for a morphism $f\colon X \to Y$, $\mathscr{P}f\colon \mathscr{P}X \to \mathscr{P}Y$ takes a subset $S \subseteq X$ to its direct image $f[S]$. Coalgebras $\alpha\colon X \to \mathscr{P}X$ may be identified with directed graphs on the set X of vertices, and the coalgebra structure α describes the edges: $b \in \alpha(a)$ means that there is an edge $a \to b$ in the graph.

(2) Let Σ be a signature, i.e. a set of operation symbols, each with a finite arity. The *polynomial functor* H_Σ associated to Σ assigns to a set X the set

$$H_\Sigma X = \coprod_{n\in\mathbb{N}} \Sigma_n \times X^n,$$

where Σ_n is the set of operation symbols of arity n. This may be identified with the set of all terms $\sigma(x_1,\dots,x_n)$, for $\sigma \in \Sigma_n$, and $x_1,\dots,x_n \in X$. Algebras for H_Σ are the usual Σ-algebras.

(3) Deterministic automata over an input alphabet Σ are coalgebras for the functor $FX = \{0,1\} \times X^\Sigma$. Indeed, given a set S of states, a next-state map $S \times \Sigma \to S$ may be curried to $\delta\colon S \to S^\Sigma$. The set of final states yields the acceptance predicate $a\colon S \to \{0,1\}$. So an automaton may be regarded as a coalgebra $\langle a,\delta\rangle\colon S \to \{0,1\} \times S^\Sigma$.

(4) Labelled transitions systems are coalgebras for $FX = \mathscr{P}(\Sigma \times X)$.

(5) To describe linear weighted automata, i.e. weighted automata over the input alphabet Σ with weights in a field K, as coalgebras, one works with the category Vec_K of vector spaces over K. A linear weighted automaton is then a coalgebra for $FX = K \times X^\Sigma$.

2.2 Preservation Properties. Recall that an intersection of two subobjects $s_i\colon S_i \rightarrowtail A$ $(i = 1,2)$ of a given object A is given by their pullback. Analogously, (general) intersections are given by wide pullbacks. Furthermore, the inverse image of a subobject $s\colon S \rightarrowtail B$ under a morphism $f\colon A \to B$ is the subobject $t\colon T \rightarrowtail A$ obtained by a pullback of s along f.

All of the 'usual' set functors preserve intersections and inverse images:

Example 2.2. (1) Every polynomial functor preserves intersections and inverse images.

(2) The power-set functor $\mathscr{P}$ preserves intersections and inverse images.

(3) Intersection-preserving set functors are closed under taking coproducts, products and composition. Similarly, for inverse images.

(4) Consider next the set functor R defined by $RX = \{(x,y) \in X \times X\colon x \neq y\} + \{d\}$ for sets X. For a function $f\colon X \to Y$ put $Rf(x,y) = (f(x),f(y))$ if $f(x) \neq f(y)$, and d otherwise. R preserves intersections but not inverse images.

Proposition 2.3 [27]. *For every set functor F there exists an essentially unique set functor $\bar{F}$ which coincides with F on nonempty sets and functions and preserves finite intersections (whence monomorphisms).*

Remark 2.4. (1) In fact, Trnková gave a construction of $\bar{F}$: she defined $\bar{F}\emptyset$ as the set of all natural transformations $C_{01} \to F$, where C_{01} is the set functor with $C_{01}\emptyset = \emptyset$ and $C_{01}X = 1$ for all nonempty sets X. For the empty map $e\colon \emptyset \to X$ with $X \neq \emptyset$, $\bar{F}e$ maps a natural transformation $\tau\colon C_{01} \to F$ to the element given by $\tau_X\colon 1 \to FX$.

(2) The above functor $\bar{F}$ is called the *Trnková hull* of F. It allows us to achieve preservation of intersections for all *finitary* set functors. Intuitively, a functor on

sets is finitary if its behavior is completely determined by its action on *finite* sets and functions. For a general functor, this intuition is captured by requiring that the functor preserves filtered colimits [8]. For a set functor F this is equivalent to being *finitely bounded*, which is the following condition: for each element $x \in FX$ there exists a finite subset $M \subseteq X$ such that $x \in Fi[FM]$, where $i: M \hookrightarrow X$ is the inclusion map [7, Rem. 3.14].

Proposition 2.5 [4, p. 66]. *The Trnková hull of a finitary set functor preserves all intersections.*

2.3 Factorizations. Recall that an epimorphism $e: A \to B$ is called *strong* if it satisfies the following *diagonal fill-in property*: given a monomorphism $m: C \rightarrowtail D$ and morphisms $f: A \to C$ and $g: B \to D$ such that $m \cdot f = g \cdot e$ then there exists a unique $d: B \to C$ such that $f = d \cdot e$ and $g = m \cdot d$.

Every complete and well-powered category has factorizations of morphisms: every morphism f may be written as $f = m \cdot e$, where e is a strong epimorphism and m is a monomorphism [9, Prop. 4.4.3]. We call the subobject m the *image* of f. It follows from a result in Kurz' thesis [16, Prop. 1.3.6] that factorizations of morphisms lift to coalgebras:

Proposition 2.6 ($\mathsf{Coalg}\,F$ inherits factorizations from $\mathscr{A}$). *Suppose that F preserves monomorphisms. Then the category $\mathsf{Coalg}\,F$ has factorizations of homomorphisms f as $f = m \cdot e$, where e is carried by a strong epimorphism and m by a monomorphism in $\mathscr{A}$. The diagonal fill-in property holds in $\mathsf{Coalg}\,F$.*

Remark 2.7. By a *subcoalgebra* of a coalgebra (A, α) we mean a subobject in $\mathsf{Coalg}\,F$ represented by a homomorphism $m: (B, \beta) \rightarrowtail (A, \alpha)$, where m is monic in $\mathscr{A}$. Similarly, by a *strong quotient* of a coalgebra (A, α) we mean one represented by a homomorphism $e: (A, \alpha) \twoheadrightarrow (C, \gamma)$ with e strongly epic in $\mathscr{A}$.

2.4 Chains. By a *transfinite chain* in a category $\mathscr{A}$ we understand a functor from the ordered class Ord of all ordinals into $\mathscr{A}$. Moreover, for an ordinal λ, a λ-*chain* in $\mathscr{A}$ is a functor from λ to $\mathscr{A}$. A category *has colimits of chains* if for every ordinal λ it has a colimit of every λ-chain. This includes the initial object 0 (the case $\lambda = 0$).

Definition 2.8. (1) A category $\mathscr{A}$ has *smooth monomorphisms* if for every λ-chain C of monomorphisms a colimit exists, its colimit cocone is formed by monomorphisms, and for every cone of C formed by monomorphisms, the factorizing morphism from $\mathsf{colim}\, C$ is monic. In particuar, every morphism from 0 is monic.

(2) $\mathscr{A}$ has *universally smooth monomorphisms* if $\mathscr{A}$ also has pullbacks, and for every morphism $f: X \to \mathsf{colim}\, C$, the functor $\mathscr{A}/\mathsf{colim}\, C \to \mathscr{A}/X$ forming pullbacks along f preserves the colimit of C. This implies that initial object 0 is *strict*, i.e. every morphism $f: X \to 0$ is an isomorphism. Indeed, consider the empty chain ($\lambda = 0$).

Example 2.9. (1) Set has universally smooth monomorphisms.

(2) Vec_K has smooth monomorphisms, but not universally so because the initial object is not strict.

(3) Categories in which colimits of chains and pullbacks are formed "set-like" have universally smooth monomorphisms. These include the categories of posets, graphs, topological spaces, presheaf categories, and many varieties, such as monoids, groups, and unary algebras.

(4) Every locally finitely presentable category $\mathscr{A}$ with a strict initial object (see Remark 2.12(1)) has smooth monomorphisms. This follows from [8, Prop. 1.62]. Moreover, since pullbacks commute with colimits of chains, it is easy to prove that colimits of chains are universal using the strictness of 0.

(5) The category CPO of complete partial orders does not have smooth monomorphisms. Indeed, consider the ω-chain of linearly ordered sets $A_n = \{0, \dots, n\} + \{\top\}$ ($\top$ a top element) with inclusion maps $A_n \to A_{n+1}$. Its colimit is the linearly ordered set $\mathbb{N} + \{\top, \top'\}$ of natural numbers with two added top elements $\top' < \top$. For the sub-cpo $\mathbb{N} + \{\top\}$, the inclusions of A_n are monic and form a cocone. But the unique factorizing morphism from the colimit is not monic.

Notation 2.10. For every object A we denote by $\mathsf{Sub}(A)$ the poset of all subobjects of A (represented by monomorphisms $s\colon S \rightarrowtail A$), where $s \leq s'$ if there exists i with $s = s' \cdot i$. If $\mathscr{A}$ has pullbacks we have, for every morphism $f\colon A \to B$, the *inverse image operator*, viz. the monotone map $\overleftarrow{f} : \mathsf{Sub}(B) \to \mathsf{Sub}(A)$ assigning to a subobject $s\colon S \rightarrowtail A$ the subobject of B obtained by forming the inverse image of s under f, i.e. the pullback of s along f.

Lemma 2.11. *If $\mathscr{A}$ is complete and well-powered, then $\overleftarrow{f}$ has a left adjoint given by the* (direct) image operator $\overrightarrow{f} : \mathsf{Sub}(A) \to \mathsf{Sub}(B)$. *It maps a subobject $t\colon T \rightarrowtail B$ to the subobject of A given by the image of $f \cdot t$; in symbols we have $\overrightarrow{f}(t) \leq s$ iff $t \leq \overleftarrow{f}(s)$.*

Remark 2.12. If $\mathscr{A}$ is a complete and well-powered category, then $\mathsf{Sub}(A)$ is a complete lattice. Now suppose that $\mathscr{A}$ has smooth monomorphisms.

(1) In this setting, the unique morphism $\perp_A\colon 0 \to A$ is a monomorphism and therefore is the bottom element of the poset $\mathsf{Sub}(A)$.

(2) Furthermore, a join of a chain in $\mathsf{Sub}(A)$ is obtained by forming a colimit, in the obvious way.

(3) If $\mathscr{A}$ has universally smooth monomorphisms, then for every morphism $f\colon A \to B$, the operator $\overleftarrow{f} : \mathsf{Sub}(B) \to \mathsf{Sub}(A)$ preserves unions of chains.

Remark 2.13. Recall [1] that every endofunctor F yields the *initial-algebra chain*, viz. a transfinite chain formed by the objects $F^i 0$ of $\mathscr{A}$, as follows: $F^0 0 = 0$, the initial object; $F^{i+1}0 = F(F^i 0)$, and for a limit ordinal i we take the colimit of the chain $(F^j 0)_{j<i}$. The connecting morphisms $w_{i,j}\colon F^i 0 \to F^j 0$ are defined by a similar transfinite recursion.

3 Recursive Coalgebras

Assumption 3.1. We work with a standard set theory (e.g. Zermelo-Fraenkel), assuming the Axiom of Choice. In particular, we use transfinite induction on several occasions. (We are not concerned with constructive foundations in this paper.)

Throughout this paper we assume that $\mathscr{A}$ is a complete and well-powered category $\mathscr{A}$ and that $F\colon \mathscr{A} \to \mathscr{A}$ preserves monomorphisms.

For $\mathscr{A} = \mathsf{Set}$ the condition that F preserves monomorphisms may be dropped. In fact, preservation of non-empty monomorphism is sufficient in general (for a suitable notion of non-empty monomorphism) [21, Lemma 2.5], and this holds for every set functor.

The following definition of recursive coalgebras was first given by Osius [22]. Taylor [24] speaks of *coalgebras obeying the recursion scheme*. Capretta et al. [10] extended the concept to *parametrically recursive* coalgebra by dualizing completely iterative algebras [20].

Definition 3.2. A coalgebra $\alpha\colon A \to FA$ is called *recursive* if for every algebra $e\colon FX \to X$ there exists a unique coalgebra-to-algebra morphism $e^\dagger\colon A \to X$, i.e. a unique morphism such that the square on the left below commutes:

$$\begin{array}{ccc} A & \xrightarrow{e^\dagger} & X \\ {\scriptstyle\alpha}\big\downarrow & & \big\uparrow{\scriptstyle e} \\ FA & \xrightarrow{Fe^\dagger} & FX \end{array} \qquad\qquad \begin{array}{ccc} A & \xrightarrow{\quad e^\dagger \quad} & X \\ {\scriptstyle\langle\alpha, A\rangle}\big\downarrow & & \big\uparrow{\scriptstyle e} \\ FA \times A & \xrightarrow{Fe^\dagger \times A} & FX \times A \end{array}$$

(A, α) is called *parametrically recursive* if for every morphism $e\colon FX \times A \to X$ there is a unique morphism $e^\dagger\colon A \to X$ such that the square on the right above commutes.

Example 3.3. (1) A graph regarded as a coalgebra for $\mathscr{P}$ is recursive iff it has no infinite path. This is an immediate consequence of the General Recursion Theorem (see Corollary 5.6 and Example 4.5(2)).

(2) Let $\iota\colon F(\mu F) \to \mu F$ be an initial algebra. By Lambek's Lemma, ι is an isomorphism. So we have a coalgebra $\iota^{-1}\colon \mu F \to F(\mu F)$. This algebra is (parametrically) recursive. By [20, Thm. 2.8], in dual form, this is precisely the same as the terminal parametrically recursive coalgebra (see also [10, Prop. 7]).

(3) The initial coalgebra $0 \to F0$ is recursive.

(4) If (C, γ) is recursive so is $(FC, F\gamma)$, see [10, Prop. 6].

(5) Colimits of recursive coalgebras in $\mathsf{Coalg}\,F$ are recursive. This is easy to prove, using that colimits of coalgebras are formed on the level of the underlying category.

(6) It follows from items (3)–(5) that in the initial-algebra chain from Remark 2.13 all coalgebras $w_{i,i+1}\colon F^i0 \to F^{i+1}0$, $i \in \mathsf{Ord}$, are recursive.

(7) Every parametrically recursive coalgebra is recursive. (To see this, form for a given $e\colon FX \to X$ the morphism $e' = e \cdot \pi$, where $\pi\colon FX \times A \to FX$ is the projection.) In Corollaries 5.6 and 5.9 we will see that the converse often holds.

Here is an example where the converse fails [3]. Let $R\colon \mathsf{Set} \to \mathsf{Set}$ be the functor defined in Example 2.2(4). Also, let $C = \{0, 1\}$, and define $\gamma\colon C \to RC$ by $\gamma(0) = \gamma(1) = (0, 1)$. Then (C, γ) is a recursive coalgebra. Indeed, for every algebra $\alpha\colon RA \to A$ the constant map $h\colon C \to A$ with $h(0) = h(1) = \alpha(d)$ is the unique coalgebra-to-algebra morphism.

However, (C, γ) is not parametrically recursive. To see this, consider any morphism $e\colon RX \times \{0, 1\} \to X$ such that RX contains more than one pair (x_0, x_1), $x_0 \neq x_1$ with $e((x_0, x_1), i) = x_i$ for $i = 0, 1$. Then each such pair yields $h\colon C \to X$ with $h(i) = x_i$ making the appropriate square commutative. Thus, (C, γ) is not parametrically recursive.

(8) Capretta et al. [11] showed that recursivity semantically models divide-and-conquer programs, as demonstrated by the example of Quicksort. For every linearly ordered set A (of data elements), Quicksort is usually defined as the recursive function $q\colon A^* \to A^*$ given by

$$q(\varepsilon) = \varepsilon \qquad \text{and} \qquad q(aw) = q(w_{\leq a}) \star (aq(w_{>a})),$$

where A^* is the set of all lists on A, ε is the empty list, $\star$ is the concatenation of lists and $w_{\leq a}$ denotes the list of those elements of w which are less than or equal than a; analogously for $w_{>a}$.

Now consider the functor $FX = 1 + A \times X \times X$ on Set, where $1 = \{\bullet\}$, and form the coalgebra $s\colon A^* \to 1 + A \times A^* \times A^*$ given by

$$s(\varepsilon) = \bullet \qquad \text{and} \qquad s(aw) = (a, w_{\leq a}, w_{>a}) \qquad \text{for } a \in A \text{ and } w \in A^*.$$

We shall see that this coalgebra is recursive in Example 5.3. Thus, for the F-algebra $m : 1 + A \times A^* \times A^* \to A^*$ given by

$$m(\bullet) = \varepsilon \qquad \text{and} \qquad m(a, w, v) = w \star (av)$$

there exists a unique function q on A^* such that $q = m \cdot Fq \cdot s$. Notice that the last equation reflects the idea that Quicksort is a divide-and-conquer algorithm. The coalgebra structure s divides a list into two parts $w_{\leq a}$ and $w_{>a}$. Then Fq sorts these two smaller lists, and finally in the combine- (or conquer-) step, the algebra structure m merges the two sorted parts to obtain the desired whole sorted list.

Jeannin et al. [15, Sec. 4] provide a number of recursive functions arising in programming that are determined by recursivity of a coalgebra, e.g. the gcd of integers, the Ackermann function, and the Towers of Hanoi.

4 The Next Time Operator and Well-Founded Coalgebras

As we have mentioned in the Introduction, the main issue of this paper is the relationship between two concepts pertaining to coalgebras: recursiveness and well-foundedness. The concept of well-foundedness is well-known for directed graphs $(G, \to)$: it means that there are no infinite directed paths $g_0 \to g_1 \to \cdots$. For a set X with a relation R, well-foundedness means that there are no *backwards* sequences $\cdots R\, x_2\, R\, x_1\, R\, x_0$, i.e. the converse of the relation is well-founded as a graph. Taylor [24, Def. 6.2.3] gave a more general category theoretic formulation of well-foundedness. We observe here that his definition can be presented in a compact way, by using an operator that generalizes the way one thinks of the semantics of the 'next time' operator of temporal logics for non-deterministic (or even probabilistic) automata and transitions systems. It is also strongly related to the algebraic semantics of modal logic, where one passes from a graph G to a function on $\mathscr{P}G$. Jacobs [14] defined and studied the 'next time' operator on coalgebras for Kripke polynomial set functors. This can be generalized to arbitrary functors as follows.

Recall that $\mathsf{Sub}(A)$ denotes the complete lattice of subobjects of A.

Definition 4.1 [4, Def. 8.9]. Every coalgebra $\alpha\colon A \to FA$ induces an endofunction on $\mathsf{Sub}(A)$, called the *next time operator*

$$\bigcirc\colon \mathsf{Sub}(A) \to \mathsf{Sub}(A), \qquad \bigcirc(s) = \overleftarrow{\alpha}(Fs) \quad \text{for } s \in \mathsf{Sub}(A).$$

In more detail: we define $\bigcirc s$ and $\alpha(s)$ by the pullback in (4.1). (Being a pullback is indicated by the "corner" symbol.) In words, $\bigcirc$ assigns to each subobject $s\colon S \rightarrowtail A$ the inverse image of Fs under α. Since Fs is a monomorphism, $\bigcirc s$ is a monomorphism and $\alpha(s)$ is (for every representation $\bigcirc s$ of that subobject of A) uniquely determined.

$$\begin{array}{ccc} \bigcirc S & \xrightarrow{\alpha(s)} & FS \\ {\scriptstyle \bigcirc s}\big\downarrow & \lrcorner & \big\downarrow{\scriptstyle Fs} \\ A & \xrightarrow{\alpha} & FA \end{array} \tag{4.1}$$

Example 4.2. (1) Let A be a graph, considered as a coalgebra for $\mathscr{P}\colon \mathsf{Set} \to \mathsf{Set}$. If $S \subseteq A$ is a set of vertices, then $\bigcirc S$ is the set of vertices all of whose successors belong to S.

(2) For the set functor $FX = \mathscr{P}(\Sigma \times X)$ expressing labelled transition systems the operator $\bigcirc$ for a coalgebra $\alpha\colon A \to \mathscr{P}(\Sigma \times A)$ is the semantic counterpart of the next time operator of classical linear temporal logic, see e.g. Manna and Pnüeli [18]. In fact, for a subset $S \hookrightarrow A$ we have that $\bigcirc S$ consists of those states all of whose next states lie in S, in symbols:

$$\bigcirc S = \{x \in A \mid (s, y) \in \alpha(x) \text{ implies } y \in S, \text{ for all } s \in \Sigma\}.$$

The next time operator allows a compact definition of well-foundedness as characterized by Taylor [24, Exercise VI.17] (see also [6, Corollary 2.19]):

Definition 4.3. A coalgebra is *well-founded* if id_A is the only fixed point of its next time operator.

Remark 4.4. (1) Let us call a subcoalgebra $m\colon (B, \beta) \rightarrowtail (A, \alpha)$ *cartesian* provided that the square (4.2) is a pullback. Then (A, α) is well-founded iff it has no proper cartesian subcoalgebra. That is, if $m\colon (B, \beta) \rightarrowtail (A, \alpha)$ is a cartesian subcoalgebra, then m is an isomorphism. Indeed, the fixed points of next time are precisely the

$$\begin{array}{ccc} B & \xrightarrow{\beta} & FB \\ {\scriptstyle m}\big\downarrow & \lrcorner & \big\downarrow{\scriptstyle Fm} \\ A & \xrightarrow{\alpha} & FA \end{array} \tag{4.2}$$

cartesian subcoalgebras.

(2) A coalgebra is well-founded iff $\bigcirc$ has a unique pre-fixed point $\bigcirc m \leq m$. Indeed, since $\mathsf{Sub}(A)$ is a complete lattice, the least fixed point of a monotone map is its least pre-fixed point. Taylor's definition [24, Def. 6.3.2] uses that property: he calls a coalgebra well-founded iff $\bigcirc$ has no proper subobject as a pre-fixed point.

Example 4.5. (1) Consider a graph as a coalgebra $\alpha\colon A \to \mathscr{P}A$ for the power-set functor (see Example 2.1). A subcoalgebra is a subset $m\colon B \hookrightarrow A$ such that with every vertex v it contains all neighbors of v. The coalgebra structure $\beta\colon B \to \mathscr{P}B$ is then the domain-codomain restriction of α. To say that B is a cartesian subcoalgebra means that whenever a vertex of A has all neighbors in B, it also lies in B. It follows that (A, α) is well-founded iff it has no infinite directed path, see [24, Example 6.3.3].

(2) If μF exists, then as a coalgebra it is well-founded. Indeed, in every pull-back (4.2), since ι^{-1} (as α) is invertible, so is β. The unique algebra homomorph-ism from μF to the algebra $\beta^{-1}\colon FB \to B$ is clearly inverse to m.

(3) If a set functor F fulfils $F\emptyset = \emptyset$, then the only well-founded coalgebra is the empty one. Indeed, this follows from the fact that the empty coalgebra is a fixed point of $\bigcirc$. For example, a deterministic automaton over the input alphabet Σ, as a coalgebra for $FX = \{0,1\} \times X^{\Sigma}$, is well-founded iff it is empty.

(4) A non-deterministic automaton may be considered as a coalgebra for the set functor $FX = \{0,1\} \times (\mathscr{P}X)^{\Sigma}$. It is well-founded iff the state transition graph is well-founded (i.e. has no infinite path). This follows from Corollary 4.10 below.

(5) A linear weighted automaton, i.e. a coalgebra for $FX = K \times X^{\Sigma}$ on Vec_K, is well-founded iff every path in its state transition graph eventually leads to 0. This means that every path starting in a given state leads to the state 0 after finitely many steps (where it stays).

Notation 4.6. Given a set functor F, we define for every set X the map $\tau_X\colon FX \to \mathscr{P}X$ assigning to every element $x \in FX$ the intersection of all subsets $m\colon M \hookrightarrow X$ such that x lies in the image of Fm:

$$\tau_X(x) = \bigcap\{m \mid m\colon M \hookrightarrow X \text{ satisfies } x \in Fm[FM]\}. \tag{4.3}$$

Recall that a functor *preserves intersections* if it preserves (wide) pullbacks of families of monomorphisms.

Gumm [13, Thm. 7.3] observed that for a set functor preserving intersections, the maps $\tau_X\colon FX \to \mathscr{P}X$ in (4.3) form a "subnatural" transformation from F to the power-set functor $\mathscr{P}$. Subnaturality means that (although these maps do not form a natural transformation in general) for every monomorphism $i\colon X \to Y$ we have a commutative square:

$$\begin{array}{ccc} FX & \xrightarrow{\tau_X} & \mathscr{P}X \\ {\scriptstyle Fi}\downarrow & & \downarrow{\scriptstyle \mathscr{P}i} \\ FY & \xrightarrow{\tau_Y} & \mathscr{P}Y \end{array} \tag{4.4}$$

Remark 4.7. As shown in [13, Thm. 7.4] and [23, Prop. 7.5], a set functor F preserves intersections iff the squares in (4.4) above are pullbacks. Moreover, *loc. cit.* and [13, Thm. 8.1] prove that $\tau\colon F \to \mathscr{P}$ is a natural transformation, provided F preserves inverse images and intersections.

Definition 4.8. Let F be a set functor. For every coalgebra $\alpha\colon A \to FA$ its *canonical graph* is the following coalgebra for $\mathscr{P}$: $A \xrightarrow{\alpha} FA \xrightarrow{\tau_A} \mathscr{P}A$.

Thanks to the subnaturality of τ one obtains the following results.

Proposition 4.9. *For every set functor F preserving intersections, the next time operator of a coalgebra (A, α) coincides with that of its canonical graph.*

Corollary 4.10 [24, Rem. 6.3.4]. *A coalgebra for a set functor preserving intersections is well-founded iff its canonical graph is well-founded.*

Example 4.11. (1) For a (deterministic or non-deterministic) automaton, the canonical graph has an edge from s to t iff there is a transition from s to t for some input letter. Thus, we obtain the characterization of well-foundedness as stated in Example 4.5(3) and (4).

(2) Every polynomial functor $H_\Sigma\colon \mathsf{Set} \to \mathsf{Set}$ preserves intersections. Thus, a coalgebra (A, α) is well-founded if there are no infinite paths in its canonical graph. The canonical graph of A has an edge from a to b if $\alpha(a)$ is of the form $\sigma(c_1, \ldots, c_n)$ for some $\sigma \in \Sigma_n$ and if b is one of the c_i's.

(3) Thus, for the functor $FX = 1 + A \times X \times X$, the coalgebra (A^*, s) of Example 3.3(8) is easily seen to be well-founded via its canonical graph. Indeed, this graph has for every list w one outgoing edge to the list $w_{\leq a}$ and one to $w_{>a}$ for every $a \in A$. Hence, this is a well-founded graph.

Lemma 4.12. *The next time operator is monotone: if $m \leq n$, then $\bigcirc m \leq \bigcirc n$.*

Lemma 4.13. *Let $\alpha\colon A \to FA$ be a coalgebra and $m\colon B \rightarrowtail A$ a subobject.*

(1) There is a coalgebra structure $\beta\colon B \to FB$ for which m gives a subcoalgebra of (A, α) iff $m \leq \bigcirc m$.

(2) There is a coalgebra structure $\beta\colon B \to FB$ for which m gives a cartesian subcoalgebra of (A, α) iff $m = \bigcirc m$.

Lemma 4.14. *For every coalgebra homomorphism $f\colon (B, \beta) \to (A, \alpha)$ we have*

$$\bigcirc_\beta \cdot \overleftarrow{f} \leq \overleftarrow{f} \cdot \bigcirc_\alpha,$$

where $\bigcirc_\alpha$ and $\bigcirc_\beta$ denote the next time operators of the coalgebras (A, α) and (B, β), respectively, and $\leq$ is the pointwise order.

Corollary 4.15. *For every coalgebra homomorphism $f\colon (B, \beta) \to (A, \alpha)$ we have $\bigcirc_\beta \cdot \overleftarrow{f} = \overleftarrow{f} \cdot \bigcirc_\alpha$, provided that either*

(1) *f is a monomorphism in $\mathscr{A}$ and F preserves finite intersections, or*

(2) *F preserves inverse images.*

Definition 4.16 [4]. The *well-founded part* of a coalgebra is its largest well-founded subcoalgebra.

The well-founded part of a coalgebra always exists and is the coreflection in the category of well-founded coalgebras [6, Prop. 2.27]. We provide a new, shorter proof of this fact. The well-founded part is obtained by the following:

Construction 4.17 [6, Not. 2.22]. Let $\alpha\colon A \to FA$ be a coalgebra. We know that $\mathsf{Sub}(A)$ is a complete lattice and that the next time operator $\bigcirc$ is monotone (see Lemma 4.12). Hence, by the Knaster-Tarski fixed point theorem, $\bigcirc$ has a least fixed point, which we denote by $a^*\colon A^* \rightarrowtail A$.

By Lemma 4.13(2), we know that there is a coalgebra structure $\alpha^*\colon A^* \to FA^*$ so that $a^*\colon (A^*, \alpha^*) \rightarrowtail (A, \alpha)$ is the smallest cartesian subcoalgebra of (A, α).

Proposition 4.18. *For every coalgebra (A, α), the coalgebra (A^*, α^*) is well-founded.*

Proof. Let $m\colon (B, \beta) \rightarrowtail (A^*, \alpha^*)$ be a cartesian subcoalgebra. By Lemma 4.13, $a^* \cdot m\colon B \to A$ is a fixed point of $\bigcirc$. Since a^* is the least fixed point, we have $a^* \le a^* \cdot m$, i.e. $a^* = a^* \cdot m \cdot x$ for some $x\colon A^* \rightarrowtail B$. Since a^* is monic, we thus have $m \cdot x = id_{A^*}$. So m is a monomorphism and a split epimorphism, whence an isomorphism. □

Proposition 4.19. *The full subcategory of* $\mathsf{Coalg}\,F$ *given by well-founded coalgebras is coreflective. In fact, the well-founded coreflection of a coalgebra (A, α) is its well-founded part $a^*\colon (A^*, \alpha^*) \rightarrowtail (A, \alpha)$.*

Proof. We are to prove that for every coalgebra homomorphism $f\colon (B, \beta) \to (A, \alpha)$, where (B, β) is well-founded, there exists a coalgebra homomorphism $f^\sharp\colon (B, \beta) \to (A^*, \alpha^*)$ such that $a^* \cdot f^\sharp = f$. The uniqueness is easy.

For the existence of $f^\sharp$, we first observe that $\overleftarrow{f}(a^*)$ is a pre-fixed point of $\bigcirc_\beta$: indeed, using Lemma 4.14 we have $\bigcirc_\beta(\overleftarrow{f}(a^*)) \le \overleftarrow{f}(\bigcirc_\alpha(a^*)) = \overleftarrow{f}(a^*)$. By Remark 4.4(2), we therefore have $id_B = b^* \le \overleftarrow{f}(a^*)$ in $\mathsf{Sub}(B)$. Using the adjunction of Lemma 2.11, we have $\overrightarrow{f}(id_B) \le a^*$ in $\mathsf{Sub}(A)$. Now factorize f as $B \stackrel{e}{\twoheadrightarrow} C \stackrel{m}{\rightarrowtail} A$. We have $\overrightarrow{f}(id_B) = m$, and we then obtain $m = \overrightarrow{f}(id_B) \le a^*$, i.e. there exists a morphism $h\colon C \rightarrowtail A^*$ such that $a^* \cdot h = m$. Thus, $f^\sharp = h \cdot e\colon B \to A^*$ is a morphism satisfying $a^* \cdot f^\sharp = a^* \cdot h \cdot e = m \cdot e = f$. It follows that $f^\sharp$ is a coalgebra homomorphism from (B, β) to (A^*, α^*) since f and a^* are and F preserves monomorphisms. □

Construction 4.20 [6, Not. 2.22]. Let (A, α) be a coalgebra. We obtain a^*, the least fixed point of $\bigcirc$, as the join of the following transfinite chain of subobjects $a_i\colon A_i \rightarrowtail A$, $i \in \mathsf{Ord}$. First, put $a_0 = \bot_A$, the least subobject of A. Given $a_i\colon A_i \rightarrowtail A$, put $a_{i+1} = \bigcirc a_i\colon A_{i+1} = \bigcirc A_i \rightarrowtail A$. For every limit ordinal j, put $a_j = \bigvee_{i<j} a_i$. Since $\mathsf{Sub}(A)$ is a set, there exists an ordinal i such that $a_i = a^*\colon A^* \rightarrowtail A$.

Remark 4.21. Note that, whenever monomorphisms are smooth, we have $A_0 = 0$ and the above join a_j is obtained as the colimit of the chain of the subobject $a_i \colon A_i \rightarrowtail A$, $i < j$ (see Remark 2.12).

If F is a finitary functor on a locally finitely presentable category, then the least ordinal i with $a^* = a_i$ is at most ω, but in general one needs transfinite iteration to reach a fixed point.

Example 4.22. Let (A, α) be a graph regarded as a coalgebra for $\mathscr{P}$ (see Example 2.1). Then $A_0 = \emptyset$, A_1 is formed by all leaves; i.e. those nodes with no neighbors, A_2 by all leaves and all nodes such that every neighbor is a leaf, etc. We see that a node x lies in A_{i+1} iff every path starting in x has length at most i. Hence $A^* = A_\omega$ is the set of all nodes from which no infinite paths start.

We close with a general fact on well-founded parts of *fixed points* (i.e. (co)algebras whose structure is invertible). The following result generalizes [15, Cor. 3.4], and it also appeared before for functors preserving finite intersections [4, Theorem 8.16 and Remark 8.18]. Here we lift the latter assumption (see [5, Theorem 7.6] for the new proof):

Theorem 4.23. *Let $\mathscr{A}$ be a complete and well-powered category with smooth monomorphisms. For F preserving monomorphisms, the well-founded part of every fixed point is an initial algebra. In particular, the only well-founded fixed point is the initial algebra.*

Example 4.24. We illustrate that for a set functor F preserving monomorphisms, the well-founded part of the terminal coalgebra is the initial algebra. Consider $FX = A \times X + 1$. The terminal coalgebra is the set $A^\infty \cup A^*$ of finite and infinite sequences from the set A. The initial algebra is A^*. It is easy to check that A^* is the well-founded part of $A^\infty \cup A^*$.

5 The General Recursion Theorem and its Converse

The main consequence of well-foundedness is parametric recursivity. This is Taylor's General Recursion Theorem [24, Theorem 6.3.13]. Taylor assumed that F preserves inverse images. We present a new proof for which it is sufficient that F preserves monomorphisms, assuming those are smooth.

Theorem 5.1 (General Recursion Theorem). *Let $\mathscr{A}$ be a complete and wellpowered category with smooth monomorphisms. For $F \colon \mathscr{A} \to \mathscr{A}$ preserving monomorphisms, every well-founded coalgebra is parametrically recursive.*

Proof sketch. (1) Let (A, α) be well-founded. We first prove that it is recursive. We use the subobjects $a_i \colon A_i \rightarrowtail A$ of Construction 4.20[4], the corresponding

[4] One might object to this use of transfinite recursion, since Theorem 5.1 itself could be used as a justification for transfinite recursion. Let us emphasize that we are not presenting Theorem 5.1 as a foundational contribution. We are building on the classical theory of transfinite recursion.

morphisms $\alpha(a_i)\colon A_{i+1} = \bigcirc A_i \to FA_i$ (cf. Definition 4.3), and the recursive coalgebras $(F^i0, w_{i,i+1})$ of Example 3.3(6). We obtain a natural transformation h from the chain (A_i) in Construction 4.20 to the initial-algebra chain (F^i0) (see Remark 2.13) by transfinite recursion.

Now for every algebra $e\colon FX \to X$, we obtain a unique coalgebra-to-algebra morphism $f_i\colon F^i0 \to X$, i.e. we have that $f_i = e \cdot Ff_i \cdot w_{i,i+1}$. Since (A, α) is well-founded, we know that $\alpha = \alpha^* = \alpha(a_i)$ for some i. From this it is not difficult to prove that $f_i \cdot h_i$ is a coalgebra-to-algebra morphism from (A, α) to (X, e).

In order to prove uniqueness, we prove by transfinite induction that for any given coalgebra-to-algebra homomorphism $e^\dagger$, one has $e^\dagger \cdot a_j = f_j \cdot h_j \cdot a_j$ for every ordinal number j. Then for the above ordinal number i with $a_i = id_A$, we have $e^\dagger = f_i \cdot h_i$, as desired. This shows that (A, α) is recursive.

(2) We prove that (A, α) is parametrically recursive. Consider the coalgebra $\langle \alpha, id_A \rangle\colon A \to FA \times A$ for $F(-) \times A$. This functor preserves monomorphisms since F does and monomorphisms are closed under products. The next time operator $\bigcirc$ on $\mathsf{Sub}(A)$ is the same for both coalgebras since the square (4.1) is a pullback if and only if the square on the right below is one.

$$\begin{array}{ccc} \bigcirc S & \xrightarrow{\langle \alpha(m), \bigcirc m \rangle} & FS \times A \\ {\scriptstyle \bigcirc m}\downarrow & \lrcorner & \downarrow{\scriptstyle Fm \times A} \\ A & \xrightarrow{\langle \alpha, A \rangle} & FA \times A \end{array}$$

Since id_A is the unique fixed point of $\bigcirc$ w.r.t. F (see Definition 4.3), it is also the unique fixed point of $\bigcirc$ w.r.t. $F(-) \times A$. Thus, $(A, \langle \alpha, id_A \rangle)$ is a well-founded coalgebra for $F(-) \times A$. By the previous argument, this coalgebra is thus recursive for $F(-) \times A$; equivalently, (A, α) is parametrically recursive for F. □

Theorem 5.2. *For every endofunctor on* Set *or* Vec_K *(vector spaces and linear maps), every well-founded coalgebra is parametrically recursive.*

Proof sketch. For Set, we apply Theorem 5.1 to the Trnková hull $\bar{F}$ (see Proposition 2.3), noting that F and $\bar{F}$ have the same (non-empty) coalgebras. Moreover, one can show that every well-founded (or recursive) F-coalgebra is a well-founded (recursive, resp.) $\bar{F}$-coalgebra. For Vec_K, observe that monomorphisms split and are therefore preserved by every endofunctor F. □

Example 5.3. We saw in Example 4.11(3) that for $FX = 1 + A \times X \times X$ the coalgebra (A, s) from Example 3.3(8) is well-founded, and therefore it is (parametrically) recursive.

Example 5.4. Well-founded coalgebras need not be recursive when F does not preserve monomorphisms. We take $\mathscr{A}$ to be the category of *sets with a predicate*, i.e. pairs (X, A), where $A \subseteq X$. Morphisms $f\colon (X, A) \to (Y, B)$ satisfy $f[A] \subseteq B$. Denote by $\mathbb{1}$ the terminal object $(1, 1)$. We define an endofunctor F by $F(X, \emptyset) = (X + 1, \emptyset)$, and for $A \neq \emptyset$, $F(X, A) = \mathbb{1}$. For a morphism $f\colon (X, A) \to (Y, B)$, put $Ff = f + id$ if $A = \emptyset$; if $A \neq \emptyset$, then also $B \neq \emptyset$ and Ff is $id\colon \mathbb{1} \to \mathbb{1}$.

The terminal coalgebra is $id\colon \mathbb{1} \to \mathbb{1}$, and it is easy to see that it is well-founded. But it is not recursive: there are no coalgebra-to-algebra morphisms into an algebra of the form $F(X, \emptyset) \to (X, \emptyset)$.

We next prove a converse to Theorem 5.1: "recursive $\Longrightarrow$ well-founded". Related results appear in Taylor [23, 24], Adámek et al. [3] and Jeannin et al. [15].

Recall universally smooth monomorphisms from Definition 2.8(2). A *pre-fixed point* of F is a monic algebra $\alpha\colon FA \rightarrowtail A$.

Theorem 5.5. *Let $\mathscr{A}$ be a complete and wellpowered category with universally smooth monomorphisms, and suppose that $F\colon \mathscr{A} \to \mathscr{A}$ preserves inverse images and has a pre-fixed point. Then every recursive coalgebra is well-founded.*

Proof. (1) We first observe that an initial algebra exists. This follows from results by Trnková et al. [25] as we now briefly recall. Recall the initial-algebra chain from Remark 2.13. Let $\beta\colon FB \rightarrowtail B$ be a pre-fixed point. Then there is a unique cocone $\beta_i\colon F^i 0 \to B$ satisfying $\beta_{i+1} = \beta \cdot F\beta_i$. Moreover, each β_i is monomorphic. Since B has only a set of subobjects, there is some λ such that for every $i > \lambda$, all of the morphisms β_i represent the same subobject of B. Consequently, $w_{\lambda,\lambda+1}$ of Remark 2.13 is an isomorphism, due to $\beta_\lambda = \beta_{\lambda+1} \cdot w_{\lambda,\lambda+1}$. Then $\mu F = F^\lambda 0$ with the structure $\iota = w_{\lambda,\lambda+1}^{-1}\colon F(\mu F) \to \mu F$ is an initial algebra.
(2) Now suppose that (A, α) is a recursive coalgebra. Then there exists a unique coalgebra homomorphism $h\colon (A, \alpha) \to (\mu F, \iota^{-1})$. Let us abbreviate $w_{i\lambda}$ by $c_i\colon F^i 0 \rightarrowtail \mu F$, and recall the subobjects $a_i\colon A_i \rightarrowtail A$ from Construction 4.20. We will prove by transfinite induction that a_i is the inverse image of c_i under h; in symbols: $a_i = \overleftarrow{h}(c_i)$ for all ordinals i. Then it follows that a_λ is an isomorphism, since so is c_λ, whence (A, α) is well-founded.

In the base case $i = 0$ this is clear since $A_0 = W_0 = 0$ is a strict initial object.

For the isolated step we compute the pullback of $c_{i+1}\colon W_{i+1} \to \mu F$ along h using the following diagram:

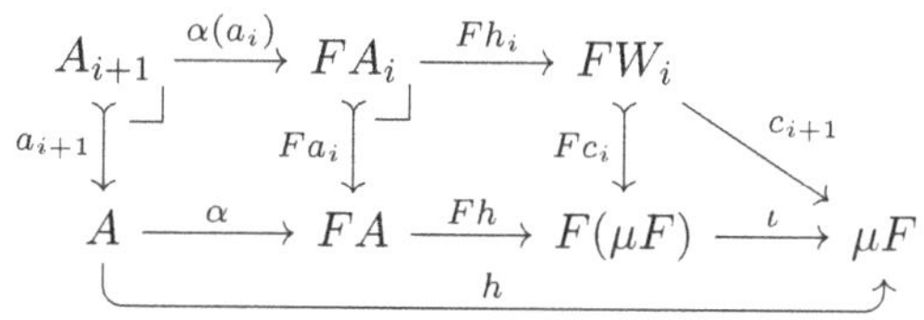

$$
\begin{array}{ccccccc}
A_{i+1} & \xrightarrow{\alpha(a_i)} & FA_i & \xrightarrow{Fh_i} & FW_i & & \\
{\scriptstyle a_{i+1}}\big\downarrow & \lrcorner & {\scriptstyle Fa_i}\big\downarrow & \lrcorner & {\scriptstyle Fc_i}\big\downarrow & \searrow^{c_{i+1}} & \\
A & \xrightarrow{\alpha} & FA & \xrightarrow{Fh} & F(\mu F) & \xrightarrow{\iota} & \mu F
\end{array}
$$
(with $h\colon A \to \mu F$ the bottom composite)

By the induction hypothesis and since F preserves inverse images, the middle square above is a pullback. Since the structure map ι of the initial algebra is an isomorphism, it follows that the middle square pasted with the right-hand triangle is also a pullback. Finally, the left-hand square is a pullback by the definition of a_{i+1}. Thus, the outside of the above diagram is a pullback, as required.

For a limit ordinal j, we know that $a_j = \bigvee_{i<j} a_i$ and similarly, $c_j = \bigvee_{i<j} c_i$ since $W_j = \mathsf{colim}_{i<j} W_j$ and monomorphisms are smooth (see Remark 2.12(2)). Using Remark 2.12(3) and the induction hypothesis we thus obtain $\overleftarrow{h}(c_j) = \overleftarrow{h}\left(\bigvee_{i<j} c_i\right) = \bigvee_{i<j} \overleftarrow{h}(c_i) = \bigvee_{i<j} a_i = a_j$. $\square$

Corollary 5.6. *Let $\mathscr{A}$ and F satisfy the assumptions of Theorem 5.5. Then the following properties of a coalgebra are equivalent:*

(1) *well-foundedness,*

(2) *parametric recursiveness,*

(3) *recursiveness,*

(4) *existence of a homomorphism into $(\mu F, \iota^{-1})$,*

(5) *existence of a homomorphism into a well-founded coalgebra.*

Proof sketch. We already know (1) $\Rightarrow$ (2) $\Rightarrow$ (3). Since F has an initial algebra (as proved in Theorem 5.5), the implication (3) $\Rightarrow$ (4) follows from Example 3.3(2). In Theorem 5.5 we also proved (4) $\Rightarrow$ (1). The implication (4) $\Rightarrow$ (5) follows from Example 4.5(2). Finally, it follows from [6, Remark 2.40] that $(\mu F, \iota^{-1})$ is a terminal well-founded coalgebra, whence (5) $\Rightarrow$ (4). □

Example 5.7. (1) The category of many-sorted sets satisfies the assumptions of Theorem 5.5, and polynomial endofunctors on that category preserve inverse images. Thus, we obtain Jeannin et al.'s result [15, Thm. 3.3] that (1)–(4) in Corollary 5.6 are equivalent as a special instance.

(2) The implication (4) $\Rightarrow$ (3) in Corollary 5.6 does not hold for vector spaces. In fact, for the identity functor on Vec_K we have $\mu Id = (0, id)$. Hence, every coalgebra has a homomorphism into μId. However, not every coalgebra is recursive, e.g. the coalgebra (K, id) admits many coalgebra-to-algebra morphisms to the algebra (K, id). Similarly, the implication (4) $\Rightarrow$ (1) does not hold.

We also wish to mention a result due to Taylor [23, Rem. 3.8]. It uses the concept of a *subobject classifier* originating in [17] and prominent in topos theory. This is an object Ω with a subobject $t\colon 1 \rightarrowtail \Omega$ such that for every subobject $b\colon B \rightarrowtail A$ there is a unique $\hat{b}\colon A \to \Omega$ such that b is the inverse image of t under $\hat{b}$. By definition, every elementary topos has a subobject classifier, in particular every category $\mathsf{Set}^{\mathscr{C}}$ with $\mathscr{C}$ small.

Our standing assumption that $\mathscr{A}$ is a complete and well-powered category is not needed for the next result: finite limits are sufficient.

Theorem 5.8 (Taylor [23]). *Let F be an endofunctor preserving inverse images on a finitely complete category with a subobject classifier. Then every recursive coalgebra is well-founded.*

Corollary 5.9. *For every set functor preserving inverse images, the following properties of a coalgebra are equivalent:*

$$\textit{well-foundedness} \iff \textit{parametric recursiveness} \iff \textit{recursiveness}.$$

Example 5.10. The hypothesis in Theorems 5.5 and 5.8 that the functor preserves inverse images cannot be lifted. In order to see this, we consider the functor $R\colon \mathsf{Set} \to \mathsf{Set}$ of Example 2.2(4). It preserves monomorphisms but not inverse images. The coalgebra $A = \{0, 1\}$ with the structure α constant to $(0, 1)$ is recursive: given an algebra $\beta\colon RB \to B$, the unique coalgebra-to-algebra homomorphism $h\colon \{0,1\} \to B$ is given by $h(0) = h(1) = \beta(d)$. But A is not well-founded: $\emptyset$ is a cartesian subcoalgebra.

Recall that an initial algebra $(\mu F, \iota)$ is also considered as a coalgebra $(\mu F, \iota^{-1})$. Taylor [23, Cor. 9.9] showed that, for functors preserving inverse images, the terminal well-founded coalgebra is the initial algebra. Surprisingly, this result is true for *all* set functors.

Theorem 5.11 [6, Thm. 2.46]. *For every set functor, a terminal well-founded coalgebra is precisely an initial algebra.*

Theorem 5.12. *For every functor on* Vec_K *preserving inverse images, the following properties of a coalgebra are equivalent:*

$$\textit{well-foundedness} \iff \textit{parametric recursiveness} \iff \textit{recursiveness}.$$

6 Closure Properties of Well-founded Coalgebras

In this section we will see that strong quotients and subcoalgebras (see Remark 2.7) of well-founded coalgebras are well-founded again. We mention the following corollary to Proposition 4.19. For endofunctors on sets preserving inverse images this was stated by Taylor [24, Exercise VI.16]:

Proposition 6.1. *The subcategory of* $\mathsf{Coalg}\,F$ *formed by all well-founded coalgebras is closed under strong quotients and coproducts in* $\mathsf{Coalg}\,F$.

This follows from a general result on coreflective subcategories [2, Thm. 16.8]: the category $\mathsf{Coalg}\,F$ has the factorization system of Proposition 2.6, and its full subcategory of well-founded coalgebras is coreflective with monomorphic coreflections (see Proposition 4.19). Consequently, it is closed under strong quotients and colimits.

We prove next that, for an endofunctor preserving finite intersections, well-founded coalgebras are closed under subcoalgebras provided that the complete lattice $\mathsf{Sub}(A)$ is a *frame*. This means that for every subobject $m\colon B \rightarrowtail A$ and every family m_i $(i \in I)$ of subobjects of A we have $m \wedge \bigvee_{i\in I} m_i = \bigvee_{i \in I}(m \wedge m_i)$. Equivalently, $\overleftarrow{m}\colon \mathsf{Sub}(A) \to \mathsf{Sub}(B)$ (see Notation 2.10) has a right adjoint $m_*\colon \mathsf{Sub}(B) \to \mathsf{Sub}(A)$.

This property holds for Set as well as for the categories of posets, graphs, topological spaces, and presheaf categories $\mathsf{Set}^{\mathscr{C}}$, $\mathscr{C}$ small. Moreover, it holds for every Grothendieck topos. The categories of complete partial orders and Vec_K do not satisfy this requirement.

Proposition 6.2. *Suppose that F preserves finite intersections, and let (A, α) be a well-founded coalgebra such that* $\mathsf{Sub}(A)$ *a frame. Then every subcoalgebra of (A, α) is well-founded.*

Proof. Let $m\colon (B,\beta) \rightarrowtail (A,\alpha)$ be a subcoalgebra. We will show that the only pre-fixed point of $\bigcirc_\beta$ is id_B (cf. Remark 4.4(2)). Suppose $s\colon S \rightarrowtail B$ fulfils $\bigcirc_\beta(s) \leq s$. Since F preserves finite intersections, we have $\overleftarrow{m} \cdot \bigcirc_\alpha = \bigcirc_\beta \cdot \overleftarrow{m}$ by Corollary 4.15(1). The counit of the above adjunction $\overleftarrow{m} \dashv m_*$ yields $\overleftarrow{m}(m_*(s)) \leq s$, so that we obtain $\overleftarrow{m}(\bigcirc_\alpha(m_*(s))) = \bigcirc_\beta(\overleftarrow{m}(m_*(s))) \leq \bigcirc_\beta(s) \leq s$. Using again the adjunction $\overleftarrow{m} \dashv m_*$, we have equivalently that $\bigcirc_\alpha(m_*(s)) \leq m_*(s)$; i.e. $m_*(s)$ is a pre-fixed point of $\bigcirc_\alpha$. Since (A,α) is well-founded, Corollary 4.15(1) implies that $m_*(s) = id_A$. Since $\overleftarrow{m}$ is also a right adjoint and therefore preserves the top element of $\mathsf{Sub}(B)$, we thus obtain $id_B = \overleftarrow{m}(id_A) = \overleftarrow{m}(m_*(s)) \leq s$. □

Remark 6.3. Given a set functor F preserving inverse images, a much better result was proved by Taylor [24, Corollary 6.3.6]: for every coalgebra homomorphism $f\colon (B,\beta) \to (A,\alpha)$ with (A,α) well-founded so is (B,β). In fact, our proof above is essentially Taylor's.

Corollary 6.4. *If a set functor preserves finite intersections, then subcoalgebras of well-founded coalgebras are well-founded.*

Trnková [26] proved that every set functor preserves all *nonempty* finite intersections. However, this does not suffice for Corollary 6.4:

Example 6.5. A well-founded coalgebra for a set functor can have non-well-founded subcoalgebras. Let $F\emptyset = 1$ and $FX = 1+1$ for all nonempty sets X, and let $Ff = \mathsf{inl}\colon 1 \to 1+1$ be the left-hand injection for all maps $f\colon \emptyset \to X$ with X nonempty. The coalgebra $\mathsf{inr}\colon 1 \to F1$ is not well-founded because its empty subcoalgebra is cartesian. However, this is a subcoalgebra of $id\colon 1+1 \to 1+1$ (via the embedding inr), and the latter is well-founded.

The fact that subcoalgebras of a well-founded coalgebra are well-founded does not necessarily need the assumption that $\mathsf{Sub}(A)$ is a frame. Instead, one may assume that the class of morphisms is universally smooth:

Theorem 6.6. *If $\mathscr{A}$ has universally smooth monomorphisms and F preserves finite intersections, every subcoalgebra of a well-founded coalgebra is well-founded.*

7 Conclusions

Well-founded coalgebras introduced by Taylor [24] have a compact definition based on an extension of Jacobs' 'next time' operator. Our main contribution is a new proof of Taylor's General Recursion Theorem that every well-founded coalgebra is recursive, generalizing this result to all endofunctors preserving monomorphisms on a complete and well-powered category with smooth monomorphisms. For functors preserving inverse images, we also have seen two variants of the converse implication "recursive ⇒ well-founded", under additional hypothesis: one due to Taylor for categories with a subobject classifier, and the second one provided that the category has universally smooth monomorphisms and the functor has a pre-fixed point. Various counterexamples demonstrate that all our hypotheses are necessary.

References

1. Adámek, J.: Free algebras and automata realizations in the language of categories. Comment. Math. Univ. Carolin. 15, 589–602 (1974)
2. Adámek, J., Herrlich, H., Strecker, G.E.: Abstract and Concrete Categories: The Joy of Cats. Dover Publications, 3rd edn. (2009)
3. Adámek, J., Lücke, D., Milius, S.: Recursive coalgebras of finitary functors. Theor. Inform. Appl. 41(4), 447–462 (2007)
4. Adámek, J., Milius, S., Moss, L.S.: Fixed points of functors. J. Log. Algebr. Methods Program. 95, 41–81 (2018)
5. Adámek, J., Milius, S., Moss, L.S.: On well-founded and recursive coalgebras (2019), full version; available online at http://arxiv.org/abs/1910.09401
6. Adámek, J., Milius, S., Moss, L.S., Sousa, L.: Well-pointed coalgebras. Log. Methods Comput. Sci. 9(2), 1–51 (2014)
7. Adámek, J., Milius, S., Sousa, L., Wißmann, T.: On finitary functors. Theor. Appl. Categ. 34, 1134–1164 (2019). available online at https://arxiv.org/abs/1902.05788
8. Adámek, J., Rosický, J.: Locally Presentable and Accessible Categories. Cambridge University Press (1994)
9. Borceux, F.: Handbook of Categorical Algebra: Volume 1, Basic Category Theory. Encyclopedia of Mathematics and its Applications, Cambridge University Press (1994)
10. Capretta, V., Uustalu, T., Vene, V.: Recursive coalgebras from comonads. Inform. and Comput. 204, 437–468 (2006)
11. Capretta, V., Uustalu, T., Vene, V.: Corecursive algebras: A study of general structured corecursion. In: Oliveira, M., Woodcock, J. (eds.) Formal Methods: Foundations and Applications, Lecture Notes in Computer Science, vol. 5902, pp. 84–100. Springer Berlin Heidelberg (2009)
12. Eppendahl, A.: Coalgebra-to-algebra morphisms. In: Proc. Category Theory and Computer Science (CTCS). Electron. Notes Theor. Comput. Sci., vol. 29, pp. 42–49 (1999)
13. Gumm, H.: From T-coalgebras to filter structures and transition systems. In: Fiadeiro, J.L., Harman, N., Roggenbach, M., Rutten, J. (eds.) Algebra and Coalgebra in Computer Science, Lecture Notes in Computer Science, vol. 3629, pp. 194–212. Springer Berlin Heidelberg (2005)
14. Jacobs, B.: The temporal logic of coalgebras via Galois algebras. Math. Structures Comput. Sci. 12(6), 875–903 (2002)
15. Jeannin, J.B., Kozen, D., Silva, A.: Well-founded coalgebras, revisited. Math. Structures Comput. Sci. 27, 1111–1131 (2017)
16. Kurz, A.: Logics for Coalgebras and Applications to Computer Science. Ph.D. thesis, Ludwig-Maximilians-Universität München (2000)
17. Lawvere, W.F.: Quantifiers and sheaves. Actes Congès Intern. Math. 1, 329–334 (1970)
18. Manna, Z., Pnüeli, A.: The Temporal Logic of Reactive and Concurrent Systems: Specification. Springer-Verlag (1992)
19. Meseguer, J., Goguen, J.A.: Initiality, induction, and computability. In: Algebraic methods in semantics (Fontainebleau, 1982), pp. 459–541. Cambridge Univ. Press, Cambridge (1985)
20. Milius, S.: Completely iterative algebras and completely iterative monads. Inform. and Comput. 196, 1–41 (2005)

21. Milius, S., Pattinson, D., Wißmann, T.: A new foundation for finitary corecursion and iterative algebras. Inform. and Comput. 217 (2020), available online at https://doi.org/10.1016/j.ic.2019.104456.
22. Osius, G.: Categorical set theory: a characterization of the category of sets. J. Pure Appl. Algebra 4(79–119) (1974)
23. Taylor, P.: Towards a unified treatment of induction I: the general recursion theorem (1995–6), preprint, available at www.paultaylor.eu/ordinals/#towuti
24. Taylor, P.: Practical Foundations of Mathematics. Cambridge University Press (1999)
25. Trnková, V., Adámek, J., Koubek, V., Reiterman, J.: Free algebras, input processes and free monads. Comment. Math. Univ. Carolin. 16, 339–351 (1975)
26. Trnková, V.: Some properties of set functors. Comment. Math. Univ. Carolin. 10, 323–352 (1969)
27. Trnková, V.: On a descriptive classification of set functors I. Comment. Math. Univ. Carolin. 12, 143–174 (1971)

Timed Negotiations*

S. Akshay[1(✉)], Blaise Genest[2], Loïc Hélouët[3], and Sharvik Mital[1]

[1] IIT Bombay, Mumbai, India {akshayss,sharky}@cse.iitb.ac.in
[2] Univ Rennes, CNRS, IRISA, Rennes, France blaise.genest@irisa.fr
[3] Univ Rennes, Inria, Rennes, France loic.helouet@inria.fr

Abstract. Negotiations were introduced in [6] as a model for concurrent systems with multiparty decisions. What is very appealing with negotiations is that it is one of the very few non-trivial concurrent models where several interesting problems, such as soundness, i.e. absence of deadlocks, can be solved in PTIME [3]. In this paper, we introduce the model of timed negotiations and consider the problem of computing the minimum and the maximum execution times of a negotiation. The latter can be solved using the algorithm of [10] computing costs in negotiations, but surprisingly minimum execution time cannot.
This paper proposes new algorithms to compute both minimum and maximum execution time, that work in much more general classes of negotiations than [10], that only considered sound and deterministic negotiations. Further, we uncover the precise complexities of these questions, ranging from PTIME to Δ_2^P-complete. In particular, we show that computing the minimum execution time is more complex than computing the maximum execution time in most classes of negotiations we consider.

1 Introduction

Distributed systems are notoriously difficult to analyze, mainly due to the explosion of the number of configurations that have to be considered to answer even simple questions. A challenging task is then to propose models on which analysis can be performed with tractable complexities, preferably within polynomial time. Free choice Petri nets are a classical model of distributed systems that allow for efficient verification, in particular when the nets are 1-safe [4, 5].

Recently, [6] introduced a new model called *negotiations* for workflows and business processes. A negotiation describes how processes interact in a distributed system: a subset of processes in a node of the system take a synchronous decisions among several *outcomes*. The effect of this outcome sends contributing processes to a new set of nodes. The execution of a negotiation ends when processes reach a *final configuration*. Negotiations can be deterministic (once an outcome is fixed, each process knows its unique successor node) or not.

Negotiations are an interesting model since several properties can be decided with a reasonable complexity. The question of *soundness*, i.e., deadlock-freedom:

* Supported by DST/CEFIPRA/INRIA Associated team EQuaVE and DST/SERB Matrices grant MTR/2018/000744.

J. Goubault-Larrecq and B. König (Eds.): FOSSACS 2020, LNCS 12077, pp. 37–56, 2020.
https://doi.org/10.1007/978-3-030-45231-5_3

whether from every reachable configuration one can reach a final configuration, is PSPACE-complete. However, for deterministic negotiations, it can be decided in PTIME [7]. The decision procedure uses reduction rules. Reduction techniques were originally proposed for Petri nets [2,8,11,16]. The main idea is to define transformations rules that produce a model of smaller size w.r.t. the original model, while preserving the property under analysis. In the context of negotiations, [7,3] proposed a sound and complete set of soundness-preserving reduction rules and algorithms to apply these rules efficiently. The question of soundness for deterministic negotiations was revisited in [9] and showed NLOGSPACE-complete using anti patterns instead of reduction rules. Further, they show that the PTIME result holds even when relaxing determinism [9]. Negotiation games have also been considered to decide whether one particular process can force termination of a negotiation. While this question is EXPTIME-complete in general, for sound and deterministic negotiations, it becomes PTIME [12].

While it is natural to consider cost or time in negotiations (e.g. think of the Brexit negotiation where time is of the essence, and which we model as running example in this paper), the original model of negotiations proposed by [6] is only qualitative. Recently, [10] has proposed a framework to associate costs to the executions of negotiations, and adapt a static analysis technique based on reduction rules to compute end-to-end cost functions that are not sensitive to scheduling of concurrent nodes. For sound *and* deterministic negotiations, the end-to-end cost can be computed in $O(n.(C+n))$, where n is the size of the negotiation and C the time needed to compute the cost of an execution. Requiring soundness or determinism seems perfectly reasonable, but asking sound *and* deterministic negotiations is too restrictive: it prevents a process from waiting for decisions of other processes to know how to proceed.

In this paper, we revisit time in negotiations. We attach time intervals to outcomes of nodes. We want to compute maximal and minimal executions times, for negotiations that are not necessarily sound and deterministic. Since we are interested in minimal and maximal execution time, cycles in negotiations can be either bypassed or lead to infinite maximal time. Hence, we restrict this study to acyclic negotiations. Notice that time can be modeled as a cost, following [10], and the maximal execution time of a sound and deterministic negotiation can be computed in PTIME using the algorithm from [10]. Surprisingly however, we give an example (Example 3) for which the minimal execution time cannot be computed in PTIME by this algorithm.

The first contribution of the paper shows that reachability (whether at least one run of a negotiation terminates) is NP-complete, already for (untimed) deterministic acyclic negotiations. This implies that computing minimal or maximal execution time for deterministic (but unsound) acyclic negotiations cannot be done in PTIME (unless NP=PTIME). We characterize precisely the complexities of different decision variants (threshold, equality, etc.), with complexities ranging from (co-)NP-complete to Δ_2^P.

We thus turn to negotiations that are sound but not necessarily deterministic. Our second contribution is a new algorithm, not based on reduction rules, to compute the maximal execution time in PTIME for sound negotiations. It is based on computing the maximal execution time of critical paths in the negotiations. However, we show that *minimal* execution time cannot be computed in PTIME for sound negotiations (unless NP=PTIME): deciding whether the minimal execution time is lower than T is NP-complete, even for T given in unary, using a reduction from a Bin packing problem. This shows that minimal execution time is harder to compute than maximal execution time.

Our third contribution consists in defining a class in which the minimal execution time can be computed in (pseudo) PTIME. To do so, we define the class of k-layered negotiations, for k fixed, that is negotiations where nodes can be organized into layers of at most k nodes at the same depth. These negotiations can be executed without remembering more than k nodes at a time. In this case, we show that computing the maximal execution time is PTIME, even if the negotiation is neither deterministic nor sound. The algorithm, not based on reduction rules, uses the k-layer restriction in order to navigate in the negotiation while considering only a polynomial number of configurations. For minimal execution time, we provide a pseudo PTIME algorithm, that is PTIME if constants are given in unary. Finally, we show that the size of constants do matter: deciding whether the minimal execution time of a k-layered negotiation is less than T is NP-complete, when T is given in binary. We show this by reducing from a Knapsack problem, yet again emphasizing that the minimal execution time of a negotiation is harder to compute than its maximal execution time.

This paper is organized as follows. Section 2 introduces the key ingredients of negotiations, determinism and soundness, known results in the untimed setting, and provides our running example modeling the Brexit negotiation. Section 3 introduces time in negotiations, gives a semantics to this new model, and formalizes several decision problems on maximal and minimal durations of runs in timed negotiations. We recall the main results of the paper in Section 4. Then, Section 5 considers timed execution problems for deterministic negotiations, Section 6 for sound negotiations, and section 7 for layered negotiations. Proof details for the last three sections are given in an extended version of this paper [1].

2 Negotiations: Definitions and Brexit example

In this section, we recall the definition of negotiations, of some subclasses (acyclic and deterministic), as well as important problems (soundness and reachability).

Definition 1 (Negotiation [6,10]). *A* negotiation *over a finite set of processes P is a tuple $\mathcal{N} = (N, n_0, n_f, \mathcal{X})$, where:*

- *N is a finite set of nodes. Each node is a pair $n = (P_n, R_n)$ where $P_n \subseteq P$ is a non empty set of processes participating in node n, and R_n is a finite set of outcomes of node n (also called results), with $R_{n_f} = \{r_f\}$. We denote by R the union of all outcomes of nodes in N.*
- *n_0 is the first node of the negotiation and n_f is the final node. Every process in P participates in both n_0 and n_f.*

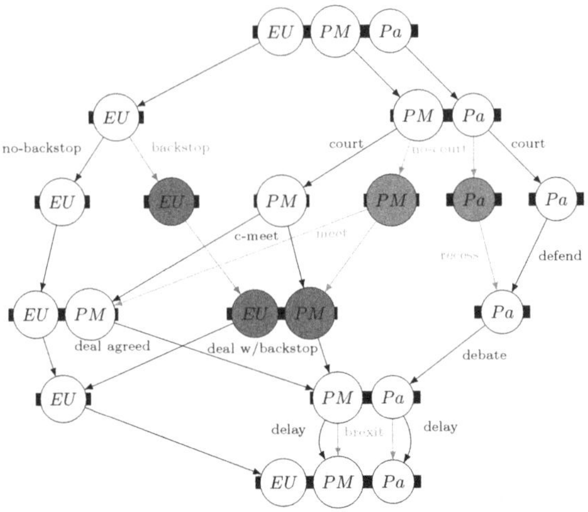

Fig. 1. A (sound but non-deterministic) negotiation modeling Brexit.

– *For all* $n \in N$, $\mathcal{X}_n : P_n \times R_n \to 2^N$ *is a map defining the transition relation from node* n, *with* $\mathcal{X}_n(p, r) = \emptyset$ *iff* $n = n_f, r = r_f$. *We denote* $\mathcal{X} : N \times P \times R \to 2^N$ *the partial map defined on* $\bigcup_{n \in N}(\{n\} \times P_n \times R_n)$, *with* $\mathcal{X}(n, p, a) = \mathcal{X}_n(p, a)$ *for all* p, a.

Intuitively, at a node $n = (P_n, R_n)$ in a negotiation, all processes of P_n have to agree on a common outcome r chosen from R_n. Once this outcome r is chosen, every process $p \in P_n$ is ready to move to any node prescribed by $\mathcal{X}(n, p, r)$. A new node m can only start when all processes of P_m are ready to move to m.

Example 1. We illustrate negotiations by considering a simplified model of the Brexit negotiation, see Figure 1. There are 3 processes, $P = \{EU, PM, Pa\}$. At first EU decides whether or not to enforce a backstop in any deal (outcome backstop) or not (outcome no-backstop). In the meantime, PM decides to proroge Pa, and Pa can choose or not to appeal to court (outcome court/no court). If it goes to court, then PM and Pa will take some time in court (c-meet, defend), before PM can meet EU to agree on a deal. Otherwise, Pa goes to recess, and PM can meet EU directly. Once EU and PM agreed on a deal, PM tries to convince Pa to vote the deal. The final outcome is whether the deal is voted, or whether Brexit is delayed.

Definition 2 (Deterministic negotiations). *A process* $p \in P$ *is* deterministic *iff, for every* $n \in N$ *and every outcome* r *of* n, $\mathcal{X}(n, p, r)$ *is a singleton. A negotiation is* deterministic *iff all its processes are deterministic. It is* weakly non-deterministic *[9] (called weakly deterministic in [3]) iff, for every node* n, *one of the processes in* P_n *is deterministic. Last, it is* very weakly non-deterministic *[9] (called weakly deterministic in [6]) iff, for every* n, *every* $p \in P_n$ *and every outcome* r *of* n, *there exists a deterministic process* q *such that* $q \in P_{n'}$ *for every* $n' \in \mathcal{X}(n, p, r)$.

In deterministic negotiations, once an outcome is chosen, each process knows the next node it will be involved in. In (very-)weakly non-deterministic negotiations, the next node might depend upon the outcome chosen in other nodes by other processes. However, once the outcomes have been chosen for all current nodes, there is only one next node possible for each process. Observe that the class of deterministic negotiations is isomorphic to the class of free choice workflow nets [10]. In Example 1, the Brexit negotiation is non-deterministic, because process PM is non-deterministic. Indeed, consider outcomes *c-meet*: it allows two nodes, according to whether the backstop is enforced or not, which is a decision taken by process EU.

Semantics: A *configuration* [3] of a negotiation is a mapping $M : P \to 2^N$. Intuitively, it tells for each process p the set $M(p)$ of nodes p is ready to engage in. The semantics of a negotiation is defined in terms of moves from a configuration to the next one. The *initial* M_0 and *final* M_f configurations, are given by $M_0(p) = \{n_0\}$ and $M_f(p) = \emptyset$ respectively for every process $p \in P$. A configuration M *enables* node n if $n \in M(p)$ for every $p \in P_n$. When n is enabled, a decision at node n can occur, and the participants at this node choose an outcome $r \in R_n$. The occurrence of (n, r) produces the configuration M' given by $M'(p) = \mathcal{X}(n, p, r)$ for every $p \in P_n$ and $M'(p) = M(p)$ for remaining processes in $P \setminus P_n$. Moving from M to M' after choosing (n, r) is called a *step*, denoted $M \xrightarrow{n,r} M'$. A *run* of $\mathcal{N}$ is a sequence $(n_1, r_1), (n_2, r_2)...(n_k, r_k)$ such that there is a sequence of configurations $M_0, M_1, \ldots, M_k$ and every (n_i, r_i) is a step between M_{i-1} and M_i. A run starting from the initial configuration and ending in the final configuration is called a *final run*. By definition, its last step is (n_f, r_f).

An important class of negotiations in the context of timed negotiations is acyclic negotiations, where infinite sequence of steps is impossible:

Definition 3 (Acyclic negotiations). *The* graph *of a negotiation* $\mathcal{N}$ *is the labeled graph* $G_{\mathcal{N}} = (V, E)$ *where* $V = N$, *and* $E = \{((n, (p, r), n') \mid n' \in \mathcal{X}(n, p, r)\}$, *with pairs of the form* (p, r) *being the labels. A negotiation is* acyclic *iff its graph is acyclic. We denote by* $Paths(G_{\mathcal{N}})$ *the set of paths in the graph of a negotiation. These paths are of form* $\pi = (n_0, (p_0, r_0), n_1) \ldots (n_{k-1}, (p_k, r_k), n_k)$.

The Brexit negotiation of Fig.1 is an example of acyclic negotiation. Despite their apparent simplicity, negotiations may express involved behaviors as shown with the Brexit example. Indeed two important questions in this setting are whether there is some way to reach a final node in the negotiation from (i) the initial node and (ii) any reachable node in the negotiation.

Definition 4 (Soundness and Reachability).

1. *A negotiation is* sound *iff every run from the initial configuration can be extended to a final run. The problem of soundness is to check if a given negotiation is sound.*
2. *The problem of reachability asks if a given negotiation has a final run.*

Notice that the Brexit negotiation of Fig.1 is sound (but not deterministic). It seems hard to preserve the important features of this negotiation while being both sound *and* deterministic. The problem of soundness has received considerable attention. We summarize the results about soudness in the next theorem:

Theorem 1. *Determining whether a negotiation is sound is PSPACE-Complete. For (very-)weakly non-deterministic negotiations, it is co-NP-complete [9]. For acyclic negotiations, it is in DP and co-NP-Hard [6]. Determining whether an acyclic weakly non-deterministic negotiation is sound is in PTIME [3, 9]. Finally, deciding soundness for deterministic negotiations is NLOGSPACE-complete [9*

Checking reachability is NP-complete, even for deterministic acyclic negotiations (surprisingly, we did not find this result stated before in the literature):

Proposition 1. *Reachability is NP-complete for acyclic negotiations, even if the negotiation is deterministic.*

Proof (sketch). One can guess a run of size $\leq |\mathcal{N}|$ in polynomial time, and verify if it reaches n_f, which gives the inclusion in NP. The hardness part comes from a reduction from 3-CNF-SAT that can be found in the proof of Theorem 3. □

k-Layered Acyclic Negotiations

We introduce a new class of negotiations which has good algorithmic properties, namely k-layered acyclic negotiations, for k fixed. Roughly speaking, nodes of a k-layered acyclic negotiations can be arranged in layers, and these layers contain at most k nodes. Before giving a formal definition, we need to define the depth of nodes in $\mathcal{N}$.

First, a *path* in a negotiation is a sequence of nodes $n_0 \dots n_\ell$ such that for all $i \in \{1, \dots, \ell - 1\}$, there exists p_i, r_i with $n_{i+1} \in \mathcal{X}(n_i, p_i, r_i)$. The *length* of a path $n_0, \dots, n_\ell$ is ℓ. The *depth* $\text{depth}(n)$ of a node n is the maximal length of a path from n_0 to n (recall that $\mathcal{N}$ is acyclic, so this number is always finite).

Definition 5. *An acyclic negotiation is* layered *if for all node n, every path reaching n has length depth(n). An acyclic negotiation is k-*layered *if it is layered, and for all $\ell \in \mathbb{N}$, there are at most k nodes at depth ℓ.*

The Brexit example of Fig. 1 is 6-layered. Notice that a layered negotiation is necessarily k-layered for some $k \leq |\mathcal{N}| - 2$. Note also that we can always transform an acyclic negotiation $\mathcal{N}$ into a layered acyclic negotiation $\mathcal{N}'$, by adding dummy nodes: for every node $m \in \mathcal{X}(n, p, r)$ with $\text{depth}(m) > \text{depth}(n) + 1$, we can add several nodes $n_1, \dots n_\ell$ with $\ell = \text{depth}(m) - (\text{depth}(n) + 1)$, and processes $P_{n_i} = \{p\}$. We compute a new relation $\mathcal{X}'$ such that $\mathcal{X}'(n, p, r) = \{n_1\}$, $\mathcal{X}(n_\ell, p, r) = \{m\}$ and for every $i \in 1..\ell - 1$, $\mathcal{X}(n_i, p, r) = n_{i+1}$. This transformation is polynomial: the resulting negotiation is of size up to $|\mathcal{N}| \times |\mathcal{X}| \times |P|$. The proof of the following Theorem can be found in [1].

Theorem 2. *Let $k \in \mathbb{N}^+$. Checking reachability or soundness for a k-layered acyclic negotiation $\mathcal{N}$ can be done in PTIME.*

3 Timed Negotiations

In many negotiations, time is an important feature to take into account. For instance, in the Brexit example, with an initial node starting at the begining of September 2019, there are 9 weeks to pass a deal till the 31st October deadline.

We extend negotiations by introducing timing constraints on outcomes of nodes, inspired by timed Petri nets [14] and by the notion of negotiations with costs [10]. We use time intervals to specify lower and upper bounds for the duration of negotiations. More precisely, we attach time intervals to pairs (n, r) where n is a node and r an outcome. In the rest of the paper, we denote by $\mathcal{I}$ the set of intervals with endpoints that are non-negative integers or ∞. For convenience we only use closed intervals in this paper (except for ∞), but the results we show can also be extended to open intervals with some notational overhead. Intuitively, outcome r can be taken at a node n with associated time interval $[a, b]$ only after a time units have elapsed from the time all processes contributing to n are ready to engage in n, and at most b time units later.

Definition 6. *A* timed negotiation *is a pair $(\mathcal{N}, \gamma)$ where $\mathcal{N}$ is a negotiation, and $\gamma : N \times R \to \mathcal{I}$ associates an interval to each pair (n, r) of node and outcome such that $r \in R_n$. For a given node n and outcome r, we denote by $\gamma^-(n, r)$ (resp. $\gamma^+(n, r)$) the lower bound (resp. the upper bound) of $\gamma(n, r)$.*

Example 2. In the Brexit example, we define the following timed constraints γ. We only specify the outcome names, as the timing only depends upon them. Backstop and no-backstop both take between 1 and 2 weeks: $\gamma(\text{backstop}) = \gamma(\text{no-backstop}) = [1, 2]$. In case of no-court, recess takes 5 weeks $\gamma(\text{recess}) = [5, 5]$, and PM can meet EU immediatly $\gamma(\text{meet}) = [0, 0]$. In case of court action, PM needs to spend 2 weeks in court $\gamma(\text{c-meet}) = [2, 2]$, and depending on the court delay and decision, Pa needs between 3 (court overules recess) to 5 (court confirms recess) weeks, $\gamma(\text{defend}) = [3, 5]$. Agreeing on a deal can take anywhere from 2 weeks to 2 years (104 weeks): $\gamma(\text{deal agreed}) = [2, 104]$—some would say infinite time is even possible! It needs more time with the backstop, $\gamma(\text{deal w/backstop}) = [5, 104]$. All other outcomes are assumed to be immediate, i.e., associated with $[0, 0]$.

Semantics: A *timed valuation* is a map $\mu : P \to \mathbb{R}^{\geq 0}$ that associates a non-negative real value to every process. A *timed configuration* is a pair (M, μ) where M is a configuration and μ a timed valuation. There is a *timed step* from (M, μ) to (M', μ'), denoted $(M, \mu) \xrightarrow{(n,r)} (M', \mu')$, if (i) $M \xrightarrow{(n,r)} M'$, (ii) $p \notin P_n$ implies $\mu'(p) = \mu(p)$ (iii) $\exists d \in \gamma(n, r)$ such that $\forall p \in P_n$, we have $\mu'(p) = \max_{p' \in P_n} \mu(p') + d$ (d is the duration of node n).

Intuitively a timed step $(M, \mu) \xrightarrow{(n,r)} (M', \mu')$ depicts a decision taken at node n, and how long each process of P_n waited in that node before taking decision (n, r). The last process engaged in n must wait for a duration contained in $\gamma(n, r)$. However, other processes may spend a time greater than $\gamma^+(n, r)$.

A *timed run* is a sequence of steps $\rho = (M_0, \mu_0) \xrightarrow{e_1} (M_1, \mu_1) \ldots (M_k, \mu_k)$ where M_0 is the initial configuration, $\mu_0(p) = 0$ for every $p \in P$, and each $(M_i, \mu_i) \xrightarrow{e_i} (M_{i+1}, \mu_{i+1})$ is a timed step. It is *final* if $M_k = M_f$. Its *execution time* $\delta(\rho)$ is defined as $\delta(\rho) = \max_{p \in P} \mu_k(p)$.

Notice that we only attached timing to processes, not to individual steps. With our definition of runs, timing on steps may not be monotonous (i.e., non-decreasing) along the run, while timing on processes is. Viewed by the lens of concurrent systems, the timing is monotonous on the partial orders of the system rather than the linearization. It is not hard to restrict paths, if necessary, to have a monotonous timing on steps as well. In this paper, we are only interested in execution time, which does not depend on the linearization considered.

Given a timed negotiation $\mathcal{N}$, we can now define the minimum and maximum execution time, which correspond to optimistic or pessimistic views:

Definition 7. *Let $\mathcal{N}$ be a timed negotiation. Its* minimum execution time, *denoted mintime($\mathcal{N}$) is the minimal $\delta(\rho)$ over all final timed run ρ of $\mathcal{N}$. We define the* maximal execution time *maxtime($\mathcal{N}$) of $\mathcal{N}$ similarly.*

Given $T \in \mathbb{N}$, the main problems we consider in this paper are the following:

- The mintime problem, i.e., do we have $mintime(\mathcal{N}) \leq T$?.
  In other words, does there exist a final timed run ρ with $\delta(\rho) \leq T$?
- The maxtime problem, i.e., do we have $maxtime(\mathcal{N}) \leq T$?.
  In other words, does $\delta(\rho) \leq T$ for every final timed run ρ?

These questions have a practical interest : in the Brexit example, the question "is there a way to have a vote on a deal within 9 weeks ?" is indeed a minimum execution time problem. We also address the equality variant of these decision problems, i.e., $mintime(\mathcal{N}) = T$: is there a final run of $\mathcal{N}$ that terminates in exactly T time units and no other final run takes less than T time units? Similarly for $maxtime(\mathcal{N}) = T$.

Example 3. We use Fig. 1 to show that it is not easy to compute the minimal execution time, and in particular one cannot use the algorithm from [10] to compute it. Consider the node n with $P_n = \{PM, Pa\}$ and $R_n = \{\text{court}, \text{no\_court}\}$. If the outcome is court, then PM needs 2 weeks before (s)he can talk to EU and Pa needs at least 3 weeks before he can debate. However, if the outcome is no_court, then PM need not wait before (s)he can talk to EU, but Pa wastes 5 weeks in recess. This means that one needs to remember different alternatives which could be faster in the end, depending on the future. On the other hand, the algorithm from [10] attaches one minimal time to process Pa, and one minimal time to process PM. No matter the choices (0 or 2 for PM and 3 or 5 for Pa), there will be futures in which the chosen number will over or underapproximate the real minimal execution time (this choice is not explicit in [10])[4].

[4] the authors of [10] acknowledged the issue with their algorithm for mintime.

For maximum execution time, it is not an issue to attach to each node a unique maximal execution time. The reason for the asymmetry between minimal and maximal execution times of a negotiation is that the execution time of a path is $\max_{p \in P} \mu_k(p)$, for μ_k the last timed valuation, which breaks the symmetry between min and max.

4 High level view of the main results

In this section, we give a high-level description of our main results. Formal statements can be found in the sections where they are proved. We gather in Fig. 2 the precise complexities for the minimal and the maximal execution time problems for 3 classes of negotiations that we describe in the following. Since we are interested in minimum and maximum execution time, cycles in negotiations can be either bypassed or lead to infinite maximal time. Hence, while we define timed negotiations in general, we always restrict to acyclic negotiations (such as Brexit) while stating and proving results.

In [10], a PTIME algorithm is given to compute different costs for negotiations that are both sound *and* deterministic. One limitation of this result is that it cannot compute the minimum execution time, as explained in Example 3. A second limitation is that the class of sound and deterministic negotiations is quite restrictive: it cannot model situations where the next node a process participates in depends on the outcome from another process, as in the Brexit example. We thus consider classes where one of these restrictions is dropped.

We first consider (Section 5) negotiations that are deterministic, but without the soundness restriction. We show that for this class, no timed problem we consider can be solved in PTIME (unless NP=PTIME). Further, we show that the equality problems ($maxtime/mintime(\mathcal{N}) = T$), are complete for the complexity class DP, i.e., at the second level of the Boolean Hierarchy [15].

We then consider (Section 6) the class of negotiations that are sound, but not necessarily deterministic. We show that maximum execution time can be solved in PTIME, and propose a new algorithm. However, the minimum execution time cannot be computed in PTIME (unless NP=PTIME). Again for the mintime equality problem we have a matching DP-completeness result.

	Deterministic	Sound	k-layered
Max $\leq T$	co-NP-complete (Thm. 3)	PTIME (Prop. 3)	PTIME (Thm. 6)
Max $= T$	DP-complete (Prop. 2)		
Min $\leq T$	NP-complete (Thm. 3)	NP-complete* (Thm. 5)	pseudo-PTIME (Thm. 8) NP-complete** (Thm. 7)
Min $= T$	DP-complete (Prop. 2)	DP-complete* (Prop. 4)	pseudo-PTIME (Thm. 8)

Fig. 2. Results for acyclic timed negotiations. *DP* refers to the complexity class, Difference Polynomial time [15], the second level of the Boolean Hierarchy.
* hardness holds even for very weakly non-deterministic negotiations, and T in unary.
** hardness holds even for sound and very weakly non-deterministic negotiations.

Finally, in order to obtain a polytime algorithm to compute the minimum execution time, we consider the class of k-layered negotiations (see Section 7): Given $k \in \mathbb{N}$, we can show that $maxtime(\mathcal{N})$ can be computed in PTIME for k-layered negotiations. We also show that while the $mintime(\mathcal{N}) \leq T$? problem is weakly NP-complete for k-layered negotiations, we can compute $mintime(\mathcal{N})$ in pseudo-PTIME, i.e. in PTIME if constants are given in unary.

5 Deterministic Negotiations

We start by considering the class of deterministic acyclic negotiations. We show that both maximal and minimal execution times cannot be computed in PTIME (unless NP=PTIME), as the threshold problems are (co-)NP-complete.

Theorem 3. *The $mintime(\mathcal{N}) \leq T$ decision problem is NP complete, and the $maxtime(\mathcal{N}) \leq T$ decision problem is co-NP-complete for acyclic deterministic timed negotiations.*

Proof. For $mintime(\mathcal{N}) \leq T$, containment in NP is easy: we just need to guess a run ρ (of polynomial size as $\mathcal{N}$ is acyclic), consider the associated timed run ρ^- where all decisions are taken at their earliest possible dates, and check whether $\delta(\rho^-) \leq T$, which can be done in time $O(|\mathcal{N}|+\log T)$.

For the hardness, we give the proof in two steps. First, we start with a proof of Proposition 1 that reachability problem is NP-hard using reduction of 3-CNF SAT, i.e., given a formula ϕ, we build a deterministic negotiation $\mathcal{N}_\phi$ s.t. ϕ is satisfiable iff $\mathcal{N}_\phi$ has a final run. In a second step, we introduce timings on this negotiation and show that $mintime(\mathcal{N}_\phi) \leq T$ iff ϕ is satisfiable.

Step 1: Reducing 3-CNF-SAT to Reachability problem.

Given a Boolean formula ϕ with variables $v_i, 1 \leq i \leq n$ and clauses $c_j, 1 \leq j \leq m$, for each variable v_i we define the sets of clauses $S_{i,\mathtt{t}} = \{c_j \mid v_i \text{ is present in } c_j\}$ and $S_{i,\mathtt{f}} = \{c_j \mid \neg v_i \text{ is present in } c_j\}$. Clauses in $S_{i,\mathtt{t}}$ and $S_{i,\mathtt{f}}$ are naturally ordered: $c_i < c_j$ iff $i < j$. We denote these elements $S_{i,\mathtt{t}}(1) < S_{i,\mathtt{t}}(2) < \dots$. Similarly for set $S_{i,\mathtt{f}}$.

Now, we construct a negotiation $\mathcal{N}_\phi$ (as depicted in Figure 3) with a process V_i for each variable v_i and a process C_j for each clause c_j:

- Initial node n_0 has a single outcome r taking each process C_j to node $Lone_{c_j}$, and each process V_i to node $Lone_{v_i}$.
- $Lone_{c_j}$ has three outcomes: if literal $v_i \in c_j$, then t_i is an outcome, taking C_j to $Pair_{c_j,v_i}$, and if literal $\neg v_i \in c_j$, then f_i is an outcome, taking C_j to $Pair_{c_j,\neg v_i}$.
- The outcomes of $Lone_{v_i}$ are `true` and `false`. Outcome `true` brings V_i to node $Tlone_{v_i,1}$ and outcome `false` brings V_i to node $Flone_{v_i,1}$.
- We have a node $Tlone_{v_i,j}$ for each $j \leq |S_{i,\mathtt{t}}|$ and $Flone_{v_i,j}$ for each $j \leq |S_{i,\mathtt{f}}|$, with V_i as only process. Let $c_r = S_{i,\mathtt{t}}(j)$. Node $Tlone_{v_i,j}$ has two outcomes $vton$ bringing V_i to $Tlone_{v_i,j+1}$ (or n_f if $j = |S_{i,\mathtt{t}}|$), and $vtoc_{i,r}$ bringing V_i to $Pair_{c_r,v_i}$. The two outcomes from $Flone_{v_i,j}$ are similar.

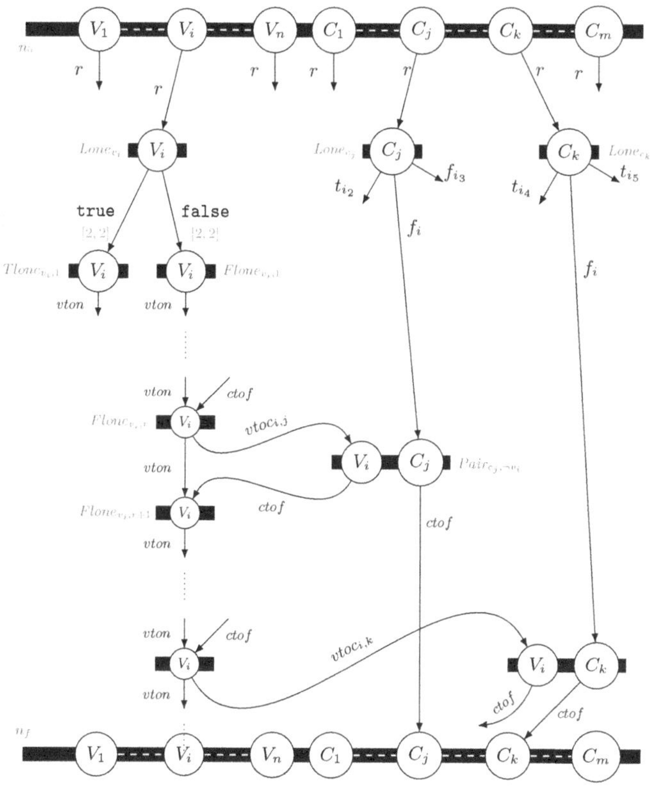

Fig. 3. A part of $\mathcal{N}_\phi$ where clause c_j is $(i_2 \vee \neg i \vee \neg i_3)$ and clause c_k is $(i_4 \vee \neg i \vee i_5)$. Timing is $[0, 0]$ whereever not mentioned

- Node $Pair_{c_r,v_i}$ has V_i and C_r as its processes and one outcome $ctof$ which takes process C_r to final node n_f and process V_i to $Tlone_{v_i,j+1}$ (with $c_r = S_{i,\mathtt{t}}(j)$), or to n_f if $j = |S_{i,\mathtt{t}}|$. Node $Pair_{c_r,\neg v_i}$ is defined in the same way from $Flone_{v_i,j}$.

With this we claim that $\mathcal{N}_\phi$ has a final run iff ϕ is satisfiable which completes the first step of the proof. We give a formal proof of this claim in Appendix A of [1]. Observe that the negotiation $\mathcal{N}_\phi$ constructed is deterministic and acyclic (but it is not sound).

Step 2: Before we introduce timing on $\mathcal{N}_\phi$, we introduce a new outcome r' at n_0 which takes all processes to n_f. Now, the timing function γ associated with $\mathcal{N}_\phi$ is: $\gamma(n_0, r) = [2, 2]$ and $\gamma(n_0, r') = [3, 3]$ and $\gamma(n, r) = [0, 0]$, for all node $n \neq n_0$ and all $r \in R_n$. Then, $mintime(\mathcal{N}_\phi) \leq 2$ iff ϕ has a satisfiable assignment: if $mintime(\mathcal{N}_\phi) \leq 2$, there is a run with decision r taken at n_0 which is final. But existence of any such final run implies satisfiability of ϕ. For reverse implication, if ϕ is satisfiable, then the corresponding run for satisfying assignment takes 2 time units, which means that $mintime(\mathcal{N}_\phi) \leq 2$.

Similarly, we can prove that the MaxTime problem is co-NP complete by changing $\gamma(n_0, r^{'}) = [1, 1]$ and asking if $maxtime(\mathcal{N}_\phi) > 1$ for the new $\mathcal{N}_\phi$. The answer will be yes iff ϕ is satisfiable. □

We now consider the related problem of checking if $mintime(\mathcal{N}) = T$ (or if $maxtime(\mathcal{N}) = T$). These problems are harder than their threshold variant under usual complexity assumptions: they are DP-complete (Difference Polynomial time class, i.e., second level of the Boolean Hierarchy, defined as intersection of a problem in NP and one in co-NP [15]).

Proposition 2. *The $mintime(\mathcal{N}) = T$ and $maxtime(\mathcal{N}) = T$ decision problems are DP-complete for acyclic deterministic negotiations.*

Proof. We only give the proof for *mintime* (the proof for *maxtime* is given in Appendix A of [1]). Indeed, it is easy to see that this problem is in DP, as it can be written as $mintime(\mathcal{N}) \leq T$ which is in NP and $\neg(mintime(\mathcal{N}) \leq T - 1))$, which is in co-NP. To show hardness, we use the negotiation constructed in the above proof as a gadget, and show a reduction from the SAT-UNSAT problem (a standard DP-complete problem).

The SAT-UNSAT Problem asks given two Boolean expressions ϕ and $\phi^{'}$, both in CNF forms with three literals per clause, is it true that ϕ is satisfiable and $\phi^{'}$ is unsatisfiable? SAT-UNSAT is known to be DP-complete [15]. We reduce this problem to $mintime(\mathcal{N}) = T$.

Given ϕ, $\phi^{'}$, we first make the corresponding negotiations $\mathcal{N}_\phi$ and $\mathcal{N}_{\phi^{'}}$ as in the previous proof. Let n_0 and n_f be the initial and final nodes of $\mathcal{N}_\phi$ and $n_0^{'}$ and $n_f^{'}$ be the initial and final nodes of $\mathcal{N}_{\phi^{'}}$. (Similarly, for other nodes we write $'$ above the nodes to signify they belong to $\mathcal{N}_{\phi^{'}}$.)

In the negotiation $\mathcal{N}_{\phi^{'}}$, we introduce a new node n_{all}, in which all the processes participate (see Figure 4). The node n_{all} has a single outcome $r_{all}^{'}$ which sends all the processes to n_f. Also, for node $n_0^{'}$, apart from the outcome r which sends all processes to different nodes, there is another outcome r_{all} which sends all the processes to n_{all}. Now we merge the nodes n_f and $n_0^{'}$ and call the merged node n_{sep}. Also nodes n_0 and $n_f^{'}$ now have all the processes of $\mathcal{N}_\phi$ and $\mathcal{N}_{\phi^{'}}$ participating in them. This merged process gives us a new negotiation $\mathcal{N}_{\phi,\phi^{'}}$ in which the structure above n_{sep} is same as $\mathcal{N}_\phi$ while below it is same as $\mathcal{N}_{\phi^{'}}$. Node n_{sep} now has all the processes of $\mathcal{N}_\phi$ and $\mathcal{N}_{\phi^{'}}$ participating in it. The outcomes of n_{sep} will be same as that of n_0' (r_{all}, r). For both the outcomes of n_{sep} the processes corresponding to $\mathcal{N}_\phi$ directly go to n_f of the $\mathcal{N}_{\phi,\phi^{'}}$. Similarly n_0 of $\mathcal{N}_{\phi,\phi^{'}}$ which is same n_0 of $\mathcal{N}_\phi$, sends processes corresponding to $\mathcal{N}_{\phi^{'}}$ directly to n_{sep} for all its outcomes. We now define timing function γ for $\mathcal{N}_{\phi,\phi^{'}}$ which is as follows: $\gamma(Lone_{v_i}^{'}, r) = [1, 1]$ for all $v_i \in \phi^{'}$ and $r \in \{\texttt{true, false}\}$, $\gamma(n_{all}, r_{all}^{'}) = [2, 2]$ and $\gamma(n, r) = [0, 0]$ for all other outcomes of nodes. With this construction, one can conclude that $mintime(\mathcal{N}_{\phi,\phi^{'}}) = 2$ iff ϕ is satisfiable and $\phi^{'}$ is unsatisfiable (see [1] for details). This completes the reduction and hence proves DP-hardness. □

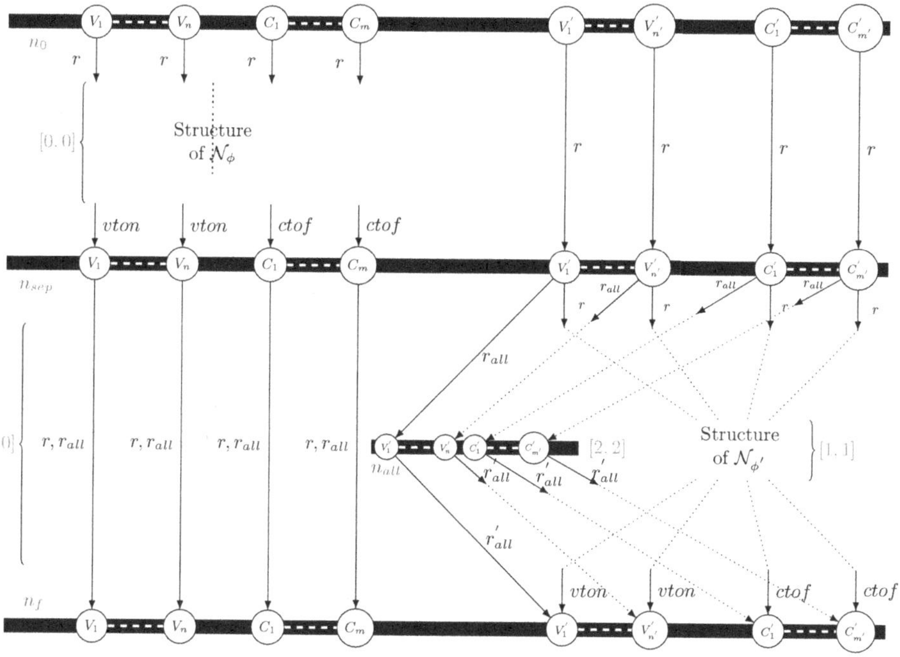

Fig. 4. Structure of $\mathcal{N}_{\phi,\phi'}$

Finally, we consider a related problem of computing the min and max time. To consider the decision variant, we rephrase this problem as checking whether an arbitrary bit of the minimum execution time is 1. Perhaps surprisingly, we obtain that this problem goes even beyond DP, the second level of the Boolean Hierarchy and is in fact hard for Δ_2^P (second level of the *polynomial* hierarchy), which contains the entire Boolean Hierarchy. Formally,

Theorem 4. *Given an acyclic deterministic timed negotiation and a positive integer k,computing the kth bit of the maximum/minimum execution time is Δ_2^P-complete.*

Finally, we remark that if we were interested in the optimization variant and not the decision variant of the problem, the above proof can be adapted to show that these variants are OptP-complete (as defined in [13]). But as optimization is not the focus of this paper, we avoid formal details of this proof.

6 Sound Negotiations

Sound negotiations are negotiations in which every run can be extended to a final run, as in Fig. 1. In this section, we show that $maxtime(\mathcal{N})$ can be computed in PTIME for sound negotiations, hence giving PTIME complexities for the $maxtime(\mathcal{N}) \leq T$? and $maxtime(\mathcal{N}) = T$? questions. However, we show that $mintime(\mathcal{N}) \leq T$ is NP-complete for sound negotiations, and that $mintime(\mathcal{N}) = T$ is DP-complete, even if T is given in unary.

Consider the graph $G_{\mathcal{N}}$ of a negotiation $\mathcal{N}$. Let $\pi = (n_0, (p_0, r_0), n_1) \cdots (n_k, (p_k, r_k), n_{k+1})$ be a path of $G_{\mathcal{N}}$. We define the *maximal execution time* of a path π as the value $\delta^+(\pi) = \sum_{i \in 0..k} \gamma^+(n_i, r_i)$. We say that a path $\pi = (n_0, (p_0, r_0), n_1) \cdots (n_\ell, (p_\ell, r_\ell), n_{\ell+1})$ is a path of some run $\rho = (M_1, \mu_1) \xrightarrow{(n_1, r_1')} \cdots (M_k, \mu_k)$ if $r_0, \dots, r_\ell$ is a subword of $r_1', \dots, r_k'$.

Lemma 1. *Let $\mathcal{N}$ be an acyclic and sound timed negotiation. Then $maxtime(\mathcal{N}) = \max_{\pi \in Paths(G_{\mathcal{N}})} \delta^+(\pi) + \gamma^+(n_f, r_f)$.*

Proof. Let us first prove that $maxtime(\mathcal{N}) \geq \max_{\pi \in Paths(G_{\mathcal{N}})} \delta^+(\pi) + \gamma^+(n_f, r_f)$. Consider any path π of $G_{\mathcal{N}}$, ending in some node n. First, as $\mathcal{N}$ is sound, we can compute a run ρ_π such that π is a path of ρ_π, and ρ_π ends in a configuration in which n is enabled. We associate with ρ_π the timed run ρ_π^+ which associates to every node the latest possible execution date. We have easily $\delta(\rho_\pi^+) \geq \delta^+(\pi)$, and then we obtain $\max_{\pi \in Paths(G_{\mathcal{N}})} \delta(\rho_\pi^+) \geq \max_{\pi \in Paths(G_{\mathcal{N}})} \delta^+(\pi)$. As $maxtime(\mathcal{N})$ is the maximal duration over all runs, it is hence necessarily greater than $\max_{\pi \in Paths(G_{\mathcal{N}})} \delta(\rho_\pi^+) + \gamma^+(n_f, r_f)$.

We now prove that $maxtime(\mathcal{N}) \leq \max_{\pi \in Paths(G_{\mathcal{N}})} \delta^+(\pi) + \gamma^+(n_f, r_f)$. Take any timed run $\rho = (M_1, \mu_1) \xrightarrow{(n_1, r_1)} \cdots (M_k, \mu_k)$ of $\mathcal{N}$ with a unique maximal node n_k. We show that there exists a path π of ρ such that $\delta(\rho) \leq \delta^+(\pi)$ by induction on the length k of ρ. The initialization is trivial for $k = 1$. Let $k \in \mathbb{N}$. Because n_k is the unique maximal node of ρ, we have $\delta^+(\rho) = \max_{p \in P_{n_k}} \mu_{k-1}(p) + \gamma^+(n_k, r_k)$. We choose one p_{k-1} maximizing $\mu_{k-1}(p)$. Let $\ell < k$ be the maximal index of a decision involving process p_{k-1} (i.e. $p_{k-1} \in P_{n_\ell}$). Now, consider the timed run ρ' subword of ρ, but with n_ℓ as unique maximal node (that is, it is ρ where nodes $n_i, i > \ell$ has been removed, but also where some nodes $n_i, i < \ell$ have been removed if they are not causally before n_ℓ (in particular, $P_{n_i} \cap P_{n_\ell} = \emptyset$).)

By definition, we have that $\delta^+(\rho) = \delta^+(\rho') + \gamma^+(n_\ell, r_\ell) + \gamma^+(n_k, r_k)$. We apply the induction hypothesis on ρ', and obtain a path π' of ρ' ending in n_ℓ such that $\delta^+(\rho') + \gamma^+(n_\ell, r_\ell) \leq \delta^+(\pi')$. It suffices to consider path $\pi = \pi'.(n_\ell, (p_{k-1}, r_\ell), n_k)$ to prove the inductive step $\delta^+(\rho) \leq \delta^+(\pi) + \gamma^+(n_k, r_k)$.

Thus $maxtime(\mathcal{N}) = \max \delta^+(\rho) \leq \max_{\pi \in Paths(G_{\mathcal{N}})} \delta^+(\pi) + \gamma^+(n_f, r_f)$. □

Lemma 1 gives a way to evaluate the maximal execution time. This amounts to finding a path of maximal weight in an acyclic graph, which is a standard PTIME problem that can be solved using standard max-cost calculation.

Proposition 3. *Computing the maximal execution time for an acyclic sound negotiation $\mathcal{N} = (N, n_0, n_f, \mathcal{X})$ can be done in time $O(|N| + |\mathcal{X}|)$.*

A direct consequence is that $maxtime(\mathcal{N}) \leq T$ and $maxtime(\mathcal{N}) = T$ problems can be solved in polynomial time when $\mathcal{N}$ is sound. Notice that if $\mathcal{N}$ is deterministic but not sound, then Lemma 1 does not hold: we only have an inequality.

We now turn to $mintime(\mathcal{N})$. We show that it is strictly harder to compute for sound negotiations than $maxtime(\mathcal{N})$.

Theorem 5. *$mintime(\mathcal{N}) \leq T$ is NP-complete in the strong sense for sound acyclic negotiations, even if $\mathcal{N}$ is very weakly non-deterministic.*

Proof (sketch). First, we can decide $mintime(\mathcal{N}) \leq T$ in NP. Indeed, one can guess a final (untimed) run ρ of size $\leq |N|$, consider ρ^- the timed run corresponding to ρ where all outcomes are taken at the earliest possible dates, and compute in linear time $\delta(\rho^-)$, and check that $\delta(\rho^-) \leq T$.

The hardness part is obtained by reduction from the **Bin Packing** problem. The reduction is similar to Knapsack, that we will present in Thm. 7. The difference is that we use ℓ bins in parallel, rather than 2 processes, one for the weight and one for the value. The hardness is thus strong, but the negotiation is not k-layered for a bounded k (it is $2\ell + 1$ bounded, with ℓ depending on the input). A detailed proof is given in Appendix B of [1]. □

We show that $mintime(\mathcal{N}) = T$ is harder to decide than $mintime(\mathcal{N}) \leq T$, with a proof similar to Prop. 2.

Proposition 4. *The $mintime(\mathcal{N}) = T$? decision problem is DP-complete for sound acyclic negotiations, even if it is very weakly non-deterministic.*

An open question is whether the minimal execution time can be computed in PTIME if the negotiation is both sound and deterministic. The reduction from Bin Packing does not work with deterministic (and sound) negotiations.

7 k-Layered Negotiations

In this section, we consider k-layeredness, a syntactic property that can be efficiently verified (see Section 2).

7.1 Algorithmic properties

Let k be a fixed integer. We first show that the maximum execution time can be computed in PTIME for k-layered negotiations. Let N_i be the set of nodes at layer i. We define for every layer i the set S_i of subsets of nodes $X \subseteq N_i$ which can be jointly enabled and such that for every process p, there is exactly one node $n(X,p)$ in X with $p \in n(X,p)$. An element X in S_i is a subset of nodes that can be selected by solving all non-determnism with an appropriate choice of outcomes. Formally, we define S_i inductively. We start with $S_0 = \{n_0\}$. We then define S_{i+1} from the contents of layer S_i: we have $Y \in S_{i+1}$ iff $\bigcup_{n \in Y} P_n = P$ and there exist $X \in S_i$ and an outcome $r_m \in R_m$ for every $m \in X$, such that $n \in \mathcal{X}(n(X,p), p, r_m)$ for each $n \in Y$ and $p \in P_n$.

Theorem 6. *Let $k \in \mathbb{N}^+$. Computing the maximum execution time for a k-layered acyclic negotiation $\mathcal{N}$ can be done in PTIME. More precisely, the worst-case time complexity is $O(|P| \cdot |\mathcal{N}|^{k+1})$.*

Proof (Sketch). The first step is to compute S_i layer by layer, by following its inductive definition. The set S_i is of size at most 2^k, as $|N_i| < k$ by definition of k-layeredness. Knowing S_i, it is easy to build S_{i+1} by induction. This takes time in $O(|P||\mathcal{N}|^{k+1})$: We need to consider all k-uples of outcomes for each layer. There can be $|\mathcal{N}|^k$ such tuples. We need to do that for all processes ($|P|$), and for all layers (at most $|\mathcal{N}|$).

We then keep for each subset $X \in S_i$ and each node $n \in X$, the maximal time $f_i(n, X) \in \mathbb{N}$ associated with n and X. From S_{i+1} and f_i, we inductively compute f_{i+1} in the following way: for all $X \in S_i$ with successor $Y \in S_{i+1}$ for outcomes $(r_p)_{p \in P}$, we denote $f_{i+1}(Y, n, X) = \max_{p \in P(n)} f_i(X, n(X,p)) + \gamma^+(n(X,p), r_p)$. If there are several choices of $(r_p)_{p \in P}$ leading to the same Y, we take r_p with the maximal $f_i(X, n(X,p)) + \gamma^+(n(X,p), r_p)$. We then define $f_{i+1}(Y, n) = \max_{X \in S_i} f_{i+1}(Y, n, X)$. Again, the initialization is trivial, with $f_0(\{n_0\}, n_0) = 0$. The maximal execution time of $\mathcal{N}$ is $f(\{n_f\}, n_f)$. □

We can bound the complexity precisely by $O(d(\mathcal{N}) \cdot C(\mathcal{N}) \cdot ||R||^{k^*})$, with:

- $d(\mathcal{N}) \leq |\mathcal{N}|$ the depth of n_f, that is the number of layers of $\mathcal{N}$, and $||R||$ is the maximum number of outcomes of a node,
- $C(\mathcal{N}) = \max_i |S_i| \leq 2^k$, which we will call the *number of contexts of* $\mathcal{N}$, and which is often much smaller than 2^k.
- $k^* = \max_{X \in \bigcup_i S_i} |X| \leq k$. We say that $\mathcal{N}$ is k^**-thread bounded*, meaning that there cannot be more that k^* nodes in the same context X of any layer. Usually, k^* is strictly smaller than $k = \max_i |N_i|$, as $N_i = \bigcup_{X \in S_i} X$.

Consider again the Brexit example Figure 1. We have $(k+1) = 7$, while we have the depth $d(\mathcal{N}) = 6$, the negotiation is $k^* = 3$-thread bounded (k^* is bounded by the number of processes), $||R|| = 2$, and the number of contexts is at most $C(\mathcal{N}) = 4$ (EU chooses to enforce backstop or not, and Pa chooses to go to court or not).

7.2 Minimal Execution Time

As with sound negotiations, computing minimal time is much harder than computing the maximal time for k-layered negotiations:

Theorem 7. *Let $k \geq 6$. The $Min \leq T$ problem is NP-Complete for k-layered acyclic negotiations, even if the negotiation is sound and very weakly non-deterministic.*

Proof. One can guess in polynomial time a final run of size $\leq |\mathcal{N}|$. If the execution time of this final run is smaller than T then we have found a final run witnessing $mintime(\mathcal{N}) \leq T$. Hence the problem is in NP.

Let us now show that the problem is NP-hard. We proceed by reduction from the **Knapsack** decision problem. Let us consider a set of items $U = \{u_1, \dots u_n\}$ of respective values $v_1, \dots v_n$ and weight $w_1, \dots, w_n$ and a knapsack of maximal capacity W. The knapsack problem asks, given a value V whether there exists a subset of items $U' \subseteq U$ such that $\sum_{u_i \in U'} v_i \geq V$ and such that $\sum_{u_i \in U'} w_i \leq W$.

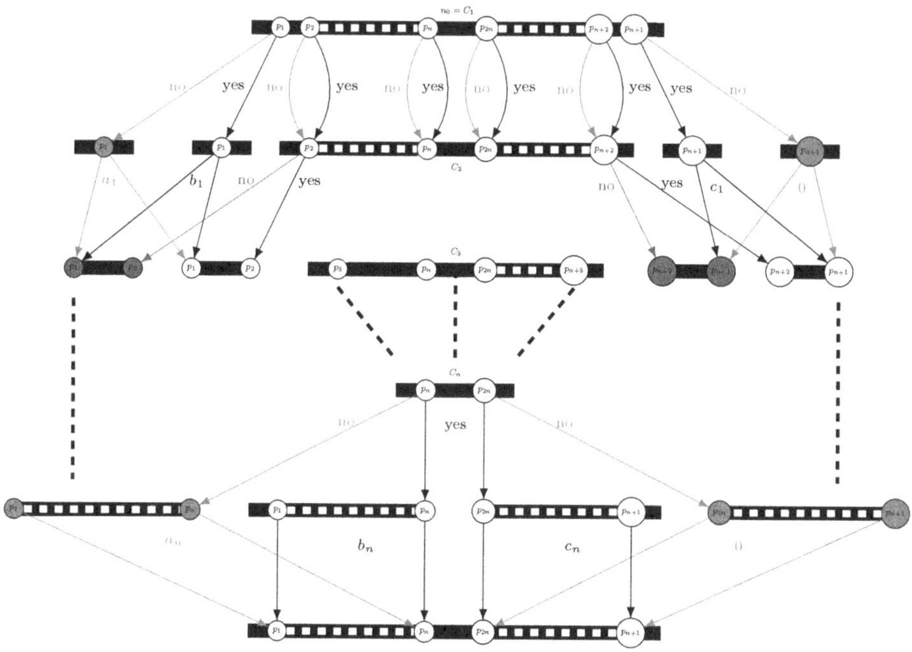

Fig. 5. The negotiation encoding Knapsack

We build a negotiation with $2n$ processes $P = \{p_1, \ldots p_{2n}\}$, as shown in Fig. 5. Intuitively, $p_i, i \leq n$ will serve to encode the value of selected items as timing, while $p_i, i > n$ will serve to encode the weight of selected items as timing.

Concerning timing constraints for outcomes we do the following: Outcomes 0, yes and no are associated with $[0, 0]$. Outcome c_i is associated with $[w_i, w_i]$, the weight of u_i. Last, outcome b_i is associated with a more complex function, such that $\sum_i b_i \leq W$ iff $\sum_i v_i \geq V$. For that, we set $[\frac{(v_{max}-v_i)W}{n \cdot v_{max}-V}, \frac{v_{max}W}{n \cdot v_{max}-v_i}]$ for outcome b_i, where v_{max} is the largest value of an item, and V is the total value we want to reach at least. Also, we set $[\frac{(v_{max})W}{n \cdot v_{max}-V}, \frac{v_{max}W}{n \cdot v_{max}-v_i}]$ for outcome a_i. We set $T = W$, the maximal weight of the knapsack.

Now, consider a final run ρ in $\mathcal{N}$. The only choices in ρ are outcomes *yes* or *no* from $C_1, \ldots, C_n$. Let I be the set of indices such that *yes* is the outcome from all C_i in this path. We obtain $\delta(\rho) = \max(\sum_{i \notin I} a_i + \sum_{i \in I} b_i, \sum_{i \in I} c_i)$. We have $\delta(\rho) \leq T = W$ iff $\sum_{i \in I} w_i \leq W$, that is the sum of the weights is lower than W, and $\sum_{i \notin I} \frac{(v_{max})W}{n \cdot v_{max}-V} + \sum_{i \in I} \frac{(v_{max}-v_i)W}{n \cdot v_{max}-V} \leq W$. That is, $n \cdot v_{max} - \sum_{i \in I} v_i \leq n \cdot v_{max} - V$, i.e. $\sum_{i \in I} v_i \geq V$. Hence, there exists a path ρ with $\delta(\rho) \leq T = W$ iff there exists a set of items of weight less than W and of value more than V. □

It is well known that Knapsack is weakly NP-hard, that is, it is NP-hard only when weights/values are given in binary. This means that Thm. 7 shows that minimum execution time $\leq T$ is NP-hard only when T is given in binary. We can actually show that for k-layered negotiations, the $mintime(\mathcal{N}) \leq T$ problem can be decided in PTIME if T is given in unary (i.e. if T is not too large):

Theorem 8. *Let $k \in \mathbb{N}$. Given a k-layered negotiation $\mathcal{N}$ and T written in unary, one can decide in PTIME whether the minimum execution time of $\mathcal{N}$ is $\leq T$. The worst-case time complexity is $O(|\mathcal{N}| \cdot |P| \cdot (T \cdot |\mathcal{N}|)^k)$.*

Proof. We will remember for each layer i a set $\mathcal{T}_i$ of functions τ from nodes N_i of layer i to a value in $\{1, \ldots, T, \bot\}$. Basically, we have $\tau \in \mathcal{T}_i$ if there exists a path ρ reaching $X = \{n \in N_i \mid \tau(n) \neq \bot\}$, and this path reaches node $n \in X$ after $\tau(n)$ time units. As for S_i, for all p, we should have a unique node $n(\tau, p)$ such that $p \in n(\tau, p)$ and $\tau(n(\tau, p)) \neq \bot$. Again, it is easy to initialize $\mathcal{T}_0 = \{\tau_0\}$, with $\tau_0(n_0) = 0$, and $\tau_0(n) = \bot$ for all $n \neq n_0$.

Inductively, we build $\mathcal{T}_{i+1}$ in the following way: $\tau_{i+1} \in \mathcal{T}_{i+1}$ iff there exists a $\tau_i \in \mathcal{T}_i$ and $r_p \in R_{n(\tau_i,p)}$ for all $p \in P$ such that for all n with $\tau_{i+1}(n) \neq \bot$, we have $\tau_{i+1}(n) = \max_p \tau_i^-(n(\tau_i, p)) + \gamma(n(\tau_i, p), r_p)$.

We have that the minimum execution time for $\mathcal{N}$ is $\min_{\tau \in \mathcal{T}_n} \tau(n_\tau)$, for n the depth of n_f. There are at most T^k functions τ in any $\mathcal{T}_i$, and there are at most $|\mathcal{N}|$ layers to consider, giving the complexity. □

As with Thm. 6, we can more accurately state the complexity as $O(d(\mathcal{N}) \cdot C(\mathcal{N}) \cdot ||R||^{k^*} \cdot T^{k^*-1})$. The $k^* - 1$ is because we only need to remember minimal functions $\tau \in \mathcal{T}_i$: if $\tau'(n) \geq \tau(n)$ for all n, then we do not need to keep τ' in $\mathcal{T}_i$. In particular, for the knapsack encoding in the proof of Thm. 7, we have $k^* = 3$, $||R|| = 2$ and $C(\mathcal{N}) = 4$. Notice that if k is part of the input, then the problem is strongly NP-hard, even if T is given in unary, as e.g. encoding bin packing with ℓ bins result to a $2\ell + 1$-layered negotiations.

8 Conclusion

In this paper, we considered timed negotiations. We believe that time is of the essence in negotiations, as examplified by the Brexit negotiation. It is thus important to be able to compute in a tractable way the minimal and maximal execution time of negotiations. We showed that we can compute in PTIME the maximal execution time for acyclic negotiations that are either sound or k-layered, for k fixed. We showed that we cannot compute in PTIME the maximal execution time for negotiations that are not sound nor k-layered, even if they are deterministic and acyclic (unless NP=PTIME). We also showed that surprisingly, computing the minimal execution time is much harder, with strong NP-hardness results in most of the classes of negotiations, contradicting a claim in [10]. We came up with a new reasonable class of negotiations, namely k-layered negotiations, which enjoys a pseudo PTIME algorithm to compute the minimal execution time. That is, the algorithm is PTIME when the timing constants are given in unary. We showed that this restriction is necessary, as the problem becomes NP-hard for constants given in binary, even when the negotiation is sound and very weakly non-deterministic. The problem to know whether the minimal execution time can be computed in PTIME for deterministic and sound negotiation remains open.

References

1. S. Akshay, B. Genest, L. Hélouët, and S. Mital. Timed Negotiations (extended version). In *Research report, https://hal.inria.fr/hal-02337887*, 2020.
2. J. Desel. Reduction and Design of Well-behaved Concurrent Systems. In *CONCUR '90, Theories of Concurrency: Unification and Extension, Amsterdam, The Netherlands, August 27-30, 1990, Proceedings*, volume 458 of *Lecture Notes in Computer Science*, pages 166–181. Springer, 1990.
3. J. Desel, J. Esparza, and P. Hoffmann. Negotiation as Concurrency Primitive. *Acta Inf.*, 56(2):93–159, 2019.
4. J. Esparza. Decidability and Complexity of Petri Net Problems - An Introduction. In *Lectures on Petri Nets I: Basic Models, Advances in Petri Nets, Dagstuhl, September 1996*, volume 1491 of *Lecture Notes in Computer Science*, pages 374–428. Springer, 1998.
5. J. Esparza and J. Desel. *Free Choice Petri Nets.* Cambridge University Press, 1995.
6. J. Esparza and J. Desel. On Negotiation as Concurrency Primitive. In *CONCUR 2013 - Concurrency Theory - 24th International Conference, CONCUR 2013, Buenos Aires, Argentina, August 27-30, 2013. Proceedings*, volume 8052 of *Lecture Notes in Computer Science*, pages 440–454. Springer, 2013.
7. J. Esparza and J. Desel. On Negotiation as Concurrency Primitive II: Deterministic Cyclic Negotiations. In *FOSSACS'14*, volume 8412 of *Lecture Notes in Computer Science*, pages 258–273. Springer, 2014.
8. J. Esparza and P. Hoffmann. Reduction Rules for Colored Workflow Nets. In *Fundamental Approaches to Software Engineering - 19th International Conference, FASE 2016, Held as Part of the European Joint Conferences on Theory and Practice of Software, ETAPS 2016, Eindhoven, The Netherlands, April 2-8, 2016, Proceedings*, volume 9633 of *Lecture Notes in Computer Science*, pages 342–358. Springer, 2016.
9. J. Esparza, D. Kuperberg, A. Muscholl, and I. Walukiewicz. Soundness in Negotiations. *Logical Methods in Computer Science*, 14(1), 2018.
10. J. Esparza, A. Muscholl, and I. Walukiewicz. Static Analysis of Deterministic Negotiations. In *32nd Annual ACM/IEEE Symposium on Logic in Computer Science, LICS 2017, Reykjavik, Iceland, June 20-23, 2017*, pages 1–12, 2017.
11. S. Haddad. A Reduction Theory for Coloured Nets. In *Advances in Petri Nets 1989*, volume 424 of *Lecture Notes in Computer Science*, pages 209–235. Springer, 1990.
12. P. Hoffmann. Negotiation Games. In Javier Esparza and Enrico Tronci, editors, *Proceedings Sixth International Symposium on Games, Automata, Logics and Formal Verification, GandALF 2015, Genoa, Italy, 21-22nd September 2015.*, volume 193 of *EPTCS*, pages 31–42, 2015.
13. M. W. Krentel. The Complexity of Optimization Problems. *Journal of computer and system sciences*, 36(3):490–509, 1988.
14. P.M. Merlin. *A Study of the Recoverability of Computing Systems.* PhD thesis, University of California, Irvine, CA, USA, 1974.
15. C. H. Papadimitriou and M. Yannakakis. The Complexity of Facets (and Some Facets of Complexity). In *Proceedings of the Fourteenth Annual ACM Symposium on Theory of Computing*, STOC '82, pages 255–260, New York, NY, USA, 1982. ACM.

16. R.H. Sloan and U.A. Buy. Reduction Rules for Time Petri Nets. *Acta Inf.*, 33(7):687–706, 1996.

Cartesian Difference Categories

Mario Alvarez-Picallo[1] and Jean-Simon Pacaud Lemay (✉)[2*]

[1] Department of Computer Science, University of Oxford, Oxford, UK
mario.alvarez-picallo@cs.ox.ac.uk
[2] Department of Computer Science, University of Oxford, Oxford, UK
jean-simon.lemay@kellogg.ox.ac.uk

Abstract. Cartesian differential categories are categories equipped with a differential combinator which axiomatizes the directional derivative. Important models of Cartesian differential categories include classical differential calculus of smooth functions and categorical models of the differential λ-calculus. However, Cartesian differential categories cannot account for other interesting notions of differentiation such as the calculus of finite differences or the Boolean differential calculus. On the other hand, change action models have been shown to capture these examples as well as more "exotic" examples of differentiation. However, change action models are very general and do not share the nice properties of a Cartesian differential category. In this paper, we introduce Cartesian difference categories as a bridge between Cartesian differential categories and change action models. We show that every Cartesian differential category is a Cartesian difference category, and how certain well-behaved change action models are Cartesian difference categories. In particular, Cartesian difference categories model both the differential calculus of smooth functions and the calculus of finite differences. Furthermore, every Cartesian difference category comes equipped with a tangent bundle monad whose Kleisli category is again a Cartesian difference category.

Keywords: Cartesian Difference Categories · Cartesian Differential Categories · Change Actions · Calculus Of Finite Differences · Stream Calculus.

1 Introduction

In the early 2000s, Ehrhard and Regnier introduced the differential λ-calculus [10], an extension of the λ-calculus equipped with a differential combinator capable of taking the derivative of arbitrary higher-order functions. This development, based on models of linear logic equipped with a natural notion of "derivative" [11], sparked a wave of research into categorical models of differentiation.

One of the most notable developments in the area is the introduction of Cartesian differential categories [4] by Blute, Cockett and Seely, which provide an abstract categorical axiomatization of the directional derivative from differential

* The second author is financially supported by Kellogg College, the Oxford-Google Deep Mind Graduate Scholarship, and the Clarendon Fund.

J. Goubault-Larrecq and B. König (Eds.): FOSSACS 2020, LNCS 12077, pp. 57–76, 2020.
https://doi.org/10.1007/978-3-030-45231-5_4

calculus. The relevance of Cartesian differential categories lies in their ability to model both "classical" differential calculus (with the canonical example being the category of Euclidean spaces and smooth functions between) and the differential λ-calculus (as every categorical model for it gives rise to a Cartesian differential category [14]). However, while Cartesian differential categories have proven to be an immensely successful formalism, they have, by design, some limitations. Firstly, they cannot account for certain "exotic" notions of derivative, such as the difference operator from the calculus of finite differences [16] or the Boolean differential calculus [19]. This is because the axioms of a Cartesian differential category stipulate that derivatives should be linear in their second argument (in the same way that the directional derivative is), whereas these aforementioned discrete sorts of derivative need not be. Additionally, every Cartesian differential category is equipped with a tangent bundle monad [7, 15] whose Kleisli category can be intuitively understood as a category of generalized vector fields. This Kleisli category has an obvious differentiation operator which comes close to making it a Cartesian differential category, but again fails the requirement of being linear in its second argument.

More recently, discrete derivatives have been suggested as a semantic framework for understanding incremental computation. This led to the development of change structures [6] and change actions [2]. Change action models have been successfully used to provide a model for incrementalizing Datalog programs [1], but have also been shown to model the calculus of finite differences as well as the Kleisli category of the tangent bundle monad of a Cartesian differential category. Change action models, however, are very general, lacking many of the nice properties of Cartesian differential categories (for example, addition in a change action model is not required to be commutative), even though they are verified in most change action models. As a consequence of this generality, the tangent bundle endofunctor in a change action model can fail to be a monad.

In this work, we introduce Cartesian difference categories (Section 4.2), whose key ingredients are an infinitesimal extension operator and a difference combinator, whose axioms are a generalization of the differential combinator axioms of a Cartesian differential category. In Section 4.3, we show that every Cartesian differential category is, in fact, a Cartesian difference category whose infinitesimal extension operator is zero, and conversely how every Cartesian difference category admits a full subcategory which is a Cartesian differential category. In Section 4.4, we show that every Cartesian difference category is a change action model, and conversely how a full subcategory of suitably well-behaved objects of a change action model is a Cartesian difference category. In Section 6, we show that every Cartesian difference category comes equipped with a monad whose Kleisli category again a Cartesian difference category. Finally, in Section 5 we provide some examples of Cartesian difference categories; notably, the calculus of finite differences and the stream calculus.

2 Cartesian Differential Categories

In this section, we briefly review Cartesian differential categories, so that the reader may compare Cartesian differential categories with the new notion of Cartesian *difference* categories which we introduce in the next section. For a full detailed introduction on Cartesian differential categories, we refer the reader to the original paper [4].

2.1 Cartesian Left Additive Categories

Here we recall the definition of Cartesian left additive categories [4] – where "additive" is meant being skew enriched over commutative monoids, which in particular means that we do not assume the existence of additive inverses, i.e., "negative elements". By a Cartesian category we mean a category $\mathbb{X}$ with chosen finite products where we denote the binary product of objects A and B by $A \times B$ with projection maps $\pi_0 : A \times B \to A$ and $\pi_1 : A \times B \to B$ and pairing operation $\langle -, - \rangle$, and the chosen terminal object as $\top$ with unique terminal maps $!_A : A \to \top$.

Definition 1. *A **left additive category** [4] is a category $\mathbb{X}$ such that each hom-set $\mathbb{X}(A, B)$ is a commutative monoid with addition operation $+ : \mathbb{X}(A, B) \times \mathbb{X}(A, B) \to \mathbb{X}(A, B)$ and zero element (called the zero map) $0 \in \mathbb{X}(A, B)$, such that pre-composition preserves the additive structure: $(f + g) \circ h = f \circ h + g \circ h$ and $0 \circ f = 0$. A map k in a left additive category is **additive** if post-composition by k preserves the additive structure: $k \circ (f + g) = k \circ f + k \circ g$ and $k \circ 0 = 0$. A **Cartesian left additive category** [4] is a Cartesian category $\mathbb{X}$ which is also a left additive category such all projection maps $\pi_0 : A \times B \to A$ and $\pi_1 : A \times B \to B$ are additive.*

We note that the definition given here of a Cartesian left additive category is slightly different from the one found in [4], but it is indeed equivalent. By [4, Proposition 1.2.2], an equivalent axiomatization is of a Cartesian left additive category is that of a Cartesian category where every object comes equipped with a commutative monoid structure such that the projection maps are monoid morphisms. This will be important later in Section 4.2.

2.2 Cartesian Differential Categories

Definition 2. *A **Cartesian differential category** [4] is a Cartesian left additive category equipped with a **differential combinator** D of the form*

$$\frac{f : A \to B}{\mathsf{D}[f] : A \times A \to B}$$

verifying the following coherence conditions:

[CD.1] $\mathsf{D}[f + g] = \mathsf{D}[f] + \mathsf{D}[g]$ *and* $\mathsf{D}[0] = 0$

[CD.2] $\mathsf{D}[f] \circ \langle x, y+z \rangle = \mathsf{D}[f] \circ \langle x, y \rangle + \mathsf{D}[f] \circ \langle x, z \rangle$ *and* $\mathsf{D}[f] \circ \langle x, 0 \rangle = 0$
[CD.3] $\mathsf{D}[1_A] = \pi_1$ *and* $\mathsf{D}[\pi_0] = \pi_0 \circ \pi_1$ *and* $\mathsf{D}[\pi_1] = \pi_1 \circ \pi_1$
[CD.4] $\mathsf{D}[\langle f, g \rangle] = \langle \mathsf{D}[f], \mathsf{D}[g] \rangle$ *and* $\mathsf{D}[!_A] = !_{A \times A}$
[CD.5] $\mathsf{D}[g \circ f] = \mathsf{D}[g] \circ \langle f \circ \pi_0, \mathsf{D}[f] \rangle$
[CD.6] $\mathsf{D}[\mathsf{D}[f]] \circ \langle \langle x, y \rangle, \langle 0, z \rangle \rangle = \mathsf{D}[f] \circ \langle x, z \rangle$
[CD.7] $\mathsf{D}[\mathsf{D}[f]] \circ \langle \langle x, y \rangle, \langle z, 0 \rangle \rangle = \mathsf{D}[\mathsf{D}[f]] \circ \langle \langle x, z \rangle, \langle y, 0 \rangle \rangle$

Note that here, following the more recent work on Cartesian differential categories, we've flipped the convention found in [4], so that the linear argument is in the second argument rather than in the first argument.

We highlight that by [7, Proposition 4.2], the last two axioms **[CD.6]** and **[CD.7]** have an equivalent alternative expression.

Lemma 1. *In the presence of the other axioms,* **[CD.6]** *and* **[CD.7]** *are equivalent to:*

[CD.6.a] $\mathsf{D}[\mathsf{D}[f]] \circ \langle \langle x, 0 \rangle, \langle 0, y \rangle \rangle = \mathsf{D}[f] \circ \langle x, y \rangle$
[CD.7.a] $\mathsf{D}[\mathsf{D}[f]] \circ \langle \langle x, y \rangle, \langle z, w \rangle \rangle = \mathsf{D}[\mathsf{D}[f]] \circ \langle \langle x, z \rangle, \langle y, w \rangle \rangle$

As a Cartesian difference category is a generalization of a Cartesian differential category, we leave the discussion of the intuition of these axioms for later in Section 4.2 below. We also refer to [4, Section 4] for a term calculus which may help better understand the axioms of a Cartesian differential category. The canonical example of a Cartesian differential category is the category of real smooth functions, which we will discuss in Section 5.1. Other interesting examples of can be found throughout the literature such as categorical models of the differential λ-calculus [10, 14], the subcategory of differential objects of a tangent category [7], and the coKleisli category of a differential category [3, 4].

3 Change Action Models

Change actions [1, 2] have recently been proposed as a setting for reasoning about higher-order incremental computation, based on a discrete notion of differentiation. Together with Cartesian differential categories, they provide the core ideas behind Cartesian difference categories. In this section, we quickly review change actions and change action models, in particular, to highlight where some of the axioms of a Cartesian difference category come from. For more details on change actions, we invite readers to see the original paper [2].

3.1 Change Actions

Definition 3. *A* ***change action*** $\overline{A}$ *in a Cartesian category* $\mathbb{X}$ *is a quintuple* $\overline{A} \equiv (A, \Delta A, \oplus_A, +_A, 0_A)$ *consisting of two objects* A *and* ΔA*, and three maps:*

$$\oplus_A : A \times \Delta A \to A \qquad +_A : \Delta A \times \Delta A \to \Delta A \qquad 0_A : \top \to \Delta A$$

such that $(\Delta A, +_A, 0_A)$ *is a monoid and* $\oplus_A : A \times \Delta A \to A$ *is an action of* ΔA *on* A*, that is, the following equalities hold:*

$$\oplus_A \circ \langle 1_A, 0_A \circ !_A \rangle = 1_A \qquad \oplus_A \circ (1_A \times +_A) = \oplus_A \circ (\oplus_A \times 1_{\Delta A})$$

For a change action $\overline{A}$ and given a pair of maps $f : C \to A$ and $g : C \to \Delta A$, we define $f \oplus_{\overline{A}} g : C \to A$ as $f \oplus_{\overline{A}} g = \oplus_A \circ \langle f, g \rangle$. Similarly, for maps $h : C \to \Delta A$ and $k : C \to \Delta A$, define $h +_{\overline{A}} k = +_A \circ \langle h, k \rangle$. Therefore, that $\oplus_A$ is an action of ΔA on A can be rewritten as:

$$1_A \oplus_{\overline{A}} 0_A = 1_A \qquad 1_A \oplus_{\overline{A}} (1_{\Delta A} +_{\overline{A}} 1_{\Delta A}) = (1_A \oplus_{\overline{A}} 1_{\Delta A}) \oplus_{\overline{A}} 1_{\Delta A}$$

The intuition behind the above definition is that the monoid ΔA is a type of possible "changes" or "updates" that might be applied to A, with the monoid structure on ΔA representing the capability to compose updates.

Change actions give rise to a notion of derivative, with a distinctly "discrete" flavour. Given change actions on objects A and B, a map $f : A \to B$ can be said to be differentiable when changes to the input (in the sense of elements of ΔA) are mapped to changes to the output (that is, elements of ΔB). In the setting of incremental computation, this is precisely what it means for f to be incrementalizable, with the derivative of f corresponding to an incremental version of f.

Definition 4. *Let $\overline{A} \equiv (A, \Delta A, \oplus_A, +_A, 0_A)$ and $\overline{B} \equiv (B, \Delta B, \oplus_B, +_B, 0_B)$ be change actions. For a map $f : A \to B$, a map $\boldsymbol{\partial}[f] : A \times \Delta A \to \Delta B$ is a* ***derivative*** *of f whenever the following equalities hold:*

[CAD.1] $f \circ (x \oplus_{\overline{A}} y) = f \circ x \oplus_{\overline{B}} (\boldsymbol{\partial}[f] \circ \langle x, y \rangle)$
[CAD.2] $\boldsymbol{\partial}[f] \circ \langle x, y +_{\overline{A}} z \rangle = (\boldsymbol{\partial}[f] \circ \langle x, y \rangle) +_{\overline{B}} (\boldsymbol{\partial}[f] \circ \langle x \oplus_{\overline{A}} y, z \rangle)$ *and* $\boldsymbol{\partial}[f] \circ \langle x, 0_B \circ !_B \rangle = 0_B \circ !_{A \times \Delta A}$

The intuition for these axioms will be explained in more detail in Section 4.2 when we explain the axioms of a Cartesian difference category. Note that although there is nothing in the above definition guaranteeing that any given map has at most a single derivative, the chain rule does hold. As a corollary, differentiation is compositional and therefore the change actions in $\mathbb{X}$ form a category.

Lemma 2. *Whenever $\boldsymbol{\partial}[f]$ and $\boldsymbol{\partial}[g]$ are derivatives for composable maps f and g respectively, then $\boldsymbol{\partial}[g] \circ \langle f \circ \pi_0, \boldsymbol{\partial}[f] \rangle$ is a derivative for $g \circ f$.*

3.2 Change Action Models

Definition 5. *Given a Cartesian category $\mathbb{X}$, define its change actions category $\mathsf{CAct}(\mathbb{X})$ as the category whose objects are change actions in $\mathbb{X}$ and whose arrows $\overline{f} : \overline{A} \to \overline{B}$ are the pairs $(f, \boldsymbol{\partial}[f])$, where $f : A \to B$ is an arrow in $\mathbb{X}$ and $\boldsymbol{\partial}[f] : A \times \Delta A \to \Delta B$ is a derivative for f. The identity is $(1_A, \pi_1)$, while composition of $(f, \boldsymbol{\partial}[f])$ and $(g, \boldsymbol{\partial}[g])$ is $(g \circ f, \boldsymbol{\partial}[g] \circ \langle f \circ \pi_0, \boldsymbol{\partial}[f] \rangle)$.*

There is an obvious product-preserving forgetful functor $\mathcal{E} : \mathsf{CAct}(\mathbb{X}) \to \mathbb{X}$ sending every change action $(A, \Delta A, \oplus, +, 0)$ to its base object A and every map $(f, \boldsymbol{\partial}[f])$ to the underlying map f. As a setting for studying differentiation, the category $\mathsf{CAct}(\mathbb{X})$ is rather lacklustre, since there is no notion of higher derivatives, so we will instead work with change action models. Informally, a change action model consists of a rule which for every object A of $\mathbb{X}$ associates a change action over it, and for every map a choice of a derivative.

Definition 6. *A **change action model** is a Cartesian category $\mathbb{X}$ is a product-preserving functor $\alpha : \mathbb{X} \to \mathsf{CAct}(\mathbb{X})$ that is a section of the forgetful functor $\mathcal{E}$.*

For brevity, when A is an object of a change action model, we will write ΔA, $\oplus_A$, $+_A$, and 0_A to refer to the components of the corresponding change action $\alpha(A)$. Examples of change action models can be found in [2]. In particular, we highlight that a Cartesian differential category always provides a change model action. We will generalize this result, and show in Section 4.4 that a Cartesian difference category also always provides a change action model.

4 Cartesian Difference Categories

In this section, we introduce *Cartesian difference categories*, which are generalizations of Cartesian differential categories. Examples of Cartesian difference categories can be found in Section 5.

4.1 Infinitesimal Extensions in Left Additive Categories

We first introduce infinitesimal extensions, which is an operator that turns a map into an "infinitesimal" version of itself – in the sense that every map coincides with its Taylor approximation on infinitesimal elements.

Definition 7. *A Cartesian left additive category $\mathbb{X}$ is said to have an **infinitesimal extension** ε if every homset $\mathbb{X}(A, B)$ comes equipped with a monoid morphism $\varepsilon : \mathbb{X}(A, B) \to \mathbb{X}(A, B)$, that is, $\varepsilon(f + g) = \varepsilon(f) + \varepsilon(g)$ and $\varepsilon(0) = 0$, and such that $\varepsilon(g \circ f) = \varepsilon(g) \circ f$ and $\varepsilon(\pi_0) = \pi_0 \circ \varepsilon(1_{A\times B})$ and $\varepsilon(\pi_1) = \pi_1 \circ \varepsilon(1_{A\times B})$.*

Note that since $\varepsilon(g \circ f) = \varepsilon(g) \circ f$, it follows that $\varepsilon(f) = \varepsilon(1_B) \circ f$ and $\varepsilon(1_A) : A \to A$ is an additive map (Definition 1). In light of this, it turns out that infinitesimal extensions can equivalently be described as a class of additive maps $\varepsilon_A : A \to A$ such that $\varepsilon_{A\times B} = \varepsilon_A \times \varepsilon_B$. The equivalence is given by setting $\varepsilon(f) = \varepsilon_B \circ f$ and $\varepsilon_A = \varepsilon(1_A)$. Furthermore, infinitesimal extensions equipped each object with a canonical change action structure:

Lemma 3. *Let $\mathbb{X}$ be a Cartesian left additive category with infinitesimal extension ε. For every object A, define the maps $\oplus_A : A \times A \to A$ as $\oplus_A = \pi_0 + \varepsilon(\pi_1)$, $+_A : A \times A \to A$ as $\pi_0 + \pi_1$, and $0_A : \top \to A$ as $0_A = 0$. Then $(A, A, \oplus_A, +_A, 0_A)$ is a change action in $\mathbb{X}$.*

Proof. As mentioned earlier, that $(A, +_A, 0_A)$ is a commutative monoid was shown in [4]. On the other hand, that $\oplus_A$ is a change action follows from the fact that ε preserves the addition. ∎

Setting $\overline{A} \equiv (A, A, \oplus_A, +_A, 0_A)$, we note that $f \oplus_{\overline{A}} g = f + \varepsilon(g)$ and $f +_{\overline{A}} g = f + g$, and so in particular $+_{\overline{A}} = +$. Therefore, from now on we will omit the subscripts and simply write $\oplus$ and $+$.

For every Cartesian left additive category, there are always at least two possible infinitesimal extensions:

Lemma 4. *For any Cartesian left additive category* $\mathbb{X}$,

1. *Setting* $\varepsilon(f) = 0$ *defines an infinitesimal extension on* $\mathbb{X}$ *and therefore in this case,* $\oplus_A = \pi_0$ *and* $f \oplus g = f$.
2. *Setting* $\varepsilon(f) = f$ *defines an infinitesimal extension on* $\mathbb{X}$ *and therefore in this case,* $\oplus_A = +_A$ *and* $f \oplus g = f + g$.

We note that while these examples of infinitesimal extensions may seem trivial, they are both very important as they will give rise to key examples of Cartesian difference categories.

4.2 Cartesian Difference Categories

Definition 8. *A* ***Cartesian difference category*** *is a Cartesian left additive category with an infinitesimal extension* ε *which is equipped with a* ***difference combinator*** ∂ *of the form:*

$$\frac{f : A \to B}{\partial[f] : A \times A \to B}$$

verifying the following coherence conditions:

[C∂.0] $f \circ (x + \varepsilon(y)) = f \circ x + \varepsilon\left(\partial[f] \circ \langle x, y \rangle\right)$
[C∂.1] $\partial[f + g] = \partial[f] + \partial[g]$, $\partial[0] = 0$, *and* $\partial[\varepsilon(f)] = \varepsilon(\partial[f])$
[C∂.2] $\partial[f] \circ \langle x, y + z \rangle = \partial[f] \circ \langle x, y \rangle + \partial[f] \circ \langle x + \varepsilon(y), z \rangle$ *and* $\partial[f] \circ \langle x, 0 \rangle = 0$
[C∂.3] $\partial[1_A] = \pi_1$ *and* $\partial[\pi_0] = \pi_1; \pi_0$ *and* $\partial[\pi_1] = \pi_1; \pi_0$
[C∂.4] $\partial[\langle f, g \rangle] = \langle \partial[f], \partial[g] \rangle$ *and* $\partial[!_A] = !_{A \times A}$
[C∂.5] $\partial[g \circ f] = \partial[g] \circ \langle f \circ \pi_0, \partial[f] \rangle$
[C∂.6] $\partial\left[\partial[f]\right] \circ \langle \langle x, y \rangle, \langle 0, z \rangle \rangle = \partial[f] \circ \langle x + \varepsilon(y), z \rangle$
[C∂.7] $\partial\left[\partial[f]\right] \circ \langle \langle x, y \rangle, \langle z, 0 \rangle \rangle = \partial\left[\partial[f]\right] \circ \langle \langle x, z \rangle, \langle y, 0 \rangle \rangle$

Before giving some intuition on the axioms **[C∂.0]** to **[C∂.7]**, we first observe that one could have used change action notation to express **[C∂.0]**, **[C∂.2]**, and **[C∂.6]** which would then be written as:

[C∂.0] $f \circ (x \oplus y) = (f \circ x) \oplus (\partial[f] \circ \langle x, y \rangle)$
[C∂.2] $\partial[f] \circ \langle x, y + z \rangle = \partial[f] \circ \langle x, y \rangle + \partial[f] \circ \langle x \oplus y, z \rangle$ and $\partial[f] \circ \langle x, 0 \rangle = 0$
[C∂.6] $\partial\left[\partial[f]\right] \circ \langle \langle x, y \rangle, \langle 0, z \rangle \rangle = \partial[f] \circ \langle x \oplus y, z \rangle$

And also, just like Cartesian differential categories, **[C∂.6]** and **[C∂.7]** have alternative equivalent expressions.

Lemma 5. *In the presence of the other axioms,* **[C∂.6]** *and* **[C∂.7]** *are equivalent to:*

[C∂.6.a] $\partial\left[\partial[f]\right] \circ \langle\langle x, 0\rangle, \langle 0, y\rangle\rangle = \partial[f] \circ \langle x, y\rangle$
[C∂.7.a] $\partial\left[\partial[f]\right] \circ \langle\langle x, y\rangle, \langle z, w\rangle\rangle = \partial\left[\partial[f]\right] \circ \langle\langle x, z\rangle, \langle y, w\rangle\rangle$

Proof. The proof is essentially the same as [7, Proposition 4.2]. ∎

The keen eyed reader will notice that the axioms of a Cartesian difference category are very similar to the axioms of a Cartesian differential category. Indeed, **[C∂.1]**, **[C∂.3]**, **[C∂.4]**, **[C∂.5]**, and **[C∂.7]** are the same as their Cartesian differential category counterpart. The axioms which are different are **[C∂.2]** and **[C∂.6]** where the infinitesimal extension ε is now included, and also there is the new extra axiom **[C∂.0]**. On the other hand, interestingly enough, **[C∂.6.a]** is the same as **[CD.6.a]**. We also point out that writing out **[C∂.0]** and **[C∂.2]** using change action notion, we see that these axioms are precisely **[CAD.1]** and **[CAD.2]** respectively. To better understand **[C∂.0]** to **[C∂.7]** it may be useful to write them out using element-like notation. In element-like notation, **[C∂.0]** is written as:

$$f(x + \varepsilon(y)) = f(x) + \varepsilon\left(\partial[f](x, y)\right)$$

This condition can be read as a generalization of the Kock-Lawvere axiom that characterizes the derivative in from synthetic differential geometry [13]. Broadly speaking, the Kock-Lawvere axiom states that, for any map $f : \mathcal{R} \to \mathcal{R}$ and any $x \in \mathcal{R}$ and $d \in \mathcal{D}$, there exists a unique $f'(x) \in \mathcal{R}$ verifying

$$f(x + d) = f(x) + d \cdot f'(x)$$

where $\mathcal{D}$ is the subset of $\mathcal{R}$ consisting of infinitesimal elements. It is by analogy with the Kock-Lawvere axiom that we refer to ε as an "infinitesimal extension" as it can be thought of as embedding the space A into a subspace $\varepsilon(A)$ of infinitesimal elements.

[C∂.1] states that the differential of a sum of maps is the sum of differentials, and similarly for zero maps and the infinitesimal extension of a map. **[C∂.2]** is the first crucial difference between a Cartesian difference category and a Cartesian differential category. In a Cartesian differential category, the differential of a map is assumed to be additive in its second argument. In a Cartesian difference category, just as derivatives for change actions, while the differential is still required to preserve zeros in its second argument, it is only additive "up to a small perturbation", that is:

$$\partial[f](x, y + z) = \partial[f](x, y) + \partial[f](x + \varepsilon(y), z)$$

[C∂.3] tells us what the differential of the identity and projection maps are, while **[C∂.4]** says that the differential of a pairing of maps is the pairing of their differentials. **[C∂.5]** is the chain rule which expresses what the differential of a composition of maps is:

$$\partial[g \circ f](x, y) = \partial[g](f(x), \partial[f](x, y))$$

[C∂.6] and **[C∂.7]** tell us how to work with second order differentials. **[C∂.6]** is expressed as follows:

$$\partial\left[\partial[f]\right](x, y, 0, z) = \partial[f](x + \varepsilon(y), z)$$

and finally **[C$\boldsymbol{\partial}$.7]** is expressed as:

$$\boldsymbol{\partial}\left[\boldsymbol{\partial}[f]\right](x,y,z,0) = \boldsymbol{\partial}\left[\boldsymbol{\partial}[f]\right](x,z,y,0)$$

It is interesting to note that while **[C$\boldsymbol{\partial}$.6]** is different from **[CD.6]**, its alternative version **[C$\boldsymbol{\partial}$.6.a]** is the same as **[CD.6.a]**.

$$\boldsymbol{\partial}\left[\boldsymbol{\partial}[f]\right]((x,0),(0,y)) = \boldsymbol{\partial}[f](x,z)$$

4.3 Another look at Cartesian Differential Categories

Here we explain how a Cartesian differential category is a Cartesian difference category where the infinitesimal extension is given by zero.

Proposition 1. *Every Cartesian differential category $\mathbb{X}$ with differential combinator D is a Cartesian difference category where the infinitesimal extension is defined as $\varepsilon(f) = 0$ and the difference combinator is defined to be the differential combinator, $\boldsymbol{\partial} = \mathsf{D}$.*

Proof. As noted before, the first two parts of the **[C$\boldsymbol{\partial}$.1]**, the second part of **[C$\boldsymbol{\partial}$.2]**, **[C$\boldsymbol{\partial}$.3]**, **[C$\boldsymbol{\partial}$.4]**, **[C$\boldsymbol{\partial}$.5]**, and **[C$\boldsymbol{\partial}$.7]** are precisely the same as their Cartesian differential axiom counterparts. On the other hand, since $\varepsilon(f) = 0$, **[C$\boldsymbol{\partial}$.0]** and the third part of **[C$\boldsymbol{\partial}$.1]** trivial state that $0 = 0$, while the first part of **[C$\boldsymbol{\partial}$.2]** and **[C$\boldsymbol{\partial}$.6]** end up being precisely the first part of **[CD.2]** and **[CD.6]**. Therefore, the differential combinator satisfies the Cartesian difference axioms and we conclude that a Cartesian differential category is a Cartesian difference category. ■

Conversely, one can always build a Cartesian differential category from a Cartesian difference category by considering the objects for which the infinitesimal extension is the zero map.

Proposition 2. *For a Cartesian difference category $\mathbb{X}$ with infinitesimal extension ε and difference combinator $\boldsymbol{\partial}$, then $\mathbb{X}_0$, the full subcategory of objects A such that $\varepsilon(1_A) = 0$, is a Cartesian differential category where the differential combinator is defined to be the difference combinator, $\mathsf{D} = \boldsymbol{\partial}$.*

Proof. First note that if $\varepsilon(1_A) = 0$ and $\varepsilon(1_B) = 0$, then by definition it also follows that $\varepsilon(1_{A\times B}) = 0$, and also that for the terminal object $\varepsilon(1_\mathsf{T}) = 0$ by uniqueness of maps into the terminal object. Thus $\mathbb{X}_0$ is closed under finite products and is therefore a Cartesian left additive category. Furthermore, we again note that since $\varepsilon(f) = 0$, this implies that for maps between such objects the Cartesian difference axioms are precisely the Cartesian differential axioms. Therefore, the difference combinator is a differential combinator for this subcategory, and so $\mathbb{X}_0$ is a Cartesian differential category. ■

In any Cartesian difference category $\mathbb{X}$, the terminal object $\top$ always satisfies that $\varepsilon(1_\top) = 0$, and so therefore, $\mathbb{X}_0$ is never empty. On the other hand, applying Proposition 2 to a Cartesian differential category results in the entire category. It is also important to note that the above two propositions do not imply that if a difference combinator is a differential combinator then the infinitesimal extension must be zero. In Section 5.3, we provide such an example of a Cartesian differential category that comes equipped with a non-zero infinitesimal extension such that the differential combinator is a difference combinator with respect to this non-zero infinitesimal extension.

4.4 Cartesian Difference Categories as Change Action Models

In this section, we show how every Cartesian difference category is a particularly well-behaved change action model, and conversely how every change action model contains a Cartesian difference category.

Proposition 3. *Let $\mathbb{X}$ be a Cartesian difference category with infinitesimal extension ε and difference combinator $\boldsymbol{\partial}$. Define the functor $\alpha : \mathbb{X} \to \mathsf{CAct}(\mathbb{X})$ as $\alpha(A) = (A, A, \oplus_A, +_A, 0_A)$ (as defined in Lemma 3) and $\alpha(f) = (f, \boldsymbol{\partial}[f])$. Then $(\mathbb{X}, \alpha : \mathbb{X} \to \mathsf{CAct}(\mathbb{X}))$ is a change action model.*

Proof. By Lemma 3, $(A, A, \oplus_A, +_A, 0_A)$ is a change action and so α is well-defined on objects. While for a map f, $\boldsymbol{\partial}[f]$ is a derivative of f in the change action sense since **[C∂.0]** and **[C∂.2]** are precisely **[CAD.1]** and **[CAD.2]**, and so α is well-defined on maps. That α preserves identities and composition follows from **[C∂.3]** and **[C∂.5]** respectively, and so α is a functor. That α preserves finite products will follow from **[C∂.3]** and **[C∂.4]**. Lastly, it is clear that α section of the forgetful functor, and therefore we conclude that $(\mathbb{X}, \alpha)$ is a change action model. ∎

It is clear that not every change action model is a Cartesian difference category. For example, change action models do not require the addition to be commutative. On the other hand, it can be shown that every change action model contains a Cartesian difference category as a full subcategory.

Definition 9. *Let $(\mathbb{X}, \alpha : \mathbb{X} \to \mathsf{CAct}(\mathbb{X}))$ be a change action model. An object A is **flat** whenever the following hold:*

[F.1] $\Delta A = A$
[F.2] $\alpha(\oplus_A) = (\oplus_A, \oplus_A \circ \pi_1)$
[F.3] $0 \oplus_A (0 \oplus_A f) = 0 \oplus_A f$ *for any* $f : U \to A$.
[F.4] $\oplus_A$ *is right-injective, that is, if* $\oplus_A \circ \langle f, g\rangle = \oplus_A \circ \langle f, h\rangle$ *then* $g = h$.

We would like to show that for any change action model $(\mathbb{X}, \alpha)$, its full subcategory of flat objects, Flat_α is a Cartesian difference category. Starting with the finite product structure, since α preserves finite products, it is straightforward to see that $\top$ is Euclidean and if A and B are flat then so is $A \times B$. The sum of maps $f : A \to B$ and $g : A \to B$ in Flat_α is defined using the change action structure $f +_B g$, while the zero map $0 : A \to B$ is $0 = 0_B \circ !_A$. And so we obtain that:

Lemma 6. Flat_α *is a Cartesian left additive category.*

Proof. Most of the Cartesian left additive structure is straightforward. However, since the addition is not required to be commutative for arbitrary change actions, we will show that the addition is commutative for Euclidean objects. Using that $\oplus_B$ is an action, that by **[F.2]** we have that $\oplus_B \circ \pi_1$ is a derivative for $\oplus_B$, and **[CAD.1]**, we obtain that:

$$0_B \oplus_B (f +_B g) = (0_B \oplus_B f) \oplus_B g = (0_B \oplus_B g) \oplus_B f = 0_B \oplus_B (g +_B f)$$

By **[F.4]**, $\oplus_B$ is right-injective and we conclude that $f + g = g + f$. ∎

As an immediate consequence We note that for any change action model $(\mathbb{X}, \alpha)$, since the terminal object is always flat, Flat_α is never empty.

We use the action of the change action structure to define the infinitesimal extension. So for a map $f : A \to B$ in Flat_α, define $\varepsilon(f) : A \to B$ as follows:

$$\varepsilon(f) = \oplus_B \circ \langle 0_B \circ !_A, f \rangle = 0 \oplus_B f$$

Lemma 7. ε *is an infinitesimal extension for* Flat_α.

Proof. We show that ε preserve the addition. Following the same idea as in the proof of Lemma 6, we obtain the following:

$$\begin{aligned}
&0_B \oplus_B \varepsilon(f +_B g) = 0_B \oplus_B (0_B \oplus_B (f +_B g)) \\
&= (0_B \oplus_B 0_B) \oplus_B ((0_B \oplus_B f) \oplus_B g) = (0_B \oplus_B (0_B \oplus_B f)) \oplus_B (0_B \oplus_B g) \\
&= (0_B \oplus_B \varepsilon(f)) \oplus_B \varepsilon(g) = 0_B \oplus_B (\varepsilon(f) +_B \varepsilon(g))
\end{aligned}$$

Then by **[F.3]**, it follows that $\varepsilon(f+g) = \varepsilon(f) + \varepsilon(g)$. The remaining infinitesimal extension axioms are proven in a similar fashion. ∎

Lastly, the difference combinator for Flat_α is defined in the obvious way, that is, $\boldsymbol{\partial}[f]$ is defined as the second component of $\alpha(f)$.

Proposition 4. *Let* $(\mathbb{X}, \alpha : \mathbb{X} \to \mathsf{CAct}(\mathbb{X}))$ *be a change action model. Then* Flat_α *is a Cartesian difference category.*

Proof (Sketch). The full calculations will appear in an upcoming extended journal version of this paper, but we give an informal explanation. **[C∂.0]** and **[C∂.2]** are a straightforward consequences of **[CAD.1]** and **[CAD.2]**. **[C∂.3]** and **[C∂.4]** follow trivially from the fact that α preserves finite products and from the structure of products in $\mathsf{CAct}(\mathbb{X})$, while **[C∂.5]** follows from composition in $\mathsf{CAct}(\mathbb{X})$. **[C∂.1]**, **[C∂.6]** and **[C∂.7]** are obtained by mechanical calculation in the spirit of Lemma 6. Note that every axiom except for **[C∂.6]** can be proven without using **[F.3]** ∎

4.5 Linear Maps and ε-Linear Maps

An important subclass of maps in a Cartesian differential category is the subclass of *linear maps* [4, Definition 2.2.1]. One can also define linear maps in a Cartesian difference category by using the same definition.

Definition 10. *In a Cartesian difference category, a map f is* **linear** *if the following equality holds: $\partial[f] = f \circ \pi_1$.*

Using element-like notation, a map f is linear if $\partial[f](x, y) = f(y)$. Linear maps in a Cartesian difference category satisfy many of the same properties found in [4, Lemma 2.2.2].

Lemma 8. *In a Cartesian difference category,*

1. *If $f : A \to B$ is linear then $\varepsilon(f) = f \circ \varepsilon(1_A)$;*
2. *If $f : A \to B$ is linear, then f is additive (Definition 1);*
3. *Identity maps, projection maps, and zero maps are linear;*
4. *The composite, sum, and pairing of linear maps is linear;*
5. *If $f : A \to B$ and $k : C \to D$ are linear, then for any map $g : B \to C$, the following equality holds: $\partial[k \circ g \circ f] = k \circ \partial[g] \circ (f \times f)$;*
6. *If an isomorphism is linear, then its inverse is linear;*
7. *For any object A, $\oplus_A$ and $+_A$ are linear.*

Using element-like notation, the first point of the above lemma says that if f is linear then $f(\varepsilon(x)) = \varepsilon(f(x))$. And while all linear maps are additive, the converse is not necessarily true, see [4, Corollary 2.3.4]. However, an immediate consequence of the above lemma is that the subcategory of linear maps of a Cartesian difference category has finite biproducts.

Another interesting subclass of maps is the subclass of ε-linear maps, which are maps whose infinitesimal extension is linear.

Definition 11. *In a Cartesian difference category, a map f is* **ε-linear** *if $\varepsilon(f)$ is linear.*

Lemma 9. *In a Cartesian difference category,*

1. *If $f : A \to B$ is ε-linear then $f \circ (x + \varepsilon(y)) = f \circ x + \varepsilon(f) \circ y$;*
2. *Every linear map is ε-linear;*
3. *The composite, sum, and pairing of ε-linear maps is ε-linear;*
4. *If an isomorphism is ε-linear, then its inverse is again ε-linear.*

Using element-like notation, the first point of the above lemma says that if f is ε-linear then $f(x + \varepsilon(y)) = f(x) + \varepsilon(f(y))$. So ε-linear maps are additive on "infinitesimal" elements (i.e. those of the form $\varepsilon(y)$).

For a Cartesian differential category, linear maps in the Cartesian difference category sense are precisely the same as the Cartesian differential category sense [4, Definition 2.2.1], while every map is ε-linear since $\varepsilon = 0$.

5 Examples of Cartesian Difference Categories

5.1 Smooth Functions

Every Cartesian differential category is a Cartesian difference category where the infinitesimal extension is zero. As a particular example, we consider the category of real smooth functions, which as mentioned above, can be considered to be the canonical (and motivating) example of a Cartesian differential category.

Let $\mathbb{R}$ be the set of real numbers and let SMOOTH be the category whose objects are Euclidean spaces $\mathbb{R}^n$ (including the point $\mathbb{R}^0 = \{*\}$), and whose maps are smooth functions $F : \mathbb{R}^n \to \mathbb{R}^m$. SMOOTH is a Cartesian left additive category where the product structure is given by the standard Cartesian product of Euclidean spaces and where the additive structure is defined by point-wise addition, $(F + G)(\boldsymbol{x}) = F(\boldsymbol{x}) + G(\boldsymbol{x})$ and $0(\boldsymbol{x}) = (0, \dots, 0)$, where $\boldsymbol{x} \in \mathbb{R}^n$. SMOOTH is a Cartesian differential category where the differential combinator is defined by the directional derivative of smooth functions. Explicitly, for a smooth function $F : \mathbb{R}^n \to \mathbb{R}^m$, which is in fact a tuple of smooth functions $F = (f_1, \dots, f_n)$ where $f_i : \mathbb{R}^n \to \mathbb{R}$, $\mathsf{D}[F] : \mathbb{R}^n \times \mathbb{R}^n \to \mathbb{R}^m$ is defined as follows:

$$\mathsf{D}[F](\boldsymbol{x}, \boldsymbol{y}) := \left(\sum_{i=1}^{n} \frac{\partial f_1}{\partial u_i}(\boldsymbol{x}) y_i, \dots, \sum_{i=1}^{n} \frac{\partial f_n}{\partial u_i}(\boldsymbol{x}) y_i \right)$$

where $\boldsymbol{x} = (x_1, \dots, x_n), \boldsymbol{y} = (y_1, \dots, y_n) \in \mathbb{R}^n$. Alternatively, $\mathsf{D}[F]$ can also be defined in terms of the Jacobian matrix of F. Therefore SMOOTH is a Cartesian difference category with infinitesimal extesion $\varepsilon = 0$ and with difference combinator D. Since $\varepsilon = 0$, the induced action is simply $\boldsymbol{x} \oplus_{\mathbb{R}^n} \boldsymbol{y} = \boldsymbol{x}$. Also a smooth function is linear in the Cartesian difference category sense precisely if it is $\mathbb{R}$-linear in the classical sense, and every smooth function is ε-linear.

5.2 Calculus of Finite Differences

Here we explain how the difference operator from the calculus of finite differences gives an example of a Cartesian difference category but *not* a Cartesian differential category. This example was the main motivating example for developing Cartesian difference categories. The calculus of finite differences is captured by the category of abelian groups and arbitrary set functions between them.

Let $\overline{\mathsf{Ab}}$ be the category whose objects are abelian groups G (where we use additive notation for group structure) and where a map $f : G \to H$ is simply an arbitrary function between them (and therefore does not necessarily preserve the group structure). $\overline{\mathsf{Ab}}$ is a Cartesian left additive category where the product structure is given by the standard Cartesian product of sets and where the additive structure is again given by point-wise addition, $(f + g)(x) = f(x) + g(x)$ and $0(x) = 0$. $\overline{\mathsf{Ab}}$ is a Cartesian difference category where the infinitesimal extension is simply given by the identity, that is, $\varepsilon(f) = f$, and and where the difference combinator $\boldsymbol{\partial}$ is defined as follows for a map $f : G \to H$:

$$\boldsymbol{\partial}[f](x, y) = f(x + y) - f(x)$$

On the other hand, ∂ is not a differential combinator for $\overline{\mathsf{Ab}}$ since it does not satisfy **[CD.6]** and part of **[CD.2]**. Thanks to the addition of the infinitesimal extension, ∂ does satisfy **[C∂.2]** and **[C∂.6]**, as well as **[C∂.0]**. However, as noted in [5], it is interesting to note that this ∂ does satisfy **[CD.1]**, the second part of **[CD.2]**, **[CD.3]**, **[CD.4]**, **[CD.5]**, **[CD.7]**, and **[CD.6.a]**. It is worth noting that in [5], the goal was to drop the addition and develop a "non-additive" version of Cartesian differential categories.

In $\overline{\mathsf{Ab}}$, since the infinitesimal operator is given by the identity, the induced action is simply the addition, $x \oplus_G y = x + y$. On the other hand, the linear maps in $\overline{\mathsf{Ab}}$ are precisely the group homomorphisms. Indeed, f is linear if $\partial[f](x, y) = f(y)$. But by **[C∂.0]** and **[C∂.2]**, we get that:

$$f(x + y) = f(x) + \partial[f](x, y) = f(x) + f(y) \qquad f(0) = \partial[f](x, 0) = 0$$

So f is a group homomorphism. Conversely, if f is a group homomorphism:

$$\partial[f](x, y) = f(x + y) - f(x) = f(x) + f(y) - f(x) = f(y)$$

So f is linear. Since $\varepsilon(f) = f$, the ε-linear maps are precisely the linear maps.

5.3 Module Morphisms

Here we provide a simple example of a Cartesian difference category whose difference combinator is also a differential combinator, but where the infinitesimal extension is neither zero nor the identity.

Let R be a commutative semiring and let MOD_R be the category of R-modules and R-linear maps between them. MOD_R has finite biproducts and is therefore a Cartesian left additive category where every map is additive. Every $r \in R$ induces an infinitesimal extension ε^r defined by scalar multiplication, $\varepsilon^r(f)(m) = rf(m)$. Then MOD_R is a Cartesian difference category with the infinitesimal extension ε^r for any $r \in R$ and difference combinator ∂ defined as:

$$\partial[f](m, n) = f(n)$$

R-linearity of f assures that **[C∂.0]** holds, while the remaining Cartesian difference axioms hold trivially. In fact, ∂ is also a differential combinator and therefore MOD_R is also a Cartesian differential category. The induced action is given by $m \oplus_M n = m + rn$. By definition of ∂, every map in MOD_R is linear, and by definition of ε^r and R-linearity, every map is also ε-linear.

5.4 Stream calculus

Here we show how one can extend the calculus of finite differences example to stream calculus. The differential calculus of causal functions and interesting applications have recently been studying in [17, 18].

For a set A, let A^ω denote the set of infinite sequences of elements of A, where we write $[a_i]$ for the infinite sequence $[a_i] = (a_1, a_2, a_3, \ldots)$ and $a_{i:j}$ for the (finite) subsequence $(a_i, a_{i+1}, \ldots, a_j)$. A function $f : A^\omega \to B^\omega$ is **causal** whenever the n-th element $f([a_i])_n$ of the output sequence only depends on the first n elements of $[a_i]$, that is, f is causal if and only if whenever $a_{0:n} = b_{0:n}$ then $f([a_i])_{0:n} = f([b_i])_{0:n}$. We now consider streams over abelian groups, so let $\overline{\mathsf{Ab}}^\omega$ be the category whose objects are all the Abelian groups and whose morphisms are causal maps from G^ω to H^ω. $\overline{\mathsf{Ab}}^\omega$ is a Cartesian left-additive category, where the product is given by the standard product of abelian groups and where the additive structure is lifted point-wise from the structure of $\overline{\mathsf{Ab}}$, that is, $(f+g)([a_i])_n = f([a_i])_n + g([a_i])_n$ and $0([a_i])_n = 0$. In order to define the infinitesimal extension, we first need to define the truncation operator $\mathbf{z}$. So let G be an abelian group and $[a_i] \in G^\omega$, then define the sequence $\mathbf{z}([a_i])$ as:

$$\mathbf{z}([a_i])_0 = 0 \qquad\qquad \mathbf{z}([a_i])_{n+1} = a_{n+1}$$

The category $\overline{\mathsf{Ab}}^\omega$ is a Cartesian difference category where the infinitesimal extension is given by the truncation operator, $\varepsilon(f)([a_i]) = \mathbf{z}(f([a_i]))$,
and where the difference combinator $\boldsymbol{\partial}$ is defined as follows:

$$\boldsymbol{\partial}[f]([a_i], [b_i])_0 = f([a_i] + [b_i])_0 - f([a_i])_0$$
$$\boldsymbol{\partial}[f]([a_i], [b_i])_{n+1} = f([a_i] + \mathbf{z}([b_i]))_{n+1} - f([a_i])_{n+1}$$

Note the similarities between the difference combinator on $\overline{\mathsf{Ab}}$ and that on $\overline{\mathsf{Ab}}^\omega$. The induced action is computed out to be:

$$([a_i] \oplus [b_i])_0 = a_0 \qquad\qquad ([a_i] \oplus [b_i])_{n+1} = a_{n+1} + b_{n+1}$$

A causal map is linear (in the Cartesian difference category sense) if and only if it is a group homomorphism. While a causal map f is ε-linear if and only if it is a group homomorphism which does not the depend on the 0-th term of its input, that is, $f([a_i]) = f(\mathbf{z}([a_i]))$.

6 Tangent Bundles in Cartesian Difference Categories

In this section, we show that the difference combinator of a Cartesian difference category induces a monad, called the *tangent monad*, whose Kleisli category is again a Cartesian difference category. This construction is a generalization of the tangent monad for Cartesian differential categories [7, 15]. However, the Kleisli category of the tangent monad of a Cartesian differential category is *not* a Cartesian differential category, but rather a Cartesian difference category.

6.1 The Tangent Bundle Monad

Let $\mathbb{X}$ be a Cartesian difference category with infinitesimal extension ε and difference combinator $\boldsymbol{\partial}$. Define the functor $\mathsf{T} : \mathbb{X} \to \mathbb{X}$ as follows:

$$\mathsf{T}(A) = A \times A \qquad\qquad \mathsf{T}(f) = \langle f \circ \pi_0, \boldsymbol{\partial}[f] \rangle$$

and define the natural transformations $\eta : 1_{\mathbb{X}} \Rightarrow \mathsf{T}$ and $\mu : \mathsf{T}^2 \Rightarrow \mathsf{T}$ as follows:

$$\eta_A := \langle 1_A, 0 \rangle \qquad \mu_A := \langle \pi_0 \circ \pi_0, \pi_1 \circ \pi_0 + \pi_0 \circ \pi_1 + \varepsilon(\pi_1 \circ \pi_1) \rangle$$

Proposition 5. (T, μ, η) *is a monad.*

Proof. Functoriality of T will follow from **[C∂.3]** and the chain rule **[C∂.5]**. Naturality of η and μ and the monad identities will follow from the remaining difference combinator axioms. The full lengthy brute force calculations will appear in an upcoming extended journal version of this paper. ■

When $\mathbb{X}$ is a Cartesian differential category with the difference structure arising from setting $\varepsilon = 0$, this tangent bundle monad coincides with the standard tangent monad corresponding to its tangent category structure [7, 15].

6.2 The Kleisli Category of T

Recall that the Kleisli category of the monad (T, μ, η) is defined as the category $\mathbb{X}_\mathsf{T}$ whose objects are the objects of $\mathbb{X}$, and where a map $A \to B$ in $\mathbb{X}_\mathsf{T}$ is a map $f : A \to \mathsf{T}(B)$ in $\mathbb{X}$, which would be a pair $f = \langle f_0, f_1 \rangle$ where $f_j : A \to B$. The identity map in $\mathbb{X}_\mathsf{T}$ is the monad unit $\eta_A : A \to \mathsf{T}(A)$, while composition of Kleisli maps $f : A \to \mathsf{T}(B)$ and $g : B \to \mathsf{T}(C)$ is defined as the composite $\mu_C \circ \mathsf{T}(g) \circ f$. To distinguish between composition in $\mathbb{X}$ and $\mathbb{X}_\mathsf{T}$, we denote Kleisli composition as $g \circ^\mathsf{T} f = \mu_C \circ \mathsf{T}(g) \circ f$. If $f = \langle f_0, f_1 \rangle$ and $g = \langle g_0, g_1 \rangle$, then their Kleisli composition can be explicitly computed out to be:

$$g \circ^\mathsf{T} f = \langle g_0, g_1 \rangle \circ^\mathsf{T} \langle f_0, f_1 \rangle = \langle g_0 \circ f_0, \boldsymbol{\partial}[g_0] \circ \langle f_0, f_1 \rangle + g_1 \circ (f_0 + \varepsilon(f_1)) \rangle$$

Kleisli maps can be understood as "generalized" vector fields. Indeed, $\mathsf{T}(A)$ should be thought of as the tangent bundle over A, and therefore a vector field would be a map $\langle 1, f \rangle : A \to \mathsf{T}(A)$, which is of course also a Kleisli map. For more details on the intuition behind this Kleisli category see [7]. We now wish to explain how the Kleisli category is again a Cartesian difference category.

We begin by exhibiting the Cartesian left additive structure of the Kleisli category. The product of objects in $\mathbb{X}_\mathsf{T}$ is defined as $A \times B$ with projections $\pi_0^\mathsf{T} : A \times B \to \mathsf{T}(A)$ and $\pi_1^\mathsf{T} : A \times B \to \mathsf{T}(B)$ defined respectively as $\pi_0^\mathsf{T} = \langle \pi_0, 0 \rangle$ and $\pi_1^\mathsf{T} = \langle \pi_1, 0 \rangle$. The pairing of Kleisli maps $f = \langle f_0, f_1 \rangle$ and $g = \langle , g_0, g_1 \rangle$ is defined as $\langle f, g \rangle^\mathsf{T} = \langle \langle f_0, g_0 \rangle, \langle f_1, g_1 \rangle \rangle$. The terminal object is again $\top$ and where the unique map to the terminal object is $!_A^\mathsf{T} = 0$. The sum of Kleisli maps f Kleisli maps $f = \langle f_0, f_1 \rangle$ and $g = \langle , g_0, g_1 \rangle$ is defined as $f +^\mathsf{T} g = f + g = \langle f_0 + g_0, f_1 + g_1 \rangle$, and the zero Kleisli maps is simply $0^\mathsf{T} = 0 = \langle 0, 0 \rangle$. Therefore we conclude that the Kleisli category of the tangent monad is a Cartesian left additive category.

Lemma 10. $\mathbb{X}_\mathsf{T}$ *is a Cartesian left additive category.*

The infinitesimal extension ε^T for the Kleisli category is defined as follows for a Kleisli map $f = \langle f_0, f_1 \rangle$:

$$\varepsilon^\mathsf{T}(f) = \langle 0, f_0 + \varepsilon(f_1) \rangle$$

Lemma 11. *ε^{T} is an infinitesimal extension on $\mathbb{X}_{\mathsf{T}}$.*

It is interesting to point out that for an object A the induced action $\oplus_A^{\mathsf{T}}$ can be computed out to be:

$$\oplus_A^{\mathsf{T}} = \pi_0^{\mathsf{T}} +^{\mathsf{T}} \varepsilon^{\mathsf{T}}(\pi_1) = \langle \pi_0, 0 \rangle + \langle 0, \pi_1 \rangle = \langle \pi_0, \pi_1 \rangle = 1_{\mathsf{T}(A)}$$

and we stress that this is the identity of $\mathsf{T}(A)$ in the base category $\mathbb{X}$ (but not in the Kleisli category).

To define the difference combinator for the Kleisli category, first note that difference combinators by definition do not change the codomain. That is, if $f : A \to \mathsf{T}(B)$ is a Kleisli arrow, then the type of its derivative *qua* Kleisli arrow should be $A \times A \to \mathsf{T}(B) \times \mathsf{T}(B)$, which coincides with the type of its derivative in $\mathbb{X}$. Therefore, the difference combinator $\boldsymbol{\partial}^{\mathsf{T}}$ for the Kleisli category can be defined to be the difference combinator of the base category, that is, for a Kleisli map $f = \langle f_0, f_1 \rangle$:

$$\boldsymbol{\partial}^{\mathsf{T}}[f] = \boldsymbol{\partial}[f] = \langle \boldsymbol{\partial}[f_0], \boldsymbol{\partial}[f_1] \rangle$$

Proposition 6. *For a Cartesian difference category $\mathbb{X}$, the Kleisli category $\mathbb{X}_{\mathsf{T}}$ is a Cartesian difference category with infinitesimal extension ε^{T} and difference combinator $\boldsymbol{\partial}^{\mathsf{T}}$.*

Proof. The full lengthy brute force calculations will appear in an upcoming extended journal version of this paper. We do note that a crucial identity for this proof is that for any map f in $\mathbb{X}$, the following equality holds:

$$\mathsf{T}(\boldsymbol{\partial}[f]) = \boldsymbol{\partial}\left[\mathsf{T}(f)\right] \circ \langle \pi_0 \times \pi_0, \pi_1 \times \pi_1 \rangle$$

This helps simplify many of the calculations for the difference combinator axioms since $\mathsf{T}(\boldsymbol{\partial}[f])$ appears everywhere due to the definition of Kleisli composition. ∎

As a result, the Kleisli category of a Cartesian difference category is again a Cartesian difference category, whose infinitesimal extension is neither the identity or the zero map. This allows one to build numerous examples of interesting and exotic Cartesian difference categories, such as the Kleisli category of Cartesian differential categories (or iterating this process, taking the Kleisli category of the Kleisli category). We highlight the importance of this construction in the Cartesian differential case as it does not in general result in a Cartesian differential category. Indeed, even if $\varepsilon = 0$, it is always the case that $\varepsilon^{\mathsf{T}} \neq 0$. We conclude this section by taking a look at the linear maps and the ε^{T}-linear maps in the Kleisli category. A Kleisli map $f = \langle f_0, f_1 \rangle$ is linear in the Kleisli category if $\boldsymbol{\partial}^{\mathsf{T}}[f] = f \circ^{\mathsf{T}} \pi_1^{\mathsf{T}}$, which amounts to requiring that:

$$\langle \boldsymbol{\partial}[f_0], \boldsymbol{\partial}[f_1] \rangle = \langle f_0 \circ \pi_1, f_1 \circ \pi_1 \rangle$$

Therefore a Kleisli map is linear in the Kleisli category if and only if it is the pairing of maps which are linear in the base category. On the other hand, f is ε^{T}-linear if $\varepsilon^{T}(f) = \langle 0, f_0 + \varepsilon(f_1) \rangle$ is linear in the Kleisli category, which in this case amounts to requiring that $f_0 + \varepsilon(f_1)$ is linear. Therefore, if f_0 is linear and f_1 is ε-linear, then f is ε^{T}-linear.

7 Conclusions and Future Work

We have presented Cartesian difference categories, which generalize Cartesian differential categories to account for more discrete definitions of derivatives while providing an additional structure that is absent in change action models. We have also exhibited important examples and shown that Cartesian difference categories arise quite naturally from considering tangent bundles in any Cartesian differential category. We claim that Cartesian difference categories can facilitate the exploration of differentiation in discrete spaces, by generalizing techniques and ideas from the study of their differential counterparts. For example, Cartesian differential categories can be extended to allow objects whose tangent space is not necessarily isomorphic to the object itself [9]. The same generalization could be applied to Cartesian difference categories – with some caveats: for example, the equation defining a linear map (Definition 10) becomes ill-typed, but the notion of ε-linear map remains meaningful.

Another relevant path to consider is developing the analogue of the "tensor" story for Cartesian difference categories. Indeed, an important source of examples of Cartesian differential categories are the coKleisli categories of a tensor differential category [3, 4]. A similar result likely holds for a hypothetical "tensor difference category", but it is not clear how these should be defined: **[C∂.2]** implies that derivatives in the difference sense are non-linear and therefore their interplay with the tensor structure will be much different.

A further generalization of Cartesian differential categories, categories with tangent structure [7] are defined directly in terms of a tangent bundle functor rather than requiring that every tangent bundle be trivial (that is, in a tangent category it may not be the case that $\mathsf{T}A = A \times A$). Some preliminary research on change actions has already shown that, when generalized in this way, change actions are precisely internal categories, but the consequences of this for change action models (and, *a fortiori*, Cartesian difference categories) are not understood. More recently, some work has emerged about differential equations using the language of tangent categories [8]. We believe similar techniques can be applied in a straightforward way to Cartesian difference categories, where they might be of use to give an abstract formalization of discrete dynamical systems and difference equations.

An important open question is whether Cartesian difference categories (or a similar notion) admit an internal language. It is well-known that the differential λ-calculus can be interpreted in Cartesian closed differential categories [14]. Given their similarities, we believe there will be a very similar "difference λ-calculus" which could potentially have applications to automatic differentiation (change structures, a notion similar to change actions, have already been proposed as models of forward-mode automatic differentiation [12], although work on the area seems to have stagnated).

Lastly, we should mention that there are adjunctions between the categories of Cartesian difference categories, change action models, and Cartesian differential categories given by Proposition 1, 2, 3, and 4. These adjunctions will be explored in detail in the upcoming journal version of this paper.

References

1. Alvarez-Picallo, M., Eyers-Taylor, A., Jones, M.P., Ong, C.H.L.: Fixing incremental computation. In: European Symposium on Programming. pp. 525–552. Springer (2019)
2. Alvarez-Picallo, M., Ong, C.H.L.: Change actions: models of generalised differentiation. In: International Conference on Foundations of Software Science and Computation Structures. pp. 45–61. Springer (2019)
3. Blute, R.F., Cockett, J.R.B., Seely, R.A.G.: Differential categories. Mathematical structures in computer science **16**(06), 1049–1083 (2006)
4. Blute, R.F., Cockett, J.R.B., Seely, R.A.G.: Cartesian differential categories. Theory and Applications of Categories **22**(23), 622–672 (2009)
5. Bradet-Legris, J., Reid, H.: Differential forms in non-linear cartesian differential categories (2018), Foundational Methods in Computer Science
6. Cai, Y., Giarrusso, P.G., Rendel, T., Ostermann, K.: A theory of changes for higher-order languages: Incrementalizing λ-calculi by static differentiation. In: ACM SIGPLAN Notices. vol. 49, pp. 145–155. ACM (2014)
7. Cockett, J.R.B., Cruttwell, G.S.H.: Differential structure, tangent structure, and sdg. Applied Categorical Structures **22**(2), 331–417 (2014)
8. Cockett, J., Cruttwell, G.: Connections in tangent categories. Theory and Applications of Categories **32**(26), 835–888 (2017)
9. Cruttwell, G.S.: Cartesian differential categories revisited. Mathematical Structures in Computer Science **27**(1), 70–91 (2017)
10. Ehrhard, T., Regnier, L.: The differential lambda-calculus. Theoretical Computer Science **309**(1), 1–41 (2003)
11. Ehrhard, T.: An introduction to differential linear logic: proof-nets, models and antiderivatives. Mathematical Structures in Computer Science **28**(7), 995–1060 (2018)
12. Kelly, R., Pearlmutter, B.A., Siskind, J.M.: Evolving the incremental {\lambda} calculus into a model of forward automatic differentiation (ad). arXiv preprint arXiv:1611.03429 (2016)
13. Kock, A.: Synthetic differential geometry, vol. 333. Cambridge University Press (2006)
14. Manzonetto, G.: What is a categorical model of the differential and the resource λ-calculi? Mathematical Structures in Computer Science **22**(3), 451–520 (2012)
15. Manzyuk, O.: Tangent bundles in differential lambda-categories. arXiv preprint arXiv:1202.0411 (2012)
16. Richardson, C.H.: An introduction to the calculus of finite differences. Van Nostrand (1954)
17. Sprunger, D., Jacobs, B.: The differential calculus of causal functions. arXiv preprint arXiv:1904.10611 (2019)
18. Sprunger, D., Katsumata, S.y.: Differentiable causal computations via delayed trace. In: 2019 34th Annual ACM/IEEE Symposium on Logic in Computer Science (LICS). pp. 1–12. IEEE (2019)
19. Steinbach, B., Posthoff, C.: Boolean differential calculus. In: Logic Functions and Equations, pp. 75–103. Springer (2009)

Contextual Equivalence for Signal Flow Graphs

Filippo Bonchi[1], Robin Piedeleu[2⋆], Paweł Sobociński[3⋆⋆], and Fabio Zanasi[2⋆](✉)

[1] Università di Pisa, Italy
[2] University College London, UK, {r.piedeleu, f.zanasi}@ucl.ac.uk
[3] Tallinn University of Technology, Estonia

Abstract. We extend the signal flow calculus—a compositional account of the classical signal flow graph model of computation—to encompass affine behaviour, and furnish it with a novel operational semantics. The increased expressive power allows us to define a canonical notion of contextual equivalence, which we show to coincide with denotational equality. Finally, we characterise the realisable fragment of the calculus: those terms that express the computations of (affine) signal flow graphs.

Keywords: signal flow graphs · affine relations · full abstraction · contextual equivalence · string diagrams

1 Introduction

Compositional accounts of models of computation often lead one to consider *relational* models because a decomposition of an input-output system might consist of internal parts where flow and causality are not always easy to assign. These insights led Willems [33] to introduce a new current of control theory, called *behavioural* control: roughly speaking, behaviours and observations are of prime concern, notions such as state, inputs or outputs are secondary. Independently, programming language theory converged on similar ideas, with *contextual equivalence* [25,28] often considered as *the* equivalence: programs are judged to be different if we can find some context in which one behaves differently from the other, and what is observed about "behaviour" is often something quite canonical and simple, such as termination. Hoare [17] and Milner [23] discovered that these programming language theory innovations also bore fruit in the non-deterministic context of concurrency. Here again, research converged on studying simple and canonical contextual equivalences [24,18].

This paper brings together all of the above threads. The model of computation of interest for us is that of signal flow graphs [32,21], which are feedback systems well known in control theory [21] and widely used in the modelling of linear dynamical systems (in continuous time) and signal processing circuits (in

⋆ Supported by EPSRC grant EP/R020604/1.
⋆⋆ Supported by the ESF funded Estonian IT Academy research measure (project 2014-2020.4.05.19-0001)

J. Goubault-Larrecq and B. König (Eds.): FOSSACS 2020, LNCS 12077, pp. 77–96, 2020.
https://doi.org/10.1007/978-3-030-45231-5_5

discrete time). The *signal flow calculus* [10,9] is a syntactic presentation with an underlying compositional denotational semantics in terms of linear relations. Armed with *string diagrams* [31] as a syntax, the tools and concepts of programming language theory and concurrency theory can be put to work and the calculus can be equipped with a structural operational semantics. However, while in previous work [9] a connection was made between operational equivalence (essentially trace equivalence) and denotational equality, the signal flow calculus was not quite expressive enough for contextual equivalence to be a useful notion.

The crucial step turns out to be moving from *linear* relations to *affine* relations, i.e. linear subspaces translated by a vector. In recent work [6], we showed that they can be used to study important physical phenomena, such as current and voltage sources in electrical engineering, as well as fundamental synchronisation primitives in concurrency, such as mutual exclusion. Here we show that, in addition to yielding compelling mathematical domains, affinity proves to be the magic ingredient that ties the different components of the story of signal flow graphs together: it provides us with a canonical and simple notion of observation to use for the *definition* of contextual equivalence, and gives us the expressive power to prove a bona fide full abstraction result that relates contextual equivalence with denotational equality.

To obtain the above result, we extend the signal flow calculus to handle affine behaviour. While the denotational semantics and axiomatic theory appeared in [6], the operational account appears here for the first time and requires some technical innovations: instead of traces, we consider *trajectories*, which are infinite traces that may start in the past. To record the time, states of our transition system have a runtime environment that keeps track of the global clock.

Because the affine signal flow calculus is oblivious to flow directionality, some terms exhibit pathological operational behaviour. We illustrate these phenomena with several examples. Nevertheless, for the linear sub-calculus, it is known [9] that every term is denotationally equal to an executable realisation: one that is in a form where a consistent flow can be identified, like the classical notion of signal flow graph. We show that the question has a more subtle answer in the affine extension: not all terms are realisable as (affine) signal flow graphs. However, we are able to characterise the class of diagrams for which this is true.

Related work. Several authors studied signal flow graphs by exploiting concepts and techniques of programming language semantics, see e.g. [4,22,29,2]. The most relevant for this paper is [2], which, independently from [10], proposed the same syntax and axiomatisation for the ordinary signal flow calculus and shares with our contribution the same methodology: the use of *string diagrams* as a mathematical playground for the compositional study of different sorts of systems. The idea is common to diverse, cross-disciplinary research programmes, including Categorical Quantum Mechanics [1,11,12], Categorical Network Theory [3], Monoidal Computer [26,27] and the analysis of (a)synchronous circuits [14,15].

Outline In Section 2 we recall the affine signal flow calculus. Section 3 introduces the operational semantics for the calculus. Section 4 defines contextual equivalence and proves full abstraction. Section 5 introduces a well-behaved class of circuits, that denotes functional input-output systems, laying the groundwork for Section 6, in which the concept of realisability is introduced before a characterisation of which circuit diagrams are realisable. Missing proofs can be found in the extended version of this paper [7].

2 Background: the Affine Signal Flow Calculus

The *Affine Signal Flow Calculus* extends the signal flow calculus [9] with an extra generator ⊢— that allows to express affine relations. In this section, we first recall its syntax and denotational semantics from [6] and then we highlight two key properties for proving full abstraction that are enabled by the affine extension. The operational semantics is delayed to the next section.

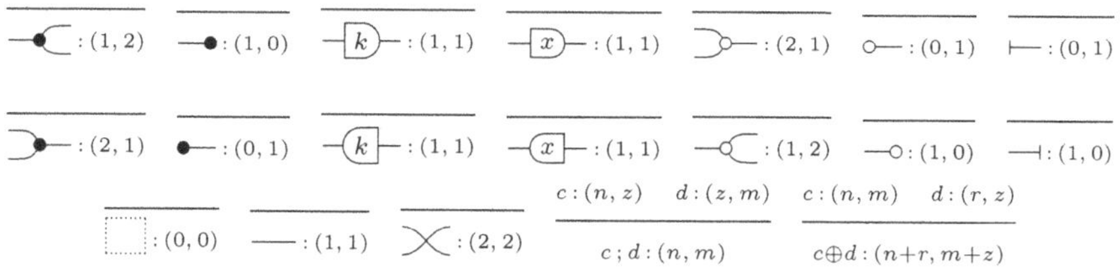

Fig. 1. Sort inference rules.

2.1 Syntax

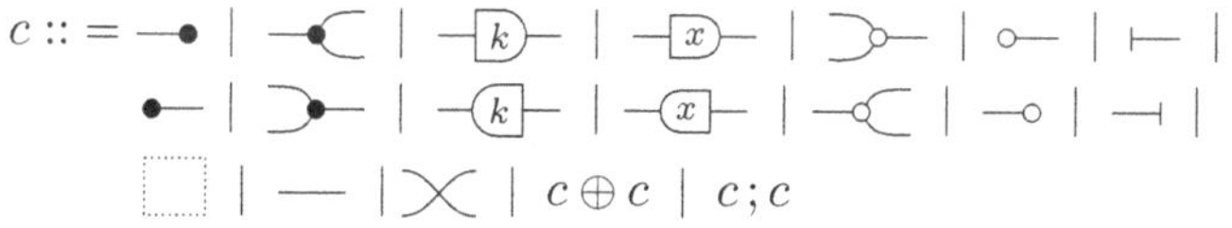

$$c ::= \; \multimap\!\!\bullet \; | \; \multimap\!\!\mathrel{\text{C}} \; | \; \multimap\!\boxed{k}\!\multimap \; | \; \multimap\!\boxed{x}\!\multimap \; | \; \mathrel{\text{D}}\!\multimap \; | \; \circ\!\multimap \; | \; \vdash\!\multimap \; | \tag{1}$$

$$\bullet\!\multimap \; | \; \mathrel{\text{D}}\!\multimap \; | \; \multimap\!\boxed{k}\!\vdash \; | \; \multimap\!\boxed{x}\!\vdash \; | \; \multimap\!\mathrel{\text{C}} \; | \; \multimap\!\circ \; | \; \multimap\!\dashv \; | \tag{2}$$

$$\boxed{} \; | \; \text{---} \; | \; \times \; | \; c \oplus c \; | \; c \, ; c \tag{3}$$

The syntax of the calculus, generated by the grammar above, is parametrised over a given field k, with k ranging over k. We refer to the constants in rows (1)-(2) as *generators*. Terms are constructed from generators, $\boxed{}$, —, $\times$, and the two binary operations in (3). We will only consider those terms that are *sortable*, i.e. they can be associated with a pair (n, m), with $n, m \in \mathbb{N}$. Sortable terms are called *circuits*: intuitively, a circuit with sort (n, m) has n ports on the left and m on the right. The sorting discipline is given in Fig. 1. We delay discussion of computational intuitions to Section 3 but, for the time being, we observe that the generators of row (2) are those of row (1) "reflected about the y-axis".

2.2 String Diagrams

It is convenient to consider circuits as the arrows of a symmetric monoidal category ACirc (for Affine Circuits). Objects of ACirc are natural numbers (thus

ACirc is a *prop* [19]) and morphisms $n \to m$ are the circuits of sort (n, m), quotiented by the laws of symmetric monoidal categories [20,31][4]. The circuit grammar yields the symmetric monoidal structure of ACirc: sequential composition is given by $c\,;d$, the monoidal product is given by $c \oplus d$, and identities and symmetries are built by pasting together —— and $\times$ in the obvious way. We will adopt the usual convention of writing morphisms of ACirc as *string diagrams*,

meaning that $c\,;c'$ is drawn $\boxed{c}\boxed{c'}$ and $c \oplus c'$ is drawn $\begin{smallmatrix}\boxed{c}\\\boxed{c'}\end{smallmatrix}$. More suc-

cinctly, ACirc is the free prop on generators (1)-(2). The free prop on (1)-(2) sans $\vdash$ and $\dashv$, hereafter called Circ, is the signal flow calculus from [9].

Example 1. The diagram represents the circuit

$$((\bullet\!-\!\!;\,-\!\!\bigcirc)\oplus\!-\!\!)\,;(-\!\!\oplus(\bigcirc\!-\!\!;\,-\!\!\bigcirc))\,;(((-\!\!\oplus\!\boxed{x}\!-\!\!)\oplus\!-\!\!)\,;((\bigcirc\!\!-\!\!;\,-\!\!\bullet)\oplus\!-\!\!)).$$

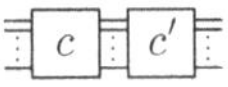

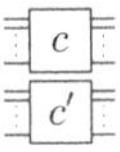

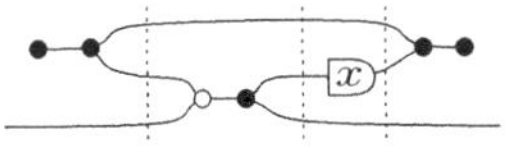

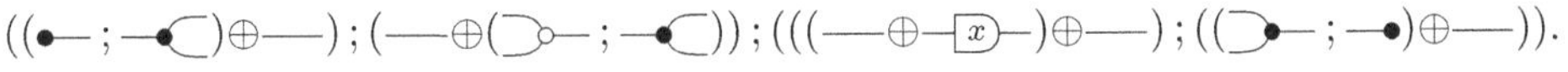

2.3   Denotational Semantics and Axiomatisation

The semantics of circuits can be given denotationally by means of affine relations.

Definition 1. *Let* k *be a field. An affine subspace of* k^d *is a subset* $V \subseteq \mathsf{k}^d$ *that is either empty or for which there exists a vector* $a \in \mathsf{k}^d$ *and a linear subspace* L *of* k^d *such that* $V = \{a + v \mid v \in L\}$. *A* k-affine relation *of type* $n \to m$ *is an affine subspace of* $\mathsf{k}^n \times \mathsf{k}^m$, *considered as a* k-*vector space.*

Note that every linear subspace is affine, taking a above to be the zero vector. Affine relations can be organised into a prop:

Definition 2. *Let* k *be a field. Let* ARel$_k$ *be the following prop:*

- *arrows* $n \to m$ *are* k-*affine relations.*
- *composition is relational: given* $G = \{(u,v) \mid u \in \mathsf{k}^n, v \in \mathsf{k}^m\}$ *and* $H = \{(v,w) \mid v \in \mathsf{k}^m, w \in \mathsf{k}^l\}$, *their composition is* $G\,;H := \{(u,w) \mid \exists v.(u,v) \in G \wedge (v,w) \in H\}$.
- *monoidal product given by* $G\oplus H = \left\{ \left(\binom{u}{u'}, \binom{v}{v'}\right) \mid (u,v) \in G, (u',v') \in H\right\}$.

In order to give semantics to ACirc, we use the prop of affine relations over the field $\mathsf{k}(x)$ of fractions of polynomials in x with coefficients from k. Elements $q \in \mathsf{k}(x)$ are a fractions $\frac{k_0 + k_1 \cdot x^1 + k_2 \cdot x^2 + \cdots + k_n \cdot x^n}{l_0 + l_1 \cdot x^1 + l_2 \cdot x^2 + \cdots + l_m \cdot l^m}$ for some $n, m \in \mathbb{N}$ and $k_i, l_i \in \mathsf{k}$. Sum, product, 0 and 1 in $\mathsf{k}(x)$ are defined as usual.

---

[4] This quotient is harmless: both the denotational semantics from [6] and the operational semantics we introduce in this paper satisfy those axioms on the nose.

Definition 3. *The prop morphism* $[\![\cdot]\!] \colon \mathsf{ACirc} \to \mathsf{ARel}_{\mathsf{k}(x)}$ *is inductively defined on circuits as follows. For the generators in* (1)

$$-\!\!\bullet\!\!\!< \;\longmapsto\; \left\{\left(p, \begin{pmatrix} p \\ p \end{pmatrix}\right) \mid p \in \mathsf{k}(x)\right\} \qquad >\!\!\circ\!- \;\longmapsto\; \left\{\left(\begin{pmatrix} p \\ q \end{pmatrix}, p+q\right) \mid p, q \in \mathsf{k}(x)\right\}$$

$$-\!\bullet \;\longmapsto\; \{(p, \bullet) \mid p \in \mathsf{k}(x)\} \qquad \circ\!- \;\longmapsto\; \{(\bullet, 0)\} \qquad \vdash\!- \;\longmapsto\; \{(\bullet, 1)\}$$

$$-\!\boxed{r}\!\!>\!- \;\longmapsto\; \{(p, p \cdot r) \mid p \in \mathsf{k}(x)\} \qquad -\!\boxed{x}\!\!>\!- \;\longmapsto\; \{(p, p \cdot x) \mid p \in \mathsf{k}(x)\}$$

where $\bullet$ *is the only element of* $\mathsf{k}(x)^0$. *The semantics of components in* (2) *is symmetric, e.g.* $\bullet\!-$ *is mapped to* $\{(p, \bullet) \mid p \in \mathsf{k}(x)\}$. *For* (3)

$$— \;\longmapsto\; \{(p, p) \mid p \in \mathsf{k}(x)\} \qquad \times \;\longmapsto\; \left\{\left(\begin{pmatrix} p \\ q \end{pmatrix}, \begin{pmatrix} q \\ p \end{pmatrix}\right) \mid p, q \in \mathsf{k}(x)\right\}$$

$$\square \;\longmapsto\; \{(\bullet, \bullet)\} \qquad c_1 \oplus c_2 \;\longmapsto\; [\![c_1]\!] \oplus [\![c_2]\!] \qquad c_1 \,;\, c_2 \;\longmapsto\; [\![c_1]\!] \,;\, [\![c_2]\!]$$

The reader can easily check that the pair of 1-dimensional vectors $\left(1, \frac{1}{1-x}\right) \in \mathsf{k}(x)^1 \times \mathsf{k}(x)^1$ belongs to the denotation of the circuit in Example 1.

The denotational semantics enjoys a sound and complete axiomatisation. The axioms involve only basic interactions between the generators (1)-(2). The resulting theory is that of *Affine Interacting Hopf Algebras* ($\mathbb{aIH}$).The generators in (1) form a Hopf algebra, those in (2) form another Hopf algebra, and the interaction of the two give rise to two Frobenius algebras. We refer the reader to [6] for the full set of equations and all further details.

Proposition 1. *For all* c, d *in* ACirc, $[\![c]\!] = [\![d]\!]$ *if and only if* $c \stackrel{\mathbb{aIH}}{=} d$.

2.4 Affine vs Linear Circuits

It is important to highlight the differences between ACirc and Circ. The latter is the purely linear fragment: circuit diagrams of Circ denote exactly the *linear* relations over $\mathsf{k}(x)$ [8], while those of ACirc denote the *affine* relations over $\mathsf{k}(x)$.

The additional expressivity afforded by affine circuits is essential for our development. One crucial property is that every polynomial fraction can be expressed as an affine circuit of sort (0, 1).

Lemma 1. *For all* $p \in \mathsf{k}(x)$, *there is* $c_p \in \mathsf{ACirc}[0, 1]$ *with* $[\![c_p]\!] = \{(\bullet, p)\}$.

Proof. For each $p \in \mathsf{k}(x)$, let P be the linear subspace generated by the pair of 1-dimensional vectors $(1, p)$. By fullness of the denotational semantics of Circ [8], there exists a circuit c in Circ such that $[\![c]\!] = P$. Then, $[\![\vdash\!- \,;\, c]\!] = \{(\bullet, p)\}$. □

The above observation yields the following:

Proposition 2. *Let* $(u, v) \in \mathsf{k}(x)^n \times \mathsf{k}(x)^m$. *There exist circuits* $c_u \in \mathsf{ACirc}[0, n]$ *and* $c_v \in \mathsf{ACirc}[m, 0]$ *such that* $[\![c_u]\!] = \{(\bullet, u)\}$ *and* $[\![c_v]\!] = \{(v, \bullet)\}$.

Proof. Let $u = \begin{pmatrix} p_1 \\ \vdots \\ p_n \end{pmatrix}$ and $v = \begin{pmatrix} q_1 \\ \vdots \\ q_m \end{pmatrix}$. By Lemma 1, for each p_i, there exists a circuit c_{p_i} such that $[\![c_{p_i}]\!] = \{(\bullet, p_i)\}$. Let $c_u = c_{p_1} \oplus \ldots \oplus c_{p_n}$. Then $[\![c_u]\!] = \{(\bullet, u)\}$. For c_v, it is enough to see that Proposition 1 also holds with 0 and 1 switched, then use the argument above. □

Proposition 2 asserts that any behaviour (u, v) occurring in the denotation of some circuit c, i.e., such that $(u, v) \in [\![c]\!]$, can be expressed by a pair of circuits (c_u, c_v). We will, in due course, think of such a pair as a *context*, namely an environment with which a circuit can interact. Observe that this is not possible with the linear fragment Circ, since the only singleton linear subspace is 0.

Another difference between linear and affine concerns circuits of sort (0, 0). Indeed $\mathsf{k}(x)^0 = \{\bullet\}$, and the only linear relation over $\mathsf{k}(x)^0 \times \mathsf{k}(x)^0$ is the singleton $\{(\bullet, \bullet)\}$, which is id_0 in $\mathsf{ARel}_{\mathsf{k}(x)}$. But there is another affine relation, namely the *empty relation* $\emptyset \in \mathsf{k}(x)^0 \times \mathsf{k}(x)^0$. This can be represented by ⊢–○, for instance, since $[\![$⊢–○$]\!] = \{(\bullet, 1)\} ; \{(0, \bullet)\} = \emptyset$.

Proposition 3. *Let* $c \in \mathsf{ACirc}[0, 0]$. *Then* $[\![c]\!]$ *is either* id_0 *or* $\emptyset$.

3 Operational Semantics for Affine Circuits

Here we give the structural operational semantics of affine circuits, building on previous work [9] that considered only the core linear fragment, Circ. We consider circuits to be *programs* that have an observable behaviour. Observations are possible interactions at the circuit's interface. Since there are two interfaces: a left and a right, each transition has two labels.

In a transition $t \triangleright c \xrightarrow[w]{v} t' \triangleright c'$, c and c' are *states*, that is, circuits augmented with information about which values $k \in \mathsf{k}$ are stored in each register (–[x]– and –[x]–) at that instant of the computation. When transitioning to c', the v above the arrow is a vector of values with which c synchronises on the left, and the w below the arrow accounts for the synchronisation on the right. States are decorated with runtime contexts: t and t' are (possibly negative) integers that—intuitively—indicate the time when the transition happens. Indeed, in Fig. 2, every rule advances time by 1 unit. "Negative time" is important: as we shall see in Example 3, some executions must start in the past.

The rules in the top section of Fig. 2 provide the semantics for the generators in (1): –•< is a *copier*, duplicating the signal arriving on the left; –• accepts any signal on the left and discards it, producing nothing on the right; >○– is an *adder* that takes two signals on the left and emits their sum on the right, ○– emits the constant 0 signal on the right; –[k]– is an *amplifier*, multiplying the signal on the left by the scalar $k \in \mathsf{k}$. All the generators described so far are stateless. State is provided by –[x]– with l above it, which is a *register*; a synchronous one place buffer with the value l stored. When it receives some value k on the left, it emits l on the right and stores k. The behaviour of the affine generator ⊢–

$$t \triangleright \!-\!\!\mathsf{C} \quad \xrightarrow[k\,k]{k} \quad t+1 \triangleright \!-\!\!\mathsf{C} \qquad t \triangleright \!-\!\!\bullet \quad \xrightarrow[\bullet]{k} \quad t+1 \triangleright \!-\!\!\bullet$$

$$t \triangleright \mathsf{\supset}\!\!- \quad \xrightarrow[k+l]{k\,l} \quad t+1 \triangleright \mathsf{\supset}\!\!- \qquad t \triangleright \mathsf{o}\!\!- \quad \xrightarrow[0]{\bullet} \quad t+1 \triangleright \mathsf{o}\!\!-$$

$$t \triangleright \!-\!\boxed{x}\!\!-\!\! \quad \xrightarrow[l]{k} \quad t+1 \triangleright \!-\!\boxed{x}\!\!-\!\! \qquad t \triangleright \!-\!\boxed{r}\!\!-\!\! \quad \xrightarrow[rl]{l} \quad t+1 \triangleright \!-\!\boxed{r}\!\!-\!\!$$

$$0 \triangleright \!\vdash \quad \xrightarrow[\bullet]{\bullet} \quad 1 \triangleright \!\vdash \qquad t \triangleright \!\vdash \quad \xrightarrow[0]{\bullet} \quad t+1 \triangleright \!\vdash \quad (t \neq 0)$$

$$t \triangleright \mathsf{\supset}\!\!\bullet\!- \quad \xrightarrow[k]{k\,k} \quad t+1 \triangleright \mathsf{\supset}\!\!\bullet\!- \qquad t \triangleright \!\bullet\!- \quad \xrightarrow[k]{\bullet} \quad t+1 \triangleright \!\bullet\!-$$

$$t \triangleright \!-\!\!\mathsf{C}\!\!\circ \quad \xrightarrow[k\,l]{k+l} \quad t+1 \triangleright \!-\!\!\mathsf{C}\!\!\circ \qquad t \triangleright \!-\!\!\circ \quad \xrightarrow[\bullet]{0} \quad t+1 \triangleright \!-\!\!\circ$$

$$t \triangleright \!-\!\boxed{x}\!\!- \quad \xrightarrow[k]{l} \quad t+1 \triangleright \!-\!\boxed{x}\!\!- \qquad t \triangleright \!-\!\boxed{r}\!\!- \quad \xrightarrow[l]{rl} \quad t+1 \triangleright \!-\!\boxed{r}\!\!-$$

$$0 \triangleright \!-\!\!\dashv \quad \xrightarrow[\bullet]{1} \quad 1 \triangleright \!-\!\!\dashv \qquad t \triangleright \!-\!\!\dashv \quad \xrightarrow[\bullet]{0} \quad t+1 \triangleright \!-\!\!\dashv \quad (t \neq 0)$$

$$t \triangleright \!-\!\!- \quad \xrightarrow[k]{k} \quad t+1 \triangleright \!-\!\!- \qquad t \triangleright \!\!\times\!\! \quad \xrightarrow[l\,k]{k\,l} \quad t+1 \triangleright \!\!\times\!\! \qquad t \triangleright \square \quad \xrightarrow[\bullet]{\bullet} \quad t+1 \triangleright \square$$

$$\frac{t \triangleright c \xrightarrow[v]{u} t+1 \triangleright c' \qquad t \triangleright d \xrightarrow[w]{v} t+1 \triangleright d'}{t \triangleright c\,;d \xrightarrow[w]{u} t+1 \triangleright c'\,;d'}$$

$$\frac{t \triangleright c \xrightarrow[v_1]{u_1} t+1 \triangleright c' \qquad t \triangleright d \xrightarrow[v_2]{u_2} t+1 \triangleright d'}{t \triangleright c \oplus d \xrightarrow[v_1\,v_2]{u_1\,u_2} t+1 \triangleright c' \oplus d'}$$

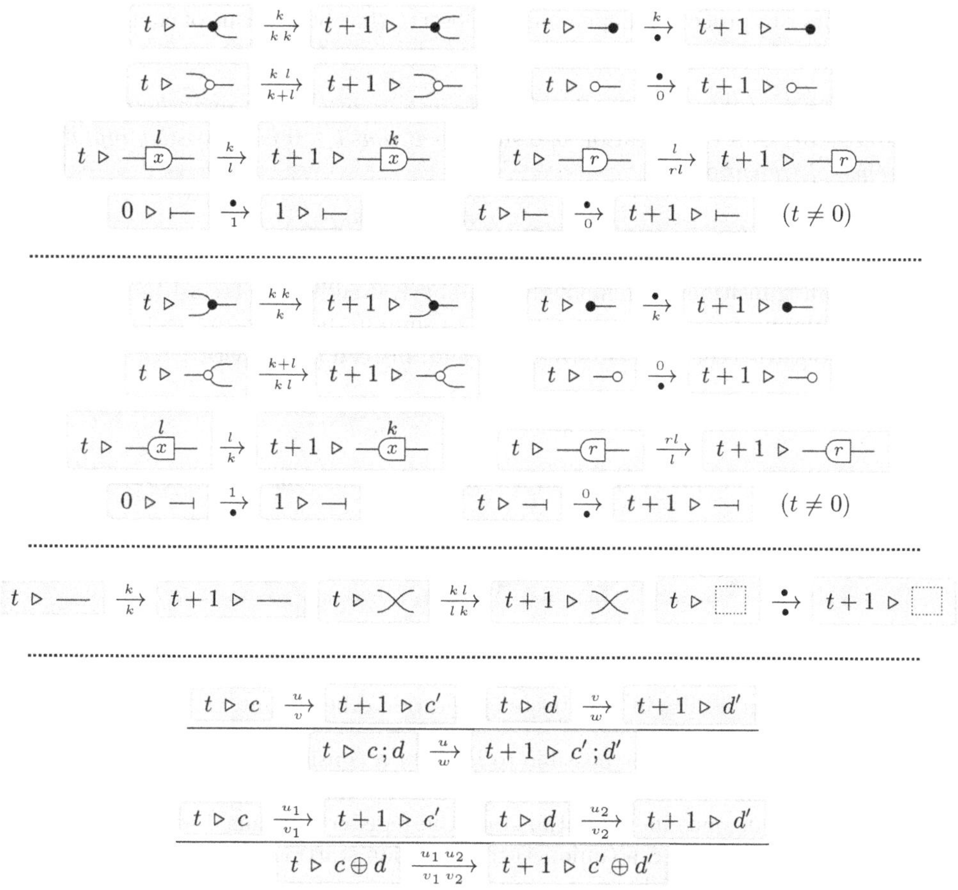

Fig. 2. Structural rules for operational semantics, with $p \in \mathbb{Z}$, k,l ranging over k and u,v,w vectors of elements of k of the appropriate size. The only vector of k^0 is written as $\bullet$ (as in Definition 3), while a vector $(k_1 \ \dots \ k_n)^T \in \mathsf{k}^n$ as $k_1 \dots k_n$.

depends on the time: when $t = 0$, it emits 1, otherwise it emits 0. Observe that the behaviour of all other generators is time-independent.

So far, we described the behaviour of the components in (1) using the intuition that signal flows from left to right: in a transition $\xrightarrow[w]{v}$, the signal v on the left is thought as trigger and w as effect. For the generators in (2), whose behaviour is defined by the rules in the second section of Fig. 2, the behaviour is symmetric—indeed, here it is helpful to think of signals as flowing from right to left. The next section of Fig. 2 specifies the behaviours of the structural connectors of (3): $\times$ is a *twist*, swapping two signals, $\square$ is the empty circuit and $\!-\!$ is the *identity* wire: the signals on the left and on the right ports are equal. Finally, the rule for sequential ; composition forces the two components to have the same value v on the shared interface, while for parallel $\oplus$ composition, components can proceed independently. Observe that both forms of composition require component transitions to happen at the same time.

Definition 4. *Let $c \in$ ACirc. The* initial state c_0 *of c is the one where all the registers store* 0*. A* computation *of c starting at time $t \leq 0$ is a (possibly infinite) sequence of transitions*

$$t \triangleright c_0 \xrightarrow[w_t]{v_t} t+1 \triangleright c_1 \xrightarrow[w_{t+1}]{v_{t+1}} t+2 \triangleright c_2 \xrightarrow[w_{t+2}]{v_{t+2}} \cdots \tag{4}$$

Since all transitions increment the time by 1, it suffices to record the time at which a computation starts. As a result, to simplify notation, we will omit the runtime context after the first transition and, instead of (4), write

$$t \triangleright c_0 \xrightarrow[w_t]{v_t} c_1 \xrightarrow[w_{t+1}]{v_{t+1}} c_2 \xrightarrow[w_{t+2}]{v_{t+2}} \cdots$$

Example 2. The circuit in Example 1 can perform the following computation.

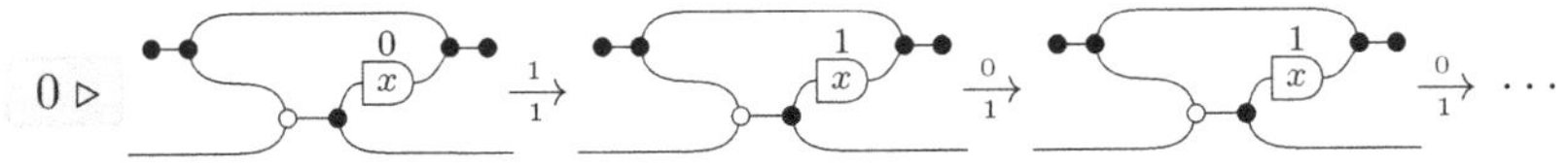

In the example above, the flow has a clear left-to-right orientation, albeit with a feedback loop. For arbitrary circuits of ACirc this is not always the case, which sometimes results in unexpected operational behaviour.

Example 3. In $\vdash\!\!\!-\!(x)\!\!-$ is not possible to identify a consistent flow: $\vdash\!\!-$ goes from left to right, while $-(x)\!\!-$ from right to left. Observe that there is no computation starting at $t = 0$, since in the initial state the register contains 0 while $\vdash\!\!-$ must emit 1. There is, however, a (unique!) computation starting at time $t = -1$, that loads the register with 1 before $\vdash\!\!-$ can also emit 1 at time $t = 0$.

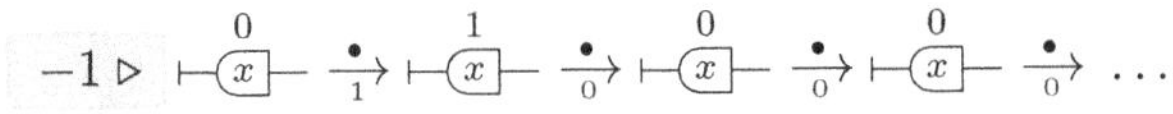

Similarly, $\vdash\!\!\!-\!(x)\!\!-\!(x)\!\!-$ features a unique computation starting at time $t = -2$.

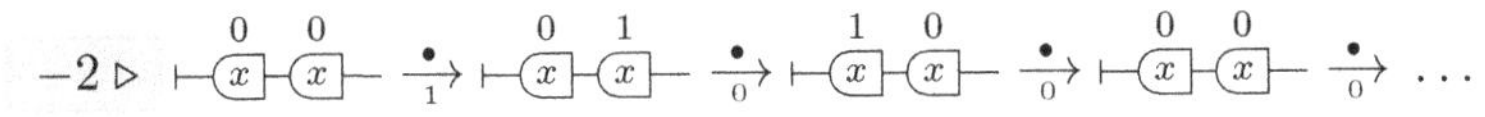

It is worthwhile clarifying the reason why, in the affine calculus, some computations start in the past. As we have already mentioned, in the linear fragment the semantics of all generators is time-independent. It follows easily that time-independence is a property enjoyed by all purely linear circuits. The behaviour of $\vdash\!\!-$, however, enforces a particular action to occur at time 0. Considering this in conjunction with a right-to-left register results in $\vdash\!\!\!-\!(x)\!\!-$, and the effect is to anticipate that action by one step to time -1, as shown in Example 3. It is obvious that this construction can be iterated, and it follows that the presence of a single time-dependent generator results in a calculus in which the computation of some terms must start at a finite, but unbounded time in the past.

Example 4. Another circuit with conflicting flow is $\vdash\!\!\circ$. Here there is no possible transition at $t = 0$, since at that time $\vdash$ must emit a 1 and $-\!\!\circ$ can only synchronise on a 0. Instead, the circuit $\boxed{}$ can always perform an infinite computation $t \triangleright \boxed{} \xrightarrow[\bullet]{\bullet} \boxed{} \xrightarrow[\bullet]{\bullet} \dots$, for any $t \leq 0$. Roughly speaking, the computations of these two (0, 0) circuits are operational mirror images of the two possible denotations of Proposition 3. This intuition will be made formal in Section 4. For now, it is worth observing that for all c, $\boxed{} \oplus c$ can perform the same computations of c, while $\vdash\!\!\circ \oplus c$ cannot ever make a transition at time 0.

Example 5. Consider the circuit $-\!(x]\!\!-\!\![x)\!-$, which again features conflicting flow. Our equational theory equates it with $—$, but the computations involved are subtly different. Indeed, for any sequence $a_i \in \mathsf{k}$, it is obvious that $—$ admits the computation

$$0 \triangleright — \xrightarrow[a_0]{a_0} — \xrightarrow[a_1]{a_1} — \xrightarrow[a_2]{a_2} \dots \tag{5}$$

The circuit $-\!(x]\!\!-\!\![x)\!-$ admits a similar computation, but we must begin at time $t = -1$ in order to first "load" the registers with a_0:

$$-1 \triangleright \overset{0\;\;0}{-\!(x]\!\!-\!\![x)\!-} \xrightarrow[0]{0} \overset{a_0\;\;a_0}{-\!(x]\!\!-\!\![x)\!-} \xrightarrow[a_0]{a_0} \overset{a_1\;\;a_1}{-\!(x]\!\!-\!\![x)\!-} \xrightarrow[a_1]{a_1} \overset{a_2\;\;a_2}{-\!(x]\!\!-\!\![x)\!-} \xrightarrow[a_2]{a_2} \dots \tag{6}$$

The circuit $-\![x)\!\!-\!\!(x]\!-$, which again is equated with $—$ by the equational theory, is more tricky. Although every computation of $—$ can be reproduced, $-\![x)\!\!-\!\!(x]\!-$ admits additional, problematic computations. Indeed, consider

$$0 \triangleright \overset{0\;\;0}{-\![x)\!\!-\!\!(x]\!-} \xrightarrow[1]{0} \overset{0\;\;1}{-\![x)\!\!-\!\!(x]\!-} \tag{7}$$

at which point no further transition is possible—the circuit can deadlock.

The following lemma is an easy consequence of the rules of Fig. 2 and follows by structural induction. It states that all circuits can stay idle *in the past.*

Lemma 2. *Let* $c \in \mathsf{ACirc}[n, m]$ *with initial state* c_0. *Then* $t \triangleright c_0 \xrightarrow[0]{0} c_0$ *if* $t < 0$.

3.1 Trajectories

For the non-affine version of the signal flow calculus, we studied in [9] *traces* arising from computations. For the affine extension, this is not possible since, as explained above, we must also consider computations that start in the past. In this paper, rather than traces we adopt a common control theoretic notion.

Definition 5. *An* (n, m)-trajectory σ *is a* $\mathbb{Z}$-*indexed sequence* $\sigma : \mathbb{Z} \to \mathsf{k}^n \times \mathsf{k}^m$ *that is finite in the past, i.e., for which* $\exists j \in \mathbb{Z}$ *such that* $\sigma(i) = (0, 0)$ *for* $i \leq j$.

By the universal property of the product we can identify $\sigma : \mathbb{Z} \to \mathsf{k}^n \times \mathsf{k}^m$ with the pairing $\langle \sigma_l, \sigma_r \rangle$ of $\sigma_l : \mathbb{Z} \to \mathsf{k}^n$ and $\sigma_r : \mathbb{Z} \to \mathsf{k}^m$. A (k, m)-trajectory σ and (m, n)-trajectory τ are *compatible* if $\sigma_r = \tau_l$. In this case, we can define their composite, a (k,n)-trajectory $\sigma\,;\tau$ by $\sigma\,;\tau := \langle \sigma_l, \tau_r \rangle$. Given an (n_1, m_1)-trajectory σ_1, and an (n_2, m_2)-trajectory σ_2, their product, an $(n_1{+}n_2, m_1{+}m_2)$-trajectory $\sigma_1 \oplus \sigma_2$, is defined $(\sigma_1 \oplus \sigma_2)(i) := \begin{pmatrix} \sigma(i) \\ \tau(i) \end{pmatrix}$. Using these two operations we can organise *sets* of trajectories into a prop.

Definition 6. *The composition of two sets of trajectories is defined as* $S\,;T := \{\sigma\,;\tau \mid \sigma \in S, \tau \in T \text{ are compatible}\}$. *The product of sets of trajectories is defined as* $S_1 \oplus S_2 := \{\sigma_1 \oplus \sigma_2 \mid \sigma_1 \in S_1, \sigma_2 \in S_2\}$.

Clearly both operations are strictly associative. The unit for $\oplus$ is the singleton with the unique $(0,0)$-trajectory. Also $;$ has a two sided identity, given by sets of "copycat" (n,n)-trajectories. Indeed, we have that:

Proposition 4. *Sets of (n,m)-trajectories are the arrows $n \to m$ of a prop* Traj *with composition and monoidal product given as in Definition 6.*

Traj serves for us as the domain for operational semantics: given a circuit c and an *infinite* computation

$$t \rhd c_0 \xrightarrow[v_t]{u_t} c_1 \xrightarrow[v_{t+1}]{u_{t+1}} c_2 \xrightarrow[v_{t+2}]{u_{t+2}} \cdots$$

its associated trajectory σ is

$$\sigma(i) = \begin{cases} (u_i, v_i) & \text{if } i \geq t, \\ (0,0) & \text{otherwise.} \end{cases} \tag{8}$$

Definition 7. *For a circuit c, $\langle c \rangle$ is the set of trajectories given by its infinite computations, following the translation* (8) *above.*

The assignment $c \mapsto \langle c \rangle$ is compositional, that is:

Theorem 1. $\langle \cdot \rangle \colon \mathsf{ACirc} \to \mathsf{Traj}$ *is a morphism of props.*

Example 6. Consider the computations (5) and (6) from Example 5. According to (8) both are translated into the trajectory σ mapping $i \geq 0$ into (a_i, a_i) and $i < 0$ into $(0,0)$. The reader can easily verify that, more generally, it holds that $\langle \text{—} \rangle = \langle \text{—}(x)\text{—}[x]\text{—} \rangle$. At this point it is worth to remark that the two circuits would be distinguished when looking at their traces: the trace of computation (5) is different from the trace of (6). Indeed, the full abstraction result in [9] does not hold for all circuits, but only for those of a certain kind. The affine extension obliges us to consider computations that starts in the past and, in turn, this drives us toward a stronger full abstraction result, shown in the next section.

Before concluding, it is important to emphasise that $\langle \text{—} \rangle = \langle \text{—}[x]\text{—}(x)\text{—} \rangle$ also holds. Indeed, problematic computations, like (7), are all finite and, by definition, do not give rise to any trajectory. The reader should note that the use of trajectories is not a semantic device to get rid of problematic computations. In fact, trajectories do not appear in the statement of our full abstraction result; they are merely a convenient tool to prove it. Another result (Proposition 9) independently takes care of ruling out problematic computations.

4 Contextual Equivalence and Full Abstraction

This section contains the main contribution of the paper: a traditional full abstraction result asserting that contextual equivalence agrees with denotational equivalence. It is not a coincidence that we prove this result in the affine setting: affinity plays a crucial role, both in its statement and proof. In particular, Proposition 3 gives us two possibilities for the denotation of (0, 0) circuits: *(i)* $\emptyset$—which, roughly speaking, means that there is a problem (see e.g. Example 4) and no infinite computation is possible—or *(ii)* id_0, in which case infinite computations are possible. This provides us with a basic notion of observation, akin to observing termination vs non-termination in the λ-calculus.

Definition 8. *For a circuit* $c \in \mathsf{ACirc}[0,0]$ *we write* $c \uparrow$ *if* c *can perform an infinite computation and* $c \not\uparrow$ *otherwise. For instance* ⬚ $\uparrow$, *while* ⊢○ $\not\uparrow$.

To be able to make observations about arbitrary circuits we need to introduce an appropriate notion of context. Roughly speaking, contexts for us are (0, 0)-circuits with a hole into which we can plug another circuit. Since ours is a variable-free presentation, "dangling wires" assume the role of free variables [16]: restricting to (0, 0) contexts is therefore analogous to considering *ground* contexts—i.e. contexts with no free variables—a standard concept of programming language theory.

To define contexts formally, we extend the syntax of Section 2.1 with an extra generator "$-$" of sort (n, m). A (0, 0)-circuit of this extended syntax is a *context* when "$-$" occurs exactly once. Given an (n, m)-circuit c and a context $C[-]$, we write $C[c]$ for the circuit obtained by replacing the unique occurrence of "$-$" by c.

With this setup, given an (n, m)-circuit c, we can insert it into a context $C[-]$ and observe the possible outcome: either $C[c] \uparrow$ or $C[c] \not\uparrow$. This naturally leads us to contextual equivalence and the statement of our main result.

Definition 9. *Given* $c, d \in \mathsf{ACirc}[n,m]$*, we say that they are* contextually equivalent, *written* $c \equiv d$*, if for all contexts* $C[-]$*,*

$$C[c] \uparrow \text{ iff } C[d] \uparrow .$$

Example 7. Recall from Example 5, the circuits —— and —[x]—(x]—. Take the context $C[-] = c_\sigma \, ; \; - \; ; c_\tau$ for $c_\sigma \in \mathsf{ACirc}[0,1]$ and $c_\tau \in \mathsf{ACirc}[1,0]$. Assume that c_σ and c_τ have a single infinite computation. Call σ and τ the corresponding trajectories. If $\sigma = \tau$, both C[——] and C[—[x]—(x]—] would be able to perform an infinite computation. Instead if $\sigma \neq \tau$, none of them would perform any infinite computation: —— would stop at time t, for t the first moment such that $\sigma(t) \neq \tau(t)$, while C[—[x]—(x]—] would stop at time $t + 1$.

Now take as context $C[-] =$ •— $; \; - \; ;$ —•. In contrast to c_σ and c_τ, •— and —• can perform more than one single computation: at any time they can nondeterministically emit any value. Thus every computation of C[——] $=$ •—•

can *always* be extended to an infinite one, forcing synchronisation of $\bullet\!-$ and $-\!\bullet$ at each step. For $C[-\boxed{x}\!-\!(x)-] = \bullet\!-\boxed{x}\!-\!(x)-\!\bullet$, $\bullet\!-$ and $-\!\bullet$ may emit different values at time t, but the computation will get stuck at $t+1$. However, our definition of $\uparrow$ only cares about whether $C[-\boxed{x}\!-\!(x)-]$ *can* perform an infinite computation. Indeed it can, as long as $\bullet\!-$ and $-\!\bullet$ consistently emit the same value at each time step.

If we think of contexts as tests, and say that a circuit c passes test $C[-]$ if $C[c]$ perform an infinite computation, then our notion of contextual equivalence is *may-testing* equivalence [13]. From this perspective, $-$ and $-\boxed{x}\!-\!(x)-$ are not *must equivalent*, since the former must pass the test $\bullet\!- \,;\, - \,;\, -\!\bullet$ while $-\boxed{x}\!-\!(x)-$ may not. It is worth to remark here that the distinction between may and must testing will cease to make sense in Section 5 where we identify a certain class of circuits equipped with a proper flow directionality and thus a deterministic, input-output, behaviour.

Theorem 2 (Full abstraction). *$c \equiv d$ iff $c \stackrel{ctx}{=} d$*

The remainder of this section is devoted to the proof of Theorem 2. We will start by clarifying the relationship between fractions of polynomials (the denotational domain) and trajectories (the operational domain).

4.1 From Polynomial Fractions to Trajectories

The missing link between polynomial fractions and trajectories are *(formal) Laurent series*: we now recall this notion. Formally, a Laurent series is a function $\sigma\colon \mathbb{Z} \to \mathsf{k}$ for which there exists $j \in \mathbb{Z}$ such that $\sigma(i) = 0$ for all $i < j$. We write σ as $\dots, \sigma(-1), \underline{\sigma(0)}, \sigma(1), \dots$ with position 0 underlined, or as formal sum $\sum_{i=d}^{\infty} \sigma(i)x^i$. Each Laurent series σ has then a *degree* $d \in \mathbb{Z}$, which is the first non-zero element. Laurent series form a field $\mathsf{k}((x))$: sum is pointwise, product is by convolution, and the inverse σ^{-1} of σ with degree d is defined as:

$$\sigma^{-1}(i) = \begin{cases} 0 & \text{if } i < -d \\ \sigma(d)^{-1} & \text{if } i = -d \\ \frac{\sum_{i=1}^{n} \left(\sigma(d+i)\cdot\sigma^{-1}(-d+n-i)\right)}{-\sigma(d)} & \text{if } i = -d+n \text{ for } n > 0 \end{cases} \tag{9}$$

Note (formal) power series, which form 'just' a ring $\mathsf{k}[[x]]$, are a particular case of Laurent series, namely those σs for which $d \geq 0$. What is most interesting for our purposes is how polynomials and fractions of polynomials relate to $\mathsf{k}((x))$ and $\mathsf{k}[[x]]$. First, the ring $\mathsf{k}[x]$ of polynomials embeds into $\mathsf{k}[[x]]$, and thus into $\mathsf{k}((x))$: a polynomial $p_0 + p_1x + \cdots + p_nx^n$ can also be regarded as the power series $\sum_{i=0}^{\infty} p_ix^i$ with $p_i = 0$ for all $i > n$. Because Laurent series are closed under division, this immediately gives also an embedding of the field of polynomial fractions $\mathsf{k}(x)$ into $\mathsf{k}((x))$. Note that the full expressiveness of $\mathsf{k}((x))$ is required: for instance, the fraction $\frac{1}{x}$ is represented as the Laurent series $\dots, 0, 1, \underline{0}, 0, \dots$, which is not a power series, because a non-zero value appears before position 0. In fact, fractions that are expressible as power series are precisely the *rational* fractions, i.e. of the form $\frac{k_0+k_1x+k_2x^2\cdots+k_nx^n}{l_0+l_1x+l_2x^2\cdots+l_nx^n}$ where $l_0 \neq 0$.

Rational fractions form a ring $\mathsf{k}\langle x\rangle$ which, differently from the full field $\mathsf{k}(x)$, embeds into $\mathsf{k}[[x]]$. Indeed, whenever $l_0 \neq 0$, the inverse of $l_0 + l_1x + l_2x^2 \cdots + l_nx^n$ is, by (9), a *bona fide* power series. The commutative diagram on the right is a summary.

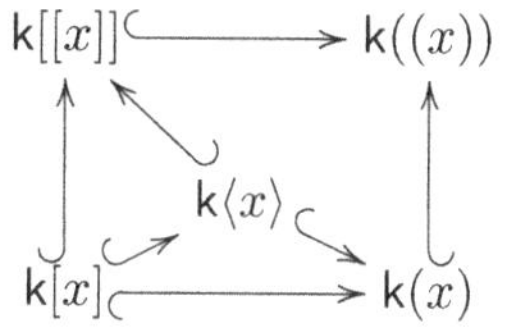

Relations between $\mathsf{k}((x))$-vectors organise themselves into a prop $\mathsf{ARel}_{\mathsf{k}((x))}$ (see Definition 2). There is an evident prop morphism $\iota\colon \mathsf{ARel}_{\mathsf{k}(x)} \to \mathsf{ARel}_{\mathsf{k}((x))}$: it maps the empty affine relation on $\mathsf{k}(x)$ to the one on $\mathsf{k}((x))$, and otherwise applies pointwise the embedding of $\mathsf{k}(x)$ into $\mathsf{k}((x))$. For the next step, observe that trajectories are in fact rearrangements of Laurent series: each pair of vectors $(u, v) \in \mathsf{k}((x))^n \times \mathsf{k}((x))^m$, as on the left below, yields the trajectory $\kappa(u, v)$ defined for all $i \in \mathbb{Z}$ as on the right below.

$$(u,v) = \left(\begin{pmatrix} \alpha^1 \\ \vdots \\ \alpha^n \end{pmatrix}, \begin{pmatrix} \beta^1 \\ \vdots \\ \beta^m \end{pmatrix} \right) \qquad \kappa(u,v)(i) = \left(\begin{pmatrix} \alpha^1(i) \\ \vdots \\ \alpha^n(i) \end{pmatrix}, \begin{pmatrix} \beta^1(i) \\ \vdots \\ \beta^m(i) \end{pmatrix} \right)$$

Similarly to ι, the assignment κ extends to sets of vectors, and also to a prop morphism from $\mathsf{ARel}_{\mathsf{k}((x))}$ to Traj. Together, κ and ι provide the desired link between operational and denotational semantics.

Theorem 3. $\langle\cdot\rangle = \kappa \circ \iota \circ \llbracket\cdot\rrbracket$

Proof. Since both are symmetric monoidal functors from a free prop, it is enough to check the statement for the generators of ACirc. We show, as an example, the case of —●⊂. By Definition 3, $\llbracket$—●⊂$\rrbracket$ $= \left\{ \left(p, \begin{pmatrix} p \\ p \end{pmatrix} \right) \mid p \in \mathsf{k}(x) \right\}$. This is mapped by ι to $\left\{ \left(\alpha, \begin{pmatrix} \alpha \\ \alpha \end{pmatrix} \right) \mid \alpha \in \mathsf{k}((x)) \right\}$. Now, to see that $\kappa(\iota(\llbracket$—●⊂$\rrbracket)) = \langle$—●⊂$\rangle$, it is enough to observe that a trajectory σ is in $\kappa(\iota(\llbracket$—●⊂$\rrbracket))$ precisely when, for all i, there exists some $k_i \in \mathsf{k}$ such that $\sigma(i) = \left(k_i, \begin{pmatrix} k_i \\ k_i \end{pmatrix} \right)$. □

4.2 Proof of Full Abstraction

We now have the ingredients to prove Theorem 2. First, we prove an adequacy result for (0, 0) circuits.

Proposition 5. *Let* $c \in \mathsf{ACirc}[0,0]$. *Then* $\llbracket c \rrbracket = id_0$ *if and only if* $c\uparrow$.

Proof. By Proposition 3, either $\llbracket c \rrbracket = id_0$ or $\llbracket c \rrbracket = \emptyset$, which, combined with Theorem 3, means that $\langle c \rangle = \kappa \circ \iota(id_0)$ or $\langle c \rangle = \kappa \circ \iota(\emptyset)$. By definition of ι this implies that either $\langle c \rangle$ contains a trajectory or not. In the first case $c\uparrow$; in the second $c\not\uparrow$. □

Next we obtain a result that relates denotational equality in all contexts to equality in $\mathbb{aIH}$. Note that it is not trivial: since we consider ground contexts it does not make sense to merely consider "identity" contexts. Instead, it is at this point that we make another crucial use of affinity, taking advantage of the increased expressivity of affine circuits, as showcased by Proposition 2.

Proposition 6. *If* $[\![C[c]]\!] = [\![C[d]]\!]$ *for all contexts* $C[-]$, *then* $c \overset{\mathbb{aIH}}{=} d$.

Proof. Suppose that $c \overset{\mathbb{aIH}}{\neq} d$. Then $[\![c]\!] \neq [\![d]\!]$. Since both $[\![c]\!]$ and $[\![d]\!]$ are affine relations over $\mathsf{k}(x)$, there exists a pair of vectors $(u, v) \in \mathsf{k}(x)^n \times \mathsf{k}(x)^m$ that is in one of $[\![c]\!]$ and $[\![d]\!]$, but not both. Assume wlog that $(u, v) \in [\![c]\!]$ and $(u, v) \notin [\![d]\!]$. By Proposition 2, there exists c_u and c_v such that $[\![c_u\,; c\,; c_v]\!] = [\![c_u]\!]\,; [\![c]\!]\,; [\![c_v]\!] = \{(\bullet, u)\}\,; [\![c]\!]\,; \{(v, \bullet)\}$. Since $(u, v) \in [\![c]\!]$, then $[\![c_u\,; c\,; c_v]\!] = \{(\bullet, \bullet)\}$. Instead, since $(u, v) \notin [\![d]\!]$, we have that $[\![c_u\,; d\,; c_v]\!] = \emptyset$. Therefore, for the context $C[-] = c_u\,; -\,; c_v$, we have that $[\![C[c]]\!] \neq [\![C[d]]\!]$. □

The proof of our main result is now straightforward.

Proof of Theorem 2. Let us first suppose that $c \overset{\mathbb{aIH}}{=} d$. Then $[\![C[c]]\!] = [\![C[d]]\!]$ for all contexts $C[-]$, since $[\![\cdot]\!]$ is a morphism of props. By Corollary 5, it follows immediately that $C[c] \uparrow$ if and only if $C[d] \uparrow$, namely $c \equiv d$.

Conversely, suppose that, for all $C[-]$, $C[c] \uparrow$ iff $C[d] \uparrow$. Again by Corollary 5, we have that $[\![C[c]]\!] = [\![C[d]]\!]$. We conclude by invoking Proposition 6. □

5 Functional Behaviour and Signal Flow Graphs

There is a sub-prop SF of Circ of classical *signal flow graphs* (see *e.g.* [21]). Here signal flows left-to-right, possibly featuring *feedback loops*, provided that these go through at least one register. Feedback can be captured algebraically via an operation $\mathsf{Tr}(\cdot)\colon \mathsf{Circ}[n+1, m+1] \to \mathsf{Circ}[n, m]$ taking $c\colon n+1 \to m+1$ to:

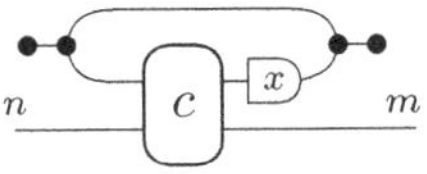

Following [9], let us call $\overrightarrow{\mathsf{Circ}}$ the free sub-prop of Circ of circuits built from (3) and the generators of (1), without ⊢—. Then SF is defined as the closure of $\overrightarrow{\mathsf{Circ}}$ under $\mathsf{Tr}(\cdot)$. For instance, the circuit of Example 2 is in SF.

Signal flow graphs are intimately connected to the executability of circuits. In general, the rules of Figure 2 do not assume a fixed flow orientation. As a result, some circuits in Circ are not executable as *functional input-output* systems, as we have demonstrated with 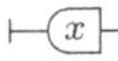,  and 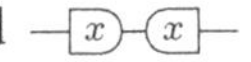 of Examples 3-5. Notice that none of these are signal flow graphs. In fact, the circuits of SF do not have pathological behaviour, as we shall state more precisely in Proposition 9.

At the denotational level, signal flow graphs correspond precisely to *rational* functional behaviours, that is, matrices whose coefficients are in the ring $\mathsf{k}\langle x\rangle$

of *rational fractions* (see Section 4.1). We call such matrices, rational matrices. One may check that the semantics of a signal flow graph $c\colon (n, m)$ is always of the form $[\![c]\!] = \{(v, A \cdot v) \mid v \in \mathsf{k}(x)^n\}$, for some $m \times n$ rational matrix A. Conversely, all relations that are the graph of rational matrices can be expressed as signal flow graphs.

Proposition 7. *Given $c\colon (n, m)$, we have $[\![c]\!] = \{(p, A \cdot p) \mid p \in \mathsf{k}(x)^n\}$ for some rational $m \times n$ matrix A iff there exists a signal flow graph f, i.e., a circuit $f\colon (n, m)$ of* SF*, such that $[\![f]\!] = [\![c]\!]$.*

Proof. This is a folklore result in control theory which can be found in [30]. The details of the translation between rational matrices and circuits of SF can be found in [10, Section 7]. □

The following gives an alternative characterisation of rational matrices—and therefore, by Proposition 7, of the behaviour of signal flow graphs—that clarifies their role as realisations of circuits.

Proposition 8. *An $m \times n$ matrix is rational iff $A \cdot r \in \mathsf{k}\langle x\rangle^m$ for all $r \in \mathsf{k}\langle x\rangle^n$.*

Proposition 8 is another guarantee of good behaviour—it justifies the name of inputs (resp. outputs) for the left (resp. right) ports of signal flow graphs. Recall from Section 4.1 that rational fractions can be mapped to Laurent series of nonnegative degree, i.e., to plain power series. Operationally, these correspond to trajectories that start after $t = 0$. Proposition 8 guarantees that any trajectory of a signal flow graph whose first nonzero value on the left appears at time $t = 0$, will not have nonzero values on the right starting before time $t = 0$. In other words, signal flow graphs can be seen as processing a stream of values from left to right. As a result, their ports can be clearly partitioned into inputs and outputs.

But the circuits of SF are too restrictive for our purposes. For example, 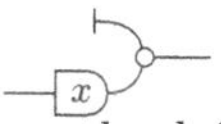 can also be seen to realise a functional behaviour transforming inputs on the left into outputs on the right yet it is not in SF. Its behaviour is no longer linear, but affine. Hence, we need to extend signal flow graphs to include functional affine behaviour. The following definition does just that.

Definition 10. *Let* ASF *be the sub-prop of* ACirc *obtained from* all *the generators in* (1)*, closed under* $\mathsf{Tr}(\cdot)$*. Its circuits are called* affine signal flow graphs.

As before, none of 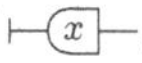, 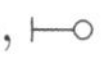 and 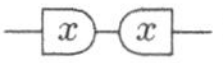 from Examples 3-5 are affine signal flow graphs. In fact, ASF rules out pathological behaviour: all computations can be extended to be infinite, or in other words, do not get stuck.

Proposition 9. *Given an affine signal flow graph f, for every computation*

$$t \triangleright f_0 \xrightarrow[v_t]{u_t} f_1 \xrightarrow[v_{p+1}]{u_{t+1}} \cdots f_n$$

there exists a trajectory $\sigma \in \langle c\rangle$ such that $\sigma(i) = (u_i, v_i)$ for $t \le i \le t + n$.

Proof. By induction on the structure of affine signal flow graphs. □

If SF circuits correspond precisely to $\mathsf{k}\langle x\rangle$-matrices, those of ASF correspond precisely to $\mathsf{k}\langle x\rangle$-affine transformations.

Definition 11. *A map* $f\colon \mathsf{k}(x)^n \to \mathsf{k}(x)^m$ *is an* affine map *if there exists an* $m \times n$ *matrix* A *and* $b \in \mathsf{k}(x)^m$ *such that* $f(p) = A \cdot p + b$ *for all* $p \in \mathsf{k}(x)^n$. *We call the pair* (A, b) *the representation of* f.

The notion of rational affine map is a straightforward extension of the linear case and so is the characterisation in terms of rational input-output behaviour.

Definition 12. *An affine map* $f\colon p \mapsto A \cdot p + b$ *is* rational *if* A *and* b *have coefficients in* $\mathsf{k}\langle x\rangle$.

Proposition 10. *An affine map* $f\colon \mathsf{k}(x)^n \to \mathsf{k}(x)^m$ *is rational iff* $f(r) \in \mathsf{k}\langle x\rangle^m$ *for all* $r \in \mathsf{k}\langle x\rangle^n$.

The following extends the correspondence of Proposition 7, showing that ASF is the rightful affine heir of SF.

Proposition 11. *Given* $c\colon (n, m)$, *we have* $[\![c]\!] = \{(p, f(p)) \mid p \in \mathsf{k}(x)^n\}$ *for some rational affine map* f *iff there exists an affine signal flow graph* g, *i.e., a circuit* $g\colon (n, m)$ *of* ASF, *such that* $[\![g]\!] = [\![c]\!]$.

Proof. Let f be given by $p \mapsto Ap + b$ for some rational $m \times n$ matrix A and vector $b \in \mathsf{k}\langle x\rangle^m$. By Proposition 7, we can find a circuit c_A of SF such that $[\![c_A]\!] = \{(p, A \cdot p) \mid p \in \mathsf{k}(x)\}$. Similarly, we can represent b as a signal flow graph c_b of sort $(1, m)$. Then, the circuit on the right is clearly in ASF and verifies $[\![c]\!] = \{(p, Ap + b) \mid p \in \mathsf{k}(x)\}$ as required.

$c :=$

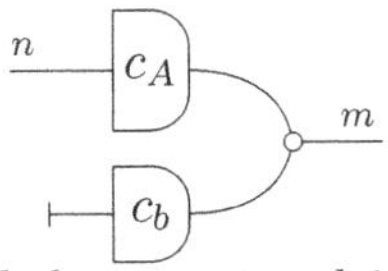

For the converse direction it is straightforward to check by structural induction that the denotation of affine signal flow graphs is the graph (in the set-theoretic sense of pairs of values) of some rational affine map. □

6 Realisability

In the previous section we gave a restricted class of morphisms with good behavioural properties. We may wonder how much of ACirc we can capture with this restricted class. The answer is, in a precise sense: most of it.

Surprisingly, the behaviours realisable in Circ—the purely linear fragment—are not more expressive. In fact, from an operational (or denotational, by full abstraction) point of view, Circ is nothing more than jumbled up version of SF. Indeed, it turns out that Circ enjoys a *realisability* theorem: any circuit c of Circ can be associated with one of SF, that implements or realises the behaviour of c into an executable form.

But the corresponding realisation may not flow neatly from left to right like signal flow graphs do—its inputs and outputs may have been moved from one side to the other. Consider for example, the circuit on the right

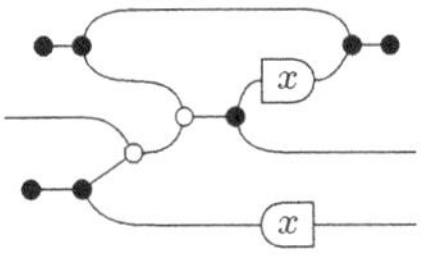

It does not belong to SF but it can be read as a signal flow graph with an input that has been bent and moved to the bottom right. The behaviour it realises can therefore executed by rewiring this port to obtain a signal flow graph:

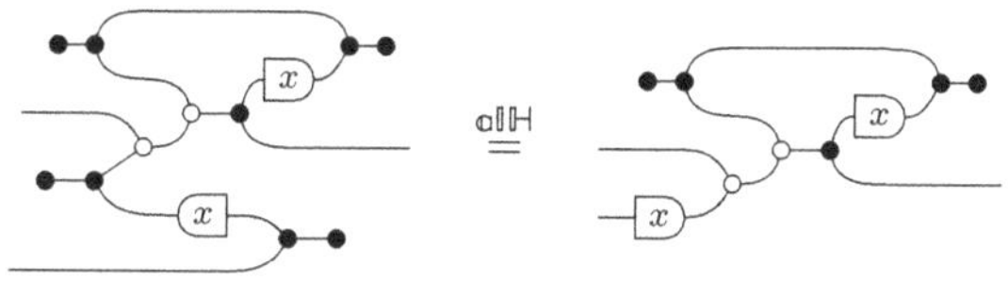

We will not make this notion of rewiring precise here but refer the reader to [9] for the details. The intuition is simply that a rewiring partitions the ports of a circuit into two sets—that we call inputs and outputs—and uses ●-◖ or ◗-● to bend input ports to the left and and output ports to the right. The realisability theorem then states that we can always recover a (not necessarily unique) signal flow graph from any circuit by performing these operations.

Theorem 4. *[9, Theorem 5] Every circuit in* Circ *is equivalent to the rewiring of a signal flow graph, called its* realisation.

This theorem allows us to extend the notion of inputs and outputs to all circuits of Circ.

Definition 13. *A port of a circuit c of* Circ *is an* input *(resp.* output*) port, if there exists a realisation for which it is an input (resp. output).*

Note that, since realisations are not necessarily unique, the same port can be both an input and an output. Then, the realisability theorem (Theorem 4) says that every port is always an input, an output or both (but never neither).

An output-only port is an output port that is not an input port. Similarly an input-only port in an input port that is not an output port.

Example 8. The left port of the register –[x⟩– is input-only whereas its right port is output-only. In the identity wire, both ports are input and output ports. The single port of ○— is output-only ; that of —● is input-only.

While in the purely linear case, all behaviours are realisable, the general case of ACirc is a bit more subtle. To make this precise, we can extend our definition of realisability to include affine signal flow graphs.

Definition 14. *A circuit of* ACirc *is* realisable *if its ports can be rewired so that it is equivalent to a circuit of* ASF.

Example 9. ⊢— is realisable; ⊢⟨x]– is not.

Notice that Proposition 11, gives the following equivalent semantic criterion for realisability. Realisable behaviours are precisely those that map rationals to rationals.

Theorem 5. *A circuit c is realisable iff its ports can be partitioned into two sets, that we call inputs and outputs, such that the corresponding rewiring of c is an affine rational map from inputs to outputs.*

We offer another perspective on realisability below: realisable behaviours correspond precisely to those for which the $\vdash$ constants are connected to inputs of the underlying Circ-circuit. First, notice that, since

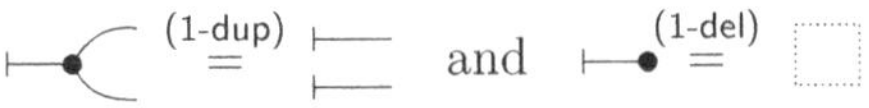

$$\vdash\!\!\!\!\prec \overset{(1\text{-}dup)}{=} \begin{matrix}\vdash\\\vdash\end{matrix} \quad \text{and} \quad \vdash\!\!\bullet \overset{(1\text{-}del)}{=} \square$$

in $\circ\!\!\parallel\!\!\mathbb{H}$, we can assume without loss of generality that each circuit contains exactly one $\vdash$.

Proposition 12. *Every circuit c of* ACirc *is equivalent to one with precisely one* $\vdash$ *and no* $\dashv$.

For $c\colon (n, m)$ a circuit of ACirc, we will call $\hat{c}$ the circuit of Circ of sort $(n + 1, m)$ that one obtains by first transforming c into an equivalent circuit with a single $\vdash$ and no $\dashv$ as above, then removing this $\vdash$, and replacing it by an identity wire that extends to the left boundary.

Theorem 6. *A circuit c is realisable iff* $\vdash$ *is connected to an input port of* $\hat{c}$.

7　Conclusion and Future Work

We introduced the operational semantics of the *affine* extension of the signal flow calculus and proved that contextual equivalence coincides with denotational equality, previously introduced and axiomatised in [6]. We have observed that, at the denotational level, affinity provides two key properties (Propositions 2 and 3) for the proof of full abstraction. However, at the operational level, affinity forces us to consider computations starting in the *past* (Example 3) as the syntax allows terms lacking a proper flow directionality. This leads to circuits that might deadlock ($\vdash\!\!\circ$ in Example 4) or perform some problematic computations ($-\boxed{x}\!\!-\!\!\boxed{x}\!\!\vdash$ in Example 5). We have identified a proper subclass of circuits, called affine signal flow graphs (Definition 10), that possess an inherent flow directionality: in these circuits, the same pathological behaviours do not arise (Proposition 9). This class is not too restrictive as it captures all desirable behaviours: a realisability result (Theorem 5) states that all and only the circuits that do not need computations to start in the past are equivalent to (the rewiring of) an affine signal flow graph.

The reader may be wondering why we do not restrict the syntax to affine signal flow graphs. The reason is that, like in the behavioural approach to control theory [33], the lack of flow direction is what allows the (affine) signal flow calculus to achieve a strong form of compositionality and a complete axiomatisation (see [9] for a deeper discussion).

We expect that similar methods and results can be extended to other models of computation. Our next step is to tackle Petri nets, which, as shown in [5], can be regarded as terms of the signal flow calculus, but over $\mathbb{N}$ rather than a field.

References

1. Abramsky, S., Coecke, B.: A categorical semantics of quantum protocols. In: Proceedings of the 19th Annual IEEE Symposium on Logic in Computer Science (LICS), 2004. pp. 415–425. IEEE (2004)
2. Baez, J., Erbele, J.: Categories in control. Theory and Applications of Categories **30**, 836–881 (2015)
3. Baez, J.C.: Network theory (2014), `http://math.ucr.edu/home/baez/networks/`, website (retrieved 15/04/2014)
4. Basold, H., Bonsangue, M., Hansen, H., Rutten, J.: (Co)Algebraic characterizations of signal flow graphs. In: van Breugel, F., Kashefi, E., Palamidessi, C., Rutten, J. (eds.) Horizons of the Mind. A Tribute to Prakash Panangaden, Lecture Notes in Computer Science, vol. 8464, pp. 124–145. Springer International Publishing (2014)
5. Bonchi, F., Holland, J., Piedeleu, R., Sobociński, P., Zanasi, F.: Diagrammatic algebra: from linear to concurrent systems. Proceedings of the 46th ACM SIGPLAN Symposium on Principles of Programming Languages (POPL) **3**, 1–28 (2019)
6. Bonchi, F., Piedeleu, R., Sobociński, P., Zanasi, F.: Graphical affine algebra. In: Proceedings of the 34th Annual ACM/IEEE Symposium on Logic in Computer Science (LICS). pp. 1–12 (2019)
7. Bonchi, F., Piedeleu, R., Sobociński, P., Zanasi, F.: Contextual equivalence for signal flow graphs (2020), `https://arxiv.org/abs/2002.08874`
8. Bonchi, F., Sobociński, P., Zanasi, F.: A categorical semantics of signal flow graphs. In: Proceedings of the 25th International Conference on Concurrency Theory (CONCUR). pp. 435–450. Springer (2014)
9. Bonchi, F., Sobocinski, P., Zanasi, F.: Full abstraction for signal flow graphs. In: Proceedings of the 42nd Annual ACM SIGPLAN Symposium on Principles of Programming Languages (POPL). pp. 515–526 (2015)
10. Bonchi, F., Sobocinski, P., Zanasi, F.: The calculus of signal flow diagrams I: linear relations on streams. Information and Computation **252**, 2–29 (2017)
11. Coecke, B., Duncan, R.: Interacting quantum observables. In: Proceedings of the 35th international colloquium on Automata, Languages and Programming (ICALP), Part II. pp. 298–310 (2008)
12. Coecke, B., Kissinger, A.: Picturing Quantum Processes - A first course in Quantum Theory and Diagrammatic Reasoning. Cambridge University Press (2017)
13. De Nicola, R., Hennessy, M.C.: Testing equivalences for processes. Theoretical Computer Science **34**(1-2), 83–133 (1984)
14. Ghica, D.R.: Diagrammatic reasoning for delay-insensitive asynchronous circuits. In: Computation, Logic, Games, and Quantum Foundations. The Many Facets of Samson Abramsky, pp. 52–68. Springer (2013)
15. Ghica, D.R., Jung, A.: Categorical semantics of digital circuits. In: Proceedings of the 16th Conference on Formal Methods in Computer-Aided Design (FMCAD). pp. 41–48 (2016)
16. Ghica, D.R., Lopez, A.: A structural and nominal syntax for diagrams. In: Proceedings 14th International Conference on Quantum Physics and Logic (QPL). pp. 71–83 (2017)
17. Hoare, C.A.R.: Communicating Sequential Processes. Prentice Hall (1985)
18. Honda, K., Yoshida, N.: On reduction-based process semantics. Theoretical Computer Science **152**(2), 437–486 (1995)

19. Mac Lane, S.: Categorical algebra. Bulletin of the American Mathematical Society **71**, 40–106 (1965)
20. Mac Lane, S.: Categories for the Working Mathematician. Springer (1998)
21. Mason, S.J.: Feedback Theory: I. Some Properties of Signal Flow Graphs. MIT Research Laboratory of Electronics (1953)
22. Milius, S.: A sound and complete calculus for finite stream circuits. In: Proceedings of the 2010 25th Annual IEEE Symposium on Logic in Computer Science (LICS). pp. 421–430 (2010)
23. Milner, R.: A Calculus of Communicating Systems, Lecture Notes in Computer Science, vol. 92. Springer (1980)
24. Milner, R., Sangiorgi, D.: Barbed bisimulation. In: Proceedings of the 19th International Colloquium on Automata, Languages and Programming (ICALP). pp. 685–695 (1992)
25. Morris Jr, J.H.: Lambda-calculus models of programming languages. Ph.D. thesis, Massachusetts Institute of Technology (1969)
26. Pavlovic, D.: Monoidal computer I: Basic computability by string diagrams. Information and Computation **226**, 94–116 (2013)
27. Pavlovic, D.: Monoidal computer II: Normal complexity by string diagrams. arXiv:1402.5687 (2014)
28. Plotkin, G.D.: Call-by-name, call-by-value and the λ-calculus. Theoretical Computer Science **1**(2), 125–159 (1975)
29. Rutten, J.J.M.M.: A tutorial on coinductive stream calculus and signal flow graphs. Theoretical Computer Science **343**(3), 443–481 (2005)
30. Rutten, J.J.M.M.: Rational streams coalgebraically. Logical Methods in Computer Science **4**(3) (2008)
31. Selinger, P.: A survey of graphical languages for monoidal categories. Springer Lecture Notes in Physics **13**(813), 289–355 (2011)
32. Shannon, C.E.: The theory and design of linear differential equation machines. Tech. rep., National Defence Research Council (1942)
33. Willems, J.C.: The behavioural approach to open and interconnected systems. IEEE Control Systems Magazine **27**, 46–99 (2007)

Parameterized Synthesis for Fragments of First-Order Logic over Data Words*

Béatrice Bérard[1], Benedikt Bollig[2], Mathieu Lehaut[1] (✉), and Nathalie Sznajder[1]

[1] Sorbonne Université, CNRS, LIP6, F-75005 Paris, France
[2] CNRS, LSV & ENS Paris-Saclay, Université Paris-Saclay, Cachan, France

Abstract. We study the synthesis problem for systems with a parameterized number of processes. As in the classical case due to Church, the system selects actions depending on the program run so far, with the aim of fulfilling a given specification. The difficulty is that, at the same time, the environment executes actions that the system cannot control. In contrast to the case of fixed, finite alphabets, here we consider the case of parameterized alphabets. An alphabet reflects the number of processes, which is static but unknown. The synthesis problem then asks whether there is a finite number of processes for which the system can satisfy the specification. This variant is already undecidable for very limited logics. Therefore, we consider a first-order logic without the order on word positions. We show that even in this restricted case synthesis is undecidable if both the system and the environment have access to all processes. On the other hand, we prove that the problem is decidable if the environment only has access to a bounded number of processes. In that case, there is even a cutoff meaning that it is enough to examine a bounded number of process architectures to solve the synthesis problem.

1 Introduction

Synthesis deals with the problem of automatically generating a program that satisfies a given specification. The problem goes back to Church [9], who formulated it as follows: The environment and the system alternately select an input symbol and an output symbol from a finite alphabet, respectively, and in this way generate an infinite sequence. The question now is whether the system has a *winning strategy*, which guarantees that the resulting infinite run is contained in a given (ω)-regular language representing the specification, no matter how the environment behaves. This problem is decidable and very well understood [8,37], and it has been extended in several different ways (e.g., [24,26,28,36,43]).

In this paper, we consider a variant of the synthesis problem that allows us to model programs with a variable number of processes. As we then deal with an unbounded number of process identifiers, a fixed finite alphabet is not suitable anymore. It is more appropriate to use an infinite alphabet, in which every

* Partly supported by ANR FREDDA (ANR-17-CE40-0013).

J. Goubault-Larrecq and B. König (Eds.): FOSSACS 2020, LNCS 12077, pp. 97–118, 2020.
https://doi.org/10.1007/978-3-030-45231-5_6

letter contains a process identifier and a program action. One can distinguish two cases here. In [16], a potentially infinite number of data values are involved in an infinite program run (e.g. by dynamic process generation). In a *parameterized* system [4, 13], on the other hand, one has an unknown but *static* number of processes so that, along each run, the number of processes is finite. In this paper, we are interested in the latter, i.e., parameterized case. Parameterized programs are ubiquitous and occur, e.g., in distributed algorithms, ad-hoc networks, telecommunication protocols, cache-coherence protocols, swarm robotics, and biological systems. The synthesis question asks whether the system has a winning strategy for some number of processes (existential version) or no matter how many processes there are (universal version).

Over infinite alphabets, there are a variety of different specification languages (e.g., [5, 11, 12, 19, 29, 33, 40]). Unlike in the case of finite alphabets, there is no canonical definition of regular languages. In fact, the synthesis problem has been studied for N-memory automata [7], the Logic of Repeating Values [16], and register automata [15,30,31]. Though there is no agreement on a "regular" automata model, first-order (FO) logic over data words can be considered as a canonical logic, and this is the specification language we consider here. In addition to classical FO logic on words over finite alphabets, it provides a predicate $x \sim y$ to express that two events x and y are triggered by the same process. Its two-variable fragment FO^2 has a decidable emptiness and universality problem [5] and is, therefore, a promising candidate for the synthesis problem.

Previous generalizations of Church's synthesis problem to infinite alphabets were generally *synchronous* in the sense that the system and the environment perform their actions in strictly alternating order. This assumption was made, e.g., in the above-mentioned recent papers [7, 15, 16, 30, 31]. If there are several processes, however, it is realistic to relax this condition, which leads us to an *asynchronous* setting in which the system has no influence on when the environment acts. Like in [21], where the asynchronous case for a fixed number of processes was considered, we only make the reasonable fairness assumption that the system is not blocked forever.

In summary, the synthesis problem over infinite alphabets can be classified as (*i*) parameterized vs. dynamic, (*ii*) synchronous vs. asynchronous, and (*iii*) according to the specification language (register automata, Logic of Repeating Values, FO logic, etc.). As explained above, we consider here the *parameterized asynchronous case for specifications written in FO logic.* To the best of our knowledge, this combination has not been considered before. For flexible modeling, we also distinguish between three types of processes: those that can only be controlled by the system; those that can only be controlled by the environment; and finally those that can be triggered by both. A partition into system and environment processes is also made in [3, 18], but for a fixed number of processes and in the presence of an arena in terms of a Petri net.

Let us briefly describe our results. We show that the general case of the synthesis problem is undecidable for FO^2 logic. This follows from an adaptation of an undecidability result from [16, 17] for a fragment of the Logic of Repeating

Values [11]. We therefore concentrate on an orthogonal logic, namely FO without the order on the word positions. First, we show that this logic can essentially count processes and actions of a given process up to some threshold. Though it has limited expressive power (albeit orthogonal to that of FO^2), it leads to intricate behaviors in the presence of an uncontrollable environment. In fact, we show that the synthesis problem is still undecidable. Due to the lack of the order relation, the proof requires a subtle reduction from the reachability problem in 2-counter Minsky machines. However, it turns out that the synthesis problem is decidable if the number of processes that are controllable by the environment is bounded, while the number of system processes remains unbounded. In this case, there is even a cutoff k, an important measure for parameterized systems (cf. [4] for an overview): If the system has a winning strategy for k processes, then it has one for any number of processes greater than k, and the same applies to the environment. The proofs of both main results rely on a reduction of the synthesis problem to turn-based *parameterized vector games*, in which, similar to Petri nets, tokens corresponding to processes are moved around between states.

The paper is structured as follows. In Section 2, we define FO logic (especially FO without word order), and in Section 3, we present the parameterized synthesis problem. In Section 4, we transform a given formula into a normal form and finally into a parameterized vector game. Based on this reduction, we investigate cutoff properties and show our (un)decidability results in Section 5. We conclude in Section 6. Some proof details can be found in the long version of this paper [2]

2 Preliminaries

For a finite or infinite alphabet Σ, let Σ^* and Σ^ω denote the sets of finite and, respectively, infinite words over Σ. The empty word is ε. Given $w \in \Sigma^* \cup \Sigma^\omega$, let $|w|$ denote the length of w and $Pos(w)$ its set of positions: $|w| = n$ and $Pos(w) = \{1, \ldots, n\}$ if $w = \sigma_1\sigma_2 \ldots \sigma_n \in \Sigma^*$, and $|w| = \omega$ and $Pos(w) = \{1, 2, \ldots\}$ if $w \in \Sigma^\omega$. Let $w[i]$ be the i-th letter of w for all $i \in Pos(w)$.

Executions. We consider programs involving a finite (but not fixed) number of processes. Processes are controlled by antagonistic protagonists, System and Environment. Accordingly, each process has a *type* among $\mathbb{T} = \{\mathsf{s}, \mathsf{e}, \mathsf{se}\}$, and we let $\mathbb{P}_\mathsf{s}$, $\mathbb{P}_\mathsf{e}$, and $\mathbb{P}_\mathsf{se}$ denote the pairwise disjoint finite sets of processes controlled by System, by Environment, and by both System and Environment, respectively. We let $\mathbb{P}$ denote the triple $(\mathbb{P}_\mathsf{s}, \mathbb{P}_\mathsf{e}, \mathbb{P}_\mathsf{se})$. Abusing notation, we sometimes refer to $\mathbb{P}$ as the disjoint union $\mathbb{P}_\mathsf{s} \cup \mathbb{P}_\mathsf{e} \cup \mathbb{P}_\mathsf{se}$.

Given any set S, vectors $s \in S^\mathbb{T}$ are usually referred to as triples $s = (s_\mathsf{s}, s_\mathsf{e}, s_\mathsf{se})$. Moreover, for $s, s' \in \mathbb{N}^\mathbb{T}$, we write $s \leq s'$ if $s_\theta \leq s'_\theta$ for all $\theta \in \mathbb{T}$. Finally, let $s + s' = (s_\mathsf{s} + s'_\mathsf{s},\ s_\mathsf{e} + s'_\mathsf{e},\ s_\mathsf{se} + s'_\mathsf{se})$.

Processes can execute actions from a finite alphabet A. Whenever an action is executed, we would like to know whether it was triggered by System or by Environment. Therefore, A is partitioned into $A = A_\mathsf{s} \uplus A_\mathsf{e}$. Let $\Sigma_\mathsf{s} = A_\mathsf{s} \times (\mathbb{P}_\mathsf{s} \cup \mathbb{P}_\mathsf{se})$ and $\Sigma_\mathsf{e} = A_\mathsf{e} \times (\mathbb{P}_\mathsf{e} \cup \mathbb{P}_\mathsf{se})$. Their union $\Sigma = \Sigma_\mathsf{s} \cup \Sigma_\mathsf{e}$ is the set of *events*. A word $w \in \Sigma^* \cup \Sigma^\omega$ is called a $\mathbb{P}$-*execution*.

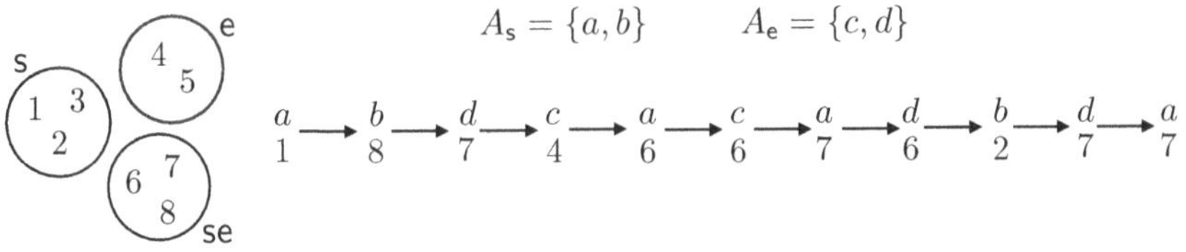

Fig. 1. Representation of $\mathbb{P}$-execution as a mathematical structure

Logic. Formulas of our logic are evaluated over $\mathbb{P}$-executions. We fix an infinite supply $\mathcal{V} = \{x, y, z, \ldots\}$ of variables, which are interpreted as processes from $\mathbb{P}$ or positions of the execution. The logic $\text{FO}_A[\sim, <, +1]$ is given by the grammar

$$\varphi ::= \theta(x) \mid a(x) \mid x = y \mid x \sim y \mid x < y \mid +1(x, y) \mid \neg\varphi \mid \varphi \vee \varphi \mid \exists x.\varphi$$

where $x, y \in \mathcal{V}$, $\theta \in \mathbb{T}$, and $a \in A$. Conjunction ($\wedge$), universal quantification ($\forall$), implication ($\Longrightarrow$), *true*, and *false* are obtained as abbreviations as usual.

Let $\varphi \in \text{FO}_A[\sim, <, +1]$. By $\mathit{Free}(\varphi) \subseteq \mathcal{V}$, we denote the set of variables that occur free in φ. If $\mathit{Free}(\varphi) = \emptyset$, then we call φ a *sentence*. We sometimes write $\varphi(x_1, \ldots, x_n)$ to emphasize the fact that $\mathit{Free}(\varphi) \subseteq \{x_1, \ldots, x_n\}$.

To evaluate φ over a $\mathbb{P}$-execution $w = (a_1, p_1)(a_2, p_2) \ldots$, we consider $(\mathbb{P}, w)$ as a structure $\mathcal{S}_{(\mathbb{P},w)} = (\mathbb{P} \uplus \mathit{Pos}(w), \mathbb{P}_{\mathsf{s}}, \mathbb{P}_{\mathsf{e}}, \mathbb{P}_{\mathsf{se}}, (R_a)_{a \in A}, \sim, <, +1)$ where $\mathbb{P} \uplus \mathit{Pos}(w)$ is the universe, $\mathbb{P}_{\mathsf{s}}$ $\mathbb{P}_{\mathsf{e}}$, and $\mathbb{P}_{\mathsf{se}}$ are interpreted as unary relations, R_a is the unary relation $\{i \in \mathit{Pos}(w) \mid a_i = a\}$, $< = \{(i, j) \in \mathit{Pos}(w) \times \mathit{Pos}(w) \mid i < j\}$, $+1 = \{(i, i+1) \mid 1 \leq i < |w|\}$, and $\sim$ is the smallest equivalence relation over $\mathbb{P} \uplus \mathit{Pos}(w)$ containing

- (p, i) for all $p \in \mathbb{P}$ and $i \in \mathit{Pos}(w)$ such that $p = p_i$, and
- (i, j) for all $(i, j) \in \mathit{Pos}(w) \times \mathit{Pos}(w)$ such that $p_i = p_j$.

An equivalence class of $\sim$ is often simply referred to as a *class*. Note that it contains exactly one process.

Example 1. Suppose $A_{\mathsf{s}} = \{a, b\}$ and $A_{\mathsf{e}} = \{c, d\}$. Let the set of processes $\mathbb{P}$ be given by $\mathbb{P}_{\mathsf{s}} = \{1, 2, 3\}$, $\mathbb{P}_{\mathsf{e}} = \{4, 5\}$, and $\mathbb{P}_{\mathsf{se}} = \{6, 7, 8\}$. Moreover, let $w = (a, 1)(b, 8)(d, 7)(c, 4)(a, 6)(c, 6)(a, 7)(d, 6)(b, 2)(d, 7)(a, 7) \in \Sigma^*$. Figure 1 illustrates $\mathcal{S}_{(\mathbb{P},w)}$. The edge relation represents $+1$, its transitive closure is $<$. $\lhd$

An *interpretation* for $(\mathbb{P}, w)$ is a partial mapping $I : \mathcal{V} \to \mathbb{P} \cup \mathit{Pos}(w)$. Suppose $\varphi \in \text{FO}_A[\sim, <, +1]$ such that $\mathit{Free}(\varphi) \subseteq \text{dom}(I)$. The satisfaction relation $(\mathbb{P}, w), I \models \varphi$ is then defined as expected, based on the structure $\mathcal{S}_{(\mathbb{P},w)}$ and interpreting free variables according to I. For example, let $w = (a_1, p_1)(a_2, p_2) \ldots$ and $i \in \mathit{Pos}(w)$. Then, for $I(x) = i$, we have $(\mathbb{P}, w), I \models a(x)$ if $a_i = a$.

We identify some fragments of $\text{FO}_A[\sim, <, +1]$. For $R \subseteq \{\sim, <, +1\}$, let $\text{FO}_A[R]$ denote the set of formulas that do not use symbols in $\{\sim, <, +1\} \setminus R$. Moreover, $\text{FO}_A^2[R]$ denotes the fragment of $\text{FO}_A[R]$ that uses only two (reusable) variables.

Let $\varphi(x_1, \ldots, x_n, y) \in \mathrm{FO}_A[\sim, <, +1]$ be a formula and $m \in \mathbb{N}$. We use $\exists^{\geq m} y.\varphi(x_1, \ldots, x_n, y)$ as an abbreviation for

$$\exists y_1 \ldots \exists y_m. \bigwedge_{1 \leq i < j \leq m} \neg(y_i = y_j) \wedge \bigwedge_{1 \leq i \leq m} \varphi(x_1, \ldots, x_n, y_i),$$

if $m > 0$, and $\exists^{\geq 0} y.\varphi(x_1, \ldots, x_n, y) = \mathit{true}$. Thus, $\exists^{\geq m} y.\varphi$ says that there are at least m distinct elements that verify φ. We also use $\exists^{=m} y.\varphi$ as an abbreviation for $\exists^{\geq m} y.\varphi \wedge \neg \exists^{\geq m+1} y.\varphi$. Note that $\varphi \in \mathrm{FO}_A[R]$ implies that $\exists^{\geq m} y.\varphi \in \mathrm{FO}_A[R]$ and $\exists^{=m} y.\varphi \in \mathrm{FO}_A[R]$.

Example 2. Let A, $\mathbb{P}$, and w be like in Example 1 and Figure 1.

- $\varphi_1 = \forall x.((\mathsf{s}(x) \vee \mathsf{se}(x)) \implies \exists y.(x \sim y \wedge (a(y) \vee b(y))))$ says that each process that System can control executes at least one system action. We have $\varphi_1 \in \mathrm{FO}_A^2[\sim]$ and $(\mathbb{P}, w) \not\models \varphi_1$, as process 3 is idle.
- $\varphi_2 = \forall x.(d(x) \implies \exists y.(x \sim y \wedge a(y)))$ says that, for every d, there is an a on the same process. We have $\varphi_2 \in \mathrm{FO}_A^2[\sim]$ and $(\mathbb{P}, w) \models \varphi_2$.
- $\varphi_3 = \forall x.(d(x) \implies \exists y.(x \sim y \wedge x < y \wedge a(y)))$ says that every d is *eventually* followed by an a executed by the same process. We have $\varphi_3 \in \mathrm{FO}_A^2[\sim, <]$ and $(\mathbb{P}, w) \not\models \varphi_3$: The event $(d, 6)$ is not followed by some $(a, 6)$.
- $\varphi_4 = \forall x.((\exists^{=2} y.(x \sim y \wedge a(y))) \iff (\exists^{=2} y.(x \sim y \wedge d(y))))$ says that each class contains exactly two occurrences of a iff it contains exactly two occurrences of d. Moreover, $\varphi_4 \in \mathrm{FO}_A[\sim]$ and $(\mathbb{P}, w) \models \varphi_4$. Note that $\varphi_4 \notin \mathrm{FO}_A^2[\sim]$, as $\exists^{=2} y$ requires the use of three different variable names. ◁

3 Parameterized Synthesis Problem

We define an asynchronous synthesis problem. A $\mathbb{P}$-*strategy* (for System) is a mapping $f : \Sigma^* \to \Sigma_\mathsf{s} \cup \{\varepsilon\}$. A $\mathbb{P}$-execution $w = \sigma_1 \sigma_2 \ldots \in \Sigma^* \cup \Sigma^\omega$ is f-*compatible* if, for all $i \in \mathit{Pos}(w)$ such that $\sigma_i \in \Sigma_\mathsf{s}$, we have $f(\sigma_1 \ldots \sigma_{i-1}) = \sigma_i$. We call w f-*fair* if the following hold: (i) If w is finite, then $f(w) = \varepsilon$, and (ii) if w is infinite and $f(\sigma_1 \ldots \sigma_{i-1}) \neq \varepsilon$ for infinitely many $i \geq 1$, then $\sigma_j \in \Sigma_\mathsf{s}$ for infinitely many $j \geq 1$.

Let $\varphi \in \mathrm{FO}_A[\sim, <, +1]$ be a sentence. We say that f is $\mathbb{P}$-*winning* for φ if, for every $\mathbb{P}$-execution w that is f-compatible and f-fair, we have $(\mathbb{P}, w) \models \varphi$.

The existence of a $\mathbb{P}$-strategy that is $\mathbb{P}$-winning for a given formula does not depend on the concrete process identities but only on the cardinality of the sets $\mathbb{P}_\mathsf{s}$, $\mathbb{P}_\mathsf{e}$, and $\mathbb{P}_\mathsf{se}$. This motivates the following definition of winning triples for a formula. Given φ, let $\mathit{Win}(\varphi)$ be the set of triples $(k_\mathsf{s}, k_\mathsf{e}, k_\mathsf{se}) \in \mathbb{N}^\mathbb{T}$ for which there is $\mathbb{P} = (\mathbb{P}_\mathsf{s}, \mathbb{P}_\mathsf{e}, \mathbb{P}_\mathsf{se})$ such that $|\mathbb{P}_\theta| = k_\theta$ for all $\theta \in \mathbb{T}$ and there is a $\mathbb{P}$-strategy that is $\mathbb{P}$-winning for φ.

Let $\mathbb{0} = \{0\}$ and $k_\mathsf{e}, k_\mathsf{se} \in \mathbb{N}$. In this paper, we focus on the intersection of $\mathit{Win}(\varphi)$ with the sets $\mathbb{N} \times \mathbb{0} \times \mathbb{0}$ (which corresponds to the usual satisfiability problem); $\mathbb{N} \times \{k_\mathsf{e}\} \times \{k_\mathsf{se}\}$ (there is a constant number of environment and mixed processes); $\mathbb{N} \times \mathbb{N} \times \{k_\mathsf{se}\}$ (there is a constant number of mixed processes); $\mathbb{0} \times \mathbb{0} \times \mathbb{N}$ (each process is controlled by both System and Environment).

Definition 3 (synthesis problem). *For fixed* $\mathfrak{F} \in \{\text{FO}, \text{FO}^2\}$*, set of relation symbols* $R \subseteq \{\sim, <, +1\}$*, and* $\mathcal{N}_{\mathsf{s}}, \mathcal{N}_{\mathsf{e}}, \mathcal{N}_{\mathsf{se}} \subseteq \mathbb{N}$*, the (parameterized) synthesis problem is given as follows:*

$\textsc{Synth}(\mathfrak{F}[R], \mathcal{N}_{\mathsf{s}}, \mathcal{N}_{\mathsf{e}}, \mathcal{N}_{\mathsf{se}})$	
Input:	$A = A_{\mathsf{s}} \uplus A_{\mathsf{e}}$ *and a sentence* $\varphi \in \mathfrak{F}_A[R]$
Question:	$\mathit{Win}(\varphi) \cap (\mathcal{N}_{\mathsf{s}} \times \mathcal{N}_{\mathsf{e}} \times \mathcal{N}_{\mathsf{se}}) \neq \emptyset$?

The satisfiability problem *for* $\mathfrak{F}[R]$ *is defined as* $\textsc{Synth}(\mathfrak{F}[R], \mathbb{N}, \mathbb{0}, \mathbb{0})$.

Example 4. Suppose $A_{\mathsf{s}} = \{a, b\}$ and $A_{\mathsf{e}} = \{c, d\}$, and consider the formulas φ_1–φ_4 from Example 2.

First, we have $\mathit{Win}(\varphi_1) = \mathbb{N}^{\mathbb{T}}$. Given an arbitrary $\mathbb{P}$ and any total order $\sqsubseteq$ over $\mathbb{P}_{\mathsf{s}} \cup \mathbb{P}_{\mathsf{se}}$, a possible $\mathbb{P}$-strategy f that is $\mathbb{P}$-winning for φ_1 maps $w \in \Sigma^*$ to (a, p) if p is the smallest process from $\mathbb{P}_{\mathsf{s}} \cup \mathbb{P}_{\mathsf{se}}$ wrt. $\sqsubseteq$ that does not occur in w, and that returns ε for w if all processes from $\mathbb{P}_{\mathsf{s}} \cup \mathbb{P}_{\mathsf{se}}$ already occur in w.

For the three formulas φ_2, φ_3, and φ_4, observe that, since d is an environment action, if there is at least one process that is exclusively controlled by Environment, then there is no winning strategy. Hence we must have $\mathbb{P}_{\mathsf{e}} = \emptyset$. In fact, this condition is sufficient in the three cases and the strategies described below show that all three sets $\mathit{Win}(\varphi_2)$, $\mathit{Win}(\varphi_3)$, and $\mathit{Win}(\varphi_4)$ are equal to $\mathbb{N} \times \mathbb{0} \times \mathbb{N}$.

- For φ_2, the very same strategy as for φ_1 also works in this case, producing an a for every process in $\mathbb{P}_{\mathsf{s}} \cup \mathbb{P}_{\mathsf{se}}$, whether there is a d or not.
- For φ_3, a winning strategy f will apply the previous mechanism iteratively, performing (a, p) for $p \in \mathbb{P}_{\mathsf{se}} = \{p_0, \ldots, p_{n-1}\}$ over and over again: $f(w) = (a, p_i)$ where i is the number of occurrences of letters from Σ_{s} modulo n. By the fairness assumption, this guarantees satisfaction of φ_3. A more "economical" winning strategy f' may organize pending requests in terms of d in a queue and acknowledge them successively. More precisely, given $u \in \mathbb{P}^*$ and $\sigma \in \Sigma$, we define another word $u \odot \sigma \in \mathbb{P}^*$ by $u \odot (d, p) = u \cdot p$ (inserting p in the queue) and $(p \cdot u) \odot (a, p) = u$ (deleting it). In all other cases, $u \odot \sigma = u$. Let $w = \sigma_1 \ldots \sigma_n \in \Sigma^*$, with queue $((\varepsilon \odot \sigma_1) \odot \sigma_2 \ldots) \odot \sigma_n = p_1 \ldots p_k$. We let $f'(w) = \varepsilon$ if $k = 0$, and $f'(w) = (a, p_1)$ if $k \geq 1$.
- For φ_4, the strategy f' for φ_3 ensures that every d has a *corresponding* a so that, in the long run, there are as many a's as d's in every class. ◁

Another interesting question is whether System (or Environment) has a winning strategy as soon as the number of processes is big enough. This leads to the notion of a cutoff (cf. [4] for an overview): Let $\mathcal{N}_{\mathsf{s}}, \mathcal{N}_{\mathsf{e}}, \mathcal{N}_{\mathsf{se}} \subseteq \mathbb{N}$ and $W \subseteq \mathbb{N}^{\mathbb{T}}$. We call $\boldsymbol{k_0} \in \mathbb{N}^{\mathbb{T}}$ a *cutoff* of W wrt. $(\mathcal{N}_{\mathsf{s}}, \mathcal{N}_{\mathsf{e}}, \mathcal{N}_{\mathsf{se}})$ if $\boldsymbol{k_0} \in \mathcal{N}_{\mathsf{s}} \times \mathcal{N}_{\mathsf{e}} \times \mathcal{N}_{\mathsf{se}}$ and either

- for all $\boldsymbol{k} \in \mathcal{N}_{\mathsf{s}} \times \mathcal{N}_{\mathsf{e}} \times \mathcal{N}_{\mathsf{se}}$ such that $\boldsymbol{k} \geq \boldsymbol{k_0}$, we have $\boldsymbol{k} \in W$, or
- for all $\boldsymbol{k} \in \mathcal{N}_{\mathsf{s}} \times \mathcal{N}_{\mathsf{e}} \times \mathcal{N}_{\mathsf{se}}$ such that $\boldsymbol{k} \geq \boldsymbol{k_0}$, we have $\boldsymbol{k} \notin W$.

Let $\mathfrak{F} \in \{\text{FO}, \text{FO}^2\}$ and $R \subseteq \{\sim, <, +1\}$. If, for every alphabet $A = A_{\mathsf{s}} \uplus A_{\mathsf{e}}$ and every sentence $\varphi \in \mathfrak{F}_A[R]$, the set $\mathit{Win}(\varphi)$ has a computable cutoff wrt.

Table 1. Summary of results. Our contributions are highlighted in **bold**.

Synthesis	$(\mathbb{N}, \mathbb{0}, \mathbb{0})$	$(\mathbb{N}, \{k_{\mathsf{e}}\}, \{k_{\mathsf{se}}\})$	$(\mathbb{N}, \mathbb{N}, \mathbb{0})$	$(\mathbb{0}, \mathbb{0}, \mathbb{N})$
$\mathrm{FO}^2[\sim, <, +1]$	decidable [5]	?	?	**undecidable**
$\mathrm{FO}^2[\sim, <]$	NEXPTIME-c. [5]	?	?	?
$\mathrm{FO}[\sim]$	**decidable**	**decidable**	?*	**undecidable**

*We show, however, that there is no cutoff.

$(\mathcal{N}_{\mathsf{s}}, \mathcal{N}_{\mathsf{e}}, \mathcal{N}_{\mathsf{se}})$, then we know that $\mathrm{SYNTH}(\mathfrak{F}[R], \mathcal{N}_{\mathsf{s}}, \mathcal{N}_{\mathsf{e}}, \mathcal{N}_{\mathsf{se}})$ is decidable, as it can be reduced to a finite number of simple synthesis problems over a finite alphabet. The latter can be solved, e.g., using attractor-based backward search (cf. [42]). This is how we will show decidability of $\mathrm{SYNTH}(\mathrm{FO}[\sim], \mathbb{N}, \{k_{\mathsf{e}}\}, \{k_{\mathsf{se}}\})$ for all $k_{\mathsf{e}}, k_{\mathsf{se}} \in \mathbb{N}$.

Our contributions are summarized in Table 1. Note that known satisfiability results for data logic apply to our logic, as processes can be simulated by treating every $\theta \in \mathbb{T}$ as an ordinary letter. Let us first state undecidability of the general synthesis problem, which motivates the study of other FO fragments.

Theorem 5. *The problem* $\mathrm{SYNTH}(\mathrm{FO}^2[\sim, <, +1], \mathbb{0}, \mathbb{0}, \mathbb{N})$ *is undecidable.*

Proof (sketch). We adapt the proof from [16, 17] reducing the halting problem for 2-counter machines. We show that their encoding can be expressed in our logic, even if we restrict it to two variables, and can also be adapted to the asynchronous setting. □

4 FO[∼] and Parameterized Vector Games

Due to the undecidability result of Theorem 5, one has to switch to other fragments of first-order logic. We will henceforth focus on the logic $\mathrm{FO}[\sim]$ and establish some important properties, such as a normal form, that will allow us to deduce a couple of results, both positive and negative.

4.1 Satisfiability and Normal Form for FO[∼]

We first show that $\mathrm{FO}[\sim]$ logic essentially allows one to count letters in a class up to some threshold, and to count such classes up to some other threshold. Let $B \in \mathbb{N}$ and $\ell \in \{0, \dots, B\}^A$. Intuitively, $\ell(a)$ imposes a constraint on the number of occurrences of a in a class. We first define an $\mathrm{FO}_A[\sim]$-formula $\psi_{B,\ell}(y)$ verifying that, in the class defined by y, the number of occurrences of each letter $a \in A$, counted up to B, is $\ell(a)$:

$$\psi_{B,\ell}(y) \;=\; \bigwedge_{\substack{a \in A \,\mid\\ \ell(a) < B}} \exists^{=\ell(a)} z.(y \sim z \wedge a(z)) \wedge \bigwedge_{\substack{a \in A \,\mid\\ \ell(a) = B}} \exists^{\geq \ell(a)} z.(y \sim z \wedge a(z))$$

Theorem 6 (normal form for FO[$\sim$]). *Let $\varphi \in \mathrm{FO}_A[\sim]$ be a sentence. There is a computable $B \in \mathbb{N}$ such that φ is effectively equivalent to a disjunction of conjunctions of formulas of the form $\exists^{\bowtie m} y.(\theta(y) \wedge \psi_{B,\ell}(y))$ where $\bowtie \in \{\geq, =\}$, $m \in \mathbb{N}$, $\theta \in \mathbb{T}$, and $\ell \in \{0, \ldots, B\}^A$.*

The normal form can be obtained using known normal-form constructions [23,41] for general FO logic [2], or using Ehrenfeucht-Fraïssé games [39], or using a direct inductive transformation in the spirit of [23].

Example 7. Recall the formula $\varphi_4 = \forall x.\big((\exists^{=2} y.(x \sim y \wedge a(y))) \Longleftrightarrow (\exists^{=2} y.(x \sim y \wedge d(y)))\big) \in \mathrm{FO}_A[\sim]$ from Example 2, over $A_\mathsf{s} = \{a, b\}$ and $A_\mathsf{e} = \{c, d\}$. An equivalent formula in normal form is $\varphi_4' = \bigwedge_{\theta \in \mathbb{T},\, \ell \in Z} \exists^{=0} y.(\theta(y) \wedge \psi_{3,\ell}(y))$ where Z is the set of vectors $\ell \in \{0, \ldots, 3\}^A$ such that $\ell(a) = 2 \neq \ell(d)$ or $\ell(d) = 2 \neq \ell(a)$. The formula indeed says that there is no class with $=2$ occurrences of a and $\neq 2$ occurrences of d or vice versa, which is equivalent to φ_4. ◁

Thanks to the normal form, it is sufficient to test finitely many structures to determine whether a given formula is satisfiable:

Corollary 8. *The satisfiability problem for* FO[$\sim$] *over data words is decidable. Moreover, every satisfiable* $\mathrm{FO}_A[\sim]$ *formula has a finite model.*

Note that the satisfiability problem for $\mathrm{FO}^2[\sim]$ is already NEXPTIME-hard, due to NEXPTIME-hardness for two-variable logic with unary relations only [14, 20,22]. In fact, it is NEXPTIME-complete due to the upper bound for $\mathrm{FO}^2[\sim, <]$ [5]. It is worth mentioning that two-variable logic with one equivalence relation on arbitrary structures also has the finite-model property [32].

4.2 From Synthesis to Parameterized Vector Games

Exploiting the normal form for $\mathrm{FO}_A[\sim]$, we now present a reduction of the synthesis problem to a strictly turn-based two-player game. This game is conceptually simpler and easier to reason about. The reduction works in both directions, which will allow us to derive both decidability and undecidability results.

Note that, given a formula $\varphi \in \mathrm{FO}_A[\sim]$ (which we suppose to be in normal form with threshold B), the order of letters in an execution does not matter. Thus, given some $\mathbb{P}$, a reasonable strategy for Environment would be to just "wait and see". More precisely, it does not put Environment into a worse position if, given the current execution $w \in \Sigma^*$, it lets the System execute as many actions as it wants in terms of a word $u \in \Sigma_\mathsf{s}^*$. Due to the fairness assumption, System would be able to execute all the letters from u anyway. Environment can even require System to play a word u such that $(\mathbb{P}, wu) \models \varphi$. If System is not able to produce such a word, Environment can just sit back and do nothing. Conversely, upon wu satisfying φ, Environment has to be able to come up with a word $v \in \Sigma_\mathsf{e}^*$ such that $(\mathbb{P}, wuv) \not\models \varphi$. This leads to a turn-based game in which System and Environment play in strictly alternate order and have to provide a satisfying and, respectively, falsifying execution.

In a second step, we can get rid of process identifiers: According to our normal form, all we are interested in is the *number* of processes that agree on their letters counted up to threshold B. That is, a finite execution can be abstracted as a *configuration* $C : L \to \mathbb{N}^{\mathbb{T}}$ where $L = \{0, \ldots, B\}^A$. For $\ell \in L$ and $C(\ell) = (n_{\mathsf{s}}, n_{\mathsf{e}}, n_{\mathsf{se}})$, n_θ is the number of processes of type θ whose letter count up to threshold B corresponds to ℓ. We can also say that ℓ contains n_θ tokens of type θ. If it is System's turn, it will pick some pairs (ℓ, ℓ') and move some tokens of type $\theta \in \{\mathsf{s}, \mathsf{se}\}$ from ℓ to ℓ', provided $\ell(a) \leq \ell'(a)$ for all $a \in A_{\mathsf{s}}$ and $\ell(a) = \ell'(a)$ for all $a \in A_{\mathsf{e}}$. This actually corresponds to adding more system letters in the corresponding processes. The Environment proceeds analogously.

Finally, the formula φ naturally translates to an acceptance condition $\mathcal{F} \subseteq \mathfrak{C}^L$ over configurations, where $\mathfrak{C}$ is the set of *local acceptance conditions*, which are of the form $(\bowtie_{\mathsf{s}} n_{\mathsf{s}}\,,\, \bowtie_{\mathsf{e}} n_{\mathsf{e}}\,,\, \bowtie_{\mathsf{se}} n_{\mathsf{se}})$ where $\bowtie_{\mathsf{s}}, \bowtie_{\mathsf{e}}, \bowtie_{\mathsf{se}} \in \{=, \geq\}$ and $n_{\mathsf{s}}, n_{\mathsf{e}}, n_{\mathsf{se}} \in \mathbb{N}$.

We end up with a turn-based game in which, similarly to a VASS game [1,6,10,27,38], System and Environment move tokens along vectors from L. Note that, however, our games have a very particular structure so that undecidability for VASS games does not carry over to our setting. Moreover, existing decidability results do not allow us to infer our cutoff results below.

In the following, we will formalize *parameterized vector games.*

Definition 9. *A* parameterized vector game *(or simply* game*) is given by a triple $\mathcal{G} = (A, B, \mathcal{F})$ where $A = A_{\mathsf{s}} \uplus A_{\mathsf{e}}$ is the finite alphabet, $B \in \mathbb{N}$ is a bound, and, letting $L = \{0, \ldots, B\}^A$, $\mathcal{F} \subseteq \mathfrak{C}^L$ is a finite set called* acceptance condition.

Locations. Let ℓ_0 be the location such that $\ell_0(a) = 0$ for all $a \in A$. For $\ell \in L$ and $a \in A$, we define $\ell + a$ by $(\ell + a)(b) = \ell(b)$ for $b \neq a$ and $(\ell + a)(b) = \max\{\ell(a) + 1, B\}$ otherwise. This is extended for all $u \in A^*$ and $a \in A$ by $\ell + \varepsilon = \ell$ and $\ell + ua = (\ell + u) + a$. By $\langle\!\langle w \rangle\!\rangle$, we denote the location $\ell_0 + w$.

Configurations. As explained above, a *configuration* of $\mathcal{G}$ is a mapping $C : L \to \mathbb{N}^{\mathbb{T}}$. Suppose that, for $\ell \in L$ and $\theta \in \mathbb{T}$, we have $C(\ell) = (n_{\mathsf{s}}, n_{\mathsf{e}}, n_{\mathsf{se}})$. Then, we let $C(\ell, \theta)$ refer to n_θ. By *Conf*, we denote the set of all configurations.

Transitions. A *system transition* (respectively *environment transition*) is a mapping $\tau : L \times L \to (\mathbb{N} \times \{0\} \times \mathbb{N})$ (respectively $\tau : L \times L \to (\{0\} \times \mathbb{N} \times \mathbb{N})$) such that, for all $(\ell, \ell') \in L \times L$ with $\tau(\ell, \ell') \neq (0, 0, 0)$, there is a word $w \in A_{\mathsf{s}}^*$ (respectively $w \in A_{\mathsf{e}}^*$) such that $\ell' = \ell + w$. Let T_{s} denote the set of system transitions, T_{e} the set of environment transitions, and $T = T_{\mathsf{s}} \cup T_{\mathsf{e}}$ the set of all transitions.

For $\tau \in T$, let the mappings $out_\tau, in_\tau : L \to \mathbb{N}^{\mathbb{T}}$ be defined by $out_\tau(\ell) = \sum_{\ell' \in L} \tau(\ell, \ell')$ and $in_\tau(\ell) = \sum_{\ell' \in L} \tau(\ell', \ell)$ (recall that sum is component-wise). We say that $\tau \in T$ is *applicable* at $C \in Conf$ if, for all $\ell \in L$, we have $out_\tau(\ell) \leq C(\ell)$ (component-wise). Abusing notation, we let $\tau(C)$ denote the configuration C' defined by $C'(\ell) = C(\ell) - out_\tau(\ell) + in_\tau(\ell)$ for all $\ell \in L$. Moreover, for $\tau(\ell, \ell') = (n_{\mathsf{s}}, n_{\mathsf{e}}, n_{\mathsf{se}})$ and $\theta \in \mathbb{T}$, we let $\tau(\ell, \ell', \theta)$ refer to n_θ.

Plays. Let $C \in Conf$. We write $C \models \mathcal{F}$ if there is $\kappa \in \mathcal{F}$ such that, for all $\ell \in L$, we have $C(\ell) \models \kappa(\ell)$ (in the expected manner). A *C-play*, or simply *play*, is a finite sequence $\pi = C_0\tau_1C_1\tau_2C_2 \ldots \tau_nC_n$ alternating between configurations and transitions (with $n \geq 0$) such that $C_0 = C$ and, for all $i \in \{1, \ldots, n\}$, $C_i = \tau_i(C_{i-1})$ and

- if i is odd, then $\tau_i \in T_s$ and $C_i \models \mathcal{F}$ (System's move),
- if i is even, then $\tau_i \in T_e$ and $C_i \not\models \mathcal{F}$ (Environment's move).

The set of all C-plays is denoted by $Plays_C$.

Strategies. A *C-strategy* for System is a partial mapping $f : Plays_C \to T_s$ such that $f(C)$ is defined and, for all $\pi = C_0\tau_1C_1 \ldots \tau_iC_i \in Plays_C$ with $\tau = f(\pi)$ defined, we have that τ is applicable at C_i and $\tau(C_i) \models \mathcal{F}$. Play $\pi = C_0\tau_1C_1 \ldots \tau_nC_n$ is

- *f-compatible* if, for all odd $i \in \{1, \ldots, n\}$, $\tau_i = f(C_0\tau_1C_1 \ldots \tau_{i-1}C_{i-1})$,
- *f-maximal* if it is not the strict prefix of an f-compatible play,
- *winning* if $C_n \models \mathcal{F}$.

We say that f is *winning* for System (from C) if all f-compatible f-maximal C-plays are winning. Finally, C is *winning* if there is a C-strategy that is winning. Note that, given an initial configuration C, we deal with an acyclic finite reachability game so that, if there is a winning C-strategy, then there is a positional one, which only depends on the last configuration.

For $\boldsymbol{k} \in \mathbb{N}^{\mathbb{T}}$, let $C_{\boldsymbol{k}}$ denote the configuration that maps ℓ_0 to $\boldsymbol{k}$ and all other locations to $(0, 0, 0)$. We set $Win(\mathcal{G}) = \{\boldsymbol{k} \in \mathbb{N}^{\mathbb{T}} \mid C_{\boldsymbol{k}} \text{ is winning for System}\}$.

Definition 10 (game problem). *For sets $\mathcal{N}_s, \mathcal{N}_e, \mathcal{N}_{se} \subseteq \mathbb{N}$, the game problem is given as follows:*

GAME($\mathcal{N}_s, \mathcal{N}_e, \mathcal{N}_{se}$)	
Input:	*Parameterized vector game* $\mathcal{G}$
Question:	$Win(\mathcal{G}) \cap (\mathcal{N}_s \times \mathcal{N}_e \times \mathcal{N}_{se}) \neq \emptyset$?

One can show that parameterized vector games are equivalent to the synthesis problem in the following sense:

Lemma 11. *For every sentence $\varphi \in \mathrm{FO}_A[\sim]$, there is a parameterized vector game $\mathcal{G} = (A, B, \mathcal{F})$ such that $Win(\varphi) = Win(\mathcal{G})$. Conversely, for every parameterized vector game $\mathcal{G} = (A, B, \mathcal{F})$, there is a sentence $\varphi \in \mathrm{FO}_A[\sim]$ such that $Win(\mathcal{G}) = Win(\varphi)$. Both directions are effective.*

Example 12. To illustrate parameterized vector games and the reduction from the synthesis problem, consider the formula $\varphi_4' = \bigwedge_{\theta \in \mathbb{T},\, \ell \in Z} \exists^{=0} y.(\theta(y) \wedge \psi_{3,\ell}(y))$ in normal form from Example 7. For simplicity, we assume that $A_s = \{a\}$ and $A_e = \{d\}$. That is, Z is the set of vectors $\langle\!\langle a^i d^j \rangle\!\rangle \in L = \{0, \ldots, 3\}^{\{a,d\}}$ such that $i = 2 \neq j$ or $j = 2 \neq i$. Figure 2 illustrates a couple of configurations $C_0, \ldots, C_5 : L \to \mathbb{N}^{\mathbb{T}}$. The leftmost location in a configuration is ℓ_0, the rightmost

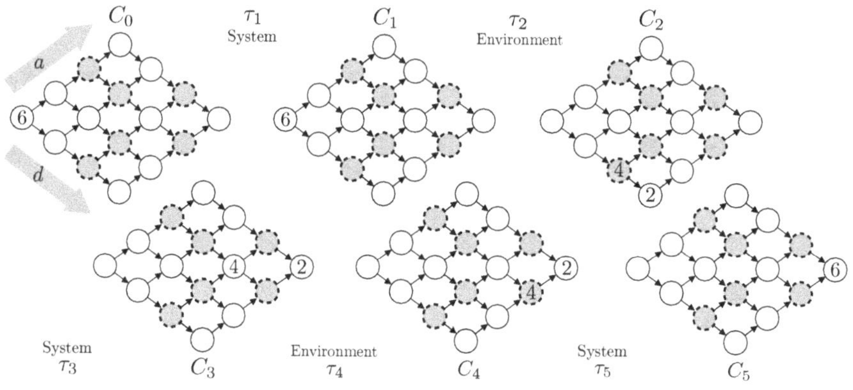

Fig. 2. A play of a parameterized vector game

location $\langle\!\langle a^3d^3\rangle\!\rangle$, the topmost one $\langle\!\langle a^3\rangle\!\rangle$, and the one at the bottom $\langle\!\langle d^3\rangle\!\rangle$. Self-loops have been omitted, and locations from Z have gray background and a dashed border.

Towards an equivalent game $\mathcal{G} = (A, 3, \mathcal{F})$, it remains to determine the acceptance condition $\mathcal{F}$. Recall that φ_4' says that every class contains two occurrences of a iff it contains two occurrences of d. This is reflected by the acceptance condition $\mathcal{F} = \{\kappa\}$ where $\kappa(\ell) = (=0\,, =0\,, =0)$ for all $\ell \in Z$ and $\kappa(\ell) = (\geq 0\,, \geq 0\,, \geq 0)$ for all $\ell \in L \setminus Z$. With this, a configuration is accepting iff no token is on a location from Z (a gray location).

We can verify that $Win(\mathcal{G}) = Win(\varphi_4') = \mathbb{N} \times \mathbb{0} \times \mathbb{N}$. In $\mathcal{G}$, a uniform winning strategy f for System that works for all $\mathbb{P}$ with $\mathbb{P}_{\mathsf{e}} = \emptyset$ proceeds as follows: System first awaits an Environment's move and then moves each token upwards as many locations as Environment has moved it downwards. Figure 2 illustrates an f-maximal $C_{(6,0,0)}$-play that is winning for System. We note that f is a "compressed" version of the winning strategy presented in Example 4, as System makes her moves only when really needed. ◁

5 Results for FO[∼] via Parameterized Vector Games

In this section, we present our results for the synthesis problem for FO[∼], which we obtain showing corresponding results for parameterized vector games. In particular, we show that $(\mathrm{FO}[\sim], \mathbb{0}, \mathbb{0}, \mathbb{N})$ and $(\mathrm{FO}[\sim], \mathbb{N}, \mathbb{N}, \mathbb{0})$ do not have a cutoff, whereas $(\mathrm{FO}[\sim], \mathbb{N}, \{k_{\mathsf{e}}\}, \{k_{\mathsf{se}}\})$ has a cutoff for all $k_{\mathsf{e}}, k_{\mathsf{se}} \in \mathbb{N}$. Finally, we prove that $\textsc{Synth}(\mathrm{FO}[\sim], \mathbb{0}, \mathbb{0}, \mathbb{N})$ is, in fact, undecidable.

Lemma 13. *There is a game $\mathcal{G} = (A, B, \mathcal{F})$ such that $Win(\mathcal{G})$ does not have a cutoff wrt. $(\mathbb{0}, \mathbb{0}, \mathbb{N})$.*

Proof. We let $A_{\mathsf{s}} = \{a\}$ and $A_{\mathsf{e}} = \{b\}$, as well as $B = 2$. For $k \in \{0, 1, 2\}$, define the local acceptance conditions ${}^{=}k = (=0\,, =0\,, =k)$ and ${}^{\geq}k = (=0\,, =0\,, \geq k)$. Set

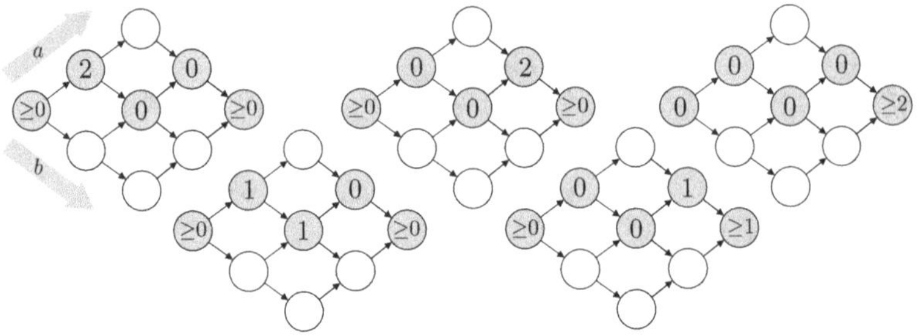

Fig. 3. Acceptance conditions for a game with no cutoff wrt. $(\mathbb{0}, \mathbb{0}, \mathbb{N})$

$\ell_1 = \langle\!\langle a \rangle\!\rangle, \ell_2 = \langle\!\langle ab \rangle\!\rangle, \ell_3 = \langle\!\langle a^2 b \rangle\!\rangle$, and $\ell_4 = \langle\!\langle a^2 b^2 \rangle\!\rangle$. For $k_0, \ldots, k_4 \in \{0, 1, 2\}$ and $\bowtie_0, \ldots, \bowtie_4 \in \{=, \geq\}$, let $[{}^{\bowtie_0}k_0\,, {}^{\bowtie_1}k_1\,, {}^{\bowtie_2}k_2\,, {}^{\bowtie_3}k_3\,, {}^{\bowtie_4}k_4]$ denote $\kappa \in \mathfrak{C}^L$ where $\kappa(\ell_i) = ({}^{\bowtie_i}k_i)$ for all $i \in \{0, \ldots, 4\}$ and $\kappa(\ell') = ({}^{=}0)$ for $\ell' \notin \{\ell_0, \ldots, \ell_4\}$. Finally,

$$\mathcal{F} = \left\{ \begin{matrix} [{}^{\geq}0\,, {}^{=}2\,, {}^{=}0\,, {}^{=}0\,, {}^{\geq}0] & [{}^{\geq}0\,, {}^{=}0\,, {}^{=}0\,, {}^{=}2\,, {}^{\geq}0] & [{}^{=}0\,, {}^{=}0\,, {}^{=}0\,, {}^{=}0\,, {}^{\geq}2] \\ [{}^{\geq}0\,, {}^{=}1\,, {}^{=}1\,, {}^{=}0\,, {}^{\geq}0] & [{}^{\geq}0\,, {}^{=}0\,, {}^{=}0\,, {}^{=}1\,, {}^{\geq}1] & \end{matrix} \right\} \cup K_{\mathsf{e}}$$

where $K_{\mathsf{e}} = \{\kappa_\ell \mid \ell \in L$ such that $\ell(b) > \ell(a)\}$ with $\kappa_\ell(\ell') = ({}^{\geq}1)$ if $\ell' = \ell$, and $\kappa_\ell(\ell') = ({}^{\geq}0)$ otherwise. This is illustrated in Figure 3.

There is a winning strategy for System from any initial configuration of size $2n$: Move two tokens from ℓ_0 to ℓ_1, wait until Environment sends them both to ℓ_2, then move them to ℓ_3, wait until they are moved to ℓ_4, then repeat with two new tokens from ℓ_0 until all the tokens are removed from ℓ_0, and Environment cannot escape $\mathcal{F}$ anymore. However, one can check that there is no winning strategy for initial configurations of odd size. □

Lemma 14. *There is a game $\mathcal{G} = (A, B, \mathcal{F})$ such that $Win(\mathcal{G})$ does not have a cutoff wrt.* $(\mathbb{N}, \mathbb{N}, \mathbb{0})$.

Proof. We define $\mathcal{G}$ such that System wins only if she has at least as many processes as Environment. Let $A_{\mathsf{s}} = \{a\}$, $A_{\mathsf{e}} = \{b\}$, and $B = 2$. As there are no shared processes, we can safely ignore locations with a letter from both System and Environment. We set $\mathcal{F} = \{\kappa_1, \kappa_2, \kappa_3, \kappa_4\}$ where

$$\begin{array}{lll} \kappa_1(\langle\!\langle a \rangle\!\rangle) = (=1\,, =0\,, =0) & \kappa_2(\langle\!\langle a \rangle\!\rangle) = (=1\,, =0\,, =0) & \kappa_3(\langle\!\langle a \rangle\!\rangle) = (=0\,, =0\,, =0) \\ \kappa_1(\langle\!\langle b \rangle\!\rangle) = (=0\,, =0\,, =0) & \kappa_2(\langle\!\langle b \rangle\!\rangle) = (=0\,, \geq 2\,, =0) & \kappa_3(\langle\!\langle b \rangle\!\rangle) = (=0\,, \geq 1\,, =0)\,, \end{array}$$

$\kappa_4(\ell_0) = (=0\,, =0\,, =0)$, and $\kappa_i(\ell') = (\geq 0\,, \geq 0\,, =0)$ for all other $\ell' \in L$ and $i \in \{1, 2, 3, 4\}$. □

We now turn to the case where the number of processes that can be triggered by Environment is bounded. Note that similar restrictions are imposed in other settings to get decidability, such as limiting the environment to a finite (Boolean) domain [16] or restricting to one environment process [3, 18]. We obtain decidability of the synthesis problem via a cutoff construction:

Theorem 15. *Given $k_\mathsf{e}, k_\mathsf{se} \in \mathbb{N}$, every game $\mathcal{G} = (A, B, \mathcal{F})$ has a cutoff wrt. $(\mathbb{N}, \{k_\mathsf{e}\}, \{k_\mathsf{se}\})$. More precisely: Let K be the largest constant that occurs in $\mathcal{F}$. Moreover, let $Max = (k_\mathsf{e} + k_\mathsf{se}) \cdot |A_\mathsf{e}| \cdot B$ and $\hat{N} = |L|^{Max+1} \cdot K$. Then, $(\hat{N}, k_\mathsf{e}, k_\mathsf{se})$ is a cutoff of $Win(\mathcal{G})$ wrt. $(\mathbb{N}, \{k_\mathsf{e}\}, \{k_\mathsf{se}\})$.*

Proof. We will show that, for all $N \geq \hat{N}$,

$$(N, k_\mathsf{e}, k_\mathsf{se}) \in Win(\mathcal{G}) \iff (N+1, k_\mathsf{e}, k_\mathsf{se}) \in Win(\mathcal{G}).$$

The main observation is that, when C contains more than K tokens in a given $\ell \in L$, adding more tokens in ℓ will not change whether $C \models \mathcal{F}$. Given $C, C' \in Conf$, we write $C <_\mathsf{e} C$ if $C \neq C'$ and there is $\tau \in T_\mathsf{e}$ such that $\tau(C) = C'$. Note that the length d of a chain $C_0 <_\mathsf{e} C_1 <_\mathsf{e} \ldots <_\mathsf{e} C_d$ is bounded by Max. In other words, Max is the maximal number of transitions that Environment can do in a play. For all $d \in \{0, \ldots, Max\}$, let $Conf_d$ be the set of configurations $C \in Conf$ such that the longest chain in $(Conf, <_\mathsf{e})$ starting from C has length d.

Claim. Suppose that $C \in Conf_d$ and $\ell \in L$ such that $C(\ell) = (N, n_\mathsf{e}, n_\mathsf{se})$ with $N \geq |L|^{d+1} \cdot K$ and $n_\mathsf{e}, n_\mathsf{se} \in \mathbb{N}$. Set $D = C[\ell \mapsto (N+1, n_\mathsf{e}, n_\mathsf{se})]$. Then,

$$C \text{ is winning for System} \iff D \text{ is winning for System.}$$

To show the claim, we proceed by induction on $d \in \mathbb{N}$, which is illustrated in Figure 4. In each implication, we distinguish the cases $d = 0$ and $d \geq 1$. For the latter, we assume that equivalence holds for all values strictly smaller than d.

For $\tau \in T_\mathsf{s}$ and $\ell, \ell' \in L$, we let $\tau[(\ell, \ell', \mathsf{s}){+}{+}]$ denote the transition $\eta \in T_\mathsf{s}$ given by $\eta(\ell_1, \ell_2, \mathsf{e}) = \tau(\ell_1, \ell_2, \mathsf{e}) = 0$, $\eta(\ell_1, \ell_2, \mathsf{se}) = \tau(\ell_1, \ell_2, \mathsf{se})$, $\eta(\ell_1, \ell_2, \mathsf{s}) = \tau(\ell_1, \ell_2, \mathsf{s}) + 1$ if $(\ell_1, \ell_2) = (\ell, \ell')$, and $\eta(\ell_1, \ell_2, \mathsf{s}) = \tau(\ell_1, \ell_2, \mathsf{s})$ if $(\ell_1, \ell_2) \neq (\ell, \ell')$. We define $\tau[(\ell, \ell', \mathsf{s}){-}{-}]$ similarly (provided $\tau(\ell, \ell', \mathsf{s}) \geq 1$).

$\Longrightarrow$: Let f be a winning strategy for System from $C \in Conf_d$. Let $\tau' = f(C)$ and $C' = \tau'(C)$. Note that $C' \models \mathcal{F}$. Since $C(\ell, \mathsf{s}) = N \geq |L|^{d+1} \cdot K$, there is $\ell' \in L$ such that $\ell + w = \ell'$ for some $w \in A_\mathsf{s}^*$ and $C'(\ell', \mathsf{s}) = N' \geq |L|^d \cdot K$.

We show that $D = C[\ell \mapsto (N{+}1, n_\mathsf{e}, n_\mathsf{se})]$ is winning for System by exhibiting a corresponding winning strategy g from D that will carefully control the position of the additional token. First, set $g(D) = \eta'$ where $\eta' = \tau'[(\ell, \ell', \mathsf{s}){+}{+}]$. Let $D' = \eta'(D)$. We obtain $D'(\ell', \mathsf{s}) = N' + 1$. Note that, since $N' \geq K$, the acceptance condition $\mathcal{F}$ cannot distinguish between C' and D'. Thus, we have $D' \models \mathcal{F}$.

Case $d = 0$: As, for all transitions $\eta'' \in T_\mathsf{e}$, we have $\eta''(D') = D' \models \mathcal{F}$, we reached a maximal play that is winning for System. We deduce that D is winning for System.

Case $d \geq 1$: Take any $\eta'' \in T_\mathsf{e}$ and D'' such that $D'' = \eta''(D') \not\models \mathcal{F}$. Let $\tau'' = \eta''$ and $C'' = \tau''(C')$. Note that $D'' = C''[(\ell', \mathsf{s}) \mapsto N+1]$, $C'' = D''[(\ell', \mathsf{s}) \mapsto N]$, and $C'', D'' \in Conf_{d^-}$ for some $d^- < d$. As f is a winning strategy for System from C, we have that C'' is winning for System. By induction hypothesis, D'' is winning for System, say by winning strategy g''. We let $g(D\,\eta'\,D'\,\eta''\,\pi) = g''(\pi)$ for all D''-plays π. For all unspecified plays, let g return any applicable system transition. Altogether, for any choice of η'', we have that g'' is winning from D''. Thus, g is a winning strategy from D.

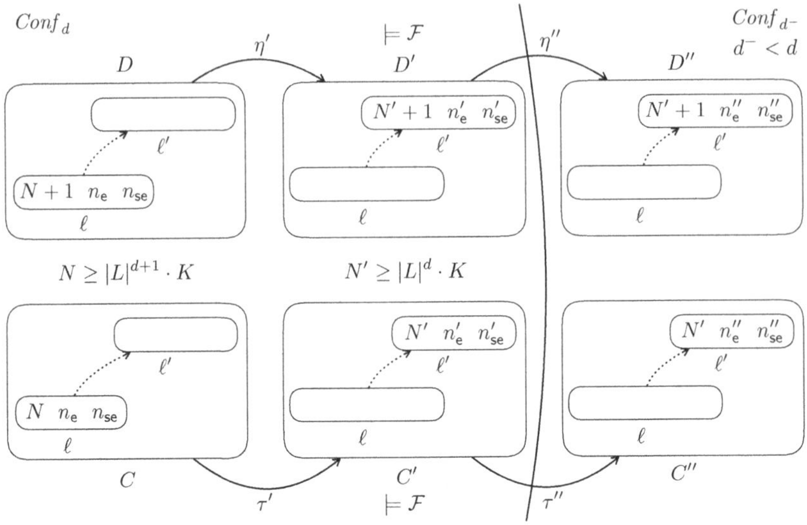

Fig. 4. Induction step in the cutoff construction

$\Longleftarrow$: Suppose g is a winning strategy for System from D. Thus, for $\eta' = g(D)$ and $D' = \eta'(D)$, we have $D' \models \mathcal{F}$. Recall that $D(\ell, \mathsf{s}) \geq (|L|^{d+1} \cdot K) + 1$. We distinguish two cases:

1. Suppose there is $\ell' \in L$ such that $\ell \neq \ell'$, $D'(\ell', \mathsf{s}) = N' + 1$ for some $N' \geq |L|^d \cdot K$, and $\eta'(\ell, \ell', \mathsf{s}) \geq 1$. Then, we set $\tau' = \eta'[(\ell, \ell', \mathsf{s})\text{--}]$.
2. Otherwise, we have $D'(\ell, \mathsf{s}) \geq (|L|^d \cdot K) + 1$, and we set $\tau' = \eta'$ (as well as $\ell' = \ell$ and $N' = N$).

Let $C' = \tau'(C)$. Since $D' \models \mathcal{F}$, one obtains $C' \models \mathcal{F}$.

Case $d = 0$: For all transitions $\tau'' \in T_{\mathsf{e}}$, we have $\tau''(C') = C' \models \mathcal{F}$. Thus, we reached a maximal play that is winning for System. We deduce that C is winning for System.

Case $d \geq 1$: Take any $\tau'' \in T_{\mathsf{e}}$ such that $C'' = \tau''(C') \not\models \mathcal{F}$. Let $\eta'' = \tau''$ and $D'' = \eta''(D')$. We have $C'' = D''[(\ell', \mathsf{s}) \mapsto N']$, $D'' = C''[(\ell', \mathsf{s}) \mapsto N' + 1]$, and $C'', D'' \in \mathit{Conf}_{d^-}$ for some $d^- < d$. As D'' is winning for System, by induction hypothesis, C'' is winning for System, say by winning strategy f''. We let $f(C\,\tau'\,C'\,\tau''\,\pi) = f''(\pi)$ for all C''-plays π. For all unspecified plays, let f return an arbitrary applicable system transition. Again, for any choice of τ'', f'' is winning from C''. Thus, f is a winning strategy from C.

This concludes the proof of the claim and, therefore, of Theorem 15. □

Corollary 16. *Let $k_{\mathsf{e}}, k_{\mathsf{se}} \in \mathbb{N}$ be the number of environment and the number of mixed processes, respectively. The problems* $\textsc{Game}(\mathbb{N}, \{k_{\mathsf{e}}\}, \{k_{\mathsf{se}}\})$ *and* $\textsc{Synth}(\mathrm{FO}[\sim], \mathbb{N}, \{k_{\mathsf{e}}\}, \{k_{\mathsf{se}}\})$ *are decidable.*

In particular, by Theorem 15, the game problem can be reduced to an exponential number of acyclic finite-state games whose size (and hence the time complexity for determining the winner) is exponential in the cutoff and, therefore, doubly exponential in the size of the alphabet, the bound B, and the fixed number of processes that are controllable by the environment.

Theorem 17. GAME($\mathbb{0}, \mathbb{0}, \mathbb{N}$) *and* SYNTH(FO[$\sim$], $\mathbb{0}, \mathbb{0}, \mathbb{N}$) *are undecidable.*

Proof. We provide a reduction from the halting problem for 2-counter machines (2CM) to GAME($\mathbb{0}, \mathbb{0}, \mathbb{N}$). A 2CM $M = (Q, \Delta, \mathsf{c}_1, \mathsf{c}_2, q_0, q_h)$ has two counters, c_1 and c_2, a finite set of states Q, and a set of transitions $\Delta \subseteq Q \times \mathsf{Op} \times Q$ where $\mathsf{Op} = \{\mathsf{c}_i\texttt{++}, \mathsf{c}_i\texttt{--}, \mathsf{c}_i\texttt{==}0 \mid i \in \{1,2\}\}$. Moreover, we have an initial state $q_0 \in Q$ and a halting state $q_h \in Q$. A configuration of M is a triple $\gamma = (q, \nu_1, \nu_2) \in Q \times \mathbb{N} \times \mathbb{N}$ giving the current state and the current respective counter values. The initial configuration is $\gamma_0 = (q_0, 0, 0)$ and the set of halting configurations is $F = \{q_h\} \times \mathbb{N} \times \mathbb{N}$. For $t \in \Delta$, configuration (q', ν_1', ν_2') is a (t-)successor of (q, ν_1, ν_2), written $(q, \nu_1, \nu_2) \vdash_t (q', \nu_1', \nu_2')$, if there is $i \in \{1,2\}$ such that $\nu_{3-i}' = \nu_{3-i}$ and one of the following holds: (i) $t = (q, \mathsf{c}_i\texttt{++}, q')$ and $\nu_i' = \nu_i + 1$, or (ii) $t = (q, \mathsf{c}_i\texttt{--}, q')$ and $\nu_i' = \nu_i - 1$, or (iii) $t = (q, \mathsf{c}_i\texttt{==}0, q')$ and $\nu_i = \nu_i' = 0$. A run of M is a (finite or infinite) sequence $\gamma_0 \vdash_{t_1} \gamma_1 \vdash_{t_2} \dots$. The 2CM halting problem asks whether there is a run reaching a configuration in F. It is known to be undecidable [34].

We fix a 2CM $M = (Q, \Delta, \mathsf{c}_1, \mathsf{c}_2, q_0, q_h)$. Let $A_\mathsf{s} = Q \cup \Delta \cup \{a_1, a_2\}$ and $A_\mathsf{e} = \{b\}$ with a_1, a_2, and b three fresh symbols. We consider the game $\mathcal{G} = (A, B, \mathcal{F})$ with $A = A_\mathsf{s} \uplus A_\mathsf{e}$, $B = 4$, and $\mathcal{F}$ defined below. Let $L = \{0, \dots, B\}^A$. Since there are only processes shared by System and Environment, we alleviate notation and consider that a configuration is simply a mapping $C : L \to \mathbb{N}$. From now on, to avoid confusion, we refer to configurations of the 2CM M as M-configurations, and to configurations of $\mathcal{G}$ as $\mathcal{G}$-configurations.

Intuitively, every valid run of M will be encoded as a play in $\mathcal{G}$, and the acceptance condition will enforce that, if a player in $\mathcal{G}$ deviates from a valid play, then she will lose immediately. At any point in the play, there will be at most one process with only a letter from Q played, which will represent the current state in the simulated 2CM run. Similarly, there will be at most one process with only a letter from Δ to represent what transition will be taken next. Finally, the value of counter c_i will be encoded by the number of processes with exactly two occurrences of a_i and two occurrences of b (i.e., $C(\langle\!\langle a_i^2 b^2 \rangle\!\rangle)$).

To increase counter c_i, the players will move a new token to $\langle\!\langle a_i^2 b^2 \rangle\!\rangle$, and to decrease it, they will move, together, a token from $\langle\!\langle a_i^2 b^2 \rangle\!\rangle$ to $\langle\!\langle a_i^4 b^4 \rangle\!\rangle$. Observe that, if c_i has value 0, then $C(\langle\!\langle a_i^2 b^2 \rangle\!\rangle) = 0$ in the corresponding configuration of the game. As expected, it is then impossible to simulate the decrement of c_i. Environment's only role is to acknowledge System's actions by playing its (only) letter when System simulates a valid run. If System tries to cheat, she loses immediately.

Encoding an M-configuration. Let us be more formal. Suppose $\gamma = (q, \nu_1, \nu_2)$ is an M-configuration and C a $\mathcal{G}$-configuration. We say that C *encodes* γ if

- $C(\langle\!\langle q \rangle\!\rangle) = 1$, $C(\langle\!\langle a_1^2 b^2 \rangle\!\rangle) = \nu_1$, $C(\langle\!\langle a_2^2 b^2 \rangle\!\rangle) = \nu_2$,
- $C(\ell) \geq 0$ for all $\ell \in \{\ell_0\} \cup \{\langle\!\langle \hat{q}^2 b^2 \rangle\!\rangle, \langle\!\langle t^2 b^2 \rangle\!\rangle, \langle\!\langle a_i^4 b^4 \rangle\!\rangle \mid \hat{q} \in Q, t \in \Delta, i \in \{1,2\}\}$,
- $C(\ell) = 0$ for all other $\ell \in L$.

We then write $\gamma = \mathsf{m}(C)$. Let $\mathbb{C}(\gamma)$ be the set of $\mathcal{G}$-configurations C that encode γ. We say that a $\mathcal{G}$-configuration C is *valid* if $C \in \mathbb{C}(\gamma)$ for some γ.

Simulating a transition of M. Let us explain how we go from a $\mathcal{G}$-configuration encoding γ to a $\mathcal{G}$-configuration encoding a successor M-configuration γ'. Observe that System cannot change by herself the M-configuration encoded. If, for instance, she tries to change the current state q, she might move one process from ℓ_0 to $\langle\!\langle q' \rangle\!\rangle$, but then the $\mathcal{G}$-configuration is not valid anymore. We need to move the process in $\langle\!\langle q \rangle\!\rangle$ into $\langle\!\langle q^2 b^2 \rangle\!\rangle$ and this requires the cooperation of Environment.

Assume that the game is in configuration C encoding $\gamma = (q, \nu_1, \nu_2)$. System will pick a transition t starting in state q, say, $t = (q, \mathsf{c}_1{+}{+}, q')$. From configuration C, System will go to the configuration C_1 defined by $C_1(\langle\!\langle t \rangle\!\rangle) = 1$, $C_1(\langle\!\langle a_1 \rangle\!\rangle) = 1$, and $C_1(\ell) = C(\ell)$ for all other $\ell \in L$.

If the transition t is correctly chosen, Environment will go to a configuration C_2 defined by $C_2(\langle\!\langle q \rangle\!\rangle) = 0$, $C_2(\langle\!\langle qb \rangle\!\rangle) = 1$, $C_2(\langle\!\langle t \rangle\!\rangle) = 0$, $C_2(\langle\!\langle tb \rangle\!\rangle) = 1$, $C_2(\langle\!\langle a_1 \rangle\!\rangle) = 0$, $C_2(\langle\!\langle a_1 b \rangle\!\rangle) = 1$ and, for all other $\ell \in L$, $C_2(\ell) = C_1(\ell)$. This means that Environment moves processes in locations $\langle\!\langle t \rangle\!\rangle$, $\langle\!\langle q \rangle\!\rangle$, $\langle\!\langle a_1 \rangle\!\rangle$ to locations $\langle\!\langle tb \rangle\!\rangle$, $\langle\!\langle qb \rangle\!\rangle$, $\langle\!\langle a_1 b \rangle\!\rangle$, respectively.

To finish the transition, System will now move a process to the destination state q' of t, and go to configuration C_3 defined by $C_3(\langle\!\langle q' \rangle\!\rangle) = 1$, $C_3(\langle\!\langle tb \rangle\!\rangle) = 0$, $C_3(\langle\!\langle t^2 b \rangle\!\rangle) = 1$, $C_3(\langle\!\langle qb \rangle\!\rangle) = 0$, $C_3(\langle\!\langle q^2 b \rangle\!\rangle) = 1$, $C_3(\langle\!\langle a_1 b \rangle\!\rangle) = 0$, $C_3(\langle\!\langle a_1^2 b \rangle\!\rangle) = 1$, and $C_3(\ell) = C_2(\ell)$ for all other $\ell \in L$.

Finally, Environment moves to configuration C_4 given by $C_4(\langle\!\langle t^2 b \rangle\!\rangle) = 0$, $C_4(\langle\!\langle t^2 b^2 \rangle\!\rangle) = C_3(\langle\!\langle t^2 b^2 \rangle\!\rangle) + 1$, $C_4(\langle\!\langle q^2 b \rangle\!\rangle) = 0$, $C_4(\langle\!\langle q^2 b^2 \rangle\!\rangle) = C_3(\langle\!\langle q^2 b^2 \rangle\!\rangle) + 1$, $C_4(\langle\!\langle a_1^2 b \rangle\!\rangle) = 0$, $C_4(\langle\!\langle a_1^2 b^2 \rangle\!\rangle) = C_3(\langle\!\langle a_1^2 b^2 \rangle\!\rangle) + 1$, and $C_4(\ell) = C_3(\ell)$ for all other $\ell \in L$. Observe that $C_4 \in \mathbb{C}((q', \nu_1 + 1, \nu_2))$.

Other types of transitions will be simulated similarly. To force System to start the simulation in γ_0, and not in any M-configuration, the configurations C such that $C(\langle\!\langle q_0^2 b^2 \rangle\!\rangle) = 0$ and $C(\langle\!\langle q \rangle\!\rangle) = 1$ for $q \neq q_0$ are not valid, and will be losing for System.

Acceptance condition. It remains to define $\mathcal{F}$ in a way that enforces the above sequence of $\mathcal{G}$-configurations. Let $L_{\checkmark} = \{\ell_0\} \cup \{\langle\!\langle a_i^2 b^2 \rangle\!\rangle, \langle\!\langle a_i^4 b^4 \rangle\!\rangle \mid i \in \{1,2\}\} \cup \{\langle\!\langle q^2 b^2 \rangle\!\rangle \mid q \in Q\} \cup \{\langle\!\langle t^2 b^2 \rangle\!\rangle \mid t \in \Delta\}$ be the set of elements in L whose values do not affect the acceptance of the configuration. By $[\ell_1 \bowtie_1 n_1, \ldots, \ell_k \bowtie_k n_k]$, we denote $\kappa \in \mathfrak{C}^L$ such that $\kappa(\ell_i) = (\bowtie_i n_i)$ for $i \in \{1, \ldots, k\}$ and $\kappa(\ell) = (=0)$ for all $\ell \in L \setminus \{\ell_1, \ldots, \ell_k\}$. Moreover, for a set of locations $\hat{L} \subseteq L$, we let $\hat{L} \geq 0$ stand for "$(\ell \geq 0)$ for all $\ell \in \hat{L}$".

First, we force Environment to play only in response to System by making System win as soon as there is a process where Environment has played more letters than System (see Condition (d) in Table 2).

If γ is not halting, the configurations in $\mathbb{C}(\gamma)$ will not be winning for System. Hence, System will have to move to win (Condition (a)).

Table 2. Acceptance conditions for the game simulating a 2CM

Requirements for System

(a) For all $t = (q, \mathsf{op}, q') \in Q$:

$$\mathcal{F}_{(q,t)} = \bigcup_{\hat{q}\in Q}\{[\langle\!\langle q\rangle\!\rangle = 1,\ \langle\!\langle t\rangle\!\rangle = 1,\ \langle\!\langle a_i\rangle\!\rangle = 1,\ \langle\!\langle \hat{q}^2b^2\rangle\!\rangle \geq 1,\ (L_{\checkmark} \setminus \{\langle\!\langle \hat{q}^2b^2\rangle\!\rangle\}) \geq 0]\} \quad \text{if } \mathsf{op} = \mathsf{c}_i{+}{+}$$

$$\mathcal{F}_{(q,t)} = \bigcup_{\hat{q}\in Q}\{[\langle\!\langle q\rangle\!\rangle = 1,\ \langle\!\langle t\rangle\!\rangle = 1,\ \langle\!\langle a_i^3b^2\rangle\!\rangle = 1,\ \langle\!\langle \hat{q}^2b^2\rangle\!\rangle \geq 1,\ (L_{\checkmark} \setminus \{\langle\!\langle \hat{q}^2b^2\rangle\!\rangle\}) \geq 0]\} \quad \text{if } \mathsf{op} = \mathsf{c}_i{-}{-}$$

$$\mathcal{F}_{(q,t)} = \bigcup_{\hat{q}\in Q}\{[\langle\!\langle q\rangle\!\rangle = 1,\ \langle\!\langle t\rangle\!\rangle = 1,\ \langle\!\langle a_i^2b^2\rangle\!\rangle = 0,\ \langle\!\langle \hat{q}^2b^2\rangle\!\rangle \geq 1,\ (L_{\checkmark} \setminus \{\langle\!\langle \hat{q}^2b^2\rangle\!\rangle, \langle\!\langle a_i^2b^2\rangle\!\rangle\}) \geq 0]\} \quad \text{if } \mathsf{op} = \mathsf{c}_i{=}{=}0$$

(b) For all $t = (q_0, \mathsf{op}, q') \in Q$ such that $\mathsf{op} \in \{\mathsf{c}_i{+}{+}, \mathsf{c}_i{=}{=}0\}$:

$$\mathcal{F}_t = \{[\langle\!\langle q_0\rangle\!\rangle = 1,\ \langle\!\langle t\rangle\!\rangle = 1,\ \langle\!\langle a_i\rangle\!\rangle = 1,\ \ell_0 \geq 0]\} \quad \text{if } \mathsf{op} = \mathsf{c}_i{+}{+}$$

$$\mathcal{F}_t = \{[\langle\!\langle q_0\rangle\!\rangle = 1,\ \langle\!\langle t\rangle\!\rangle = 1,\ \ell_0 \geq 0]\} \quad \text{if } \mathsf{op} = \mathsf{c}_i{=}{=}0$$

(c) For all $t = (q, \mathsf{op}, q') \in Q$:

$$\mathcal{F}_{(q,t,q')} = \{[\langle\!\langle q^2b\rangle\!\rangle = 1,\ \langle\!\langle t^2b\rangle\!\rangle = 1,\ \langle\!\langle a_i^2b\rangle\!\rangle = 1,\ \langle\!\langle q'\rangle\!\rangle = 1,\ L_{\checkmark} \geq 0]\} \quad \text{if } \mathsf{op} = \mathsf{c}_i{+}{+}$$

$$\mathcal{F}_{(q,t,q')} = \{[\langle\!\langle q^2b\rangle\!\rangle = 1,\ \langle\!\langle t^2b\rangle\!\rangle = 1,\ \langle\!\langle a_i^4b^3\rangle\!\rangle = 1,\ \langle\!\langle q'\rangle\!\rangle = 1,\ L_{\checkmark} \geq 0]\} \quad \text{if } \mathsf{op} = \mathsf{c}_i{-}{-}$$

$$\mathcal{F}_{(q,t,q')} = \{[\langle\!\langle q^2b\rangle\!\rangle = 1,\ \langle\!\langle t^2b\rangle\!\rangle = 1,\ L_{\checkmark} \geq 0]\} \quad \text{if } \mathsf{op} = \mathsf{c}_i{=}{=}0$$

Requirements for Environment

(d) Let $L_{\mathsf{s}<\mathsf{e}} = \{\ell \in L \mid (\sum_{\alpha\in A_\mathsf{s}} \ell(\alpha)) < \ell(b)\}$. For all $\ell \in L_{\mathsf{s}<\mathsf{e}}$: $\mathcal{F}_\ell = [\ell \geq 1, (L \setminus \{\ell\}) \geq 0]$

(e) For all $t = (q, \mathsf{op}, q') \in Q$:

$$\mathcal{F}^{\mathsf{e}}_{(q,t)} = \left\{\begin{array}{l}[\langle\!\langle qb\rangle\!\rangle = 1,\ \langle\!\langle t\rangle\!\rangle = 1,\ \langle\!\langle a_i\rangle\!\rangle = 1,\ L_{\checkmark} \geq 0],\ [\langle\!\langle q\rangle\!\rangle = 1,\ \langle\!\langle tb\rangle\!\rangle = 1,\ \langle\!\langle a_i\rangle\!\rangle = 1,\ L_{\checkmark} \geq 0],\\ [\langle\!\langle q\rangle\!\rangle = 1,\ \langle\!\langle t\rangle\!\rangle = 1,\ \langle\!\langle a_ib\rangle\!\rangle = 1,\ L_{\checkmark} \geq 0],\ [\langle\!\langle qb\rangle\!\rangle = 1,\ \langle\!\langle tb\rangle\!\rangle = 1,\ \langle\!\langle a_i\rangle\!\rangle = 1,\ L_{\checkmark} \geq 0],\\ [\langle\!\langle qb\rangle\!\rangle = 1,\ \langle\!\langle t\rangle\!\rangle = 1,\ \langle\!\langle a_ib\rangle\!\rangle = 1,\ L_{\checkmark} \geq 0],\ [\langle\!\langle q\rangle\!\rangle = 1,\ \langle\!\langle tb\rangle\!\rangle = 1,\ \langle\!\langle a_ib\rangle\!\rangle = 1,\ L_{\checkmark} \geq 0]\end{array}\right\} \quad \text{if } \mathsf{op} = \mathsf{c}_i{+}{+}$$

$$\mathcal{F}^{\mathsf{e}}_{(q,t)} = \left\{\begin{array}{l}[\langle\!\langle qb\rangle\!\rangle = 1,\ \langle\!\langle t\rangle\!\rangle = 1,\ \langle\!\langle a_i^3b^2\rangle\!\rangle = 1,\ L_{\checkmark} \geq 0],\ [\langle\!\langle q\rangle\!\rangle = 1,\ \langle\!\langle tb\rangle\!\rangle = 1,\ \langle\!\langle a_i^3b^2\rangle\!\rangle = 1,\ L_{\checkmark} \geq 0],\\ [\langle\!\langle q\rangle\!\rangle = 1,\ \langle\!\langle t\rangle\!\rangle = 1,\ \langle\!\langle a_i^3b^3\rangle\!\rangle = 1,\ L_{\checkmark} \geq 0],\ [\langle\!\langle qb\rangle\!\rangle = 1,\ \langle\!\langle tb\rangle\!\rangle = 1,\ \langle\!\langle a_i^3b^2\rangle\!\rangle = 1,\ L_{\checkmark} \geq 0],\\ [\langle\!\langle qb\rangle\!\rangle = 1,\ \langle\!\langle t\rangle\!\rangle = 1,\ \langle\!\langle a_i^3b^3\rangle\!\rangle = 1,\ L_{\checkmark} \geq 0],\ [\langle\!\langle q\rangle\!\rangle = 1,\ \langle\!\langle tb\rangle\!\rangle = 1,\ \langle\!\langle a_i^3b^3\rangle\!\rangle = 1,\ L_{\checkmark} \geq 0]\end{array}\right\} \quad \text{if } \mathsf{op} = \mathsf{c}_i{-}{-}$$

$$\mathcal{F}^{\mathsf{e}}_{(q,t)} = \{[\langle\!\langle qb\rangle\!\rangle = 1,\ \langle\!\langle t\rangle\!\rangle = 1,\ L_{\checkmark} \geq 0],\ [\langle\!\langle q\rangle\!\rangle = 1,\ \langle\!\langle tb\rangle\!\rangle = 1,\ L_{\checkmark} \geq 0]\} \quad \text{if } \mathsf{op} = \mathsf{c}_i{=}{=}0$$

(f) For all $t = (q, \mathsf{op}, q') \in Q$:

$$\mathcal{F}^{\mathsf{e}}_{(q,t,q')} = \left\{\begin{array}{l}[\langle\!\langle q'\rangle\!\rangle = 1,\ \langle\!\langle q^2b\rangle\!\rangle = 1,\ \langle\!\langle t^2b\rangle\!\rangle \geq 0,\ \langle\!\langle a_i^2b\rangle\!\rangle \geq 0,\ L_{\checkmark} \geq 0],\\ [\langle\!\langle q'\rangle\!\rangle = 1,\ \langle\!\langle q^2b\rangle\!\rangle \geq 0,\ \langle\!\langle t^2b\rangle\!\rangle = 1,\ \langle\!\langle a_i^2b\rangle\!\rangle \geq 0,\ L_{\checkmark} \geq 0],\\ [\langle\!\langle q'\rangle\!\rangle = 1,\ \langle\!\langle q^2b\rangle\!\rangle \geq 0,\ \langle\!\langle t^2b\rangle\!\rangle \geq 0,\ \langle\!\langle a_i^2b\rangle\!\rangle = 1,\ L_{\checkmark} \geq 0],\\ [\langle\!\langle q'b\rangle\!\rangle = 1,\ \langle\!\langle q^2b\rangle\!\rangle \geq 0,\ \langle\!\langle t^2b\rangle\!\rangle \geq 0,\ \langle\!\langle a_i^2b\rangle\!\rangle \geq 0,\ L_{\checkmark} \geq 0]\end{array}\right\} \quad \text{if } \mathsf{op} = \mathsf{c}_i{+}{+}$$

$$\mathcal{F}^{\mathsf{e}}_{(q,t,q')} = \left\{\begin{array}{l}[\langle\!\langle q'\rangle\!\rangle = 1,\ \langle\!\langle q^2b\rangle\!\rangle = 1,\ \langle\!\langle t^2b\rangle\!\rangle \geq 0,\ \langle\!\langle a_i^4b^3\rangle\!\rangle \geq 0,\ L_{\checkmark} \geq 0],\\ [\langle\!\langle q'\rangle\!\rangle = 1,\ \langle\!\langle q^2b\rangle\!\rangle \geq 0,\ \langle\!\langle t^2b\rangle\!\rangle = 1,\ \langle\!\langle a_i^4b^3\rangle\!\rangle \geq 0,\ L_{\checkmark} \geq 0],\\ [\langle\!\langle q'\rangle\!\rangle = 1,\ \langle\!\langle q^2b\rangle\!\rangle \geq 0,\ \langle\!\langle t^2b\rangle\!\rangle \geq 0,\ \langle\!\langle a_i^4b^3\rangle\!\rangle = 1,\ L_{\checkmark} \geq 0],\\ [\langle\!\langle q'b\rangle\!\rangle = 1,\ \langle\!\langle q^2b\rangle\!\rangle \geq 0,\ \langle\!\langle t^2b\rangle\!\rangle \geq 0,\ \langle\!\langle a_i^4b^3\rangle\!\rangle \geq 0,\ L_{\checkmark} \geq 0]\end{array}\right\} \quad \text{if } \mathsf{op} = \mathsf{c}_i{-}{-}$$

$$\mathcal{F}^{\mathsf{e}}_{(q,t,q')} = \left\{\begin{array}{l}[\langle\!\langle q'\rangle\!\rangle = 1,\ \langle\!\langle q^2b\rangle\!\rangle = 1,\ \langle\!\langle t^2b\rangle\!\rangle \geq 0,\ L_{\checkmark} \geq 0],\\ [\langle\!\langle q'\rangle\!\rangle = 1,\ \langle\!\langle q^2b\rangle\!\rangle \geq 0,\ \langle\!\langle t^2b\rangle\!\rangle = 1,\ L_{\checkmark} \geq 0],\\ [\langle\!\langle q'b\rangle\!\rangle = 1,\ \langle\!\langle q^2b\rangle\!\rangle \geq 0,\ \langle\!\langle t^2b\rangle\!\rangle \geq 0,\ \langle\!\langle a_i^4b^3\rangle\!\rangle \geq 0,\ L_{\checkmark} \geq 0]\end{array}\right\} \quad \text{if } \mathsf{op} = \mathsf{c}_i{=}{=}0$$

The first transition chosen by System must start from the initial state of M. This is enforced by Condition (b).

Once System has moved, Environment will move other processes to leave accepting configurations. The only possible move for her is to add b on a process in locations $\langle\!\langle q \rangle\!\rangle$, $\langle\!\langle t \rangle\!\rangle$, and $\langle\!\langle a_i \rangle\!\rangle$, if t is a transition incrementing counter c_i (respectively $\langle\!\langle a_i^3 b^2 \rangle\!\rangle$ if t is a transition decrementing counter c_i). All other $\mathcal{G}$-configurations accessible by Environment from already defined accepting configurations are winning for System, as established in Condition (e).

System can now encode the successor configuration of M, according to the chosen transition, by moving a process to the destination state of the transition (see Condition (c)).

Finally, Environment makes the necessary transitions for the configuration to be a valid $\mathcal{G}$-configuration. If she deviates, System wins (see Condition (f)).

If Environment reaches a configuration in $\mathbb{C}(\gamma)$ for $\gamma \in F$, System can win by moving the process in $\langle\!\langle q_h \rangle\!\rangle$ to $\langle\!\langle q_h^2 \rangle\!\rangle$. From there, all the configurations reachable by Environment are also winning for System:

$$\mathcal{F}_F = \left\{ [\langle\!\langle q_h^2 \rangle\!\rangle = 1, L_{\checkmark} \geq 0] \ , \ [\langle\!\langle q_h^2 b \rangle\!\rangle = 1, L_{\checkmark} \geq 0] \ , \ [\langle\!\langle q_h^2 b^2 \rangle\!\rangle = 1, L_{\checkmark} \geq 0] \right\} .$$

Finally, the acceptance condition is given by

$$\mathcal{F} = \bigcup_{\ell \in L_{\mathsf{s}<\mathsf{e}}} \mathcal{F}_\ell \cup \bigcup_{t=(q_0,\mathsf{op},q') \in \Delta} \mathcal{F}_t \cup \bigcup_{t=(q,\mathsf{op},q') \in \Delta} (\mathcal{F}_{(q,t)} \cup \mathcal{F}^{\mathsf{e}}_{(q,t)} \cup \mathcal{F}_{(q,t,q')} \cup \mathcal{F}^{\mathsf{e}}_{(q,t,q')}) \cup \mathcal{F}_F .$$

Note that a correct play can end in three different ways: either there is a process in $\langle\!\langle q_h \rangle\!\rangle$ and System moves it to $\langle\!\langle q_h^2 \rangle\!\rangle$, or System has no transition to pick, or there are not enough processes in ℓ_0 for System to simulate a new transition. Only the first kind is winning for System.

We can show that there is an accepting run in M iff there is some k such that System has a winning $C_{(0,0,k)}$-strategy for $\mathcal{G}$. □

6 Conclusion

There are several questions that we left open and that are interesting in their own right due to their fundamental character. Moreover, in the decidable cases, it will be worthwhile to provide tight bounds on cutoffs and the algorithmic complexity of the decision problem. Like in [7,15,16,30,31], our strategies allow the system to have a global view of the whole program run executed so far. However, it is also perfectly natural to consider uniform local strategies where each process only sees its own actions and possibly those that are revealed according to some causal dependencies. This is, e.g., the setting considered in [3,18] for a fixed number of processes and in [25] for parameterized systems over ring architectures.

Moreover, we would like to study a parameterized version of the control problem [35] where, in addition to a specification, a program in terms of an arena is already given but has to be controlled in a way such that the specification is satisfied. Finally, our synthesis results crucially rely on the fact that the number of processes in each execution is finite. It would be interesting to consider the case with potentially infinitely many processes.

References

1. P. A. Abdulla, R. Mayr, A. Sangnier, and J. Sproston. Solving parity games on integer vectors. In P. R. D'Argenio and H. C. Melgratti, editors, *CONCUR 2013 - Concurrency Theory - 24th International Conference, CONCUR 2013, Buenos Aires, Argentina, August 27-30, 2013. Proceedings*, volume 8052 of *Lecture Notes in Computer Science*, pages 106–120. Springer, 2013.
2. B. Bérard, B. Bollig, M. Lehaut, and N. Sznajder. Parameterized synthesis for fragments of first-order logic over data words. *CoRR*, abs/1910.14294, 2019.
3. R. Beutner, B. Finkbeiner, and J. Hecking-Harbusch. Translating Asynchronous Games for Distributed Synthesis. In W. Fokkink and R. van Glabbeek, editors, *30th International Conference on Concurrency Theory (CONCUR 2019)*, volume 140 of *Leibniz International Proceedings in Informatics (LIPIcs)*, pages 26:1–26:16, Dagstuhl, Germany, 2019. Schloss Dagstuhl–Leibniz-Zentrum fuer Informatik.
4. R. Bloem, S. Jacobs, A. Khalimov, I. Konnov, S. Rubin, H. Veith, and J. Widder. *Decidability of Parameterized Verification.* Morgan & Claypool Publishers, 2015.
5. M. Bojanczyk, C. David, A. Muscholl, T. Schwentick, and L. Segoufin. Two-variable logic on data words. *ACM Trans. Comput. Log.*, 12(4):27, 2011.
6. T. Brázdil, P. Jancar, and A. Kucera. Reachability games on extended vector addition systems with states. In *ICALP'10, Part II*, volume 6199 of *LNCS*, pages 478–489. Springer, 2010.
7. B. Brütsch and W. Thomas. Playing games in the Baire space. In *Proc. Cassting Workshop on Games for the Synthesis of Complex Systems and 3rd Int. Workshop on Synthesis of Complex Parameters*, volume 220 of *EPTCS*, pages 13–25, 2016.
8. J. R. Büchi and L. H. Landweber. Solving sequential conditions by finite-state strategies. *Transactions of the American Mathematical Society*, 138:295–311, Apr. 1969.
9. A. Church. Applications of recursive arithmetic to the problem of circuit synthesis. In *Summaries of the Summer Institute of Symbolic Logic – Volume 1*, pages 3–50. Institute for Defense Analyses, 1957.
10. J. Courtois and S. Schmitz. Alternating vector addition systems with states. In E. Csuhaj-Varjú, M. Dietzfelbinger, and Z. Ésik, editors, *Mathematical Foundations of Computer Science 2014 - 39th International Symposium, MFCS 2014, Budapest, Hungary, August 25-29, 2014. Proceedings, Part I*, volume 8634 of *Lecture Notes in Computer Science*, pages 220–231. Springer, 2014.
11. S. Demri, D. D'Souza, and R. Gascon. Temporal logics of repeating values. *J. Log. Comput.*, 22(5):1059–1096, 2012.
12. S. Demri and R. Lazić. LTL with the freeze quantifier and register automata. *ACM Transactions on Computational Logic*, 10(3), 2009.
13. J. Esparza. Keeping a crowd safe: On the complexity of parameterized verification. In *STACS'14*, volume 25 of *Leibniz International Proceedings in Informatics*, pages 1–10. Leibniz-Zentrum für Informatik, 2014.
14. K. Etessami, M. Y. Vardi, and T. Wilke. First-order logic with two variables and unary temporal logic. *Inf. Comput.*, 179(2):279–295, 2002.
15. L. Exibard, E. Filiot, and P.-A. Reynier. Synthesis of Data Word Transducers. In W. Fokkink and R. van Glabbeek, editors, *30th International Conference on Concurrency Theory (CONCUR 2019)*, volume 140 of *Leibniz International Proceedings in Informatics (LIPIcs)*, pages 24:1–24:15, Dagstuhl, Germany, 2019. Schloss Dagstuhl–Leibniz-Zentrum fuer Informatik.

16. D. Figueira and M. Praveen. Playing with repetitions in data words using energy games. In A. Dawar and E. Grädel, editors, *Proceedings of the 33rd Annual ACM/IEEE Symposium on Logic in Computer Science, LICS 2018, Oxford, UK, July 09-12, 2018*, pages 404–413. ACM, 2018.
17. D. Figueira and M. Praveen. Playing with repetitions in data words using energy games. *arXiv preprint arXiv:1802.07435*, 2018.
18. B. Finkbeiner and E. Olderog. Petri games: Synthesis of distributed systems with causal memory. *Inf. Comput.*, 253:181–203, 2017.
19. H. Frenkel, O. Grumberg, and S. Sheinvald. An automata-theoretic approach to model-checking systems and specifications over infinite data domains. *J. Autom. Reasoning*, 63(4):1077–1101, 2019.
20. M. Fürer. The computational complexity of the unconstrained limited domino problem (with implications for logical decision problems). In E. Börger, G. Hasenjaeger, and D. Rödding, editors, *Logic and Machines: Decision Problems and Complexity, Proceedings of the Symposium "Rekursive Kombinatorik" held from May 23-28, 1983 at the Institut für Mathematische Logik und Grundlagenforschung der Universität Münster/Westfalen*, volume 171 of *Lecture Notes in Computer Science*, pages 312–319. Springer, 1983.
21. P. Gastin and N. Sznajder. Fair synthesis for asynchronous distributed systems. *ACM Transactions on Computational Logic*, 14(2:9), 2013.
22. E. Grädel, P. G. Kolaitis, and M. Y. Vardi. On the decision problem for two-variable first-order logic. *Bulletin of Symbolic Logic*, 3(1):53–69, 1997.
23. W. Hanf. Model-theoretic methods in the study of elementary logic. In J. W. Addison, L. Henkin, and A. Tarski, editors, *The Theory of Models*. North-Holland, Amsterdam, 1965.
24. F. Horn, W. Thomas, N. Wallmeier, and M. Zimmermann. Optimal strategy synthesis for request-response games. *RAIRO - Theor. Inf. and Applic.*, 49(3):179–203, 2015.
25. S. Jacobs and R. Bloem. Parameterized synthesis. *Logical Methods in Computer Science*, 10(1), 2014.
26. S. Jacobs, L. Tentrup, and M. Zimmermann. Distributed synthesis for parameterized temporal logics. *Inf. Comput.*, 262(Part):311–328, 2018.
27. P. Jancar. On reachability-related games on vector addition systems with states. In *RP'15*, volume 9328 of *LNCS*, pages 50–62. Springer, 2015.
28. M. Jenkins, J. Ouaknine, A. Rabinovich, and J. Worrell. The church synthesis problem with metric. In M. Bezem, editor, *Computer Science Logic, 25th International Workshop / 20th Annual Conference of the EACSL, CSL 2011, September 12-15, 2011, Bergen, Norway, Proceedings*, volume 12 of *LIPIcs*, pages 307–321. Schloss Dagstuhl - Leibniz-Zentrum fuer Informatik, 2011.
29. M. Kaminski and N. Francez. Finite-memory automata. *Theoretical Computer Science*, 134(2):329–363, 1994.
30. A. Khalimov and O. Kupferman. Register-Bounded Synthesis. In W. Fokkink and R. van Glabbeek, editors, *30th International Conference on Concurrency Theory (CONCUR 2019)*, volume 140 of *Leibniz International Proceedings in Informatics (LIPIcs)*, pages 25:1–25:16, Dagstuhl, Germany, 2019. Schloss Dagstuhl–Leibniz-Zentrum fuer Informatik.
31. A. Khalimov, B. Maderbacher, and R. Bloem. Bounded synthesis of register transducers. In S. K. Lahiri and C. Wang, editors, *Automated Technology for Verification and Analysis - 16th International Symposium, ATVA 2018, Los Angeles, CA, USA, October 7-10, 2018, Proceedings*, volume 11138 of *Lecture Notes in Computer Science*, pages 494–510. Springer, 2018.

32. E. Kieronski and M. Otto. Small substructures and decidability issues for first-order logic with two variables. *J. Symb. Log.*, 77(3):729–765, 2012.
33. L. Libkin, T. Tan, and D. Vrgoc. Regular expressions for data words. *J. Comput. Syst. Sci.*, 81(7):1278–1297, 2015.
34. M. L. Minsky. *Computation: Finite and Infinite Machines.* Prentice Hall, Upper Saddle River, NJ, USA, 1967.
35. A. Muscholl. Automated synthesis of distributed controllers. In M. M. Halldórsson, K. Iwama, N. Kobayashi, and B. Speckmann, editors, *Automata, Languages, and Programming - 42nd International Colloquium, ICALP 2015, Kyoto, Japan, July 6-10, 2015, Proceedings, Part II*, volume 9135 of *Lecture Notes in Computer Science*, pages 11–27. Springer, 2015.
36. A. Pnueli and R. Rosner. Distributed reactive systems are hard to synthesize. In *31st Annual Symposium on Foundations of Computer Science, St. Louis, Missouri, USA, October 22-24, 1990, Volume II*, pages 746–757. IEEE Computer Society, 1990.
37. M. O. Rabin. *Automata on infinite objects and Church's problem.* Number 13 in Regional Conference Series in Mathematics. American Mathematical Soc., 1972.
38. J. Raskin, M. Samuelides, and L. V. Begin. Games for counting abstractions. *Electr. Notes Theor. Comput. Sci.*, 128(6):69–85, 2005.
39. A. Sangnier and O. Stietel. Private communication, 2020.
40. L. Schröder, D. Kozen, S. Milius, and T. Wißmann. Nominal automata with name binding. In J. Esparza and A. S. Murawski, editors, *Foundations of Software Science and Computation Structures - 20th International Conference, FOSSACS 2017, Held as Part of the European Joint Conferences on Theory and Practice of Software, ETAPS 2017, Uppsala, Sweden, April 22-29, 2017, Proceedings*, volume 10203 of *Lecture Notes in Computer Science*, pages 124–142, 2017.
41. T. Schwentick and K. Barthelmann. Local normal forms for first-order logic with applications to games and automata. In *Annual Symposium on Theoretical Aspects of Computer Science*, pages 444–454. Springer, 1998.
42. W. Thomas. Church's problem and a tour through automata theory. In *Pillars of Computer Science, Essays Dedicated to Boris (Boaz) Trakhtenbrot on the Occasion of His 85th Birthday*, volume 4800 of *Lecture Notes in Computer Science*, pages 635–655. Springer, 2008.
43. Y. Velner and A. Rabinovich. Church synthesis problem for noisy input. In M. Hofmann, editor, *Foundations of Software Science and Computational Structures - 14th International Conference, FOSSACS 2011, Held as Part of the Joint European Conferences on Theory and Practice of Software, ETAPS 2011, Saarbrücken, Germany, March 26-April 3, 2011. Proceedings*, volume 6604 of *Lecture Notes in Computer Science*, pages 275–289. Springer, 2011.

Controlling a random population*

Thomas Colcombet[1], Nathanaël Fijalkow[2,3](✉), and Pierre Ohlmann[1]

[1] Université de Paris, IRIF, CNRS, Paris, France
{thomas.colcombet,pierre.ohlmann}@irif.fr
[2] CNRS, LaBRI, Bordeaux, France
nathanael.fijalkow@labri.fr
[3] The Alan Turing Institute of data science, London, United Kingdom

Abstract. Bertrand et al. introduced a model of parameterised systems, where each agent is represented by a finite state system, and studied the following control problem: for any number of agents, does there exist a controller able to bring all agents to a target state? They showed that the problem is decidable and **EXPTIME**-complete in the adversarial setting, and posed as an open problem the stochastic setting, where the agent is represented by a Markov decision process. In this paper, we show that the stochastic control problem is decidable. Our solution makes significant uses of well quasi orders, of the max-flow min-cut theorem, and of the theory of regular cost functions.

1 Introduction

The control problem for populations of identical agents. The model we study was introduced in [3] (see also the journal version [4]): a population of agents are controlled uniformly, meaning that the controller applies the same action to every agent. The agents are represented by a finite state system, the same for every agent. The key difficulty is that there is an arbitrary large number of agents: the control problem is whether for every $n \in \mathbb{N}$, there exists a controller able to bring all n agents synchronously to a target state.

The technical contribution of [3,4] is to prove that in the adversarial setting where an opponent chooses the evolution of the agents, the (adversarial) control problem is **EXPTIME**-complete.

In this paper, we study the stochastic setting, where each agent evolves independently according to a probabilistic distribution, *i.e.* the finite state system modelling an agent is a Markov decision process. The control problem becomes whether for every $n \in \mathbb{N}$, there exists a controller able to bring all n agents synchronously to a target state with probability one.

* The authors are committed to making professional choices acknowledging the climate emergency. We submitted this work to FoSSaCS for its excellence and because its location induces for us a low carbon footprint. This work was supported by the European Research Council (ERC) under the European Union's Horizon 2020 research and innovation programme (grant agreement No.670624), and by the DeLTA ANR project (ANR-16-CE40-0007).

J. Goubault-Larrecq and B. König (Eds.): FOSSACS 2020, LNCS 12077, pp. 119–135, 2020.
https://doi.org/10.1007/978-3-030-45231-5_7

Our main technical result is that the stochastic control problem is decidable. In the next paragraphs we discuss four motivations for studying this problem: control of biological systems, parameterised verification and control, distributed computing, and automata theory.

Modelling biological systems. The original motivation for studying this model was for controlling population of yeasts ([21]). In this application, the concentration of some molecule is monitored through fluorescence level. Controlling the frequency and duration of injections of a sorbitol solution influences the concentration of the target molecule, triggering different chemical reactions which can be modelled by a finite state system. The objective is to control the population to reach a predetermined fluorescence state. As discussed in the conclusions of [3,4], the stochastic semantics is more satisfactory than the adversarial one for representing the behaviours of the chemical reactions, so our decidability result is a step towards a better understanding of the modelling of biological systems as populations of arbitrarily many agents represented by finite state systems.

From parameterised verification to parameterised control. Parameterised verification was introduced in [12]: it is the verification of a system composed of an arbitrary number of identical components. The control problem we study here and introduced in [3,4] is the first step towards *parameterised control*: the goal is control a system composed of many identical components in order to ensure a given property. To the best of our knowledge, the contributions of [3,4] are the first results on parameterised control; by extension, we present the first results on parameterised control in a stochastic setting.

Distributed computing. Our model resembles two models introduced for the study of distributed computing. The first and most widely studied is population protocols, introduced in [2]: the agents are modelled by finite state systems and interact by pairs drawn at random. The mode of interaction is the key difference with the model we study here: in a time step, all of our agents perform simultaneously and independently the same action. This brings us closer to broadcast protocols as studied for instance in [8], in which one action involves an arbitrary number of agents. As explained in [3,4], our model can be seen as a subclass of (stochastic) broadcast protocols, but key differences exist in the semantics, making the two bodies of work technically independent.

The focus of the distributed computing community when studying population or broadcast protocols is to construct the most efficient protocols for a given task, such as (prominently) electing a leader. A growing literature from the verification community focusses on checking the correctness of a given protocol against a given specification; we refer to the recent survey [7] for an overview. We concentrate on the control problem, which can then be seen as a first result in the control of distributed systems in a stochastic setting.

Alternative semantics for probabilistic automata. It is very tempting to consider the limit case of infinitely many agents: the parameterised control question becomes the value 1 problem for probabilistic automata, which was proved undecidable in [13], and even in very restricted cases ([10]). Hence abstracting continuous distributions by a discrete population of arbitrary size can be seen as an approximation technique for probabilistic automata. Using n agents correponds to using numerical approximation up to 2^{-n} with random rounding; in this sense the control problem considers arbitrarily fine approximations. The plague of undecidability results on probabilistic automata (see *e.g.* [9]) is nicely contrasted by our positive result, which is one of the few decidability results on probabilistic automata not making structural assumptions on the underlying graph.

Our results. We prove decidability of the stochastic control problem. The first insight is given by the theory of well quasi orders, which motivates the introduction of a new problem called the sequential flow problem. The first step of our solution is to reduce the stochastic control problem to (many instances of) the sequential flow problem. The second insight comes from the theory of regular cost functions, providing us with a set of tools for addressing the key difficulty of the problem, namely the fact that there are arbitarily many agents. Our key technical contribution is to show the computability of the sequential flow problem by reducing it to a boundedness question expressed in the cost monadic second order logic using the max-flow min-cut theorem.

Related work. The notion of decisive Markov chains was introduced in [1] as a unifying property for studying infinite-state Markov chains with finite-like properties. A typical example of decisive Markov chains is lossy channel systems where tokens can be lost anytime inducing monotonicity properties. Our situation is the exact opposite as we are considering (using the Petri nets terminology) safe Petri nets where the number of tokens along a run is constant. So it is not clear whether the underlying argument in both cases can be unified using decisiveness.

Organisation of the paper. We define the stochastic control problem in Section 2, and the sequential flow problem in Section 3. We construct a reduction from the former to (many instances of) the latter in Section 4, and show the decidability of the sequential flow problem in Section 5.

2 The stochastic control problem

Definition 1. *A* Markov decision process *(*MDP *for short) consists of*

- *a finite set of* states $\mathcal{Q}$*,*
- *a finite set of* actions $\mathcal{A}$*,*
- *a stochastic* transition table $\rho : \mathcal{Q} \times \mathcal{A} \to \mathcal{D}(\mathcal{Q})$.

The interpretation of the transition table is that from the state p under action a, the probability to transition to q is $\rho(p, a)(q)$. The *transition relation* Δ is defined by

$$\Delta = \{(p, a, q) \in \mathcal{Q} \times \mathcal{A} \times \mathcal{Q} : \rho(p, a)(q) > 0\}.$$

We also use Δ_a given by $\{(p, q) \in Q \times Q : (p, a, q) \in \Delta\}$.

We refer to [17] for the usual notions related to MDPs; it turns out that very little probability theory will be needed in this paper, so we restrict ourselves to mentioning only the relevant objects. In an MDP $\mathcal{M}$, a strategy is a function $\sigma : \mathcal{Q} \to \mathcal{A}$; note that we consider only pure and positional strategies, as they will be sufficient for our purposes.

Given a *source* $s \in \mathcal{Q}$ and a *target* $t \in \mathcal{Q}$, we say that the strategy σ *almost surely* reaches t if the probability that a path starting from s and consistent with σ eventually leads to t is 1. As we shall recall in Section 4, whether there exists a strategy ensuring to reach t almost surely from s, called the *almost sure reachability problem* for MDP can be reduced to solving a two player Büchi game, and in particular does not depend upon the exact probabilities. In other words, the only relevant information for each $(p, a, q) \in \mathcal{Q} \times \mathcal{A} \times \mathcal{Q}$ is whether $\rho(p, a)(q) > 0$ or not. Since the same will be true for the stochastic control problem we study in this paper, in our examples we do not specify the exact probabilities, and an edge from p to q labelled a means that $\rho(p, a)(q) > 0$.

Let us now fix an MDP $\mathcal{M}$ and consider a population of n *tokens* (we use tokens to represent the agents). Each token evolves in an independent copy of the MDP $\mathcal{M}$. The controller acts through a *strategy* $\sigma : \mathcal{Q}^n \to \mathcal{A}$, meaning that given the state each of the n tokens is in, the controller chooses *one* action to be performed by all tokens independently. Formally, we are considering the product MDP $\mathcal{M}^n$ whose set of states is $\mathcal{Q}^n$, set of actions is $\mathcal{A}$, and transition table is $\rho^n(u, a)(v) = \prod_{i=1}^n \rho(u_i, a)(v_i)$, where $u, v \in \mathcal{Q}^n$ and u_i, v_i are the ith components of u and v.

Let $s, t \in \mathcal{Q}$ be the source and target states, we write s^n and t^n for the constant n-tuples where all components are s and t. For a fixed value of n, whether there exists a strategy ensuring to reach t^n almost surely from s^n can be reduced to solving a two player Büchi game in the same way as above for a single MDP, replacing $\mathcal{M}$ by $\mathcal{M}^n$. The stochastic control problem asks whether this is true for arbitrary values of n:

Problem 1 (Stochastic control problem*)*. The inputs are an MDP $\mathcal{M}$, a source state $s \in \mathcal{Q}$ and a target state $t \in \mathcal{Q}$. The question is whether for all $n \in \mathbb{N}$, there exists a strategy ensuring to reach t^n almost surely from s^n.

Our main result is the following.

Theorem 1. *The stochastic control problem is decidable.*

The fact that the problem is co-recursively enumerable is easy to see: if the answer is "no", there exists $n \in \mathbb{N}$ such that there exist no strategy ensuring to reach t^n almost surely from s^n. Enumerating the values of n and solving the almost sure reachability problem for $\mathcal{M}^n$ eventually finds this out. However, it is not clear whether one can place an upper bound on such a witness n, which would yield a simple (yet inefficient!) algorithm. As a corollary of our analysis we can indeed derive such an upper bound, but it is non elementary in the size of the MDP.

In the remainder of this section we present a few interesting examples.

Example 1 Let us consider the MDP represented in Figure 1. We show that for this MDP, for any $n \in \mathbb{N}$, the controller has an almost sure strategy to reach t^n from s^n. Starting with n tokens on s, we iterate the following strategy:

- Repeatedly play action a until all tokens are in q;
- Play action b.

The first step is eventually successful with probability one, since at each iteration there is a positive probability that the number of tokens in state q increases. In the second step, with non zero probability at least one token goes to t, while the rest go back to s. It follows that each iteration of this strategy increases with non zero probability the number of tokens in t. Hence, all tokens are eventually transferred to t^n almost surely.

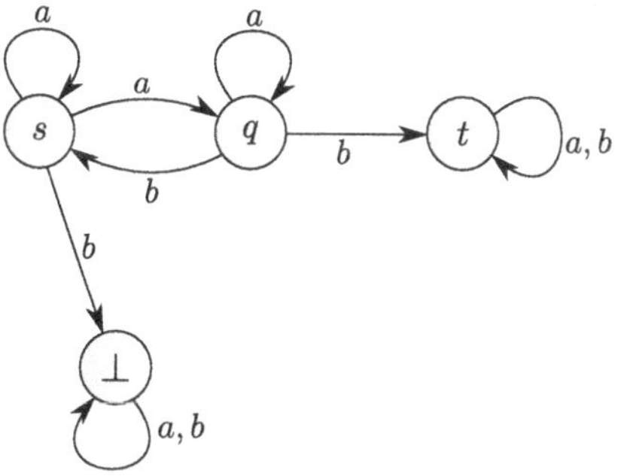

Fig. 1. The controller can almost surely reach t^n from s^n, for any $n \in \mathbb{N}$.

Example 2 We now consider the MDP represented in Figure 2. By convention, if from a state some action does not have any outgoing transition (for instance the action u from s), then it goes to the sink state $\bot$.

We show that there exists a controller ensuring to transfer seven tokens from s to t, but that the same does not hold for eight tokens. For the first assertion, we present the following strategy:

- Play a. One of the states $q_1^{i_1}$ for $i_1 \in \{u, d\}$ receives at least 4 tokens.
- Play $i_1 \in \{u, d\}$. At least 4 tokens go to t while at most 3 go to q_1.
- Play a. One of the states $q_2^{i_2}$ for $i_2 \in \{u, d\}$ receives at least 2 tokens.
- Play $i_2 \in \{u, d\}$. At least 2 tokens go to t while at most 1 token goes to q_2.
- Play a. The token (if any) goes to q_3^i for $i_3 \in \{u, d\}$.

– Play $i_3 \in \{u, d\}$. The remaining token (if any) goes to t.

Now assume that there are 8 tokens or more on s. The only choices for a strategy are to play u or d on the second, fourth, and sixth move. First, with non zero probability at least 4 tokens are in each of q_1^i for $i \in \{u, d\}$. Then, whatever the choice of action $i \in \{u, d\}$, there are at least 4 tokens in q_1 after the next step. Proceeding likewise, there are at least 2 tokens in q_2 with non zero probability two steps later. Then again two steps later, at least 1 token falls in the sink with non zero probability.

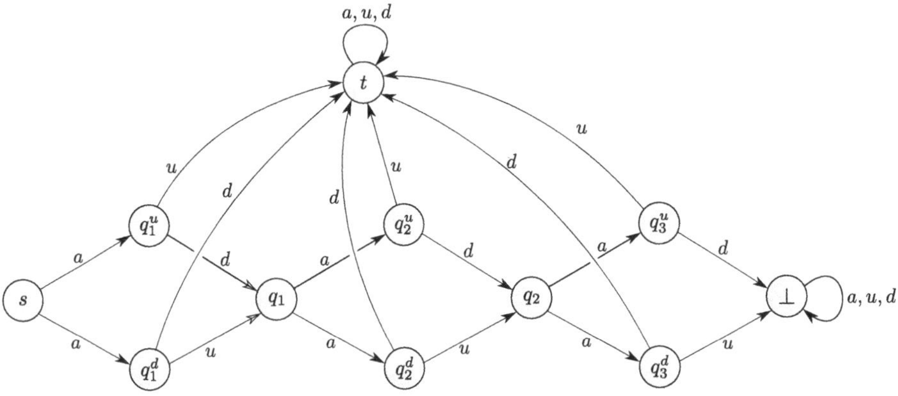

Fig. 2. The controller can synchronise up to 7 tokens on the target state t almost surely, but not more.

Generalising this example shows that if the answer to the stochastic control problem is "no", the smallest number of tokens n for which there exist no almost surely strategy for reaching t^n from s^n may be exponential in $|Q|$. This can further extended to show a doubly exponential in Q lower bound, as done in [3,4]; the example produced there holds for both the adversarial and the stochastic setting. Interestingly, for the adversarial setting this doubly exponential lower bound is tight. Our proof for the stochastic setting yields a non-elementary bound, leaving a very large gap.

Example 3 We consider the MDP represented in Figure 3. For any $n \in \mathbb{N}$, there exists a strategy almost surely reaching t^n from s^n. However, this strategy has to pass tokens one by one through q_1. We iterate the following strategy:

– Repeatedly play action a until exactly 1 token is in q_1.
– Play action b. The token goes to q_i for some $i \in \{l, r\}$.
– Play action $i \in \{l, r\}$, which moves the token to t.

Note that the first step may take a very long time (the expectation of the number of as to be played until this happens is exponential in the number of tokens), but it is eventually successful with probability one. This very slow strategy is necessary: if q_1 contains at least two tokens, then action b should not be played: with non zero probability, at least one token ends up in each of q_l, q_r, so at the next step some token ends up in $\bot$. It follows that any strategy almost surely reaching t^n has to be able to detect the presence of at most 1 token in q_1. This is a key example for understanding the difficulty of the stochastic control problem.

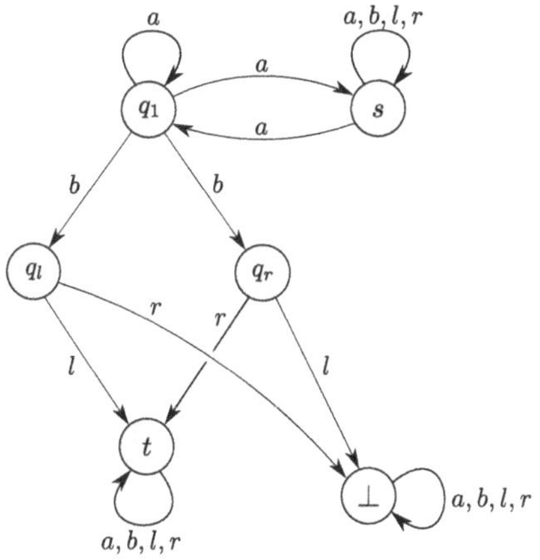

Fig. 3. The controller can synchronise any number of tokens almost surely on the target state t, but they have to go one by one.

3 The sequential flow problem

We let Q be a finite set of states. We call *configuration* an element of $\mathbb{N}^Q$ and *flow* an element of $f \in \mathbb{N}^{Q \times Q}$. A flow f induces two configurations $\text{pre}(f)$ and $\text{post}(f)$ defined by

$$\text{pre}(f)(p) = \sum_{q \in Q} f(p, q) \qquad \text{and} \qquad \text{post}(f)(q) = \sum_{p \in Q} f(p, q).$$

Given c, c' two configurations and f a flow, we say that c *goes to* c' using f and write $c \rightarrow^f c'$, if $c = \text{pre}(f)$ and $c' = \text{post}(f)$.

A *flow word* is $f = f_1 \dots f_\ell$ where each f_i is a flow. We write $c \rightsquigarrow^f c'$ if there exists a sequence of configurations $c = c_0, c_1, \dots, c_\ell = c'$ such that $c_{i-1} \rightarrow^{f_i} c_i$ for all $i \in \{1, \dots, \ell\}$. In this case, we say that c goes to c' using the flow word f.

We now recall some classical definitions related to well quasi orders ([15,16], see [19] for an exposition of recent results). Let $(E, \leqslant)$ be a quasi ordered set (*i.e.* $\leqslant$ is reflexive and transitive), it is a *well quasi ordered set* (WQO) if any infinite sequence contains an increasing pair. We say that $S \subseteq E$ is *downward closed* if for any $x \in S$, if $y \leqslant x$ then $y \in S$. An *ideal* is a non-empty downward closed set $I \subseteq E$ such that for all $x, y \in I$, there exists some $z \in I$ satisfying both $x \leqslant z$ and $y \leqslant z$.

Lemma 1.

- *Any infinite sequence of decreasing downward closed sets in a WQO is eventually constant.*
- *A subset is downward closed if and only if it is a finite union of incomparable ideals. We call it its* decomposition into ideals *(or simply, its* decomposition*), which is unique (up to permutation).*
- *An ideal is included in a downward closed set if and only if it is included in one of the ideals of its decomposition.*

We equip the set of configurations $\mathbb{N}^Q$ and the set of flows $\mathbb{N}^{Q\times Q}$ with the quasi order $\leqslant$ defined component wise, yielding thanks to Dickson's Lemma [6] two WQOs.

Lemma 2. *Let X be a finite set. A subset of $\mathbb{N}^X$ is an ideal if and only if it is of the form*

$$a{\downarrow} = \{c \in \mathbb{N}^X \mid c \leqslant a\},$$

for some $a \in (\mathbb{N} \cup \{\omega\})^X$ (in which ω is larger than all integers).

We represent downward closed sets of configurations and flows using their decomposition into finitely many ideals of the form $a\downarrow$ for $a \in (\mathbb{N} \cup \{\omega\})^Q$ or $a \in (\mathbb{N} \cup \{\omega\})^{Q\times Q}$.

*Problem 2 (*Sequential flow problem*).* Let Q be a finite set of states. Given a downward closed set of flows $\mathit{Flows} \subseteq \mathbb{N}^{Q\times Q}$ and a downward closed set of final configurations $F \subseteq \mathbb{N}^Q$, compute the downward closed set

$$\mathrm{Pre}^*(\mathit{Flows}, F) = \{c \in \mathbb{N}^Q \mid c \rightsquigarrow^f c' \in F,\ f \in \mathit{Flows}^*\}\ ,$$

i.e. the configurations from which one may reach F using only flows from *Flows*.

4 Reduction of the stochastic control problem to the sequential flow problem

Let us consider an MDP $\mathcal{M}$ and a target $t \in Q$. We first recall a folklore result reducing the almost sure reachability question for MDPs to solving a two player Büchi game (we refer to [14] for the definitions and notations of Büchi games). The Büchi game is played between *Eve* and *Adam* as follows. From a state p:

1. Eve chooses an action a and a transition $(p, q) \in \Delta_a$;
2. Adam can either choose to
   agree and the game continues from q, or
   interrupt and choose another transition $(p, q') \in \Delta_a$, the game continues from q'.

The Büchi objective is satisfied (meaning Eve wins) if either the target state t is reached or Adam interrupts infinitely many times.

Lemma 3. *There exists a strategy ensuring almost surely to reach t from s if and only if Eve has a winning strategy from s in the above Büchi game.*

We now explain how this reduction can be extended to the stochastic control problem. Let us consider an MDP $\mathcal{M}$ and a target $t \in \mathcal{Q}$. We now define an infinite Büchi game $\mathcal{G}_{\mathcal{M}}$. The set of vertices is the set of configurations $\mathbb{N}^{\mathcal{Q}}$. For a flow f, we write $\text{supp}(f) = \{(p,q) \in \mathcal{Q}^2 : f(p,q) > 0\}$. The game is played as follows from a configuration c:

1. Eve chooses an action a and a flow f such that $\text{pre}(f) = c$ and $\text{supp}(f) \subseteq \Delta_a$.
2. Adam can either choose to

   agree and the game continues from $c' = \text{post}(f)$

   interrupt and choose a flow f' such that $\text{pre}(f') = c$ and $\text{supp}(f') \subseteq \Delta_a$, and the game continues from $c'' = \text{post}(f')$.

Note that Eve choosing a flow f is equivalent to choosing for each token a transition $(p,q) \in \Delta_a$, inducing the configuration c', and simiarly for Adam should he decide to interrupt.

Eve wins if either all tokens are in the target state, or if Adam interrupts infinitely many times.

Note that although the game is infinite, it is actually a disjoint union of finite games. Indeed, along a play the number of tokens is fixed, so each play is included in $\mathcal{Q}^n$ for some $n \in \mathbb{N}$.

Lemma 4. *Let c be a configuration with n tokens in total, the following are equivalent:*

- *There exists a strategy almost surely reaching t^n from c,*
- *Eve has a winning strategy in the Büchi game $\mathcal{G}_{\mathcal{M}}$ starting from c.*

Lemma 4 follows from applying Lemma 3 on the product MDP $\mathcal{M}^n$.

We also consider the game $\mathcal{G}_{\mathcal{M}}^{(i)}$ for $i \in \mathbb{N}$, which is defined just as $\mathcal{G}_{\mathcal{M}}$ except for the winning objective: Eve wins in $\mathcal{G}_{\mathcal{M}}^{(i)}$ if either all tokens are in the target state, or if Adam interrupts more than i times. It is clear that if Eve has a winning strategy in $\mathcal{G}_{\mathcal{M}}$ then she has a winning strategy in $\mathcal{G}_{\mathcal{M}}^{(i)}$. Conversely, the following result states that $\mathcal{G}_{\mathcal{M}}^{(i)}$ is equivalent to $\mathcal{G}_{\mathcal{M}}$ for some i.

Lemma 5. *There exists $i \in \mathbb{N}$ such that from any configuration $c \in \mathbb{N}^{\mathcal{Q}}$, Eve has a winning strategy in $\mathcal{G}_{\mathcal{M}}$ if and only if Eve has a winning strategy in $\mathcal{G}_{\mathcal{M}}^{(i)}$.*

Proof: Let $X^{(i)} \subseteq \mathbb{N}^{\mathcal{Q}}$ be the winning region for Eve in $\mathcal{G}_{\mathcal{M}}^{(i)}$. We first argue that $X = \bigcap_i X^{(i)}$ is the winning region in $\mathcal{G}_{\mathcal{M}}$. It is clear that X is contained in the winning region: if Eve has a strategy to ensure that either all tokens are in the target state, or that Adam interrupts infinitely many times, then it particular this is true for Adam interrupting more than i times for any i. The converse inclusion holds because $\mathcal{G}_{\mathcal{M}}$ is a disjoint union of finite Büchi games. Indeed, in a finite Büchi game, since Adam can restrict himself to playing a memoryless winning strategy, if Eve can ensure that he interrupts a certain number of times (larger than the size of the game), then by a simple pumping argument this implies that Adam will interrupt infinitely many times.

To conclude, we note that each $X^{(i)}$ is downward closed: indeed, a winning strategy from a configuration c can be used from a configuration c' where there are fewer tokens in each state. It follows that $(X^{(i)})_{i \geq 0}$ is a decreasing sequence of downward closed sets in $\mathbb{N}^{\mathcal{Q}}$, hence it stabilises thanks to Lemma 1, *i.e.* there exists $i_0 \in \mathbb{N}$ such that $X^{(i_0)} = \bigcap_i X^{(i)}$, which concludes. □

Note that Lemma 4 and Lemma 5 substantiate the claims made in Section 2: pure positional strategies are enough and the answer to the stochastic control problem does not depend upon the exact probabilities in the MDP. Indeed, the construction of the Büchi games do not depend on them, and the answer to the former is equivalent to determining whether Eve has a winning strategy in each of them.

We are now fully equipped to show that a solution to the sequential flow problem yields the decidability of the stochastic control problem.

Let F be the set of configurations for which all tokens are in state t. we let $X^{(i)} \subseteq \mathbb{N}^{\mathcal{Q}}$ denote the winning region for Eve in the game $\mathcal{G}_{\mathcal{M}}^{(i)}$. Note first that $X^{(0)} = \mathrm{Pre}^*(\mathit{Flows}^0, F)$ where

$$\mathit{Flows}^0 = \{f \in \mathbb{N}^{\mathcal{Q} \times \mathcal{Q}} : \exists a \in \mathcal{A},\ \mathrm{supp}(f) \subseteq \Delta_a\}.$$

Indeed, in the game $\mathcal{G}_{\mathcal{M}}^{(0)}$ Adam cannot interrupt as this would make him lose immediately. Hence, the winning region for Eve in $\mathcal{G}_{\mathcal{M}}^{(0)}$ is $\mathrm{Pre}^*(\mathit{Flows}^0, F)$.

We generalise this by setting Flows^i for all $i > 0$ to be the set of flows $f \in \mathbb{N}^{\mathcal{Q} \times \mathcal{Q}}$ such that for some action $a \in \mathcal{A}$,

- $\mathrm{supp}(f) \subseteq \Delta_a$, and
- for f' with $\mathrm{pre}(f') = \mathrm{pre}(f)$ and $\mathrm{supp}(f') \subseteq \Delta_a$, we have $\mathrm{post}(f') \in X^{(i-1)}$.

Equivalently, this is the set of flows for which, when played in the game $\mathcal{G}_{\mathcal{M}}$ by Eve, Adam cannot use an interrupt move and force the configuration outside of $X^{(i-1)}$.

We now claim that

$$X^{(i)} = \mathrm{Pre}^*(\mathit{Flows}^i, F)$$

for all $i \geq 0$.

We note that this means that for each i computing $X^{(i)}$ reduces to solving one instance of the sequential flow problem. This induces an algorithm for solving the stochastic control problem: compute the sequence $(X^{(i)})_{i\geq 0}$ until it stabilises, which is ensured by Lemma 5 and yields the winning region of $\mathcal{G}_{\mathcal{M}}$. The answer to the stochastic control problem is then whether the initial configuration where all tokens are in s belongs to the winning region of $\mathcal{G}_{\mathcal{M}}$.

Let us prove the claim by induction on i.

Let c be a configuration in $\mathrm{Pre}^*(\mathit{Flows}^i, F)$. This means that there exists a flow word $f = f_1 \cdots f_\ell$ such that $f_k \in \mathit{Flows}^i$ for all k, and $c \rightsquigarrow^f c' \in F$. Expanding the definition, there exist $c_0 = c, \dots, c_\ell = c'$ such that $c_{k-1} \rightarrow^{f_k} c_k$ for all k.

Let us now describe a strategy for Eve in $\mathcal{G}_{\mathcal{M}}^{(i)}$ starting from c. As long as Adam agrees, Eve successively chooses the sequence of flows $f_1, f_2, \dots$ and the corresponding configurations $c_1, c_2, \dots$. If Adam never interrupts, then the game reaches the configuration $c' \in F$, and Eve wins. Otherwise, as soon as Adam interrupts, by definition of Flows^i, we reach a configuration $d \in X^{(i-1)}$. By induction hypothesis, Eve has a strategy which ensures from d to either reach F or that Adam interrupts at least $i-1$ times. In the latter case, adding the interrupt move leading to d yields i interrupts, so this is a winning strategy for Eve in $\mathcal{G}_{\mathcal{M}}^{(i)}$, witnessing that $c \in X^{(i)}$.

Conversely, assume that there is a winning strategy σ of Eve in $\mathcal{G}_{\mathcal{M}}^{(i)}$ from a configuration c. Consider a play consistent with σ, it either reaches F or Adam interrupts. Let us denote by $f = f_1, f_2, \dots, f_\ell$ the sequence of flows until then. We argue that $f_k \in \mathit{Flows}^i$ for $k \in \{1, \dots, \ell\}$. Let $f = f_k$ for some k, by definition of the game $\mathrm{supp}(f) \subseteq \Delta_a$ for some action a. Let f' such that $\mathrm{pre}(f') = \mathrm{pre}(f)$ and $\mathrm{supp}(f') \subsetneq \Delta_a$. In the game $\mathcal{G}_{\mathcal{M}}$ after Eve played f_k, Adam has the possibility to interrupt and choose f'. From this configuration onward the strategy σ is winning in $\mathcal{G}_{\mathcal{M}}^{(i-1)}$, implying that $f \in \mathit{Flows}^i$. Thus $f = f_1 f_2 \dots f_\ell$ is a witness that $c \in X^{(i)}$.

5 Computability of the sequential flow problem

Let Q be a finite set of states, $\mathit{Flows} \subseteq \mathbb{N}^{Q \times Q}$ a downward closed set of flows and $F \subseteq \mathbb{N}^Q$ a downward closed set of configurations, the sequential flow problem is to compute the downward closed set Pre^* defined by

$$\mathrm{Pre}^*(\mathit{Flows}, F) = \{c \in \mathbb{N}^Q \mid c \rightsquigarrow^f c' \in F,\ f \in \mathit{Flows}^*\}\ ,$$

i.e. the configurations from which one may reach F using only flows from Flows.

The following classical result of [22] allows us to further reduce our problem.

Lemma 6. *The task of computing a downward closed set can be reduced to the task of deciding whether a given ideal is included in a downward closed set.*

Thanks to Lemma 6, it is sufficient for solving the sequential flow problem to establish the following result.

Lemma 7. *Let I be an ideal of the form $a\downarrow$ for $a \in (\mathbb{N} \cup \{\omega\})^Q$, and $Flows \subseteq \mathbb{N}^{Q \times Q}$ be a downward closed set of flows. It is decidable whether F can be reached from all configurations of I using only flows from $Flows$.*

We call a vector $a \in (\mathbb{N} \cup \{\omega\})^{Q \times Q}$ a *capacity*. A *capacity word* is a finite sequence of capacities. For two capacity words w, w' of the same length, we write $w \leq w'$ to mean that $w_i \leq w'_i$ for each i. Since flows are particular cases of capacities, we can compare flows with capacities in the same way.

Before proving Lemma 7 let us give an example and some notations.

Given a state q, we write $q \in \mathbb{N}^Q$ for the vector which has value 1 on the q component and 0 elsewhere. More generally we let αq for $\alpha \in \mathbb{N} \cup \{\omega\}$ denote the vector with value α on the q component and 0 elsewhere. We use similar notations for flows. For instance, $\omega q_1 + q_2$ has value ω in the q_1 component, 1 in the q_2 component, and 0 elsewhere.

In the instance of the sequential flow problem represented in Figure 4, we ask the following question: can F be reached from any configuration of $I = (\omega q_2)\downarrow$? The answer is yes: the capacity word $w = (ac^{n-1}b)^n$ is such that $nq_2 \leadsto^f nq_4 \in F$ for a flow word $f \leqslant w$, the begining of which is described in Figure 5.

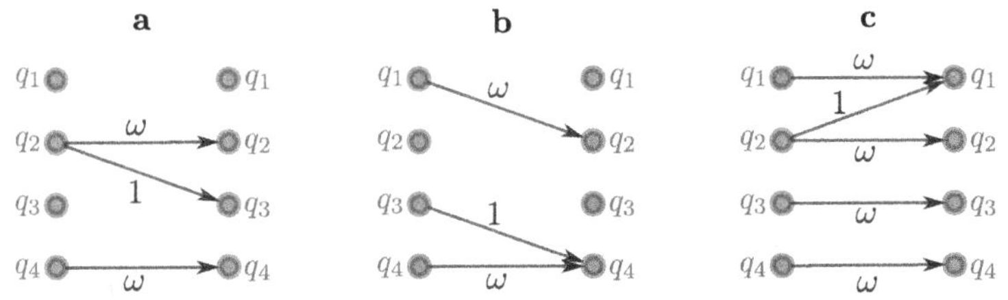

Fig. 4. An instance of the sequential flow problem. We let $Flows = a\downarrow \cup\, b\downarrow \cup\, c\downarrow$ where $a = \omega(q_2, q_2) + (q_2, q_3) + \omega(q_4, q_4)$, $b = \omega(q_1, q_2) + (q_3, q_4) + \omega(q_4, q_4)$, and $c = \omega(q_1, q_1) + (q_2, q_1) + \omega(q_2, q_2) + \omega(q_3, q_3) + \omega(q_4, q_4)$. Set also $F = (\omega q_4)\downarrow$.

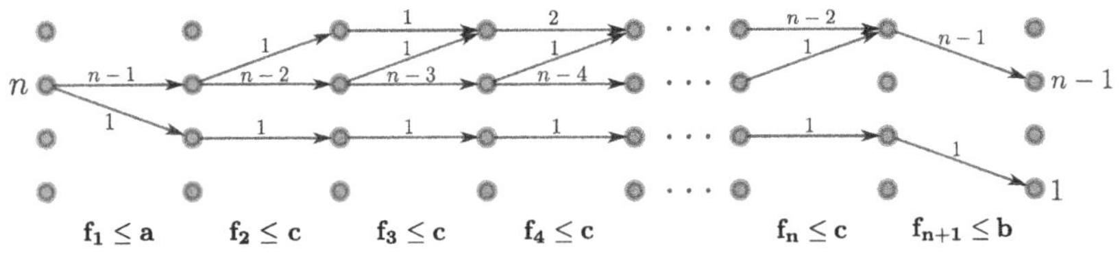

Fig. 5. A flow word $f = f_1 f_2 \ldots f_{n+1} \leqslant ac^{n-1}b$ such that nq_2 goes to $(n-1)q_1 + q_4$ using f. This construction can be extended to $f \leqslant w$ such that nq_2 goes to nq_4 using f.

We write $a[\omega \leftarrow n]$ for the configuration obtained from a by replacing all ωs by n.

The key idea for solving the sequential flow problem is to rephrase it using *regular cost functions* (a set of tools for solving boundedness questions). Indeed, whether F can be reached from all configurations of $I = a\downarrow$ using only flows from *Flows* can be equivalently phrased as a boundedness question, as follows:

> does there exist a bound on the values of $n \in \mathbb{N}$ such that $a[\omega \leftarrow n] \rightsquigarrow^{f} c$ for some $c \in F$ and $f \in \mathit{Flows}^*$?

We show that this boundedness question can be formulated as a boundedness question for a formula of *cost monadic logic*, a formalism that we introduce now. We assume that the reader is familiar with *monadic second order logic* (MSO) over finite words, and refer to [20] for the definitions. The syntax of cost monadic logic (cost MSO for short) extends MSO with the construct $|X| \leq N$, where X is a second order variable and N is a bounding variable. The semantics is defined as usual: $w, n \models \varphi$ for a word $w \in A^*$, with $n \in \mathbb{N}$ specifying the bound N. We assume that there is at most one bounding variable, and that the construct $|X| \leq N$ appears positively, *i.e.* under an even number of negations. This ensures that the larger N, the more true the formula is: if $w, n \models \varphi$, then $w, n' \models \varphi$ for all $n' \geq n$. The semantics of a formula φ of cost MSO induces a function $A^* \to \mathbb{N} \cup \{\infty\}$ defined by $\varphi(w) = \inf\{n \in \mathbb{N} \mid w, n \models \varphi\}$.

The *boundedness problem* for cost monadic logic is the following problem: given a cost MSO formula φ over A^*, is it true that the function $A^* \to \mathbb{N} \cup \{\infty\}$ is bounded, *i.e.*:

$$\exists n \in \mathbb{N},\ \forall w \in A^*,\ w, n \models \varphi?$$

The decidability of the boundedness problem is a central result in the theory of regular cost functions ([5]). Since in the theory of regular cost functions, when considering functions we are only interested in whether they are bounded or not, we will consider functions "up to boundedness properties". Concretely, this means that a *cost function* is an equivalence class of functions $A^* \to \mathbb{N} \cup \{\infty\}$, with the equivalence being $f \approx g$ if there exists $\alpha : \mathbb{N} \to \mathbb{N}$ such that $f(w)$ is finite if and only if $g(w)$ is finite, and in this case, $f(w) \leqslant \alpha(g(w))$ and $g(w) \leqslant \alpha(f(w))$. This is equivalent to stating that for all $X \subseteq A^*$, if f is bounded over X if and only if g is bounded over X.

Let us now establish Lemma 7.

Proof: Let $T = \{q \in Q \mid a(q) = \omega\}$. Note that for n sufficiently large, we have $a[\omega \leftarrow n]\downarrow = I \cap \{0, 1, \ldots, n\}$. We let $\mathscr{C} \subseteq (\mathbb{N} \cup \{\omega\})^{Q \times Q}$ be the decomposition of *Flows* into ideals, that is, $\mathscr{C}$ is the minimal finite set such that

$$\mathit{Flows} = \bigcup_{b \in \mathscr{C}} b\downarrow .$$

We let k denote the largest finite value that appears in the definition of $\mathscr{C}$, that is, $k = \max\{b(q, q') : b \in \mathscr{C}, q, q' \in Q, b(q, q') \neq \omega\}$.

Let us define the function

$$\begin{aligned} \Phi : \mathscr{C}^* &\longrightarrow \mathbb{N} \cup \{\omega\} \\ w &\longmapsto \sup\{n \in \mathbb{N} : \exists f \leqslant w, a[\omega \leftarrow n] \rightsquigarrow^{f} F\}. \end{aligned}$$

By definition Φ is unbounded if and only if F can be reached from all configurations of I. Since boundedness of cost MSO is decidable, it suffices to construct a formula in cost monadic logic for Φ to obtain the decidability of our problem. Our approach will be to additively decompose the capacity word w into a finitary part $w^{(\mathrm{fin})}$ (which is handled using a regular language), and several unbounded parts $w^{(s)}$ for each $s \in T$. The unbounded parts require a more careful analysis which notably goes through the use of the max-flow min-cut theorem.

Note that $a[\omega \leftarrow n]$ decomposes as the sum of its finite part $a_{\mathrm{fin}} = a[\omega \leftarrow 0]$ and $\sum_{s \in T} ns$. Since flows are additive, it holds that $f \leqslant w = w_1 \dots w_l$ is a flow from c_n to F if and only if the capacity word w may be decomposed into $(w^{(s)})_{s\in T} = (w_1^{(s)} \dots w_l^{(s)})_{s \in T}$ and $w^{(\mathrm{fin})} = w_1^{(\mathrm{fin})} \dots w_l^{(\mathrm{fin})}$ such that

- all the numbers appearing in the $w_i^{(s)}$ capacities are bounded by k,
- for all $i \in \{1, \dots, l\}, w_i = \sum_{s \in T \cup \{fin\}} w_i^{(s)}$,
- for all $s \in T$, $ns \rightsquigarrow^f F$ for some flow word $f \leqslant w^{(s)}$,
- and $a_{\mathrm{fin}} \rightsquigarrow^f F$ for some flow word $f \leqslant w^{(\mathrm{fin})}$.

In order to encode such capacity words in cost MSO we use monadic variables $W^{(s)}_{q,q',p}$ where $q, q' \in \mathcal{Q}$, $p \in \{0, \dots, k, \omega\}$ and $s \in T \cup \{\mathrm{fin}\}$. They are meant to satisfy that $i \in W^{(s)}_{q,q',p,s}$ if and only if $w_i^{(s)}(q, q') = p$. We use bold $\boldsymbol{W}$ to denote the tuple $(W^{(s)}_{q,q',p})_{q,q',p,s}$, and $\boldsymbol{W}^{(s)}$ for $(W^{(s)}_{q,q',p})_{q,q',p}$ when $s \in T \cup \{\omega\}$ is fixed. The MSO formula $\texttt{IsDecomp}(\boldsymbol{W}, w)$ states that a decomposition $(w^{(s)})_{s \in T \cup \{\omega\}}$ is semantically valid and sums to w:

$$\forall i, \quad \left[\bigwedge_{q,q',s} \bigvee_{p \in \{0,\dots,k,\omega\}} \left(i \in W^{(s)}_{q,q',p} \wedge \bigwedge_{p' \neq p} i \notin W^{(s)}_{q,q',p}\right)\right]$$
$$\wedge \left[\left(\bigwedge_{q,q' p} w_i(q,q') = p\right) \implies \bigvee_{\substack{(p_s)_{s \in T \cup \{\mathrm{fin}\}} \\ \sum p_s = p}} \bigwedge_{s \in T \cup \{\mathrm{fin}\}} i \in W^{(s)}_{q,q',p_s}\right]$$

For $s \in T$, we now consider the function

$$\begin{aligned} \Psi^{(s)} : (\{0, 1, \dots, k, \omega\}^{\mathcal{Q} \times \mathcal{Q}})^* &\longrightarrow \mathbb{N} \cup \{\omega\} \\ w^{(s)} &\longmapsto \sup\{n \in \mathbb{N} \mid \exists f \leqslant w^{(s)},\ ns \overset{f}{\rightsquigarrow} F\}. \end{aligned}$$

We also define $\Psi^{(\mathrm{fin})} \subseteq (\{0, \dots, k, \omega\})^{\mathcal{Q} \times \mathcal{Q}}$ to be the language of capacity words $w^{(\mathrm{fin})}$ such that there exists a flow $f \leqslant w^{(\mathrm{fin})}$ with $a_{\mathrm{fin}} \rightsquigarrow^f F$. Note that $\Psi^{(\mathrm{fin})}$ is a regular language since it is recognized by a finite automaton over $\{0, 1, \dots, k|Q|\}^{\mathcal{Q}}$ that may update the current bounded configuration only with flows smaller than the current letter of $w^{(\mathrm{fin})}$.

We have

$$\Phi(w) = \sup_n \left[\exists \boldsymbol{W}, \texttt{IsDecomp}(\boldsymbol{W}, w) \wedge \left(\bigwedge_{s \in T} \Psi^{(s)}(\boldsymbol{W}^{(s)}) \geq n\right) \wedge \boldsymbol{W}^{(\mathrm{fin})} \in \Psi^{(\mathrm{fin})}\right].$$

Hence, it is sufficient to prove that for each $s \in T$, $\Psi^{(s)}$ is definable in cost MSO.

Let us fix s and a capacity word $w \in \{0, \ldots, k, \omega\}^{Q \times Q}$ of length $|w| = \ell$. Consider the finite graph G with vertex set $Q \times \{0, 1, \ldots, \ell\}$ and for all $i \geq 1$, an edge from $(q, i-1)$ to (q', i) labelled by $w_i(q, q')$. Then $\Psi^{(s)}(w)$ is the maximal flow from $(s, 0)$ to (t, ℓ) in G. We recall that a *cut* in a graph with distinguished source s and target t is a set of edges such that removing them disconnects s and t. The *cost of a cut* is the sum of the weight of its edges. The *max-flow min-cut theorem* states that the maximal flow in a graph is exactly the minimal cost of a cut ([11]).

We now define a cost MSO formula $\tilde{\Psi}^{(s)}$ which is equivalent (in terms of cost functions) to the minimal cost of cut in the previous graph G and thus to $\Psi^{(s)}$. In the following formula, $\boldsymbol{X} = (X_{q,q'})_{q,q' \in Q}$ represents a cut in the graph: $i \in X_{q,q'}$ means that edge $((q, i-1), (q', i))$ belongs to the cut. Likewise, $\boldsymbol{P} = (P_{q,q'})_{q,q' \in Q}$ represents paths in the graph. Let $\tilde{\Psi}^{(s)}(w)$ be defined by

$$\inf_n \left\{ \exists \boldsymbol{X} \left[\bigwedge_{q,q'} n \geq |X_{q,q'}| \right] \wedge \Big(\forall i, i \in X_{q,q'} \implies w_i(q,q') < \omega \Big) \ \wedge \ \mathrm{Disc}_{s,t}(\boldsymbol{X}, w) \right\},$$

where $\mathrm{Disc}_{s,t}(\boldsymbol{X}, w)$ expresses that $\boldsymbol{X}$ disconnects $(s, 0)$ and (t, ℓ) in G. For instance $\mathrm{Disc}_{s,t}(\boldsymbol{X}, w)$ is defined by

$$\forall \boldsymbol{P}, \left[\Big(\forall i, \bigwedge_{q,q'} i \in P_{q,q'} \implies w_i(q,q') > 0 \Big) \wedge \Big(\bigvee_{q'} 0 \in P_{s,q'} \Big) \wedge \Big(\bigvee_{q} \ell \in P_{q,t} \Big) \wedge \right.$$

$$\left. \forall i \geq 1, \bigwedge_{q,q'} i \in P_{q,q'} \implies \Big(\bigvee_{q''} i - 1 \in P_{q'',q} \Big) \right] \implies \exists i, \bigvee_{q,q'} \big(i \in X_{q,q'} \wedge i \in P_{q,q'} \big).$$

Now $\tilde{\Psi}^{(s)}(w)$ does not exactly define the minimal total weight $\Phi^{(s)}(w)$ of a cut, but rather the minimal value over all cuts of the minimum over $(q, q') \in Q^2$ of how many edges are of the form $((q, i-1), (q', i))$. This is good enough for our purposes since these two values are related by

$$\tilde{\Psi}^{(s)}(w) \leqslant \Phi^{(s)}(w) \leqslant k|Q|^2 \tilde{\Psi}^{(s)}(w),$$

implying that the functions $\tilde{\Psi}^{(s)}$ and $\Phi^{(s)}$ define the same cost function. In particular, $\Phi^{(s)}$ is definable in cost MSO. □

6 Conclusions

We showed the decidability of the stochastic control problem. Our approach uses well quasi orders and the sequential flow problem, which is then solved using the theory of regular cost functions.

Together with the original result of [3,4] in the adversarial setting, our result contributes to the theoretical foundations of parameterised control. We return to the first application of this model, control of biological systems. As we discussed the stochastic setting is perhaps more satisfactory than the adversarial one, although as we saw very complicated behaviours emerge in the stochastic setting involving single agents, which are arguably not pertinent for modelling biological systems.

We thus pose two open questions. The first is to settle the complexity status of the stochastic control problem. Very recently [18] proved the **EXPTIME**-hardness of the problem, which is interesting because the underlying phenomena involved in this hardness result are specific to the stochastic setting (and do not apply to the adversarial setting). Our algorithm does not even yield elementary upper bounds, leaving a very large complexity gap. The second question is towards more accurately modelling biological systems: can we refine the stochastic control problem by taking into account the synchronising time of the controller, and restrict it to reasonable bounds?

Acknowledgements

We thank Nathalie Bertrand and Blaise Genest for introducing us to this fascinating problem, and the preliminary discussions at the Simons Institute for the Theory of Computing in Fall 2015.

References

1. Abdulla, P.A., Henda, N.B., Mayr, R.: Decisive Markov chains. Logical Methods in Computer Science **3**(4) (2007). https://doi.org/10.2168/LMCS-3(4:7)2007
2. Angluin, D., Aspnes, J., Diamadi, Z., Fischer, M.J., Peralta, R.: Computation in networks of passively mobile finite-state sensors. Distributed Computing **18**(4), 235–253 (2006). https://doi.org/10.1007/s00446-005-0138-3
3. Bertrand, N., Dewaskar, M., Genest, B., Gimbert, H.: Controlling a population. In: CONCUR. pp. 12:1–12:16 (2017). https://doi.org/10.4230/LIPIcs.CONCUR.2017.12
4. Bertrand, N., Dewaskar, M., Genest, B., Gimbert, H., Godbole, A.A.: Controlling a population. Logical Methods in Computer Science **15**(3) (2019), https://lmcs.episciences.org/5647
5. Colcombet, T.: Regular cost functions, part I: logic and algebra over words. Logical Methods in Computer Science **9**(3) (2013). https://doi.org/10.2168/LMCS-9(3:3)2013
6. Dickson, L.E.: Finiteness of the odd perfect and primitive abundant numbers with n distinct prime factors. American Journal of Mathematics **35**(4), 413–422 (1913), http://www.jstor.org/stable/2370405
7. Esparza, J.: Parameterized verification of crowds of anonymous processes. In: Dependable Software Systems Engineering, pp. 59–71. IOS Press (2016). https://doi.org/10.3233/978-1-61499-627-9-59
8. Esparza, J., Finkel, A., Mayr, R.: On the verification of broadcast protocols. In: LICS. pp. 352–359 (1999). https://doi.org/10.1109/LICS.1999.782630
9. Fijalkow, N.: Undecidability results for probabilistic automata. SIGLOG News **4**(4), 10–17 (2017), https://dl.acm.org/citation.cfm?id=3157833

10. Fijalkow, N., Gimbert, H., Horn, F., Oualhadj, Y.: Two recursively inseparable problems for probabilistic automata. In: MFCS. pp. 267–278 (2014). https://doi.org/10.1007/978-3-662-44522-8_23
11. Ford, L.R., Fulkerson, D.R.: Maximal flow through a network. Canadian Journal of Mathematics **8**, 399–404 (1956). https://doi.org/10.4153/CJM-1956-045-5
12. German, S.M., Sistla, A.P.: Reasoning about systems with many processes. Journal of the ACM **39**(3), 675–735 (1992)
13. Gimbert, H., Oualhadj, Y.: Probabilistic automata on finite words: Decidable and undecidable problems. In: ICALP. pp. 527–538 (2010). https://doi.org/10.1007/978-3-642-14162-1_44
14. Grädel, E., Thomas, W., Wilke, T. (eds.): Automata, Logics, and Infinite Games, LNCS, vol. 2500. Springer (2002)
15. Higman, G.: Ordering by divisibility in abstract algebras. Proceedings of the London Mathematical Society **s3-2**(1), 326–336 (1952). https://doi.org/10.1112/plms/s3-2.1.326
16. Kruskal, J.B.: The theory of well-quasi-ordering: A frequently discovered concept. J. Comb. Theory, Ser. A **13**(3), 297–305 (1972). https://doi.org/10.1016/0097-3165(72)90063-5
17. Kučera, A.: Turn-Based Stochastic Games. Lectures in Game Theory for Computer Scientists, Cambridge University Press (2011)
18. Mascle, C., Shirmohammadi, M., Totzke, P.: Controlling a random population is EXPTIME-hard. CoRR (2019), http://arxiv.org/abs/1909.06420
19. Schmitz, S.: Algorithmic Complexity of Well-Quasi-Orders. Habilitation à diriger des recherches, École normale supérieure Paris-Saclay (Nov 2017), https://tel.archives-ouvertes.fr/tel-01663266
20. Thomas, W.: Languages, automata, and logic. In: Handbook of Formal Language Theory, vol. III, pp. 389–455. Springer (1997)
21. Uhlendorf, J., Miermont, A., Delaveau, T., Charvin, G., Fages, F., Bottani, S., Hersen, P., Batt, G.: In silico control of biomolecular processes. Computational Methods in Synthetic Biology **13**, 277–285 (2015)
22. Valk, R., Jantzen, M.: The residue of vector sets with applications to decidability problems in Petri nets. Acta Informatica **21**, 643–674 (03 1985). https://doi.org/10.1007/BF00289715

Decomposing Probabilistic Lambda-Calculi

Ugo Dal Lago[1], Giulio Guerrieri[2(✉)], and Willem Heijltjes[2]

[1] Dipartimento di Informatica - Scienza e Ingegneria
Università di Bologna, Bologna, Italy
ugo.dallago@unibo.it

[2] Department of Computer Science
University of Bath, Bath, UK
{w.b.heijltjes,g.guerrieri}@bath.ac.uk

Abstract. A notion of probabilistic lambda-calculus usually comes with a prescribed reduction strategy, typically call-by-name or call-by-value, as the calculus is non-confluent and these strategies yield different results. This is a break with one of the main advantages of lambda-calculus: confluence, which means that results are independent from the choice of strategy. We present a probabilistic lambda-calculus where the probabilistic operator is decomposed into two syntactic constructs: a generator, which represents a probabilistic event; and a consumer, which acts on the term depending on a given event. The resulting calculus, the Probabilistic Event Lambda-Calculus, is confluent, and interprets the call-by-name and call-by-value strategies through different interpretations of the probabilistic operator into our generator and consumer constructs. We present two notions of reduction, one via fine-grained local rewrite steps, and one by generation and consumption of probabilistic events. Simple types for the calculus are essentially standard, and they convey strong normalization. We demonstrate how we can encode call-by-name and call-by-value probabilistic evaluation.

1 Introduction

Probabilistic lambda-calculi [24,22,17,11,18,9,15] extend the standard lambda-calculus with a probabilistic choice operator $N \oplus_p M$, which chooses N with probability p and M with probability $1-p$ (throughout this paper, we let p be $1/2$ and will omit it). Duplication of $N \oplus M$, as is wont to happen in lambda-calculus, raises a fundamental question about its semantics: do the duplicate occurrences represent *the same* probabilistic event, or *different* ones with the same probability? For example, take the term $\top \oplus \bot$ that represents a coin flip between boolean values *true* $\top$ and *false* $\bot$. If we duplicate this term, do the copies represent two distinct coin flips with possibly distinct outcomes, or do these represent a single coin flip that determines the outcome for both copies? Put differently again, when we duplicate $\top \oplus \bot$, do we duplicate the *event*, or only its *outcome*?

In probabilistic lambda-calculus, these two interpretations are captured by the evaluation strategies of call-by-name ($\rightarrow_{\mathsf{cbn}}$), which duplicates events, and

J. Goubault-Larrecq and B. König (Eds.): FOSSACS 2020, LNCS 12077, pp. 136–156, 2020.
https://doi.org/10.1007/978-3-030-45231-5_8

call-by-value ($\to_{\mathsf{cbv}}$), which evaluates any probabilistic choice before it is duplicated, and thus only duplicates outcomes. Consider the following example, where $=$ tests equality of boolean values.

$$\top \;\; {}_{\mathsf{cbv}}\twoheadleftarrow \;\; (\lambda x.\, x = x)(\top \oplus \bot) \;\; \twoheadrightarrow_{\mathsf{cbn}} \;\; \top \oplus \bot$$

This situation is not ideal, for several, related reasons. Firstly, it demonstrates how probabilistic lambda-calculus is non-confluent, negating one of the central properties of the lambda-calculus, and one of the main reasons why it is the prominent model of computation that it is. Secondly, it means that a probabilistic lambda-calculus must derive its semantics from a prescribed reduction strategy, and its terms only have meaning in the context of that strategy. Thirdly, combining different kinds of probabilities becomes highly involved [15], as it would require specialized reduction strategies. These issues present themselves even in a more general setting, namely that of commutative (algebraic) effects, which in general do not commute with copying.

We address these issues by a decomposition of the probabilistic operator into a *generator* $\boxed{a}$ and a *choice* $\overset{a}{\oplus}$, as follows.

$$N \oplus M \quad \overset{\Delta}{=} \quad \boxed{a}.\, N \overset{a}{\oplus} M$$

Semantically, $\boxed{a}$ represents a probabilistic event, that generates a boolean value recorded as a. The choice $N \overset{a}{\oplus} M$ is simply a conditional on a, choosing N if a is false and M if a is true. Syntactically, a is a boolean variable with an occurrence in $\overset{a}{\oplus}$, and $\boxed{a}$ acts as a probabilistic quantifier, binding all occurrences in its scope. (To capture a non-equal chance, one would attach a probability p to a generator, as $\boxed{a}_p$, though we will not do so in this paper.)

The resulting *probabilistic event lambda-calculus* Λ_{PE}, which we present in this paper, is confluent. Our decomposition allows us to separate duplicating an *event*, represented by the generator $\boxed{a}$, from duplicating only its *outcome* a, through having multiple choice operators $\overset{a}{\oplus}$. In this way our calculus may interpret both original strategies, call-by-name and call-by-value, by different translations of standard probabilistic terms into Λ_{PE}: call-by-name by the above decomposition (see also Section 2), and call-by-value by a different one (see Section 7). For our initial example, we get the following translations and reductions.

$$\mathsf{cbn}: \quad (\lambda x.\, x{=}x)(\boxed{a}.\, \top \overset{a}{\oplus} \bot) \;\; \to_\beta \;\; (\boxed{a}.\, \top \overset{a}{\oplus} \bot){=}(\boxed{b}.\, \top \overset{b}{\oplus} \bot) \;\; \twoheadrightarrow \;\; \top \oplus \bot \qquad (1)$$

$$\mathsf{cbv}: \quad \boxed{a}.\, (\lambda x.\, x{=}x)(\top \overset{a}{\oplus} \bot) \;\; \to_\beta \;\; \boxed{a}.\, (\top \overset{a}{\oplus} \bot){=}(\top \overset{a}{\oplus} \bot) \;\; \twoheadrightarrow \;\; \top \qquad (2)$$

We present two reduction relations for our probabilistic constructs, both independent of beta-reduction. Our main focus will be on *permutative* reduction (Sections 2, 3), a small-step local rewrite relation which is computationally inefficient but gives a natural and very fine-grained operational semantics. *Projective* reduction (Section 6) is a more standard reduction, following the intuition that $\boxed{a}$ generates a coin flip to evaluate $\overset{a}{\oplus}$, and is coarser but more efficient.

We further prove confluence (Section 4), and we give a system of simple types and prove strong normalization for typed terms by reducibility (Section 5). Omitted proofs can be found in [7], the long version of this paper.

1.1 Related Work

Probabilistic λ-calculi are a topic of study since the pioneering work by Saheb-Djaromi [24], the first to give the syntax and operational semantics of a λ-calculus with binary probabilistic choice. Giving well-behaved denotational models for probabilistic λ-calculi has proved to be challenging, as witnessed by the many contributions spanning the last thirty years: from Jones and Plotkin's early study of the probabilistic powerdomain [17], to Jung and Tix's remarkable (and mostly negative) observations [18], to the very recent encouraging results by Goubault-Larrecq [16]. A particularly well-behaved model for probabilistic λ-calculus can be obtained by taking a probabilistic variation of Girard's coherent spaces [10], this way getting full abstraction [13].

On the operational side, one could mention a study about the various ways the operational semantics of a calculus with binary probabilistic choice can be specified, namely by small-step or big-step semantics, or by inductively or coinductively defined sets of rules [9]. Termination and complexity analysis of higher-order probabilistic programs seen as λ-terms have been studied by way of type systems in a series of recent results about size [6], intersection [4], and refinement type disciplines [1]. Contextual equivalence on probabilistic λ-calculi has been studied, and compared with equational theories induced by Böhm Trees [19], applicative bisimilarity [8], or environmental bisimilarity [25].

In all the aforementioned works, probabilistic λ-calculi have been taken as implicitly endowed with either call-by-name or call-by-value strategies, for the reasons outlined above. There are only a few exceptions, namely some works on Geometry of Interaction [5], Probabilistic Coherent Spaces [14], and Standardization [15], which achieve, in different contexts, a certain degree of independence from the underlying strategy, thus accommodating both call-by-name and call-by-value evaluation. The way this is achieved, however, invariably relies on Linear Logic or related concepts. This is deeply different from what we do here.

Some words of comparison with Faggian and Ronchi Della Rocca's work on confluence and standardization [15] are also in order. The main difference between their approach and the one we pursue here is that the operator ! in their calculus $\Lambda^!_\oplus$ plays *both* the roles of a marker for duplicability and of a checkpoint for any probabilistic choice "flowing out" of the term (*i.e.* being fired). In our calculus, we do not control duplication, but we definitely make use of checkpoints. Saying it another way, Faggian and Ronchi Della Rocca's work is inspired by linear logic, while our approach is inspired by deep inference, even though this is, on purpose, not evident in the design of our calculus.

Probabilistic λ-calculi can also be seen as vehicles for expressing probabilistic models in the sense of bayesian programming [23,3]. This, however, requires an operator for modeling conditioning, which complicates the metatheory considerably, and that we do not consider here.

Our permutative reduction is a refinement of that for the call-by-name probabilistic λ-calculus [20], and is an implementation of the equational theory of *(ordered) binary decision trees* via rewriting [27]. Probabilistic decision trees have been proposed with a primitive binary probabilistic operator [22], but not with a decomposition as we explore here.

2 The Probabilistic Event λ-Calculus Λ_{PE}

Definition 1. The *probabilistic event λ-calculus* (Λ_{PE}) is given by the following grammar, with from left to right: a *variable* (denoted by $x, y, z, \ldots$), an *abstraction*, an *application*, a *(labeled) choice*, and a *(probabilistic) generator.*

$$M, N \quad ::= \quad x \mid \lambda x.N \mid NM \mid N \overset{a}{\oplus} M \mid \boxed{a}.\,N$$

In a term $\lambda x.\,M$ the abstraction λx binds the free occurrences of the variable x in its scope M, and in $\boxed{a}.\,N$ the generator $\boxed{a}$ binds the *label* a in N. The calculus features a decomposition of the usual probabilistic sum $\oplus$, as follows.

$$N \oplus M \quad \triangleq \quad \boxed{a}.\,N \overset{a}{\oplus} M \tag{3}$$

The generator $\boxed{a}$ represents a probabilistic *event*, whose outcome, a binary value $\{0, 1\}$ represented by the label a, is used by the choice operator $\overset{a}{\oplus}$. That is, $\boxed{a}$ flips a coin setting a to 0 (resp. 1), and depending on this $N \overset{a}{\oplus} M$ reduces to N (resp. M). We will use the unlabeled choice $\oplus$ as in (3). This convention also gives the translation from a *call-by-name* probabilistic λ-calculus into Λ_{PE} (the interpretation of a *call-by-value* probabilistic λ-calculus is in Section 7).

Reduction. Reduction in Λ_{PE} will consist of standard β-reduction $\to_\beta$ plus an evaluation mechanism for generators and choice operators, which implements probabilistic choice. We will present two such mechanisms: *projective* reduction $\to_\pi$ and *permutative* reduction $\to_{\mathsf{p}}$. While projective reduction implements the given intuition for the generator and choice operator, we relegate it to Section 6 and make permutative reduction our main evaluation mechanism, for the reason that it is more fine-grained, and thus more general.

Permutative reduction is based on the idea that any operator distributes over the labeled choice operator (see the reduction steps in Figure 1), even other choice operators, as below.

$$(N \overset{a}{\oplus} M) \overset{b}{\oplus} P \;\sim\; (N \overset{b}{\oplus} P) \overset{a}{\oplus} (M \overset{b}{\oplus} P)$$

To orient this as a rewrite rule, we need to give priority to one label over another. Fortunately, the relative position of the associated generators $\boxed{a}$ and $\boxed{b}$ provides just that. Then to define $\to_{\mathsf{p}}$, we will want every choice to belong to some generator, and make the order of generators explicit.

Definition 2. The set $\mathsf{fl}(N)$ of *free labels* of a term N is defined inductively by:

$$\mathsf{fl}(x) = \emptyset \qquad \mathsf{fl}(MN) = \mathsf{fl}(M) \cup \mathsf{fl}(N) \qquad \mathsf{fl}(\lambda x.\,M) = \mathsf{fl}(M)$$
$$\mathsf{fl}(\boxed{a}.\,M) = \mathsf{fl}(M) \setminus \{a\} \quad \mathsf{fl}(M \overset{a}{\oplus} N) = \mathsf{fl}(M) \cup \mathsf{fl}(N) \cup \{a\}$$

A term M is *label-closed* if $\mathsf{fl}(M) = \emptyset$.

$$(\lambda x.N)M \to_\beta N[M/x] \qquad (\beta)$$

$$N \overset{a}{\oplus} N \to_{\mathsf{p}} N \qquad (\mathsf{i})$$

$$(N \overset{a}{\oplus} M) \overset{a}{\oplus} P \to_{\mathsf{p}} N \overset{a}{\oplus} P \qquad (\mathsf{c}_1)$$

$$N \overset{a}{\oplus} (M \overset{a}{\oplus} P) \to_{\mathsf{p}} N \overset{a}{\oplus} P \qquad (\mathsf{c}_2)$$

$$\lambda x.\,(N \overset{a}{\oplus} M) \to_{\mathsf{p}} (\lambda x.\,N) \overset{a}{\oplus} (\lambda x.\,M) \qquad (\oplus\lambda)$$

$$(N \overset{a}{\oplus} M)P \to_{\mathsf{p}} (NP) \overset{a}{\oplus} (MP) \qquad (\oplus\mathsf{f})$$

$$N(M \overset{a}{\oplus} P) \to_{\mathsf{p}} (NM) \overset{a}{\oplus} (NP) \qquad (\oplus\mathsf{a})$$

$$(N \overset{a}{\oplus} M) \overset{b}{\oplus} P \to_{\mathsf{p}} (N \overset{b}{\oplus} P) \overset{a}{\oplus} (M \overset{b}{\oplus} P) \qquad (\text{if } a < b) \qquad (\oplus\oplus_1)$$

$$N \overset{b}{\oplus} (M \overset{a}{\oplus} P) \to_{\mathsf{p}} (N \overset{b}{\oplus} M) \overset{a}{\oplus} (N \overset{b}{\oplus} P) \qquad (\text{if } a < b) \qquad (\oplus\oplus_2)$$

$$\boxed{b}.\,(N \overset{a}{\oplus} M) \to_{\mathsf{p}} (\boxed{b}.\,N) \overset{a}{\oplus} (\boxed{b}.\,M) \qquad (\text{if } a \neq b) \qquad (\oplus\Box)$$

$$\boxed{a}.\,N \to_{\mathsf{p}} N \qquad (\text{if } a \notin \mathsf{fl}(N)) \qquad (\not\Box)$$

$$\lambda x.\boxed{a}.\,N \to_{\mathsf{p}} \boxed{a}.\,\lambda x.\,N \qquad (\Box\lambda)$$

$$(\boxed{a}.\,N)M \to_{\mathsf{p}} \boxed{a}.\,(NM) \qquad (\text{if } a \notin \mathsf{fl}(M)) \qquad (\Box\mathsf{f})$$

Fig. 1. Reduction Rules for β-reduction and p-reduction.

From here on, we consider only label-closed terms (we implicitly assume this, unless otherwise stated). All terms are identified up to renaming of their bound variables and labels. Given some terms M and N and a variable x, $M[N/x]$ is the capture-avoiding (for both variables and labels) substitution of N for the free occurrences of x in M. We speak of a *representative* M of a term when M is not considered up to such a renaming. A representative M of a term is *well-labeled* if for every occurrence of $\boxed{a}$ in M there is no $\boxed{a}$ occurring in its scope.

Definition 3 (Order for labels). Let M be a well-labeled representative of a term. We define an *order* $<_M$ for the labels occurring in M as follows: $a <_M b$ if and only if $\boxed{b}$ occurs in the scope of $\boxed{a}$.

For a well-labeled and label-closed representative M, $<_M$ is a finite tree order.

Definition 4. *Reduction* $\to = \to_\beta \cup \to_{\mathsf{p}}$ in Λ_{PE} consists of *β-reduction* $\to_\beta$ and *permutative* or p*-reduction* $\to_{\mathsf{p}}$, both defined as the contextual closure of the rules given in Figure 1. We write $\twoheadrightarrow$ for the reflexive–transitive closure of $\to$, and $\twoheadrightarrow\!\!\!|$ for reduction to normal form; similarly for $\to_\beta$ and $\to_{\mathsf{p}}$. We write $=_{\mathsf{p}}$ for the symmetric and reflexive–transitive closure of $\to_{\mathsf{p}}$.

$$
\begin{array}{llll}
\boxed{a}.\,(\lambda x.\,x\!=\!x)(\top \overset{a}{\oplus} \bot) & \rightarrow_{\mathsf{p}} & \boxed{a}.\,(\lambda x.\,x\!=\!x)\top \overset{a}{\oplus} (\lambda x.\,x\!=\!x)\bot & (\oplus\mathsf{a}) \\
& \twoheadrightarrow_{\beta} & \boxed{a}.\,(\top = \top) \overset{a}{\oplus} (\bot = \bot) & \\
& = & \boxed{a}.\,\top \overset{a}{\oplus} \top \;\rightarrow_{\mathsf{p}}\; \boxed{a}.\,\top \;\rightarrow_{\mathsf{p}}\; \top & (\mathsf{i},\not\emptyset)
\end{array}
$$

Fig. 2. Example Reduction of the cbv-translation of the Term on p. 137.

Two example reductions are (1)-(2) on p. 137; a third, complete reduction is in Figure 2. The crucial feature of p-reduction is that a choice $\overset{a}{\oplus}$ *does* permute out of the argument position of an application, but a generator $\boxed{a}$ does *not*, as below. Since the argument of a redex may be duplicated, this is how we characterize the difference between the *outcome* of a probabilistic event, whose duplicates may be identified, and the event itself, whose duplicates may yield different outcomes.

$$
N\,(M \overset{a}{\oplus} P) \rightarrow_{\mathsf{p}} (NM) \overset{a}{\oplus} (NP) \qquad\qquad N\,(\boxed{a}.\,M) \not\rightarrow_{\mathsf{p}} \boxed{a}.\,N\,M
$$

By inspection of the rewrite rules in Figure 1, we can then characterize the normal forms of $\rightarrow_{\mathsf{p}}$ and $\rightarrow$ as follows.

Proposition 5 (Normal forms). *The normal forms P_0 of $\rightarrow_{\mathsf{p}}$, respectively N_0 of $\rightarrow$, are characterized by the following grammars.*

$$
\begin{array}{lll}
P_0 ::= P_1 \mid P_0 \oplus P_0' \\
P_1 ::= x \mid \lambda x.P_1 \mid P_1\,P_0
\end{array}
\qquad\qquad
\begin{array}{lll}
N_0 ::= N_1 \mid N_0 \oplus N_0' \\
N_1 ::= N_2 \mid \lambda x.N_1 \\
N_2 ::= x \mid N_2\,N_0
\end{array}
$$

3 Properties of Permutative Reduction

We will prove strong normalization and confluence of $\rightarrow_{\mathsf{p}}$. For strong normalization, the obstacle is the interaction between different choice operators, which may duplicate each other, creating super-exponential growth.[3] Fortunately, Dershowitz's *recursive path orders* [12] seem tailor-made for our situation.

Observe that the set Λ_{PE} endowed with $\rightarrow_{\mathsf{p}}$ is a first-order term rewriting system over a countably infinite set of variables and the signature Σ given by:

- the binary function symbol $\overset{a}{\oplus}$, for any label a;
- the unary function symbol $\boxed{a}$, for any label a;
- the unary function symbol λx, for any variable x;
- the binary function symbol @, letting $@(M,N)$ stand for MN.

---

[3] This was inferred only from a simple simulation; we would be interested to know a rigorous complexity result.

Definition 6. Let M be a well-labeled representative of a label-closed term, and let Σ_M be the set of signature symbols occurring in M. We define $\prec_M$ as the (strict) partial order on Σ_M generated by the following rules.

$$\begin{array}{ll} \overset{a}{\oplus} \prec_M \overset{b}{\oplus} & \text{if } a <_M b \\ \overset{a}{\oplus} \prec_M \boxed{b} & \text{for any labels } a, b \\ \boxed{b} \prec_M @, \lambda x & \text{for any label } b \end{array}$$

Lemma 7. *The reduction* $\rightarrow_{\mathsf{p}}$ *is strongly normalizing.*

Proof. For the first-order term rewriting system $(\Lambda_{\mathsf{PE}}, \rightarrow_{\mathsf{p}})$ we derive a well-founded recursive path ordering $<$ from $\prec_M$ following [12, p. 289]. Let f and g range over function symbols, let $[N_1, \ldots, N_n]$ denote a multiset and extend $<$ to multisets by the standard multiset ordering, and let $N = f(N_1, \ldots, N_n)$ and $M = g(M_1, \ldots, M_m)$; then

$$N < M \iff \begin{cases} [N_1, \ldots, N_n] < [M_1, \ldots, M_m] & \text{if } f = g \\ [N_1, \ldots, N_n] < [M] & \text{if } f \prec_M g \\ [N] \leq [M_1, \ldots, M_m] & \text{if } f \not\preceq_M g\,. \end{cases}$$

While $\prec_M$ is defined only relative to Σ_M, reduction may only reduce the signature. Inspection of Figure 1 then shows that $M \rightarrow_{\mathsf{p}} N$ implies $N < M$. □

Confluence of Permutative Reduction. With strong normalization, confluence of $\rightarrow_{\mathsf{p}}$ requires only local confluence. We reduce the number of cases to consider, by casting the permutations of $\overset{a}{\oplus}$ as instances of a common shape.

Definition 8. We define a *context* $C[\,]$ (with exactly one hole $[\,]$) as follows, and let $C[N]$ represent $C[\,]$ with the hole $[\,]$ replaced by N.

$$C[\,] ::= [\,] \mid \lambda x.C[\,] \mid C[\,]M \mid NC[\,] \mid C[\,]\overset{a}{\oplus}M \mid N\overset{a}{\oplus}C[\,] \mid \boxed{a}.C[\,]$$

Observe that the six reduction rules $\oplus\lambda$ through $\oplus\square$ in Figure 1 are all of the following form. We refer to these collectively as $\oplus\star$.

$$C[N\overset{a}{\oplus}M] \rightarrow_{\mathsf{p}} C[N]\overset{a}{\oplus}C[M] \tag{$\oplus\star$}$$

Lemma 9 (Confluence of $\rightarrow_{\mathsf{p}}$). *Reduction* $\rightarrow_{\mathsf{p}}$ *is confluent.*

Proof. By Newman's lemma and strong normalization of $\rightarrow_{\mathsf{p}}$ (Lemma 7), confluence follows from local confluence. The proof of local confluence consists of joining all critical pairs given by $\rightarrow_{\mathsf{p}}$. Details are in the Appendix of [7]. □

Definition 10. We denote the unique p-normal form of a term N by N_{p}.

4 Confluence

We aim to prove that $\twoheadrightarrow = \twoheadrightarrow_\beta \cup \twoheadrightarrow_{\mathsf{p}}$ is confluent. We will use the standard technique of *parallel* β-reduction [26], a simultaneous reduction step on a number of β-redexes, which we define via a labeling of the redexes to be reduced. The central point is to find a notion of reduction that is *diamond*, *i.e.* every critical pair can be closed in one (or zero) steps. This will be our *complete* reduction, which consists of parallel β-reduction followed by p-reduction to normal form.

Definition 11. A *labeled* term $P^\bullet$ is a term P with chosen β-redexes annotated as $(\lambda x.\,N)^\bullet M$. The unique *labeled β-step* $P^\bullet \Rightarrow_\beta P_\bullet$ from $P^\bullet$ to the *labeled reduct* $P_\bullet$ reduces every labeled redex, and is defined inductively as follows.

$$(\lambda x.\,N^\bullet)^\bullet M^\bullet \Rightarrow_\beta N_\bullet[M_\bullet/x] \qquad N^\bullet M^\bullet \Rightarrow_\beta N_\bullet M_\bullet$$
$$x \Rightarrow_\beta x \qquad N^\bullet \overset{a}{\oplus} M^\bullet \Rightarrow_\beta N_\bullet \overset{a}{\oplus} M_\bullet$$
$$\lambda x.\,N^\bullet \Rightarrow_\beta \lambda x.\,N_\bullet \qquad \boxed{a}.\,N^\bullet \Rightarrow_\beta \boxed{a}.\,N_\bullet$$

A *parallel β-step* $P \Rightarrow_\beta P_\bullet$ is a labeled step $P^\bullet \Rightarrow_\beta P_\bullet$ for some labeling $P^\bullet$.

Note that $P_\bullet$ is an unlabeled term, since all labels are removed in the reduction. For the empty labeling, $P^\bullet = P_\bullet = P$, so parallel reduction is reflexive: $P \Rightarrow_\beta P$.

Lemma 12. *A parallel β-step $P \Rightarrow_\beta P_\bullet$ is a β-reduction $P \twoheadrightarrow_\beta P_\bullet$.*

Proof. By induction on the labeled term $P^\bullet$ generating $P \Rightarrow_\beta P_\bullet$. □

Lemma 13. *Parallel β-reduction is diamond.*

Proof. Let $P^\bullet \Rightarrow_\beta P_\bullet$ and $P^\circ \Rightarrow_\beta P_\circ$ be two labeled reduction steps on a term P. We annotate each step with the label of the other, preserved by reduction, to give the span from the doubly labeled term $P^{\bullet\circ} = P^{\circ\bullet}$ below left. Reducing the remaining labels will close the diagram, as below right.

$$P^\circ_\bullet \;{}_\beta\!\Leftarrow\; P^{\bullet\circ} = P^{\circ\bullet} \Rightarrow_\beta P^\bullet_\circ \qquad\qquad P^\circ_\bullet \Rightarrow_\beta P_{\bullet\circ} = P_{\circ\bullet} \;{}_\beta\!\Leftarrow\; P^\bullet_\circ$$

This is proved by induction on $P^{\bullet\circ}$, where only two cases are not immediate: those where a redex carries one but not the other label. One case follows by the below diagram; the other case is symmetric. Below, for the step top right, induction on $N^\bullet$ shows that $N^\bullet[M^\bullet/x] \Rightarrow_\beta N_\bullet[M_\bullet/x]$.

$$\begin{array}{ccccc}(\lambda x.\,N^{\circ\bullet})^\circ M^{\circ\bullet} & \Rightarrow_\beta & N^\bullet_\circ[M^\bullet_\circ/x] & \Rightarrow_\beta & N_{\circ\bullet}[M_{\circ\bullet}/x] \\ = & & & & = \\ (\lambda x.\,N^{\bullet\circ})^\circ M^{\bullet\circ} & \Rightarrow_\beta & (\lambda x.\,N^\circ_\bullet)^\circ M^\circ_\bullet & \Rightarrow_\beta & N_{\bullet\circ}[M_{\bullet\circ}/x]\end{array}$$
□

4.1 Parallel Reduction and Permutative Reduction

For the commutation of (parallel) β-reduction with p-reduction, we run into the minor issue that a permuting generator or choice operator may block a redex: in both cases below, before $\to_{\mathsf{p}}$ the term has a redex, but after $\to_{\mathsf{p}}$ it is blocked.

$$(\lambda x.\, N \overset{a}{\oplus} M)\, P \to_{\mathsf{p}} ((\lambda x.\, N) \overset{a}{\oplus} (\lambda x.\, M))\, P \qquad (\lambda x.\, \boxed{a}.\, N)\, M \to_{\mathsf{p}} (\boxed{a}.\, \lambda x.\, N)\, M$$

We address this by an adaptation $\mathrel{-\!\!\blacktriangleright}_{\mathsf{p}}$ of p-reduction on labeled terms, which is a strategy in $\twoheadrightarrow_{\mathsf{p}}$ that permutes past a labeled redex in one step.

Definition 14. A *labeled* p-reduction $N^\bullet \mathrel{-\!\!\blacktriangleright}_{\mathsf{p}} M^\bullet$ on labeled terms is a p-reduction of one of the forms

$$(\lambda x.\, N^\bullet \overset{a}{\oplus} M^\bullet)^\bullet P^\bullet \twoheadrightarrow_{\mathsf{p}} (\lambda x.\, N^\bullet)^\bullet P^\bullet \overset{a}{\oplus} (\lambda x.\, M^\bullet)^\bullet P^\bullet$$

$$(\lambda x.\, \boxed{a}.\, N^\bullet)^\bullet M^\bullet \twoheadrightarrow_{\mathsf{p}} \boxed{a}.\, (\lambda x.\, N^\bullet)^\bullet M^\bullet$$

or a single p-step $\to_{\mathsf{p}}$ on unlabeled constructors in $N^\bullet$.

Lemma 15. *Reduction to normal form in* $\mathrel{-\!\!\blacktriangleright}_{\mathsf{p}}$ *is equal to* $\twoheadrightarrow\!|_{\mathsf{p}}$ *(on labeled terms).*

Proof. Observe that $\to_{\mathsf{p}}$ and $\mathrel{-\!\!\blacktriangleright}_{\mathsf{p}}$ have the same normal forms. Then in one direction, since $\mathrel{-\!\!\blacktriangleright}_{\mathsf{p}} \subseteq \twoheadrightarrow_{\mathsf{p}}$ we have $\mathrel{-\!\!\blacktriangleright\!\!\blacktriangleright\!|}_{\mathsf{p}} \subseteq \twoheadrightarrow\!|_{\mathsf{p}}$. Conversely, let $N \twoheadrightarrow\!|_{\mathsf{p}} M$. On this reduction, let $P \to_{\mathsf{p}} Q$ be the first step such that $P \not\mathrel{-\!\!\blacktriangleright}_{\mathsf{p}} Q$. Then there is an R such that $P \mathrel{-\!\!\blacktriangleright}_{\mathsf{p}} R$ and $Q \to_{\mathsf{p}} R$. Note that we have $N \mathrel{-\!\!\blacktriangleright\!\!\blacktriangleright}_{\mathsf{p}} R$. By confluence, $R \twoheadrightarrow\!|_{\mathsf{p}} M$, and by induction on the sum length of paths in $\to_{\mathsf{p}}$ from R (smaller than from N) we have $R \mathrel{-\!\!\blacktriangleright\!\!\blacktriangleright\!|}_{\mathsf{p}} M$, and hence $N \mathrel{-\!\!\blacktriangleright\!\!\blacktriangleright\!|}_{\mathsf{p}} M$. □

The following lemmata then give the required commutation properties of the relations $\mathrel{-\!\!\blacktriangleright}_{\mathsf{p}}$, $\twoheadrightarrow\!|_{\mathsf{p}}$, and $\Rightarrow_{\beta}$. Figure 3 illustrates these by commuting diagrams.

Lemma 16. *If* $N^\bullet \mathrel{-\!\!\blacktriangleright}_{\mathsf{p}} M^\bullet$ *then* $N_\bullet =_{\mathsf{p}} M_\bullet$.

Proof. By induction on the rewrite step $\mathrel{-\!\!\blacktriangleright}_{\mathsf{p}}$. The two interesting cases are:

$$\begin{array}{ccc} (\lambda x.\, M^\bullet)^\bullet (N^\bullet \overset{a}{\oplus} P^\bullet) & \xrightarrow{\ \mathsf{p}\ } & ((\lambda x.\, M^\bullet)^\bullet N^\bullet) \overset{a}{\oplus} ((\lambda x.\, M^\bullet)^\bullet P^\bullet) \\ \beta \Big\Downarrow & & \Big\Downarrow \beta \\ M_\bullet[(N_\bullet \overset{a}{\oplus} P_\bullet)/x] & \overset{}{\underset{\mathsf{p}}{\dashrightarrow\!\!\!\!\rightarrow}} & M_\bullet[N_\bullet/x] \overset{a}{\oplus} M_\bullet[P_\bullet/x] \end{array} \qquad (x \in \mathsf{fv}(M))$$

$$\begin{array}{ccc} (\lambda x.\, M^\bullet)^\bullet (N^\bullet \overset{a}{\oplus} P^\bullet) & \xrightarrow{\ \mathsf{p}\ } & ((\lambda x.\, M^\bullet)^\bullet N^\bullet) \overset{a}{\oplus} ((\lambda x.\, M^\bullet)^\bullet P^\bullet) \\ \beta \Big\Downarrow & & \Big\Downarrow \beta \\ M_\bullet & \underset{\mathsf{p}}{\dashleftarrow} & M_\bullet \overset{a}{\oplus} M_\bullet \end{array} \qquad (x \notin \mathsf{fv}(M))$$

□

How the critical pairs in the above diagrams are joined shows that we cannot use the Hindley-Rosen Lemma [2, Prop. 3.3.5] to prove confluence of $\rightarrow_\beta \cup \rightarrow_{\mathsf{p}}$.

Lemma 17. $N_\bullet =_{\mathsf{p}} N_{\mathsf{p}\bullet}$.

Proof. Using Lemma 15 we decompose $N^\bullet \twoheadrightarrow\!\!|_{\mathsf{p}} N^\bullet_{\mathsf{p}}$ as

$$N^\bullet = N^\bullet_1 \rightarrow_{\mathsf{p}} N^\bullet_2 \rightarrow_{\mathsf{p}} \cdots \rightarrow_{\mathsf{p}} N^\bullet_n = N^\bullet_{\mathsf{p}}$$

where $(N_i)_\bullet =_{\mathsf{p}} (N_{i+1})_\bullet$ by Lemma 16. □

4.2 Complete Reduction

To obtain a reduction strategy with the diamond property for $\rightarrow$, we combine parallel reduction $\Rightarrow_\beta$ with permutative reduction to normal form $\twoheadrightarrow\!\!|_{\mathsf{p}}$ into a notion of *complete reduction* $\Rrightarrow$. We will show that it is diamond (Lemma 19), and that any step in $\rightarrow$ maps onto a complete step of p-normal forms (Lemma 20). Confluence of $\rightarrow$ (Theorem 21) then follows: any two paths $\twoheadrightarrow$ map onto complete paths $\Rrightarrow\!\!\!\!\Rrightarrow$ on p-normal forms, which then converge by the diamond property.

Definition 18. A *complete* reduction step $N \Rrightarrow N_{\bullet\mathsf{p}}$ is a parallel β-step followed by p-reduction to normal form:

$$N \Rrightarrow N_{\bullet\mathsf{p}} \quad := \quad N \Rightarrow_\beta N_\bullet \twoheadrightarrow\!\!|_{\mathsf{p}} N_{\bullet\mathsf{p}} \ .$$

Lemma 19 (Complete reduction is diamond). *If $P \Lleftarrow N \Rrightarrow M$ then for some Q, $P \Rrightarrow Q \Lleftarrow M$.*

Proof. By the following diagram, where $M = N_{\circ\mathsf{p}}$ and $P = N_{\bullet\mathsf{p}}$, and $Q = N_{\circ\bullet\mathsf{p}}$. The square top left is by Lemma 13, top right and bottom left are by Lemma 17, and bottom right is by confluence and strong normalization of p-reduction.

$$\begin{array}{ccccc}
N^{\circ\bullet} & \xRightarrow{\ \beta\ } & N^\bullet_\circ & \xrightarrow{\ \mathsf{p}\ }\!\!| & N^\bullet_{\circ\mathsf{p}} \\
\Downarrow_\beta & & \Downarrow_\beta & & \Downarrow_\beta \\
N^\circ_\bullet & \xRightarrow{\ \beta\ } & N_{\circ\bullet} & =_{\mathsf{p}} & N_{\circ\mathsf{p}\bullet} \\
\downarrow\!\!\downarrow_{\mathsf{p}} & & =_{\mathsf{p}} & & \downarrow\!\!\downarrow_{\mathsf{p}} \\
N^\circ_{\bullet\mathsf{p}} & \xRightarrow{\ \beta\ } & N_{\bullet\mathsf{p}\circ} & \xrightarrow{\ \mathsf{p}\ }\!\!| & N_{\circ\bullet\mathsf{p}}
\end{array}$$

□

Lemma 20 (p-Normalization maps reduction to complete reduction). *If $N \rightarrow M$ then $N_{\mathsf{p}} \Rrightarrow M_{\mathsf{p}}$.*

Proof. For a p-step $N \rightarrow_{\mathsf{p}} M$ we have $N_{\mathsf{p}} = M_{\mathsf{p}}$ while $\Rightarrow_\beta$ is reflexive. For a β-step $N \rightarrow_\beta M$ we label the reduced redex in N to get $N^\bullet \Rightarrow_\beta N_\bullet = M$. Then Lemma 17 gives $N_{\mathsf{p}\bullet} =_{\mathsf{p}} M$, and hence $N_{\mathsf{p}} \Rightarrow_\beta N_{\mathsf{p}\bullet} \twoheadrightarrow\!\!|_{\mathsf{p}} M_{\mathsf{p}}$. □

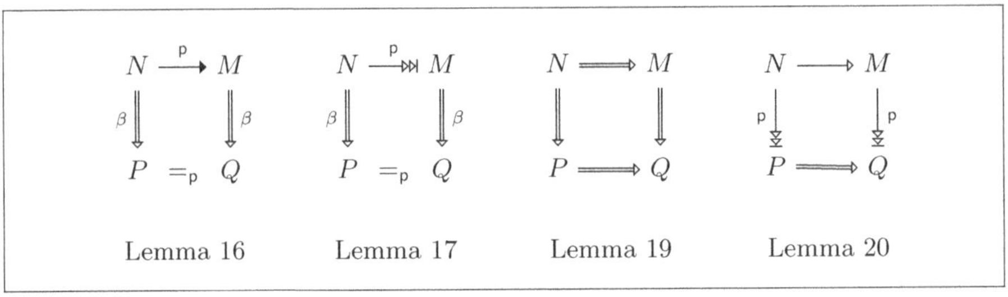

Fig. 3. Diagrams for the Lemmata Leading up to Confluence

Theorem 21. *Reduction $\rightarrow$ is confluent.*

Proof. By the following diagram. For the top and left areas, by Lemma 20 any reduction path $N \twoheadrightarrow M$ maps onto one $N_{\mathsf{p}} \Rightarrow\!\!\!\!\!\Rightarrow M_{\mathsf{p}}$. The main square follows by the diamond property of complete reduction, Lemma 19.

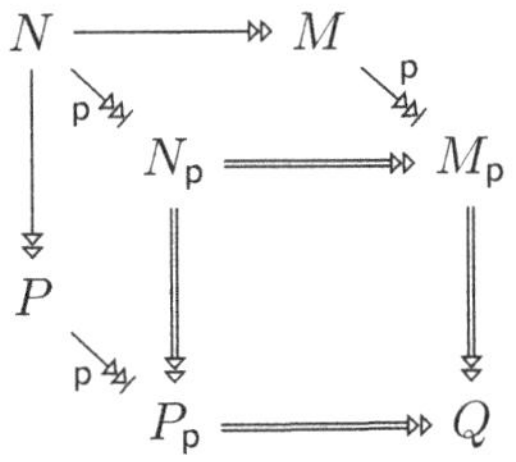

□

5 Strong Normalization for Simply-Typed Terms

In this section, we prove that the relation $\rightarrow$ enjoys strong normalization in *simply typed* terms. Our proof of strong normalization is based on the classic reducibility technique, and inherently has to deal with label-open terms. It thus make great sense to turn the order $<_M$ from Definition 3 into something more formal, at the same time allowing terms to be label-*open*. This is in Figure 4. It is easy to realize that, of course modulo label α-equivalence, for every term M there is at least one θ such that $\theta \vdash_L M$. An easy fact to check is that if $\theta \vdash_L M$ and $M \rightarrow N$, then $\theta \vdash_L N$. It thus makes sense to parametrize $\rightarrow$ on a sequence of labels θ, *i.e.*, one can define a family of reduction relations $\rightarrow^{\theta}$ on pairs in the form (M, θ). The set of strongly normalizable terms, and the number of steps to normal forms become themselves parametric:

- The set SN^{θ} of those terms M such that $\theta \vdash_L M$ and (M, θ) is strongly normalizing modulo $\rightarrow^{\theta}$;
- The function sn^{θ} assigning to any term in SN^{θ} the maximal number of $\rightarrow^{\theta}$ steps to normal form.

Label Sequences: $\theta \quad ::= \quad \varepsilon \mid a \cdot \theta$

Label Judgments: $\xi \quad ::= \quad \theta \vdash_L M$

Label Rules:

$$\frac{}{\theta \vdash_L x} \qquad \frac{\theta \vdash_L M}{\theta \vdash_L \lambda x.M} \qquad \frac{a \cdot \theta \vdash_L M}{\theta \vdash_L \boxed{a}.\, M}$$

$$\frac{\theta \vdash_L M \quad \theta \vdash_L N}{\theta \vdash_L MN} \qquad \frac{\theta \vdash_L M \quad \theta \vdash_L N \quad a \in \theta}{\theta \vdash_L M \overset{a}{\oplus} N}$$

Fig. 4. Labeling Terms

Types: $\tau \quad ::= \quad \alpha \mid \tau \Rightarrow \rho$

Environments: $\Gamma \quad ::= \quad x_1 : \tau_1, \ldots, x_n : \tau_n$

Judgments: $\pi \quad ::= \quad \Gamma \vdash M : \tau$

Typing Rules:

$$\frac{}{\Gamma, x : \tau \vdash x : \tau} \qquad \frac{\Gamma, x : \tau \vdash M : \rho}{\Gamma \vdash \lambda x.M : \tau \Rightarrow \rho} \qquad \frac{\Gamma \vdash M : \tau}{\Gamma \vdash \boxed{a}.\, M : \tau}$$

$$\frac{\Gamma \vdash M : \tau \Rightarrow \rho \quad \Gamma \vdash N : \tau}{\Gamma \vdash MN : \rho} \qquad \frac{\Gamma \vdash M : \tau \quad \Gamma \vdash N : \tau}{\Gamma \vdash M \overset{a}{\oplus} N : \tau}$$

Fig. 5. Types, Environments, Judgments, and Rules

$$\frac{L_1 \in SN^\theta \quad \cdots \quad L_m \in SN^\theta}{x L_1 \ldots L_m \in SN^\theta} \qquad \frac{M L_1 \ldots L_m \in SN^\theta \quad N L_1 \ldots L_m \in SN^\theta \quad a \in \theta}{M \overset{a}{\oplus} N L_1 \ldots L_m \in SN^\theta}$$

$$\frac{M[L_0/x] L_1 \ldots L_m \in SN^\theta \quad L_0 \in SN^\theta}{(\lambda x.\, M) L_0 \ldots L_m \in SN^\theta} \qquad \frac{M L_1 \ldots L_m \in SN^{a \cdot \theta} \quad \forall i. a \notin L_i}{(\boxed{a}.\, M) L_1 \ldots L_m \in SN^\theta}$$

Fig. 6. Closure Rules for Sets SN^θ

We can now define types, environments, judgments, and typing rules in Figure 5.

Please notice that the type structure is precisely the one of the usual, vanilla, simply-typed λ-calculus (although terms are of course different), and we can thus reuse most of the usual proof of strong normalization, for example in the version given by Ralph Loader's notes [21], page 17.

Lemma 22. *The closure rules in Figure 6 are all sound.*

Since the structure of the type system is the one of plain, simple types, the definition of reducibility sets is the classic one:

$$\begin{aligned} Red_\alpha &= \{(\Gamma, \theta, M) \mid M \in SN^\theta \wedge \Gamma \vdash M : \alpha\}; \\ Red_{\tau \Rightarrow \rho} &= \{(\Gamma, \theta, M) \mid (\Gamma \vdash M : \tau \Rightarrow \rho) \wedge (\theta \vdash_L M) \wedge \\ &\qquad \forall (\Gamma\Delta, \theta, N) \in Red_\tau . (\Gamma\Delta, \theta, MN) \in Red_\rho\}. \end{aligned}$$

Before proving that all terms are reducible, we need some auxiliary results.

Lemma 23. *1. If $(\Gamma, \theta, M) \in Red_\tau$, then $M \in SN^\theta$.*
2. If $\Gamma \vdash xL_1 \ldots L_m : \tau$ and $L_1, \ldots, L_m \in SN^\theta$, then $(\Gamma, \theta, xL_1 \ldots L_m) \in Red_\tau$.
3. If $(\Gamma, \theta, M[L_0/x]L_1 \ldots L_m) \in Red_\tau$ with $\Gamma \vdash L_0 : \rho$ and $L_0 \in SN^\theta$, then $(\Gamma, \theta, (\lambda x. M)L_0 \ldots L_m) \in Red_\tau$.
4. If $(\Gamma, \theta, ML_1 \ldots L_m) \in Red_\tau$ with $(\Gamma, \theta, NL_1 \ldots L_m) \in Red_\tau$ and $a \in \theta$, then $(\Gamma, \theta, (M \overset{a}{\oplus} N)L_1 \ldots L_m) \in Red_\tau$.
5. If $(\Gamma, a \cdot \theta, ML_1 \ldots L_m) \in Red_\tau$ and $a \notin L_i$ for all i, then $(\Gamma, \theta, (\boxed{a}. M)L_1 \ldots L_m) \in Red_\tau$.

Proof. The proof is an induction on τ: If τ is an atom α, then Point 1 follows by definition, while points 2 to 5 come from Lemma 22. If τ is $\rho \Rightarrow \mu$, Points 2 to 5 come directly from the induction hypothesis, while Point 1 can be proved by observing that M is in SN^θ if Mx is itself SN^θ, where x is a fresh variable. By induction hypothesis (on Point 2), we can say that $(\Gamma(x : \rho), \theta, x) \in Red_\rho$, and conclude that $(\Gamma(x : \rho), \theta, Mx) \in Red_\mu$. □

The following is the so-called Main Lemma:

Proposition 24. *Suppose $y_1 : \tau_1, \ldots, y_n : \tau_n \vdash M : \rho$ and $\theta \vdash_L M$, with $(\Gamma, \theta, N_j) \in Red_{\tau_j}$ for all $1 \leq j \leq n$. Then $(\Gamma, \theta, M[N_1/y_1, \ldots, N_n/y_n]) \in Red_\rho$.*

Proof. This is an induction on the structure of the term M:
- If M is a variable, necessarily one among $y_1, \ldots, y_n$, then the result is trivial.
- If M is an application LP, then there exists a type ξ such that $y_1 : \tau_1, \ldots, y_n : \tau_n \vdash L : \xi \Rightarrow \rho$ and $y_1 : \tau_1, \ldots, y_n : \tau_n \vdash P : \xi$. Moreover, $\theta \vdash_L L$ and $\theta \vdash_L P$ we can then safely apply the induction hypothesis and conclude that

  $$(\Gamma, \theta, L[\overline{N}/\overline{y}]) \in Red_{\xi \Rightarrow \rho} \qquad (\Gamma, \theta, P[\overline{N}/\overline{y}]) \in Red_\xi \ .$$

  By definition, we get

  $$(\Gamma, \theta, (LP)[\overline{N}/\overline{y}]) \in Red_\rho \ .$$

- If M is an abstraction $\lambda x. L$, then ρ is an arrow type $\xi \Rightarrow \mu$ and $y_1 : \tau_1, \ldots, y_n : \tau_n, x : \xi \vdash L : \mu$. Now, consider any $(\Gamma\Delta, \theta, P) \in Red_\xi$. Our objective is to prove with this hypothesis that $(\Gamma\Delta, \theta, (\lambda x.L[\overline{N}/\overline{y}])P) \in Red_\mu$. By induction hypothesis, since $(\Gamma\Delta, N_i) \in Red_{\tau_i}$, we get that $(\Gamma\Delta, \theta, L[\overline{N}/\overline{y}, P/x]) \in Red_\mu$. The thesis follows from Lemma 23.

- If M is a sum $L \overset{a}{\oplus} P$, we can make use of Lemma 23 and the induction hypothesis, and conclude.
- If M is a generator $\boxed{a}. P$, we can make use of Lemma 23 and the induction hypothesis. We should however observe that $a \cdot \theta \vdash_L P$, since $\theta \vdash_L M$. □

We now have all the ingredients for our proof of strong normalization:

Theorem 25. *If $\Gamma \vdash M : \tau$ and $\theta \vdash_L M$, then $M \in SN^\theta$.*

Proof. Suppose that $x_1 : \rho_1, \ldots, x_n : \rho_n \vdash M : \tau$. Since $x_1 : \rho_1, \ldots, x_n : \rho_n \vdash x_i : \rho_i$ for all i, and clearly $\theta \vdash_L x_i$ for every i, we can apply Lemma 24 and obtain that $(\Gamma, \theta, M[\overline{x}/\overline{x}]) \in Red_\tau$ from which, via Lemma 23, one gets the thesis. □

6 Projective Reduction

Permutative reduction $\rightarrow_{\mathsf{p}}$ evaluates probabilistic sums purely by rewriting. Here we look at a more standard *projective* notion of reduction, which conforms more closely to the intuition that $\boxed{a}$ generates a probabilistic event to determine the choice $\overset{a}{\oplus}$. Using $+$ for an external probabilistic sum, we expect to reduce $\boxed{a}. N$ to $N_0 + N_1$ where each N_i is obtained from N by projecting every subterm $M_0 \overset{a}{\oplus} M_1$ to M_i. The question is, in what context should we admit this reduction? We first limit ourselves to reducing in *head* position.

Definition 26. The *a-projections* $\pi_0^a(N)$ and $\pi_1^a(N)$ are defined as follows:

$$\begin{aligned}
\pi_0^a(N \overset{a}{\oplus} M) &= \pi_0^a(N) & \pi_i^a(\lambda x. N) &= \lambda x. \pi_i^a(N) & \\
\pi_1^a(N \overset{a}{\oplus} M) &= \pi_1^a(M) & \pi_i^a(NM) &= \pi_i^a(N)\, \pi_i^a(M) & \\
\pi_i^a(\boxed{a}. N) &= \boxed{a}. N & \pi_i^a(N \overset{b}{\oplus} M) &= \pi_i^a(N) \overset{b}{\oplus} \pi_i^a(M) & \text{if } a \neq b \\
\pi_i^a(x) &= x & \pi_i^a(\boxed{b}. N) &= \boxed{b}. \pi_i^a(N) & \text{if } a \neq b.
\end{aligned}$$

Definition 27. A *head context* $H[\,]$ is given by the following grammar.

$$H[\,] ::= [\,] \mid \lambda x. H[\,] \mid H[\,]N$$

Definition 28. *Projective head reduction* $\rightarrow_{\pi\mathsf{h}}$ is given by

$$H[\boxed{a}. N] \rightarrow_{\pi\mathsf{h}} H[\pi_0^a(N)] + H[\pi_1^a(N)]\ .$$

We can simulate $\rightarrow_{\pi\mathsf{h}}$ by permutative reduction if we interpret the external sum $+$ by an outermost $\oplus$ (taking special care if the label does not occur).

Proposition 29. *Permutative reduction simulates projective head reduction:*

$$H[\boxed{a}. N] \twoheadrightarrow_{\mathsf{p}} \begin{cases} H[N] & \textit{if } a \notin \mathsf{fl}(N) \\ H[\pi_0^a(N)] \oplus H[\pi_1^a(N)] & \textit{otherwise.} \end{cases}$$

Proof. The case $a \notin \mathsf{fl}(N)$ is immediate by a $\not\boxed{}$ step. For the other case, observe that $H[\boxed{a}.\,N] \twoheadrightarrow_{\mathsf{p}} \boxed{a}.\,H[N]$ by $\Box\lambda$ and $\Box\mathsf{f}$ steps, and since a does not occur in $H[\,]$, that $H[\pi_i^a(N)] = \pi_i^a(H[N])$. By induction on N, if a is minimal in N (*i.e.* $a \in \mathsf{fl}(N)$ and $a \leq b$ for all $b \in \mathsf{fl}(N)$) then $N \twoheadrightarrow_{\mathsf{p}} \pi_0^a(N) \overset{a}{\oplus} \pi_1^a(N)$. As required,

$$H[\boxed{a}.\,N] \twoheadrightarrow_{\mathsf{p}} \boxed{a}.\ H[\pi_0^a(N)] \overset{a}{\oplus} H[\pi_1^a(N)] \quad \text{if } a \in \mathsf{fl}(N)\ . \qquad \Box$$

A gap remains between which generators will not be duplicated, which we *should* be able to reduce, and which generators projective head reduction *does* reduce. In particular, to interpret call-by-value probabilistic reduction in Section 7, we would like to reduce under other generators. However, permutative reduction does not permit exchanging generators, and so only simulates reducing in head position. While (independent) probabilistic events are generally considered interchangeable, it is a question whether the below equivalence is desirable.

$$\boxed{a}.\boxed{b}.\,N \ \overset{?}{\sim}\ \boxed{b}.\boxed{a}.\,N \tag{4}$$

We elide the issue by externalizing probabilistic events, and reducing with reference to a predetermined binary stream $s \in \{0,1\}^{\mathbb{N}}$ representing their outcomes. In this way, we will preserve the intuitions of both permutative and projective reduction: we obtain a qualified version of the equivalence (4) (see (5) below), and will be able to reduce any generator on the *spine* of a term: under (other) generators and choices as well as under abstractions and in function position.

Definition 30. The set of *streams* is $\mathbb{S} = \{0,1\}^{\mathbb{N}}$, ranged over by r, s, t, and $i \cdot s$ denotes a stream with $i \in \{0,1\}$ as first element and s as the remainder.

Definition 31. The *stream labeling* N^s of a term N with a stream $s \in \mathbb{S}$, which annotates generators as $\boxed{a}^i$ with $i \in \{0,1\}$ and variables as x^s with a stream s, is given inductively below. We lift β-reduction to stream-labeled terms by introducing a substitution case for stream-labeled variables: $x^s[M/x] = M^s$.

$$(\lambda x.\,N)^s = \lambda x.\,N^s \qquad\qquad (\boxed{a}.\,N)^{i\cdot s} = \boxed{a}^i.\,N^s$$

$$(N\,M)^s = N^s\,M \qquad\qquad (N \overset{a}{\oplus} M)^s = N^s \overset{a}{\oplus} M^s$$

Definition 32. *Projective reduction* $\rightarrow_\pi$ on stream-labeled terms is the rewrite relation given by

$$\boxed{a}^i.\,N \ \rightarrow_\pi\ \pi_i^a(N)\ .$$

Observe that in N^s a generator that occurs under n other generators on the spine of N, is labeled with the element of s at position $n+1$. Generators in argument position remain unlabeled, until a β-step places them on the spine, in which case they become labeled by the new substitution case. We allow to annotate a term with a finite prefix of a stream, *e.g.* N^i with a singleton i, so that only part of the spine is labeled. Subsequent labeling of a partly labeled term is then by $(N^r)^s = N^{r\cdot s}$ (abusing notation). To introduce streams via the external probabilistic sum, and to ignore an unused remaining stream after completing a probabilistic computation, we adopt the following equation.

$$N = N^0 + N^1$$

Proposition 33. *Projective reduction generalizes projective head reduction:*

$$H[\boxed{a}.N] \;=\; H[\boxed{a}^0.N] + H[\boxed{a}^1.N] \;\to_\pi\; H[\pi_0^a(N)] + H[\pi_1^a(N)] \;.$$

Returning to the interchangeability of probabilistic events, we refine (4) by exchanging the corresponding elements of the annotating streams:

$$
\begin{array}{ccccc}
(\boxed{a}.\boxed{b}.N)^{i\cdot j\cdot s} & = & \boxed{a}^i.\boxed{b}^j.N^s & \xrightarrow{\pi} & \pi_i^a(\pi_j^b(N^s)) \\
\sim & & & & = \\
(\boxed{b}.\boxed{a}.N)^{j\cdot i\cdot s} & = & \boxed{b}^j.\boxed{a}^i.N^s & \xrightarrow{\pi} & \pi_j^b(\pi_i^a(N^s))
\end{array}
\tag{5}
$$

Stream-labeling externalizes all probabilities, making reduction deterministic. This is expressed by the following proposition, that stream-labeling commutes with reduction: if a generator remains unlabeled in M and becomes labeled after a reduction step $M \to N$, what label it receives is predetermined. The deep reason is that stream labeling assigns an outcome to each generator in a way that corresponds to a call-by-name strategy for probabilistic reduction.

Proposition 34. *If $M \to N$ by a step other than $\not{\boxtimes}$ then $M^s \to N^s$.*

Remark 35. The statement is false for the $\not{\boxtimes}$ rule $\boxed{a}.N \to_\mathsf{p} N$ $(a \notin \mathsf{fl}(N))$, as it removes a generator but not an element from the stream. Arguably, for this reason the rule should be excluded from the calculus. On the other hand, the rule is necessary to implement idempotence of $\oplus$, rather than just $\overset{a}{\oplus}$, as follows.

$$N \oplus N \;=\; \boxed{a}.N \overset{a}{\oplus} N \;\to_\mathsf{p}\; \boxed{a}.N \;\to_\mathsf{p}\; N \qquad \text{where } a \notin \mathsf{fl}(N)$$

The below proposition then expresses that projective reduction is an *invariant* for permutative reduction. If $N \to_\mathsf{p} M$ by a step (that is not $\not{\boxtimes}$) on a labeled generator $\boxed{a}^i$ or a corresponding choice $\overset{a}{\oplus}$, then N and M reduce to a common term, $N \to_\pi P \;_\pi\!\!\leftarrow M$, by the projective steps evaluating $\boxed{a}^i$.

Proposition 36. *Projective reduction is an invariant for permutative reduction, as follows (with a case for c_2 symmetric to c_1, and where $D[\,]$ is a context).*

$$
\boxed{a}^i.C[N \overset{a}{\oplus} N] \xrightarrow{\;\mathsf{p}\;} \boxed{a}^i.C[N]
\qquad\qquad
\boxed{a}^i.C[(N_0 \overset{a}{\oplus} M) \overset{a}{\oplus} N_1] \xrightarrow{\;\mathsf{p}\;} \boxed{a}^i.C[N_0 \overset{a}{\oplus} N_1]
$$

$$
\begin{array}{ccc}
 & \searrow_\pi \quad i \quad \swarrow_\pi & \\
 & \pi_i^a(C[N]) &
\end{array}
\qquad\qquad
\begin{array}{ccc}
 & \searrow_\pi \quad \mathsf{c}_1 \quad \swarrow_\pi & \\
 & \pi_i^a(C[N_i]) &
\end{array}
$$

$$
\boxed{a}^i.C[D[N_0 \overset{a}{\oplus} N_1]] \xrightarrow{\;\mathsf{p}\;} \boxed{a}^i.C[D[N_0] \overset{a}{\oplus} D[N_1]]
$$

$$
\begin{array}{ccc}
 & \searrow_\pi \quad \oplus\star \quad \swarrow_\pi & \\
 & \pi_i^a(C[D[N_i]]) &
\end{array}
$$

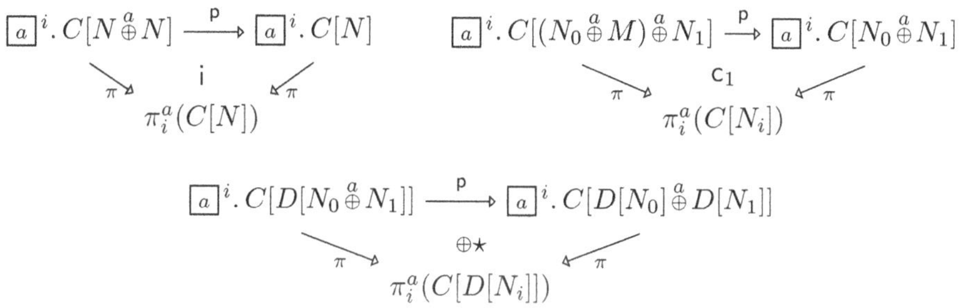

$$\begin{array}{ccc} \lambda x.\,\boxed{a}^{i}.\,N & \xrightarrow{\;\mathsf{p}\;} & \boxed{a}^{i}.\,\lambda x.\,N \\ \pi\downarrow & \square\lambda & \downarrow\pi \\ \lambda x.\,\pi_i^a(N) & = & \pi_i^a(\lambda x.\,N) \end{array} \qquad \begin{array}{ccc} (\boxed{a}^{i}.\,N)M & \xrightarrow{\;\mathsf{p}\;} & \boxed{a}^{i}.\,NM \\ \pi\downarrow & \square\mathsf{f} & \downarrow\pi \\ \pi_i^a(N)\,M & = & \pi_i^a(N\,M) \end{array}$$

7 Call-by-value Interpretation

We consider the interpretation of a call-by-value probabilistic λ-calculus. For simplicity we will allow duplicating (or deleting) β-redexes, and only restrict duplicating probabilities; our *values* V are then just deterministic—*i.e.* without choices—terms, possibly applications and not necessarily β-normal (so that our $\twoheadrightarrow_{\beta\mathsf{v}}$ is actually β-reduction on deterministic terms, unlike [9]). We evaluate the internal probabilistic choice $\oplus_{\mathsf{v}}$ to an external probabilistic choice $+$.

$$N ::= x \mid \lambda x.N \mid MN \mid M \oplus_{\mathsf{v}} N \qquad\qquad (\lambda x.N)V \twoheadrightarrow_{\beta\mathsf{v}} N[V/x]$$

$$V, W ::= x \mid \lambda x.V \mid VW \qquad\qquad M \oplus_{\mathsf{v}} N \twoheadrightarrow_{\mathsf{v}} M + N$$

The interpretation $[\![N]\!]_{\mathsf{v}}$ of a call-by-value term N into Λ_{PE} is given as follows. First, we translate N to a label-open term $[\![N]\!]_{\mathsf{open}} = \theta \vdash_L P$ by replacing each choice $\oplus_{\mathsf{v}}$ with one $\overset{a}{\oplus}$ with a unique label, where the label-context θ collects the labels used. Then $[\![N]\!]_{\mathsf{v}}$ is the *label closure* $[\![N]\!]_{\mathsf{v}} = \lfloor \theta \vdash_L P \rfloor$, which prefixes P with a generator $\boxed{a}$ for every a in θ.

Definition 37. (Call-by-value interpretation) The *open interpretation* $[\![N]\!]_{\mathsf{open}}$ of a call-by-value term N is as follows, where all labels are fresh, and inductively $[\![N_i]\!]_{\mathsf{open}} = \theta_i \vdash_L P_i$ for $i \in \{1, 2\}$.

$$\begin{array}{rclrcl} [\![x]\!]_{\mathsf{open}} & = & \vdash_L x & [\![N_1 N_2]\!]_{\mathsf{open}} & = & \theta_2 \cdot \theta_1 \vdash_L P_1 P_2 \\ [\![\lambda x.N_1]\!]_{\mathsf{open}} & = & \theta_1 \vdash_L \lambda x.P_1 & [\![N_1 \oplus_{\mathsf{v}} N_2]\!]_{\mathsf{open}} & = & \theta_2 \cdot \theta_1 \cdot a \vdash_L P_1 \overset{a}{\oplus} P_2 \end{array}$$

The *label closure* $\lfloor \theta \vdash_L P \rfloor$ is given inductively as follows.

$$\lfloor \vdash_L P \rfloor = P \qquad \lfloor a \cdot \theta \vdash_L P \rfloor = \lfloor \theta \vdash_L \boxed{a}.\,P \rfloor$$

The *call-by-value interpretation* of N is $[\![N]\!]_{\mathsf{v}} = \lfloor [\![N]\!]_{\mathsf{open}} \rfloor$.

Our call-by-value reduction may choose an arbitrary order in which to evaluate the choices $\oplus_{\mathsf{v}}$ in a term N, but the order of generators in the interpretation $[\![N]\!]_{\mathsf{v}}$ is necessarily fixed. Then to simulate a call-by-value reduction, we cannot choose a fixed context stream a priori; all we can say is that for every reduction, there is some stream that allows us to simulate it. Specifically, a reduction step $C[N_0 \oplus_{\mathsf{v}} N_1] \twoheadrightarrow_{\mathsf{v}} C[N_j]$ where $C[\,]$ is a call-by-value term context is simulated by the following projective step.

$$\ldots \boxed{a}^{i}.\,\boxed{b}^{j}.\,\boxed{c}^{k} \ldots D[P_0 \overset{b}{\oplus} P_1] \twoheadrightarrow_{\pi} \ldots \boxed{a}^{i}.\,\boxed{c}^{k} \ldots D[P_j]$$

Here, $[\![C[N_0 \oplus_{\mathsf{v}} N_1]]\!]_{\mathsf{open}} = \theta \vdash_L D[P_0 \overset{b}{\oplus} P_1]$ with $D[\,]$ a Λ_{PE}-context, and θ giving rise to the sequence of generators $\ldots \boxed{a}.\boxed{b}.\boxed{c}\ldots$ in the call-by-value translation. To simulate the reduction step, if b occupies the n-th position in θ, then the n-th position in the context stream s must be the element j. Since β-reduction survives the translation and labeling process intact, we may simulate call-by-value probabilistic reduction by projective and β-reduction.

Theorem 38. *If* $N \twoheadrightarrow_{\mathsf{v},\beta\mathsf{v}} V$ *then* $[\![N]\!]^s_{\mathsf{v}} \twoheadrightarrow_{\pi,\beta} [\![V]\!]_{\mathsf{v}}$ *for some stream* $s \in \mathbb{S}$.

8 Conclusions and Future Work

We believe our decomposition of probabilistic choice in λ-calculus to be an elegant and compelling way of restoring confluence, one of the core properties of the λ-calculus. Our probabilistic event λ-calculus captures traditional call-by-name and call-by-value probabilistic reduction, and offers finer control beyond those strategies. Permutative reduction implements a natural and fine-grained equivalence on probabilistic terms as internal rewriting, while projective reduction provides a complementary and more traditional external perspective.

There are a few immediate areas for future work. Firstly, within probabilistic λ-calculus, it is worth exploring if our decomposition opens up new avenues in semantics. Secondly, our approach might apply to probabilistic reasoning more widely, outside the λ-calculus. Most importantly, we will explore if our approach can be extended to other computational effects. Our use of streams interprets probabilistic choice as a *read* operation from an external source, which means other read operations can be treated similarly. A complementary treatment of *write* operations would allow us to express a considerable range of effects, including input/output and state.

Acknowledgments

This work was supported by EPSRC Project EP/R029121/1 *Typed Lambda-Calculi with Sharing and Unsharing.* The first author is partially supported by the ANR project 19CE480014 PPS, the ERC Consolidator Grant 818616 DIAPASoN, and the MIUR PRIN 201784YSZ5 ASPRA. We thank the referees for their diligence and their helpful comments. We are grateful to Chris Barrett and—indirectly—Anupam Das for pointing us to Zantema and Van de Pol's work [27].

References

1. Avanzini, M., Dal Lago, U., Ghyselen, A.: Type-based complexity analysis of probabilistic functional programs. In: 34th Annual ACM/IEEE Symposium on Logic in Computer Science, LICS 2019. pp. 1–13. IEEE Computer Society (2019). https://doi.org/10.1109/LICS.2019.8785725
2. Barendregt, H.P.: The Lambda Calculus – Its Syntax and Semantics, Studies in logic and the foundations of mathematics, vol. 103. North-Holland (1984)

3. Borgström, J., Dal Lago, U., Gordon, A.D., Szymczak, M.: A lambda-calculus foundation for universal probabilistic programming. In: 21st ACM SIGPLAN International Conference on Functional Programming, ICFP 2016. pp. 33–46. ACM (2016). https://doi.org/10.1145/2951913.2951942
4. Breuvart, F., Dal Lago, U.: On intersection types and probabilistic lambda calculi. In: roceedings of the 20th International Symposium on Principles and Practice of Declarative Programming, PPDP 2018. pp. 8:1–8:13. ACM (2018). https://doi.org/10.1145/3236950.3236968
5. Dal Lago, U., Faggian, C., Valiron, B., Yoshimizu, A.: The geometry of parallelism: classical, probabilistic, and quantum effects. In: Proceedings of the 44th ACM SIGPLAN Symposium on Principles of Programming Languages, POPL 2017. pp. 833–845. ACM (2017). https://doi.org/10.1145/3009837
6. Dal Lago, U., Grellois, C.: Probabilistic termination by monadic affine sized typing. ACM Transactions on Programming Languages and Systems **41**(2), 10:1–10:65 (2019). https://doi.org/10.1145/3293605
7. Dal Lago, U., Guerrieri, G., Heijltjes, W.: Decomposing probabilistic lambda-calculi (long version) (2020), https://arxiv.org/abs/2002.08392
8. Dal Lago, U., Sangiorgi, D., Alberti, M.: On coinductive equivalences for higher-order probabilistic functional programs. In: The 41st Annual ACM SIGPLAN-SIGACT Symposium on Principles of Programming Languages, POPL '14. pp. 297–308. ACM (2014). https://doi.org/10.1145/2535838.2535872
9. Dal Lago, U., Zorzi, M.: Probabilistic operational semantics for the lambda calculus. RAIRO - Theoretical Informatics and Applications **46**(3), 413–450 (2012). https://doi.org/10.1051/ita/2012012
10. Danos, V., Ehrhard, T.: Probabilistic coherence spaces as a model of higher-order probabilistic computation. Information and Compututation **209**(6), 966–991 (2011). https://doi.org/10.1016/j.ic.2011.02.001
11. de'Liguoro, U., Piperno, A.: Non deterministic extensions of untyped lambda-calculus. Information and Computation **122**(2), 149–177 (1995). https://doi.org/10.1006/inco.1995.1145
12. Dershowitz, N.: Orderings for term-rewriting systems. Theoretical Computer Science **17**, 279–301 (1982). https://doi.org/10.1016/0304-3975(82)90026-3
13. Ehrhard, T., Pagani, M., Tasson, C.: Full abstraction for probabilistic PCF. Journal of the ACM **65**(4), 23:1–23:44 (2018). https://doi.org/10.1145/3164540
14. Ehrhard, T., Tasson, C.: Probabilistic call by push value. Logical Methods in Computer Science **15**(1) (2019). https://doi.org/10.23638/LMCS-15(1:3)2019
15. Faggian, C., Ronchi Della Rocca, S.: Lambda calculus and probabilistic computation. In: 34th Annual ACM/IEEE Symposium on Logic in Computer Science, LICS 2019. pp. 1–13. IEEE Computer Society (2019). https://doi.org/10.1109/LICS.2019.8785699
16. Goubault-Larrecq, J.: A probabilistic and non-deterministic call-by-push-value language. In: 34th Annual ACM/IEEE Symposium on Logic in Computer Science, LICS 2019. pp. 1–13. IEEE Computer Society (2019). https://doi.org/10.1109/LICS.2019.8785809
17. Jones, C., Plotkin, G.D.: A probabilistic powerdomain of evaluations. In: Proceedings of the Fourth Annual Symposium on Logic in Computer Science (LICS '89). pp. 186–195. IEEE Computer Society (1989). https://doi.org/10.1109/LICS.1989.39173
18. Jung, A., Tix, R.: The troublesome probabilistic powerdomain. Electronic Notes in Theoretical Computer Science **13**, 70–91 (1998). https://doi.org/10.1016/S1571-0661(05)80216-6

19. Leventis, T.: Probabilistic Böhm trees and probabilistic separation. In: Proceedings of the 33rd Annual ACM/IEEE Symposium on Logic in Computer Science, LICS 2018. pp. 649–658. IEEE Computer Society (2018). https://doi.org/10.1145/3209108.3209126
20. Leventis, T.: A deterministic rewrite system for the probabilistic λ-calculus. Mathematical Structures in Computer Science **29**(10), 1479–1512 (2019). https://doi.org/10.1017/S0960129519000045
21. Loader, R.: Notes on simply typed lambda calculus. Reports of the laboratory for foundations of computer science ECS-LFCS-98-381, University of Edinburgh, Edinburgh (1998), http://www.lfcs.inf.ed.ac.uk/reports/98/ECS-LFCS-98-381/
22. Manber, U., Tompa, M.: Probabilistic, nondeterministic, and alternating decision trees. In: 14th Annual ACM Symposium on Theory of Computing. pp. 234–244 (1982). https://doi.org/10.1145/800070.802197
23. Ramsey, N., Pfeffer, A.: Stochastic lambda calculus and monads of probability distributions. In: Conference Record of POPL 2002: The 29th SIGPLAN-SIGACT Symposium on Principles of Programming Languages. pp. 154–165. POPL '02 (2002). https://doi.org/10.1145/503272.503288
24. Saheb-Djahromi, N.: Probabilistic LCF. In: Mathematical Foundations of Computer Science 1978, Proceedings, 7th Symposium. Lecture Notes in Computer Science, vol. 64, pp. 442–451. Springer (1978). https://doi.org/10.1007/3-540-08921-7_92
25. Sangiorgi, D., Vignudelli, V.: Environmental bisimulations for probabilistic higher-order languages. In: Proceedings of the 43rd Annual ACM SIGPLAN-SIGACT Symposium on Principles of Programming Languages, POPL 2016. pp. 595–607 (2016). https://doi.org/10.1145/2837614.2837651
26. Takahashi, M.: Parallel reductions in lambda-calculus. Information and Computation **118**(1), 120–127 (1995). https://doi.org/10.1006/inco.1995.1057
27. Zantema, H., van de Pol, J.: A rewriting approach to binary decision diagrams. The Journal of Logic and Algebraic Programming **49**(1-2), 61–86 (2001). https://doi.org/10.1016/S1567-8326(01)00013-3

On the k-synchronizability of Systems

Cinzia Di Giusto (✉), Laetitia Laversa, and Etienne Lozes

Université Côte d'Azur, CNRS, I3S, Sophia Antipolis, France
{cinzia.di-giusto,laetitia.laversa,etienne.lozes}@univ-cotedazur.fr

Abstract. We study k-synchronizability: a system is k-synchronizable if any of its executions, up to reordering causally independent actions, can be divided into a succession of k-bounded interaction phases. We show two results (both for mailbox and peer-to-peer automata): first, the reachability problem is decidable for k-synchronizable systems; second, the membership problem (whether a given system is k-synchronizable) is decidable as well. Our proofs fix several important issues in previous attempts to prove these two results for mailbox automata.

Keywords: Verification · Communicating Automata · A/Synchronous communication.

1 Introduction

Asynchronous message-passing is ubiquitous in communication-centric systems; these include high-performance computing, distributed memory management, event-driven programming, or web services orchestration. One of the parameters that play an important role in these systems is whether the number of pending sent messages can be bounded in a predictable fashion, or whether the buffering capacity offered by the communication layer should be unlimited. Clearly, when considering implementation, testing, or verification, bounded asynchrony is preferred over unbounded asynchrony. Indeed, for bounded systems, reachability analysis and invariants inference can be solved by regular model-checking [5]. Unfortunately and even if designing a new system in this setting is easier, this is not the case when considering that the buffering capacity is unbounded, or that the bound is not known a priori . Thus, a question that arises naturally is how can we bound the "behaviour" of a system so that it operates as one with unbounded buffers? In a recent work [4], Bouajjani *et al.* introduced the notion of k-synchronizable system of finite state machines communicating through mailboxes and showed that the reachability problem is decidable for such systems. Intuitively, a system is k-synchronizable if any of its executions, up to reordering causally independent actions, can be chopped into a succession of k-bounded interaction phases. Each of these phases starts with at most k send actions that are followed by at most k receptions. Notice that, a system may be k-synchronizable even if some of its executions require buffers of unbounded capacity.

As explained in the present paper, this result, although valid, is surprisingly non-trivial, mostly due to complications introduced by the mailbox semantics of

J. Goubault-Larrecq and B. König (Eds.): FOSSACS 2020, LNCS 12077, pp. 157–176, 2020.
https://doi.org/10.1007/978-3-030-45231-5_9

communications. Some of these complications were missed by Bouajjani *et al.* and the algorithm for the reachability problem in [4] suffers from false positives. Another problem is the membership problem for the subclass of k-synchronizable systems: for a given k and a given system of communicating finite state machines, is this system k-synchronizable? The main result in [4] is that this problem is decidable. However, again, the proof of this result contains an important flaw at the very first step that breaks all subsequent developments; as a consequence, the algorithm given in [4] produces both false positives and false negatives.

In this work, we present a new proof of the decidability of the reachability problem together with a new proof of the decidability of the membership problem. Quite surprisingly, the reachability problem is more demanding in terms of causality analysis, whereas the membership problem, although rather intricate, builds on a simpler dependency analysis. We also extend both decidability results to the case of peer-to-peer communication.

Outline. Next section recalls the definition of communicating systems and related notions. In Section 3 we introduce k-synchronizability and we give a graphical characterisation of this property. This characterisation corrects Theorem 1 in [4] and highlights the flaw in the proof of the membership problem. Next, in Section 4, we establish the decidability of the reachability problem, which is the core of our contribution and departs considerably from [4]. In Section 5, we show the decidability of the membership problem. Section 6 extends previous results to the peer-to-peer setting. Finally Section 7 concludes the paper discussing other related works. Proofs and some additional material are available at https://hal.archives-ouvertes.fr/hal-02272347.

2 Preliminaries

A communicating system is a set of finite state machines that exchange messages: automata have transitions labelled with either send or receive actions. The paper mainly considers as communication architecture, mailboxes: i.e., messages await to be received in FIFO buffers that store all messages sent to a same automaton, regardless of their senders. Section 6, instead, treats peer-to-peer systems, their introduction is therefore delayed to that point.

Let $\mathbb{V}$ be a finite set of messages and $\mathbb{P}$ a finite set of processes. A send action, denoted $send(p, q, \mathbf{v})$, designates the sending of message $\mathbf{v}$ from process p to process q. Similarly a receive action $rec(p, q, \mathbf{v})$ expresses that process q is receiving message $\mathbf{v}$ from p. We write a to denote a send or receive action. Let $S = \{send(p, q, \mathbf{v}) \mid p, q \in \mathbb{P}, \mathbf{v} \in \mathbb{V}\}$ be the set of send actions and $R = \{rec(p, q, \mathbf{v}) \mid p, q \in \mathbb{P}, \mathbf{v} \in \mathbb{V}\}$ the set of receive actions. S_p and R_p stand for the set of sends and receives of process p respectively. Each process is encoded by an automaton and by abuse of notation we say that a *system* is the parallel composition of processes.

Definition 1 (System). *A system is a tuple* $\mathfrak{S} = \big((L_p, \delta_p, l_p^0) \mid p \in \mathbb{P}\big)$ *where, for each process* p, L_p *is a finite set of local control states,* $\delta_p \subseteq (L_p \times (S_p \cup R_p) \times L_p)$ *is the transition relation (also denoted* $l \xrightarrow{a}_p l'$*) and* l_p^0 *is the initial state.*

Definition 2 (Configuration). *Let* $\mathfrak{S} = ((L_p, \delta_p, l_p^0) \mid p \in \mathbb{P})$*, a configuration is a pair* $(\vec{l}, \texttt{Buf})$ *where* $\vec{l} = (l_p)_{p \in \mathbb{P}} \in \Pi_{p \in \mathbb{P}} L_p$ *is a global control state of* $\mathfrak{S}$ *(a local control state for each automaton), and* $\texttt{Buf} = (b_p)_{p \in \mathbb{P}} \in (\mathbb{V}^*)^{\mathbb{P}}$ *is a vector of buffers, each* b_p *being a word over* $\mathbb{V}$*.*

We write $\vec{l_0}$ to denote the vector of initial states of all processes $p \in \mathbb{P}$, and $\texttt{Buf}_0$ stands for the vector of empty buffers. The semantics of a system is defined by the two rules below.

[SEND]

$$\frac{l_p \xrightarrow{send(p,q,\mathbf{v})}_p l'_p \quad b'_q = b_q \cdot \mathbf{v}}{(\vec{l}, \texttt{Buf}) \xrightarrow{send(p,q,\mathbf{v})} (\vec{l}[l'_p/l_p], \texttt{Buf}[b'_q/b_q])}$$

[RECEIVE]

$$\frac{l_q \xrightarrow{rec(p,q,\mathbf{v})}_q l'_q \quad b_q = \mathbf{v} \cdot b'_q}{(\vec{l}, \texttt{Buf}) \xrightarrow{rec(p,q,\mathbf{v})} (\vec{l}[l'_q/l_q], \texttt{Buf}[b'_q/b_q])}$$

A send action adds a message in the buffer b of the receiver, and a receive action pops the message from this buffer. An execution $e = a_1 \cdots a_n$ is a sequence of actions in $S \cup R$ such that $(\vec{l_0}, \texttt{Buf}_0) \xrightarrow{a_1} \cdots \xrightarrow{a_n} (\vec{l}, \texttt{Buf})$ for some $\vec{l}$ and $\texttt{Buf}$. As usual $\stackrel{e}{\Rightarrow}$ stands for $\xrightarrow{a_1} \cdots \xrightarrow{a_n}$. We write $asEx(\mathfrak{S})$ to denote the set of asynchronous executions of a system $\mathfrak{S}$. In a sequence of actions $e = a_1 \cdots a_n$, a send action $a_i = send(p, q, \mathbf{v})$ is *matched* by a reception $a_j = rec(p', q', \mathbf{v}')$ (denoted by $a_i \vdash\!\dashv a_j$) if $i < j$, $p = p'$, $q = q'$, $\mathbf{v} = \mathbf{v}'$, and there is $\ell \geq 1$ such that a_i and a_j are the ℓth actions of e with these properties respectively. A send action a_i is *unmatched* if there is no matching reception in e. A *message exchange* of a sequence of actions e is a set either of the form $v = \{a_i, a_j\}$ with $a_i \vdash\!\dashv a_j$ or of the form $v = \{a_i\}$ with a_i unmatched. For a message $\mathbf{v}_i$, we will note v_i the corresponding message exchange. When v is either an unmatched $send(p, q, \mathbf{v})$ or a pair of matched actions $\{send(p, q, \mathbf{v}), rec(p, q, \mathbf{v})\}$, we write $\mathsf{proc}_S(v)$ for p and $\mathsf{proc}_R(v)$ for q. Note that $\mathsf{proc}_R(v)$ is defined even if v is unmatched. Finally, we write $\mathsf{procs}(v)$ for $\{p\}$ in the case of an unmatched send and $\{p, q\}$ in the case of a matched send.

An execution imposes a total order on the actions. We are interested in stressing the causal dependencies between messages. We thus make use of message sequence charts (MSCs) that only impose an order between matched pairs of actions and between the actions of a same process. Informally, an MSC will be depicted with vertical timelines (one for each process) where time goes from top to bottom, that carry some events (points) representing send and receive actions of this process (see Fig. 1). An arc is drawn between two matched events. We will also draw a dashed arc to depict an unmatched send event. An MSC is, thus, a partially ordered set of events, each corresponding to a send or receive action.

Definition 3 (MSC). *A message sequence chart is a tuple* $(Ev, \lambda, \prec)$*, where*

- *Ev is a finite set of events,*
- *$\lambda : Ev \to S \cup R$ tags each event with an action,*
- *$\prec = (\prec_{po} \cup \prec_{src})^+$ is the transitive closure of $\prec_{po}$ and $\prec_{src}$ where:*
    - *$\prec_{po}$ is a partial order on Ev such that, for all process p, $\prec_{po}$ induces a total order on the set of events of process p, i.e., on $\lambda^{-1}(S_p \cup R_p)$*

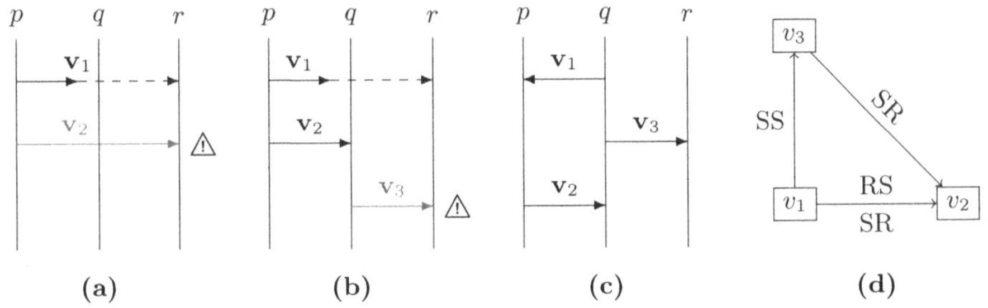

Fig. 1: (a) and (b): two MSCs that violate causal delivery. (c) and (d): an MSC and its conflict graph

- $\prec_{src}$ *is a binary relation that relates each receive event to its preceding send event :*
  - $*$ *for all events* $r \in \lambda^{-1}(R)$*, there is exactly one events* s *such that* $s \prec_{src} r$
  - $*$ *for all events* $s \in \lambda^{-1}(S)$*, there is at most one event* r *such that* $s \prec_{src} r$
  - $*$ *for any two events* s, r *such that* $s \prec_{src} r$*, there are* $p, q, \mathbf{v}$ *such that* $\lambda(s) = send(p, q, \mathbf{v})$ *and* $\lambda(r) = rec(p, q, \mathbf{v})$*.*

We identify MSCs up to graph isomorphism (i.e., we view an MSC as a labeled graph). For a given *well-formed* (i.e., each reception is matched) sequence of actions $e = a_1 \dots a_n$, we let $msc(e)$ be the MSC where $Ev = [1..n]$, $\prec_{po}$ is the set of pairs of indices (i, j) such that $i < j$ and $\{a_i, a_j\} \subseteq S_p \cup R_p$ for some $p \in \mathbb{P}$ (i.e., a_i and a_j are actions of a same process), and $\prec_{src}$ is the set of pairs of indices (i, j) such that $a_i \vdash\!\dashv a_j$. We say that $e = a_1 \dots a_n$ is a *linearisation* of $msc(e)$, and we write $asTr(\mathfrak{S})$ to denote $\{msc(e) \mid e \in asEx(\mathfrak{S})\}$ the set of MSCs of system $\mathfrak{S}$.

Mailbox communication imposes a number of constraints on what and when messages can be read. The precise definition is given below, we now discuss some of the possible scenarios. For instance: if two messages are sent to a same process, they will be received in the same order as they have been sent. As another example, unmatched messages also impose some constraints: if a process p sends an unmatched message to r, it will not be able to send matched messages to r afterwards (Fig. 1a); or similarly, if a process p sends an unmatched message to r, any process q that receives subsequent messages from p will not be able to send matched messages to r afterwards (Fig. 1b). When an MSC satisfies the constraint imposed by mailbox communication, we say that it satisfies causal delivery. Notice that, by construction, all executions satisfy causal delivery.

Definition 4 (Causal Delivery). *Let* $(Ev, \lambda, \prec)$ *be an MSC. We say that it satisfies causal delivery if the MSC has a linearisation* $e = a_1 \dots a_n$ *such that for any two events* $i \prec j$ *such that* $a_i = send(p, q, \mathbf{v})$ *and* $a_j = send(p', q, \mathbf{v}')$*, either* a_j *is unmatched, or there are* i', j' *such that* $a_i \vdash\!\dashv a_{i'}$*,* $a_j \vdash\!\dashv a_{j'}$*, and* $i' \prec j'$*.*

Our definition enforces the following intuitive property.

Proposition 1. *An MSC msc satisfies causal delivery if and only if there is a system* $\mathfrak{S}$ *and an execution* $e \in asEx(\mathfrak{S})$ *such that* $msc = msc(e)$.

We now recall from [4] the definition of *conflict graph* depicting the causal dependencies between message exchanges. Intuitively, we have a dependency whenever two messages have a process in common. For instance an $\xrightarrow{SS}$ dependency between message exchanges v and v' expresses the fact that v' has been sent after v, by the same process.

Definition 5 (Conflict Graph). *The conflict graph* $\mathsf{CG}(e)$ *of a sequence of actions* $e = a_1 \cdots a_n$ *is the labeled graph* $(V, \{\xrightarrow{XY}\}_{X,Y \in \{R,S\}})$ *where* V *is the set of message exchanges of* e, *and for all* $X, Y \in \{S, R\}$, *for all* $v, v' \in V$, *there is a* XY *dependency edge* $v \xrightarrow{XY} v'$ *between* v *and* v' *if there are* $i < j$ *such that* $\{a_i\} = v \cap X$, $\{a_j\} = v' \cap Y$, *and* $\mathsf{proc}_X(v) = \mathsf{proc}_Y(v')$.

Notice that each linearisation e of an MSC will have the same conflict graph. We can thus talk about an MSC and the associated conflict graph. (As an example see Figs. 1c and 1d.)

We write $v \to v'$ if $v \xrightarrow{XY} v'$ for some $X, Y \in \{R, S\}$, and $v \to^* v'$ if there is a (possibly empty) path from v to v'.

3 k-synchronizable Systems

In this section, we define k-synchronizable systems. The main contribution of this part is a new characterisation of k-synchronizable executions that corrects the one given in [4].

In the rest of the paper, k denotes a given integer $k \geq 1$. A k-exchange denotes a sequence of actions starting with at most k sends and followed by at most k receives matching some of the sends. An MSC is *k-synchronous* if there exists a linearisation that is breakable into a sequence of *k-exchanges*, such that a message sent during a k-exchange cannot be received during a subsequent one: either it is received during the same k-exchange, or it remains orphan forever.

Definition 6 (k-synchronous). *An MSC msc is k-synchronous if:*

1. *there exists a linearisation of msc* $e = e_1 \cdot e_2 \cdots e_n$ *where for all* $i \in [1..n]$, $e_i \in S^{\leq k} \cdot R^{\leq k}$,
2. *msc satisfies causal delivery,*
3. *for all* j, j' *such that* $a_j \mapsto a_{j'}$ *holds in* e, $a_j \mapsto a_{j'}$ *holds in some* e_i.

An execution e is k-synchronizable if msc(e) is k-synchronous.

We write $sTr_k(\mathfrak{S})$ to denote the set $\{msc(e) \mid e \in asEx(\mathfrak{S})$ and $msc(e)$ is k-synchronous$\}$.

Example 1 (k-synchronous MSCs and k-synchronizable Executions).

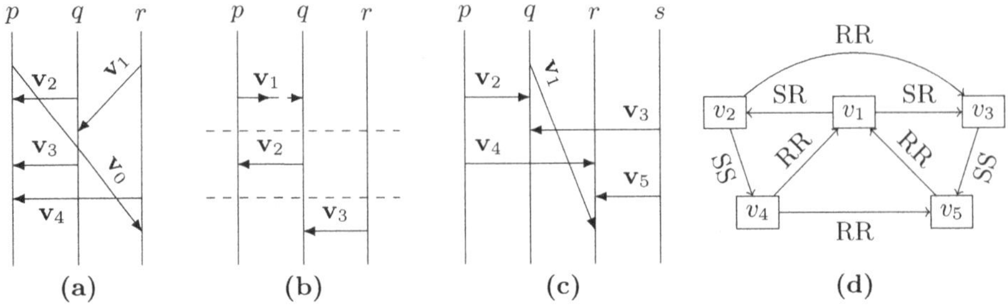

Fig. 2: (a) the MSC of Example 1.1. (b) the MSC of Example 1.2. (c) the MSC of Example 2 and (d) its conflict graph.

1. There is no k such that the MSC in Fig. 2a is k-synchronous. All messages must be grouped in the same k-exchange, but it is not possible to schedule all the sends first, because the reception of $\mathbf{v}_1$ happens before the sending of $\mathbf{v}_3$. Still, this MSC satisfies causal delivery.
2. Let $e_1 = send(r, q, \mathbf{v}_3) \cdot send(q, p, \mathbf{v}_2) \cdot send(p, q, \mathbf{v}_1) \cdot rec(q, p, \mathbf{v}_2) \cdot rec(r, q, \mathbf{v}_3)$ be an execution. Its MSC, $msc(e_1)$ depicted in Fig. 2b satisfies causal delivery. Notice that e_1 can not be divided in 1-exchanges. However, if we consider the alternative linearisation of $msc(e_1)$: $e_2 = send(p, q, \mathbf{v}_1) \cdot send(q, p, \mathbf{v}_2) \cdot rec(q, p, \mathbf{v}_2) \cdot send(r, q, \mathbf{v}_3) \cdot rec(r, q, \mathbf{v}_3)$, we have that e_2 is breakable into 1-exchanges in which each matched send is in a 1-exchange with its reception. Therefore, $msc(e_1)$ is 1-synchronous and e_1 is 1-synchronizable. Remark that e_2 is not an execution and there exists no execution that can be divided into 1-exchanges. A k-synchronous MSC highlights dependencies between messages but does not impose an order for the execution.

Comparison with [4]. In [4], the authors define set $sEx_k(\mathfrak{S})$ as the set of k-synchronous executions of system $\mathfrak{S}$ in the k-synchronous semantics. Nonetheless as remarked in Example 1.2 not all executions of a system can be divided into k-exchanges even if they are k-synchronizable. Thus, in order not to lose any executions, we have decided to reason only on MSCs (called traces in [4]).

Following standard terminology, we say that a set $U \subseteq V$ of vertices is a *strongly connected component* (SCC) of a given graph $(V, \rightarrow)$ if between any two vertices $v, v' \in U$, there exist two oriented paths $v \rightarrow^* v'$ and $v' \rightarrow^* v$. The statement below fixes some issues with Theorem 1 in [4].

Theorem 1 (Graph Characterisation of k-synchronous MSCs). *Let msc be a causal delivery MSC. msc is k-synchronous iff every SCC in its conflict graph is of size at most k and if no RS edge occurs on any cyclic path.*

Example 2 (A 5-synchronous MSC). Fig. 2c depicts a 5-synchronous MSC, that is not 4-synchronous. Indeed, its conflict graph (Fig. 2d) contains a SCC of size 5 (all vertices are on the same SCC).

Comparison with [4]. Bouajjani *et al.* give a characterisation of k-synchronous executions similar to ours, but they use the word *cycle* instead of SCC, and the subsequent developments of the paper suggest that they intended to say *Hamiltonian cycle* (i.e., a cyclic path that does not go twice through the same vertex). It is not the case that a MSC is k-synchronous if and only if every Hamiltonian cycle in its conflict graph is of size at most k and if no RS edge occurs on any cyclic path. Indeed, consider again Example 2. This graph is not Hamiltonian, and the largest Hamiltonian cycle indeed is of size 4 only. But as we already discussed in Example 2, the corresponding MSC is not 4-synchronous.

As a consequence, the algorithm that is presented in [4] for deciding whether a system is k-synchronizable is not correct as well: the MSC of Fig. 2c would be considered 4-synchronous according to this algorithm, but it is not.

4 Decidability of Reachability for k-synchronizable Systems

We show that the reachability problem is decidable for k-synchronizable systems. While proving this result, we have to face several non-trivial aspects of causal delivery that were missed in [4] and that require a completely new approach.

Definition 7 (k-synchronizable System). *A system $\mathfrak{S}$ is* k-synchronizable *if all its executions are k-synchronizable, i.e., $sTr_k(\mathfrak{S}) = asTr(\mathfrak{S})$.*

In other words, a system $\mathfrak{S}$ is k-synchronizable if for every execution e of $\mathfrak{S}$, $msc(e)$ may be divided into k-exchanges.

Remark 1. In particular, a system may be k-synchronizable even if some of its executions fill the buffers with more than k messages. For instance, the only linearisation of the 1-synchronous MSC Fig. 2b that is an execution of the system needs buffers of size 2.

For a k-synchronizable system, the reachability problem reduces to the reachability through a k-synchronizable execution. To show that k-synchronous reachability is decidable, we establish that the set of k-synchronous MSCs is regular. More precisely, we want to define a finite state automaton that accepts a sequence $e_1 \cdot e_2 \cdots e_n$ of k-exchanges if and only if they satisfy causal delivery.

We start by giving a graph-theoretic characterisation of causal delivery. For this, we define the *extended edges* $v \overset{XY}{\dashrightarrow} v'$ of a given conflict graph. The relation $\overset{XY}{\dashrightarrow}$ is defined in Fig. 3 with $X, Y \in \{S, R\}$. Intuitively, $v \overset{XY}{\dashrightarrow} v'$ expresses that event X of v must happen before event Y of v' due to either their order on the same machine (Rule 1), or the fact that a send happens before its matching receive (Rule 2), or due to the mailbox semantics (Rules 3 and 4), or because of a chain of such dependencies (Rule 5). We observe that in the *extended conflict graph*, obtained applying such rules, a cyclic dependency appears whenever causal delivery is not satisfied.

$$\text{(Rule 1)}\ \frac{v_1 \xrightarrow{XY} v_2}{v_1 \overset{XY}{\dashrightarrow} v_2} \qquad \text{(Rule 2)}\ \frac{v \cap R \neq \emptyset}{v \overset{SR}{\dashrightarrow} v} \qquad \text{(Rule 3)}\ \frac{v_1 \xrightarrow{RR} v_2}{v_1 \overset{SS}{\dashrightarrow} v_2}$$

$$\text{(Rule 4)}\ \frac{v_1 \cap R \neq \emptyset \qquad v_2 \cap R = \emptyset \qquad \mathsf{proc}_R(v_1) = \mathsf{proc}_R(v_2)}{v_1 \overset{SS}{\dashrightarrow} v_2} \qquad \text{(Rule 5)}\ \frac{v_1 \overset{XY}{\dashrightarrow}\overset{YZ}{\dashrightarrow} v_2}{v_1 \overset{XZ}{\dashrightarrow} v_2}$$

Fig. 3: Deduction rules for extended dependency edges of the conflict graph

Example 3. Fig. 5a and 5b depict an MSC and its associated conflict graph with some extended edges. This MSC violates causal delivery and there is a cyclic dependency $v_1 \overset{SS}{\dashrightarrow} v_1$.

Theorem 2 (Graph-theoretic Characterisation of Causal Delivery). *An MSC satisfies causal delivery iff there is no cyclic causal dependency of the form $v \overset{SS}{\dashrightarrow} v$ for some vertex v of its extended conflict graph.*

Let us now come back to our initial problem: we want to recognise with finite memory the sequences $e_1, e_2 \ldots e_n$ of k-exchanges that composed give an MSC that satisfies causal delivery. We proceed by reading each k-exchange one by one in sequence. This entails that, at each step, we have only a partial view of the global conflict graph. Still, we want to determine whether the acyclicity condition of Theorem 2 is satisfied in the global conflict graph. The crucial observation is that only the edges generated by Rule 4 may "go back in time". This means that we have to remember enough information from the previously examined k-exchanges to determine whether the current k-exchange contains a vertex v that shares an edge with some unmatched vertex v' seen in a previous k-exchange and whether this could participate in a cycle. This is achieved by computing two sets of processes $C_{S,p}$ and $C_{R,p}$ that collect the following information: a process q is in $C_{S,p}$ if it performs a send action causally after an unmatched send to p, or it is the sender of the unmatched send; a process q belongs to $C_{R,p}$ if it receives a message that was sent after some unmatched message directed to p. More precisely, we have:

$$C_{S,p} = \{\mathsf{proc}_S(v) \mid v' \overset{SS}{\dashrightarrow} v \ \&\ v' \text{ is unmatched } \&\ \mathsf{proc}_R(v') = p\}$$

$$C_{R,p} = \{\mathsf{proc}_R(v) \mid v' \overset{SS}{\dashrightarrow} v \ \&\ v' \text{ is unmatched } \&\ \mathsf{proc}_R(v') = p \ \&\ v \cap R \neq \emptyset\}$$

These sets abstract and carry from one k-exchange to another the necessary information to detect violations of causal delivery. We compute them in any local conflict graph of a k-exchange incrementally, i.e., knowing what they were at the end of the previous k-exchange, we compute them at the end of the current one. More precisely, let $e = s_1 \cdots s_m \cdot r_1 \cdots r_{m'}$ be a k-exchange, $\mathsf{CG}(e) = (V, E)$ its conflict graph and $B : \mathbb{P} \to (2^{\mathbb{P}} \times 2^{\mathbb{P}})$ the function that associates to each $p \in \mathbb{P}$ the two sets $B(p) = (C_{S,p}, C_{R,p})$. Then, the conflict graph $\mathsf{CG}(e, B)$ is the graph (V', E') with $V' = V \cup \{\psi_p \mid p \in \mathbb{P}\}$ and $E' \supseteq E$ as defined below. For each process $p \in \mathbb{P}$, the "summary node" ψ_p shall account for all past unmatched

$$\frac{\begin{array}{c} e = s_1 \cdots s_m \cdot r_1 \cdots r_{m'} \quad s_1 \cdots s_m \in S^* \quad r_1 \cdots r_{m'} \in R^* \quad 0 \le m' \le m \le k \\ (\vec{l}, \mathtt{Buf}_0) \overset{e}{\Rightarrow} (\vec{l'}, \mathtt{Buf}) \text{ for some } \mathtt{Buf} \\ \text{for all } p \in \mathbb{P} \quad B(p) = (C_{S,p}, C_{R,p}) \text{ and } B'(p) = (C'_{S,p}, C'_{R,p}), \\ \mathsf{Unm}_p = \{\psi_p\} \cup \{v \mid v \text{ is unmatched}, \mathsf{proc}_R(v) = p\} \\ C'_{X,p} = C_{X,p} \cup \{p \mid p \in C_{X,q}, v \overset{SS}{\dashrightarrow} \psi_q, (\mathsf{proc}_R(v) = p \text{ or } v = \psi_p)\} \cup \\ \{\mathsf{proc}_X(v) \mid v \in \mathsf{Unm}_p \cap V, X = S\} \cup \{\mathsf{proc}_X(v') \mid v \overset{SS}{\dashrightarrow} v', v \in \mathsf{Unm}_p, v \cap X \neq \emptyset\} \\ \text{for all } p \in \mathbb{P}, p \notin C'_{R,p} \end{array}}{(\vec{l}, B) \xRightarrow[\mathsf{cd}]{e,k} (\vec{l'}, B')}$$

Fig. 4: Definition of the relation $\xRightarrow[\mathsf{cd}]{e,k}$

messages sent to p that occurred in some k-exchange before e. E' is the set E of edges $\xrightarrow{XY}$ among message exchanges of e, as in Definition 5, augmented with the following set of extra edges that takes into account summary nodes.

$$\{\psi_p \xrightarrow{SX} v \mid \mathsf{proc}_X(v) \in C_{S,p} \;\&\; v \cap X \neq \emptyset \text{ for some } X \in \{S, R\}\} \quad (1)$$
$$\cup \{\psi_p \xrightarrow{SS} v \mid \mathsf{proc}_X(v) \in C_{R,p} \;\&\; v \cap R \neq \emptyset \text{ for some } X \in \{S, R\}\} \quad (2)$$
$$\cup \{\psi_p \xrightarrow{SS} v \mid \mathsf{proc}_R(v) \in C_{R,p} \;\&\; v \text{ is unmatched}\} \quad (3)$$
$$\cup \{v \xrightarrow{SS} \psi_p \mid \mathsf{proc}_R(v) = p \;\&\; v \cap R \neq \emptyset\} \cup \{\psi_q \xrightarrow{SS} \psi_p \mid p \in C_{R,q}\} \quad (4)$$

These extra edges summarise/abstract the connections to and from previous k-exchanges. Equation (1) considers connections $\xrightarrow{SS}$ and $\xrightarrow{SR}$ that are due to two sends messages or, respectively, a send and a receive on the same process. Equations (2) and (3) considers connections $\xrightarrow{RR}$ and $\xrightarrow{RS}$ that are due to two received messages or, respectively, a receive and a subsequent send on the same process. Notice how the rules in Fig. 3 would then imply the existence of a connection $\overset{SS}{\dashrightarrow}$, in particular Equation (3) abstract the existence of an edge $\overset{SS}{\dashrightarrow}$ built because of Rule 4. Equations in (4) abstract edges that would connect the current k-exchange to previous ones. As before those edges in the global conflict graph would correspond to extended edges added because of Rule 4 in Fig. 3. Once we have this enriched local view of the conflict graph, we take its extended version. Let $\overset{XY}{\dashrightarrow}$ denote the edges of the extended conflict graph as defined from rules in Fig. 3 taking into account the new vertices ψ_p and their edges.

Finally, let $\mathfrak{S}$ be a system and $\xRightarrow[\mathsf{cd}]{e,k}$ be the transition relation given in Fig. 4 among abstract configurations of the form $(\vec{l}, B)$. $\vec{l}$ is a global control state of $\mathfrak{S}$ and $B : \mathbb{P} \to (2^{\mathbb{P}} \times 2^{\mathbb{P}})$ is the function defined above that associates to each process p a pair of sets of processes $B(p) = (C_{S,p}, C_{R,p})$. Transition $\xRightarrow[\mathsf{cd}]{e,k}$ updates these sets with respect to the current k-exchange e. Causal delivery is verified by checking that for all $p \in \mathbb{P}, p \notin C'_{R,p}$ meaning that there is no cyclic dependency

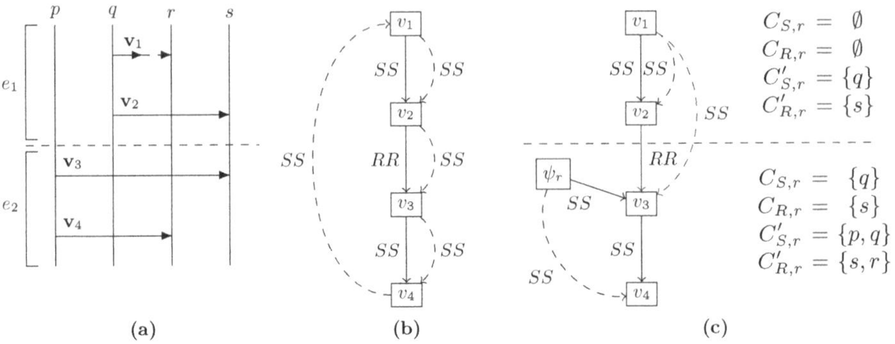

Fig. 5: (a) an MSC (b) its associated global conflict graph, (c) the conflict graphs of its k-exchanges

as stated in Theorem 2. The initial state is $(\vec{l_0}, B_0)$, where $B_0 : \mathbb{P} \to (2^{\mathbb{P}} \times 2^{\mathbb{P}})$ denotes the function such that $B_0(p) = (\emptyset, \emptyset)$ for all $p \in \mathbb{P}$.

Example 4 (An Invalid Execution). Let $e = e_1 \cdot e_2$ with e_1 and e_2 the two 2-exchanges of this execution. such that $e_1 = send(q, r, \mathbf{v}_1) \cdot send(q, s, \mathbf{v}_2) \cdot rec(q, s, \mathbf{v}_2)$ and $e_2 = send(p, s, \mathbf{v}_3) \cdot rec(p, s, \mathbf{v}_3) \cdot send(p, r, \mathbf{v}_4) \cdot rec(p, r, \mathbf{v}_4)$. Fig. 5a and 5c show the MSC and corresponding conflict graph of each of the 2-exchanges. Note that two edges of the global graph (in blue) "go across" k-exchanges. These edges do not belong to the local conflict graphs and are mimicked by the incoming and outgoing edges of summary nodes. The values of sets $C_{S,r}$ and $C_{R,r}$ at the beginning and at the end of the k-exchange are given on the right. All other sets $C_{S,p}$ and $C_{R,p}$ for $p \neq r$ are empty, since there is only one unmatched message to process r. Notice how at the end of the second k-exchange, $r \in C'_{R,r}$ signalling that message v_4 violates causal delivery.

Comparison with [4]. In [4] the authors define $\xRightarrow[\mathsf{cd}]{e,k}$ in a rather different way: they do not explicitly give a graph-theoretic characterisation of causal delivery; instead they compute, for every process p, the set $B(p)$ of processes that either sent an unmatched message to p or received a message from a process in $B(p)$. They then make sure that any message sent to p by a process $q \in B(p)$ is unmatched. According to that definition, the MSC of Fig. 5b would satisfy causal delivery and would be 1-synchronous. However, this is not the case (this MSC does not satisfy causal delivery) as we have shown in Example 3. Due to to the above errors, we had to propose a considerably different approach. The extended edges of the conflict graph, and the graph-theoretic characterisation of causal delivery as well as summary nodes, have no equivalent in [4].

Next lemma proves that Fig. 4 properly characterises causal delivery.

Lemma 1. *An MSC msc is k-synchronous iff there is* $e = e_1 \cdots e_n$ *a linearisation such that* $(\vec{l_0}, B_0) \xRightarrow[\mathsf{cd}]{e_1,k} \cdots \xRightarrow[\mathsf{cd}]{e_n,k} (\vec{l'}, B')$ *for some global state* $\vec{l'}$ *and some* $B' : \mathbb{P} \to (2^{\mathbb{P}} \times 2^{\mathbb{P}})$.

Note that there are only finitely many abstract configurations of the form $(\vec{l}, B)$ with $\vec{l}$ a tuple of control states and $B : \mathbb{P} \to (2^{\mathbb{P}} \times 2^{\mathbb{P}})$. Moreover, since $\mathbb{V}$ is finite, the alphabet over the possible k-exchange for a given k is also finite. Therefore $\xRightarrow[\mathsf{cd}]{e,k}$ is a relation on a finite set, and the set $sTr_k(\mathfrak{S})$ of k-synchronous MSCs of a system $\mathfrak{S}$ forms a regular language. It follows that it is decidable whether a given abstract configuration of the form $(\vec{l}, B)$ is reachable from the initial configuration following a k-synchronizable execution.

Theorem 3. *Let* $\mathfrak{S}$ *be a k-synchronizable system and* $\vec{l}$ *a global control state of* $\mathfrak{S}$. *The problem whether there exists* $e \in asEx(\mathfrak{S})$ *and* $\mathtt{Buf}$ *such that* $(\vec{l_0}, \mathtt{Buf}_0) \overset{e}{\Rightarrow} (\vec{l}, \mathtt{Buf})$ *is decidable.*

Remark 2. Deadlock-freedom, unspecified receptions, and absence of orphan messages are other properties that become decidable for a k-synchronizable system because of the regularity of the set of k-synchronous MSCs.

5 Decidability of k-synchronizability for Mailbox Systems

We establish the decidability of k-synchronizability; our approach is similar to the one of [4] based on the notion of borderline violation, but we adjust it to adapt to the new characterisation of k-synchronizable executions (Theorem 1).

Definition 8 (Borderline Violation). *A non k-synchronizable execution e is a borderline violation if* $e = e' \cdot r$, r *is a reception and* e' *is k-synchronizable.*

Note that a system $\mathfrak{S}$ that is not k-synchronizable always admits at least one borderline violation $e' \cdot r \in asEx(\mathfrak{S})$ with $r \in R$: indeed, there is at least one execution $e \in asEx(\mathfrak{S})$ which contains a unique minimal prefix of the form $e' \cdot r$ that is not k-synchronizable; moreover since e' is k-synchronizable, r cannot be a k-exchange of just one send action, therefore it must be a receive action. In order to find such a borderline violation, Bouajjani *et al.* introduced an instrumented system $\mathfrak{S}'$ that behaves like $\mathfrak{S}$, except that it contains an extra process π, and such that a non-deterministically chosen message that should have been sent from a process p to a process q may now be sent from p to π, and later forwarded by π to q. In $\mathfrak{S}'$, each process p has the possibility, instead of sending a message $\mathbf{v}$ to q, to deviate this message to π; if it does so, p continues its execution as if it really had sent it to q. Note also that the message sent to π get tagged with the original destination process q. Similarly, for each possible reception, a process has the possibility to receive a given message not from the initial sender but from π. The process π has an initial state from which it can receive any messages from the system. Each reception makes it go into a different state. From this state, it is able to send the message back to the original recipient. Once a message is forwarded, π reaches its final state and remains idle. The following example illustrates how the instrumented system works.

Example 5 (A Deviated Message).
Let e_1, e_2 be two executions of a system $\mathfrak{S}$ with MSCs respectively $msc(e_1)$ and $msc(e_2)$. e_1 is not 1-synchronizable. It is borderline in $\mathfrak{S}$. If we delete the last reception, it becomes indeed 1-synchronizable. $msc(e_2)$ is the MSC obtained from the instrumented system $\mathfrak{S}'$ where the message $\mathbf{v}_1$ is first deviated to π and then sent back to q from π.
Note that $msc(e_2)$ is 1-synchronous. In this case, the instrumented system $\mathfrak{S}'$ in the 1-synchronous semantics "reveals" the existence of a borderline violation of $\mathfrak{S}$.

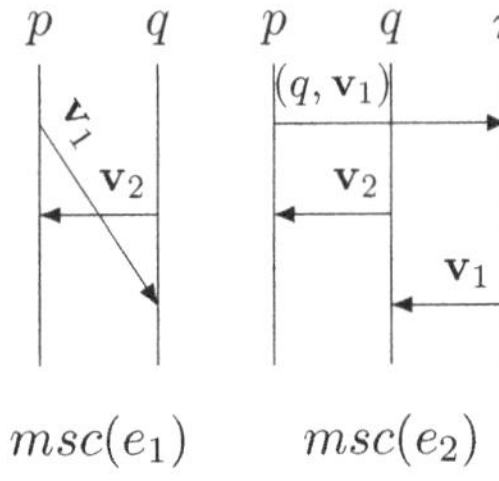

For each execution $e \cdot r \in asEx(\mathfrak{S})$ that ends with a reception, there exists an execution $\mathsf{deviate}(e \cdot r) \in asEx(\mathfrak{S}')$ where the message exchange associated with the reception r has been deviated to π; formally, if $e \cdot r = e_1 \cdot s \cdot e_2 \cdot r$ with $r = rec(p, q, \mathbf{v})$ and $s \vdash\!\dashv r$, then

$$\mathsf{deviate}(e{\cdot}r) = e_1{\cdot}send(p, \pi, (q, \mathbf{v})){\cdot}rec(p, \pi, (q, \mathbf{v})){\cdot}e_2{\cdot}send(\pi, q, (\mathbf{v})){\cdot}rec(\pi, q, \mathbf{v}).$$

Definition 9 (Feasible Execution, Bad Execution). *A k-synchronizable execution e' of $\mathfrak{S}'$ is* feasible *if there is an execution $e \cdot r \in asEx(\mathfrak{S})$ such that $\mathsf{deviate}(e{\cdot}r) = e'$. A feasible execution $e' = \mathsf{deviate}(e{\cdot}r)$ of $\mathfrak{S}'$ is* bad *if execution $e \cdot r$ is not k-synchronizable in $\mathfrak{S}$.*

Example 6 (A Non-feasible Execution).
Let e' be an execution such that $msc(e')$ is as depicted on the right. Clearly, this MSC satisfies causal delivery and could be the execution of some instrumented system $\mathfrak{S}'$. However, the sequence $e{\cdot}r$ such that $\mathsf{deviate}(e{\cdot}r) = e'$ does not satisfy causal delivery, therefore it cannot be an execution of the original system $\mathfrak{S}$. In other words, the execution e' is not feasible.

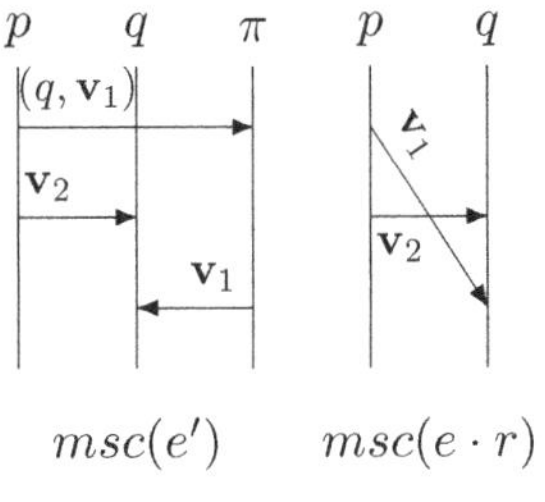

Lemma 2. *A system $\mathfrak{S}$ is not k-synchronizable iff there is a k-synchronizable execution e' of $\mathfrak{S}'$ that is feasible and bad.*

As we have already noted, the set of k-synchronous MSCs of $\mathfrak{S}'$ is regular. The decision procedure for k-synchronizability follows from the fact that the set of MSCs that have as linearisation a feasible bad execution as we will see, is regular as well, and that it can be recognised by an (effectively computable) non-deterministic finite state automaton. The decidability of k-synchronizability follows then from Lemma 2 and the decidability of the emptiness problem for non-deterministic finite state automata.

Recognition of Feasible Executions. We start with the automaton that recognises feasible executions; for this, we revisit the construction we just used for recognising sequences of k-exchanges that satisfy causal delivery.

In the remainder, we assume an execution $e' \in asEx(\mathfrak{S}')$ that contains exactly one send of the form $send(p, \pi, (q, \mathbf{v}))$ and one reception of the form $rec(\pi, q, \mathbf{v})$, this reception being the last action of e'. Let $(V, \{\xrightarrow{XY}\}_{X,Y \in \{R,S\}})$ be the conflict graph of e'. There are two uniquely determined vertices $v_{\mathsf{start}}, v_{\mathsf{stop}} \in V$ such that $\mathsf{proc}_R(v_{\mathsf{start}}) = \pi$ and $\mathsf{proc}_S(v_{\mathsf{stop}}) = \pi$ that correspond, respectively, to the first and last message exchanges of the deviation. The conflict graph of $e \cdot r$ is then obtained by merging these two nodes.

Lemma 3. *The execution e' is not feasible iff there is a vertex v in the conflict graph of e' such that $v_{\mathsf{start}} \overset{SS}{\dashrightarrow} v \xrightarrow{RR} v_{\mathsf{stop}}$.*

In order to decide whether an execution e' is feasible, we want to forbid that a send action $send(p', q, \mathbf{v}')$ that happens causally after v_{start} is matched by a receive $rec(p', q, \mathbf{v}')$ that happens causally before the reception v_{stop}. As a matter of fact, this boils down to deal with the deviated send action as an unmatched send. So we will consider sets of processes C_S^π and C_R^π similar to the ones used for $\xRightarrow[\mathsf{cd}]{e,k}$, but with the goal of computing which actions happen causally after the send to π. We also introduce a summary node ψ_{start} and the extra edges following the same principles as in the previous section. Formally, let $B : \mathbb{P} \to (2^{\mathbb{P}} \times 2^{\mathbb{P}})$, $C_S^\pi, C_R^\pi \subseteq \mathbb{P}$ and $e \in S^{\leq k}R^{\leq k}$ be fixed, and let $\mathsf{CG}(e, B) = (V', E')$ be the constraint graph with summary nodes for unmatched sent messages as defined in the previous section. The local constraint graph $\mathsf{CG}(e, B, C_S^\pi, C_R^\pi)$ is defined as the graph (V'', E'') where $V'' = V' \cup \{\psi_{\mathsf{start}}\}$ and E'' is E' augmented with

$$\{\psi_{\mathsf{start}} \xrightarrow{SX} v \mid \mathsf{proc}_X(v) \in C_S^\pi \;\&\; v \cap X \neq \emptyset \text{ for some } X \in \{S, R\}\}$$
$$\cup\ \{\psi_{\mathsf{start}} \xrightarrow{SS} v \mid \mathsf{proc}_X(v) \in C_R^\pi \;\&\; v \cap R \neq \emptyset \text{ for some } X \in \{S, R\}\}$$
$$\cup\ \{\psi_{\mathsf{start}} \xrightarrow{SS} v \mid \mathsf{proc}_R(v) \in C_R^\pi \;\&\; v \text{ is unmatched}\} \ \cup\ \{\psi_{\mathsf{start}} \xrightarrow{SS} \psi_p \mid p \in C_R^\pi\}$$

As before, we consider the "closure" $\overset{XY}{\dashrightarrow}$ of these edges by the rules of Fig. 3. The transition relation $\xRightarrow[\mathsf{feas}]{e,k}$ is defined in Fig. 6. It relates abstract configurations of the form $(\vec{l}, B, \vec{C}, \mathtt{dest}_\pi)$ with $\vec{C} = (C_{S,\pi}, C_{R,\pi})$ and $\mathtt{dest}_\pi \in \mathbb{P} \cup \{\bot\}$ storing to whom the message deviated to π was supposed to be delivered. Thus, the initial abstract configuration is $(l_0, B_0, (\emptyset, \emptyset), \bot)$, where $\bot$ means that the processus $\mathtt{dest}_\pi$ has not been determined yet. It will be set as soon as the send to process π is encountered.

Lemma 4. *Let e' be an execution of $\mathfrak{S}'$. Then e' is a k-synchronizable feasible execution iff there are $e'' = e_1 \cdots e_n \cdot send(\pi, q, \mathbf{v}) \cdot rec(\pi, q, \mathbf{v})$ with $e_1, \ldots, e_n \in S^{\leq k}R^{\leq k}$, $B' : \mathbb{P} \to 2^{\mathbb{P}}$, $\vec{C'} \in (2^{\mathbb{P}})^2$, and a tuple of control states $\vec{l'}$ such that $msc(e') = msc(e'')$, $\pi \notin C_{R,q}$ (with $B'(q) = (C_{S,q}, C_{R,q})$), and*

$$(\vec{l_0}, B_0, (\emptyset, \emptyset), \bot) \xRightarrow[\mathsf{feas}]{e_1,k} \ldots \xRightarrow[\mathsf{feas}]{e_n,k} (\vec{l'}, B', \vec{C'}, q).$$

$$\frac{\begin{array}{c}(\vec{l}, B) \xRightarrow[\mathsf{cd}]{e,k} (\vec{l'}, B') \qquad e = a_1 \cdots a_n \qquad (\forall v)\ \mathsf{proc}_S(v) \neq \pi \\ (\forall v, v')\ \mathsf{proc}_R(v) = \mathsf{proc}_R(v') = \pi \implies v = v' \wedge \mathtt{dest}_\pi = \bot \\ (\forall v)\ v \ni send(p, \pi, (q, \mathbf{v})) \implies \mathtt{dest}'_\pi = q \quad \mathtt{dest}_\pi \neq \bot \implies \mathtt{dest}'_\pi = \mathtt{dest}_\pi \\ C_X^{\pi\prime} = C_X^\pi \cup \{\mathsf{proc}_X(v') \mid v \overset{SS}{\dashrightarrow} v'\ \&\ v' \cap X \neq \emptyset\ \&\ (\mathsf{proc}_R(v) = \pi \text{ or } v = \psi_{\mathsf{start}})\} \\ \cup\ \{\mathsf{proc}_S(v) \mid \mathsf{proc}_R(v) = \pi\ \&\ X = S\} \\ \cup\ \{p \mid p \in C_{X,q}\ \&\ v \overset{SS}{\dashrightarrow} \psi_q\ \&\ (\mathsf{proc}_R(v) = \pi \text{ or } v = \psi_{\mathsf{start}})\} \\ \mathtt{dest}'_\pi \notin C_R^{\pi\prime}\end{array}}{(\vec{l}, B, C_S^\pi, C_R^\pi, \mathtt{dest}_\pi) \xRightarrow[\mathsf{feas}]{e,k} (\vec{l'}, B', C_S^{\pi\prime}, C_R^{\pi\prime}, \mathtt{dest}'_\pi)}$$

Fig. 6: Definition of the relation $\xRightarrow[\mathsf{feas}]{e,k}$

Comparison with [4]. In [4] the authors verify that an execution is feasible with a *monitor* which reviews the actions of the execution and adds processes that no longer are allowed to send a message to the receiver of π. Unfortunately, we have here a similar problem that the one mentioned in the previous comparison paragraph. According to their monitor, the following execution $e' = \mathsf{deviate}(e \cdot r)$ is feasible, i.e., is runnable in $\mathfrak{S}'$ and $e \cdot r$ is runnable in $\mathfrak{S}$.

$$\begin{aligned} e' = \ & send(q, \pi, (r, \mathbf{v}_1)) \cdot rec(q, \pi, (r, \mathbf{v}_1)) \cdot send(q, s, \mathbf{v}_2) \cdot rec(q, s, \mathbf{v}_2) \cdot \\ & send(p, s, \mathbf{v}_3) \cdot rec(p, s, \mathbf{v}_3) \cdot send(p, r, \mathbf{v}_4) \cdot rec(p, r, \mathbf{v}_4) \cdot \\ & send(\pi, r, \mathbf{v}_1) \cdot rec(\pi, r, \mathbf{v}_4) \end{aligned}$$

However, this execution is not feasible because there is a causal dependency between $\mathbf{v}_1$ and $\mathbf{v}_3$. In [4] this execution would then be considered as feasible and therefore would belong to set $sTr_k(\mathfrak{S}')$. Yet there is no corresponding execution in $asTr(\mathfrak{S})$, the comparison and therefore the k-synchronizability, could be distorted and appear as a false negative.

Recognition of Bad Executions. Finally, we define a non-deterministic finite state automaton that recognizes MSCs of bad executions, i.e., feasible executions $e' = \mathsf{deviate}(e \cdot r)$ such that $e \cdot r$ is not k-synchronizable. We come back to the "non-extended" conflict graph, without edges of the form $\overset{XY}{\dashrightarrow}$. Let $\mathsf{Post}^*(v) = \{v' \in V \mid v \to^* v'\}$ be the set of vertices reachable from v, and let $\mathsf{Pre}^*(v) = \{v' \in V \mid v' \to^* v\}$ be the set of vertices co-reachable from v. For a set of vertices $U \subseteq V$, let $\mathsf{Post}^*(U) = \bigcup\{\mathsf{Post}^*(v) \mid v \in U\}$, and $\mathsf{Pre}^*(U) = \bigcup\{\mathsf{Pre}^*(v) \mid v \in U\}$.

Lemma 5. *The feasible execution e' is bad iff one of the two holds*

1. $v_{\mathsf{start}} \longrightarrow^* \xrightarrow{RS} \longrightarrow^* v_{\mathsf{stop}}$, *or*
2. *the size of the set* $\mathsf{Post}^*(v_{\mathsf{start}}) \cap \mathsf{Pre}^*(v_{\mathsf{stop}})$ *is greater or equal to* $k + 2$.

In order to determine whether a given message exchange v of $\mathsf{CG}(e')$ should be counted as reachable (resp. co-reachable), we will compute at the entry and exit of every k-exchange of e' which processes are "reachable" or "co-reachable".

Example 7. (Reachable and Co-reachable Processes)

Consider the MSC on the right made of five 1-exchanges. While sending message $(s, \mathbf{v}_0)$ that corresponds to v_{start}, process r becomes "reachable": any subsequent message exchange that involves r corresponds to a vertex of the conflict graph that is reachable from v_{start}. While sending $\mathbf{v}_2$, process s becomes "reachable", because process r will be reachable when it will receive message $\mathbf{v}_2$. Similary, q becomes reachable after receiving $\mathbf{v}_3$ because r was reachable when it sent $\mathbf{v}_3$, and p becomes reachable after receiving $\mathbf{v}_4$ because q was reachable when it sent $\mathbf{v}_4$. Co-reachability works similarly, but reasoning backwards on the timelines. For instance, process s stops being "co-reachable" while it receives $\mathbf{v}_0$, process r stops being co-reachable after it receives $\mathbf{v}_2$, and process p stops being co-reachable by sending $\mathbf{v}_1$. The only message that is sent by a process being both reachable and co-reachable at the instant of the sending is $\mathbf{v}_2$, therefore it is the only message that will be counted as contributing to the SCC.

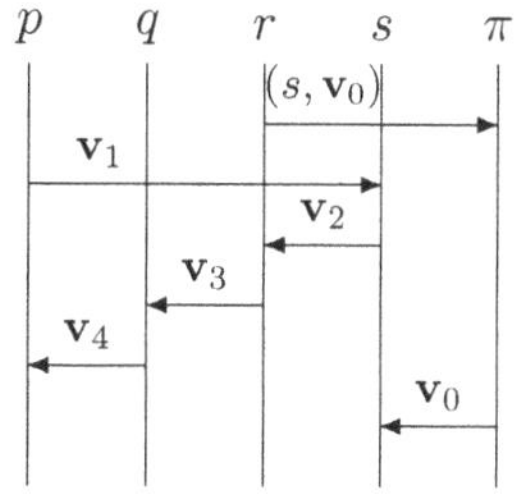

$msc(e)$

More formally, let e be sequence of actions, $\mathsf{CG}(e)$ its conflict graph and P, Q two sets of processes, $\mathsf{Post}_e(P) = \mathsf{Post}^*\Big(\{v \mid \mathsf{procs}(v) \cap P \neq \emptyset\}\Big)$ and $\mathsf{Pre}_e(Q) = \mathsf{Pre}^*\Big(\{v \mid \mathsf{procs}(v) \cap Q \neq \emptyset\}\Big)$ are introduced to represent the local view through k-exchanges of $\mathsf{Post}^*(v_{\mathsf{start}})$ and $\mathsf{Pre}^*(v_{\mathsf{stop}})$. For instance, for e as in Example 7, we get $\mathsf{Post}_e(\{\pi\}) = \{(s, \mathbf{v}_0), \mathbf{v}_2, \mathbf{v}_3, \mathbf{v}_4, \mathbf{v}_0\}$ and $\mathsf{Pre}_e(\{\pi\}) = \{\mathbf{v}_0, \mathbf{v}_2, \mathbf{v}_1, (s, \mathbf{v}_0)\}$. In each k-exchange e_i the size of the intersection between $\mathsf{Post}_{e_i}(P)$ and $\mathsf{Pre}_{e_i}(Q)$ will give the local contribution of the current k-exchange to the calculation of the size of the global SCC. In the transition relation $\xrightarrow[\mathsf{bad}]{e,k}$ this value is stored in variable `cnt`. The last ingredient to consider is to recognise if an edge RS belongs to the SCC. To this aim, we use a function `lastisRec` : $\mathbb{P} \to \{\mathsf{True}, \mathsf{False}\}$ that for each process stores the information whether the last action in the previous k-exchange was a reception or not. Then depending on the value of this variable and if a node is in the current SCC or not the value of `sawRS` is set accordingly.

The transition relation $\xrightarrow[\mathsf{bad}]{e,k}$ defined in Fig. 7 deals with abstract configurations of the form $(P, Q, \mathtt{cnt}, \mathtt{sawRS}, \mathtt{lastisRec}')$ where $P, Q \subseteq \mathbb{P}$, `sawRS` is a boolean value, and `cnt` is a counter bounded by $k+2$. We denote by $\mathtt{lastisRec}_0$ the function where all $\mathtt{lastisRec}(p) = \mathsf{False}$ for all $p \in \mathbb{P}$.

Lemma 6. *Let e' be a feasible k-synchronizable execution of $\mathfrak{S}'$. Then e' is a bad execution iff there are $e'' = e_1 \cdots e_n \cdot send(\pi, q, \mathbf{v}) \cdot rec(\pi, q, \mathbf{v})$ with $e_1, \ldots, e_n \in S^{\leq k} R^{\leq k}$ and $msc(e') = msc(e'')$, $P', Q \subseteq \mathbb{P}$, $\mathtt{sawRS} \in \{\mathsf{True}, \mathsf{False}\}$, $\mathtt{cnt} \in \{0, \ldots, k+2\}$, such that*

$$(\{\pi\}, Q, 0, \mathsf{False}, \mathtt{lastisRec}_0) \xrightarrow[\mathsf{bad}]{e_1,k} \cdots \xrightarrow[\mathsf{bad}]{e_n,k} (P', \{\pi\}, \mathtt{cnt}, \mathtt{sawRS}, \mathtt{lastisRec})$$

$$\frac{\begin{array}{c} P' = \mathsf{procs}(\mathsf{Post}_e(P)) \qquad Q = \mathsf{procs}(\mathsf{Pre}_e(Q')) \\ SCC_e = \mathsf{Post}_e(P) \cap \mathsf{Pre}_e(Q') \\ \mathtt{cnt}' = \mathsf{min}(k+2, \mathtt{cnt}+n) \quad \text{where } n = |SCC_e| \\ \mathtt{lastisRec}'(q) \Leftrightarrow (\exists v \in SCC_e.\mathsf{proc}_R(v) = q \wedge v \cap R \neq \emptyset) \vee \\ (\mathtt{lastisRec}(q) \wedge \nexists v \in V.\mathsf{proc}_S(v) = q) \\ \mathtt{sawRS}' = \mathtt{sawRS} \vee \\ (\exists v \in SCC_e)(\exists p \in \mathbb{P} \setminus \{\pi\})\ \mathsf{proc}_S(v) = p \wedge \mathtt{lastisRec}(p) \wedge p \in P \cap Q \end{array}}{(P, Q, \mathtt{cnt}, \mathtt{sawRS}, \mathtt{lastisRec}) \xrightarrow[\mathsf{bad}]{e,k} (P', Q', \mathtt{cnt}', \mathtt{sawRS}', \mathtt{lastisRec}')}$$

Fig. 7: Definition of the relation $\xRightarrow[\mathsf{bad}]{e,k}$

and at least one of the two holds: either $\mathtt{sawRS} = \mathsf{True}$*, or* $\mathtt{cnt} = k + 2$.

Comparison with [4]. As for the notion of feasibility, to determine if an execution is bad, in [4] the authors use a monitor that builds a path between the send to process π and the send from π. In addition to the problems related to the wrong characterisation of k-synchronizability, this monitor not only can detect an RS edge when there should be none, but also it can miss them when they exist. In general, the problem arises because the path is constructed by considering only an endpoint at the time.

We can finally conclude that:

Theorem 4. *The k-synchronizability of a system $\mathfrak{S}$ is decidable for $k \geq 1$.*

6 k-synchronizability for Peer-to-Peer Systems

In this section, we will apply k-synchronizability to peer-to-peer systems. A peer-to-peer system is a composition of communicating automata where each pair of machines exchange messages via two private FIFO buffers, one per direction of communication. Here we only give an insight on what changes with respect to the mailbox setting.

Causal delivery reveals the order imposed by FIFO buffers. Definition 4 must then be adapted to account for peer-to-peer communication. For instance, two messages that are sent to a same process p by two different processes can be received by p in any order, regardless of any causal dependency between the two sends. Thus, checking causal delivery in peer-to-peer systems is easier than in the mailbox setting, as we do not have to carry information on causal dependencies.

Within a peer-to-peer architecture, MSCs and conflict graphs are defined as within a mailbox communication. Indeed, they represents dependencies over machines, i.e., the order in which the actions can be done on a given machine, and over the send and the reception of a same message, and they do not depend on the type of communication. The notion of k-exchange remains also unchanged.

Decidability of Reachability for k-synchronizable Peer-to-Peer Systems. To establish the decidability of reachability for k-synchronizable peer-to-peer systems, we define a transition relation $\overset{e,k}{\underset{\mathsf{cd}}{\Longrightarrow}}^{\mathsf{p2p}}$ for a sequence of action e describing a k-exchange. As for mailbox systems, if a send action is unmatched in the current k-exchange, it will stay orphan forever. Moreover, after a process p sent an orphan message to a process q, p is forbidden to send any matched message to q. Nonetheless, as a consequence of the simpler definition of causal delivery, , we no longer need to work on the conflict graph. Summary nodes and extended edges are not needed and all the necessary information is in function B that solely contains all the forbidden senders for process p.

The characterisation of a k-synchronizable execution is the same as for mailbox systems as the type of communication is not relevant. We can thus conclude, as within mailbox communication, that reachability is decidable.

Theorem 5. *Let $\mathfrak{S}$ be a k-synchronizable system and $\vec{l}$ a global control state of $\mathfrak{S}$. The problem whether there exists $e \in asEx(\mathfrak{S})$ and* $\mathtt{Buf}$ *such that $(\vec{l_0}, \mathtt{Buf}_0) \overset{e}{\Rightarrow} (\vec{l}, \mathtt{Buf})$ is decidable.*

Decidability of k-synchronizability for Peer-to-Peer Systems. As in mailbox system, the detection of a borderline execution determines whether a system is k-synchronizable.

The relation transition $\overset{e,k}{\underset{\mathsf{feas}}{\Longrightarrow}}^{\mathsf{p2p}}$ allows to obtain feasible executions. Differently from the mailbox setting, we need to save not only the recipient $\mathtt{dest}_\pi$ but also the sender of the delayed message (information stored in variable $\mathtt{exp}_\pi$). The transition rule then checks that there is no message that is violating causal delivery, i.e., there is no message sent by $\mathtt{exp}_\pi$ to $\mathtt{dest}_\pi$ after the deviation. Finally the recognition of bad execution, works in the same way as for mailbox systems. The characterisation of a bad execution and the definition of $\overset{e,k}{\underset{\mathsf{bad}}{\Longrightarrow}}^{\mathsf{p2p}}$ are, therefore, the same.

As for mailbox systems, we can, thus, conclude that for a given k, k-synchronizability is decidable.

Theorem 6. *The k-synchronizability of a system $\mathfrak{S}$ is decidable for $k \geq 1$.*

7 Concluding Remarks and Related works

In this paper we have studied k-synchronizability for mailbox and peer-to-peer systems. We have corrected the reachability and decidability proofs given in [4]. The flaws in [4] concern fundamental points and we had to propose a considerably different approach. The extended edges of the conflict graph, and the graph-theoretic characterisation of causal delivery as well as summary nodes, have no equivalent in [4]. Transition relations $\overset{e,k}{\underset{\mathsf{feas}}{\Longrightarrow}}$ and $\overset{e,k}{\underset{\mathsf{bad}}{\Longrightarrow}}$ building on the graph-theoretic characterisations of causal delivery and k-synchronizability, depart considerably from the proposal in [4].

We conclude by commenting on some other related works. The idea of "communication layers" is present in the early works of Elrad and Francez [8] or Chou and Gafni [7]. More recently, Chaouch-Saad *et al.* [6] verified some consensus algorithms using the Heard-Of Model that proceeds by "communication-closed rounds". The concept that an asynchronous system may have an "equivalent" synchronous counterpart has also been widely studied. Lipton's reduction [14] reschedules an execution so as to move the receive actions as close as possible from their corresponding send. Reduction recently received an increasing interest for verification purpose, e.g. by Kragl *et al.* [12], or Gleissenthal *et al.* [11].

Existentially bounded communication systems have been studied by Genest *et al.* [10,15]: a system is existentially k-bounded if any execution can be rescheduled in order to become k-bounded. This approach targets a broader class of systems than k-synchronizability, because it does not require that the execution can be chopped in communication-closed rounds. In the perspective of the current work, an interesting result is the decidability of existential k-boundedness for deadlock-free systems of communicating machines with peer-to-peer channels. Despite the more general definition, these older results are incomparable with the present ones, that deal with systems communicating with mailboxes, and not peer-to-peer channels.

Basu and Bultan studied a notion they also called synchronizability, but it differs from the notion studied in the present work; synchronizability and k-synchronizability define incomparable classes of communicating systems. The proofs of the decidability of synchronizability [3,2] were shown to have flaws by Finkel and Lozes [9]. A question left open in their paper is whether synchronizability is decidable for mailbox communications, as originally claimed by Basu and Bultan. Akroun and Salaün defined also a property they called stability [1] and that shares many similarities with the synchronizability notion in [2].

Context-bounded model-checking is yet another approach for the automatic verification of concurrent systems. La Torre *et al.* studied systems of communicating machines extended with a calling stack, and showed that under some conditions on the interplay between stack actions and communications, context-bounded reachability was decidable [13]. A context-switch is found in an execution each time two consecutive actions are performed by a different participant. Thus, while k-synchronizability limits the number of consecutive sendings, bounded context-switch analysis limits the number of times two consecutive actions are performed by two different processes.

As for future work, it would be interesting to explore how both context-boundedness and communication-closed rounds could be composed. Moreover refinements of the definition of k-synchronizability can also be considered. For instance, we conjecture that the current development can be greatly simplified if we forbid linearisation that do not correspond to actual executions.

References

1. Akroun, L., Salaün, G.: Automated verification of automata communicating via FIFO and bag buffers. Formal Methods in System Design **52**(3), 260–276 (2018). https://doi.org/10.1007/s10703-017-0285-8
2. Basu, S., Bultan, T.: On deciding synchronizability for asynchronously communicating systems. Theor. Comput. Sci. **656**, 60–75 (2016). https://doi.org/10.1016/j.tcs.2016.09.023
3. Basu, S., Bultan, T., Ouederni, M.: Synchronizability for verification of asynchronously communicating systems. In: Kuncak, V., Rybalchenko, A. (eds.) Verification, Model Checking, and Abstract Interpretation - 13th International Conference, VMCAI 2012, Philadelphia, PA, USA, January 22-24, 2012. Proceedings. Lecture Notes in Computer Science, vol. 7148, pp. 56–71. Springer (2012). https://doi.org/10.1007/978-3-642-27940-9_5
4. Bouajjani, A., Enea, C., Ji, K., Qadeer, S.: On the completeness of verifying message passing programs under bounded asynchrony. In: Chockler, H., Weissenbacher, G. (eds.) Computer Aided Verification - 30th International Conference, CAV 2018, Held as Part of the Federated Logic Conference, FloC 2018, Oxford, UK, July 14-17, 2018, Proceedings, Part II. Lecture Notes in Computer Science, vol. 10982, pp. 372–391. Springer (2018). https://doi.org/10.1007/978-3-319-96142-2_23
5. Bouajjani, A., Habermehl, P., Vojnar, T.: Abstract regular model checking. In: Alur, R., Peled, D.A. (eds.) Computer Aided Verification, 16th International Conference, CAV 2004, Boston, MA, USA, July 13-17, 2004, Proceedings. Lecture Notes in Computer Science, vol. 3114, pp. 372–386. Springer (2004). https://doi.org/10.1007/978-3-540-27813-9_29
6. Chaouch-Saad, M., Charron-Bost, B., Merz, S.: A reduction theorem for the verification of round-based distributed algorithms. In: Bournez, O., Potapov, I. (eds.) Reachability Problems, 3rd International Workshop, RP 2009, Palaiseau, France, September 23-25, 2009. Proceedings. Lecture Notes in Computer Science, vol. 5797, pp. 93–106. Springer (2009). https://doi.org/10.1007/978-3-642-04420-5_10
7. Chou, C., Gafni, E.: Understanding and verifying distributed algorithms using stratified decomposition. In: Dolev, D. (ed.) Proceedings of the Seventh Annual ACM Symposium on Principles of Distributed Computing, Toronto, Ontario, Canada, August 15-17, 1988. pp. 44–65. ACM (1988). https://doi.org/10.1145/62546.62556
8. Elrad, T., Francez, N.: Decomposition of distributed programs into communication-closed layers. Sci. Comput. Program. **2**(3), 155–173 (1982). https://doi.org/10.1016/0167-6423(83)90013-8
9. Finkel, A., Lozes, É.: Synchronizability of communicating finite state machines is not decidable. In: Chatzigiannakis, I., Indyk, P., Kuhn, F., Muscholl, A. (eds.) 44th International Colloquium on Automata, Languages, and Programming, ICALP 2017, July 10-14, 2017, Warsaw, Poland. LIPIcs, vol. 80, pp. 122:1–122:14. Schloss Dagstuhl - Leibniz-Zentrum fuer Informatik (2017). https://doi.org/10.4230/LIPIcs.ICALP.2017.122, http://www.dagstuhl.de/dagpub/978-3-95977-041-5
10. Genest, B., Kuske, D., Muscholl, A.: On communicating automata with bounded channels. Fundam. Inform. **80**(1-3), 147–167 (2007), http://content.iospress.com/articles/fundamenta-informaticae/fi80-1-3-09
11. von Gleissenthall, K., Kici, R.G., Bakst, A., Stefan, D., Jhala, R.: Pretend synchrony: synchronous verification of asynchronous distributed programs. PACMPL **3**(POPL), 59:1–59:30 (2019). https://doi.org/10.1145/3290372

12. Kragl, B., Qadeer, S., Henzinger, T.A.: Synchronizing the asynchronous. In: Schewe, S., Zhang, L. (eds.) 29th International Conference on Concurrency Theory, CONCUR 2018, September 4-7, 2018, Beijing, China. LIPIcs, vol. 118, pp. 21:1–21:17. Schloss Dagstuhl - Leibniz-Zentrum fuer Informatik (2018). https://doi.org/10.4230/LIPIcs.CONCUR.2018.21
13. La Torre, S., Madhusudan, P., Parlato, G.: Context-bounded analysis of concurrent queue systems. In: Ramakrishnan, C.R., Rehof, J. (eds.) Tools and Algorithms for the Construction and Analysis of Systems, 14th International Conference, TACAS 2008, Held as Part of the Joint European Conferences on Theory and Practice of Software, ETAPS 2008, Budapest, Hungary, March 29-April 6, 2008. Proceedings. Lecture Notes in Computer Science, vol. 4963, pp. 299–314. Springer (2008). https://doi.org/10.1007/978-3-540-78800-3_21
14. Lipton, R.J.: Reduction: A method of proving properties of parallel programs. Commun. ACM **18**(12), 717–721 (1975). https://doi.org/10.1145/361227.361234
15. Muscholl, A.: Analysis of communicating automata. In: Dediu, A., Fernau, H., Martín-Vide, C. (eds.) Language and Automata Theory and Applications, 4th International Conference, LATA 2010, Trier, Germany, May 24-28, 2010. Proceedings. Lecture Notes in Computer Science, vol. 6031, pp. 50–57. Springer (2010). https://doi.org/10.1007/978-3-642-13089-2_4

General Supervised Learning as Change Propagation with Delta Lenses

Zinovy Diskin(✉)

McMaster University, Hamilton, Canada
diskinz@mcmaster.ca

Abstract. Delta lenses are an established mathematical framework for modelling and designing bidirectional model transformations (Bx). Following the recent observations by Fong et al, the paper extends the delta lens framework with a a new ingredient: learning over a parameterized space of model transformations seen as functors. We will define a notion of an asymmetric learning delta lens with amendment (ala-lens), and show how ala-lenses can be organized into a symmetric monoidal (sm) category. We also show that sequential and parallel composition of well-behaved (wb) ala-lenses are also wb so that wb ala-lenses constitute a full sm-subcategory of ala-lenses.

1 Introduction

The goal of the paper is to develop a formal model of *supervised learning* in a very general context of *bidirectional model transformation* or *Bx*, i.e., synchronization of two arbitrary complex structures (called *models*) related by a transformation.[1] Rather than learning parameterized functions between Euclidean spaces as is typical for machine learning (ML), we will consider learning mappings between model spaces and formalize them as parameterized functors between categories, $f\colon P\times\mathbf{A} \to \mathbf{B}$, with P being a parameter space. The basic ML-notion of a *training pair* $(A, B') \in \mathbf{A}_0 \times \mathbf{B}_0$ will be considered as an inconsistency between models caused by a change (*delta*) $v\colon B \to B'$ of the target model $B = f(p, A)$, $p \in P$, that was first consistent with A w.r.t. the transformation (functor) $f(p, \_)$. An inconsistency is repaired by an appropriate change of the source structure, $u\colon A \to A'$, changing the parameter p to p', and an *amendment* of the target structure $v^@\colon B' \to B^@$ so that $f(p', A') = B^@$ is a consistent state of the parameterized two-model system.

The setting above without parameterization and learning (i.e., $p' = p$ always holds), and without amendment ($v^@ = \mathsf{id}_{B'}$ always holds), is well known in the Bx literature under the name of *delta lenses*— mathematical structures, in

[1]Term *Bx* refers to a wide area including file synchronization, data exchange in databases, and model synchronization in Model-Driven software Engineering (MDE), see [7] for a survey. In the present paper, Bx will mainly refer to Bx in the MDE context.

J. Goubault-Larrecq and B. König (Eds.): FOSSACS 2020, LNCS 12077, pp. 177–197, 2020.
https://doi.org/10.1007/978-3-030-45231-5_10

which consistency restoration via change propagation is modelled by functorial-like algebraic operations over categories [12,6]. There are several types of delta lenses tailored for modelling different synchronization tasks and scenarios, particularly, symmetric and asymmetric. In the paper, we only consider asymmetric delta lenses and will often omit explicit mentioning these attributes. Despite their extra-generality, (delta) lenses have been proved useful in the design and implementation of practical model synchronization systems with triple graph grammars (TGG) [5,2]; enriching lenses with amendment is a recent extension of the framework motivated and formalized in [11]. A major advantage of the lens framework for synchronization is its compositionality: a lens satisfying several equational laws specifying basic synchronization requirements is called *well-behaved (wb)*, and basic lens theorems state that sequential and parallel composition of wb lenses is again wb. In practical applications, it allows the designer of a complex synchronizer to avoid integration testing: if elementary synchronizers are tested and proved to be wb, their composition is automatically wb too.

The present paper makes the following contributions to the delta lens framework for Bx. a) We motivate model synchronization enriched with learning and, moreover, with *categorical* learning, in which the parameter space is a category, and introduce the notion of a *wb asymmetric learning (delta) lens* with *amendment* (a *wb ala-lens*) (this is the content of Sect. 3). b) We prove compositionality of wb ala-lenses and show how their universe can be organized into a symmetric monoidal (sm) category (Theorems 1-3 in Sect. 4). All proofs (rather straightforward but notationally laborious) can be found in the long version of the paper [9]. One more compositional result is c) a definition of a *compositional bidirectional transformation language* (Def. 6) that formalizes an important requirement to model synchronization tools, which (surprisingly) is missing from the Bx literature. Background Sect. 2 provides a simple example demonstrating main concepts of Bx and delta lenses in the MDE context. Section 5 briefly surveys related work, and Sect. 6 concludes.

Notation. Given a category $\mathbf{A}$, its objects are denoted by capital letters A, A', etc. to recall that in MDE applications, objects are complex structures, which themselves have elements $a, a', ...$; the collection of all objects of category $\mathbf{A}$ is denoted by $\mathbf{A}_0$. An arrow with domain $A \in \mathbf{A}_0$ is written as $u\colon A \to \_$ or $u \in \mathbf{A}(A, \_)$; we also write $\mathsf{dom}(u) = A$ (and sometimes $u^{\mathsf{dom}} = A$ to shorten formulas). Similarly, formula $u\colon \_ \to A'$ denotes an arrow with codomain $u.\mathsf{cod} = A'$. Given a functor $f\colon \mathbf{A} \to \mathbf{B}$, its object function is denoted by $f_0\colon \mathbf{A}_0 \to \mathbf{B}_0$.

A subcategory $\mathbf{B} \subset \mathbf{A}$ is called *wide* if it has the same objects. All categories we consider in the paper are small.

2 Background: Update propagation and delta lenses

Although Bx ideas work well only in domains conforming to the slogan *any implementation satisfying the specification is good enough* such as code generation (see [10] for discussion), and have limited applications in databases (only so called updatable views can be treated in the Bx-way), we will employ a simple database example: it allows demonstrating the core ideas without any special domain knowledge required by typical Bx-amenable areas. The presentation will be semi-formal as our goal is to motivate the delta lens formalism that abstracts the details away rather than formalize the example as such.

2.1 Why deltas.

Bx-lenses first appeared in the work on file synchronization, and if we have two sets of strings, say, $B = \{\text{John}, \text{Mary}\}$ and $B' = \{\text{Jon}, \text{Mary}\}$, we can readily see the difference: John $\neq$ Jon but Mary $=$ Mary. We thus have a structure in-between B and B' (which maybe rather complex if B and B' are big files), but this structure can be recovered by string matching and thus updates can be identified with pairs. The situation dramatically changes if B and B' are object structures, e.g., $B = \{o_1, o_2\}$ with $\mathsf{Name}(o_1) = \text{John}$, $\mathsf{Name}(o_2) = \text{Mary}$ and similarly $B' = \{o'_1, o'_2\}$ with $\mathsf{Name}(o'_1) = \text{Jon}$, $\mathsf{Name}(o'_2) = \text{Mary}$. Now string matching does not say too much: it may happen that o_1 and o'_1 are the same object (think of a typo in the dataset), while o_2 and o'_2 are different (although equally named) objects. Of course, for better matching we could use full names or ID numbers or something similar (called, in the database parlance, primary keys), but absolutely reliable keys are rare, and typos and bugs can compromise them anyway. Thus, for object structures that Bx needs to keep in sync, deltas between models need to be independently specified, e.g., by specifying a *sameness relation* $u \subset B \times B'$ between models. For example, $u = \{o_1, o'_1\}$ says that John@B and Jon@B' are the same person while Mary@B and Mary@B' are not. Hence, model spaces in Bx are categories (objects are models and arrows are update/delta specifications) rather than sets (codiscrete categories).

2.2 Consistency restoration via update propagation: An Example

Figure 1 presents a simple example of delta propagation for consistency restoration. Models consist of objects (in the sense of OO programming) with attributes (a.k.a. labelled records), e.g., the source model A consists of three objects identified by their oids (object identifiers) #A, #J, #M (think about employees of some company) with attribute values as shown in the table: attribute Expr. refers to Experience measured by a number of years, and Depart. is the column of department names. The schema of the table, i.e., the triple $S_{\mathbf{A}}$ of attributes (Name, Expr., Depart.) with their domains of values ***String***, ***Integer***, ***String*** resp., determines a model space $\mathbf{A}$. A model $X \in \mathbf{A}$ is given by its set of objects OID^X together with three functions Name^X, $\mathsf{Expr.}^X$, $\mathsf{Depart.}^X$ from the same domain OID^X to targets ***String***, ***Integer***, ***String*** resp., which are compactly specified by tables as shown for model A. The target model space $\mathbf{B}$ is given by a similar schema $S_{\mathbf{B}}$ consisting of two attributes. The $\mathbf{B}$-view $\mathsf{get}(X)$ of an $\mathbf{A}$-model X is computed by selecting those oids $\#O \in \mathsf{OID}^X$ for which $\mathsf{Depart.}^X(\#O)$ is an *IT-department*, i.e., an element of the set $IT \stackrel{\text{def}}{=} \{\text{ML}, \text{DB}\}$. For example, the upper part of the figure shows the IT-view B of model A.

We assume that all column names in schemas $S_{\mathbf{A}}$, and $S_{\mathbf{B}}$ are qualified by schema names, e.g., $\mathsf{OID}@S_{\mathbf{A}}$, $\mathsf{OID}@S_{\mathbf{B}}$ etc, so that schemas are disjoint except elementary domains like ***String*** etc. Also disjoint are OID-values, e.g., $\#\mathrm{J}@A$ and $\#\mathrm{J}@B$ are different elements, but constants like John and Mary are elements of set ***String*** shared by both schemas. To shorten long expressions in the diagrams, we will often omit qualifiers and write $\#J = \#J$ meaning $\#J@A = \#J@B$ or $\#J@B = \#J@B'$ depending on the context given by the diagram; often we will also write $\#J$ and $\#J'$ for such OIDs. Also, when we write $\#J = \#J$ inside block arrows denoting updates, we actually mean a pair, e.g., $(\#J@B, \#J@B')$.

Given two models over the same schema, say, B and B' over $S_{\mathbf{B}}$, an update $v\colon B \to B'$ is a relation $v \subset \mathsf{OID}^B \times \mathsf{OID}^{B'}$; if a schema contains several nodes, an update should provide a relation v_N for each node N in the schema.

Note an essential difference between the two parallel updates $v_1, v_2\colon B \to B'$ specified in the figure. Update v_1 says that John's name was changed to Jon (think of fixing a typo), and the experience data for Mary were also corrected (either because of a typo or, e.g., because the department started to use a new ML method for which Mary has a longer experience). Update v_2 specifies the same story for John but a new story for Mary: it says that Mary $\#M$ left the IT-view and Mary $\#M'$ is a new employee in one of IT-departments.

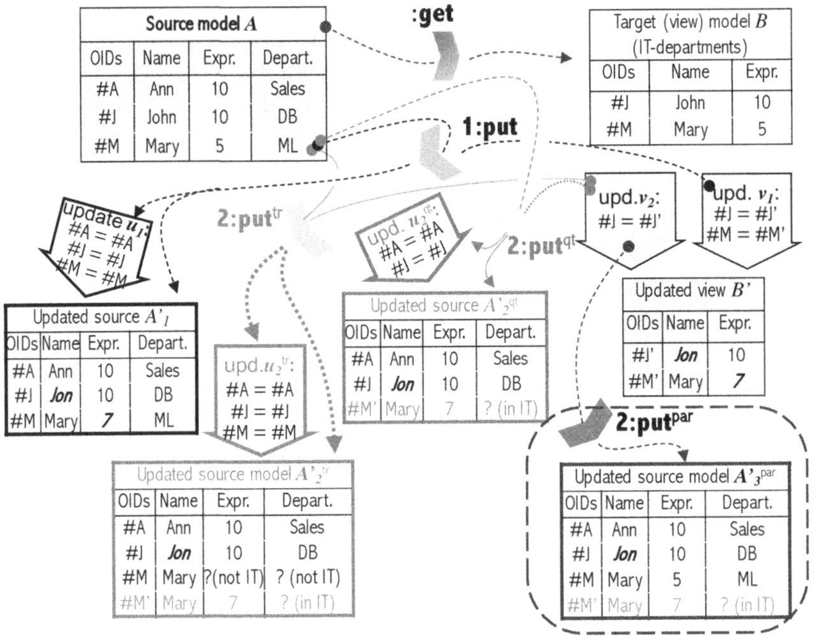

Fig. 1: Example of update propagation

2.3 Update propagation and update policies

The updated view B' is inconsistent with the source A and the latter is to be updated accordingly — we say that update v is to be propagated back to A. Propagation of v_1 is easy: we just update accordingly the values of the attributes as shown in the figure in the block arrow u_1: $A \to A'_1$ (of black colour). Importantly, propagation needs two pieces of data: the view update v_1 and the original state A of the source as shown in the figure by two data-flow lines into the chevron 1:put; the latter denotes invocation of the backward propagation operation put (read "put view update back to the source"). The quadruple $1 = (v_1, A,\ u_1, A')$ can be seen as an *instance* of operation put, hence the notation 1:put (borrowed from the UML).

Propagation of update v_2 is more challenging: Mary can disappear from the IT-view because a) she quit the company, b) she transitioned to a non-IT department, and c) the view definition has changed, e.g., the new view must only show employees with experience more than 5 years. Choosing between these possibilities is often called choosing an *(update) policy*. We will consider the case of changing the view in Sect. 3, and in the current section discuss policies a) and b) (ignore for a while the propagation scenario shown in blue in the right lower corner of the figure that shows policy c)).

For policy a), further referred to as *quiting* and briefly denoted by qt, the result of update propagation is shown in the figure with green colour: notice the update (block) arrow u_2^{qt} and its result, model $A_2'^{\mathsf{qt}}$, produced by invoking operation $\mathsf{put}^{\mathsf{qt}}$. Note that while we know the new employee Mary works in one of IT departments, we do not know in which one. This is specified with a special value '?' (a.k.a. labelled null in the database parlance).

For policy b), further referred to as *transition* and denoted tr, the result of update propagation is shown in the figure with orange colour: notice update arrow u_2^{tr} and its result, model $A_2'^{\mathsf{tr}}$ produced by $\mathsf{put}^{\mathsf{tr}}$. Mary #M is the old employee who transitioned to a new non-IT department, for which her expertize is unknown. Mary #M' is a new employee in one of IT-departments (we assume that the set of departments is not exhausted by those appearing in a particular state $A \in \mathbf{A}$). There are also updates whose backward propagation is uniquely defined and does not need a policy, e.g., update v_1 is such.

An important property of update propagations we have considered is that they restore consistency: the view of the updated source equals to the updated view initiated the update: $\mathsf{get}(A') = B'$; moreover, this equality extends for update arrows: $\mathsf{get}(u_i) = v_i$, $i = 1, 2$. Such extensions can be derived from view definitions if the latter are determined by so called monotonic queries (which encompass a wide class of practically useful queries including the Select-Project-Join class). For views defined by non-monotonic queries, in order to obtain get's action on source updates u: $A \to A'$, a suitable policy is to be added to the view definition (see [1,14,12] for details and discussion). Moreover, normally get preserves identity updates, $\mathsf{get}(\mathsf{id}_A) = \mathsf{id}_{\mathsf{get}(A)}$, and update composition: for any u: $A \to A'$ and u': $A' \to A''$, equality $\mathsf{get}(u; u') = \mathsf{get}(u); \mathsf{get}(u')$ holds.

2.4 Delta lenses

Our discussion of the example can be summarized in the following algebraic terms. We have two categories of *models* and *updates*, $\mathbf{A}$ and $\mathbf{B}$, and a functor $\mathsf{get}\colon \mathbf{A} \to \mathbf{B}$ incrementally computing $\mathbf{B}$-views of $\mathbf{A}$-models (we will often write $A.\mathsf{get}$ for $\mathsf{get}(A)$). We also suppose that for a chosen update policy, we have worked out precise procedures for how to propagate any view update backwards. This gives us a family of operations $\mathsf{put}_A : \mathbf{A}(A, \_) \leftarrow \mathbf{B}(A.\mathsf{get}, \_)$ indexed by $\mathbf{A}$-objects, $A \in \mathbf{A}_0$, for which we write $\mathsf{put}_A.v$ or $\mathsf{put}_A(v)$ interchangeably.

Definition 1 (Delta Lenses ([12])) Let $\mathbf{A}$, $\mathbf{B}$ be two categories. An *(asymmetric delta) lens* from $\mathbf{A}$ (the source of the lens) to $\mathbf{B}$ (the target) is a pair $\ell = (\mathsf{get}, \mathsf{put})$, where $\mathsf{get}\colon \mathbf{A} \to \mathbf{B}$ is a functor and put is a family of operations $\mathsf{put}_A : \mathbf{A}(A, \_) \leftarrow \mathbf{B}(A.\mathsf{get}, \_)$ indexed by objects of $\mathbf{A}$, $A \in \mathbf{A}_0$. Given A, operation put_A maps any arrow $v\colon A.\mathsf{get} \to B'$ to an arrow $u\colon A \to A'$ such that $A'.\mathsf{get} = B'$. The last condition is called (co)discrete Putget law:

$$(\mathsf{Putget})_0 \quad (\mathsf{put}_A.v).\mathsf{cod}.\mathsf{get}_0 = v.\mathsf{cod} \text{ for all } A \in \mathbf{A}_0 \text{ and } v \in \mathbf{B}(A.\mathsf{get}, \_)$$

where get_0 denotes the object function of functor get. We will write a lens as an arrow $\ell\colon \mathbf{A} \to \mathbf{B}$ going in the direction of get.

Note that family put corresponds to a chosen update policy, e.g., in terms of the example above, for the same view functor get, we have two families of put-operations, $\mathsf{put}^{\mathsf{qt}}$ and $\mathsf{put}^{\mathsf{tr}}$, corresponding to the two updated policies we discussed. These two policies determine two lenses $\ell^{\mathsf{qt}} = (\mathsf{get}, \mathsf{put}^{\mathsf{qt}})$ and $\ell^{\mathsf{tr}} = (\mathsf{get}, \mathsf{put}^{\mathsf{tr}})$ sharing the same get.

Definition 2 (Well-behavedness) A *(lens) equational law* is an equation to hold for all values of two variables: $A \in \mathbf{A}_0$ and $v\colon A.\mathsf{get} \to T'$. A lens is called *well-behaved (wb)* if the following two laws hold:

$$(\mathsf{Stability}) \quad \mathsf{id}_A = \mathsf{put}_A.\mathsf{id}_{A.\mathsf{get}} \text{ for all } A \in \mathbf{A}_0$$

$$(\mathsf{Putget}) \quad (\mathsf{put}_A.v).\mathsf{get} = v \text{ for all } A \in \mathbf{A}_0 \text{ and all } v \in \mathbf{B}(A.\mathsf{get}, \_)$$

Remark 1. Stability law says that a wb lens does nothing if nothing happens on the target side (no actions without triggers). Putget requires consistency after the backward propagation is finished. Note the distinction between the Putget$_0$ condition included into the very definition of a lens, and the full Putget law required for the wb lenses. The former is needed to ensure smooth tiling of put-squares (i.e., arrow squares describing application of put to a view update and its result) both horizontally (for sequential composition) and vertically (not considered in the paper). The full Putget assures true consistency as considering a state B' alone does not say much about the real update and elements of B' cannot be properly interpreted. The real story is specified by delta $v\colon B \to B'$, and consistency restoration needs the full (PutGet) law as above. [2]

A more detailed trailer of lenses can be found in the long version [9].

[2] As shown in [6], the Putget$_0$ condition is needed if we want to define operations put separately from the functor get: then we still need a function $\mathsf{get}_0\colon \mathbf{A}_0 \to \mathbf{B}_0$ and the codiscrete Putget law to ensure a reasonable behaviour of put.

3 Asymmetric Learning Lenses with Amendments

We will begin with a brief motivating discussion, and then proceed with formal definitions

3.1 Does Bx need categorical learning?

Enriching delta lenses with learning capabilities has a clear practical sense for Bx. Having a lens $(\mathsf{get}, \mathsf{put})$: $\mathbf{A} \to \mathbf{B}$ and inconsistency $A.\mathsf{get} \neq B'$, the idea of learning extends the notion of the search space and allows us to update the transformation itself so that the final consistency is achieved for a new transformation get': $A.\mathsf{get}' = B'$. For example, in the case shown in Fig. 1, disappearance of Mary #M in the updated view B' can be caused by changing the view definition, which now requires to show only those employees whose experience is more than 5 years and hence Mary #M is to be removed from the view, while Mary #M' is a new IT-employee whose experience satisfies the new definition. Then the update v_2 can be propagated as shown in the bottom right corner of Fig. 1, where index par indicates a new update policy allowing for view definition (parameter) change.

To manage the extended search possibilities, we parameterize the space of transformations as a family of mappings get_p: $\mathbf{A} \to \mathbf{B}$ indexed over some parameter space $p \in \mathbf{P}$. For example, we may define the IT-view to be parameterized by the experience of employees shown in the view (including *any* experience as a special parameter value). Then we have two interrelated propagation operations that map an update $B \rightsquigarrow B'$ to a parameter update $p \rightsquigarrow p'$ and a source update $A \rightsquigarrow A'$. Thus, the extended search space allows for new update policies that look for updating the parameter as an update propagation possibility. The possibility to update the transformation appears to be very natural in at least two important Bx scenarios: a) model transformation design and b) model transformation evolution (cf. [21]), which necessitates the enrichment of the delta lens framework with parameterization and learning. Note that all transformations get_p, $p \in \mathbf{P}$ are to be elements of the same lens, and operations put are *not* indexed by p, hence, formalization of learning by considering a family of ordinary lenses would *not* do the job.

Categorical vs. codiscrete learning Suppose that the parameter p is itself a set, e.g., the set of departments forming a view can vary depending on some context. Then an update from p to p' has a relational structure as discussed above, i.e., e: $p \to p'$ is a relation $e \subset p \times p'$ specifying which departments disappeared from the view and which are freshly added. This is a general phenomenon: as soon as parameters are structures (sets of objects or graphs of objects and attributes), a parameter change becomes a structured delta and the space of parameters gives rise to a category $\mathbf{P}$. The search/propagation procedure returns an arrow e: $p \to p'$ in this category, which updates the parameter value from p to p'. Hence, a general model of supervised learning should assume $\mathbf{P}$ to be a category (and we say that learning is *categorical*). The case of the parameter space being a set is captured by considering a codiscrete category $\mathbf{P}$ whose only arrows are pairs of its objects; we call such learning *codiscrete*.

3.2 Ala-lenses

The notion of a *parameterized functor (p-functor)* is fundamental for ala-lenses, but is not a lens notion per se and is thus placed into Appendix Sect. A.1. We will work with its exponential (rather than equivalent product-based) formulation but will do uncurrying and currying back if necessary, and often using the same symbol for an arrow f and its uncurried version $\check{f}$.

Definition 3 (ala-lenses) Let $\mathbf{A}$ and $\mathbf{B}$ be categories. An *ala-lens* from $\mathbf{A}$ (the *source* of the lens) to $\mathbf{B}$ (the *target*) is a pair $\ell = (\mathsf{get}, \mathsf{put})$ whose first component is a p-functor $\mathsf{get}\colon \mathbf{A} \xrightarrow{\mathbf{P}} \mathbf{B}$ and the second component is a triple of (families of) operations $\mathsf{put} = (\mathsf{put}^{\mathsf{upd}}_{p,A}, \mathsf{put}^{\mathsf{req}}_{p,A}, \mathsf{put}^{\mathsf{self}}_{p,A})$ indexed by pairs $p \in \mathbf{P}_0$, $A \in \mathbf{A}_0$; arities of the operations are specified below after we introduce some notation. Names req (for 'request') and upd (for 'update') are chosen to match the terminology in [17].

Categories $\mathbf{A}$, $\mathbf{B}$ are called *model spaces*, their objects are *models* and their arrows are *(model) updates* or *deltas*. Objects of $\mathbf{P}$ are called *parameters* and are denoted by small letters $p, p', ..$ rather than capital ones to avoid confusion with [17], in which capital P is used for the entire parameter set. Arrows of $\mathbf{P}$ are called *parameter deltas*. For a parameter $p \in \mathbf{P}_0$, we write get_p for the functor $\mathsf{get}(p)\colon \mathbf{A} \to \mathbf{B}$ (read "get $\mathbf{B}$-views of $\mathbf{A}$"), and if $A \in \mathbf{A}_0$ is a source model, its get_p-view is denoted by $\mathsf{get}_p(A)$ or $A.\mathsf{get}_p$ or even A_p (so that $\__p$ becomes yet another notation for functor get_p). Given a parameter delta $e\colon p \to p'$ and a source model $A \in \mathbf{A}_0$, the model delta $\mathsf{get}(e)\colon \mathsf{get}_p(A) \to \mathsf{get}_{p'}(A)$ will be denoted by $\mathsf{get}_e(A)$ or e_S (rather than A_e as we would like to keep capital letters for objects only). In the uncurried version, $\mathsf{get}_e(A)$ is nothing but $\check{\mathsf{get}}(e, \mathsf{id}_S)$

Since get_e is a natural transformation, for any delta $u\colon A \to A'$ we have a commutative square $e_S; u_{p'} = u_p; e_{A'}$ (whose diagonal is $\check{\mathsf{get}}(e, u)$). We will denote the diagonal of this square by $u.\mathsf{get}_e$ or $u_e\colon A_p \to A'_{p'}$. Thus, we use notation

$$(1)\quad \begin{array}{l} A_p \overset{\text{def}}{=} A.\mathsf{get}_p \overset{\text{def}}{=} \mathsf{get}_p(A) \overset{\text{def}}{=} \mathsf{get}(p)(A) \\ u_e \overset{\text{def}}{=} u.\mathsf{get}_e \overset{\text{def}}{=} \mathsf{get}_e(u) \overset{\text{def}}{=} \mathsf{get}(e)(u) \overset{\text{def}}{=} e_S; u_{p'} \overset{\text{nat}}{=} u_p; e_{A'}\colon A_p \to A'_{p'} \end{array}$$

Now we describe operations put. They all have the same indexing set $\mathbf{P}_0 \times \mathbf{A}_0$, and the same domain: for any index p, A and any model delta $v\colon A_p \to B'$ in $\mathbf{B}$, the value $\mathsf{put}^{\mathbf{x}}_{p,A}(p, A)$, $\mathbf{x} \in \{\mathsf{req}, \mathsf{upd}, \mathsf{self}\}$ is defined and unique:

$$(2)\quad \begin{array}{ll} \mathsf{put}^{\mathsf{upd}}_{p,A}\colon p \to p' & \text{is a parameter delta from } p, \\ \mathsf{put}^{\mathsf{req}}_{p,A}\colon A \to A' & \text{is a model delta from } A, \\ \mathsf{put}^{\mathsf{self}}_{p,A}\colon B' \to A'_{p'} & \text{is a model delta from } B' \\ & \text{called the \textit{amendment} and denoted by } v^{@}. \end{array}$$

Note that the definition of $\mathsf{put}^{\mathsf{self}}$ involves an equational dependency between all three operations: for all $A \in \mathbf{A}_0$, $v \in \mathbf{B}(A.\mathsf{get}, \_)$, we require

$$(\mathsf{Putget})_0 \quad (\mathsf{put}^{\mathsf{req}}_A.v).\mathsf{cod}.\mathsf{get}_{p'} = (v; \mathsf{put}^{\mathsf{self}}_A).\mathsf{cod} \text{ where } p' = (\mathsf{put}^{\mathsf{upd}}_A.v).\mathsf{cod}$$

We will write an ala-lens as an arrow $\ell = (\mathsf{get}, \mathsf{put})\colon \mathbf{A} \xrightarrow{\mathbf{P}} \mathbf{B}$.

A lens is called *(twice) codiscrete* if categories $\mathbf{A}$, $\mathbf{B}$, $\mathbf{P}$ are codiscrete and thus $\mathsf{get}\colon \mathbf{A} \xrightarrow{\mathbf{P}} \mathbf{B}$ is a parameterized function. If only $\mathbf{P}$ is codiscrete, we call ℓ a *codiscretely learning* delta lens, while if only model spaces are codiscrete, we call ℓ a *categorically learning* codiscrete lens.

Diagram in Fig. 2 shows how a lens' operations are interrelated. The upper part shows an arrow $e\colon p \to p'$ in category $\mathbf{P}$ and two corresponding functors from $\mathbf{A}$ to $\mathbf{B}$. The lower part is to be seen as a 3D-prism with visible front face $AA_{p'}A'_{p'}A'$ and visible upper face $AA_pA_{p'}$, the bottom and two back faces are invisible, and the corresponding arrows are dashed. The prism denotes an algebraic term: given elements are shown with black fill and white font while derived elements are blue (recalls being mechanically computed) and blank (double-body arrows are considered as "blank"). The two pairs of arrows originating from A and A' are not blank because they denote pairs of nodes (the UML says *links*) rather than mappings/deltas between nodes.

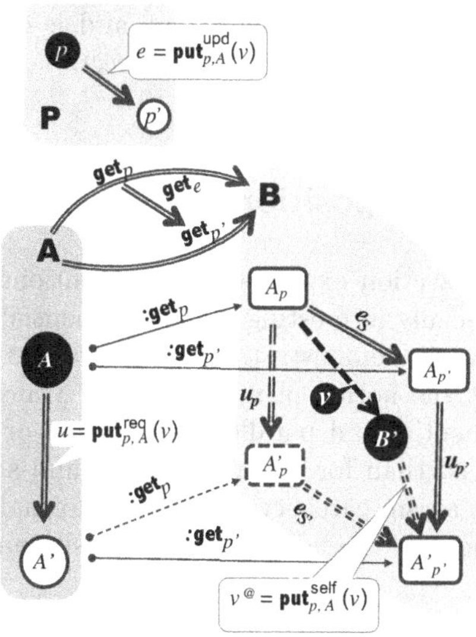

Fig. 2: Ala-lens operations

Equational definitions of deltas $e, u, v^@$ are written up in the three callouts near them. The right back face of the prism is formed by the two vertical derived deltas $u_p = u.\mathsf{get}_p$ and $u_{p'} = u.\mathsf{get}_{p'}$, and the two matching them horizontal derived deltas $e_S = \mathsf{get}_e(A)$ and $e_{A'} = \mathsf{get}_e(A')$; together they form a commutative square due to the naturality of $\mathsf{get}(e)$ as explained earlier.

Definition 4 (Well-behavedness) An ala-lens is called *well-behaved (wb)* if the following two laws hold for all $p \in \mathbf{P}_0$, $A \in \mathbf{A}_0$ and $v\colon A_p \to B'$:

(Stability) if $v = \mathsf{id}_{A_p}$ then all three propagated updates $e, u, v^@$ are identities:
$$\mathsf{put}^{\mathsf{upd}}_{p,A}(\mathsf{id}_{A_p}) = \mathsf{id}_p, \quad \mathsf{put}^{\mathsf{req}}_{p,A}(\mathsf{id}_{A_p}) = \mathsf{id}_S, \quad \mathsf{put}^{\mathsf{self}}_{p,A}(\mathsf{id}_{A_p}) = \mathsf{id}_{A_p}$$

(Putget) $(\mathsf{put}^{\mathsf{req}}_{p,A}.v).\mathsf{get}_e = v; v^@$ where $e = \mathsf{put}^{\mathsf{upd}}_{p,A}(v)$ and $v^@ = \mathsf{put}^{\mathsf{self}}_{p,A}(v)$

Remark 2. Note that Remark 1 about the Putget law is again applicable.

Example 1 (Identity lenses). Any category $\mathbf{A}$ gives rise to an ala-lens $id_{\mathbf{A}}$ with the following components. The source and target spaces are equal to $\mathbf{A}$, and the parameter space is **1**. Functor get is the identity functor and all puts are identities. Obviously, this lens is wb.

Example 2 (Iso-lenses). Let ι: $\mathbf{A} \to \mathbf{B}$ be an isomorphism between model spaces. It gives rise to a wb ala-lens $\ell(\iota)$: $\mathbf{A} \to \mathbf{B}$ with $\mathbf{P}^{\ell(\iota)} = \mathbf{1} = \{*\}$ as follows. Given any A in $\mathbf{A}$ and v: $\iota(A) \to B'$ in $\mathbf{B}$, we define $\mathsf{put}^{\ell(\iota).\mathsf{req}}_{*,A}(v) = \iota^{-1}(v)$ while the two other put operations map v to identities.

Example 3 (Bx lenses). Examples of wb aa-lenses modelling a Bx can be found in [11]: they all can be considered as ala-lenses with a trivial parameter space **1**.

Example 4 (Learners). Learners defined in [17] are codiscretely learning codiscrete lenses with amendment, and as such satisfy (the amended) Putget (Remark 1). Looking at the opposite direction, ala-lenses are a categorification of learners as detailed in Fig. 8 on p. 194.

4 Compositionality of ala-lenses

This section explores the compositional structure of the universe of ala-lenses; especially interesting is their sequential composition. We will begin with a small example demonstrating sequential composition of ordinary lenses and showing that the notion of update policy transcends individual lenses. Then we define sequential and parallel composition of ala-lenses (the former is much more involved than for ordinary lenses) and show that wb ala-lenses can be organized into an sm-category. Finally, we formalize the notion of a compositional update policy via the notion of a compositional bidirectional language.

4.1 Compositionality of update policies: An example

Fig. 3 extends the example in Fig. 1 with a new model space **C** whose schema consists of the only attribute Name, and a view of the IT-view, in which only employees of the ML department are to be shown. Thus, we now have two functors, get1: $\mathbf{A} \to \mathbf{B}$ and get2: $\mathbf{B} \to \mathbf{C}$, and their composition Get: $\mathbf{A} \to \mathbf{C}$ (referred to as the *long* get). The top part of Fig. 3 shows how it works for model A considered above.

Each of the two policies, policy qt (green) and policy tr (orange), in which person's disappearance from the view are interpreted, resp., as quiting the company and transitioning to a department not included into the view, is applicable to the new view mappings get2 and Get, thus giving us six lenses shown in Fig. 4 with solid arrows; amongst them, lenses, $\mathcal{L}^{\mathsf{qt}}$ and $\mathcal{L}^{\mathsf{tr}}$ are obtained by applying policy *pol* to the (long) functor Get;, and we will refer to them *long* lenses. In addition, we can compose lenses of the same colour as shown in Fig. 4 by dashed arrows (and we can also compose lenses of different colours (ℓ_1^{qt} with ℓ_2^{tr} and ℓ_1^{tr} with ℓ_2^{qt}) but we do not need them). Now an important question is how long and composed lenses are related: whether $\mathcal{L}^{pol}$ and $\ell_1^{pol};\ell_2^{pol}$ for $pol \in \{\mathsf{qt},\mathsf{tr}\}$, are equal (perhaps up to some equivalence) or different?

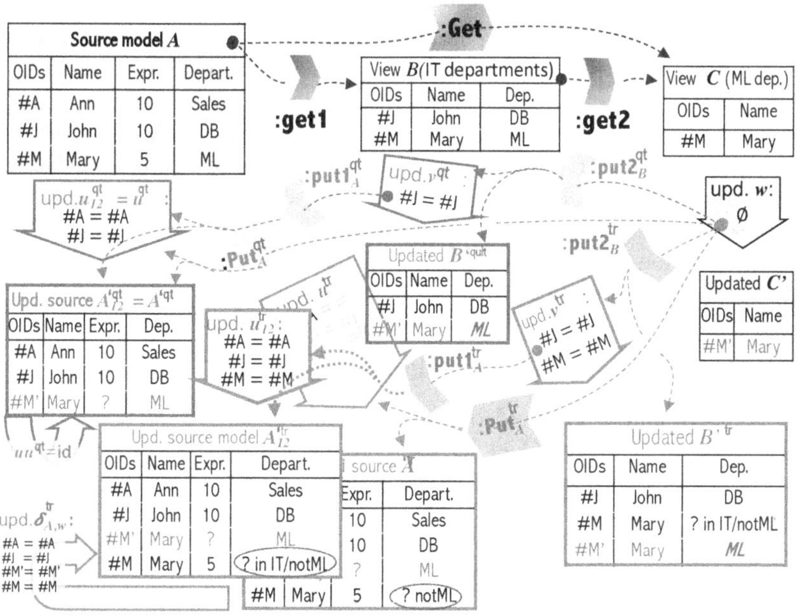

Fig. 3: Example cont'd: functoriality of update policies

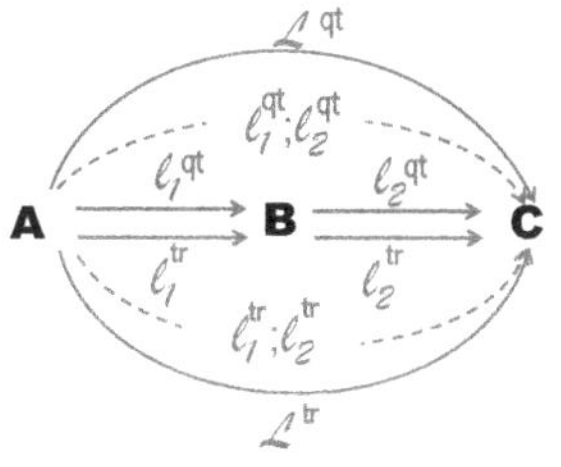

Fig. 4: Lens combination schemas for Fig. 3

Fig. 3 demonstrates how the mechanisms work with a simple example. We begin with an update w of the view C that says that Mary $\#M$ left the ML department, and a new Mary $\#M'$ was hired for ML. Policy qt interprets Mary's disappearance as quiting the company, and hence this Mary doesn't appear in view B'^{qt} produced by $\mathsf{put2}^{\mathsf{qt}}$ nor in view A'^{qt}_{12} produced from B'^{qt} by $\mathsf{put1}^{\mathsf{qt}}$, and updates v^{qt} and u^{qt}_{12} are written accordingly. Obviously, Mary also does not appear in view A'^{qt} produced by the long lens's $\mathsf{Put}^{\mathsf{qt}}$. Thus, $\mathsf{put1}^{\mathsf{qt}}_A(\mathsf{put2}^{\mathsf{qt}}_A(w)) = \mathsf{Put}^{\mathsf{qt}}_A(w)$, and it is easy to understand that such equality will hold for any source model A and any update $w\colon C \to C'$ due to the nature of our two views $\mathsf{get1}$ and $\mathsf{get2}$. Hence, $\mathcal{L}^{\mathsf{qt}} = \ell^{\mathsf{qt}}_1;\ell^{\mathsf{qt}}_2$ where $\mathcal{L}^{\mathsf{qt}} = (\mathsf{Get}, \mathsf{Put}^{\mathsf{qt}})$ and $\ell^{\mathsf{qt}}_i = (\mathsf{get}i, \mathsf{put}i^{\mathsf{qt}})$.

The situation with policy tr is more interesting. Model A'^{tr}_{12} produced by the composed lens $\ell^{\mathsf{tr}}_1;\ell^{\mathsf{tr}}_2$, and model A'^{tr} produced by the long lens $\mathcal{L}^{\mathsf{tr}} = (\mathsf{Get}, \mathsf{Put}^{\mathsf{tr}})$ are different as shown in the figure (notice the two different values for Mary's department framed with red ovals in the models). Indeed, the composed lens has more information about the old employee Mary—it knows that Mary was in the IT view, and hence can propagate the update more accurately. The comparison update $\delta^{\mathsf{tr}}_{A,w}\colon A'^{\mathsf{tr}} \to A'^{\mathsf{tr}}_{12}$ adds this missing information so that equality $u^{\mathsf{tr}};\delta^{\mathsf{tr}}_{A,w} = u^{\mathsf{tr}}_{12}$ holds. This is a general phenomenon: functor composition looses information and, in general, functor $\mathsf{Get} = \mathsf{get1};\mathsf{get2}$ knows less than the pair $(\mathsf{get1}, \mathsf{get2})$. Hence, operation Put back-propagating updates over Get (we will also say *inverting* Get) will, in general, result in less certain models than composition $\mathsf{put1} \circ \mathsf{put2}$ that inverts the composition $\mathsf{get1};\mathsf{get2}$ (a discussion and examples of this phenomenon in the context of vertical composition of updates can be found in [8]). Hence, comparison updates such as $\delta^{\mathrm{tr}}_{A,w}$ should exist for any A and any $w\colon A.\mathsf{Get} \to C'$, and together they should give rise to something like a natural transformation between lenses, $\delta^{\mathrm{tr}}_{\mathbf{A},\mathbf{B},\mathbf{C}}\colon \mathscr{L}^{\mathrm{tr}} \Rightarrow \ell^{\mathrm{tr}}_1;\ell^{\mathrm{tr}}_2$. To make this notion precise, we need a notion of natural transformation between "functors" put, which we leave for future work. In the present paper, we will consider policies like qt, for which strict equality holds.

4.2 Sequential composition of ala-lenses

Let $\mathscr{k}\colon \mathbf{A} \to \mathbf{B}$ and $\ell\colon \mathbf{B} \to \mathbf{C}$ be two ala-lenses with parameterized functors $\mathsf{get}^{\mathscr{k}}\colon \mathbf{P} \to [\mathbf{A},\mathbf{B}]$ and $\mathsf{get}^{\ell}\colon \mathbf{Q} \to [\mathbf{B},\mathbf{C}]$ resp. Their *composition* is the following ala-lens $\mathscr{k};\ell$. Its parameter space is the product $\mathbf{P} \times \mathbf{Q}$, and the get-family is defined as follows. For any pair of parameters (p,q) (we will write pq), $\mathsf{get}^{\mathscr{k};\ell}_{pq} = \mathsf{get}^{\mathscr{k}}_{p};\mathsf{get}^{\ell}_{q}\colon \mathbf{A} \to \mathbf{C}$. Given a pair of parameter deltas, $e\colon p \to p'$ in $\mathbf{P}$ and $h\colon q \to q'$ in $\mathbf{Q}$, their $\mathsf{get}^{\mathscr{k};\ell}$-image is the Godement product $*$ of natural transformations, $\mathsf{get}^{\mathscr{k};\ell}(eh) = \mathsf{get}^{\mathscr{k}}(e) * \mathsf{get}^{\ell}(h)$ (we will also write $\mathsf{get}^{\mathscr{k}}_{e} \,||\, \mathsf{get}^{\ell}_{h}$)

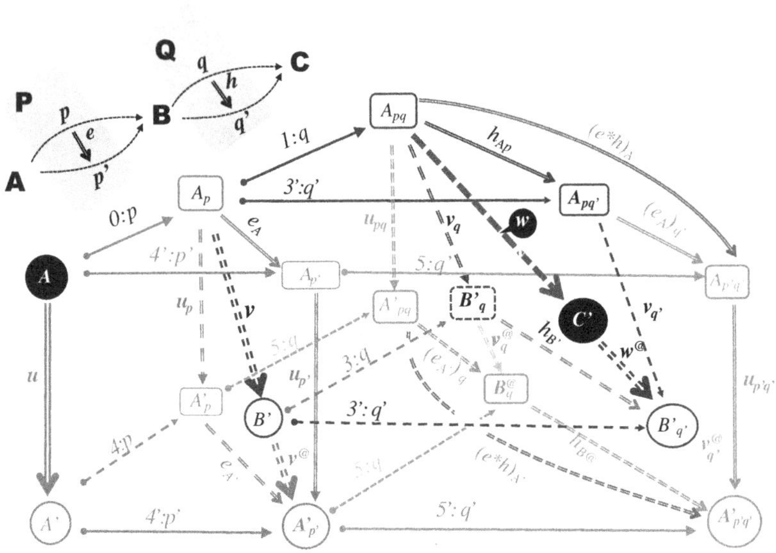

Fig. 5: Sequential composition of apa-lenses

Now we define $\mathscr{k};\ell$'s propagation operations puts. Let (A, pq, A_{pq}) with $A \in \mathbf{A}_0$, $pq \in (\mathbf{P}\times\mathbf{Q})_0$, $A.\mathsf{get}^{\mathscr{k}}_{p}.\mathsf{get}^{\ell}_{q} = A_{pq} \in \mathbf{C}_0$ be a state of lens $\mathscr{k};\ell$, and $w\colon A_{pq} \to C'$ is a target update as shown in Fig. 3. For the first propagation step, we run lens ℓ as shown in Fig. 3 with the blue colour for derived elements: this is just an

instantiation of the pattern of Fig. 2 with the source object being $A_p = A.\mathsf{get}_p$ and parameter q. The results are deltas
(3)
$$h = \mathsf{put}^{\ell.\mathsf{upd}}_{q,A_p}(w)\colon q \to q', v = \mathsf{put}^{\ell.\mathsf{req}}_{q,A_p}(w)\colon A_p \to B', w^{@} = \mathsf{put}^{\ell.\mathsf{self}}_{q,A_p}(w)\colon C' \to B'_{q'}.$$

Next we run lens k at state (p, A) and the target update v produced by lens ℓ; it is yet another instantiation of pattern in Fig. 2 (this time with the green colour for derived elements), which produces three deltas
(4)
$$e = \mathsf{put}^{k.\mathsf{upd}}_{p,A}(v)\colon p \to p', u = \mathsf{put}^{k.\mathsf{req}}_{p,A}(v)\colon A \to A', v^{@} = \mathsf{put}^{k.\mathsf{self}}_{p,A}(v)\colon B' \to A'_{p'}.$$

These data specify the green prism adjoint to the blue prism: the edge v of the latter is the "first half" of the right back face diagonal $A_pA'_{p'}$ of the former. In order to make an instance of the pattern in Fig. 2 for lens $k;\ell$, we need to extend the blue-green diagram to a triangle prism by filling-in the corresponding "empty space". These filling-in arrows are provided by functors get^{ℓ} and get^{k} and shown in orange (where we have chosen one of the two equivalent ways of forming the Godement product – note two curve brown arrows). In this way we obtain yet another instantiation of the pattern in Fig. 2 denoted by $k;\ell$:

$$(5)\qquad \mathsf{put}^{(k;\ell)\mathsf{upd}}_{A,pq}(w) = (e, h), \quad \mathsf{put}^{(k;\ell)\mathsf{req}}_{A,pq}(w) = u, \quad \mathsf{put}^{(k;\ell)\mathsf{self}}_{A,pq}(w) = w^{@}; v^{@}_{q'}$$

where $v^{@}_{q'}$ denotes $v^{@}.\mathsf{get}_{q'}$. Thus, we built an ala-lens $k;\ell$, which satisfies equation Putget_0 by construction.

Theorem 1 (Sequential composition and lens laws). *Given ala-lenses $k\colon \mathbf{A} \to \mathbf{B}$ and $\ell\colon \mathbf{B} \to \mathbf{C}$, let lens $k;\ell\colon \mathbf{A} \to \mathbf{C}$ be their sequential composition as defined above. Then the lens $k;\ell$ is wb as soon as lenses k and ℓ are such.*

See [9, Appendix A.3] for a proof.

4.3 Parallel composition of ala-lenses

Let $\ell_i\colon \mathbf{A}_i \to \mathbf{B}_i$, $i = 1, 2$ be two ala-lenses with parameter spaces $\mathbf{P}_i$. The lens $\ell_1\|\ell_2\colon \mathbf{A}_1{\times}\mathbf{A}_2 \to \mathbf{B}_1{\times}\mathbf{B}_2$ is defined as follows. Parameter space $\ell_1\|\ell_2.\mathbf{P} = \mathbf{P}_1 \times \mathbf{P}_2$. For any pair $p_1\|p_2 \in (\mathbf{P}_1{\times}\mathbf{P}_2)_0$, define $\mathsf{get}^{\ell_1\|\ell_2}_{p_1\|p_2} = \mathsf{get}^{\ell_1}_{p_1} \times \mathsf{get}^{\ell_2}_{p_2}$ (we denote pairs of parameters by $p_1\|p_2$ rather than $p_1 \otimes p_2$ to shorten long formulas going beyond the page width). Further, for any pair of models $A_1\|A_2 \in (\mathbf{A}_1 \times \mathbf{A}_2)_0$ and deltas $v_1\|v_2\colon (A_1\|A_2).\mathsf{get}^{\ell_1\|\ell_2}_{p_1\|p_2} \to B'_1\|B'_2$, we define componentwise

$$e = \mathsf{put}^{(\ell_1\|\ell_2)\mathsf{upd}}_{p_1\|p_2,A_1\|A_2}(v_1\|v_2)\colon p_1\|p_2 \to p'_1\|p'_2$$

by setting $e = e_1\|e_2$ where $e_i = \mathsf{put}^{\ell_i}_{p_i,S_i}(v_i)$, $i = 1, 2$ and similarly for $\mathsf{put}^{(\ell_1\|\ell_2)\mathsf{req}}_{p_1\|p_2,A_1\|A_2}$ and $\mathsf{put}^{(\ell_1\|\ell_2)\mathsf{self}}_{p_1\|p_2,A_1\|A_2}$ The following result is obvious.

Theorem 2 (Parallel composition and lens laws). *Lens $\ell_1\|\ell_2$ is wb as soon as lenses ℓ_1 and ℓ_2 are such.*

4.4 Symmetric monoidal structure over ala-lenses

Our goal is to organize ala-lenses into an sm-category. To make sequential composition of ala-lenses associative, we need to consider them up to some equivalence (indeed, Cartesian product is not strictly associative).

Definition 5 (Ala-lens Equivalence) Two parallel ala-lenses $\ell, \hat{\ell}\colon \mathbf{A} \to \mathbf{B}$ are called *equivalent* if their parameter spaces are isomorphic via a functor $\iota\colon \mathbf{P} \to \hat{\mathbf{P}}$ such that for any $A \in \mathbf{A}_0$, $e\colon p \to p' \in \mathbf{P}$ and $v\colon (A.\mathsf{get}_p) \to T'$ the following holds (for $\mathbf{x}{\in}\{\mathsf{req}, \mathsf{self}\}$):

$$A.\mathsf{get}_e = A.\widehat{\mathsf{get}}_{\iota(e)},\ \iota(\mathsf{put}^{\mathsf{upd}}_{p,A}(v)) = \widehat{\mathsf{put}}_{\iota(p),A}(v),\ \text{and}\ \mathsf{put}^{\mathbf{x}}_{p,A}(v) = \widehat{\mathsf{put}}^{\mathbf{x}}_{\iota(p),A}(v)$$

Remark 3. It would be more categorical to require delta isomorphisms (i.e., commutative squares whose horizontal edges are isomorphisms) rather than equalities as above. However, model spaces appearing in Bx-practice are skeletal categories (and even stronger than skeletal in the sense that all isos, including iso loops, are identities), for which isos become equalities so that the generality would degenerate into equality anyway.

It is easy to see that operations of lens' sequential and parallel composition are compatible with lens' equivalence and hence are well-defined for equivalence classes. Below we identify lenses with their equivalence classes by default.

Theorem 3 (Ala-lenses form an sm-category). *Operations of sequential and parallel composition of ala-lenses defined above give rise to an sm-category* ***aLaLens****, whose objects are model spaces (= categories) and arrows are (equivalence classes of) ala-lenses.* See [9, p.17 and Appendix A.4] for a proof.

4.5 Functoriality of learning in the *delta* lens setting

As example in Sect. 4.1 shows, the notion of update policy transcends individual lenses. Hence, its proper formalization needs considering the entire category of ala-lenses and functoriality of a suitable mapping.

Definition 6 (Bx-transformation language) A *compositional bidirectional model transformation language* $\mathcal{L}_{\mathsf{bx}}$ is given by **(i)** an sm-category ***pGet***$(\mathcal{L}_{\mathsf{bx}})$ whose objects are *($\mathcal{L}_{\mathsf{bx}}$-)model spaces* and arrows are *($\mathcal{L}_{\mathsf{bx}}$-)transformations* which is supplied with forgetful functor into ***pCat***, and **(ii)** an sm-functor $L_{\mathcal{L}_{\mathsf{bx}}}\colon$ ***pGet***$(\mathcal{L}_{\mathsf{bx}}) \to$ ***aLaLens*** such that the lower triangle in the inset diagram commutes. (Forgetful functors in this diagram are named "$-X$" with X referring to the structure to be forgotten.)

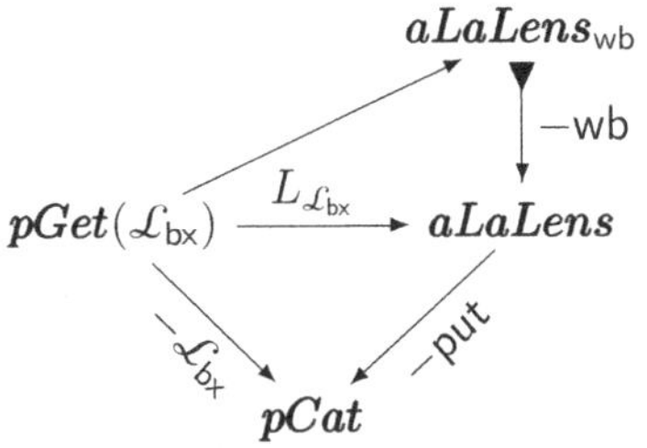

An $\mathcal{L}_{\mathsf{bx}}$-language is *well-behaved (wb)* if functor $L_{\mathcal{L}_{\mathsf{bx}}}$ factorizes as shown by the upper triangle of the diagram.

Example. A major compositionality result of Fong *et al* [17] states the existence of an sm-functor from the category of Euclidean spaces and parameterized differentiable functions (pd-functions) ***Para*** into the category ***Learn*** of learning algorithms (*learners*) as shown by the inset commutative diagram. (The functor is itself parameterized by a *step size* $0 < \varepsilon \in \mathbb{R}$ and an *error function* err: $\mathbb{R}\times\mathbb{R} \to \mathbb{R}$ needed to specify the gradient descent procedure.) However, learners are nothing but codiscrete ala-lenses (see Sect. A.2), and thus the inset diagram is a codiscrete specialization of the diagram in Def. 6 above. That is, the category of Euclidean spaces and pd-functions, and the gradient descent method for back propagation, give rise to a (codiscrete) compositional bx-transformation language (over ***pSet*** rather than ***pCat***).

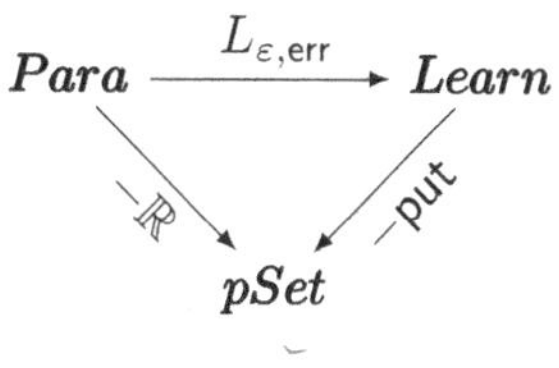

Finding a specifically Bx instance of Def. 6 (e.g., checking whether it holds for concrete languages and tools such as EMOFLON [23] or GROUNDTRAM [22]) is laborious and left for future work.

5 Related work

Figure 6 on the right is a simplified version of Fig. 8 on p. 194 convenient for our discussion here: immediate related work should be found in areas located at points (0,1) (codiscrete learning lenses) and (1,0) (delta lenses) of the plane. For the point (0,1), the paper [17] by Fong, Spivak and Tuyéras is fundamental: they defined the notion of a codiscrete learning lens (called a learner), proved a fundamental results about sm-functoriality of the gradient descent approach to ML, and thus laid a foundation for the compositional approach to change propagation with learning. One follow-up of that work is paper [16] by Fong and Johnson, in which they build an sm-functor ***Learn*** $\to$ ***sLens*** which maps learners to so called symmetric lenses. That paper is probably the first one where the terms 'lens' and 'learner' are met, but the initial observation that a learner whose parameter set is a singleton is actually a lens is due to Jules Hedges, see [16].

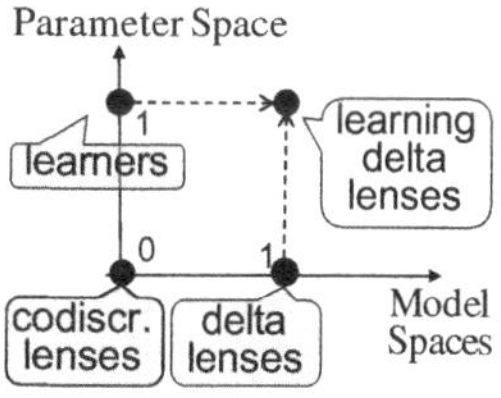

Fig. 6

There are conceptual and technical distinctions between [16] and the present paper. On the conceptual level, by encoding learners as symmetric lenses, they "hide" learning inside the lens framework and make it a technical rather than conceptual idea. In contrast, we consider parameterization and supervised learning as a fundamental idea and a first-class citizen for the lens framework, which grants creation of a new species of lenses. Moreover, while an ordinary lens is a way to invert a functor, a learning lens is a way to invert a parameterized functor so that learning lenses appear as an extension of the parameterization idea from functors to lenses. (This approach can probably be specified formally by treating parameterization as a suitably defined functorial construction.) Besides technical advantages (working with asymmetric lenses is simpler), our asymmetric model seems more adequate to the problem of learning functions rather than relations. On the technical level, the lens framework we develop in the paper is much more general than in [16]: we categorificated both the parameter space and model spaces, and we work with lenses with amendment (which allows us to relax the Putget law if needed).

As for the delta lens roots (the point (1,0) in the figure), delta lenses were motivated and formally defined in [12] (the asymmetric case) and [13] (the symmetric one). Categorical foundations for the delta lens theory were developed by Johnson and Rosebrugh in a series of papers (see [20] for references); this line is continued in Clarke's work [6]. The notion of a delta lens with amendments (in both asymmetric and symmetric variants) was defined in [11], and several composition results were proved. Another extensive body of work within the delta-based area is modelling and implementing model transformations with triple-graph grammars (TGG) [4,23]. TGG provide an implementation framework for delta lenses as is shown and discussed in [5,19,2], and thus inevitably consider change propagation on a much more concrete level than lenses. The author is not aware of any work considering functoriality of update policies developed within the TGG framework.

The present paper is probably the first one at the intersection (1,1) of the plane. The preliminary results have recently been reported at ACT'19 in Oxford to a representative lens community, and no references besides [17], [16] mentioned above were provided.

6 Conclusion

The perspective on Bx presented in the paper is an example of a fruitful interaction between two domains—ML and Bx. In order to be ported to Bx, the compositional approach to ML developed in [17] is to be categorificated as shown in Fig. 8 on p. 194. This opens a whole new program for Bx: checking that currently existing Bx languages and tools are compositional (and well-behaved) in the sense of Def. 6 p. 190. The wb compositionality is an important practical requirement as it allows for modular design and testing of bidirectional transformations. Surprisingly, but this important requirement has been missing from the agenda of the Bx community, e.g., the recent endeavour of developing an effective benchmark for Bx-tools [3] does not discuss it.

In a wider context, the main message of the paper is that the learning idea transcends its applications in ML: it is applicable and usable in many domains in which lenses are applicable such as model transformations, data migration, and open games [18]. Moreover, the categorificated learning may perhaps find useful applications in ML itself. In the current ML setting, the object to be learnt is a function $f\colon \mathbb{R}^m \to \mathbb{R}^n$ that, in the OO class modelling perspective, is a very simple structure: it can be seen as one object with a (huge) amount of attributes, or, perhaps, a predefined set of objects, which is not allowed to be changed during the search — only attribute values may be changed. In the delta lens view, such changes constitute a rather narrow class of updates and thus unjustifiably narrow the search space. Learning with the possibility to change dimensions m, n may be an appropriate option in several contexts. On the other hand, while categorification of model spaces extends the search space, categorification of the parameter space would narrow the search space as we are allowed to replace a parameter p by parameter p' only if there is a suitable arrow $e\colon p \to p'$ in category $\mathbf{P}$. This narrowing may, perhaps, improve performance. All in all, the interaction between ML and Bx could be bidirectional!

A Appendices

A.1 Category of parameterized functors *pCat*

Category ***pCat*** has all small categories as objects. ***pCat***-arrows $\mathbf{A} \to \mathbf{B}$ are *parameterized* functors (*p-functors*) i.e., functors $f\colon \mathbf{P} \to [\mathbf{A}, \mathbf{B}]$ with $\mathbf{P}$ a small category of *parameters* and $[\mathbf{A}, \mathbf{B}]$ the category of functors from $\mathbf{A}$ to $\mathbf{B}$ and their natural transformations. For an object p and an arrow $e\colon p \to p'$ in $\mathbf{P}$, we write f_p for the functor $f(p)\colon \mathbf{A} \to \mathbf{B}$ and f_e for the natural transformation $f(e)\colon f_p \Rightarrow f_{p'}$. We will write p-functors as labelled arrows $f\colon \mathbf{A} \xrightarrow{\mathbf{P}} \mathbf{B}$. As ***Cat*** is Cartesian closed, we have a natural isomorphism between $\boldsymbol{Cat}(\mathbf{P}, [\mathbf{A}, \mathbf{B}])$ and $\boldsymbol{Cat}(\mathbf{P}\times\mathbf{A}, \mathbf{B})$ and can reformulate the above definition in an equivalent way with functors $\mathbf{P}\times\mathbf{A} \to \mathbf{B}$. We prefer the former formulation as it corresponds to the notation $f\colon \mathbf{A} \xrightarrow{\mathbf{P}} \mathbf{B}$ visualizing $\mathbf{P}$ as a hidden state of the transformation, which seems adequate to the intuition of parameterized in our context. (If some technicalities may perhaps be easier to see with the product formulation, we will switch to the product view thus doing currying and uncurrying without special mentioning.) Sequential composition of of $f\colon \mathbf{A} \xrightarrow{\mathbf{P}} \mathbf{B}$ and $g\colon \mathbf{B} \xrightarrow{\mathbf{Q}} \mathbf{C}$ is $f.g\colon \mathbf{A} \xrightarrow{\mathbf{P}\times\mathbf{Q}} \mathbf{C}$ given by $(f.g)_{pq} \stackrel{\text{def}}{=} f_p.g_q$ for objects, i.e., pairs $p{\in}\mathbf{P}$, $q{\in}\mathbf{Q}$, and by the Godement product of natural transformations for arrows in $\mathbf{P}\times\mathbf{Q}$. That is, given a pair $e\colon p \to p'$ in $\mathbf{P}$ and $h\colon q \to q'$ in $\mathbf{Q}$, we define the transformation $(f.g)_{eh}\colon f_p.g_q \Rightarrow f_{p'}.g_{q'}$ to be the Godement product $f_e * g_h$.

Any category $\mathbf{A}$ gives rise to a p-functor $\mathsf{Id}_{\mathbf{A}}\colon \mathbf{A} \xrightarrow{\mathbf{1}} \mathbf{A}$, whose parameter space is a singleton category $\mathbf{1}$ with the only object $*$, $\mathsf{Id}_{\mathbf{A}}(*) = \mathsf{id}_{\mathbf{A}}$ and $\mathsf{Id}_A(\mathsf{id}_*)\colon \mathsf{id}_{\mathbf{A}} \Rightarrow \mathsf{id}_{\mathbf{A}}$ is the identity transformation. It's easy to see that p-functors $\mathsf{Id}_{\_}$ are units of the sequential composition. To ensure associativity we need to consider p-functors up to an equivalence of their parameter spaces. Two parallel p-functors $f\colon \mathbf{A} \xrightarrow{\mathbf{P}} \mathbf{B}$ and $\hat{f}\colon \mathbf{A} \xrightarrow{\hat{\mathbf{P}}} \mathbf{B}$, are *equivalent* if there is an isomorphism $\alpha\colon \mathbf{P} \to \hat{\mathbf{P}}$ such that two parallel functors $f\colon \mathbf{P} \to [\mathbf{A}, \mathbf{B}]$ and $\alpha; \hat{f}\colon \mathbf{P} \to [\mathbf{A}, \mathbf{B}]$ are naturally isomorphic; then we write $f \approx_\alpha \hat{f}$. It's easy to see that if $f \approx_\alpha \hat{f}\colon \mathbf{A} \to \mathbf{B}$ and $g \approx_\beta \hat{g}\colon \mathbf{B} \to \mathbf{C}$, then $f; g \approx_{\alpha\times\beta} \hat{f}; \hat{g}\colon \mathbf{A} \to \mathbf{C}$, i.e., sequential composition is stable under equivalence. Below we will identify p-functors and their equivalence classes. Using a natural isomorphism $(\mathbf{P}\times\mathbf{Q})\times\mathbf{R} \cong \mathbf{P}\times(\mathbf{Q}\times\mathbf{R})$, strict associativity of the functor composition and strict associativity of the Godement product, we conclude that sequential composition of (equivalence classes of) p-functors is strictly associative. Hence, ***pCat*** is a category.

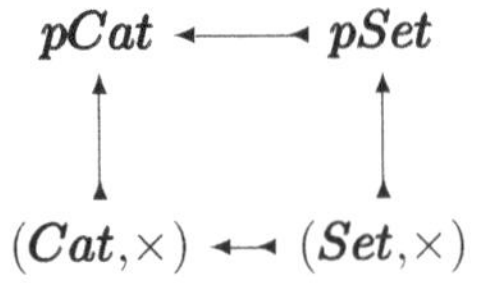

Fig. 7

Our next goal is to supply it with a monoidal structure. We borrow the latter from the sm-category $(\boldsymbol{Cat},\times)$, whose tensor is given by the product. There is an identical on objects embedding $(\boldsymbol{Cat},\times) \rightarrowtail \boldsymbol{pCat}$ that maps a functor $f\colon \mathbf{A} \to \mathbf{B}$ to a p-functor $\bar{f}\colon \mathbf{A} \xrightarrow{\mathbf{1}} \mathbf{B}$ whose parameter space is the singleton category $\mathbf{1}$. Moreover, as this embedding is a functor, the coherence equations for the associators and unitors that hold in $(\boldsymbol{Cat},\times)$ hold in ***pCat*** as well (this proof idea is borrowed from [17]). In this way, ***pCat*** becomes an sm-category. In a similar way, we define the sm-category ***pSet*** of small sets and parametrized functions between them — the codiscrete version of ***pCat***. The diagram in Fig. 7 shows how these categories are related.

A.2 Ala-lenses as categorification of ML-learners

Figure 8 shows a discrete two-dimensional plane with each axis having three points: a space is a singleton, a set, a category encoded by coordinates 0,1,2 resp. Each of the points x_{ij} is then the location of a corresponding sm-category of

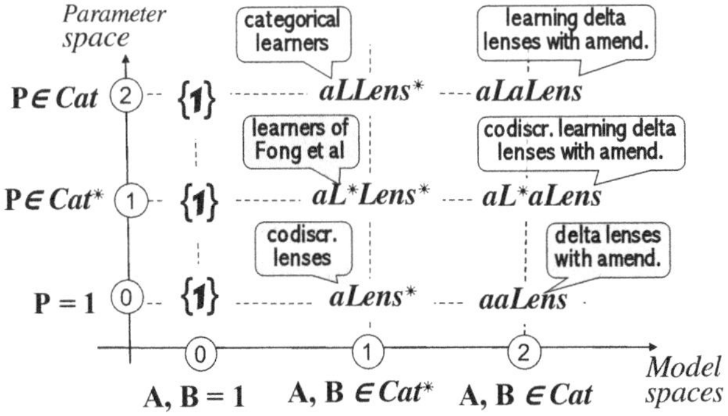

Fig. 8: The universe of categories of learning delta lenses

(asymmetric) learning (delta) lenses. Category $\{\boldsymbol{1}\}$ is a terminal category whose only arrow is the identity lens $\boldsymbol{1} = (\mathsf{id}_1, \mathsf{id}_1)\colon \mathbf{1} \to \mathbf{1}$ propagating from a terminal category $\mathbf{1}$ to itself. Label $*$ refers to the codiscrete specialization of the construct being labelled: $\boldsymbol{L}^*$ means codiscrete learning (i.e., the parameter space $\mathbf{P}$ is a set considered as a codiscrete category) and $\boldsymbol{aLens}^*$ refers to codiscrete model spaces. The category of learners defined in [17] is located at point (1,1), and the category of learning delta lenses with amendments defined in the present paper is located at (2,2). There are also two semi-categorificated species of learning lenses: categorical learners at point (1,2) and codiscretely learning delta lenses at (2,1), which are special cases of ala-lenses.

References

1. Abiteboul, S., McHugh, J., Rys, M., Vassalos, V., J.Wiener: Incremental Maintenance for Materialized Views over Semistructured Data. In: Gupta, A., Shmueli, O., Widom, J. (eds.) VLDB. Morgan Kaufmann (1998)
2. Anjorin, A.: An introduction to triple graph grammars as an implementation of the delta-lens framework. In: Gibbons, J., Stevens, P. (eds.) Bidirectional Transformations - International Summer School, Oxford, UK, July 25-29, 2016, Tutorial Lectures. Lecture Notes in Computer Science, vol. 9715, pp. 29–72. Springer (2016). https://doi.org/10.1007/978-3-319-79108-1
3. Anjorin, A., Diskin, Z., Jouault, F., Ko, H., Leblebici, E., Westfechtel, B.: Benchmarx reloaded: A practical benchmark framework for bidirectional transformations. In: Eramo and Johnson [15], pp. 15–30, http://ceur-ws.org/Vol-1827/paper6.pdf
4. Anjorin, A., Leblebici, E., Schürr, A.: 20 years of triple graph grammars: A roadmap for future research. ECEASST **73** (2015). https://doi.org/10.14279/tuj.eceasst.73.1031
5. Anjorin, A., Rose, S., Deckwerth, F., Schürr, A.: Efficient model synchronization with view triple graph grammars. In: Modelling Foundations and Applications - 10th European Conference, ECMFA 2014, York, UK, July 21-25, 2014. Proceedings. Lecture Notes in Computer Science, vol. 8569, pp. 1–17. Springer (2014). https://doi.org/10.1007/978-3-319-09195-2_1
6. Clarke, B.: Internal lenses as functors and cofunctors. In: Pre-proceedings of ACT'19, Oxford, 2019. http://www.cs.ox.ac.uk/ACT2019/preproceedings/BryceClarke.pdf
7. Czarnecki, K., Foster, J.N., Hu, Z., Lämmel, R., Schürr, A., Terwilliger, J.F.: Bidirectional transformations: A cross-discipline perspective. In: Theory and Practice of Model Transformations, pp. 260–283. Springer (2009)
8. Diskin, Z.: Compositionality of update propagation: Lax putput. In: Eramo and Johnson [15], pp. 74–89, http://ceur-ws.org/Vol-1827/paper12.pdf
9. Diskin, Z.: General supervised learning as change propagation with delta lenses. CoRR **abs/1911.12904** (2019), http://arxiv.org/abs/1911.12904
10. Diskin, Z., Gholizadeh, H., Wider, A., Czarnecki, K.: A three-dimensional taxonomy for bidirectional model synchronization. Journal of System and Software **111**, 298–322 (2016). https://doi.org/10.1016/j.jss.2015.06.003
11. Diskin, Z., König, H., Lawford, M.: Multiple model synchronization with multiary delta lenses with amendment and K-Putput. Formal Asp. Comput. **31**(5), 611–640 (2019). https://doi.org/10.1007/s00165-019-00493-0, (Sect.7.1 of the paper is unreadable and can be found in http://arxiv.org/abs/1911.11302)
12. Diskin, Z., Xiong, Y., Czarnecki, K.: From State- to Delta-Based Bidirectional Model Transformations: the Asymmetric Case. Journal of Object Technology **10**, 6: 1–25 (2011)
13. Diskin, Z., Xiong, Y., Czarnecki, K., Ehrig, H., Hermann, F., Orejas, F.: From state-to delta-based bidirectional model transformations: the symmetric case. In: MODELS, pp. 304–318. Springer (2011)
14. El-Sayed, M., Rundensteiner, E.A., Mani, M.: Incremental Maintenance of Materialized XQuery Views. In: Liu, L., Reuter, A., Whang, K.Y., Zhang, J. (eds.) ICDE. p. 129. IEEE Computer Society (2006). https://doi.org/10.1109/ICDE.2006.80
15. Eramo, R., Johnson, M. (eds.): Proceedings of the 6th International Workshop on Bidirectional Transformations co-located with The European Joint Conferences

on Theory and Practice of Software, Bx@ETAPS 2017, Uppsala, Sweden, April 29, 2017, CEUR Workshop Proceedings, vol. 1827. CEUR-WS.org (2017), http://ceur-ws.org/Vol-1827
16. Fong, B., Johnson, M.: Lenses and learners. In: Cheney, J., Ko, H. (eds.) Proceedings of the 8th International Workshop on Bidirectional Transformations co-located with the Philadelphia Logic Week, Bx@PLW 2019, Philadelphia, PA, USA, June 4, 2019. CEUR Workshop Proceedings, vol. 2355, pp. 16–29. CEUR-WS.org (2019), http://ceur-ws.org/Vol-2355/paper2.pdf
17. Fong, B., Spivak, D.I., Tuyéras, R.: Backprop as functor: A compositional perspective on supervised learning. In: The 34th Annual ACM/IEEE Symposium on Logic in Computer Science, LICS 2019, Vancouver, BC, Canada, June 24-27, 2019. pp. 1–13. IEEE (2019). https://doi.org/10.1109/LICS.2019.8785665
18. Hedges, J.: From open learners to open games. CoRR **abs/1902.08666** (2019), http://arxiv.org/abs/1902.08666
19. Hermann, F., Ehrig, H., Orejas, F., Czarnecki, K., Diskin, Z., Xiong, Y., Gottmann, S., Engel, T.: Model synchronization based on triple graph grammars: correctness, completeness and invertibility. Software and System Modeling **14**(1), 241–269 (2015). https://doi.org/10.1007/s10270-012-0309-1
20. Johnson, M., Rosebrugh, R.D.: Unifying set-based, delta-based and edit-based lenses. In: The 5th International Workshop on Bidirectional Transformations, Bx 2016. pp. 1–13 (2016), http://ceur-ws.org/Vol-1571/paper_13.pdf
21. Kappel, G., Langer, P., Retschitzegger, W., Schwinger, W., Wimmer, M.: Model transformation by-example: A survey of the first wave. In: Conceptual Modelling and Its Theoretical Foundations - Essays Dedicated to Bernhard Thalheim on the Occasion of His 60th Birthday. pp. 197–215 (2012). https://doi.org/10.1007/978-3-642-28279-9_15
22. Sasano, I., Hu, Z., Hidaka, S., Inaba, K., Kato, H., Nakano, K.: Toward bidirectionalization of ATL with GRoundTram. In: Theory and Practice of Model Transformations - 4th International Conference, ICMT 2011, Zurich, Switzerland, June 27-28, 2011. Proceedings. Lecture Notes in Computer Science, vol. 6707, pp. 138–151. Springer (2011). https://doi.org/10.1007/978-3-642-21732-6_10
23. Weidmann, N., Anjorin, A., Fritsche, L., Varró, G., Schürr, A., Leblebici, E.: Incremental bidirectional model transformation with emoflon: Ibex. In: The 8th International Workshop on Bidirectional Transformations co-located with the Philadelphia Logic Week, Bx@PLW 2019, Philadelphia, PA, USA, June 4, 2019. CEUR Workshop Proceedings, vol. 2355, pp. 45–55. CEUR-WS.org (2019), http://ceur-ws.org/Vol-2355/paper4.pdf

Non-idempotent intersection types in logical form*

Thomas Ehrhard [✉]

Université de Paris, IRIF, CNRS, F-75013 Paris, France
ehrhard@irif.fr
https://www.irif.fr/ ehrhard/

Abstract. Intersection types are an essential tool in the analysis of operational and denotational properties of lambda-terms and functional programs. Among them, non-idempotent intersection types provide precise quantitative information about the evaluation of terms and programs. However, unlike simple or second-order types, intersection types cannot be considered as a logical system because the application rule (or the intersection rule, depending on the presentation of the system) involves a condition stipulating that the proofs of premises must have the same structure. Using earlier work introducing an indexed version of Linear Logic, we show that non-idempotent typing can be given a logical form in a system where formulas represent hereditarily indexed families of intersection types.

Keywords: Lambda Calculus · Denotational Semantics · Intersection Types · Linear Logic

Introduction

Intersection types, introduced in the work of Coppo and Dezani [4,5] and developed since then by many authors, are still a very active research topic. As quite clearly explained in [13], the Coppo and Dezani intersection type system $D\Omega$ can be understood as a syntactic presentation of the denotational interpretation of λ-terms in the Engeler's model, which is a model of the pure λ-calculus in the cartesian closed category of prime-algebraic complete lattices and Scott continuous functions.

Intersection types can be considered as formulas of the propositional calculus with implication $\Rightarrow$ and conjunction $\wedge$ as connectives. However, as pointed out by Hindley [12], intersection types deduction rules depart drastically from the standard logical rules of intuitionistic logic (and of any standard logical system) by the fact that, in the $\wedge$-introduction rule, it is assumed that the proofs of the two premises are typings of the *same* λ-term, which means that, in some sense made precise by the typing system itself, they have the same structure. Such requirements on *proofs* premises, and not only on formulas proven in premises,

* Partially supported by the project ANR-19-CE48-0014 PPS.

J. Goubault-Larrecq and B. König (Eds.): FOSSACS 2020, LNCS 12077, pp. 198–216, 2020.
https://doi.org/10.1007/978-3-030-45231-5_11

are absent from standard (intuitionistic or classical) logical systems where the proofs of premises are completely independent from each other. Many authors have addressed this issue, we refer to [14] for a discussion on several solutions which mainly focus on the design of *à la Church* presentations of intersection typing systems, thus enriching λ-terms with additional structures. Among the most recent and convincing contributions to this line of research we should certainly mention [15].

In our "new" approach to this problem — not so new actually since it dates back to [3] —, we change formulas instead of changing terms. It is based on a specific model of Linear Logic (and thus of the λ-calculus): the *relational model*. It is fair to credit Girard for the introduction of this model since it appears at least implicitly in [11]. It was probably known by many people in the Linear Logic community as a piece of folklore since the early 1990's and is presented formally in [3]. In this quite simple and canonical denotational model, types are interpreted as sets (without any additional structure) and a closed term of type σ is interpreted as a subset of the interpretation of σ. It is quite easy to define, in this semantic framework, analogues of the usual models of the pure λ-calculus such as Scott's D_∞ or Engeler's model, which in some sense are simpler than the original ones since the sets interpreting types need not to be pre-ordered. As explained in the work of De Carvalho [6,7], the intersection type counterpart of this semantics is a typing system where "intersection" is non-idempotent (in sharp contrast with the original systems introduced by Coppo and Dezani), sometimes called *system R*. Notice that the precise connection between the idempotent and non-idempotent approaches is analyzed in [8], in a quite general Linear Logic setting by means of an extensional collapse.

In order to explain our approach, we restrict first to simple types, interpreted as follows in the relational model: a basic type α is interpreted as a given set $[\![\alpha]\!]$ and the type $\sigma \Rightarrow \tau$ is interpreted as the set $\mathcal{M}_{\text{fin}}([\![\sigma]\!]) \times [\![\tau]\!]$ (where $\mathcal{M}_{\text{fin}}(E)$ is the set of finite multisets of elements of E). Remember indeed that intersection types can be considered as a syntactic presentation of denotational semantics, so it makes sense to define intersection types relative to simple types (in the spirit of [10]) as we do in Section 3: an intersection type relative to the base type α is an element of $[\![\alpha]\!]$ and an intersection type relative to $\sigma \Rightarrow \tau$ is a pair $([\,a_1, \dots, a_n\,], b)$ where the a_is are intersection types relative to σ and b is an intersection type relative to τ; with more usual notations[1] $([\,a_1, \dots, a_n\,], b)$ would be written $(a_1 \wedge \cdots \wedge a_n) \to b$. Then, given a type σ, the main idea consists in representing an indexed family of elements of $[\![\sigma]\!]$ as a formula of a new logical system. If $\sigma = (\varphi \Rightarrow \psi)$ then the family can be written[2] $([\,a_k \mid k \in K \text{ and } u(k) = j\,], b_j)_{j \in J}$ where J and K are indexing sets, $u : K \to J$ is a function such that $f^{-1}(\{j\})$ is finite for all $j \in J$, $(b_j)_{j \in J}$ is a family of elements of $[\![\psi]\!]$ (represented by a formula B) and $(a_k)_{k \in K}$ is a family of elements of $[\![\varphi]\!]$ (represented by a formula A): in that case we introduce the implicative formula $(A \Rightarrow_u B)$ to represent the family

[1] That we prefer not to use for avoiding confusions between these two levels of typing.

[2] We use $[\cdots]$ for denoting multisets much as one uses $\{\cdots\}$ for denoting sets, the only difference is that multiplicities are taken into account.

$([\, a_k \mid k \in K \text{ and } u(k) = j \,], b_j)_{j \in J}$. It is clear that a family of simple types has generally infinitely many representations as such formulas; this huge redundancy makes it possible to establish a tight link between inhabitation of intersection types with provability of formulas representing them (in an indexed version $\mathsf{LJ}(I)$ of intuitionistic logic). Such a correspondence is exhibited in Section 3 in the simply typed setting and the idea is quite simple:

> given a type σ, a family $(a_j)_{j \in J}$ of elements of $[\![\sigma]\!]$, and a closed λ-term of type σ, it is equivalent to say that $\vdash M : a_j$ holds for all j and to say that some (and actually any) formula A representing $(a_j)_{j \in J}$ has an $\mathsf{LJ}(I)$ proof[3] whose underlying λ-term is M.

In Section 4 we extend this approach to the untyped λ-calculus taking as underlying model of the pure λ-calculus our relational version R_∞ of Scott's D_∞. We define an adapted version of $\mathsf{LJ}(I)$ and establish a similar correspondence, with some slight modifications due to the specificities of R_∞.

1 Notations and preliminary definitions

If E is a set, a *finite multiset of elements of* E is a function $m : E \to \mathbb{N}$ such that the set $\{a \in E \mid m(a) \neq 0\}$ (called the *domain* of m) is finite. The cardinal of such a multiset m is $\# m = \sum_{a \in E} m(a)$. We use $+$ for the obvious addition operation on multisets, and if $a_1, \dots, a_n$ are elements of E, we use $[\, a_1, \dots, a_n \,]$ for the corresponding multiset (taking multiplicities into account); for instance $[\, 0, 1, 0, 2, 1 \,]$ is the multiset m of elements of $\mathbb{N}$ such that $m(0) = 2$, $m(1) = 2$, $m(2) = 1$ and $m(i) = 0$ for $i > 2$. If $(a_i)_{i \in I}$ is a family of elements of E and if J is a finite subset of I, we use $[\, a_i \mid i \in J \,]$ for the multiset of elements of E which maps $a \in E$ to the number of elements $i \in J$ such that $a_i = a$ (which is finite since J is). We use $\mathcal{M}_{\mathrm{fin}}(E)$ for the set of finite multisets of elements of E.

We use $+$ to denote set union when we we want to stress the fact that the involved sets are disjoint. A function $u : J \to K$ is *almost injective* if $\# u^{-1}\{k\}$ is finite for each $k \in K$ (equivalently, the inverse image of any finite subset of K under u is finite). If $s = (a_1, \dots, a_n)$ is a sequence of elements of E and $i \in \{1, \dots, n\}$, we use $(s) \setminus i$ for the sequence $(a_1, \dots, a_{i-1}, a_{i+1}, \dots, a_n)$. Given sets E and F, we use F^E for the set of function from E to F. The elements of F^E are sometimes considered as functions u (with a functional notation $u(e)$ for application) and sometimes as indexed families a (with index notations a_e for application) especially when E is countable.

If $i \in \{1, \dots, n\}$ and $j \in \{1, \dots, n-1\}$, we define $\mathsf{s}(j, i) \in \{1, \dots, n\}$ as follows: $\mathsf{s}(j, i) = j$ if $j < i$ and $\mathsf{s}(j, i) = j + 1$ if $j \geq i$.

[3] Any such proof can be stripped from its indexing data giving rise to a proof of σ in intuitionistic logic.

2 The relational model of the λ-calculus

Let $\mathbf{Rel}_!$ the category whose objects are sets[4] and $\mathbf{Rel}_!(X,Y) = \mathcal{P}(\mathcal{M}_{\mathrm{fin}}(X) \times Y)$ with $\mathsf{Id}_X = \{([a],a) \mid a \in X\}$ and composition of $s \in \mathbf{Rel}_!(X,Y)$ and $t \in \mathbf{Rel}_!(Y,Z)$ given by

$$t \circ s = \{(m_1 + \cdots + m_k, c) \mid \\ \exists b_1,\dots,b_k \in Y \ ([b_1,\dots,b_k], c) \in t \text{ and } \forall j\, (m_j, b_j) \in s\}\,.$$

It is easily checked that this composition law is associative and that Id is neutral for composition[5]. This category has all countable products: let $(X_j)_{j\in J}$ be a countable family of sets, their product is $X = \&_{j\in J} X_j = \bigcup_{j\in J}\{j\} \times X_j$ and projections $(\mathsf{pr}_j)_{j\in J}$ given by $\mathsf{pr}_j = \{([(j,a)],a) \mid a \in X_j\} \in \mathbf{Rel}_!(X, X_j)$ and if $(s_j)_{j\in J}$ is a family of morphisms $s_j \in \mathbf{Rel}_!(Y, X_j)$ then their tupling is $\langle s_j\rangle_{j\in J} = \{([a],(j,b))) \mid j \in J \text{ and } ([a],b) \in s_j\} \in \mathbf{Rel}_!(Y,X)$.

The category $\mathbf{Rel}_!$ is cartesian closed with object of morphisms from X to Y the set $(X \Rightarrow Y) = \mathcal{M}_{\mathrm{fin}}(X) \times Y$ and evaluation morphism $\mathsf{Ev} \in \mathbf{Rel}_!((X \Rightarrow Y)\ \&\ X, Y)$ is given by $\mathsf{Ev} = \{([(1,[a_1,\dots,a_k],b),(2,a_1),\dots,(2,a_k)],b) \mid a_1,\dots,a_k \in X \text{ and } b \in Y\}$. The transpose (or curryfication) of $s \in \mathbf{Rel}_!(Z\ \&\ X, Y)$ is $\mathsf{Cur}(s) \in \mathbf{Rel}_!(Z, X \Rightarrow Y)$ given by $\mathsf{Cur}(s) = \{([c_1,\dots,c_n],([a_1,\dots,a_k],b)) \mid ([(1,c_1),\dots,(1,c_n),(2,a_1),\dots,(2,a_k)],c) \in s\}$.

Relational D_∞. Let R_∞ be the least set such that $(m_0, m_1, \dots) \in \mathsf{R}_\infty$ as soon as $m_0, m_1 \dots$ are finite multisets of elements of R_∞ which are almost all equal to $[\,]$. Notice in particular that $\mathsf{e} = ([\,],[\,],\dots) \in \mathsf{R}_\infty$ and satisfies $\mathsf{e} = ([\,],\mathsf{e})$. By construction we have $\mathsf{R}_\infty = \mathcal{M}_{\mathrm{fin}}(\mathsf{R}_\infty) \times \mathsf{R}_\infty$, that is $\mathsf{R}_\infty = (\mathsf{R}_\infty \Rightarrow \mathsf{R}_\infty)$ and hence R_∞ is a model of the pure λ-calculus in $\mathbf{Rel}_!$ which also satisfies the η-rule. See [1] for general facts on this kind of model.

3 The simply typed case

We assume to be given a set of type atoms $\alpha, \beta, \dots$ and of variables $x, y, \dots$; types and terms are given as usual by $\sigma, \tau, \dots := \alpha \mid \sigma \Rightarrow \tau$ and $M, N, \dots := x \mid (M)\,N \mid \lambda x^\sigma\, N$.

With any type atom we associate a set $[\![\alpha]\!]$. This interpretation is extended to all types by $[\![\sigma \Rightarrow \tau]\!] = [\![\sigma]\!] \Rightarrow [\![\tau]\!] = \mathcal{M}_{\mathrm{fin}}([\![\sigma]\!]) \times [\![\tau]\!]$. The relational semantics of this λ-calculus can be described as a non-idempotent intersection type system, with judgments of shape $x_1 : m_1 : \sigma_1, \dots, x_n : m_n : \sigma_n \vdash M : a : \sigma$ where the x_i's are pairwise distinct variables, M is a term, $a \in [\![\sigma]\!]$ and $m_i \in \mathcal{M}_{\mathrm{fin}}([\![\sigma_i]\!])$ for each i. Here are the typing rules:

$$\frac{j \neq i \Rightarrow m_j = [\,] \text{ and } m_i = [a]}{(x_i : m_i : \sigma_i)_{i=1}^n \vdash x_i : a : \sigma} \qquad \frac{\Phi, x : m : \sigma \vdash M : b : \tau}{\Phi \vdash \lambda x^\sigma\, M : (m,b) : \sigma \Rightarrow \tau}$$

[4] We can restrict to countable sets.

[5] This results from the fact that $\mathbf{Rel}_!$ arises as the Kleisli category of the LL model of sets and relations, see [3] for instance.

$$\frac{\Phi \vdash M : ([a_1, \ldots, a_k], b) : \sigma \Rightarrow \tau \qquad (\Phi_l \vdash N : a_l : \sigma)_{l=1}^{k}}{\Psi \vdash (M)\,N : b : \tau}$$

where $\Phi = (x_i : m_i : \sigma_i)_{i=1}^n$, $\Phi_l = (x_i : m_i^l : \sigma_i)_{i=1}^n$ for $l = 1, \ldots, k$ and $\Psi = (x_i : m_i + \sum_{l=1}^k m_i^l : \sigma_i)_{i=1}^n$.

3.1 Why do we need another system?

The trouble with this deduction system is that it cannot be considered as the term decorated version of an underlying "logical system for intersection types" allowing to prove sequents of shape $m_1 : \sigma_1, \ldots, m_n : \sigma_n \vdash a : \sigma$ (where non-idempotent intersection types m_i and a are considered as logical formulas, the ordinary types σ_i playing the role of "kinds") because, in the application rule above, it is required that all the proofs of the k right hand side premises have the same shape given by the λ-term N. We propose now a "logical system" derived from [3] which, in some sense, solves this issue. The main idea is quite simple and relies on three principles: (1) replace *hereditarily* multisets with indexed families in intersection types, (2) instead of proving single types, prove indexed families of hereditarily indexed types and (3) represent syntactically such families (of hereditarily indexed types) as formulas of a new system of *indexed logic.*

3.2 Minimal $\mathsf{LJ}(I)$

We define now the syntax of indexed formulas. Assume to be given an infinite countable set I of indices. Then we define indexed types A; with each such type we associate an underlying type $\underline{A}$, a set $\mathsf{d}(A)$ and a family $\langle A\rangle \in [\![\underline{A}]\!]^{\mathsf{d}(A)}$. These formulas are given by the following inductive definition:

- if $J \subseteq I$ and $f : J \to [\![\alpha]\!]$ is a function then $\alpha[f]$ is a formula with $\underline{\alpha[f]} = \alpha$, $\mathsf{d}(\alpha[f]) = J$ and $\langle \alpha[f]\rangle = f$
- and if A and B are formulas and $u : \mathsf{d}(A) \to \mathsf{d}(B)$ is almost injective then $A \Rightarrow_u B$ is a formula with $\underline{A \Rightarrow_u B} = \underline{A} \Rightarrow \underline{B}$, $\mathsf{d}(A \Rightarrow_u B) = \mathsf{d}(B)$ and, for $k \in \mathsf{d}(B)$, $\langle A \Rightarrow_u B\rangle_k = ([\, \langle A\rangle_j \mid j \in \mathsf{d}(A) \text{ and } u(j) = k\,], \langle B\rangle_k)$.

Proposition 1. *Let σ be a type, J be a subset of I and $f \in [\![\sigma]\!]^J$. There is a formula A such that $\underline{A} = \sigma$, $\mathsf{d}(A) = J$ and $\langle A\rangle = f$ (actually, there are infinitely many such A's as soon as σ is not an atom and $J \neq \emptyset$).*

Proof. The proof is by induction on σ. If σ is an atom α then we take $A = \alpha[f]$. Assume that $\sigma = (\rho \Rightarrow \tau)$ so that $f(j) = (m_j, b_j)$ with $m_j \in \mathcal{M}_{\mathsf{fin}}([\![\rho]\!])$ and $b_j \in [\![\tau]\!]$. Since each m_j is finite and I is infinite, we can find a family $(K_j)_{j\in J}$ of pairwise disjoint finite subsets of I such that $\#K_j = \#m_j$. Let $K = \bigcup_{j\in J} K_j$, there is a function $g : K \to [\![\rho]\!]$ such that $m_j = [\, g(k) \mid k \in K_j\,]$ for each $j \in J$ (choose first an enumeration $g_j : K_j \to [\![\rho]\!]$ of m_j for each j and then define $g(k) = g_j(k)$ where j is the unique element of J such that $k \in K_j$). Let $u : K \to J$ be the unique function such that $k \in K_{u(k)}$ for all $k \in K$; since each K_j is finite, this function u is almost injective. By inductive hypothesis there is a formula A such that $\underline{A} = \rho$, $\mathsf{d}(A) = K$ and $\langle A \rangle = g$, and there is a formula B such that $\underline{B} = \tau$, $\mathsf{d}(B) = J$ and $\langle B \rangle = (b_j)_{j \in J}$. Then the formula $A \Rightarrow_u B$ is well formed (since u is an almost injective function $\mathsf{d}(A) = K \to \mathsf{d}(B) = J$) and satisfies $\underline{A \Rightarrow_u B} = \sigma$, $\mathsf{d}(A \Rightarrow_u B) = J$ and $\langle A \Rightarrow_u B \rangle = f$ as contended. □

As a consequence, for any type σ and any element a of $[\![\sigma]\!]$ (so a is a non-idempotent intersection type of kind σ), one can find a formula A such that $\underline{A} = \sigma$, $\mathsf{d}(A) = \{j\}$ (where j is an arbitrary element of I) and $\langle A \rangle_j = a$. In other word, any intersection type can be represented as a formula (in infinitely many different ways in general of course, but up to renaming of indices, that is, up to "hereditary α-equivalence", this representation is unique).

For any formula A and $J \subseteq I$, we define a formula $A \restriction_J$ such that $\underline{A \restriction_J} = \underline{A}$, $\mathsf{d}(A \restriction_J) = \mathsf{d}(A) \cap J$ and $\langle A \restriction_J \rangle = \langle A \rangle \restriction_J$. The definition is by induction on A.

- $\alpha[f] \restriction_J = \alpha[f \restriction_J]$
- $(A \Rightarrow_u B) \restriction_J = (A \restriction_K \Rightarrow_v B \restriction_J)$ where $K = u^{-1}(\mathsf{d}(B) \cap J)$ and $v = u \restriction_K$.

Let $u : \mathsf{d}(A) \to J$ be a *bijection* (so that $u(\mathsf{d}(A)) = J$), we define a formula $u_*(A)$ such that $\underline{u_*(A)} = \underline{A}$, $\mathsf{d}(u_*(A)) = u(\mathsf{d}(A))$ and $\langle u_*(A) \rangle_j = \langle A \rangle_{u^{-1}(j)}$. The definition is by induction on A:

- $u_*(\alpha[f]) = \alpha[f \circ u^{-1}]$
- $u_*(A \Rightarrow_v B) = (A \Rightarrow_{u \circ v} u_*(B))$.

Using these two auxiliary notions, we can give a set of three deduction rules for a minimal natural deduction allowing to prove formulas in this indexed intuitionistic logic. This logical system allows to derive sequents which are of shape

$$A_1^{u_1}, \ldots, A_n^{u_n} \vdash B \tag{1}$$

where for each $i = 1, \ldots, n$, the function $u_i : \mathsf{d}(A_i) \to \mathsf{d}(B)$ is almost injective (it is not required that $\mathsf{d}(B) = \bigcup_{i=1}^n u_i(\mathsf{d}(A_i))$). Notice that the expressions $A_i^{u_i}$ are not formulas; this construction A^u is part of the syntax of sequents, just as the "," separating these pseudo-formulas. Given a formula A and $u : \mathsf{d}(A) \to J$ almost injective, it is nevertheless convenient to define $\langle A^u \rangle \in \mathcal{M}_{\text{fin}}([\![\underline{A}]\!])^J$ by $\langle A^u \rangle_j = [\, \langle A \rangle_k \mid u(k) = j \,]$. In particular, when u is a bijection, $\langle A^u \rangle_j = [\, \langle A \rangle_{u^{-1}(j)} \,]$.

The crucial point here is that such a sequent (1) involves no λ-term.

The main difference between the original system $\mathsf{LL}(I)$ of [3] and the present system is the way axioms are dealt with. In $\mathsf{LL}(I)$ there is no explicit identity axiom and only "atomic axioms" restricted to the basic constants of LL; indeed it is well-known that in LL all identity axioms can be η-expanded, leading to proofs using only such atomic axioms. In the λ-calculus, and especially in the untyped λ-calculus we want to deal with in next sections, such η-expansions are hard to handle so we prefer to use explicit identity axioms.

The axiom is

$$\frac{j \neq i \Rightarrow \mathsf{d}(A_j) = \emptyset \text{ and } u_i \text{ is a bijection}}{A_1^{u_1}, \ldots, A_n^{u_n} \vdash u_{i*}(A_i)}$$

so that for $j \neq i$, the function u_j is empty. A special case is

$$\frac{j \neq i \Rightarrow \mathsf{d}(A_j) = \emptyset \text{ and } u_i \text{ is the identity function}}{A_1^{u_1}, \dots, A_n^{u_n} \vdash A_i}$$

which may look more familiar, but the general axiom rule, allowing to "delocalize" the proven formula A_i by an arbitrary bijection u_i, is required as we shall see. The $\Rightarrow$ introduction rule is quite simple

$$\frac{A_1^{u_1}, \dots, A_n^{u_n}, A^u \vdash B}{A_1^{u_1}, \dots, A_n^{u_n} \vdash A \Rightarrow_u B}$$

Last the $\Rightarrow$ elimination rule is more complicated (from a Linear Logic point of view, this is due to the fact that it combines 3 LL logical rules: $\multimap$ elimination, contraction and promotion). We have the deduction

$$\frac{C_1^{u_1}, \dots, C_n^{u_n} \vdash A \Rightarrow_u B \qquad D_1^{v_1}, \dots, D_n^{v_n} \vdash A}{E_1^{w_1}, \dots, E_n^{w_n} \vdash B}$$

under the following conditions, to be satisfied by the involved formulas and functions: for each $i = 1, \dots, n$ one has $\mathsf{d}(C_i) \cap \mathsf{d}(D_i) = \emptyset$, $\mathsf{d}(E_i) = \mathsf{d}(C_i) + \mathsf{d}(D_i)$, $C_i = E_i \lceil_{\mathsf{d}(C_i)}$, $D_i = E_i \lceil_{\mathsf{d}(D_i)}$, $w_i \lceil_{\mathsf{d}(C_i)} = u_i$, and $w_i \lceil_{\mathsf{d}(D_i)} = u \circ v_i$.

Let π be a deduction tree of the sequent $A_1^{u_1}, \dots, A_n^{u_n} \vdash B$ in this system. By dropping all index information we obtain a derivation tree $\underline{\pi}$ of $\underline{A_1}, \dots, \underline{A_n} \vdash \underline{B}$, and, upon choosing a sequence $\vec{x}$ of n pairwise distinct variables, we can associate with this derivation tree a simply typed λ-term $\underline{\pi}_{\vec{x}}$ which satisfies $x_1 : \underline{A_1}, \dots, x_n : \underline{A_n} \vdash \underline{\pi}_{\vec{x}} : \underline{B}$.

3.3 Basic properties of $\mathsf{LJ}(I)$

We prove some basic properties of this logical system. This is also the opportunity to get some acquaintance with it. Notice that in many places we drop the type annotations of variables in λ-terms, first because they are easy to recover, and second because the very same results and proofs are also valid in the untyped setting of Section 4.

Lemma 1 (Weakening). *Assume that $\Phi \vdash A$ is provable by a proof π and let B be a formula such that $\mathsf{d}(B) = \emptyset$. Then $\Phi' \vdash A$ is provable by a proof π', where Φ' is obtained by inserting $B^{0_{\mathsf{d}(A)}}$ at any place in Φ. Moreover $\underline{\pi}_{\vec{x}} = \underline{\pi'}_{\vec{x}'}$ (where $\vec{x}'$ is obtained from $\vec{x}$ by inserting a dummy variable at the same place).*

The proof is an easy induction on the proof of $\Phi \vdash A$.

Lemma 2 (Relocation). *Let π be a proof of $(A_i^{u_i})_{i=1}^n \vdash A$ let $u : \mathsf{d}(A) \to J$ be a bijection, there is a proof π' of $(A_i^{u \circ u_i})_{i=1}^n \vdash u_*(A)$ such that $\underline{\pi'}_{\vec{x}} = \underline{\pi}_{\vec{x}}$.*

The proof is a straightforward induction on π.

Lemma 3 (Restriction). *Let π be a proof of $(A_i^{u_i})_{i=1}^n \vdash A$ and let $J \subseteq \mathsf{d}(A)$. For $i = 1, \dots, n$, let $K_i = u_i{}^{-1}(J) \subseteq \mathsf{d}(A_i)$ and $u_i' = u_i \lceil_{K_i} : K_i \to J$. Then the sequent $((A_i \lceil_{K_i})^{u_i'})_{i=1}^n \vdash A \lceil_J$ has a proof π' such that $\underline{\pi'}_{\vec{x}} = \underline{\pi}_{\vec{x}}$.*

Proof. By induction on π. Assume that π consists of an axiom $(A_j^{u_j})_{j=1}^n \vdash u_{i*}(A_i)$ with $\mathsf{d}(A_j) = \emptyset$ if $j \neq i$, and u_i a bijection. With the notations of the lemma, $K_j = \emptyset$ for $j \neq i$ and u'_i is a bijection $K_i \to J$. Moreover $u'_{i*}(A_i \upharpoonright_{K_i}) = u_{i*}(A_i) \upharpoonright_J$ so that $((A_i \upharpoonright_{K_i})^{u'_i})_{i=1}^n \vdash A \upharpoonright_J$ is obtained by an axiom π' with $\underline{\pi'}_{\vec{x}} = x_i = \underline{\pi}_{\vec{x}}$.

Assume that π ends with a $\Rightarrow$-introduction rule:

$$\frac{\begin{array}{c}\rho \\ (A_i^{u_i})_{i=1}^{n+1} \vdash B\end{array}}{(A_i^{u_i})_{i=1}^n \vdash A_{n+1} \Rightarrow_{u_{n+1}} B}$$

with $A = (A_{n+1} \Rightarrow_{u_{n+1}} B)$, and we have $\underline{\pi}_{\vec{x}} = \lambda x_{n+1}\, \underline{\rho}_{\vec{x}, x_{n+1}}$. With the notations of the lemma we have $A \upharpoonright_J = (A_{n+1} \upharpoonright_{K_{n+1}} \Rightarrow_{u'_{n+1}} B \upharpoonright_J)$. By inductive hypothesis there is a proof ρ' of $(A_i \upharpoonright_{K_i}^{u'_i})_{i=1}^{n+1} \vdash B \upharpoonright_J$ such that $\underline{\rho'}_{\vec{x}, x_{n+1}} = \underline{\rho}_{\vec{x}, x_{n+1}}$ and hence we have a proof π' of $(A_i \upharpoonright_{K_i}^{u'_i})_{i=1}^n \vdash A \upharpoonright_J$ with $\underline{\pi'}_{\vec{x}} = \lambda x_{n+1}\, \underline{\rho'}_{\vec{x}, x_{n+1}} = \underline{\pi}_{\vec{x}}$ as contended.

Assume last that π ends with a $\Rightarrow$-elimination rule:

$$\frac{\begin{array}{c}\mu \\ (B_i^{v_i})_{i=1}^n \vdash B \Rightarrow_v A\end{array} \qquad \begin{array}{c}\rho \\ (C_i^{w_i})_{i=1}^n \vdash B\end{array}}{(A_i^{u_i})_{i=1}^n \vdash A}$$

with $\mathsf{d}(A_i) = \mathsf{d}(B_i) + \mathsf{d}(C_i)$, $B_i = A_i \upharpoonright_{\mathsf{d}(B_i)}$ and $C_i = A_i \upharpoonright_{\mathsf{d}(C_i)}$, $u_i \upharpoonright_{\mathsf{d}(B_i)} = v_i$ and $u_i \upharpoonright_{\mathsf{d}(C_i)} = v \circ w_i$ for $i = 1, \dots, n$, and of course $\underline{\pi}_{\vec{x}} = \left(\underline{\mu}_{\vec{x}}\right) \underline{\rho}_{\vec{x}}$. Let $L = v^{-1}(J) \subseteq \mathsf{d}(B)$. Let $L_i = {v_i}^{-1}(J)$ and $R_i = {w_i}^{-1}(L)$ for $i = 1, \dots, n$ (we also set $v'_i = v_i \upharpoonright_{L_i}$, $w'_i = w_i \upharpoonright_{R_i}$ and $v' = v \upharpoonright_L$). By inductive hypothesis, we have a proof μ' of $(B_i \upharpoonright_{L_i}^{v'_i})_{i=1}^n \vdash B \upharpoonright_L \Rightarrow_{v'} A \upharpoonright_J$ such that $\underline{\mu'}_{\vec{x}} = \underline{\mu}_{\vec{x}}$ and a proof ρ' of $(C_i \upharpoonright_{R_i}^{w'_i})_{i=1}^n \vdash B \upharpoonright_L$ such that $\underline{\rho'}_{\vec{x}} = \underline{\rho}_{\vec{x}}$. Now, setting $K_i = {u_i}^{-1}(K)$, observe that

- $\mathsf{d}(B_i) \cap K_i = L_i = \mathsf{d}(B_i \upharpoonright_{L_i})$ and $u_i \upharpoonright_{L_i} = v'_i$ since $u_i \upharpoonright_{\mathsf{d}(B_i)} = v_i$
- $\mathsf{d}(C_i) \cap K_i = R_i = \mathsf{d}(C_i) \cap {w_i}^{-1}(L)$ since $u_i \upharpoonright_{\mathsf{d}(C_i)} = v \circ w_i$ and $L = v^{-1}(J)$, hence $\mathsf{d}(C_i) \cap K_i = \mathsf{d}(C_i \upharpoonright_{R_i})$, and also $u_i \upharpoonright_{L_i} = v' \circ w'_i$.

It follows that $\mathsf{d}(A_i \upharpoonright_{K_i}) = L_i + R_i$, and, setting $u'_i = u_i \upharpoonright_{K_i}$, we have $u'_i \upharpoonright_{L_i} = v'_i$ and $u'_i \upharpoonright_{R_i} = v' \circ w'_i$. Hence we have a proof π' of $(A_i \upharpoonright_{K_i}^{u'_i})_{i=1}^n \vdash A \upharpoonright_J$ such that $\underline{\pi'}_{\vec{x}} = \left(\underline{\mu'}_{\vec{x}}\right) \underline{\rho'}_{\vec{x}} = \left(\underline{\mu}_{\vec{x}}\right) \underline{\rho}_{\vec{x}} = \underline{\pi}_{\vec{x}}$ as contended. $\square$

Though substitution lemmas are usually trivial, the $\mathsf{LJ}(I)$ substitution lemma requires some care in its statement and proof[6].

Lemma 4 (Substitution). *Assume that* $(A_j^{u_j})_{j=1}^n \vdash A$ *with a proof* μ *and that, for some* $i \in \{1, \dots, n\}$, $(B_j^{v_j})_{j=1}^{n-1} \vdash A_i$ *with a proof* ρ. *Then there is a proof* π *of* $(C_j^{w_j})_{j=1}^{n-1} \vdash A$ *such that* $\underline{\pi}_{(\vec{x})\setminus i} = \underline{\mu}_{\vec{x}} \left[\underline{\rho}_{(\vec{x})\setminus i}/x_i\right]$ *as soon as for each* $j = 1, \dots, n-1$, $\mathsf{d}(C_j) = \mathsf{d}(A_{\mathsf{s}(j,i)}) + \mathsf{d}(B_j)$ *for each* $j = 1, \dots, n-1$ *(remember that this requires also that* $\mathsf{d}(A_{\mathsf{s}(j,i)}) \cap \mathsf{d}(B_j) = \emptyset$*) with:*

[6] We use notations introduced in Section 1, especially for $\mathsf{s}(j,i)$.

– $C_j \upharpoonright_{\mathsf{d}(A_{\mathsf{s}(j,i)})} = A_{\mathsf{s}(j,i)}$ *and* $w_j \upharpoonright_{\mathsf{d}(A_{\mathsf{s}(j,i)})} = u_{\mathsf{s}(j,i)}$
– $C_j \upharpoonright_{\mathsf{d}(B_j)} = B_j$ *and* $w_j \upharpoonright_{\mathsf{d}(B_j)} = u_i \circ v_j$.

Proof. By induction on the proof μ. Assume that μ is an axiom, so that there is a $k \in \{1, \ldots, n\}$ such that $A = u_{k*}(A_k)$, u_k is a bijection and $\mathsf{d}(A_j) = \emptyset$ for all $j \neq k$. In that case we have $\underline{\mu}_{\vec{x}} = x_k$. There are two subcases to consider. Assume first that $k = i$. By Lemma 2 there is a proof ρ' of $(B_j^{u_i \circ v_j})_{j=1}^{n-1} \vdash u_{i*}(A_i)$ such that $\underline{\rho'}_{(\vec{x})\setminus i} = \underline{\rho}_{(\vec{x})\setminus i}$. We have $C_j = B_j$ and $w_j = u_i \circ v_j$ for $j = 1, \ldots, n-1$, so that ρ' is a proof of $(C_j^{w_j})_{j=1}^{n-1} \vdash A$, so we take $\pi = \rho'$ and equation $\underline{\pi}_{(\vec{x})\setminus i} = \underline{\mu}_{\vec{x}} \left[\underline{\rho}_{(\vec{x})\setminus i}/x_i\right]$ holds since $\underline{\mu}_{\vec{x}} = x_i$. Assume next that $k \neq i$, then $\mathsf{d}(A_i) = \emptyset$ and hence $\mathsf{d}(B_j) = \emptyset$ (and $v_j = 0_\emptyset$) for $j = 1, \ldots, n-1$. Therefore $C_j = A_{\mathsf{s}(j,i)}$ and $w_j = v_{\mathsf{s}(j,i)}$ for $j = 1, \ldots, n-1$. So our target sequent $(C_j^{w_j})_{j=1}^{n-1} \vdash A$ can also be written $(A_{\mathsf{s}(j,i)}^{u_{\mathsf{s}(j,i)}})_{j=1}^{n-1} \vdash u_{k*}(A_k)$ and is provable by a proof π such that $\underline{\pi}_{(\vec{x})\setminus i} = x_k$ as contended.

Assume now that μ is a $\Rightarrow$-intro, that is $A = (A_{n+1} \Rightarrow_{u_{n+1}} A')$ and μ is

$$\dfrac{\begin{array}{c}\theta\\ (A_j^{u_j})_{j=1}^{n+1} \vdash A'\end{array}}{(A_j^{u_j})_{j=1}^{n} \vdash A}$$

We set $B_n = A_{n+1} \upharpoonright_\emptyset$ and of course $v_{n+1} = 0_{\mathsf{d}(A)}$. Then we have a proof ρ' of $(B_j^{v_j})_{j=1}^{n} \vdash A_i$ such that $\underline{\rho'}_{(\vec{x})\setminus i, x_{n+1}} = \underline{\rho}_{(\vec{x})\setminus i}$ by Lemma 1. We set $C_n = A_{n+1}$ and $w_n = u_{n+1}$. Then by inductive hypothesis applied to θ we have a proof π^0 of $(C_j^{w_j})_{j=1}^{n} \vdash A'$ which satisfies $\underline{\pi^0}_{(\vec{x})\setminus i, x_{n+1}} = \underline{\theta}_{\vec{x}, x_{n+1}} \left[\underline{\rho}_{(\vec{x})\setminus i}/x_i\right]$ and applying a $\Rightarrow$-introduction rule we get a proof π of $(C_j^{w_j})_{j=1}^{n-1} \vdash A$ such that $\underline{\pi}_{(\vec{x})\setminus i} = \lambda x_{n+1} \left(\underline{\theta}_{\vec{x}, x_{n+1}} \left[\underline{\rho}_{(\vec{x})\setminus i}/x_i\right]\right) = \underline{\mu}_{\vec{x}} \left[\underline{\rho}_{(\vec{x})\setminus i}/x_i\right]$ as expected.

Assume last that the proof μ ends with

$$\dfrac{\begin{array}{c}\varphi\\ (E_j^{s_j})_{j=1}^{n} \vdash E \Rightarrow_s A\end{array} \qquad \begin{array}{c}\psi\\ (F_j^{t_j})_{j=1}^{n} \vdash E\end{array}}{(A_j^{u_j})_{j=1}^{n} \vdash A}$$

with $\mathsf{d}(A_j) = \mathsf{d}(E_j) + \mathsf{d}(F_j)$, $A_j \upharpoonright_{\mathsf{d}(E_j)} = E_j$, $A_j \upharpoonright_{\mathsf{d}(F_j)} = F_j$, $u_j \upharpoonright_{\mathsf{d}(E_j)} = s_j$ and $u_j \upharpoonright_{\mathsf{d}(F_j)} = s \circ t_j$, for $j = 1, \ldots, n$. And we have $\underline{\mu}_{\vec{x}} = \left(\underline{\varphi}_{\vec{x}}\right) \underline{\psi}_{\vec{x}}$. The idea is to "share" the substituting proof ρ of $(B_j^{v_j})_{j=1}^{n} \vdash A_i$ among φ and ψ according to what they need, as specified by the formulas E_i and F_i. So we write $\mathsf{d}(B_j) = L_j + R_j$ where $L_j = {v_j}^{-1}(\mathsf{d}(E_i))$ and $R_j = {v_j}^{-1}(\mathsf{d}(F_i))$ and by Lemma 3 we have two proofs ρ^L of $(B_j \upharpoonright_{L_j}^{v_j^L})_{j=1}^{n-1} \vdash E_i$ and $(B_j \upharpoonright_{R_j}^{v_j^R})_{j=1}^{n-1} \vdash F_i$ where we set $v_j^L = v_j \upharpoonright_{L_j}$ and $v_j^R = v_j \upharpoonright_{R_j}$, obtained from ρ by restriction. These proofs satisfy $\underline{\rho^L}_{(\vec{x})\setminus i} = \underline{\rho^R}_{(\vec{x})\setminus i} = \underline{\rho}_{(\vec{x})\setminus i}$.

Now we want to apply the inductive hypothesis to φ and ρ^L, in order to get a proof of the sequent $(G_j^{w_j^L})_{j=1}^{n-1} \vdash E \Rightarrow_s A$ where $G_j = C_j\lceil_{\mathsf{d}(E_{\mathsf{s}(j,i)})+L_j}$ (observe indeed that $\mathsf{d}(E_{\mathsf{s}(j,i)}) \subseteq \mathsf{d}(A_{\mathsf{s}(j,i)})$ and $L_j \subseteq \mathsf{d}(B_j)$ and hence are disjoint by our assumption that $\mathsf{d}(C_j) = \mathsf{d}(A_{\mathsf{s}(j,i)}) + \mathsf{d}(B_j)$) and $w_j^L = w_j\lceil_{\mathsf{d}(E_{\mathsf{s}(j,i)})+L_j}$. With these definitions, and by our assumptions about C_j and w_j, we have for all $j = 1, \ldots, n-1$

$$\begin{aligned}
G_j\lceil_{\mathsf{d}(E_{\mathsf{s}(j,i)})} &= C_j\lceil_{\mathsf{d}(A_{\mathsf{s}(j,i)})}\lceil_{\mathsf{d}(E_{\mathsf{s}(j,i)})} = A_{\mathsf{s}(j,i)}\lceil_{\mathsf{d}(E_{\mathsf{s}(j,i)})} = E_{\mathsf{s}(j,i)}\\
w_j^L\lceil_{\mathsf{d}(E_{\mathsf{s}(j,i)})} &= w_j\lceil_{\mathsf{d}(A_{\mathsf{s}(j,i)})}\lceil_{\mathsf{d}(E_{\mathsf{s}(j,i)})} = u_{\mathsf{s}(j,i)}\lceil_{\mathsf{d}(E_{\mathsf{s}(j,i)})} = s_{\mathsf{s}(j,i)}\\
G_j\lceil_{L_j} &= C_j\lceil_{\mathsf{d}(B_j)}\lceil_{L_j} = B_j\lceil_{L_j}\\
w_j^L\lceil_{L_j} &= w_j\lceil_{\mathsf{d}(B_j)}\lceil_{L_j} = (u_i \circ v_j)\lceil_{L_j} = u_i\lceil_{\mathsf{d}(E_i)} \circ v_j^L = s_i \circ v_j^L\,.
\end{aligned}$$

Therefore the inductive hypothesis applies yielding a proof φ' of $(G_j^{w_j^L})_{j=1}^{n-1} \vdash E \Rightarrow_s A$ such that $\underline{\varphi'}_{(\vec{x})\setminus i} = \underline{\varphi}_{\vec{x}}\left[\underline{\rho^L}_{(\vec{x})\setminus i}/x_i\right] = \underline{\varphi}_{\vec{x}}\left[\underline{\rho}_{(\vec{x})\setminus i}/x_i\right]$.

Next we want to apply the inductive hypothesis to ψ and ρ^R, in order to get a proof of the sequent $(H_j^{r_j})_{j=1}^{n-1} \vdash E$ where, for $j = 1, \ldots, n-1$, $H_j = C_j\lceil_{\mathsf{d}(F_{\mathsf{s}(j,i)})+R_j}$ (again $\mathsf{d}(F_{\mathsf{s}(j,i)}) \subseteq \mathsf{d}(A_{\mathsf{s}(j,i)})$ and $R_j \subseteq \mathsf{d}(B_j)$ are disjoint by our assumption that $\mathsf{d}(C_j) = \mathsf{d}(A_{\mathsf{s}(j,i)}) + \mathsf{d}(B_j)$) and r_j is defined by $r_j\lceil_{\mathsf{d}(F_{\mathsf{s}(j,i)})} = t_{\mathsf{s}(j,i)}$ and $r_j\lceil_{R_j} = t_i \circ v_j^R$. Remember indeed that $v_j^R : R_j \to \mathsf{d}(F_i)$ and $t_i : \mathsf{d}(F_i) \to \mathsf{d}(E)$. We have

$$\begin{aligned}
H_j\lceil_{\mathsf{d}(F_{\mathsf{s}(j,i)})} &= C_j\lceil_{\mathsf{d}(A_{\mathsf{s}(j,i)})}\lceil_{\mathsf{d}(F_{\mathsf{s}(j,i)})} = A_{\mathsf{s}(j,i)}\lceil_{\mathsf{d}(F_{\mathsf{s}(j,i)})} = F_{\mathsf{s}(j,i)}\\
H_j\lceil_{R_j} &= C_j\lceil_{\mathsf{d}(B_j)}\lceil_{R_j} = B_j\lceil_{R_j}
\end{aligned}$$

and hence by inductive hypothesis there is a proof ψ' of $(H_j^{r_j})_{j=1}^{n-1} \vdash E$ such that $\underline{\psi'}_{(\vec{x})\setminus i} = \underline{\psi}_{\vec{x}}\left[\underline{\rho^R}_{(\vec{x})\setminus i}/x_i\right] = \underline{\psi}_{\vec{x}}\left[\underline{\rho}_{(\vec{x})\setminus i}/x_i\right]$.

To end the proof of the lemma, it will be sufficient to prove that we can apply a $\Rightarrow$-elimination rule to the sequents $(G_j^{w_j^L})_{j=1}^{n-1} \vdash E \Rightarrow_s A$ and $(H_j^{r_j})_{j=1}^{n-1} \vdash E$ in order to get a proof π of the sequent $(C_j^{w_j})_{j=1}^{n-1} \vdash A$. Indeed, the proof π obtained in that way will satisfy $\underline{\pi}_{(\vec{x})\setminus i} = \left(\underline{\varphi'}_{(\vec{x})\setminus i}\right)\underline{\psi'}_{(\vec{x})\setminus i} = \underline{\mu}_{\vec{x}}\left[\underline{\rho}_{(\vec{x})\setminus i}/x_i\right]$. Let $j \in \{1, \ldots, n-1\}$. We have $C_j\lceil_{\mathsf{d}(G_j)} = G_j$ and $C_j\lceil_{\mathsf{d}(H_j)} = H_j$ simply because G_j and H_j are defined by restricting C_j. Moreover $\mathsf{d}(G_j) = \mathsf{d}(E_{\mathsf{s}(j,i)}) + L_j$ and $\mathsf{d}(H_j) = \mathsf{d}(F_{\mathsf{s}(j,i)}) + R_j$. Therefore $\mathsf{d}(G_j) \cap \mathsf{d}(H_j) = \emptyset$ and

$$\mathsf{d}(C_j) = \mathsf{d}(A_{\mathsf{s}(j,i)}) + \mathsf{d}(B_j) = \mathsf{d}(E_{\mathsf{s}(j,i)}) + \mathsf{d}(F_{\mathsf{s}(j,i)}) + L_j + R_j = \mathsf{d}(G_j) + \mathsf{d}(H_j)\,.$$

We have $w_j\lceil_{\mathsf{d}(G_j)} = w_j^L$ by definition of w_j^L as $w_j\lceil_{\mathsf{d}(E_{\mathsf{s}(j,i)})+L_j}$. We have

$$\begin{aligned}
w_j\lceil_{\mathsf{d}(H_j)}\lceil_{\mathsf{d}(F_{\mathsf{s}(j,i)})} &= w_j\lceil_{\mathsf{d}(A_{\mathsf{s}(j,i)})}\lceil_{\mathsf{d}(F_{\mathsf{s}(j,i)})} = u_{\mathsf{s}(j,i)}\lceil_{\mathsf{d}(F_{\mathsf{s}(j,i)})}\\
&= s \circ t_{\mathsf{s}(j,i)} = (s \circ r_j)\lceil_{\mathsf{d}(F_{\mathsf{s}(j,i)})}\\
w_j\lceil_{\mathsf{d}(H_j)}\lceil_{R_j} &= w_j\lceil_{\mathsf{d}(B_j)}\lceil_{R_j} = (u_i \circ v_j)\lceil_{R_j}\\
&= u_i\lceil_{\mathsf{d}(F_i)} \circ v_j^R = s \circ t_i \circ v_j^R = s \circ r_j\lceil_{R_j} = (s \circ r_j)\lceil_{R_j}
\end{aligned}$$

and therefore $w_j \restriction_{\mathsf{d}(H_j)} = s \circ r_j$ as required. □

We shall often use the two following consequences of the Substitution Lemma.

Lemma 5. *Given a proof μ of $(A_j^{u_j})_{j=1}^n \vdash A$ and a proof ρ of $B^v \vdash A_i$ (for some $i \in \{1, \ldots, n\}$), there is a proof π of $(A_j^{u_j})_{j=1}^{i-1}, B^{u_i \circ v}, (A_j^{u_j})_{j=i+1}^n \vdash A$ such that $\underline{\pi}_{\vec{x}} = \underline{\mu}_{\vec{x}} \left[\underline{\rho}_{x_i}/x_i\right]$*

Proof. By weakening we have a proof μ' of $(A_j^{u_j})_{j=1}^i, B\restriction_{\emptyset}^{0_{\mathsf{d}(A)}}, (A_j^{u_j})_{j=i+1}^n \vdash A$ such that $\underline{\mu'}_{\vec{x}} = \underline{\mu}_{(\vec{x})\setminus i+1}$ (where $\vec{x}$ is a list of pairwise distinct variables of length $n+1$), as well as a proof ρ' of $(A_j \restriction_{\emptyset}^{0_{\mathsf{d}(A_i)}})_{j=1}^i, B^v, (A_j \restriction_{\emptyset}^{0_{\mathsf{d}(A_i)}})_{j=i+1}^n \vdash A_i$ such that $\underline{\rho'}_{\vec{x}} = \underline{\rho}_{x_{i+1}}$. By Lemma 4, we have a proof π' of $(A_j^{u_j})_{j=1}^{i-1}, B^{u_i \circ v}, (A_j^{u_j})_{j=i+1}^n \vdash A$ which satisfies $\underline{\pi'}_{(\vec{x})\setminus i} = \underline{\mu'}_{\vec{x}} \left[\underline{\rho'}_{(\vec{x})\setminus i}/x_i\right] = \underline{\mu}_{\vec{x}} \left[\underline{\rho}_{x_i}/x_i\right]$. □

Lemma 6. *Given a proof μ of $A^v \vdash B$ and a proof ρ of $(A_j^{u_j})_{j=1}^n \vdash A$, there is a proof π of $(A_j^{v \circ u_j})_{j=1}^n \vdash B$ such that $\underline{\pi}_{\vec{x}} = \underline{\mu}_x \left[\underline{\rho}_{\vec{x}}/x\right]$.*

The proof is similar to the previous one.

If A and B are formulas such that $\underline{A} = \underline{B}$, $\mathsf{d}(A) = \mathsf{d}(B)$ and $\langle A \rangle = \langle B \rangle$, we say that A and B are similar and we write $A \sim B$. One fundamental property of our deduction system is that two formulas which represent the same family of intersection types are logically equivalent.

Theorem 1. *If $A \sim B$ then $A^{\mathsf{Id}} \vdash B$ with a proof π such that $\underline{\pi}_x \sim_\eta x$.*

Proof. Assume that $A = \alpha[f]$, then we have $B = A$ and $A^{\mathsf{Id}} \vdash B$ is an axiom.

Assume that $A = (C \Rightarrow_u D)$ and $B = (E \Rightarrow_v F)$. We have $D \sim F$ and hence $D^{\mathsf{Id}} \vdash F$ with a proof ρ such that $\underline{\rho}_x \sim_\eta x$. And there is a bijection $w : \mathsf{d}(E) \to \mathsf{d}(C)$ such that $w_*(E) \sim C$ and $u \circ w = v$. By inductive hypothesis we have a proof μ of $w_*(E)^{\mathsf{Id}} \vdash C$ such that $\underline{\mu}_y \sim_\eta y$, and hence using the axiom $E^w \vdash w_*(E)$ and Lemma 5 we have a proof μ' of $E^w \vdash C$ such that $\underline{\mu'}_x = \underline{\mu}_x$.

There is a proof π^1 of $(C \Rightarrow_u D)^{\mathsf{Id}}, C^u \vdash D$ such that $\underline{\pi^1}_{x,y} = (x)\, y$ (consider the two axioms $(C \Rightarrow_u D)^{\mathsf{Id}}, C\restriction_{\emptyset}^{0_{\mathsf{d}(D)}} \vdash C \Rightarrow_u D$ and $(C \Rightarrow_u D)\restriction_{\emptyset}^{0_{\mathsf{d}(C)}}, C^{\mathsf{Id}} \vdash C$ and use a $\Rightarrow$-elimination rule). So by Lemma 5 there is a proof π^2 of $(C \Rightarrow_u D)^{\mathsf{Id}}, E^{u \circ w} \vdash D$, that is of $(C \Rightarrow_u D)^{\mathsf{Id}}, E^v \vdash D$, such that $\underline{\pi^2}_{x,y} = (x)\, \underline{\mu}_y$. Applying Lemma 6 we get a proof π^3 of $(C \Rightarrow_u D)^{\mathsf{Id}}, E^v \vdash F$ such that $\underline{\pi^3}_{x,y} = \underline{\rho}_z \left[(x)\, \underline{\mu}_y/z\right]$. We get the expected proof π by a $\Rightarrow$-introduction rule so that $\underline{\pi}_x = \lambda y\, \underline{\rho}_z \left[(x)\, \underline{\mu}_y/z\right]$. By inductive hypothesis $\underline{\pi}_x \sim_\eta x$. □

3.4 Relation between intersection types and $\mathsf{LJ}(I)$

Now we explain the precise connection between non-idempotent intersection types and our logical system $\mathsf{LJ}(I)$. This connection consists of two statements:

- the first one means that any proof of $\mathsf{LJ}(I)$ can be seen as a typing derivation in non-idempotent intersection types (soundness)
- and the second one means that any non-idempotent intersection typing can be seen as a derivation in $\mathsf{LJ}(I)$ (completeness).

Theorem 2 (Soundness). *Let π be a deduction tree of the sequent $(A_i^{u_i})_{i=1}^n \vdash B$ and $\vec{x}$ a sequence of n pairwise distinct variables. Then the λ-term $\underline{\pi}_{\vec{x}}$ satisfies $(x_i : \langle A_i^{u_i}\rangle_j : \underline{A_i})_{i=1}^n \vdash \underline{\pi}_{\vec{x}} : \langle B\rangle_j : \underline{B}$ in the intersection type system, for each $j \in \mathsf{d}(B)$.*

Proof. We prove the first part by induction on π (in the course of this induction, we recall the precise definition of $\underline{\pi}_{\vec{x}}$). If π is the proof

$$\frac{q \neq i \Rightarrow \mathsf{d}(A_q) = \emptyset \text{ and } u_i \text{ is a bijection}}{(A_q^{u_q})_{q=1}^n \vdash u_{i*}(A_i)}$$

(so that $B = u_{i*}(A_i)$) then $\underline{\pi}_{\vec{x}} = x_i$. We have $\langle A_q^{u_q}\rangle_j = [\,]$ if $q \neq i$, $\langle A_i^{u_i}\rangle_j = [\,\langle A_i\rangle_{u_i^{-1}(j)}\,]$ and $\langle u_{i*}(A_i)\rangle_j = \langle A_i\rangle_{u_i^{-1}(j)}$. It follows that $(x_q : \langle A_q^{u_q}\rangle_j : \underline{A_q})_{q=1}^n \vdash x_i : \langle B\rangle_j : \underline{B}$ is a valid axiom in the intersection type system.

Assume that π is the proof

$$\frac{\begin{array}{c}\pi^0\\ A_1^{u_1},\dots,A_n^{u_n},A^u \vdash B\end{array}}{A_1^{u_1},\dots,A_n^{u_n} \vdash A \Rightarrow_u B}$$

where π^0 is the proof of the premise of the last rule of π. By inductive hypothesis the λ-term $\underline{\pi^0}_{\vec{x},x}$ satisfies $(x_i : \langle A_i^{u_i}\rangle_j : \underline{A_i})_{i=1}^n, x : \langle A^u\rangle_j : \underline{A} \vdash \underline{\pi^0}_{\vec{x},x} : \langle B\rangle_j : \underline{B}$ from which we deduce $(x_i : \langle A_i^{u_i}\rangle_j : \underline{A_i})_{i=1}^n \vdash \lambda x^{\underline{A}}\,\underline{\pi^0}_{\vec{x},x} : (\langle A^u\rangle_j, \langle B\rangle_j) : \underline{A} \Rightarrow \underline{B}$ which is the required judgment since $\underline{\pi}_{\vec{x}} = \lambda x^{\underline{A}}\,\underline{\pi^0}_{\vec{x},x}$ and $(\langle A_i^{u_i}\rangle_j, \langle B\rangle_j) = \langle A \Rightarrow_u B\rangle_j$ as easily checked.

Assume last that π ends with

$$\frac{\begin{array}{c}\pi^1\\ C_1^{u_1},\dots,C_n^{u_n} \vdash A \Rightarrow_u B\end{array} \qquad \begin{array}{c}\pi^2\\ D_1^{v_1},\dots,D_n^{v_n} \vdash A\end{array}}{E_1^{w_1},\dots,E_n^{w_n} \vdash B}$$

with: for each $i = 1,\dots,n$ there are two disjoint sets L_i and R_i such that $\mathsf{d}(E_i) = L_i + R_i$, $C_i = E_i\lceil_{L_i}$, $D_i = E_i\lceil_{R_i}$, $w_i\lceil_{L_i} = u_i$, and $w_i\lceil_{R_i} = u \circ v_i$.

Let $j \in \mathsf{d}(B)$. By inductive hypothesis, the judgment $(x_i : \langle C_i^{u_i}\rangle_j : \underline{C_i})_{i=1}^n \vdash \underline{\pi^1}_{\vec{x}} : \langle A \Rightarrow_u B\rangle_j : \underline{A} \Rightarrow \underline{B}$ is derivable in the intersection type system. Let $K_j = u^{-1}(\{j\})$, which is a finite subset of $\mathsf{d}(A)$. By inductive hypothesis again, for each $k \in K_j$ we have $(x_i : \langle D_i^{u_i}\rangle_k : \underline{D_i})_{i=1}^n \vdash \underline{\pi^2}_{\vec{x}} : \langle A\rangle_k : \underline{A}$. Now observe that $\langle A \Rightarrow_u B\rangle_j = ([\,\langle A\rangle_k \mid k \in K_j\,], \langle B\rangle_j)$ so that

$$(x_i : \langle C_i^{u_i}\rangle_j + \sum_{k\in K_j} \langle D_i^{u_i}\rangle_k : \underline{E_i})_{i=1}^n \vdash (\underline{\pi^1}_{\vec{x}})\,\underline{\pi^2}_{\vec{x}} : \langle B\rangle_j : \underline{B}$$

is derivable in intersection types (remember that $\underline{C_i} = \underline{D_i} = \underline{E_i}$). Since $\underline{\pi}_{\vec{x}} = (\underline{\pi^1}_{\vec{x}})\,\underline{\pi^2}_{\vec{x}}$ it will be sufficient to prove that

$$\langle E_i^{w_i}\rangle_j = \langle C_i^{u_i}\rangle_j + \sum_{k\in K_j} \langle D_i^{v_i}\rangle_k\,. \tag{2}$$

For this, since $\langle E_i^{w_i}\rangle_j = [\,\langle E_i\rangle_l \mid w_i(l) = j\,]$, consider an element l of $\mathsf{d}(E_i)$ such that $w_i(l) = j$. There are two possibilities: (1) either $l \in L_i$ and in that case we know that $\langle E_i\rangle_l = \langle C_i\rangle_l$ since $E_i\!\upharpoonright_{L_i} = C_i$ and moreover we have $u_i(l) = w_i(l) = j$ (2) or $l \in R_i$. In that case we have $\langle E_i\rangle_l = \langle D_i\rangle_l$ since $E_i\!\upharpoonright_{R_i} = D_i$. Moreover $u(v_i(l)) = w_i(l) = j$ and hence $v_i(l) \in K_j$. Therefore

$$[\,\langle E_i\rangle_l \mid l \in L_i \text{ and } w_i(l) = j\,] = [\,\langle C_i\rangle_l \mid u_i(l) = j\,] = \langle C_i^{u_i}\rangle_j$$
$$[\,\langle E_i\rangle_l \mid l \in R_i \text{ and } w_i(l) = j\,] = [\,\langle D_i\rangle_l \mid v_i(l) \in K_j\,] = \sum_{k\in K_j} \langle D_i^{v_i}\rangle_k$$

and (2) follows. □

Theorem 3 (Completeness). *Let $J \subseteq I$. Let M be a λ-term and $x_1, \dots, x_n$ be pairwise distinct variables, such that $(x_i : m_i^j : \sigma_i)_{i=1}^n \vdash M : b_j : \tau$ in the intersection type system for all $j \in J$. Let $A_1, \dots, A_n$ and B be formulas and let $u_1, \dots, u_n$ be almost injective functions such that $u_i : \mathsf{d}(A_i) \to J = \mathsf{d}(B)$. Assume also that $\underline{A_i} = \sigma_i$ for each $i = 1, \dots, n$ and that $\underline{B} = \tau$. Last assume that, for all $j \in J$, one has $\langle B\rangle_j = b_j$ and $\langle A_i^{u_i}\rangle_j = m_i^j$ for $i = 1, \dots, n$. Then the judgment $(A_i^{u_i})_{i=1}^n \vdash B$ has a proof π such that $\underline{\pi}_{\vec{x}} \sim_\eta M$.*

Proof. By induction on M. Assume first that $M = x_i$ for some $i \in \{1, \dots, n\}$. Then we must have $\tau = \sigma_i$, $m_q^j = [\,]$ for $q \neq i$ and $m_i^j = [b_j]$ for all $j \in J$. Therefore $\mathsf{d}(A_q) = \emptyset$ and u_q is the empty function for $q \neq i$, u_i is a bijection $\mathsf{d}(A_i) \to J$ and $\forall k \in \mathsf{d}(A_i)$ $\langle A_i\rangle_k = b_{u_i(k)}$, in other words $u_{i*}(A_i) \sim B$. By Theorem 1 we know that the judgment $(u_{i*}(A_i))^{\mathsf{Id}} \vdash B$ is provable in $\mathsf{LJ}(I)$ with a proof ρ such that $\underline{\rho}_x \sim_\eta x$. We have a proof θ of $(A_i^{u_i})_{i=1}^n \vdash u_{i*}(A_i)$ which consists of an axiom so that $\underline{\theta}_{\vec{x}} = x_i$ and hence by Lemma 6 we have a proof π of $(A_i^{u_i})_{i=1}^n \vdash B$ such that $\underline{\pi}_{\vec{x}} = \underline{\rho}_x\,[\underline{\theta}_{\vec{x}}/x] \sim_\eta x_i$.

Assume that $M = \lambda x^\sigma N$, that $\tau = (\sigma \Rightarrow \varphi)$ and that we have a family of deductions (for $j \in J$) of $(x_i : m_i^j : \sigma_i)_{i=1}^n \vdash M : (m^j, c_j) : \sigma \Rightarrow \varphi$ with $b_j = (m^j, c_j)$ and the premise of this conclusion in each of these deductions is $(x_i : m_i^j : \sigma_i)_{i=1}^n, x : m^j : \sigma \vdash N : c_j : \varphi$. We must have $B = (C \Rightarrow_u D)$ with $\underline{D} = \varphi$, $\underline{C} = \sigma$, $\mathsf{d}(D) = J$, $u : \mathsf{d}(C) \to \mathsf{d}(D)$ almost injective, $\langle D\rangle_j = c_j$ and

$[\langle C\rangle_k \mid k \in \mathsf{d}(C) \text{ and } u(k) = j\,] = m^j$, that is $\langle C^u\rangle_j = m^j$, for each $j \in J$. By inductive hypothesis we have a proof ρ of $(A_i^{u_i})_{i=1}^n, C^u \vdash D$ such that $\underline{\rho}_{\vec{x},x} \sim_\eta N$ from which we obtain a proof π of $(A_i^{u_i})_{i=1}^n \vdash C \Rightarrow_u D$ such that $\underline{\pi}_{\vec{x}} = \lambda x^\sigma\, \underline{\rho}_{\vec{x},x} \sim_\eta M$ as expected.

Assume last that $M = (N)\,P$ and that we have a J-indexed family of deductions $(x_i : m_i^j : \sigma_i)_{i=1}^n \vdash M : b_j : \tau$. Let $A_1,\dots,A_n$, $u_1,\dots,u_n$ and B be $\mathsf{LJ}(I)$ formulas and almost injective functions as in the statement of the theorem.

Let $j \in J$. There is a finite set $L_j \subseteq I$ and multisets $m_i^{j,0}$, $(m_i^{j,l})_{l\in L_j}$ such that we have deductions[7] of $(x_i : m_i^{j,0} : \sigma_i)_{i=1}^n \vdash N : ([\,a_l^j \mid l \in L_j\,], b_j) : \sigma \Rightarrow \tau$ and, for each $l \in L_j$, of $(x_i : m_i^{j,l} : \sigma_i)_{i=1}^n \vdash P : a_l^j : \sigma$ with

$$m_i^j = m_i^{j,0} + \sum_{l\in L_j} m_i^{j,l}\,. \tag{3}$$

We assume the finite sets L_j to be pairwise disjoint (this is possible because I is infinite) and we use L for their union. Let $u : L \to J$ be the function which maps $l \in L$ to the unique j such that $l \in L_j$, this function is almost injective. Let A be an $\mathsf{LL}(J)$ formula such that $\underline{A} = \sigma$, $\mathsf{d}(A) = L$ and $\langle A\rangle_l = a_l^{u(l)}$; such a formula exists by Proposition 1.

Let $i \in \{1,\dots,n\}$. For each $j \in J$ we know that

$$[\,\langle A_i\rangle_r \mid r \in \mathsf{d}(A_i) \text{ and } u_i(r) = j\,] = m_i^j = m_i^{j,0} + \sum_{l\in L_j} m_i^{j,l}$$

and hence we can split the set $\mathsf{d}(A_i) \cap {u_i}^{-1}(\{j\})$ into disjoint subsets $R_i^{j,0}$ and $(R_i^{j,l})_{l\in L_j}$ in such a way that

$$[\,\langle A_i\rangle_r \mid r \in R_i^{j,0}\,] = m_i^{j,0} \quad \text{and} \quad \forall l \in L_j\ [\,\langle A_i\rangle_r \mid r \in R_i^{j,l}\,] = m_i^{j,l}\,.$$

We set $R_i^0 = \bigcup_{j\in J} R_i^{j,0}$; observe that this is a disjoint union because $R_i^{j,0} \subseteq {u_i}^{-1}(\{j\})$. Similarly we define $R_i^1 = \bigcup_{l\in L} R_i^{u(l),l}$ which is a disjoint union for the following reason: if $l, l' \in L$ satisfy $u(l) = u(l') = j$ then $R_i^{j,l}$ and $R_i^{j,l'}$ have been chosen disjoint and if $u(l) = j$ and $u(l') = j'$ with $j \neq j'$ we have $R_i^{j,l} \subseteq {u_i}^{-1}\{j\}$ and $R_i^{j',l'} \subseteq {u_i}^{-1}(\{j'\})$. Let $v_i : R_i^1 \to L$ be defined by: $v_i(r)$ is the unique $l \in L$ such that $r \in R_i^{u(l),l}$. Since each $R_i^{j,l}$ is finite the function v_i is almost injective. Moreover $u \circ v_i = u_i\lceil_{R_i^1}$.

We use u_i' for the restriction of u_i to R_i^0 so that $u_i' : R_i^0 \to J$. By inductive hypothesis we have $((A_i\lceil_{R_i^0})^{u_i'})_{i=1}^n \vdash A \Rightarrow_u B$ with a proof μ such that $\underline{\mu}_{\vec{x}} \sim_\eta N$. Indeed $[\,\langle A_i\lceil_{R_i^0}\rangle_r \mid r \in R_i^0 \text{ and } u_i'(r) = j\,] = m_i^{j,0}$ and $\langle A \Rightarrow_u B\rangle_j = ([\,a_l^j \mid u(l) = j\,], b_j)$ for each $j \in J$. For the same reason we have $((A_i\lceil_{R_i^1})^{v_i})_{i=1}^n \vdash A$ with a proof ρ such that $\underline{\rho}_{\vec{x}} \sim_\eta P$. Indeed for each $l \in L = \mathsf{d}(A)$ we have

[7] Notice that our λ-calculus is in *Church style* and hence the type σ is uniquely determined by the sub-term N of M.

$[\langle A_i \upharpoonright_{R_i^1} \rangle_r \mid v_i(r) = l] = m_i^{j,l}$ and $\langle A \rangle_l = a_l^j$ where $j = u(l)$. By an application rule we get a proof π of $(A_i^{u_i})_{i=1}^n \vdash B$ such that $\underline{\pi}_{\overrightarrow{x}} = \left(\underline{\mu}_{\overrightarrow{x}}\right) \underline{\rho}_{\overrightarrow{x}} \sim_\eta (N) P = M$ as contended. □

4 The untyped Scott case

Since intersection types usually apply to the pure λ-calculus, we move now to this setting by choosing in $\mathbf{Rel}_!$ the set R_∞ as model of the pure λ-calculus. The R_∞ intersection typing system has the elements of R_∞ as types, and the typing rules involve sequents of shape $(x_i : m_i)_{i=1}^n \vdash M : a$ where $m_i \in \mathcal{M}_{\mathrm{fin}}(\mathsf{R}_\infty)$ and $a \in \mathsf{R}_\infty$.

We use Λ for the set of terms of the pure λ-calculus, and Λ_Ω as the pure λ-calculus extended with a constant Ω subject to the two following $\rightsquigarrow_\omega$ reduction rules: $\lambda x\, \Omega \rightsquigarrow_\omega \Omega$ and $(\Omega)\, M \rightsquigarrow_\omega \Omega$. We use $\sim_{\eta\omega}$ for the least congruence on Λ_Ω which contains $\rightsquigarrow_\eta$ and $\rightsquigarrow_\omega$ and similarly for $\sim_{\beta\eta\omega}$. We define a family $(\mathcal{H}(x))_{x\in\mathcal{V}}$ of subsets of Λ_Ω minimal such that, for any sequence $\overrightarrow{x} = (x_1, \dots, x_n)$ and $\overrightarrow{y} = (y_1, \dots, y_k)$ such that $\overrightarrow{x}, \overrightarrow{y}$ is repetition-free, and for any terms $M_i \in \mathcal{H}(x_i)$ (for $i = 1, \dots, n$), one has $\lambda \overrightarrow{x}\, \lambda \overrightarrow{y}\, (x)\, M_1 \cdots M_n\, O_1 \cdots O_l \in \mathcal{H}(x)$ where $O_j \sim_\omega \Omega$ for $j = 1, \dots, l$. Notice that $x \in \mathcal{H}(x)$.

The typing rules of R_∞ are

$$\frac{}{x_1 : [\,], \dots, x_i : [a], \dots, x_n : [\,] \vdash x_i : a} \qquad \frac{\Phi, x : m \vdash M : a}{\Phi \vdash \lambda x\, M : (m, a)}$$

$$\frac{\Phi \vdash M : ([a_1, \dots, a_k], b) \qquad (\Phi_j \vdash N : a_j)_{j=1}^k}{\Phi + \sum_{j=1}^k \Phi_j \vdash (M)\, N : b}$$

where we use the following convention: when we write $\Phi + \Psi$ it is assumed that Φ is of shape $(x_i : m_i)_{i=1}^n$ and Ψ is of shape $(x_i : p_i)_{i=1}^n$, and then $\Phi + \Psi$ is $(x_i : m_i + p_i)_{i=1}^n$. This typing system is just a "proof-theoretic" rephrasing of the denotational semantics of the terms of Λ_Ω in R_∞.

Proposition 2. *Let $M, M' \in \Lambda_\Omega$ and $\overrightarrow{x} = (x_1, \dots, x_n)$ be a list of pairwise distinct variables containing all the free variables of M and M'. Let $m_i \in \mathcal{M}_{\mathrm{fin}}(\mathsf{R}_\infty)$ for $i = 1, \dots, n$ and $b \in \mathsf{R}_\infty$. If $M \sim_{\beta\eta\omega} M'$ then $(x_i : m_i)_{i=1}^n \vdash M : b$ iff $(x_i : m_i)_{i=1}^n \vdash M' : b$.*

4.1 Formulas

We define the associated formulas as follows, each formula A being given together with $\mathsf{d}(A) \subseteq I$ and $\langle A \rangle \in \mathsf{R}_\infty^{\mathsf{d}(A)}$.

- If $J \subseteq I$ then ε_J is a formula with $\mathsf{d}(\varepsilon_J) = J$ and $\langle \varepsilon_J \rangle_j = \mathsf{e}$ for $j \in J$
- and if A and B are formulas and $u : \mathsf{d}(A) \to \mathsf{d}(B)$ is almost injective then $A \Rightarrow_u B$ is a formula with $\mathsf{d}(A \Rightarrow_u B) = \mathsf{d}(B)$ and $\langle A \Rightarrow_u B \rangle_j = ([\langle A \rangle_k \mid u(k) = j], \langle B \rangle_j) \in \mathsf{R}_\infty$.

We can consider that there is a type o of pure λ-terms interpreted as R_∞ in $\mathbf{Rel}_!$, such that $(\mathsf{o} \Rightarrow \mathsf{o}) = \mathsf{o}$, and then for any formula A we have $\underline{A} = \mathsf{o}$.

Operations of restriction and relocation of formulas are the same as in Section 3 (setting $\varepsilon_J \upharpoonright_K = \varepsilon_{J \cap K}$) and satisfy the same properties, for instance $\langle A \upharpoonright_K \rangle = \langle A \rangle \upharpoonright_K$ and one sets $u_*(\varepsilon_J) = \varepsilon_K$ if $u : J \to K$ is a bijection.

The deduction rules are exactly the same as those of Section 3, plus the axiom $\vdash \varepsilon_\emptyset$. With any deduction π of $(A_i^{u_i})_{i=1}^n \vdash B$ and sequence $\overrightarrow{x} = (x_1, \ldots, x_n)$ of pairwise distinct variables, we can associate a *pure* $\underline{\pi}_{\overrightarrow{x}} \in \Lambda_\Omega$ defined exactly as in Section 3 (just drop the types associated with variables in abstractions). If π consists of an instance of the additional axiom, we set $\underline{\pi}_{\overrightarrow{x}} = \Omega$.

Lemma 7. *Let $A, A_1, \ldots, A_n$ be a formula such that $\mathsf{d}(A) = \mathsf{d}(A_i) = \emptyset$. Then $(A_i^{0_\emptyset})_{i=1}^n \vdash A$ is provable by a proof π which satisfies $\underline{\pi}_{x_1,\ldots,x_k} \sim_\omega \Omega$.*

The proof is a straightforward induction on A using the additional axiom, Lemma 1 and the observations that if $\mathsf{d}(B \Rightarrow_u C) = \emptyset$ then $u = 0_\emptyset$.

One can easily define a size function $\mathsf{sz} : \mathsf{R}_\infty \to \mathbb{N}$ such that $\mathsf{sz}(\mathsf{e}) = 0$ and $\mathsf{sz}([a_1, \ldots, a_k], a) = \mathsf{sz}(a) + \sum_{i=1}^k (1 + \mathsf{sz}(a_i))$. First we have to prove an adapted version of Proposition 1; here it will be restricted to finite sets.

Proposition 3. *Let J be a finite subset of I and $f \in \mathsf{R}_\infty^J$. There is a formula A such that $\mathsf{d}(A) = J$ and $\langle A \rangle = f$.*

Proof. Observe that, since J is finite, there is an $N \in \mathbb{N}$ such that $\forall j \in J\ \forall q \in \mathbb{N}\ q \geq N \Rightarrow f(j)_q = [\,]$ (remember that $f(j) \in \mathcal{M}_{\mathrm{fin}}(\mathsf{R}_\infty)^{\mathbb{N}}$). Let $N(f)$ be the least such N. We set $\mathsf{sz}(f) = \sum_{j \in J} \mathsf{sz}(f(j))$ and the proof is by induction on $(\mathsf{sz}(f), N(f))$ lexicographically.

If $\mathsf{sz}(f) = 0$ this means that $f(j) = \mathsf{e}$ for all $j \in J$ and hence we can take $A = \varepsilon_J$. Assume that $\mathsf{sz}(f) > 0$, one can write[8] $f(j) = (m_j, a_j)$ with $m_j \in \mathcal{M}_{\mathrm{fin}}(\mathsf{R}_\infty)$ and $a_j \in \mathsf{R}_\infty$ for each $j \in J$. Just as in the proof of Proposition 1 we choose a set K, a function $g : K \to \mathsf{R}_\infty$ and an almost injective function $u : K \to J$ such that $m_j = [\, g(k) \mid u(k) = j \,]$. The set K is finite since J is and we have $\mathsf{sz}(g) < \mathsf{sz}(f)$ because $\mathsf{sz}(f) > 0$. Therefore by inductive hypothesis there is a formula B such that $\mathsf{d}(B) = K$ and $\langle B \rangle = g$. Let $f' : J \to \mathsf{R}_\infty$ defined by $f'(j) = a_j$, we have $\mathsf{sz}(f') \leq \mathsf{sz}(f)$ and $N(f') < N(f)$ and hence by inductive hypothesis there is a formula C such that $\langle C \rangle = f$. We set $A = (B \Rightarrow_u C)$ which satisfies $\langle A \rangle = f$ as required. □

Theorem 1 still holds up to some mild adaptation. First notice that $A \sim B$ simply means now that $\mathsf{d}(A) = \mathsf{d}(B)$ and $\langle A \rangle = \langle B \rangle$.

Theorem 4. *If A and B are such that $A \sim B$ then $A^{\mathsf{Id}} \vdash B$ with a proof π which satisfies $\underline{\pi}_x \in \mathcal{H}(x)$.*

[8] This is also possible if $\mathsf{sz}(f) = 0$ actually.

Proof. By induction on the sum of the sizes of A and B. Assume that $A = \varepsilon_J$ so that $\mathsf{d}(B) = J$ and $\forall j \in J\ \langle B \rangle_j = \mathsf{e}$. There are two cases as to B. In the first case B is of shape ε_K but then we must have $K = J$ and we can take for π an axiom so that $\underline{\pi}_x = x \in \mathcal{H}(x)$. Otherwise we have $B = (C \Rightarrow_u D)$ with $\mathsf{d}(D) = J$, $\forall j \in J\ \langle D \rangle_j = \mathsf{e}$ and $\mathsf{d}(C) = \emptyset$, so that $u = 0_J$. We have $A \sim D$ and hence by inductive hypothesis we have a proof ρ of $A^{\mathsf{Id}} \vdash D$ such that $\underline{\rho}_x \in \mathcal{H}(x)$. By weakening and $\Rightarrow$-introduction we get a proof π of $A^{\mathsf{Id}} \vdash B$ which satisfies $\underline{\pi}_x = \lambda y\, \underline{\rho}_x \in \mathcal{H}(x)$.

Assume that $A = (C \Rightarrow_u D)$. If $B = \varepsilon_J$ then we must have $\mathsf{d}(C) = \emptyset$, $u = 0_J$ and $D \sim B$ and hence by inductive hypothesis we have a proof ρ of $D^{\mathsf{Id}} \vdash B$ such that $\underline{\rho}_x \in \mathcal{H}(x)$. By Lemma 7 there is a proof θ of $\vdash C$ such that $\underline{\theta} \sim_\omega \Omega$. Hence there is a proof π of $A^{\mathsf{Id}} \vdash B$ such that $\underline{\pi}_x = \underline{\rho}_y\,[(x)\,\underline{\theta}/y] \in \mathcal{H}(x)$.

Assume last that $B = (E \Rightarrow_v F)$, then we must have $D \sim F$ and there must be a bijection $w : \mathsf{d}(E) \to \mathsf{d}(C)$ such that $u \circ w = v$ and $w_*(E) \sim C$. We reason as in the proof of Lemma 1: by inductive hypothesis we have a proof ρ of $D^{\mathsf{Id}} \vdash F$ and a proof μ of $w_*(E)^{\mathsf{Id}} \vdash C$ from which we build a proof π of $A^{\mathsf{Id}} \vdash B$ such that $\underline{\pi}_x = \lambda y\, \underline{\rho}_z \left[(x)\, \underline{\mu}_y / z\right] \in \mathcal{H}(x)$ by inductive hypothesis. □

Theorem 5 (Soundness). *Let π be a deduction tree of $A_1^{u_1}, \dots, A_n^{u_n} \vdash B$ and $\vec{x}$ a sequence of n pairwise distinct variables. Then the λ-term $\underline{\pi}_{\vec{x}} \in \Lambda_\Omega$ satisfies $(x_i : \langle A_i^{u_i} \rangle_j)_{i=1}^n \vdash \underline{\pi}_{\vec{x}} : \langle B \rangle_j$ in the R_∞ intersection type system, for each $j \in \mathsf{d}(B)$.*

The proof is exactly the same as that of Theorem 2, dropping all simple types.

For all λ-term $M \in \Lambda$, we define $\mathcal{H}_\Omega(M)$ as the least subset of element of Λ_Ω such that:

- if $O \in \Lambda_\Omega$ and $O \sim_\omega \Omega$ then $O \in \mathcal{H}_\Omega(M)$ for all $M \in \Lambda$
- if $M = x$ then $\mathcal{H}(x) \subseteq \mathcal{H}_\Omega(M)$
- if $M = \lambda y\, N$ and $N' \in \mathcal{H}_\Omega(N)$ then $\lambda y\, N' \in \mathcal{H}_\Omega(M)$
- if $M = (N)\, P$, $N' \in \mathcal{H}_\Omega(N)$ and $P' \in \mathcal{H}_\Omega(P)$ then $(N')\, P' \in \mathcal{H}_\Omega(M)$.

The elements of $\mathcal{H}_\Omega(M)$ can probably be seen as approximates of M.

Theorem 6 (Completeness). *Let $J \subseteq I$ be finite. Let $M \in \Lambda_\Omega$ and $x_1, \dots, x_n$ be pairwise distinct variables, such that $(x_i : m_i^j)_{i=1}^n \vdash M : b_j$ in the R_∞ intersection type system for all $j \in J$. Let $A_1, \dots, A_n$ and B be formulas and let $u_1, \dots, u_n$ be almost injective functions such that $u_i : \mathsf{d}(A_i) \to J = \mathsf{d}(B)$. Assume also that, for all $j \in J$, one has $\langle B \rangle_j = b_j$ and $\langle A_i^{u_i} \rangle_j = m_i^j$ for $i = 1, \dots, n$. Then the judgment $A_1^{u_1}, \dots, A_n^{u_n} \vdash B$ has a proof π such that $\underline{\pi}_{\vec{x}} \in \mathcal{H}_\Omega(M)$.*

The proof is very similar to that of Theorem 3.

5 Concluding remarks and acknowledgments

The results presented in this paper show that, at least in non-idempotent intersection types, the problem of knowing whether all elements of a given family of intersection types $(a_j)_{j\in J}$ are inhabited by a common λ-term can be reformulated logically: is it true that one (or equivalently, any) of the indexed formulas A such that $\mathsf{d}(A) = J$ and $\forall j \in \langle A\rangle_j = a_j$ is provable in $\mathsf{LJ}(I)$? Such a strong connection between intersection and Indexed Linear Logic was already mentioned in the introduction of [2], but we never made it more explicit until now.

To conclude we propose a typed λ-calculus *à la Church* to denote proofs of the $\mathsf{LJ}(I)$ system of Section 4. The syntax of *pre-terms* is given by $s, t \ldots :=$ $x[J] \mid \lambda x : A^u\, s \mid (s)\, t$ where in $x[J]$, x is a variable and $J \subseteq I$ and, in $\lambda x : A^u\, s$, u is an almost injective function from $\mathsf{d}(A)$ to a set $J \subseteq I$. Given a pre-term s and a variable x, the *domain of x in s* is the subset $\mathsf{dom}(x, s)$ of I given by $\mathsf{dom}(x, x[J]) = J$, $\mathsf{dom}(x, y[J]) = \emptyset$ if $y \neq x$, $\mathsf{dom}(x, \lambda y : A^u\, s) = \mathsf{dom}(x, s)$ (assuming of course $y \neq x$) and $\mathsf{dom}(x, (s)\, t) = \mathsf{dom}(x, s) \cup \mathsf{dom}(x, t)$. Then a pre-term s is a term if any subterm of t which is of shape $(s_1)\, s_2$ satisfies $\mathsf{dom}(x, s_1) \cap \mathsf{dom}(x, s_2) = \emptyset$ for all variable x. A typing judgment is an expression $(x_i : A_i^{u_i})_{i=1}^n \vdash s : B$ where the x_i's are pairwise distinct variables, s is a term and each u_i is an almost injective function $\mathsf{d}(A_i) \to \mathsf{d}(B)$. The following typing rules exactly mimic the logical rules of $\mathsf{LJ}(I)$:

$$\frac{\mathsf{d}(A) = \emptyset}{((x_i : A_i^{0_\emptyset})_{i=1}^n) \vdash \Omega : A}$$

$$\frac{q \neq i \Rightarrow \mathsf{d}(A_i) = \emptyset \text{ and } u_i \text{ bijection}}{(x_q : A_q^{u_q})_{q=1}^n \vdash x_i[\mathsf{d}(A_i)] : u_{i*}(A_i)} \qquad \frac{(x_i : A_i^{u_i})_{i=1}^n, x : A^u \vdash s : B}{(x_i : A_i^{u_i})_{i=1}^n \vdash \lambda x : A^u\, s : A \Rightarrow_u B}$$

$$\frac{(x_i : A_i\lceil^{v_i}_{\mathsf{dom}(x_i,s)})_{i=1}^n \vdash s : A \Rightarrow_u B \qquad (x_i : A_i\lceil^{w_i}_{\mathsf{dom}(x_i,t)})_{i=1}^n \vdash t : A}{(x_i : A_i^{v_i + (u \circ w_i)})_{i=1}^n \vdash (s)\, t : B}$$

The properties of this calculus, and more specifically of its β-reduction, and its connections with the resource calculus of [9] will be explored in further work.

Another major objective will be to better understand the meaning of $\mathsf{LJ}(I)$ formulas, using ideas developed in [3] where a *phase semantics* is introduced and related to (non-uniform) coherence space semantics. In the intuitionistic present setting, it is tempting to look for Kripke-like interpretations with the hope of generalizing indexed logic beyond the (perhaps too) specific relational setting we started from.

Last, we would like to thank Luigi Liquori and Claude Stolze for many helpful discussions on intersection types and the referees for their careful reading and insightful comments and suggestions.

References

1. F. Breuvart, G. Manzonetto, and D. Ruoppolo. Relational graph models at work. *Logical Methods in Computer Science*, 14(3), 2018.
2. A. Bucciarelli and T. Ehrhard. On phase semantics and denotational semantics in multiplicative-additive linear logic. *Annals of Pure and Applied Logic*, 102(3):247–282, 2000.

3. A. Bucciarelli and T. Ehrhard. On phase semantics and denotational semantics: the exponentials. *Annals of Pure and Applied Logic*, 109(3):205–241, 2001.
4. M. Coppo and M. Dezani-Ciancaglini. An extension of the basic functionality theory for the λ-calculus. *Notre Dame Journal of Formal Logic*, 21(4):685–693, 1980.
5. M. Coppo, M. Dezani-Ciancaglini, and B. Venneri. Functional characters of solvable terms. *Mathematical Logic Quarterly*, 27(2-6):45–58, 1981.
6. D. de Carvalho. Execution time of lambda-terms via denotational semantics and intersection types. *CoRR*, abs/0905.4251, 2009.
7. D. de Carvalho. Execution time of λ-terms via denotational semantics and intersection types. *MSCS*, 28(7):1169–1203, 2018.
8. T. Ehrhard. The Scott model of linear logic is the extensional collapse of its relational model. *Theoretical Computer Science*, 424:20–45, 2012.
9. T. Ehrhard and L. Regnier. Uniformity and the Taylor expansion of ordinary lambda-terms. *Theoretical Computer Science*, 403(2-3):347–372, 2008.
10. T. S. Freeman and F. Pfenning. Refinement Types for ML. In D. S. Wise, editor, *Proceedings of the ACM SIGPLAN'91 Conference on Programming Language Design and Implementation (PLDI), Toronto, Ontario, Canada, June 26-28, 1991*, pages 268–277. ACM, 1991.
11. J.-Y. Girard. Normal functors, power series and the λ-calculus. *Annals of Pure and Applied Logic*, 37:129–177, 1988.
12. J. R. Hindley. Coppo-dezani types do not correspond to propositional logic. *Theoretical Computer Science*, 28:235–236, 1984.
13. J.-L. Krivine. *Lambda-Calculus, Types and Models.* Ellis Horwood Series in Computers and Their Applications. Ellis Horwood, 1993. Translation by René Cori from French 1990 edition (Masson).
14. L. Liquori and S. R. D. Rocca. Intersection-types à la Church. *Information and Computation*, 205(9):1371–1386, 2007.
15. L. Liquori and C. Stolze. The Delta-calculus: Syntax and Types. In H. Geuvers, editor, *4th International Conference on Formal Structures for Computation and Deduction, FSCD 2019, June 24-30, 2019, Dortmund, Germany.*, volume 131 of *LIPIcs*, pages 28:1–28:20. Schloss Dagstuhl - Leibniz-Zentrum fuer Informatik, 2019.

On Computability of Data Word Functions Defined by Transducers*

Léo Exibard[1,2**(✉)], Emmanuel Filiot[1***], and Pierre-Alain Reynier[2†]

[1] Université Libre de Bruxelles, Brussels, Belgium
leo.exibard@ulb.ac.be
[2] Aix Marseille Univ, Université de Toulon, CNRS, LIS, Marseille, France

Abstract. In this paper, we investigate the problem of synthesizing computable functions of infinite words over an infinite alphabet (data ω-words). The notion of computability is defined through Turing machines with infinite inputs which can produce the corresponding infinite outputs in the limit. We use non-deterministic transducers equipped with registers, an extension of register automata with outputs, to specify functions. Such transducers may not define functions but more generally relations of data ω-words, and we show that it is PSPACE-complete to test whether a given transducer defines a function. Then, given a function defined by some register transducer, we show that it is decidable (and again, PSPACE-c) whether such function is computable. As for the known finite alphabet case, we show that computability and continuity coincide for functions defined by register transducers, and show how to decide continuity. We also define a subclass for which those problems are PTIME.

Keywords: Data Words · Register Automata · Register Transducers · Functionality · Continuity · Computability.

1 Introduction

Context Program synthesis aims at deriving, in an automatic way, a program that fulfils a given specification. Such setting is very appealing when for instance the specification describes, in some abstract formalism (an automaton or ideally a logic), important properties that the program must satisfy. The synthesised program is then *correct-by-construction* with regards to those properties. It is particularly important and desirable for the design of safety-critical systems with hard dependability constraints, which are notoriously hard to design correctly.

Program synthesis is hard to realise for general-purpose programming languages but important progress has been made recently in the automatic synthesis

* A version with full proofs can be found at https://arxiv.org/abs/2002.08203.
** Funded by a FRIA fellowship from the F.R.S.-FNRS.
*** Research associate of F.R.S.-FNRS. Supported by the ARC Project Transform Fédération Wallonie-Bruxelles and the FNRS CDR J013116F; MIS F451019F projects.
† Partly funded by the ANR projects DeLTA (ANR-16-CE40-0007) and Ticktac (ANR-18-CE40-0015).

J. Goubault-Larrecq and B. König (Eds.): FOSSACS 2020, LNCS 12077, pp. 217–236, 2020.
https://doi.org/10.1007/978-3-030-45231-5_12

of *reactive systems*. In this context, the system continuously receives input signals to which it must react by producing output signals. Such systems are not assumed to terminate and their executions are usually modelled as infinite words over the alphabets of input and output signals. A specification is thus a set of pairs (in,out), where in and out are infinite words, such that out is a legitimate output for in. Most methods for reactive system synthesis only work for *synchronous* systems over *finite* sets of input and output signals Σ and Γ. In this synchronous setting, input and output signals alternate, and thus *implementations* of such a specification are defined by means of *synchronous* transducers, which are Büchi automata with transitions of the form (q, σ, γ, q'), expressing that in state q, when getting input $\sigma \in \Sigma$, output $\gamma \in \Gamma$ is produced and the machine moves to state q'. We aim at building *deterministic* implementations, in the sense that the output γ and state q' uniquely depend on q and σ. The realisability problem of specifications given as synchronous non-deterministic transducers, by implementations defined by synchronous deterministic transducers is known to be decidable [14,20]. In this paper, we are interested in the *asynchronous* setting, in which transducers can produce none or several outputs at once every time some input is read, i.e., transitions are of the form (q, σ, w, q') where $w \in \Gamma^*$. However, such generalisation makes the realisability problem undecidable [2,9].

Synthesis of Transducers with Registers In the setting we just described, the set of signals is considered to be finite. This assumption is not realistic in general, as signals may come with unbounded information (e.g. process ids) that we call here *data*. To address this limitation, recent works have considered the synthesis of reactive systems processing *data words* [17,6,16,7]. Data words are infinite words over an alphabet $\Sigma \times \mathcal{D}$, where Σ is a finite set and $\mathcal{D}$ is a possibly infinite countable set. To handle data words, just as automata have been extended to *register automata*, transducers have been extended to *register transducers*. Such transducers are equipped with a finite set of registers in which they can store data and with which they can compare data for equality or inequality. While the realisability problem of specifications given as synchronous non-deterministic register transducers ($\mathsf{NRT_{syn}}$) by implementation defined by synchronous deterministic register transducers ($\mathsf{DRT_{syn}}$) is undecidable, decidability is recovered for specifications defined by universal register transducers and by giving as input the number of registers the implementation must have [7,17].

Computable Implementations In the previously mentioned works, both for finite or infinite alphabets, implementations are considered to be deterministic transducers. Such an implementation is guaranteed to use only a constant amount of memory (assuming data have size $O(1)$). While it makes sense with regards to memory-efficiency, some problems turn out to be undecidable, as already mentioned: realisability of $\mathsf{NRT_{syn}}$ specifications by $\mathsf{DRT_{syn}}$, or, in the finite alphabet setting, when both the specification and implementation are asynchronous. In this paper, we propose to study computable implementations, in the sense of (partial) functions f of data ω-words computable by some Turing machine M that has an infinite input $x \in \mathrm{dom}(f)$, and produces longer and longer prefixes of the output

$f(x)$ as it reads longer and longer prefixes of the input x. Therefore, such a machine produces the output $f(x)$ in the limit. We denote by TM the class of Turing machines computing functions in this sense. As an example, consider the function f that takes as input any data ω-word $u = (\sigma_1, d_1)(\sigma_2, d_2)\dots$ and outputs $(\sigma_1, d_1)^\omega$ if d_1 occurs at least twice in u, and otherwise outputs u. This function is not computable, as an hypothetic machine could not output anything as long as d_1 is not met a second time. However, the following function g is computable. It is defined only on words $(\sigma_1, d_1)(\sigma_2, d_2)\dots$ such that $\sigma_1\sigma_2\dots \in ((a+b)c^*)^\omega$, and transforms any (σ_i, d_i) by (σ_i, d_1) if the next symbol in $\{a, b\}$ is an a, otherwise it keeps (σ_i, d_i) unchanged. To compute it, a TM would need to store d_1, and then wait until the next symbol in $\{a, b\}$ is met before outputting something. Since the finite input labels are necessarily in $((a+b)c^*)^\omega$, this machine will produce the whole output in the limit. Note that g cannot be defined by any deterministic register transducer, as it needs unbounded memory to be implemented.

However, already in the finite alphabet setting, the problem of deciding if a specification given as some non-deterministic synchronous transducer is realisable by some computable function is open. The particular case of realisability by computable functions of universal domain (the set of all ω-words) is known to be decidable [12]. In the asynchronous setting, the undecidability proof of [2] can be easily adapted to show the undecidability of realisability of specifications given by non-deterministic (asynchronous) transducers by computable functions.

Functional Specifications As said before, a specification is in general a relation from inputs to outputs. If this relation is a function, we call it functional. Due to the negative results just mentioned about the synthesis of computable functions from non-functional specifications, we instead here focus on the case of functional specifications and address the following general question: given the specification of a function of data ω-words, is this function "implementable", where we define "implementable" as "being computable by some Turing machine". Moreover, if it is implementable, then we want a procedure to automatically generate an algorithm that computes it. This raises another important question: how to decide whether a specification is functional ? We investigate these questions for asynchronous register transducers, here called register transducers. This asynchrony allows for much more expressive power, but is a source of technical challenge.

Contributions In this paper, we solve the questions mentioned before for the class of (asynchronous) non-deterministic register transducers (NRT). We also give fundamental results on this class. In particular, we prove that:

1. deciding whether an NRT defines a function is PSPACE-complete,
2. deciding whether two functions defined by NRT are equal on the intersection of their domains is PSPACE-complete,
3. the class of functions defined by NRT is effectively closed under composition,
4. computability and continuity are equivalent notions for functions defined by NRT, where continuity is defined using the classical Cantor distance,
5. deciding whether a function given as an NRT is computable is PSPACE-c,

6. those problems are in PTime for a subclass of NRT, called test-free NRT.

Finally, we also mention that considering the class of deterministic register transducers (DRT for short) instead of computable functions as a yardstick for the notion of being "implementable" for a function would yield undecidability. Indeed, given a function defined by some NRT, it is in general undecidable to check whether this function is realisable by some DRT, by a simple reduction from the universality problem of non-deterministic register automata [19].

Related Work The notion of continuity with regards to Cantor distance is not new, and for rational functions over finite alphabets, it was already known to be decidable [21]. Its connection with computability for functions of ω-words over a finite alphabet has recently been investigated in [3] for one-way and two-way transducers. Our results lift some of theirs to the setting of data words. The model of test-free NRT can be seen as a one-way non-deterministic version of a model of two-way transducers considered in [5].

2 Data Words and Register Transducers

For a (possibly infinite) set S, we denote by S^* (resp. S^ω) the set of finite (resp. infinite) words over this alphabet, and we let $S^\infty = S^* \cup S^\omega$. For a word $u = u_1 \dots u_n$, we denote $\|u\| = n$ its length, and, by convention, for $u \in S^\omega$, $\|u\| = \infty$. The empty word is denoted ε. For $1 \leq i \leq j \leq \|u\|$, we let $u[i{:}j] = u_i u_{i+1} \dots u_j$ and $u[i] = u[i{:}i]$ the ith letter of u. For $u, v \in S^\infty$, we say that u is a prefix of v, written $u \preceq v$, if there exists $w \in S^\infty$ such that $v = uw$. In this case, we define $u^{-1}v = w$. For $u, v \in S^\infty$, we say that u and v *mismatch*, written $\mathsf{mismatch}(u, v)$, when there exists a position i such that $1 \leq i \leq \|u\|$, $1 \leq i \leq \|v\|$ and $u[i] \neq v[i]$. Finally, for $u, v \in S^\infty$, we denote by $u \wedge v$ their longest common prefix, i.e. the longest word $w \in S^\infty$ such that $w \preceq u$ and $w \preceq v$.

Data Words In this paper, Σ and Γ are two finite alphabets and $\mathcal{D}$ is a countably infinite set of *data*. We use letter σ (resp. γ, d) to denote elements of Σ (resp. Γ, $\mathcal{D}$). We also distinguish an arbitrary data value $\mathsf{d}_0 \in \mathcal{D}$. Given a set R, let τ_0^R be the constant function defined by $\tau_0^R(r) = \mathsf{d}_0$ for all $r \in R$. Given a finite alphabet A, a *labelled data* is a pair $x = (a, d) \in A \times \mathcal{D}$, where a is the *label* and d the *data*. We define the projections $\mathsf{lab}(x) = a$ and $\mathsf{dt}(x) = d$. A *data word* over A and $\mathcal{D}$ is an infinite sequence of labelled data, i.e. a word $w \in (A \times \mathcal{D})^\omega$. We extend the projections lab and dt to data words naturally, i.e. $\mathsf{lab}(w) \in A^\omega$ and $\mathsf{dt}(w) \in \mathcal{D}^\omega$. A *data word language* is a subset $L \subseteq (A \times \mathcal{D})^\omega$. Note that here, data words are infinite, otherwise they are called *finite data words*.

2.1 Register Transducers

Register transducers are transducers recognising data word relations. They are an extension of finite transducers to data word relations, in the same way register automata [15] are an extension of finite automata to data word languages. Here, we define them over infinite data words with a Büchi acceptance condition, and allow multiple registers to contain the same data, with a syntax close to [18]. The current data can be compared for equality with the register contents via tests, which are symbolic and defined via Boolean formulas of the following form. Given R a set of registers, a *test* is a formula ϕ satisfying the following syntax:

$$\phi ::= \top \mid \bot \mid r^{=} \mid r^{\neq} \mid \phi \wedge \phi \mid \phi \vee \phi \mid \neg \phi$$

where $r \in R$. Given a valuation $\tau : R \to \mathcal{D}$, a test ϕ and a data d, we denote by $\tau, d \models \phi$ the satisfiability of ϕ by d in valuation τ, defined as $\tau, d \models r^{=}$ if $\tau(r) = d$ and $\tau, d \models r^{\neq}$ if $\tau(r) \neq d$. The Boolean combinators behave as usual. We denote by Tst_R the set of (symbolic) tests over R.

Definition 1. *A non-deterministic register transducer (*NRT*) is a tuple* $T = (Q, R, i_0, F, \Delta)$*, where* Q *is a finite set of* states*,* $i_0 \in Q$ *is the* initial *state,* $F \subseteq Q$ *is the set of* accepting *states,* R *is a finite set of* registers *and* $\Delta \subseteq Q \times \Sigma \times \mathsf{Tst}_R \times 2^R \times (\Gamma \times R)^* \times Q$ *is a finite set of* transitions*. We write* $q \xrightarrow[T]{\sigma,\phi|\mathsf{asgn},o} q'$ *for* $(q, \sigma, \phi, \mathsf{asgn}, o, q') \in \Delta$ *(*T *is sometimes omitted).*

The semantics of a register transducer is given by a labelled transition system: we define $L_T = (C, \Lambda, \to)$, where $C = Q \times (R \to \mathcal{D})$ is the set of configurations, $\Lambda = (\Sigma \times \mathcal{D}) \times (\Gamma \times \mathcal{D})^*$ is the set of labels, and we have, for all $(q, \tau), (q', \tau') \in C$ and for all $(l, w) \in \Lambda$, that $(q, \tau) \xrightarrow{(l,w)} (q', \tau')$ whenever there exists a transition $q \xrightarrow[T]{\sigma,\phi|\mathsf{asgn},o} q'$ such that, by writing $l = (\sigma', d)$ and $w = (\gamma'_1, d_1) \dots (\gamma'_n, d_n)$:

- (Matching labels) $\sigma = \sigma'$
- (Compatibility) d satisfies the test $\phi \in \mathsf{Tst}_R$, i.e. $\tau, d \models \phi$.
- (Update) τ' is the successor register configuration of τ with regards to d and asgn: $\tau'(r) = d$ if $r \in \mathsf{asgn}$, and $\tau'(r) = \tau(r)$ otherwise
- (Output) By writing $o = (\gamma_1, r_1) \dots (\gamma_m, r_m)$, we have that $m = n$ and for all $1 \leq i \leq n$, $\gamma_i = \gamma'_i$ and $d_i = \tau'(r_i)$.

Then, a *run* of T is an infinite sequence of configurations and transitions $\rho = (q_0, \tau_0) \xrightarrow[L_T]{(u_1,v_1)} (q_1, \tau_1) \xrightarrow[L_T]{(u_2,v_2)} \cdots$. Its input is $\mathsf{in}(\rho) = u_1 u_2 \dots$, its output is $\mathsf{out}(\rho) = v_1 \cdot v_2 \dots$. We also define its sequence of states $\mathsf{st}(\rho) = q_0 q_1 \dots$, and its *trace* $\mathsf{tr}(\rho) = u_1 \cdot v_1 \cdot u_2 \cdot v_2 \dots$. Such run is *initial* if $(q_0, \tau_0) = (i_0, \tau_0^R)$. It is *final* if it satisfies the Büchi condition, i.e. $\inf(\mathsf{st}) \cap F \neq \varnothing$, where $\inf(\mathsf{st}) = \{q \in Q \mid q = q_i$ for infinitely many $i\}$. Finally, it is *accepting* if it is both initial and final. We then write $(q_0, \tau_0) \xrightarrow[T]{u|v}$ to express that there is a final run ρ of T starting from (q_0, τ_0) such that $\mathsf{in}(\rho) = u$ and $\mathsf{out}(\rho) = v$. In the whole paper, and unless stated otherwise, we always assume that the output of an accepting run is infinite ($v \in (\Gamma \times \mathcal{D})^\omega$), which can be ensured by a Büchi condition.

A *partial run* is a finite prefix of a run. The notions of input, output and states are extended by taking the corresponding prefixes. We then write $(q_0, \tau_0) \xrightarrow[T]{u|v}$

(q_n, τ_n) to express that there is a partial run ρ of T starting from configuration (q_0, τ_0) and ending in configuration (q_n, τ_n) such that $\mathsf{in}(\rho) = u$ and $\mathsf{out}(\rho) = v$.

Finally, the relation represented by a transducer T is:

$$[\![T]\!] = \{(u, v) \in (\Sigma \times \mathcal{D})^\omega \times (\Gamma \times \mathcal{D})^\omega \mid \text{there exists an accepting run } \rho \text{ of } T \\ \text{such that } \mathsf{in}(\rho) = u \text{ and } \mathsf{out}(\rho) = v\}$$

Example 2. As an example, consider the register transducer T_{rename} depicted in Figure 1. It realises the following transformation: consider a setting in which we deal with logs of communications between a set of clients. Such a log is an infinite sequence of pairs consisting of a tag, chosen in some finite alphabet Σ, and the identifier of the client delivering this tag, chosen in some infinite set of data values. The transformation should modify the log as follows: for a given client that needs to be modified, each of its messages should now be associated with some new identifier. The transformation has to verify that this new identifier is indeed free, *i.e.* never used in the log. Before treating the log, the transformation receives as input the id of the client that needs to be modified (associated with the tag del), and then a sequence of identifiers (associated with the tag ch), ending with #. The transducer is non-deterministic as it has to guess which of these identifiers it can choose to replace the one of the client. In particular, observe that it may associate multiple output words to a same input if two such free identifiers exist.

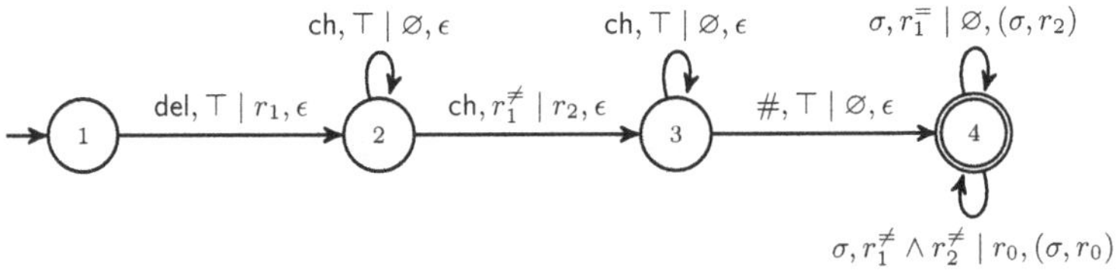

Fig. 1. A register transducer T_{rename}. It has three registers r_1, r_2 and r_0 and four states. σ denotes any letter in Σ, r_1 stores the id of del and r_2 the chosen id of ch, while r_0 is used to output the last data value read as input. As we only assign data to single registers, we write r_i for the singleton assignment set $\{r_i\}$.

Finite Transducers Since we reduce the decision of continuity and functionality of NRT to the one of finite transducers, let us introduce them: a finite transducer (NFT for short) is an NRT with 0 registers (i.e. $R = \varnothing$). Thus, its transition relation can be represented as $\Delta \subseteq Q \times \Sigma \times \Gamma^* \times Q$. A direct extension of the construction of [15, Proposition 1] allows to show that:

Proposition 3. *Let T be an* NRT *with k registers, and let $X \subset_f \mathcal{D}$ be a finite subset of data. Then, $[\![T]\!] \cap (\Sigma \times X)^\omega \times (\Gamma \times X)^\omega$ is recognised by an* NFT *of exponential size, more precisely with $O(|Q| \times |X|^{|R|})$ states.*

2.2 Technical Properties of Register Automata

Although automata are simpler machines than transducers, we only use them as tools in our proofs, which is why we define them from transducers, and not the other way around. A non-deterministic register automaton, denoted NRA, is a transducer without outputs: its transition relation is $\Delta \subseteq Q \times \Sigma \times \mathsf{Tst}_R \times 2^R \times \{\varepsilon\} \times Q$ (simply represented as $\Delta \subseteq Q \times \Sigma \times \mathsf{Tst}_R \times 2^R \times Q$). The semantics are the same, except that now we lift the condition that the output v is infinite since there is no output. For A an NRA, we denote $L(A) = \{u \in (\Sigma \times \mathcal{D})^\omega \mid$ there exists an accepting run ρ of A over $u\}$. Necessarily the output of an accepting run is ε. In this section, we establish technical properties about NRA.

Proposition 4, the so-called "indistinguishability property", was shown in the seminal paper by Kaminski and Francez [15, Proposition 1]. Their model differs in that they do not allow distinct registers to contain the same data, and in the corresponding test syntax, but their result easily carries to our setting. It states that if an NRA accepts a data word, then such data word can be relabelled with data from any set containing d_0 and with at least $k+1$ elements. Indeed, at any point of time, the automaton can only store at most k data in its registers, so its notion of "freshness" is a local one, and forgotten data can thus be reused as fresh ones. Moreover, as the automaton only tests data for equality, their actual value does not matter, except for d_0 which is initially contained in the registers.

Such "small-witness" property is fundamental to NRA, and will be paramount in establishing decidability of functionality (Section 3) and computability (Section 4). We use it jointly with Lemma 5, which states that the interleaving of the traces of runs of an NRT can be recognised with an NRA, and Lemma 6, which expresses that an NRA can check whether interleaved words coincide on some bounded prefix, and/or mismatch before some given position.

Proposition 4 ([15]). *Let A be an* NRA *with k registers. If $L(A) \neq \varnothing$, then, for any $X \subseteq \mathcal{D}$ of size $|X| \geq k+1$ such that $\mathsf{d}_0 \in X$, $L(A) \cap (\Sigma \times X)^\omega \neq \varnothing$.*

The runs of a register transducer T can be flattened to their traces, so as to be recognised by an NRA. Those traces can then be interleaved, in order to be compared. The proofs of the following properties are straightforward.

Let $\rho_1 = (q_0, \tau_0) \xrightarrow[L_T]{(u_1, u_1')} (q_1, \tau_1) \ldots$ and $\rho_2 = (p_0, \mu_0) \xrightarrow[L_T]{(v_1, v_1')} (p_1, \mu_1) \ldots$ be two runs of a transducer T. Then, we define their *interleaving* $\rho_1 \otimes \rho_2 = u_1 \cdot u_1' \cdot v_1 \cdot v_1' \cdot u_2 \cdot u_2' \cdot v_2 \cdot v_2' \ldots$ and $L_\otimes(T) = \{\rho_1 \otimes \rho_2 \mid \rho_1$ and ρ_2 are accepting runs of $T\}$.

Lemma 5. *If T has k registers, then $L_\otimes(T)$ is recognised by an* NRA *with $2k$ registers.*

Lemma 6. *Let $i, j \in \mathbb{N} \cup \{\infty\}$. We define $M_j^i = \{u_1 u_1' v_1 v_1' \cdots \mid \forall k \geq 1, u_k, v_k \in (\Sigma \times \mathcal{D}), u_k', v_k' \in (\Gamma \times \mathcal{D})^*, \forall 1 \leq k \leq j, v_k = u_k$ and $\|u_1' \cdot u_2' \cdots \wedge v_1' \cdot v_2' \ldots\| \leq i\}$. Then, M_j^i is recognisable by an* NRA *with 2 registers and with 1 register if $i = \infty$.*

3 Functionality, Equivalence and Composition of NRT

In general, since they are non-deterministic, NRT may not define functions but relations, as illustrated by Example 2. In this section, we first show that deciding whether a given NRT defines a function is PSpace-complete, in which case we call it *functional.* We show, as a consequence, that testing whether two functional NRT define two functions which coincide on their common domain is PSpace-complete. Finally, we show that functions defined by NRT are closed under composition. This is an appealing property in transducer theory, as it allows to define complex functions by composing simple ones.

Example 7. As explained before, the transducer T_{rename} described in Example 2 is not functional. To gain functionality, one can reinforce the specification by considering that one gets at the beginning a list of k possible identifiers, and that one has to select the first one which is free, for some fixed k. This transformation is realised by the register transducer T_{rename2} depicted in Figure 2 (for $k = 2$).

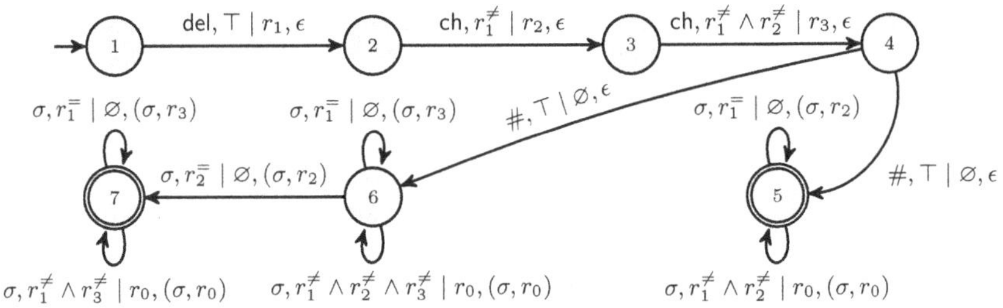

Fig. 2. A NRT T_{rename2}, with four registers r_1, r_2, r_3 and r_0 (the latter being used, as in Figure 1, to output the last read data). After reading the # symbol, it guesses whether the value of register r_2 appears in the suffix of the input word. If not, it goes to state 5, and replaces occurrences of r_1 by r_2. Otherwise, it moves to state 6, waiting for an occurrence of r_2, and replaces occurrences of r_1 by r_3.

Let us start with the functionality problem in the data-free case. It is already known that checking whether an NFT over ω-words is functional is decidable [13,11]. By relying on the pattern logic of [10] designed for transducers of *finite* words, it can be shown that it is decidable in NLogSpace.

Proposition 8. *Deciding whether an* NFT *is functional is in* NLogSpace.

The following theorem shows that a relation between data-words defined by an NRT with k registers is a function iff its restriction to a set of data with at most $2k + 3$ data is a function. As a consequence, functionality is decidable as it reduces to the functionality problem of transducers over a finite alphabet.

Theorem 9. *Let T be an* NRT *with k registers. Then, for all $X \subseteq \mathcal{D}$ of size $|X| \geq 2k + 3$ such that $\mathsf{d}_0 \in X$, we have that T is functional if and only if $[\![T]\!] \cap ((\Sigma \times X)^\omega \times (\Gamma \times X)^\omega)$ is functional.*

Proof. The left-to-right direction is trivial. Now, assume T is not functional. Let $x \in (\Sigma \times \mathcal{D})^\omega$ be such that there exists $y, z \in (\Gamma \times \mathcal{D})^\omega$ such that $y \neq z$ and $(x, y), (x, z) \in [\![T]\!]$. Let $i = \|y \wedge z\|$. Then, consider the language $L = \{\rho_1 \otimes \rho_2 \mid \rho_1$ and ρ_2 are accepting runs of $T, \mathsf{in}(\rho_1) = \mathsf{in}(\rho_2)$ and $\|\mathsf{out}(\rho_1) \wedge \mathsf{out}(\rho_2)\| \leq i\}$. Since, by Lemma 5, $L_\otimes(T)$ is recognised by an NRA with $2k$ registers and, by Lemma 6, M^i_∞ is recognised by an NRA with 2 registers, we get that $L = L_\otimes(T) \cap M^i_\infty$ is recognised by an NRA with $2k+2$ registers.

Now, $L \neq \varnothing$, since, by letting ρ_1 and ρ_2 be the runs of T both with input x and with respective outputs y and z, we have that $w = \rho_1 \otimes \rho_2 \in L$. Let $X \subseteq \mathcal{D}$ such that $|X| \geq 2k+3$ and $\mathsf{d}_0 \in X$. By Proposition 4, we get that $L \cap (\Sigma \times X)^\omega \neq \varnothing$. By letting $w' = \rho'_1 \otimes \rho'_2 \in L \cap (\Sigma \times X)^\omega$, and $x' = \mathsf{in}(\rho'_1) = \mathsf{in}(\rho'_2)$, $y' = \mathsf{out}(\rho'_1)$ and $z' = \mathsf{out}(\rho'_2)$, we have that $(x', y'), (x', z') \in [\![T]\!] \cap ((\Sigma \times X)^\omega \times (\Gamma \times X)^\omega)$ and $\|y' \wedge z'\| \leq i$, so, in particular, $y' \neq z'$ (since both are infinite words). Thus, $[\![T]\!] \cap ((\Sigma \times X)^\omega \times (\Gamma \times X)^\omega)$ is not functional. □

As a consequence of Proposition 8 and Theorem 9, we obtain the following result. The lower bound is obtained by encoding non-emptiness of register automata, which is PSPACE-complete [4].

Corollary 10. *Deciding whether an* NRT T *is functional is* PSPACE*-complete.*

Hence, the following problem on the equivalence of NRT is decidable:

Theorem 11. *The problem of deciding, given two functions f, g defined by* NRT*, whether for all $x \in \mathrm{dom}(f) \cap \mathrm{dom}(g)$, $f(x) = g(x)$, is* PSPACE*-complete.*

Proof. The formula $\forall x \in \mathrm{dom}(f) \cap \mathrm{dom}(g) \cdot f(x) = g(x)$ is true iff the relation $f \cup g = \{(x, y) \mid y = f(x) \vee y = g(x)\}$ is a function. The latter can be decided by testing whether the disjoint union of the transducers defining f and g defines a function, which is in PSPACE by Corollary 10. To show the hardness, we similarly reduce the emptiness problem of NRA A over finite words, just as in the proof of Corollary 10. In particular, the functions f_1 and f_2 defined in this proof (which have the same domain) are equal iff $L(A) = \varnothing$. □

Note that under the promise that f and g have the same domain, the latter theorem implies that it is decidable to check whether the two functions are equal. However, checking $\mathrm{dom}(f) = \mathrm{dom}(g)$ is undecidable, as the language-equivalence problem for non-deterministic register automata is undecidable, since, in particular, universality is undecidable [19].

Closure under composition is a desirable property for transducers, which holds in the data-free setting [1]. We show that it also holds for functional NRT.

Theorem 12. *Let f, g be two functions defined by* NRT*. Then, their composition $f \circ g$ is (effectively) definable by some* NRT.

Proof (Sketch). By $f \circ g$ we mean $f \circ g : x \mapsto f(g(x))$. Assume f and g are defined by $T_f = (Q_f, R_f, q_0, F_f, \Delta_f)$ and $T_g = (Q_g, R_g, p_0, F_g, \Delta_g)$ respectively. Wlog we assume that the input and output finite alphabets of T_f and T_g are all equal to Σ, and that R_f and R_g are disjoint. We construct T such that $[\![T]\!] = f \circ g$. The proof is similar to the data-free case where the composition is shown via a product construction which simulates both transducers in parallel, executing the second on the output of the first. Assume T_g has some transition

$p \xrightarrow{\sigma,\phi|\{r\},o} q$ where $o \in (\Sigma \times R_g)^*$. Then T has to be able to execute transitions of T_f while processing o, even though o does not contain any concrete data values (it is here the main important difference with the data-free setting). However, if T knows the equality types between R_f and R_g, then it is able to trigger the transitions of T_f. For example, assume that $o = (a, r_g)$ and assume that the content of r_g is equal to the content of r_f, r_f being a register of T_f, then if T_f has some transition of the form $p' \xrightarrow{a,r_f^{=}|\{r'_f\},o'} q'$ then T can trigger the transition $(p,q) \xrightarrow{\sigma,\phi|\{r\}\cup\{r'_f:=r_g\},o'} (p',q')$ where the operation $r'_f := r_g$ is a syntactic sugar on top of NRT that intuitively means "put the content of r_g into r'_f". □

Remark 13. The proof of Theorem 12 does not use the hypothesis that f and g are functions, and actually shows a stronger result, namely that relations defined by NRT are closed under composition.

4 Computability and Continuity

We equip the set of (finite or infinite) data words with the usual distance: for $u, v \in (\Sigma \times \mathcal{D})^\omega$, $d(u,v) = 0$ if $u = v$ and $d(u,v) = 2^{-\|u \wedge v\|}$ otherwise. A sequence of (finite or infinite) data words $(x_n)_{n\in\mathbb{N}}$ converges to some infinite data word x if for all $\epsilon > 0$, there exists $N \geq 0$ such that for all $n \geq N$, $d(x_n, x) \leq \epsilon$.

In order to reason with computability, we assume in the sequel that the infinite set of data values $\mathcal{D}$ we are dealing with has an effective representation. For instance, this is the case when $\mathcal{D} = \mathbb{N}$.

We now define how a Turing machine can compute a function of data words. We consider deterministic Turing machines, which three tapes: a read-only one-way input tape (containing the infinite input data word), a two-way working tape, and a write-only one-way output tape (on which it writes the infinite output data word). Consider some input data word $x \in (\Sigma \times \mathcal{D})^\omega$. For any integer $k \in \mathbb{N}$, we let $M(x,k)$ denote the output written by M on its output tape after having read the k first cells of the input tape. Observe that as the output tape is write-only, the sequence of data words $(M(x,k))_{k\geq 0}$ is non-decreasing.

Definition 14 (Computability). *A function $f : (\Sigma \times \mathcal{D})^\omega \to (\Gamma \times \mathcal{D})^\omega$ is computable if there exists a deterministic multi-tape machine M such that for all $x \in \mathrm{dom}(f)$, the sequence $(M(x,k))_{k\geq 0}$ converges to $f(x)$.*

Definition 15 (Continuity). *A function $f : (\Sigma \times \mathcal{D})^\omega \to (\Gamma \times \mathcal{D})^\omega$ is* continuous *at $x \in \mathrm{dom}(f)$ if (equivalently):*

(a) for all sequences of data words $(x_n)_{n\in\mathbb{N}}$ converging towards x, where for all $i \in \mathbb{N}$, $x_i \in \mathrm{dom}(f)$, we have that $(f(x_n))_{n\in\mathbb{N}}$ converges to $f(x)$.
(b) $\forall i \geq 0, \exists j \geq 0, \forall y \in \mathrm{dom}(f), \|x \wedge y\| \geq j \Rightarrow \|f(x) \wedge f(y)\| \geq i$.

Then, f is continuous if and only if it is continuous at each $x \in \mathrm{dom}(f)$. Finally, a functional NRT *T is* continuous *when $[\![T]\!]$ is continuous.*

Example 16. We give an example of a non-continuous function f. The finite input and output alphabets are unary, and are therefore ignored in the description of f. Such function associates with every sequence $s = d_1 d_2 \cdots \in \mathcal{D}^\omega$ the word $f(s) = d_1^\omega$ if d_1 occurs infinitely many times in s, otherwise $f(s) = s$ itself.

The function f is not continuous. Indeed, by taking $d \neq d'$, the sequence of data words $d(d')^n d^\omega$ converges to $d(d')^\omega$, while $f(d(d')^n d^\omega) = d^\omega$ converges to $d^\omega \neq f(d(d')^\omega) = d(d')^\omega$.

Moreover, f is realisable by some NRT which non-deterministically guesses whether d_1 repeats infinitely many times or not. It needs only one register r in which to store d_1. In the first case, it checks whether the current data d is equal the content r infinitely often, and in the second case, it checks that this test succeeds finitely many times, using Büchi conditions.

One can show that the register transducer T_{rename2} considered in Example 7 also realises a function which is not continuous, as the value stored in register r_2 may appear arbitrarily far in the input word. One could modify the specification to obtain a continuous function as follows. Instead of considering an infinite log, one considers now an infinite sequence of finite logs, separated by \$ symbols. The register transducer T_{rename3}, depicted in Figure 3, defines such a function.

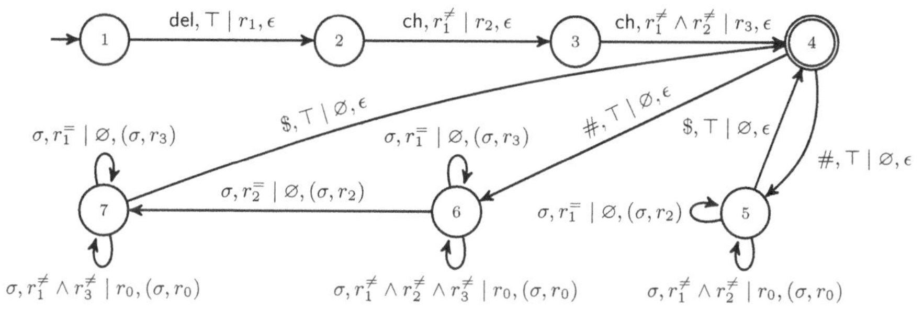

Fig. 3. A register transducer T_{rename3}. This transducer is non-deterministic, yet it defines a continuous function.

We now prove the equivalence between continuity and computability for functions defined by NRT. One direction, namely the fact that computability implies continuity, is easy, almost by definition. For the other direction, we rely on the following lemma which states that it is decidable whether a word v can be safely output, only knowing a prefix u of the input. In particular, given a function f, we let $\hat{f}$ be the function defined over all finite prefixes u of words in $\mathrm{dom}(f)$ by $\hat{f}(u) = \bigwedge(f(uy) \mid uy \in \mathrm{dom}(f))$, the longest common prefix of all outputs of continuations of u by f. Then, we have the following decidability result:

Lemma 17. *The following problem is decidable. Given an* NRT T *defining a function* f*, two finite data words* $u \in (\Sigma \times \mathcal{D})^*$ *and* $v \in (\Gamma \times \mathcal{D})^*$*, decide whether* $v \preceq \hat{f}(u)$.

Theorem 18. *Let f be a function defined by some* NRT T. *Then f is continuous iff f is computable.*

Proof. $\Leftarrow$ Assuming $f = [\![T]\!]$ is computable by some Turing machine M, we show that f is continuous. Indeed, consider some $x \in \text{dom}(f)$, and some $i \geq 0$. As the sequence of finite words $(M(x,k))_{k\in\mathbb{N}}$ converges to $f(x)$ and these words have non-decreasing lengths, there exists $j \geq 0$ such that $|M(x,j)| \geq i$. Hence, for any data word $y \in \text{dom}(f)$ such that $|x \wedge y| \geq j$, the behaviour of M on y is the same during the first j steps, as M is deterministic, and thus $|f(x) \wedge f(y)| \geq i$, showing that f is continuous at x.

$\Rightarrow$ Assume that f is continuous. We describe a Turing machine computing f; the corresponding algorithm is formalised as Algorithm 1. When reading a finite prefix $x[{:}j]$ of its input $x \in \text{dom}(f)$, it computes the set P_j of all configurations (q,τ) reached by T on $x[{:}j]$. This set is updated along taking increasing values of j. It also keeps in memory the finite output word o_j that has been output so far. For any j, if $\mathsf{dt}(x[{:}j])$ denotes the data that appear in x, the algorithm then decides, for each input $(\sigma,d) \in \Sigma \times (\mathsf{dt}(x[{:}j]) \cup \{\mathsf{d}_0\})$ whether (σ,d) can safely be output, i.e., whether all accepting runs on words of the form $x[{:}j]y$, for an infinite word y, outputs at least $o_j(\sigma,d)$. The latter can be decided, given T, o_j and $x[{:}j]$, by Lemma 17. Note that it suffices to look at data in $\mathsf{dt}(x[{:}j]) \cup \{\mathsf{d}_0\}$ only since, by definition of NRT, any data that is output is necessarily stored in some register, and therefore appears in $x[{:}j]$ or is equal to d_0. Let us show that

Algorithm 1: Algorithm describing the machine M_f computing f.

Data: $x \in \text{dom}(f)$

1 $o := \epsilon$;
2 **for** $j = 0$ **to** ∞ **do**
3   **for** $(\sigma,d) \in \Sigma \times (dt(x[{:}j]) \cup \{d_0\})$ **do**
4   **if** $o.(\sigma,d) \preceq \hat{f}(x[{:}j])$ **then** `// such test is decidable by Lemma 17`
5             $o := o.(\sigma,d)$;
6   output (σ,d);
7   **end**
8   **end**
9 **end**

M_f actually computes f. Let $x \in \text{dom}(f)$. We have to show that the sequence $(M_f(x,j))_j$ converges to $f(x)$. Let o_j be the content of variable o of M_f when exiting the inner loop at line 8, when the outer loop (line 2) has been executed j times (hence j input symbols have been read). Note that $o_j = M_f(x,j)$. We have $o_1 \preceq o_2 \preceq \dots$ and $o_j \preceq \hat{f}(x[{:}j])$ for all $j \geq 0$. Hence, $o_j \preceq f(x)$ for all $j \geq 0$. To show that $(o_j)_j$ converges to $f(x)$, it remains to show that $(o_j)_j$ is non-stabilising, i.e. $o_{i_1} \prec o_{i_2} \prec \dots$ for some infinite subsequence $i_1 < i_2 < \dots$. First, note that f being continuous is equivalent to the sequence $(\hat{f}(x[{:}k]))_k$ converging to $f(x)$. Therefore we have that $f(x) \wedge \hat{f}(x[{:}k])$ can be arbitrarily long, for sufficiently large k. Let $j \geq 0$ and $(\sigma, d) = f(x)[|o_j|+1]$. By the latter property and the fact that $o_j.(\sigma, d) \preceq f(x)$, necessarily, there exists some $k > j$ such that $o_j.(\sigma, d) \preceq \hat{f}(x[:k])$. Moreover, by definition of NRT, d is necessarily a data that appears in some prefix of x, therefore there exists $k' \geq k$ such that d appears in $x[:k']$ and $o_j.(\sigma, d) \preceq \hat{f}(x[:k] \preceq \hat{f}(x[:k'])$. This entails that $o_j.(\sigma, d) \preceq o_{k'}$. So, we have shown that for all for all j, there exists $k' > j$ such that $o_j \prec o_{k'}$, which concludes the proof. □

Now that we have shown that computability is equivalent with continuity for functions defined by NRT, we exhibit a pattern which allows to decide continuity. Such pattern generalises the one of [3] to the setting of data words, the difficulty lying in showing that our pattern can be restricted to a finite number of data.

Theorem 19. *Let T be a functional* NRT *with k registers. Then, for all $X \subseteq \mathcal{D}$ such that $|X| \geq 2k+3$ and $\mathsf{d}_0 \in X$, T is not continuous at some $x \in (\Sigma \times \mathcal{D})^\omega$ if and only if T is not continuous at some $z \in (\Sigma \times X)^\omega$.*

Proof. The right-to-left direction is trivial. Now, let T be a functional NRT with k registers which is not continuous at some $x \in (\Sigma \times \mathcal{D})^\omega$. Let $f : \mathrm{dom}(\llbracket T \rrbracket) \to (\Gamma \times \mathcal{D})^\omega$ be the function defined by T, as: for all $u \in \mathrm{dom}(\llbracket T \rrbracket)$, $f(u) = v$ where $v \in (\Gamma \times \mathcal{D})^\omega$ is the unique data word such that $(u, v) \in \llbracket T \rrbracket$.

Now, let $X \subseteq \mathcal{D}$ be such that $|X| \geq 2k+3$ and $\mathsf{d}_0 \in X$. We need to build two words u and v labelled over X which coincide on a sufficiently long prefix to allow for pumping, hence yielding a converging sequence of input data words whose images do not converge, witnessing non-continuity. To that end, we use a similar proof technique as for Theorem 9: we show that the language of interleaved runs whose inputs coincide on a sufficiently long prefix while their respective outputs mismatch before a given position is recognisable by an NRA, allowing us to use the indistinguishability property. We also ask that one run presents sufficiently many occurrences of a final state q_f, so that we can ensure that there exists a pair of configurations containing q_f which repeats in both runs.

On reading such u and v, the automaton behaves as a finite automaton, since the number of data is finite ([15, Proposition 1]). By analysing the respective runs, we can, using pumping arguments, bound the position on which the mismatch appears in u, then show the existence of a synchronised loop over u and v after such position, allowing us to build the sought witness for non-continuity.

Relabel over X Thus, assume T is not continuous at some point $x \in (\Sigma \times \mathcal{D})^\omega$. Let ρ be an accepting run of T over x, and let $q_f \in \inf(\mathsf{st}(\rho)) \cap F$ be an accepting state repeating infinitely often in ρ. Then, let $i \geq 0$ be such that for all $j \geq 0$, there exists $y \in \mathrm{dom}(f)$ such that $\|x \wedge y\| \geq j$ but $\|f(x) \wedge f(y)\| \leq i$. Now, define $K = |Q| \times (2k+3)^{2k}$ and let $m = (2i+3) \times (K+1)$. Finally, pick j such that $\rho[1{:}j]$ contains at least m occurrences of q_f. Consider the language:

$$L = \big\{\rho_1 \otimes \rho_2 \big| \|\mathsf{in}(\rho_1) \wedge \mathsf{in}(\rho_2)\| \geq j, \|\mathsf{out}(\rho_1) \wedge \mathsf{out}(\rho_2)\| \leq i \text{ and}$$
$$\text{there are at least } m \text{ occurrences of } q_f \text{ in } \rho_1[1{:}j]\big\}$$

By Lemma 5, $L_{\otimes}(T)$ is recognised by an NRA with $2k$ registers. Additionnally, by Lemma 6, M^i_j is recognised by an NRA with 2 registers. Thus, $L = L_{\otimes}(T) \cap O^{q_f}_{m,j} \cap M^i_j$, where $O^{q_f}_{m,j}$ checks there are at least m occurrences of q_f in $\rho_1[1{:}j]$ (this is easily doable from the automaton recognising $L_{\otimes}(T)$ by adding an m-bounded counter), is recognisable by an NRA with $2k+2$ registers.

Choose $y \in \mathrm{dom}(f)$ such that $\|x \wedge y\| \geq j$ but $\|f(x) \wedge f(y)\| \leq i$. By letting ρ_1 (resp. ρ_2) be an accepting run of T over x (resp. y) we have $\rho_1 \otimes \rho_2 \in L$, so $L \neq \varnothing$. By Proposition 4, $L \cap ((\Sigma \times X)^\omega \times (\Gamma \times X)^\omega) \neq \varnothing$. Let $w = \rho'_1 \otimes \rho'_2 \in L \cap ((\Sigma \times X)^\omega \times (\Gamma \times X)^\omega)$, $u = \mathsf{in}(\rho'_1)$ and $v = \mathsf{in}(\rho'_2)$. Then, $\|u \wedge v\| \geq j$, $\|f(u) \wedge f(v)\| \leq i$ and there are at least m occurrences of q_f in $\rho_1[1{:}j]$.

Now, we depict ρ'_1 and ρ'_2 in Figure 4, where we decompose u as $u = u_1 \dots u_m \cdot s$ and v as $v = u_1 \dots u_m \cdot t$; their corresponding images being respectively $u' = u'_1 \dots u'_m \cdot s'$ and $u'' = u''_1 \dots u''_m t''$. We also let $l = (i+1)(K+1)$ and $l' = 2(i+1)(K+1)$. Since the data of u, v and w belong to X, we know that $\tau_i, \mu_i : R \to X$.

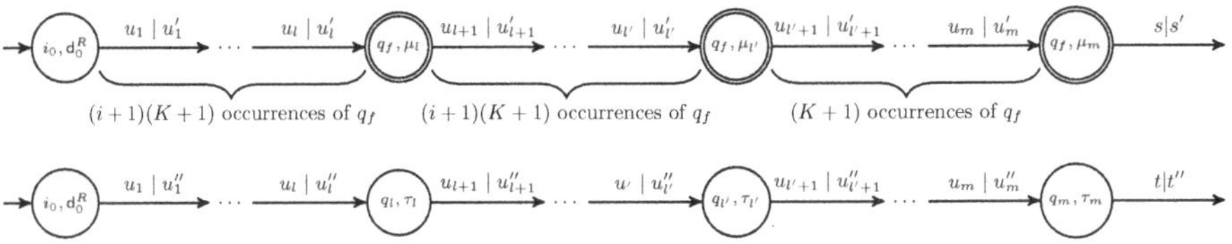

Fig. 4. Runs of f over $u = u_1 \dots u_m \cdot s$ and $v = u_1 \dots u_m \cdot t$.

Repeating configurations First, let us observe that in a partial run of ρ'_1 containing more than $|Q| \times |X|^k$ occurrences of q_f, there is at least one productive transition, i.e. a transition whose output is $o \neq \varepsilon$. Otherwise, by the pigeonhole principle, there exists a configuration $\mu : R \to X$ such that (q_f, μ) occurs at least twice in the partial run. Since all transitions are improductive, it would mean that, by writing w the corresponding part of input, we have $(q_f, \mu) \xrightarrow[T]{w|\varepsilon} (q_f, \mu)$. This partial run is part of ρ'_1, so, in particular, (q_f, μ) is accessible, hence by taking w_0 such that $(i_0, \tau_0) \xrightarrow[T]{w_0|w'_0} (q_f, \mu)$, we have that $f(w_0 w^\omega) = w'_0$, which is a finite word, contradicting our assumption that all accepting runs produce an infinite output. This implies that, for any $n \geq |Q| \times |X|^k$ (in particular for $n = l$), $\|u'_1 \dots u'_n\| \geq i+1$.

Locate the mismatch Again, upon reading $u_{l+1} \dots u_{l'}$, there are $(i+1)(K+1)$ occurrences of q_f. There are two cases:

(a) There are at least $i+1$ productive transitions in ρ'_2. Then, we obtain that $\|u''_1 \dots u''_{l'}\| > i$, so $\mathsf{mismatch}(u'_1 \dots u'_{l'}, u''_1 \dots u''_{l'})$, since we know $\|f(u) \wedge f(v)\| \leq i$ and they are respectively prefixes of $f(u)$ and $f(v)$, both of length at least $i+1$. Afterwards, upon reading $u_{l'+1} \dots u_m$, there are $K+1 > |Q| \times |X|^{2k}$ occurrences of q_f, so, by the pigeonhole principle, there is a repeating pair: there exist indices p and p' such that $l' \leq p < p' \leq m$ and $(q_f, \mu_p) = (q_f, \mu_{p'})$, $(q_p, \tau_p) = (q_{p'}, \tau_{p'})$. Thus, let $z_P = u_1 \dots u_p$, $z_R = u_{p+1} \dots u_{p'}$ and $z_C = u_{p'+1} \dots u_m \cdot t$ (P stands for *prefix*, R for *repeat* and C for *continuation*; we use capital letters to avoid confusion with indices). By denoting $z'_P = u'_1 \dots u'_p$, $z'_R = u'_{p+1} \dots u'_{p'}$, $z''_P = u''_1 \dots u''_p$, $z''_R = u''_{p+1} \dots u''_{p'}$ and $z''_C = u''_{p'+1} \dots u''_m \cdot t''$ the corresponding images, $z = z_P \cdot {z_R}^\omega$ is a point of discontinuity. Indeed, define $(z_n)_{n \in \mathbb{N}}$ as, for all $n \in \mathbb{N}$, $z_n = z_P \cdot z_R^n \cdot z_C$. Then, $(z_n)_{n \in \mathbb{N}}$ converges towards z, but, since for all $n \in \mathbb{N}$, $f(z_n) = z''_P \cdot {z''_L}^n \cdot z''_C$, we have that $f(z_n) \not\xrightarrow[n\infty]{} f(z) = z'_P \cdot {z'_L}^\omega$, since $\mathsf{mismatch}(z'_P, z''_P)$.

(b) Otherwise, by the same reasoning as above, it means there exists a repeating pair with only improductive transitions in between: there exist indices p and p' such that $l \leq p < p' \leq l'$, $(q_f, \mu_p) = (q_f, \mu_{p'})$, $(q_p, \tau_p) = (q_{p'}, \tau_{p'})$, and $(q_f, \mu_p) \xrightarrow{u_{p+1} \dots u_{p'} | \varepsilon} (q_f, \mu_{p'})$, $(q_p, \tau_p) \xrightarrow{u_{p+1} \dots u_{p'} | \varepsilon} (q_{p'}, \tau_{p'})$. Then, by taking $z_P = u_1 \dots u_p$, $z_R = u_{p+1} \dots u_{p'}$ and $z_C = u_{p'+1} \dots u_m \cdot t$, we have, by letting $z'_P = u'_1 \dots u'_p$, $z'_R = u'_{p+1} \dots u'_{p'}$, $z''_P = u''_1 \dots u''_p$, $z''_R = \varepsilon$ and $z''_C = u''_{n'+1} \dots u''_m \cdot t''$, that $z = z_P \cdot {z_R}^\omega$ is a point of discontinuity. Indeed, define $(z_n)_{n \in \mathbb{N}}$ as, for all $n \in \mathbb{N}$, $z_n = z_P \cdot z_R^n \cdot z_C$. Then, $(z_n)_{n \in \mathbb{N}}$ indeed converges towards z, but, since for all $n \in \mathbb{N}$, $f(z_n) = z''_P \cdot z''_C$, we have that $f(z_n) \not\xrightarrow[n\infty]{} f(z) = z'_P \cdot {z'_R}^\omega$, since $\mathsf{mismatch}(z'_P, z''_P \cdot z''_C)$ (the mismatch necessarily lies in z'_P, since $\|z'_P\| \geq i+1$). □

Corollary 20. *Deciding whether an* NRT *defines a continuous function is* PSPACE*-complete.*

Proof. Let $X \subseteq \mathcal{D}$ be a set of size $2k+3$ containing d_0. By Theorem 19, T is not continuous iff it is not continuous at some $z \in (\Sigma \times X)^\omega$, iff $[\![T]\!] \cap \big((\Sigma \times X)^\omega \times (\Gamma \times X)^\omega\big)$ is not continuous. By Proposition 3, such relation is recognisable by a finite transducer T_X with $O(|Q| \times |X|^{|R|})$ states, which can be built on-the-fly. By [3], the continuity of functions defined by NFT is decidable in NLOGSPACE, which yields a PSPACE procedure.

For the hardness, we reduce again from the emptiness problem of register automata, which is PSPACE-complete [4]. Let A be a register automaton over some alphabet $\Sigma \times \mathcal{D}$. We construct a transducer T which defines a continuous function iff $L(A) = \varnothing$ iff the domain of T is empty. Let f be a non-continous function realised by some NRT H (it exists by Example 16). Then, let $\# \notin \Sigma$ be a fresh symbol, and define the function g as the function mapping any data word of the form $w(\#, d)w'$ to $w(\#, d)f(w')$ if $w \in L(A)$. The function g is realised by an NRT which simulates A and copies its inputs on the output to implement the identity, until it sees #. If it was in some accepting state of A before seeing #, it branches to some initial state of H and proceeds executing H. If there is some $w_0 \in L(A)$, then the subfunction g_{w_0} mapping words of the form $w_0(\#, d)w'$ to $w_0(\#, d)f(w')$ is not continuous, since f is not. Hence g is not continuous. Conversely, if $L(A) = \varnothing$, then $\mathrm{dom}(g) = \varnothing$, so g is continuous. □

In [3], non-continuity is characterised by a specific pattern (Lemma 21, Figure 1), i.e. the existence of some particular sequence of transitions. By applying this characterisation to the finite transducer recognising $[\![T]\!] \cap ((\Sigma \times X)^\omega \times (\Gamma \times X)^\omega)$, as constructed in Proposition 3, we can characterise non-continuity by a similar pattern, which will prove useful to decide (non-)continuity of test-free NRT in NLogSpace (cf Section 5):

Corollary 21 ([3]). *Let T be an* NRT *with k registers. Then, for all $X \subseteq \mathcal{D}$ such that $|X| \geq 2k+3$ and $\mathsf{d}_0 \in X$, T is not continuous at some $x \in (\Sigma \times \mathcal{D})^\omega$ if and only if it has the pattern of Figure 5.*

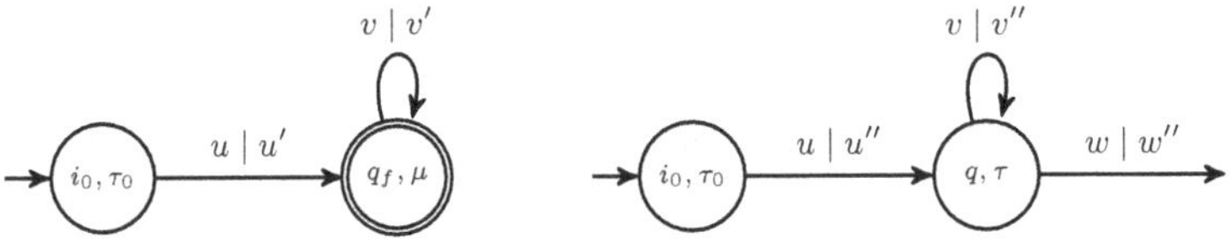

Fig. 5. A pattern characterising non-continuity of functions definable by an NRT: we ask that there exist configurations (q_f, μ) and (q, τ), where q_f is accepting, as well as finite input data words u, v, finite output data words u', v', u'', v'', and an infinite input data word w admitting an accepting run from configuration (q, τ) producing output w'', such that $\mathsf{mismatch}(u', u'') \vee (v'' = \varepsilon \wedge \mathsf{mismatch}(u', u''w''))$.

5 Test-free Register Transducers

In [7], we introduced a restriction which allows to recover decidability of the bounded synthesis problem for specifications expressed as non-deterministic register automata. Applied to transducers, such restriction also yields polynomial complexities when considering the functionality and computability problems.

An NRT T is *test-free* when its transition function does not depend on the tests conducted over the input data. Formally, we say that T is *test-free* if for all transitions $q \xrightarrow[T]{\sigma, \phi | \mathsf{asgn}, o} q'$ we have $\phi = \top$. Thus, we can omit the tests altogether and its transition relation can be represented as $\Delta' \subseteq Q \times \Sigma \times 2^R \times (\Gamma \times R)^* \times Q$.

Example 22. Consider the function $f : (\Sigma \times \mathcal{D})^\omega \to (\Gamma \times \mathcal{D})^\omega$ associating, to $x = (\sigma_1, d_1)(\sigma_2, d_2) \ldots$, the value $(\sigma_1, d_1)(\sigma_2, d_1)(\sigma_3, d_1) \ldots$ if there are infinitely many a in x, and $(\sigma_1, d_2)(\sigma_2, d_2)(\sigma_3, d_2) \ldots$ otherwise.

f can be implemented using a test-free NRT with one register: it initially guesses whether there are infinitely many a in x, if it is the case, it stores d_1 in the single register r, otherwise it waits for the next input to get d_2 and stores it in r. Then, it outputs the content of r along with each σ_i. f is not continuous, as even outputting the first data requires reading an infinite prefix when $d_1 \neq d_2$.

Note that when a transducer is test-free, the existence of an accepting run over a given input x only depends on its finite labels. Hence, the existence of two outputs y and z which mismatch over data can be characterised by a simple pattern (Figure 6), which allows to decide functionality in polynomial time:

Theorem 23. *Deciding whether a test-free* NRT *is functional is in* PTIME.

Proof. Let T be a test-free NRT such that T is not functional. Then, there exists $x \in (\Sigma \times \mathcal{D})^\omega$, $y, z \in (\Gamma \times \mathcal{D})^\omega$ such that $(x, y), (x, z) \in [\![T]\!]$ and $y \neq z$. Then, let i be such that $y[i] \neq z[i]$. There are two cases. Either $\mathsf{lab}(y[i]) \neq \mathsf{lab}(z[i])$, which means that the finite transducer T' obtained by ignoring the registers of T is not functional. By Proposition 8, such property can be decided in NLOGSPACE, so let us focus on the second case: $\mathsf{dt}(y[i]) \neq \mathsf{dt}(z[i])$.

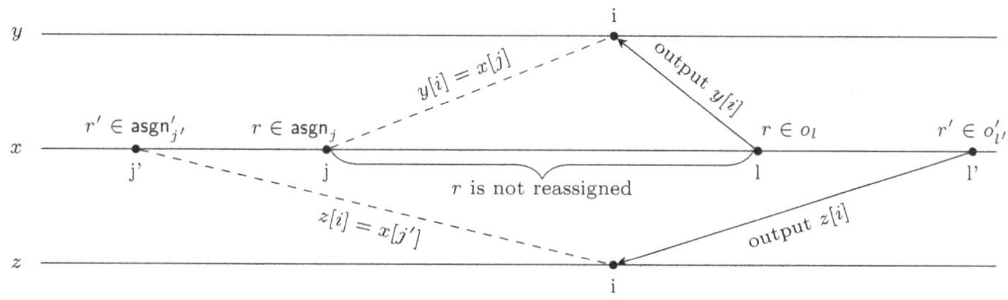

Fig. 6. A situation characterising the existence of a mismatch over data. Since acceptance does not depend on data, we can always choose x such that $\mathsf{dt}(x[j]) \neq \mathsf{dt}(x[j'])$. Here, we assume that the labels of x, y and z range over a unary alphabet; in particular $y[i] = x[j]$ iff $\mathsf{dt}(y[i]) = \mathsf{dt}(x[j])$. Finally, for readability, we did not write that r' should not be reassigned between j' and l'. Note that the position of i with regards to j, j', l and l' does not matter; nor does the position of l w.r.t. l'.

We here give a sketch of the proof: observe that an input x admits two outputs which mismatch over data if and only if it admits two runs which respectively store $x[j]$ and $x[j']$ such that $x[j] \neq x[j']$ and output them later at the same output position i; the outputs y and z are then such that $\mathsf{dt}(y[i]) \neq \mathsf{dt}(z[i])$. Since T is test-free, the existence of two runs over the same input x only depends on its finite labels. Then, the registers containing respectively $x[j]$ and $x[j']$ should not be reassigned before being output, and should indeed output their content at the same position i (cf Figure 6). Besides, again because of test-freeness, we can always assume that x is such that $x[j] \neq x[j']$. Overall, such pattern can be checked by a 2-counter Parikh automaton, whose emptiness is decidable in PTIME [8] (under conditions that are satisfied here). □

Now, let us move to the case of continuity. Here again, the fact that test-free NRT conduct no test over the input data allows to focus on the only two registers that are responsible for the mismatch, the existence of an accepting run being only determined by finite labels.

Theorem 24. *Deciding whether a test-free* NRT *defines a continuous function is in* PTIME.

Proof. Let T be a test-free NRT. First, it can be shown that T is continuous if and only if T has the pattern of Figure 7, where r is coaccessible (since acceptance only depends on finite labels, T can be trimmed[3] in polynomial time).

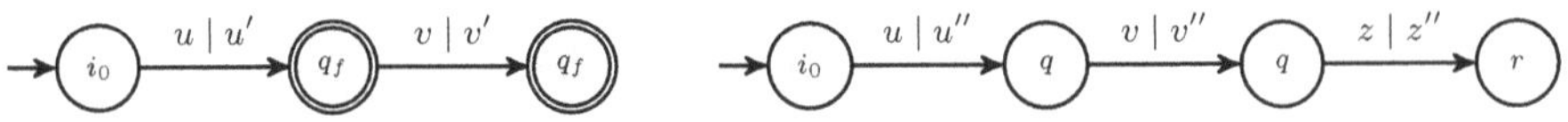

Fig. 7. A pattern characterising non-continuity of functions defined by NRT, where we ask that there exist some states q_f, q and r, where q_f is accepting, as well as finite input data words u, v, z and finite output data words u', v', u'', v'', z'' such that $\mathsf{mismatch}(u', u'') \vee (v'' = \varepsilon \wedge \mathsf{mismatch}(u', u''z''))$. Register assignments are not depicted, as there are no conditions on them. We unrolled the loops to highlight the fact that they do not necessarily loop back to the same configuration.

Now, it remains to show that such simpler pattern can be checked in PTIME. We treat each part of the disjunction separately:

(a) there exists u, u', u'', v, v', v'' s.t. $i_0 \xrightarrow{u|u'} q_f \xrightarrow{v|v'} q_f$ and $i_0 \xrightarrow{u|u''} q \xrightarrow{v|v''} q$, where $q_f \in F$ and $\mathsf{mismatch}(u', u'')$. Then, as shown in the proof of Theorem 23, there exists a mismatch between some u' and u'' produced by the same input u if and only if there exists two runs and two registers r and r' assigned at two distinct positions, and later on output at the same position. Such pattern can similarly be checked by a 2-counter Parikh automaton; the only difference is that here, instead of checking that the two end states are coaccessible with a common ω-word, we only need to check that $q_f \in F$ and that there is a synchronised loop over q_f and q, which are regular properties that can be checked by the Parikh automaton with only a polynomial increase.

(b) there exists $u, u', u'', v, v', z, z''$ s.t. $i_0 \xrightarrow{u|u'} q_f \xrightarrow{v|v'} q_f$ and $i_0 \xrightarrow{u|u''} q \xrightarrow{v|\varepsilon} q \xrightarrow{z|z''} r$, where $q_f \in F$ and $\mathsf{mismatch}(u', u''z'')$. By examining again the proof of Theorem 23, it can be shown that to obtain a mismatch, it suffices that the input is the same for both runs only up to position $\max(j, j')$. More precisely, there is a mismatch between u' and $u''z''$ if and only if there exists two registers r and r' and two positions $j, j' \in \{1, \dots, \|u\|\}$ such that $j \neq j'$, r is stored at position j, r' is stored at position j', r and r' are respectively output at input positions $l \in \{1, \dots, \|u\|\}$ and $l' \in \{1, \dots, \|uz\|\}$ and they are not reassigned in the meantime. Again, such property, along with the fact that $q_f \in F$ and the existence of a synchronised loop can be checked by a 2-counter Parikh automaton of polynomial size.

Overall, deciding whether a test-free NRT is continuous is in PTIME. □

[3] We say that T is trim when all its states are both accessible and coaccessible.

References

1. Berstel, J.: Transductions and Context-free Languages. Teubner Verlag (1979), http://www-igm.univ-mlv.fr/~berstel/LivreTransductions/LivreTransductions.html
2. Carayol, A., Löding, C.: Uniformization in Automata Theory. In: Proceedings of the 14th Congress of Logic, Methodology and Philosophy of Science, Nancy, July 19-26, 2011. pp. 153–178. London: College Publications (2014), https://hal.archives-ouvertes.fr/hal-01806575
3. Dave, V., Filiot, E., Krishna, S.N., Lhote, N.: Deciding the computability of regular functions over infinite words. CoRR **abs/1906.04199** (2019), http://arxiv.org/abs/1906.04199
4. Demri, S., Lazic, R.: LTL with the freeze quantifier and register automata. ACM Trans. Comput. Log. **10**(3), 16:1–16:30 (2009). https://doi.org/10.1145/1507244.1507246
5. Durand-Gasselin, A., Habermehl, P.: Regular transformations of data words through origin information. In: Foundations of Software Science and Computation Structures - 19th International Conference, FOSSACS 2016, Held as Part of the European Joint Conferences on Theory and Practice of Software, ETAPS 2016, Eindhoven, The Netherlands, April 2-8, Proceedings. pp. 285–300 (2016). https://doi.org/10.1007/978-3-662-49630-5_17
6. Ehlers, R., Seshia, S.A., Kress-Gazit, H.: Synthesis with identifiers. In: Proceedings of the 15th International Conference on Verification, Model Checking, and Abstract Interpretation - Volume 8318. pp. 415–433. VMCAI 2014 (2014). https://doi.org/10.1007/978-3-642-54013-4_23
7. Exibard, L., Filiot, E., Reynier, P.: Synthesis of data word transducers. In: 30th International Conference on Concurrency Theory, CONCUR 2019, August 27-30, Amsterdam, the Netherlands. pp. 24:1–24:15 (2019). https://doi.org/10.4230/LIPIcs.CONCUR.2019.24
8. Figueira, D., Libkin, L.: Path logics for querying graphs: Combining expressiveness and efficiency. In: 30th Annual ACM/IEEE Symposium on Logic in Computer Science, LICS 2015, Kyoto, Japan, July 6-10. pp. 329–340 (2015). https://doi.org/10.1109/LICS.2015.39
9. Filiot, E., Jecker, I., Löding, C., Winter, S.: On equivalence and uniformisation problems for finite transducers. In: 43rd International Colloquium on Automata, Languages, and Programming, ICALP 2016, July 11-15, Rome, Italy. pp. 125:1–125:14 (2016). https://doi.org/10.4230/LIPIcs.ICALP.2016.125
10. Filiot, E., Mazzocchi, N., Raskin, J.: A pattern logic for automata with outputs. In: Developments in Language Theory - 22nd International Conference, DLT 2018, Tokyo, Japan, September 10-14, Proceedings. pp. 304–317 (2018). https://doi.org/10.1007/978-3-319-98654-8_25
11. Gire, F.: Two decidability problems for infinite words. Inf. Process. Lett. **22**(3), 135–140 (1986). https://doi.org/10.1016/0020-0190(86)90058-X
12. Holtmann, M., Kaiser, L., Thomas, W.: Degrees of lookahead in regular infinite games. Logical Methods in Computer Science **8**(3) (2012). https://doi.org/10.2168/LMCS-8(3:24)2012
13. II, K.C., Pachl, J.K.: Equivalence problems for mappings on infinite strings. Information and Control **49**(1), 52–63 (1981). https://doi.org/10.1016/S0019-9958(81)90444-7
14. J.R. Büchi, L.H. Landweber: Solving sequential conditions finite-state strategies. Transactions of the American Mathematical Society **138**, 295–311 (1969). https://doi.org/10.2307/1994916

15. Kaminski, M., Francez, N.: Finite-memory automata. Theor. Comput. Sci. **134**(2), 329–363 (Nov 1994). https://doi.org/10.1016/0304-3975(94)90242-9
16. Khalimov, A., Kupferman, O.: Register-bounded synthesis. In: 30th International Conference on Concurrency Theory, CONCUR 2019, August 27-30, Amsterdam, the Netherlands. pp. 25:1–25:16 (2019). https://doi.org/10.4230/LIPIcs.CONCUR.2019.25
17. Khalimov, A., Maderbacher, B., Bloem, R.: Bounded synthesis of register transducers. In: Automated Technology for Verification and Analysis, 16th International Symposium, ATVA 2018, Los Angeles, October 7-10. Proceedings (2018). https://doi.org/10.1007/978-3-030-01090-4_29
18. Libkin, L., Tan, T., Vrgoc, D.: Regular expressions for data words. J. Comput. Syst. Sci. **81**(7), 1278–1297 (2015). https://doi.org/10.1016/j.jcss.2015.03.005
19. Neven, F., Schwentick, T., Vianu, V.: Finite state machines for strings over infinite alphabets. ACM Trans. Comput. Logic **5**(3), 403–435 (Jul 2004). https://doi.org/10.1145/1013560.1013562
20. Pnueli, A., Rosner, R.: On the synthesis of a reactive module. In: ACM Symposium on Principles of Programming Languages, POPL. ACM (1989). https://doi.org/10.1145/75277.75293
21. Prieur, C.: How to decide continuity of rational functions on infinite words. Theor. Comput. Sci. **276**(1-2), 445–447 (2002). https://doi.org/10.1016/S0304-3975(01)00307-3

Minimal Coverability Tree Construction Made Complete and Efficient *

Alain Finkel[1,3], Serge Haddad[1,2], and Igor Khmelnitsky[1,2](✉)

[1] LSV, ENS Paris-Saclay, CNRS, Université Paris-Saclay, Cachan, France
{finkel,haddad,khmelnitsky}@lsv.fr
[2] Inria, France
[3] Institut Universitaire de France, France

Abstract. Downward closures of Petri net reachability sets can be finitely represented by their set of maximal elements called the minimal coverability set or Clover. Many properties (coverability, boundedness, ...) can be decided using Clover, in a time proportional to the size of Clover. So it is crucial to design algorithms that compute it efficiently. We present a simple modification of the original but incomplete Minimal Coverability Tree algorithm (MCT), computing Clover, which makes it complete: it memorizes accelerations and fires them as ordinary transitions. Contrary to the other alternative algorithms for which no bound on the size of the required additional memory is known, we establish that the additional space of our algorithm is at most doubly exponential. Furthermore we have implemented a prototype `MinCov` which is already very competitive: on benchmarks it uses less space than all the other tools and its execution time is close to the one of the fastest tool.

Keywords: Petri nets · Karp-Miller tree algorithm · Coverability · Minimal coverability set · Clover · Minimal coverability tree.

1 Introduction

Coverability and coverability set in Petri nets. Petri nets are iconic as an infinite-state model used for verifying concurrent systems. Coverability, in Petri nets, is the most studied property for several reasons: (1) many properties like mutual exclusion, safety, control-state reachability reduce to coverability, (2) the coverability problem is EXPSPACE-complete (while reachability is non elementary), and (3) there exist efficient prototypes and numerous case studies. To solve the coverability problem, there are backward and forward algorithms. But these algorithms do not address relevant problems like the repeated coverability problem, the LTL model-checking, the boundedness problem and regularity of the traces.

However these problems are EXPSPACE-complete [4, 1] and are also decidable using the Karp-Miller tree algorithm (KMT) [11] that computes a finite tree

* The work was carried out in the framework of ReLaX, UMI2000 and also supported by ANR-17-CE40-0028 project BRAVAS.

J. Goubault-Larrecq and B. König (Eds.): FOSSACS 2020, LNCS 12077, pp. 237–256, 2020.
https://doi.org/10.1007/978-3-030-45231-5_13

labeled by a set of *ω-markings* $C \subseteq \mathbb{N}_\omega^P$ (where $\mathbb{N}_\omega$ is the set of naturals enlarged with an upper bound ω and P is the set of places) such that the reachability set and the finite set C have the same downward closure in $\mathbb{N}^P$. Thus a marking $\mathbf{m}$ is coverable if there exists some $\mathbf{m}' \geq \mathbf{m}$ with $\mathbf{m}' \in C$. Hence, C can be seen as *one* among all the possible finite representations of the infinite downward closure of the reachability set. This set C allows, for instance, to solve multiple instances of coverability in linear time linear w.r.t. the size of C avoiding to call many times a costly algorithm. Informally the KMT algorithm builds a reachability tree but, in order to ensure termination, substitutes ω to some finite components of a marking of a vertex when some marking of an ancestor is smaller.

Unfortunately C may contain comparable markings while only the maximal elements are important. The set of maximal elements of C can be defined independently of the KMT algorithm and was called the *minimal coverability set (MCS)* in [6] and abbreviated as the *Clover* in the more general framework of Well Structured Transition Systems (WSTS) [7].

The minimal coverability tree algorithm. So in [5,6] the author computes the minimal coverability set by modifying the KMT algorithm in such a way that at each step of the algorithm, the set of ω-markings labelling vertices is an antichain. But this aggressive strategy, implemented by the so-called Minimal Coverability Tree algorithm (MCT), contains a subtle bug and it may compute a strict under-approximation of Clover as shown in [8,10].

Alternative minimal coverability set algorithms. Since the discovery of this bug, three algorithms (with variants) [10,14,13] have been designed for computing the minimal coverability set without building the full Karp-Miller tree. In [10] the authors proposed a minimal coverability set algorithm (called `CovProc`) that is not based on the Karp-Miller tree algorithm but uses a similar but restricted introduction of ω's. In [14], Reynier and Servais proposed a modification of the MCT, called the Monotone-Pruning algorithm (called `MP`), that keeps but "deactivates" vertices labeled with smaller ω-markings while MCT would have deleted them. Recently in [15], the authors simplified their original proof of correctness. In [16], Valmari and Hansen proposed another algorithm (denoted below as `VH`) for constructing the minimal coverability set without deleting vertices. Their algorithm builds a graph and not a tree as usual. In [13], Piipponen and Valmari improved this algorithm by designing appropriate data structures and heuristics for exploration strategy that may significantly decrease the size of the graph.

Our contributions.

1. We introduce the concept of *abstraction* as an ω-transition that mimics the effect of an infinite family of firing sequences of markings w.r.t. coverability. As a consequence adding abstractions to the net does not modify its coverability set. Moreover, the classical Karp-Miller *acceleration* can be formalized as an abstraction whose incidence on places is either ω or null. The set of accelerations of a net is upward closed and well-ordered. Hence there exists a finite subset of minimal accelerations and we show that the size of all minimal acceleration is bounded by a double exponential.

2. Despite the current opinion that *"The flaw is intricate and we do not see an easy way to get rid of it.... Thus, from our point of view, fixing the bug of the MCT algorithm seems to be a difficult task"* [10], we have found a *simple* modification of MCT which makes it correct. It mainly consists in memorizing discovered accelerations and using them as ordinary transitions.
3. Contrary to *all* existing minimal coverability set algorithms that use an *unknown additional memory* that could be non primitive recursive, we show, by applying a recent result of Leroux [12], that the additional memory required for accelerations, is at most doubly exponential.
4. We have developed a prototype in order to also empirically evaluate the efficiency of our algorithm and the benchmarks (either from the literature or random ones) have confirmed that our algorithm requires significantly less memory than the other algorithms and is close to the fastest tool w.r.t. the execution time.

Organization. Section 2 introduces abstractions and accelerations and studies their properties. Section 3 presents our algorithm and establishes its correctness. Section 4 describes our tool and discusses the results of the benchmarks. We conclude and give some perspectives to this work in Section 5. One can find all the missing proofs and an illustration of the behavior of the algorithm in [9].

2 Covering abstractions

2.1 Petri nets: reachability and covering

Here we define Petri nets differently from the usual way but in an equivalent manner. i.e. based on the backward incidence matrix **Pre** and the incidence matrix **C**. The forward incidence matrix is implicitly defined by $\mathbf{C} + \mathbf{Pre}$. Such a choice is motivated by the introduction of abstractions in section 2.2.

Definition 1. *A Petri net (PN) is a tuple* $\mathcal{N} = \langle P, T, \mathbf{Pre}, \mathbf{C} \rangle$ *where:*

- P *is a finite set of* places;
- T *is a finite set of* transitions, *with* $P \cap T = \emptyset$;
- $\mathbf{Pre} \in \mathbb{N}^{P \times T}$ *is the* backward incidence matrix;
- $\mathbf{C} \in \mathbb{Z}^{P \times T}$ *is the* incidence matrix *which fulfills: for all* $p \in P$ *and* $t \in T$, $\mathbf{C}(p,t) + \mathbf{Pre}(p,t) \geq 0$.

A marked *Petri net* $(\mathcal{N}, \mathbf{m}_0)$ *is a Petri net* $\mathcal{N}$ *equipped with an initial marking* $\mathbf{m}_0 \in \mathbb{N}^P$.

The column vector of matrix **Pre** (resp. **C**) indexed by $t \in T$ is denoted $\mathbf{Pre}(t)$ (resp. $\mathbf{C}(t)$). A transition $t \in T$ is *fireable* from a marking $\mathbf{m} \in \mathbb{N}^P$ if $\mathbf{m} \geq \mathbf{Pre}(t)$. When t is fireable from $\mathbf{m}$, its *firing* leads to marking $\mathbf{m}' \stackrel{\text{def}}{=} \mathbf{m} + \mathbf{C}(t)$, denoted by $\mathbf{m} \xrightarrow{t} \mathbf{m}'$. One extends fireability and firing to a sequence $\sigma \in T^*$ by recurrence on its length. The empty sequence ε is always fireable and let the marking unchanged. Let $\sigma = t\sigma'$ be a sequence with $t \in T$ and $\sigma' \in T^*$. Then σ

is fireable from $\mathbf{m}$ if $\mathbf{m} \xrightarrow{t} \mathbf{m}'$ and σ' is fireable from $\mathbf{m}'$. The firing of σ from $\mathbf{m}$ leads to the marking $\mathbf{m}''$ reached by σ' from $\mathbf{m}'$. One also denotes this firing by $\mathbf{m} \xrightarrow{\sigma} \mathbf{m}''$.

Definition 2. *Let* $(\mathcal{N}, \mathbf{m}_0)$ *be a marked net. The* reachability set $Reach(\mathcal{N}, \mathbf{m}_0)$ *is defined by:*

$$Reach(\mathcal{N}, \mathbf{m}_0) = \{\mathbf{m} \mid \exists\sigma \in T^* \ \mathbf{m}_0 \xrightarrow{\sigma} \mathbf{m}\}$$

In order to introduce the coverability set of a Petri net, let us recall some definitions and results related to ordered sets. Let $(X, \leq)$ be an ordered set. The downward (resp. upward) *closure* of a subset $E \subseteq X$ is denoted by $\downarrow E$ (resp. $\uparrow E$) and defined by:

$$\downarrow E = \{x \in X \mid \exists y \in E \ y \geq x\} \qquad \text{(resp. } \uparrow E = \{x \in X \mid \exists y \in E \ y \leq x\})$$

A subset $E \subseteq X$ is downward (resp. upward) *closed* if $E = \downarrow E$ (resp. $E = \uparrow E$).

An *antichain* E is a set which fulfills: $\forall x \neq y \in E \ \neg(x \leq y \vee y \leq x)$. X is said *FAC* (for Finite AntiChains) if all its antichains are finite. A non empty set $E \subseteq X$ is *directed* if for all $x, y \in E$ there exists $z \in E$ such that $x \leq z$ and $y \leq z$. An *ideal* is a set which is downward closed and directed. There exists an equivalent characterization of FAC sets which provides a finite description of any downward closed set: a set is FAC if and only if every downward closed set admits a finite decomposition in ideals (a proof of this well-known result can be found in [3]).

X is *well founded* if all its (strictly) decreasing sequences are finite. X is *well ordered* if it is FAC and well founded. There are many equivalent characterizations of well order. For instance, a set X is well ordered if and only if for all sequence $(x_n)_{n\in\mathbb{N}}$ in X, there exists a non decreasing infinite subsequence. This characterization allows to design algorithms that computes trees whose finiteness is ensured by well order. Let us recall that $(\mathbb{N}, \leq)$ and $(\mathbb{N}^P, \leq)$ are well ordered sets.

We are now ready to introduce the *cover* (also called the coverability set) of a net and to state some of its properties.

Definition 3. *Let* $(\mathcal{N}, \mathbf{m}_0)$ *be a marked Petri net.* $Cover(\mathcal{N}, \mathbf{m}_0)$, *its* coverability set, *is defined by:*

$$Cover(\mathcal{N}, \mathbf{m}_0) = \downarrow Reach(\mathcal{N}, \mathbf{m}_0)$$

Since the coverability set is downward closed and $\mathbb{N}^P$ is FAC, it admits a finite decomposition in ideals. The ideals of $\mathbb{N}^P$ can be defined in an elegant way as follows. One first extends the sets of naturals and integers: $\mathbb{N}_\omega = \mathbb{N} \cup \{\omega\}$ et $\mathbb{Z}_\omega = \mathbb{Z} \cup \{\omega\}$. Then one extends the order relation and the addition to $\mathbb{Z}_\omega$: for all $n \in \mathbb{Z}$, $\omega > n$ and for all $n \in \mathbb{Z}_\omega$, $n + \omega = \omega + n = \omega$. $\mathbb{N}_\omega^P$ is also a well ordered set and its members are called ω*-markings*. There is a one-to-one mapping between ideals of $\mathbb{N}^P$ and ω-markings. Let $\mathbf{m} \in \mathbb{N}_\omega^P$. Define $[\![\mathbf{m}]\!]$ by:

$$[\![\mathbf{m}]\!] = \{\mathbf{m}' \in \mathbb{N}^P \mid \mathbf{m}' \leq \mathbf{m}\}$$

$[\![\mathbf{m}]\!]$ is an ideal of $\mathbb{N}^P$ (and all ideal can be defined in such a way). Let Ω be a set of ω-markings, $[\![\Omega]\!]$ denotes the set $\bigcup_{\mathbf{m}\in\Omega}[\![\mathbf{m}]\!]$. Due to the above properties, there exists a unique finite set with minimal size $Clover(\mathcal{N}, \mathbf{m}_0) \subseteq \mathbb{N}^p_\omega$ such that:

$$Cover(\mathcal{N}, \mathbf{m}_0) = [\![Clover(\mathcal{N}, \mathbf{m}_0)]\!]$$

A more general result can be found in [3] for well structured transition systems.

Example 1. The marked net of Figure 1 is unbounded. Its Clover is the following set:

$$\{p_i, p_{bk} + p_m, p_l + p_m + \omega p_{ba}, p_l + p_{bk} + \omega p_{ba} + \omega p_c\}$$

For instance, the marking $p_l + p_{bk} + \alpha p_{ba} + \beta p_c$ is reached thus covered by sequence $t_1 t_5^{\alpha+\beta} t_6^{\beta}$.

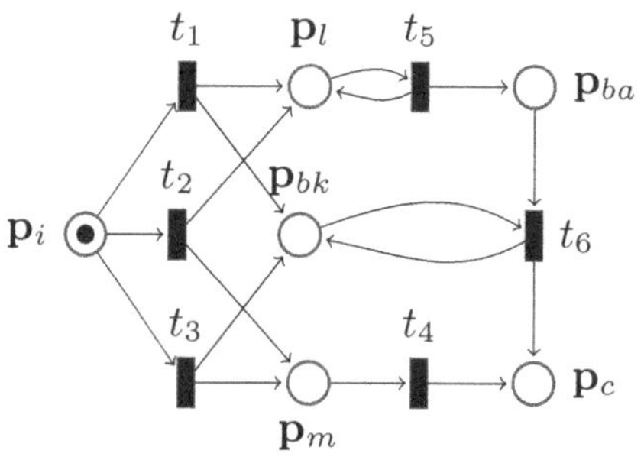

Fig. 1. An unbounded Petri net

2.2 Abstraction and acceleration

In order to introduce abstractions and accelerations, we generalize the transitions to allow the capability to mark a place with ω tokens.

Definition 4. *Let P be a set of places. An ω-transition* $\mathbf{a}$ *is defined by:*

- $\mathbf{Pre}(\mathbf{a}) \in \mathbb{N}^P_\omega$ *its* backward incidence*;*
- $\mathbf{C}(\mathbf{a}) \in \mathbb{Z}^P_\omega$ *its* incidence *with* $\mathbf{Pre}(\mathbf{a}) + \mathbf{C}(\mathbf{a}) \geq 0$.

For sake of homogeneity, one denotes $\mathbf{Pre}(\mathbf{a})(p)$ (resp. $\mathbf{C}(\mathbf{a})(p)$) by $\mathbf{Pre}(p, \mathbf{a})$ (resp. $\mathbf{C}(p, \mathbf{a})$). An ω-transition $\mathbf{a}$ is fireable from an ω-marking $\mathbf{m} \in \mathbb{N}^P_\omega$ if $\mathbf{m} \geq \mathbf{Pre}(\mathbf{a})$. When $\mathbf{a}$ is fireable from $\mathbf{m}$, its firing leads to the ω-marking $\mathbf{m}' \stackrel{\text{def}}{=} \mathbf{m} + \mathbf{C}(\mathbf{a})$, denoted as previously $\mathbf{m} \xrightarrow{\mathbf{a}} \mathbf{m}'$. One observes that if $\mathbf{Pre}(p, \mathbf{a}) = \omega$ then for all values of $\mathbf{C}(p, \mathbf{a})$, $\mathbf{m}'(\mathbf{a}) = \omega$. So without loss of generality, one assumes that for all ω-transition $\mathbf{a}$, $\mathbf{Pre}(p, \mathbf{a}) = \omega$ implies $\mathbf{C}(p, \mathbf{a}) = \omega$.

In order to define abstractions, we first define the incidences of a sequence σ of ω-transitions by recurrence on its length. As previously, we denote $\mathbf{Pre}(p, \sigma) \stackrel{\text{def}}{=}$

$\mathbf{Pre}(\sigma)(p)$ and $\mathbf{C}(p,\sigma) \stackrel{\text{def}}{=} \mathbf{C}(\sigma)(p)$. The base case corresponds to the definition of an ω-transition. Let $\sigma = t\sigma'$, with t an ω-transition and σ' a sequence of ω-transitions, then:

- $\mathbf{C}(\sigma) = \mathbf{C}(t) + \mathbf{C}(\sigma')$;
- for all $p \in P$
  - if $\mathbf{C}(p,t) = \omega$ then $\mathbf{Pre}(p,\sigma) = \mathbf{Pre}(p,t)$;
  - else $\mathbf{Pre}(p,\sigma) = \max(\mathbf{Pre}(p,t), \mathbf{Pre}(p,\sigma') - \mathbf{C}(p,t))$.

One checks by recurrence that σ is firable from $\mathbf{m}$ if and only if $\mathbf{m} \geq \mathbf{Pre}(\sigma)$ and in this case, $\mathbf{m} \xrightarrow{\sigma} \mathbf{m} + \mathbf{C}(\sigma)$.

An *abstraction* of a net is an ω-transition which concisely expresses the behaviour of the net w.r.t. covering (see Proposition 1). One will observe that a transition t of a net is by construction (with $\sigma_n = t$ for all n) an abstraction.

Definition 5. *Let $\mathcal{N} = \langle P, T, \mathbf{Pre}, \mathbf{C}\rangle$ be a Petri net and $\mathbf{a}$ be an ω-transition. $\mathbf{a}$ is an* abstraction *if for all $n \geq 0$, there exists $\sigma_n \in T^*$ such that for all $p \in P$ with $\mathbf{Pre}(p,\mathbf{a}) \in \mathbb{N}$:*

1. *$\mathbf{Pre}(p,\sigma_n) \leq \mathbf{Pre}(p,\mathbf{a})$;*
2. *If $\mathbf{C}(p,\mathbf{a}) \in \mathbb{Z}$ then $\mathbf{C}(p,\sigma_n) \geq \mathbf{C}(p,\mathbf{a})$;*
3. *If $\mathbf{C}(p,\mathbf{a}) = \omega$ then $\mathbf{C}(p,\sigma_n) \geq n$.*

The following proposition justifies the interest of abstractions.

Proposition 1. *Let $(\mathcal{N}, \mathbf{m}_0)$ be a marked Petri net, $\mathbf{a}$ be an abstraction and $\mathbf{m}$ be an ω-marking such that: $[\![\mathbf{m}]\!] \subseteq Cover(\mathcal{N}, \mathbf{m}_0)$ and $\mathbf{m} \xrightarrow{\mathbf{a}} \mathbf{m}'$. Then $[\![\mathbf{m}']\!] \subseteq Cover(\mathcal{N}, \mathbf{m}_0)$.*

Proof. Pick some $\mathbf{m}^* \in [\![\mathbf{m}']\!]$. Denote $n = \max(\mathbf{m}^*(p) \mid \mathbf{m}'(p) = \omega)$ and $\ell = \max(\mathbf{Pre}(p,\sigma_n), n - \mathbf{C}(p,\sigma_n) \mid \mathbf{m}(p) = \omega)$. Let us define $\mathbf{m}^\sharp \in [\![\mathbf{m}]\!]$ by:

- If $\mathbf{m}(p) < \omega$ then $\mathbf{m}^\sharp(p) = \mathbf{m}(p)$;
- Else $\mathbf{m}^\sharp(p) = \ell$.

Let us check that σ_n is fireable from $\mathbf{m}^\sharp$. Let $p \in P$,

- If $\mathbf{m}(p) < \omega$ then $\mathbf{m}^\sharp(p) = \mathbf{m}(p) \geq \mathbf{Pre}(p,\mathbf{a}) \geq \mathbf{Pre}(p,\sigma_n)$;
- Else $\mathbf{m}^\sharp(p) = \ell \geq \mathbf{Pre}(p,\sigma_n)$.

Let us show that $\mathbf{m}^\sharp + \mathbf{C}(\sigma_n) \geq \mathbf{m}^*$. Let $p \in P$,

- If $\mathbf{m}(p) < \omega$ and $\mathbf{C}(p,\mathbf{a}) < \omega$ then $\mathbf{m}^\sharp(p) + \mathbf{C}(p,\sigma_n) \geq \mathbf{m}(p) + \mathbf{C}(p,\mathbf{a}) = \mathbf{m}'(p) \geq \mathbf{m}^*(p)$;
- If $\mathbf{m}(p) < \omega$ and $\mathbf{C}(p,\mathbf{a}) = \omega$ then $\mathbf{m}^\sharp(p) + \mathbf{C}(p,\sigma_n) \geq \mathbf{C}(p,\sigma_n) \geq n \geq \mathbf{m}^*(p)$;
- If $\mathbf{m}(p) = \omega$ then $\mathbf{m}^\sharp(p) + \mathbf{C}(p,\sigma_n) \geq n - \mathbf{C}(p,\sigma_n) + \mathbf{C}(p,\sigma_n) = n \geq \mathbf{m}^*(p)$.

■

An easy way to build new abstractions consists in concatenating them.

Proposition 2. *Let $\mathcal{N} = \langle P, T, \mathbf{Pre}, \mathbf{C} \rangle$ be a Petri net and σ be a sequence of abstractions. Then the ω-transition $\mathbf{a}$ defined by $\mathbf{Pre}(\mathbf{a}) = \mathbf{Pre}(\sigma)$ and $\mathbf{C}(\mathbf{a}) = \mathbf{C}(\sigma)$ is an abstraction.*

We now introduce the underlying concept of the Karp and Miller construction.

Definition 6. *Let $\mathcal{N} = \langle P, T, \mathbf{Pre}, \mathbf{C} \rangle$ be a Petri net. One says that $\mathbf{a}$ is an* acceleration *if $\mathbf{a}$ is an abstraction such that $\mathbf{C}(\mathbf{a}) \in \{0, \omega\}^P$.*

The following proposition provides a way to get an acceleration from an arbitrary abstraction.

Proposition 3. *Let $\mathcal{N} = \langle P, T, \mathbf{Pre}, \mathbf{C} \rangle$ be a Petri net and $\mathbf{a}$ be an abstraction. Define $\mathbf{a}'$ an ω-transition as follows. For all $p \in P$:*

- *If $\mathbf{C}(p, \mathbf{a}) < 0$ then $\mathbf{Pre}(p, \mathbf{a}') = \mathbf{C}(p, \mathbf{a}') = \omega$;*
- *If $\mathbf{C}(p, \mathbf{a}) = 0$ then $\mathbf{Pre}(p, \mathbf{a}') = \mathbf{Pre}(p, \mathbf{a})$ and $\mathbf{C}(p, \mathbf{a}') = 0$;*
- *If $\mathbf{C}(p, \mathbf{a}) > 0$ then $\mathbf{Pre}(p, \mathbf{a}') = \mathbf{Pre}(p, \mathbf{a})$ and $\mathbf{C}(p, \mathbf{a}') = \omega$.*

Then $\mathbf{a}'$ is an acceleration.

Let us study more deeply the set of accelerations. First we equip the set of ω-transitions with a"natural" order w.r.t. covering.

Definition 7. *Let P be a set of places and two ω-transitions $\mathbf{a}$ and $\mathbf{a}'$.*

$$\mathbf{a} \leq \mathbf{a}' \text{ if and only if } \mathbf{Pre}(\mathbf{a}) \leq \mathbf{Pre}(\mathbf{a}') \wedge \mathbf{C}(\mathbf{a}) \geq \mathbf{C}(\mathbf{a}')$$

In other words, $\mathbf{a} \leq \mathbf{a}'$ if given any ω-marking $\mathbf{m}$, if $\mathbf{a}'$ is fireable from $\mathbf{m}$ then $\mathbf{a}$ is also fireable and its firing leads to a marking greater or equal that the one reached by the firing of $\mathbf{a}'$.

Proposition 4. *Let $\mathcal{N}$ be a Petri net. Then the set of abstractions of $\mathcal{N}$ is upward closed. Similarly, the set of accelerations is upward closed in the set of ω-transitions whose incidence belongs to $\{0, \omega\}^P$.*

Proposition 5. *The set of accelerations of a Petri net is well ordered.*

Proof. The set of accelerations is a subset of $\mathbb{N}^P \times \{0, \omega\}^P$ (where P is the set of places) with the order obtained by iterating cartesian products of sets $(\mathbb{N}, \leq)$ and $(\{0, \omega\}, \geq)$. These sets are well ordered and the cartesian product preserves this property. So we are done. ∎

Since the set of accelerations is well ordered and it is upward closed, it is equal to the upward closure of the finite set of *minimal* accelerations. Let us study the size of a minimal acceleration. Given some Petri net, one denotes $d = |P|$ and $e = \max_{p,t}(\max(\mathbf{Pre}(p, t), \mathbf{Pre}(p, t) + \mathbf{C}(p, t))$.

We are going to use the following result of Jérôme Leroux (published on HAL in June 2019) which provides a bound for the lengths of shortest sequences between two markings $\mathbf{m}_1$ and $\mathbf{m}_2$ mutually reachable.

Theorem 1. *(Theorem 2, [12]) Let $\mathcal{N}$ be a Petri net, $\mathbf{m}_1, \mathbf{m}_2$ be markings, σ_1, σ_2 be sequences of transitions such that $\mathbf{m}_1 \xrightarrow{\sigma_1} \mathbf{m}_2 \xrightarrow{\sigma_2} \mathbf{m}_1$. Then there exist σ'_1, σ'_2 such that $\mathbf{m}_1 \xrightarrow{\sigma'_1} \mathbf{m}_2 \xrightarrow{\sigma'_2} \mathbf{m}_1$ fulfilling:*

$$|\sigma'_1\sigma'_2| \leq ||\mathbf{m}_1 - \mathbf{m}_2||_\infty (3de)^{(d+1)^{2d+4}}$$

One deduces an upper bound on the size of minimal accelerations.
Let $\mathbf{v} \in \mathbb{N}_\omega^P$. One denotes $||\mathbf{v}||_\infty = \max(\mathbf{v}(p) \mid \mathbf{v}(p) \in \mathbb{N})$.

Proposition 6. *Let $\mathcal{N}$ be a Petri net and $\mathbf{a}$ be a minimal acceleration. Then $||\mathbf{Pre}(\mathbf{a})||_\infty \leq e(3de)^{(d+1)^{2d+4}}$.*

Proof. Let us consider the net $\mathcal{N}' = \langle P', T', \mathbf{Pre}', \mathbf{C}' \rangle$ obtained from $\mathcal{N}$ by deleting the set of places $\{p \mid \mathbf{Pre}(p, \mathbf{a}) = \omega\}$ and adding the set of transitions $T_1 = \{t_p \mid p \in P'\}$ with $\mathbf{Pre}(t_p) = p$ et $\mathbf{C}(t_p) = -p$. Observe that $d' \leq d$ and $e' = e$.
One denotes $P_1 = \{p \mid \mathbf{Pre}(p, \mathbf{a}) < \omega = \mathbf{C}(p, \mathbf{a})\}$. One introduces $\mathbf{m}_1$ the marking obtained by restricting $\mathbf{Pre}(\mathbf{a})$ to P' and $\mathbf{m}_2 = \mathbf{m}_1 + \sum_{p \in P_1} p$.
Let $\{\sigma_n\}_{n \in \mathbb{N}}$ be a family of sequences associated with $\mathbf{a}$. Let $n^* = ||\mathbf{Pre}(\mathbf{a})||_\infty + 1$. Then σ_{n^*} is fireable in $\mathcal{N}'$ from $\mathbf{m}_1$ and its firing leads to a marking that covers $\mathbf{m}_2$. By concatenating some occurrences of transitions of T_1, one gets a firing sequence in $\mathcal{N}'$ $\mathbf{m}_1 \xrightarrow{\sigma_1} \mathbf{m}_2$. Using the same process, one gets a firing sequence $\mathbf{m}_2 \xrightarrow{\sigma_2} \mathbf{m}_1$.
Let us apply Theorem 1. There exists a sequence σ'_1 with $\mathbf{m}_1 \xrightarrow{\sigma'_1} \mathbf{m}_2$ and $|\sigma'_1| \leq (3de)^{(d+1)^{2d+4}}$ since $||\mathbf{m}_1 - \mathbf{m}_2||_\infty = 1$. By deleting the transitions of T_1 occurring in σ'_1, one gets a sequence $\sigma''_1 \in T^*$ such that $\mathbf{m}_1 \xrightarrow{\sigma''_1} \mathbf{m}'_2 \geq \mathbf{m}_2$ with $|\sigma''_1| \leq (3de)^{(d+1)^{2d+4}}$.
The ω-transition $\mathbf{a}'$, defined by $\mathbf{Pre}(p, \mathbf{a}') = \mathbf{Pre}(p, \sigma''_1)$ for all $p \in P'$, $\mathbf{Pre}(p, \mathbf{a}') = \omega$ for all $p \in P \setminus P'$ and $\mathbf{C}(\mathbf{a}') = \mathbf{C}(\mathbf{a})$, is an acceleration whose associated family is $\{\sigma_1''^n\}_{n \in \mathbb{N}}$. By definition of $\mathbf{m}_1$, $\mathbf{a}' \leq \mathbf{a}$. Since $\mathbf{a}$ is minimal, $\mathbf{a}' = \mathbf{a}$. Observing that $|\sigma''_1| \leq (3de)^{(d+1)^{2d+4}}$, one gets $||\mathbf{Pre}(\mathbf{a})||_\infty = ||\mathbf{Pre}(\mathbf{a}')||_\infty \leq e(3de)^{(d+1)^{2d+4}}$. ■

Thus given any acceleration, one can easily obtain a smaller acceleration whose (representation) size is exponential.

Proposition 7. *Let $\mathcal{N}$ be a Petri net and $\mathbf{a}$ be an acceleration. Then the ω-transition trunc($\mathbf{a}$) defined by:*

- $\mathbf{C}(trunc(\mathbf{a})) = \mathbf{C}(\mathbf{a})$;
- *for all p such that $\mathbf{Pre}(p, \mathbf{a}) \neq \omega$,* $\mathbf{Pre}(p, trunc(\mathbf{a})) = \min(\mathbf{Pre}(p, \mathbf{a}), e(3de)^{(d+1)^{2d+4}})$;
- *for all p such that $\mathbf{Pre}(p, \mathbf{a}) = \omega$, $\mathbf{Pre}(p, trunc(\mathbf{a})) = \omega$.*

is an acceleration.

Proof. Let $\mathbf{a}' \leq \mathbf{a}$, be a minimal acceleration. For all p such that $\mathbf{Pre}(p, \mathbf{a}) \neq \omega$, $Pre(p, \mathbf{a}') \leq e(3de)^{(d+1)^{2d+4}}$. So $\mathbf{a}' \leq trunc(\mathbf{a})$. Since the set of accelerations is upward closed, one gets that $trunc(\mathbf{a})$ is an acceleration. ■

3 A coverability tree algorithm

3.1 Specification and illustration

As discussed in the introduction, to compute the clover of a Petri net, most algorithms build coverability trees (or graphs), which are variants of the Karp and Miller tree with the aim of reducing the peak memory during the execution. The seminal algorithm [6] is characterized by a main difference with the KMT construction: when finding that the marking associated with the current vertex strictly covers the marking of another vertex, it deletes the subtree issued from this vertex, and when the current vertex belonged to the removed subtree it substitutes it to the root of the deleted subtree. This operation drastically reduces the peak memory but as shown in [8] entails incompleteness of the algorithm.

Like the previous algorithms that ensure completeness with deletions, our algorithm also needs additional memory. However unlike the other algorithms, it memorizes accelerations instead of ω-markings. This approach has two advantages. First, we are able to exhibit a theoretical upper bound on the additional memory which is doubly exponential, while the other algorithms do not have such a bound. Furthermore, accelerations are reused in the construction and thus may even shorten the execution time and peak space w.r.t. the algorithm in [6].

Before we delve into a high level description of this algorithm, let us present some of the variables, functions, and definitions used by the algorithm. Algorithm 1, denoted from now on as MinCov takes as an input a marked net $(\mathcal{N}, \mathbf{m}_0)$ and constructs a directed labeled tree $CT = (V, E, \lambda, \delta)$, and a set Acc of ω-transitions (which by Lemma 2 are accelerations). Each $v \in V$ is labeled by an ω-marking, $\lambda(v) \in \mathbb{N}_\omega^P$. Since CT is a directed tree, every vertex $v \in V$, has a predecessor (except the root r) denoted by $prd(v)$ and a set of descendants denoted by $Des(v)$. By convention, $prd(r) = r$. Each edge $e \in E$ is labeled by a firing sequence $\delta(e) \in T_o \cdot \mathsf{Acc}^*$, consisting of an ordinary transition followed by a sequence of accelerations (which by Lemma 1 fulfills $\lambda(prd(v)) \xrightarrow{\delta(prd(v),v)} \lambda(v)$). In addition, again by Lemma 1, $\mathbf{m}_0 \xrightarrow{\delta(r,r)} \lambda(r)$. Let $\gamma = e_1 e_2 \dots e_k \in E^*$ be a path in the tree, we denote by $\delta(\gamma) := \delta(e_1)\delta(e_2)\dots\delta(e_k) \in (T \cup \mathsf{Acc})^*$. The subset $\mathsf{Front} \subset V$ is the set of vertices 'to be processed'.

MinCov may call function `Delete`(v) that removes from V a leaf v of CT and function `Prune`(v) that removes from V all descendants of $v \in V$ except v itself as illustrated in the following figure:

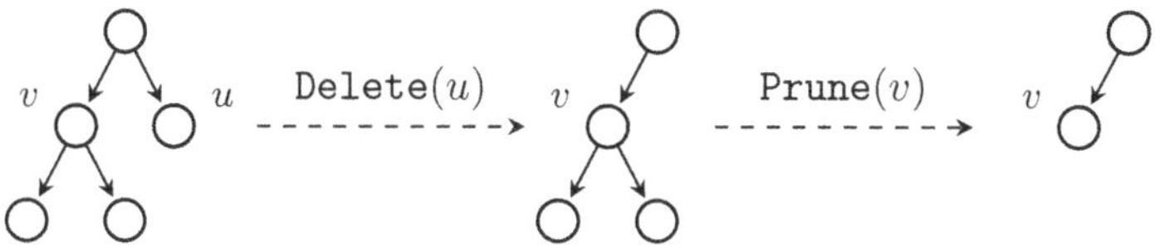

First MinCov does some initializations, and sets the tree CT to be a single vertex r with marking $\lambda(r) = \mathbf{m}_0$ and $\mathsf{Front} = \{r\}$. Afterwards the main loop builds the tree, where each iteration consists in processing some vertex in Front as follows.

MinCov picks a vertex $u \in$ Front (line 3). From $\lambda(u)$, MinCov fires a sequence $\sigma \in Acc^*$ reaching some $\mathbf{m}_u$ that maximizes the number of ω produced, i.e. $|\{p \in P \mid \lambda(u)(p) \neq \omega \wedge \mathbf{m}_u(p) = \omega\}|$. Thus in σ, no acceleration occurs twice and its length is bounded by $|P|$. Then MinCov updates $\lambda(u)$ with $\mathbf{m}_u$ (line 5) and the label of the edge incoming to u by concatenating σ. Afterwards it performs one of the following actions according to the marking $\lambda(u)$:

- **Cleaning** (line 7): If there exists $u' \in V \setminus$ Front with $\lambda(u') \geq \lambda(u)$. The vertex u is redundant and MinCov calls `Delete`(u)
- **Accelerating** (lines 8-16): If there exists u', an ancestor of u with $\lambda(u') < \lambda(u)$ then an acceleration can be computed. The acceleration $\mathbf{a}$ is deduced from the firing sequence labeling the path from u' to u. MinCov inserts $\mathbf{a}$ into Acc, calls `Prune`(u') and pushes back u' in Front.
- **Exploring** (lines 18 - 25): Otherwise MinCov calls `Prune`(u') followed by `Delete`(u') for all $u' \in V$ with $\lambda(u') < \lambda(u)$ since they are redundant. Afterwards, it removes u from Front and for all fireable transition $t \in T$ from $\lambda(u)$, it creates a new child for u in CT and inserts it into Front.

For a detailed example of a run of the algorithm see Example 2 in [9].

3.2 Correctness Proof

We now establish the correctness of Algorithm 1 by proving the following properties (where for all $W \subseteq V$, $\lambda(W)$ denotes $\bigcup_{v \in W} \lambda(v)$):

- its termination;
- the incomparability of ω-markings associated with vertices in V: $\lambda(V)$ is an antichain;
- its consistency: $[\![\lambda(V)]\!] \subseteq Cover(\mathcal{N}, \mathbf{m}_0)$;
- its completeness: $Cover(\mathcal{N}, \mathbf{m}_0) \subseteq [\![\lambda(V)]\!]$.

We get termination by using the well order of $\mathbb{N}_\omega^P$ and Koenig Lemma.

Proposition 8. *MinCov terminates.*

Proof. Consider the following variation of the algorithm.

Instead of deleting the current vertex when its marking is smaller or equal than the marking of a vertex, one marks it as 'cut' and extract it from Front.

Instead of cutting a subtree when the marking of the current vertex v is greater than the marking of a vertex which is not an ancestor of v, one marks them as 'cut' and extract from Front those who are inside.

Instead of cutting a subtree when the marking of the current vertex v is greater than the marking of a vertex which is an ancestor of v, say v^*, one marks those on the path from v^* to v (except v) as 'accelerated', one marks the other vertices

Algorithm 1: Computing the minimal coverability set

MinCov($\mathcal{N}, \mathbf{m}_0$)

Input: A marked Petri net $(\mathcal{N}, \mathbf{m}_0)$

Data: V set of vertices; $E \subseteq V \times V$; $\mathsf{Front} \subseteq V$; $\lambda : V \to \mathbb{N}^p_\omega$; $\delta : E \to T_o\mathsf{Acc}^*$; $CT = (V, E, \lambda, \delta)$ a labeled tree; Acc a set of ω-transitions;

Output: A labeled tree $CT = (V, E, \lambda, \delta)$

$V \leftarrow \{r\}$; $E \leftarrow \emptyset$; $\mathsf{Front} \leftarrow \{r\}$; $\lambda(r) \leftarrow \mathbf{m}_0$; $\mathsf{Acc} \leftarrow \emptyset$; $\delta(r, r) \leftarrow \varepsilon$
while $Front \neq \emptyset$ **do**
    Select $u \in \mathsf{Front}$
    Let $\sigma \in \mathsf{Acc}^*$ a maximal fireable sequence of accelerations from $\lambda(u)$
    // Maximal w.r.t. the number of ω's produced
    $\lambda(u) \leftarrow \lambda(u) + \mathbf{C}(\sigma)$
    $\delta((prd(u), u)) \leftarrow \delta((prd(u), u)) \cdot \sigma$
    if $\exists u' \in V \setminus \mathsf{Front}$ $s.t.$ $\lambda(u') \geq \lambda(u)$ **then** Delete(u) // $\lambda(u)$ is covered
    else if $\exists u' \in \mathsf{Anc}(V)$ $s.t.$ $\lambda(u) > \lambda(u')$ **then**
        // An acceleration was found between u and one of u's ancestors
        Let $\gamma \in E^*$ the path from u' to u in CT
        $\mathbf{a} \leftarrow$ NewAcceleration()
        foreach $p \in P$ **do**
            if $\mathbf{C}(p, \delta(\gamma)) < 0$ **then** $\mathbf{Pre}(p, \mathbf{a}) \leftarrow \omega$; $\mathbf{C}(p, \mathbf{a}) \leftarrow \omega$
            if $\mathbf{C}(p, \delta(\gamma)) = 0$ **then** $\mathbf{Pre}(p, \mathbf{a}) \leftarrow \mathbf{Pre}(p, \delta(\gamma))$; $\mathbf{C}(p, \mathbf{a}) \leftarrow 0$
            if $\mathbf{C}(p, \delta(\gamma)) > 0$ **then** $\mathbf{Pre}(p, \mathbf{a}) \leftarrow \mathbf{Pre}(p, \delta(\gamma))$; $\mathbf{C}(p, \mathbf{a}) \leftarrow \omega$
        end
        $\mathbf{a} \leftarrow trunc(\mathbf{a})$; $\mathsf{Acc} \leftarrow \mathsf{Acc} \cup \{\mathbf{a}\}$; Prune($u'$); $\mathsf{Front} = \mathsf{Front} \cup \{u'\}$;
    else
        for $u' \in V$ **do**
            // Remove vertices labeled by markings covered by $\lambda(u)$
            if $\lambda(u') < \lambda(u)$ **then** Prune(u'); Delete(u')
        end
        $\mathsf{Front} \leftarrow \mathsf{Front} \setminus \{u\}$
        foreach $t \in T \wedge \lambda(u) \geq \mathbf{Pre}(t)$ **do**
            // Add the children of u
            $u' \leftarrow$ NewNode(); $V \leftarrow V \cup \{u'\}$; $\mathsf{Front} \leftarrow \mathsf{Front} \cup \{u'\}$); $E \leftarrow E \cup \{(u, u')\}$
            $\lambda(u') \leftarrow \lambda(u) + \mathbf{C}(t)$; $\delta((u, u')) \leftarrow t$
        end
    end
end
return CT

of the subtree as 'cut' and inserts v again in Front with the marking of v^*. All the markings of the subtree in Front are extracted from it.

All the vertices marked as 'cut' or 'accelerated' are ignored for comparisons and discovering accelerations. This alternative algorithm behaves as the original one except that the size of the tree never decreases and so if the algorithm does not terminate the tree is infinite. Since this tree is finitely branching, due to Koenig Lemma it contains an infinite path. On this infinite path, no vertex can be marked as 'cut' since it would belong to a finite subtree. Observe that the marking labelling the vertex following an accelerated subpath has at least one more ω than the marking of the first vertex of this subpath. So there is an infinite subpath with unmarked vertices in V. But $\mathbb{N}_\omega^P$ is well-ordered, so there should be two vertices v and v', where v' is a descendant of v with $\lambda(v') \geq \lambda(v)$, which contradicts the behaviour of the algorithm. ∎

Since we are going to use recurrence on the number of iterations of the main loop of Algorithm 1, we introduce the following notations: $CT_n = (V_n, E_n, \lambda_n, \delta_n)$, Front_n, and Acc_n are the the values of variables CT, Front, and Acc at line 2 when n iterations have been executed.

Proposition 9. *For all $n \in \mathbb{N}$, $\lambda(V_n \setminus \mathsf{Front}_n)$ is an antichain. Thus on termination, $\lambda(V)$ is an antichain.*

Proof. Let us introduce $V' := V \setminus \mathsf{Front}$ and $V'_n := V_n \setminus \mathsf{Front}_n$. We are going to prove by induction on the number n of iterations of the while-loop that V'_n is an antichain. MinCov initializes variables V and Front at line 1. So $V_0 = \{r\}$ and $\mathsf{Front}_0 = \{r\}$, therefore $V'_0 = V_0 \setminus \mathsf{Front}_0 = \emptyset$ is an antichain.
Assume that $V'_n = V_n \setminus \mathsf{Front}_n$ is an antichain. Modifying V'_n can be done by *adding* or *removing* vertices from V_n and *removing* vertices from Front_n while keeping them in V_n. The actions that MinCov may perform in order to modify the sets V and Front are: `Delete` (lines 7 and 19), `Prune` (lines 16 and 19), adding vertices to V (line 23), adding vertices to Front (lines 16 and 23), and removing vertices from Front (line 21).

- Both `Delete` and `Prune` do not add new vertices to V'. Thus the antichain feature is preserved.
- MinCov may add vertices to V only at line 23 where it simultaneously adds them to Front and therefore does not add new vertices to V'. Thus the antichain feature is preserved.
- Adding vertices to Front may only remove vertices from V'_n. Thus the antichain feature is preserved.
- MinCov can only add a vertex to V' when it removes it from Front while keeping it in V. This is done only at line 21. There the only vertex MinCov may remove (line 21) is the working vertex u. However if (in the iteration) MinCov reaches line 21 then it did not reach line 7 hence, (1) all markings of $\lambda(V'_n) \subseteq \lambda(V_n)$ are either smaller or incomparable to $\lambda_{n+1}(u)$. Moreover, MinCov has also reached line 18-20, where (2) it performs `Delete` on all vertices $u' \in V'_n \subseteq V_n$ with $\lambda_n(u') < \lambda_{n+1}(u)$. Let us denote by $V''_n \subseteq V'_n$ the set V' at the end of line

20. Due to (1) and (2), marking $\lambda_{n+1}(u)$ is incomparable to any marking in $\lambda_{n+1}(V''_n)$. Since $V''_n \subseteq V'_n$, $\lambda_{n+1}(V''_n)$ is an antichain. Combining this fact with the incomparability between $\lambda_{n+1}(u)$ and any marking in $\lambda_{n+1}(V''_n)$, we conclude that the set $\lambda_{n+1}(V'_{n+1}) = \lambda_{n+1}(V''_n) \cup \{\lambda_{n+1}(u)\}$ is an antichain.

■

In order to establish consistency, we prove that the labelling of vertices and edges is compatible with the firing rule and that *Acc* is a set of accelerations.

Lemma 1. *For all $n \in \mathbb{N}$, for all $u \in V_n \setminus \{r\}$, $\lambda_n(prd(u)) \xrightarrow{\delta(prd(u),u)} \lambda_n(u)$ and $\mathbf{m}_0 \xrightarrow{\delta(r,r)} \lambda_n(r)$.*

Proof. Let us prove by induction on the number n of iterations of the main loop that for all $v \in V_n$, the assertions of the lemma hold. Initially, $V_0 = \{r\}$ and $\lambda_0(r) = \mathbf{m}_0$. Since $\mathbf{m}_0 \xrightarrow{\varepsilon} \mathbf{m}_0 = \lambda_0(r)$ the base case is established.
Assume that the assertions hold for CT_n. Observe that MinCov may change the labeling function λ and/or add new vertices in exactly two places: at lines 4-6 and at lines 22-25. Therefore in order to prove the assertion, we show that after each group of lines it still holds.

- After lines 4-6: MinCov computes (1) a maximal fireable sequence $\sigma \in \mathsf{Acc}_n^*$ from $\lambda_n(u)$ (line 4), and updates u's marking to $\mathbf{m}_u = \lambda_n(u) + \mathbf{C}(\sigma)$ (line 5). Since the assertions hold for CT_n, (2) if $u \neq r$, $\lambda_n(prd(u)) \xrightarrow{\delta(prd(u),u)} \lambda_n(u)$ else $\mathbf{m}_0 \xrightarrow{\delta(r,r)} \lambda_n(r)$. By concatenation, we get $\lambda_n(prd(u)) \xrightarrow{\delta(prd(u),u)\sigma} \mathbf{m}_u$ if $u \neq r$ and otherwise $\mathbf{m}_0 \xrightarrow{\delta(r,r)\sigma} \mathbf{m}_u$ which establishes that the assertions hold after line 6.
- After lines 22-25: The vertices for which λ is updated at these lines are the children of u that are added to the tree. For every fireable transition $t \in T$ from $\lambda(u)$, MinCov creates a child v_t for u (lines 22-23). The marking of any child v_t is set to $\mathbf{m}_{n+1}(v) := \mathbf{m}_{n+1}(u) + \mathbf{C}(t)$ (line 24). Therefore since $\lambda_{n+1}(u) \xrightarrow{t} \lambda_{n+1}(v_t)$, the assertions hold.

■

Lemma 2. *At any execution point of MinCov,* Acc *is a set of accelerations.*

Proof. At most one acceleration is added per iteration. Let us prove by induction on the number n of iterations of the main loop that Acc_n is a set of accelerations. Since $\mathsf{Acc}_0 = \emptyset$, the base case is straightforward.
Assume that Acc_n is a set of accelerations and consider Acc_{n+1}. In an iteration, MinCov may add an ω-transition $\mathbf{a}$ to Acc. Due to the inductive hypothesis, $\delta(\gamma)$ is a sequence of abstractions where γ is defined at line 9. Consider b, the ω-transition defined by $\mathbf{Pre}(b) = \mathbf{Pre}(\delta(\gamma))$ and $\mathbf{C}(b) = \mathbf{C}(\delta(\gamma))$. Due to Proposition 2, b is an abstraction. Due to Proposition 3, the loop of lines 11-15 transforms b into an acceleration $\mathbf{a}$. Due to Proposition 7, after truncation at line 16, $\mathbf{a}$ is still an acceleration. ■

Proposition 10. $[\![\lambda(V)]\!] \subseteq Cover(\mathcal{N}, \mathbf{m}_0)$.

Proof. Let $v \in V$. Consider the path $u_0, \ldots, u_k$ of CT from the root $r = u_0$ to $u_k = v$. Let $\sigma \in (T \cup \mathsf{Acc})^*$ denote $\delta(prd(u_0), u_0) \cdots \delta(prd(u_k), u_k)$. Due to Lemma 1, $m_0 \xrightarrow{\sigma} \lambda(v)$. Due to Lemma 2, σ is a sequence of abstractions. Due to Proposition 2, the ω-transition $\mathbf{a}$ defined by $\mathbf{Pre}(\mathbf{a}) = \mathbf{Pre}(\sigma)$ and $\mathbf{C}(\mathbf{a}) = \mathbf{C}(\sigma)$ is an abstraction. Due to Proposition 1, $[\![\lambda(v)]\!] \subseteq Cover(\mathcal{N}, \mathbf{m}_0)$. ■

The following definitions are related to an arbitrary execution point of $\mathtt{M}inCov$ and are introduced to establish its completeness.

Definition 8. *Let $\sigma = \sigma_0 t_1 \sigma_1 \ldots t_k \sigma_k$ with for all i, $t_i \in T$ and $\sigma_i \in \mathsf{Acc}^*$. Then the firing sequence $\mathbf{m} \xrightarrow{\sigma} \mathbf{m}'$ is an* exploring sequence *if:*

- *There exists $v \in \mathsf{Front}$ with $\lambda(v) = \mathbf{m}$*
- *For all $0 \leq i \leq k$, there does not exist $v' \in V \setminus \mathsf{Front}$ with $\mathbf{m} + \mathbf{C}(\sigma_0 t_1 \sigma_1 \ldots t_i \sigma_i) \leq \lambda(v')$.*

Definition 9. *Let $\widehat{\mathbf{m}}$ be a marking. Then $\widehat{\mathbf{m}}$ is* quasi-covered *if:*

- *either there exists $v \in V \setminus \mathsf{Front}$ with $\lambda(v) \geq \widehat{\mathbf{m}}$;*
- *or there exists an exploring sequence $\mathbf{m} \xrightarrow{\sigma} \mathbf{m}' \geq \widehat{\mathbf{m}}$.*

In order to prove completeness of the algorithm, we want to prove that at the beginning of every iteration, any $\mathbf{m} \in Cover(\mathcal{N}, \mathbf{m}_0)$ is quasi-covered. To establish this assertion, we introduce several lemmas showing that this assertion is preserved by some actions of the algorithm with some prerequisites. More precisely, Lemma 3 corresponds to the deletion of the current vertex, Lemma 4 to the discovery of an acceleration, Lemma 5 to the deletion of a subtree whose marking of the root is smaller than the marking of the current vertex and Lemma 6 to the creation of the children of the current vertex.

Lemma 3. *Let CT, Front and Acc be the values of corresponding variables at some execution point of $\mathtt{MinCov}$ and $u \in V$ be a leaf in CT such that the following items hold:*

1. *All $\mathbf{m} \in Cover(\mathcal{N}, \mathbf{m}_0)$ are quasi-covered;*
2. *$\lambda(V \setminus \mathsf{Front})$ is an antichain;*
3. *For all $\mathbf{a} \in \mathsf{Acc}$ fireable from $\lambda(u)$, $\lambda(u) = \lambda(u) + \mathbf{C}(\mathbf{a})$;*
4. *There exists $v \in V \setminus \{u\}$ such that $\lambda(v) \geq \lambda(u)$.*

Then all $\mathbf{m} \in Cover(\mathcal{N}, \mathbf{m}_0)$ are quasi-covered after performing $\mathtt{Delete}(u)$.

Lemma 4. *Let CT, Front and Acc be the values of corresponding variables at some execution point of $\mathtt{MinCov}$. and $u \in V$ such that the following items hold:*

1. *All $\mathbf{m} \in Cover(\mathcal{N}, \mathbf{m}_0)$ are quasi-covered;*
2. *$\lambda(V \setminus \mathsf{Front})$ is an antichain;*
3. *For all $v \in V \setminus \{r\}$, $\lambda(prd(v)) \xrightarrow{\delta(prd(v),v)} \lambda(v)$.*

Then all $\mathbf{m} \in Cover(\mathcal{N}, \mathbf{m}_0)$ are quasi-covered after performing $\mathtt{Prune}(u)$ and then adding u to Front.

Lemma 5. *Let CT, Front and Acc be the values of corresponding variables at some execution point of MinCov, $u \in \mathsf{Front}$ and $u' \in V$ such that the following items hold:*

1. *All $\mathbf{m} \in Cover(\mathcal{N}, \mathbf{m}_0)$ are quasi-covered;*
2. *$\lambda(V \setminus \mathsf{Front})$ is an antichain;*
3. *For all $v \in V \setminus \{r\}$, $\lambda(prd(v)) \xrightarrow{\delta(prd(v),v)} \lambda(v)$;*
4. *$\lambda(u') < \lambda(u)$ and u is not a descendant of u'.*

Then after performing $\mathtt{Prune}(u'); \mathtt{Delete}(u')$,

1. *All $\mathbf{m} \in Cover(\mathcal{N}, \mathbf{m}_0)$ are quasi-covered;*
2. *$\lambda(V \setminus \mathsf{Front})$ is an antichain;*
3. *For all $v \in V \setminus \{r\}$, $\lambda(prd(v)) \xrightarrow{\delta(prd(v),v)} \lambda(v)$.*

Lemma 6. *Let CT, Front and Acc be the values of corresponding variables at some execution point of MinCov. and $u \in \mathsf{Front}$ such that the following items hold:*

1. *All $\mathbf{m} \in Cover(\mathcal{N}, \mathbf{m}_0)$ are quasi-covered;*
2. *$\lambda(V \setminus \mathsf{Front}) \cup \{\lambda(u)\}$ is an antichain;*
3. *For all $\mathbf{a} \in \mathsf{Acc}$ fireable from $\lambda(u)$, $\lambda(u) = \lambda(u) + \mathbf{C}(\mathbf{a})$.*

Then after removing u from Front and for all $t \in T$ fireable from $\lambda(u)$, adding a child v_t to u in Front with marking of v_t defined by $\lambda_u(v_t) = \lambda(u) + \mathbf{C}(t)$, all $\mathbf{m} \in Cover(\mathcal{N}, \mathbf{m}_0)$ are quasi-covered.

Proposition 11. *At the beginning of every iteration, all $\mathbf{m} \in Cover(\mathcal{N}, \mathbf{m}_0)$ are quasi-covered.*

Proof. Let us prove by induction on the number of iterations that all $\mathbf{m} \in Cover(\mathcal{N}, \mathbf{m}_0)$ are quasi-covered.
Let us consider the base case. MinCov initializes V and Front to $\{r\}$ and $\lambda(r)$ to $\mathbf{m}_0$. By definition, for all $\mathbf{m} \in Cov(\mathcal{N}, \mathbf{m}_0)$ there exists $\sigma = t_1 t_2 \cdots t_k \in T^*$ such that $\mathbf{m}_0 \xrightarrow{\sigma} \mathbf{m}' \geq \mathbf{m}$. Since $V \setminus \mathsf{Front} = \emptyset$, this firing sequence is an exploring sequence.
Assume that all $\mathbf{m} \in Cover(\mathcal{N}, \mathbf{m}_0)$ are quasi-covered at the beginning of some iteration. Let us examine what may happen during the iteration. In lines 4-6, MinCov computes the maximal fireable sequence $\sigma \in \mathsf{Acc}_n^*$ from $\lambda_n(u)$ (line 4) and sets u's marking to $\mathbf{m}_u := \lambda_n(u) + \mathbf{C}(\sigma)$ (line 5). Afterwards, there are three possible cases: (1) either $\mathbf{m}_u$ is covered by some marking associated with a vertex out of Front, (2) either an acceleration is found, (3) or MinCov computes the successors of u and removes u from Front.

Line 7. MinCov calls $\mathtt{Delete}(u)$. So CT_{n+1} is obtained by deleting u. Moreover, $\lambda(u') \geq \mathbf{m}_u$. Let us check the hypotheses of Lemma 3. Assertion 1 follows from induction since (1) the only change in the data is the increasing of $\lambda(u)$ by firing some accelerations and (2) u belongs to Front so cannot cover intermediate markings of exploring sequences. Assertion 2 follows from Proposition 9 since $V \setminus \mathsf{Front}$ is unchanged. Assertion 3 follows immediately from lines 4-6. Assertion 4 follows with $v = u'$. Thus using this lemma the induction is proved in this case.

Lines 8-16. Let us check the hypotheses of Lemma 4. Assertions 1 and 2 are established as in the previous case. Assertion 3 holds due to Lemma 1, and the fact that no edge has been added since the beginning of iteration. Thus using this lemma the induction is proved in this case.

Lines 18-25. We first show that the hypotheses of Lemma 6 hold before line 21. Let us denote the values of CT and Front after line 20 by $\widehat{CT}_n$ and $\widehat{\mathsf{Front}}_n$. Observe that for all iteration of Line 19 in the inner loop, the hypotheses of Lemma 5 are satisfied. Therefore, in order to apply Lemma 6 it remains only to check assertions 2 and 3 of this lemma. Assertion 2 holds since (1) $\lambda(V \setminus \mathsf{Front})$ is an antichain, (2) due to Line 7 there is no $w \in V \setminus \mathsf{Front}$ such that $\lambda(w) \geq \lambda(u)$, and (3) by iteration of Line 19 all $w \in V \setminus \mathsf{Front}$ such that $\lambda(w) < \lambda(u)$ have been deleted. Assertion 3 holds due to Line 5 (all useful enabled accelerations have been fired) and Line 8 (no acceleration has been added).

Lines 21-25 correspond to the operations related to Lemma 6. Thus using this lemma, the induction is proved in this case.

■

The completeness of `MinCov` is an immediate consequence of the previous proposition.

Corollary 1. *When MinCov terminates,* $Cover(\mathcal{N}, \mathbf{m}_0) \subseteq [\![\lambda(V)]\!]$.

Proof. By Proposition 11 all $\mathbf{m} \in Cover(\mathcal{N}, \mathbf{m}_0)$ are quasi-covered. Since on termination, Front is empty for all $\mathbf{m} \in Cover(\mathcal{N}, \mathbf{m}_0)$, there exists $v \in V$ such that $\mathbf{m} \leq \lambda(v)$. ■

4 Tool and benchmarks

In order to empirically evaluate our algorithm, we have implemented a prototype tool which computes the clover and solves the coverability problem. This tool is developed in the programming language Python, using the Numpy library. It can be found on GitHub[3]. All benchmarks were performed on a computer equipped by Intel i5-8250U CPU with 4 cores, 16GB of memory and Ubuntu Linux 18.03.

Minimal coverability set. We compare `MinCov` with the tool `MP` [14], the tool `VH` [16], and the tool `CovProc` [10]. We have also implemented the (incomplete) minimal coverability tree algorithm denoted by `AF` in order to measure the additional memory needed for the (complete) tools. Both `MP` and `VH` tools were sent to us by the courtesy of the authors. The tool `MP` has an implementation

[3] https://github.com/IgorKhm/MinCov

in Python and another in C++. For comparison we selected the Python one to avoid biases due to programming language.

We ran two kinds of benchmarks: (1) 123 standard benchmarks from the literature in Table 1, (which were taken from [2]), (2) 100 randomly generated Petri nets also in Table 1, since the benchmarks from the literature do not present all the features that lead to infinite state systems. These random Petri nets have the following properties: (1) $50 < |P|, |T| < 100$, (2) the number of places connected of each transition is bounded by 10, and (3) they are not structurally bounded. The execution time of the tools was limited to 900 seconds.

Table 1 contains a summary of all the instances of the benchmarks. The first column shows the number of instances on which the tool timed out. The time column consists of the total time on instances that did not time out plus 900 seconds for any instance that led to a time out. The #Nodes column consists of the peak number of nodes in instances that did not time out on any of the tools (except `CovProc` which does not provide this number). For `MinCov` we take the peak number of nodes plus accelerations. In the benchmarks from the literature

Table 1. Benchmarks for clover

123 benchmarks from the literature				**100 random benchmarks**			
	T/O	Time	#Nodes		T/O	Time	#Nodes
`MinCov`	16	18127	**48218**	`MinCov`	**14**	13989	**61164**
`VH`	**15**	**14873**	75225	`VH`	15	**13692**	208134
`MP`	24	23904	478681	`MP`	21	21726	755129
`CovProc`	49	47081	N/A	`CovProc`	80	74767	N/A
`AF`	19	19223	45660	`AF`	16	15888	63275

we observed that the instances that timed out from `MinCov` are included in those of `AF` and `MP`. However there were instances the timed out on `VH` but did not time out on `MinCov` and vice versa. `MinCov` is the second fastest tool, and compared to `VH` it is 1.2 times slower. A possible explanation would be that `VH` is implemented in C++. As could be expected, w.r.t. memory requirements `MinCov` has the least number of nodes. In the benchmarks from the literature `MinCov` has approximately 10 times less nodes then `MP` and 1.6 times less then `VH`. In the random benchmarks these ratio are significantly higher.

Coverability. We compare `MinCov` to the tool `qCover` [2] on the set of benchmarks from the literature in Table 2. In [2], `qCover` is compared to the most competitive tools for coverability and achieves a score of 142 solved instances while the second best tool achieves a score of 122. We split the results into safe instances (not coverable) and unsafe ones (coverable). In both categories we counted the number of instances on which the tools failed (columns T/O) and the total time (columns Time) as in Table 1.

We observed that the tools are complementary, i.e. `qCover` is faster at proving that an instance is safe and `MinCov` is faster at proving that an instance is unsafe.

Table 2. Benchmarks for the coverability problem (60 unsafe and 115 safe)

	Time Unsafe	T/O Unsafe	Time safe	T/O safe	T/O	Time
`MinCov`	**1754**	1	51323	53	54	53077
`qCover`	26467	26	**11865**	**11**	37	38332
`MinCov` ∥ `qCover`	1841	2	13493	11	**13**	**15334**

Therefore, by splitting the processing time between them we get better results. The third row of Table 2 represents a parallel execution of the tools, where the time for each instance is computed as follows:

$$\text{Time}(\texttt{MinCov} \parallel \texttt{qCover}) = 2\min\left(\text{Time}(\texttt{MinCov}), \text{Time}(\texttt{qCover})\right).$$

Combining both tools is 2.5 times faster than `qCover` and 3.5 times faster than `MinCov`. This confirms the above statement. We could still get better results by dynamically deciding which ratio of CPU to share between the tools depending on some predicted status of the instance.

5 Conclusion

We have proposed a simple and efficient modification of the incomplete minimal coverability tree algorithm for building the clover of a net. Our algorithm is based on the introduction of the concepts of covering abstractions and accelerations. Compared to the alternative algorithms previously designed, we have theoretically bounded the size of the additional space. Furthermore we have implemented a prototype which is already very competitive.

From a theoretical point of view, we plan to study how abstractions and accelerations, could be defined in the more general context of well structured transition systems. From an experimental point of view, we will follow three directions in order to increase the performance of our tool. First as in [13], we have to select appropriate data structures to minimize the number of comparisons between ω-markings. Then we want to precompute a set of accelerations using linear programming as the correctness of the algorithm is preserved and the efficiency could be significantly improved. Last we want to take advantage of parallelism in a more general way than simultaneously running several tools.

References

1. Blockelet, M., Schmitz, S.: Model checking coverability graphs of vector addition systems. In: Proceedings of MFCS 2011. LNCS, vol. 6907, pp. 108–119 (2011)
2. Blondin, M., Finkel, A., Haase, C., Haddad, S.: Approaching the coverability problem continuously. In: Proceedings of TACAS 2016. LNCS, vol. 9636, pp. 480–496. Springer (2016)
3. Blondin, M., Finkel, A., McKenzie, P.: Well behaved transition systems. Logical Methods in Computer Science **13**(3), 1–19 (2017)
4. Demri, S.: On selective unboundedness of VASS. J. Comput. Syst. Sci. **79**(5), 689–713 (2013)
5. Finkel, A.: Reduction and covering of infinite reachability trees. Information and Computation **89**(2), 144–179 (1990)
6. Finkel, A.: The minimal coverability graph for Petri nets. In: Advances in Petri Nets. LNCS, vol. 674, pp. 210–243 (1993)
7. Finkel, A., Goubault-Larrecq, J.: Forward analysis for WSTS, part II: Complete WSTS. Logical Methods in Computer Science **8**(4), 1–35 (2012)
8. Finkel, A., Geeraerts, G., Raskin, J.F., Van Begin, L.: A counter-example to the minimal coverability tree algorithm. Tech. rep., Université Libre de Bruxelles, Belgium (2005), http://www.lsv.fr/Publis/PAPERS/PDF/FGRV-ulb05.pdf
9. Finkel, A., Haddad, S., Khmelnitsky, I.: Minimal coverability tree construction made complete and efficient (2020), https://hal.inria.fr/hal-02479879
10. Geeraerts, G., Raskin, J.F., Van Begin, L.: On the efficient computation of the minimal coverability set of Petri nets. International Journal of Fundamental Computer Science **21**(2), 135–165 (2010)
11. Karp, R.M., Miller, R.E.: Parallel program schemata. J. Comput. Syst. Sci. **3**(2), 147–195 (1969)
12. Leroux, J.: Distance between mutually reachable Petri net configurations (Jun 2019), https://hal.archives-ouvertes.fr/hal-02156549, preprint
13. Piipponen, A., Valmari, A.: Constructing minimal coverability sets. Fundamenta Informaticae **143**(3–4), 393–414 (2016)
14. Reynier, P.A., Servais, F.: Minimal coverability set for Petri nets: Karp and Miller algorithm with pruning. Fundamenta Informaticae **122**(1–2), 1–30 (2013)
15. Reynier, P.A., Servais, F.: On the computation of the minimal coverability set of Petri nets. In: Proceedings of Reachability Problems 2019. LNCS, vol. 11674, pp. 164–177 (2019)
16. Valmari, A., Hansen, H.: Old and new algorithms for minimal coverability sets. Fundamenta Informaticae **131**(1), 1–25 (2014)

Constructing Infinitary Quotient-Inductive Types

Marcelo P. Fiore, Andrew M. Pitts, and S. C. Steenkamp(✉)

Department of Computer Science and Technology
University of Cambridge, Cambridge CB3 0FD, UK
s.c.steenkamp@cl.cam.ac.uk

Abstract This paper introduces an expressive class of quotient-inductive types, called QW-types. We show that in dependent type theory with uniqueness of identity proofs, even the infinitary case of QW-types can be encoded using the combination of inductive-inductive definitions involving strictly positive occurrences of Hofmann-style quotient types, and Abel's size types. The latter, which provide a convenient constructive abstraction of what classically would be accomplished with transfinite ordinals, are used to prove termination of the recursive definitions of the elimination and computation properties of our encoding of QW-types. The development is formalized using the Agda theorem prover.

Keywords: dependent type theory · higher inductive types · inductive-inductive definitions · quotient types · sized types · category theory

1 Introduction

One of the key features of proof assistants based on dependent type theory such as Agda, Coq and Lean is their support for inductive definitions of families of types. Homotopy Type Theory [29] introduces a potentially very useful extension of the notion of inductive definition, the *higher inductive types* (HITs). To define an ordinary inductive type one declares how its elements are constructed. To define a HIT one not only declares element constructors, but also declares equality constructors in identity types (possibly iterated ones), specifying how the constructed elements and identities are to be equated. In this paper we work in a dependent type theory satisfying uniqueness of identity proofs (UIP), so that identity types are trivial in dimensions higher than one. Nevertheless, as Altenkirch and Kaposi [5] point out, HITs are still useful in such a one-dimensional setting. They introduce the term *quotient inductive type* (QIT) for this truncated form of HIT.

Figure 1 gives two examples of QITs, using Agda-style notation for dependent type theory; in particular, Set denotes a universe of types and $\equiv$ denotes the identity type. The first example specifies the element and equality constructors for the type $\mathsf{Bag}\,X$ of finite multisets of elements from a type X. The second example, adapted from [5], specifies the element and equality constructors for the type $\omega\mathsf{Tree}\,X$ of trees whose nodes are labelled with elements of X and that have unordered countably infinite branching. Both examples illustrate the nice feature

J. Goubault-Larrecq and B. König (Eds.): FOSSACS 2020, LNCS 12077, pp. 257–276, 2020.
https://doi.org/10.1007/978-3-030-45231-5_14

Finite multisets:

$$
\begin{aligned}
&\mathsf{data}\ \mathsf{Bag}(X : \mathsf{Set}) : \mathsf{Set}\ \mathsf{where}\\
&\quad [] : \mathsf{Bag}\, X\\
&\quad \_{::}\_ : X \to \mathsf{Bag}\, X \to \mathsf{Bag}\, X\\
&\quad \mathsf{swap} : (x\ y : X)(ys : \mathsf{Bag}\, X) \to x :: y :: ys \equiv y :: x :: ys
\end{aligned}
$$

Unordered countably branching trees (elements of $\mathsf{isIso}\, f$ witness that f is a bijection):

$$
\begin{aligned}
&\mathsf{data}\ \omega\mathsf{Tree}(X : \mathsf{Set}) : \mathsf{Set}\ \mathsf{where}\\
&\quad \mathsf{leaf} : \omega\mathsf{Tree}\, X\\
&\quad \mathsf{node} : X \to (\mathbb{N} \to \omega\mathsf{Tree}\, X) \to \omega\mathsf{Tree}\, X\\
&\quad \mathsf{perm} : (x : X)(f : \mathbb{N} \to \mathbb{N})(\_ : \mathsf{isIso}\, f)(g : \mathbb{N} \to \omega\mathsf{Tree}\, X) \to\\
&\qquad\qquad\qquad \mathsf{node}\, x\, g \equiv \mathsf{node}\, x\, (g \circ f)
\end{aligned}
$$

Figure 1. Two examples of QITs

of QITs that users only have to specify the particular identifications between data needed for their applications. Thus the standard property of equality that it is an equivalence relation respecting the constructors is inherited by construction from the usual properties of identity types, without the need to say so in the declaration of the QIT.

The second example also illustrates a more technical aspect of QITs, that they enable constructive versions of structures that classically use non-constructive choice principles. The first example in Figure 1 only involves element constructors of finite arity ($[]$ is nullary and $x :: \_$ is unary) and consequently $\mathsf{Bag}\, X$ is isomorphic to the type obtained from the ordinary inductive type of finite lists over X by quotienting by the congruence generated by swap. Of course this assumes, as we do in this paper, that the type theory comes with Hofmann-style *quotient types* [18, Section 3.2.6.1]. By contrast, the second example in the figure involves an element constructor with countably infinite arity. So if one first forms the ordinary inductive type of *ordered* countably branching trees (by dropping the equality constructor perm from the declaration) and then quotients by a suitable relation to get the equalities specified by perm, one needs the axiom of countable choice to be able to lift the node element constructor to the quotient; see [5, Section 2.2] for a detailed discussion. The construction of the Cauchy reals as a higher inductive-inductive type [29, Section 11.3] provides a similar, but more complicated example where use of countable choice is avoided. Such examples have led to the folklore that as far as constructive type theories go, infinitary QITs are more expressive than the combination of ordinary inductive (or inductive-recursive, or inductive-inductive) types with quotient types. In this paper we use Abel's *sized types* [2] to show that, for a wide class of QITs, this view is not justified. Thus we make two main contributions:

First we define a family of QITs called *QW-types* and give elimination and computation rules for them (Section 2). The usual W-types of Martin-Löf [22] are inductive types giving the algebraic terms over a possibly infinitary signature.

One specifies a QW-type by giving a family of equations between such terms. So such QITs give initial algebras for possibly infinitary algebraic theories. As we indicate in Section 3, they can encode a very wide range of examples of possibly infinitary quotient-inductive types, namely those that do not involve constructors taking previously constructed equalities as arguments (so do not cover the infinitary extension of the very general scheme considered by Dybjer and Moeneclaey [12]). In set theory with the Axiom of Choice (AC), QW-types can be constructed simply as Quotients of the underlying W-type—hence the name.

Secondly, we prove that contrary to expectation, without AC it is still possible to construct QW-types using quotients, but not simply by quotienting a W-type. Instead, the type to be quotiented and the relation by which to quotient are given simultaneously by definitions that refer to each other. Thus our construction (in Section 4) involves *inductive-inductive* definitions [15]. The elimination and computation functions which witness that the quotiented type correctly represents the required QW-type are defined recursively. In order to prove that our recursive definitions terminate we combine the use of inductive definitions involving strictly positive occurrences of quotient types with sized types (currently, we do not know whether it is possible to avoid sizing in favour of, say, a suitable well-founded termination ordering). Sized types provide a convenient constructive abstraction of what classically would be accomplished with sequences of transfinite ordinal length.

The type theory in which we work

To present our results we need a version of Martin-Löf Type Theory with (1) uniqueness of identity proofs, (2) quotient types and hence also function extensionality, (3) inductive-inductive datatypes (with strictly positive occurrences of quotient types) and (4) sized types. Lean 3 provides (1) and (2) out of the box, but also the Axiom of Choice, unfortunately. Neither it, nor Coq provide (3) and (4). Agda provides (1) via unrestricted dependent pattern-matching, (2) via a combination of postulates and the rewriting mechanism of Cockx and Abel [8], (3) via its very liberal mechanism for mutual definitions and (4) thanks to the work of Abel [2]. Therefore we make use of the type theory implemented by Agda (version 2.6.0.1) to give formal proofs of our results. The Agda code can be found at DOI: 10.17863/CAM.48187. In this paper we describe the results informally, using Agda-style notation for dependent type theory. In particular we use Set to denote the universe at the lowest level of a countable hierarchy of (Russell-style) universes. We also use Agda's convention that an implicit argument of an operation can be made explicit by enclosing it in {braces}.

Acknowledgement We would like to acknowledge the contribution Ian Orton made to the initial development of the work described here. He and the first author supervised the third author's Master's dissertation *Quotient Inductive Types: A Schema, Encoding and Interpretation*, in which the notion of QW-type (there called a W^+-type) was introduced.

2 QW-types

We begin by recalling some facts about types of well-founded trees, the W-types of Martin-Löf [22]. We take *signatures* to be elements of the dependent product

$$\mathsf{Sig} = \textstyle\sum A : \mathsf{Set}, (A \to \mathsf{Set}) \tag{1}$$

So a signature is given by a pair $\Sigma = (A, B)$ consisting of a type $A : \mathsf{Set}$ and a family of types $B : A \to \mathsf{Set}$. Each such signature determines a polynomial endofunctor [1, 16] $\mathsf{S}\{\Sigma\} : \mathsf{Set} \to \mathsf{Set}$ whose value at $X : \mathsf{Set}$ is the following dependent product

$$\mathsf{S}\{\Sigma\}X = \textstyle\sum a : A, (B\,a \to X) \tag{2}$$

An S-*algebra* is by definition an element of the dependent product

$$\mathsf{Alg}\{\Sigma\} = \textstyle\sum X : \mathsf{Set}, (\mathsf{S}\,X \to X) \tag{3}$$

S-algebra morphisms $(X, s) \to (X', s')$ are given by functions $h : X \to X'$ together with an element of the type

$$\mathsf{isHom}\,h = (a : A)(b : B\,a \to X) \to s'(a, h \circ b) \equiv h(s(a, b)) \tag{4}$$

Then the W-type $\mathsf{W}\{\Sigma\}$ determined by Σ is the underlying type of an initial S-algebra. More generally, Dybjer [11] shows that the initial algebra of any non-nested, strictly positive endofunctor on Set is given by a W-type; and Abbott, Altenkirch, and Ghani [1] extend this to the case with nested uses of W-types as part of their work on containers. (These proofs take place in extensional type theory [22], but work just as well in the intensional type theory with uniqueness of identity proofs and function extensionality that we are using here.)

More concretely, given a signature $\Sigma = (A, B)$, if one thinks of elements $a : A$ as names of operation symbols whose (not necessarily finite) arity is given by the type $B\,a : \mathsf{Set}$, then the elements of $\mathsf{W}\{\Sigma\}$ represent the closed algebraic terms (i.e. well-founded trees) over the signature. From this point of view it is natural to consider not only closed terms solely built up from operations, but also open terms additionally built up with variables drawn from some type X. As well as allowing operators of possibly infinite arity, we also allow terms involving possibly infinitely many variables (the second example in Figure 1 involves such terms). Categorically, the type $\mathsf{T}\{\Sigma\}X$ of such open terms is the free S-algebra on X and is another W-type, for the signature obtained from Σ by adding the elements of X as nullary operations. Nevertheless, it is convenient to give a direct inductive definition:

$$\begin{aligned}
&\mathsf{data} : \mathsf{T}\{\Sigma : \mathsf{Sig}\}(X : \mathsf{Set}) : \mathsf{Set}\ \mathsf{where}\\
&\quad \eta : X \to \mathsf{T}\,X\\
&\quad \sigma : \mathsf{S}(\mathsf{T}\,X) \to \mathsf{T}\,X
\end{aligned} \tag{5}$$

Given an S-algebra $(Y, s) : \mathsf{Alg}\{\Sigma\}$ and a function $f : X \to Y$, the unique morphism of S-algebras from the free S-algebra $(\mathsf{T}\,X, \sigma)$ on X to (Y, s) has underlying function $\mathsf{T}\,X \to Y$ mapping each $t : \mathsf{T}\,X$ to the element $t \gg\!= f$ in Y defined[1] by recursion on the structure of t:

$$\begin{aligned} \eta\, x \gg\!= f &= f\, x \\ \sigma(a,b) \gg\!= f &= s(a, \lambda x \to b\, x \gg\!= f) \end{aligned} \tag{6}$$

As the notation suggests, $\gg\!=$ is the Kleisli lifting operation ("bind") for a monad structure on T; indeed, it is the free monad on the endofunctor S.

The notion of "QW-type" that we introduce in this section is obtained from that of W-type by considering not only the algebraic terms over a given signature, but also equations between terms. To code equations we use a type-theoretic rendering of a categorical notion of equational system introduced by Fiore and Hur, referred to as *term equational system* [14, Section 2] and as *monadic equational system* [13, Section 5], here instantiated to free monads on signatures.

Definition 1. *A* system of equations *over a signature* $\Sigma : \mathsf{Sig}$ *is specified by*

- *a type* $E : \mathsf{Set}$ *(whose elements* $e : E$ *name the equations)*
- *a family of types* $V : E \to \mathsf{Set}$ *(*$V\,e : \mathsf{Set}$ *contains the variables used in the equation named* $e : E$*)*
- *for each* $e : E$*, elements* $l\,e$ *and* $r\,e$ *of type* $\mathsf{T}(V\,e)$*, the free* S*-algebra on* $V\,e$ *(the terms with variables from* $V\,e$ *that are equated by the equation named* e*).*

Thus a system of equations over Σ *is an element of the dependent product*

$$\begin{aligned} \mathsf{Syseq}\{\Sigma\} = \textstyle\sum E : \mathsf{Set}, \sum V : (E \to \mathsf{Set}),& \\ ((e : E) \to \mathsf{T}(V\,e)) \times ((e : E) \to \mathsf{T}(V\,e))& \end{aligned} \tag{7}$$

An $\mathsf{S}\{\Sigma\}$*-algebra* $\mathsf{S}\,X \to X$ satisfies *the system of equations* $\varepsilon = (E, V, l, r) : \mathsf{Syseq}\{\Sigma\}$ *if there is an element of type*

$$\mathsf{Sat}\{\varepsilon\}X = (e : E)(\rho : V\,e \to X) \to ((l\,e) \gg\!= \rho) \equiv ((r\,e) \gg\!= \rho) \tag{8}$$

The category-theoretic view of QW-types is that they are simply S-algebras that are initial among those satisfying a given system of equations:

Definition 2. *A* QW-type *for a signature* $\Sigma = (A, B) : \mathsf{Sig}$ *and system of equations* $\varepsilon = (E, V, l, r) : \mathsf{Syseq}\{\Sigma\}$ *is given by a type* $\mathsf{QW}\{\Sigma\}\{\varepsilon\} : \mathsf{Set}$ *equipped with an* S*-algebra structure and a proof that it satisfies the equations*

$$\mathsf{qwintro} : \mathsf{S}(\mathsf{QW}) \to \mathsf{QW} \tag{9}$$

$$\mathsf{qwequ} : \mathsf{Sat}\{\varepsilon\}(\mathsf{QW}) \tag{10}$$

together with functions that witness that it is the initial such algebra:

$$\mathsf{qwrec} : (X : \mathsf{Set})(s : \mathsf{S}\,X \to X) \to \mathsf{Sat}\,X \to \mathsf{QW} \to X \tag{11}$$

$$\mathsf{qwrechom} : (X : \mathsf{Set})(s : \mathsf{S}\,X \to X)(p : \mathsf{Sat}\,X) \to \mathsf{isHom}(\mathsf{qwrec}\,X\,s\,p) \tag{12}$$

$$\begin{aligned} \mathsf{qwuniq} : (X : \mathsf{Set})(s : \mathsf{S}\,X \to X)(p : \mathsf{Sat}\,X)(f : \mathsf{QW} \to X) \to& \\ \mathsf{isHom}\,f \to \mathsf{qwrec}\,X\,s\,p \equiv f& \end{aligned} \tag{13}$$

[1] Note that the definition of $\gg\!=$ depends on the S-algebra structure s; in Agda we use *instance arguments* to hide this dependence.

Given the definitions of $\mathsf{S}\{\Sigma\}$ in (2) and $\mathsf{Sat}\{\varepsilon\}$ in (8), properties (9) and (10) suggest that a QW-type is an instance of the notion of quotient-inductive type [5] with element constructor qwintro and equality constructor qwequ. For this to be so, $\mathsf{QW}\{\Sigma\}\{\varepsilon\}$ needs to have the requisite dependently-typed elimination and computation[2] properties for these element and equality constructors. As Proposition 1 below shows, these follow from (11)–(13), because we are working in a type theory with function extensionality (by virtue of assuming quotient types). To state the proposition we need a dependent version of (6). For each

$$\begin{aligned} &P : \mathsf{QW} \to \mathsf{Set} \\ &p : (a : A)(b : B\,a \to \mathsf{QW}) \to ((x : B\,a) \to P(b\,x)) \to P(\mathsf{qwintro}(a, b)) \end{aligned} \tag{14}$$

type X : Set, function $f : X \to \sum x : \mathsf{QW}, P\,x$ and term $t : \mathsf{T}(X)$, we get an element $\mathsf{lift}\,P\,p\,f\,t : P(t \ggg= \mathsf{fst} \circ f)$ defined by recursion on the structure of t:

$$\begin{aligned} &\mathsf{lift}\,P\,p\,f\,(\eta\,x) &&= \mathsf{snd}(f\,x) \\ &\mathsf{lift}\,P\,p\,f\,(\sigma(a,b)) &&= p\,a\,(\lambda x \to b\,x \ggg= (\mathsf{fst} \circ f))(\mathsf{lift}\,P\,p\,f \circ b) \end{aligned} \tag{15}$$

Proposition 1. *For a QW-type as in the above definition, given P and p as in* (14) *and a term of type*

$$(e : E)(f : V\,e \to \textstyle\sum x : \mathsf{QW}, P\,x) \to \mathsf{lift}\,P\,p\,f\,(l\,e) \equiv\equiv \mathsf{lift}\,P\,p\,f\,(r\,e) \tag{16}$$

there are elimination and computation terms:

$$\begin{aligned} &\mathsf{qwelim} : (x : \mathsf{QW}) \to P\,x \\ &\mathsf{qwcomp} : (a : A)(b : B\,a \to \mathsf{QW}) \to \mathsf{qwelim}(\mathsf{qwintro}(a,b)) \equiv p\,a\,b\,(\mathsf{qwelim} \circ b) \end{aligned}$$

(Note that (16) *uses McBride's* heterogeneous *equality type [23], which we denote by* $\equiv\equiv$, *because* $\mathsf{lift}\,P\,p\,f\,(l\,e)$ *and* $\mathsf{lift}\,P\,p\,f\,(r\,e)$ *inhabit different types, namely* $P(l\,e \ggg= \mathsf{fst} \circ f)$ *and* $P(r\,e \ggg= \mathsf{fst} \circ f)$ *respectively.)* □

The proof of the proposition can be found in the accompanying Agda code (DOI: 10.17863/CAM.48187).

So QW-types are in particular quotient-inductive types (QITs). Conversely, in the next section we show that a wide range of QITs can be encoded as QW-types. Then in Section 4 we prove:

Theorem 1. *In constructive dependent type theory with uniqueness of identity proofs (or equivalently the Axiom K of Streicher [27]) and universes with inductive-inductive datatypes [15] permitting strictly positive occurrences of quotient types [18] and sized types [2], for every signature and system of equations (Definition 1) there is a QW-type as in Definition 2.*

[2] We only establish the computation property up to propositional rather than definitional equality; so, using the terminology of Shulman [25], these are *typal* quotient-inductive types.

Remark 1 (Free algebras). Definition 2 defines QW-types as *initial* algebras. A corollary of Theorem 1 is that *free-algebras* also exist. In other words, given a signature Σ and a type $X : \mathsf{Set}$, there is an S-algebra

$$(\mathsf{F}\{\Sigma\}\{\varepsilon\}X\,,\ \mathsf{S}\{\Sigma\}(\mathsf{F}\{\Sigma\}\{\varepsilon\}X) \to \mathsf{F}\{\Sigma\}\{\varepsilon\}X)$$

satisfying a system of equations ε and equipped with a function $X \to \mathsf{F}\{\Sigma\}\{\varepsilon\}X$, and which is universal among such S-algebras. Thus $\mathsf{QW}\{\Sigma\}\{\varepsilon\}$ is isomorphic to $\mathsf{F}\{\Sigma\}\{\varepsilon\}\varnothing$, where $\varnothing$ is the empty datatype.

To see that such free algebras can be constructed as QW-types, given a signature $\Sigma = (A, B)$, let Σ_X be the signature $(X \uplus A, B')$, where $X \uplus A$ is the coproduct datatype (with constructors $\mathsf{inl} : X \to X \uplus A$ and $\mathsf{inr} : A \to X \uplus A$) and where $B' : X \uplus A \to \mathsf{Set}$ maps each $\mathsf{inl}\, x$ to $\varnothing$ and each $\mathsf{inr}\, a$ to $B\, a$. Given a system of equations $\varepsilon = (E, V, l, r)$, let ε_X be the system (E, V, l_X, r_X) where for each $e : E$, $l_X\, e = l\, e \gg\!= \eta$ and $r_X\, e = r\, e \gg\!= \eta$ (using $\eta : V\, e \to \mathsf{T}\{\Sigma_X\}(V\, e)$ as in (5) and the $\mathsf{S}\{\Sigma\}$-algebra structure s on $\mathsf{T}\{\Sigma_X\}(V\, e)$ given by $s(a, b) = \sigma(\mathsf{inr}\, a, b)$). Then one can show that the QW-type $\mathsf{QW}\{\Sigma_X\}\{\varepsilon_X\}$ is the free algebra $\mathsf{F}\{\Sigma\}\{\varepsilon\}X$, with the function $X \to \mathsf{F}\{\Sigma\}\{\varepsilon\}X$ sending each $x : X$ to $\mathsf{qwintro}(\mathsf{inl}\, x, \_) : \mathsf{QW}\{\Sigma_X\}\{\varepsilon_X\}$, and the $\mathsf{S}\{\Sigma\}$-algebra structure on $\mathsf{F}\{\Sigma\}\{\varepsilon\}X$ being given by the function sending $(a, b) : \mathsf{S}(\mathsf{QW}\{\Sigma_X\}\{\varepsilon_X\})$ to $\mathsf{qwintro}(\mathsf{inr}\, a, b)$.

Remark 2 (Strictly positive equational systems). A very general, categorical notion of equational system was introduced by Fiore and Hur [14, Section 3]. They regard any endofunctor $S : \mathsf{Set} \to \mathsf{Set}$ as a *functorial signature.* A *functorial term* over such a signature, $S \rhd G \vdash L$, is specified by another functorial signature $G : \mathsf{Set} \to \mathsf{Set}$ (the term's context) together with a functor L from S-algebras to G-algebras that commutes with the forgetful functors to Set. Then an *equational system* is given by a pair of such terms in the same context, $S \rhd G \vdash L$ and $S \rhd G \vdash R$ say. An S-algebra $s : S\, X \to X$ satisfies the equational system if $L(X, s)$ and $R(X, s)$ are equal G-algebras.

Taking the *strictly positive* endofunctors $\mathsf{Set} \to \mathsf{Set}$ to be the smallest collection containing the identity and constant endofunctors and closed under forming dependent products and dependent functions over fixed types then, as in [11] (and also in the type theory in which we work), up to isomorphism every such endofunctor is of the form $\mathsf{S}\{\Sigma\}$ for some signature $\Sigma : \mathsf{Sig}$. If we restrict attention to equational systems $S \rhd G \vdash L, R$ with S and G strictly positive, then it turns out that such equational systems are in bijection with the systems of equations from Definition 1, and the two notions of satisfaction for an algebra coincide in that case. (See our Agda development for a proof of this.) So Dybjer's characterisation of W-types as initial algebras for strictly positive endofunctors generalises to the fact that *QW-types are initial among the algebras satisfying strictly positive equational systems in the sense of Fiore and Hur.*

3 Quotient-inductive types

Higher inductive types (HITs) are originally motivated by their use in homotopy type theory to construct homotopical cell complexes, such as spheres, tori, and so on [29]. Intuitively, a higher inductive type is an inductive type with point constructors also allowing for path constructors, surface constructors, etc., which are represented as elements of (iterated) identity types. For example, the sphere is given by the HIT[3]:

$$\begin{aligned}
&\mathsf{data}\ \mathsf{S}^2 : \mathsf{Set}\ \mathsf{where} \\
&\quad \mathsf{base} : \mathsf{S}^2 \\
&\quad \mathsf{surf} : \mathsf{refl} \equiv_{\mathsf{base} \equiv_{\mathsf{S}^2} \mathsf{base}} \mathsf{refl}
\end{aligned} \tag{17}$$

In the presence of the UIP axiom we will refer to HITs as *quotient inductive types* (QITs) [5], since all paths beyond the first level are trivial and any HIT is truncated to an h-set. We use the terms *element constructor* and *equality constructor* to refer to the point constructors and the only non-trivial level of path constructors.

We believe that QW-types can be used to encode a wide range of QITs: see Conjecture 1 below. As evidence, we give several examples of QITs encoded as QW-types, beginning with the two examples of QITs in Figure 1, giving the corresponding signature (A, B) and system of equations (E, V, l, r) as in Definition 2.

Example 1 (Finite multisets). The element constructors for finite multisets are encoded exactly as with a W-type: the constructors are $[]$ and $x :: \_$ for each $x : X$. So we take A to be $\mathbb{1} \uplus X$, the coproduct of the unit type $\mathbb{1}$ (whose single constructor is denoted tt) with X. The arity of $[]$ is zero, and the arity of each $x :: \_$ is one, represented by the empty type $\varnothing$ and unit type $\mathbb{1}$ respectively; so we take $B : A \to \mathsf{Set}$ to be the function $[\lambda\_ \to \mathbb{0} \mid \lambda\_ \to \mathbb{1}] : \mathbb{1} \uplus X \to \mathsf{Set}$ mapping $\mathsf{inl}\,\mathsf{tt}$ to $\varnothing$ and each $\mathsf{inr}\,x$ to $\mathbb{1}$.

The swap equality constructor is parameterised by elements of $E = X \times X$. For each $(x, y) : E$, $\mathsf{swap}\,x\,y$ yields an equation involving a single free variable (called $ys : \mathsf{Bag}\,X$ in Figure 1); so we take $V : E \to \mathsf{Set}$ to be $\lambda\_ \to \mathbb{1}$. Each side of the equation named by $\mathsf{swap}\,x\,y$ is coded by an element of $\mathsf{T}\{\Sigma\}(V(x, y)) = \mathsf{T}\{\Sigma\}(\mathbb{1})$. Recalling the definition of T from (5), the single free variable corresponds to $\eta\,\mathsf{tt} : \mathsf{T}\{\Sigma\}(\mathbb{1})$ and then the left-hand side of the equation is $\sigma(\mathsf{inr}\,x, (\lambda\_ \to \sigma(\mathsf{inr}\,y, (\lambda\_ \to \eta\,\mathsf{tt}))))$ and the right-hand side is $\sigma(\mathsf{inr}\,y, (\lambda\_ \to \sigma(\mathsf{inr}\,x, (\lambda\_ \to \eta\,\mathsf{tt}))))$.

So, altogether, the signature and system of equations for the QW-type corresponding to the first example in Figure 1 is:

$$\begin{aligned}
A &= \mathbb{1} \uplus X & E &= X \times X \\
B &= [\lambda_ \to \varnothing \mid \lambda_ \to \mathbb{1}] & V &= \lambda_ \to \mathbb{1} \\
l &= \lambda\,(x, y) \to \sigma(\mathsf{inr}\,x, (\lambda_ \to \sigma(\mathsf{inr}\,y, (\lambda_ \to \eta\,\mathsf{tt})))) \\
r &= \lambda\,(x, y) \to \sigma(\mathsf{inr}\,y, (\lambda_ \to \sigma(\mathsf{inr}\,x, (\lambda_ \to \eta\,\mathsf{tt}))))
\end{aligned}$$

[3] The subscript on $\equiv$ will be treated as an implicit argument and omitted when clear.

Example 2 (Unordered countably-branching trees). Here the element constructors are leaf of arity zero and, for each $x : X$, $\mathsf{node}\,x$ of arity $\mathbb{N}$. So we use the signature with $A = \mathbb{1} \uplus X$ and $B = [\lambda\_\to\varnothing \mid \lambda\_\to\mathbb{N}]$.

The perm equality constructor is parameterised by elements of

$$E = X \times \sum f : (\mathbb{N} \to \mathbb{N}),\ \mathsf{isIso}\,f$$

For each element (x, f, i) of that type, $\mathsf{perm}\,x\,f\,i$ yields an equation involving an $\mathbb{N}$-indexed family of variables (called $g : \mathbb{N} \to \omega\mathsf{Tree}\,X$ in Figure 1); so we take $V : E \to \mathsf{Set}$ to be $\lambda\_\to\mathbb{N}$. Each side of the equation named by $\mathsf{perm}\,x\,f\,i$ is coded by an element of $\mathsf{T}\{\Sigma\}(V(x, f, i)) = \mathsf{T}\{\Sigma\}(\mathbb{N})$. The $\mathbb{N}$-indexed family of variables is represented by the function $\eta : \mathbb{N} \to \mathsf{T}\{\Sigma\}(\mathbb{N})$ and its permuted version by $\eta \circ f$. Thus the left- and right-hand sides of the equation named by $\mathsf{perm}\,x\,f\,i$ are coded respectively by the elements $\sigma(\mathsf{inr}\,x, \eta)$ and $\sigma(\mathsf{inr}\,x, \eta \circ f)$ of $\mathsf{T}\{\Sigma\}(\mathbb{N})$.

So, altogether, the signature and system of equations for the QW-type corresponding to the second example in Figure 1 is:

$$\begin{aligned} A &= \mathbb{1} \uplus X & E &= X \times \textstyle\sum f : (\mathbb{N} \to \mathbb{N}),\ \mathsf{isIso}\,f \\ B &= [\lambda\_\to\varnothing \mid \lambda\_\to\mathbb{N}] & V &= \lambda\_\to\mathbb{N} \\ l &= \lambda\,(x, \_, \_) \to \sigma(\mathsf{inr}\,x, \eta) \\ r &= \lambda\,(x, f, \_) \to \sigma(\mathsf{inr}\,x, \eta \circ f) \end{aligned}$$

That unordered countably-branching trees are a QW-type is significant since no previous work on various subclasses of QITs (or indeed QIITs [19, 10]) supports infinitary QITs [6, 26, 28, 12, 19, 10]. See Example 5 for another, more substantial infinitary QW-type. So this extension represents one of our main contributions. QW-types generalise prior developments; the internal encodings for particular subclasses of 1-HITs given by Sojakova [26] and Swan [28] are direct instances of QW-types, as the next two examples show.

Example 3. W-suspensions [26] are an instance of QW-types. The data for a W-suspension is: $A', C' : \mathsf{Set}$, a type family $B' : A' \to \mathsf{Set}$ and functions $l', r' : C' \to A'$. The equivalent QW-type is:

$$\begin{aligned} A &= A' & E &= C' & l &= \lambda\,c \to \sigma((l'\,c), \eta) \\ B &= B' & V &= \lambda\,c \to (B'\,(l'\,c)) \times (B'\,(r'\,c)) & r &= \lambda\,c \to \sigma((r'\,c), \eta) \end{aligned}$$

Example 4. The non-indexed case of *W-types with reductions* [28] are QW-types. The data of such a type is: $Y : \mathsf{Set}$, $X : Y \to \mathsf{Set}$ and a reindexing map $R : (y : Y) \to X y$. The reindexing map identifies a term $\sigma\ (y, \alpha)$ with some $\alpha\ (R\,y)$ used to construct it. The equivalent QW-type is given by:

$$\begin{aligned} A &= Y & E &= Y & l &= \lambda y \to \sigma\ (y, \eta) \\ B &= X & V &= X & r &= \lambda y \to \eta\ (R\ i) \end{aligned}$$

Example 5. Lumsdaine and Shulman [21, Section 9] give an example of a HIT not constructible in type theory from only pushouts and $\mathbb{N}$. Their HIT F can be thought of as a set of notations for countable ordinals. It consists of three point constructors: $0 : F$, $S : F \to F$, and $\mathsf{sup} : (\mathbb{N} \to F) \to F$, and five path constructors which are omitted here for brevity. It is inspired by the infinitary algebraic theory of Blass [7, Section 9] and hence it is not surprising that it can be encoded by a QW-type; the details can be found in our Agda code.

3.1 General QIT schemas

Basold, Geuvers, and van der Weide [6] present a schema (though not a model) for infinitary QITs that do not support conditional path equations. Constructors are defined by arbitrary polynomial endofunctors built up using (non-dependent) products and sums, which means in particular that parameters and arguments can occur in any order. They require constructors to be in uncurried form.

Dybjer and Moeneclaey [12, Sections 3.1 and 3.2] present a schema for finitary QITs that supports *conditional* path equations, where constructors are allowed to take inductive arguments not just of the datatype being declared, but also of its identity type. This schema can be generalised to infinitary QITs with conditional path equations. We believe this extension of their schema to be the most general schema for QITs. The schema requires all parameters to appear before all arguments, whereas the schema for regular inductive types in Agda is more flexible, allowing parameters and arguments in any order.

We wish to combine the schema for infinitary QITs of Basold, Geuvers, and van der Weide [6] with the schema for QITs with conditional path equations of Dybjer and Moeneclaey [12] to provide a general schema. Moreover, we would like to combine the arbitrarily ordered parameters and arguments of the former with the curried constructors of the latter in order to support flexible pattern matching.

For consistency with the definition of inductive types in Agda [9, equation (25) and figure 1] we will define strictly positive (i.e. polynomial) endofunctors in terms of strictly positive telescopes.

A telescope is given by the grammar:

$$\begin{array}{llr} \Delta ::= \epsilon & & \text{empty telescope} \\ \quad\mid\ (x : A)\Delta & (x \notin \mathrm{dom}(\Delta)) & \text{non-empty telescope} \end{array} \tag{18}$$

A telescope extension $(x : A)\Delta$ binds (free) occurrences of x inside the tail Δ. The type A may contain free variables that are later bound by further telescope extensions on the left. A telescope can also exist in a context which binds any free variables not already bound in the telescope. Such a context is implicit in the following definitions. A function type $\Delta \to C$ from a telescope Δ to a type C is defined as an iterated dependent function type by:

$$\begin{aligned} \epsilon \to C &\stackrel{\text{def}}{=} C \\ (x : A)\Delta \to C &\stackrel{\text{def}}{=} (x : A) \to (\Delta \to C) \end{aligned} \tag{19}$$

A *strictly positive* endofunctor on a variable Y is presented by a strictly positive telescope

$$\Delta = (x_1 : \Phi_1(Y))(x_2 : \Phi_2(Y)) \cdots (x_n : \Phi_n(Y))\epsilon \tag{20}$$

where each type scheme Φ_i is described by a expression on Y made up of Π-types, Σ-types, and any (previously defined "constant") types A not containing Y, according to the grammar:

$$\Phi(Y), \Psi(Y) ::= \quad (y : A) \to \Phi(Y) \quad | \quad \Sigma p : \Phi(Y), \Psi(Y) \quad | \quad A \quad | \quad Y \tag{21}$$

For example, $\Delta \stackrel{\text{def}}{=} (x : X)(f : \mathbb{N} \to Y)\epsilon$ is the strictly positive telescope for the node constructor in Figure 1. In this instance, reordering x and f is permitted by exchange. Note that the variable Y can never appear in the argument position of a Π-type.

Now it is possible to define the form of the endpoints of an equality (within the context of a strictly positive telescope), corresponding to the notion of an abstract syntax tree with free variables. With this intuition in mind, we can take the definition in Dybjer and Moeneclaey's presentation [12] of endpoints given by *point constructor patterns*:

$$l, r, p ::= \quad c_i\ k \quad | \quad y \tag{22}$$

Where $y : Y$ is in the context of the telescope for the equality constructor, and k is a term built without any rule for Y, but which may use other point constructor patterns $p : Y$. (That is, any sub-term of type Y must either be a variable $y : Y$ found in the telescope, or a constructor for Y applied to further point constructor patterns and earlier defined constants. It could not, for instance, use the function application rule for Y with some function $g : M \to Y$, not least since such functions cannot be defined before defining Y.) Note that this exactly matches the type T in (5).

Basold, Geuvers, and van der Weide's presentation has a sightly more general notion of *constructor term* [6, Definition 6] (Dybjer and Moeneclaey's presentation [12] has more restricted telescopes). It is defined by rules which operate in the context of a strictly positive (polynomial) telescope and permit use of its bound variables, and the use of constructors c_i, but not any other rules for Y. We take the dependent form of their rules for products and functions. Note that these rules do not allow the use of terms of type $\equiv_Y$ in the endpoints.

As with inductive types, the element constructors of QITs are specified by strictly positive telescopes. The equality constructors also permit *conditions* to appear in strictly positive positions, where l and r are constructor terms according to grammar (22):

$$\Phi(Y), \Psi(Y) ::= (\textit{same grammar as in } (21)) \mid l \equiv_Y r \tag{23}$$

Definition 3. *A* QIT *is defined by a list of named element constructors and equality constructors:*

$$\begin{aligned}
&\mathsf{data}\ \mathsf{Y} : \mathsf{Set}\ \mathsf{where}\\
&\quad \mathsf{c}_1 : \Delta_1 \to \mathsf{Y}\\
&\quad \vdots\\
&\quad \mathsf{c}_n : \Delta_n \to \mathsf{Y}\\
&\quad \mathsf{p}_1 : \Theta_1 \to l_1 \equiv_{\mathsf{Y}} r_1\\
&\quad \vdots\\
&\quad \mathsf{p}_m : \Theta_m \to l_m \equiv_{\mathsf{Y}} r_m
\end{aligned}$$

where Δ_i *are strictly positive telescopes on* Y *according to* (21)*, and* Θ_j *are strictly positive telescopes on* Y *and* $\equiv_{\mathsf{Y}}$ *in which conditions may also occur in strictly positive positions according to* (23)*.*

QITs without equality constructors are inductive types. If none of the equality constructors contain Y in an argument position then it is called *non-recursive*, otherwise it is called *recursive* [6]. If none of the equality constructors contain an equality in Y then we call it a *non-conditional*, or *equational*, QIT, otherwise it is called a *conditional* [12], or *quasi-equational*, QIT. If all of the constant types A in any of the constructors are finite (isomorphic to $\mathsf{Fin}\ n$ for $n : \mathbb{N}$) then it is called a *finitary* QIT [12]. Otherwise, it is called a *generalised* [12], or *infinitary*, QIT. We are not aware of any existing examples in the literature of HITs which allow the point constructors to be conditional (though it is not difficult to imagine), nor any schemes for HITs that allow such definitions. However, we do believe this is worth investigating further.

Conjecture 1. Any equational QIT can be encoded as a QW-type.

We believe this can be proved analogously to the approach of Dybjer [11] for inductive types, though the endpoints still need to be considered and we have not yet translated the schema in definition 3 into Agda.

Remark 3. Assuming Conjecture 1, Basold, Geuvers, and van der Weide's schema [6], being an equational (non-conditional) instance of Definition 3, can be encoded as a QW-type.

4 Construction of QW-types

In Section 2 we defined a QW-type to be initial among algebras over a given (possibly infinitary) signature satisfying a given systems of equations (Definition 2). If one interprets these notions in classical Zermelo-Fraenkel set theory with the axiom of Choice (ZFC), one regains the usual notion from universal algebra of initial algebras for infinitary equational theories. Since in the set-theoretic interpretation there is an upper bound on the cardinality of arities of operators in a given signature Σ, the ordinal-indexed sequence $\mathsf{S}^{\alpha}(\varnothing)$ of iterations of the functor in (2) starting from the empty set eventually becomes stationary; and so the sequence has a small colimit, namely the set $\mathsf{W}\{\Sigma\}$ of well-founded trees over Σ. A system of equations ε (Definition 1) over Σ generates a Σ-congruence relation $\sim$ on $\mathsf{W}\{\Sigma\}$. The quotient set $\mathsf{W}\{\Sigma\}/{\sim}$ yields the desired initial algebra for (Σ, ε) provided the S-algebra structure on $\mathsf{W}\{\Sigma\}$ induces one on the quotient set. It does so, because for each operator, using AC one can pick representatives of the (possibly infinitely many) equivalence classes that are the arguments of the operator, apply the interpretation of the operator in $\mathsf{W}\{\Sigma\}$ and then take the equivalence class of that. So the set-theoretic model of type theory in ZFC models QW-types.

Is this use of choice really necessary? Blass [7, Section 9] shows that if one drops AC and just works in ZF, then provided a certain large cardinal axiom is consistent with ZFC, it is consistent with ZF that there is an infinitary equational theory with no initial algebra. He shows this by first exhibiting a countably presented equational theory whose initial algebra has to be an uncountable regular cardinal; and secondly appealing to the construction of Gitik [17] of a model of ZF with no uncountable regular cardinals (assuming a certain large cardinal axiom). Lumsdaine and Shulman [21] turn the infinitary equational theory of Blass into a higher-inductive type that cannot be proved to exist in ZF (and hence cannot be constructed in type theory just using pushouts and the natural numbers). We noted in Example 5 that this higher inductive type can be presented as a QW-type.

So one cannot hope to construct QW-types using a type theory which is interpretable in just ZF. However, the type theory in which we work, with its universes closed under inductive-inductive definitions, already requires going beyond ZF to be able to give it a naive, classical set-theoretic interpretation (by assuming the existence of enough strongly inaccessible cardinals, for example). So the above considerations about initial algebras for infinitary equational theories in classical set theory do not rule out the construction of QW-types in the type theory in which we work. However, something more than just quotienting a W-type is needed in order to prove Theorem 1.

Figure 2 gives a first attempt to do this (which later we will modify using sized types to get around a termination problem). The definition is relative to a given signature $\Sigma : \mathsf{Sig}$ and system of equations $\varepsilon = (E, V, l, r) : \mathsf{Syseq}\,\Sigma$. It makes use of quotient types, which we add to Agda via postulates, as shown in Figure 3.[4] The REWRITE pragma makes $\mathsf{elim}\,R\,B\,f\,e\,(\mathsf{mk}\,R\,x)$ definitionally equal to $f\,x$ and is not merely a computational convenience—this is what allows function extensionality to be proved from these postulated quotient types. The POLARITY pragmas enable the postulated quotients to be used in datatype declarations at positions that Adga deems to be strictly positive; a case in point being the definitions of Q_0 and Q_1 in Figure 2. Agda's test for strict positivity is sound with respect to a set-theoretic semantics of inductively defined datatypes that are built up using strictly positive uses of dependent functions; the semantics of such datatypes uses initial algebras for endofunctors possessing a rank. Here we

[4] The actual implementation is polymorphic in universe levels, but for simplicity here we just give the level-zero version.

```
mutual
  data Q₀ : Set where
    sq : T Q → Q₀

  data Q₁ : Q₀ → Q₀ → Set where
    sqeq : (e : E)(ρ : V e → Q) → Q₁ (sq(T'ρ (l e))) (sq(T'ρ (r e)))
    sqη : (x : Q₀) → Q₁ (sq(η(qu x))) x
    sqσ : (s : S(T Q)) → Q₁ (sq(σ s)) (sq(ι(S'(qu ∘ sq) s)))

  Q : Set
  Q = Q₀/Q₁

  qu : Q₀ → Q
  qu = quot.mk Q₁

QW{Σ}{ε} = Q
```

Figure 2. First attempt at constructing QW-types

are allowing the inductively defined datatypes to be built up using quotients as well, but this is semantically unproblematic, since quotienting does not increase rank. (Later we need to combine the use of POLARITY with sized types; the semantics of this has been studied for System F_ω [3], but needs to be explored further for Agda.)

We build up the underlying inductive type Q_0 to be quotiented using a constructor sq that takes well-founded trees $T(Q_0/Q_1)$ of whole equivalence classes with respect to a relation Q_1 that is mutually inductively defined with Q_0—an instance of an inductive-inductive definition [15]. The definition of Q_1 makes use of the actions on functions of the signature endofunctor S and its associated free monad T (Section 2); those actions are defined as follows:

$$\begin{aligned} &\mathsf{S}' : \{X\ Y : \mathsf{Set}\} \to (X \to Y) \to \mathsf{S}\,X \to \mathsf{S}\,Y \\ &\mathsf{S}'\,f\,(a,b) = (a, f \circ b) \end{aligned} \tag{24}$$

$$\begin{aligned} &\mathsf{T}' : \{X\ Y : \mathsf{Set}\} \to (X \to Y) \to \mathsf{T}\,X \to \mathsf{T}\,Y \\ &\mathsf{T}'\,f\,t = t \ggg\!= (\eta \circ f) \end{aligned} \tag{25}$$

The definition of Q_1 also uses the natural transformation $\iota : \{X : \mathsf{Set}\} \to \mathsf{S}\,X \to \mathsf{T}\,X$ defined by $\iota = \sigma \circ \mathsf{S}'\,\eta$.

Turning to the proof of Theorem 1 using the definitions in Figure 2, the S-algebra structure (9) is easy to define without using any form of choice, because of the type of Q_0's constructor sq. Indeed, we can just take qwintro to be $\mathsf{qu} \circ \mathsf{sq} \circ \iota : \mathsf{S}(\mathsf{QW}) \to \mathsf{QW}$.[5] The first constructor sqeq of the data type Q_1 ensures that the quotient Q_0/Q_1 satisfies the equations in ε, so that we get qwequ as in (10); and the other two constructors, sqη and sqσ make identifications that

[5] The use of the free monad $\mathsf{T}\{\Sigma\}$ in the domain of sq, rather than just $\mathsf{S}\{\Sigma\}$, seems necessary in order to define Q_1 with the properties needed for (10)–(13).

```
module quot where
  postulate
    ty : {A : Set}(R : A → A → Set) → Set
    mk : {A : Set}(R : A → A → Set) → A → ty R
    eq : {A : Set}(R : A → A → Set){x y : A} → R x y → mk R x ≡ mk R y
    elim : {A : Set}(R : A → A → Set)(B : ty R → Set)(f : (x : A) → B(mk R x))
           (e : {x y : A} → R x y → f x ≡≡ f y)(z : ty R) → B z
    comp : {A : Set}(R : A → A → Set)(B : ty R → Set)(f : (x : A) → B(mk R x))
           (e : {x y : A} → R x y → f x ≡≡ f y)(x : A) → elim R B f e (mk R x) ≡ f x
{-# REWRITE comp -#}
{-# POLARITY ty ++ ++ -#}
{-# POLARITY mk _ _ * -#}

_/_ : (A : Set)(R : A → A → Set) → Set
A/R = quot.ty R
```

Figure 3. Quotient types

enable the construction of functions qwrec, qwrechom and qwuniq as in (11)–(13). However, there is a problem. Given X : Set, s : SX → X and e : Sat X, for qwrec $X\,s\,e$ we have to construct a function r : Q → X. Since Q $=$ Q$_0$/Q$_1$ is a quotient, we will have to use the eliminator quot.elim from Figure 3 to define r. The following is an obvious candidate definition

```
mutual                                                        (26)
  r : Q → X
  r = quot.elim Q₁ (λ_ → X) r₀ r₁

  r₀ : Q₀ → X
  r₀(sq t) = t >>= r

  r₁ : {x y : Q₀} → Q₁ x y → r₀ x ≡ r₀ y
  r₁ = ···
```

(where we have elided the details of the invariance proof r$_1$). The problem with this mutually recursive definition is that it is not clear to us (and certainly not to Agda) whether it gives totally defined functions: although the value of r$_0$ at a typical element sq t is explained in terms of the structurally smaller element t, the explanation involves r, whose definition uses the whole function r$_0$ rather than some application of it at a structurally smaller argument. Agda's termination checker rejects the definition.

We get around this problem by using a type-based termination method, namely Agda's implementation of sized types [2]. Intuitively, this provides a type Size of "sizes" which give a constructive abstraction of features of ordinals in ZF when they are used to index sequences of sets that eventually become stationary, such as in various transfinite constructions of free algebras [20, 14]. In Agda, the type Size comes equipped with various relations and functions: given sizes

$$
\begin{array}{l}
\text{mutual} \\
\quad \text{data } \mathsf{Q}_0 (i : \mathsf{Size}) : \mathsf{Set} \text{ where} \\
\quad\quad \mathsf{sq} : \{j : \mathsf{Size}{<}\, i\} \to \mathsf{T}(\mathsf{Q}\, j) \to \mathsf{Q}_0\, i \\[1ex]
\quad \text{data } \mathsf{Q}_1 (i : \mathsf{Size}) : \mathsf{Q}_0\, i \to \mathsf{Q}_0\, i \to \mathsf{Set} \text{ where} \\
\quad\quad \mathsf{sqeq} : \{j : \mathsf{Size}{<}\, i\}(e : E)(\rho : V\, e \to \mathsf{Q}\, j) \to \mathsf{Q}_1\, i\, (\mathsf{sq}(\mathsf{T'}\rho\,(l\, e)))\, (\mathsf{sq}(\mathsf{T'}\rho\,(r\, e))) \\
\quad\quad \mathsf{sq}\eta : \{j : \mathsf{Size}{<}\, i\}(x : \mathsf{Q}_0\, j) \to \mathsf{Q}_1\, i\, (\mathsf{sq}(\eta(\mathsf{qu}\, j\, x)))\, (\phi_0\, i\, x) \\
\quad\quad \mathsf{sq}\sigma : \{j : \mathsf{Size}{<}\, i\}\{k : \mathsf{Size}{<}\, j\}(s : \mathsf{S}(\mathsf{T}(\mathsf{Q}\, k))) \to \\
\quad\quad\quad\quad\quad\quad \mathsf{Q}_1\, i\, (\mathsf{sq}(\sigma\, s))\, (\mathsf{sq}(\iota(\mathsf{S'}(\mathsf{qu}\, j \circ \mathsf{sq})\, s))) \\
\quad \mathsf{Q} : \mathsf{Size} \to \mathsf{Set} \\
\quad \mathsf{Q}\, i = (\mathsf{Q}_0\, i)/\mathsf{Q}_1\, i \\[1ex]
\quad \mathsf{qu} : (i : \mathsf{Size}) \to \mathsf{Q}_0\, i \to \mathsf{Q}\, i \\
\quad \mathsf{qu}\, i = \mathsf{quot.mk}\, (\mathsf{Q}_1\, i) \\[1ex]
\quad \phi_0 : (i : \mathsf{Size})\{j : \mathsf{Size}{<}\, i\} \to \mathsf{Q}_0\, j \to \mathsf{Q}_0\, i \\
\quad \phi_0\, i\, (\mathsf{sq}\, z) = \mathsf{sq}\, z \\[1ex]
\mathsf{QW}\{\Sigma\}\{\varepsilon\} = \mathsf{Q}\, \infty
\end{array}
$$

Figure 4. Construction of QW-types using sized types

i, j : Size, there is a relation i : Size< j to indicate strictly increasing size (so the type Size< j is treated as a subtype of Size); there is a successor operation $\uparrow$: Size $\to$ Size (and also a join operation $\_\sqcup^s\_$: Size $\to$ Size $\to$ Size, but we do not need it here); and a size ∞ : Size to indicate where a sequence becomes stationary. Thus we construct the QW-type $\mathsf{QW}\{\Sigma\}\{\varepsilon\}$ as $\mathsf{Q}\,\infty$ for a suitable size-indexed sequence of types $\mathsf{Q} : \mathsf{Size} \to \mathsf{Set}$, shown in Figure 4.

For each size i : Size, the type $\mathsf{Q}\, i$ is a quotient $\mathsf{Q}_0\, i/\mathsf{Q}_1\, i$, where the constructors of the data types $\mathsf{Q}_0\, i$ and $\mathsf{Q}_1\, i$ take arguments of smaller sizes j : Size< i. Consequently in the following sized version of (26)

$$
\begin{array}{l}
\text{mutual} \\
\quad \mathsf{r} : \{i : \mathsf{Size}\} \to \mathsf{Q}\, i \to X \\
\quad \mathsf{r}\{i\} = \mathsf{quot.elim}\, (\mathsf{Q}_1\, i)\, (\lambda\_ \to X)\, (\mathsf{r}_0\, \{i\})\, (\mathsf{r}_1\, \{i\}) \\[1ex]
\quad \mathsf{r}_0 : \{i : \mathsf{Size}\} \to \mathsf{Q}_0\, i \to X \\
\quad \mathsf{r}_0\{i\}(\mathsf{sq}\, \{j\}\, t) = t \gg\!= \mathsf{r}\, \{j\} \\[1ex]
\quad \mathsf{r}_1 : \{i : \mathsf{Size}\}\{x\, y : \mathsf{Q}_0\, i\} \to \mathsf{Q}_1\, i\, x\, y \to \mathsf{r}_0\, x \equiv \mathsf{r}_0\, y \\
\quad \mathsf{r}_1 = \cdots
\end{array}
\tag{27}
$$

the definition of $\mathsf{r}_0\{i\}$ involves a recursive call via r to the whole function r_0, but at a size j which is smaller than i. So now Agda accepts that the definition of $\mathsf{qwrec}\, X\, s\, e$ as $\mathsf{r}\,\infty$, with r as in (27), is terminating.

Thus we get a function qwrec for (11). We still have (9), but now with $\mathsf{qwintro} = \mathsf{qu}\,\infty \circ \mathsf{sq}\,\{\infty\} \circ \iota$; and as before, the constructor sqeq of Q_1 in Figure 4 ensures that $\mathsf{QW} = (\mathsf{Q}_0\,\infty)/\mathsf{Q}_1\,\infty$ satisfies the equations ε. With these definitions it turns out that each $\mathsf{qwrec}\, X\, s\, e$ is an S-algebra morphism up to definitional equality, so that the function qwrechom needed for (12) is straightforward to define. Finally, the function qwuniq needed for (13) can be constructed via a sequence of lemmas making use of the other two constructors of the data type Q_1, namely $\mathsf{sq}\eta$, which makes use of an auxiliary function for coercing between different size instances of Q_0, and $\mathsf{sq}\sigma$. We refer the reader to the accompanying Agda code (DOI: 10.17863/CAM.48187) for the details of the construction of qwuniq. Altogether, the sized definitions in Figure 4 allow us to complete a proof of Theorem 1.

5 Conclusion

QW-types are a general form of QIT that capture many examples, including simple 1-cell complexes and non-recursive QITs [6], non-structural QITs [26], W-types with reductions [28], and also infinitary QITs (e.g. unordered infinitely branching trees [5], and ordinals [21]). They also capture the notion of initial (and free) algebras for strictly positive equational systems [14], analogously to how W-types capture the notion of initial (and free) algebras for strictly positive endofunctors (see Remark 2). Using Agda to formalise our results, we have shown that it is possible to construct any QW-type, even infinitary ones, in intensional type theory satisfying UIP, using inductive-inductive definitions permitting strictly positive occurrences of quotients and sized types (see Theorem 1 and Section 4). We conclude by mentioning related work and some possible directions for future work.

Quotients of monads. In view of Remark 2, Section 4 gives a construction of initial algebras for equational systems [14] on the *free* monad $\mathsf{T}\{\Sigma\}$ generated by a signature Σ. By a suitable change of signature (see Remark 1) this extends to a construction of free algebras, rather than just initial ones. We can show that the construction works for an arbitrary strictly positive monad and not just for free ones. Given such a construction one gets a quotient monad morphism from the base monad to the quotient monad. This contravariantly induces a forgetful functor from the algebras of the latter to that of the former. Using the adjoint triangle theorem, one should be able to construct a left adjoint. This would then cover examples such as the free group over a monoid, free ring over a group, etc.

Quotient inductive-inductive types. The notion of QW-type generalises to *indexed* QW-types, analogously to the generalisation of W-types to Petersson-Synek trees for inductively defined indexed families of types [24, Chapter 16], and we will consider it in subsequent work. More generally, we wonder whether our analysis of QITs using quotients, inductive-inductive and sized types can be extended to cover the notion of *quotient inductive-inductive* type (QIIT) [4, 19]. Dijkstra [10] studies such types in depth and in Chapter 6 of his thesis gives a construction for finitary ones in terms of countable colimits, and hence in terms of countable coproducts and quotients. One could hope to pass to the infinitary case by using sized types as we have done, provided an analogue for QIITs can be found of the monadic construction in Section 4 for our class of QITs, the QW-types. Kaposi, Kovács, and Altenkirch [19] give a specification of finitary QIITs using a domain-specific type theory called the *theory of signatures* and prove existence of QIITs matching this specification. It might be possible to encode their theory of signatures using QW-types (it can already be encoded as a QIIT), or to extend QW-types making this possible. This would allow infinitary QIITs.

Schemas for QITs. We have shown by example that QW-types can encode a wide range of QITs. However, we have yet to extend this to a proof of Conjecture 1 that every instance of the schema for QITs considered in Section 3 can be so encoded.

Conditional path equations. In Section 3 we mentioned the fact that Dybjer and Moeneclaey [12] give a model for finitary 1-HITs and 2-HITs in which constructors are allowed to take arguments involving the identity type of the datatype being declared. On the face of it, QW-types are not able to encode such *conditional* QITs. We plan to consider whether it is possible to extend the notion of QW-type to allow encoding of infinitary QITs with such conditional equations.

Homotopy Type Theory (HoTT). Our development makes use of UIP (and heterogeneous equality), which is well-known to be incompatible with the Univalence Axiom [29, Example 3.1.9]. Given the interest in HoTT, it is certainly worth investigating whether a result like Theorem 1 holds in univalent foundations for a suitably coherent version of QW-types. We are currently investigating this using set-truncation.

Pattern matching for QITs and HITs. Our reduction of QITs to induction-induction, strictly positive quotients and sized types is of theoretical interest, but in practice one could wish for more direct support in systems like Agda, Lean and Coq for the very useful notion of quotient inductive types (or more generally, for higher inductive types). Even having better support for the special case of quotient types would be welcome. It is not hard to envisage the addition of a general schema for declaring QITs; but when it comes to defining functions on them, having to do that with eliminator forms rapidly becomes cumbersome (for example, for functions of several QIT arguments). Some extension of dependently typed pattern matching to cover equality constructors as well as element constructors is needed and the third author has begun work on that based on the approach of Cockx and Abel [9].[6]

[6] In this context it is worth mentioning that the `cubical` features of recent versions of Agda give access to cubical type theory [30]. This allows for easy declaration of HITs and hence in particular QITs (and quotients avoiding the need for POLARITY pragmas) and a certain amount of pattern matching when it comes to defining functions on them: the value of a function on a path constructor can be specified by using generic elements of the interval type in point-level patterns; but currently the user is given little mechanised assistance to solve the definitional equality constraints on end-points of paths that are generated by this method.

References

1. Abbott, M., Altenkirch, T., Ghani, N.: Containers: Constructing strictly positive types. Theoretical Computer Science *vol. 342*(1), 3–27 (2005). DOI: 10.1016/j.tcs.2005.06.002.
2. Abel, A.: Type-Based Termination, Inflationary Fixed-Points, and Mixed Inductive-Coinductive Types. Electronic Proceedings in Theoretical Computer Science *vol. 77*, 1–11 (2012). DOI: 10.4204/EPTCS.77.1.
3. Abel, A., Pientka, B.: Well-Founded Recursion with Copatterns and Sized Types. J. Funct. Prog. *vol. 26*, e2 (2016). DOI: 10.1017/S0956796816000022.
4. Altenkirch, T., Capriotti, P., Dijkstra, G., Kraus, N., Nordvall Forsberg, F.: Quotient Inductive-Inductive Types. In: Baier, C., Dal Lago, U. (eds.) Foundations of Software Science and Computation Structures, FoSSaCS 2018, LNCS, vol. 10803, pp. 293–310. Springer, Heidelberg (2018).
5. Altenkirch, T., Kaposi, A.: Type Theory in Type Theory Using Quotient Inductive Types. In: Proceedings of the 43rd Annual ACM SIGPLAN-SIGACT Symposium on Principles of Programming Languages - POPL 2016, pp. 18–29. ACM Press, St. Petersburg, FL, USA (2016). DOI: 10.1145/2837614.2837638.
6. Basold, H., Geuvers, H., van der Weide, N.: Higher Inductive Types in Programming. Journal of Universal Computer Science *vol. 23*(1), 27 (2017). DOI: 10.3217/jucs-023-01-0063.
7. Blass, A.: Words, Free Algebras, and Coequalizers. Fundamenta Mathematicae *vol. 117*(2), 117–160 (1983).
8. Cockx, J., Abel, A.: "Sprinkles of Extensionality for Your Vanilla Type Theory". Abstract for the 22nd International Conference on Types for Proofs and Programs (TYPES 2016), Novi Sad, Serbia.
9. Cockx, J., Abel, A.: Elaborating Dependent (Co)Pattern Matching. Proceedings of the ACM on Programming Languages *vol. 2*, 1–30 (2018). DOI: 10.1145/3236770.
10. Dijkstra, G.: Quotient Inductive-Inductive Definitions. PhD thesis, University of Nottingham (2017), URL: http://eprints.nottingham.ac.uk/42317/1/thesis.pdf.
11. Dybjer, P.: Representing Inductively Defined Sets by Wellorderings in Martin-Löf's Type Theory. Theoretical Computer Science *vol. 176*(1-2), 329–335 (1997). DOI: 10.1016/S0304-3975(96)00145-4.
12. Dybjer, P., Moeneclaey, H.: Finitary Higher Inductive Types in the Groupoid Model. Electronic Notes in Theoretical Computer Science *vol. 336*, 119–134 (2018). DOI: 10.1016/j.entcs.2018.03.019.
13. Fiore, M.: An Equational Metalogic for Monadic Equational Systems. Theory and Applications of Categories *vol. 27*(18), 464–492 (2013). URL: https://emis.de/journals/TAC/volumes/27/18/27-18.pdf.
14. Fiore, M., Hur, C.-K.: On the Construction of Free Algebras for Equational Systems. Theoretical Computer Science *vol. 410*(18), 1704–1729 (2009). DOI: 10.1016/j.tcs.2008.12.052.
15. Forsberg, F.N., Setzer, A.: A Finite Axiomatisation of Inductive-Inductive Definitions. In: Berger, U., Diener, H., Schuster, P., Seisenberger, M. (eds.) Logic, Construction, Computation, Ontos mathematical logic, pp. 259–287. De Gruyter (2012). DOI: 10.1515/9783110324921.259.
16. Gambino, N., Kock, J.: Polynomial Functors and Polynomial Monads. Math. Proc. Camb. Phil. Soc. *vol. 154*(1), 153–192 (2013). DOI: 10.1017/S0305004112000394.
17. Gitik, M.: All Uncountable Cardinals Can Be Singular. Israel J. Math. *vol. 35*(1–2), 61–88 (1980).

18. Hofmann, M.: Extensional Concepts in Intensional Type Theory. PhD thesis, University of Edinburgh (1995).
19. Kaposi, A., Kovács, A., Altenkirch, T.: Constructing Quotient Inductive-Inductive Types. Proc. ACM Program. Lang. *vol. 3*, 1–24 (2019). DOI: 10.1145/3290315.
20. Kelly, M.: A Unified Treatment of Transfinite Constructions for Free Algebras, Free Monoids, Colimits, Associated Sheaves, and so on. Bull. Austral. Math. Soc. *vol. 22*, 1–83 (1980).
21. Lumsdaine, P.L., Shulman, M.: Semantics of Higher Inductive Types. Math. Proc. Camb. Phil. Soc. (2019). DOI: 10.1017/S030500411900015X.
22. Martin-Löf, P.: Constructive Mathematics and Computer Programming. In: Cohen, L.J., Łoś, J., Pfeiffer, H., Podewski, K.-P. (eds.) Studies in Logic and the Foundations of Mathematics, pp. 153–175. Elsevier (1982). DOI: 10.1016/S0049-237X(09)70189-2.
23. McBride, C.: Dependently Typed Functional Programs and their Proofs. PhD thesis, University of Edinburgh (1999).
24. Nordström, B., Petersson, K., Smith, J.M.: *Programming in Martin-Löf's Type Theory*. Oxford University Press (1990).
25. Shulman, M.: Brouwer's Fixed-Point Theorem in Real-Cohesive Homotopy Type Theory. Mathematical Structures in Computer Science *vol. 28*, 856–941 (2018).
26. Sojakova, K.: Higher Inductive Types as Homotopy-Initial Algebras. In: Proceedings of the 42nd Annual ACM SIGPLAN-SIGACT Symposium on Principles of Programming Languages - POPL '15, pp. 31–42. ACM Press, Mumbai, India (2015). DOI: 10.1145/2676726.2676983.
27. Streicher, T.: Investigations into Intensional Type Theory. Habilitation Thesis, Ludwig Maximilian University (1993).
28. Swan, A.: W-Types with Reductions and the Small Object Argument. (2018). arXiv:1802.07588 [math].
29. The Univalent Foundations Program, *Homotopy Type Theory: Univalent Foundations for Mathematics*. http://homotopytypetheory.org/book, Institute for Advanced Study (2013).
30. Vezzosi, A., Mörtberg, A., Abel, A.: Cubical Agda: A Dependently Typed Programming Language with Univalence and Higher Inductive Types. Proc. ACM Program. Lang. *vol. 3*(ICFP), 87:1–87:29 (2019). DOI: 10.1145/3341691.

Relative full completeness for bicategorical cartesian closed structure

Marcelo Fiore[1] and Philip Saville[(✉)2]

[1] Department of Computer Science and Technology, University of Cambridge, UK
marcelo.fiore@cl.cam.ac.uk
[2] School of Informatics, University of Edinburgh, UK
philip.saville@ed.ac.uk

Abstract. The glueing construction, defined as a certain comma category, is an important tool for reasoning about type theories, logics, and programming languages. Here we extend the construction to accommodate '2-dimensional theories' of types, terms between types, and rewrites between terms. Taking bicategories as the semantic framework for such systems, we define the glueing bicategory and establish a bicategorical version of the well-known construction of cartesian closed structure on a glueing category. As an application, we show that free finite-product bicategories are fully complete relative to free cartesian closed bicategories, thereby establishing that the higher-order equational theory of rewriting in the simply-typed lambda calculus is a conservative extension of the algebraic equational theory of rewriting in the fragment with finite products only.

Keywords: glueing, bicategories, cartesian closure, relative full completeness, rewriting, type theory, conservative extension

1 Introduction

Relative full completeness for cartesian closed structure. Every small category $\mathbb{C}$ can be viewed as an algebraic theory. This has sorts the objects of $\mathbb{C}$ with unary operators for each morphism of $\mathbb{C}$ and equations determined by the equalities in $\mathbb{C}$. Suppose one freely extends $\mathbb{C}$ with finite products. Categorically, one obtains the free cartesian category $\mathbb{F}^{\times}[\mathbb{C}]$ on $\mathbb{C}$. From the well-known construction of $\mathbb{F}^{\times}[\mathbb{C}]$ (see *e.g.* [12] and [46, §8]) it is direct that the universal functor $\mathbb{C} \to \mathbb{F}^{\times}[\mathbb{C}]$ is fully-faithful, a property we will refer to as the *relative full completeness* (*c.f.* [2,16]) of $\mathbb{C}$ in $\mathbb{F}^{\times}[\mathbb{C}]$. Type theoretically, $\mathbb{F}^{\times}[\mathbb{C}]$ corresponds to the Simply-Typed Product Calculus (STPC) over the algebraic theory of $\mathbb{C}$, given by taking the fragment of the Simply-Typed Lambda Calculus (STLC) consisting of just the types, rules, and equational theory for products. Relative full completeness corresponds to the STPC being a *conservative extension*.

Consider now the free cartesian closed category $\mathbb{F}^{\times,\to}[\mathbb{C}]$ on $\mathbb{C}$, type-theoretically corresponding to the STLC over the algebraic theory of $\mathbb{C}$. Does the relative full completeness property, and hence conservativity, still hold for either $\mathbb{C}$ in $\mathbb{F}^{\times,\to}[\mathbb{C}]$

J. Goubault-Larrecq and B. König (Eds.): FOSSACS 2020, LNCS 12077, pp. 277–298, 2020.
https://doi.org/10.1007/978-3-030-45231-5_15

or for $\mathbb{F}^{\times}[\mathbb{C}]$ in $\mathbb{F}^{\times,\rightarrow}[\mathbb{C}]$? Precisely, is either the universal functor $\mathbb{C} \rightarrow \mathbb{F}^{\times,\rightarrow}[\mathbb{C}]$ or its universal cartesian extension $\mathbb{F}^{\times}[\mathbb{C}] \rightarrow \mathbb{F}^{\times,\rightarrow}[\mathbb{C}]$ full and faithful? The answer is affirmative, but the proof is non-trivial. One must either reason proof-theoretically (*e.g.* in the style of [63, Chapter 8]) or employ semantic techniques such as glueing [39, Annexe C].

In this paper we consider the question of relative full completeness in the bicategorical setting. This corresponds to the question of conservativity for 2-dimensional theories of types, terms between types, and rewrites between terms (see [32,20]). We focus on the particular case of the STLC with invertible rewrites given by β-reductions and η-expansions, and its STPC fragment. By identifying these two systems with cartesian closed, resp. finite product, structure 'up to isomorphism' one recovers a conservative extension result for rewrites akin to that for terms.

2-dimensional categories and rewriting. It has been known since the 1980s that one may consider 2-dimensional categories as abstract reduction systems (*e.g.* [54,51]): if sorts are 0-cells (objects) and terms are 1-cells (morphisms), then rewrites between terms ought to be 2-cells. Indeed, every sesquicategory (of which 2-categories are a special class) generates a rewriting relation $\rightsquigarrow$ on its 1-cells defined by $f \rightsquigarrow g$ if and only if there exists a 2-cell $f \Rightarrow g$ (*e.g.* [60,58]). Invertible 2-cells may be then thought of as equality witnesses.

The rewriting rules of the STLC arise naturally in this framework: Seely [56] observed that β-reduction and η-expansion may be respectively interpreted as the counit and unit of the adjunctions corresponding to lax (directed) products and exponentials in a 2-category (*c.f.* also [34,27]). This approach was taken up by Hilken [32], who developed a '2-dimensional λ-calculus' with strict products and lax exponentials to study the proof theory of rewriting in the STLC (*c.f.* also [33]).

Our concern here is with equational theories of rewriting, and we follow Seely in viewing weak categorical structure as a semantic model of rewriting modulo an equational theory. We are not aware of non-syntactic examples of 2-dimensional cartesian closed structure that are lax but not pseudo (*i.e.* up to isomorphism) and so adopt *cartesian closed bicategories* as our semantic framework.

From the perspective of rewriting, a sesquicategory embodies the rewriting of terms modulo the monoid laws for identities and composition, while a bicategory embodies the rewriting of terms modulo the equational theory on rewrites given by the triangle and pentagon laws of a monoidal category. Cartesian closed bicategories further embody the usual β-reductions and η-expansions of STLC modulo an equational theory on rewrites; for instance, this identifies the composite rewrite $\langle t_1, t_2 \rangle \Rightarrow \langle \pi_1(\langle t_1, t_2 \rangle), \pi_2(\langle t_1, t_2 \rangle) \rangle \Rightarrow \langle t_1, t_2 \rangle$ with the identity rewrite. Indeed, in the free cartesian closed bicategory over a signature of base types and constant terms, the quotient of 1-cells by the isomorphism relation provided by 2-cells is in bijection with $\alpha\beta\eta$-equivalence classes of STLC-terms (*c.f.* [55, Chapter 5]).

Bicategorical relative full completeness. The bicategorical notion of relative full completeness arises by generalising from functors that are fully-faithful to pseudofunctors $F : \mathcal{B} \to \mathcal{C}$ that are *locally an equivalence*, that is, for which every hom-functor $F_{X,Y} : \mathcal{B}(X,Y) \to \mathcal{C}(FX,FY)$ is an equivalence of categories. Interpreted in the context of rewriting, this amounts to the *conservativity of rewriting theories*. First, the equational theory of rewriting in $\mathcal{C}$ is conservative over that in $\mathcal{B}$: the hom-functors do not identify distinct rewrites. Second, the reduction relation in $\mathcal{C}(FX,FY)$ is conservative over that in $\mathcal{B}(X,Y)$: whenever $Ff \rightsquigarrow Fg$ in $\mathcal{C}$ then already $f \rightsquigarrow g$ in $\mathcal{B}$. Third, the term structure in $\mathcal{B}$ gets copied by F in $\mathcal{C}$: modulo the equational theory of rewrites, there are no new terms between types in the image of F.

Contributions. This paper makes two main contributions.

Our first contribution, in Section 3, is to introduce the *bicategorical glueing* construction and, in Section 4, to initiate the development of its theory. As well as providing an assurance that our notion is the right one, this establishes the basic framework for applications. Importantly, we bicategorify the fundamental folklore result (*e.g.* [40,12,62]) establishing mild conditions under which a glued bicategory is cartesian closed.

Our second contribution, in Section 5, is to employ bicategorical glueing to show that for a bicategory $\mathcal{B}$ with finite-product completion $\mathcal{F}^{\times}[\mathcal{B}]$ and cartesian-closed completion $\mathcal{F}^{\times,\to}[\mathcal{B}]$, the universal pseudofunctor $\mathcal{B} \to \mathcal{F}^{\times,\to}[\mathcal{B}]$ and its universal finite-product-preserving extension $\mathcal{F}^{\times}[\mathcal{B}] \to \mathcal{F}^{\times,\to}[\mathcal{B}]$ are both locally an equivalence. Since one may directly observe that the universal pseudofunctor $\mathcal{B} \to \mathcal{F}^{\times}[\mathcal{B}]$ is locally an equivalence, we obtain *relative full completeness results for bicategorical cartesian closed structure* mirroring those of the categorical setting. Establishing this proof-theoretically would require the development of a 2-dimensional proof theory. Given the complexities already present at the categorical level this seems a serious and interesting undertaking. Here, once the basic bicategorical theory has been established, the proof is relatively compact. This highlights the effectiveness of our approach for the application.

The result may also be expressed type-theoretically. For instance, in terms of the type theories of [20], the type theory $\Lambda_{\text{ps}}^{\times,\to}$ for cartesian closed bicategories is a conservative extension of the type theory $\Lambda_{\text{ps}}^{\times}$ for finite-product bicategories. It follows that, modulo the equational theory of bicategorical products and exponentials, any rewrite between STPC-terms constructed using the $\beta\eta$-rewrites for both products and exponentials may be equally presented as constructed from just the $\beta\eta$-rewrites for products (see [21,55]).

Further work. We view the foundational theory presented here as the starting point for future work. For instance, we plan to incorporate further type structure into the development, such as coproducts (*c.f.* [22,16,4]) and monoidal structure (*c.f.* [31]).

On the other hand, the importance of glueing in the categorical setting suggests that its bicategorical counterpart will find a range of applications. A case in point, which has already been developed, is the proof of a 2-dimensional normalisation property for the type theory $\Lambda_{\text{ps}}^{\times,\to}$ for cartesian closed bicategories of [20] that entails a corresponding bicategorical coherence theorem [21,55]. There are also a variety of syntactic constructions in programming languages and type theory that naturally come with a 2-dimensional semantics (see *e.g.* the use of 2-categorical constructions in [23,14,6,61,35]). In such scenarios, bicategorical glueing may prove useful for establishing properties corresponding to the notions of adequacy and/or canonicity, or for proving further conservativity properties.

2 Cartesian closed bicategories

We begin by briefly recapitulating the basic theory of bicategories, including the definition of cartesian closure. A summary of the key definitions is in [41]; for a more extensive introduction see *e.g.* [5,7].

2.1 Bicategories

Bicategories axiomatise structures in which the associativity and unit laws of composition only hold up to coherent isomorphism, for instance when composition is defined by a universal property. They are rife in mathematics and theoretical computer science, appearing in the semantics of computation [29,11,49], datatype models [1,13], categorical logic [26], and categorical algebra [19,25,18].

Definition 1 ([5]). *A* bicategory $\mathcal{B}$ *consists of*

1. *A class of objects* $ob(\mathcal{B})$*,*
2. *For every* $X, Y \in ob(\mathcal{B})$ *a* hom-category $\big(\mathcal{B}(X,Y), \bullet, \mathrm{id}\big)$ *with objects* 1-cells $f : X \to Y$ *and morphisms* 2-cells $\alpha : f \Rightarrow f' : X \to Y$*; composition of 2-cells is called* vertical composition,
3. *For every* $X, Y, Z \in ob(\mathcal{B})$ *an* identity *functor* $\mathrm{Id}_X : \mathbf{1} \to \mathcal{B}(X,X)$ *(for* $\mathbf{1}$ *the terminal category) and a* horizontal composition *functor* $\circ_{X,Y,Z}$: $\mathcal{B}(Y,Z) \times \mathcal{B}(X,Y) \to \mathcal{B}(X,Z)$*,*
4. *Invertible 2-cells*

$$\mathbf{a}_{h,g,f} : (h \circ g) \circ f \Rightarrow h \circ (g \circ f) : W \to Z$$
$$\mathbf{l}_f : \mathrm{Id}_X \circ f \Rightarrow f : W \to X$$
$$\mathbf{r}_g : g \circ \mathrm{Id}_X \Rightarrow g : X \to Y$$

for every $f : W \to X$*,* $g : X \to Y$ *and* $h : Y \to Z$*, natural in each of their parameters and satisfying a* triangle law *and a* pentagon law *analogous to those for monoidal categories.*

A bicategory is said to be locally small *if every hom-category is small.*

Example 1. 1. Every 2-category is a bicategory in which the structural isomorphisms are all the identity.
2. For any category $\mathbb{C}$ with pullbacks there exists a *bicategory of spans* over $\mathbb{C}$ [5]. The objects are those of $\mathbb{C}$, 1-cells $A \rightsquigarrow B$ are spans $(A \leftarrow X \to B)$, and 2-cells $(A \leftarrow X \to B) \to (A \leftarrow X' \to B)$ are morphisms $X \to X'$ making the expected diagram commute. Composition is defined using chosen pullbacks.

A bicategory has three notions of 'opposite', depending on whether one reverses 1-cells, 2-cells, or both (see *e.g.* [37, §1.6]). We shall only require the following.

Definition 2. *The* opposite *of a bicategory $\mathcal{B}$, denoted $\mathcal{B}^{\mathrm{op}}$, is obtained by setting $\mathcal{B}^{\mathrm{op}}(X,Y) := \mathcal{B}(Y,X)$ for all $X, Y \in \mathcal{B}$.*

A morphism of bicategories is called a *pseudofunctor* (or *homomorphism*) [5]. It is a mapping on objects, 1-cells and 2-cells that preserves horizontal composition up to isomorphism. Vertical composition is preserved strictly.

Definition 3. *A* pseudofunctor *$(F, \phi, \psi) : \mathcal{B} \to \mathcal{C}$ between bicategories $\mathcal{B}$ and $\mathcal{C}$ consists of*

1. *A mapping $F : ob(\mathcal{B}) \to ob(\mathcal{C})$,*
2. *A functor $F_{X,Y} : \mathcal{B}(X,Y) \to \mathcal{C}(FX, FY)$ for every $X, Y \in ob(\mathcal{B})$,*
3. *An invertible 2-cell $\psi_X : \mathrm{Id}_{FX} \Rightarrow F(\mathrm{Id}_X)$ for every $X \in ob(\mathcal{B})$,*
4. *An invertible 2-cell $\phi_{f,g} : F(f) \circ F(g) \Rightarrow F(f \circ g)$ for every $g : X \to Y$ and $f : Y \to Z$, natural in f and g,*

subject to two unit laws and an associativity law. A pseudofunctor for which ϕ and ψ are both the identity is called strict. *A pseudofunctor is called* locally P *if every functor $F_{X,Y}$ satisfies the property P.*

Example 2. A monoidal category is equivalently a one-object bicategory; a monoidal functor is equivalently a pseudofunctor between one-object bicategories.

Pseudofunctors $F, G : \mathcal{B} \to \mathcal{C}$ are related by *pseudonatural transformations.* A pseudonatural transformation $(\mathsf{k}, \overline{\mathsf{k}}) : F \Rightarrow G$ consists of a family of 1-cells $(\mathsf{k}_X : FX \to GX)_{X \in \mathcal{B}}$ and, for every $f : X \to Y$, an invertible 2-cell $\overline{\mathsf{k}}_f : \mathsf{k}_Y \circ Ff \Rightarrow Gf \circ \mathsf{k}_X$ witnessing naturality. The 2-cells $\overline{\mathsf{k}}_f$ are required to be natural in f and satisfy two coherence axioms. A morphism of pseudonatural transformations is called a *modification*, and may be thought of as a coherent family of 2-cells.

Notation 1. For bicategories $\mathcal{B}$ and $\mathcal{C}$ we write $\mathbf{Bicat}(\mathcal{B}, \mathcal{C})$ for the (possibly large) bicategory of pseudofunctors, pseudonatural transformations, and modifications (see *e.g.* [41]). If $\mathcal{C}$ is a 2-category, then so is $\mathbf{Bicat}(\mathcal{B}, \mathcal{C})$. We write $\mathbf{Cat}$ for the 2-category of small categories and think of the 2-category $\mathbf{Bicat}(\mathcal{B}^{\mathrm{op}}, \mathbf{Cat})$ as a bicategorical version of the presheaf category $\mathrm{Set}^{\mathbb{C}^{\mathrm{op}}}$. As for presheaf categories, one must take care to avoid size issues. We therefore adopt the convention that when considering $\mathbf{Bicat}(\mathcal{B}^{\mathrm{op}}, \mathbf{Cat})$ the bicategory $\mathcal{B}$ is small or locally small as appropriate.

Example 3. For every bicategory $\mathcal{B}$ and $X \in \mathcal{B}$ there exists the *representable pseudofunctor* $\mathrm{Y}X : \mathcal{B}^{\mathrm{op}} \to \mathbf{Cat}$, defined by $\mathrm{Y}X := \mathcal{B}(-, X)$. The 2-cells ϕ and ψ are structural isomorphisms.

The notion of equivalence between bicategories is called biequivalence. A *biequivalence* $\mathcal{B} \simeq \mathcal{C}$ consists of a pair of pseudofunctors $F : \mathcal{B} \leftrightarrows G : \mathcal{C}$ together with equivalences $FG \simeq \mathrm{id}_{\mathcal{C}}$ and $GF \simeq \mathrm{id}_{\mathcal{B}}$ in $\mathbf{Bicat}(\mathcal{C}, \mathcal{C})$ and $\mathbf{Bicat}(\mathcal{B}, \mathcal{B})$ respectively. Equivalences in an arbitrary bicategory are defined by analogy with equivalences of categories, see *e.g.* [42, pp. 28].

Remark 1. The coherence theorem for monoidal categories [44, Chapter VII] generalises to bicategories: any bicategory is biequivalent to a 2-category [45] (see [42] for a readable summary of the argument). We are therefore justified in writing simply $\cong$ for composites of $\mathbf{a}, \mathbf{l}$ and $\mathbf{r}$.

As a rule of thumb, a category-theoretic proposition lifts to a bicategorical proposition so long as one takes care to weaken isomorphisms to equivalences and sprinkle the prefixes 'pseudo' and 'bi' in appropriate places. For instance, bicategorical adjoints are called *biadjoints* and bicategorical limits are called *bilimits* [59]. The latter may be thought of as limits in which every cone is filled by a coherent choice of invertible 2-cell. Bilimits are preserved by representable pseudofunctors and by right biadjoints. The *bicategorical Yoneda lemma* [59, §1.9] says that for any pseudofunctor $P : \mathcal{B}^{\mathrm{op}} \to \mathbf{Cat}$, evaluation at the identity determines a pseudonatural family of equivalences $\mathbf{Bicat}(\mathcal{B}^{\mathrm{op}}, \mathbf{Cat})(\mathrm{Y}X, P) \simeq PX$. One may then deduce that the *Yoneda pseudofunctor* $\mathrm{Y} : \mathcal{B} \to \mathbf{Bicat}(\mathcal{B}^{\mathrm{op}}, \mathbf{Cat}) : X \mapsto \mathrm{Y}X$ is locally an equivalence. Another 'bicategorified' lemma is the following, which we shall employ in Section 5.

Lemma 1. *1. For pseudofunctors $F, G : \mathcal{B} \to \mathcal{C}$, if $F \simeq G$ and G is locally an equivalence, then so is F.*
2. For pseudofunctors $F : \mathcal{A} \to \mathcal{B}$, $G : \mathcal{B} \to \mathcal{C}$, $H : \mathcal{C} \to \mathcal{D}$, if $G \circ F$ and $H \circ G$ are local equivalences, then so is F.

2.2 fp-Bicategories

It is convenient to directly consider all finite products, as this reduces the need to deal with the equivalent objects given by re-bracketing binary products. To avoid confusion with the 'cartesian bicategories' of Carboni and Walters [10,8], we call a bicategory with all finite products an *fp-bicategory.*

Definition 4. *An* fp-bicategory $(\mathcal{B}, \Pi_n(-))$ *is a bicategory $\mathcal{B}$ equipped with the following data for every $A_1, \ldots, A_n \in \mathcal{B}$ $(n \in \mathbb{N})$:*

1. *A chosen object $\prod_n(A_1, \ldots, A_n)$,*
2. *Chosen arrows $\pi_k : \prod_n(A_1, \ldots, A_n) \to A_k$ $(k = 1, \ldots, n)$, called* projections,
3. *For every $X \in \mathcal{B}$ an adjoint equivalence*

$$\mathcal{B}\left(X, \prod_n(A_1, \ldots, A_n)\right) \underset{\langle -, \ldots, = \rangle}{\overset{(\pi_1 \circ -, \ldots, \pi_n \circ -)}{\perp\simeq}} \prod_{i=1}^{n} \mathcal{B}(X, A_i) \tag{1}$$

specified by choosing a family of universal arrows (see e.g. *[44, Theorem IV.2]) with components $\varpi^{(i)}_{f_1, \ldots, f_n} : \pi_i \circ \langle f_1, \ldots, f_n \rangle \Rightarrow f_i$ for $i = 1, \ldots, n$.*

We call the right adjoint $\langle -, \dots, = \rangle$ *the* n-ary tupling.

Explicitly, the universal property of $\varpi = (\varpi^{(1)}, \dots, \varpi^{(n)})$ is the following. For any finite family of 2-cells $(\alpha_i : \pi_i \circ g \Rightarrow f_i : X \to A_i)_{i=1,\dots,n}$, there exists a 2-cell $\mathsf{p}^\dagger(\alpha_1, \dots, \alpha_n) : g \Rightarrow \langle f_1, \dots, f_n \rangle : X \to \prod_n(A_1, \dots, A_n)$, unique such that

$$\varpi^{(k)}_{f_1,\dots,f_n} \bullet \left(\pi_k \circ \mathsf{p}^\dagger(\alpha_1, \dots, \alpha_n)\right) = \alpha_k : \pi_k \circ g \Rightarrow f_k$$

for $k = 1, \dots, n$. One thereby obtains a functor $\langle -, \dots, = \rangle$ and an adjunction as in (1) with counit $\varpi = (\varpi^{(1)}, \dots, \varpi^{(n)})$ and unit $\varsigma_g := \mathsf{p}^\dagger(\mathrm{id}_{\pi_1 \circ g}, \dots, \mathrm{id}_{\pi_n \circ g}) : g \Rightarrow \langle \pi_1 \circ g, \dots, \pi_n \circ g \rangle$. This defines a *lax* n-ary product structure: one merely obtains an adjunction in (1). One turns it into a bicategorical (*pseudo*) product by further requiring the unit and counit to be invertible. The *terminal object* $\mathbf{1}$ arises as $\prod_0()$. We adopt the same notation as for categorical products, for example by writing $\prod_{i=1}^n A_i$ for $\prod_n(A_1, \dots, A_n)$ and $\prod_{i=1}^n f_i$ for $\langle f_1 \circ \pi_1, \dots, f_n \circ \pi_n \rangle$.

Example 4. The bicategory of spans over a *lextensive category* [9] has finite products; such a bicategory is biequivalent to its opposite, so these are in fact biproducts [38, Theorem 6.2]. Biproduct structure arises using the coproduct structure of the underlying category (*c.f.* the biproduct structure of the category of relations).

*Remark 2 (*c.f. *Remark 1).* fp-Bicategories satisfy the following coherence theorem: every fp-bicategory is biequivalent to a 2-category with 2-categorical products [52, Theorem 4.1]. Thus, we shall sometimes simply write $\cong$ in diagrams for composites of 2-cells arising from either the bicategorical or product structure. In pasting diagrams we shall omit such 2-cells completely (*c.f.* [30, Remark 3.1.16]; for a detailed exposition, see [64, Appendix A]).

One may think of bicategorical product structure as an intensional version of the familiar categorical structure, except the usual equations (*e.g.* [28]) are now witnessed by natural families of invertible 2-cells. It is useful to introduce explicit names for these 2-cells.

Notation 2. In the following, and throughout, we write $A_\bullet$ for a finite sequence $\langle A_1, \dots, A_n \rangle$.

Lemma 2. *For any fp-bicategory* $(\mathcal{B}, \Pi_n(-))$ *there exist canonical choices for the following natural families of invertible 2-cells:*

1. *For every* $(h_i : Y \to A_i)_{i=1,\dots,n}$ *and* $g : X \to Y$*, a 2-cell* $\mathsf{post}(h_\bullet; g) : \langle h_1, \dots, h_n \rangle \circ g \Rightarrow \langle h_1 \circ g, \dots, h_n \circ g \rangle$*,*
2. *For every* $(h_i : A_i \to B_i)_{i=1,\dots,n}$ *and* $(g_i : X \to A_i)_{i=1,\dots,n}$*, a 2-cell* $\mathsf{fuse}(h_\bullet; g_\bullet) : (\prod_{i=1}^n h_i) \circ \langle g_1, \dots, g_n \rangle \Rightarrow \langle h_1 \circ g_1, \dots, h_n \circ g_n \rangle$.

In particular, it follows from Lemma 2(2) that there exists a canonical natural family of invertible 2-cells $\Phi_{h_\bullet, g_\bullet} : (\prod_{i=1}^n h_i) \circ (\prod_{i=1}^n g_i) \Rightarrow \prod_{i=1}^n (h_i \circ g_i)$ for any $(h_i : A_i \to B_i)_{i=1,\dots,n}$ and $(g_j : X_j \to A_j)_{j=1,\dots,n}$.

In the categorical setting, a cartesian functor preserves products up to isomorphism. An fp-pseudofunctor preserves bicategorical products up to equivalence.

Definition 5. *An* fp-pseudofunctor $(F, \mathrm{q}^\times)$ *between fp-bicategories* $(\mathcal{B}, \Pi_n(-))$ *and* $(\mathcal{C}, \Pi_n(-))$ *is a pseudofunctor* $F : \mathcal{B} \to \mathcal{C}$ *equipped with specified equivalences*

$$\langle F\pi_1, \ldots, F\pi_n \rangle : F(\prod_{i=1}^n A_i) \leftrightarrows \prod_{i=1}^n (FA_i) : \mathrm{q}^\times_{A_\bullet}$$

for every $A_1, \ldots, A_n \in \mathcal{B}$ $(n \in \mathbb{N})$. *We denote the 2-cells witnessing these equivalences by* $\mathsf{u}^\times_{A_\bullet} : \mathrm{Id}_{(\prod_i FA_i)} \Rightarrow \langle F\pi_1, \ldots, F\pi_n \rangle \circ \mathrm{q}^\times_{A_\bullet}$ *and* $\mathsf{c}^\times_{A_\bullet} : \mathrm{q}^\times_{A_\bullet} \circ \langle F\pi_1, \ldots, F\pi_n \rangle \Rightarrow \mathrm{Id}_{(F\Pi_i A_i)}$. *We call* $(F, \mathrm{q}^\times)$ strict *if* F *is strict and satisfies*

$$F(\prod_n(A_1, \ldots, A_n)) = \prod_n(FA_1, \ldots, FA_n)$$

$$F(\pi_i^{A_1,\ldots,A_n}) = \pi_i^{FA_1,\ldots,FA_n} \qquad F\varpi^{(i)}_{t_1,\ldots,t_n} = \varpi^{(i)}_{Ft_1,\ldots,Ft_n}$$

$$F\langle t_1, \ldots, t_n \rangle = \langle Ft_1, \ldots, Ft_n \rangle \qquad \mathrm{q}^\times_{A_1,\ldots,A_n} = \mathrm{Id}_{\Pi_n(FA_1,\ldots,FA_n)}$$

with equivalences given by the 2-cells $\mathsf{p}^\dagger(\mathbf{r}_{\pi_1}, \ldots, \mathbf{r}_{\pi_n}) : \mathrm{Id} \overset{\cong}{\Rightarrow} \langle \pi_1, \ldots, \pi_n \rangle$.

Notation 3. For fp-bicategories $\mathcal{B}$ and $\mathcal{C}$ we write **fp-Bicat**$(\mathcal{B}, \mathcal{C})$ for the bicategory of fp-pseudofunctors, pseudonatural transformations and modifications.[3]

We define two further families of 2-cells to witness standard properties of cartesian functors. The first witnesses the fact that any fp-pseudofunctor commutes with the $\prod_n(-, \ldots, =)$ operation. The second witnesses the equality $\langle F\pi_1, \ldots, F\pi_n \rangle \circ F\langle f_1, \ldots, f_n \rangle = \langle Ff_1, \ldots, Ff_n \rangle$ 'unpacking' an n-ary tupling from inside F.

Lemma 3. *Let* $(F, \mathrm{q}^\times) : (\mathcal{B}, \Pi_n(-)) \to (\mathcal{C}, \Pi_n(-))$ *be an fp-pseudofunctor.*

1. *For any finite family of 1-cells* $(f_i : A_i \to A'_i)_{i=1,\ldots,n}$ *in* $\mathcal{B}$, *there exists an invertible 2-cell* $\mathsf{nat}_{f_\bullet} : \mathrm{q}^\times_{A'_\bullet} \circ \prod_{i=1}^n Ff_i \Rightarrow F(\prod_{i=1}^n f_i) \circ \mathrm{q}^\times_{A_\bullet}$ *such that the pair* $(\mathrm{q}^\times, \mathsf{nat})$ *forms a a pseudonatural transformation*
$$\prod_{i=1}^n (F(-), \ldots, F(=)) \Rightarrow (F \circ \prod_{i=1}^n)(-, \ldots, =)$$
2. *For any finite family of 1-cells* $(f_i : X \to B_i)_{i=1,\ldots,n}$ *in* $\mathcal{B}$, *there exists a canonical choice of naturally invertible 2-cell* $\mathsf{unpack}_{f_\bullet} : \langle F\pi_1, \ldots, F\pi_n \rangle \circ F\langle f_1, \ldots, f_n \rangle \Rightarrow \langle Ff_1, \ldots, Ff_n \rangle : FX \to \prod_{i=1}^n FB_i$.

2.3 Cartesian closed bicategories

A cartesian closed bicategory is an fp-bicategory $(\mathcal{B}, \Pi_n(-))$ equipped with a biadjunction $(-) \times A \dashv (A \Rightarrow -)$ for every $A \in \mathcal{B}$. Examples include the bicategory of generalised species [17], bicategories of concurrent games [49], and bicategories of operads [26].

[3] In the categorical setting, every natural transformation between cartesian functors is monoidal with respect to the cartesian structure and a similar fact is true bicategorically: every pseudonatural transformation is canonically compatible with the product structure, see [55, § 4.1.1].

Definition 6. *A* cartesian closed bicategory *or* cc-bicategory *is an fp-bicategory* $(\mathcal{B}, \Pi_n(-))$ *equipped with the following data for every* $A, B \in \mathcal{B}$*:*

1. *A chosen object* $(A \Rightarrow B)$,
2. *A specified 1-cell* $\mathrm{eval}_{A,B} : (A \Rightarrow B) \times A \to B$,
3. *For every* $X \in \mathcal{B}$, *an adjoint equivalence*

$$\mathcal{B}(X, A \Rightarrow B) \underset{\lambda}{\overset{\mathrm{eval}_{A,B} \circ (- \times A)}{\rightleftarrows}} \perp\simeq \ \mathcal{B}(X \times A, B)$$

specified by a choice of universal arrow $\varepsilon_f : \mathrm{eval}_{A,B} \circ (\lambda f \times A) \overset{\cong}{\Rightarrow} f$.

We call the functor $\lambda(-)$ currying *and refer to* λf *as the* currying of f.

Explicitly, the counit ε satisfies the following universal property. For every 1-cell $g : X \to (A \Rightarrow B)$ and 2-cell $\alpha : \mathrm{eval}_{A,B} \circ (g \times A) \Rightarrow f$ there exists a unique 2-cell $\mathsf{e}^{\dagger}(\alpha) : g \Rightarrow \lambda f$ such that $\varepsilon_f \bullet \big(\mathrm{eval}_{A,B} \circ (\mathsf{e}^{\dagger}(\alpha) \times A)\big) = \alpha$. This defines a *lax* exponential structure. One obtains a *pseudo* (bicategorical) exponential structure by further requiring that ε and the unit $\eta_t := \mathsf{e}^{\dagger}(\mathrm{id}_{\mathrm{eval}_{A,B} \circ (t \times A)})$ are invertible.

Example 5. Every 'presheaf' 2-category $\mathbf{Bicat}(\mathcal{B}^{\mathrm{op}}, \mathbf{Cat})$ has all bicategorical limits [52, Proposition 3.6], given pointwise, and is cartesian closed with $(P \Rightarrow Q)X := \mathbf{Bicat}(\mathcal{B}^{\mathrm{op}}, \mathbf{Cat})(\mathrm{Y}X \times P, Q)$ [55, Chapter 6].

As for products, we adopt the notational conventions that are standard in the categorical setting, for example by writing $(f \Rightarrow g) : (A \Rightarrow B) \to (A' \Rightarrow B')$ for the currying of $(g \circ \mathrm{eval}_{A,B}) \circ (\mathrm{Id}_{A \Rightarrow B} \times f)$.

Just as fp-pseudofunctors preserve products up to equivalence, cartesian closed pseudofunctors preserve products and exponentials up to equivalence.

Definition 7. *A* cartesian closed pseudofunctor *or* cc-pseudofunctor *between cc-bicategories* $(\mathcal{B}, \Pi_n(-), \Rightarrow)$ *and* $(\mathcal{C}, \Pi_n(-), \Rightarrow)$ *is an fp-pseudofunctor* $(F, \mathrm{q}^{\times})$ *equipped with specified equivalences* $m_{A,B} : F(A \Rightarrow B) \leftrightarrows (FA \Rightarrow FB) : \mathrm{q}^{\Rightarrow}_{A,B}$ *for every* $A, B \in \mathcal{B}$, *where* $m_{A,B} : F(A \Rightarrow B) \to (FA \Rightarrow FB)$ *is the currying of* $F(\mathrm{eval}_{A,B}) \circ \mathrm{q}^{\times}_{A \Rightarrow B, A}$. *A cc-pseudofunctor* $(F, \mathrm{q}^{\times}, \mathrm{q}^{\Rightarrow})$ *is* strict *if* $(F, \mathrm{q}^{\times})$ *is a strict fp-pseudofunctor such that*

$$F(A \Rightarrow B) = (FA \Rightarrow FB)$$
$$F(\mathrm{eval}_{A,B}) = \mathrm{eval}_{FA,FB} \qquad F(\varepsilon_t) = \varepsilon_{Ft}$$
$$F(\lambda t) = \lambda(Ft) \qquad \mathrm{q}^{\Rightarrow}_{A,B} = \mathrm{Id}_{FA \Rightarrow FB}$$

with equivalences given by the 2-cells

$$\mathsf{e}^{\dagger}(\mathrm{eval}_{FA,FB} \circ \kappa) : \mathrm{Id}_{(FA \Rightarrow FB)} \overset{\cong}{\Rightarrow} \lambda(\mathrm{eval}_{FA,FB} \circ \mathrm{Id}_{(FA \Rightarrow FB) \times FA})$$

where κ *is the canonical isomorphism* $\mathrm{Id}_{FA \Rightarrow FB} \times FA \cong \mathrm{Id}_{(FA \Rightarrow FB) \times FA}$.

Remark 3. As is well-known in the case of **Cat** (*e.g.* [44, IV.2]), every equivalence $X \simeq Y$ in a bicategory gives rise to an *adjoint equivalence* between X and Y with the same 1-cells (see *e.g.* [42, pp. 28–29]). Thus, one may assume without loss of generality that all the equivalences in the preceding definition are adjoint equivalences. The same observation applies to the definition of fp-pseudofunctors.

Notation 4. For cc-bicategories $\mathcal{B}$ and $\mathcal{C}$ we write $\mathbf{cc\text{-}Bicat}(\mathcal{B}, \mathcal{C})$ for the bicategory of cc-pseudofunctors, pseudonatural transformations and modifications (*c.f.* Notation 3).

3 Bicategorical glueing

The glueing construction has been discovered in various forms, with correspondingly various names: the notions of logical relation [50,57], sconing [24], Freyd covers, and glueing (*e.g.* [40]) are all closely related (see *e.g.* [47] for an overview of the connections). Originally presented set-theoretically, the technique was quickly given categorical expression [43,47] and is now a standard component of the armoury for studying type theories (*e.g.* [40,12]).

The *glueing* $\mathrm{gl}(F)$ of categories $\mathbb{C}$ and $\mathbb{D}$ along a functor $F : \mathbb{C} \to \mathbb{D}$ may be defined as the comma category $(\mathrm{id}_{\mathbb{D}} \downarrow F)$. We define bicategorical glueing analogously.

Definition 8.

1. *Let $F : \mathcal{A} \to \mathcal{C}$ and $G : \mathcal{B} \to \mathcal{C}$ be pseudofunctors of bicategories. The* comma bicategory $(F \downarrow G)$ *has objects triples $(A \in \mathcal{A}, f : FA \to GB, B \in \mathcal{B})$. The 1-cells $(A, f, B) \to (A', f', B')$ are triples (p, α, q), where $p : A \to A'$ and $q : B \to B'$ are 1-cells and α is an invertible 2-cell $\alpha : f' \circ Fp \Rightarrow Gq \circ f$. The 2-cells $(p, \alpha, q) \Rightarrow (p', \alpha', q')$ are pairs of 2-cells $(\sigma : p \Rightarrow p', \tau : q \Rightarrow q')$ such that the following diagram commutes:*

$$\begin{array}{ccc} f' \circ F(p) & \xrightarrow{f' \circ F(\sigma)} & f' \circ F(p') \\ {\scriptstyle \alpha}\downarrow & & \downarrow{\scriptstyle \alpha'} \\ G(q) \circ f & \xrightarrow[G(\tau) \circ f]{} & G(q') \circ f \end{array} \tag{2}$$

Identities and horizontal composition are given by the following pasting diagrams.

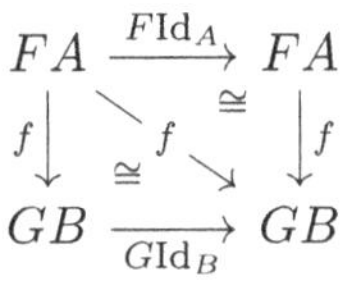

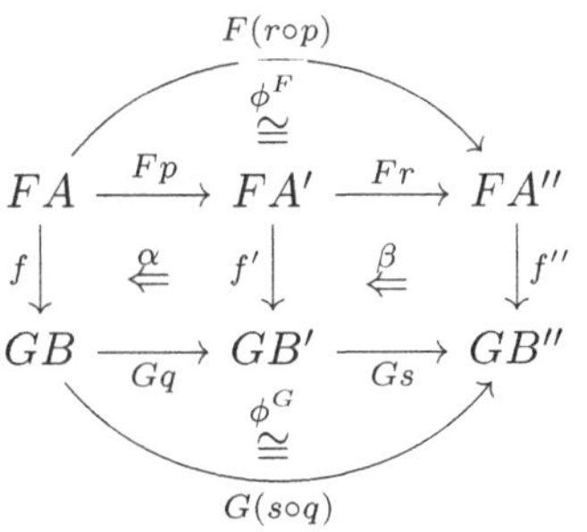

Vertical composition, the identity 2-cell, and the structural isomorphisms are given component-wise.

2. *The* glueing bicategory $\mathrm{gl}(\mathfrak{J})$ *of bicategories* $\mathcal{B}$ *and* $\mathcal{C}$ *along a pseudofunctor* $\mathfrak{J} : \mathcal{B} \to \mathcal{C}$ *is the comma bicategory* $(\mathrm{id}_{\mathcal{C}} \downarrow \mathfrak{J})$.

We call axiom (2) the *cylinder condition* due to its shape when viewed as a (3-dimensional) pasting diagram. Note that one directly obtains projection pseudofunctors $\mathcal{B} \xleftarrow{\pi_{\mathrm{dom}}} \mathrm{gl}(\mathfrak{J}) \xrightarrow{\pi_{\mathrm{cod}}} \mathcal{C}$.

We develop some basic theory of glueing bicategories, which we shall put to use in Section 5. We follow the terminology of [15].

Definition 9. *Let* $\mathfrak{J} : \mathcal{B} \to \mathcal{X}$ *be a pseudofunctor. The* relative hom-pseudofunctor $\langle\mathfrak{J}\rangle : \mathcal{X} \to \mathbf{Bicat}(\mathcal{B}^{\mathrm{op}}, \mathbf{Cat})$ *is defined by* $\langle\mathfrak{J}\rangle X := \mathcal{X}(\mathfrak{J}(-), X)$.

Following [15], one might call the glueing bicategory $\mathrm{gl}(\langle\mathfrak{J}\rangle)$ associated to a relative hom-pseudofunctor the bicategory of *$\mathcal{B}$-intensional Kripke relations of arity* $\mathfrak{J}$, and view it as an intensional, bicategorical, version of the category of Kripke relations.

The relative hom-pseudofunctor preserves all bilimits that exist in its domain. For products, this may be described explicitly.

Lemma 4. *For any fp-bicategory* $(\mathcal{X}, \Pi_n(-))$ *and pseudofunctor* $\mathfrak{J} : \mathcal{B} \to \mathcal{X}$*, the relative hom-pseudofunctor* $\langle\mathfrak{J}\rangle$ *extends canonically to an fp-pseudofunctor.*

Proof. *Take* $\mathrm{q}^{\times}_{X_\bullet}$ *to be the n-ary tupling* $\prod_{i=1}^{n} \mathcal{X}(\mathfrak{J}(-), X_i) \xrightarrow{\simeq} \mathcal{X}(\mathfrak{J}(-), \prod_{i=1}^{n} X_i)$. *This forms a pseudonatural transformation with naturality witnessed by* post.

For any pseudofunctor $\mathfrak{J} : \mathcal{B} \to \mathcal{X}$ there exists a pseudonatural transformation $(l, \bar{l}) : \mathrm{Y} \Rightarrow \langle\mathfrak{J}\rangle \circ \mathfrak{J} : \mathcal{B} \to \mathbf{Bicat}(\mathcal{B}^{\mathrm{op}}, \mathbf{Cat})$ given by the functorial action of $\mathfrak{J}$ on hom-categories. One may therefore define the following.

Definition 10. *For any pseudofunctor* $\mathfrak{J} : \mathcal{B} \to \mathcal{X}$*, define the* extended Yoneda pseudofunctor $\underline{\mathrm{Y}} : \mathcal{B} \to \mathrm{gl}(\langle\mathfrak{J}\rangle)$ *by setting* $\underline{\mathrm{Y}}B := \big(\mathrm{Y}B, (l, \bar{l})_{(-,B)}, \mathfrak{J}B\big)$, $\underline{\mathrm{Y}}f := (\mathrm{Y}f, (\phi^{\mathfrak{J}}_{-,f})^{-1}, \mathfrak{J}f)$, *and* $\underline{\mathrm{Y}}(\tau : f \Rightarrow f' : B \to B') := (\mathrm{Y}\tau, \mathfrak{J}\tau)$. *The cylinder condition holds by the naturality of* $\phi^{\mathfrak{J}}$*, and the 2-cells* $\phi^{\underline{\mathrm{Y}}}$ *and* $\psi^{\underline{\mathrm{Y}}}$ *are* $(\phi^{\mathrm{Y}}, \phi^{\mathfrak{J}})$ *and* $(\psi^{\mathrm{Y}}, \psi^{\mathfrak{J}})$*, respectively.*

The extended Yoneda pseudofunctor satisfies a corresponding 'extended Yoneda lemma' (*c.f.* [15, pp. 33]).

Lemma 5. *For any pseudofunctor* $\mathfrak{J} : \mathcal{B} \to \mathcal{X}$ *and* $\underline{P} = (P, (\mathsf{k}, \bar{\mathsf{k}}), X) \in \mathrm{gl}(\langle\mathfrak{J}\rangle)$ *there exists an equivalence of pseudofunctors* $\mathrm{gl}(\langle\mathfrak{J}\rangle)(\underline{\mathrm{Y}}(-), \underline{P}) \simeq P$ *and an invertible modification as in the diagram below. Hence* $\underline{\mathrm{Y}}$ *is locally an equivalence.*

$$\begin{array}{ccc} \mathrm{gl}(\langle\mathfrak{J}\rangle)(\underline{\mathrm{Y}}(-), \underline{P}) & \xrightarrow{\qquad\simeq\qquad} & P \\ {\scriptstyle \pi_{\mathrm{dom}}} \searrow & \overset{\cong}{\Leftarrow} & \swarrow {\scriptstyle (\mathsf{k}, \bar{\mathsf{k}})} \\ & \mathcal{X}(\mathfrak{J}(-), X) & \end{array}$$

Proof. The arrow marked $\simeq$ *is the composite of a projection and the equivalence arising from the Yoneda lemma. Its pseudo-inverse is the composite*

$$P \xrightarrow{\simeq} \mathbf{Bicat}(\mathcal{B}^{op}, \mathbf{Cat})(\mathrm{Y}(-), P) \to \mathrm{gl}(\langle\mathfrak{J}\rangle)(\underline{\mathrm{Y}}(-), \underline{P}) \tag{3}$$

in which the equivalence arises from the Yoneda lemma and the unlabelled pseudofunctor takes a pseudonatural transformation $(\mathsf{j}, \bar{\mathsf{j}}) : \mathrm{Y}B \Rightarrow P$ *to the triple with first component* $(\mathsf{j}, \bar{\mathsf{j}})$*, third component* $\mathsf{j}_B(\mathsf{k}_B(\mathrm{Id}_B)) : \mathfrak{J}B \to X$ *and second component defined using* $\bar{\mathsf{k}}$ *and* $\bar{\mathsf{j}}$*. Chasing the definitions through and evaluating at* $A, B \in \mathcal{B}$*, one sees that when* $\underline{P} := \underline{\mathrm{Y}}B$ *the composite (3) is equivalent to* $\underline{\mathrm{Y}}_{A,B}$*. Since (3) is locally an equivalence, Lemma 1(1) completes the proof.*

4 Cartesian closed structure on the glueing bicategory

It is well-known that, if $\mathbb{C}$ and $\mathbb{D}$ are cartesian closed categories, $\mathbb{D}$ has pullbacks, and $F : \mathbb{C} \to \mathbb{D}$ is cartesian, then $\mathrm{gl}(F)$ is cartesian closed (*e.g.* [40,12]). In this section we prove a corresponding result for the glueing bicategory. We shall be guided by the categorical proof, for which see *e.g.* [43, Proposition 2].

4.1 Finite products in $\mathrm{gl}(\mathfrak{J})$

Proposition 1. *Let* $(\mathcal{B}, \Pi_n(-))$ *and* $(\mathcal{C}, \Pi_n(-))$ *be fp-bicategories and* $(\mathfrak{J}, \mathrm{q}^\times) : \mathcal{B} \to \mathcal{C}$ *be an fp-pseudofunctor. Then* $\mathrm{gl}(\mathfrak{J})$ *is an fp-bicategory with both projection pseudofunctors* π_{dom} *and* π_{cod} *strictly preserving products.*

For a family of objects $(C_i, c_i, B_i)_{i=1,\dots,n}$, the n-ary product $\prod_{i=1}^n (C_i, c_i, B_i)$ is defined to be the tuple $\big(\prod_{i=1}^n C_i, \mathrm{q}^\times_{B_\bullet} \circ \prod_{i=1}^n c_i, \prod_{i=1}^n B_i\big)$. The kth projection $\underline{\pi_k}$ is (π_k, μ_k, π_k), where μ_k is defined by commutativity of the following diagram:

$$\begin{array}{ccc}
c_k \circ \pi_k & \xrightarrow{\quad\mu_k\quad} & \mathfrak{J}(\pi_k) \circ \big(\mathrm{q}^\times_{B_\bullet} \circ \prod_i c_i\big) \\
\big\downarrow{\scriptstyle (\varpi^{(k)})^{-1}} & & \big\uparrow{\scriptstyle \cong} \\
\pi_k \circ \prod_i c_i & & (\mathfrak{J}\pi_k \circ \mathrm{q}^\times_{B_\bullet}) \circ \prod_i c_i \\
\big\downarrow{\scriptstyle \cong} & & \big\uparrow{\scriptstyle (\varpi^{(k)} \circ \mathrm{q}^\times_{B_\bullet}) \circ \Pi_i c_i} \\
(\pi_k \circ \mathrm{Id}_{(\prod_i \mathfrak{J}B_i)}) \circ \prod_i c_i & & ((\pi_k \circ \langle \mathfrak{J}\pi_1, \dots, \mathfrak{J}\pi_n \rangle) \circ \mathrm{q}^\times_{B_\bullet}) \circ \prod_i c_i \\
& & \big\uparrow{\scriptstyle \cong} \\
& \xrightarrow{(\pi_k \circ \mathrm{u}^\times_{B_\bullet}) \circ \Pi_i c_i} & (\pi_k \circ (\langle \mathfrak{J}\pi_1, \dots, \mathfrak{J}\pi_n \rangle \circ \mathrm{q}^\times_{B_\bullet})) \circ \prod_i c_i
\end{array}$$

For an n-ary family of 1-cells $(g_i, \alpha_i, f_i) : (Y, y, X) \to (C_i, c_i, B_i)\,(i = 1, \dots, n)$, the n-ary tupling is $(\langle g_1, \dots, g_n \rangle, \{\alpha_1, \dots, \alpha_n\}, \langle f_1, \dots, f_n \rangle)$, where $\{\alpha_1, \dots, \alpha_n\}$

is the composite

$$\begin{array}{ccc}
(\mathrm{q}^{\times}_{B_\bullet} \circ \prod_i c_i) \circ \langle g_1, \dots, g_n \rangle & \xrightarrow{\{\alpha_1,\dots,\alpha_n\}} & \mathfrak{J}(\langle f_1, \dots, f_n \rangle) \circ y \\
\cong \big\downarrow & & \big\uparrow \cong \\
\mathrm{q}^{\times}_{B_\bullet} \circ (\prod_i c_i \circ \langle g_1, \dots, g_n \rangle) & & \mathrm{Id}_{\mathfrak{J}(\prod B_i)} \circ (\mathfrak{J}\langle f_1, \dots, f_n \rangle \circ y) \\
\mathrm{q}^{\times}_{B_\bullet} \circ \mathsf{fuse} \big\downarrow & & \big\uparrow (\mathsf{c}^{\times}_{B_\bullet} \circ \mathfrak{J}\langle f_1,\dots,f_n\rangle) \circ y \\
\mathrm{q}^{\times}_{B_\bullet} \circ \langle c_1 \circ g_1, \dots, c_n \circ g_n \rangle & & (\mathrm{q}^{\times}_{B_\bullet} \circ \langle \mathfrak{J}\pi_1, \dots, \mathfrak{J}\pi_n \rangle) \circ (\mathfrak{J}\langle f_1, \dots, f_n \rangle \circ y) \\
\mathrm{q}^{\times}_{B_\bullet} \circ \langle \alpha_1, \dots, \alpha_n \rangle \big\downarrow & & \big\uparrow \cong \\
\mathrm{q}^{\times}_{B_\bullet} \circ \langle \mathfrak{J}f_1 \circ y, \dots, \mathfrak{J}f_n \circ y \rangle & & \mathrm{q}^{\times}_{B_\bullet} \circ (((\langle \mathfrak{J}\pi_1, \dots, \mathfrak{J}\pi_n \rangle \circ \mathfrak{J}\langle f_1, \dots, f_n \rangle) \circ y) \\
 & & \big\uparrow \mathrm{q}^{\times}_{B_\bullet} \circ (\mathsf{unpack}^{-1}_{f_\bullet} \circ y) \\
 & \xrightarrow{\mathrm{q}^{\times}_{B_\bullet} \circ \mathsf{post}^{-1}} & \mathrm{q}^{\times}_{B_\bullet} \circ (\langle \mathfrak{J}f_1, \dots, \mathfrak{J}f_n \rangle \circ y)
\end{array}$$

Finally, for every family of 1-cells $(g_i, \alpha_i, f_i) : (Y, y, X) \to (C_i, c_i, B_i)$ $(i = 1, \dots, n)$ we require a glued 2-cell $\underline{\pi}_k \circ (\langle g_1, \dots, g_n \rangle, \{\alpha_1, \dots, \alpha_n\}, \langle f_1, \dots, f_n \rangle) \Rightarrow (g_k, \alpha_k, f_k)$ to act as the counit. We take simply $(\varpi^{(k)}_{g_\bullet}, \varpi^{(k)}_{f_\bullet})$. This pair forms a 2-cell in $\mathrm{gl}(\mathfrak{J})$, and the required universal property holds pointwise.

Remark 4. If $(\mathfrak{J}, \mathrm{q}^{\times}) : \mathcal{B} \to \mathcal{X}$ is an fp-pseudofunctor, then $\underline{\mathrm{Y}} : \mathcal{B} \to \mathrm{gl}(\langle \mathfrak{J} \rangle)$ canonically extends to an fp-pseudofunctor. The pseudoinverse to $\langle \underline{\mathrm{Y}}\pi_1, \dots, \underline{\mathrm{Y}}\pi_n \rangle$ is $(\langle -, \dots, = \rangle, \cong, \mathrm{q}^{\times})$, where the component of the isomorphism at $(f_i : X \to B_i)_{i=1,\dots,n}$ is $F\langle f_\bullet \rangle \overset{\cong}{\Rightarrow} \mathrm{Id}_{F(\prod_i B_i)} \circ F\langle f_\bullet \rangle \xRightarrow{(\mathsf{c}^{\times}_{B_\bullet})^{-1} \circ F\langle f_\bullet \rangle} \mathrm{q}^{\times}_{B_\bullet} \circ \langle F\pi_\bullet \rangle \circ F\langle f_\bullet \rangle \xRightarrow{\mathrm{q}^{\times}_{B_\bullet} \circ \mathsf{unpack}} \mathrm{q}^{\times}_{B_\bullet} \circ \langle F f_\bullet \rangle$.

4.2 Exponentials in $\mathrm{gl}(\mathfrak{J})$

As in the 1-categorical case, the definition of currying in $\mathrm{gl}(\mathfrak{J})$ employs pullbacks. A *pullback* of the cospan $(X_1 \to X_0 \leftarrow X_2)$ in a bicategory $\mathcal{B}$ is a bilimit for the strict pseudofunctor $X : (1 \to 0 \leftarrow 2) \to \mathcal{B}$ determined by the cospan. We state the universal property in the form that will be most useful for our applications.

Lemma 6. *The pullback of a cospan $(X_1 \xrightarrow{f_1} X_0 \xleftarrow{f_2} X_2)$ in a bicategory $\mathcal{B}$ is determined, up to equivalence, by the following data and properties: a span $(X_1 \xleftarrow{\gamma_1} P \xrightarrow{\gamma_2} X_2)$ in $\mathcal{B}$ and an invertible 2-cell filling the diagram on the left below*

$$\begin{array}{ccccc}
 & & P & & \\
 & \swarrow^{\gamma_1} & & \searrow^{\gamma_2} & \\
X_1 & & \overset{\overline{\gamma}}{\underset{\cong}{\Leftarrow}} & & X_2 \\
 & \searrow_{f_1} & & \swarrow_{f_2} & \\
 & & X_0 & &
\end{array}
\qquad\qquad
\begin{array}{ccccc}
 & & Q & & \\
 & \swarrow^{\mu_1} & & \searrow^{\mu_2} & \\
X_1 & & \overset{\overline{\mu}}{\underset{\cong}{\Leftarrow}} & & X_2 \\
 & \searrow_{f_1} & & \swarrow_{f_2} & \\
 & & X_0 & &
\end{array}$$

such that

1. *for any other diagram as on the right above there exists a* fill-in (u, Ξ_1, Ξ_2), *namely a 1-cell* $u : Q \to P$ *and invertible 2-cells* $\Xi_i : \gamma_i \circ u \Rightarrow \mu_i$ $(i = 1, 2)$ *satisfying*

$$\begin{array}{ccccc}
(f_2 \circ \gamma_2) \circ u & \xrightarrow{\cong} & f_2 \circ (\gamma_2 \circ u) & \xrightarrow{f_2 \circ \Xi_2} & f_2 \circ \mu_2 \\
{\scriptstyle \overline{\gamma} \circ u}\downarrow & & & & \downarrow{\scriptstyle \overline{\mu}} \\
(f_1 \circ \gamma_1) \circ u & \xrightarrow[\cong]{} & f_1 \circ (\gamma_1 \circ u) & \xrightarrow[f_1 \circ \Xi_1]{} & f_1 \circ \mu_1
\end{array}$$

2. *for any 1-cells* $v, w : Q \to P$ *and 2-cells* $\Psi_i : \gamma_i \circ v \Rightarrow \gamma_i \circ w$ $(i = 1, 2)$ *satisfying*

$$\begin{array}{ccccccc}
(f_2 \circ \gamma_2) \circ v & \xrightarrow{\cong} & f_2 \circ (\gamma_2 \circ v) & \xrightarrow{f_2 \circ \Psi_2} & f_2 \circ (\gamma_2 \circ w) & \xrightarrow{\cong} & (f_2 \circ \gamma_2) \circ w \\
{\scriptstyle \overline{\gamma} \circ v}\downarrow & & & & & & \downarrow{\scriptstyle \overline{\gamma} \circ w} \\
(f_1 \circ \gamma_1) \circ v & \xrightarrow[\cong]{} & f_1 \circ (\gamma_1 \circ v) & \xrightarrow[f_1 \circ \Psi_1]{} & f_1 \circ (\gamma_1 \circ w) & \xrightarrow{\cong} & (f_1 \circ \gamma_1) \circ w
\end{array}$$

there exists a unique 2-cell $\Psi : v \Rightarrow w$ *such that* $\Psi_i = \gamma_i \circ \Psi$ $(i = 1, 2)$.

Example 6. 1. In **Cat**, the pullback of a cospan $(\mathcal{B} \xrightarrow{F} \mathcal{X} \xleftarrow{G} \mathcal{C})$ is the full subcategory of the comma category $(F \downarrow G)$ consisting of objects of the form (B, f, C) for which $f : FB \to GC$ is an isomorphism. Note that this differs from the strict (2-)categorical pullback in **Cat**, in which every f is required to be an identity (*c.f.* [65, Example 2.1]).

2. Like any bilimit, pullbacks in the bicategory $\mathbf{Bicat}(\mathcal{B}^{\mathrm{op}}, \mathbf{Cat})$ are computed pointwise (see [53, Proposition 3.6]).

We now define exponentials in the glueing bicategory. Precisely, we extend Proposition 1 to the following.

Theorem 5. *Let* $(\mathcal{B}, \Pi_n(-), \Rightarrow)$ *and* $(\mathcal{C}, \Pi_n(-), \Rightarrow)$ *be cc-bicategories such that* $\mathcal{C}$ *has pullbacks. For any fp-pseudofunctor* $(\mathfrak{J}, \mathrm{q}^{\times}) : (\mathcal{B}, \Pi_n(-)) \to (\mathcal{C}, \Pi_n(-))$, *the glueing bicategory* $\mathrm{gl}(\mathfrak{J})$ *has a cartesian closed structure with forgetful pseudofunctor* $\pi_{\mathrm{dom}} : \mathrm{gl}(\mathfrak{J}) \to \mathcal{B}$ *strictly preserving products and exponentials.*

The evaluation map. We begin by defining the mapping $(-) \Rightarrow (=)$ and the evaluation 1-cell $\underline{\mathrm{eval}}$. For $\underline{C} := (C, c, B), \underline{C'} := (C', c', B') \in \mathrm{gl}(\mathfrak{J})$ we set $\underline{C} \Rightarrow \underline{C'}$ to be the left-hand vertical leg of the following pullback diagram, in which we write $m_{B,B'} := \lambda(\mathfrak{J}(\mathrm{eval}_{B,B'}) \circ \mathrm{q}^{\times}_{B \Rightarrow B', B})$.

$$\begin{array}{ccccc}
C \supset C' & & \xrightarrow{\quad q_{c,c'} \quad} & & (C \Rightarrow C') \\
{\scriptstyle p_{c,c'}}\downarrow & & \overset{\omega_{c,c'}}{\Leftarrow} & & \downarrow{\scriptstyle \lambda(c' \circ \mathrm{eval}_{C,C'})} \\
\mathfrak{J}(B \Rightarrow B') & \xrightarrow[m_{B,B'}]{} & (\mathfrak{J}B \Rightarrow \mathfrak{J}B') & \xrightarrow[\lambda(\mathrm{eval}_{\mathfrak{J}B,\mathfrak{J}B'} \circ ((\mathfrak{J}B \Rightarrow \mathfrak{J}B') \times c))]{} & (C \Rightarrow \mathfrak{J}B')
\end{array}$$

with the bottom composite $\lambda(\mathrm{eval}_{\mathfrak{J}B,\mathfrak{J}B'} \circ ((\mathfrak{J}B \Rightarrow \mathfrak{J}B') \times c)) \circ m_{B,B'}$ from $\mathfrak{J}(B \Rightarrow B')$ to $(C \Rightarrow \mathfrak{J}B')$.

(4)

Example 7. The pullback (4) generalises the well-known definition of a *logical relation of varying arity* [36]. Indeed, where $\mathfrak{J} := \langle \mathfrak{K} \rangle$ is the relative hom-pseudofunctor for an fp-pseudofunctor $(\mathfrak{K}, \mathrm{q}^{\times}) : \mathcal{B} \to \mathcal{X}$ between cc-bicategories, $A \in \mathcal{B}$ and $X, X' \in \mathcal{X}$, the functor $m_{X,X'}(A)$ takes a 1-cell $f : \mathfrak{K}A \to (X \Rightarrow X')$ in $\mathcal{X}$ to the pseudonatural transformation $\mathrm{Y}A \times \mathcal{X}(\mathfrak{K}(-), X) \Rightarrow \mathcal{X}(\mathfrak{K}(-), X')$ with components $\lambda B \,.\, \lambda(\rho : B \to A, u : \mathfrak{K}B \to X) \,.\, \mathrm{eval}_{X,X'} \circ \langle f \circ \mathfrak{K}(\rho), u \rangle$. Intuitively, therefore, the pullback enforces the usual closure condition defining a logical relation at exponential type, while also tracking the isomorphism witnessing that this condition holds (*c.f.* [36,3,15]).

Notation 6. For reasons of space—particularly in pasting diagrams—we will sometimes write $\widetilde{c} := \mathrm{eval}_{\mathfrak{J}B,\mathfrak{J}B'} \circ ((\mathfrak{J}B \Rightarrow \mathfrak{J}B') \times c) : (\mathfrak{J}B \Rightarrow \mathfrak{J}B') \times C \to \mathfrak{J}B'$ when $c : C \to \mathfrak{J}B$ in $\mathcal{C}$.

The evaluation map $\underline{\mathrm{eval}}_{\underline{C},\underline{C'}}$ is defined to be $(\mathrm{eval}_{C,C'} \circ (q_{c,c'} \times C), \mathrm{E}_{\underline{C},\underline{C'}}, \mathrm{eval}_{B,B'})$, where the witnessing 2-cell $\mathrm{E}_{\underline{C},\underline{C'}}$ is given by the pasting diagram below, in which the unlabelled arrow is $\mathrm{q}^{\times}_{(B \Rightarrow B', B)} \circ (p_{c,c'} \times c)$.

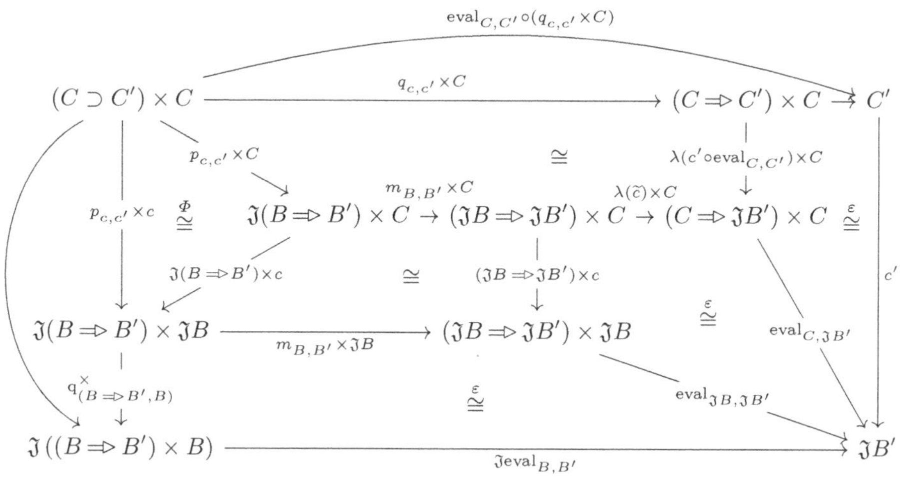

Here the bottom $\cong$ denotes a composite of Φ, structural isomorphisms and Φ^{-1}, and the top $\cong$ denotes a composite of $\omega_{c,c'} \times C$ with instances of Φ, Φ^{-1}, and the structural isomorphisms.

The currying operation. Let $\underline{R} := (R, r, Q)$, $\underline{C} := (C, c, B)$ and $\underline{C'} := (C', c', B')$ and suppose given a 1-cell $(t, \alpha, s) : \underline{R} \times \underline{C} \to \underline{C'}$. We construct $\underline{\lambda}(t, \alpha, s)$ using the universal property (4) of the pullback. To this end, we define invertible composites U_α and T_α as in the following two diagrams and set $\mathrm{L}_\alpha := \eta^{-1} \bullet \mathsf{e}^{\dagger}(\mathrm{U}_\alpha^{-1} \circ \alpha \circ \mathrm{T}_\alpha) : \lambda(c' \circ \mathrm{eval}_{C,C'}) \circ \lambda t \Rightarrow (\lambda(\widetilde{c}) \circ m_{B,B'}) \circ (\mathfrak{J}(\lambda s) \circ r)$.

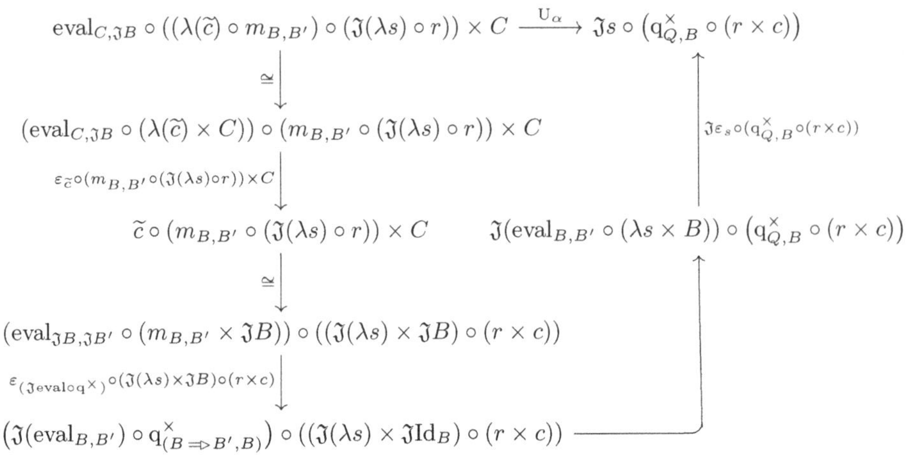

$\mathrm{eval}_{C,\mathfrak{J}B} \circ ((\lambda(\widetilde{c}) \circ m_{B,B'}) \circ (\mathfrak{J}(\lambda s) \circ r)) \times C$
U_α
$\mathfrak{J}s \circ (\mathrm{q}^{\times}_{Q,B} \circ (r \times c))$
$\cong$
$(\mathrm{eval}_{C,\mathfrak{J}B} \circ (\lambda(\widetilde{c}) \times C)) \circ (m_{B,B'} \circ (\mathfrak{J}(\lambda s) \circ r)) \times C$
$\mathfrak{J}\varepsilon_s \circ (\mathrm{q}^{\times}_{Q,B} \circ (r \times c))$
$\varepsilon_{\widetilde{c}} \circ (m_{B,B'} \circ (\mathfrak{J}(\lambda s) \circ r)) \times C$
$\widetilde{c} \circ (m_{B,B'} \circ (\mathfrak{J}(\lambda s) \circ r)) \times C$
$\mathfrak{J}(\mathrm{eval}_{B,B'} \circ (\lambda s \times B)) \circ (\mathrm{q}^{\times}_{Q,B} \circ (r \times c))$
$\cong$
$(\mathrm{eval}_{\mathfrak{J}B,\mathfrak{J}B'} \circ (m_{B,B'} \times \mathfrak{J}B)) \circ ((\mathfrak{J}(\lambda s) \times \mathfrak{J}B) \circ (r \times c))$
$\varepsilon_{(\mathfrak{J}\mathrm{eval} \circ \mathrm{q}^{\times})} \circ (\mathfrak{J}(\lambda s) \times \mathfrak{J}B) \circ (r \times c)$
$(\mathfrak{J}(\mathrm{eval}_{B,B'}) \circ \mathrm{q}^{\times}_{(B \Rightarrow B', B)}) \circ ((\mathfrak{J}(\lambda s) \times \mathfrak{J}\mathrm{Id}_B) \circ (r \times c))$

The unlabelled arrow is the canonical composite of $\mathsf{nat}_{\lambda s, \mathrm{id}_B}$ with $\phi^{\mathfrak{J}}_{\mathrm{eval}, \lambda(s) \times B}$ and structural isomorphisms. T_α is then defined using U_α:

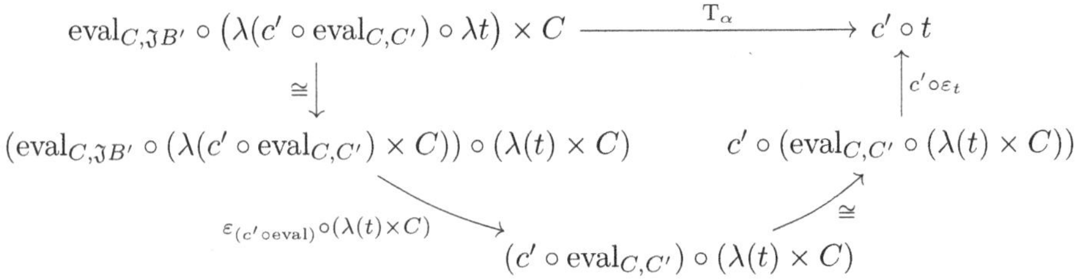

$\mathrm{eval}_{C,\mathfrak{J}B'} \circ (\lambda(c' \circ \mathrm{eval}_{C,C'}) \circ \lambda t) \times C$
T_α
$c' \circ t$
$\cong$
$c' \circ \varepsilon_t$
$(\mathrm{eval}_{C,\mathfrak{J}B'} \circ (\lambda(c' \circ \mathrm{eval}_{C,C'}) \times C)) \circ (\lambda(t) \times C)$
$c' \circ (\mathrm{eval}_{C,C'} \circ (\lambda(t) \times C))$
$\varepsilon_{(c' \circ \mathrm{eval})} \circ (\lambda(t) \times C)$
$\cong$
$(c' \circ \mathrm{eval}_{C,C'}) \circ (\lambda(t) \times C)$

Applying the universal property of the pullback (4) to L_α, one obtains a 1-cell $\underline{\mathsf{lam}}(t)$ and a pair of invertible 2-cells $\varGamma_{c,c'}$ and $\varDelta_{c,c'}$ filling the diagram

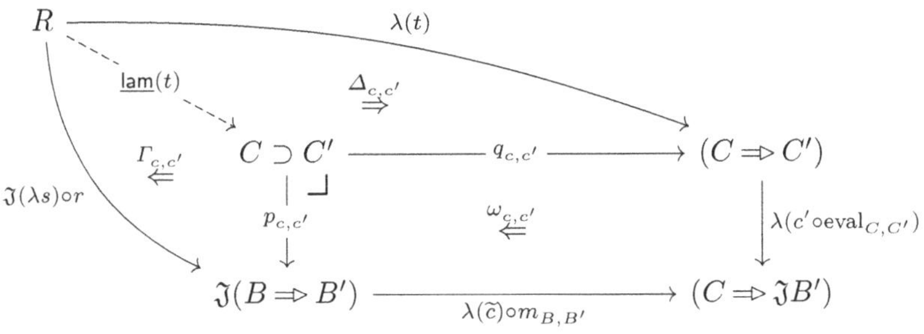

R
$\lambda(t)$
$\underline{\mathsf{lam}}(t)$
$\varDelta_{c,c'}$
$\varGamma_{c,c'}$
$C \supset C'$
$q_{c,c'}$
$(C \Rightarrow C')$
$\mathfrak{J}(\lambda s) \circ r$
$p_{c,c'}$
$\omega_{c,c'}$
$\lambda(c' \circ \mathrm{eval}_{C,C'})$
$\mathfrak{J}(B \Rightarrow B')$
$(C \Rightarrow \mathfrak{J}B')$
$\lambda(\widetilde{c}) \circ m_{B,B'}$

We define $\underline{\lambda}(t, \alpha, s) := (\underline{\mathsf{lam}}(t), \varGamma_{c,c'}, \lambda s)$.

The counit 2-cell. Finally we come to the counit. For a 1-cell $\underline{t} := (t, \alpha, s) : (R, r, Q) \times (C, c, B) \to (C', c', B')$ the 1-cell $\underline{\mathrm{eval}} \circ (\underline{\lambda}(t, \alpha, s) \times (C, c, B))$ unwinds to the pasting diagram below, in which the unlabelled arrow is $\mathrm{q}^{\times}_{Q,B} \circ (r \times c)$:

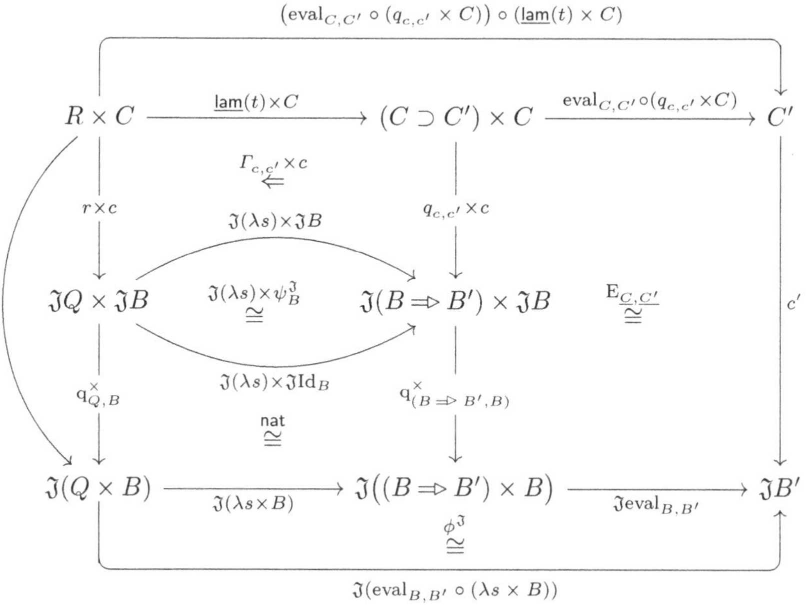

For the counit $\underline{\varepsilon_t}$ we take the 2-cell with first component $\underline{\mathrm{e}}_t$ defined by

$$\begin{array}{ccc}
(\mathrm{eval}_{C,C'} \circ (q_{c,c'} \times C)) \circ (\underline{\mathsf{lam}}(t) \times C) & \xrightarrow{\underline{\mathrm{e}}_t} & t \\
\cong \downarrow & & \uparrow \varepsilon_t \\
\mathrm{eval}_{C,C'} \circ ((q_{c,c'} \circ \underline{\mathsf{lam}}(t)) \times C) & \xrightarrow[\mathrm{eval}_{C,C'} \circ (\Delta_{c,c'} \times C)]{} & \mathrm{eval}_{C,C'} \circ (\lambda(t) \times C)
\end{array}$$

and second component simply $\varepsilon_s : \mathrm{eval}_{B,B'} \circ (\lambda(s) \times B) \Rightarrow s$. This pair forms an invertible 2-cell in $\mathrm{gl}(\mathfrak{J})$. One checks this satisfies the required universal property in a manner analogous to the 1-categorical case (see [55] for the full details). This completes the proof of Theorem 5.

5 Relative full completeness

We apply the theory developed in the preceding two sections to prove the relative full completeness result. As outlined in the introduction, this corresponds to a proof of *conservativity of the theory of rewriting* for the higher-order equational theory of rewriting in STLC over the algebraic equational theory of rewriting in STPC. We adapt 'Lafont's argument' [39, Annexe C] from the form presented in [16], for which we require bicategorical versions of the free cartesian category $\mathbb{F}^{\times}[\mathbb{C}]$ and free cartesian closed category $\mathbb{F}^{\times,\rightarrow}[\mathbb{C}]$ over a category $\mathbb{C}$. In line with the strategy for the STLC (*c.f.* [12, pp. 173–4]), we deal with the contravariance of the pseudofunctor $(- \Rightarrow =)$ by restricting to a bicategory of cc-pseudofunctors, pseudonatural equivalences (that is, pseudonatural transformations for which each component is a given equivalence), and invertible modifications. We denote this with the subscript $\simeq, \cong$.

Lemma 7. *For any bicategory $\mathcal{B}$, fp-bicategory $(\mathcal{C}, \Pi_n(-))$ and cc-bicategory $(\mathcal{D}, \Pi_n(-), \Rightarrow)$:*

1. *There exists an fp-bicategory $\mathcal{F}^{\times}[\mathcal{B}]$ and a pseudofunctor $\eta^{\times} : \mathcal{B} \to \mathcal{F}^{\times}[\mathcal{B}]$ such that composition with $\eta^{\times}$ induces a biequivalence*
$$\textbf{fp-Bicat}(\mathcal{F}^{\times}[\mathcal{B}], \mathcal{C}) \xrightarrow{\simeq} \textbf{Bicat}(\mathcal{B}, \mathcal{C})$$
2. *There exists a cc-bicategory $\mathcal{F}^{\times,\to}[\mathcal{B}]$ and a pseudofunctor $\eta^{\Rightarrow} : \mathcal{B} \to \mathcal{F}^{\times,\to}[\mathcal{B}]$ such that composition with $\eta^{\Rightarrow}$ induces a biequivalence*
$$\textbf{cc-Bicat}_{\simeq,\cong}(\mathcal{F}^{\times,\to}[\mathcal{B}], \mathcal{D}) \xrightarrow{\simeq} \textbf{Bicat}(\mathcal{B}, \mathcal{D})$$

Proof (sketch). A syntactic construction suffices: one defines formal products and exponentials and then quotients by the axioms (see [48, p. 79] or [55]).

Thus, for any bicategory $\mathcal{B}$, fp-bicategory $(\mathcal{C}, \Pi_n(-))$, and pseudofunctor $F : \mathcal{B} \to \mathcal{C}$ there exists an fp-pseudofunctor $F^{\#} : \mathcal{F}^{\times}[\mathcal{B}] \to \mathcal{C}$ and an equivalence $F^{\#} \circ \eta^{\times} \simeq F$. Moreover, for any fp-pseudofunctor $G : \mathcal{F}^{\times}[\mathcal{B}] \to \mathcal{C}$ such that $G \circ \eta^{\times} \simeq F$ one has $G \simeq F^{\#}$. A corresponding result holds for cc-bicategories and cc-pseudofunctors.

Theorem 7. *For any bicategory $\mathcal{B}$ the universal fp-pseudofunctor $\iota : \mathcal{F}^{\times}[\mathcal{B}] \to \mathcal{F}^{\times,\to}[\mathcal{B}]$ extending $\eta^{\Rightarrow}$ is locally an equivalence. Hence $\eta^{\Rightarrow} : \mathcal{B} \to \mathcal{F}^{\times,\to}[\mathcal{B}]$ is locally an equivalence.*

Proof. Since ι preserves finite products, the bicategory $\mathrm{gl}(\langle\iota\rangle)$ is cartesian closed (Theorem 5). The composite $\mathrm{K} := \underline{\mathrm{Y}} \circ \eta^{\times} : \mathcal{B} \to \mathrm{gl}(\langle\iota\rangle)$ therefore induces a cc-pseudofunctor $\mathrm{K}^{\#} : \mathcal{F}^{\times,\to}[\mathcal{B}] \to \mathrm{gl}(\langle\iota\rangle)$.

First observe that $(\mathrm{K}^{\#} \circ \iota) \circ \eta^{\times} \simeq \mathrm{K}^{\#} \circ \eta^{\Rightarrow} \simeq \mathrm{K} = \underline{\mathrm{Y}} \circ \eta^{\times}$. Since $\underline{\mathrm{Y}}$ is canonically an fp-pseudofunctor (Remark 4), it follows that $\mathrm{K}^{\#} \circ \iota \simeq \underline{\mathrm{Y}}$. Since $\underline{\mathrm{Y}}$ is locally an equivalence (Lemma 5), Lemma 1(1) entails that $\mathrm{K}^{\#} \circ \iota$ is locally an equivalence.

Next, examining the definition of $\underline{\mathrm{Y}}$ one sees that $\pi_{\mathrm{dom}} \circ \underline{\mathrm{Y}} = \iota$, and so
$$(\pi_{\mathrm{dom}} \circ \mathrm{K}^{\#}) \circ \eta^{\Rightarrow} \simeq (\pi_{\mathrm{dom}} \circ \underline{\mathrm{Y}}) \circ \eta^{\times} \simeq \iota \circ \eta^{\times} \simeq \eta^{\Rightarrow}$$
It follows that $\pi_{\mathrm{dom}} \circ \mathrm{K}^{\#} \simeq \mathrm{id}_{\mathcal{F}^{\times,\to}[\mathcal{B}]}$, and hence that $\pi_{\mathrm{dom}} \circ \mathrm{K}^{\#}$ is also locally an equivalence.

Now consider the composite $\mathcal{F}^{\times}[\mathcal{B}] \xrightarrow{\iota} \mathcal{F}^{\times,\to}[\mathcal{B}] \xrightarrow{\mathrm{K}^{\#}} \mathrm{gl}(\langle\iota\rangle) \xrightarrow{\pi_{\mathrm{dom}}} \mathcal{F}^{\times,\to}[\mathcal{B}]$. By Lemma 1(2) and the preceding, ι is locally an equivalence. Finally, it is direct from the construction of $\mathcal{F}^{\times}[\mathcal{B}]$ that $\eta^{\times}$ is locally an equivalence; thus, so are $\iota \circ \eta^{\times} \simeq \eta^{\Rightarrow}$.

Acknowledgements. We thank all the anonymous reviewers for their comments: these improved the paper substantially. We are especially grateful to the reviewer who pointed out an oversight in the original formulation of Lemma 1(2), which consequently affected the argument in Theorem 7, and provided the elegant fix therein.

The second author was supported by a Royal Society University Research Fellow Enhancement Award.

References

1. Abbott, M.G.: Categories of containers. Ph.D. thesis, University of Leicester (2003)
2. Abramsky, S., Jagadeesan, R.: Games and full completeness for multiplicative linear logic. Journal of Symbolic Logic **59**(2), 543–574 (1994). https://doi.org/10.2307/2275407
3. Alimohamed, M.: A characterization of lambda definability in categorical models of implicit polymorphism. Theoretical Computer Science **146**(1-2), 5–23 (1995). https://doi.org/10.1016/0304-3975(94)00283-O
4. Balat, V., Di Cosmo, R., Fiore, M.: Extensional normalisation and typed-directed partial evaluation for typed lambda calculus with sums. In: Proceedings of the 31st Annual ACM SIGPLAN-SIGACT Symposium on Principles of Programming Languages. pp. 64–76 (2004)
5. Bénabou, J.: Introduction to bicategories. In: Reports of the Midwest Category Seminar. pp. 1–77. Springer Berlin Heidelberg, Berlin, Heidelberg (1967)
6. Bloom, S.L., Ésik, Z., Labella, A., Manes, E.G.: Iteration 2-theories. Applied Categorical Structures **9**(2), 173–216 (2001). https://doi.org/10.1023/a:1008708924144
7. Borceux, F.: Bicategories and distributors, Encyclopedia of Mathematics and its Applications, vol. 1, pp. 281–324. Cambridge University Press (1994). https://doi.org/10.1017/CBO9780511525858.009
8. Carboni, A., Kelly, G.M., Walters, R.F.C., Wood, R.J.: Cartesian bicategories II. Theory and Applications of Categories **19**(6), 93–124 (2008), http://www.tac.mta.ca/tac/volumes/19/6/19-06abs.html
9. Carboni, A., Lack, S., Walters, R.F.C.: Introduction to extensive and distributive categories. Journal of Pure and Applied Algebra **84**(2), 145–158 (1993). https://doi.org/10.1016/0022-4049(93)90035-r
10. Carboni, A., Walters, R.F.C.: Cartesian bicategories I. Journal of Pure and Applied Algebra **49**(1), 11–32 (1987). https://doi.org/10.1016/0022-4049(87)90121-6
11. Castellan, S., Clairambault, P., Rideau, S., Winskel, G.: Games and strategies as event structures. Logical Methods in Computer Science **13** (2017)
12. Crole, R.L.: Categories for Types. Cambridge University Press (1994). https://doi.org/10.1017/CBO9781139172707
13. Dagand, P.E., McBride, C.: A categorical treatment of ornaments. In: Proceedings of the 28th Annual ACM/IEEE Symposium on Logic in Computer Science. pp. 530–539. IEEE Computer Society, Washington, DC, USA (2013). https://doi.org/10.1109/LICS.2013.60
14. Fiore, M.: Axiomatic Domain Theory in Categories of Partial Maps. Distinguished Dissertations in Computer Science, Cambridge University Press (1996)
15. Fiore, M.: Semantic analysis of normalisation by evaluation for typed lambda calculus. In: Proceedings of the 4th ACM SIGPLAN International Conference on Principles and Practice of Declarative Programming. pp. 26–37. ACM, New York, NY, USA (2002). https://doi.org/10.1145/571157.571161
16. Fiore, M., Di Cosmo, R., Balat, V.: Remarks on isomorphisms in typed lambda calculi with empty and sum types. In: Proceedings of the 28th Annual IEEE Symposium on Logic in Computer Science. pp. 147–156. IEEE Computer Society Press (2002). https://doi.org/10.1109/LICS.2002.1029824
17. Fiore, M., Gambino, N., Hyland, M., Winskel, G.: The cartesian closed bicategory of generalised species of structures. Journal of the London Mathematical Society **77**(1), 203–220 (2007). https://doi.org/10.1112/jlms/jdm096

18. Fiore, M., Gambino, N., Hyland, M., Winskel, G.: Relative pseudomonads, Kleisli bicategories, and substitution monoidal structures. Selecta Mathematica New Series (2017)
19. Fiore, M., Joyal, A.: Theory of para-toposes. Talk at the Category Theory 2015 Conference. Departamento de Matematica, Universidade de Aveiro (Portugal)
20. Fiore, M., Saville, P.: A type theory for cartesian closed bicategories. In: Proceedings of the 34th Annual ACM/IEEE Symposium on Logic in Computer Science (2019). https://doi.org/10.1109/LICS.2019.8785708
21. Fiore, M., Saville, P.: Coherence and normalisation-by-evaluation for bicategorical cartesian closed structure. Preprint (2020)
22. Fiore, M., Simpson, A.: Lambda definability with sums via Grothendieck logical relations. In: Girard, J.Y. (ed.) Typed lambda calculi and applications: 4th international conference. pp. 147–161. Springer Berlin Heidelberg, Berlin, Heidelberg (1999)
23. Freyd, P.: Algebraically complete categories. In: Lecture Notes in Mathematics, pp. 95–104. Springer Berlin Heidelberg (1991). https://doi.org/10.1007/bfb0084215
24. Freyd, P.J., Scedrov, A.: Categories, Allegories. Elsevier North Holland (1990)
25. Gambino, N., Joyal, A.: On operads, bimodules and analytic functors. Memoirs of the American Mathematical Society **249**(1184), 153–192 (2017)
26. Gambino, N., Kock, J.: Polynomial functors and polynomial monads. Mathematical Proceedings of the Cambridge Philosophical Society **154**(1), 153–192 (2013). https://doi.org/10.1017/S0305004112000394
27. Ghani, N.: Adjoint rewriting. Ph.D. thesis, University of Edinburgh (1995)
28. Gibbons, J.: Conditionals in distributive categories. Tech. rep., University of Oxford (1997)
29. G.L. Cattani, Fiore, M., Winskel, G.: A theory of recursive domains with applications to concurrency. In: Proceedings of the 13th Annual IEEE Symposium on Logic in Computer Science. pp. 214–225. IEEE Computer Society (1998)
30. Gurski, N.: An Algebraic Theory of Tricategories. University of Chicago, Department of Mathematics (2006)
31. Hasegawa, M.: Logical predicates for intuitionistic linear type theories. In: Girard, J.Y. (ed.) Typed lambda calculi and applications: 4th international conference. pp. 198–213. Springer Berlin Heidelberg, Berlin, Heidelberg (1999)
32. Hilken, B.: Towards a proof theory of rewriting: the simply typed 2λ-calculus. Theoretical Computer Science **170**(1), 407–444 (1996). https://doi.org/10.1016/S0304-3975(96)80713-4
33. Hirschowitz, T.: Cartesian closed 2-categories and permutation equivalence in higher-order rewriting. Logical Methods in Computer Science **9**, 1–22 (2013)
34. Jay, C.B., Ghani, N.: The virtues of eta-expansion. Journal of Functional Programming **5**(2), 135–154 (1995). https://doi.org/10.1017/S0956796800001301
35. Johann, P., Polonsky, P.: Higher-kinded data types: Syntax and semantics. In: 34th Annual ACM/IEEE Symposium on Logic in Computer Science. IEEE (2019). https://doi.org/10.1109/lics.2019.8785657
36. Jung, A., Tiuryn, J.: A new characterization of lambda definability. In: Bezem, M., Groote, J.F. (eds.) Typed Lambda Calculi and Applications. pp. 245–257. Springer Berlin Heidelberg, Berlin, Heidelberg (1993)
37. Lack, S.: A 2-Categories Companion, pp. 105–191. Springer New York, New York, NY (2010)
38. Lack, S., Walters, R.F.C., Wood, R.J.: Bicategories of spans as cartesian bicategories. Theory and Applications of Categories **24**(1), 1–24 (2010)

39. Lafont, Y.: Logiques, catégories et machines. Ph.D. thesis, Université Paris VII (1987)
40. Lambek, J., Scott, P.J.: Introduction to Higher Order Categorical Logic. Cambridge University Press, New York, NY, USA (1986)
41. Leinster, T.: Basic bicategories (May 1998), https://arxiv.org/pdf/math/9810017.pdf
42. Leinster, T.: Higher operads, higher categories. No. 298 in London Mathematical Society Lecture Note Series, Cambridge University Press (2004)
43. Ma, Q.M., Reynolds, J.C.: Types, abstraction, and parametric polymorphism, part 2. In: Brookes, S., Main, M., Melton, A., Mislove, M., Schmidt, D. (eds.) Mathematical Foundations of Programming Semantics. pp. 1–40. Springer Berlin Heidelberg, Berlin, Heidelberg (1992)
44. Mac Lane, S.: Categories for the Working Mathematician, Graduate Texts in Mathematics, vol. 5. Springer-Verlag New York, second edn. (1998). https://doi.org/10.1007/978-1-4757-4721-8
45. Mac Lane, S., Paré, R.: Coherence for bicategories and indexed categories. Journal of Pure and Applied Algebra **37**, 59–80 (1985). https://doi.org/10.1016/0022-4049(85)90087-8
46. Marmolejo, F., Wood, R.J.: Kan extensions and lax idempotent pseudomonads. Theory and Applications of Categories **26**(1), 1–29 (2012)
47. Mitchell, J.C., Scedrov, A.: Notes on sconing and relators. In: Börger, E., J., G., Kleine Büning, H., Martini, S., Richter, M.M. (eds.) Computer Science Logic. pp. 352–378. Springer Berlin Heidelberg, Berlin, Heidelberg (1993)
48. Ouaknine, J.: A two-dimensional extension of Lambek's categorical proof theory. Master's thesis, McGill University (1997)
49. Paquet, H.: Probabilistic concurrent game semantics. Ph.D. thesis, University of Cambridge (2020)
50. Plotkin, G.D.: Lambda-definability and logical relations. Tech. rep., University of Edinburgh School of Artificial Intelligence (1973), memorandum SAI-RM-4
51. Power, A.J.: An abstract formulation for rewrite systems. In: Pitt, D.H., Rydeheard, D.E., Dybjer, P., Pitts, A.M., Poigné, A. (eds.) Category Theory and Computer Science. pp. 300–312. Springer Berlin Heidelberg, Berlin, Heidelberg (1989)
52. Power, A.J.: Coherence for bicategories with finite bilimits I. In: Gray, J.W., Scedrov, A. (eds.) Categories in Computer Science and Logic: Proceedings of the AMS-IMS-SIAM Joint Summer Research Conference, vol. 92, pp. 341–349. AMS (1989)
53. Power, A.J.: A general coherence result. Journal of Pure and Applied Algebra **57**(2), 165–173 (1989). https://doi.org/https://doi.org/10.1016/0022-4049(89)90113-8
54. Rydeheard, D.E., Stell, J.G.: Foundations of equational deduction: A categorical treatment of equational proofs and unification algorithms. In: Pitt, D.H., Poigné, A., Rydeheard, D.E. (eds.) Category Theory and Computer Science. pp. 114–139. Springer Berlin Heidelberg, Berlin, Heidelberg (1987)
55. Saville, P.: Cartesian closed bicategories: type theory and coherence. Ph.D. thesis, University of Cambridge (Submitted)
56. Seely, R.A.G.: Modelling computations: A 2-categorical framework. In: Gries, D. (ed.) Proceedings of the 2nd Annual IEEE Symposium on Logic in Computer Science. pp. 65–71. IEEE Computer Society Press (June 1987)
57. Statman, R.: Logical relations and the typed λ-calculus. Information and Control **65**, 85–97 (1985)
58. Stell, J.: Modelling term rewriting systems by sesqui-categories. In: Proc. Catégories, Algèbres, Esquisses et Néo-Esquisses (1994)

59. Street, R.: Fibrations in bicategories. Cahiers de Topologie et Géométrie Différentielle Catégoriques **21**(2), 111–160 (1980), https://eudml.org/doc/91227
60. Street, R.: Categorical structures. In: Hazewinkel, M. (ed.) Handbook of Algebra, vol. 1, chap. 15, pp. 529–577. Elsevier (1995)
61. Tabareau, N.: Aspect oriented programming: A language for 2-categories. In: Proceedings of the 10th International Workshop on Foundations of Aspect-oriented Languages. pp. 13–17. ACM, New York, NY, USA (2011). https://doi.org/10.1145/1960510.1960514
62. Taylor, P.: Practical Foundations of Mathematics, Cambridge Studies in Advanced Mathematics, vol. 59. Cambridge University Press (1999)
63. Troelstra, A.S., Schwichtenberg, H.: Basic proof theory. No. 43 in Cambridge Tracts in Theoretical Computer Science, Cambridge University Press, second edn. (2000)
64. Verity, D.: Enriched categories, internal categories and change of base. Ph.D. thesis, University of Cambridge (1992), TAC reprint available at http://www.tac.mta.ca/tac/reprints/articles/20/tr20abs.html
65. Weber, M.: Yoneda structures from 2-toposes. Applied Categorical Structures **15**(3), 259–323 (2007). https://doi.org/10.1007/s10485-007-9079-2

A duality theoretic view on limits of finite structures*

Mai Gehrke[1], Tomáš Jakl[1], and Luca Reggio[2](✉)

[1] CNRS and Université Côte d'Azur, Nice, France
{mgehrke,tomas.jakl}@unice.fr
[2] Institute of Computer Science of the Czech Academy of Sciences, Prague, Czech Republic and Mathematical Institute, University of Bern, Switzerland
luca.reggio@math.unibe.ch

Abstract. A systematic theory of *structural limits* for finite models has been developed by Nešetřil and Ossona de Mendez. It is based on the insight that the collection of finite structures can be embedded, via a map they call the *Stone pairing*, in a space of measures, where the desired limits can be computed. We show that a closely related but finer grained space of measures arises — via Stone-Priestley duality and the notion of types from model theory — by enriching the expressive power of first-order logic with certain "probabilistic operators". We provide a sound and complete calculus for this extended logic and expose the functorial nature of this construction.

The consequences are two-fold. On the one hand, we identify the logical gist of the theory of structural limits. On the other hand, our construction shows that the duality-theoretic variant of the Stone pairing captures the adding of a layer of quantifiers, thus making a strong link to recent work on semiring quantifiers in logic on words. In the process, we identify the model theoretic notion of *types* as the unifying concept behind this link. These results contribute to bridging the strands of logic in computer science which focus on semantics and on more algorithmic and complexity related areas, respectively.

Keywords: Stone duality · finitely additive measures · structural limits · finite model theory · formal languages · logic on words

1 Introduction

While topology plays an important role, via Stone duality, in many parts of semantics, topological methods in more algorithmic and complexity oriented areas of theoretical computer science are not so common. One of the few examples,

* This project has been supported by the European Research Council (ERC) under the European Union's Horizon 2020 research and innovation program (grant agreement No.670624). Luca Reggio has received an individual support under the grants GA17-04630S of the Czech Science Foundation, and No.184693 of the Swiss National Science Foundation.

J. Goubault-Larrecq and B. König (Eds.): FOSSACS 2020, LNCS 12077, pp. 299–318, 2020.
https://doi.org/10.1007/978-3-030-45231-5_16

the one we want to consider here, is the study of limits of finite relational structures. We will focus on the *structural limits* introduced by Nešetřil and Ossona de Mendez [15,17]. These provide a common generalisation of various notions of limits of finite structures studied in probability theory, random graphs, structural graph theory, and finite model theory. The basic construction in this work is the so-called *Stone pairing*. Given a relational signature σ and a first-order formula φ in the signature σ with free variables $v_1, \ldots, v_n$, define

$$\langle \varphi, A \rangle = \frac{|\{\overline{a} \in A^n \mid A \models \varphi(\overline{a})\}|}{|A|^n} \qquad \textit{(the probability that a random assignment in } A \textit{ satisfies } \varphi\textit{).} \tag{1}$$

Nešetřil and Ossona de Mendez view the map $A \mapsto \langle \text{-}, A \rangle$ as an embedding of the finite σ-structures into the space of probability measures over the Stone space dual to the Lindenbaum-Tarski algebra of all first-order formulas in the signature σ. This space is complete and thus provides the desired limit objects for all sequences of finite structures which embed as Cauchy sequences.

Another example of topological methods in an algorithmically oriented area of computer science is the use of profinite monoids in automata theory. In this setting, profinite monoids are the subject of the extensive theory, based on theorems by Eilenberg and Reiterman, and used, among others, to settle decidability questions [18]. In [4], it was shown that this theory may be understood as an application of Stone duality, thus making a bridge between semantics and more algorithmically oriented work. Bridging this semantics-versus-algorithmics gap in theoretical computer science has since gained quite some momentum, notably with the recent strand of research by Abramsky, Dawar and co-workers [2,3]. In this spirit, a natural question is whether the structural limits of Nešetřil and Ossona de Mendez also can be understood semantically, and in particular whether the topological component may be seen as an application of Stone duality.

More precisely, recent work on understanding quantifiers in the setting of logic on finite words [5] has shown that adding a layer of certain quantifiers (such as classical and modular quantifiers) corresponds dually to measure space constructions. The measures involved are not classical but only finitely additive and they take values in finite semirings rather than in the unit interval. Nevertheless, this appearance of *measures as duals of quantifiers* begs the further question whether the measure spaces in the theory of structural limits may be obtained via Stone duality from a semantic addition of certain quantifiers to classical first-order logic.

The purpose of this paper is to address this question. Our main result is that the Stone pairing of Nešetřil and Ossona de Mendez is related by a retraction to a Stone space of measures, which is dual to the Lindenbaum-Tarski algebra of a logic fragment obtained from first-order logic by adding one layer of probabilistic quantifiers, and which arises in exactly the same way as the spaces of semiring-valued measures in logic on words. That is, the Stone pairing, although originating from other considerations, may be seen as arising by duality from a semantic construction.

A foreseeable hurdle is that spaces of classical measures are valued in the unit interval $[0, 1]$ which is not zero-dimensional and hence outside the scope of Stone duality. This is well-known to cause problems e.g. in attempts to combine non-determinism and probability in domain theory [12]. However, in the structural limits of Nešetřil and Ossona de Mendez, at the base, one only needs to talk about finite models equipped with normal distributions and thus only the finite intervals $I_n = \{0, \frac{1}{n}, \frac{2}{n}, \ldots, 1\}$ are involved. A careful duality-theoretic analysis identifies a codirected diagram (i.e. an inverse limit system) based on these intervals compatible with the Stone pairing. The resulting inverse limit, which we denote $\mathbf{\Gamma}$, is a Priestley space. It comes equipped with an algebra-like structure, which allows us to reformulate many aspects of the theory of structural limits in terms of $\mathbf{\Gamma}$-valued measures as opposed to $[0, 1]$-valued measures.

The analysis justifying the structure of $\mathbf{\Gamma}$ is based on duality theory for double quasi-operator algebras [7,8]. In the presentation, we have tried to compromise between giving interesting topo-relational insights into why $\mathbf{\Gamma}$ is as it is, and not overburdening the reader with technical details. Some interesting features of $\mathbf{\Gamma}$, dictated by the nature of the Stone pairing and the ensuing codirected diagram, are that

- $\mathbf{\Gamma}$ is based on a version of $[0, 1]$ in which the rationals are doubled;
- $\mathbf{\Gamma}$ comes with section-retraction maps $[0, 1] \overset{\iota}{\hookrightarrow} \mathbf{\Gamma} \overset{\gamma}{\twoheadrightarrow} [0, 1]$;
- the map ι is lower semicontinuous while the map γ is continuous.

These features are a consequence of general theory and precisely allow us to witness continuous phenomena relative to $[0, 1]$ in the setting of $\mathbf{\Gamma}$.

Our contribution

We show that the ambient measure space for the structural limits of Nešetřil and Ossona de Mendez can be obtained via *"adding a layer of quantifiers"* in a suitable enrichment of first-order logic. The conceptual framework for seeing this is that of *types* from classical model theory. More precisely, we will see that a variant of the Stone pairing is a map into a space of measures with values in a Priestley space $\mathbf{\Gamma}$. Further, we show that this map is in fact the embedding of the finite structures into the space of (0-)types of an extension of first-order logic, which we axiomatise. On the other hand, $\mathbf{\Gamma}$-valued measures and $[0, 1]$-valued measures are tightly related by a retraction-section pair which allows the transfer of properties. These results identify the logical gist of the theory of structural limits and provide a new interesting connection between logic on words and the theory of structural limits in finite model theory.

Outline of the paper. In section 2 we briefly recall Stone-Priestley duality, its application in logic via spaces of types, and the particular instance of logic on words (needed only to show the similarity of the constructions). In Section 3 we introduce the Priestley space $\mathbf{\Gamma}$ with its additional operations, and show that it admits $[0, 1]$ as a retract. The spaces of $\mathbf{\Gamma}$-valued measures are introduced in

Section 4, and the retraction of $\mathbf{\Gamma}$ onto $[0,1]$ is lifted to the appropriate spaces of measures. In Section 5 we introduce the $\mathbf{\Gamma}$-valued Stone pairing and make the link with logic on words. Further, we compare convergence in the space of $\mathbf{\Gamma}$-valued measures with the one considered by Nešetřil and Ossona de Mendez. Finally, in Section 6 we show that constructing the space of $\mathbf{\Gamma}$-valued measures dually corresponds to enriching the logic with probabilistic operators.

2 Preliminaries

Notation. Throughout this paper, if $X \xrightarrow{f} Y \xrightarrow{g} Z$ are functions, their composition is denoted $g \cdot f$. For a subset $S \subseteq X$, $f_{\restriction S} \colon S \to Y$ is the obvious restriction. Given any set T, $\wp(T)$ denotes its power-set. Further, for a poset P, P^{∂} is the poset obtained by turning the order of P upside down.

2.1 Stone-Priestley duality

In this paper, we will need Stone duality for bounded distributive lattices in the order topological form due to Priestley [19]. It is a powerful and well established tool in the study of propositional logic and semantics of programming languages, see e.g. [9,1] for major landmarks. We briefly recall how this duality works.

A *compact ordered space* is a pair $(X, \leq)$ where X is a compact space and $\leq$ is a partial order on X which is closed in the product topology of $X \times X$. (Note that such a space is automatically Hausdorff). A compact ordered space is a *Priestley space* provided it is *totally order-disconnected.* That is, for all $x, y \in X$ such that $x \nleq y$, there is a *clopen* (i.e. simultaneously closed and open) $C \subseteq X$ which is an up-set for $\leq$, and satisfies $x \in C$ but $y \notin C$. We recall the construction of the Priestley space of a distributive lattice D.[3]

A non-empty proper subset $F \subset D$ is a *prime filter* if it is *(i)* upward closed (in the natural order of D), *(ii)* closed under finite meets, and *(iii)* if $a \vee b \in F$, either $a \in F$ or $b \in F$. Denote by X_D the set of all prime filters of D. By Stone's Prime Filter Theorem, the map

$$[\![\text{-}]\!] \colon D \to \wp(X_D), \quad a \mapsto [\![a]\!] = \{F \in X_D \mid a \in F\}$$

is an embedding. Priestley's insight was that D can be recovered from X_D, if the latter is equipped with the inclusion order and the topology generated by the sets of the form $[\![a]\!]$ and their complements. This makes X_D into a Priestley space — the *dual space* of D — and the map $[\![\text{-}]\!]$ is an isomorphism between D and the lattice of clopen up-sets of X_D. Conversely, any Priestley space X is the dual space of the lattice of its clopen up-sets. We call the latter the *dual lattice* of X. This correspondence extends to morphisms. In fact, Priestley duality states that the category of distributive lattices with homomorphisms is dually equivalent to the category of Priestley spaces and continuous monotone maps.

[3] We assume all distributive lattices are bounded, with the bottom and top denoted by 0 and 1, respectively. The bounds need to be preserved by homomorphisms.

When restricting to Boolean algebras, we recover the celebrated Stone duality restricted to Boolean algebras and *Boolean spaces*, i.e. compact Hausdorff spaces in which the clopen subsets form a basis.

2.2 Stone duality and logic: type spaces

The *theory of types* is an important tool for first-order logic. We briefly recall the concept as it is closely related to, and provides the link between, two otherwise unrelated occurrences of topological methods in theoretical computer science.

Consider a signature σ and a first-order theory T in this signature. For each $n \in \mathbb{N}$, let Fm_n denote the set of first-order formulas whose free variables are among $\overline{v} = \{v_1, \ldots, v_n\}$, and let $\mathrm{Mod}_n(T)$ denote the class of all pairs (A, α) where A is a model of T and α is an interpretation of $\overline{v}$ in A. Then the satisfaction relation, $(A, \alpha) \models \varphi$, is a binary relation from Mod_n to Fm_n. It induces the equivalence relations of elementary equivalence $\equiv$ and logical equivalence $\approx$ on these sets, respectively. The quotient $\mathrm{FO}_n(T) = \mathrm{Fm}_n/{\approx}$ carries a natural Boolean algebra structure and is known as the *n-th Lindenbaum-Tarski algebra* of T. Its dual space is $\mathrm{Typ}_n(T)$, the *space of n-types* of T, whose points can be identified with elements of $\mathrm{Mod}_n(T)/{\equiv}$. The Boolean algebra $\mathrm{FO}(T)$ of *all* first-order formulas modulo logical equivalence over T is the directed colimit of the $\mathrm{FO}_n(T)$ for $n \in \mathbb{N}$ while its dual space, $\mathrm{Typ}(T)$, is the codirected limit of the $\mathrm{Typ}_n(T)$ for $n \in \mathbb{N}$ and consists of models equipped with interpretations of the full set of variables.

If we want to study finite models, there are two equivalent approaches: e.g. at the level of sentences, we can either consider the theory T_{fin} of finite T-models, or the closure of the collection of all finite T-models in the space $\mathrm{Typ}_0(T)$. This closure yields a space, which should tell us about finite T-structures. Indeed, it is equal to $\mathrm{Typ}_0(T_{\mathit{fin}})$, the space of pseudofinite T-structures. For an application of this, see [10]. Below, we will see an application in finite model theory of the case $T = \emptyset$ (in this case we write $\mathrm{FO}(\sigma)$ and $\mathrm{Typ}(\sigma)$ instead of $\mathrm{FO}(\emptyset)$ and $\mathrm{Typ}(\emptyset)$).

In light of the theory of types as exposed above, the Stone pairing of Nešetřil and Ossona de Mendez (see equation (1)) can be regarded as an embedding of finite structures into the space of probability measures on $\mathrm{Typ}(\sigma)$, which set-theoretically are finitely additive functions $\mathrm{FO}(\sigma) \to [0, 1]$.

2.3 Duality and logic on words

As mentioned in the introduction, spaces of measures arise via duality in *logic on words* [5]. Logic on words, as introduced by Büchi, see e.g. [14] for a recent survey, is a variation and specialisation of finite model theory where only models based on words are considered. I.e., a word $w \in A^*$ is seen as a relational structure on $\{1, \ldots, |w|\}$, where $|w|$ is the length of w, equipped with a unary relation P_a, for each $a \in A$, singling out the positions in the word where the letter a appears. Each sentence φ in a language interpretable over these structures yields a language $L_\varphi \subseteq A^*$ consisting of the words satisfying φ. Thus, logic fragments are considered modulo the theory of finite words and the Lindenbaum-Tarski algebras are subalgebras of $\wp(A^*)$ consisting of the appropriate L_φ's, cf. [10] for a treatment of first-order logic on words.

For lack of logical completeness, the duals of the Lindenbaum-Tarski algebras have more points than those given by models. Nevertheless, the dual spaces of types, which act as compactifications and completions of the collections of models, provide a powerful tool for studying logic fragments by topological means. The central notion is that of *recognition*, in which, a Boolean subalgebra $\mathcal{B} \subseteq \wp(A^*)$ is studied by means of the dual map $\eta\colon \beta(A^*) \to X_{\mathcal{B}}$. Here $\beta(A^*)$ is the Stone dual of $\wp(A^*)$, also known in topology as the Čech-Stone compactification of the discrete space A^*, and $X_{\mathcal{B}}$ is the Stone dual of $\mathcal{B}$. The set A^* embeds in $\beta(A^*)$, and η is uniquely determined by its restriction $\eta_0\colon A^* \to X_{\mathcal{B}}$. Now, Stone duality implies that $L \subseteq A^*$ is in $\mathcal{B}$ iff there is a clopen subset $V \subseteq X_{\mathcal{B}}$ so that $\eta_0^{-1}(V) = L$. Anytime the latter is true for a map η and a language L as above, one says that η *recognises* L.[4]

When studying logic fragments via recognition, the following inductive step is central: given a notion of quantifier and a recogniser for a Boolean algebra of formulas with a free variable, construct a recogniser for the Boolean algebra generated by the formulas obtained by applying the quantifier. This problem was solved in [5], using duality theory, in a general setting of *semiring quantifiers*. The latter are defined as follows: let $(S, +, \cdot, 0_S, 1_S)$ be a semiring, and $k \in S$. Given a formula $\psi(v)$, the formula $\exists_{S,k} v.\psi(v)$ is true of a word $w \in A^*$ iff $k = 1_S + \cdots + 1_S$, m times, where m is the number of assignments of the variable v in w satisfying $\psi(v)$. If $S = \mathbb{Z}/q\mathbb{Z}$, we obtain the so-called *modular quantifiers*, and for S the two-element lattice we recover the existential quantifier $\exists$.

To deal with formulas with a free variable, one considers maps of the form $f\colon \beta((A \times 2)^*) \to X$ (the extra bit in $A \times 2$ is used to mark the interpretation of the free variable). In [5] (see also [6]), it was shown that $L_{\psi(v)}$ is recognised by f iff for every $k \in S$ the language $L_{\exists_{S,k} v.\psi(v)}$ is recognised by the composite

$$\xi\colon A^* \xrightarrow{\ R\ } \widehat{\mathbf{S}}(\beta((A \times 2)^*)) \xrightarrow{\widehat{\mathbf{S}}(f)} \widehat{\mathbf{S}}(X), \tag{2}$$

where $\widehat{\mathbf{S}}(X)$ is the space of finitely additive S-valued measures on X, and R maps $w \in A^*$ to the measure $\mu_w\colon \wp((A \times 2)^*) \to S$ sending $K \subseteq (A \times 2)^*$ to the sum $1_S + \cdots + 1_S$, $n_{w,K}$ times. Here, $n_{w,K}$ is the number of interpretations α of the free variable v in w such that the pair (w, α), seen as an element of $(A \times 2)^*$, belongs to K. Finally, $\widehat{\mathbf{S}}(f)$ sends a measure to its pushforward along f.

3 The space Γ

Central to our results is a Priestley space $\mathbf{\Gamma}$ closely related to $[0, 1]$, in which our measures will take values. Its construction comes from the insight that the range

[4] Here, being beyond the scope of this paper, we are ignoring the important role of the monoid structure available on the spaces (in the form of profinite monoids or BiMs, cf. [10,5]).

of the Stone pairing $\langle \text{-}, A\rangle$, for a finite structure A and formulas restricted to a fixed number of free variables, can be confined to a chain $I_n = \{0, \frac{1}{n}, \frac{2}{n}, \ldots, 1\}$. Moreover, the floor functions $f_{mn,n}\colon I_{mn} \twoheadrightarrow I_n$ are monotone surjections. The ensuing system $\{f_{mn,n}\colon I_{mn} \twoheadrightarrow I_n \mid m, n \in \mathbb{N}\}$ can thus be seen as a codirected diagram of finite discrete posets and monotone maps. Let us define $\mathbf{\Gamma}$ to be the limit of this diagram. Then, $\mathbf{\Gamma}$ is naturally equipped with a structure of Priestley space, see e.g. [11, Corollary VI.3.3], and can be represented as based on the set

$$\{r^- \mid r \in (0,1]\} \cup \{q^\circ \mid q \in \mathbb{Q} \cap [0,1]\}.$$

The order of $\mathbf{\Gamma}$ is the unique total order which has 0° as bottom element, satisfies $r^* < s^*$ if and only if $r < s$ for $* \in \{-, \circ\}$, and such that q° is a cover of q^- for every rational $q \in (0,1]$ (i.e. $q^- < q^\circ$, and there is no element strictly in between). In a sense, the values q^- represent approximations of the values of the form q°. Cf. Figure 1. The topology of $\mathbf{\Gamma}$ is generated by the sets of the form

$$\uparrow p^\circ = \{x \in \mathbf{\Gamma} \mid p^\circ \leq x\} \quad \text{and} \quad \downarrow q^- = \{x \in \mathbf{\Gamma} \mid x \leq q^-\}$$

for $p, q \in \mathbb{Q} \cap [0,1]$ such that $q \neq 0$. The distributive lattice dual to $\mathbf{\Gamma}$, denoted by $\mathbf{L}$, is given by

$$\mathbf{L} = \{\bot\} \cup (\mathbb{Q} \cap [0,1])^\partial, \text{ with } \bot <_{\mathbf{L}} q \text{ and } q \leq_{\mathbf{L}} p \text{ for every } p \leq q \text{ in } \mathbb{Q} \cap [0,1].$$

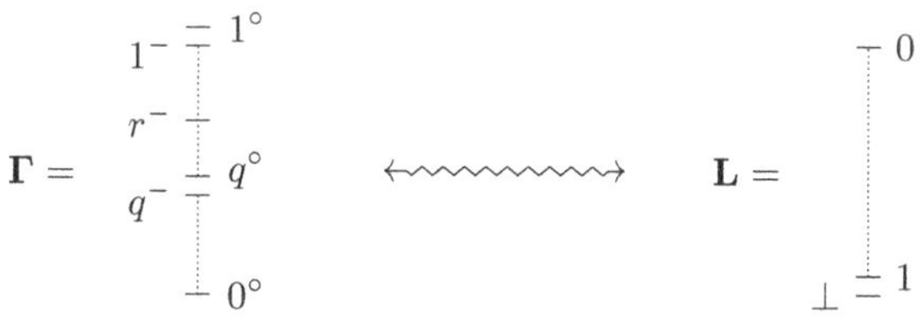

Fig. 1. The Priestley space $\mathbf{\Gamma}$ and its dual lattice $\mathbf{L}$

3.1 The algebraic structure on $\mathbf{\Gamma}$

When defining measures we need an algebraic structure available on the space of values. The space $\mathbf{\Gamma}$ fulfils this requirement as it comes equipped with a partial operation $-\colon \mathrm{dom}(-) \to \mathbf{\Gamma}$, where $\mathrm{dom}(-) = \{(x,y) \in \mathbf{\Gamma} \times \mathbf{\Gamma} \mid y \leq x\}$ and

$$\begin{aligned} r^\circ - s^\circ &= (r-s)^\circ \\ r^- - s^\circ &= (r-s)^- \end{aligned} \qquad \left.\begin{aligned} r^\circ - s^- \\ r^- - s^- \end{aligned}\right\} = \begin{cases} (r-s)^\circ & \text{if } r - s \in \mathbb{Q} \\ (r-s)^- & \text{otherwise.} \end{cases}$$

In fact, this (partial) operation is dual to the truncated addition on the lattice $\mathbf{L}$. However, explaining this would require us to delve into extended Priestley duality for lattices with operations, which is beyond the scope of this paper. See [9] and also [7,8] for details. It also follows from the general theory that there exists another partial operation definable from $-$, namely:

$$\sim\colon \mathrm{dom}(-) \to \mathbf{\Gamma}, \quad x \sim y = \bigvee \{x - q^\circ \mid y < q^\circ \leq x\}.$$

Next, we collect some basic properties of $-$ and $\sim$, needed in Section 4, which follow from the general theory of [7,8]. First, recall that a map into an ordered topological space is *lower* (resp. *upper*) *semicontinuous* provided the preimage of any open down-set (resp. open up-set) is open.

Lemma 1. *If* $\mathrm{dom}(-)$ *is seen as a subspace of* $\mathbf{\Gamma} \times \mathbf{\Gamma}^{\partial}$*, the following hold:*

1. $\mathrm{dom}(-)$ *is a closed up-set in* $\mathbf{\Gamma} \times \mathbf{\Gamma}^{\partial}$*;*
2. *both* $-\colon \mathrm{dom}(-) \to \mathbf{\Gamma}$ *and* $\sim\colon \mathrm{dom}(-) \to \mathbf{\Gamma}$ *are monotone in the first coordinate, and antitone in the second;*
3. $-\colon \mathrm{dom}(-) \to \mathbf{\Gamma}$ *is lower semicontinuous;*
4. $\sim\colon \mathrm{dom}(-) \to \mathbf{\Gamma}$ *is upper semicontinuous.*

3.2 The retraction $\mathbf{\Gamma} \twoheadrightarrow [0,1]$

In this section we show that, with respect to appropriate topologies, the unit interval $[0,1]$ can be obtained as a topological retract of $\mathbf{\Gamma}$, in a way which is compatible with the operation $-$. This will be important in Sections 4 and 5, where we need to move between [0,1]-valued and $\mathbf{\Gamma}$-valued measures. Let us define the monotone surjection given by collapsing the doubled elements:

$$\gamma\colon \mathbf{\Gamma} \to [0,1],\ r^-, r^\circ \mapsto r. \tag{3}$$

The map γ has a right adjoint, given by

$$\iota\colon [0,1] \to \mathbf{\Gamma},\ r \mapsto \begin{cases} r^\circ & \text{if } r \in \mathbb{Q} \\ r^- & \text{otherwise.} \end{cases} \tag{4}$$

Indeed, it is readily seen that $\gamma(y) \leq x$ iff $y \leq \iota(x)$, for all $y \in \mathbf{\Gamma}$ and $x \in [0,1]$. The composition $\gamma \cdot \iota$ coincides with the identity on $[0,1]$, i.e. ι is a section of γ. Moreover, this retraction lifts to a topological retract provided we equip $\mathbf{\Gamma}$ and $[0,1]$ with the topologies consisting of the open down-sets:

Lemma 2. *The map* $\gamma\colon \mathbf{\Gamma} \to [0,1]$ *is continuous and the map* $\iota\colon [0,1] \to \mathbf{\Gamma}$ *is lower semicontinuous.*

Proof. To check continuity of γ observe that, for a rational $q \in (0,1)$, $\gamma^{-1}(q,1]$ and $\gamma^{-1}[0,q)$ coincide, respectively, with the open sets

$$\bigcup\{\uparrow p^\circ \mid p \in \mathbb{Q} \cap [0,1] \text{ and } q < p\} \ \text{ and } \ \bigcup\{\downarrow p^- \mid p \in \mathbb{Q} \cap (0,1] \text{ and } p < q\}.$$

Also, ι is lower semicontinuous, for $\iota^{-1}(\downarrow q^-) = [0,q)$ whenever $q \in \mathbb{Q} \cap (0,1]$. □

It is easy to see that both γ and ι preserve the minus structure available on $\mathbf{\Gamma}$ and [0,1] (the unit interval is equipped with the usual minus operation $x - y$ defined whenever $y \leq x$), that is,

- $\gamma(x - y) = \gamma(x \sim y) = \gamma(x) - \gamma(y)$ whenever $y \leq x$ in $\mathbf{\Gamma}$, and
- $\iota(x - y) = \iota(x) - \iota(y)$ whenever $y \leq x$ in [0,1].

Remark. $\iota\colon [0,1] \to \mathbf{\Gamma}$ is not upper semicontinuous because, for every $q \in \mathbb{Q} \cap [0,1]$, $\iota^{-1}(\uparrow q^\circ) = \{x \in [0,1] \mid q^\circ \leq \iota(x)\} = \{x \in [0,1] \mid \gamma(q^\circ) \leq x\} = [q,1]$.

4 Spaces of measures valued in $\mathbf{\Gamma}$ and in $[0,1]$

The aim of this section is to replace $[0,1]$-valued measures by $\mathbf{\Gamma}$-valued measures. The reason for doing this is two-fold. First, the space of $\mathbf{\Gamma}$-valued measures is Priestley (Proposition 4), and thus amenable to a duality theoretic treatment and a dual logic interpretation (cf. Section 6). Second, it retains more topological information than the space of $[0,1]$-valued measures. Indeed, the former retracts onto the latter (Theorem 10).

Let D be a distributive lattice. Recall that, classically, a monotone function $m\colon D \to [0,1]$ is a (finitely additive, probability) measure provided $m(0)=0$, $m(1)=1$, and $m(a)+m(b)=m(a\vee b)+m(a\wedge b)$ for every $a,b\in D$. The latter property is equivalently expressed as

$$\forall a,b\in D,\ m(a)-m(a\wedge b)=m(a\vee b)-m(b). \tag{5}$$

We write $\mathcal{M}_{\mathrm{I}}(D)$ for the set of all measures $D\to[0,1]$, and regard it as an ordered topological space, with the structure induced by the product order and product topology of $[0,1]^D$. The notion of (finitely additive, probability) $\mathbf{\Gamma}$-valued measure is analogous to the classical one, except that the finite additivity property (5) splits into two conditions, involving $-$ and $\sim$.

Definition 3. *Let D be a distributive lattice. A* $\mathbf{\Gamma}$-valued measure *(or simply a* measure*) on D is a function $\mu\colon D\to\mathbf{\Gamma}$ such that*

1. *$\mu(0)=0^\circ$ and $\mu(1)=1^\circ$,*
2. *μ is monotone, and*
3. *for all $a,b\in D$,*

$$\mu(a)\sim\mu(a\wedge b)\le\mu(a\vee b)-\mu(b) \quad \text{and} \quad \mu(a)-\mu(a\wedge b)\ge\mu(a\vee b)\sim\mu(b).$$

We denote by $\mathcal{M}_\Gamma(D)$ the subspace of $\mathbf{\Gamma}^D$ consisting of the measures $\mu\colon D\to\mathbf{\Gamma}$.

Since $\mathbf{\Gamma}$ is a Priestley space, so is $\mathbf{\Gamma}^D$ equipped with the product order and topology. Hence, we regard $\mathcal{M}_\Gamma(D)$ as an ordered topological space, whose topology and order are induced by those of $\mathbf{\Gamma}^D$. In fact $\mathcal{M}_\Gamma(D)$ is a Priestley space:

Proposition 4. *For any distributive lattice D, $\mathcal{M}_\Gamma(D)$ is a Priestley space.*

Proof. It suffices to show that $\mathcal{M}_\Gamma(D)$ is a closed subspace of $\mathbf{\Gamma}^D$. Let

$$C_{1,2}=\{f\in\mathbf{\Gamma}^D\mid f(0)=0^\circ\}\cap\{f\in\mathbf{\Gamma}^D\mid f(1)=1^\circ\}\cap\bigcap_{a\le b}\{f\in\mathbf{\Gamma}^D\mid f(a)\le f(b)\}.$$

Note that the evaluation maps $\mathrm{ev}_a\colon\mathbf{\Gamma}^D\to\mathbf{\Gamma}$, $f\mapsto f(a)$, are continuous for every $a\in D$. Thus, the first set in the intersection defining $C_{1,2}$ is closed because it is the equaliser of the evaluation map ev_0 and the constant map of value 0°. Similarly, for the set $\{f\in\mathbf{\Gamma}^D\mid f(1)=1^\circ\}$. The last one is the intersection of the sets of the form $\langle\mathrm{ev}_a,\mathrm{ev}_b\rangle^{-1}(\le)$, which are closed because $\le$ is closed in $\mathbf{\Gamma}\times\mathbf{\Gamma}$. Whence, $C_{1,2}$ is a closed subset of $\mathbf{\Gamma}^D$. Moreover,

$$\mathcal{M}_\Gamma(D)=\bigcap_{a,b\in D}\{f\in C_{1,2}\mid f(a)\sim f(a\wedge b)\le f(a\vee b)-f(b)\}$$

$$\cap \bigcap_{a,b \in D} \{f \in C_{1,2} \mid f(a) - f(a \wedge b) \geq f(a \vee b) \sim f(b)\}.$$

From semicontinuity of $-$ and $\sim$ (Lemma 1) and the following well-known fact in order-topology we conclude that $\mathcal{M}_\Gamma(D)$ is closed in $\mathbf{\Gamma}^D$.

Fact. Let X, Y be compact ordered spaces, $f\colon X \to Y$ a lower semicontinuous function and $g\colon X \to Y$ an upper semicontinuous function. If X' is a closed subset of X, then so is $E = \{x \in X' \mid g(x) \leq f(x)\}$. □

Next, we prove a property which is very useful when approximating a fragment of a logic by smaller fragments (see, e.g., Section 5.1). Let us denote by **DLat** the category of distributive lattices and homomorphisms, and by **Pries** the category of Priestley spaces and continuous monotone maps.

Proposition 5. *The assignment $D \mapsto \mathcal{M}_\Gamma(D)$ yields a contravariant functor $\mathcal{M}_\Gamma$:* **DLat** $\to$ **Pries** *which sends directed colimits to codirected limits.*

Proof. If $h\colon D \to E$ is a lattice homomorphism and $\mu\colon E \to \mathbf{\Gamma}$ is a measure, it is not difficult to see that $\mathcal{M}_\Gamma(h)(\mu) = \mu \cdot h\colon D \to \mathbf{\Gamma}$ is a measure. The mapping $\mathcal{M}_\Gamma(h)\colon \mathcal{M}_\Gamma(E) \to \mathcal{M}_\Gamma(D)$ is clearly monotone. For continuity, recall that the topology of $\mathcal{M}_\Gamma(D)$ is generated by the sets $[\![a < q]\!] = \{\nu\colon D \to \mathbf{\Gamma} \mid \nu(a) < q^\circ\}$ and $[\![a \geq q]\!] = \{\nu\colon D \to \mathbf{\Gamma} \mid \nu(a) \geq q^\circ\}$, with $a \in D$ and $q \in \mathbb{Q} \cap [0,1]$. We have

$$\mathcal{M}_\Gamma(h)^{-1}([\![a < q]\!]) = \{\mu\colon E \to \mathbf{\Gamma} \mid \mu(h(a)) < q^\circ\} = [\![h(a) < q]\!]$$

which is open in $\mathcal{M}_\Gamma(E)$. Similarly, $\mathcal{M}_\Gamma(h)^{-1}([\![a \geq q]\!]) = [\![h(a) \geq q]\!]$, showing that $\mathcal{M}_\Gamma(h)$ is continuous. Thus, $\mathcal{M}_\Gamma$ is a contravariant functor.

The rest of the proof is a routine verification. □

Remark 6. We work with the contravariant functor $\mathcal{M}_\Gamma$: **DLat** $\to$ **Pries** because $\mathcal{M}_\Gamma$ is concretely defined on the lattice side. However, by Priestley duality, **DLat** is dually equivalent to **Pries**, so we can think of $\mathcal{M}_\Gamma$ as a covariant functor **Pries** $\to$ **Pries** (this is the perspective traditionally adopted in analysis, and also in the works of Nešetřil and Ossona de Mendez). From this viewpoint, Section 6 provides a description of the endofunctor on **DLat** dual to $\mathcal{M}_\Gamma$: **Pries** $\to$ **Pries**.

Recall the maps $\gamma\colon \mathbf{\Gamma} \to [0,1]$ and $\iota\colon [0,1] \to \mathbf{\Gamma}$ from equations (3)–(4). In Section 3.2 we showed that this is a retraction-section pair. In Theorem 10 this retraction is lifted to the spaces of measures. We start with an easy observation:

Lemma 7. *Let D be a distributive lattice. The following statements hold:*

1. *for every $\mu \in \mathcal{M}_\Gamma(D)$, $\gamma \cdot \mu \in \mathcal{M}_\mathrm{I}(D)$,*
2. *for every $m \in \mathcal{M}_\mathrm{I}(D)$, $\iota \cdot m \in \mathcal{M}_\Gamma(D)$.*

Proof. 1. The only non-trivial condition to verify is finite additivity. In view of the discussion after Lemma 2, the map γ preserves both minus operations on $\mathbf{\Gamma}$. Hence, for every $a, b \in D$, the inequalities $\mu(a) \sim \mu(a \wedge b) \leq \mu(a \vee b) - \mu(b)$ and $\mu(a) - \mu(a \wedge b) \geq \mu(a \vee b) \sim \mu(b)$ imply that $\gamma \cdot \mu(a) - \gamma \cdot \mu(a \wedge b) = \gamma \cdot \mu(a \vee b) - \gamma \cdot \mu(b)$.

2. The first two conditions in Definition 3 are immediate. The third condition follows from the fact that $\iota(r-s) = \iota(r) - \iota(s)$ whenever $s \leq r$ in [0,1], and $x \sim y \leq x - y$ for every $(x, y) \in \mathrm{dom}(-)$. □

In view of the previous lemma, there are well-defined functions

$$\gamma^{\#}\colon \mathcal{M}_{\Gamma}(D) \to \mathcal{M}_{\mathrm{I}}(D),\ \mu \mapsto \gamma \cdot \mu \quad \text{and} \quad \iota^{\#}\colon \mathcal{M}_{\mathrm{I}}(D) \to \mathcal{M}_{\Gamma}(D),\ m \mapsto \iota \cdot m.$$

Lemma 8. *$\gamma^{\#}\colon \mathcal{M}_{\Gamma}(D) \to \mathcal{M}_{\mathrm{I}}(D)$ is a continuous and monotone map.*

Proof. The topology of $\mathcal{M}_{\mathrm{I}}(D)$ is generated by the sets of the form $\{m \in \mathcal{M}_{\mathrm{I}}(D) \mid m(a) \in O\}$, for $a \in D$ and O an open subset of $[0,1]$. In turn,

$$(\gamma^{\#})^{-1}\{m \in \mathcal{M}_{\mathrm{I}}(D) \mid m(a) \in O\} = \{\mu \in \mathcal{M}_{\Gamma}(D) \mid \mu(a) \in \gamma^{-1}(O)\}$$

is open in $\mathcal{M}_{\Gamma}(D)$ because $\gamma\colon \mathbf{\Gamma} \to [0,1]$ is continuous by Lemma 2. This shows that $\gamma^{\#}\colon \mathcal{M}_{\Gamma}(D) \to \mathcal{M}_{\mathrm{I}}(D)$ is continuous. Monotonicity is immediate. □

Note that $\gamma^{\#}\colon \mathcal{M}_{\Gamma}(D) \to \mathcal{M}_{\mathrm{I}}(D)$ is surjective, since it admits $\iota^{\#}$ as a (set-theoretic) section. It follows that $\mathcal{M}_{\mathrm{I}}(D)$ is a compact ordered space:

Corollary 9. *For each distributive lattice D, $\mathcal{M}_{\mathrm{I}}(D)$ is a compact ordered space.*

Proof. The surjection $\gamma^{\#}\colon \mathcal{M}_{\Gamma}(D) \to \mathcal{M}_{\mathrm{I}}(D)$ is continuous (Lemma 8). Since $\mathcal{M}_{\Gamma}(D)$ is compact by Proposition 4, so is $\mathcal{M}_{\mathrm{I}}(D)$. The order of $\mathcal{M}_{\mathrm{I}}(D)$ is clearly closed in the product topology, thus $\mathcal{M}_{\mathrm{I}}(D)$ is a compact ordered space. □

Finally, we see that the set-theoretic retraction of $\mathcal{M}_{\Gamma}(D)$ onto $\mathcal{M}_{\mathrm{I}}(D)$ lifts to the topological setting, provided we restrict to the down-set topologies. If $(X, \leq)$ is a partially ordered topological space, write $X^{\downarrow}$ for the space with the same underlying set as X and whose topology consists of the open down-sets of X.

Theorem 10. *The maps $\gamma^{\#}\colon \mathcal{M}_{\Gamma}(D)^{\downarrow} \to \mathcal{M}_{\mathrm{I}}(D)^{\downarrow}$ and $\iota^{\#}\colon \mathcal{M}_{\mathrm{I}}(D)^{\downarrow} \to \mathcal{M}_{\Gamma}(D)^{\downarrow}$ are a retraction-section pair of topological spaces.*

Proof. It suffices to show that $\gamma^{\#}$ and $\iota^{\#}$ are continuous. It is not difficult to see, using Lemma 8, that $\gamma^{\#}\colon \mathcal{M}_{\Gamma}(D)^{\downarrow} \to \mathcal{M}_{\mathrm{I}}(D)^{\downarrow}$ is continuous. For the continuity of $\iota^{\#}$, note that the topology of $\mathcal{M}_{\Gamma}(D)^{\downarrow}$ is generated by the sets of the form $\{\mu \in \mathcal{M}_{\Gamma}(D) \mid \mu(a) \leq q^{-}\}$, for $a \in D$ and $q \in \mathbb{Q} \cap (0,1]$. We have

$$\begin{aligned}(\iota^{\#})^{-1}\{\mu \in \mathcal{M}_{\Gamma}(D) \mid \mu(a) \leq q^{-}\} &= \{m \in \mathcal{M}_{\mathrm{I}}(D) \mid m(a) \in \iota^{-1}(\downarrow q^{-})\} \\ &= \{m \in \mathcal{M}_{\mathrm{I}}(D) \mid m(a) < q\},\end{aligned}$$

which is an open set in $\mathcal{M}_{\mathrm{I}}(D)^{\downarrow}$. This concludes the proof. □

5 The Γ-valued Stone pairing and limits of finite structures

In the work of Nešetřil and Ossona de Mendez, the Stone pairing $\langle \text{-}, A\rangle$ is $[0,1]$-valued, i.e. an element of $\mathcal{M}_{\mathrm{I}}(\mathrm{FO}(\sigma))$. In this section, we show that basically the same construction for the recognisers arising from the application of a layer of semiring quantifiers in logic on words (cf. Section 2.3) provides an embedding of finite σ-structures into the space of $\mathbf{\Gamma}$-valued measures. It turns out that this embedding is a $\mathbf{\Gamma}$-valued version of the Stone pairing. Hereafter we make a notational difference, writing $\langle -, - \rangle_{\mathrm{I}}$ for the (classical) [0,1]-valued Stone pairing.

The main ingredient of the construction are the $\mathbf{\Gamma}$-valued finitely supported functions. To start with, we point out that the partial operation $-$ on $\mathbf{\Gamma}$ uniquely determines a partial "plus" operation on $\mathbf{\Gamma}$. Define

$$+\colon \mathrm{dom}(+) \to \mathbf{\Gamma}, \quad \text{where} \quad \mathrm{dom}(+) = \{(x,y) \mid x \leq 1^\circ - y\},$$

by the following rules (whenever the expressions make sense):

$$r^\circ + s^\circ = (r+s)^\circ, \; r^- + s^\circ = (r+s)^-, \; r^\circ + s^- = (r+s)^-, \; \text{and} \; r^- + s^- = (r+s)^-.$$

Then, for every $y \in \mathbf{\Gamma}$, the function $(-) + y$ sending x to $x + y$ is left adjoint to the function $(-) - y$ sending x to $x - y$.

Definition 11. *For any set X, $\mathcal{F}(X)$ is the set of all functions $f\colon X \to \mathbf{\Gamma}$ s.t.*

1. *the set* $\mathrm{supp}(f) = \{x \in X \mid f(x) \neq 0^\circ\}$ *is finite, and*
2. *$f(x_1) + \cdots + f(x_n)$ is defined and equal to 1°, where* $\mathrm{supp}(f) = \{x_1, \ldots, x_n\}$.

To improve readability, if the sum $y_1 + \cdots + y_m$ exists in $\mathbf{\Gamma}$, we denote it $\sum_{i=1}^m y_i$. Finitely supported functions in the above sense always determine measures over the power-set algebra (the proof is an easy verification and is omitted):

Lemma 12. *Let X be any set. There is a well-defined mapping $\int\colon \mathcal{F}(X) \to \mathcal{M}_\Gamma(\wp(X))$, assigning to every $f \in \mathcal{F}(X)$ the measure*

$$\textstyle\int f\colon M \mapsto \int_M f = \sum\{f(x) \mid x \in M \cap \mathrm{supp}(f)\}.$$

5.1 The $\mathbf{\Gamma}$-valued Stone pairing and logic on words

Fix a countably infinite set of variables $\{v_1, v_2, \ldots\}$. Recall that $\mathrm{FO}_n(\sigma)$ is the Lindenbaum-Tarski algebra of first-order formulas with free variables among $\{v_1, \ldots, v_n\}$. The dual space of $\mathrm{FO}_n(\sigma)$ is the space of n-types $\mathrm{Typ}_n(\sigma)$. Its points are the equivalence classes of pairs (A, α), where A is a σ-structure and $\alpha\colon \{v_1, \ldots, v_n\} \to A$ is an interpretation of the variables. Write $\mathrm{Fin}(\sigma)$ for the set of all finite σ-structures and define a map $\mathrm{Fin}(\sigma) \to \mathcal{F}(\mathrm{Typ}_n(\sigma))$ as $A \mapsto f_n^A$, where f_n^A is the function which sends an equivalence class $E \in \mathrm{Typ}_n(\sigma)$ to

$$f_n^A(E) = \sum_{(A,\alpha)\in E} \left(\frac{1}{|A|^n}\right)^\circ \qquad \textit{(Add } \tfrac{1}{|A|^n} \textit{ for every interpretation } \alpha \textit{ of the free variables s.t. } (A,\alpha) \textit{ is in the equivalence class).}$$

By Lemma 12, we get a measure $\int f_n^A\colon \wp(\mathrm{Typ}_n(\sigma)) \to \mathbf{\Gamma}$. Now, for each $\varphi \in \mathrm{FO}_n(\sigma)$, let $[\![\varphi]\!]_n \subseteq \mathrm{Typ}_n(\sigma)$ be the set of (equivalence classes of) σ-structures with interpretations satisfying φ. By Stone duality we obtain an embedding $[\![-]\!]_n\colon \mathrm{FO}_n(\sigma) \hookrightarrow \wp(\mathrm{Typ}_n(\sigma))$. Restricting $\int f_n^A$ to $\mathrm{FO}_n(\sigma)$, we get a measure

$$\mu_n^A\colon \mathrm{FO}_n(\sigma) \to \mathbf{\Gamma}, \quad \varphi \mapsto \textstyle\int_{[\![\varphi]\!]_n} f_n^A.$$

Summing up, we have the composite map

$$\mathrm{Fin}(\sigma) \to \mathcal{M}_{\Gamma}(\wp(\mathrm{Typ}_n(\sigma))) \to \mathcal{M}_{\Gamma}(\mathrm{FO}_n(\sigma)), \quad A \mapsto \textstyle\int f_n^A \mapsto \mu_n^A. \tag{6}$$

Essentially the same construction is featured in logic on words, cf. equation (2):

- The set of finite σ-structures $\mathrm{Fin}(\sigma)$ corresponds to the set of finite words A^*.
- The collection $\mathrm{Typ}_n(\sigma)$ of (equivalence classes of) σ-structures with interpretations corresponds to $(A \times 2)^*$ or, interchangeably, $\beta(A \times 2)^*$ (in the case of one free variable).
- The fragment $\mathrm{FO}_n(\sigma)$ of first-order logic corresponds to the Boolean algebra of languages, defined by formulas with a free variable, dual to the Boolean space X appearing in (2).
- The first map in the composite (6) sends a finite structure A to the measure $\int f_n^A$ which, evaluated on $K \subseteq \mathrm{Typ}_n(\sigma)$, counts the (proportion of) interpretations $\alpha\colon \{v_1, \ldots, v_n\} \to A$ such that $(A, \alpha) \in K$, similarly to R from (2).
- Finally, the second map in (6) sends a measure in $\mathcal{M}_{\Gamma}(\wp(\mathrm{Typ}_n(\sigma)))$ to its pushforward along $[\![\text{-}]\!]_n\colon \mathrm{FO}_n(\sigma) \hookrightarrow \wp(\mathrm{Typ}_n(\sigma))$. This is the second map in the composition (2).

On the other hand, the assignment $A \mapsto \mu_n^A$ defined in (6) is also closely related to the classical Stone pairing. Indeed, for every formula φ in $\mathrm{FO}_n(\sigma)$,

$$\begin{aligned}\mu_n^A(\varphi) &= \sum_{E \in [\![\varphi]\!]_n} f_n^A(E) = \sum_{E \in [\![\varphi]\!]_n} \sum_{(A,\alpha) \in E} \left(\frac{1}{|A|^n}\right)^{\circ} \\ &= \left(\frac{|\{\bar{a} \in A^n \mid A \models \varphi(\bar{a})|}{|A|^n}\right)^{\circ} = (\langle \varphi, A \rangle_{\mathrm{I}})^{\circ}.\end{aligned} \tag{7}$$

In this sense, μ_n^A can be regarded as a $\mathbf{\Gamma}$-valued Stone pairing, relative to the fragment $\mathrm{FO}_n(\sigma)$. Next, we show how to extend this to the full first-order logic $\mathrm{FO}(\sigma)$. First, we observe that the construction is invariant under extensions of the set of free variables (the proof is the same as in the classical case).

Lemma 13. *Given $m, n \in \mathbb{N}$ and $A \in \mathrm{Fin}(\sigma)$, if $m \geq n$ then $(\mu_m^A)_{\restriction \mathrm{FO}_n(\sigma)} = \mu_n^A$.*

The Lindenbaum-Tarski algebra of all first-order formulas $\mathrm{FO}(\sigma)$ is the directed colimit of the Boolean subalgebras $\mathrm{FO}_n(\sigma)$, for $n \in \mathbb{N}$. Since the functor $\mathcal{M}_{\Gamma}$ turns directed colimits into codirected limits (Proposition 5), the Priestley space $\mathcal{M}_{\Gamma}(\mathrm{FO}(\sigma))$ is the limit of the diagram

$$\left\{\mathcal{M}_{\Gamma}(\mathrm{FO}_n(\sigma)) \xleftarrow{q_{n,m}} \mathcal{M}_{\Gamma}(\mathrm{FO}_m(\sigma)) \mid m, n \in \mathbb{N},\ m \geq n\right\}$$

where, for any $\mu\colon \mathrm{FO}_m(\sigma) \to \mathbf{\Gamma}$ in $\mathcal{M}_{\Gamma}(\mathrm{FO}_m(\sigma))$, the measure $q_{n,m}(\mu)$ is the restriction of μ to $\mathrm{FO}_n(\sigma)$. In view of Lemma 13, for every $A \in \mathrm{Fin}(\sigma)$, the tuple $(\mu_n^A)_{n \in \mathbb{N}}$ is compatible with the restriction maps. Thus, recalling that limits in the category of Priestley spaces are computed as in sets, by universality of the limit construction, this tuple yields a measure

$$\langle \text{-}, A \rangle_{\Gamma} : \mathrm{FO}(\sigma) \to \mathbf{\Gamma}$$

in the space $\mathcal{M}_\Gamma(\mathrm{FO}(\sigma))$. This we call the **Γ**-*valued Stone pairing* associated with A. As in the classical case, it is not difficult to see that the mapping $A \mapsto \langle \text{-}, A\rangle_\Gamma$ gives an embedding $\langle \text{-}, \text{-}\rangle_\Gamma : \mathrm{Fin}(\sigma) \hookrightarrow \mathcal{M}_\Gamma(\mathrm{FO}(\sigma))$. The following theorem illustrates the relation between the classical Stone pairing $\langle \text{-}, \text{-}\rangle_\mathrm{I} : \mathrm{Fin}(\sigma) \hookrightarrow \mathcal{M}_\mathrm{I}(\mathrm{FO}(\sigma))$, and the **Γ**-valued one.

Theorem 14. *The following diagram commutes:*

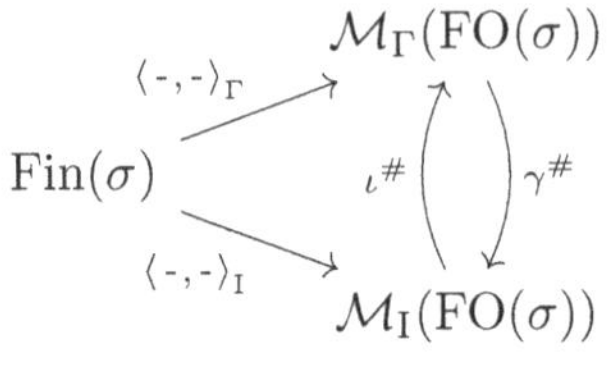

Proof. Fix an arbitrary finite structure $A \in \mathrm{Fin}(\sigma)$. Let φ be a formula in $\mathrm{FO}(\sigma)$ with free variables among $\{v_1, \ldots, v_n\}$, for some $n \in \mathbb{N}$. By construction, $\langle \varphi, A\rangle_\Gamma = \mu_n^A(\varphi)$. Therefore, by equation (7), $\langle \varphi, A\rangle_\Gamma = (\langle \varphi, A\rangle_\mathrm{I})^\circ$. The statement then follows at once. □

Remark. The construction in this section works also for proper fragments, i.e. for sublattices $D \subseteq \mathrm{FO}(\sigma)$. This corresponds to composing the embedding $\mathrm{Fin}(\sigma) \hookrightarrow \mathcal{M}_\Gamma(\mathrm{FO}(\sigma))$ with the restriction map $\mathcal{M}_\Gamma(\mathrm{FO}(\sigma)) \to \mathcal{M}_\Gamma(D)$ sending $\mu \colon \mathrm{FO}(\sigma) \to \mathbf{\Gamma}$ to $\mu_{\restriction D} \colon D \to \mathbf{\Gamma}$. The only difference is that the ensuing map $\mathrm{Fin}(\sigma) \to \mathcal{M}_\Gamma(D)$ need not be injective, in general.

5.2 Limits in the spaces of measures

By Theorem 14 the **Γ**-valued Stone pairing $\langle \text{-}, \text{-}\rangle_\Gamma$ and the classical Stone pairing $\langle \text{-}, \text{-}\rangle_\mathrm{I}$ determine each other. However, the notions of convergence associated with the spaces $\mathcal{M}_\Gamma(\mathrm{FO}(\sigma))$ and $\mathcal{M}_\mathrm{I}(\mathrm{FO}(\sigma))$ are different: since the topology of $\mathcal{M}_\Gamma(\mathrm{FO}(\sigma))$ is richer, there are "fewer" convergent sequences. Recall from Lemma 8 that $\gamma^\# \colon \mathcal{M}_\Gamma(\mathrm{FO}(\sigma)) \to \mathcal{M}_\mathrm{I}(\mathrm{FO}(\sigma))$ is continuous. Also, $\gamma^\#(\langle \text{-}, A\rangle_\Gamma) = \langle \text{-}, A\rangle_\mathrm{I}$ by Theorem 14. Thus, for any sequence of finite structures $(A_n)_{n\in\mathbb{N}}$, if

$$\langle \text{-}, A_n\rangle_\Gamma \text{ converges to a measure } \mu \text{ in } \mathcal{M}_\Gamma(\mathrm{FO}(\sigma))$$

then

$$\langle \text{-}, A_n\rangle_\mathrm{I} \text{ converges to the measure } \gamma^\#(\mu) \text{ in } \mathcal{M}_\mathrm{I}(\mathrm{FO}(\sigma)).$$

The converse is not true. For example, consider the signature $\sigma = \{<\}$ consisting of a single binary relation symbol, and let $(A_n)_{n\in\mathbb{N}}$ be the sequence of finite posets displayed in the picture below.

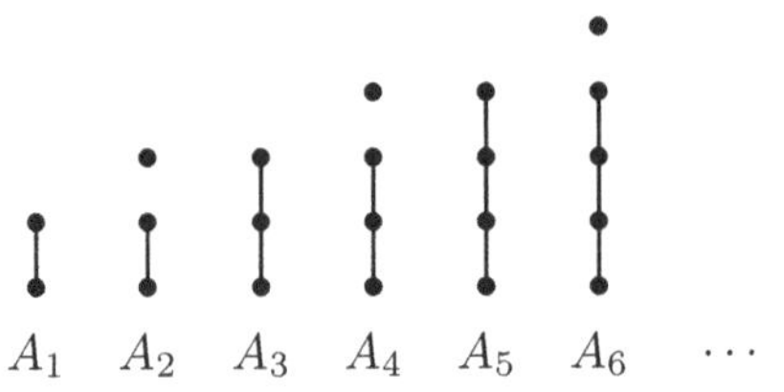

Let $\psi(x) \approx \forall y \neg(x < y) \wedge \exists z \neg(z < x) \wedge \neg(z = x)$ be the formula stating that x is maximal but not the maximum in the order given by $<$. Then, for the sublattice $D = \{\mathbf{f}, \psi, \mathbf{t}\}$ of $\mathrm{FO}(\sigma)$, the sequences $\langle \text{-}, A_n \rangle_\Gamma$ and $\langle \text{-}, A_n \rangle_\mathrm{I}$ converge in $\mathcal{M}_\Gamma(D)$ and $\mathcal{M}_\mathrm{I}(D)$, respectively. However, if we consider the Boolean algebra $B = \{\mathbf{f}, \psi, \neg\psi, \mathbf{t}\}$, then the $\langle \text{-}, A_n \rangle_\mathrm{I}$'s still converge whereas the $\langle \text{-}, A_n \rangle_\Gamma$'s do not. Indeed, the following sequence does not converge in $\mathbf{\Gamma}$:

$$(\langle \neg\psi, A_n \rangle_\Gamma)_n = (1^\circ, (\tfrac{1}{3})^\circ, 1^\circ, (\tfrac{2}{4})^\circ, 1^\circ, (\tfrac{3}{5})^\circ, \ldots),$$

because the odd terms converge to 1°, while the even terms converge to 1^-. However, there is a sequence $\langle \text{-}, B_n \rangle_\Gamma$ whose image under $\gamma^\#$ coincides with the limit of the $\langle \text{-}, A_n \rangle_\mathrm{I}$'s (e.g., take the subsequence of even terms of $(A_n)_{n\in\mathbb{N}}$). In the next theorem, we will see that this is a general fact.

Identify $\mathrm{Fin}(\sigma)$ with a subset of $\mathcal{M}_\Gamma(\mathrm{FO}(\sigma))$ (resp. $\mathcal{M}_\mathrm{I}(\mathrm{FO}(\sigma))$) through $\langle \text{-}, \text{-} \rangle_\Gamma$ (resp. $\langle \text{-}, \text{-} \rangle_\mathrm{I}$). A central question in the theory of structural limits, cf. [16], is to determine the closure of $\mathrm{Fin}(\sigma)$ in $\mathcal{M}_\mathrm{I}(\mathrm{FO}(\sigma))$, which consists precisely of the limits of sequences of finite structures. The following theorem gives an answer to this question in terms of the corresponding question for $\mathcal{M}_\Gamma(\mathrm{FO}(\sigma))$.

Theorem 15. *Let* $\overline{\mathrm{Fin}(\sigma)}$ *denote the closure of* $\mathrm{Fin}(\sigma)$ *in* $\mathcal{M}_\Gamma(\mathrm{FO}(\sigma))$. *Then the set* $\gamma^\#(\overline{\mathrm{Fin}(\sigma)})$ *coincides with the closure of* $\mathrm{Fin}(\sigma)$ *in* $\mathcal{M}_\mathrm{I}(\mathrm{FO}(\sigma))$.

Proof. Write U for the image of $\langle \text{-}, \text{-} \rangle_\Gamma : \mathrm{Fin}(\sigma) \hookrightarrow \mathcal{M}_\Gamma(\mathrm{FO}(\sigma))$, and V for the image of $\langle \text{-}, \text{-} \rangle_\mathrm{I} : \mathrm{Fin}(\sigma) \hookrightarrow \mathcal{M}_\mathrm{I}(\mathrm{FO}(\sigma))$. We must prove that $\gamma^\#(\overline{U}) = \overline{V}$. By Theorem 14, $\gamma^\#(U) = V$. The map $\gamma^\# \colon \mathcal{M}_\Gamma(\mathrm{FO}(\sigma)) \to \mathcal{M}_\mathrm{I}(\mathrm{FO}(\sigma))$ is continuous (Lemma 8), and the spaces $\mathcal{M}_\Gamma(\mathrm{FO}(\sigma))$ and $\mathcal{M}_\mathrm{I}(\mathrm{FO}(\sigma))$ are compact Hausdorff (Proposition 4 and Corollary 9). Since continuous maps between compact Hausdorff spaces are closed, $\gamma^\#(\overline{U}) = \overline{\gamma^\#(U)} = \overline{V}$. □

6 The logic of measures

Let D be a distributive lattice. We know from Proposition 4 that the space $\mathcal{M}_\Gamma(D)$ of $\mathbf{\Gamma}$-valued measures on D is a Priestley space, whence it has a dual distributive lattice $\mathbf{P}(D)$. In this section we show that $\mathbf{P}(D)$ can be represented as the Lindenbaum-Tarski algebra for a propositional logic $P\mathcal{L}_D$ obtained from D by adding probabilistic quantifiers. Since we adopt a logical perspective, we write $\mathbf{f}$ and $\mathbf{t}$ for the bottom and top elements of D, respectively.

The set of propositional variables of $P\mathcal{L}_D$ consists of the symbols $\mathbb{P}_{\geq p}\, a$, for every $a \in D$ and $p \in \mathbb{Q} \cap [0,1]$. For every measure $\mu \in \mathcal{M}_\Gamma(D)$, we set

$$\mu \models \mathbb{P}_{\geq p}\, a \;\Leftrightarrow\; \mu(a) \geq p^\circ. \tag{8}$$

This satisfaction relation extends in the obvious way to the closure under finite conjunctions and finite disjunctions of the set of propositional variables. Define

$$\varphi \models \psi \quad \text{if,} \quad \forall \mu \in \mathcal{M}_\Gamma(D), \quad \mu \models \varphi \text{ implies } \mu \models \psi.$$

Also, write $\models \varphi$ if $\mu \models \varphi$ for every $\mu \in \mathcal{M}_\Gamma(D)$, and $\varphi \models$ if there is no $\mu \in \mathcal{M}_\Gamma(D)$ with $\mu \models \varphi$.

Consider the following conditions, for any $p, q, r \in \mathbb{Q} \cap [0,1]$ and $a, b \in D$.

(L1) $\mathbb{P}_{\geq q}\, a \models \mathbb{P}_{\geq p}\, a$ whenever $p \leq q$
(L2) $\mathbb{P}_{\geq p}\, \mathbf{f} \models$ whenever $p > 0$, $\models \mathbb{P}_{\geq 0}\, \mathbf{f}$ and $\models \mathbb{P}_{\geq q}\, \mathbf{t}$
(L3) $\mathbb{P}_{\geq q}\, a \models \mathbb{P}_{\geq q}\, b$ whenever $a \leq b$
(L4) $\mathbb{P}_{\geq p}\, a \wedge \mathbb{P}_{\geq q}\, b \models \mathbb{P}_{\geq p+q-r}\,(a \vee b) \vee \mathbb{P}_{\geq r}\,(a \wedge b)$ whenever $0 \leq p+q-r \leq 1$
(L5) $\mathbb{P}_{\geq p+q-r}\,(a \vee b) \wedge \mathbb{P}_{\geq r}\,(a \wedge b) \models \mathbb{P}_{\geq p}\, a \vee \mathbb{P}_{\geq q}\, b$ whenever $0 \leq p+q-r \leq 1$

It is not hard to see that the interpretation in (8) validates these conditions:

Lemma 16. *The conditions (L1)–(L5) are satisfied in* $\mathcal{M}_\Gamma(D)$.

Write $\mathbf{P}(D)$ for the quotient of the free distributive lattice on the set

$$\{\mathbb{P}_{\geq p}\, a \mid p \in \mathbb{Q} \cap [0,1],\ a \in D\}$$

with respect to the congruence generated by the conditions (L1)–(L5).

Proposition 17. *Let $F \subseteq \mathbf{P}(D)$ be a prime filter. The assignment*

$$a \mapsto \bigvee \{q^\circ \mid \mathbb{P}_{\geq q}\, a \in F\} \quad \text{defines a measure } \mu_F \colon D \to \mathbf{\Gamma}.$$

Proof. Items (L2) and (L3) take care of the first two conditions defining $\mathbf{\Gamma}$-valued measures (cf. Definition 3). We prove the first half of the third condition, as the other half is proved in a similar fashion. We must show that, for every $a, b \in D$,

$$\mu_F(a) \sim \mu_F(a \wedge b) \leq \mu_F(a \vee b) - \mu_F(b). \tag{9}$$

It is not hard to show that $\mu_F(a) - r^\circ = \bigvee\{p_1^\circ - r^\circ \mid r^\circ \leq p_1^\circ \leq \mu_F(a)\}$, and $x - (\text{-})$ transforms non-empty joins into meets (this follows by Scott continuity of $x - (\text{-})$ seen as a map $[0^\circ, x] \to \mathbf{\Gamma}^\partial$). Hence, equation (9) is equivalent to

$$\bigvee \{p^\circ - r^\circ \mid \mu_F(a \wedge b) < r^\circ \leq p^\circ \leq \mu_F(a)\} \leq \bigwedge \{\mu_F(a \vee b) - q^\circ \mid q^\circ \leq \mu_F(b)\}.$$

To settle this inequality it is enough to show that, provided $\mu_F(a \wedge b) < r^\circ \leq p^\circ \leq \mu_F(a)$ and $q^\circ \leq \mu_F(b)$, we have $(p - r)^\circ \leq \mu_F(a \vee b) - q^\circ$. The latter inequality is equivalent to $(p + q - r)^\circ \leq \mu_F(a \vee b)$. In turn, using (L4) and the fact that F is a prime filter, $\mathbb{P}_{\geq p}\, a, \mathbb{P}_{\geq q}\, b \in F$ and $\mathbb{P}_{\geq r}\,(a \wedge b) \notin F$ entail $\mathbb{P}_{\geq p+q-r}\,(a \vee b) \in F$. Whence,

$$\mu_F(a \vee b) = \bigvee \{s^\circ \mid \mathbb{P}_{\geq s}\,(a \vee b) \in F\} \geq (p + q - r)^\circ. \qquad \square$$

We can now describe the dual lattice of $\mathcal{M}_\Gamma(D)$ as the Lindenbaum-Tarski algebra for the logic $P\mathcal{L}_D$, built from the propositional variables $\mathbb{P}_{\geq p}\, a$ by imposing the laws (L1)–(L5).

Theorem 18. *Let D be a distributive lattice. Then the lattice $\mathbf{P}(D)$ is isomorphic to the distributive lattice dual to the Priestley space $\mathcal{M}_\Gamma(D)$.*

Proof. Let $X_{\mathbf{P}(D)}$ be the space dual to $\mathbf{P}(D)$. By Proposition 17 there is a map $\vartheta \colon X_{\mathbf{P}(D)} \to \mathcal{M}_\Gamma(D)$, $F \mapsto \mu_F$. We claim that ϑ is an isomorphism of Priestley space. Clearly, ϑ is monotone. If $\mu_{F_1}(a) \nleq \mu_{F_2}(a)$ for some $a \in D$, we have

$$\bigvee \{q^\circ \mid \mathbb{P}_{\geq q}\, a \in F_1\} = \mu_{F_1}(a) \nleq \mu_{F_2}(a) = \bigwedge \{p^- \mid \mathbb{P}_{\geq p}\, a \notin F_2\}. \tag{10}$$

Equation (10) implies the existence of p, q satisfying $\mathbb{P}_{\geq q}\, a \in F_1$, $\mathbb{P}_{\geq p}\, a \notin F_2$ and $q \geq p$. It follows by (L1) that $\mathbb{P}_{\geq p}\, a \in F_1$. We conclude that $\mathbb{P}_{\geq p}\, a \in F_1 \setminus F_2$, whence $F_1 \nsubseteq F_2$. This shows that ϑ is an order embedding, whence injective.

We prove that ϑ is surjective, thus a bijection. Fix a measure $\mu \in \mathcal{M}_\Gamma(D)$. It is not hard to see, using Lemma 16, that the filter $F_\mu \subseteq \mathbf{P}(D)$ generated by

$$\{\mathbb{P}_{\geq q}\, a \mid a \in D,\ q \in \mathbb{Q} \cap [0,1],\ \mu(a) \geq q^\circ\}$$

is prime. Further, $\vartheta(F_\mu)(a) = \bigvee \{q^\circ \mid \mathbb{P}_{\geq q}\, a \in F_\mu\} = \bigvee \{q^\circ \mid \mu(a) \geq q^\circ\} = \mu(a)$ for every $a \in D$. Hence, $\vartheta(F_\mu) = \mu$ and ϑ is surjective.

To settle the theorem it remains to show that ϑ is continuous. Note that for a basic clopen of the form $C = \{\mu \in \mathcal{M}_\Gamma(D) \mid \mu(a) \geq p^\circ\}$ where $a \in D$ and $p \in \mathbb{Q} \cap [0,1]$, the preimage $\vartheta^{-1}(C) = \{F \subseteq \mathbf{P}(D) \mid \mu_F(a) \geq p^\circ\}$ is equal to

$$\{F \in X_{\mathbf{P}(D)} \mid \bigvee \{q^\circ \mid \mathbb{P}_{\geq q}\, a \in F\} \geq p^\circ\} = \{F \in X_{\mathbf{P}(D)} \mid \mathbb{P}_{\geq p}\, a \in F\},$$

which is a clopen of $X_{\mathbf{P}(D)}$. Similarly, if $C = \{\mu \in \mathcal{M}_\Gamma(D) \mid \mu(a) \leq q^-\}$ for some $a \in D$ and $q \in \mathbb{Q} \cap (0,1]$, by the claim above $\vartheta^{-1}(C) = \{F \in X_{\mathbf{P}(D)} \mid \mathbb{P}_{\geq q}\, a \notin F\}$, which is again a clopen of $X_{\mathbf{P}(D)}$. □

By Theorem 18, for any distributive lattice D, the lattice of clopen up-sets of $\mathcal{M}_\Gamma(D)$ is isomorphic to the Lindenbaum-Tarski algebra $\mathbf{P}(D)$ of our *positive* propositional logic $P\mathcal{L}_D$. Moving from the lattice of clopen up-sets to the Boolean algebra of all clopens logically corresponds to adding negation to the logic. The logic obtained this way can be presented as follows. Introduce a new propositional variable $\mathbb{P}_{<q}\, a$, for each $a \in D$ and $q \in \mathbb{Q} \cap [0,1]$. For a measure $\mu \in \mathcal{M}_\Gamma(D)$, set

$$\mu \models \mathbb{P}_{<q}\, a \ \Leftrightarrow\ \mu(a) < q^\circ.$$

We also add a new rule, stating that $\mathbb{P}_{<q}\, a$ is the negation of $\mathbb{P}_{\geq q}\, a$:

(L6) $\mathbb{P}_{<q}\, a \wedge \mathbb{P}_{\geq q}\, a \models$ and $\models \mathbb{P}_{<q}\, a \vee \mathbb{P}_{\geq q}\, a$

Clearly, (L6) is satisfied in $\mathcal{M}_\Gamma(D)$. Moreover, the Boolean algebra of *all* clopens of $\mathcal{M}_\Gamma(D)$ is isomorphic to the quotient of the free distributive lattice on

$$\{\mathbb{P}_{\geq p}\, a \mid p \in \mathbb{Q} \cap [0,1],\ a \in D\} \cup \{\mathbb{P}_{<q}\, b \mid q \in \mathbb{Q} \cap [0,1],\ b \in D\}$$

with respect to the congruence generated by the conditions (L1)–(L6).

Specialising to FO(σ). Let us briefly discuss what happens when we instantiate D with the full first-order logic FO(σ). For a formula $\varphi \in \mathrm{FO}(\sigma)$ with free variables $v_1, \ldots, v_n$ and a $q \in \mathbb{Q} \cap [0,1]$, we have two new sentences $\mathbb{P}_{\geq q}\, \varphi$ and $\mathbb{P}_{<q}\, \varphi$. For a finite σ-structure A identified with its $\mathbf{\Gamma}$-valued Stone pairing $\langle \text{-}, A \rangle_\Gamma$,

$$A \models \mathbb{P}_{\geq q}\, \varphi \ \ (\text{resp. } A \models \mathbb{P}_{<q}\, \varphi) \quad \text{iff} \quad \langle \varphi, A \rangle_\Gamma \geq q^\circ \ \ (\text{resp. } \langle \varphi, A \rangle_\Gamma < q^\circ).$$

That is, $\mathbb{P}_{\geq q}\, \varphi$ is true in A if a random assignment of the variables $v_1, \ldots, v_n$ in A satisfies φ with probability at least q. Similarly for $\mathbb{P}_{<q}\, \varphi$. If we regard $\mathbb{P}_{\geq q}$ and $\mathbb{P}_{<q}$ as probabilistic quantifiers that bind all free variables of a given formula, the Stone pairing $\langle \text{-}, \text{-} \rangle_\Gamma : \mathrm{Fin} \to \mathcal{M}_\Gamma(\mathrm{FO}(\sigma))$ can be seen as the embedding of finite structures into the space of types for the logic $P\mathcal{L}_{\mathrm{FO}(\sigma)}$.

Conclusion

Types are points of the dual space of a logic (viewed as a Boolean algebra). In classical first-order logic, 0-types are just the models modulo elementary equivalence. But when there are not 'enough' models, as in finite model theory, the spaces of types provide completions of the sets of models.

In [5], it was shown that for logic on words and various quantifiers we have that, given a Boolean algebra of formulas with a free variable, the space of types of the Boolean algebra generated by the formulas obtained by quantification is given by a measure space construction. Here we have shown that a suitable enrichment of first-order logic gives rise to a space of measures $\mathcal{M}_{\Gamma}(\mathrm{FO}(\sigma))$ closely related to the space $\mathcal{M}_{\mathrm{I}}(\mathrm{FO}(\sigma))$ used in the theory of structural limits. Indeed, Theorem 14 tells us that the ensuing Stone pairings interdetermine each other. Further, the Stone pairing for $\mathcal{M}_{\Gamma}(\mathrm{FO}(\sigma))$ is just the embedding of the finite models in the completion/compactification provided by the space of types of the enriched logic.

These results identify the logical gist of the theory of structural limits, and provide a new and interesting connection between logic on words and the theory of structural limits in finite model theory. But we also expect that it may prove a useful tool in its own right. Thus, for structural limits, it is an open problem to characterise the closure of the image of the $[0,1]$-valued Stone pairing [16]. Reasoning in the $\mathbf{\Gamma}$-valued setting, native to logic and where we can use duality, one would expect that this is the subspace $\mathcal{M}_{\Gamma}(\mathrm{Th}(\mathrm{Fin}))$ of $\mathcal{M}_{\Gamma}(\mathrm{FO}(\sigma))$ given by the quotient $\mathrm{FO}(\sigma) \twoheadrightarrow \mathrm{Th}(\mathrm{Fin})$ onto the theory of pseudofinite structures. The purpose of such a characterisation would be to understand the points of the closure as "generalised models". Another subject that we would like to investigate is that of zero-one laws. The zero-one law for first-order logic states that the sequence of measures for which the nth measure, on a sentence ψ, yields the proportion of n-element structures satisfying ψ, converges to a $\{0,1\}$-valued measure. Over $\mathbf{\Gamma}$ this will no longer be true as 1 is split into its 'limiting' and 'achieved' personae. Yet, we expect the above sequence to converge also in this setting and, by Theorem 14, it will converge to a $\{0^{\circ}, 1^{-}, 1^{\circ}\}$-valued measure. Understanding this more fine-grained measure may yield useful information about the zero-one law.

Further, it would be interesting to investigate whether the limits for schema mappings introduced by Kolaitis *et al.* [13] may be seen also as a type-theoretic construction. Finally, we would want to explore the connections with other semantically inspired approaches to finite model theory, such as those recently put forward by Abramsky, Dawar *et al.* [2,3].

References

1. Abramsky, S.: Domain theory in logical form. Ann. Pure Appl. Logic **51**, 1–77 (1991)
2. Abramsky, S., Dawar, A., Wang, P.: The pebbling comonad in finite model theory. In: 32nd Annual ACM/IEEE Symposium on Logic in Computer Science, LICS. pp. 1–12 (2017)
3. Abramsky, S., Shah, N.: Relating Structure and Power: Comonadic semantics for computational resources. In: 27th EACSL Annual Conference on Computer Science Logic, CSL. pp. 2:1–2:17 (2018)
4. Gehrke, M., Grigorieff, S., Pin, J.-É.: Duality and equational theory of regular languages. In: Automata, languages and programming II, LNCS, vol. 5126, pp. 246–257. Springer, Berlin (2008)
5. Gehrke, M., Petrişan, D., Reggio, L.: Quantifiers on languages and codensity monads. In: 32nd Annual ACM/IEEE Symposium on Logic in Computer Science, LICS. pp. 1–12 (2017)
6. Gehrke, M., Petrişan, D., Reggio, L.: Quantifiers on languages and codensity monads (2019), extended version. Submitted. Preprint available at https://arxiv.org/abs/1702.08841
7. Gehrke, M., Priestley, H.A.: Canonical extensions of double quasioperator algebras: an algebraic perspective on duality for certain algebras with binary operations. J. Pure Appl. Algebra **209**(1), 269–290 (2007)
8. Gehrke, M., Priestley, H.A.: Duality for double quasioperator algebras via their canonical extensions. Studia Logica **86**(1), 31–68 (2007)
9. Goldblatt, R.: Varieties of complex algebras. Ann. Pure Appl. Logic **44**(3), 173–242 (1989)
10. van Gool, S.J., Steinberg, B.: Pro-aperiodic monoids via saturated models. In: 34th Symposium on Theoretical Aspects of Computer Science, STACS. pp. 39:1–39:14 (2017)
11. Johnstone, P.T.: Stone spaces, Cambridge Studies in Advanced Mathematics, vol. 3. Cambridge University Press (1986), reprint of the 1982 edition
12. Jung, A.: Continuous domain theory in logical form. In: Coecke, B., Ong, L., Panangaden, P. (eds.) Computation, Logic, Games, and Quantum Foundations, Lecture Notes in Computer Science, vol. 7860, pp. 166–177. Springer Verlag (2013)
13. Kolaitis, P.G., Pichler, R., Sallinger, E., Savenkov, V.: Limits of schema mappings. Theory of Computing Systems **62**(4), 899–940 (2018)
14. Matz, O., Schweikardt, N.: Expressive power of monadic logics on words, trees, pictures, and graphs. In: Logic and Automata: History and Perspectives. pp. 531–552 (2008)
15. Nešetřil, J., Ossona de Mendez, P.: A model theory approach to structural limits. Commentationes Mathematicae Universitatis Carolinae **53**(4), 581–603 (2012)
16. Nešetřil, J., Ossona de Mendez, P.: First-order limits, an analytical perspective. European Journal of Combinatorics **52**, 368–388 (2016)
17. Nešetřil, J., Ossona de Mendez, P.: A unified approach to structural limits and limits of graphs with bounded tree-depth (2020), to appear in *Memoirs of the American Mathematical Society*
18. Pin, J.-É.: Profinite methods in automata theory. In: 26th Symposium on Theoretical Aspects of Computer Science, STACS. pp. 31–50 (2009)
19. Priestley, H.A.: Representation of distributive lattices by means of ordered Stone spaces. Bull. London Math. Soc. **2**, 186–190 (1970)

Correctness of Automatic Differentiation via Diffeologies and Categorical Gluing

Mathieu Huot ✉[1*], Sam Staton[1*], and Matthijs Vákár[2*]

[1] University of Oxford, UK
[2] Utrecht University, The Netherlands
*Equal contribution mathieu.huot@stx.ox.ac.uk

Abstract. We present semantic correctness proofs of Automatic Differentiation (AD). We consider a forward-mode AD method on a higher order language with algebraic data types, and we characterise it as the unique structure preserving macro given a choice of derivatives for basic operations. We describe a rich semantics for differentiable programming, based on diffeological spaces. We show that it interprets our language, and we phrase what it means for the AD method to be correct with respect to this semantics. We show that our characterisation of AD gives rise to an elegant semantic proof of its correctness based on a gluing construction on diffeological spaces. We explain how this is, in essence, a logical relations argument. Finally, we sketch how the analysis extends to other AD methods by considering a continuation-based method.

1 Introduction

Automatic differentiation (AD), loosely speaking, is the process of taking a program describing a function, and building the derivative of that function by applying the chain rule across the program code. As gradients play a central role in many aspects of machine learning, so too do automatic differentiation systems such as TensorFlow [1] or Stan [6].

Differentiation has a well developed mathematical theory in terms of differential geometry. The aim of this paper is to formalize this connection between differential geometry and the syntactic operations of AD. In this way we achieve two things: (1) a compositional, denotational understanding of differentiable programming and AD; (2) an explanation of the correctness of AD.

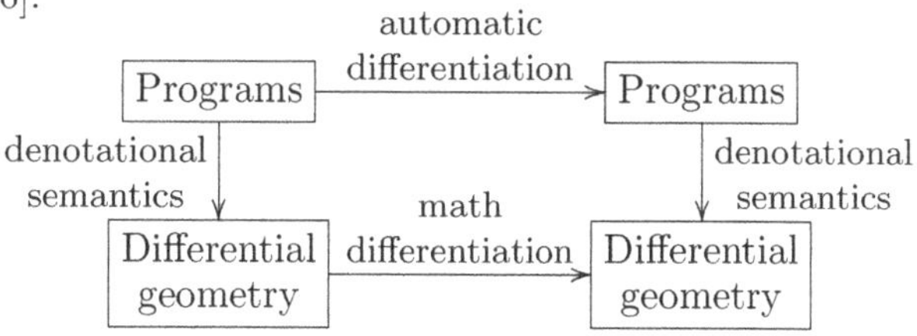

Fig. 1. Overview of semantics/correctness of AD.

This intuitive correspondence (summarized in Fig. 1) is in fact rather complicated. In this paper we focus on resolving the following problem: higher order functions play a key role in programming, and yet they have no counterpart in traditional differential geometry. Moreover, we resolve this problem while retaining the compositionality of denotational semantics.

J. Goubault-Larrecq and B. König (Eds.): FOSSACS 2020, LNCS 12077, pp. 319–338, 2020.
https://doi.org/10.1007/978-3-030-45231-5_17

Higher order functions and differentiation. A major application of higher order functions is to support disciplined code reuse. Code reuse is particularly acute in machine learning. For example, a multi-layer neural network might be built of millions of near-identical neurons, as follows.

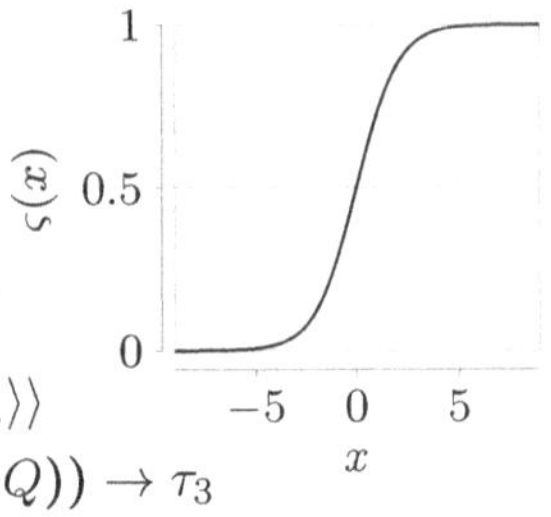

$$\begin{aligned}
&\mathrm{neuron}_n : (\mathbf{real}^n * (\mathbf{real}^n * \mathbf{real})) \to \mathbf{real}\\
&\mathrm{neuron}_n \stackrel{\mathrm{def}}{=} \lambda\langle x, \langle w, b\rangle\rangle.\, \varsigma(w \cdot x + b)\\
&\mathrm{layer}_n : ((\tau_1 * P) \to \tau_2) \to (\tau_1 * P^n) \to \tau_2^n\\
&\mathrm{layer}_n \stackrel{\mathrm{def}}{=} \lambda f.\, \lambda\langle x, \langle p_1, \dots, p_n\rangle\rangle.\, \langle f\langle x, p_1\rangle, \dots, f\langle x, p_n\rangle\rangle\\
&\mathrm{comp} : (((\tau_1 * P) \to \tau_2) * ((\tau_2 * Q) \to \tau_3)) \to (\tau_1 * (P * Q)) \to \tau_3\\
&\mathrm{comp} \stackrel{\mathrm{def}}{=} \lambda\langle f, g\rangle.\, \lambda\langle x, (p, q)\rangle.\, g\langle f\langle x, p\rangle, q\rangle
\end{aligned}$$

(Here $\varsigma(x) \stackrel{\mathrm{def}}{=} \frac{1}{1+e^{-x}}$ is the sigmoid function, as illustrated.) We can use these functions to build a network as follows (see also Fig. 2):

$$\mathrm{comp}\langle \mathrm{layer}_m(\mathrm{neuron}_k), \mathrm{comp}\langle \mathrm{layer}_n(\mathrm{neuron}_m), \mathrm{neuron}_n\rangle\rangle : (\mathbf{real}^k * P) \to \mathbf{real} \tag{1}$$

Here $P \cong \mathbf{real}^p$ with $p = (m(k+1)+n(m+1)+n+1)$. This program (1) describes a smooth (infinitely differentiable) function. The goal of automatic differentiation is to find its derivative.

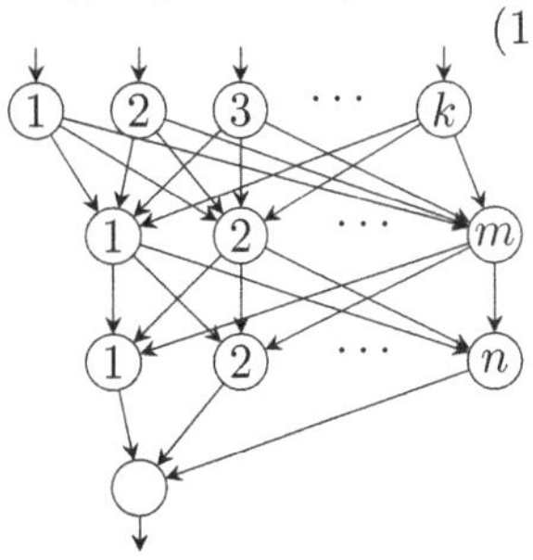

Fig. 2. The network in (1) with k inputs and two hidden layers.

If we β-reduce all the λ's, we end up with a very long function expression just built from the sigmoid function and linear algebra. We can then find a program for calculating its derivative by applying the chain rule. However, automatic differentiation can also be expressed without first β-reducing, in a compositional way, by explaining how higher order functions like (layer) and (comp) propagate derivatives. This paper is a semantic analysis of this compositional approach.

The general idea of denotational semantics is to interpret types as spaces and programs as functions between the spaces. In this paper, we propose to use diffeological spaces and smooth functions [32, 16] to this end. These satisfy the following three desiderata:

- $\mathbb{R}$ is a space, and the smooth functions $\mathbb{R} \to \mathbb{R}$ are exactly the functions that are infinitely differentiable;
- The set of smooth functions $X \to Y$ between spaces again forms a space, so we can interpret function types.
- The disjoint union of a sequence of spaces again forms a space, so we can interpret variant types and inductive types.

We emphasise that the most standard formulation of differential geometry, using manifolds, does not support spaces of functions. Diffeological spaces seem to us the simplest notion of space that satisfies these conditions, but there are other candidates [3,33]. A diffeological space is in particular a set X equipped with a chosen set of curves $C_X \subseteq X^{\mathbb{R}}$ and a smooth map $f : X \to Y$ must be such that if $\gamma \in C_X$ then $\gamma; f \in C_Y$. This is remiscent of the method of logical relations.

From smoothness to automatic derivatives at higher types. Our denotational semantics in diffeological spaces guarantees that all definable functions are smooth. But we need more than just to know that a definable function happens to have a mathematical derivative: we need to be able to find that derivative.

In this paper we focus on a simple, forward mode automatic differentiation method, which is a macro translation on syntax (called $\overrightarrow{\mathcal{D}}$ in §2). We are able to show that it is correct, using our denotational semantics.

Here there is one subtle point that is central to our development. Although differential geometry provides established derivatives for first order functions (such as neuron above), there is no canonical notion of derivative for higher order functions (such as layer and comp) in the theory of diffeological spaces (e.g. [7]). We propose a new way to resolve this, by interpreting types as triples (X, X', S) where, intuitively, X is a space of inhabitants of the type, X' is a space serving as a chosen bundle of tangents over X, and $S \subseteq X^{\mathbb{R}} \times X'^{\mathbb{R}}$ is a binary relation between curves, informally relating curves in X with their tangent curves in X'. This new model gives a denotational semantics for automatic differentiation.

In §3 we boil this new approach down to a straightforward and elementary logical relations argument for the correctness of automatic differentiation. The approach is explained in detail in §5.

Related work and context. AD has a long history and has many implementations. AD was perhaps first phrased in a functional setting in [26], and there are now a number of teams working on AD in the functional setting (e.g. [34, 31,12]), some providing efficient implementations. Although that work does not involve formal semantics, it is inspired by intuitions from differential geometry and category theory.

This paper adds to a very recent body of work on verified automatic differentiation. Much of this is concurrent with and independent from the work in this article. In the first order setting, there are recent accounts based on denotational semantics in manifolds [13] and based on synthetic differential geometry [9], as well as work making a categorical abstraction [8] and work connecting operational semantics with denotational semantics [2,28]. Recently there has also been significant progress at higher types. The work of Brunel et al. gives formal correctness proofs for reverse-mode derivatives on computation graphs [5]. The work of Barthe et al. [4] provides a general discussion of some new syntactic logical relations arguments including one very similar to our syntactic proof of Theorem 1. We understand that the authors of [9] are working on higher types.

The differential λ-calculus [11] is related to AD, and explicit connections are made in [22,23]. One difference is that the differential λ-calculus allows addition of terms at all types, and hence vector space models are suitable, but this appears peculiar with the variant and inductive types that we consider here.

Finally we emphasise that we have chosen the neural network (1) as our running example mainly for its simplicity. There are many other examples of AD outside the neural networks literature: AD is useful whenever derivatives need to be calculated on high dimensional spaces. This includes optimization problems more generally, where the derivative is passed to a gradient descent method (e.g. [30, 18, 29, 19, 10, 21]). Other applications of AD are in advanced *integration* methods, since derivatives play a role in Hamiltonian Monte Carlo [25, 14] and variational inference [20].

Summary of contributions. We have provided a semantic analysis of automatic differentiation. Our syntactic starting point is a well-known forward-mode AD macro on a typed higher order language (e.g. [31, 34]). We recall this in §2 for function types, and in §4 we extend it to inductive types and variants. The main contributions of this paper are as follows.

- We give a denotational semantics for the language in diffeological spaces, showing that every definable expression is smooth (§3).
- We show correctness of the AD macro by a logical relations argument (Th. 1).
- We give a categorical analysis of this correctness argument with two parts: canonicity of the macro in terms of syntactic categories, and a new notion of glued space that abstracts the logical relation (§5).
- We then use this analysis to state and prove a correctness argument at all first order types (Th. 2).
- We show that our method is not specific to one particular AD macro, by also considering a continuation-based AD method (§6).

2 A simple forward-mode AD translation

Rudiments of differentiation and dual numbers. Recall that the derivative of a function $f : \mathbb{R} \to \mathbb{R}$, if it exists, is a function $\nabla f : \mathbb{R} \to \mathbb{R}$ such that $\nabla f(x_0) = \frac{\mathrm{d}f(x)}{\mathrm{d}x}(x_0)$ is the gradient of f at x_0.

To find ∇f in a compositional way, two generalizations are reasonable:

- We need both f and ∇f when calculating $\nabla(f; g)$ of a composition $f; g$, using the chain rule, so we are really interested in the pair $(f, \nabla f) : \mathbb{R} \to \mathbb{R} \times \mathbb{R}$;
- In building f we will need to consider functions of multiple arguments, such as $+ : \mathbb{R}^2 \to \mathbb{R}$, and these functions should propagate derivatives.

Thus we are more generally interested in transforming a function $g : \mathbb{R}^n \to \mathbb{R}$ into a function $h : (\mathbb{R}\times\mathbb{R})^n \to \mathbb{R}\times\mathbb{R}$ in such a way that for any $f_1 \dots f_n : \mathbb{R} \to \mathbb{R}$,

$$(f_1, \nabla f_1, \dots, f_n, \nabla f_n); h = ((f_1, \dots, f_n); g, \nabla((f_1, \dots, f_n); g)). \tag{2}$$

An intuition for h is often given in terms of dual numbers. The transformed function operates on pairs of numbers, (x, x'), and it is common to think of such a pair as $x + x'\epsilon$ for an 'infinitesimal' ϵ. But while this is a helpful intuition, the formalization of infinitesimals can be intricate, and the development in this paper is focussed on the elementary formulation in (2).

The reader may also notice that h encodes all the partial derivatives of g. For example, if $g\colon \mathbb{R}^2 \to \mathbb{R}$, then with $f_1(x) \stackrel{\text{def}}{=} x$ and $f_2(x) \stackrel{\text{def}}{=} x_2$, by applying (2) to x_1 we obtain $h(x_1, 1, x_2, 0) = (g(x_1, x_2), \frac{\partial g(x, x_2)}{\partial x}(x_1))$ and similarly

$h(x_1, 0, x_2, 1) = (g(x_1, x_2), \frac{\partial g(x_1,x)}{\partial x}(x_2))$. And conversely, if g is differentiable in each argument, then a unique h satisfying (2) can be found by taking linear combinations of partial derivatives:

$$h(x_1, x_1', x_2, x_2') = (g(x_1, x_2), x_1' \cdot \frac{\partial g(x,x_2)}{\partial x}(x_1) + x_2' \cdot \frac{\partial g(x_1,x)}{\partial x}(x_2)).$$

In summary, the idea of differentiation with dual numbers is to transform a differentiable function $g : \mathbb{R}^n \to \mathbb{R}$ to a function $h : \mathbb{R}^{2n} \to \mathbb{R}^2$ which captures g and all its partial derivatives. We packaged this up in (2) as a sort-of invariant which is useful for building derivatives of compound functions $\mathbb{R} \to \mathbb{R}$ in a compositional way. The idea of forward mode automatic differentiation is to perform this transformation at the source code level.

A simple language of smooth functions. We consider a standard higher order typed language with a first order type **real** of real numbers. The types (τ, σ) and terms (t, s) are as follows.

τ, σ, ρ ::= types | $(\tau_1 * \ldots * \tau_n)$ finite product
| **real** real numbers | $\tau \to \sigma$ function

t, s, r ::= terms
x variable
| $\underline{c}$ | $t + s$ | $t * s$ | $\varsigma(t)$ operations/constants
| $\langle t_1, \ldots, t_n \rangle$ | $\mathbf{case}\, t\, \mathbf{of}\, \langle x_1, \ldots, x_n \rangle \to s$ tuples/pattern matching
| $\lambda x.t$ | $t\, s$ function abstraction/app.

The typing rules are in Figure 3. We have included a minimal set of operations for the sake of illustration, but it is not difficult to add further operations. We add some simple syntactic sugar $t - u \stackrel{\text{def}}{=} t + \underline{(-1)} * u$. We intend ς to stand for the sigmoid function, $\varsigma(x) \stackrel{\text{def}}{=} \frac{1}{1+e^{-x}}$. We further include syntactic sugar $\mathbf{let}\, x = t\, \mathbf{in}\, s$ for $(\lambda x.s)\, t$ and $\lambda\langle x_1, \ldots, x_n \rangle.t$ for $\lambda x.\mathbf{case}\, x\, \mathbf{of}\, \langle x_1, \ldots, x_n \rangle \to t$.

Syntactic automatic differentiation: a functorial macro. The aim of forward mode AD is to find the dual numbers representation of a function by syntactic manipulations. For our simple language, we implement this as the following inductively defined macro $\overrightarrow{\mathcal{D}}$ on both types and terms (see also [34, 31]):

$\overrightarrow{\mathcal{D}}(\mathbf{real}) \stackrel{\text{def}}{=} (\mathbf{real}{*}\mathbf{real})$ $\qquad$ $\overrightarrow{\mathcal{D}}(\tau \to \sigma) \stackrel{\text{def}}{=} \overrightarrow{\mathcal{D}}(\tau) \to \overrightarrow{\mathcal{D}}(\sigma)$

$\overrightarrow{\mathcal{D}}((\tau_1 * \cdots * \tau_n)) \stackrel{\text{def}}{=} (\overrightarrow{\mathcal{D}}(\tau_1) * \cdots * \overrightarrow{\mathcal{D}}(\tau_n))$

$$\frac{}{\Gamma \vdash \underline{c} : \mathbf{real}}(c \in \mathbb{R}) \qquad \frac{\Gamma \vdash t : \mathbf{real} \quad \Gamma \vdash s : \mathbf{real}}{\Gamma \vdash t + s : \mathbf{real}} \qquad \frac{\Gamma \vdash t : \mathbf{real} \quad \Gamma \vdash s : \mathbf{real}}{\Gamma \vdash t * s : \mathbf{real}} \qquad \frac{\Gamma \vdash t : \mathbf{real}}{\Gamma \vdash \varsigma(t) : \mathbf{real}}$$

$$\frac{\Gamma \vdash t_1 : \tau_1 \quad \ldots \quad \Gamma \vdash t_n : \tau_n}{\Gamma \vdash \langle t_1, \ldots, t_n \rangle : (\tau_1 * \ldots * \tau_n)} \qquad \frac{\Gamma \vdash t : (\sigma_1 * \ldots * \sigma_n) \quad \Gamma, x_1 : \sigma_1, \ldots, x_n : \sigma_n \vdash s : \tau}{\Gamma \vdash \mathbf{case}\, t\, \mathbf{of}\, \langle x_1, \ldots, x_n \rangle \to s : \tau}$$

$$\frac{}{\Gamma \vdash x : \tau}((x : \tau) \in \Gamma) \qquad \frac{\Gamma, x : \tau \vdash t : \sigma}{\Gamma \vdash \lambda x : \tau.t : \tau \to \sigma} \qquad \frac{\Gamma \vdash t : \sigma \to \tau \quad \Gamma \vdash s : \sigma}{\Gamma \vdash t\, s : \tau}$$

Fig. 3. Typing rules for the simple language.

$$\overrightarrow{\mathcal{D}}(x) \stackrel{\text{def}}{=} x \qquad\qquad \overrightarrow{\mathcal{D}}(\underline{c}) \stackrel{\text{def}}{=} \langle \underline{c}, 0\rangle$$
$$\overrightarrow{\mathcal{D}}(t+s) \stackrel{\text{def}}{=} \mathbf{case}\,\overrightarrow{\mathcal{D}}(t)\,\mathbf{of}\,\langle x, x'\rangle \to \mathbf{case}\,\overrightarrow{\mathcal{D}}(s)\,\mathbf{of}\,\langle y, y'\rangle \to \langle x+y, x'+y'\rangle$$
$$\overrightarrow{\mathcal{D}}(t*s) \stackrel{\text{def}}{=} \mathbf{case}\,\overrightarrow{\mathcal{D}}(t)\,\mathbf{of}\,\langle x, x'\rangle \to \mathbf{case}\,\overrightarrow{\mathcal{D}}(s)\,\mathbf{of}\,\langle y, y'\rangle \to \langle x*y, x*y'+x'*y\rangle$$
$$\overrightarrow{\mathcal{D}}(\varsigma(t)) \stackrel{\text{def}}{=} \mathbf{case}\,\overrightarrow{\mathcal{D}}(t)\,\mathbf{of}\,\langle x, x'\rangle \to \mathbf{let}\,y = \varsigma(x)\,\mathbf{in}\,\langle y, x'*y*(1-y)\rangle$$
$$\overrightarrow{\mathcal{D}}(\lambda x.t) \stackrel{\text{def}}{=} \lambda x.\overrightarrow{\mathcal{D}}(t) \quad \overrightarrow{\mathcal{D}}(t\,s) \stackrel{\text{def}}{=} \overrightarrow{\mathcal{D}}(t)\,\overrightarrow{\mathcal{D}}(s) \quad \overrightarrow{\mathcal{D}}(\langle t_1,\ldots,t_n\rangle) \stackrel{\text{def}}{=} \langle \overrightarrow{\mathcal{D}}(t_1),\ldots,\overrightarrow{\mathcal{D}}(t_n)\rangle$$
$$\overrightarrow{\mathcal{D}}(\mathbf{case}\,t\,\mathbf{of}\,\langle x_1,\ldots,x_n\rangle \to s) \stackrel{\text{def}}{=} \mathbf{case}\,\overrightarrow{\mathcal{D}}(t)\,\mathbf{of}\,\langle x_1,\ldots,x_n\rangle \to \overrightarrow{\mathcal{D}}(s)$$

We extend $\overrightarrow{\mathcal{D}}$ to contexts: $\overrightarrow{\mathcal{D}}(\{x_1{:}\tau_1,\ldots,x_n{:}\tau_n\}) \stackrel{\text{def}}{=} \{x_1{:}\overrightarrow{\mathcal{D}}(\tau_1),\ldots,x_n{:}\overrightarrow{\mathcal{D}}(\tau_n)\}$. This turns $\overrightarrow{\mathcal{D}}$ into a well-typed, functorial macro in the following sense.

Lemma 1 (Functorial macro). *If $\Gamma \vdash t : \tau$ then $\overrightarrow{\mathcal{D}}(\Gamma) \vdash \overrightarrow{\mathcal{D}}(t) : \overrightarrow{\mathcal{D}}(\tau)$.*
If $\Gamma, x : \sigma \vdash t : \tau$ and $\Gamma \vdash s : \sigma$ then $\overrightarrow{\mathcal{D}}(\Gamma) \vdash \overrightarrow{\mathcal{D}}(t[{}^{s}/_{x}]) = \overrightarrow{\mathcal{D}}(t)[{}^{\overrightarrow{\mathcal{D}}(s)}/_{x}]$.

Example 1 (Inner products). Let us write τ^n for the n-fold product $(\tau * \ldots * \tau)$. Then, given $\Gamma \vdash t, s : \mathbf{real}^n$ we can define their inner product

$$\Gamma \vdash t \cdot_n s \stackrel{\text{def}}{=} \mathbf{case}\,t\,\mathbf{of}\,\langle z_1,\ldots,z_n\rangle \to \mathbf{case}\,s\,\mathbf{of}\,\langle y_1,\ldots,y_n\rangle \to z_1 * y_1 + \cdots + z_n * y_n : \mathbf{real}$$

To illustrate the calculation of $\overrightarrow{\mathcal{D}}$, let us expand (and β-reduce) $\overrightarrow{\mathcal{D}}(t \cdot_2 s)$:

$$\mathbf{case}\,\overrightarrow{\mathcal{D}}(t)\,\mathbf{of}\,\langle z_1, z_2\rangle \to \mathbf{case}\,\overrightarrow{\mathcal{D}}(s)\,\mathbf{of}\,\langle y_1, y_2\rangle \to \mathbf{case}\,z_1\,\mathbf{of}\,\langle z_{1,1}, z_{1,2}\rangle \to \mathbf{case}\,y_1\,\mathbf{of}\,\langle y_{1,1}, y_{1,2}\rangle \to \mathbf{case}\,z_2\,\mathbf{of}\,\langle z_{2,1}, z_{2,2}\rangle \to \mathbf{case}\,y_2\,\mathbf{of}\,\langle y_{2,1}, y_{2,2}\rangle \to \langle z_{1,1}*y_{1,1} + z_{2,1}*y_{2,1}\,,\; z_{1,1}*y_{1,2} + z_{1,2}*y_{1,1} + z_{2,1}*y_{2,2} + z_{2,2}*y_{2,1}\rangle$$

Example 2 (Neural networks). In our introduction (1), we provided a program in our language to build a neural network out of expressions neuron, layer, comp; this program makes use of the inner product of Ex. 1. We can similarly calculate $\overrightarrow{\mathcal{D}}$ of such deep neural nets by mechanically applying the macro.

3 Semantics of differentiation

Consider for a moment the first order fragment of the language in § 2, with only one type, **real**, and no λ's or pairs. This has a simple semantics in the category of cartesian spaces and smooth maps. Indeed, a term $x_1 \ldots x_n : \mathbf{real} \vdash t : \mathbf{real}$ has a natural reading as a function $[\![t]\!] : \mathbb{R}^n \to \mathbb{R}$ by interpreting our operation symbols by the well-known operations on $\mathbb{R}^n \to \mathbb{R}$ with the corresponding name. In fact, the functions that are definable in this first order fragment are smooth, which means that they are continuous, differentiable, and their derivatives are continuous, differentiable, and so on. Let us write **CartSp** for this category of cartesian spaces ($\mathbb{R}^n$ for some n) and smooth functions.

The category **CartSp** has cartesian products, and so we can also interpret product types, tupling and pattern matching, giving us a useful syntax for constructing functions into and out of products of $\mathbb{R}$. For example, the interpretation of (neuron_n) in (1) becomes

$$\mathbb{R}^n \times \mathbb{R}^n \times \mathbb{R} \xrightarrow{[\![\cdot_n]\!] \times \mathrm{id}_{\mathbb{R}}} \mathbb{R} \times \mathbb{R} \xrightarrow{[\![+]\!]} \mathbb{R} \xrightarrow{[\![\varsigma]\!]} \mathbb{R}.$$

where $[\![\cdot_n]\!]$, $[\![+]\!]$ and $[\![\varsigma]\!]$ are the usual inner product, addition and the sigmoid function on $\mathbb{R}$, respectively.

Inside this category, we can straightforwardly study the first order language without λ's, and automatic differentiation. In fact, we can prove the following by plain induction on the syntax:
The interpretation of the (syntactic) forward AD $\overrightarrow{\mathcal{D}}(t)$ of a first-order term t equals the usual (semantic) derivative of the interpretation of t as a smooth function.

However, as is well known, the category **CartSp** does not support function spaces. To see this, notice that we have polynomial terms

$$x_1, \ldots, x_d : \mathbf{real} \vdash \lambda y.\ \sum_{n=1}^{d} x_n y^n : \mathbf{real} \to \mathbf{real}$$

for each d, and so if we could interpret ($\mathbf{real} \to \mathbf{real}$) as a Euclidean space $\mathbb{R}^p$ then, by interpreting these polynomial expressions, we would be able to find continuous injections $\mathbb{R}^d \to \mathbb{R}^p$ for every d, which is topologically impossible for any p, for example as a consequence of the Borsuk-Ulam theorem (see [15], Appx. A).

This means that we cannot interpret the functions (layer) and (comp) from (1) in **CartSp**, as they are higher order functions, even though they are very useful and innocent building blocks for differential programming! Clearly, we could define neural nets such as (1) directly as smooth functions without any higher order subcomponents, though that would quickly become cumbersome for deep networks. A problematic consequence of the lack of a semantics for higher order differential programs is that we have no obvious way of establishing compositional semantic correctness of $\overrightarrow{\mathcal{D}}$ for the given implementation of (1).

Diffeological spaces. This motivates us to turn to a more general notion of differential geometry for our semantics, based on *diffeological spaces* [16]. The key idea will be that a higher order function is called smooth if it sends smooth functions to smooth functions, meaning that we can never use it to build first order functions that are not smooth. For example, (comp) in (1) has this property.

Definition 1. *A* diffeological space $(X, \mathcal{P}_X)$ *consists of a set X together with, for each n and each open subset U of $\mathbb{R}^n$, a set $\mathcal{P}_X^U \subseteq [U \to X]$ of functions, called* plots, *such that*
- *all constant functions are plots;*
- *if $f : V \to U$ is a smooth function and $p \in \mathcal{P}_X^U$, then $f;p \in \mathcal{P}_X^V$;*
- *if $\left(p_i \in \mathcal{P}_X^{U_i}\right)_{i \in I}$ is a compatible family of plots ($x \in U_i \cap U_j \Rightarrow p_i(x) = p_j(x)$) and $(U_i)_{i \in I}$ covers U, then the gluing $p : U \to X : x \in U_i \mapsto p_i(x)$ is a plot.*

We call a function $f : X \to Y$ between diffeological spaces *smooth* if, for all plots $p \in \mathcal{P}_X^U$, we have that $p; f \in \mathcal{P}_Y^U$. We write $\mathbf{Diff}(X, Y)$ for the set of smooth maps from X to Y. Smooth functions compose, and so we have a category **Diff** of diffeological spaces and smooth functions.

A diffeological space is thus a set equipped with structure. Many constructions of sets carry over straightforwardly to diffeological spaces.

Example 3 (Cartesian diffeologies). Each open subset U of $\mathbb{R}^n$ can be given the structure of a diffeological space by taking all the smooth functions $V \to U$

as $\mathcal{P}_U^V$. It is easily seen that smooth functions from $V \to U$ in the traditional sense coincide with smooth functions in the sense of diffeological spaces. Thus diffeological spaces have a profound relationship with ordinary calculus.

In categorical terms, this gives a full embedding of **CartSp** in **Diff**.

Example 4 (Product diffeologies). Given a family $(X_i)_{i\in I}$ of diffeological spaces, we can equip the product $\prod_{i\in I} X_i$ of sets with the *product diffeology* in which U-plots are precisely the functions of the form $(p_i)_{i\in I}$ for $p_i \in \mathcal{P}_{X_i}^U$.

This gives us the categorical product in **Diff**.

Example 5 (Functional diffeology). We can equip the set $\mathbf{Diff}(X, Y)$ of smooth functions between diffeological spaces with the *functional diffeology* in which U-plots consist of functions $f : U \to \mathbf{Diff}(X, Y)$ such that $(u, x) \mapsto f(u)(x)$ is an element of $\mathbf{Diff}(U \times X, Y)$.

This specifies the categorical function object in **Diff**.

Semantics and correctness of AD. We can now give a denotational semantics to our language from § 2. We interpret each type τ as a set $[\![\tau]\!]$ equipped with the relevant diffeology, by induction on the structure of types:

$$[\![\mathbf{real}]\!] \stackrel{\text{def}}{=} \mathbb{R} \qquad [\![(\tau_1 * \ldots * \tau_n)]\!] \stackrel{\text{def}}{=} \prod_{i=1}^n [\![\tau_i]\!] \qquad [\![\tau \to \sigma]\!] \stackrel{\text{def}}{=} \mathbf{Diff}([\![\tau]\!], [\![\sigma]\!])$$

A context $\Gamma = (x_1 : \tau_1 \ldots x_n : \tau_n)$ is interpreted as a diffeological space $[\![\Gamma]\!] \stackrel{\text{def}}{=} \prod_{i=1}^n [\![\tau_i]\!]$. Now well typed terms $\Gamma \vdash t : \tau$ are interpreted as smooth functions $[\![t]\!] : [\![\Gamma]\!] \to [\![\tau]\!]$, giving a meaning for t for every valuation of the context. This is routinely defined by induction on the structure of typing derivations. Constants $\underline{c}$: **real** are interpreted as constant functions; and the first order operations $(+, *, \varsigma)$ are interpreted by composing with the corresponding functions, which are smooth. For example, $[\![\varsigma(t)]\!](\rho) \stackrel{\text{def}}{=} \varsigma([\![t]\!](\rho))$, where $\rho \in [\![\Gamma]\!]$. Variables are interpreted as $[\![x_i]\!](\rho) \stackrel{\text{def}}{=} \rho_i$. The remaining constructs are interpreted as follows, and it is straightforward to show that smoothness is preserved.

$$[\![\langle t_1, \ldots, t_n \rangle]\!](\rho) \stackrel{\text{def}}{=} ([\![t_1]\!](\rho), \ldots, [\![t_n]\!](\rho)) \quad [\![\lambda x{:}\tau.t]\!](\rho)(a) \stackrel{\text{def}}{=} [\![t]\!](\rho, a)\ (a \in [\![\tau]\!])$$

$$[\![\mathbf{case}\, t\, \mathbf{of}\, \langle \ldots \rangle \to s]\!](\rho) \stackrel{\text{def}}{=} [\![s]\!](\rho, [\![t]\!](\rho)) \qquad [\![t\, s]\!](\rho) \stackrel{\text{def}}{=} [\![t]\!](\rho)([\![s]\!](\rho))$$

Notice that a term x_1: **real**, $\ldots$, x_n: **real** $\vdash t$: **real** is interpreted as a smooth function $[\![t]\!] : \mathbb{R}^n \to \mathbb{R}$, even if t involves higher order functions (like (1)). Moreover the macro differentiation $\overrightarrow{\mathcal{D}}(t)$ is a function $[\![\overrightarrow{\mathcal{D}}(t)]\!] : (\mathbb{R} \times \mathbb{R})^n \to (\mathbb{R} \times \mathbb{R})$. This enables us to state a limited version of our main correctness theorem:

Theorem 1 (Semantic correctness of $\overrightarrow{\mathcal{D}}$ (limited)). *For any term* x_1: **real**, $\ldots$, x_n: **real** $\vdash t$: **real**, *the function* $[\![\overrightarrow{\mathcal{D}}(t)]\!]$ *is the dual numbers representation* (2) *of* $[\![t]\!]$. *In detail: for any smooth functions* $f_1 \ldots f_n : \mathbb{R} \to \mathbb{R}$,

$$(f_1, \nabla f_1, \ldots, f_n, \nabla f_n); [\![\overrightarrow{\mathcal{D}}(t)]\!] = \big((f_1 \ldots f_n); [\![t]\!], \nabla((f_1 \ldots f_n); [\![t]\!])\big).$$

(For instance, if $n = 2$, then $[\![\overrightarrow{\mathcal{D}}(t)]\!](x_1, 1, x_2, 0) = ([\![t]\!](x_1, x_2), \frac{\partial [\![t]\!](x, x_2)}{\partial x}(x_1))$.)

Proof. We prove this by logical relations. Although the following proof is elementary, we found it by using the categorical methods in § 5.

For each type τ, we define a binary relation S_τ between curves in $[\![\tau]\!]$ and curves in $[\![\overrightarrow{\mathcal{D}}(\tau)]\!]$, i.e. $S_\tau \subseteq \mathcal{P}^{\mathbb{R}}_{[\![\tau]\!]} \times \mathcal{P}^{\mathbb{R}}_{[\![\overrightarrow{\mathcal{D}}(\tau)]\!]}$, by induction on τ:

- $S_{\mathbf{real}} \stackrel{\text{def}}{=} \{(f, (f, \nabla f)) \mid f : \mathbb{R} \to \mathbb{R} \text{ smooth}\}$;
- $S_{(\tau * \sigma)} \stackrel{\text{def}}{=} \{((f_1, g_1), (f_2, g_2)) \mid (f_1, f_2) \in S_\tau, (g_1, g_2) \in S_\sigma\}$;
- $S_{\tau \to \sigma} \stackrel{\text{def}}{=} \{(f_1, f_2) \mid \forall (g_1, g_2) \in S_\tau.(x \mapsto f_1(x)(g_1(x)), x \mapsto f_2(x)(g_2(x))) \in S_\sigma\}$.

Then, we establish the following 'fundamental lemma':

> If $x_1{:}\tau_1, ..., x_n{:}\tau_n \vdash t : \sigma$ and, for all $1 \le i \le n$, $y_1...y_m : \mathbf{real} \vdash s_i : \tau_i$ is such that $((f_1, \ldots, f_m); [\![s_i]\!], (f_1, \nabla f_1), ..., f_m, \nabla f_m); [\![\overrightarrow{\mathcal{D}}(s_i)]\!]) \in S_{\tau_i}$ for all smooth $f_i : \mathbb{R} \to \mathbb{R}$, then
> $\big((f_1, ..., f_m); [\![t[^{s_1}/_{x_1}, ..., ^{s_n}/_{x_n}]]\!], (f_1, \nabla f_1, ..., f_m, \nabla f_m); [\![\overrightarrow{\mathcal{D}}(t[^{s_1}/_{x_1}, ..., ^{s_n}/_{x_n}])]\!]\big)$
> is in S_σ for all smooth $f_i : \mathbb{R} \to \mathbb{R}$.

This is proved routinely by induction on the typing derivation of t. The case for $*$ relies on the precise definition of $\overrightarrow{\mathcal{D}}(t * s)$, and similarly for $+, \varsigma$.

We conclude the theorem from the fundamental lemma by considering the case where $\tau_i = \sigma = \mathbf{real}$, $m = n$ and $s_i = y_i$. □

4 Extending the language: variant and inductive types

In this section, we show that the definition of forward AD and the semantics generalize if we extend the language of §2 with variants and inductive types. As an example of inductive types, we consider lists. This specific choice is only for expository purposes and the whole development works at the level of generality of arbitrary algebraic data types generated as initial algebras of (polynomial) type constructors formed by finite products and variants.

Similarly, our choice of operations is for expository purposes. More generally, assume given a family of operations $(\mathsf{Op}_n)_{n \in \mathbb{N}}$ indexed by their arity n. Further assume that each $\mathsf{op} \in \mathsf{Op}_n$ has type $\mathbf{real}^n \to \mathbf{real}$. We then ask for a certain closure of these operations under differentiation, that is we define

$$\overrightarrow{\mathcal{D}}(\mathsf{op}(t_1, \ldots, t_n)) \stackrel{\text{def}}{=} \mathbf{case}\, \overrightarrow{\mathcal{D}}(t_1)\, \mathbf{of}\, \langle x_1, x_1' \rangle \to \ldots \to \mathbf{case}\, \overrightarrow{\mathcal{D}}(t_n)\, \mathbf{of}\, \langle x_n, x_n' \rangle \to \langle \mathsf{op}(x_1, \ldots, x_n), \textstyle\sum_{i=1}^{n} x_i' * \partial_i \mathsf{op}(x_1, \ldots, x_n) \rangle$$

where $\partial_i \mathsf{op}(x_1, \ldots, x_n)$ is some chosen term in the language, involving free variables from $x_1, \ldots, x_n$, which we think of as implementing the partial derivative of op with respect to its i-th argument. For constructing the semantics, every op must be interpreted by some smooth function, and, to establish correctness, the semantics of $\partial_i \mathsf{op}(x_1, \ldots, x_n)$ must be the semantic i-th partial derivative of the semantics of $\mathsf{op}(x_1, \ldots, x_n)$.

Language. We additionally consider the following types and terms:

τ, σ, ρ ::=		types	| $\mathbf{list}(\tau)$	list
	| $\{\ell_1\, \tau_1 \mid \ldots \mid \ell_n\, \tau_n\}$	variant		

t, s, r ::=		terms
	$\mid \tau.\ell\, t$	variant constructor
	$\mid []\mid t :: s$	empty list and cons
	$\mid \textbf{case}\, t\, \textbf{of}\, \{\ell_1\, x_1 \to s_1 \mid \cdots \mid \ell_n\, x_n \to s_n\}$	pattern matching: variants
	$\mid \textbf{fold}\, (x_1, x_2).t\, \textbf{over}\, s\, \textbf{from}\, r$	list fold

We extend the type system according to:

$$\frac{\Gamma \vdash t : \tau_i}{\Gamma \vdash \tau.\ell_i\, t : \tau}((\ell_i\, \tau_i) \in \tau) \qquad \frac{}{\Gamma \vdash [] : \textbf{list}(\tau)} \qquad \frac{\Gamma \vdash t : \tau \qquad \Gamma \vdash s : \textbf{list}(\tau)}{\Gamma \vdash t :: s : \textbf{list}(\tau)}$$

$$\frac{\Gamma \vdash t : \{\ell_1\, \tau_1 \mid \ldots \mid \ell_n\, \tau_n\} \qquad \text{for each } 1 \leq i \leq n\text{: } \Gamma, x_i : \tau_i \vdash s_i : \tau}{\Gamma \vdash \textbf{case}\, t\, \textbf{of}\, \{\ell_1\, x_1 \to s_1 \mid \cdots \mid \ell_n\, x_n \to s_n\} : \tau}$$

$$\frac{\Gamma \vdash s : \textbf{list}(\tau) \qquad \Gamma \vdash r : \sigma \qquad \Gamma, x_1 : \tau, x_2 : \sigma \vdash t : \sigma}{\Gamma \vdash \textbf{fold}\, (x_1, x_2).t\, \textbf{over}\, s\, \textbf{from}\, r : \sigma}$$

We can then extend $\overrightarrow{\mathcal{D}}$ to our new types and terms by

$$\overrightarrow{\mathcal{D}}(\{\ell_1\, \tau_1 \mid \ldots \mid \ell_n\, \tau_n\}) \stackrel{\text{def}}{=} \{\ell_1\, \overrightarrow{\mathcal{D}}(\tau_1) \mid \ldots \mid \ell_n\, \overrightarrow{\mathcal{D}}(\tau_n)\} \qquad \overrightarrow{\mathcal{D}}(\textbf{list}(\tau)) \stackrel{\text{def}}{=} \textbf{list}(\overrightarrow{\mathcal{D}}(\tau))$$

$$\overrightarrow{\mathcal{D}}(\tau.\ell\, t) \stackrel{\text{def}}{=} \overrightarrow{\mathcal{D}}(\tau).\ell\, \overrightarrow{\mathcal{D}}(t) \qquad \overrightarrow{\mathcal{D}}([]) \stackrel{\text{def}}{=} [] \qquad \overrightarrow{\mathcal{D}}(t :: s) \stackrel{\text{def}}{=} \overrightarrow{\mathcal{D}}(t) :: \overrightarrow{\mathcal{D}}(s)$$

$$\overrightarrow{\mathcal{D}}(\textbf{case}\, t\, \textbf{of}\, \{\ell_1\, x_1 \to s_1 \mid \cdots \mid \ell_n\, x_n \to s_n\}) \stackrel{\text{def}}{=}$$
$$\textbf{case}\, \overrightarrow{\mathcal{D}}(t)\, \textbf{of}\, \{\ell_1\, x_1 \to \overrightarrow{\mathcal{D}}(s_1) \mid \cdots \mid \ell_n\, x_n \to \overrightarrow{\mathcal{D}}(s_n)\}$$

$$\overrightarrow{\mathcal{D}}(\textbf{fold}\, (x_1, x_2).t\, \textbf{over}\, s\, \textbf{from}\, r) \stackrel{\text{def}}{=} \textbf{fold}\, (x_1, x_2).\overrightarrow{\mathcal{D}}(t)\, \textbf{over}\, \overrightarrow{\mathcal{D}}(s)\, \textbf{from}\, \overrightarrow{\mathcal{D}}(r)$$

To demonstrate the practical use of expressive type systems for differential programming, we consider the following two examples.

Example 6 (Lists of inputs for neural nets). Usually, we run a neural network on a large data set, the size of which might be determined at runtime. To evaluate a neural network on multiple inputs, in practice, one often sums the outcomes. This can be coded in our extended language as follows. Suppose that we have a network $f : (\textbf{real}^n * P) \to \textbf{real}$ that operates on single input vectors. We can construct one that operates on lists of inputs as follows:

$$g \stackrel{\text{def}}{=} \lambda\langle l, w\rangle.\textbf{fold}\, (x_1, x_2).f\langle x_1, w\rangle + x_2\, \textbf{over}\, l\, \textbf{from}\, \underline{0} : (\textbf{list}(\textbf{real}^n) * P) \to \textbf{real}$$

Example 7 (Missing data). In practically every application of statistics and machine learning, we face the problem of *missing data*: for some observations, only partial information is available. In an expressive typed programming language like we consider, we can model missing data conveniently using the data type $\textbf{maybe}(\tau) = \{\mathsf{Nothing}\, () \mid \mathsf{Just}\, \tau\}$. In the context of a neural network, one might use it as follows. First, define some helper functions

$$\text{fromMaybe} \stackrel{\text{def}}{=} \lambda x.\lambda m.\textbf{case}\, m\, \textbf{of}\, \{\mathsf{Nothing}\, \_ \to x \mid \mathsf{Just}\, x' \to x'\}$$

$$\text{fromMaybe}^n \stackrel{\text{def}}{=} \lambda\langle x_1, ..., x_n\rangle.\lambda\langle m_1, ..., m_n\rangle.\langle \text{fromMaybe}\, x_1\, m_1, ..., \text{fromMaybe}\, x_n\, m_n\rangle$$
$$: (\textbf{maybe}(\textbf{real}))^n \to \textbf{real}^n \to \textbf{real}^n$$

$$\text{map} \stackrel{\text{def}}{=} \lambda f.\lambda l.\textbf{fold}\, (x_1, x_2).f\, x_1 :: x_2\, \textbf{over}\, l\, \textbf{from}\, [] : (\tau \to \sigma) \to \textbf{list}(\tau) \to \textbf{list}(\sigma)$$

Given a neural network $f : (\mathbf{list}(\mathbf{real}^k)*P) \to \mathbf{real}$, we can build a new one that operates on on a data set for which some covariates (features) are missing, by passing in default values to replace the missing covariates:

$$\lambda\langle l, \langle m, w\rangle\rangle.f\,\langle \mathrm{map}\,(\mathrm{fromMaybe}^k\ m)\,l, w\rangle$$
$$: (\mathbf{list}((\mathbf{maybe}(\mathbf{real}))^k)*(\mathbf{real}^k*P)) \to \mathbf{real}$$

Then, given a data set l with missing covariates, we can perform automatic differentiation on this network to optimize, simultaneously, the ordinary network parameters w *and* the default values for missing covariates m.

Semantics. In § 3 we gave a denotational semantics for the simple language in diffeological spaces. This extends to the language in this section, as follows. As before, each type τ is interpreted as a diffeological space, which is a set equipped with a family of plots:

- A variant type $\{\ell_1\,\tau_1 \mid \ldots \mid \ell_n\,\tau_n\}$ is inductively interpreted as the disjoint union $[\![\{\ell_1\,\tau_1 \mid \cdots \mid \ell_n\,\tau_n\}]\!] \stackrel{\text{def}}{=} \biguplus_{i=1}^n [\![\tau_i]\!]$ with U-plots
$$\mathcal{P}^U_{[\![\{\ell_1\,\tau_1|\ldots|\ell_n\,\tau_n\}]\!]} \stackrel{\text{def}}{=} \left\{ \left[U_j \xrightarrow{f_j} [\![\tau_j]\!] \to \biguplus_{i=1}^n [\![\tau_i]\!] \right]_{j=1}^n \;\middle|\; U = \biguplus_{j=1}^n U_j,\ f_j \in \mathcal{P}^{U_j}_{[\![\tau_j]\!]} \right\}.$$
- A list type $\mathbf{list}(\tau)$ is interpreted as the set of lists, $[\![\mathbf{list}(\tau)]\!] \stackrel{\text{def}}{=} \biguplus_{i=1}^\infty [\![\tau]\!]^i$ with U-plots
$$\mathcal{P}^U_{[\![\mathbf{list}(\tau)]\!]} \stackrel{\text{def}}{=} \left\{ \left[U_j \xrightarrow{f_j} [\![\tau]\!]^j \to \biguplus_{i=1}^\infty [\![\tau]\!]^i \right]_{j=1}^\infty \;\middle|\; U = \biguplus_{j=1}^\infty U_j,\ f_j \in \mathcal{P}^{U_j}_{[\![\tau]\!]^j} \right\}.$$

The constructors and destructors for variants and lists are interpreted as in the usual set theoretic semantics. It is routine to show inductively that these interpretations are smooth. Thus every term $\Gamma \vdash t : \tau$ in the extended language is interpreted as a smooth function $[\![t]\!] : [\![\Gamma]\!] \to [\![\tau]\!]$ between diffeological spaces.

(In this section we focused on a language with lists, but other inductive types are easily interpreted in the category of diffeological spaces in much the same way; the categorically minded reader may regard this as a consequence of **Diff** being a concrete Grothendieck quasitopos, e.g. [3].)

5 Categorical analysis of forward AD and its correctness

This section has three parts. First, we give a categorical account of the functoriality of AD (Ex. 8). Then we introduce our gluing construction, and relate it to the correctness of AD (dgm. (3)). Finally, we state and prove a correctness theorem for all first order types by considering a category of manifolds (Th. 2).

Syntactic categories. Our language induces a syntactic category as follows.

Definition 2. *Let* **Syn** *be the category whose objects are types, and where a morphism $\tau \to \sigma$ is a term in context $x : \tau \vdash t : \sigma$ modulo the $\beta\eta$-laws (Fig. 4). Composition is by substitution.*

For simplicity, we do not impose arithmetic identities such as $x + y = y + x$ in **Syn**. As is standard, this category has the following universal property.

Lemma 2 (e.g. [27]). *For every bicartesian closed category $\mathcal{C}$ with list objects, and every object $F(\mathbf{real}) \in \mathcal{C}$ and morphisms $F(\underline{c}) \in \mathcal{C}(1, F(\mathbf{real}))$, $F(+), F(*) \in \mathcal{C}(F(\mathbf{real}) \times F(\mathbf{real}), F(\mathbf{real}))$, $F(\varsigma) \in \mathbf{Syn}(F(\mathbf{real}), F(\mathbf{real}))$ in $\mathcal{C}$, there is a unique functor $F : \mathbf{Syn} \to \mathcal{C}$ respecting the interpretation and preserving the bicartesian closed structure as well as list objects.*

Proof (notes). The functor $F : \mathbf{Syn} \to \mathcal{C}$ is a canonical denotational semantics for the language, interpreting types as objects of $\mathcal{C}$ and terms as morphisms. For instance, $F(\tau \to \sigma) \stackrel{\text{def}}{=} (F\tau \to F\sigma)$, the function space in the category $\mathcal{C}$, and $F(t\,s) \stackrel{\text{def}}{=}$ is the composite $(Ft, Fs); eval$. When $\mathcal{C} = \mathbf{Diff}$, the denotational semantics of the language in diffeological spaces (§3,4) can be understood as the unique structure preserving functor $[\![-]\!] : \mathbf{Syn} \to \mathbf{Diff}$ satisfying $[\![\mathbf{real}]\!] = \mathbb{R}$, $[\![\varsigma]\!] = \varsigma$ and so on. □

Example 8 (Canonical definition forward AD). The forward AD macro $\overrightarrow{\mathcal{D}}$ (§2,4) arises as a canonical cartesian closed functor on **Syn**. Consider the unique cartesian closed functor $F : \mathbf{Syn} \to \mathbf{Syn}$ such that $F(\mathbf{real}) = \mathbf{real}{*}\mathbf{real}$, $F(\underline{c}) = \overrightarrow{\mathcal{D}}(\underline{c})$, $F(\varsigma) = \overrightarrow{\mathcal{D}}(\varsigma(x))$, and

$F(+) = z : F(\mathbf{real}){*}F(\mathbf{real}) \vdash \mathbf{case}\, z \,\mathbf{of}\, \langle x, y\rangle \to \overrightarrow{\mathcal{D}}(x + y) : F(\mathbf{real})$

$F(*) = z : F(\mathbf{real}){*}F(\mathbf{real}) \vdash \mathbf{case}\, z \,\mathbf{of}\, \langle x, y\rangle \to \overrightarrow{\mathcal{D}}(x * y) : F(\mathbf{real})$

Then for any type τ, $F(\tau) = \overrightarrow{\mathcal{D}}(\tau)$, and for any term $x : \tau \vdash t : \sigma$, $F(t) = \overrightarrow{\mathcal{D}}(t)$ as morphisms $F(\tau) \to F(\sigma)$ in the syntactic category.

Categorical gluing and logical relations. Gluing is a method for building new categorical models which has been used for many purposes, including logical relations and realizability [24]. Our logical relations argument in the proof of Th. 1 can be understood in this setting. In this subsection, for the categorically minded, we explain this, and in doing so we quickly recover a correctness result for the more general language in § 4 and for arbitrary first order types.

We define a category $\mathbf{Gl}_U$ whose objects are triples (X, X', S) where X and X' are diffeological spaces and $S \subseteq \mathcal{P}^U_X \times \mathcal{P}^U_{X'}$ is a relation between their U-plots. A morphism $(X, X', S) \to (Y, Y', T)$ is a pair of smooth functions

$\mathbf{case}\, \langle t_1, \ldots, t_n\rangle \,\mathbf{of}\, \langle x_1, \ldots, x_n\rangle \to s = s[{}^{t_1}/_{x_1}, \ldots, {}^{t_n}/_{x_n}]$

$s[{}^{t}/_{y}] \stackrel{\#x_1,\ldots,x_n}{=} \mathbf{case}\, t \,\mathbf{of}\, \langle x_1, \ldots, x_n\rangle \to s[{}^{\langle x_1,\ldots,x_n\rangle}/_{y}]$

$\mathbf{case}\, \ell_i\, t \,\mathbf{of}\, \{\ell_1\, x_1 \to s_1 \mid \cdots \mid \ell_n\, x_n \to s_n\} = s_i[{}^{t}/_{x_i}]$

$s[{}^{t}/_{y}] \stackrel{\#x_1,\ldots,x_n}{=} \mathbf{case}\, t \,\mathbf{of}\, \{\ell_1\, x_1 \to s[{}^{\ell_1\, x_1}/_{y}] \mid \cdots \mid \ell_n\, x_n \to s[{}^{\ell_n\, x_n}/_{y}]\}$

$\mathbf{fold}\, (x_1, x_2).t \,\mathbf{over}\, [\,] \,\mathbf{from}\, r = r$

$\mathbf{fold}\, (x_1, x_2).t \,\mathbf{over}\, s_1 :: s_2 \,\mathbf{from}\, r = t[{}^{s_1}/_{x_1}, {}^{\mathbf{fold}\,(x_1,x_2).t\,\mathbf{over}\,s_2\,\mathbf{from}\,r}/_{x_2}]$

$u = s[{}^{[\,]}/_{y}], r[{}^{s}/_{x_2}] = s[{}^{x_1::y}/_{y}] \Rightarrow s[{}^{t}/_{y}] \stackrel{\#x_1,x_2}{=} \mathbf{fold}\, (x_1, x_2).r \,\mathbf{over}\, t \,\mathbf{from}\, u$

$(\lambda x.t)\, s = t[{}^{s}/_{x}]$

$t \stackrel{\#x}{=} \lambda x.t\, x$

We write $\stackrel{\#x_1,\ldots,x_n}{=}$ to indicate that the variables are free in the left hand side.

Fig. 4. Standard $\beta\eta$-laws (e.g. [27]) for products, functions, variants and lists.

$f: X \to Y$, $f': X' \to Y'$, such that if $(g, g') \in S$ then $(g; f, g'; f') \in T$. The idea is that this is a semantic domain where we can simultaneously interpret the language and its automatic derivatives.

Proposition 1. *The category* $\mathbf{Gl}_U$ *is bicartesian closed, has list objects, and the projection functor* $\mathrm{proj} : \mathbf{Gl}_U \to \mathbf{Diff} \times \mathbf{Diff}$ *preserves this structure.*

Proof (notes). The category $\mathbf{Gl}_U$ is a full subcategory of the comma category $\mathrm{id}_{\mathbf{Set}} \downarrow \mathbf{Diff}(U, -) \times \mathbf{Diff}(U, -)$. The result thus follows by the general theory of categorical gluing (e.g. [17, Lemma 15]). □

We give a semantics $\llparenthesis - \rrparenthesis = (\llparenthesis - \rrparenthesis_0, \llparenthesis - \rrparenthesis_1, S_-)$ for the language in $\mathbf{Gl}_{\mathbb{R}}$, interpreting types τ as objects $(\llparenthesis \tau \rrparenthesis_0, \llparenthesis \tau \rrparenthesis_1, S_\tau)$, and terms as morphisms. We let $\llparenthesis \mathbf{real} \rrparenthesis_0 \stackrel{\text{def}}{=} \mathbb{R}$ and $\llparenthesis \mathbf{real} \rrparenthesis_1 \stackrel{\text{def}}{=} \mathbb{R}^2$, with the relation $S_{\mathbf{real}} \stackrel{\text{def}}{=} \{(f, (f, \nabla f)) \mid f : \mathbb{R} \to \mathbb{R} \text{ smooth}\}$. We interpret the constants $\underline{c}$ as pairs $\llparenthesis \underline{c} \rrparenthesis_0 \stackrel{\text{def}}{=} \underline{c}$ and $\llparenthesis \underline{c} \rrparenthesis_1 \stackrel{\text{def}}{=} (\underline{c}, 0)$, and we interpret $+, \times, \varsigma$ in the standard way (meaning, like $[\![-]\!]$) in $\llparenthesis - \rrparenthesis_0$, but according to the derivatives in $\llparenthesis - \rrparenthesis_1$, for instance, $\llparenthesis * \rrparenthesis_1 : \mathbb{R}^2 \times \mathbb{R}^2 \to \mathbb{R}^2$ is

$$\llparenthesis * \rrparenthesis_1((x, x'), (y, y')) \stackrel{\text{def}}{=} (xy, xy' + x'y).$$

At this point one checks that these interpretations are indeed morphisms in $\mathbf{Gl}_{\mathbb{R}}$. This amounts to checking that these interpretations are dual numbers representations in the sense of (2). The remaining constructions of the language are interpreted using the categorical structure of $\mathbf{Gl}_{\mathbb{R}}$, following Lem. 2.

Notice that the diagram below commutes. One can check this by hand or note that it follows from the initiality of $\mathbf{Syn}$ (Lem. 2): all the functors preserve all the structure.

$$\begin{array}{ccc} \mathbf{Syn} & \xrightarrow{(\mathrm{id}, \overrightarrow{\mathcal{D}}(-))} & \mathbf{Syn} \times \mathbf{Syn} \\ {\scriptstyle \llparenthesis - \rrparenthesis} \downarrow & & \downarrow {\scriptstyle [\![-]\!] \times [\![-]\!]} \\ \mathbf{Gl}_{\mathbb{R}} & \xrightarrow[\mathrm{proj}]{} & \mathbf{Diff} \times \mathbf{Diff} \end{array} \tag{3}$$

We thus arrive at a restatement of the correctness theorem (Th. 1), which holds even for the extended language with variants and lists, because for any $x_1 ... x_n : \mathbf{real} \vdash t : \mathbf{real}$, the interpretations $([\![t]\!], [\![\overrightarrow{\mathcal{D}}(t)]\!])$ are in the image of the projection $\mathbf{Gl}_{\mathbb{R}} \to \mathbf{Diff} \times \mathbf{Diff}$, and hence $[\![\overrightarrow{\mathcal{D}}(t)]\!]$ is a dual numbers encoding of $[\![t]\!]$.

Correctness at all first order types, via manifolds. We now generalize Theorem 1 to hold at all first order types, not just the reals. To do this, we need to define the derivative of a smooth map between the interpretations of first order types. We do this by recalling the well known theory of manifolds and tangent bundles.

For our purposes, a smooth manifold M is a second-countable Hausdorff topological space together with a smooth atlas: an open cover $\mathcal{U}$ together with homeomorphisms $\left(\phi_U : U \to \mathbb{R}^{n(U)}\right)_{U \in \mathcal{U}}$ (called charts) such that $\phi_U^{-1}; \phi_V$ is smooth on its domain of definition for all $U, V \in \mathcal{U}$. A function $f : M \to N$ between manifolds is smooth if $\phi_U^{-1}; f; \psi_V$ is smooth for all charts ϕ_U and ψ_V of M and N, respectively. Let us write **Man** for this category.

Our manifolds are slightly unusual because different charts in an atlas may have different finite dimension $n(U)$. Thus we consider manifolds with dimensions that are potentially unbounded, albeit locally finite. This does not affect the theory of differential geometry as far as we need it here.

Each open subset of $\mathbb{R}^n$ can be regarded as a manifold. This lets us regard the category of manifolds **Man** as a full subcategory of the category of diffeological spaces. We consider a manifold $(X, \{\phi_U\}_U)$ as a diffeological space with the same carrier set X and where the plots $\mathcal{P}_X^U$ are the smooth functions in $\mathbf{Man}(U, X)$. A function $X \to Y$ is smooth in the sense of manifolds if and only if it is smooth in the sense of diffeological spaces [16]. For the categorically minded reader, this means that we have a full embedding of **Man** into **Diff**. Moreover, the natural interpretation of the first order fragment of our language in **Man** coincides with that in **Diff**. That is, the embedding of **Man** into **Diff** preserves finite products and countable coproducts (hence initial algebras of polynomial endofunctors).

Proposition 2. *Suppose that a type τ is first order, i.e. it is just built from reals, products, variants, and lists (or, again, arbitrary inductive types), and not function types. Then the diffeological space $[\![\tau]\!]$ is a manifold.*

Proof (notes). This is proved by induction on the structure of types. In fact, one may show that every such $[\![\tau]\!]$ is isomorphic to a manifold of the form $\biguplus_{i=1}^{n} \mathbb{R}^{d_i}$ where the bound n is either finite or ∞, but this isomorphism is typically not an identity function. □

The constraint to first order types is necessary because, e.g. the space $[\![\mathbf{real} \to \mathbf{real}]\!]$ is not a manifold, because of a Borsuk-Ulam argument (see [15], Appx. A).

We recall that the derivative of any morphism $f : M \to N$ of manifolds is given as follows. For each point x in a manifold M, define the *tangent space* $\mathcal{T}_x M$ to be the set $\{\gamma \in \mathbf{Man}(\mathbb{R}, M) \mid \gamma(0) = x\}/\sim$ of equivalence classes $[\gamma]$ of smooth curves γ in M based at x, where we identify $\gamma_1 \sim \gamma_2$ iff $\nabla(\gamma_1; f)(0) = \nabla(\gamma_2; f)(0)$ for all smooth $f : M \to \mathbb{R}$. The *tangent bundle* of M is the set $\mathcal{T}(M) \stackrel{\text{def}}{=} \biguplus_{x \in M} \mathcal{T}_x(M)$. The charts of M equip $\mathcal{T}(M)$ with a canonical manifold structure. Then for smooth $f : M \to N$, the derivative $\mathcal{T}(f) : \mathcal{T}(M) \to \mathcal{T}(N)$ is defined as $\mathcal{T}(f)(x, [\gamma]) \stackrel{\text{def}}{=} (f(x), [\gamma; f])$. All told, the derivative is a functor $\mathcal{T} : \mathbf{Man} \to \mathbf{Man}$.

As is standard, we can understand the tangent bundle of a composite space in terms of that of its parts.

Lemma 3. *There are canonical isomorphisms $\mathcal{T}(\biguplus_{i=1}^{\infty} M_i) \cong \biguplus_{i=1}^{\infty} \mathcal{T}(M_i)$ and $\mathcal{T}(M_1 \times \ldots \times M_n) \cong \mathcal{T}(M_1) \times \ldots \times \mathcal{T}(M_n)$.*

We define a canonical isomorphism $\phi_\tau^{\overrightarrow{\mathcal{D}}\mathcal{T}} : [\![\overrightarrow{\mathcal{D}}(\tau)]\!] \to \mathcal{T}([\![\tau]\!])$ for every type τ, by induction on the structure of types. We let $\phi_{\mathbf{real}}^{\overrightarrow{\mathcal{D}}\mathcal{T}} : [\![\overrightarrow{\mathcal{D}}(\mathbf{real})]\!] \to \mathcal{T}([\![\mathbf{real}]\!])$ be given by $\phi^{\overrightarrow{\mathcal{D}}\mathcal{T}}_{\mathbf{real}}(x,x') \stackrel{\text{def}}{=} (x,[t\mapsto x+x't])$. For the other types, we use Lemma 3. We can now phrase correctness at all first order types.

Theorem 2 (Semantic correctness of $\overrightarrow{\mathcal{D}}$ (full)). *For any ground τ, any first order context Γ and any term $\Gamma\vdash t:\tau$, the syntactic translation $\overrightarrow{\mathcal{D}}$ coincides with the tangent bundle functor, modulo these canonical isomorphisms:*

$$\begin{array}{ccc} [\![\overrightarrow{\mathcal{D}}(\Gamma)]\!] & \xrightarrow{[\![\overrightarrow{\mathcal{D}}(t)]\!]} & [\![\overrightarrow{\mathcal{D}}(\tau)]\!] \\ {\scriptstyle \phi^{\overrightarrow{\mathcal{D}}\mathcal{T}}_{\Gamma}}\downarrow\cong & & \cong\downarrow{\scriptstyle \phi^{\overrightarrow{\mathcal{D}}\mathcal{T}}_{\tau}} \\ \mathcal{T}([\![\Gamma]\!]) & \xrightarrow[\mathcal{T}([\![t]\!])]{} & \mathcal{T}([\![\tau]\!]) \end{array}$$

Proof (notes). For any curve $\gamma\in\mathbf{Man}(\mathbb{R},M)$, let $\bar\gamma\in\mathbf{Man}(\mathbb{R},\mathcal{T}(M))$ be the tangent curve, given by $\bar\gamma(x)=(\gamma(x),[t\mapsto\gamma(x+t)])$. First, we note that a smooth map $h:\mathcal{T}(M)\to\mathcal{T}(N)$ is of the form $\mathcal{T}(g)$ for some $g:M\to N$ if for all smooth curves $\gamma:\mathbb{R}\to M$ we have $\bar\gamma;h=\overline{(\gamma;g)}:\mathbb{R}\to\mathcal{T}(N)$. This generalizes (2). Second, for any first order type τ, $S_{[\![\tau]\!]}=\{(f,\tilde f)\mid \tilde f;\phi^{\overrightarrow{\mathcal{D}}\mathcal{T}}_{\tau}=\bar f\}$. This is shown by induction on the structure of types. We conclude the theorem from diagram (3), by putting these two observations together. □

6 A continuation-based AD algorithm

We now illustrate the flexibility of our framework by briefly describing an alternative syntactic translation $\overleftarrow{\mathcal{D}}_\rho$. This alternative translation uses aspects of continuation passing style, inspired by recent developments in reverse mode AD [34, 5]. In brief, $\overleftarrow{\mathcal{D}}_\rho$ works by $\overleftarrow{\mathcal{D}}_\rho(\mathbf{real})=(\mathbf{real}*(\mathbf{real}\to\rho))$. Thus instead of using a pair of a number and its tangent, we use a pair of a number and a continuation. The answer type $\rho=\mathbf{real}^k$ needs to have the structure of a vector space, and the continuations that we consider will turn out to be linear maps. Because we work in continuation passing style, the chain rule is applied contravariantly. If the reader is familiar with reverse-mode AD algorithms, they may think of the dimension k as the number of memory cells used to store the result.

Computing the whole gradient of a term $x_1:\mathbf{real},...,x_k:\mathbf{real}\vdash t:\mathbf{real}$ at once is then achieved by running $\overleftarrow{\mathcal{D}}_k(t)$ on a k-tuple of basis vectors for $\mathbf{real}^k$.

We define the continuation-based AD macro $\overleftarrow{\mathcal{D}}_k$ on types and terms as the unique structure preserving functor $\mathbf{Syn}\to\mathbf{Syn}$ with $\overleftarrow{\mathcal{D}}_k(\mathbf{real})=(\mathbf{real}*(\mathbf{real}\to\mathbf{real}^k))$ and

$$\overleftarrow{\mathcal{D}}_k(\underline{c}) \stackrel{\text{def}}{=} \langle \underline{c}, \lambda z.\langle \underline{0},\ldots,\underline{0}\rangle\rangle$$
$$\overleftarrow{\mathcal{D}}_k(t+s) \stackrel{\text{def}}{=} \mathbf{case}\,\overleftarrow{\mathcal{D}}_k(t)\,\mathbf{of}\,\langle x,x'\rangle\to\mathbf{case}\,\overleftarrow{\mathcal{D}}_k(s)\,\mathbf{of}\,\langle y,y'\rangle\to\langle x+y,\lambda z.x'\,z+y'\,z\rangle$$
$$\overleftarrow{\mathcal{D}}_k(t*s) \stackrel{\text{def}}{=} \mathbf{case}\,\overleftarrow{\mathcal{D}}_k(t)\,\mathbf{of}\,\langle x,x'\rangle\to\mathbf{case}\,\overleftarrow{\mathcal{D}}_k(s)\,\mathbf{of}\,\langle y,y'\rangle\to$$
$$\langle x*y,\lambda z.x'\,(y*z)+y'\,(x*z)\rangle$$
$$\overleftarrow{\mathcal{D}}_k(\varsigma(t)) \stackrel{\text{def}}{=} \mathbf{case}\,\overleftarrow{\mathcal{D}}_k(t)\,\mathbf{of}\,\langle x,x'\rangle\to\mathbf{let}\,y=\varsigma(x)\,\mathbf{in}\,\langle y,\lambda z.x'\,(y*(1-y)*z)\rangle.$$

Here, we use sugar $x:\mathbf{real}^k, y:\mathbf{real}^k\vdash x+y \stackrel{\text{def}}{=} \mathbf{case}\,x\,\mathbf{of}\,\langle x_1,\ldots,x_k\rangle\to$

case y **of** $\langle y_1,\ldots,y_k\rangle \to \langle x_1+y_1,\ldots,x_k+y_k\rangle$. (We could easily expand this definition by making $\overleftarrow{\mathcal{D}}_k$ preserve all other term and type formers, as we did for $\overrightarrow{\mathcal{D}}$.) Note that the corresponding scheme for an arbitrary n-ary operation op would be (c.f. the scheme for forward AD in §4)

$$\overleftarrow{\mathcal{D}}_k(\mathsf{op}(t_1,\ldots,t_n)) \stackrel{\text{def}}{=} \mathbf{case}\, \overleftarrow{\mathcal{D}}_k(t_1)\, \mathbf{of}\, \langle x_1,x_1'\rangle \to \ldots \to \mathbf{case}\, \overleftarrow{\mathcal{D}}_k(t_n)\, \mathbf{of}\, \langle x_n,x_n'\rangle \to \\ \langle \mathsf{op}(x_1,\ldots,x_n), \lambda z.\textstyle\sum_{i=1}^n x_i'(\partial_i\mathsf{op}(x_1,\ldots,x_n) * z)\rangle.$$

The idea is that $\overleftarrow{\mathcal{D}}_k(t)$ is a higher order function that simultaneously computes t (the forward pass) and defines as a continuation the reverse pass which computes the gradient. In order to actually run the algorithm, we need two auxiliary definitions

$$\mathrm{lamR}^k_{\mathbf{real}} \stackrel{\text{def}}{=} \lambda z.\,\mathbf{case}\, z\, \mathbf{of}\, \langle x,x'\rangle \to \mathbf{case}\, x'\, \mathbf{of}\, \langle x_1',\ldots,x_k'\rangle \to \\ \langle x, \lambda y.\langle x_1' * y,\ldots,x_k' * y\rangle\rangle : \overrightarrow{\mathcal{D}}_k(\mathbf{real}) \to \overleftarrow{\mathcal{D}}_k(\mathbf{real})$$

$$\mathrm{evR}^k_{\mathbf{real}} \stackrel{\text{def}}{=} \lambda z.\,\mathbf{case}\, z\, \mathbf{of}\, \langle x,x'\rangle \to \langle x, x'\,\underline{1}\rangle : \overleftarrow{\mathcal{D}}_k(\mathbf{real}) \to \overrightarrow{\mathcal{D}}_k(\mathbf{real}).$$

Here, $\overrightarrow{\mathcal{D}}_k$ is a macro on types (and terms) with exactly the same inductive definition as $\overrightarrow{\mathcal{D}}$ except for the base case $\overrightarrow{\mathcal{D}}_k(\mathbf{real}) = (\mathbf{real}{*}\mathbf{real}^k)$. By noting that both $\overrightarrow{\mathcal{D}}_k$ and $\overleftarrow{\mathcal{D}}_k$ preserve all type formers, we can extend these definitions to all first order types τ: $z : \overrightarrow{\mathcal{D}}_k(\tau) \vdash \mathrm{lamR}^k_\tau(z) : \overleftarrow{\mathcal{D}}_k(\tau)$, $z : \overleftarrow{\mathcal{D}}_k(\tau) \vdash \mathrm{evR}^k_\tau(z) : \overrightarrow{\mathcal{D}}_k(\tau)$. We can think of $\mathrm{lamR}^k_\tau(z)$ as encoding k tangent vectors $z : \overrightarrow{\mathcal{D}}_k(\tau)$ as a closure, so it is suitable for running $\overleftarrow{\mathcal{D}}_k(t)$ on, and $\mathrm{evR}^k_\tau(z)$ as actually evaluating the reverse pass defined by $z : \overleftarrow{\mathcal{D}}_k(\tau)$ and returning the result as k tangent vectors. The idea is that given some $x : \tau \vdash t : \sigma$ between first order types τ,σ, we run our continuation-based AD by running $\mathrm{evR}^k_\sigma(\overleftarrow{\mathcal{D}}_k(t)[^{\mathrm{lamR}^k_\tau(z)}/_x])$.

The correctness proof closely follows that for forward AD. In particular, one defines a binary logical relation $(\!|\mathbf{real}|\!)^{r,k} = (\mathbb{R}, \mathbb{R}\times(\mathbb{R}^k)^{\mathbb{R}}, S^{r,k}_{\mathbf{real}})$, where $S^{r,k}_{\mathbf{real}} = \left\{(f, x\mapsto(f(x), y\mapsto(\partial_1 f(x) * y,\ldots,\partial_k f(x) * y))) \mid f\in\mathcal{P}^{\mathbb{R}^k}_{\mathbb{R}}\right\}$, on the plots $\mathcal{P}^{\mathbb{R}^k}_{\mathbb{R}} \times \mathcal{P}^{\mathbb{R}^k}_{\mathbb{R}\times((\mathbb{R}^k)^{\mathbb{R}})}$ and verifies that $[\![\underline{c}]\!]\times[\![\overleftarrow{\mathcal{D}}_k(\underline{c})]\!]$, $[\![x+y]\!]\times[\![\overleftarrow{\mathcal{D}}_k(x+y)]\!]$, $[\![x*y]\!]\times[\![\overleftarrow{\mathcal{D}}_k(x*y)]\!]$ and $[\![\varsigma(x)]\!]\times[\![\overleftarrow{\mathcal{D}}_k(\varsigma(x))]\!]$ respect this logical relation. It follows that this relation extends to a functor $(\!|-|\!)^{r,k} : \mathbf{Syn}\to\mathbf{Gl}_{\mathbb{R}^k}$ such that $\mathrm{id}\times\overleftarrow{\mathcal{D}}_k$ factors over $(\!|-|\!)^{r,k}$, implying the correctness of the continuation-based AD by the following lemma.

Lemma 4. *For all first order types τ (i.e. types not involving function types), we have that* $[\![\mathrm{evR}^k_\tau(\mathrm{lamR}^k_\tau(t))]\!] = [\![t]\!]$.

Proof (notes). This follows by an induction on the structure of τ. The idea is that lamR^k_τ embeds reals into function spaces as linear maps, which is undone by evR^k_τ by evaluating the linear maps at $\underline{1}$. □

To phrase correctness, in this setting, however, we need a few definitions. Keeping in mind the canonical projection $\mathcal{T}(M)\to M$, we define $\mathcal{T}^k(M)$ as the k-fold categorical pullback (fibre product) $\mathcal{T}(M)\times_M\ldots\times_M\mathcal{T}(M)$. To be explicit, $\mathcal{T}^k_x M$ consists of k-tuples of tangent vectors at the base point x. Again, $\mathcal{T}^k$ extends to a functor $\mathbf{Man}\to\mathbf{Man}$ by defining $\mathcal{T}^k(f)(x,(v_1,\ldots,v_k)) \stackrel{\text{def}}{=} (f(x),(\mathcal{T}_x(f)(v_1),\ldots,\mathcal{T}_x(f)(v_k)))$. As $\mathcal{T}^k$ preserves countable coproducts and finite products (like $\mathcal{T}$), it follows that the isomorphisms $\phi_{\tau}^{\overrightarrow{\mathcal{D}}\mathcal{T}}$ generalize to canonical isomorphisms $\phi_{\tau,k}^{\overrightarrow{\mathcal{D}}\mathcal{T}} : [\![\overrightarrow{\mathcal{D}}_k(\tau)]\!] \to \mathcal{T}^k([\![\tau]\!])$ for first order types τ. This leads to the following correctness statement for continuation-based AD.

Theorem 3 (Semantic correctness of $\overleftarrow{\mathcal{D}}_k$). *For any ground τ, any first order context Γ and any term $\Gamma \vdash t : \tau$, syntactic translation $t \mapsto \mathrm{evR}^k_\tau(\overleftarrow{\mathcal{D}}_k(t)[^{\mathrm{lamR}^k_\Gamma(z)}/_{\ldots}])$ coincides with the tangent bundle functor, modulo these canonical isomorphisms:*

$$\begin{array}{ccc}
[\![\overrightarrow{\mathcal{D}}_k(\Gamma)]\!] & \xrightarrow{[\![\mathrm{lamR}^k_\Gamma;\overleftarrow{\mathcal{D}}_k(t);\mathrm{evR}^k_\tau]\!]} & [\![\overrightarrow{\mathcal{D}}_k(\tau)]\!] \\
\phi_{\Gamma,k}^{\overrightarrow{\mathcal{D}}\mathcal{T}} \Big\downarrow \cong & & \cong \Big\downarrow \phi_{\tau,k}^{\overrightarrow{\mathcal{D}}\mathcal{T}} \\
\mathcal{T}^k([\![\Gamma]\!]) & \xrightarrow[\mathcal{T}^k([\![t]\!])]{} & \mathcal{T}^k([\![\tau]\!])
\end{array}$$

For example, when $\tau = \mathbf{real}$ and $\Gamma = x, y : \mathbf{real}$, we can run our continuation-based AD to compute the gradient of a program $x, y : \mathbf{real} \vdash t : \mathbf{real}$ at values $x = V, y = W$ by evaluating

$$\mathrm{evR}^2_{\mathbf{real}}(\overleftarrow{\mathcal{D}}_2(t)[^{(\mathrm{lamR}^2_{x:\mathbf{real}}\, v)}/_x, ^{(\mathrm{lamR}^2_{y:\mathbf{real}}\, w)}/_y])[^{\langle V, \langle \underline{1}, \underline{0}\rangle\rangle}/_v, ^{\langle W, \langle \underline{0}, \underline{1}\rangle\rangle}/_w].$$

Indeed,

$$[\![\mathrm{evR}^2_{\mathbf{real}}(\overleftarrow{\mathcal{D}}_2(t)[^{(\mathrm{lamR}^2_{x:\mathbf{real}}\, v)}/_x, ^{(\mathrm{lamR}^2_{y:\mathbf{real}}\, w)}/_y])[^{\langle V, \langle \underline{1}, \underline{0}\rangle\rangle}/_v, ^{\langle W, \langle \underline{0}, \underline{1}\rangle\rangle}/_w]]\!] =$$
$$([\![t]\!]([\![V]\!], [\![W]\!]), \partial_1 [\![t]\!]([\![V]\!], [\![W]\!]), \partial_2 [\![t]\!]([\![V]\!], [\![W]\!])).$$

7 Discussion and future work

Summary. We have shown that diffeological spaces provide a denotational semantics for a higher order language with variants and inductive types (§3,4). We have used this to show correctness of a simple AD translation (Thm. 1, Thm. 2). But the method is not tied to this specific translation, as we illustrated in Section 6.

The structure of our elementary correctness argument for Theorem 1 is a typical logical relations proof. As explained in Section 5, this can equivalently be understood as a denotational semantics in a new kind of space obtained by categorical gluing.

Overall, then, there are two logical relations at play. One is in diffeological spaces, which ensures that all definable functions are smooth. The other is in the correctness proof (equivalently in the categorical gluing), which explicitly tracks the derivative of each function, and tracks the syntactic AD even at higher types.

Connection to the state of the art in AD implementation. As is common in denotational semantics research, we have here focused on an idealized language and simple translations to illustrate the main aspects of the method. There are a number of points where our approach is simplistic compared to the advanced current practice, as we now explain.

Representation of vectors. In our examples we have treated n-vectors as tuples of length n. This style of programming does not scale to large n. A better solution would be to use array types, following [31]. Our categorical semantics and correctness proofs straightforwardly extend to cover them, in a similar way to our treatment of lists.

Efficient forward-mode AD. For AD to be useful, it must be fast. The syntactic translation $\overrightarrow{\mathcal{D}}$ that we use is the basis of an efficient AD library [31]. However, numerous optimizations are needed, ranging from algebraic manipulations, to partial evaluations, to the use of an optimizing C compiler. A topic for future work would be to validate some of these manipulations using our semantics. The resulting implementation is performant in experiments [31].

Efficient reverse-mode AD. Our sketch of continuation-based AD is primarily intended to emphasise that our denotational approach is not tied to any specific translation $\overrightarrow{\mathcal{D}}$. Nonetheless, it is worth noting that this algorithm shares similarities with advanced reverse-mode implementations: (1) it calculates derivatives in a (contravariant) "reverse pass" in which derivatives of operations are evaluated in the reverse order compared to their order in calculating the function value; (2) it can be used to calculate the full gradient of a function $\mathbb{R}^n \to \mathbb{R}$ in a single reverse pass (while n passes of fwd AD would be necessary). However, it lacks important optimizations and the continuation scales with the size of the input n where it should scale with the size of the output. This adds an important overhead, as pointed out in [26]. Speed being the main attraction of reverse-mode AD, its implementations tend to rely on mutable state, control operators and/or staging [26, 6, 34, 5], which we have not considered here.

Other language features. The idealized languages that we considered so far do not touch on several useful language constructs. For example: the use of functions that are partial (such as division) or partly-smooth (such as RelU); phenomena such as iteration, recursion; and probabilities. There are suggestions that the denotational approach using diffeological spaces can be adapted to these features using standard categorical methods. We leave this for future work.

Acknowledgements. We have benefited from discussing this work with many people, including B. Pearlmutter, O. Kammar, C. Mak, L. Ong, G. Plotkin, A. Shaikhha, J. Sigal, and others. Our work is supported by the Royal Society and by a Facebook Research Award. In the course of this work, MV has also been employed at Oxford (EPSRC Project EP/M023974/1) and at Columbia in the Stan development team. This project has received funding from the European Union's Horizon 2020 research and innovation programme under the Marie Skłodowska-Curie grant agreement No. 895827.

References

1. Abadi, M., Barham, P., Chen, J., Chen, Z., Davis, A., Dean, J., Devin, M., Ghemawat, S., Irving, G., Isard, M., et al.: Tensorflow: A system for large-scale machine learning. In: 12th USENIX Symposium on Operating Systems Design and Implementation (OSDI 16). pp. 265–283 (2016)
2. Abadi, M., Plotkin, G.D.: A simple differentiable programming language. In: Proc. POPL 2020. ACM (2020)
3. Baez, J., Hoffnung, A.: Convenient categories of smooth spaces. Transactions of the American Mathematical Society **363**(11), 5789–5825 (2011)
4. Barthe, G., Crubillé, R., Lago, U.D., Gavazzo, F.: On the versatility of open logical relations: Continuity, automatic differentiation, and a containment theorem. In: Proc. ESOP 2020. Springer (2020), to appear
5. Brunel, A., Mazza, D., Pagani, M.: Backpropagation in the simply typed lambda-calculus with linear negation. In: Proc. POPL 2020 (2020)
6. Carpenter, B., Hoffman, M.D., Brubaker, M., Lee, D., Li, P., Betancourt, M.: The Stan math library: Reverse-mode automatic differentiation in C++. arXiv preprint arXiv:1509.07164 (2015)
7. Christensen, J.D., Wu, E.: Tangent spaces and tangent bundles for diffeological spaces. arXiv preprint arXiv:1411.5425 (2014)
8. Cockett, J.R.B., Cruttwell, G.S.H., Gallagher, J., Lemay, J.S.P., MacAdam, B., Plotkin, G.D., Pronk, D.: Reverse derivative categories. In: Proc. CSL 2020 (2020)
9. Cruttwell, G., Gallagher, J., MacAdam, B.: Towards formalizing and extending differential programming using tangent categories. In: Proc. ACT 2019 (2019)
10. Duchi, J., Hazan, E., Singer, Y.: Adaptive subgradient methods for online learning and stochastic optimization. Journal of Machine Learning Research **12**(Jul), 2121–2159 (2011)
11. Ehrhard, T., Regnier, L.: The differential lambda-calculus. Theoretical Computer Science **309**(1-3), 1–41 (2003)
12. Elliott, C.: The simple essence of automatic differentiation. Proceedings of the ACM on Programming Languages **2**(ICFP), 70 (2018)
13. Fong, B., Spivak, D., Tuyéras, R.: Backprop as functor: A compositional perspective on supervised learning. In: 2019 34th Annual ACM/IEEE Symposium on Logic in Computer Science (LICS). pp. 1–13. IEEE (2019)
14. Hoffman, M.D., Gelman, A.: The No-U-Turn sampler: adaptively setting path lengths in Hamiltonian Monte Carlo. Journal of Machine Learning Research **15**(1), 1593–1623 (2014)
15. Huot, M., Staton, S., Vákár, M.: Correctness of automatic differentiation via diffeologies and categorical gluing. Full version (2020), arxiv:2001.02209
16. Iglesias-Zemmour, P.: Diffeology. American Mathematical Soc. (2013)
17. Johnstone, P.T., Lack, S., Sobocinski, P.: Quasitoposes, quasiadhesive categories and Artin glueing. In: Proc. CALCO 2007 (2007)
18. Kiefer, J., Wolfowitz, J., et al.: Stochastic estimation of the maximum of a regression function. The Annals of Mathematical Statistics **23**(3), 462–466 (1952)
19. Kingma, D.P., Ba, J.: Adam: A method for stochastic optimization. arXiv preprint arXiv:1412.6980 (2014)
20. Kucukelbir, A., Tran, D., Ranganath, R., Gelman, A., Blei, D.M.: Automatic differentiation variational inference. The Journal of Machine Learning Research **18**(1), 430–474 (2017)

21. Liu, D.C., Nocedal, J.: On the limited memory BFGS method for large scale optimization. Mathematical programming **45**(1-3), 503–528 (1989)
22. Mak, C., Ong, L.: A differential-form pullback programming language for higher-order reverse-mode automatic differentiation (2020), arxiv:2002.08241
23. Manzyuk, O.: A simply typed λ-calculus of forward automatic differentiation. In: Proc. MFPS 2012 (2012)
24. Mitchell, J.C., Scedrov, A.: Notes on sconing and relators. In: International Workshop on Computer Science Logic. pp. 352–378. Springer (1992)
25. Neal, R.M., et al.: MCMC using Hamiltonian dynamics. Handbook of Markov Chain Monte Carlo **2**(11), 2 (2011)
26. Pearlmutter, B.A., Siskind, J.M.: Reverse-mode AD in a functional framework: Lambda the ultimate backpropagator. ACM Transactions on Programming Languages and Systems (TOPLAS) **30**(2), 7 (2008)
27. Pitts, A.M.: Categorical logic. Tech. rep., University of Cambridge, Computer Laboratory (1995)
28. Plotkin, G.D.: Some principles of differential programming languages (2018), invited talk, POPL 2018
29. Qian, N.: On the momentum term in gradient descent learning algorithms. Neural networks **12**(1), 145–151 (1999)
30. Robbins, H., Monro, S.: A stochastic approximation method. The annals of mathematical statistics pp. 400–407 (1951)
31. Shaikhha, A., Fitzgibbon, A., Vytiniotis, D., Peyton Jones, S.: Efficient differentiable programming in a functional array-processing language. Proceedings of the ACM on Programming Languages **3**(ICFP), 97 (2019)
32. Souriau, J.M.: Groupes différentiels. In: Differential geometrical methods in mathematical physics, pp. 91–128. Springer (1980)
33. Stacey, A.: Comparative smootheology. Theory Appl. Categ. **25**(4), 64–117 (2011)
34. Wang, F., Wu, X., Essertel, G., Decker, J., Rompf, T.: Demystifying differentiable programming: Shift/reset the penultimate backpropagator. Proceedings of the ACM on Programming Languages **3**(ICFP) (2019)

Deep Induction: Induction Rules for (Truly) Nested Types

Patricia Johann✉ and Andrew Polonsky
Appalachian State University, Boone, NC, USA
johannp@appstate.edu, polonskya@appstate.edu

Abstract. This paper introduces *deep induction*, and shows that it is the notion of induction most appropriate to nested types and other data types defined over, or mutually recursively with, (other) such types. Standard induction rules induct over only the top-level structure of data, leaving any data internal to the top-level structure untouched. By contrast, deep induction rules induct over *all* of the structured data present. We give a grammar generating a robust class of nested types (and thus ADTs), and develop a fundamental theory of deep induction for them using their recently defined semantics as fixed points of accessible functors on locally presentable categories. We then use our theory to derive deep induction rules for some common ADTs and nested types, and show how these rules specialize to give the standard structural induction rules for these types. We also show how deep induction specializes to solve the long-standing problem of deriving principled and practically useful structural induction rules for bushes and other *truly* nested types. Overall, deep induction opens the way to making induction principles appropriate to richly structured data types available in programming languages and proof assistants. Agda implementations of our development and examples, including two extended case studies, are available.

1 Introduction

This paper is concerned with the problem of inductive reasoning about inductive data types that are defined over, or are defined mutually recursively with, (other) such data types. Examples of such *deep data types* include, trivially, ordinary algebraic data types (ADTs), such as list and tree types; data types, such as the forest type, whose recursive occurrences appear below other type constructors; simple nested types, such as the type of perfect trees, whose recursive occurrences never appear below their own type constructors; truly[1] nested types, such as the type of bushes (also called *bootstrapped heaps* by Okasaki [16]), whose recursive occurrences do appear below their own type constructors; and GADTs. Proof assistants, including Coq and Agda, currently provide insufficient support for performing induction over deep data types. The induction rules, if any, they generate for such types induct over only their top-level structures, leaving any data internal to the top-level structure untouched. This paper develops a principle that, by contrast, inducts over *all* of the structured data present. We call this principle *deep induction*. Deep induction not only provides general support for solving problems that previously had only (usually quite painful and) *ad hoc* solutions, but also opens the way for incorporating automatic generation of useful induction rules for deep data types into proof assistants.

[1] Nested types that are defined over themselves are known as *truly nested types*.

J. Goubault-Larrecq and B. König (Eds.): FOSSACS 2020, LNCS 12077, pp. 339–358, 2020.
https://doi.org/10.1007/978-3-030-45231-5_18

To illustrate the difference between structural induction and deep induction, note that the data inside a structure of type $\texttt{List a} = \texttt{Nil} \mid \texttt{Cons a (List a)}$ is treated monolithically (i.e., ignored) by the structural induction rule for lists:

$$\begin{aligned}&\forall (\mathsf{a} : \mathsf{Set})\,(\mathtt{P} : \mathtt{List\,a} \to \mathsf{Set}) \to \mathtt{P\,Nil} \to \\ &\quad (\forall\,(\mathtt{x} : \mathtt{a})\,(\mathtt{xs} : \mathtt{List\,a}) \to \mathtt{P\,xs} \to \mathtt{P\,(Cons\,x\,xs)}) \to \forall\,(\mathtt{xs} : \mathtt{List\,a}) \to \mathtt{P\,xs}\end{aligned}$$

By contrast, the deep induction rule for lists traverses not just the outer list structure with a predicate P, but also each data element of that list with a custom predicate Q:

$$\begin{aligned}&\forall\,(\mathsf{a} : \mathsf{Set})\,(\mathtt{P} : \mathtt{List\,a} \to \mathsf{Set})\,(\mathtt{Q} : \mathtt{a} \to \mathsf{Set}) \to \\ &\quad \mathtt{P\,Nil} \to (\forall (\mathtt{x} : \mathtt{a})\,(\mathtt{xs} : \mathtt{List\,a}) \to \mathtt{Q\,x} \to \mathtt{P\,xs} \to \mathtt{P\,(Cons\,x\,xs)}) \to \\ &\quad \forall (\mathtt{xs} : \mathtt{List\,a}) \to \mathtt{List}^{\wedge}\,\mathtt{Q\,xs} \to \mathtt{P\,xs}\end{aligned}$$

Here, $\mathtt{List}^{\wedge}$ lifts its argument predicate Q on data of type a to a predicate on data of type List a asserting that Q holds for every element of its argument list. The structural induction rule for lists is, like that for any ADT, recovered by taking the custom predicate in the corresponding deep rule to be $\lambda \mathtt{x}.\,\mathtt{True}$.

A particular advantage of deep induction is that it obviates the need to reflect properties as data types. For example, although the set of primes cannot be defined by an ADT, the primeness predicate Prime on the ADT of natural numbers *can* be lifted to a predicate $\mathtt{List}^{\wedge}\,\mathtt{Prime}$ characterizing lists of primes. Properties can then be proved for lists of primes using the following deep induction rule:

$$\begin{aligned}&\forall (\mathtt{P} : \mathtt{List\,Nat} \to \mathsf{Set}) \to \mathtt{P\,Nil} \to \\ &\quad (\forall (\mathtt{x} : \mathtt{Nat})\,(\mathtt{xs} : \mathtt{List\,Nat}) \to \mathtt{Prime\,x} \to \mathtt{P\,xs} \to \mathtt{P\,(Cons\,x\,xs)}) \to \\ &\quad \forall (\mathtt{xs} : \mathtt{List\,Nat}) \to \mathtt{List}^{\wedge}\,\mathtt{Prime\,xs} \to \mathtt{P\,xs}\end{aligned}$$

As we'll see in Sections 3, 4, and 5, the extra flexibility afforded by lifting predicates like Q and Prime on data internal to a structure makes it possible to derive useful induction principles for more complex types, such as truly nested ones.

In each of the above examples, a predicate on the data is lifted to a predicate on the list. This is an example of lifting a predicate on a type in a *non-recursive* position of an ADT's definition to the entire ADT. However, the predicate to be lifted can also be on the type in a *recursive* position of a definition — i.e., on the ADT being defined itself — and this ADT can appear below another type constructor in the definition. This is exactly the situation for the ADT Forest a, which appears below the type constructor List in the definition

$$\mathtt{Forest\,a} = \mathtt{FEmpty} \mid \mathtt{FNode\,a\,(List\,(Forest\,a))}$$

The induction rule Coq generates for forests is

$$\begin{aligned}&\forall\,(\mathsf{a} : \mathsf{Set})\,(\mathtt{P} : \mathtt{Forest\,a} \to \mathsf{Set}) \to \mathtt{P\,FEmpty} \to \\ &\quad (\forall\,(\mathtt{x} : \mathtt{a})\,(\mathtt{ts} : \mathtt{List\,(Forest\,a)}) \to \mathtt{P\,(FNode\,x\,ts)}) \to \forall\,(\mathtt{x} : \mathtt{Forest\,a}) \to \mathtt{P\,x}\end{aligned}$$

However, this is neither the induction rule we intuitively expect, nor is it expressive enough to prove even basic properties of forests that ought to be amenable to inductive proof. The approach of [11,12] does give the expected rule[2]

[2] This is equivalent to the rule as classically stated in Coq/Isabelle/HOL.

$$\forall (\mathtt{a} : \mathsf{Set})\, (\mathtt{P} : \mathtt{Forest\,a} \to \mathsf{Set}) \to \mathtt{P\,FEmpty} \to$$
$$(\forall (\mathtt{x} : \mathtt{a})\, (\mathtt{ts} : \mathtt{List\,(Forest\,a)}) \to (\forall (\mathtt{k} < \mathtt{length\,ts}) \to \mathtt{P\,(ts!!k)})$$
$$\to \mathtt{P\,(FNode\,x\,ts)}) \to \forall (\mathtt{x} : \mathtt{Forest\,a}) \to \mathtt{P\,x}$$

But to derive it, a technique based on list positions is used to propagate the predicate to be proved over the list of forests that is the second argument to the data constructor `FNode`. Unfortunately, this technique does not obviously extend to other deep data types, including the type of "generalized forests" introduced in Section 4.4 below, which combines smaller generalized forests into larger ones using a type constructor `f` potentially different from `List`. Nevertheless, replacing $\forall (\mathtt{k} < \mathtt{length\,ts}) \to \mathtt{P\,(ts!!k)}$ in the expected rule with $\mathtt{List}^\wedge\,\mathtt{P\,ts}$, which is equivalent, reveals that it is nothing more than the special case for $\mathtt{Q} = \lambda\mathtt{x}.\,\mathtt{True}$ of the following deep induction rule for `Forest a`:

$$\forall (\mathtt{a} : \mathsf{Set})\, (\mathtt{P} : \mathtt{Forest\,a} \to \mathsf{Set})\, (\mathtt{Q} : \mathtt{a} \to \mathsf{Set}) \to \mathtt{P\,FEmpty} \to$$
$$(\forall (\mathtt{x} : \mathtt{a})\, (\mathtt{ts} : \mathtt{List\,(Forest\,a)}) \to \mathtt{Q\,x} \to \mathtt{List}^\wedge\,\mathtt{P\,ts} \to \mathtt{P\,(FNode\,x\,ts)}) \to$$
$$\forall (\mathtt{x} : \mathtt{Forest\,a}) \to \mathtt{Forest}^\wedge\,\mathtt{Q\,x} \to \mathtt{P\,x}$$

When types, like `Forest a` and `List (Forest a)` above, are defined by mutual recursion, their (deep) induction rules are defined by mutual recursion as well. For example, the induction rules for the ADTs

```
data Expr  = Lit Nat | Add Expr Expr | If BExpr Expr Expr
data BExpr = BLit Bool | And BExpr BExpr | Not BExpr | Equal Expr Expr
```

of integer and boolean expressions are defined by mutual recursion as

$$\forall (\mathtt{P} : \mathsf{Expr} \to \mathsf{Set})\, (\mathtt{Q} : \mathsf{BExpr} \to \mathsf{Set}) \to$$
$$(\forall (\mathtt{n} : \mathtt{Nat}) \to \mathtt{P\,(Lit\,n)}) \to$$
$$(\forall (\mathtt{e1} : \mathsf{Expr})\, (\mathtt{e2} : \mathsf{Expr}) \to \mathtt{P\,e1} \to \mathtt{P\,e2} \to \mathtt{P\,(Add\,e1\,e2)}) \to$$
$$(\forall (\mathtt{b} : \mathsf{BExpr})\, (\mathtt{e1} : \mathsf{Expr})\, (\mathtt{e2} : \mathsf{Expr}) \to \mathtt{Q\,b} \to \mathtt{P\,e1} \to \mathtt{P\,e2} \to \mathtt{P\,(If\,b\,e1\,e2)}) \to$$
$$(\forall (\mathtt{b} : \mathtt{Bool}).\, \mathtt{Q\,(BLit\,b)}) \to$$
$$(\forall (\mathtt{b1} : \mathsf{BExpr})\, (\mathtt{b2} : \mathsf{BExpr}) \to \mathtt{Q\,b1} \to \mathtt{Q\,b2} \to \mathtt{Q\,(And\,b1\,b2)}) \to$$
$$(\forall (\mathtt{b} : \mathsf{BExpr}) \to \mathtt{Q\,b} \to \mathtt{Q\,(Not\,b)}) \to$$
$$(\forall (\mathtt{e1} : \mathsf{Expr})\, (\mathtt{e2} : \mathsf{Expr}) \to \mathtt{P\,e1} \to \mathtt{P\,e2} \to \mathtt{Q\,(Equal\,e1\,e2)}) \to$$
$$(\forall (\mathtt{e} : \mathsf{Expr}) \to \mathtt{P\,e}) \times (\forall (\mathtt{b} : \mathsf{BExpr}) \to \mathtt{Q\,b})$$

2 The Key Idea

As the examples of the previous section suggest, the key to deriving deep induction rules from (deep) data type declarations is to parameterize the induction rules not just over a predicate over the top-level data type being defined, but over predicates on the types of primitive data they contain as well. These additional predicates are then lifted to predicates on any internal structures containing these data, and the resulting predicates on these internal structures are lifted to predicates on any internal structures containing structures at the previous level, and so on, until the internal structures at all levels of the data type definition, including the top level, have been so processed. Satisfaction of a predicate by the data at one level of a structure is then conditioned upon satisfaction of the appropriate predicates by *all* of the data at the preceding level.

The above deep induction rules were all obtained using this technique. For example, the deep induction rule for lists is derived by first noting that structures of type `List a` contain only data of type `a`, so that only one additional predicate parameter, which we called `Q` above, is needed. Then, since the only data structure internal to the type `List a` is `List a` itself, `Q` need only be lifted to lists containing data of type `a`. This is exactly what `List^ Q` does. Finally, the deep induction rule for lists is obtained by parameterizing the standard one over not just `P` but also `Q`, adding the additional hypothesis `Q x` to its second antecedent, and adding the additional hypothesis `List^ Q xs` to its conclusion.

The deep induction rule for forests is similarly obtained from the Coq-generated rule by first parameterizing over an additional predicate `Q` on the type `a` of data stored in the forest, then lifting `P` to a predicate on lists containing data of type `Forest a` and `Q` to forests containing data of type `a`, and, finally, adding the additional hypotheses `Q x` and `List^ P ts` to its second antecedent and the additional hypothesis `Forest^ Q x` to its conclusion.

Predicate liftings such as `List^` and `Forest^` may either be supplied as primitives, or be generated automatically from the definitions of the types themselves, as described in Section 4. For container types, lifting a predicate amounts to traversing the container and applying the argument predicate pointwise.

Our technique for deriving deep induction rules for ADTs, as well as its generalization to nested types given in Section 3, is both made precise and rigorously justified in Section 4 using the results of [13]. This paper can thus be seen as a concrete application, in the specific category Fam, of the very general semantics developed in [13]; indeed, our induction rules are computed as the interpretations of the syntax for nested types in Fam. A general methodology is extracted in Section 5. The rest of this paper can be read either as "just" describing how to generate deep induction rules in practice, or as also proving that our technique for doing so is provably correct and general. Our Agda code is at [14].

3 Extending to Nested Types

Appropriately generalizing the basic technique of Section 2 derives deep induction rules, and therefore structural induction rules, for nested types, including truly nested types and other deep nested types. Nested types generalize ADTs by allowing elements at one instance of a data type to depend on data at other instances of the same type so that, in effect, the entire family of instances is constructed simultaneously. That is, rather than defining standalone *families of inductive types*, one for each choice of types to which type constructors like `List` and `Tree` are applied, the type constructors for nested types define *inductive families of types*. The structural induction rule for a nested type must therefore account for its changing type parameters by parameterizing over an appropriately polymorphic predicate, and appropriately instantiating that predicate's type argument at each application site. For example, the structural induction rule for the nested type

$$\mathtt{PTree\,a} = \mathtt{PLeaf\,a} \mid \mathtt{PNode}\,(\mathtt{PTree}\,(\mathtt{a} \times \mathtt{a}))$$

of perfect trees is

$$\begin{aligned}
&\forall\,(\mathtt{P} : \forall\,(\mathtt{a} : \mathsf{Set}) \to \mathtt{PTree\,a} \to \mathsf{Set}) \to \\
&\quad (\forall\,(\mathtt{a} : \mathsf{Set})\,(\mathtt{x} : \mathtt{a}) \to \mathtt{P\,a\,(PLeaf\,x)}) \to \\
&\quad (\forall\,(\mathtt{a} : \mathsf{Set})\,(\mathtt{x} : \mathtt{PTree}\,(\mathtt{a} \times \mathtt{a})) \to \mathtt{P}\,(\mathtt{a} \times \mathtt{a})\,\mathtt{x} \to \mathtt{P\,a\,(PNode\,x)}) \to \\
&\quad \forall\,(\mathtt{a} : \mathsf{Set})\,(\mathtt{x} : \mathtt{PTree\,a}) \to \mathtt{P\,a\,x}
\end{aligned}$$

and the structural induction rule for the nested type

```
data Lam a where
  Var :: a → Lam a
  App :: Lam a → Lam a → Lam a
  Abs :: Lam (Maybe a) → Lam a
```

of de Bruijn encoded lambda terms [9] with variables of type a is

$$\begin{aligned}
&\forall(\mathtt{P} : \forall(\mathtt{a} : \mathsf{Set}) \to \mathtt{Lam\,a} \to \mathsf{Set}) \to \\
&\quad (\forall(\mathtt{a} : \mathsf{Set})\,(\mathtt{x} : \mathtt{a}) \to \mathtt{P\,a\,(Var\,x)}) \to \\
&\quad (\forall(\mathtt{a} : \mathsf{Set})\,(\mathtt{x} : \mathtt{Lam\,a})\,(\mathtt{y} : \mathtt{Lam\,a}) \to \mathtt{P\,a\,x} \to \mathtt{P\,a\,y} \to \mathtt{P\,a\,(App\,x\,y)}) \to \\
&\quad (\forall(\mathtt{a} : \mathsf{Set})\,(\mathtt{x} : \mathtt{Lam\,(Maybe\,a)}) \to \mathtt{P\,(Maybe\,a)\,x} \to \mathtt{P\,a\,(Abs\,x)}) \to \\
&\quad \forall(\mathtt{a} : \mathsf{Set})\,(\mathtt{x} : \mathtt{Lam\,a}) \to \mathtt{P\,a\,x}
\end{aligned}$$

Deep induction rules for nested types must similarly account for their type constructors' changing type parameters while also parameterizing over the additional predicate on the type of data they contain. Letting $\mathtt{Pair}^\wedge\,\mathtt{Q}$ be the lifting of a predicate $\mathtt{Q}$ on $\mathtt{a}$ to pairs of type $\mathtt{a} \times \mathtt{a}$, so that $\mathtt{Pair}^\wedge\,\mathtt{Q}\,(\mathtt{x},\mathtt{y}) = \mathtt{Q\,x} \times \mathtt{Q\,y}$, this gives the deep induction rule

$$\begin{aligned}
&\forall\,(\mathtt{P} : \forall\,(\mathtt{a} : \mathsf{Set}) \to (\mathtt{a} \to \mathsf{Set}) \to \mathtt{PTree\,a} \to \mathsf{Set}) \to \\
&\quad (\forall\,(\mathtt{a} : \mathsf{Set})\,(\mathtt{Q} : \mathtt{a} \to \mathsf{Set})\,(\mathtt{x} : \mathtt{a}) \to \mathtt{Q\,x} \to \mathtt{P\,a\,Q\,(PLeaf\,x)}) \to \\
&\quad (\forall\,(\mathtt{a} : \mathsf{Set})\,(\mathtt{Q} : \mathtt{a} \to \mathsf{Set})\,(\mathtt{x} : \mathtt{PTree}\,(\mathtt{a} \times \mathtt{a})) \to \mathtt{P}\,(\mathtt{a} \times \mathtt{a})\,(\mathtt{Pair}^\wedge\,\mathtt{Q})\,\mathtt{x} \to \\
&\qquad\qquad\qquad\qquad\qquad\qquad \mathtt{P\,a\,Q\,(PNode\,x)}) \to \\
&\quad \forall\,(\mathtt{a} : \mathsf{Set})\,(\mathtt{Q} : \mathtt{a} \to \mathsf{Set})\,(\mathtt{x} : \mathtt{PTree\,a}) \to \mathtt{PTree}^\wedge\,\mathtt{Q\,x} \to \mathtt{P\,a\,Q\,x}
\end{aligned}$$

for perfect trees, and the deep induction rule

$$\begin{aligned}
&\forall(\mathtt{P} : \forall(\mathtt{a} : \mathsf{Set}) \to (\mathtt{a} \to \mathsf{Set}) \to \mathtt{Lam\,a} \to \mathsf{Set}) \to \\
&\quad (\forall(\mathtt{a} : \mathsf{Set})\,(\mathtt{Q} : \mathtt{a} \to \mathsf{Set})\,(\mathtt{x} : \mathtt{a}) \to \mathtt{Q\,x} \to \mathtt{P\,a\,Q\,(Var\,x)}) \to \\
&\quad (\forall(\mathtt{a} : \mathsf{Set})\,(\mathtt{Q} : \mathtt{a} \to \mathsf{Set})\,(\mathtt{x} : \mathtt{Lam\,a})\,(\mathtt{y} : \mathtt{Lam\,a}) \to \mathtt{P\,a\,Q\,x} \to \mathtt{P\,a\,Q\,y} \to \\
&\qquad\qquad\qquad\qquad\qquad\qquad \mathtt{P\,a\,Q\,(App\,x\,y)}) \to \\
&\quad (\forall(\mathtt{a} : \mathsf{Set})\,(\mathtt{Q} : \mathtt{a} \to \mathsf{Set})\,(\mathtt{x} : \mathtt{Lam\,(Maybe\,a)}) \to \mathtt{P\,(Maybe\,a)}\,(\mathtt{Maybe}^\wedge\,\mathtt{Q})\,\mathtt{x} \to \\
&\qquad\qquad\qquad\qquad\qquad\qquad \mathtt{P\,a\,Q\,(Abs\,x)}) \to \\
&\quad \forall(\mathtt{a} : \mathsf{Set})\,(\mathtt{Q} : \mathtt{a} \to \mathsf{Set})\,(\mathtt{x} : \mathtt{Lam\,a}) \to \mathtt{Lam}^\wedge\,\mathtt{Q\,x} \to \mathtt{P\,a\,Q\,x}
\end{aligned}$$

for lambda terms. As usual, the structural induction rules for these types can be recovered by setting $\mathtt{Q} = \lambda\mathtt{x}.\,\mathtt{True}$ in their deep induction rules. Moreover, the basic technique described in Section 2 can be recovered from the more general one described in this section by noting that the type arguments to ADT data type constructors don't change, and that the internal predicate parameter to $\mathtt{P}$ can therefore be lifted to the outermost level of ADT induction rules.

We conclude this section by giving both structural and deep induction rules for the following truly nested type of bushes [8]:

$$\texttt{Bush}\,\texttt{a} = \texttt{BNil} \mid \texttt{BCons}\,\texttt{a}\,(\texttt{Bush}\,(\texttt{Bush}\,\texttt{a}))$$

(Note that this type is not even definable in Agda.) Correct and useful structural induction rules for bushes and other truly nested types have long been elusive. One recent effort to derive such rules has been recorded in [10], but the approach taken there is more *ad hoc* than not, and generates induction rules for data types *related* to the nested types of interest rather than for the original nested types themselves. To treat bushes, for example, Fu and Selinger rewrite the type $\texttt{Bush}\,\texttt{a}$ as $\texttt{NBush}\,(\texttt{Succ}\,\texttt{Zero})\,\texttt{a}$, where $\texttt{NBush} = \texttt{NTimes}\,\texttt{Bush}$ and

$$\begin{array}{ll}\texttt{NTimes} & :: (\mathsf{Set} \to \mathsf{Set}) \to \texttt{Nat} \to \mathsf{Set} \to \mathsf{Set}\\ \texttt{NTimes}\,\texttt{p}\,\texttt{Zero}\,\texttt{s} & = \texttt{s}\\ \texttt{NTimes}\,\texttt{p}\,(\texttt{Succ}\,\texttt{n})\,\texttt{s} & = \texttt{p}\,(\texttt{NTimes}\,\texttt{p}\,\texttt{n}\,\texttt{s})\end{array}$$

Their induction rule for bushes is then given in terms of these rewritten ones as

$$\begin{array}{l}\forall\,(\texttt{a} : \mathsf{Set})\,(\texttt{P} : \forall\,(\texttt{n} : \texttt{Nat}) \to \texttt{NBush}\,\texttt{n}\,\texttt{a} \to \mathsf{Set}) \to\\ \quad (\forall\,(\texttt{x} : \texttt{a}) \to \texttt{P}\,\texttt{Zero}\,\texttt{x}) \to\\ \quad (\forall\,(\texttt{n} : \texttt{Nat}) \to \texttt{P}\,(\texttt{Succ}\,\texttt{n})\,\texttt{BNil}) \to\\ \quad (\forall\,(\texttt{n} : \texttt{Nat})\,(\texttt{x} : \texttt{NBush}\,\texttt{n}\,\texttt{a})\,(\texttt{xs} : \texttt{NBush}\,(\texttt{Succ}\,(\texttt{Succ}\,\texttt{n}))\,\texttt{a}) \to\\ \qquad\qquad \texttt{P}\,\texttt{n}\,\texttt{x} \to \texttt{P}\,(\texttt{Succ}\,(\texttt{Succ}\,\texttt{n}))\,\texttt{xs} \to \texttt{P}\,(\texttt{Succ}\,\texttt{n})\,(\texttt{BCons}\,\texttt{x}\,\texttt{xs})) \to\\ \quad \forall\,(\texttt{n} : \texttt{Nat})\,(\texttt{xs} : \texttt{NBush}\,\texttt{n}\,\texttt{a}) \to \texttt{P}\,\texttt{n}\,\texttt{xs}\end{array}$$

This approach appears promising, but is not yet fully mature. The core difficulty is that, although Fu and Selinger "hint at how the construction ... can be generalized to arbitrary nested types" and "give an example of nested data type [sic] that is hopefully general enough to make it clear what one would do in the general case" in Section 5 of [10], they do not show how to derive their induction rules in a uniform and principled way even for the "reasonably arbitrary and general" nested types they consider. As a result, it is unclear what guarantees that the induction rules they derive are correct, *either* for the original nested types *or* for their rewritten versions, or whether the induction rules for the rewritten nested types are sufficiently expressive to prove all results about the original nested types that one would expect to be provable by induction. This latter point echoes the issue with Coq-derived induction rules for forests mentioned above, and has the unfortunate effect of forcing users to manually write induction (and other) rules for such types for use in that system [17].

Direct application of the general technique illustrated above and explicated in full in Section 4 below derives the following first-ever useful induction rule for bushes, respectively — a full 20 years after bushes were first introduced!

$$\begin{array}{l}\forall(\texttt{P} : \forall(\texttt{a} : \mathsf{Set}) \to \texttt{Bush}\,\texttt{a} \to \mathsf{Set}) \to\\ \quad (\forall(\texttt{a} : \mathsf{Set}) \to \texttt{P}\,\texttt{a}\,\texttt{BNil}) \to\\ \quad (\forall(\texttt{a} : \mathsf{Set})\,(\texttt{x} : \texttt{a})\,(\texttt{y} : \texttt{Bush}\,(\texttt{Bush}\,\texttt{a})) \to \texttt{P}\,(\texttt{Bush}\,\texttt{a})\,\texttt{y} \to \texttt{P}\,\texttt{a}\,(\texttt{BCons}\,\texttt{x}\,\texttt{y})) \to\\ \quad \forall(\texttt{a} : \mathsf{Set})\,(\texttt{x} : \texttt{Bush}\,\texttt{a}) \to \texttt{P}\,\texttt{a}\,\texttt{x}\end{array}$$

In the next section we will see that this rule is derivable from the following more general one:

$$
\begin{aligned}
&\forall\,(\mathtt{P} : \forall\,(\mathsf{a} : \mathsf{Set}) \to (\mathsf{a} \to \mathsf{Set}) \to \mathtt{Bush}\,\mathsf{a} \to \mathsf{Set}) \to \\
&\quad (\forall\,(\mathsf{a} : \mathsf{Set})\,(\mathtt{Q} : \mathsf{a} \to \mathsf{Set}) \to \mathtt{P}\,\mathsf{a}\,\mathtt{Q}\,\mathtt{Bnil}) \to \\
&\quad (\forall\,(\mathsf{a} : \mathsf{Set})\,(\mathtt{Q} : \mathsf{a} \to \mathsf{Set})\,(\mathtt{x} : \mathsf{a})\,(\mathtt{y} : \mathtt{Bush}\,(\mathtt{Bush}\,\mathsf{a})) \to \\
&\qquad\qquad \mathtt{Q}\,\mathtt{x} \to \mathtt{P}\,(\mathtt{Bush}\,\mathsf{a})\,(\mathtt{P}\,\mathsf{a}\,\mathtt{Q})\,\mathtt{y} \to \mathtt{P}\,\mathsf{a}\,\mathtt{Q}\,(\mathtt{BCons}\,\mathtt{x}\,\mathtt{y})) \to \\
&\quad \forall\,(\mathsf{a} : \mathsf{Set})\,(\mathtt{Q} : \mathsf{a} \to \mathsf{Set})\,(\mathtt{x} : \mathtt{Bush}\,\mathsf{a}) \to \mathtt{Bush}^{\wedge}\,\mathtt{Q}\,\mathtt{x} \to \mathtt{P}\,\mathsf{a}\,\mathtt{Q}\,\mathtt{x}
\end{aligned}
$$

4 Theoretical Foundations

This section gives a grammar generating a robust class of nested types, including ADTs and truly nested types, and recaps the semantics given in [13] for them from which we derive their deep induction rules. This entire paper can thus be read as a practical application of the abstract results of [13].

4.1 Categorical Preliminaries

We write $a : A$ if A is category and a is an object of A. We write 0_A and 1_A for the initial and terminal objects of A, and o_A and $!_A$ for the unique maps $o_A : 0_A \to A$ and $!_A : A \to 1_A$, respectively. If A is the category Set of sets and functions between them, we write 0 for 0_{Set}, i.e., for $\emptyset$, and 1 for any 1-element set, i.e., for 1_{Set}. If $a : A$ we write K_a for the constantly a-valued functor on A. The category Fam, which we will use to interpret predicates, is given by:

Definition 1. *The category* Fam *comprises the following:*

- ***Objects:*** *An object of* Fam *is a pair* (A, P) *where* $A : \mathsf{Set}$ *and* $P : A \to \mathsf{Set}$.
- ***Morphisms:*** *A morphism* $f : (A, P) \to (A', P')$ *in* Fam *is a pair* (α, β), *where* $\alpha : A \to A'$ *and* $\beta : \Pi_{a:A}\, Pa \to P'(\alpha a)$.
- ***Identities:*** *The identity morphism* $\underline{id}_{(A,P)} : (A, P) \to (A, P)$ *in* Fam *is* $(id_A, \lambda a : A.\, id_{Pa})$.
- ***Composition:*** *If* $(\alpha, \beta) : (A, P) \to (A', P')$ *and* $(\alpha', \beta') : (A', P') \to (A'', P'')$, *then the composition* $(\alpha', \beta') \,\underline{\circ}\, (\alpha, \beta) : (A, P) \to (A'', P'')$ *in* Fam *is defined by* $(\alpha', \beta') \,\underline{\circ}\, (\alpha, \beta) = (\alpha' \circ \alpha, \lambda a : A.\, \beta'(\alpha a) \circ \beta a)$.

4.2 Syntax and Semantics of ADTs

If $\mathcal{V}$ is a countable set of type variables, $V \subseteq \mathcal{V}$ is finite, $\alpha \in \mathcal{V}$, and we write V, α for $V \cup \{\alpha\}$, then the following grammar generates (representations of) all standard *polynomial ADTs* over V, i.e., all ADTs defined over data of primitive types:

$$\mathcal{A}^V \;:=\; 0 \mid 1 \mid \alpha \in V \mid \mathcal{A}^V + \mathcal{A}^V \mid \mathcal{A}^V \times \mathcal{A}^V \mid \mu\alpha.\mathcal{A}^{V,\alpha}$$

The grammar $\mathcal{A} = \bigcup_V \mathcal{A}^V$ also generates (representations of) *deep ADTs*, i.e., ADTs defined not just over data of the primitive types, but over data of other ADTs as well. For example, it generates the representation $\mathit{List}\,\alpha := \mu\beta.\, 1 + \alpha \times \beta$ of the type `List a`, the representation $\mathit{Forest}\,\alpha := \mu\beta.\, 1 + \alpha \times \mu\gamma.\, 1 + \beta \times \gamma$ of the type `Forest a`, and the representation $\mu\delta.\, 1 + (\mu\beta.\, 1 + \alpha \times \mu\gamma.\, 1 + \beta \times \gamma) \times \delta$ of the type `List (Forest a)`. Using Bekič's Lemma, it can also generate (representations of) ADTs defined by mutual recursion such as $Expr := \mu\alpha.\, s(\alpha, \mu\beta.\, t(\alpha,\beta))$ and $BExpr := \mu\beta.\, t(Expr, \beta)$, where $s(\alpha,\beta) := Nat + \alpha \times \alpha + \beta \times \alpha \times \alpha$ and $t(\alpha,\beta) := Bool + \beta \times \beta + \beta + \alpha \times \alpha$ for the ADTs of integer and boolean expressions from Section 1. ADTs with more than one type argument can be handled by tupling them into one or, equivalently, by noting that such ADTs are generated by the extension $\mathcal{N}$ of the grammar $\mathcal{A}$ given in Section 4.4. We adopt the usual conventions regarding free and bound type variables for $\mathcal{A}$.

As usual, ADTs are interpreted relative to environments.

Definition 2. *A* set environment σ *is a function from a finite subset* V *of* $\mathcal{V}$ *to* Set. *We write* $\mathsf{Env}_V^{\mathsf{Set}}$ *for the set of set environments whose domain is* V. *If* $A \in \mathsf{Set}$, $\sigma \in \mathsf{Env}_V^{\mathsf{Set}}$, *and* $\alpha \notin V$, *then* $\sigma[\alpha := A]$ *is the set environment with domain* V, α *that extends* σ *by mapping* α *to* A. *We write* $\sigma\alpha$ *in place of* $\sigma(\alpha)$ *for the image of* α *under* σ, *and* $[]$ *for the set environment with domain* $V = \emptyset$.

It is well-known that the ADTs generated by the grammar $\mathcal{A}$ have initial algebra semantics in the category Set. That is, each such ADT $\mu\alpha.\, E$ can be interpreted as the carrier μF of the initial algebra for the polynomial endofunctor F on Set that interprets its body E. In particular, the final clause of the next definition is well-defined.

Definition 3. *The interpretation function* $[\![\cdot]\!]^{\mathsf{Set}} : \mathcal{A}^V \to \mathsf{Env}_V^{\mathsf{Set}} \to \mathsf{Set}$ *is:*

$$
\begin{aligned}
[\![0]\!]^{\mathsf{Set}}\sigma &= 0 \\
[\![1]\!]^{\mathsf{Set}}\sigma &= 1 \\
[\![\alpha]\!]^{\mathsf{Set}}\sigma &= \alpha\sigma \\
[\![E_1 + E_2]\!]^{\mathsf{Set}}\sigma &= [\![E_1]\!]^{\mathsf{Set}}\sigma + [\![E_2]\!]^{\mathsf{Set}}\sigma \\
[\![E_1 \times E_2]\!]^{\mathsf{Set}}\sigma &= [\![E_1]\!]^{\mathsf{Set}}\sigma \times [\![E_2]\!]^{\mathsf{Set}}\sigma \\
[\![\mu\alpha.\, E]\!]^{\mathsf{Set}}\sigma &= \mu(A \mapsto [\![E]\!]^{\mathsf{Set}}\sigma[\alpha := A])
\end{aligned}
$$

Like Set, the category Fam has sufficient structure to interpret ADTs generated by the grammar $\mathcal{A}$. In particular, it interprets bodies of polynomial ADTs.

Definition 4. *The category* Fam *supports the following constructions:*

- ***Initial object***: *The initial object* $\underline{0}$ *of* Fam *is* $(0, K_0 : 0 \to \mathsf{Set})$. *For* $(A, P) : \mathsf{Fam}$, $(o_A, \lambda x : 0.\, o_{P(o_A x)}) : \underline{0} \to (A, P)$ *is the unique map from* $\underline{0}$ *to* (A, P).
- ***Terminal object***: *The terminal object* $\underline{1}$ *of* Fam *is* $(1, K_1 : 1 \to \mathsf{Set})$, *where* $()$ *is the unique element of the set* 1 *and* $K_1() = 1$. *For* $(A, P) : \mathsf{Fam}$, $(!_A, \lambda a : A.\, !_{Pa}) : (A, P) \to \underline{1}$ *is the unique map from* (A, P) *to* $\underline{1}$.
- ***Coproducts***: *Given* $(A, P), (A', P') : \mathsf{Fam}$, *the coproduct* $(A, P) \underline{+} (A', P') : \mathsf{Fam}$ *is* $(A + A', P + P')$, *where* $P + P' : A + A' \to \mathsf{Set}$ *is just the usual coproduct of* P *and* P' *as functions. The associated injections* $\underline{\mathsf{inL}} : (A, P) \to (A, P) \underline{+} (A', P')$ *and* $\underline{\mathsf{inR}} : (A', P') \to (A, P) \underline{+} (A', P')$ *are given by* $\underline{\mathsf{inL}} = (inL, \lambda a : A.\, id_{Pa})$ *and* $\underline{\mathsf{inR}} = (inR, \lambda a' : A'.\, id_{P'a'})$. *The coproduct* $(\alpha, \beta) \underline{+} (\alpha', \beta') : (A, P) \underline{+} (A', P') \to (B, Q)$ *of morphisms* $(\alpha, \beta) : (A, P) \to (B, Q)$

and $(\alpha',\beta') : (A',P') \to (B,Q)$ *is* $(\alpha+\alpha',\delta)$*, where* $\delta : \Pi_{x\in A+A'}(P+P')x \to Q((\alpha+\alpha')x)$ *is defined by* $\delta(\mathsf{inL}\,a) = \beta a$ *and* $\delta(\mathsf{inR}\,a') = \beta' a'$*. As expected,* $((\alpha,\beta)\,\underline{+}\,(\alpha',\beta'))\,\underline{\circ}\,\underline{\mathsf{inL}} = (\alpha,\beta)$ *and* $((\alpha,\beta)\,\underline{+}\,(\alpha',\beta'))\,\underline{\circ}\,\underline{\mathsf{inR}} = (\alpha',\beta')$.

- ***Products:*** *Given* $(A,P),(A',P') : \mathsf{Fam}$*, the product* $(A,P)\,\underline{\times}\,(A',P') : \mathsf{Fam}$ *is* $(A\times A', \lambda(a,a') : A\times A'.\, Pa \times P'a')$*. The associated projections* $\underline{\pi_1} : (A,P)\,\underline{\times}\,(A',P') \to (A,P)$ *and* $\underline{\pi_2} : (A,P)\,\underline{\times}\,(A',P') \to (A',P')$ *are given by* $\underline{\pi_1} = (\pi_1, \lambda(a,a') : A\times A'.\,\pi_1)$ *and* $\underline{\pi_2} = (\pi_2, \lambda(a,a') : A\times A'.\,\pi_2)$*. The product* $(\alpha,\beta)\,\underline{\times}\,(\alpha',\beta') : (A,P) \to (B,Q)\,\underline{\times}\,(B',Q')$ *of morphisms* $(\alpha,\beta) : (A,P) \to (B,Q)$ *and* $(\alpha',\beta') : (A,P) \to (B',Q')$ *is* $(\lambda a : A.(\alpha a, \alpha' a), \lambda a : A.\,\lambda x : Pa.\,(\beta a x, \beta' a x))$*. As expected,* $\underline{\pi_1}\,\underline{\circ}\,((\alpha,\beta)\,\underline{\times}\,(\alpha',\beta')) = (\alpha,\beta)$ *and* $\underline{\pi_2}\,\underline{\circ}\,((\alpha,\beta)\,\underline{\times}\,(\alpha',\beta')) = (\alpha',\beta')$.

To interpret ADTs generated by $\mathcal{A}$ in Fam we also need to be able to interpret expressions of the form $\mu\alpha.E$. This we do by computing the least fixed point in Fam of the functor $G : \mathsf{Fam} \to \mathsf{Fam}$ interpreting E. It is natural to try to do this using the same technique in Fam that gives its Set-interpretation, i.e., by iterating G ω-many times starting from the initial object $\underline{0}$ of Fam. This gives the least fixed point $\underline{\mu}G$ of G as the colimit $G^\omega\underline{0}$ in Fam of the sequence

$$\underline{0} \hookrightarrow G\underline{0} \hookrightarrow G^2\underline{0} \hookrightarrow \ldots \hookrightarrow G^n\underline{0} \hookrightarrow \ldots \qquad (*)$$

This approach is indeed viable, and is formally justified by [13]. There, it is shown that if λ is a regular cardinal, $\mathcal{C}$ is a locally λ-presentable category, and $G : \mathcal{C} \to \mathcal{C}$ is a λ-accessible functor drawn from a particular class of functors that goes far beyond just first-order polynomial ones, then the least fixed point $\underline{\mu}G$ of G exists in $\mathcal{C}$ and can be computed as the transfinite colimit $G^\lambda\underline{0}$ of the sequence $\underline{0} \hookrightarrow G\underline{0} \hookrightarrow G^2\underline{0} \hookrightarrow \ldots \hookrightarrow G^n\underline{0} \hookrightarrow \ldots \hookrightarrow G^\omega\underline{0} \hookrightarrow \ldots \hookrightarrow G^\alpha\underline{0} \hookrightarrow \ldots$ over all $\alpha < \lambda$. That the sequence (*) computes $\underline{\mu}G$ for all polynomial functors on Fam then follows by taking λ to be ω, noting that Fam is locally finitely presentable, and recalling that all such functors are ω-accessible. That (*) further computes $\underline{\mu}G$ for *every* functor G on Fam that interprets an expression generated by $\mathcal{A}$ now follows easily by structural induction. We record this as:

Theorem 1. *If* $G : \mathsf{Fam} \to \mathsf{Fam}$ *is a functor interpreting an expression (with a distinguished variable) generated by the grammar* $\mathcal{A}$*, then the least fixed point* $\underline{\mu}G$ *of* G *(with respect to that variable) is* $G^\omega\underline{0}$*. Concretely, the colimit* $G^\omega\underline{0}$ *can be computed as* $\varinjlim_{n\in\mathbb{N}}(A_n,P_n) = (A,P)$*, where* $A = \varinjlim_{n\in\mathbb{N}} A_n$ *with mediating morphisms* $\alpha_n : A_n \to A$*, and* P *is defined by* $P\,x = \varinjlim_{n\in\mathbb{N}, y\in\alpha_n^{-1}(x)} P_n\,y$.

To define interpretations in Fam for ADTs generated by $\mathcal{A}$ we need the following analogue of Definition 2:

Definition 5. *A* predicate environment ρ *is a function from a finite subset* V *of* $\mathcal{V}$ *to* Fam*. We write* $\mathsf{Env}_V^{\mathsf{Fam}}$ *for the set of predicate environments whose domain is* V*. If* $(A,P) \in \mathsf{Fam}$*,* $\rho \in \mathsf{Env}_V^{\mathsf{Fam}}$*, and* $\alpha \notin V$*, we write* $\rho[\alpha := (A,P)]$ *for the predicate environment with domain* V,α *that extends* ρ *by mapping* α *to* (A,P)*. We write* $\alpha\rho$ *in place of* $\rho(\alpha)$ *for the image of* α *under* ρ*.*

Let $\sigma \in \mathsf{Env}_V^{\mathsf{Set}}$*. If* $\rho \in \mathsf{Env}_V^{\mathsf{Fam}}$ *is such that* $\pi_1(\alpha\rho) = \alpha\sigma$ *for all* $\alpha \in V$ *then we say that* ρ *is a* lifting *of* σ*. We write* $\overline{\sigma}$ *for the particular lifting* ρ *of* σ *such* *that* $\alpha\rho = (\alpha\sigma, K_1)$ *for all* $\alpha \in V$*. In addition, if* $\rho \in \mathsf{Env}_V^{\mathsf{Fam}}$ *maps each* $\alpha \in V$ *to* (A_α, P_α) *then we write* $\pi_1\rho$ *for the set environment with domain* V *mapping each* $\alpha \in V$ *to* A_α*. We write* $[]$ *for the unique environment with domain* $V = \emptyset$*.*

We then have the following Fam-interpretations for ADTs generated by $\mathcal{A}$:

Definition 6. *The interpretation function* $[\![\cdot]\!]^{\mathsf{Fam}} : \mathcal{A}^V \to \mathsf{Env}_V^{\mathsf{Fam}} \to \mathsf{Fam}$ *is:*

$$\begin{aligned}
[\![0]\!]^{\mathsf{Fam}}\rho &= \underline{0}\\
[\![1]\!]^{\mathsf{Fam}}\rho &= \underline{1}\\
[\![\alpha]\!]^{\mathsf{Fam}}\rho &= \alpha\rho\\
[\![E_1 + E_2]\!]^{\mathsf{Fam}}\rho &= [\![E_1]\!]^{\mathsf{Fam}}\rho \,\underline{+}\, [\![E_2]\!]^{\mathsf{Fam}}\rho\\
[\![E_1 \times E_2]\!]^{\mathsf{Fam}}\rho &= [\![E_1]\!]^{\mathsf{Fam}}\rho \,\underline{\times}\, [\![E_2]\!]^{\mathsf{Fam}}\rho\\
[\![\mu\alpha.E]\!]^{\mathsf{Fam}}\rho &= \underline{\mu}(Z \mapsto [\![E]\!]^{\mathsf{Fam}}\rho[\alpha := Z])
\end{aligned}$$

Before showing how to derive induction rules for the ADTs generated by $\mathcal{A}$ we prove two crucial lemmas linking their Set- and Fam-interpretations.

Lemma 1. *If* $E \in \mathcal{A}^V$ *and* $\rho \in \mathsf{Env}_V^{\mathsf{Fam}}$*, then* $\pi_1([\![E]\!]^{\mathsf{Fam}}\rho) = [\![E]\!]^{\mathsf{Set}}(\pi_1\rho)$*. Furthermore, if* $\pi_2(\beta\rho) = K_1$ *for all* $\beta \in V$*, then* $\pi_2([\![E]\!]^{\mathsf{Fam}}\rho) = K_1$*.*

Proof. By induction on the structure of expressions. The only non-trivial case is for $\mu\alpha.E \in \mathcal{A}^V$. Let $\rho \in \mathsf{Env}_V^{\mathsf{Fam}}$ be given. Letting $F : \mathsf{Set} \to \mathsf{Set}$ be defined by $FA = [\![E]\!]^{\mathsf{Set}}(\pi_1\rho)[\alpha := A]$ and $G : \mathsf{Fam} \to \mathsf{Fam}$ be defined by $G(A,Q) = [\![E]\!]^{\mathsf{Fam}}\rho[\alpha := (A,Q)]$, the induction hypothesis gives

$$\pi_1(G(A,Q)) = \pi_1([\![E]\!]^{\mathsf{Fam}}\rho[\alpha := (A,Q)]) = [\![E]\!]^{\mathsf{Set}}(\pi_1\rho)[\alpha := A] = FA \qquad (\dagger)$$

and if $\pi_2(\beta\rho) = K_1$ for all $\beta \in V$ then, moreover, $\pi_2(G(A,K_1)) = K_1$. We then have $\pi_1([\![\mu\alpha.E]\!]^{\mathsf{Fam}}\rho) = \pi_1(\underline{\mu}((A,Q) \mapsto [\![E]\!]^{\mathsf{Fam}}\rho[\alpha := (A,Q)])) = \pi_1(\underline{\mu}G) = \pi_1(\varinjlim_{n\in\mathbb{N}} G^n\underline{0}) = \varinjlim_{n\in\mathbb{N}}\pi_1(G^n\underline{0}) = \varinjlim_{n\in\mathbb{N}} F^n 0 = \mu F = \mu(A \mapsto [\![E]\!]^{\mathsf{Set}}(\pi_1\rho)[\alpha := A]) = [\![\mu\alpha.E]\!]^{\mathsf{Set}}(\pi_1\rho)$. Here, the fourth equality is justified by Theorem 1, and the fifth is justified by (†) and induction on n. If $\pi_2(\beta\rho) = K_1$ for all $\beta \in V$ as well, then $\pi_2([\![\mu\alpha.E]\!]^{\mathsf{Fam}}\rho) = \pi_2(\underline{\mu}((A,Q) \mapsto [\![E]\!]^{\mathsf{Fam}}\rho[\alpha := (A,Q)])) = \pi_2(\underline{\mu}G) = \pi_2(\varinjlim_{n\in\mathbb{N}} G^n\underline{0}) = \pi_2(\varinjlim_{n\in\mathbb{N}}(F^n 0, K_1)) = \lambda x. \varinjlim_{n\in\mathbb{N}, y\in\alpha_n^{-1}x} K_1 y = K_1$. Here, the morphisms $\alpha_n : F^n 0 \to \mu F$ are the mediating morphisms for the colimit, as in Theorem 1, and the fourth equality is justified by the fact that $\pi_2(G(A,K_1)) = K_1$ and induction on n.

Corollary 1. *If* E *is closed then* $[\![E]\!]^{\mathsf{Fam}}[] = ([\![E]\!]^{\mathsf{Set}}[], K_1)$*.*

Lemma 2. *If* $\sigma \in \mathsf{Env}_V^{\mathsf{Set}}$*, and if* $F : \mathsf{Set} \to \mathsf{Set}$ *and* $G : \mathsf{Fam} \to \mathsf{Fam}$ *are given by* $FA = [\![E]\!]^{\mathsf{Set}}\sigma[\alpha := A]$ *and* $G(A,Q) = [\![E]\!]^{\mathsf{Fam}}\overline{\sigma}[\alpha := (A,Q)]$*, then* $\underline{\mu}G = (\mu F, K_1)$*.*

Proof. We have $\underline{\mu}G = \underline{\mu}((A,Q) \mapsto [\![E]\!]^{\mathsf{Fam}}\overline{\sigma}[\alpha := (A,Q)]) = [\![\mu\alpha.E]\!]^{\mathsf{Fam}}\overline{\sigma} = ([\![\mu\alpha.E]\!]^{\mathsf{Set}}\sigma, K_1) = (\mu F, K_1)$, where the third equality holds by Lemma 1.

4.3 Induction Rules for ADTs

To derive induction rules for the ADTs generated by $\mathcal{A}$, we first observe that, given an ADT $\mu\alpha.E \in \mathcal{A}^V$ and a set environment $\sigma \in \mathsf{Env}_V^{\mathsf{Set}}$ interpreting its free variables, the interpretation $[\![E]\!]^{\mathsf{Set}}\sigma$ defines a functor $F_\sigma A = [\![E]\!]^{\mathsf{Set}}\sigma[\alpha := A]$ such that $[\![\mu\alpha.E]\!]^{\mathsf{Set}}\sigma = \mu(A \mapsto [\![E]\!]^{\mathsf{Set}}\sigma[\alpha := A]) = \mu(A \mapsto F_\sigma A) = \mu F_\sigma$. We can therefore think of F_σ as representing the data type constructor associated with the ADT. Thus, as argued in [11,12], the semantic induction rule for proving predicates over the σ-instance of the ADT $\mu\alpha.E$ has the form

$$\forall(P : \mu F_\sigma \to \mathsf{Set}).\, ??? \to \forall(x : \mu F_\sigma).\, Px$$

for some appropriate hypotheses ???. We can use the Fam-interpretation of E to discover a semantic counterpart to the hypotheses ???. Reflecting the resulting semantic rule for the σ-instance of $\mu\alpha.E$ back into the programming language syntax will then derive induction rules for polynomial ADTs.

To deduce what ??? is, we first observe that the conclusion $\forall(x : \mu F_\sigma).\, Px$ of the induction rule for the σ-instance of $\mu\alpha.E$ is isomorphic to the type of the second component of a morphism in Fam from $(\mu F_\sigma, K_1)$ to $(\mu F_\sigma, P)$ whose first component is id. Lemma 1 suggests that if we can see $(\mu F_\sigma, K_1)$ as $\underline{\mu}G$ for some functor $G : \mathsf{Fam} \to \mathsf{Fam}$, then we can fold over a G-algebra on $(\mu F_\sigma, P)$ in Fam to get such a morphism, i.e., to inhabit the type that is the structural induction rule for the σ-instance of $\mu\alpha.E$. This will provide a proof $ind_{\mu\alpha.E,\overline{\sigma}}\, P$ that the property P holds for all elements of the σ-instance of $\mu\alpha.E$.

To this end, let $\rho \in \mathsf{Env}_V^{\mathsf{Fam}}$ be any lifting of σ, and consider again the functor $\hat{F}_\rho(A, Q) = [\![E]\!]^{\mathsf{Fam}}\rho[\alpha := (A, Q)]$ on Fam given in Lemma 1 (there called G). An $\hat{F}_\rho$-algebra structure on $(\mu F_\sigma, P)$ is a morphism $(k', k) : \hat{F}_\rho(\mu F_\sigma, P) \to (\mu F_\sigma, P)$ in Fam. Then $\pi_1(\hat{F}_\rho(\mu F_\sigma, P)) = \pi_1([\![E]\!]^{\mathsf{Fam}}\rho[\alpha := (\mu F_\sigma, P)]) = (\pi_1([\![E]\!]^{\mathsf{Fam}}\rho))[\alpha := \mu F_\sigma] = [\![E]\!]^{\mathsf{Set}}\sigma[\alpha := \mu F_\sigma] = F_\sigma(\mu F_\sigma)$, with the third equality holding by Lemma 1. If we take $k' = in$, then $k : \forall(x : F_\sigma(\mu F_\sigma)).\, \pi_2([\![E]\!]^{\mathsf{Fam}}\rho[\alpha := (\mu F_\sigma, P)])x \to P(in\, x)$, so that

$$\begin{aligned}
ind_{\mu\alpha.E,\rho} \quad &: \forall(P : \mu F_\sigma \to \mathsf{Set}).\\
&\qquad (\forall(x : F_\sigma(\mu F_\sigma)).\, \pi_2([\![E]\!]^{\mathsf{Fam}}\rho[\alpha := (\mu F_\sigma, P)])x \to P(in\, x))\\
&\qquad \to \forall(x : \mu F_\sigma).\, Px\\
ind_{\mu\alpha.E,\rho}\, P\, k\, x &= \pi_2\, (fold^{\mathsf{Fam}}_{\mu\alpha.E,\rho}\, (in, k))\, x\, ()
\end{aligned}$$

Here, $fold^{\mathsf{Fam}}_{\mu\alpha.\,E,\rho}(in, k)$ is the unique $\hat{F}_\rho$-algebra morphism from $\underline{in} : \hat{F}_\rho(\underline{\mu}\hat{F}_\rho) \to \underline{\mu}\hat{F}_\rho$ to (in, k) in Fam.

Taking $\rho = \overline{\sigma}$ in the above development derives the expected structural induction rules for ADTs generated by $\mathcal{A}$. But this development is actually far more flexible, since the induction rule it derives is parameterized over an *arbitrary* lifting ρ of the set environment σ, and later specialized to $\overline{\sigma}$ to obtain structural induction rules for ADTs. The non-specialized rule can therefore be used to prove properties of ADTs that are parameterized over non-trivial (i.e., non-K_1) predicates on the type parameters to the type constructors induced by those ADTs; these are precisely our deep induction rules for ADTs.

As expected, the conclusion of an ADT's deep induction rule will have an additional hypothesis involving the lifting of this predicate to that ADT. As we have seen, the ability to lift a predicate Q on a set A to a predicate T_Q on TA, where T is an ADT's type constructor, is therefore central to deep induction. Every type constructor for every ADT generated by the grammar $\mathcal{A}$ has such a lifting. Concretely, it is computed as the second component of the interpretation in Fam of that data type. For example, the lifting $List_Q : List\,A \to \mathsf{Set}$ is $\pi_2[\![\mu\beta.\,1+\alpha\times\beta]\!]^{\mathsf{Fam}}[\alpha := (A,Q)]$. This can be coded in Agda as

```
List^ : ∀ {a : Set} → (a → Set) → (List a → Set)
List^ Q Nil = ⊤
List^ Q (Cons x xs) = Q x × List^ Q xs
```

Example 1. The deep induction rule for lists can be computed as the type of $ind_{List\,\alpha,\,\rho}$ for the ADT $List\,\alpha := \mu\beta.\,1+\alpha\times\beta$ and the predicate environment $\rho = [\alpha := (A,Q)]$ for $(A,Q) \in \mathsf{Fam}$. Letting $FY = [\![1+\alpha\times\beta]\!]^{\mathsf{Set}}(\pi_1\rho)[\beta := Y] = 1 + A\times Y$ with the obviously named injections, we have that $\mu F = List\,A$. This gives the deep induction rule

$$\begin{aligned} ind_{List\,\alpha,\rho} : \forall (P : \mu F \to \mathsf{Set}).\,\forall\,(Q : A \to \mathsf{Set}).\\ (\forall (x : F(\mu F)).\,\pi_2\,([\![1+\alpha\times\beta]\!]^{\mathsf{Fam}}[\alpha := (A,Q), \beta := (\mu F, P)])x \to\\ P(in\,x)) \to \forall (x : \mu F).\, List_Q\,x \to P\,x \end{aligned}$$

Simplifying π_2's argument gives $(1, K_1)\,\underline{+}\,(A,Q)\,\underline{\times}\,(\mu F, P)$. Its predicate part, obtained by applying π_2, is $K_1 + (Q \times P)$, so the hypotheses for $ind_{List\,\alpha,\rho}$ are

$$\begin{aligned} &\forall (x : 1 + A \times List\,A).(K_1 + (Q\times P))x \to P(in\,x)\\ &= (\forall (x:1).\,1 \to P\,Nil) \times (\forall (y : A).\,\forall (ys : List\,A).\,Q\,y \to P\,ys \to P\,(Cons\,y\,ys))\\ &= P\,Nil \times (\forall (y : A).\,\forall (ys : List\,A).\,Q\,y \to P\,ys \to P\,(Cons\,y\,ys)) \end{aligned}$$

Reflecting back into syntax gives the deep induction rule from Section 1:

```
∀ (a : Set) (P : List a → Set) (Q : a → Set) →
      P Nil → (∀(y : a) (ys : List a) → Q y → P ys → P (Cons y ys)) →
      ∀(xs : List a) → List^ Q xs → P xs
```

Taking $Q = K_1$ gives the usual structural induction rule for lists from Section 1.

Example 2. Since `Forest a` and `List (Forest a)` are mutually recursively defined, the deep induction rule for forests is defined by mutual recursion with the deep induction rule for lists. It can be computed as the type of $ind_{Forest\,\alpha,\,\rho}$ for the ADT $Forest\,\alpha := \mu\beta.\,\alpha\times\mu\gamma.\,1+\beta\times\gamma$ using the same technique as in Example 1. This gives the (deep) induction rule for forests from Section 1.

Example 3. Exactly the same technique delivers the deep induction rules from Section 1 for the mutually recursive ADTs `Expr` and `BExpr` whose representations are given before Definition 2.

4.4 Syntax and Semantics of Nested Types

We can use the technique from Section 4.3 to derive induction rules for nested types as well, including truly nested types and other deep nested types. To do so we first need an extension of the grammar $\mathcal{A}$ that generates these types.

Since nested types generalize ADTs to allow elements of a nested type at one instance of a type to depend on data at other instances of that nested type, they are interpreted as least fixed points not of ordinary (first-order) functors on Fam, as ADTs are, but rather as least fixed points of higher-order such functors. Moreover, since nested types can be parameterized over any number of type arguments, the (first-order) functors interpreting them can have correspondingly arbitrary arities. For each $k \geq 0$ we therefore assume a countable set $\mathcal{F}^k$ of *functor variables of arity* k, disjoint for distinct k. We use lower case Greek letters for functor variables, write φ^k to indicate that $\varphi \in \mathcal{F}^k$, and say that φ has *arity* k in this case. Type variables are exactly functor variables of arity 0; we continue to write α, β, etc., rather than α^0, β^0, etc., for them. We write $\mathcal{F} = \bigcup_{k\geq 0} \mathcal{F}^k$. If $V \subseteq \mathcal{F}$ is finite and $\varphi \in \mathcal{F}^k$ for some k, write V, φ for $V \cup \{\varphi\}$.

Definition 7. *For a finite set V of $\mathcal{F}$, the set of* (truly) nested data types over *V is generated by the following grammar:*

$$\mathcal{N}^V \;:=\; 0 \mid 1 \mid \varphi^k\overline{\mathcal{N}^V} \mid \mathcal{N}^V + \mathcal{N}^V \mid \mathcal{N}^V \times \mathcal{N}^V \mid (\mu\varphi^k.\lambda\alpha_1...\alpha_k.\mathcal{N}^{V,\alpha_1,...,\alpha_k,\varphi})\overline{\mathcal{N}^V}$$

Here, $\varphi^k \in V$ and the lengths of the vectors of terms in $\overline{\mathcal{N}^V}$ in the third and final clauses of the above grammar are both k.

The grammar $\mathcal{N} = \bigcup_V \mathcal{N}^V$ generalizes $\mathcal{A}$ by allowing recursion not just at the level of type variables, but also at the level of functor variables. This reflects the fact that, in programming language syntax, nested types can be parameterized over both types and type constructors. For example, $\mathcal{N}^V$ generates the representation $\mathit{PTree}\,\alpha := \big(\mu\varphi^1.\lambda\beta.\beta + \varphi(\beta\times\beta)\big)\,\alpha \in \mathcal{N}^\alpha$ of the type `PTree a`, the representation $\mathit{Lam}\,\alpha := \big(\mu\varphi^1.\lambda\beta.\beta + \varphi\beta\times\varphi\beta + \varphi(\beta{+}1)\big)\,\alpha \in \mathcal{N}^\alpha$ of the type `Lam a` and the representation $\mathit{Bush}\,\alpha := \big(\mu\varphi^1.\lambda\beta.\,1 + \beta\times\varphi\,(\varphi\,\beta)\big)\,\alpha \in \mathcal{N}^\alpha$ of the type `Bush a`. But it also generates the representation $\mathit{GForest}\,\varphi\,\alpha := \mu\beta.\,1 + \alpha\times\varphi\,\beta \in \mathcal{N}^{\varphi,\alpha}$ of the following nested type of generalized forests with data of type `a`:

$$\texttt{GForest f a} \;=\; \texttt{FEmpty} \mid \texttt{FNode a (f (GForest f a))}$$

This type is higher-order in the sense that the type constructor `GForest` takes not just a type, but also a (unary) type constructor, as an argument. It therefore cannot be expressed as an element of $\mathcal{A}$, and thus demonstrates the benefit of working with the more expressive grammar $\mathcal{N}$. On the other hand, it is decidedly ADT-like, in the sense that it defines a family of inductive types rather than an inductive family of types. In fact, if `f` were a type constructor induced by a nested type generated by our grammar, then `GForest f a` and `f (GForest f a)` would be mutually recursively defined. In this case, generalizing Example 2, their structural induction rules would also be defined by mutual recursion.

It is not hard to see that $\mathcal{A} \subseteq \mathcal{N}$. Moreover, the grammar $\mathcal{N}$ allows nested types to be parameterized over (other) nested data types, just as $\mathcal{A}$ allows ADTs to be parameterized over (other) ADTs. For instance, we could have perfect trees of lists or binary trees, bushes of perfect trees, etc.

We have the following notions of functor and application in Fam:

Definition 8. *A (k-ary) lifted functor $G : \mathsf{Fam}^k \to \mathsf{Fam}$ is a pair (F, P), where $F : \mathsf{Set}^k \to \mathsf{Set}$ and $P : \forall(X_1, P_1)....(X_k, P_k).\, FX_1...X_k \to \mathsf{Set}$ is a Fam-indexed predicate. The* application *of a functor $(F, P) : \mathsf{Fam}^k \to \mathsf{Fam}$ to an object $(A_1, Q_1), ..., (A_k, Q_k)$ of Fam^k is given by*

$$(F, P)(A_1, Q_1)...(A_k, Q_k) \;=\; (FA_1...A_k,\; P(A_1, Q_1)...(A_k, Q_k))$$

We call a lifted functor $G = (F, P)$ a *lifting* of F from Set to Fam, and call P a Fam-*indexed predicate.* A Set-*indexed predicate* is a Fam-indexed predicate that does not depend on its arguments' second components. We extend the notions of *set environment* and *predicate environment* from Definitions 2 and 5 as follows:

Definition 9. *A* set environment *σ is a mapping from a finite subset $V = \{\varphi_1^{k_1}, ..., \varphi_n^{k_n}\}$ of $\mathcal{F}$ such that $\varphi_i\sigma : \mathsf{Set}^{k_i} \to \mathsf{Set}$ for $i = 1, ..., n$. We write $\mathsf{Env}_V^{\mathsf{Set}}$ for the set of set environments whose domain is V. If $F \in \mathsf{Set}^k \to \mathsf{Set}$, $\sigma \in \mathsf{Env}_V^{\mathsf{Set}}$, and $\varphi^k \notin V$, we write $\sigma[\varphi := F]$ for the set environment with domain V, φ that extends σ by mapping φ to F. Similarly, a* predicate environment *ρ is a mapping from a finite subset $V = \{\varphi_1^{k_1}, ..., \varphi_n^{k_n}\}$ of $\mathcal{F}$ such that $\varphi_i\rho : \mathsf{Fam}^{k_i} \to \mathsf{Fam}$ is a lifted functor for $i = 1, ..., n$. We write $\mathsf{Env}_V^{\mathsf{Fam}}$ for the set of predicate environments whose domain is V. If $(F, P) \in \mathsf{Fam}^k \to \mathsf{Fam}$, $\rho \in \mathsf{Env}_V^{\mathsf{Fam}}$, and $\varphi^k \notin V$, we write $\rho[\varphi := (F, P)]$ for the predicate environment with domain V, φ that extends ρ by mapping φ to (F, P).*

The notions of a predicate environment being a lifting of a set environment and the notations $\overline{\sigma}$, $\pi_1\rho$, and $[]$ are now extended in the obvious ways.

The following interpretations of nested types generated by $\mathcal{N}$ in the locally finitely presentable categories Set and Fam are shown in [13] to be well-defined:

Definition 10. *The interpretation functions $[\![\cdot]\!]^{\mathsf{Set}} : \mathcal{N}^V \to \mathsf{Env}_V^{\mathsf{Set}} \to \mathsf{Set}$ and $[\![\cdot]\!]^{\mathsf{Fam}} : \mathcal{N}^V \to \mathsf{Env}_V^{\mathsf{Fam}} \to \mathsf{Fam}$ are:*

$$\begin{aligned}
[\![0]\!]^{\mathsf{Set}}\sigma &= 0\\
[\![1]\!]^{\mathsf{Set}}\sigma &= 1\\
[\![\varphi^k E_1...E_k]\!]^{\mathsf{Set}}\sigma &= (\varphi\sigma)\overline{([\![E_i]\!]^{\mathsf{Set}}\sigma)}\\
[\![E_1 + E_2]\!]^{\mathsf{Set}}\sigma &= [\![E_1]\!]^{\mathsf{Set}}\sigma + [\![E_2]\!]^{\mathsf{Set}}\sigma\\
[\![E_1 \times E_2]\!]^{\mathsf{Set}}\sigma &= [\![E_1]\!]^{\mathsf{Set}}\sigma \times [\![E_2]\!]^{\mathsf{Set}}\sigma\\
[\![(\mu\varphi^k.\lambda\alpha_1...\alpha_k.\, E)E_1...E_k]\!]^{\mathsf{Set}}\sigma &= (\mu(F \mapsto \lambda A_1...A_k.\\
&\qquad [\![E]\!]^{\mathsf{Set}}\sigma\overline{[\alpha_i := A_i]}[\varphi := F]))\overline{([\![E_i]\!]^{\mathsf{Set}}\sigma)}
\end{aligned}$$

$$
\begin{aligned}
[\![0]\!]^{\mathsf{Fam}}\rho &= \underline{0} \\
[\![1]\!]^{\mathsf{Fam}}\rho &= \underline{1} \\
[\![\varphi^k E_1 ... E_k]\!]^{\mathsf{Fam}}\rho &= (\varphi\rho)\overline{([\![E_i]\!]^{\mathsf{Fam}}\rho)} \\
[\![E_1 + E_2]\!]^{\mathsf{Fam}}\rho &= [\![E_1]\!]^{\mathsf{Fam}}\rho \,\underline{+}\, [\![E_2]\!]^{\mathsf{Fam}}\rho \\
[\![E_1 \times E_2]\!]^{\mathsf{Fam}}\rho &= [\![E_1]\!]^{\mathsf{Fam}}\rho \,\underline{\times}\, [\![E_2]\!]^{\mathsf{Fam}}\rho \\
[\![(\mu\varphi^k.\lambda\alpha_1...\alpha_k.\, E)E_1...E_k]\!]^{\mathsf{Fam}}\rho &= (\underline{\mu}(F \mapsto \lambda Z_1...Z_k. \\
&\qquad [\![E]\!]^{\mathsf{Fam}}\rho\overline{[\alpha_i := Z_i]}[\varphi := F]))\overline{([\![E_i]\!]^{\mathsf{Fam}}\rho)}
\end{aligned}
$$

4.5 Induction Rules for Nested Types

Straightforward generalization of the analysis in Section 4.3 to $\mathcal{N}$ gives induction rules for the type constructors nested types induce. Given a nested type $(\mu\varphi^k.\lambda\alpha_1...\alpha_k.\, E)E_1...E_k \in \mathcal{N}^V$ with type constructor $T = \mu\varphi^k.\lambda\alpha_1...\alpha_k.\, E$ and a set environment $\sigma \in \mathsf{Env}_V^{\mathsf{Set}}$ interpreting its free variables, we have that

$$
[\![T\overline{E_i}]\!]^{\mathsf{Set}}\sigma = \mu(F \mapsto \lambda A_1...A_k.\, [\![E]\!]^{\mathsf{Set}}\sigma\overline{[\alpha_i := A_i]}[\varphi := F])\overline{([\![E_i]\!]^{\mathsf{Set}}\sigma)} = (\mu H_\sigma)\overline{([\![E_i]\!]^{\mathsf{Set}}\sigma)}
$$

where the higher-order functor H_σ on Set is defined by

$$
H_\sigma F A_1 ... A_k = [\![E]\!]^{\mathsf{Set}}\sigma\overline{[\alpha_i := A_i]}[\varphi := F]
$$

For any lifting ρ of σ, the predicate counterpart to H_σ is the higher-order functor $\hat{H}_\rho$ on Fam whose action on a k-ary lifted functor (F, P) is the k-ary lifted functor $\hat{H}_\rho(F, P)$ given by

$$
\hat{H}_\rho\,(F, P)\,(A_1, Q_1)....(A_k, Q_k) \;=\; [\![E]\!]^{\mathsf{Fam}}\rho\overline{[\alpha := (A_i, Q_i)]}[\varphi := (F, P)]
$$

The induction rule $ind_{T,\,\rho}$ for proving predicates over the σ-instance of the type constructor T relative to the lifting ρ is thus given by

$$
\begin{aligned}
ind_{T,\,\rho} :\; & \forall (P : \forall\overline{(A_i, Q_i)}.(\mu H_\sigma)\,\overline{A_i} \to \mathsf{Set}). \\
& \quad (\forall\overline{(A_i, Q_i)}.\, \pi_2(\hat{H}_\rho(\mu H_\sigma, P))\overline{(A_i, Q_i)} \to P\overline{(A_i, Q_i)}) \to \\
& \quad (\forall\overline{(A_i, Q_i)}.\, \pi_2(\mu\hat{H}_\rho)\overline{(A_i, Q_i)} \to P\overline{(A_i, Q_i)}) \\
= & \;\forall (P : \forall\overline{(A_i.Q_i)}.(\mu H_\sigma)\,\overline{A_i} \to \mathsf{Set}). \\
& \quad (\forall\overline{(A_i, Q_i)}.\, \forall (x : H_\sigma(\mu H_\sigma)\overline{A_i}). \\
& \qquad \pi_2(\hat{H}_\rho(\mu H_\sigma, P))\overline{(A_i, Q_i)}\, x \to P\overline{(A_i, Q_i)}(in\, x)) \to \\
& \quad (\forall\overline{(A_i, Q_i)}.\, \forall (x : (\mu H_\sigma)\overline{A_i}).\, \pi_2(\mu\hat{H}_\rho)\overline{(A_i, Q_i)}\, x \to P\overline{(A_i, Q_i)}x) \\
ind_{T,\,\rho} = & \;\lambda\; P\, k\, \overline{(A_i, Q_i)}.\, \pi_2(fold^{\mathsf{Fam}}_{T,\,\rho}\,(in, k))
\end{aligned}
$$

To get analogues for nested types of the structural induction rules for ADTs note that, since each σ-instance of the type constructor $T = \mu\varphi^k.\lambda\alpha_1...\alpha_k.\, E$ associated with a nested type $(\mu\varphi^k.\lambda\alpha_1...\alpha_k.E)E_1...E_k \in \mathcal{N}^V$ gives rise to an inductive family of types, the appropriate notion of predicate for a nested type is actually a Set-indexed predicate. By direct analogy with structural induction for ADTs, the structural induction rule for a nested type with type constructor T whose σ-instance is interpreted by μH_σ is then

$$\begin{aligned}
&\forall (P : \forall \overline{A_i}. (\mu H_\sigma)\, \overline{A_i} \to \mathsf{Set}).\\
&\quad (\forall \overline{A_i}. \forall (x : H_\sigma(\mu H_\sigma)\overline{A_i}).\, \pi_2(\hat{H}_{\overline{\sigma}}(\mu H_\sigma, \hat{P}))\overline{(A_i, K_1)}\, x \to \hat{P}\overline{(A_i, K_1)}(in\, x)) \to\\
&\quad (\forall \overline{A_i}. \forall (x : (\mu H_\sigma)\overline{A_i}).\, \pi_2(\mu \hat{H}_{\overline{\sigma}})\overline{(A_i, K_1)}\, x \to \hat{P}\overline{(A_i, K_1)} x)\\
= &\forall (P : \forall \overline{A_i}. (\mu H_\sigma)\, \overline{A_i} \to \mathsf{Set}).\\
&\quad (\forall \overline{A_i}. \forall (x : H_\sigma(\mu H_\sigma)\overline{A_i}).\, \pi_2(\hat{H}_{\overline{\sigma}}(\mu H_\sigma, \hat{P}))\overline{(A_i, K_1)}\, x \to \hat{P}\overline{(A_i, K_1)}(in\, x)) \to\\
&\quad (\forall \overline{A_i}. \forall (x : (\mu H_\sigma)\overline{A_i}).\, \hat{P}\overline{A_i} x)
\end{aligned} \tag{$\ddagger$}$$

where $\hat{P}$ is defined below. To see that the structural induction rule ($\ddagger$) is indeed a specialization of $ind_{T,\rho}$, suppose we are given a predicate $P : \forall \overline{(A_i, Q_i)}.\, (\mu H_\sigma)\overline{A_i} \to \mathsf{Set}$ for a nested type with type constructor T whose σ-instance is interpreted by μH_σ, together with induction hypotheses

$$R \;=\; \forall \overline{A_i}. \forall (x : H_\sigma(\mu H_\sigma)\overline{A_i}).\, \pi_2(\hat{H}_{\overline{\sigma}}(\mu H_\sigma, \hat{P}))\overline{(A_i, K_1)}\, x \to \hat{P}\overline{(A_i, K_1)}(in\, x)$$

Let $\hat{P} : \forall \overline{(A_i, Q_i)}.\, (\mu H_\sigma)\overline{A_i} \to \mathsf{Set}$ be the Fam-indexed predicate $\hat{P} = \lambda \overline{(A_i, Q_i)}.\, P\overline{A_i}$, and consider the instantiation $ind_{T,\overline{\sigma}}\, \hat{P}\, \hat{R}$, where the induction hypothesis $\hat{R} \;:\; \forall \overline{(A_i, Q_i)}. \forall (x : H_\sigma(\mu H_\sigma)\overline{A_i}).\, \pi_2(\hat{H}_{\overline{\sigma}}(\mu H_\sigma, \hat{P}))\overline{(A_i, Q_i)} x \to \hat{P}\overline{(A_i, Q_i)}(in\, x)$ for $ind_{T,\overline{\sigma}}$ is given by $\hat{R}\,\overline{(A_i, Q_i)}\, x\, y \;=\; R\,\overline{A_i}\, x\, (\pi_2(\hat{H}_{\overline{\sigma}}(\mu H_\sigma, \hat{P})\, t)\, x\, y)$.

5 The General Methodology

We can distill from the foundations given in Section 4 a general methodology that will derive correct deep induction rules for any nested type generated by $\mathcal{N}$. Concretely, this methodology comprises the following steps:

1. Given a nested data type definition D, translate its type constructor into an expression N in the grammar $\mathcal{N}$ (or, more simply, $\mathcal{A}$, if D defines an ADT).
2. Interpret N in Set to get a fixpoint equation defining D as μH for some (higher-order) operator H.
3. Reinterpret N in Fam to define a corresponding (higher-order) operator $\hat{H}$ on predicates whose fixed point $\mu\hat{H}$ is an inductive predicate on μH, i.e., on D.
4. Initiality of $\mu\hat{H}$ guarantees that there is a unique predicate morphism from $\mu\hat{H}$ to any other predicate P admitting an $\hat{H}$-algebra structure. This gives D's deep induction rule.

These are precisely the steps carried out in all of our examples, including those below, which illustrate the derivation for nested types given in Section 4.5.

Example 4. Since the nested type $Lam\,\alpha := \big(\mu\varphi^1.\lambda\beta.\beta + \varphi\beta \times \varphi\beta + \varphi(\beta{+}1)\big)\,\alpha$ of lambda terms is uniform in its index α, it induces a type constructor $Lam := \mu\varphi^1.\lambda\beta.\beta + \varphi\beta \times \varphi\beta + \varphi(\beta{+}1)$. Writing H for $H_{[]}$ and $\hat{H}$ for $\hat{H}_{[]}$, and letting

$$H\,F\,A \;=\; [\![\beta + \varphi\beta \times \varphi\beta + \varphi(\beta + 1)]\!]^{\mathsf{Set}}[\beta := A][\varphi := F] \;=\; A{+}FA{\times}FA{+}F(A{+}1)$$

we have that $\mu H = Lam$ and that the predicate counterpart $\hat{H}$ to H is given by

$$\begin{aligned}\hat{H}\,(F,\hat{P})\,(A,Q) &= [\![\beta + \varphi\beta \times \varphi\beta + \varphi(\beta+1)]\!]^{\mathsf{Fam}}[\beta := (A,Q)][\varphi := (F,\hat{P})] \\ &= (A,Q) \underline{+} (F,\hat{P})(A,Q) \underline{\times} (F,\hat{P})(A,Q) \underline{+} (F,\hat{P})((A,Q) \underline{+} (1,K_1)) \\ &= (A + FA \times FA + F(A+1), \\ &\qquad \pi_2((A,Q) \underline{+} (F,\hat{P})(A,Q) \underline{\times} (F,\hat{P})(A,Q) \underline{+} (F,\hat{P})((A,Q) \underline{+} (1,K_1)))\end{aligned}$$

Reflecting $\mu\hat{H}$ back into syntax gives the inductive predicate

```
Lam^ : ∀(a : Set) → (a → Set) → (Lam a → Set) where
  Var^ : ∀(a : Set) (Q : a → Set) (x : a) → Q x → Lam^ a Q (Var x)
  App^ : ∀(a : Set) (Q : a → Set) (x : Lam a) (y : Lam a) → Lam^ a Q x →
                      Lam^ a Q y → Lam^ a Q (App x y)
  Abs^ : ∀(a : Set) (Q : a → Set) (x : Lam a) → Lam^ (Maybe a) (Maybe^ a Q) x →
                      Lam^ a Q (Abs x)
```

Now, if P is any other predicate on Lam admitting an $\hat{H}$-algebra structure, then there must exist a morphism $k : \forall(x : A + Lam\,A \times Lam\,A + Lam(A+1)).\,(Q + PAQ \times PAQ + P(A+1)((+1)^\wedge Q))x \to PAQ\,(in\,x)$, i.e., $k = (k_1, k_2, k_3)$, where

$$\begin{aligned}&k_1 : \forall(x : A).\, Q\,x \to P\,A\,Q\,(Var\,x) \\ &k_2 : \forall(x : Lam\,A).\,\forall(y : Lam\,A).\, P\,A\,Q\,x \to P\,A\,Q\,y \to P\,A\,Q\,(App\,x\,y) \\ &k_3 : \forall(x : Lam\,(A+1)).\, P\,(A+1)\,((+1)^\wedge\,Q)\,x \to P\,A\,Q\,(Abs\,x)\end{aligned}$$

Since `Lam^` reflects the initial $\hat{H}$-algebra, there is a unique algebra morphism from $in : \hat{H}(\mu\hat{H}) \to \mu\hat{H}$ to the $\hat{H}$-algebra k on P, i.e., from $\mu\hat{H}$ to P. Reflecting this morphism back into syntax gives the deep induction rule for lambda terms from Section 3.

The deep induction rule for lambda terms can be used to prove, e.g., properties of lambda terms whose variables are represented by prime numbers or lambda terms over strings that can represent variable names. It can also be used to prove properties of lambda terms over lambda terms, such as the associativity laws needed to show that the functor Lam is a monad; such a proof is included as the first case study in the accompanying Agda code. The second uses deep induction rule we derive in Example 5 to prove some results about bushes.

Since truly nested types are a special case of deep nested types, our methodology can derive useful induction rules for them — including the perpetually problematic truly nested type of bushes [8,10,15] introduced in Section 3.

Example 5. Since the truly nested type $Bush\,\alpha := \big(\mu\varphi^1.\lambda\beta.\,1 + \beta \times \varphi\,(\varphi\,\beta)\big)\,\alpha \in \mathcal{N}^\alpha$ is uniform in its index α, it induces a type constructor $Bush := \mu\varphi^1.\lambda\beta.\,1 + \beta \times \varphi\,(\varphi\,\beta)$. Writing H for $H_{[]}$ and $\hat{H}$ for $\hat{H}_{[]}$, and letting

$$H\,F\,A \;=\; [\![1 + \beta \times \varphi\,(\varphi\,\beta)]\!]^{\mathsf{Set}}\sigma[\beta := A][\varphi := F] \;=\; 1 + A \times F(FA)$$

we have that $\mu H = Bush$ and the predicate counterpart $\hat{H}$ to H is given by

$$\begin{aligned}\hat{H}\,(F,P)\,(A,Q) &= [\![1 + \beta \times \varphi\,(\varphi\,\beta)]\!]^{\mathsf{Fam}}\overline{\sigma}[\beta := (A,Q)][\varphi := (F,P)] \\ &= (1,K_1) \underline{+} (A,Q) \underline{\times} (F,P)((F,P)(A,Q)) \\ &= (1 + A \times F(FA),\; K_1 + Q \times \pi_2((F,P)((F,P)(A,Q))))\end{aligned}$$

Reflecting $\mu\hat{H}$ back into syntax gives the inductive predicate

```
Bush^ : ∀(a : Set) → (a → Set) → (Bush a → Set) where
  BNil^ : ∀(a : Set) (Q : a → Set) → Bush^ a Q BNil
  BCons^ : ∀(a : Set) (Q : a → Set) (x : a) (y : Bush (Bush a)) →
                  Q x → Bush^ (Bush a) (Bush^ Q) x → Bush^ a Q (BCons x y)
```

Now, if P is any other predicate on *Bush* admitting an $\hat{H}$-algebra structure, then there must exist a morphism

$$\begin{aligned} k: \; & \forall(x : 1 + Bush\,(Bush\,A)).\\ & \quad (K_1 + Q \times \pi_2((Bush, \hat{P})((Bush, \hat{P})(A, Q))))x \to P\,A\,Q\,(in\,x)\\ = \; & \forall(x : 1 + Bush\,(Bush\,A)).\,(K_1 + Q \times P\,(Bush\,A)\,(P\,A\,Q))x \to P\,A\,Q\,(in\,x) \end{aligned}$$

i.e., (k_1, k_2), where $k_1 : \forall(x : 1).\,1 \to P\,A\,Q\,BNil$ and $k_2 : \forall(x : A).\ \forall(y : Bush\,(Bush\,A)).\,1 \to P\,(Bush\,A)\,(P\,A\,Q)\,y \to P\,A\,Q(BCons\,x\,y)$. Since `Bush^` reflects the initial $\hat{H}$-algebra, there is a unique predicate morphism from $\mu\hat{H}$ to P. Reflecting this morphism back into syntax gives the deep induction rule for bushes from Section 3.

The function `BDind⇒MBDind` in our Agda code shows that our methodology also derives a mutually recursive deep induction rule for bushes, there called `MBDind`.

Examples 4 and 5 show that when the definition of a nested type contains an instance of *another* nested type constructor `C` — e.g., `Maybe a` in the argument `Lam (Maybe a)` to `Abs` — its inductive predicate definition, and thus its deep induction rule, will involve a call to the predicate interpretation `C^` of `C`. When the definition contains an instance of the constructor for *the same* type being defined — e.g., `Bush a` in the type argument `Bush (Bush a)` to `BCons` — its inductive predicate definition, and thus its deep induction rule, will involve a recursive call to the inductive predicate being defined. The treatment of a truly nested type is thus exactly the same as the treatment of any other nested type.

Independently of deriving induction rules, even defining some nested types in Agda requires turning off its termination checks in a few tightly compartmentalized places. For example, neither Coq nor Agda currently allows the definition of the bush data type because of the non-positive occurrence of `Bush` in the type of `BCons`. The correctness of our development in those places is justified by [13]. This work suggests that the current notion of positivity should be generalized.

6 Related Work and Directions for Further Investigation

As far as we know, the phenomenon of deep induction has not previously even been identified, let alone studied. This paper treats deep induction for *nested types*, which extend ADTs by allowing higher-order recursion. Other generalizations of ADTs are also well-studied in the literature, including *(indexed) containers* [1,2], which extend ADTs by allowing type dependency. In particular, [3] defines a class of "nested" containers corresponding to inductive types whose constructors can recursively depend on the data type at different instances than the one being defined. The case of *truly* nested types is not treated there, however. We hope eventually to extend the results of this paper to derive provably correct deep induction rules for (indexed) containers, GADTs, dependent types, and other classes of more advanced data types. One interesting question is whether or not a common generalization of indexed containers and the class of nested types studied here has a rigorous initial algebra semantics as in [13].

A more recent line of investigation concerns *sized types* [5]. These are particularly well-suited to termination checking of (co)recursive definitions, and are implemented in the latest versions of Agda [6]. Although originally defined in the context of a type theory with higher-order functions [4], the current incarnation of sized types does not appear to admit definitions with true nesting. What seems to be missing is an addition operation on sizes, which would allow a constructor such as `BCons` to combine a structure with size of depth "up to α" with one of depth "up to β" to define a data element of depth "up to $\alpha + \beta$".

Tassi [17] has independently implemented a tool for deriving induction principles of data type definitions in Coq using unary parametricity. Although neither rigorous derivation nor justification is provided, his technique seems to be essentially equivalent to ours, and could perhaps be justified by our general framework. True nesting still is not permitted, however. In [7], mutually recursively defined induction and coinduction rules are derived for mutually recursive and corecursive data types. But these are still the standard structural (co)induction rules, rather than deep ones. This suggests a need for deep coinduction rules, too.

References

1. Abbott, M. G., Altenkirch, T, Ghani, N.: Containers: Constructing strictly positive types. Theoretical Computer Science 342(2), pp. 3-27 (2005)
2. Altenkirch, T, Ghani, N., Hancock, P., McBride, C., Morris, P.: Indexed containers. Journal of Functional Programming 25 (2015)
3. Abbott, M. G., Altenkirch, T., Ghani, N.: Representing nested inductive types using W-types. In: Automata, Languages and Programming, pp. 59-71 (2004)
4. Abel, A.: Type-based termination: A polymorphic lambda-calculus with sized higher-order types. Ph.D. Dissertation, Ludwig Maximilians University Munich. https://dblp.org/rec/bib/phd/de/Abel2007. 2007.
5. Abel, A.: Semi-continuous sized types and termination. Logical Methods in Computer Science 4(2) (2008)
6. Abel, A.: MiniAgda: Integrating sized and dependent types. In: Partiality and Recursion in Interactive Theorem Provers, pp. 18-32 (2010)
7. Blanchette, J. C., Hölzl, J., Lochbihler, A., Panny, L., Popescu, A., Traytel, D.: Truly Modular (Co)datatypes for Isabelle/HOL. In: Interactive Theorem Proving, pp. 99-110 (2014)
8. Bird, R., Meertens, L.: Nested datatypes. In: Mathematics of Program Construction, pp. 52-67 (1998)
9. Bird, R., Paterson, R.: De Bruijn notation as a nested datatype. Journal of Functional Programming 9(1), pp. 77-91 (1999)
10. Fu, P., Selinger, P.: Dependently typed folds for nested data types. ArXiv, https://arxiv.org/abs/1806.05230. 2018.
11. Ghani, N., Johann, P., Fumex, C.: Fibrational induction rules for initial algebras. In: Computer Science Logic, pp. 336-350 (2010)

12. Ghani, N., Johann, P., Fumex, C.: Generic fibrational induction. Logical Methods in Computer Science 8(2) (2012)
13. Johann, P., Polonsky, A.: Higher-kinded data types: Syntax and semantics. In: Logic in Computer Science, pp. 1-13 (2019)
14. Johann, P. and Polonsky, A.: Accompanying Agda code for this paper. Available at https://cs.appstate.edu/~johannp/FoSSaCS19Code.html, 2019.
15. Matthes, R.: An induction principle for nested datatypes in intensional type theory. Journal of Functional Programming 3 & 4, pp. 439-468 (2009)
16. Okasaki, C.: Purely Functional Data Structures. Cambridge University Press (1999)
17. Tassi, E.: Deriving proved equality tests in Coq-elpi: Stronger induction principles for containers in Coq. In: Interactive Theorem Proving, pp. 1-18 (2019)

Exponential Automatic Amortized Resource Analysis*

David M. Kahn [✉] and Jan Hoffmann

Carnegie Mellon University, Pittsburgh PA, USA
davidkah@cs.cmu.edu cs.cmu.edu/~davidkah
jhoffmann@cmu.edu cs.cmu.edu/~janh

Abstract. Automatic amortized resource analysis (AARA) is a type-based technique for inferring concrete (non-asymptotic) bounds on a program's resource usage. Existing work on AARA has focused on bounds that are polynomial in the sizes of the inputs. This paper presents and extension of AARA to exponential bounds that preserves the benefits of the technique, such as compositionality and efficient type inference based on linear constraint solving. A key idea is the use of the Stirling numbers of the second kind as the basis of potential functions, which play the same role as the binomial coefficients in polynomial AARA. To formalize the similarities with the existing analyses, the paper presents a general methodology for AARA that is instantiated to the polynomial version, the exponential version, and a combined system with potential functions that are formed by products of Stirling numbers and binomial coefficients. The soundness of exponential AARA is proved with respect to an operational cost semantics and the analysis of representative example programs demonstrates the effectiveness of the new analysis.

Keywords: Functional programming · Resource consumption · Quantitative analysis · Amortized analysis · Stirling numbers · Exponential

1 Introduction

"Time is money" is a phrase that also applies to executing software, most directly in domains such as on-demand cloud computing and smart contracts where execution comes with a explicit price tag. In such domains, there is an increasing interest in formally analyzing and certifying the precise resource usage of programs. However, the cost of formally verifying properties by hand is an obstacle to even getting projects off the ground. For this reason, it would be desirable if such resource analyses could be performed mostly automatically, with reduced burden on the programmer.

* This article is based on research supported by DARPA under AA Contract FA8750-18-C-0092 and by the National Science Foundation under SaTC Award 1801369, SHF Award 1812876, and CAREER Award 1845514. Any opinions, findings, and conclusions contained in this document are those of the authors and do not necessarily reflect the views of the sponsoring organizations.

J. Goubault-Larrecq and B. König (Eds.): FOSSACS 2020, LNCS 12077, pp. 359–380, 2020.
https://doi.org/10.1007/978-3-030-45231-5_19

Techniques and tools for automatic and semi-automatic resource analysis have been extensively studied. The applied methods range from deriving and analyzing recurrence relations [55, 1, 16, 2, 12, 36, 10, 37], to abstract interpretation and static analysis [18, 7, 49, 39], to type systems [11, 56, 53], to proof assistants and program logics [4, 9, 8, 48, 19, 45, 42], to term rewriting [6, 5, 47]. Many techniques focus on upper bounds on the worst-case bounds, but average-case bounds [15, 35, 43, 54] and lower-bounds have also been studied [3, 17, 44].

In this paper, we extend automatic amortized resource analysis (AARA) to cover *exponential* worst-case bounds. AARA is an effective type-based technique for deriving concrete (non-asymptotic) worst-case bounds, in particular for functional languages. It has been introduced by Hofmann and Jost [31] to derive *linear bounds* on the heap-space usage of strict first-order functional programs with lists. Subsequently, AARA has been extended to programs with recursive types and general resource metrics [34], higher-order functions [33], lazy evaluation [52], parallel evaluation [29], univariate polynomial bounds [27], multivariate polynomial bounds [23, 25], session-typed concurrency [13], and side effects [38, 46]. However, none of the aforementioned works explores exponential bounds.

The idea of AARA is to enrich types with numeric annotations that represent coefficients in a potential function in the sense of amortized analysis [51]. Bound inference is reduced to Hindley-Milner type inference extended with linear constraints for the numeric annotations. Advantages of the technique include compositionality, efficient bound inference via off-the-shelf LP solving, and the ability to derive bounds on the high-water mark for non-monotone resources like memory. A powerful innovation leveraged in polynomial AARA is the representation of potential functions as non-negative linear combinations of binomial coefficients. Their combinatorial identities yield simple and local typing rules and support a natural semantic understanding of types and bounds. Moreover, these potential functions are more expressive than non-negative linear-combinations of the standard polynomial basis.

However, polynomial potential is not always enough. Functional languages make it particularly easy to use exponentially many resources just by having two or more recursive calls. The following function `subsetSum : int list → int → bool` exemplifies this by naively solving the well-known NP-complete problem subset sum. In the worst case, it performs $3 * 2^{|nums|} - 2$ Boolean and arithmetic operations (where $|x|$ gives the length of the list x).

```
let subsetSum nums target =
    match nums with
    | [] → target = 0
    | hd::tl → subsetSum tl (target-hd) || subsetSum tl target
```

Such a function could appear in a program with polynomial resource usage if applied to arguments of logarithmic size. In this case, polynomial AARA would not be able to derive a bound. Section 6 contains a relevant example.

To handle such functions, we introduce an extension to AARA that allows working with potential functions of the form $f(n) = b^n$. This extension exploits the combinatorial properties of *Stirling numbers of the second kind* [50] in much the same way that AARA currently exploits those of binomial coefficients. Moreover, we allow both multiplicative and additive mixtures of exponential and polynomial potential functions. The techniques used in this process could easily be applied to other potential functions in the future.

The paper first details a generalized AARA type system fit for reuse between polynomial, exponential, and other potential functions. We then instantiate this system with Stirling numbers of the second kind, yielding the first AARA that can infer exponential resource bounds. Finally, we pick out the characteristics that allow for mixing different families of potential functions and maximizing the space they express, and we instantiate the general system with products of exponential and polynomial potential functions. To focus on the main contribution, we develop the system for a simple first-order language with lists in which resource usage is defined with explicit *tick* expressions. However, we are confident that the results smoothly generalize to more general resource metrics, recursive types, and higher-order functions. As in previous work, we prove the soundness of the analysis with respect to a big-step cost semantics that models the high-water mark of the resource usage.

2 Language and Cost Semantics

Abstract Syntax To begin, we define an abstract binding tree (ABT, see [20]) underlying a simple strict first-order functional language. Expressions are in let-normal form to simplify the AARA typing rules. For code examples, however, we overlay the ABT with corresponding ML-based syntax. For example, $1{::}[]$, $[1]$, and $cons(1, nil)$ all represent the same list.

A program *prog* is a collection of functions as defined in the following grammar. The symbols *lit*, *binop*, and *unop* refer to standard literal values, binary operations, and unary operations respectively, of *basic* types (*int*, *bool*, etc.). The symbols f, x, and r refer to function identifiers, variables, and rational numbers, respectively.

$$\begin{aligned} prog &::= func\{f\}(x.e)\ prog \mid \epsilon \\ e &::= lit \mid x \mid binop(x_1; x_2) \mid unop(x) \mid app\{f\}(x) \mid let(e_1; x.e_2) \\ &\mid share(x_1; x_2, x_3.e) \mid tick\{r\} \mid pair(x_1; x_2) \mid nil \mid cons(x_1; x_2) \\ &\mid cond(x; e_1; e_2) \mid pairMatch(x_1; x_2, x_3.e) \mid listMatch(x_1; e_1; x_2, x_3.e_2) \end{aligned}$$

Expressions include function applications, conditionals, and the usual introduction and elimination forms for pairs and lists. They also include two special expressions: $tick\{r\}$ and *share*. The former, $tick\{r\}$, is used to specify constant resource cost r. We allow r to be negative in the case of resources becoming available instead of being consumed. The latter, $share(x_1; x_2, x_3.e)$, provides two copies of its argument x_1 for use in e. This is useful because the affine features of the AARA type system do not allow naive variable reuse. In practice, *share* can be left implicit by automatically preceding every variable usage by *share*.

To focus on the technical novelties, we keep function identifiers and variables disjoint, that is, the types of variables do not contain arrow types and functions are first-order. Higher-order functions can be handled as in previous AARA literature [25]. As a further simplification, we only let functions take one argument;

multiple arguments can be simulated with nested pairs. Finally, the language here only supports the inductive types of lists; future work could extend this to more general types as in other AARA literature [38, 25, 30, 28].

Operational Cost Semantics To define resource usage, AARA literature uses the operational big-step judgment $V \vdash e \Downarrow v \mid (q, q')$ (see e.g. [22]) defined in Figure 1. This judgment means that, under the environment V, the expression e evaluates to the value v under some resource constraints given by the pair q, q'. The environment V maps variables to values. The resource constraints are that q is the high-water mark of resource usage, and $q - q'$ is the net amount of resources consumed during evaluation. In other words, if one started with exactly as many resources needed to evaluate e, that amount would be q, and the amount of leftover resources after evaluation would be q'. It is essential to track both of these values to model resources that might be returned after use, like space. Space usage usually has a positive high-water mark but no net resource consumption, as space could be reused.

The above big-step judgment only formalizes terminating evaluations. To deal with divergence, the additional judgment $V \vdash e \Downarrow \circ \mid q$ has been introduced [26]. This merely drops the parts of the previous judgment relevant to post-termination, focusing on partial evaluation. It means that some partial evaluation of e uses a high-water mark of q resources. Should it exist, the largest q such that $V \vdash e \Downarrow \circ \mid q$ holds would be the high-water mark of resource usage across any partial evaluation of e. For a formal definition, see [26].

3 Automatic Amortized Resource Analysis

Here we lay out a generalized version of the AARA system with the potential functions abstracted. Existing AARA literature is specialized to polynomial functions (see e.g. [27]). This existing polynomial system may be obtained as an instantiation, as may the exponential system that we introduce in Section 4.

AARA uses the *potential* (or physicist's) method to account for resource use, as is commonly used in amortized analyses. The potential method uses the physical analogy of converting between potential and actual energy that can be used to perform work. Whereas a physicist might find potential in the chemical bonds of a fuel, however, AARA places it in the constructors of lists.

To prime intuition with an example, consider paying a resource for each :: operation performed in the following code. It performs *snoc*, which is like *cons* but adds onto the back of the list rather than the front.

```
let snoc x xs =
    match xs with
    | [] → tick 1; x::[]                      (* pay 1 resource here *)
    | hd::tl → tick 1; hd::(snoc x tl)        (* pay 1 resource here *)
```

The resource consumption of *snoc* x xs as defined by the *tick* expressions is $1 + |xs|$. Using the potential method, we can justify this bound as follows. If 1 resource is initially available, then the base case of the empty list can be paid for. If there is 1 stored per element of the list then 1 resource is released in the cons case of the pattern match. This suffices to pay for the additional ::

Fig. 1. Terminating operational cost semantics rules.

$$\frac{q = max(r,0) \quad q' = max(-r,0)}{V \vdash tick\{r\} \Downarrow () \mid (q,q')}\ \textit{Tick} \qquad \frac{binop(V(x_1),V(x_2)) \mapsto v}{V \vdash binop(x_1,x_2) \Downarrow v \mid (0,0)}\ \textit{Binop}$$

$$\frac{}{V \vdash lit \Downarrow lit \mid (0,0)}\ \textit{Lit} \qquad \frac{V(x) = v}{V \vdash x \Downarrow v \mid (0,0)}\ \textit{Var} \qquad \frac{V(x_1) = v_1 \quad V(x_2) = v_2}{V \vdash pair(x_1,x_2) \Downarrow (v_1,v_2) \mid (0,0)}\ \textit{Pair}$$

$$\frac{unop(V(x)) \mapsto v}{V \vdash unop(x) \Downarrow v \mid (0,0)}\ \textit{Unop} \qquad \frac{V(x_p) = (v_1,v_2) \quad V[x_1 {\mapsto} v_1, x_2 {\mapsto} v_2] \vdash e \Downarrow v \mid (q,q')}{V \vdash pairMatch(x_p;x_1,x_2.e) \Downarrow v \mid (q,q')}\ \textit{PMat}$$

$$\frac{V \vdash e_1 \Downarrow v_1 \mid (q,q') \quad V[x \mapsto v_1] \vdash e_2 \Downarrow v_2 \mid (p,p')}{V \vdash let(e_1;x.e_2) \Downarrow v_2 \mid (q + max(p-q',0), p' + max(q'-p,0))}\ \textit{Let}$$

$$\frac{V(x_b) = true \quad V \vdash e_t \Downarrow v \mid (q,q')}{V \vdash cond(x_b;e_t;e_f) \Downarrow v \mid (q,q')}\ \textit{CondT} \qquad \frac{V(x_b) = false \quad V \vdash e_f \Downarrow v \mid (q,q')}{V \vdash cond(x_b;e_t;e_f) \Downarrow v \mid (q,q')}\ \textit{CondF}$$

$$\frac{func\{f\}(x'.e) \in prog \quad V(x) = v_x \quad V[x' \mapsto v_x] \vdash e \Downarrow v \mid (q,q')}{V \vdash app\{f\}(x) \Downarrow v \mid (q,q')}\ \textit{App}$$

$$\frac{V(x) = nil \quad V \vdash e_1 \Downarrow v \mid (q,q')}{V \vdash listMatch(x;e_1;x_h,x_t.e_2) \Downarrow v \mid (q,q')}\ \textit{LMat0} \qquad \frac{V(x_h) = v_h \quad V(x_t) = v_t}{V \vdash cons(x_h;x_t) \Downarrow v_h :: v_t \mid (0,0)}\ \textit{Cons}$$

$$\frac{V(x) = v_h :: v_t \quad V[x_h \mapsto v_h, x_t \mapsto v_t] \vdash e_2 \Downarrow v \mid (q,q')}{V \vdash listMatch(x;e_1;x_h,x_t.e_2) \Downarrow v \mid (q,q')}\ \textit{LMat1} \qquad \frac{}{V \vdash nil \Downarrow nil \mid (0,0)}\ \textit{Nil}$$

$$\frac{V[x_2 \mapsto V(x_1), x_3 \mapsto V(x_1)] \vdash e \Downarrow v \mid (q,q')}{V \vdash share(x_1;x_2,x_3.e) \Downarrow v \mid (q,q')}\ \textit{Share}$$

operation. The remaining potential on xs can be assigned to tl for the recursive call. One can sum these costs to infer that the initial potential $1 + |xs|$ covers the cost of all the :: operations. The AARA type system could describe this with the typing $L^1(\mathbb{Z})$ for xs (describing the linear potential in the superscript) and $\mathbb{Z} \times L^1(\mathbb{Z}) \overset{1/0}{\to} L^0(\mathbb{Z})$ for *snoc* (describing the initial/remaining resources above the arrow). Another valid type is $\mathbb{Z} \times L^2(\mathbb{Z}) \overset{1/0}{\to} L^1(\mathbb{Z})$, which could be used in a context where the result of *snoc* must be used to pay for additional cost.

Types The AARA system laid out here supports the types given below. The symbol F gives the types of functions, where q and q' are non-negative rationals. The symbol S gives the remaining non-function types, where *basic* stands for the basic types like *int* or *unit*, and the resource annotation P is an indexed family of rationals representing the coefficients in a linear combination of basic potential functions.

$$F ::= S \overset{q/q'}{\to} S \qquad\qquad S ::= basic \mid L^P(S) \mid S \times S$$

The typing rules for these types are given in Figure 2 and explained in the following sections. The values of these types are the usual values.

Potential To understand typing rules, it is necessary to define potential. The following potential constructs are generalized from polynomial AARA work [27].

As mentioned, $P = (p_i)_{i \in I}$ is in $\mathbb{Q}^I$ as an indexed family of rationals. Each entry represents a coefficient in a linear combination of basic potential functions. This linearity makes it natural to overload the type of P as a vector or matrix of rationals, so it is treated as such whenever the context is appropriate. Finally, let those basic potential functions be fixed as some family $(f_i)_{i \in I}$, where $f_i(0) = 0$.

We define the potential represented with P using the function ϕ where

$$\phi(n, P) = \textstyle\sum_i p_i \cdot f_i(n) \ .$$

The function ϕ yields the total potential on a list (excluding the potential of its elements) as a function of the list's size n and its potential annotation P.

We can then relate resource potential between different sizes of list with the shift operator $\lhd : \mathbb{Q}^I \to \mathbb{Q}^I$ and constant difference operator $\delta : \mathbb{Q}^I \to \mathbb{Q}$. These functions need only satisfy the following property equation.

$$\phi(n+1, P) = \delta(P) + \phi(n, \lhd P) \tag{1}$$

Though we leave open the explicit definition of these functions for generality, we only later work with instances of them that are linear operators, such that Equation 1 denotes a linear recurrence. Such a refinement leaves $\lhd P$ and $\delta(P)$ linear functions of P.

These functions come in handy for understanding the stored potential in a value of a given type, defined by the potential function Φ as follows.

$$\begin{aligned}
\Phi(v : basic) &= 0 \\
\Phi((v_1, v_2) : A_1 \times A_2) &= \Phi(v_1 : A_1) + \Phi(v_2 : A_2) \\
\Phi([] : L^P(A)) &= 0 \\
\Phi(h :: t : L^P(A)) &= \delta(P) + \Phi(h : A) + \Phi(t : L^{\lhd P}(A))
\end{aligned}$$

We often need to measure the potential across an entire evaluation context of typed values $V : \Gamma$ given by a typing context Γ and variable bindings V. We do so by extending the definition of potential Φ as follows.

$$\Phi(\emptyset) = 0 \qquad\qquad \Phi(V : (\Gamma, v : A)) = \Phi(V : \Gamma) + \Phi(v : A)$$

Finally, we can use these definitions to obtain a closed-form expression for the potential over an entire list (including its elements) with the following:

Lemma 1. *Let* $l = [a_n, ..., a_1]$ *be a list of* n *values. Then* $\Phi(l : L^P(A)) = \phi(n, P) + \sum_{i=1}^{n} \Phi(a_i : A)$

Proof. We induct over the structure of the list l.

For the empty list of length 0:

$$\Phi([] : L^P(A)) = 0 = \textstyle\sum_i p_i \cdot f_i(0) = \phi(0, P) + \sum_{i=1}^{0} \Phi(a_i : A)$$

For $l = h :: t$ of size $n + 1$:

$$\begin{aligned}
\Phi(a_{n+1} :: b : L^P(A)) &= \delta(P) + \Phi(a_{n+1} : A) + \Phi(l' : L^{\lhd P}(A)) \\
&= \delta(P) + \Phi(a_{n+1} : A) + \phi(n, \lhd P) + \textstyle\sum_{i=1}^{n} \Phi(a_i : A) \\
&= \phi(n+1, P) + \textstyle\sum_{i=1}^{n+1} \Phi(a_i : A)
\end{aligned}$$

We can apply Lemma 1 to the previously defined function *snoc* to see the change in potential between input and output. This difference in potential should bound the resources consumed. For this case, the basic potential functions (f_i) only need contain $\lambda n.n$, and we can let $\lhd(p) = p = \delta((p))$. Letting y be the result of $snoc\ x\ xs$, the type $\mathbb{Z} \times L^1(\mathbb{Z}) \overset{1/0}{\to} L^0(\mathbb{Z})$ indicates the following bound

$$\Phi(x : \mathbb{Z}, xs : L^1(\mathbb{Z})) + 1 - \Phi(y : L^0(\mathbb{Z})) = \phi(|xs|, 1) + 1 - \phi(|y|, 0) = |xs| + 1$$

This is exactly the amount of resources consumed, so the bound is tight.

In this work we only consider so-called *univariate* potential, wherein every term in the potential sum is dependent on the length of only one input list. However, different univariate potential summands may depend on different inputs, and thus univariate potential may still be multivariate. The term *multivariate potential* refers to using more general multivariate functions for potential. There is existent work on multivariate potential using polynomial functions [24]. We expect that the work here extends to multivariate potential similarly.

Typing Rules The typing rules in Figure 2 use the judgment $\Sigma; \Gamma \vdash^{q}_{q'} e : A$. In this typing judgment, Γ maps variables to types, while Σ maps function labels to sets of types. This judgment holds when, in the typing environment given by Σ and Γ, the expression e is of type A, subject to the constraints that q and q' are the amount of available resources before and after some evaluation of e. Unlike the judgment $V \vdash e \Downarrow v \mid (q, q')$, these values need not be tight.

By expressing available resources on the turnstile, and potential resources in the types given by Σ, Γ, and A, the type system is set up to formalize the reasoning of the potential method. Theorem 1 shows that it is sound with respect to the operational semantics of Section 2.

Many typing rules preserve the total resource potential they are given, consuming none of it themselves. They therefore usually either have no explicit interaction with potential (e.g. *Lit*) or pass around exactly what they are given (e.g. *Let*). All basic rules in the first block of Figure 2 fit this characterization.

The typing rules concerning functions in second block of Figure 2 are the only to make use of Σ. For each function f defined in *prog* via $func\{f\}(x.e)$, $\Sigma(f)$ refers to the set of types that its body e could be given. That we allow for sets of types is important because recursive calls to a function may not always make use of a type with the same resource annotations; this is called *resource-polymorphic recursion*. Despite these rules capturing the intuition behind typing resource-polymorphic recursion, they are not used in existing implementation, as they lead to infinite type derivations. Nonetheless there exists an effective way to type resource-polymorphic recursion with a finite derivation; see [26]. In the examples provided in this article, it usually suffices to consider only *resource-monomorphic recursion*, wherein inner and outer calls use the same annotation.

All of the rules discussed so far are simply those of existing AARA literature with their parameter for operation cost set to 0 (see e.g. [27]). This does not change their generality, as such constant cost can (and could already in prior work) be simulated using *tick*. Similarly, non-constant costs could be simulated by running helper functions using *tick* the appropriate number of times.

Fig. 2. AARA typing rules.

Basic rules:

$$\frac{}{\Sigma;\emptyset \vdash^{0}_{0} \mathit{lit} : \mathit{basic}}\;\mathit{Lit} \qquad \frac{\Sigma;\Gamma_1 \vdash^{q}_{p} e_1 : A \quad \Sigma;\Gamma_2, x : A \vdash^{p}_{q'} e_2 : B}{\Sigma;\Gamma_1,\Gamma_2 \vdash^{q}_{q'} \mathit{let}(e_1; x.e_2) : B}\;\mathit{Let}$$

$$\frac{}{\Sigma; x : \mathit{basic} \vdash^{0}_{0} \mathit{unop}(x) : \mathit{basic}'}\;\mathit{Unop} \qquad \frac{}{\Sigma; x_i : \mathit{basic} \vdash^{0}_{0} \mathit{binop}(x_1, x_2) : \mathit{basic}'}\;\mathit{Binop}$$

$$\frac{}{\Sigma; x : A \vdash^{0}_{0} x : A}\;\mathit{Var} \qquad \frac{}{\Sigma; x_1 : A_1, x_2 : A_2 \vdash^{0}_{0} \mathit{pair}(x_1, x_2) : A_1 \times A_2}\;\mathit{Pair}$$

$$\frac{\Sigma;\Gamma, x_1 : A_1, x_2 : A_2 \vdash^{q}_{q'} e : B}{\Sigma;\Gamma, x : A_1 \times A_2 \vdash^{q}_{q'} \mathit{pairMatch}(x; x_1, x_2.e) : B}\;\mathit{PMat}$$

$$\frac{\Sigma;\Gamma, x : \mathit{bool} \vdash^{q}_{q'} e_1 : A \quad \Sigma;\Gamma, x : \mathit{bool} \vdash^{q}_{q'} e_2 : A}{\Sigma;\Gamma, x : \mathit{bool} \vdash^{q}_{q'} \mathit{cond}(x; e_1; e_2) : A}\;\mathit{Cond}$$

Function rules:

$$\frac{A \xrightarrow{q/q'} B \in \Sigma(f)}{\Sigma; x : A \vdash^{q}_{q'} \mathit{app}\{f\}(x) : B}\;\mathit{App} \qquad \frac{\mathit{func}\{f\}(x.e) \in \mathit{prog} \quad \Sigma; x : A \vdash^{q}_{q'} e : B}{A \xrightarrow{q/q'} B \in \Sigma(f)}\;\mathit{Fun}$$

Potential-focused rules:

$$\frac{}{\Sigma;\Gamma \vdash^{\mathit{max}(r,0)}_{\mathit{max}(-r,0)} \mathit{tick}\{r\} : \mathit{unit}}\;\mathit{Tick} \qquad \frac{\Sigma;\Gamma \vdash^{p}_{p'} e : A \quad q \geq p \quad q - p \geq q' - p'}{\Sigma;\Gamma \vdash^{q}_{q'} e : A}\;\mathit{Relax}$$

$$\frac{\Sigma;\Gamma, x : A \vdash^{q}_{q'} e : B \quad A' <: A}{\Sigma;\Gamma, x : A' \vdash^{q}_{q'} e : B}\;\mathit{SubWeakL} \qquad \frac{\Sigma;\Gamma \vdash^{q}_{q'} e : A' \quad A' <: A}{\Sigma;\Gamma \vdash^{q}_{q'} e : A}\;\mathit{SubWeakR}$$

$$\frac{\Sigma;\Gamma, x_2 : A_2, x_3 : A_3 \vdash^{q}_{q'} e : B \quad A_1 \curlyvee (A_2, A_3)}{\Sigma;\Gamma, x_1 : A_1 \vdash^{q}_{q'} \mathit{share}(x_1; x_2, x_3.e) : B}\;\mathit{Sharing}$$

List rules:

$$\frac{}{\Sigma;\emptyset \vdash^{0}_{0} \mathit{nil} : L^{P}(A)}\;\mathit{Nil} \qquad \frac{}{\Sigma; x_h : A, x_t : L^{\lhd P}(A) \vdash^{\delta(P)}_{0} \mathit{cons}(x_h; x_t) : L^{P}(A)}\;\mathit{Cons}$$

$$\frac{\Sigma;\Gamma \vdash^{q}_{q'} e_1 : B \quad \Sigma;\Gamma, x_h : A, x_t : L^{\lhd P}(A) \vdash^{q + \delta(P)}_{q'} e_2 : B}{\Sigma;\Gamma, x : L^{P}(A) \vdash^{q}_{q'} \mathit{listMatch}(x; e_1; x_h, x_t.e_2) : B}\;\mathit{ListMatch}$$

Fig. 3. AARA subtyping and sharing judgments.

$$\frac{\forall i.p_i \geq q_i}{L^P(A) <: L^Q(A)}\ \textit{Subtype} \qquad \frac{}{basic \curlyvee (basic, basic)}\ \textit{ShareBasic}$$

$$\frac{A_1 \curlyvee (A_2, A_3) \quad B_1 \curlyvee (B_2, B_3)}{A_1 \times B_1 \curlyvee (A_2 \times B_2, A_3 \times B_3)}\ \textit{SharePair} \qquad \frac{A_1 \curlyvee (A_2, A_3) \quad P = Q + R}{L^P(A_1) \curlyvee (L^Q(A_2), L^R(A_3))}\ \textit{ShareList}$$

The remaining rules cover sharing, subtype-weakening, and the rules concerning lists. Weakening, though not listed, is also allowed.

Sharing is a form of contraction. By sharing, the rest of the typing rules can become affine, allowing only single usages of a given variable. Intuitively, sharing is meant to prevent duplicating potential across multiple usages of a variable, and instead split the potential across them. The rules for the sharing judgment, indicating how to split potential, can be found in Figure 3. Note that the rule *ShareList* adds indexed collections of rationals; this should be interpreted pointwise, as if the addends were vectors or matrices.

Subtype-weakening is a form of subtyping based on potential. It discards potential on a list, weakening the upper bound on resources it represents. This rule follows all usual subtyping rules, as well as *Subtype* from Figure 3. Relaxing behaves similarly, but loosens the bounds on the available resources instead.

The intuition for the rules concerning lists in the last block of Figure 2 is that total resources should be conserved between constructions and destructions. Because $\delta(P)$ expresses the difference in potential, it is exactly how many resource units are released after a pattern match on a list of type $L^P(A)$. For the same reason, it is also how many need to be stored when reversing the process and putting an element on a list of type $L^{\lhd P}(A)$. Finally, when a list is empty, it has no room to store potential. Every potential function f_i maps 0 to 0, so an empty list can safely be assigned any scalar of zero potential.

Soundness The soundness of the type system is expressed with the following theorem. It states that the evaluation of an expression e does not require more resources than initially present, and (should evaluation terminate) it leaves at least as many resource as dictated. The proof is a straightforward generalization of the version from [27], but we nonetheless reproduce the proof below.

Theorem 1. *Let* $\Sigma; \Gamma \vdash^{q}_{q'} e : B$ *and* V *provide the variable bindings for* Γ

1. *If* $V \vdash e \Downarrow v \mid (p, p')$ *then* $p \leq \Phi(V : \Gamma) + q$ *and* $p - p' \leq \Phi(V : \Gamma) + q - \Phi(v : B) - q'$
2. *If* $V \vdash e \Downarrow \circ \mid p$ *then* $p \leq \Phi(V : \Gamma) + q$

Proof. Assume V binds Γ's variables and perform nested induction on the type derivation and operational judgment for an expression in let-normal form. We show the induction below only for the terminating operational judgment cases, but the partial-evaluation cases are nearly identical.

(**Base Non-Cons**) Suppose the last rule applied in the typing derivation is any non-*Cons* base case, i.e., *Lit*, *Var*, *Unop*, *Binop*, *Pair*, *Nil*, or *Tick*. Then assume the appropriate terminating operational judgment rule applies. In such a case, one finds $p \leq q$, $p' \geq q'$, and $\Phi(v : B) = \Phi(V : \Gamma)$. This and the non-negativity of potential are sufficient to satisfy the desired inequalities.

(**Base Cons**) Suppose the last rule is *Cons*, so $q = \delta(P)$ and $q' = 0$. Assume the *Cons* operational judgment applies, so that $p = p' = 0$. Note $\Phi(v_h :: v_t : L^P)$ is equal to $\delta(P) + \Phi(v_h : A) + \Phi(v_t : L^P(A))$ by definition. This identity and the non-negativity of potential satisfy the desired inequalities.

(**Step Implicit Inequalities**) Suppose the last rule is one of *SubWeakL*, *SubWeakR*, *Relax*, or substructural weakening, and assume some operational judgment applies. Each typing requires a similar typing judgment as a premiss. Further, none changes any values, so the same operational judgment still applies. Thus, the inductive hypothesis applies, and gives almost the inequalities we need. Each case provides the inequalities needed to finish. For subtype-weakening, it is sufficient note that $C <: D$ entails $\Phi(v : C) \geq \Phi(v : D)$, since C is pointwise greater-then-or-equal to D. For *relax*, the premisses of the *relax* rule directly include the inequalities needed to complete the case. And we can complete the substructural weakening case by noting that the non-negativity of potential entails $\Phi(V : \Gamma, v : A) \geq \Phi(V : \Gamma)$.

(**Step Let**) Suppose the last rule is *Let*, and suppose its operational judgment applies. The premisses of the typing rule require that $\Sigma; \Gamma_1 \vdash^{q}_{r} e_1 : A$ and $\Sigma; \Gamma_2, x : A \vdash^{r}_{q'} e_2 : B$. The premisses of the operational judgment require that $V \vdash e_1 \Downarrow v_1 \mid (s, s')$ and $V[x \mapsto v_1] \vdash e_2 \Downarrow v_2 \mid (t, t')$, where $p = s + max(t - s', 0)$ and $p' = t' + max(s' - t, 0)$. Applying the inductive hypothesis to these premiss pairs and adding the resulting inequalities cancels terms to complete the case.

(**Step Sharing**) Suppose the last is *Sharing*, so that $\Gamma = \Gamma', x_1 : A_1$. It requires as a premiss that $\Sigma; \Gamma', x_2 : A_2, x_3 : A_3 \vdash^{q}_{q'} e : B$, where $A_1 \curlyvee (A_2, A_3)$. Assuming the operational judgment *Share* applies, $V[x_2 \mapsto V(x_1), x_3 \mapsto V(x_1)] \vdash e \Downarrow v \mid (p, p')$ also holds. The inductive hypothesis applies, yielding the needed inequalities, but for x_2, x_3 instead of x_1. However, the sharing relation ensures that $\Phi(v_1 : A_1) = \Phi(v_2 : A_2, v_3 : A_3)$, and this identity finishes the case.

(**Step ListMatch**) Suppose the last is *ListMatch*, so $\Gamma = \Gamma', x : L^P(A)$. There are two operational judgments which could apply: *LMat0* and *LMat1*.

Suppose the former judgment applies. It requires that $V \vdash e_1 \Downarrow v \mid (p, p')$. At the same time, the *ListMatch* rule requires as a premiss that $\Sigma; \Gamma' \vdash^{q}_{q'} e_1 : B$. The inductive hypothesis applies, yielding the needed inequalities, but for Γ' instead of Γ. However, because $\Phi(nil : L^P(A)) = 0$, we see $\Phi(V : \Gamma') = \Phi(V : \Gamma)$, and the desired inequalities result.

Suppose instead the latter judgment applies. This judgment requires as a premiss that $V[x_h \mapsto v_h, x_t \mapsto v_t] \vdash e_2 \Downarrow v \mid (p, p')$. At the same time, the *ListMatch* rule requires that $\Sigma; \Gamma', x_h : A, x_t : L^{\lhd P}(A) \vdash^{q + \delta(P)}_{q'} e_2 : B$. The inductive hypothesis applies, telling us that $p - p' \leq \Phi(V : \Gamma', v_h : A, v_t : L^{\lhd P}(A)) + q + \delta(P) - \Phi(v : B) - q'$ and $p \leq \Phi(V : \Gamma', v_h : A, v_t : L^{\lhd P}(A)) + q + \delta(P)$. By definition, $\Phi(v_h :: v_t : L^P) = \delta(P) + \Phi(v_h : A) + \Phi(v_t : L^P(A))$, and applying this identity to the inequalities yields the inequalities needed.

(**Step Cond**) Suppose the last rule is *Cond*, and that either of the *CondT* or *CondF* operational judgments apply. In either case, applying the inductive hypothesis to its premiss and the premiss of *Cond* gives the needed inequalities.

(**Step PMat**) Suppose that the last rule applied is *PMat*, so that $\Gamma = \Gamma', x : A_1 \times A_2$. This rule would require as a premiss that $\Sigma; \Gamma', x_1 : A_1, x_2 : A_2 \vdash^{q}_{q'} e' : B$, for e' the body of the match statement e. Suppose the *PMat* operational judgment applies. This judgment requires as a premiss that $V[x_1 \mapsto v_1, x_2 \mapsto v_2] \vdash e' \Downarrow v \mid (p, p')$, where the value of x is (v_1, v_2). Applying the inductive hypothesis to these premisses followed by the definitional identity $\Phi((v_1, v_2) : A_1 \times A_2) = \Phi(v_1 : A_1) + \Phi(v_2 : A_2)$ completes the case.

(**Step App**) Suppose the last rule is *App*. Note that this rule requires *Fun* as a premiss, which in turn requires $\Sigma; x : A \vdash^{q}_{q'} e' : B$ where e' is the body of the function being applied. If the *App* operational judgment applies, its premiss would require $V[x' \mapsto V(x)] \vdash e \Downarrow v \mid (p, p')$. Although e' might not be a smaller expression than e, the operational judgment derivation still shrinks. This means the inductive hypothesis applies, and it gives the exact inequalities needed.

Type Inference Type inference for the Hindley-Milner part of the type system is decidable [21, 41]. The only new barrier for automating inference in AARA is obtaining witnesses for all the coefficients in each annotation P in a derivation.

Each typing rule naturally gives a set of linear constraints on the entries of P. If the relation given by $\lhd$ and δ can likewise be expressed with linear constraints, then all such constraints are linear. So long as $|P|$ is finite, this forms a linear program. A linear program solver can then find minimal witnesses efficiently.

Existing AARA literature (see e.g. [27]), however, uses binomial coefficients as the basis functions for P, of which there are infinitely many. This nonetheless works because only a particular finite prefix of their set, $\binom{-}{1}, \ldots, \binom{-}{k}$, are used as a basis in a given analysis. Each such prefix basis also yields the same locally-definable shift operation: the linear equality $\lhd p_i = p_i + p_{i+1}$, where p_k is the coefficient of $\binom{-}{k}$ and is 0 if the function is outside the prefix. As this is a linear relation, and each prefix is finite, inference can be performed via linear program. The prefix bases of binomial coefficients thereby form an infinite family of finite bases, each of which allows automated inference of resource polynomials up to a fixed degree in the AARA system.

As a caveat, not all programs use resources in a manner compatible with the AARA system. Indeed, it is undecidable whether or not a program uses e.g. polynomial amounts of resources, as this could solve the halting problem.

4 Exponential Potential

Stirling numbers of the second kind $\left\{ {n \atop k} \right\} = \frac{1}{k!} \sum_{i=0}^{k} (-1)^i \binom{k}{i} (k-i)^n$ count the number of ways to form a k-partition of a set of n elements. These can be used to express exponential potential functions similarly to how binomial coefficients can express polynomial ones. In particular, we make use of Stirling numbers with arguments n, k offset by 1, $\left\{ {n+1 \atop k+1} \right\}$, so that $\phi(n, P) = \sum_i p_i \cdot \left\{ {n+1 \atop i+1} \right\}$. While other bases could also express exponential potential, these offset Stirling numbers have a few particularly desirable properties, which are described in this section.

Simple Shift Operation Like binomial coefficients, the prefixes of the basis of the offset Stirling numbers of the second kind form an infinite family of finite bases, each of which allows automated inference in the AARA system. However, these potential functions are exponential rather than polynomial.

Stirling numbers of the second kind satisfy the recurrence $\left\{{n+1 \atop k+1}\right\} = (k+1)\left\{{n \atop k+1}\right\} + \left\{{n \atop k}\right\}$. This recurrence allows the $\lhd$ operation to have the same local definition for every annotation entry in every prefix basis: $\lhd p_i = (i+1)p_i + p_{i+1}$, where p_k is the coefficient of $\left\{{n+1 \atop k+1}\right\}$, and is 0 if the function index is outside the chosen prefix. Given this definition for $\lhd$ and letting $\delta(P) = p_0$, we find $p_0 + \sum_i \lhd p_i \left\{{n+1 \atop i+1}\right\} = \sum_i p_i \left\{{n+2 \atop i+1}\right\}$, satisfying Equation 1.

This shift operation yields a linear relation, as the coefficient of a given p_i is a constant scalar. Thus, exactly like when using binomial coefficients, inference is automatable via linear programming. Certain other exponential bases, like Gaussian binomial coefficients, could be similarly automated.

Expressivity Because $\left\{{n+1 \atop k+1}\right\} = \frac{1}{k!}\sum_{i=0}^{k}(-1)^{k-i}\binom{k}{i}(i+1)^n \in \Theta((k+1)^n)$, the offset Stirling numbers of the second kind can form a linear basis for the space of sums of exponential functions. Each function $\lambda n.b^n$ with $b \geq 1$ can be expressed as a linear combination of the functions $\lambda n.\left\{{n+1 \atop k+1}\right\}$.

The function $\lambda n.\left\{{n+1 \atop k+1}\right\}$ is also non-negative for natural n, and non-decreasing with respect to n. These are two natural properties to require of basic potential functions, since amortized analysis requires non-negative resources, and larger inputs should not usually become cheaper to process. Further, the properties are preserved by non-negative linear (i.e. conical) combination, and by $\lhd$ when defined with a non-negative linear recurrence - the combinations given by P and $\lhd P$ always satisfy the two potential function properties.

Ensuring these properties for more general potential functions requires determining if such a function on a natural domain is always non-negative. This is non-trivial. In the existing literature on multivariate polynomials, we find this is *undecidable* in the worst case [40]. However, restricting to non-negative linear (that is, *conical*) combinations of non-negative, non-decreasing functions - as we have done here - gives simple linear constraints that ensure both desired properties. For finite bases, this is easily handled via linear programming.

When considering expressivity in this conical combination model of potential functions, one finds some otherwise-valid potential functions are not be expressible in the conical space given by the offset Stirling number functions. Nonetheless, Stirling number functions are a *maximally expressive* basis; it is not possible to express additional potential functions using a different basis without losing expressibility elsewhere. Notably, the standard exponential basis is *not* maximal in this sense. The formal statement of such maximal expressivity is generalized in the theorem below. Any finite, sequential subset of the offset Stirling number functions satisfy the prerequisites of this theorem, as do the binomial coefficient functions and other well-known functions like the Gaussian polynomials.

Theorem 2. *Let $\{f_i\}$ be a finite set of linearly independent functions on the naturals that are non-negative and non-decreasing. Let $f_i(n)$ be 0 until $n \geq i$,* *and let $i \leq j$ imply that $O(f_i) \subseteq O(f_j)$, with asymptotic equality only when $i = j$. Let L be the linear span (collection of linear combinations) of $\{f_i\}$, and let C be its conical span (collection of conical combinations).*

There does not exist another linearly independent basis $\{g_i\}$ with linear span L and conical span $D \supsetneq C$ such that each function in $\{g_i\}$ is non-negative and non-decreasing. That is, $\{f_i\}$ has a maximally expressive conical span.

Proof. Suppose there is such a basis $\{g_i\}$. We express each basis $\{f_i\}$ and $\{g_i\}$ with linear combinations of the other, and derive a contradiction.

If there is any function in the conical span D of $\{g_i\}$ that is not in C, then this is the case for some basis function g_k. Because $g_k \in L$, it can be written as a linear combination of $\{f_i\}$; let $\sum_i \alpha_i f_i = g_k$. Because $g_k \notin C$, there is at least one coefficient $\alpha_i < 0$; let it be α_m. In case there are multiple candidate elements g_k, pick g_k to be the basis function such that this index m is minimized.

We then see that $g_k(m) = \sum_i \alpha_i f_i(m) = (\sum_{i<m} \alpha_i f_i(m)) + \alpha_m f_m(m)$ because $f_i(m)$ for $i > m$ is 0. This yields two observations: First, $m < k$, as otherwise the fastest-growing term of g_k would be negative, but g_k is never negative. Second, the term $\alpha_m f_m(m)$ is negative, yet $g_k \geq 0$, so it must be that $\sum_{i<m} \alpha_i f_i(m) > 0$. Thus there exists a coefficient $\alpha_p > 0$ where $p < m$.

Now we look at representing $\{f_i\}$ with $\{g_i\}$. Because the conical span D contains C, it can represent each f_i as a conical combination. Notably, a given f_i cannot be represented only with functions outside of $\Omega(f_i)$, nor any function outside of $O(f_i)$, due to growth rates. There is therefore at least one function in $\{g_i\}$ that is $\Theta(f_i)$, for each i. Since the linear span of these corresponding g_i already has the same (finite) dimension as L, any additional functions would not be linearly independent. Due to this, we can say $g_i \in \Theta(f_i)$ uniquely for each i.

Take f_k in particular as a conical combination of $\{g_i\}$. We now consider replacing each element of $\{g_i\}$ in that conical combination with its equivalent linear combination of elements of $\{f_i\}$. Because of the above correspondence of growth rates, there must be a positive coefficient for g_k. Because g_k has positive weight α_p on f_p where $p < m < k$, another basis function g_i in the conical combination must have negative weight on f_p to cancel it out in their linear combination. However, g_k was picked such that it had the lowest index m with negative weight across all $\{g_i\}$; it is contradictory for there to be such a $p < m$.

Natural Semantics The values of $\left\{ {n+1 \atop k+1} \right\}$ count the number of ways to pick k non-empty disjoint subsets of n elements. Many programs with exponential resource use iterate over collections of subsets, so these numbers naturally arise.

Recall the naive solution to subset sum from the introduction. The algorithm iterates through all the subsets of numbers in the input list. When considering Fagin's descriptive complexity result that NP problems are precisely those expressible in existential second order logic [14], it becomes clear that naive solutions to any NP-complete problem fit this characterization: naively brute-forcing through second order terms to find an existential witness is just iterating through tuples of subsets.

Example Consider the naive solution to subset sum from the introduction. One can verify that the number of Boolean and arithmetic operations used on an input of size n is $3 * 2^n - 2$ by induction. We find the same bound here by preceding each such operation with an explicit $tick\{1\}$ operation. Thee AARA type system then verifies that the type of *subsetSum* is $L^3(\mathbb{Z}) \times \mathbb{Z} \xrightarrow{1/0} bool$.

Here is the code again, with type annotations on each line tracking the amount of $\left\{ {n+1 \atop 2} \right\}$ potential on lists, and comments tracking available constant potential. For clarity, the code is re-written in a let-normal form, and sharing locations are marked.

```
let subsetSum nums:L³(ℤ) target =                        (* 1 *)
    match nums:L³(ℤ) with
    | [] →                                               (* 1 *)
        tick 1; target = 0                               (* 0 *)
    | hd::(tl:L⁶(ℤ)) →                                   (* 4 *)
        tick 1; let newTarget = target - hd in           (* 3 *)
        (* share tl:L⁶(ℤ) as L³(ℤ), L³(ℤ) *)
        let withNum = subsetSum tl:L³(ℤ) newTarget in    (* 2 *)
        let without = subsetSum tl:L³(ℤ) target in       (* 1 *)
        tick 1; withNum || without                       (* 0 *)
```

The indicated values yield witnesses for the AARA typing rules, so we know via soundness that the difference between initial and ending potential gives an upper bound on how many operations were used. That difference is $1+3*\left\{ {n+1 \atop 2} \right\} = 3 * 2^n - 2$, where n is the size of *nums*, exactly the amount used.

Exponential terms with higher bases than 2 can come into play with more recursive calls, like in the code below enumerating the 3^n ways to put n labelled balls into 3 labelled bins.

```
let helper xs:L^{2,2}(ℤ) a b c =                              (* 1 *)
        match xs with
        | [] →                                                (* 1 *)
                tick 1; [(a,b,c)]                             (* 0 *)
        | hd::(tl:L^{6,6}(ℤ)) →                               (* 3 *)
                (* share tl:L^{6,6}(ℤ) as L^{2,2}(ℤ), L^{2,2}(ℤ), L^{2,2}(ℤ) *)
                let newA = hd::a in                           (* 3 *)
                let tmp1 = helper tl:L^{2,2}(ℤ) newA b c in   (* 2 *)
                let newB = hd::b in                           (* 2 *)
                let tmp2 = helper tl:L^{2,2}(ℤ) a newB c in   (* 1 *)
                let newC = hd::c in                           (* 1 *)
                let tmp3 = helper tl:L^{2,2}(ℤ) a b newC in   (* 0 *)
                tmp1 @ tmp2 @ tmp3                            (* 0 *)

let ballBins3 xs:L^{2,2}(ℤ) =                                 (* 1 *)
        helper xs:L^{2,2}(ℤ) [] [] []                         (* 0 *)
```

By paying a unit of resource for each such way using *tick*, we can use AARA to bound the count. It assigns a type of $L^{2,2}(\mathbb{Z}) \xrightarrow{1/0} L^{0,0}(L^{0,0}(\mathbb{Z}) \times L^{0,0}(\mathbb{Z}) \times L^{0,0}(\mathbb{Z}))$ to *ballBins*3, where the superscript tracks $\left\{ {n+1 \atop 2} \right\}$ and $\left\{ {n+1 \atop 3} \right\}$ potential, respectively. Since $2\left\{ {n+1 \atop 3} \right\} + 2\left\{ {n+1 \atop 2} \right\} + 1 = 3^n$, this bound is exact.

5 Mixed Potential

It is possible to combine the existing polynomial potential functions with these new exponential potential functions to not only conservatively extend both, but further represent potentials functions with their products. This space represents functions in $\Theta(n^k(b+1)^n)$ for naturals k, b, and does so with terms of the form $\binom{n}{k}\left\{{n+1 \atop b+1}\right\}$ so that $\phi(n, P) = \sum_{b,k} p_{b,k} \cdot \binom{n}{k}\left\{{n+1 \atop b+1}\right\}$. Note that for k or b equal to 0, the potential functions here reduce to the offset Stirling numbers or binomial coefficients, respectively.

The methods used to combine these potential functions here can easily be generalized to combine any two suitable sets.

Simple Shift Operation It is straightforward to find a linear recurrence for these products by distributing over their linear recurrences.

$$\begin{aligned}\binom{n+1}{k+1}\left\{{n+2 \atop b+2}\right\} &= \left(\binom{n}{k+1}+\binom{n}{k}\right)\left((b+2)\left\{{n+1 \atop b+2}\right\}+\left\{{n+1 \atop b+1}\right\}\right)\\ &= (b+2)\binom{n}{k+1}\left\{{n+1 \atop b+2}\right\}+(b+2)\binom{n}{k}\left\{{n+1 \atop b+2}\right\}+\binom{n}{k+1}\left\{{n+1 \atop b+1}\right\}+\binom{n}{k}\left\{{n+1 \atop b+1}\right\}\end{aligned}$$

As before, this yields a definition for δ and $\triangleleft$ with Equation 1. Letting P now be indexed by pairs b, k: $\triangleleft p_{b,k} = (b+1)p_{b,k} + (b+1)p_{b,k+1} + p_{b+1,k} + p_{b+1,k+1}$, and $\delta(P) = p_{0,1} + p_{1,0} + p_{1,1}$. Noting that these definitions are linear again yields automatability for finite (2-dimensional) prefixes of the basis.

Expressivity The product of non-negative, non-decreasing functions is still non-negative and non-decreasing, so products of valid potential functions are still valid. Soundness is preserved by letting p_0 be shorthand for the new constant function coefficient $p_{0,0}$ wherever it is used in Theorem 1. Moreover, maximality of expressivity is preserved, simply by giving index pairs the ordering relation $(i_1, i_2) \leq (j_1, j_2) \iff i_1 \leq j_1 \wedge i_2 \leq j_2$ and applying Theorem 2.

Example Consider bounding the number of Boolean and arithmetic operations in a variation of subset sum: *single-use* subset sum. Here the input may contain duplicate numbers that should be ignored, so as to treat the input as a true set. This is a trivial change to the mathematical problem, but one that real code might have to deal with, depending on the implementation of sets.

The code can be changed to handle this by removing all later duplicates of each number it reaches, so that later recursive calls will never see the number again. It is easy to create a function *remove* of type $\mathbb{Z} \times L^{a+1,b,c}(\mathbb{Z}) \overset{d/d}{\rightarrow} L^{a,b,c}(\mathbb{Z})$ to do this for any a, b, c, d, where the superscript values represent linear, $\left\{{n+1 \atop 2}\right\}$, and $n\left\{{n+1 \atop 2}\right\}$ potential, respectively.

One can prove by induction that at most $4 * 2^n - n - 3$ Boolean or arithmetic operations are required. Although this can be bounded with only exponential functions, the purely exponential potential system cannot reason about the exact (linear) cost associated with *remove*, and overestimates the bound to be in $\theta(3^n)$. This mixed system can provide a better (though still loose) bound of $n2^n + 2 * 2^n - n - 1$, giving a type of $L^{0,2,1}(\mathbb{Z}) \times \mathbb{Z} \overset{1/0}{\rightarrow} bool$ to $subSum1$. After showing this derivation, we will show how to find the exact bound with AARA.

The following is the single-use subset sum code, with comments on each line tracking the amount of available resources on each line. For clarity, we indicate sharing and subtype-weakening locations.

```
let subSum1 nums:L^{0,2,1}(ℤ) target =                           (* 1 *)
    match nums with
    | [] →                                                        (* 1 *)
        tick 1; target = 0                                        (* 0 *)
    | hd::(tl:L^{1,6,2}(ℤ)) →                                     (* 4 *)
        let otherNums:L^{0,6,2}(ℤ) = remove hd tl:L^{1,6,2}(ℤ) in (* 4 *)
        tick 1; let newTarg = target - hd in                      (* 3 *)
        (* weaken otherNums:L^{0,6,2}(ℤ) to L^{0,4,2}(ℤ) *)
        (* share otherNums:L^{0,4,2}(ℤ) as L^{0,2,1}(ℤ), L^{0,2,1}(ℤ) *)
        let withNum = subSum1 otherNums:L^{0,2,1}(ℤ) newTarg in   (* 2 *)
        let without = subSum1 otherNums:L^{0,2,1}(ℤ) target in    (* 1 *)
        tick 1; withNum || without                                (* 0 *)
```

The difference between initial and ending potential gives the upper bound of $1+2\genfrac\{\}{0pt}{}{n+1}{2}+n*\genfrac\{\}{0pt}{}{n+1}{2}=n2^n+2*2^n-n-1$ Boolean or arithmetic operations.

Note that we use the subtype-weakening rule, throwing away 2 units of $\genfrac\{\}{0pt}{}{n+1}{2}$ potential. This indicates why the bound is not tight. Next we show how to improve this bound using potential demotion.

Demotion There is one special exception to the non-negativity of potential annotations that may be added due to the particular nature of the relation between binomial coefficients and Stirling numbers. It represents the concept of *demoting* exponential potential into polynomial potential.

The relevant relation is $\genfrac\{\}{0pt}{}{n+1}{2}=2^n-1=\sum_{i=1}^{\infty}\binom{n}{i}\geq\sum_{i=1}^{k}\binom{n}{i}$. This allows a unit of $\genfrac\{\}{0pt}{}{n+1}{2}$ potential to account for one unit *each* of all non-constant binomial coefficient potentials. We can express this with the following additional subtyping rule. In this rule we interpret the 2-dimensional indexing of the potential annotation as a matrix, and we let $\overrightarrow{p}$ refer to the vector of potential entries at index coordinates $0, i$ for $i \geq 1$.

$$\frac{P=R+\begin{bmatrix}0 & \overrightarrow{p}\\ r & 0\end{bmatrix} \quad Q=R+\begin{bmatrix}0 & \overrightarrow{p}+s*\overrightarrow{1}\\ r-s & 0\end{bmatrix}}{L^P(A) <: L^Q(A)}\;\textit{Demote}$$

Theorem 3. *The demotion rule is sound.*

Proof. We need only show that $C <: D$ implies $\Phi(v : D) \leq \Phi(v : C)$ for unchanged values v. The rest of soundness then follows as in Theorem 1. To do so, it is sufficient to show for $l=[a_1,\ldots,a_n]$ we have $\Phi(a : L^Q(A)) \leq \Phi(a : L^P(A))$.

Without loss of generality, we need only consider where $R=0$.

$$\begin{aligned}\Phi(l : L^Q(A)) =& \phi(n,Q)+\textstyle\sum_{i=1}^{n}\Phi(a_i : A)\\ =&(r-s)\genfrac\{\}{0pt}{}{n+1}{2}+\textstyle\sum_{i=1}^{k}(\overrightarrow{p}_{i-1}+s)\binom{n}{i}+\sum_{i=1}^{n}\Phi(a_i : A)\end{aligned}$$

$$
\begin{aligned}
&=\sum_{i=1}^{\infty}(r-s)\tbinom{n}{i}+\sum_{i=1}^{k}(\overrightarrow{p}_{i-1}+s)\tbinom{n}{i}+\sum_{i=1}^{n}\Phi(a_i:A)\\
&\le\sum_{i=1}^{\infty}r\tbinom{n}{i}+\sum_{i=1}^{k}\overrightarrow{p}_{i-1}\tbinom{n}{i}+\sum_{i=1}^{n}\Phi(a_i:A)\\
&=r\left\{{n+1 \atop 2}\right\}+\sum_{i=1}^{k}\overrightarrow{p}_{i-1}\tbinom{n}{i}+\sum_{i=1}^{n}\Phi(a_i:A)\\
&=\phi(n,P)+\sum_{i=1}^{n}\Phi(a_i:A)=\Phi(l:L^P(A))
\end{aligned}
$$

As a corollary, this allows us to loosen the constraint that every annotation P contains only non-negative rationals. In particular, it is no longer required that $\forall i.p_{0,i}\ge 0$. Instead, we require that $\forall i.p_{0,i}+p_{1,0}\ge 0$. Each unit of $\left\{{n+1 \atop 2}\right\}$ potential may "pay" for one unit of deficit from each polynomial potential function. Because this is still a linear constraint, type inference remains automatable.

Using *Demote*, tighter bounds can be obtained. Consider the single-use subset sum solution from the previous section. Here it is again below, but this time allowing the linear potential to be paid for by $\left\{{n+1 \atop 2}\right\}$ potential. AARA can now provide a type of $L^{-1,4,0}(\mathbb{Z})\times\mathbb{Z}\overset{1/0}{\to}$ *bool* for *subSum*1, corresponding to the exact upper bound of $4*2^n-n-3$ operations. This time $n*\left\{{n+1 \atop 2}\right\}$ is elided in the annotated potentials, as it is not needed.

```
let subSum1 nums:L^{-1,4}(ℤ) target =                              (* 1 *)
  match nums with
  | [] →                                                           (* 1 *)
    tick 1; target = 0                                             (* 0 *)
  | hd::(tl:L^{-1,8}(ℤ)) →                                         (* 4 *)
    let otherNums:L^{-2,8}(ℤ) = remove hd tl:L^{-1,8}(ℤ) in        (* 4 *)
    tick 1; let newTarg = target - hd in                           (* 3 *)
    (* share otherNums:L^{-2,8}(ℤ) as L^{-1,4}(ℤ), L^{-1,4}(ℤ) *)
    let withNum = subSum1 otherNums:L^{-1,4}(ℤ) newTarg in         (* 2 *)
    let without = subSum1 otherNums:L^{-1,4}(ℤ) target in          (* 1 *)
    tick 1; withNum || without                                     (* 0 *)
```

The difference between initial and ending potential gives the upper bound of $1-n+4\left\{{n+1 \atop 2}\right\}=4*2^n-n-3$, as desired.

6 Exponentials, Polynomials, and Logarithms

The addition of exponential potential also allows for the inference of previously nonderivable polynomial-resource types for certain programs. One such way this can happen is by compacting the potential of a list into a new list logarithmic in size to the first. Performing exponential-cost operations, such as *subsetSum*, on a list of logarithmic size only has linear cost in total.

In the code below, *log* takes a list x of length n and returns a list of length roughly $log_2(n)$. If x begins with one unit of linear potential, the type system assigns the output of *log* one unit of base-2 exponential (2^n-1) potential. We show in the code below with types of the form $L^{a,b}$, where a is the linear potential, and b is the base-2 exponential potential. This lets us find that *half* can have type $L^{1,0}(\mathbb{Z})\overset{0/0}{\to}L^{2,0}(\mathbb{Z})$ and *log* has type $L^{1,0}(\mathbb{Z})\overset{0/0}{\to}L^{0,1}(\mathbb{Z})$. The typing of *log* shows the conversion from linear to exponential potential.

```
let half x: L^{1,0}(Z) =                          (* 0 *)
    match x with
    | [] →                                        (* 0 *)
        []: L^{2,0}(Z)                            (* 0 *)
    | hd::(tl: L^{1,0}(Z)) →                      (* 1 *)
        match tl with
        | [] →                                    (* 1 *)
            []: L^{2,0}(Z)                        (* 1 *)
        | hd2::(tl2: L^{1,0}(Z)) →                (* 2 *)
            let halfTail: L^{2,0}(Z) = half tl2 in (* 2 *)
            (hd::halfTail): L^{2,0}(Z)            (* 0 *)

let log x: L^{1,0}(Z) =                           (* 0 *)
    match x with
    | [] →                                        (* 0 *)
        []: L^{0,1}(Z)                            (* 0 *)
    | hd::(tl: L^{1,0}(Z)) →                      (* 1 *)
        let halfTail: L^{2,0}(Z) = half tl in     (* 1 *)
        let subSoln: L^{0,2}(Z) = log halfTail in (* 1 *)
        (hd::subSoln): L^{0,1}(Z)                 (* 0 *)
```

Typing *log* above requires resource-polymorphic recursion. However, this can be justified by noting that the above can be thought of to show *half* has type $L^{a,0}(\mathbb{Z}) \overset{0/0}{\to} L^{2a,0}(\mathbb{Z})$ and *log* has type $L^{a,0}(\mathbb{Z}) \overset{0/0}{\to} L^{0,a}(\mathbb{Z})$ for any $a \geq 0$.

Coincidentally, *log* conversion of linear to exponential potential certifies that the output list's size can be bounded by a logarithm of the input's size. Nonetheless, logarithmic *potential* is not directly compatible with the approach this work takes. Sublinear functions have negative second derivatives, and this yields negative annotation entries under $\lhd$ applications. This may not be insurmountable, as the demotion rule showed here, but new ideas are needed overall. Logarithmic potential has been explored in [32], though the approach there departs from the automatable AARA framework of linear constraint solving.

7 Conclusion and Future Work

Using Stirling numbers of the second kind allows for the automated inference of exponential resource usages via Automatic Amortized Resource Analysis. This may be combined with the existing polynomial system, allowing mixtures of polynomial and exponential functions to be inferred. Under this system, more kinds of programs can now be automatically analyzed, in particular those making use of multiple recursive calls, or logarithmically-sized lists. Finally, the framework put in place to accomplish this separates the concerns of the type system and potential functions, paving the way to allow modular addition of different potential functions. Future work could extend the work here to cover additional language features supported in polynomial AARA literature, like trees [22].

References

1. Albert, E., Arenas, P., Genaim, S., Puebla, G., Zanardini, D.: Cost Analysis of Java Bytecode. In: 16th Euro. Symp. on Prog. (ESOP'07) (2007)
2. Albert, E., Fernández, J.C., Román-Díez, G.: Non-cumulative Resource Analysis. In: Tools and Algorithms for the Construction and Analysis of Systems - 21st International Conference, (TACAS'15) (2015)
3. Albert, E., Genaim, S., Masud, A.N.: On the Inference of Resource Usage Upper and Lower Bounds. ACM Transactions on Computational Logic **14**(3) (2013)
4. Atkey, R.: Amortised Resource Analysis with Separation Logic. In: 19th Euro. Symp. on Prog. (ESOP'10) (2010)
5. Avanzini, M., Lago, U.D., Moser, G.: Analysing the Complexity of Functional Programs: Higher-Order Meets First-Order. In: 29th Int. Conf. on Functional Programming (ICFP'15) (2012)
6. Avanzini, M., Moser, G.: A Combination Framework for Complexity. In: 24th International Conference on Rewriting Techniques and Applications (RTA'13) (2013)
7. Blanc, R., Henzinger, T.A., Hottelier, T., Kovács, L.: ABC: Algebraic Bound Computation for Loops. In: Logic for Prog., AI., and Reasoning - 16th Int. Conf. (LPAR'10) (2010)
8. Carbonneaux, Q., Hoffmann, J., Reps, T., Shao, Z.: Automated Resource Analysis with Coq Proof Objects. In: 29th International Conference on Computer-Aided Verification (CAV'17) (2017)
9. Carbonneaux, Q., Hoffmann, J., Shao, Z.: Compositional Certified Resource Bounds. In: 36th Conference on Programming Language Design and Implementation (PLDI'15) (2015), artifact submitted and approved
10. Chatterjee, K., Fu, H., Goharshady, A.K.: Non-polynomial worst-case analysis of recursive programs. In: Computer Aided Verification - 29th International Conference (CAV '17). pp. 41–63 (2017)
11. Dal Lago, U., Gaboardi, M.: Linear Dependent Types and Relative Completeness. In: 26th IEEE Symp. on Logic in Computer Science (LICS'11) (2011)
12. Danner, N., Licata, D.R., Ramyaa, R.: Denotational Cost Semantics for Functional Languages with Inductive Types. In: 29th Int. Conf. on Functional Programming (ICFP'15) (2012)
13. Das, A., Hoffmann, J., Pfenning, F.: Work analysis with resource-aware session types. In: 33th ACM/IEEE Symposium on Logic in Computer Science (LICS'18) (2018)
14. Fagin, R.: Generalized First-Order Spectra, and Polynomial-Time Recognizable Sets. SIAM-AMS Proc. **7** (01 1974)
15. Flajolet, P., Salvy, B., Zimmermann, P.: Automatic Average-case Analysis of Algorithms. Theoret. Comput. Sci. **79**(1), 37–109 (1991)
16. Flores-Montoya, A., Hähnle, R.: Resource Analysis of Complex Programs with Cost Equations. In: Programming Languages and Systems - 12th Asian Symposiu (APLAS'14) (2014)
17. Frohn, F., Naaf, M., Hensel, J., Brockschmidt, M., Giesl, J.: Lower Runtime Bounds for Integer Programs. In: Automated Reasoning - 8th International Joint Conference (IJCAR'16) (2016)
18. Gulwani, S., Mehra, K.K., Chilimbi, T.M.: SPEED: Precise and Efficient Static Estimation of Program Computational Complexity. In: 36th ACM Symp. on Principles of Prog. Langs. (POPL'09) (2009)

19. Guéneau, A., Charguéraud, A., Pottier, F.: A fistful of dollars: Formalizing asymptotic complexity claims via deductive program verification. In: Ahmed, A. (ed.) European Symposium on Programming (ESOP). Lecture Notes in Computer Science, vol. 10801, pp. 533–560. Springer (Apr 2018)
20. Harper, R.: Practical Foundations for Programming Languages. Cambridge University Press (2016)
21. Hindley, R.: The Principal Type-Scheme of an Object in Combinatory Logic. Transactions of the American Mathematical Society **146**, 29–60 (1969), http://www.jstor.org/stable/1995158
22. Hoffmann, J.: Types with Potential: Polynomial Resource Bounds via Automatic Amortized Analysis. Ph.D. thesis, Ludwig-Maximilians-Universität München (2011)
23. Hoffmann, J., Aehlig, K., Hofmann, M.: Multivariate Amortized Resource Analysis. In: 38th Symposium on Principles of Programming Languages (POPL'11) (2011)
24. Hoffmann, J., Aehlig, K., Hofmann, M.: Multivariate Amortized Resource Analysis. ACM Trans. Program. Lang. Syst. (2012)
25. Hoffmann, J., Das, A., Weng, S.C.: Towards Automatic Resource Bound Analysis for OCaml. In: 44th Symposium on Principles of Programming Languages (POPL'17) (2017)
26. Hoffmann, J., Hofmann, M.: Amortized Resource Analysis with Polymorphic Recursion and Partial Big-Step Operational Semantics. In: 8th Asian Symposium on Programming Languages (APLAS'10) (2010)
27. Hoffmann, J., Hofmann, M.: Amortized Resource Analysis with Polynomial Potential. In: 19th European Symposium on Programming (ESOP'10) (2010)
28. Hoffmann, J., Shao, Z.: Type-Based Amortized Resource Analysis with Integers and Arrays. In: 12th International Symposium on Functional and Logic Programming (FLOPS'14) (2014)
29. Hoffmann, J., Shao, Z.: Automatic Static Cost Analysis for Parallel Programs. In: 24th European Symposium on Programming (ESOP'15) (2015)
30. Hoffmann, J., Shao, Z.: Type-Based Amortized Resource Analysis with Integers and Arrays. J. Funct. Program. (2015)
31. Hofmann, M., Jost, S.: Static Prediction of Heap Space Usage for First-Order Functional Programs. In: 30th ACM Symp. on Principles of Prog. Langs. (POPL'03) (2003)
32. Hofmann, M., Moser, G.: Analysis of logarithmic amortised complexity (2018)
33. Jost, S., Hammond, K., Loidl, H.W., Hofmann, M.: Static Determination of Quantitative Resource Usage for Higher-Order Programs. In: 37th ACM Symp. on Principles of Prog. Langs. (POPL'10) (2010)
34. Jost, S., Loidl, H.W., Hammond, K., Scaife, N., Hofmann, M.: Carbon Credits for Resource-Bounded Computations using Amortised Analysis. In: 16th Symp. on Form. Meth. (FM'09) (2009)
35. Kaminski, B.L., Katoen, J.P., Matheja, C., Olmedo, F.: Weakest Precondition Reasoning for Expected Run–Times of Probabilistic Programs. In: Proceedings of the European Symposium on Programming Languages and Systems (ESOP'16). Springer (2016)
36. Kincaid, Z., Breck, J., Boroujeni, A.F., Reps, T.: Compositional recurrence analysis revisited. In: Conference on Programming Language Design and Implementation (PLDI'17) (2017)
37. Kincaid, Z., Cyphert, J., Breck, J., Reps, T.: Non-linear reasoning for invariant synthesis. Proc. ACM Program. Lang. **2**(POPL), 54:1–54:33 (Dec 2017)

38. Lichtman, B., Hoffmann, J.: Arrays and References in Resource Aware ML. In: 2nd International Conference on Formal Structures for Computation and Deduction (FSCD'17) (2017)
39. Madhavan, R., Kulal, S., Kuncak, V.: Contract-based resource verification for higher-order functions with memoization. In: Proceedings of the 44th Symposium on Principles of Programming Languages (POPL'17) (2017)
40. Matiyasevich, Y.V.: The Diophantineness of Enumerable Sets. In: Doklady Akademii Nauk. vol. 191, pp. 279–282. Russian Academy of Sciences (1970)
41. Milner, R.: A Theory of Type Polymorphism in Programming. Journal of Computer and System Sciences **17**, 348–375 (1978)
42. Mével, G., Jourdan, J.H., Pottier, F.: Time credits and time receipts in Iris. In: Caires, L. (ed.) European Symposium on Programming (ESOP). Lecture Notes in Computer Science, vol. 11423, pp. 1–27. Springer (Apr 2019)
43. Ngo, V.C., Carbonneaux, Q., Hoffmann, J.: Bounded expectations: Resource analysis for probabilistic programs. In: 39th Conference on Programming Language Design and Implementation (PLDI'18) (2018)
44. Ngo, V.C., Dehesa-Azuara, M., Fredrikson, M., Hoffmann, J.: Verifying and Synthesizing Constant-Resource Implementations with Types. In: 38th IEEE Symposium on Security and Privacy (S&P '17) (2017)
45. Nipkow, T., Brinkop, H.: Amortized complexity verified. J. Autom. Reasoning **62**(3), 367–391 (2019)
46. Niu, Y., Hoffmann, J.: Automatic space bound analysis for functional programs with garbage collection. In: 22nd International Conference on Logic for Programming Artificial Intelligence and Reasoning (LPAR'18) (2018)
47. Noschinski, L., Emmes, F., Giesl, J.: Analyzing Innermost Runtime Complexity of Term Rewriting by Dependency Pairs. J. Autom. Reasoning **51**(1), 27–56 (2013)
48. Radiček, I., Barthe, G., Gaboardi, M., Garg, D., Zuleger, F.: Monadic Refinements for Relational Cost Analysis. Proc. ACM Program. Lang. **2**(POPL) (2017)
49. Sinn, M., Zuleger, F., Veith, H.: A Simple and Scalable Approach to Bound Analysis and Amortized Complexity Analysis. In: Computer Aided Verification - 26th Int. Conf. (CAV'14) (2014)
50. Stirling, J.: The Differential Method: Or, A Treatise Concerning Summation and Interpolation of Infinite Series. E. Cave (1749)
51. Tarjan, R.E.: Amortized Computational Complexity. SIAM J. Algebraic Discrete Methods **6**(2), 306–318 (1985)
52. Vasconcelos, P.B., Jost, S., Florido, M., Hammond, K.: Type-Based Allocation Analysis for Co-recursion in Lazy Functional Languages. In: 24th European Symposium on Programming (ESOP'15) (2015)
53. Wang, P., Wang, D., Chlipala, A.: TiML: A Functional Language for Practical Complexity Analysis with Invariants. In: Object-Oriented Prog., Syst., Lang., and Applications (OOPSLA'17) (2017)
54. Wang, P., Fu, H., Goharshady, A.K., Chatterjee, K., Qin, X., Shi, W.: Cost analysis of nondeterministic probabilistic programs. In: 40th ACM SIGPLAN Conference on Programming Language Design and Implementation (PLDI '19). pp. 204–220 (2019)
55. Wegbreit, B.: Mechanical Program Analysis. Commun. ACM **18**(9), 528–539 (1975)
56. Çiçek, E., Barthe, G., Gaboardi, M., Garg, D., Hoffmann, J.: Relational Cost Analysis. In: 44th Symposium on Principles of Programming Languages (POPL'17) (2017)

Concurrent Kleene Algebra with Observations: from Hypotheses to Completeness

Tobias Kappé (✉), Paul Brunet, Alexandra Silva,
Jana Wagemaker, and Fabio Zanasi

University College London, London, United Kingdom; tkappe@cs.ucl.ac.uk

Abstract. Concurrent Kleene Algebra (CKA) extends basic Kleene algebra with a parallel composition operator, which enables reasoning about concurrent programs. However, CKA fundamentally misses *tests*, which are needed to model standard programming constructs such as conditionals and while-loops. It turns out that integrating tests in CKA is subtle, due to their interaction with parallelism. In this paper we provide a solution in the form of Concurrent Kleene Algebra with Observations (CKAO). Our main contribution is a completeness theorem for CKAO. Our result resorts on a more general study of CKA "with hypotheses", of which CKAO turns out to be an instance: this analysis is of independent interest, as it can be applied to extensions of CKA other than CKAO.

Acknowledgments. This work was partially supported by the ERC Starting Grant ProFoundNet, grant code 679127. We acknowledge support from the EPSRC grants EP/S028641/1 (A. Silva); EP/R020604/1 (F. Zanasi); EP/R006865/1 (P. Brunet).

1 Introduction

Kleene algebra with tests (KAT) is a (co)algebraic framework [17,19] that allows one to study properties of imperative programs with conditional branching, i.e. if-statements and while-loops. KAT is build on Kleene algebra (KA) [6,16], the algebra of regular languages. Both KA and KAT enjoy a rich meta-theory, which makes them a suitable foundation for reasoning about program verification. In particular, it is well-known that the equational theories of KA and KAT characterise rational languages [27,21,16] and guarded rational languages [17] respectively. Efficient procedures for deciding equivalence have been studied in recent years, also in view of recent applications to network verification [3,8,28].

Concurrency is a known source of bugs and hence challenges for verification. Hoare, Struth, and collaborators [11], have proposed an extension of KA, *Concurrent Kleene Algebra* (CKA), as an algebraic foundation for concurrent programming. CKA enriches the basic language of KA with a parallel composition operator $\cdot \parallel \cdot$. Analogously to KA, CKA also has a semantic characterisation for which the equational theory is complete, in terms of rational languages of *pomsets* (words with a partial order on letters) [23,24,15].

J. Goubault-Larrecq and B. König (Eds.): FOSSACS 2020, LNCS 12077, pp. 381–400, 2020.
https://doi.org/10.1007/978-3-030-45231-5_20

The development of CKA raises a natural question, namely how tests, which were essential in KAT for the study of sequential programs, can be integrated into CKA. At first glance, the obvious answer may appear to be to merge KAT with CKA, yielding Concurrent Kleene Algebra with Tests (CKAT) — as attempted in [12]. However, as it turns out, integrating tests into CKA is quite subtle and this naive combination does not adequately capture the behaviour of concurrent programs. In particular, using the CKAT framework of [12] one can prove that for any test b and CKAT program e:

$$0 \quad \leq_{\mathsf{KAT}} \quad b \cdot e \cdot \overline{b} \quad \leq_{\mathsf{CKA}} \quad e \parallel (b \cdot \overline{b}) \quad \equiv_{\mathsf{KAT}} \quad e \parallel 0 \quad \equiv_{\mathsf{CKA}} \quad 0$$

thus $b \cdot e \cdot \overline{b} \equiv_{\mathsf{CKAT}} 0$, meaning no program e can change the outcome of any test b. Or equivalently, and undesirably, that any test is an invariant of any program!

The core issue is the identification in KAT of sequential composition $\cdot$ and Boolean conjunction $\wedge$. In the concurrent setting this is not sound as the values of variables — and hence tests — can be changed between the two tests.

In order to fix this issue, we have presented *Kleene Algebra with Observations* (KAO) in previous work [13]. Algebraically, KAO differs from KAT in that conjunction of tests $b \wedge b'$ and their sequential composition $b \cdot b'$ are distinct operations. In particular, $b \wedge b'$ expresses a single test executed *atomically*, whereas $b \cdot b'$ describes two distinct executions, occurring one after the other. As mentioned above, this distinction is crucial when moving from the sequential setting of KA to the concurrent setting of CKA, as actions from another thread that happen to be scheduled after b but before b' may as well change the outcome of b'.

This newly developed extension of KA enables a novel attempt to enrich CKA with the ability to reason about programs that also have the traditional conditionals: in this paper, we present Concurrent Kleene Algebra with Observations (CKAO) and show that it overcomes the problems present in CKAT.

The traditional plan for developing a variant of (C)KA is to define a separate syntax, semantics, and set of axioms, before establishing a formal correspondence with the base syntax, semantics and axioms of (C)KA proper, and arguing that this correspondence allows one to conclude soundness and completeness of the axioms w.r.t. the semantics, as well as decidability of equivalence in the semantics. Instead of such a tailor-made proof, however, we take a more general approach by first proposing CKA with hypotheses (CKAH) as a formalism for studying extensions of CKA, akin to how Kleene algebra with hypotheses [5,18,20,7] can be used to extend Kleene algebra. We then apply CKAH to study CKAO, but the meta-theory developed can also be applied to extensions other than CKAO.

Using the CKAH formalism, we instantiate CKAO as CKAH with a particular set of hypotheses, and we immediately obtain a syntax and semantics; we can then use the meta-theory of CKAH to argue completeness and decidability in a modular proof, which composes results about CKA [15] and KAO [13].

The technical roadmap of the paper and its contributions are as follows.

- We introduce Concurrent Kleene Algebra with Hypotheses (CKAH), a formalism for studying extensions of CKA; this is a concurrent extension of Kleene Algebra with Hypotheses (Section 4). We show how CKAH is sound with respect to rational pomset languages closed under an operation arising from the set of hypotheses. We propose techniques to argue completeness of the extended set of axioms with respect to the sound model as well as decidability of equivalence, capturing methods commonly used in literature to argue completeness and decidability for extensions of (concurrent) KA.

- We prove that CKAO can be presented as an instance of CKAH, for a certain set of hypotheses (Section 5). This gives us a sound model of CKAO 'for free'. We then prove that the axioms of CKAO are also complete for this model, and that equivalence is decidable, using the techniques developed previously.

We conclude this introduction by giving an example of how hypotheses can be added to CKA to include the meaning of primitive actions. Suppose we were designing a DSL for recipes, specifically, the steps necessary, and their order. A recipe to prepare cookies might contain the actions mix (mixing the ingredients), $\mathsf{preheat}$ (pre-heating the oven), chill (chilling the dough) and bake (baking the cookies). Using these actions, a recipe like "mix the ingredients until combined; chill the dough while pre-heating the oven; bake cookies in the oven" may be encoded as $\mathsf{mix}^* \cdot (\mathsf{chill} \parallel \mathsf{preheat}) \cdot \mathsf{bake}$. Now, imagine that we have only one oven, meaning that we cannot bake two batches of cookies concurrently. We might encode this restriction on concurrent behaviour by forcing the equation

$$(e \cdot \mathsf{bake} \cdot f) \parallel (g \cdot \mathsf{bake} \cdot h) = (e \cdot \mathsf{bake} \parallel g) \cdot (f \parallel \mathsf{bake} \cdot h) + (e \parallel g \cdot \mathsf{bake}) \cdot (\mathsf{bake} \cdot f \parallel h)$$

As a consequence of this hypothesis, one could then derive properties such as

$$\mathsf{bake} \parallel (\mathsf{bake} \cdot \mathsf{mix}) = \mathsf{bake} \cdot \mathsf{bake} \cdot \mathsf{mix} + \mathsf{bake} \cdot \mathsf{mix} \cdot \mathsf{bake}$$

In a nutshell, this paper provides an algebraic framework — CKAH — together with techniques for soundness and completeness results. The framework is flexible in that different instantiations of the hypotheses generate very different algebraic systems. We provide one instantiation — CKAO — that enables analysis of programs with both concurrency primitives and Boolean assertions. This is the first sound and complete algebraic theory to reason about such programs.

For the sake of brevity, some proofs appear in the extended version [14].

2 Preliminaries

We recall basic definitions on pomset languages, used in the semantics of CKA, which generalise languages to allow letters in words to be partially ordered. We fix a (possibly infinite) alphabet Σ. When defining sets parametrised by Σ, say $\mathsf{S}(\Sigma)$, if Σ is clear from the context we use S to refer to $\mathsf{S}(\Sigma)$.

Posets and Pomsets Pomsets [9,10] are labelled posets, up to isomorphism.

Definition 2.1 (Labellet poset). *A* labelled poset *over Σ is a tuple $\mathbf{u} = \langle S, \leq, \lambda \rangle$, where S is a finite set (the* carrier *of $\mathbf{u}$), $\leq_{\mathbf{u}}$ is a partial order on S (the* order *of $\mathbf{u}$), and $\lambda : S \to \Sigma$ is a function (the* labelling *of $\mathbf{u}$).*

We will denote labelled posets by bold lower-case letters $\mathbf{u}$, $\mathbf{v}$, etc. We write $S_{\mathbf{u}}$ for the carrier of $\mathbf{u}$, $\leq_{\mathbf{u}}$ for the order of $\mathbf{u}$, and $\lambda_{\mathbf{u}}$ for the labelling of $\mathbf{u}$. We assume that any labelled poset has a carrier that is a subset of some countably infinite set, say $\mathbb{N}$; this allows us to speak about the *set of labelled posets* over Σ. The precise contents of the carrier, however, are not important — what matters to us is the labels of the points, and the ordering between them.

Definition 2.2 (Poset isomorphism, pomset). *Let $\mathbf{u}, \mathbf{v}$ be labelled posets over Σ. We say $\mathbf{u}$ is* isomorphic *to $\mathbf{v}$, denoted $\mathbf{u} \cong \mathbf{v}$, if there exists a bijection $h : S_{\mathbf{u}} \to S_{\mathbf{v}}$ that preserves labels, and preserves and reflects ordering. More precisely, we require that $\lambda_{\mathbf{v}} \circ h = \lambda_{\mathbf{u}}$, and $s \leq_{\mathbf{u}} s'$ if and only if $h(s) \leq_{\mathbf{v}} h(s')$.*

A pomset *over Σ is an isomorphism class of labelled posets over Σ, i.e., the class $[\mathbf{v}] = \{\mathbf{u} : \mathbf{u} \cong \mathbf{v}\}$ for some labelled poset $\mathbf{v}$.*

We write $\mathsf{Pom}(\Sigma)$ for the set of pomsets over Σ, and 1 for the empty pomset. As long as we have countably many pomsets in scope, the above allows us to assume w.l.o.g. that those pomsets are represented by labelled posets with pairwise disjoint carriers; we tacitly make this assumption throughout this paper.

Pomsets can be concatenated, creating a new pomset that contains all events of the operands, with the same label, but which orders all events of the left operand before those of the right one. We can also compose pomsets in parallel, where events of the operands are juxtaposed without any ordering between them.

Definition 2.3 (Pomset composition). *Let $U = [\mathbf{u}]$ and $V = [\mathbf{v}]$ be pomsets over Σ. We write $U \parallel V$ for the* parallel composition *of U and V, which is the pomset over Σ represented by the labelled poset $\mathbf{u} \parallel \mathbf{v}$, where*

$$S_{\mathbf{u}\parallel\mathbf{v}} = S_{\mathbf{u}} \cup S_{\mathbf{v}} \qquad \leq_{\mathbf{u}\parallel\mathbf{v}} = \leq_{\mathbf{u}} \cup \leq_{\mathbf{v}} \qquad \lambda_{\mathbf{u}\parallel\mathbf{v}}(x) = \begin{cases} \lambda_{\mathbf{u}}(x) & x \in S_{\mathbf{u}} \\ \lambda_{\mathbf{v}}(x) & x \in S_{\mathbf{v}} \end{cases}$$

Similarly, we write $U \cdot V$ for the sequential composition *of U and V, that is, the pomset represented by the labelled poset $\mathbf{u} \cdot \mathbf{v}$, where*

$$S_{\mathbf{u}\cdot\mathbf{v}} = S_{\mathbf{u}\parallel\mathbf{v}} \qquad \leq_{\mathbf{u}\cdot\mathbf{v}} = \leq_{\mathbf{u}} \cup \leq_{\mathbf{v}} \cup (S_{\mathbf{u}} \times S_{\mathbf{v}}) \qquad \lambda_{\mathbf{u}\cdot\mathbf{v}} = \lambda_{\mathbf{u}\parallel\mathbf{v}}$$

Just like words are built up from the empty word and letters using concatenation, we can build a particular set of pomsets using only sequential and parallel composition; this will be the primary type of pomset that we will use.

Definition 2.4 (Series-parallel). *The set of* series-parallel pomsets *(*sp-pomsets*) over Σ, denoted $\mathsf{SP}(\Sigma)$, is the smallest set s.t. $1 \in \mathsf{SP}(\Sigma)$, $\mathsf{a} \in \mathsf{SP}(\Sigma)$ for every $\mathsf{a} \in \Sigma$, and it is closed under parallel and sequential composition.*

The following characterisation of SP is very useful in proofs.

Theorem 2.5 (Gischer [9]). *Let $U = [\mathbf{u}] \in \mathsf{Pom}$. Then $U \in \mathsf{SP}$ if and only if U is* N-free, *which is to say that if there exist no distinct $s_0, s_1, s_2, s_3 \in S_{\mathbf{u}}$ such that $s_0 \leq_{\mathbf{u}} s_1$ and $s_2 \leq_{\mathbf{u}} s_3$ and $s_0 \leq_{\mathbf{u}} s_3$, with no other relation between them.*

One way of comparing pomsets is to see whether they have the same events and labels, except that one is "more sequential" in the sense that more events are ordered. This is captured by the notion of *subsumption* [9], defined as follows.

Definition 2.6 (Subsumption). *Let* $U = [\mathbf{u}]$ *and* $V = [\mathbf{v}]$. *We say* U *is subsumed by* V, *written* $U \sqsubseteq V$, *if there exists a label- and order-preserving bijection* $h : S_{\mathbf{v}} \to S_{\mathbf{u}}$. *That is,* $\lambda_{\mathbf{u}} \circ h = \lambda_{\mathbf{v}}$ *and if* $s \leq_{\mathbf{v}} s'$, *then* $h(s) \leq_{\mathbf{u}} h(s')$.

Subsumption between sp-pomsets can be characterised as follows [9].

Lemma 2.7. *Let* $\sqsubseteq^{\mathsf{sp}}$ *be* $\sqsubseteq$ *restricted to* SP. *Then* $\sqsubseteq^{\mathsf{sp}}$ *is the smallest precongruence (preorder monotone w.r.t. the operators) such that for all* $U, V, W, X \in \mathsf{SP}$*:*

$$(U \parallel V) \cdot (W \parallel X) \sqsubseteq^{\mathsf{sp}} (U \cdot W) \parallel (V \cdot X)$$

CKA: syntax and semantics. CKA terms are generated by the grammar

$$e, f \in \mathcal{T}(\Sigma) ::= 0 \mid 1 \mid \mathtt{a} \in \Sigma \mid e + f \mid e \cdot f \mid e \parallel f \mid e^*$$

Semantics of CKA is given in terms of *pomset languages*, that is subsets of SP, which we simply denote by 2^{SP}. Formally, the function $[\![-]\!] : \mathcal{T} \to 2^{\mathsf{SP}}$ assigning languages to CKA terms is defined as follows:

$$[\![0]\!] = \emptyset \qquad [\![1]\!] = \{1\} \qquad [\![e+f]\!] = [\![e]\!] \cup [\![f]\!] \qquad [\![e \cdot f]\!] = [\![e]\!] \cdot [\![f]\!]$$
$$[\![e^*]\!] = [\![e]\!]^* \qquad [\![\mathtt{a}]\!] = \{\mathtt{a}\} \qquad [\![e \parallel f]\!] = [\![e]\!] \parallel [\![f]\!]$$

Here, we use the pointwise lifting of sequential and parallel composition from pomsets to pomset languages, i.e., when $\mathcal{U}, \mathcal{V} \subseteq \mathsf{SP}(\Sigma)$, we define

$$\mathcal{U} \cdot \mathcal{V} = \{U \cdot V : U \in \mathcal{U}, V \in \mathcal{V}\} \qquad \mathcal{U} \parallel \mathcal{V} = \{U \parallel V : U \in \mathcal{U}, V \in \mathcal{V}\}$$

Furthermore, the Kleene star of a pomset language $\mathcal{U}$ is defined as $\mathcal{U}^* = \bigcup_{n \in \mathbb{N}} \mathcal{U}^n$, where $\mathcal{U}^0 = \{1\}$ and $\mathcal{U}^{n+1} = \mathcal{U}^n \cdot \mathcal{U}$.

Equivalence of CKA terms can be axiomatised in the style of Kleene algebra. The relation $\equiv$ is the smallest congruence on $\mathcal{T}$ (with respect to all operators) such that for all $e, f, g \in \mathcal{T}$:

$$e + 0 \equiv e \qquad e + e \equiv e \qquad e + f \equiv f + e \qquad e + (f + g) \equiv (f + g) + h$$

$$e \cdot (f \cdot g) \equiv (e \cdot f) \cdot g \qquad e \cdot (f + g) \equiv e \cdot f + e \cdot h \qquad (e + f) \cdot g \equiv e \cdot g + f \cdot g$$

$$e \cdot 1 \equiv e \equiv 1 \cdot e \qquad e \cdot 0 \equiv 0 \equiv 0 \cdot e \qquad e \parallel f \equiv f \parallel e \qquad e \parallel 1 \equiv e \qquad e \parallel 0 \equiv 0$$

$$e \parallel (f \parallel g) \equiv (e \parallel f) \parallel g \qquad e \parallel (f + g) \equiv e \parallel f + e \parallel g \qquad 1 + e \cdot e^* \equiv e^* \equiv 1 + e^* \cdot e$$

$$e + f \cdot g \leqq g \implies f^* \cdot e \leqq g \qquad e + f \cdot g \leqq f \implies e \cdot g^* \leqq f$$

in which $e \leqq f$ is the natural order $e + f \equiv f$. The final (conditional) axioms are referred to as the *least fixpoint axioms.*

Laurence and Struth [23] proved this axiomatisation to be sound and complete. A decision procedure was proposed in [4].

Theorem 2.8 (Soundness, completeness, decidability). *Let $e, f \in \mathcal{T}$. We have: $e \equiv f$ if and only if $[\![e]\!] = [\![f]\!]$, and it is decidable whether $[\![e]\!] = [\![f]\!]$.*

Readers familiar with CKA will notice that the algebra defined here is not in fact CKA as defined in [11]. Indeed the signature axiom of CKA, the exchange law, has been omitted. However, as we show in Section 4.2, the standard definition of CKA, as well as its completeness proof [15], may be recovered using hypotheses.

3 Pomset contexts

The linear one-dimensional structure of words makes it straightforward to define occurrences of subwords: if one wants to state that a word w appears in another word v, one can simply say that $v = xwy$ for some x and y. Due to the two-dimensional nature of pomsets, it is not straightforward to define when a pomset occurs inside another pomset, because the pomset could appear below a parallel, which is nested in a sequential, which is in a parallel, etc. In what follows we define *pomset contexts*, that will enable us to talk about pomset factorisations in a similar fashion as we do for words, and prove some useful properties for these.

Definition 3.1. *Let $*$ be a symbol not occurring in Σ. A* pomset context *is a pomset over $\Sigma \cup \{*\}$ with exactly* one *node labelled by $*$. More precisely, C is a pomset context if $C = [\mathbf{c}]$ with exactly one $s_* \in S_\mathbf{c}$ with $\lambda_\mathbf{c}(s_*) = *$.*

Intuitively, $*$ is a placeholder or gap where another pomset can be inserted. We write $\mathsf{PC}(\Sigma)$ for the set of pomset contexts over Σ, and $\mathsf{PC}^{\mathsf{sp}}(\Sigma)$ for the series-parallel pomset contexts over Σ.

Given a $C \in \mathsf{PC}$ and $U \in \mathsf{Pom}$, we can "plug" U into the gap left in C to obtain the pomset $C[U] \in \mathsf{Pom}$. More precisely, let $U = [\mathbf{u}]$ and $C = [\mathbf{c}]$ with $\mathbf{u}$ disjoint from $\mathbf{c}$. We write $C[U]$ for the pomset represented by $\mathbf{c}[\mathbf{u}]$, where $S_{\mathbf{c}[\mathbf{u}]} = S_\mathbf{u} \cup S_\mathbf{c} - \{*\}$ and $\lambda_{\mathbf{c}[\mathbf{u}]}(s)$ is given by $\lambda_\mathbf{c}(s)$ if $s \in S_\mathbf{c} - \{*\}$, and $\lambda_\mathbf{u}(s)$ when $s \in S_\mathbf{u}$; lastly, $\leq_{\mathbf{c}[\mathbf{u}]}$ is the smallest relation on $S_{\mathbf{c}[\mathbf{u}]}$ satisfying

$$\frac{s \leq_\mathbf{u} s'}{s \leq_{\mathbf{c}[\mathbf{u}]} s'} \qquad \frac{s \leq_\mathbf{c} s'}{s \leq_{\mathbf{c}[\mathbf{u}]} s'} \qquad \frac{s_* \leq_\mathbf{c} s \qquad s' \in S_\mathbf{u}}{s' \leq_{\mathbf{c}[\mathbf{u}]} s} \qquad \frac{s' \in S_\mathbf{u} \qquad s \leq_\mathbf{c} s_*}{s \leq_{\mathbf{c}[\mathbf{u}]} s'}$$

It follows easily that $\leq_{\mathbf{c}[\mathbf{u}]}$ is a partial order. We may also apply contexts to languages: if $L \subseteq \mathsf{Pom}$ and $C \in \mathsf{PC}$, the language $C[L]$ is defined as $\{C[U] : U \in L\}$.

We now prove some properties of contexts that will be useful later in our technical development. First, we note that pomset contexts respect subsumption.

Lemma 3.2. *Let $C, D \in \mathsf{PC}$, $U \in \mathsf{Pom}$. If $C \sqsubseteq D$, then $C[U] \sqsubseteq D[U]$.*

Series-parallel pomset contexts can be given an inductive characterisation.

Lemma 3.3. *$\mathsf{PC}^{\mathsf{sp}}$ is the smallest pomset language L satisfying*

$$\frac{}{* \in L} \qquad \frac{U \in \mathsf{SP} \qquad C \in L}{U \cdot C \in L} \qquad \frac{C \in L \qquad V \in \mathsf{SP}}{C \cdot V \in L} \qquad \frac{U \in \mathsf{SP} \qquad C \in L}{U \parallel C \in L}$$

We will identify *totally ordered pomsets* with words, i.e., $\Sigma^* \subseteq \mathsf{SP}$. If the pomset U inserted in a context C is a non-empty word, and the resulting pomset is a parallel pomset, then we can infer how to factorise C.

Lemma 3.4. *Let $C \in \mathsf{PC}^{\mathsf{sp}}$ be a pomset context, let $V, W \in \mathsf{Pom}$, and let $U \in \Sigma^*$ be non-empty. If $C[U] = V \parallel W$, then there exists a $C' \in \mathsf{PC}^{\mathsf{sp}}$ such that either $C = C' \parallel W$ and $C'[U] = V$, or $C = V \parallel C'$ and $C'[U] = W$.*

Application of series-parallel contexts preserves series-parallel pomsets.

Lemma 3.5. *Let $C \in \mathsf{PC}^{\mathsf{sp}}$. If $U \in \mathsf{SP}$, then $C[U] \in \mathsf{SP}$ as well.*

If we plug the empty pomset into a context, then any subsumed pomset can be obtained by plugging the empty pomset into a subsumed context. If the subsumed pomset is series-parallel, then so is the subsumed context.

Lemma 3.6. *Let $C \in \mathsf{PC}$ and $V \in \mathsf{Pom}$ with $V \sqsubseteq C[1]$. We can construct $C' \in \mathsf{PC}$ such that $C' \sqsubseteq C$ and $C'[1] = V$. Moreover, if $V \in \mathsf{SP}$, then $C' \in \mathsf{PC}^{\mathsf{sp}}$.*

An analogue to the previous lemma can be obtained if instead of the empty pomset one inserts a single letter pomset a.

Lemma 3.7. *Let $C \in \mathsf{PC}$, $V \in \mathsf{Pom}$ and $\mathsf{a} \in \Sigma$ with $V \sqsubseteq C[\mathsf{a}]$. We can construct $C' \in \mathsf{PC}$ s.t. $C' \sqsubseteq C$ and $C'[\mathsf{a}] = V$. Moreover, if $V \in \mathsf{SP}$, then $C' \in \mathsf{PC}^{\mathsf{sp}}$.*

4 Concurrent Kleene Algebra with Hypotheses

Kleene algebra has basic axioms about how program composition operators should work in general, and hence does not make any assumptions about how these operators work on specific programs. When reasoning about equivalence in a programming language, however, it makes sense to embed domain-specific truths about the operators into the axioms. For instance, if a programming language includes assignments to variables, then subsequent assignments to the same variable could be merged into one, giving rise to an equation such as

$$x \leftarrow m \leq x \leftarrow n \cdot x \leftarrow m, \tag{1}$$

which says that the behaviour of first assigning n, then m to x (on the right) includes the behaviour of simply assigning m to x directly (on the left).

Kleene algebra with hypotheses (KAH) [5,18,20,7] enables the addition of extra axioms, called *hypotheses*, to the axioms of KA. The appeal of KAH is that it allows a wide range of such hypotheses about programs to be added to the equational theory, while retaining the theoretical boilerplate of KA. In particular, it turns out that we can derive a sound model for any set of hypotheses, using the language model that is sound for KA proper [7]. Moreover, the completeness and decidability results that hold for KA can be leveraged to obtain completeness and decidability results for some specific types of hypotheses [5,20,7]; in general, equivalence under other hypotheses may turn out to be undecidable [18].

In this section, we propose a generalisation of so-called Kleene algebra with hypotheses to a concurrent setting, showing how one can obtain a sound (pomset language) model for any set of hypotheses. We then discuss a number of techniques that allow one to prove completeness and decidability of the resulting system for a large set of hypotheses, by relying on analogous results about CKA.

Definition 4.1. *A* hypothesis *is an inequation* $e \leq f$ *where* $e, f \in \mathcal{T}$*. When* H *is a set of hypotheses, we write* $\equiv^H$ *for the smallest congruence on* $\mathcal{T}$ *generated by the hypotheses in* H *as well as the axioms and implications that build* $\equiv$*. More concretely, whenever* $e \leq f \in H$*, also* $e \leqq^H f$*.*

A hypothesis that declares two programs to be equivalent, such as in (1), can be encoded by including both $e \leq f$ and $f \leq e$ in H.

Example 4.2. Suppose the set of primitive actions Σ includes the increments of the form $\mathtt{incr}\,x$, as well as a statement `print`, which writes the complete state of the machine (including variables) on the standard output. Since we would like to depict the state consistently, the state should not change while the output is rendered; hence, `print` cannot be executed concurrently with any other action. Instead, when a program containing `print` is scheduled to run in parallel with an assignment, it must be interleaved such that the assignment runs either entirely before or after `print`. To encode this, we can include in H the hypotheses

$$\mathtt{incr}\,x \parallel \mathtt{print} = \mathtt{incr}\,x \cdot \mathtt{print} + \mathtt{print} \cdot \mathtt{incr}\,x$$

for all variables x. This allows us to prove, for instance, that

$$\mathtt{print} \cdot \mathtt{incr}\,x \cdot \mathtt{incr}\,x \cdot \mathtt{print} \leqq^H (\mathtt{incr}\,x \parallel \mathtt{print})^*$$

That is, if we run some number of increments and `print` statements in parallel, it is possible that x is incremented twice between print statements.

To obtain a model of CKAH, it is not enough to use $[\![-]\!]$, as some programs equated by the hypotheses might have different semantics. To get around this, we adapt the method from [7]: take $[\![-]\!]$ as a base semantics, and adapt the resulting language using hypotheses, such that the pomsets that could be obtained by rearranging the term using the hypotheses are also present in the language:

Definition 4.3. *Let* $L \subseteq \mathsf{Pom}$*. We define the* H*-closure of* L*, written* $L{\downarrow}^H$*, as the smallest language containing* L *such that for all* $e \leq f \in H$ *and* $C \in \mathsf{PC}^{\mathsf{sp}}$*, if* $C[[\![f]\!]] \subseteq L{\downarrow}^H$*, then* $C[[\![e]\!]] \subseteq L{\downarrow}^H$*. Formally,* $L{\downarrow}^H$ *may be described as the smallest language satisfying the following inference rules:*

$$\frac{}{L \subseteq L{\downarrow}^H} \qquad \frac{e \leq f \in H \qquad C \in \mathsf{PC}^{\mathsf{sp}} \qquad C[[\![f]\!]] \subseteq L{\downarrow}^H}{C[[\![e]\!]] \subseteq L{\downarrow}^H}$$

Example 4.4. Continuing with H and Σ as in the previous examples, note that if $L = [\![\mathtt{incr}\,x \parallel \mathtt{print}]\!]$, then $\mathtt{incr}\,x \parallel \mathtt{print} \in L{\downarrow}^H$. Choose $C = *$; we have $C[\mathtt{incr}\,x \cdot \mathtt{print}] = \mathtt{incr}\,x \cdot \mathtt{print}$. Because $\mathtt{incr}\,x \cdot \mathtt{print} + \mathtt{print} \cdot \mathtt{incr}\,x \leq \mathtt{incr}\,x \parallel \mathtt{print} \in H$ and for all $U \in [\![\mathtt{incr}\,x \parallel \mathtt{print}]\!]$ we have $C[U] \in L \subseteq L{\downarrow}^H$, we get $C[\mathtt{incr}\,x \cdot \mathtt{print}] \in L{\downarrow}^H$ and therefore $\mathtt{incr}\,x \cdot \mathtt{print} \in L{\downarrow}^H$.

We observe the following useful properties about the interaction between closure and other operators on pomset languages.

Lemma 4.5. *Let $L, K \subseteq \mathsf{Pom}$ and $C \in \mathsf{PC}^{\mathsf{sp}}$. The following hold.*

1. $L \subseteq K{\downarrow}^H$ *iff* $L{\downarrow}^H \subseteq K{\downarrow}^H$.
2. *If* $L \subseteq K$, *then* $L{\downarrow}^H \subseteq K{\downarrow}^H$.
3. $(L \cup K){\downarrow}^H = (L{\downarrow}^H \cup K{\downarrow}^H){\downarrow}^H$
4. $(L \cdot K){\downarrow}^H = (L{\downarrow}^H \cdot K{\downarrow}^H){\downarrow}^H$
5. $(L \parallel K){\downarrow}^H = (L{\downarrow}^H \parallel K{\downarrow}^H){\downarrow}^H$
6. $(L^*){\downarrow}^H = ((L{\downarrow}^H)^*){\downarrow}^H$
7. *If* $L{\downarrow}^H \subseteq K{\downarrow}^H$, *then* $C[L]{\downarrow}^H \subseteq C[K]{\downarrow}^H$.
8. *If* $L \subseteq \mathsf{SP}$, *then* $L{\downarrow}^H \subseteq \mathsf{SP}$.

Remark 4.6. Property (1) states that $-{\downarrow}^H$ is a closure operator. However, it is not in general a Kuratowski closure operator [22], since it fails to commute with union. For instance, let $\mathsf{a}, \mathsf{b}, \mathsf{c} \in \Sigma$ and $H = \{\mathsf{a} \leq \mathsf{b} + \mathsf{c}\}$; then $\{\mathsf{b}\}{\downarrow}^H \cup \{\mathsf{c}\}{\downarrow}^H = \{\mathsf{b}, \mathsf{c}\}$, while $\mathsf{a} \in (\{\mathsf{b}\} \cup \{\mathsf{c}\}){\downarrow}^H$.

Using Lemma 4.5, we can show that, if we combine the semantics from $[\![-]\!]$ with H-closure, we obtain a sound semantics for CKA with hypotheses H.

Lemma 4.7 (Soundness). *If $e \equiv^H f$, then $[\![e]\!]{\downarrow}^H = [\![f]\!]{\downarrow}^H$.*

The converse of the above, where semantic equivalence is sufficient to establish axiomatic equivalence, is called *completeness*. Similarly, we may also be interested in *deciding* whether $[\![e]\!]{\downarrow}^H$ and $[\![f]\!]{\downarrow}^H$ coincide.

Definition 4.8. *Let $e, f \in \mathcal{T}$.*

(i) If $[\![e]\!]{\downarrow}^H = [\![f]\!]{\downarrow}^H$ implies $e \equiv^H f$, then H is called complete.
(ii) If $[\![e]\!]{\downarrow}^H = [\![f]\!]{\downarrow}^H$ is decidable, then H is said to be decidable.

Note that, in the special case where $H = \emptyset$, we know that H is complete and decidable by Theorem 2.8. One method to find out whether H is complete or decidable is to reduce the problem to this special case. More concretely, suppose we know $[\![e]\!]{\downarrow}^H = [\![f]\!]{\downarrow}^H$, and want to establish that $e \equiv^H f$. If we could find a set of hypotheses H' that is complete, and we could map e and f to terms $r(e)$ and $r(f)$ such that $[\![r(e)]\!]{\downarrow}^{H'} = [\![r(f)]\!]{\downarrow}^{H'}$, then we would have $r(e) \equiv^{H'} r(f)$. If we could then "lift" that equivalence to prove $e \equiv^H f$, we are done. Similarly, if we would know that $[\![r(e)]\!]{\downarrow}^{H'} = [\![r(f)]\!]{\downarrow}^{H'}$ is equivalent to $[\![e]\!]{\downarrow}^H = [\![f]\!]{\downarrow}^H$, we could decide the latter. To formalise this intuition, we first need the following.

Definition 4.9. *We say that H* implies *H' if we can use the hypotheses in H to prove those of H', i.e., if for every hypothesis $e \leq f \in H'$ it holds that $e \leqq^H f$.*

Implication relates to equivalence and closure as follows.

Lemma 4.10. *Let H and H' be sets of hypotheses such that H implies H'.*

(i) If $e, f \in \mathcal{T}$ with $e \equiv^{H'} f$, then $e \equiv^H f$.
(ii) If $L \subseteq \mathsf{Pom}$, then $L{\downarrow}^{H'} \subseteq L{\downarrow}^H$.
(iii) If $L \subseteq \mathsf{Pom}$, then $(L{\downarrow}^{H'}){\downarrow}^H = L{\downarrow}^H$.

If H implies H' and vice versa, then H is complete (resp. decidable) precisely when H' is. In general, however, this is not very helpful; we need something more asymmetrical, in order to get from a complicated set of hypotheses H to a simpler set of hypotheses H', where completeness or decidability might be easier to prove. Ideally, we would like to reduce to $H' = \emptyset$, which is complete and decidable.

One idea to formalise this idea of a reduction is as follows.

Definition 4.11. *Let H and H' be sets of hypotheses such that H implies H'. A map $r : \mathcal{T} \to \mathcal{T}$ is a* reduction *from H to H' when both of the following are true:*

(i) for $e \in \mathcal{T}$, it holds that $e \equiv^H r(e)$, and
(ii) for $e, f \in \mathcal{T}$, if $[\![e]\!]{\downarrow}^H = [\![f]\!]{\downarrow}^H$, then $[\![r(e)]\!]{\downarrow}^{H'} = [\![r(f)]\!]{\downarrow}^{H'}$.

We call H reducible *to H' if there exists a reduction from H to H'.*

It is straightforward to show that reductions do indeed carry over completeness and decidability results, in the following sense.

Lemma 4.12. *Suppose H is reducible to H'. If H' is complete (respectively decidable), then so is H.*

Example 4.13. Let $\Sigma = \{\mathsf{a}, \mathsf{b}\}$. Let $H = \{\mathsf{a} \leq \mathsf{b}\}$. We can define for $e \in \mathcal{T}$ the term $r(e) \in \mathcal{T}$, which is e but with every occurrence of b replaced by $\mathsf{a} + \mathsf{b}$. For instance, $r(\mathsf{a} \cdot \mathsf{b}^* \parallel \mathsf{c}) = \mathsf{a} \cdot (\mathsf{a} + \mathsf{b})^* \parallel \mathsf{c}$. An inductive argument on the structure of e shows that r reduces H to $\emptyset$, and hence H is complete and decidable.

It is not very hard to show that reductions can be chained, as follows.

Lemma 4.14. *If H reduces to H', which reduces to H'', then H reduces to H''.*

Another way of reducing H is to find two sets of hypotheses H_0 and H_1, and reduce each of those to another set of hypotheses H' [7]. The idea is that a proof of $e \equiv^H f$ can be split up in a phase where we find $e', f' \in \mathcal{T}$ such that $e \equiv^{H_0} e'$ and $f \equiv^{H_0} f'$, after which we find $e'', f'' \in \mathcal{T}$ with $e' \equiv^{H_1} e''$ and $f' \equiv^{H_1} f''$. Finally, we establish that $e'' \equiv^{H'} f''$, before lifting those equivalences to H, concluding

$$e \equiv^H e' \equiv^H e'' \equiv^H f'' \equiv^H f' \equiv^H f$$

One way of achieving this is as follows.

Definition 4.15. *We say that H* factorises *into H_0 and H_1 if H implies both H_0 and H_1, and for all $L \subseteq \mathsf{SP}$ we have that $L{\downarrow}^H = (L{\downarrow}^{H_0}){\downarrow}^{H_1}$.*

In order to use factorisation to compose simpler reductions into more complicated ones, we need a slightly stronger notion of reduction, as follows.

Definition 4.16. *We say that r is a* strong reduction *from H to H' if it is a reduction such that for $e \in \mathcal{T}$, it holds that $[\![e]\!]{\downarrow}^H = [\![r(e)]\!]{\downarrow}^{H'}$.*

Note that this additional condition essentially strengthens the second condition in Definition 4.11. Factorisation then lets us compose strong reductions.

Lemma 4.17. *Suppose H factorises into H_0 and H_1, and both H_0 and H_1 strongly reduce to H'. Then H strongly reduces to H'.*

The remainder of this section is devoted to developing techniques that can be used to design reductions, based on the properties of the sets of hypotheses under consideration. Using the lemmas we have established so far, these techniques may then be leveraged to obtain completeness and decidability results.

4.1 Reification

It can happen that the hypotheses in H impose an algebraic structure on the letters in Σ; for instance, as we will see later on, the letters in H could be propositional terms, whose equivalence is mediated by the axioms of Boolean algebra. In order to peel away this layer of axioms and reduce to a smaller H', we can try to reduce to terms over a smaller alphabet, making the algebraic structure on the letters irrelevant to equivalence. In a sense, performing this kind of reduction is like showing that the equivalences between letters from the hypotheses can already be guaranteed by replacing them with the right terms.

Example 4.18. Let Σ be the set of group terms over a (finite) alphabet Λ, that is, Σ consists of the terms generated by the grammar $g, h ::= u \mid \mathsf{a} \in \Lambda \mid g \circ h \mid \overline{g}$. Furthermore, let $\equiv_G$ be the smallest congruence generated by the group axioms, i.e., for all $g, h, i \in \Lambda$ it holds that

$$g \circ (h \circ i) \equiv_G (g \circ h) \circ i \qquad g \circ u \equiv_G g \equiv_G u \circ g \qquad \overline{g} \circ g \equiv_G u \equiv_G g \circ \overline{g}$$

Lastly, let $\mathsf{group} = \{g \leq h : g \equiv_G h\}$. We can then define a reduction from group to $\emptyset$ by replacing every letter (group term) in a term e with its reduced form, that is, with the (unique) equivalent group term of minimum size. For instance, if $\Lambda = \{\mathsf{a}, \mathsf{b}, \mathsf{c}\}$, then we send the term $\mathsf{a} \circ \overline{\mathsf{a}} \parallel \mathsf{b} \circ \mathsf{c} \circ \overline{\mathsf{c}}$ to the term $u \parallel \mathsf{b}$.

For the remainder of this section, we fix a subalphabet $\Gamma \subseteq \Sigma$. When $r : \Sigma \to \mathcal{T}(\Gamma)$, we extend r to a map from $\mathcal{T}(\Sigma)$ to $\mathcal{T}(\Gamma)$, by inductively applying r to terms. We can also apply r to a series-parallel pomset, obtaining a pomset language. More precisely, when U is a pomset, we define $r(U)$ as follows:

$$r(1) = \{1\} \quad r(U \cdot V) = r(U) \cdot r(V) \quad r(\mathsf{a}) = [\![r(\mathsf{a})]\!] \quad r(U \parallel V) = r(U) \parallel r(V)$$

Lastly, when $L \subseteq \mathsf{SP}$, we write $r(L)$ for the set $\bigcup\{r(U) : U \in L\}$.

The following then formalises the idea of reducing by replacing letters.

Definition 4.19. *A map $r : \Sigma \to \mathcal{T}(\Gamma)$ is a* reification *from H to H' if*

(i) For all $\mathsf{a} \in \Sigma$, it holds that $r(\mathsf{a}) \equiv^H \mathsf{a}$.
(ii) r is expansive on Γ, i.e., for all $\mathsf{a} \in \Gamma$, $\mathsf{a} \leqq r(\mathsf{a})$.
(iii) H'-closure preserves Γ, i.e., for all $L \subseteq \mathsf{SP}(\Gamma)$, also $L{\downarrow}^{H'} \subseteq \mathsf{SP}(\Gamma)$.
(iv) For all $e \leq f \in H$, it holds that $r(e) \leqq^{H'} r(f)$.

Example 4.20. Continuing with the previous example, let r be the map that sends a group term to its reduced form; we claim that r is a reification from group to $\emptyset$. By definition, we then know that for a group term $g \in \Sigma$, we have $r(g) \equiv_G g$, and hence $r(g) \equiv^{\mathsf{group}} g$. Furthermore, the reduction of a reduced term is that term itself; hence, the second condition is satisfied. The third condition holds trivially. Lastly, if $e \leq f \in \mathsf{group}$, then $e, f \in \Sigma$ such that $e \equiv_G f$. Since reductions are unique, we then know that $r(e) = r(f)$, and hence $r(e) \leqq^{\emptyset} r(f)$.

We have the following general properties of a map r, which we will use in demonstrating how to obtain a reduction from a reification.

Lemma 4.21. *Let $r : \Sigma \to \mathcal{T}$ be some map.*

(i) For all $C \in \mathsf{PC}^{\mathsf{sp}}$, we have $r(C) \subseteq \mathsf{PC}^{\mathsf{sp}}$.
(ii) For all $L \subseteq \mathsf{SP}$ and $C \in \mathsf{PC}^{\mathsf{sp}}$, we have $r(C[L]) = \bigcup_{D \in r(C)} D[r(L)]$.
(iii) For all $e \in \mathcal{T}$, it holds that $r(\llbracket e \rrbracket) = \llbracket r(e) \rrbracket$.

The following technical lemma is a consequence of property (iv).

Lemma 4.22. *If r is a reification and $L \subseteq \mathsf{SP}(\Sigma)$, then $r(L{\downarrow}^H) \subseteq r(L){\downarrow}^{H'}$.*

Using this, we can then show how to obtain a reduction from a reification.

Lemma 4.23. *If H implies H' and r is a reification from H to H', then r is a reduction from H to H'.*

Proof. The first condition, i.e., that for $e \in \mathcal{T}$ we have $e \equiv^H r(e)$, can be checked using the first property of reification by induction on the structure of e. It thus remains to check the second condition; we do this by proving that for all $e \in \mathcal{T}(\Sigma)$ we have $r\left(\llbracket e \rrbracket{\downarrow}^H\right) = \llbracket r(e) \rrbracket{\downarrow}^{H'}$. To this end, we derive as follows:

$$
\begin{aligned}
r(\llbracket e \rrbracket{\downarrow}^H) &\subseteq r(\llbracket e \rrbracket){\downarrow}^{H'} && \text{(Lemma 4.22)}\\
&= \llbracket r(e) \rrbracket{\downarrow}^{H'} && \text{(Lemma 4.21(iii))}\\
&\subseteq r(\llbracket r(e) \rrbracket{\downarrow}^{H'}) && \text{(property (ii))}\\
&\subseteq r(\llbracket r(e) \rrbracket{\downarrow}^{H}) && \text{(Lemma 4.10(ii))}\\
&= r(\llbracket e \rrbracket{\downarrow}^{H}) && \text{(property (i), soundness)}
\end{aligned}
$$

Specifically, in the third step, property (ii) ensures that for $L \subseteq \mathsf{SP}(\Gamma)$ we have $L \subseteq r(L)$. We can use this property because H'-closure preserves the Γ-language by property (iii). This completes the proof. □

4.2 Factoring the exchange law

In the basic axioms that generate $\equiv$, there is no interaction between sequential and parallel composition. One sensible way of adding that kind of interaction is, as suggested by Hoare, Struth and collaborators [11], by adding an axiom of the form $(e \parallel f) \cdot (g \parallel h) \leqq (e \cdot g) \parallel (f \cdot h)$, known as the *exchange law*. Essentially, this axiom encodes the possibility of (partial) interleaving: when $e \cdot g$ runs in parallel with $f \cdot h$, one possible behaviour is that, first e runs in parallel with f, and then g runs in parallel with h. The core observation of this section is that the exchange law can be treated as another set of hypotheses, as we show below, and this can then be used to recover the completeness result of CKA [15].

Definition 4.24. *We write* exch *for the set*

$$\{(e \parallel f) \cdot (g \parallel h) \leq (e \cdot g) \parallel (f \cdot h) : e, f, g, h \in \mathcal{T}\}$$

The semantic effect of adding exch to our hypotheses is that, if U is a pomset in a series-parallel language L, and V is a series-parallel pomset subsumed by U, then V is in the exch-closure of L. Intuitively, the exch-closure adds pomsets that are more sequential, i.e., have more ordering, than the ones already in L. Indeed, exch-closure coincides with the downward closure w.r.t. $\sqsubseteq^{\mathsf{sp}}$.

Lemma 4.25. *Let* $L \subseteq \mathsf{SP}$ *and* $U \in \mathsf{SP}$. *Now* $U \in L{\downarrow}^{\mathsf{exch}}$ *if and only if there exists a* $V \in L$ *such that* $U \sqsubseteq^{\mathsf{sp}} V$.

We have previously shown that exch is complete [15]; as a matter of fact, the pivotal result from op. cit. can be presented as follows.

Theorem 4.26. *The set of hypotheses* exch *is strongly reducible to* $\emptyset$.

When exch is contained in our hypotheses, it is not immediately clear whether those hypotheses can be reduced. What we can do is try to factorise our hypotheses into exch and some residual set of hypotheses, and prove strong reducibility for that residual set. To this end, we first note that, in some circumstances, the H-closure of the exch-closure remains downward-closed w.r.t. $\sqsubseteq^{\mathsf{sp}}$.

Lemma 4.27. *Suppose that for each* $e \leq f \in H$ *we have that* $e = 1$ *or* $e = \mathtt{a}$ *for some* $\mathtt{a} \in \Sigma$, *and let* $L \subseteq \mathsf{SP}$. *If* $U, V \in \mathsf{SP}$ *such that* $U \sqsubseteq^{\mathsf{sp}} V$ *and* $V \in (L{\downarrow}^{\mathsf{exch}}){\downarrow}^{H}$, *then* $U \in (L{\downarrow}^{\mathsf{exch}}){\downarrow}^{H}$.

Using this fact, we can now show that, under the same precondition, $\mathsf{exch} \cup H$ factors into exch and H. This factorisation is what we were looking for: it tells us that whenever H strongly reduces to $\emptyset$, so does $H \cup \mathsf{exch}$.

Lemma 4.28. *Suppose that for each* $e \leq f \in H$ *we have that* $e = 1$, *or* $e = \mathtt{a}$ *for some* $\mathtt{a} \in \Sigma$. *Then* $H \cup \mathsf{exch}$ *factorises into* exch *and* H.

Proof. Since $H, \mathsf{exch} \subseteq H \cup \mathsf{exch}$, it should be obvious that $H \cup \mathsf{exch}$ implies both H and exch. It remains to show that, if $L \subseteq \mathsf{SP}$, then $(L{\downarrow}^{\mathsf{exch}}){\downarrow}^{H} = L{\downarrow}^{H \cup \mathsf{exch}}$. The inclusion from left to right is a consequence of Lemma 4.10(ii)–(iii).

For the other inclusion, we show that if $A \subseteq L{\downarrow}^{H \cup \mathsf{exch}}$, then $A \subseteq (L{\downarrow}^{\mathsf{exch}}){\downarrow}^{H}$. The proof proceeds by induction on the construction of $A \subseteq L{\downarrow}^{H \cup \mathsf{exch}}$. In the base, we have that $A \subseteq L{\downarrow}^{H \cup \mathsf{exch}}$ because $A = L$; in that case, $A \subseteq L{\downarrow}^{\mathsf{exch}} \subseteq (L{\downarrow}^{\mathsf{exch}}){\downarrow}^{H}$.

For the inductive step, $A \subseteq L{\downarrow}^{H \cup \mathsf{exch}}$ because there exist $e \leq f \in H \cup \mathsf{exch}$ and $C \in \mathsf{PC}^{\mathsf{sp}}$ such that $A = C[\llbracket e \rrbracket]$, and $C[\llbracket f \rrbracket] \subseteq L{\downarrow}^{H \cup \mathsf{exch}}$. By induction, we then know that $C[\llbracket f \rrbracket] \subseteq \left(L{\downarrow}^{\mathsf{exch}}\right){\downarrow}^{H}$. On the one hand, if $e \leq f \in H$, then $A = C[\llbracket e \rrbracket] \subseteq \left(L{\downarrow}^{\mathsf{exch}}\right){\downarrow}^{H}$ immediately. On the other hand, if $e \leq f \in \mathsf{exch}$, then $\llbracket e \rrbracket \sqsubseteq^{\mathsf{sp}} \llbracket f \rrbracket$, and hence $C[\llbracket e \rrbracket] \sqsubseteq^{\mathsf{sp}} C[\llbracket f \rrbracket]$ by Lemma 3.2. By Lemma 3.5 and Lemma 4.27, it then follows that $A = C[\llbracket e \rrbracket] \subseteq (L{\downarrow}^{\mathsf{exch}}){\downarrow}^{H}$. □

4.3 Lifting

A number of reduction procedures already exist at the level of Kleene algebra [20,7]; ideally, one would like to lift those procedures to CKA.

Example 4.29. The reductions in Example 4.13 and Example 4.18 worked out for terms without $\parallel$, and then extended inductively, by defining the reduction of $e \parallel f$ to be the parallel composition of the reductions of e and f respectively.

As a non-example, consider $H = \{\mathsf{a} \leq 1\}$. Even though this hypothesis can be reduced to $\emptyset$ within Kleene algebra [5], it is not obvious how this would work for pomset languages. In particular, if $1 \in L$, then $1 \parallel \cdots \parallel 1 \in L$ for any number of 1's, and hence $\mathsf{a} \parallel \cdots \parallel \mathsf{a} \in L{\downarrow}^H$ for any number of a's. This precludes the possibility of a strong reduction to $\emptyset$, because $[\![1]\!]{\downarrow}^H$ is a pomset language of unbounded (parallel) width, which cannot be expressed by any $e \in \mathcal{T}$ [25].

We now establish a set of sufficient conditions for such a lifting to work. To this end, we first formally define Kleene algebra syntax, axioms and semantics.

Definition 4.30. *Write $\mathcal{T}_{\mathsf{KA}}$ for the set of* Kleene algebra terms, *i.e., the terms in $\mathcal{T}$ that do not contain $\parallel$. Furthermore, we write $\equiv_{\mathsf{KA}}$ for the smallest congruence on $\mathcal{T}_{\mathsf{KA}}$ that is generated by the axioms of $\equiv$ that do not involve $\parallel$.*

When $e \in \mathcal{T}_{\mathsf{KA}}$, it is not hard to see that $[\![e]\!]$ contains totally ordered pomsets, i.e., words, exclusively. Using these definitions, we can now specialise the notions of hypotheses, context, and closure to the sequential setting, as follows.

Definition 4.31. *The relation $\equiv^H_{\mathsf{KA}}$ is generated from H and $\equiv_{\mathsf{KA}}$ as before.*

A context $C \in \mathsf{PC}^{\mathsf{sp}}$ is sequential *if it is totally ordered, i.e., if it is a word with one occurrence of $*$; we write $\mathsf{PC}^{\mathsf{seq}}$ for the set of sequential contexts.*

Given a set of hypotheses H and a language $L \subseteq \Sigma^*$, we define the sequential closure *of L with respect to H, written $L{\downarrow}^H_{\mathsf{seq}}$, as the least language containing L such that for all $e \leq f \in H$ and $C \in \mathsf{PC}^{\mathsf{seq}}$, if $C[[\![f]\!]] \subseteq L{\downarrow}^H_{\mathsf{seq}}$, then $C[[\![e]\!]] \subseteq L{\downarrow}^H_{\mathsf{seq}}$.*

If $\parallel$ does not occur in any hypothesis, then the definition of sequential closure coincides with the closure operator from [7]. Thus, if $L \subseteq \Sigma^*$, then $L{\downarrow}^H_{\mathsf{seq}} \subseteq \Sigma^*$.

The analogue of strong reduction for the sequential setting is as follows.

Definition 4.32. *Suppose that H implies H'. A map $r : \mathcal{T}_{\mathsf{KA}} \to \mathcal{T}_{\mathsf{KA}}$ is a* sequential reduction *from H to H' when the following hold:*

(i) for $e \in \mathcal{T}_{\mathsf{KA}}$, it holds that $e \equiv^H_{\mathsf{KA}} r(e)$, and
(ii) for $e \in \mathcal{T}_{\mathsf{KA}}$, it holds that $[\![e]\!]_{\mathsf{KA}}{\downarrow}^H_{\mathsf{seq}} = [\![r(e)]\!]_{\mathsf{KA}}{\downarrow}^{H'}_{\mathsf{seq}}$.

H sequentially reduces *to H' if there exists a sequential reduction from H to H'.*

To lift a sequential reduction to a proper reduction, the following class of hypotheses will turn out to be useful.

Definition 4.33. *A hypothesis $e \leq f$ with $e, f \in \mathcal{T}_{\mathsf{KA}}$ is called* grounded *if $[\![f]\!] = \{W\}$ for some non-empty word (totally ordered pomset) W, and $e \in \mathcal{T}_{\mathsf{KA}}$. We say that a set of hypotheses H is grounded if every $e \leq f \in H$ is grounded.*

Example 4.34. Any hypothesis of the form $e \leq \mathsf{a}_1 \cdots \mathsf{a}_n$ for $n > 0$ is grounded. On the other hand, the hypothesis $\mathsf{a} \leq 1$ that we saw in the previous example is not grounded, since the semantics of 1 contains the empty pomset.

The closure of a language of words can be expressed in terms of its sequential closure, provided that the set of hypotheses is grounded.

Lemma 4.35. *Let H be grounded. If $L \subseteq \Sigma^*$, then $L{\downarrow}^H = L{\downarrow}^H_{\mathsf{seq}}$. Moreover, for $L, L' \subseteq \mathsf{SP}$, we have that $(L \parallel L'){\downarrow}^H = L{\downarrow}^H \parallel L'{\downarrow}^H$.*

The above then allows us to turn a sequential reduction into a reduction.

Lemma 4.36. *Suppose that H sequentially reduces to H'. If H and H' are grounded, then H strongly reduces to H'.*

5 Instantiation to CKA with Observations

In this section, we will present Concurrent Kleene Algebra with Observations (CKAO), an extension of CKA with Boolean assertions that enable the specification of programs with the usual guarded conditionals and loops. We will obtain CKAO as an instance of CKAH by choosing a particular set of hypotheses. First, we define the set of propositional terms or Boolean observations.

Definition 5.1. *Fix a finite set Ω of* primitive observations. *The set of* propositional terms, *written $\mathcal{T}_{\mathsf{BA}}$, is generated by*

$$p, q ::= \bot \mid \top \mid o \in \Omega \mid p \vee q \mid p \wedge q \mid \overline{p}$$

The relation $\equiv_{\mathsf{BA}}$ is the smallest congruence on $\mathcal{T}_{\mathsf{BA}}$ s.t. for $p, q, r \in \mathcal{T}_{\mathsf{BA}}$, we have

$$p \vee \bot \equiv_{\mathsf{BA}} p \qquad p \vee q \equiv_{\mathsf{BA}} q \vee p \qquad p \vee \overline{p} \equiv_{\mathsf{BA}} \top \qquad p \vee (q \vee r) \equiv_{\mathsf{BA}} (p \vee q) \vee r$$

$$p \wedge \top \equiv_{\mathsf{BA}} p \qquad p \wedge q \equiv_{\mathsf{BA}} q \wedge p \qquad p \wedge \overline{p} \equiv_{\mathsf{BA}} \bot \qquad p \wedge (q \wedge r) \equiv_{\mathsf{BA}} (p \wedge q) \wedge r$$

$$p \vee (q \wedge r) \equiv_{\mathsf{BA}} (p \vee q) \wedge (p \vee r) \qquad p \wedge (q \vee r) \equiv_{\mathsf{BA}} (p \wedge q) \vee (p \wedge r)$$

We will write $p \leqq_{\mathsf{BA}} q$ as a shorthand for $p \vee q \equiv_{\mathsf{BA}} q$.

We write At for 2^Ω, the set of *atoms* of the Boolean algebra. It is well known that every $\alpha \in \mathsf{At}$ corresponds canonically to a Boolean term π_α, such that every Boolean term $p \in \mathcal{T}_{\mathsf{BA}}$ is equivalent to the disjunction of all π_α with $\pi_\alpha \leqq_{\mathsf{BA}} p$ [2]. To simplify notation we identify $\alpha \in \mathsf{At}$ with π_α.

We can now use $\mathcal{T}_{\mathsf{BA}}$ in defining the terms and axioms of CKAO, which will be given as a CKA over a specific alphabet with the following hypotheses:

Definition 5.2 (CKAO). *We define the* terms *of CKAO, denoted $\mathcal{T}_{\mathsf{CKAO}}$, as $\mathcal{T}(\Sigma \cup \mathcal{T}_{\mathsf{BA}})$, that is, as the CKA terms over $\mathcal{T}_{\mathsf{BA}} \cup \Sigma$. We furthermore define the following set of hypotheses over $\mathcal{T}_{\mathsf{CKAO}}$:*

$$\mathsf{bool} = \{p = q : p, q \in \mathcal{T}_{\mathsf{BA}} \text{ s.t. } p \equiv_{\mathsf{BA}} q\} \qquad \mathsf{contr} = \{p \wedge q \leq p \cdot q : p, q \in \mathcal{T}_{\mathsf{BA}}\}$$

$$\mathsf{glue} = \{0 = \bot\} \cup \{p + q = p \vee q : p, q \in \mathcal{T}_{\mathsf{BA}}\} \qquad \mathsf{obs} = \mathsf{bool} \cup \mathsf{contr} \cup \mathsf{exch} \cup \mathsf{glue}$$

The semantics *of CKAO is then given by $[\![-]\!]{\downarrow}^{\mathsf{obs}}$.*

The hypotheses bool contain the boolean identities, and glue identifies the disjunction with the union (and their respective units as well). contr specifies that if p and q hold simultaneously, then it is possible to observe them in sequence. Note that the converse inequality is not included: observing p and q in sequence has strictly more behaviour than observing p and q simultaneously, as some intervening action can happen between the two observations.

The above definition gives us the semantics of CKAO as the standard pomset language model obtained from taking the obs-closure of the semantics of CKA. As a matter of fact, we find by Lemma 4.7 that if $e, f \in \mathcal{T}_{\mathsf{CKAO}}$ with $e \equiv^{\mathsf{obs}} f$, then $[\![e]\!]{\downarrow}^{\mathsf{obs}} = [\![f]\!]{\downarrow}^{\mathsf{obs}}$; hence, we already have a sound model of CKAO.

To prove completeness, we will use the techniques from the previous section.

First step: reification. We start by using reification to rid ourselves of the hypotheses from bool and glue, and to simplify the hypotheses in contr. To this end, let contr' be the set of hypotheses given by $\{\alpha \leq \alpha \cdot \alpha : \alpha \in \mathsf{At}\}$. Let $\Gamma = \mathsf{At} \cup \Sigma \subseteq \mathcal{T}_{\mathsf{BA}} \cup \Sigma$. We define $r : \Sigma \cup \mathcal{T}_{\mathsf{BA}} \to \mathcal{T}(\Gamma)$ by setting

$$r(a) = \begin{cases} \sum_{\alpha \leqq_{\mathsf{BA}} p} \alpha & a = p \in \mathcal{T}_{\mathsf{BA}} \\ \mathtt{a} & a = \mathtt{a} \in \Sigma \end{cases}$$

Lemma 5.3. *The hypotheses* obs *reduce to* $\mathsf{exch} \cup \mathsf{contr}'$.

Proof. By Lemma 4.23, it suffices to show that r is a reification, and that obs implies $\mathsf{exch} \cup \mathsf{contr}'$. To see that r is a reification, we check the conditions.

(i): If $\mathtt{a} \in \Sigma$, then $r(\mathtt{a}) = \mathtt{a} \equiv^{\mathsf{obs}} \mathtt{a}$ immediately. Otherwise, if $p \in \mathcal{T}_{\mathsf{BA}}$, then we derive $r(p) = \sum_{\alpha \leqq_{\mathsf{BA}} p} \alpha \equiv^{\mathsf{glue}} \bigvee_{\alpha \leqq_{\mathsf{BA}} p} \alpha \equiv^{\mathsf{bool}} p$ and hence $r(p) \equiv^{\mathsf{obs}} p$.

(ii): If $\mathtt{a} \in \Sigma$, then we already know that $r(\mathtt{a}) = \mathtt{a}$. Otherwise, if $\alpha \in \mathsf{At}$, then

$$r(\alpha) = \sum_{\beta \leqq_{\mathsf{BA}} \alpha} \beta = \alpha$$

(iii): This property holds because all hypotheses in $\mathsf{exch} \cup \mathsf{contr}'$ preserve Γ-languages, i.e., if $e \leq f \in \mathsf{exch} \cup \mathsf{contr}'$ where $[\![f]\!] \subseteq \mathsf{SP}(\Gamma)$, then $[\![e]\!] \subseteq \mathsf{SP}(\Gamma)$ too. It follows that $\mathsf{exch} \cup \mathsf{contr}'$-closure must preserve Γ-languages.

(iv): We should show that if $e \leq f \in \mathsf{obs}$, then $r(e) \leqq^{\mathsf{exch} \cup \mathsf{contr}'} r(f)$. To this end, we analyse the separate sets of hypotheses that make up obs.

- Let $e \leq f \in \mathsf{exch}$, then $e = (g_{00} \parallel g_{01}) \cdot (g_{10} \parallel g_{11})$ and $f = (g_{00} \cdot g_{10}) \parallel (g_{01} \cdot g_{11})$, for some $g_{00}, g_{01}, g_{10}, g_{11} \in \mathcal{T}$. We then find that

$$r(e) = (r(g_{00}) \parallel r(g_{01})) \cdot (r(g_{10}) \parallel r(g_{11}))$$

$$r(f) = (r(g_{00}) \cdot r(g_{10})) \parallel (r(g_{01}) \cdot r(g_{11}))$$

  hence $r(e) \leq r(f) \in \mathsf{exch}$, and therefore $r(e) \leqq^{\mathsf{exch} \cup \mathsf{contr}'} r(f)$.
- Let $e \leq f \in \mathsf{bool}$, then $e = p$ and $f = q$ such that $p \equiv_{\mathsf{BA}} q$. In that case,

$$r(p) = \sum_{\alpha \leqq_{\mathsf{BA}} p} \alpha = \sum_{\alpha \leqq_{\mathsf{BA}} q} \alpha = r(q)$$

– Let $e \leq f \in \mathsf{contr}$; then $e = p \wedge q$ and $f = p \cdot q$ for $p, q \in \mathcal{T}_{\mathsf{BA}}$. Then

$$\begin{aligned} r(p \wedge q) &= \sum_{\alpha \leqq_{\mathsf{BA}} p \wedge q} \alpha \leqq^{\mathsf{contr}'} \sum_{\alpha \leqq_{\mathsf{BA}} p \wedge q} \alpha \cdot \alpha \\ &\leqq \Big(\sum_{\alpha \leqq_{\mathsf{BA}} p} \alpha \Big) \cdot \Big(\sum_{\alpha \leqq_{\mathsf{BA}} q} \alpha \Big) = r(p) \cdot r(q) = r(p \cdot q) \end{aligned}$$

– Let $e \leq f \in \mathsf{glue}$. On the one hand, if $e = p \vee q$ and $f = p + q$, then

$$r(p \vee q) = \sum_{\alpha \leqq_{\mathsf{BA}} p \vee q} \alpha \equiv \sum_{\alpha \leqq_{\mathsf{BA}} p} \alpha + \sum_{\alpha \leqq_{\mathsf{BA}} q} \alpha = r(p) + r(q) = r(p + q)$$

This also establishes the case for $f \leq e \in \mathsf{glue}$. On the other hand, if $e = 0$ and $p = \bot$, then $r(0) = 0 = \sum_{\alpha \leqq_{\mathsf{BA}} \bot} \alpha = r(\bot)$.

To see that obs implies $\mathsf{exch} \cup \mathsf{contr}'$, it suffices to show that obs implies contr'. To this end, note that if $e \leq f \in \mathsf{contr}'$, then $e = \alpha$ and $f = \alpha \cdot \alpha$ for some $\alpha \in \mathsf{At}$. We can then derive that $\alpha \equiv^{\mathsf{bool}} \alpha \wedge \alpha \leqq^{\mathsf{contr}} \alpha \cdot \alpha$, and hence $e \leqq^{\mathsf{obs}} f$. □

Second step: factorising. Since contr' satisfies the precondition of Lemma 4.28, we obtain the following.

Lemma 5.4. *The hypotheses* $\mathsf{exch} \cup \mathsf{contr}'$ *factorise into* exch *and* contr'.

This means that, by Lemma 4.17 all that remains to do is strongly reduce exch and contr' to $\emptyset$; we have already taken care of the former in Theorem 4.26.

Third step: reducing contr'. In [13], we have already shown that contr' sequentially reduces to $\emptyset$. Since contr' is grounded we find the following, by Lemma 4.36.

Lemma 5.5. *The hypotheses* contr' *strongly reduce to* $\emptyset$.

Last step: putting it all together. Using the above reductions, we can then prove completeness of $\equiv^{\mathsf{obs}}$ w.r.t. $[\![-]\!]{\downarrow}^{\mathsf{obs}}$, and decidability of semantic equivalence, too.

Theorem 5.6 (Soundness and Completeness of CKAO). *Let* $e, f \in \mathcal{T}_{\mathsf{CKAO}}$.

(i) We have $e \equiv^{\mathsf{obs}} f$ *if and only if* $[\![e]\!]{\downarrow}^{\mathsf{obs}} = [\![f]\!]{\downarrow}^{\mathsf{obs}}$.
(ii) It is decidable whether $[\![e]\!]{\downarrow}^{\mathsf{obs}} = [\![f]\!]{\downarrow}^{\mathsf{obs}}$.

Proof. For the first claim, we already knew the implication from left to right from Lemma 4.7. Conversely, and for the second claim, first note that that obs reduces to $\mathsf{exch} \cup \mathsf{contr}'$ by Lemma 5.3. By Lemma 5.4 and Lemma 4.17, the latter reduces to $\emptyset$, if we apply Theorem 4.26 and Lemma 5.5. By Lemma 4.12, we then conclude that obs is complete and decidable, hence establishing the claim. □

6 Discussion

The first contribution of this paper is to extend Kleene algebra with hypotheses [7] with a parallel operator. The resulting framework, concurrent Kleene algebra with hypotheses (CKAH), is interpreted over pomset languages, a standard model of concurrency. We start from simple axioms, known to capture equality of pomset languages [23]. CKAH allows to add custom axioms, the so-called hypotheses. These may be used to include domain-specific information in the language. We develop this framework by providing a systematic way of producing from the hypotheses a sound pomset language model. We also propose techniques that may be used to prove completeness and decidability of the resulting model.

An important instance of this framework is concurrent Kleene algebra (CKA) as presented in [11]. The only additional axiom there, known as the exchange law, may be added as a set of hypotheses. We prove that the resulting semantics coincides with the (subsumption-closed) semantics of CKA and, more interestingly, the completeness proof of [15] can be recovered as an instance of this framework.

The second contribution is a new framework to reason about programs with concurrency: concurrent Kleene algebra with observations (CKAO). CKAO is obtained as an instance of CKAH, where we add the exchange law to model concurrent behaviour, and Boolean assertions to model control flow. The Boolean assertions we consider are as in Kleene algebra with observations (KAO) [13] — in fact, CKAO is a conservative extension of KAO. Using the techniques developed earlier, we obtain a sound and complete semantics for this algebra. While CKAO is similar to concurrent Kleene algebra with tests [12], it avoids the problems of the latter by distinguishing conjunction and sequential composition. CKAO provides the first sound and complete algebraic theory that seems sensible as a framework to reason about concurrent programs with Boolean assertions.

Future work is to explore other meaningful instances of CKAH. Synchronous Kleene algebra [29,26] is a natural candidate for this. We also want to try and design domain specific languages, specifically, a concurrent variant of NetKAT [1,8].

The class of hypotheses considered in this paper for which decidability and completeness may be established systematically is somewhat restrictive; identifying larger classes of tractable hypotheses is a challenging open problem.

Because of the compositional nature of our model, the CKAO semantics of a program contains behaviours that are not possible to obtain in isolation. These behaviours are present to allow the program to interact meaningfully with its environment, i.e., when placed in a context. However, for practical purposes one might want to close the system, and only consider behaviours that are possible in isolation. Studying this semantics remains subject of future work.

In the semantics of concurrent programs with assertions, it would be natural to see atoms as partial instead of total functions. This captures the intuition that a thread might not have access to the complete machine state, but instead holds a partial view of it. Pseudo-complemented distributive lattices (PCDL) have been proposed [12] as an alternative to Boolean algebra, modelling this partiality of information. We leave it to future work to investigate the variant of CKAO obtained by replacing the Boolean algebra of observations with a PCDL.

References

1. Anderson, C.J., Foster, N., Guha, A., Jeannin, J.B., Kozen, D., Schlesinger, C., Walker, D.: NetKAT: Semantic foundations for networks. In: POPL. pp. 113–126. ACM (2014)
2. Birkhoff, G., Bartee, T.C.: Modern applied algebra. McGraw-Hill (1970)
3. Bonchi, F., Pous, D.: Checking NFA equivalence with bisimulations up to congruence. In: POPL. pp. 457–468 (2013)
4. Brunet, P., Pous, D., Struth, G.: On decidability of concurrent Kleene algebra. In: CONCUR. pp. 28:1–28:15 (2017)
5. Cohen, E.: Hypotheses in Kleene algebra. Tech. rep., Bellcore (1994)
6. Conway, J.H.: Regular Algebra and Finite Machines. Chapman and Hall, Ltd., London (1971)
7. Doumane, A., Kuperberg, D., Pous, D., Pradic, P.: Kleene algebra with hypotheses. In: FOSSACS. pp. 207–223 (2019)
8. Foster, N., Kozen, D., Milano, M., Silva, A., Thompson, L.: A coalgebraic decision procedure for NetKAT. In: POPL. pp. 343–355 (2015)
9. Gischer, J.L.: The equational theory of pomsets. Theor. Comput. Sci. 61, 199–224 (1988)
10. Grabowski, J.: On partial languages. Fundam. Inform. 4(2), 427 (1981)
11. Hoare, T., Möller, B., Struth, G., Wehrman, I.: Concurrent Kleene algebra. In: CONCUR. pp. 399–414 (2009)
12. Jipsen, P., Moshier, M.A.: Concurrent Kleene algebra with tests and branching automata. J. Log. Algebr. Meth. Program. 85(4), 637–652 (2016)
13. Kappé, T., Brunet, P., Rot, J., Silva, A., Wagemaker, J., Zanasi, F.: Kleene algebra with observations. In: CONCUR. pp. 41:1–41:16 (2019)
14. Kappé, T., Brunet, P., Silva, A., Wagemaker, J., Zanasi, F.: Concurrent Kleene algebra with observations: from hypotheses to completeness (2020), arXiv:2002.09682
15. Kappé, T., Brunet, P., Silva, A., Zanasi, F.: Concurrent Kleene algebra: Free model and completeness. In: ESOP. pp. 856–882 (2018)
16. Kozen, D.: A completeness theorem for Kleene algebras and the algebra of regular events. Inf. Comput. 110(2), 366–390 (1994)
17. Kozen, D.: Kleene algebra with tests and commutativity conditions. In: TACAS. pp. 14–33 (1996)
18. Kozen, D.: On the complexity of reasoning in Kleene algebra. Inf. Comput. 179(2), 152–162 (2002)
19. Kozen, D.: On the coalgebraic theory of Kleene algebra with tests. In: Başkent, C., Moss, L.S., Ramanujam, R. (eds.) Rohit Parikh on Logic, Language and Society, Outstanding Contributions to Logic, vol. 11, pp. 279–298. Springer (2017)
20. Kozen, D., Mamouras, K.: Kleene algebra with equations. In: ICALP. pp. 280–292 (2014)
21. Krob, D.: A complete system of B-rational identities. In: ICALP. pp. 60–73 (1990)
22. Kuratowski, C.: Sur l'opération $\bar{A}$ de l'Analysis Situs. Fundamenta Mathematicae 3(1), 182–199 (1922)
23. Laurence, M.R., Struth, G.: Completeness theorems for bi-Kleene algebras and series-parallel rational pomset languages. In: RAMiCS. pp. 65–82 (2014)
24. Laurence, M.R., Struth, G.: Completeness theorems for pomset languages and concurrent Kleene algebras (2017), arXiv:1705.05896
25. Lodaya, K., Weil, P.: Series-parallel languages and the bounded-width property. Theoretical Computer Science 237(1), 347–380 (2000)

26. Prisacariu, C.: Synchronous Kleene algebra. The Journal of Logic and Algebraic Programming 79(7), 608 – 635 (2010)
27. Salomaa, A.: Two complete axiom systems for the algebra of regular events. J. ACM 13(1), 158–169 (1966)
28. Smolka, S., Foster, N., Hsu, J., Kappé, T., Kozen, D., Silva, A.: Guarded Kleene algebra with tests: verification of uninterpreted programs in nearly linear time. In: POPL. pp. 61:1–61:28 (2020)
29. Wagemaker, J., Bonsangue, M., Kappé, T., Rot, J., Silva, A.: Completeness and incompleteness of synchronous Kleene algebra. In: MPC (2019)

Graded Algebraic Theories

Satoshi Kura[1,2]

[1] National Institute of Informatics, Tokyo, Japan
[2] The Graduate University for Advanced Studies (SOKENDAI), Kanagawa, Japan
kura@nii.ac.jp

Abstract. We provide graded extensions of algebraic theories and Lawvere theories that correspond to graded monads. We prove that graded algebraic theories, graded Lawvere theories, and finitary graded monads are equivalent via equivalence of categories, which extends the equivalence for monads. We also give sums and tensor products of graded algebraic theories to combine computational effects as an example of importing techniques based on algebraic theories to graded monads.

1 Introduction

In the field of denotational semantics of programming languages, monads have been used to express computational effects since Moggi's seminal work [18]. They have many applications from both theoretical and practical points of view.

Monads correspond to *algebraic theories* [5]. This correspondence gives natural presentations of many kinds of computational effects by operations and equations [21], which is the basis of algebraic effect [20]. The algebraic perspective of monads also provides ways of combining [9], reasoning about [22], and handling computational effects [23].

Graded monads [27] are a refinement of monads and defined as a monad-like structure indexed by a monoidal category (or a preordered monoid). The unit and multiplication of graded monads are required to respect the monoidal structure. This structure enables graded monads to express some kind of "abstraction" of effectful computations. For example, graded monads are used to give denotational semantics of effect systems [12], which are type systems designed to estimate scopes of computational effects caused by programs.

$$\frac{f \in \Sigma_{n,m} \qquad t_i \in T^{\Sigma}_{m'}(X) \text{ for each } i \in \{1, \dots, n\}}{f(t_1, \dots, t_n) \in T^{\Sigma}_{m \otimes m'}(X)}$$

Fig. 1. A rule of term formation.

This paper provides a *graded extension of algebraic theories* that corresponds to monads graded by small strict monoidal categories. This generalizes $\mathbb{N}$-graded theories in [17]. The main ideas of this extension are the following. First, we assign to each operation a *grade*, i.e., an object in a monoidal category that represents effects. Second, our extension provides a mechanism (Fig 1) to keep track of effects in the same way as graded monads. That is, if an operation f with grade m is applied to terms with grade m', then the grade of the whole term is the product $m \otimes m'$.

J. Goubault-Larrecq and B. König (Eds.): FOSSACS 2020, LNCS 12077, pp. 401–421, 2020.
https://doi.org/10.1007/978-3-030-45231-5_21

For example, graded algebraic theories enable us to estimate (an overapproximation of) the set of memory locations computations may access. The side-effects theory [21] is given by operations lookup_l and $\mathsf{update}_{l,v}$ for each location $l \in L$ and value $v \in V$ together with several equations, and each term represents a computation with side-effects. Since lookup_l and $\mathsf{update}_{l,v}$ only read from or write to the location l, we assign $\{l\} \in \mathbf{2}^L$ as the grade of the operations in the graded version of the side-effects theory where $\mathbf{2}^L$ is the join-semilattice of subsets of locations L. The grade of a term is (an overapproximation of) the set of memory locations the computations may access thanks to the rule in Fig 1.

We also provide *graded Lawvere theories* that correspond to graded algebraic theories. The intuition of a Lawvere theory is a category whose arrows are terms of an algebraic theory. We use this intuition to define graded Lawvere theories. In graded algebraic theories, each term has a grade, and substitution of terms must respect the monoidal structure of grades. To characterize this structure of "graded" terms, we consider Lawvere theories enriched in a presheaf category.

Like algebraic theories brought many concepts and techniques to the semantics of computational effects, we expect that the proposed graded algebraic theories will do the same for effect systems. We look into one example out of such possibilities: combining graded algebraic theories.

The main contributions of this paper are summarized as follows.

- We generalize ($\mathbb{N}$-)graded algebraic theories of [17] to $\mathbf{M}$-graded algebraic theories and also provide $\mathbf{M}$-graded Lawvere theories where $\mathbf{M}$ is a small strict monoidal category. We show that there exist translations between these notions and finitary graded monads, which yield equivalences of categories.
- We extend sums and tensor products of algebraic theories [9] to graded algebraic theories. We define sums in the category of $\mathbf{M}$-graded algebraic theories, and tensor products as an $\mathbf{M} \times \mathbf{M}'$-graded algebraic theory made from an $\mathbf{M}$-graded and an $\mathbf{M}'$-graded algebraic theory. We also show a few properties and examples of these constructions.

2 Preliminaries

2.1 Enriched Category Theory

We review enriched category theory and introduce notations. See [13] for details.

Let $\mathbf{V}_0 = (\mathbf{V}_0, \otimes, I)$ be a (not necessarily symmetric) monoidal category. $\mathbf{V}_0$ is *right closed* if $(-) \otimes X : \mathbf{V}_0 \to \mathbf{V}_0$ has a right adjoint $[X, -]$ for each $X \in \mathrm{ob}\mathbf{V}_0$. Similarly, $\mathbf{V}_0$ is *left closed* if $X \otimes (-)$ has a right adjoint $[\![X, -]\!]$ for each $X \in \mathrm{ob}\mathbf{V}_0$. $\mathbf{V}_0$ is *biclosed* if $\mathbf{V}_0$ is left and right closed.

Let ${\mathbf{V}_0}^{\mathrm{t}}$ denote the monoidal category $(\mathbf{V}_0, \otimes^{\mathrm{t}}, I)$ where $\otimes^{\mathrm{t}}$ is defined by $X \otimes^{\mathrm{t}} Y := Y \otimes X$. Note that ${\mathbf{V}_0}^{\mathrm{t}}$ is right closed if and only if $\mathbf{V}_0$ is left closed.

We define $\mathbf{V}_0$*-category*, $\mathbf{V}_0$*-functor* and $\mathbf{V}_0$*-natural transformation* as in [13].

If $\mathbf{V}_0$ is right closed, then $\mathbf{V}_0$ itself enriches to a $\mathbf{V}_0$-category $\mathbf{V}$ with hom-object given by $\mathbf{V}(X, Y) := [X, Y]$. We use the subscript $(-)_0$ to distinguish the enriched category $\mathbf{V}$ from its underlying category $\mathbf{V}_0$.

Assume that $\mathbf{V}_0$ is biclosed and let $\mathbf{A}$ be a $\mathbf{V}_0$-category. The *opposite category* $\mathbf{A}^{\mathrm{op}}$ is the ${\mathbf{V}_0}^{\mathrm{t}}$-category defined by $\mathbf{A}^{\mathrm{op}}(X,Y) = \mathbf{A}(Y,X)$. For any $X \in \mathrm{ob}\mathbf{A}$, $\mathbf{A}(X,-) : \mathbf{A} \to \mathbf{V}_0$ is a $\mathbf{V}_0$-functor where $\mathbf{A}(X,-)_{Y,Z} : \mathbf{A}(Y,Z) \to [\mathbf{A}(X,Y), \mathbf{A}(X,Z)]$ is defined by transposing the composition law $\overline{\circ}$ of $\mathbf{A}$. A ${\mathbf{V}_0}^{\mathrm{t}}$-functor $\mathbf{A}(-,X)$ is defined by $\mathbf{A}^{\mathrm{op}}(X,-) : \mathbf{A}^{\mathrm{op}} \to {\mathbf{V}_0}^{\mathrm{t}}$.

Let $\mathbf{A}$ be a $\mathbf{V}_0$-category. For each $X \in \mathbf{V}_0$ and $C \in \mathbf{A}$, a *tensor* $X \otimes C$ is an object in $\mathbf{A}$ together with a counit morphism $\nu : X \to \mathbf{A}(C, X \otimes C)$ such that a $\mathbf{V}_0$-natural transformation $\mathbf{A}(X \otimes C, -) \to [\![X, \mathbf{A}(C,-)]\!]$ obtained by transposing $(\overline{\circ}) \circ (\mathbf{A}(X \otimes C, B) \otimes \nu)$ is isomorphic where $\overline{\circ}$ is the composition in the $\mathbf{V}_0$-category $\mathbf{A}$. A *cotensor* $X \pitchfork C$ is a tensor in $\mathbf{A}^{\mathrm{op}}$. For example, if $\mathbf{V}_0 = \mathbf{Set}$, then tensors $X \otimes C$ are copowers $X \cdot C$, and cotensors $X \pitchfork C$ are powers C^X.

A $\mathbf{V}_0$-functor $F : \mathbf{A} \to \mathbf{B}$ is said to preserve a tensor $X \otimes C$ if $F_{C, X \otimes C} \circ \nu : X \to \mathbf{B}(FC, F(X \otimes C))$ is again a counit morphism. F preserves cotensors if F^{op} preserves tensors.

Let Φ be a collection of objects in $\mathbf{V}_0$. A $\mathbf{V}_0$-functor $F : \mathbf{A} \to \mathbf{B}$ is said to preserve Φ-(co)tensors if F preserves (co)tensors of the form $X \otimes C$ ($X \pitchfork C$) for each $X \in \Phi$ and $C \in \mathrm{ob}\mathbf{A}$.

2.2 Graded Monads

We review the notion of graded monad in [7, 12], and then define the category $\mathbf{GMnd}_{\mathbf{M}}$ of finitary $\mathbf{M}$-graded monads. Throughout this section, we fix a small strict monoidal category $\mathbf{M} = (\mathbf{M}, \otimes, I)$.

Definition 1 (graded monads). An $\mathbf{M}$-*graded monad* on $\mathbf{C}$ is a lax monoidal functor $\mathbf{M} \to [\mathbf{C}, \mathbf{C}]$ where $[\mathbf{C}, \mathbf{C}]$ is a monoidal category with composition as multiplication. That is, an $\mathbf{M}$-graded monad is a tuple $(*, \eta, \mu)$ of a functor $* : \mathbf{M} \times \mathbf{C} \to \mathbf{C}$ and natural transformations $\eta_X : X \to I * X$ and $\mu_{m_1, m_2, X} : m_1 * (m_2 * X) \to (m_1 \otimes m_2) * X$ such that the following diagrams commute.

$$\begin{array}{ccc} m * X & \xrightarrow{\eta} & I * (m * X) \\ {\scriptstyle m*\eta}\downarrow & \diagdown\!\!\diagdown & \downarrow{\scriptstyle \mu} \\ m * (I * X) & \xrightarrow[\mu]{} & m * X \end{array} \qquad \begin{array}{ccc} m_1 * (m_2 * (m_3 * X)) & \xrightarrow{m_1 * \mu} & m_1 * ((m_2 \otimes m_3) * X) \\ {\scriptstyle \mu}\downarrow & & \downarrow{\scriptstyle \mu} \\ (m_1 \otimes m_2) * (m_3 * X) & \xrightarrow[\mu]{} & (m_1 \otimes m_2 \otimes m_3) * X \end{array}$$

A *morphism of* $\mathbf{M}$-*graded monad* is a monoidal natural transformation $\alpha : (*, \eta, \mu) \to (*', \eta', \mu')$, i.e. a natural transformation $\alpha : * \to *'$ that is compatible with η and μ.

An intuition of graded monads is a refinement of monads: $m * X$ is a computation whose scope of effect is indicated by m and whose result is in X. The monoidal category $\mathbf{M}$ defines the granularity of the refinement, and a $\mathbf{1}$-graded monad is just an ordinary monad. Note that we do not assume that $\mathbf{M}$ is symmetric because some of graded monads in [12] require $\mathbf{M}$ to be nonsymmetric. We also deal with such a nonsymmetric case in Example 25.

A *finitary functor* is a functor that preserves filtered colimits. In this paper, we focus on finitary graded monads on $\mathbf{Set}$.

Definition 2. A *finitary* **M**-*graded monad on* **Set** is a lax monoidal functor $\mathbf{M} \to [\mathbf{Set}, \mathbf{Set}]_f$ where $[\mathbf{Set}, \mathbf{Set}]_f$ denotes the full subcategory of $[\mathbf{Set}, \mathbf{Set}]$ on finitary functors. Let $\mathbf{GMnd_M}$ denote the category of finitary **M**-graded monads and monoidal natural transformations between them.

A morphism in $\mathbf{GMnd_M}$ is determined by the restriction to $\aleph_0 \subseteq \mathbf{Set}$ where $\aleph_0$ is the full subcategory of **Set** on natural numbers.

Lemma 3. *Let $T = (*, \eta, \mu)$ and $T' = (*', \eta', \mu')$ be finitary* **M**-*graded monads. There exists one-to-one correspondence between the following.*

1. *Morphisms $\alpha : T \to T'$.*
2. *Natural transformations $\beta : * \circ (\mathbf{M} \times i) \to *' \circ (\mathbf{M} \times i)$ (where $i : \aleph_0 \to \mathbf{Set}$ is the inclusion functor) such that the following diagrams commute for each $n, n' \in \aleph_0$, $m_1, m_2 \in \mathbf{M}$ and $f : n \to m_2 * n'$.*

$$\begin{array}{ccc} n & \xrightarrow{\eta_n} & I * n \\ & \searrow^{\eta'_n} & \downarrow \beta \\ & & I *' n \end{array} \qquad \begin{array}{ccc} m_1 * n & \xrightarrow{\beta} & m_1 *' n \\ & & \downarrow m_1 *' f \\ \downarrow m_1 * f & & m_1 *' (m_2 * n') \\ & & \downarrow m_1 *' \beta \\ m_1 * (m_2 * n') & & m_1 *' (m_2 *' n') \\ \downarrow \mu & & \downarrow \mu' \\ (m_1 \otimes m_2) * n' & \xrightarrow{\beta} & (m_1 \otimes m_2) *' n' \end{array}$$

Proof. By the equivalence $[\mathbf{Set}, \mathbf{Set}]_f \simeq [\aleph_0, \mathbf{Set}]$ induced by restriction and the left Kan extension along the inclusion $i : \aleph_0 \to \mathbf{Set}$. □

2.3 Day Convolution

We describe a monoidal biclosed structure on the (covariant) presheaf category $[\mathbf{M}, \mathbf{Set}]_0$ where $\mathbf{M} = (\mathbf{M}, \otimes, I)$ is a small monoidal category [3]. Here, we use the subscript $(-)_0$ to indicate that $[\mathbf{M}, \mathbf{Set}]_0$ is an ordinary (not enriched) category since we also use the enriched version $[\mathbf{M}, \mathbf{Set}]$ later.

The *external tensor product* $F \boxtimes G : \mathbf{M} \times \mathbf{M} \to \mathbf{Set}$ is defined by $(F \boxtimes G)(m_1, m_2) = Fm_1 \times Gm_2$ for any $F, G : \mathbf{M} \to \mathbf{Set}$.

Definition 4. Let $F, G : \mathbf{M} \to \mathbf{Set}$ be functors. The *Day tensor product* $F \check{\otimes} G : \mathbf{M} \to \mathbf{Set}$ is the left Kan extension $\mathrm{Lan}_{\otimes}(F \boxtimes G)$ of the external tensor product $F \boxtimes G : \mathbf{M} \times \mathbf{M} \to \mathbf{Set}$ along the tensor product $\otimes : \mathbf{M} \times \mathbf{M} \to \mathbf{M}$.

Note that a natural transformation $\overline{\theta} : F \check{\otimes} G \to H$ is equivalent to a natural transformation $\theta_{m_1,m_2} : Fm_1 \times Gm_2 \to H(m_1 \otimes m_2)$ by the universal property.

The Day convolution induces a monoidal biclosed structure in $[\mathbf{M}, \mathbf{Set}]_0$ [3].

Proposition 5. *The Day tensor product makes $([\mathbf{M}, \mathbf{Set}]_0, \check{\otimes}, y(I))$ a monoidal biclosed category where $y : \mathbf{M}^{\mathrm{op}} \to [\mathbf{M}, \mathbf{Set}]_0$ is the Yoneda embedding $y(m) := \mathbf{M}(m, -)$.* □

The left and the right closed structure are given by $[\![F, G]\!]\, m = [\mathbf{M}, \mathbf{Set}]_0(F, G(m \otimes -))$ and $[F, G]\, m = [\mathbf{M}, \mathbf{Set}]_0(F, G(- \otimes m))$ for each $m \in \mathbf{M}$, respectively.

Note that since we do not assume **M** to be symmetric, neither is $[\mathbf{M}, \mathbf{Set}]_0$. Note also that the twisting and the above construction commute: there is an isomorphism ${[\mathbf{M}, \mathbf{Set}]_0}^{\mathrm{t}} \cong [\mathbf{M}^{\mathrm{t}}, \mathbf{Set}]_0$ of monoidal categories.

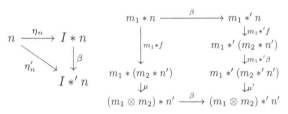

2.4 Categories Enriched in a Presheaf Category

We rephrase the definitions of $[\mathbf{M}, \mathbf{Set}]_0$-enriched category, functor and natural transformation in elementary terms. An $[\mathbf{M}, \mathbf{Set}]_0$-category is, so to say, an "$\mathbf{M}$-graded" category: each morphism has a grade $m \in \mathrm{ob}\mathbf{M}$ and the grade of the composite of two morphisms with grades m and m' is the product $m \otimes m'$ of the grades of each morphism. Likewise, $[\mathbf{M}, \mathbf{Set}]_0$-functors and $[\mathbf{M}, \mathbf{Set}]_0$-natural transformations can be also understood as an "$\mathbf{M}$-graded" version of ordinary functors and natural transformations. Specifically, the following lemma holds [2].

Lemma 6. *There is a one-to-one correspondence between (1) an $[\mathbf{M}, \mathbf{Set}]_0$-category $\mathbf{C}$ and (2) the following data satisfying the following conditions.*

- *A class of objects* $\mathrm{ob}\mathbf{C}$.
- *For each $X, Y \in \mathrm{ob}\mathbf{C}$, a hom objects $\mathbf{C}(X, Y) \in [\mathbf{M}, \mathbf{Set}]_0$.*
- *For each $X \in \mathrm{ob}\mathbf{C}$, an element $1_X \in \mathbf{C}(X, X)I$.*
- *For each $X, Y, Z \in \mathrm{ob}\mathbf{C}$, a family of morphisms $\left(\circ_{m_1,m_2} : \mathbf{C}(Y, Z)m_1 \times \mathbf{C}(X, Y)m_2 \to \mathbf{C}(X, Z)(m_1 \otimes m_2)\right)_{m_1,m_2 \in M}$ which is natural in m_1 and m_2. The subscripts m_1 and m_2 are often omitted.*

These data must satisfy the identity law $1_Y \circ f = f = f \circ 1_X$ for each $f \in \mathbf{C}(X, Y)m$ and the associativity $(h \circ g) \circ f = h \circ (g \circ f)$ for each $f \in \mathbf{C}(X, Y)m_1$, $g \in \mathbf{C}(Y, Z)m_2$ and $h \in \mathbf{C}(Z, W)m_3$.

Proof. The identity $\overline{1_X} : y(I) \to \mathbf{C}(X, X)$ in $\mathbf{C}$ corresponds to $1_X \in \mathbf{C}(X, X)I$ by the Yoneda lemma, and the composition $\overline{\circ} : \mathbf{C}(Y, Z) \check{\otimes} \mathbf{C}(X, Y) \to \mathbf{C}(X, Z)$ in $\mathbf{C}$ corresponds to the natural transformation $\circ_{m_1,m_2} : \mathbf{C}(Y, Z)m_1 \times \mathbf{C}(X, Y)m_2 \to \mathbf{C}(X, Z)(m_1 \otimes m_2)$ by the universal property of the Day convolution. The rest of the proof is easy. □

An $[\mathbf{M}, \mathbf{Set}]_0$-functor $F : \mathbf{C} \to \mathbf{D}$ consists of a mapping $X \mapsto FX$ and a natural transformation $F_{X,Y} : \mathbf{C}(X, Y) \to \mathbf{D}(FX, FY)$ (for each X, Y) that preserves identities and compositions of morphisms. An $[\mathbf{M}, \mathbf{Set}]_0$-natural transformation $\overline{\alpha} : F \to G$ is a family of elements $\left(\alpha_X \in \mathbf{D}(FX, GX)I\right)_{X \in \mathrm{ob}(C)}$ that satisfies $\alpha_Y \circ Ff = Gf \circ \alpha_X$ for each $f \in \mathbf{C}(X, Y)m$. Vertical and horizontal compositions of $[\mathbf{M}, \mathbf{Set}]_0$-natural transformations are defined as expected.

We introduce a useful construction of ${[\mathbf{M}, \mathbf{Set}]_0}^{\mathrm{t}}$-categories. Given an $\mathbf{M}$-graded monad (in other words, a lax left $\mathbf{M}$-action) on $\mathbf{C}$, we can define an ${[\mathbf{M}, \mathbf{Set}]_0}^{\mathrm{t}}$-enriched category as follows.

Definition 7. Let $T = (*, \eta, \mu)$ be an $\mathbf{M}$-graded monad on $\mathbf{C}$. An ${[\mathbf{M}, \mathbf{Set}]_0}^{\mathrm{t}}$-category $\widetilde{\mathbf{C}_T}$ is defined by $\mathrm{ob}\widetilde{\mathbf{C}_T} := \mathrm{ob}\mathbf{C}$ and $\widetilde{\mathbf{C}_T}(X, Y)m := \mathbf{C}(X, m * Y)$. The identity morphisms are the unit morphisms $\eta_X \in \widetilde{\mathbf{C}_T}(X, X)I$, and the composite of $f \in \widetilde{\mathbf{C}_T}(Y, Z)m$ and $g \in \widetilde{\mathbf{C}_T}(X, Y)m'$ is $\mu \circ (m * g) \circ f$.

The definition of $\widetilde{\mathbf{C}_T}$ is similar to the definition of the Kleisli categories for ordinary monads. Actually, $\widetilde{\mathbf{C}_T}$ can be constructed via the Kleisli category $\mathbf{C}_T$ for the graded monad T presented in [7] (although $\mathbf{C}_T$ itself is not enriched). This can be observed by $\mathbf{C}_T((I, X), (m, Y)) \cong \widetilde{\mathbf{C}_T}(X, Y)m$.

3 Graded Algebraic Theories

We explain a framework of universal algebra for graded monads, which is a natural extension of [17, 27]. The key idea of this framework is that each term is associated with not only an arity but also a "grade", which is represented by an object in a monoidal category $\mathbf{M}$. We also add coercion construct for terms that changes the grade of terms along a morphism of the monoidal category $\mathbf{M}$. Then, a mapping that takes $m \in \mathbf{M}$ and a set of variables X and returns the set of terms with grade m (modulo the equational axioms) yields a graded monad.

We fix a small strict monoidal category $\mathbf{M} = (\mathbf{M}, \otimes, I)$ throughout this section. We sometimes identify $n \in \mathbb{N}$ with $\{1, \dots, n\}$, or $\{x_1, \dots, x_n\}$ if it is used as a set of variables.

3.1 Equational Logic

A *signature* is a family of sets of symbols $\Sigma = (\Sigma_{n,m})_{n \in \mathbb{N}, m \in \mathbf{M}}$. An element $f \in \Sigma_{n,m}$ is called an operation with arity n and grade m. We define a sufficient structure to interpret operations in a category $\mathbf{C}$ as follows.

Definition 8. $\mathbf{M}$-*model condition* is defined by the following conditions on a tuple $(\mathbf{C}, (\circledast, \eta^{\circledast}, \mu^{\circledast}))$.

- $\mathbf{C}$ is a category with finite power.
- $(\circledast, \eta^{\circledast}, \mu^{\circledast})$ is a strong $\mathbf{M}^{\mathrm{t}}$-action (i.e. an $\mathbf{M}^{\mathrm{t}}$-graded monad whose unit and multiplication are invertible).
- For each $m \in \mathbf{M}$, $m \circledast (-)$ preserves finite powers: $m \circledast c^n \cong (m \circledast c)^n$.

Example 9. If $\mathbf{A}$ is a category with finite powers, then the functor category $[\mathbf{M}, \mathbf{A}]$ has strong $\mathbf{M}^{\mathrm{t}}$-action defined by $m \circledast F := F(m \otimes (-))$ and satisfies $\mathbf{M}$-model condition. Especially, $[\mathbf{M}, \mathbf{Set}]_0$ satisfies $\mathbf{M}$-model condition.

A *model* $A = (A, |\cdot|^A)$ of Σ in a category $\mathbf{C}$ satisfying $\mathbf{M}$-model condition consists of an object $A \in \mathbf{C}$ and an interpretation $|f|^A : A^n \to m \circledast A$ for each $f \in \Sigma_{n,m}$. A *homomorphism* $\alpha : A \to B$ between two models A, B is a morphism $\alpha : A \to B$ in $\mathbf{C}$ such that $(m \circledast \alpha) \circ |f|^A = |f|^B \circ \alpha^n$ for each $f \in \Sigma_{n,m}$.

Definition 10. Let X be a set of variables. The set of ($\mathbf{M}$-graded) Σ-terms $T^{\Sigma}_m(X)$ for each $m \in \mathbf{M}$ is defined inductively as follows.

$$\frac{x \in X}{x \in T^{\Sigma}_I(X)} \qquad \frac{t \in T^{\Sigma}_m(X) \qquad w : m \to m'}{c_w(t) \in T^{\Sigma}_{m'}(X)} \qquad \frac{f \in \Sigma_{n,m} \qquad \forall i \in \{1, \dots, n\},\ t_i \in T^{\Sigma}_{m'}(X)}{f(t_1, \dots, t_n) \in T^{\Sigma}_{m \otimes m'}(X)}$$

That is, we build Σ-terms from variables by applying operations in Σ and coercions c_w while keeping track of the grade of terms. When applying operations, we sometimes write $f(\lambda i \in n.t_i)$ or $f(\lambda i.t_i)$ instead of $f(t_1, \dots, t_n)$.

Definition 11. Let A be a model of a signature Σ. For each $m \in \mathbf{M}$ and $s \in T^{\Sigma}_m(n)$, the *interpretation* $|s|^A : A^n \to m \circledast A$ is defined as follows.

- For any variable x_i, $|x_i|^A = \eta^{\circledast} \circ \pi_i$ where $\pi_i : A^n \to A$ is the i-th projection.
- For each $w : m' \to m$ and $s \in T^{\Sigma}_{m'}(\{x_1, \ldots, x_n\})$, $|c_w(s)|^A = (w \circledast A) \circ |s|^A$.
- If $f \in \Sigma_{k,m'}$ and $t_i \in T^{\Sigma}_{m''}(\{x_1, \ldots, x_n\})$ for each $i \in \{1, \ldots, k\}$, then $|f(t_1, \ldots, t_k)|^A$ is defined by the following composite.

$$A^n \xrightarrow{\langle |t_1|, \ldots, |t_k| \rangle} (m''\circledast A)^k \overset{\cong}{\to} m''\circledast A^k \xrightarrow{m''\circledast |f|} m''\circledast(m'\circledast A) \overset{\mu}{\to} (m'\otimes m'')\circledast A$$

When we interpret a term $t \in T^{\Sigma}_m(X)$, we need to pick a finite set n such that $\mathrm{fv}(t) \subseteq n \subseteq X$ where $\mathrm{fv}(t)$ is the set of free variables in t, but the choice of the finite set does not matter when we consider only equality of interpretations by the following fact. If $\sigma : n \to n'$ is a renaming of variables and $\overline{\sigma} : T^{\Sigma}_m(n) \to T^{\Sigma}_m(n')$ is a mapping induced by the renaming σ, then for each $t \in T^{\Sigma}_m(n)$, $|\overline{\sigma}(t)|^A = |t|^A \circ A^{\sigma}$, which implies that equality of the interpretations of two terms s, t is preserved by renaming: $|s| = |t|$ implies $|\overline{\sigma}(s)| = |\overline{\sigma}(s)|$.

An *equational axiom* is a family of sets $E = (E_m)_{m\in\mathbf{M}}$ where E_m is a set of pairs of terms in $T^{\Sigma}_m(X)$. We sometimes identify E with its union $\bigcup_{m\in\mathbf{M}} E_m$. A *presentation of an* $\mathbf{M}$*-graded algebraic theory* (or an $\mathbf{M}$*-graded algebraic theory*) is a pair $\mathcal{T} = (\Sigma, E)$ of a signature and an equational axiom. A *model* A of (Σ, E) is a model of Σ that satisfies $|s|^A = |t|^A$ for each $(s = t) \in E$. Let $\mathrm{Mod}_{\mathcal{T}}(\mathbf{C})$ denote the category of models of $\mathcal{T}$ in $\mathbf{C}$ and homomorphisms between them.

To obtain a graded monad on **Set** from $\mathcal{T}$, we need a strict left action of $\mathbf{M}$ on $\mathrm{Mod}_{\mathcal{T}}([\mathbf{M}, \mathbf{Set}]_0)$ and an adjunction between $\mathrm{Mod}_{\mathcal{T}}([\mathbf{M}, \mathbf{Set}]_0)$ and **Set**. The former is defined by the following, while the latter is described in §3.2.

Lemma 12. *Let* $\mathbf{C}$ *be a category satisfying* $\mathbf{M}_1 \times \mathbf{M}_2$*-model condition. If* $\mathcal{T}$ *is an* $\mathbf{M}_1$*-graded algebraic theory, then* $\mathbf{C}$ *satisfies* $\mathbf{M}_1$*-model condition and* $\mathrm{Mod}_{\mathcal{T}}(\mathbf{C})$ *satisfies* $\mathbf{M}_2$*-model condition.*

Proof. An $\mathbf{M}_1^{\mathrm{t}}$-action on $\mathbf{C}$ is obtained by the composition of $\mathbf{M}_1^{\mathrm{t}} \times \mathbf{M}_2^{\mathrm{t}}$-action and the strong monoidal functor $\mathbf{M}_1^{\mathrm{t}} \to \mathbf{M}_1^{\mathrm{t}} \times \mathbf{M}_2^{\mathrm{t}}$ defined by $m \mapsto (m, I)$. Finite powers and an $\mathbf{M}_2^{\mathrm{t}}$-action for $\mathrm{Mod}_{\mathcal{T}}(\mathbf{C})$ are induced by those for $\mathbf{C}$. □

Corollary 13. $\mathrm{Mod}_{\mathcal{T}}([\mathbf{M}, \mathbf{Set}]_0)$ *has an* $\mathbf{M}$*-action, which is given by the precomposition of* $m \otimes (-)$ *like the* $\mathbf{M}$*-action of Example 9.*

Proof. $[\mathbf{M}, \mathbf{Set}]_0$ has $\mathbf{M}^{\mathrm{t}} \times \mathbf{M}$-action defined by $(m_1, m_2) * F = F(m_1 \otimes (-) \otimes m_2)$. Thus, $\mathbf{M}$-action for $\mathrm{Mod}_{\mathcal{T}}([\mathbf{M}, \mathbf{Set}]_0)$ is obtained by Lemma 12. □

Substitution $s[t_1/x_1, \ldots, t_k/x_k]$ for $\mathbf{M}$-graded Σ-terms can be defined as usual, but we have to take care of grades: given $s \in T^{\Sigma}_m(k)$ and $t_1, \ldots, t_k \in T^{\Sigma}_{m'}(n)$, the substitution $s[t_1/x_1, \ldots, t_k/x_k]$ is defined as a term in $T^{\Sigma}_{m\otimes m'}(n)$.

We obtain an equational logic for graded theories by adding some additional rules to the usual equational logic.

Definition 14. The entailment relation $\mathcal{T} \vdash s = t$ (where $s, t \in T_m(X)$) for an $\mathbf{M}$-graded theory $\mathcal{T}$ is defined by adding the following rules to the standard rules i.e. reflexivity, symmetry, transitivity, congruence, substitution and axiom in E (see e.g. [26] for the standard rules of equational logic).

$$\frac{s,t \in T^{\Sigma}_m(X) \qquad \mathcal{T} \vdash s = t \qquad w : m \to m'}{\mathcal{T} \vdash c_w(s) = c_w(t)} \qquad \frac{t \in T^{\Sigma}_m(X)}{\mathcal{T} \vdash c_{1_m}(t) = t}$$

$$\frac{t \in T^{\Sigma}_m(X) \qquad w : m \to m' \qquad w' : m' \to m''}{\mathcal{T} \vdash c_{w'}(c_w(t)) = c_{w' \circ w}(t)}$$

$$\frac{f \in \Sigma_{n,m} \qquad t_i \in T^{\Sigma}_{m'}(X) \text{ for each } i \in \{1,\ldots,n\} \qquad w : m' \to m''}{\mathcal{T} \vdash f(c_w(t_1),\ldots,c_w(t_n)) = c_{m\otimes w}(f(t_1,\ldots,t_n))}$$

Definition 15. Given a model A of $\mathcal{T}$, we denote $A \Vdash s = t$ if $s,t \in T^{\Sigma}_m(n)$ (for some n) and $|s|^A = |t|^A$. If $\mathbf{C}$ is a category satisfying $\mathbf{M}$-model condition, we denote $\mathcal{T}, \mathbf{C} \Vdash s = t$ if $A \Vdash s = t$ for any model A of $\mathcal{T}$ in $\mathbf{C}$.

It is easy to verify that the equational logic in Definition 14 is sound.

Theorem 1 (soundness). *$\mathcal{T} \vdash s = t$ implies $\mathcal{T}, \mathbf{C} \Vdash s = t$.* □

3.2 Free Models

We describe a construction of a free model $F^{\mathcal{T}}X \in \mathrm{Mod}_{\mathcal{T}}([\mathbf{M},\mathbf{Set}]_0)$ of a graded theory $\mathcal{T}$ generated by a set X, which induces an adjunction between $\mathrm{Mod}_{\mathcal{T}}([\mathbf{M},\mathbf{Set}]_0)$ and $\mathbf{Set}$. This adjunction, together with the $\mathbf{M}$-action of Corollary 13, gives a graded monad as described in [7].

Definition 16 (free model $F^{\mathcal{T}}X$). Let $\mathcal{T} = (\Sigma, E)$ be an $\mathbf{M}$-graded theory. We define a functor $F^{\mathcal{T}}X : \mathbf{M} \to \mathbf{Set}$ by $F^{\mathcal{T}}Xm := T^{\Sigma}_m(X)/{\sim_m}$ for each $m \in \mathbf{M}$ and any $X \in \mathbf{Set}$ where $s \sim_m t$ is the equivalence relation defined by $\mathcal{T} \vdash s = t$ and $F^{\mathcal{T}}Xw([t]_m) := [c_w(t)]_{m'}$ for any $w : m \to m'$ where $[t]_m$ is the equivalence class of $t \in T^{\Sigma}_m(X)$. For each $f \in \Sigma_{n,m'}$, let $|f|^{F^{\mathcal{T}}X} : (F^{\mathcal{T}}X)^n \to m' \circledast F^{\mathcal{T}}X$ be a mapping defined by $|f|^{F^{\mathcal{T}}X}_m([t_1]_m,\ldots,[t_n]_m) = [f(t_1,\ldots,t_n)]_{m'\otimes m}$ for each $m \in \mathbf{M}$. We define a model of $\mathcal{T}$ by $F^{\mathcal{T}}X = (F^{\mathcal{T}}X, |\cdot|^{F^{\mathcal{T}}X})$.

The model $F^{\mathcal{T}}X$, together with the mapping $\eta_X : X \to F^{\mathcal{T}}XI$ defined by $x \mapsto [x]_I$, has the following universal property as a free model generated by X.

Lemma 17. *For any model A in $[\mathbf{M},\mathbf{Set}]_0$ and any mapping $v : X \to AI$, there exists a unique homomorphism $\overline{v} : F^{\mathcal{T}}X \to A$ satisfying $\overline{v}_I \circ \eta_X = v$.* □

Corollary 18. *Let $U : \mathrm{Mod}_{\mathcal{T}}([\mathbf{M},\mathbf{Set}]_0) \to \mathbf{Set}$ be the forgetful functor defined by the evaluation at I, that is, $UA = A_I$ and $U\alpha = \alpha_I$. The free model functor $F^{\mathcal{T}} : \mathbf{Set} \to \mathrm{Mod}_{\mathcal{T}}([\mathbf{M},\mathbf{Set}]_0)$ is a left adjoint of U.* □

By considering the interpretation in the free model, we obtain the following completeness theorem.

Theorem 19 (completeness). *$\mathcal{T}, [\mathbf{M},\mathbf{Set}]_0 \Vdash s = t$ implies $\mathcal{T} \vdash s = t$.* □

Recall that $\mathrm{Mod}_{\mathcal{T}}([\mathbf{M},\mathbf{Set}]_0)$ has a left action (Corollary 13). Therefore the above adjunction induces an $\mathbf{M}$-graded monad as described in [7].

The relationship between $\mathrm{Mod}_{\mathcal{T}}([\mathbf{M},\mathbf{Set}]_0)$ and the Eilenberg–Moore construction is as follows. In [7], the Eilenberg–Moore category $\mathbf{C}^{\mathbf{T}}$ for any graded monad $\mathbf{T}$ on $\mathbf{C}$ is introduced together with a left action $\circledast : \mathbf{M} \times \mathbf{C}^{\mathbf{T}} \to \mathbf{C}^{\mathbf{T}}$. If $\mathbf{C} = \mathbf{Set}$ and $\mathbf{T}$ is the graded monad obtained from an $\mathbf{M}$-graded theory $\mathcal{T}$, then the Eilenberg–Moore category $\mathbf{Set}^{\mathbf{T}}$ is essentially the same as $\mathrm{Mod}_{\mathcal{T}}([\mathbf{M}, \mathbf{Set}]_0)$.

Theorem 20. *The comparison functor $K : \mathrm{Mod}_{\mathcal{T}}([\mathbf{M}, \mathbf{Set}]_0) \to \mathbf{Set}^{\mathbf{T}}$ (see [7] for the definition) where $\mathcal{T}$ is an $\mathbf{M}$-graded theory and $\mathbf{T}$ is the graded monad induced from the graded theory $\mathcal{T}$ is isomorphic. Moreover, K preserves the $\mathbf{M}$-action: $\circledast \circ (\mathbf{M} \times K) = K \circ \circledast$.* □

We define the category $\mathbf{GS}_{\mathbf{M}}$ of graded algebraic theories as follows.

Definition 21. Let $\mathcal{T} = (\Sigma, E)$ and $\mathcal{T}' = (\Sigma', E')$. A morphism $\alpha : \mathcal{T} \to \mathcal{T}'$ between graded algebraic theories is a family of mappings $\alpha_{n,m} : \Sigma_{n,m} \to F^{\mathcal{T}'} nm$ from operations in Σ to Σ'-terms such that the equations in E are preserved by α, i.e. for each $s, t \in T^{\Sigma}_{m}(X)$, $(s, t) \in E$ implies $|s|^{(F^{\mathcal{T}'}X, \alpha)} = |t|^{(F^{\mathcal{T}'}X, \alpha)}$ where $(F^{\mathcal{T}'}X, \alpha)$ is a model of $\mathcal{T}$ induced by α.

Definition 22. Given a morphism $\alpha : \mathcal{T} \to \mathcal{T}'$, let $F^{\alpha} : F^{\mathcal{T}} \to F^{\mathcal{T}'}$ be a natural transformation defined by $F^{\alpha}([t]) = |t|^{(F^{\mathcal{T}'}X, \alpha)}$ for each $t \in T^{\Sigma}_{m}(X)$.

Definition 23. We write $\mathbf{GS}_{M}$ for the category of graded algebraic theories and morphisms between them. The identity morphisms are defined by $1_{\mathcal{T}}(f) = [f(x_1, \ldots, x_n)]$ for each $f \in \Sigma_{n,m}$. The composition of $\alpha : \mathcal{T} \to \mathcal{T}'$ and $\beta : \mathcal{T}' \to \mathcal{T}''$ is defined by $\beta \circ \alpha(f) = F^{\beta}(\alpha(f))$.

3.3 Examples

Example 24 (graded modules). Let $\mathbf{M} = (\mathbb{N}, +, 0)$ where $\mathbb{N}$ is regarded as a discrete category. Given a graded ring $A = \bigoplus_{n \in \mathbb{N}} A_n$, let Σ be a set of operations which consists of the binary addition operation $+$ (arity: 2, grade: 0), the unary inverse operation $-$ (arity: 1, grade: 0), the identity element (nullary operation) 0 (arity: 0, grade: 0) and the unary scalar multiplication operation $a \cdot (-)$ (arity: 1, grade: n) for each $a \in A_n$. Let E be the equational axiom for modules.

A model $(F, |\cdot|)$ of the $\mathbf{M}$-graded theory (Σ, E) in $[\mathbf{M}, \mathbf{Set}]_0$ consists of a set F_n for each $n \in \mathbb{N}$ and functions $|+|_n : (F_n)^2 \to F_n$, $|-|_n : F_n \to F_n$, $|0|_n \in F_n$ and $|a \cdot (-)|_n : F_n \to F_{m+n}$ for each $n \in \mathbb{N}$ and each $a \in A_m$, and these interpretations satisfy E. Therefore models of (Σ, E) in $[\mathbf{M}, \mathbf{Set}]_0$ correspond one-to-one with graded modules.

Example 25 (graded exception monad [12, Example 3.4]). We give an algebraic presentation of the graded exception monad.

Let $\mathbf{M}$ and $(*, \eta, \mu)$ be a preordered monoid and the graded monad defined as follows. Let $P^{+}(X)$ denote the set of nonempty subsets of X. Let Ex be a set of exceptions and $\mathbf{M} = ((P^{+}(\mathrm{Ex} \cup \{\mathrm{Ok}\}), \subseteq), I, \otimes)$ be a preordered monoid where $I = \{\mathrm{Ok}\}$ and the multiplication $\otimes$ is defined by $m \otimes m' = (m \setminus \{\mathrm{Ok}\}) \cup m'$ if $\mathrm{Ok} \in m$ and $m \otimes m' = m$ otherwise (note that this is not commutative). The graded exception monad $(*, \eta, \mu)$ is the $\mathbf{M}$-graded monad given as follows.

$$m * X = \{\mathrm{Er}(e) \mid e \in m \setminus \{\mathrm{Ok}\}\} \cup \{\mathrm{Ok}(x) \mid x \in X \wedge \mathrm{Ok} \in m\}$$
$$\eta_X(x) = \mathrm{Ok}(x) \qquad \mu_{m_1,m_2,X}(\mathrm{Er}(e)) = \mathrm{Er}(e) \quad \mu_{m_1,m_2,X}(\mathrm{Ok}(x)) = x$$

The $\mathbf{M}$-graded theory $\mathcal{T}^{\mathrm{ex}}$ for the graded exception monad is defined by $(\Sigma^{\mathrm{ex}}, \emptyset)$ where Σ^{ex} is the set that consists of an operation raise_e (arity: 0, grade: $\{e\}$) for each $e \in \mathrm{Ex}$.

The graded monad induced by $\mathcal{T}^{\mathrm{ex}}$ coincides with the graded exception monad. Indeed, the free model functor $F^{\mathcal{T}^{\mathrm{ex}}}$ for $\mathcal{T}^{\mathrm{ex}}$ is given by $F^{\mathcal{T}^{\mathrm{ex}}}Xm = m * X$. Here, the operations raise_e are interpreted by $e \in \mathrm{Ex}$.

$$|\mathsf{raise}_e|_m^{F^{\mathcal{T}^{\mathrm{ex}}}X} = \mathrm{Er}(e) \in F^{\mathcal{T}^{\mathrm{ex}}}X(\{e\} \otimes m)$$

Example 26 (extending an ordinary monad to an M-graded monad). We consider the problem of extending an $\mathbf{M}'$-graded theory to an $\mathbf{M}$-graded theory along a lax monoidal functor of type $\mathbf{M}' \to \mathbf{M}$, but here we restrict ourselves to the case of $\mathbf{M}' = \mathbf{1}$ and the strict monoidal functor of type $\mathbf{1} \to \mathbf{M}$.

Let $\mathbf{M} = (\mathbf{M}, I, \otimes)$ be an arbitrary small strict monoidal category. Let $\mathcal{T} = (\Sigma, E)$ be a ($\mathbf{1}$-graded) theory and (T, η^T, μ^T) be the corresponding ordinary monad. Let $\mathcal{T}^{\mathbf{M}} = (\Sigma^{\mathbf{M}}, E^{\mathbf{M}})$ be the $\mathbf{M}$-graded theory obtained when we regard each operation in $\mathcal{T}$ as an operation with grade $I \in \mathbf{M}$, that is, $\Sigma^{\mathbf{M}}_{n,m} := \Sigma_n$ if $m = I$ and $\Sigma^{\mathbf{M}}_{n,m} := \emptyset$ otherwise, and $E^{\mathbf{M}} := E$.

The free model functor for $\mathcal{T}^{\mathbf{M}}$ is $F^{\mathcal{T}^{\mathbf{M}}}X = F^{\mathcal{T}}(\mathbf{M}(I, -) \times X)$ where $F^{\mathcal{T}} : \mathbf{Set} \to \mathrm{Mod}_{\mathcal{T}}(\mathbf{Set})$ is the free model functor for $\mathcal{T}$ as a $\mathbf{1}$-graded theory, and the interpretation of an operation $f \in \Sigma_n$ in $F^{\mathcal{T}^{\mathbf{M}}}X$ is defined by the interpretation in the free models of $\mathcal{T}$.

$$|f|_m^{F^{\mathcal{T}^{\mathbf{M}}}X} = |f|^{F^{\mathcal{T}}(\mathbf{M}(I,m)\times X)} : \left(F^{\mathcal{T}}(\mathbf{M}(I,m) \times X)\right)^n \to F^{\mathcal{T}}(\mathbf{M}(I,m) \times X)$$

Intuitively, this can be understood as follows. Since all the operations are of grade I, coercions c_w in a term can be moved to the innermost places where variables occur by repeatedly applying $c_w(f(t_1, \ldots, t_n)) = f(c_w(t_1), \ldots, c_w(t_n))$ (see Definition 14). Therefore, we can consider terms of $\mathcal{T}^{\mathbf{M}}$ as terms of $\mathcal{T}$ whose variables are of the form $c_w(x)$.

An $\mathbf{M}$-graded monad $(*, \eta, \mu)$ obtained from $\mathcal{T}^{\mathbf{M}}$ is as follows.

$$m * X = T(\mathbf{M}(I, m) \times X) \qquad \eta = \eta^T(1_I, -) \qquad \mu = T(\otimes \times X) \circ \mu^T \circ T\mathrm{st}$$

Here, $\otimes : \mathbf{M}(I, m_1) \times \mathbf{M}(I, m_2) \to \mathbf{M}(I, m_1 \otimes m_2)$ is induced by $\otimes$: $\mathbf{M} \times \mathbf{M} \to \mathbf{M}$ and $\mathrm{st}_{X,Y} : X \times TY \to T(X \times Y)$ is the strength for T.

4 Graded Lawvere Theories

We present a categorical formulation of graded algebraic theories of §3 in a similar fashion to ordinary Lawvere theories.

For ordinary (single-sorted) finitary algebraic theories, a *Lawvere theory* is defined as a small category $\mathbf{L}$ with finite products together with a strict finite-product preserving identity-on-objects functor $J : \aleph_0^{\mathrm{op}} \to \mathbf{L}$ where $\aleph_0$ is the full subcategory of **Set** on natural numbers. Intuitively, morphisms in the Lawvere theory $\mathbf{L}$ are terms of the corresponding algebraic theory, and objects of $\mathbf{L}$, which are exactly the objects in $\mathrm{ob}\aleph_0$, are arities.

According to the above intuition, it is expected that a graded Lawvere theory is also defined as a category whose objects are natural numbers and morphisms are graded terms. However, since terms in a graded algebraic theory are stratified by a monoidal category $\mathbf{M}$, mere sets are insufficient to express hom-objects of graded Lawvere theories. Instead, we take hom-objects from the functor category $[\mathbf{M}, \mathbf{Set}]_0$ and define graded Lawvere theories using $[\mathbf{M}, \mathbf{Set}]_0$-categories where $[\mathbf{M}, \mathbf{Set}]_0$ is equipped with the Day convolution monoidal structure. Specifically, $\aleph_0$ (in ordinary Lawvere theories) is replaced with an $[\mathbf{M}, \mathbf{Set}]_0$-category $\mathbf{N_M}$, $\mathbf{L}$ with an $[\mathbf{M}, \mathbf{Set}]_0$-category, and "finite products" with "$\mathbf{N_M^{op}}$-cotensors".

So, we first provide an enriched category $\mathbf{N_M}$ that we use as arities. Since we do not assume that $\mathbf{M}$ is symmetric, $\mathbf{N_M}$ is defined to be an ${[\mathbf{M}, \mathbf{Set}]_0}^{\mathrm{t}}$-category so that the opposite category $\mathbf{N_M^{op}}$ is an $[\mathbf{M}, \mathbf{Set}]_0$-category. Let $[\mathbf{M}, \mathbf{Set}]^{\mathrm{t}}$ be an ${[\mathbf{M}, \mathbf{Set}]_0}^{\mathrm{t}}$-category induced by the closed structure of ${[\mathbf{M}, \mathbf{Set}]_0}^{\mathrm{t}}$. That is, hom-objects of $[\mathbf{M}, \mathbf{Set}]^{\mathrm{t}}$ are given by $[\mathbf{M}, \mathbf{Set}]^{\mathrm{t}}(G, H)m = [\mathbf{M}, \mathbf{Set}]_0(G, H(- \otimes m))$.

Definition 27. An ${[\mathbf{M}, \mathbf{Set}]_0}^{\mathrm{t}}$-category $\mathbf{N_M}$ is defined by the full sub-${[\mathbf{M}, \mathbf{Set}]_0}^{\mathrm{t}}$-category of $[\mathbf{M}, \mathbf{Set}]^{\mathrm{t}}$ whose set of objects is given by $\mathrm{ob}\mathbf{N_M} = \{n \cdot y(I) \mid n \in \mathbb{N}\} \subseteq \mathrm{ob}[\mathbf{M}, \mathbf{Set}]^{\mathrm{t}}$ where $\mathbb{N}$ is the set of natural numbers and $n \cdot y(I)$ is the n-fold coproduct of $y(I)$. We sometimes identify $\mathrm{ob}\mathbf{N_M}$ with $\mathbb{N}$ via the mapping $n \mapsto \underline{n} := n \cdot y(I)$.

Lemma 28. *The $[\mathbf{M}, \mathbf{Set}]_0$-category $\mathbf{N_M^{op}}$ has $\mathbf{N_M^{op}}$-cotensors, which are given by $\underline{n} \pitchfork \underline{n'} = \underline{n \cdot n'}$ for each n and n'.* □

Proof. A cotensor $(n \cdot y(I)) \pitchfork (n' \cdot y(I))$ is a tensor $(n \cdot y(I)) \otimes^{\mathrm{t}} (n' \cdot y(I))$ in $[\mathbf{M}, \mathbf{Set}]^{\mathrm{t}}$. Since $\otimes^{\mathrm{t}}$ is biclosed, $\otimes^{\mathrm{t}}$ preserves colimits in both arguments. Therefore, $(n \cdot y(I)) \otimes^{\mathrm{t}} (n' \cdot y(I)) \cong (n \cdot n') \cdot y(I)$. □

$\mathbf{N_M^{op}}$-cotensors (i.e. $n \cdot y(I) \pitchfork C$) behave like an enriched counterpart of finite powers $(-)^n$. We show that $\mathbf{N_M^{op}}$-cotensors in a general $[\mathbf{M}, \mathbf{Set}]_0$-category $\mathbf{A}$ are characterized by projections satisfying a universal property. Given a unit morphism $\nu : \underline{n} \to \mathbf{A}(\underline{n} \pitchfork C, C)$ of the cotensor $\underline{n} \pitchfork C$, an $[\mathbf{M}, \mathbf{Set}]_0$-natural transformation $\overline{\nu} : \mathbf{A}(B, \underline{n} \pitchfork C) \to [\underline{n}, \mathbf{A}(B, C)]$ is given by $f \mapsto (x \mapsto \nu(x) \circ f)$. The condition that $\overline{\nu}$ is isomorphic can be rephrased as follows.

Lemma 29. *An $[\mathbf{M}, \mathbf{Set}]_0$-category $\mathbf{A}$ has $\mathbf{N_M^{op}}$-cotensors if and only if for any $n \in \mathbb{N}$ and $C \in \mathrm{ob}\mathbf{A}$, there exist an object $\underline{n} \pitchfork C \in \mathrm{ob}\mathbf{A}$ and $(\pi_1, \dots, \pi_n) \in (\mathbf{A}(\underline{n} \pitchfork C, C)I)^n$ such that the following condition holds: for each m, the function $f \mapsto (\pi_1 \circ f, \dots, \pi_n \circ f)$ of type $\mathbf{A}(B, \underline{n} \pitchfork C)m \to (\mathbf{A}(B, C)m)^n$ is bijective.*

An $[\mathbf{M}, \mathbf{Set}]_0$-functor $F : \mathbf{A} \to \mathbf{B}$ preserves $\mathbf{N_M^{op}}$-cotensors if and only if $(F_{\underline{n} \pitchfork C, C, I} \circ \pi_1, \dots, F_{\underline{n} \pitchfork C, C, I} \circ \pi_n) \in (\mathbf{B}(F(\underline{n} \pitchfork C), FC)I)^n$ satisfies the same condition for each n and C.

Proof. The essence of the proof is that the unit morphism $\nu : n \cdot y(I) \to \mathbf{A}(\underline{n} \pitchfork C, C)$ corresponds to elements $\pi_1, \dots, \pi_n \in \mathbf{A}(\underline{n} \pitchfork C, C)I$ by $[\mathbf{M}, \mathbf{Set}]_0(n \cdot$

$y(I), \mathbf{A}(\underline{n} \pitchfork C, C)) \cong [\mathbf{M}, \mathbf{Set}]_0(y(I), \mathbf{A}(\underline{n} \pitchfork C, C))^n \cong \left(\mathbf{A}(\underline{n} \pitchfork C, C)I\right)^n$. The $[\mathbf{M}, \mathbf{Set}]_0$-natural transformation $\overline{\nu}$ is isomorphic if and only if each component $\overline{\nu}_m : \mathbf{A}(B, \underline{n} \pitchfork C)m \to [\underline{n}, \mathbf{A}(B, C)]\, m$ of $\overline{\nu}$ is isomorphic, which is moreover equivalent to the condition that $f \mapsto (\pi_1 \circ f, \ldots, \pi_n \circ f) : \mathbf{A}(B, \underline{n} \pitchfork C)m \to (\mathbf{A}(B, C))^n$ is isomorphic since we have $[\underline{n}, \mathbf{A}(B, C)]\, m \cong (\mathbf{A}(B, C)m)^n$.

The latter part of the lemma follows from the former part. □

If $(\pi_1, \ldots, \pi_n) \in (\mathbf{A}(\underline{n} \pitchfork C, C)I)^n$ satisfies the condition in Lemma 29, we call the element $\pi_i \in \mathbf{A}(\underline{n} \pitchfork C, C)I$ the *i-th projection* of $\underline{n} \pitchfork C$. Note that the choice of projections is not necessarily unique. However, when we say that $\mathbf{A}$ is an $[\mathbf{M}, \mathbf{Set}]_0$-category with $\mathbf{N}_{\mathbf{M}}^{\mathrm{op}}$-cotensors, we implicitly assume that there are a chosen cotensor $\underline{n} \pitchfork C$ and chosen projections $(\pi_1, \ldots, \pi_n) \in (\mathbf{A}(\underline{n} \pitchfork C, C)I)^n$ for each $\underline{n} \in \mathrm{ob}\mathbf{N}_{\mathbf{M}}^{\mathrm{op}}$ and $C \in \mathrm{ob}\mathbf{A}$. We also assume that $\underline{1} \pitchfork X = X$ without loss of generality. Given n-tuple $(f_1, \ldots, f_n)$ of elements in $\mathbf{A}(B, C)m$, we denote by $\langle f_1, \ldots, f_n \rangle$ an element in $\mathbf{A}(B, \underline{n} \pitchfork C)m$ obtained by the inverse of $f \mapsto (\pi_1 \circ f, \ldots, \pi_n \circ f)$ and call this a tupling. Tuplings and projections for $\mathbf{N}_{\mathbf{M}}^{\mathrm{op}}$-cotensors behave like those for finite products.

The following proposition claims that $\mathbf{N}_{\mathbf{M}}^{\mathrm{op}}$ is a free $[\mathbf{M}, \mathbf{Set}]_0$-category with chosen $\mathbf{N}_{\mathbf{M}}^{\mathrm{op}}$-cotensors generated by one object.

Proposition 30. *Let $\mathbf{A}$ be an $[\mathbf{M}, \mathbf{Set}]_0$-category with $\mathbf{N}_{\mathbf{M}}^{\mathrm{op}}$-cotensors and C be an object in $\mathbf{A}$. Then there exists a unique $\mathbf{N}_{\mathbf{M}}^{\mathrm{op}}$-cotensor preserving $[\mathbf{M}, \mathbf{Set}]_0$-functor $F : \mathbf{N}_{\mathbf{M}}^{\mathrm{op}} \to \mathbf{A}$ such that $Fn = \underline{n} \pitchfork C$ and $F\pi_i = \pi_i$.* □

We define $\mathbf{M}$-graded Lawvere theories in a similar fashion to enriched Lawvere theories.

Definition 31. An $\mathbf{M}$-graded Lawvere theory is a tuple $(\mathbf{L}, J)$ where $\mathbf{L}$ is an $[\mathbf{M}, \mathbf{Set}]_0$-category with $\mathbf{N}_{\mathbf{M}}^{\mathrm{op}}$-cotensors and $J : \mathbf{N}_{\mathbf{M}}^{\mathrm{op}} \to \mathbf{L}$ is an identity-on-objects $\mathbf{N}_{\mathbf{M}}^{\mathrm{op}}$-cotensor preserving $[\mathbf{M}, \mathbf{Set}]_0$-functor. A morphism $F : (\mathbf{L}, J) \to (\mathbf{L}', J')$ between two graded Lawvere theories is an $[\mathbf{M}, \mathbf{Set}]_0$-functor $F : \mathbf{L} \to \mathbf{L}'$ such that $FJ = J'$. We denote the category of graded Lawvere theories and morphisms between them by $\mathbf{GLaw}_{\mathbf{M}}$.

By Proposition 30, the existence of the above $J : \mathbf{N}_{\mathbf{M}}^{\mathrm{op}} \to \mathbf{L}$ is equivalent to requiring that $\mathrm{ob}\mathbf{L} = \mathbb{N}$ and projections in $\mathbf{L}$ are chosen in some way. So, we sometimes leave J implicit and just write $\mathbf{L} \in \mathbf{GLaw}_{\mathbf{M}}$ for $(\mathbf{L}, J) \in \mathbf{GLaw}_{\mathbf{M}}$.

Definition 32. A *model* of graded Lawvere theory $\mathbf{L}$ in an $[\mathbf{M}, \mathbf{Set}]_0$-category $\mathbf{A}$ with $\mathbf{N}_{\mathbf{M}}^{\mathrm{op}}$-cotensor is an $\mathbf{N}_{\mathbf{M}}^{\mathrm{op}}$-cotensor preserving $[\mathbf{M}, \mathbf{Set}]_0$-functor of type $\mathbf{L} \to \mathbf{A}$. A morphism $\alpha : F \to G$ between two models F, G of graded Lawvere theory $\mathbf{L}$ is an $[\mathbf{M}, \mathbf{Set}]_0$-natural transformation. Let $\mathrm{Mod}(\mathbf{L}, \mathbf{A})$ be the category of models of graded Lawvere theory $\mathbf{L}$ in the $[\mathbf{M}, \mathbf{Set}]_0$-category $\mathbf{A}$.

In §3, we use a category $\mathbf{C}$ satisfying $\mathbf{M}$-model condition to define a model of graded algebraic theory. Actually, $\mathbf{M}$-model condition is sufficient to give an $[\mathbf{M}, \mathbf{Set}]_0$-category with $\mathbf{N}_{\mathbf{M}}^{\mathrm{op}}$-cotensors.

Lemma 33. *If $\mathbf{C}$ satisfies $\mathbf{M}$-model condition, then the $[\mathbf{M}, \mathbf{Set}]_0$-category $\widetilde{\mathbf{C}_T}^{\mathrm{op}}$ defined in Definition 7 has $\mathbf{N}_{\mathbf{M}}^{\mathrm{op}}$-cotensors.*

Proof. For any $X \in \widetilde{\mathbf{C}_T}^{\mathrm{op}}$ and n, the cotensor $\underline{n} \pitchfork X$ is given by finite power X^n, and the i-th projection is given by $\eta^{\circledast} \circ \pi_i \in \widetilde{\mathbf{C}_T}^{\mathrm{op}} I$ where $\pi_i : X^n \to X$ is the i-th projection of the finite power X^n. The rest of the proof is routine. □

If we apply Lemma 33 to $[\mathbf{M}, \mathbf{Set}]_0$ equipped with the $\mathbf{M}^{\mathrm{t}}$-action in Example 9 (here denoted by T), then $(\widetilde{[\mathbf{M}, \mathbf{Set}]_0)_T}^{\mathrm{op}}$ coincides with $[\mathbf{M}, \mathbf{Set}]$ (i.e. the $[\mathbf{M}, \mathbf{Set}]_0$-category obtained by the closed structure of $[\mathbf{M}, \mathbf{Set}]_0$).

5 Equivalence

We have shown three graded notions: graded algebraic theories, graded Lawvere theories and finitary graded monads, which give rise to categories $\mathbf{GS}_{\mathbf{M}}$, $\mathbf{GLaw}_{\mathbf{M}}$ and $\mathbf{GMnd}_{\mathbf{M}}$, respectively. This section is about the equivalence of these three notions. We give only a sketch of the proof of the equivalence, and the details are deferred to [14, Appendix A].

5.1 Graded Algebraic Theories and Graded Lawvere Theories

We prove that the category of graded algebraic theories $\mathbf{GS}_{\mathbf{M}}$ and the category of graded Lawvere theories $\mathbf{GLaw}_{\mathbf{M}}$ are equivalent by showing the existence of an adjoint equivalence $\mathbf{Th} \vdash U : \mathbf{GLaw}_{\mathbf{M}} \to \mathbf{GS}_{\mathbf{M}}$.

Let $\mathbf{M}$ be a small strict monoidal category and $\mathcal{T} = (\Sigma, E)$ be an $\mathbf{M}$-graded algebraic theory. We define $\mathbf{Th}\mathcal{T}$ (the object part of $\mathbf{Th}$) as an $\mathbf{M}$-graded Lawvere theory whose morphisms are terms of $\mathcal{T}$ modulo equational axioms.

Definition 34. An $[\mathbf{M}, \mathbf{Set}]_0$-category $\mathbf{Th}\mathcal{T}$ is defined by $\mathrm{ob}(\mathbf{Th}\mathcal{T}) := \mathbb{N}$ and $(\mathbf{Th}\mathcal{T})(n, n')m := (F^{\mathcal{T}} nm)^{n'}$ with composition defined by substitution.

It is easy to show that $\mathbf{Th}\mathcal{T}$ has $\mathbf{N}_{\mathbf{M}}^{\mathrm{op}}$-cotensors (by Lemma 29). Therefore, $\mathbf{Th}$ is a mapping from an object in $\mathbf{GS}_{\mathbf{M}}$ to an object in $\mathbf{GLaw}_{\mathbf{M}}$.

We define a functor $U : \mathbf{GLaw}_M \to \mathbf{GS}_M$ by taking all the morphism $f \in L(n, 1)m$ in $L \in \mathbf{GLaw}_M$ as operations and all the equations that hold in L as equational axioms.

Definition 35. A functor $U : \mathbf{GLaw}_M \to \mathbf{GS}_M$ is defined as follows.

- For each $\mathbf{L} \in \mathrm{ob}\mathbf{GLaw}_M$, $U\mathbf{L} = (\Sigma, E)$ where $\Sigma_{n,m} = \mathbf{L}(n, 1)m$, $E = \{(s, t) \mid |s|^{\mathbf{L}} = |t|^{\mathbf{L}}\}$ and $|\cdot|^{\mathbf{L}} : T_m^{\Sigma}(n) \to \mathbf{L}(n, 1)m$ is an interpretation of terms defined in the same way as Definition 11.
- Given $G : \mathbf{L} \to \mathbf{L}'$, let $UG : U\mathbf{L} \to U\mathbf{L}'$ be a functor defined by $UG(f) = [G(f)(x_1, \ldots, x_n)]$ for each $f \in \mathbf{L}(n, 1)m$.

Then, $\mathbf{Th}\mathcal{T}$ has the following universal property as a left adjoint of U.

Lemma 36. *For each $\mathcal{T}$, let $\eta_{\mathcal{T}} : \mathcal{T} \to U\mathbf{Th}\mathcal{T}$ be a family of functions $\eta_{\mathcal{T},n,m} : \Sigma_{n,m} \to F^{U\mathbf{Th}\mathcal{T}} nm$ defined by $\eta_{\mathcal{T},n,m}(f) = [[f(x_1, \ldots, x_n)](x_1, \ldots, x_n)]$. For any $\alpha : \mathcal{T} \to U\mathbf{L}$, there exists a unique morphism $\overline{\alpha} : \mathbf{Th}\mathcal{T} \to \mathbf{L}$ such that $\alpha = U\overline{\alpha} \circ \eta_{\mathcal{T}}$.* □

Moreover, the unit and the counit of $\mathbf{Th} \dashv U$ are isomorphic. Therefore:

Theorem 37. *Two categories $\mathbf{GS}_{\mathbf{M}}$ and $\mathbf{GLaw}_{\mathbf{M}}$ are equivalent.* □

We can also prove the equivalence of the categories of models.

Lemma 38. *If $\mathbf{C}$ is a category satisfying $\mathbf{M}$-model condition, then $\mathrm{Mod}_{\mathcal{T}}(\mathbf{C})$ is equivalent to $\mathrm{Mod}(\mathbf{Th}\mathcal{T}, \widetilde{\mathbf{C}_T})$ where T is the $\mathbf{M}^{\mathrm{t}}$-action on $\mathbf{C}$.* □

5.2 Graded Lawvere theories and Finitary Graded Monads

We prove that the category of graded Lawvere theories $\mathbf{GLaw}_{\mathbf{M}}$ and the category of finitary graded monads $\mathbf{GMnd}_{\mathbf{M}}$ are equivalent. Given a graded Lawvere theory, a finitary graded monad is obtained as a coend that represents the set of terms. On the other hand, given a finitary graded monad, a graded Lawvere theory is obtained from taking the full sub-$[\mathbf{M}, \mathbf{Set}]_0$-category on arities $\mathrm{ob}(\mathbf{N}_{\mathbf{M}}^{\mathrm{op}})$ of the opposite category of the Kleisli(-like) category in Definition 7. These constructions give rise to an equivalence of categories.

An $\mathbf{M}$-graded Lawvere theory yields a finitary graded monad by letting $m*X$ be the set of terms of grade m whose variables range over X.

Definition 39. Let $\mathbf{L}$ be an $\mathbf{M}$-graded Lawvere theory. We define $T_{\mathbf{L}} = (*, \eta, \mu)$ by a (finitary) $\mathbf{M}$-graded monad whose functor part is given as follows.

$$m * X := \int^{n \in \aleph_0} \mathbf{L}(\underline{n}, \underline{1})m \times X^n$$

Note that $\mathbf{L}(\underline{-}, \underline{1}) : \aleph_0 \to [\mathbf{M}, \mathbf{Set}]_0$ is a $\mathbf{Set}$-functor here.

Given a graded monad, a graded Lawvere theory is obtained as follows.

Definition 40. Let $T = (*, \eta, \mu)$ be an $\mathbf{M}$-graded monad on $\mathbf{Set}$. Let $\mathbf{L}_T$ be the full sub-$[\mathbf{M}, \mathbf{Set}]_0$-category of $(\widetilde{\mathbf{Set}_T})^{\mathrm{op}}$ with $\mathrm{ob}(\mathbf{L}_T) = \mathbb{N}$.

Since $\mathbf{L}_T$ has $\mathbf{N}_M$-cotensors $\underline{n} \pitchfork 1 = n$ whose projections are given by $\pi_i = (* \mapsto \eta(i)) \in \mathbf{Set}(1, I * n)$, $\mathbf{L}_T$ is a graded Lawvere theory.

Given a morphism $\alpha : T \to T'$ in $\mathbf{GMnd}_{\mathbf{M}}$, we define $\mathbf{L}_\alpha : \mathbf{L}_T \to \mathbf{L}_{T'}$ by $(\mathbf{L}_\alpha)_{n,n',m} = \mathbf{Set}(n', \alpha_{n,m}) : \mathbf{L}_T(n,n')m \to \mathbf{L}_{T'}(n,n')m$. It is easy to prove that $\mathbf{L}_\alpha$ is a morphism in $\mathbf{GLaw}_{\mathbf{M}}$ and $\mathbf{L}_{(-)} : \mathbf{GMnd}_{\mathbf{M}} \to \mathbf{GLaw}_{\mathbf{M}}$ is a functor.

Theorem 41. *Two categories $\mathbf{GLaw}_{\mathbf{M}}$ and $\mathbf{GMnd}_{\mathbf{M}}$ are equivalent.*

Proof. $\mathbf{L}_{(-)}$ is an essentially surjective fully faithful functor. □

6 Combining Effects

Under the correspondence to algebraic theories, combinations of computational effects can be understood as combinations of algebraic theories. In particular, sums and tensor products are well-known constructions [9]. In this section, we show that these constructions can be adapted to graded algebraic theories. By the equivalence $\mathbf{GMnd}_{\mathbf{M}} \simeq \mathbf{GLaw}_{\mathbf{M}} \simeq \mathbf{GS}_{\mathbf{M}}$ in §5, constructions like sums and tensor products in one of these categories induce those in the other two categories. So, we choose $\mathbf{GS}_{\mathbf{M}}$ and describe sums as colimits in $\mathbf{GS}_{\mathbf{M}}$ and tensor products as a mapping $\mathbf{GS}_{\mathbf{M}_1} \times \mathbf{GS}_{\mathbf{M}_2} \to \mathbf{GS}_{\mathbf{M}_1 \times \mathbf{M}_2}$.

6.1 Sums

We prove that $\mathbf{GS_M}$ has small colimits.

Lemma 42. *The category* $\mathbf{GS}_M$ *has small coproducts.*

Proof. Given a family $\{(\Sigma^{(i)}, E^{(i)})\}_{i\in I}$ of objects in $\mathbf{GS_M}$, the coproduct is obtained by the disjoint union of operations and equations: $\coprod_{i\in I}(\Sigma^{(i)}, E^{(i)}) = \left(\bigcup_{i\in I} \Sigma^{(i)}, \bigcup_{i\in I} E^{(i)}\right)$. □

Lemma 43. *The category* $\mathbf{GS}_M$ *has coequalizers.*

Proof. Let $\mathcal{T} = (\Sigma, E)$ and $\mathcal{T}' = (\Sigma', E')$ be graded algebraic theories and $\alpha, \beta : \mathcal{T} \to \mathcal{T}'$ be a morphism. The coequalizer $\mathcal{T}''$ of α and β is given by adding the set of equations induced by α and β to $\mathcal{T}'$, that is, $\mathcal{T}'' := (\Sigma', E' \cup E'')$ where $E'' = \{(s,t) \mid \exists f \in \Sigma, \alpha(f) = [s] \wedge \beta(f) = [t]\}$. □

Since a category has all small colimits if and only if it has all small coproducts and coequalizers, we obtain the following corollary.

Corollary 44. *Three equivalent categories* $\mathbf{GS_M}$, $\mathbf{GMnd_M}$ *and* $\mathbf{GLaw_M}$ *are cocomplete.* □

Example 45. It is known that the sum of an ordinary monad T and the exception monad $(-)+\mathrm{Ex}$ (where Ex is a set of exceptions) is given by $T((-)+\mathrm{Ex})$ [9, Corollary 3]. We show that a similar result holds for the graded exception monad.

Let $\mathcal{T}^{\mathrm{ex}}$ be the theory in Example 25 and $\mathbf{M}$ be the preordered monoid used there. We denote $(*^{\mathrm{ex}}, \eta^{\mathrm{ex}}, \mu^{\mathrm{ex}})$ for the graded exception monad. Let $\mathcal{T} = (\Sigma, E)$ be a ($\mathbf{1}$-graded) theory and (T, η^T, μ^T) be the corresponding ordinary monad. Let $\mathcal{T}^{\mathbf{M}} = (\Sigma^{\mathbf{M}}, E^{\mathbf{M}})$ be the $\mathbf{M}$-graded theory obtained from $\mathcal{T}$ as in Example 26. We consider a graded monad obtained as the sum of $\mathcal{T}^{\mathrm{ex}}$ and $\mathcal{T}^{\mathbf{M}}$.

A free model functor F for $\mathcal{T}^{\mathrm{ex}} + \mathcal{T}^{\mathbf{M}}$ is given by $FXm = T(m *^{\mathrm{ex}} X)$. For each n-ary operation f in $\mathcal{T}$, $|f|_m^{FX} : (T(m *^{\mathrm{ex}} X))^n \to T(m *^{\mathrm{ex}} X)$ is induced by free models of $\mathcal{T}$, and for each $e \in \mathrm{Ex}$, $|\mathsf{raise}_e|_m^{FX} : 1 \to T(\{e\} *^{\mathrm{ex}} X)$ is defined by $\eta^T_{\{e\} *^{\mathrm{ex}} X}(e) \in T(\{e\} *^{\mathrm{ex}} X)$. It is easy to see that FX defined above is indeed a model of $\mathcal{T}^{\mathrm{ex}} + \mathcal{T}^{\mathbf{M}}$. Therefore, we obtain a graded monad $m * X = T(m *^{\mathrm{ex}} X)$.

6.2 Tensor Products

The tensor product of two ordinary algebraic theories (Σ, E) and (Σ', E') is constructed as $(\Sigma \cup \Sigma', E \cup E' \cup E_\otimes)$ where $E_\otimes$ consists of $f(\lambda i.g(\lambda j.x_{ij})) = g(\lambda j.f(\lambda i.x_{ij}))$ for each $f \in \Sigma$ and $g \in \Sigma'$. However, when we extend tensor products to graded algebraic theories, the grades of the both sides are not necessarily equal. If the grade of f is m and the grade of g is m', then the grades of $f(\lambda i.g(\lambda j.x_{ij}))$ and $g(\lambda j.f(\lambda i.x_{ij}))$ are $m \otimes m'$ and $m' \otimes m$, respectively. Therefore, we have to somehow guarantee that the grade of $f \in \Sigma$ and the grade of $g \in \Sigma'$ commute. We solve this problem by taking the product of monoidal categories. That is, we define the tensor product of an $\mathbf{M}_1$-graded algebraic theory and an $\mathbf{M}_2$-graded algebraic theory as an $\mathbf{M}_1 \times \mathbf{M}_2$-graded algebraic theory.

Before defining tensor products, we consider extending an $\mathbf{M}$-graded theory to $\mathbf{M}'$-graded theory along a lax monoidal functor $G = (G, \eta^G, \mu^G) : \mathbf{M} \to \mathbf{M}'$. Given an $\mathbf{M}$-graded theory $\mathcal{T} = (\Sigma, E)$, we define the $\mathbf{M}'$-graded theory $G_*\mathcal{T} = (G_*\Sigma, G_*E)$ by $(G_*\Sigma)_{n,m'} := \{f \in \Sigma_{n,m} \mid Gm = m'\}$ and $G_*E := \{G_*(s) = G_*(t) \mid (s = t) \in E\}$ where for each term t of $\mathcal{T}$ (with grade m), $G_*(t)$ is the term of $G_*\mathcal{T}$ (with grade Gm) defined inductively as follows: if x is a variable, then $G_*(x) := c_{\eta^G}(x)$; for each $w : m \to m'$ and term t, $G_*(c_w(t)) := c_{Gw}(G_*(t))$; for each $f \in \Sigma_{n,m}$ and terms $t_1, \ldots, t_n$ with grade m', $G_*(f(t_1, \ldots, t_n)) := c_{\mu^G_{m,m'}}(f(G_*(t_1), \ldots, G_*(t_n)))$.

The tensor product of $\mathcal{T}_1 \in \mathbf{GS}_{\mathbf{M}_1}$ and $\mathcal{T}_2 \in \mathbf{GS}_{\mathbf{M}_2}$ is defined by first extending $\mathcal{T}_1$ and $\mathcal{T}_2$ to $\mathbf{M}_1 \times \mathbf{M}_2$-graded theories and then adding commutation equations.

Definition 46 (tensor product). Let $\mathcal{T}_1 = (\Sigma, E) \in \mathbf{GS}_{\mathbf{M}_1}$ and $\mathcal{T}_2 = (\Sigma', E') \in \mathbf{GS}_{\mathbf{M}_2}$. The *tensor product* $\mathcal{T}_1 \otimes \mathcal{T}_2$ is defined by $(K_*\Sigma \cup K'_*\Sigma', K_*E \cup K'_*E' \cup E_{\mathcal{T}_1 \otimes \mathcal{T}_2}) \in \mathbf{GS}_{\mathbf{M}_1 \times \mathbf{M}_2}$ where $K : \mathbf{M}_1 \to \mathbf{M}_1 \times \mathbf{M}_2$ and $K' : \mathbf{M}_2 \to \mathbf{M}_1 \times \mathbf{M}_2$ are lax monoidal functors defined by $Km_1 := (m_1, I_2)$ and $K'm_2 := (I_1, m_2)$, and

$$E_{\mathcal{T}_1 \otimes \mathcal{T}_2} := \{f(\lambda i.g(\lambda j.x_{ij})) = g(\lambda j.f(\lambda i.x_{ij})) \mid f \in (K_*\Sigma)_{n,m}, g \in (K'_*\Sigma')_{n',m'}\}.$$

That is, if f is an operation in $\mathcal{T}_1$ with grade $m_1 \in \mathbf{M}_1$, then $\mathcal{T}_1 \otimes \mathcal{T}_2$ has the operation f with grade $(m_1, I_2) \in \mathbf{M}_1 \times \mathbf{M}_2$ and similarly for operations in $\mathcal{T}_2$.

The tensor products satisfy the following fundamental property.

Proposition 47. *Let* $\mathbf{C}$ *be a category satisfying* $\mathbf{M}_1 \times \mathbf{M}_2$*-model condition. Let* $\mathcal{T}_i$ *be an* $\mathbf{M}_i$*-graded algebraic theory for* $i = 1, 2$*. Then we have an isomorphism* $\mathrm{Mod}_{\mathcal{T}_1}(\mathrm{Mod}_{\mathcal{T}_2}(\mathbf{C})) \cong \mathrm{Mod}_{\mathcal{T}_1 \otimes \mathcal{T}_2}(\mathbf{C})$.

Proof. Let $((A, |\cdot|'), |\cdot|) \in \mathrm{Mod}_{\mathcal{T}_1}(\mathrm{Mod}_{\mathcal{T}_2}(\mathbf{C}))$ be a model. For each operation f in $\mathcal{T}_1$, $|f| : (A, |\cdot|')^n \to m \circledast (A, |\cdot|')$ is a homomorphism. This condition is equivalent to satisfying the equations in $E_{\mathcal{T}_1 \otimes \mathcal{T}_2}$. □

Example 48. We exemplify the tensor product by showing a graded version of [9, Corollary 6], which claims that the L-fold tensor product of the side-effects theory in [21] with one location is the side-effects theory with L locations.

First, we consider the situation where there is only one memory cell whose value ranges over a finite set V. Let $\mathbf{2}$ the preordered monoid (join-semilattice) $(\{\bot, \top\}, \leq, \vee, \bot)$ where $\leq$ is the preorder defined by $\bot \leq \top$. Intuitively, $\bot$ represents pure computations, and $\top$ represents (possibly) stateful computations. Let $\mathcal{T}_{\mathrm{st}}$ be a $\mathbf{2}$-graded theory of two types of operations lookup (arity: V, grade: $\top$) and update_v (arity: 1, grade: $\top$) for each $v \in V$ and the four equations in [21] for the interaction of lookup and update. Note that we have to insert coercion to arrange the grade of the equation $\mathsf{lookup}(\lambda v \in V.\mathsf{update}_v(x)) = c_{\bot \leq \top}(x)$.

The graded monad $(*, \eta, \mu)$ induced by $\mathcal{T}_{\mathrm{st}}$ is as follows.

$$\bot * X = X \qquad \top * X = (V \times X)^V \qquad ((\bot \leq \top) * X)(x) = \lambda v.(v, x)$$

The middle equation can be explained as follows: any term with grade $\top$ can be presented by a canonical form $t_f := \mathsf{lookup}(\lambda v.\mathsf{update}_{f_V(v)}(f_X(v)))$ where $f = \langle f_V, f_X \rangle : V \to V \times X$ is a function, and therefore, the mapping $f \mapsto t_f$ gives a bijection between $(V \times X)^V$ and $\top * X = T^{\Sigma}_{\top}(X)/{\sim}$.

The L-fold tensor product of $\mathcal{T}_{\mathrm{st}}$, which we denote by $\mathcal{T}_{\mathrm{st}}^{\otimes L}$, is a $\mathbf{2}^L$-graded theory where $\mathbf{2}^L = (2^L, \subseteq, \cup, \emptyset)$ is the join-semilattice of subsets of L. Specifically, $\mathcal{T}_{\mathrm{st}}^{\otimes L}$ consists of operations lookup_l and $\mathsf{update}_{l,v}$ with grade $\{l\}$ for each $l \in L$ and $v \in V$ with additional three commutation equations in [21]. The induced graded monad is $L' *^{\otimes L} X = \{f : V^L \to (V^L \times X) \mid \mathrm{read}(L', f) \wedge \mathrm{write}(L', f)\}$ where $L' \subseteq L$, and $\mathrm{read}(L', f)$ and $\mathrm{write}(L', f)$ assert that f depends only on values at locations in L' and does not change values at locations outside L'. That is, $L' *^{\otimes L} X$ represents computations that touch only memory locations in L'.

$$\begin{aligned} \mathrm{read}(L', f) &:= \forall \sigma, \sigma' \in V^n, (\forall l \in L', \sigma(l) = \sigma'(l)) \implies f(\sigma) = f(\sigma') \\ \mathrm{write}(L', f) &:= \forall \sigma, \sigma' \in V^n, x \in X, (\sigma', x) = f(\sigma) \implies \forall l \notin L', \sigma(l) = \sigma'(l) \end{aligned}$$

7 Related Work

Algebraic theories for graded monads. Graded monads are introduced in [27], and notions of graded theory and graded Eilenberg–Moore algebra appear in [4,17] for coalgebraic treatment of trace semantics. However, these work only deal with $\mathbb{N}$-graded monads where $\mathbb{N}$ is regarded as a discrete monoidal category, while we deal with general monoidal categories. The Kleisli construction and the Eilenberg–Moore construction for graded monads are presented in [7] by adapting the 2-categorical argument on resolutions of monads [29].

Algebraic operations for graded monads are introduced in [12] and classified into two types, which are different in how to integrate the grades of subterms. One is operations that take terms with the same grade, and these are what we treated in this paper. The other is operations that take terms with different grades: the grade of $f(t_1, \dots, t_n)$ is determined by an *effect function* $\epsilon : \mathbf{M}^n \to \mathbf{M}$ associated to f. Although the latter type of operations is also important to give natural presentations of computational effects, we leave it for future work.

Enriched Lawvere theories. There are many variants of Lawvere theories [1, 10, 11, 15, 16, 19, 24, 25, 28], and most of them share a common pattern: they are defined as an identity-on-objects functor from a certain category (e.g., $\aleph_0^{\mathrm{op}}$) which represents arities, and the functor must preserve a certain class of products (or cotensors if enriched). Among the most relevant work to ours are enriched Lawvere theories [24] and discrete Lawvere theories [10].

For a given monoidal category $\mathbf{V}$, a Lawvere $\mathbf{V}$-theory is defined as an identity-on-objects finite cotensor (i.e. $\mathbf{V}_{\mathrm{fp}}$-cotensor) preserving $\mathbf{V}^{\mathrm{t}}$-functor $J : \mathbf{V}_{\mathrm{fp}}^{\mathrm{op}} \to \mathbf{L}$ where $\mathbf{V}_{\mathrm{fp}}$ is the full subcategory of $\mathbf{V}$ spanned by finitely presentable objects. If $\mathbf{V} = {[\mathbf{M}, \mathbf{Set}]_0}^{\mathrm{t}}$, Lawvere ${[\mathbf{M}, \mathbf{Set}]_0}^{\mathrm{t}}$-theories are analogous to our graded Lawvere theories except that we used $\mathbf{N}_{\mathbf{M}}^{\mathrm{op}}$ instead of $([\mathbf{M}, \mathbf{Set}]_0)_{\mathrm{fp}}$. Since $n \cdot y(I) \in \mathbf{N}_{\mathbf{M}}^{\mathrm{op}}$ is finitely presentable, we can say that the notion of graded Lawvere theory is obtained from enriched Lawvere theories by restricting arities to $\mathbf{N}_{\mathbf{M}}^{\mathrm{op}} \subseteq ([\mathbf{M}, \mathbf{Set}]_0)_{\mathrm{fp}}$. However, the correspondence to finitary graded monads on $\mathbf{Set}$ is an interesting point of our graded Lawvere theories compared to Lawvere $\mathbf{V}$-theories, which correspond to finitary $\mathbf{V}$-monads on $\mathbf{V}$.

Discrete Lawvere theories restrict arities of Lawvere $\mathbf{V}$-theories to $\aleph_0$, that is, a discrete Lawvere $\mathbf{V}$-theory is defined as a ($\mathbf{Set}$-enriched) finite-product preserving functor $J : \aleph_0^{\mathrm{op}} \to \mathbf{L}_0$ where $\mathbf{L}$ is a $\mathbf{V}^{\mathrm{t}}$-category. Actually, discrete Lawvere $[\mathbf{M}, \mathbf{Set}]_0{}^{\mathrm{t}}$-theories are equivalent to graded Lawvere theories because there is a finite-product preserving functor $\iota : \aleph_0^{\mathrm{op}} \to \mathbf{N}_{\mathbf{M}}^{\mathrm{op}}$ such that the composition with ι gives a bijection between graded Lawvere theories $J : \mathbf{N}_{\mathbf{M}}^{\mathrm{op}} \to \mathbf{L}$ and discrete Lawvere $[\mathbf{M}, \mathbf{Set}]_0{}^{\mathrm{t}}$-theories $J_0 \circ \iota : \aleph_0^{\mathrm{op}} \to \mathbf{L}_0$. However, we considered not only symmetric monoidal categories but also nonsymmetric ones, which cause a nontrivial problem when we define tensor products of algebraic theories. The problem is that adding commutation equations requires some kind of commutativity of monoidal categories. We solved this problem by considering product monoidal categories and defining the tensor product of an $\mathbf{M}_1$-graded theory and an $\mathbf{M}_2$-graded theory as an $\mathbf{M}_1 \times \mathbf{M}_2$-graded theory, and the use of two different monoidal categories is new to the best of our knowledge.

8 Conclusions and Future Work

To extend the correspondence between algebraic theories, Lawvere theories, and (finitary) monads, we introduced notions of graded algebraic theory and graded Lawvere theory and proved their correspondence with finitary graded monads. We also provided sums and tensor products for graded algebraic theories, which are natural extensions of those for ordinary algebraic theories. Since we do not assume monoidal categories to be symmetric, our tensor products are a bit different from the ordinary ones in that this combines two theories graded by (or enriched in) different monoidal categories. We hope that these results will lead us to apply many kinds of techniques developed for monads to graded monads.

As future work, we are interested in "change-of-effects", that is, changing the monoidal category $\mathbf{M}$ in $\mathbf{M}$-graded algebraic theory along a (lax) monoidal functor $F : \mathbf{M} \to \mathbf{M}'$. The problem already appeared in §6.2 to define tensor products, but we want to look for more properties of this operation. We are also interested in integrating a more general framework for notions of algebraic theory [6] and obtaining a graded version of the framework. Another direction is exploiting models of graded algebraic theories as modalities in the study of coalgebraic modal logic [4, 17] or weakest precondition semantics [8].

Acknowledgement. We thank Soichiro Fujii, Shin-ya Katsumata, Yuichi Nishiwaki, Yoshihiko Kakutani and the anonymous referees for useful comments. This work was supported by JST ERATO HASUO Metamathematics for Systems Design Project (No. JPMJER1603).

References

1. Berger, C., Melliès, P.A., Weber, M.: Monads with arities and their associated theories. Journal of Pure and Applied Algebra **216**(8), 2029 – 2048 (2012). https://doi.org/10.1016/j.jpaa.2012.02.039, special Issue devoted to the International Conference in Category Theory 'CT2010'

2. Curien, P., Fiore, M.P., Munch-Maccagnoni, G.: A theory of effects and resources: adjunction models and polarised calculi. In: Bodík, R., Majumdar, R. (eds.) Proceedings of the 43rd Annual ACM SIGPLAN-SIGACT Symposium on Principles of Programming Languages, POPL 2016, St. Petersburg, FL, USA, January 20 - 22, 2016. pp. 44–56. ACM (2016). https://doi.org/10.1145/2837614.2837652
3. Day, B.: On closed categories of functors. In: Reports of the Midwest Category Seminar IV. Lecture Notes in Mathematics, vol. 137, pp. 1–38. Springer, Berlin, Heidelberg (1970)
4. Dorsch, U., Milius, S., Schröder, L.: Graded monads and graded logics for the linear time - branching time spectrum. In: Fokkink, W., van Glabbeek, R. (eds.) 30th International Conference on Concurrency Theory, CONCUR 2019, August 27-30, 2019, Amsterdam, the Netherlands. LIPIcs, vol. 140, pp. 36:1–36:16. Schloss Dagstuhl - Leibniz-Zentrum für Informatik (2019). https://doi.org/10.4230/LIPIcs.CONCUR.2019.36
5. E. J. Linton, F.: Some aspects of equational categories. In: Proceedings of the Conference on Categorical Algebra. pp. 84–94 (01 1966). https://doi.org/10.1007/978-3-642-99902-4_3
6. Fujii, S.: A unified framework for notions of algebraic theory (2019), https://arxiv.org/abs/1904.08541
7. Fujii, S., Katsumata, S., Melliès, P.: Towards a formal theory of graded monads. In: Jacobs, B., Löding, C. (eds.) Foundations of Software Science and Computation Structures - 19th International Conference, FOSSACS 2016, Held as Part of the European Joint Conferences on Theory and Practice of Software, ETAPS 2016, Eindhoven, The Netherlands, April 2-8, 2016, Proceedings. Lecture Notes in Computer Science, vol. 9634, pp. 513–530. Springer (2016). https://doi.org/10.1007/978-3-662-49630-5_30
8. Hasuo, I.: Generic weakest precondition semantics from monads enriched with order. Theor. Comput. Sci. **604**, 2–29 (2015). https://doi.org/10.1016/j.tcs.2015.03.047
9. Hyland, M., Plotkin, G.D., Power, J.: Combining effects: Sum and tensor. Theor. Comput. Sci. **357**(1-3), 70–99 (2006). https://doi.org/10.1016/j.tcs.2006.03.013
10. Hyland, M., Power, J.: Discrete Lawvere theories and computational effects. Theoretical Computer Science **366**(1), 144 – 162 (2006). https://doi.org/10.1016/j.tcs.2006.07.007, algebra and Coalgebra in Computer Science
11. Hyland, M., Power, J.: The category theoretic understanding of universal algebra: Lawvere theories and monads. Electronic Notes in Theoretical Computer Science **172**, 437–458 (2007). https://doi.org/10.1016/j.entcs.2007.02.019, computation, Meaning, and Logic: Articles dedicated to Gordon Plotkin
12. Katsumata, S.: Parametric effect monads and semantics of effect systems. In: Jagannathan, S., Sewell, P. (eds.) The 41st Annual ACM SIGPLAN-SIGACT Symposium on Principles of Programming Languages, POPL '14, San Diego, CA, USA, January 20-21, 2014. pp. 633–646. ACM (2014). https://doi.org/10.1145/2535838.2535846
13. Kelly, M.: Basic Concepts of Enriched Category Theory. Lecture note series / London mathematical society, Cambridge University Press (1982)
14. Kura, S.: Graded algebraic theories (2020), https://arxiv.org/abs/2002.06784
15. Lack, S., Power, J.: Gabriel-Ulmer duality and Lawvere theories enriched over a general base. Journal of Functional Programming **19**(3-4), 265–286 (2009). https://doi.org/10.1017/S0956796809007254

16. Lucyshyn-Wright, R.B.B.: Enriched algebraic theories and monads for a system of arities. Theory and Applications of Categories **31**(5), 101–137 (2016)
17. Milius, S., Pattinson, D., Schröder, L.: Generic trace semantics and graded monads. In: Moss, L.S., Sobocinski, P. (eds.) 6th Conference on Algebra and Coalgebra in Computer Science, CALCO 2015, June 24-26, 2015, Nijmegen, The Netherlands. LIPIcs, vol. 35, pp. 253–269. Schloss Dagstuhl - Leibniz-Zentrum fuer Informatik (2015). https://doi.org/10.4230/LIPIcs.CALCO.2015.253
18. Moggi, E.: Computational lambda-calculus and monads. In: Proceedings of the Fourth Annual Symposium on Logic in Computer Science (LICS '89), Pacific Grove, California, USA, June 5-8, 1989. pp. 14–23. IEEE Computer Society (1989). https://doi.org/10.1109/LICS.1989.39155
19. Nishizawa, K., Power, J.: Lawvere theories enriched over a general base. Journal of Pure and Applied Algebra **213**(3), 377 – 386 (2009). https://doi.org/10.1016/j.jpaa.2008.07.009
20. Plotkin, G.D., Power, J.: Adequacy for algebraic effects. In: Honsell, F., Miculan, M. (eds.) Foundations of Software Science and Computation Structures, 4th International Conference, FOSSACS 2001 Held as Part of the Joint European Conferences on Theory and Practice of Software, ETAPS 2001 Genova, Italy, April 2-6, 2001, Proceedings. Lecture Notes in Computer Science, vol. 2030, pp. 1–24. Springer (2001). https://doi.org/10.1007/3-540-45315-6_1
21. Plotkin, G.D., Power, J.: Notions of computation determine monads. In: Nielsen, M., Engberg, U. (eds.) Foundations of Software Science and Computation Structures, 5th International Conference, FOSSACS 2002. Held as Part of the Joint European Conferences on Theory and Practice of Software, ETAPS 2002 Grenoble, France, April 8-12, 2002, Proceedings. Lecture Notes in Computer Science, vol. 2303, pp. 342–356. Springer (2002). https://doi.org/10.1007/3-540-45931-6_24
22. Plotkin, G.D., Pretnar, M.: A logic for algebraic effects. In: Proceedings of the Twenty-Third Annual IEEE Symposium on Logic in Computer Science, LICS 2008, 24-27 June 2008, Pittsburgh, PA, USA. pp. 118–129. IEEE Computer Society (2008). https://doi.org/10.1109/LICS.2008.45
23. Plotkin, G.D., Pretnar, M.: Handling algebraic effects. Logical Methods in Computer Science **9**(4) (2013). https://doi.org/10.2168/LMCS-9(4:23)2013
24. Power, J.: Enriched Lawvere theories. Theory and Applications of Categories **6**(7), 83–93 (1999)
25. Power, J.: Countable Lawvere theories and computational effects. Electr. Notes Theor. Comput. Sci. **161**, 59–71 (2006). https://doi.org/10.1016/j.entcs.2006.04.025
26. Sankappanavar, H.P., Burris, S.: A course in universal algebra. Springer-Verlag (1981)
27. Smirnov, A.: Graded monads and rings of polynomials. Journal of Mathematical Sciences **151**(3), 3032–3051 (2008)
28. Staton, S.: Freyd categories are enriched Lawvere theories. Electronic Notes in Theoretical Computer Science **303**, 197 – 206 (2014). https://doi.org/https://doi.org/10.1016/j.entcs.2014.02.010, proceedings of the Workshop on Algebra, Coalgebra and Topology (WACT 2013)
29. Street, R.: The formal theory of monads. Journal of Pure and Applied Algebra **2**(2), 149 – 168 (1972). https://doi.org/10.1016/0022-4049(72)90019-9

A Curry-style Semantics of Interaction: From untyped to second-order lazy $\lambda\mu$-calculus

James Laird

Department of Computer Science, University of Bath, UK

Abstract. We propose a "Curry-style" semantics of programs in which a nominal labelled transition system of types, characterizing observable behaviour, is overlaid on a nominal LTS of untyped computation. This leads to a notion of program equivalence as typed bisimulation.
Our semantics reflects the role of types as hiding operators, firstly via an axiomatic characterization of "parallel composition with hiding" which yields a general technique for establishing congruence results for typed bisimulation, and secondly via an example which captures the hiding of implementations in abstract data types: a typed bisimulation for the (Curry-style) lazy $\lambda\mu$-calculus with polymorphic types. This is built on an abstract machine for CPS evaluation of $\lambda\mu$-terms: we first give a basic typing system for this LTS which characterizes acyclicity of the environment and local control flow, and then refine this to a polymorphic typing system which uses equational constraints on instantiated type variables, inferred from observable interaction, to capture behaviour at polymorphic and abstract types.

1 Introduction

"Church-style" and "Curry-style" are used to distinguish programming languages in which the type of a term is intrinsic to its definition from those in which it is an extrinsic property. The same distinction may be applied to semantics of programming languages: in many models, type-objects are essential to the interpretation of a term — e.g. as a morphism between objects (types) in a category — but interpreting terms independently of their types (as in e.g. realizability interpretations) may have conceptual and practical advantages, particularly for describing Curry-style type systems. The aim of this semantic investigation of higher-order programs is to develop a Curry-style semantics of interaction by overlaying a labelled transition system of types onto a LTS of untyped computation, so that the observable behaviour of a typed state is restricted to the actions made available by its type. Our objective is to apply this to lazy functional programs: untyped and with Curry-style polymorphic typing systems, and to develop a theory of program equivalence — *typed bisimulation* — able to describe genericity and abstract datatypes in this setting.

Game Semantics Games models for programming languages are typically (but not invariably) given in a Church-style: terms are interpreted as strategies on

J. Goubault-Larrecq and B. König (Eds.): FOSSACS 2020, LNCS 12077, pp. 422–441, 2020.
https://doi.org/10.1007/978-3-030-45231-5_22

a specified two-player game which represents their type [2,9]. This kind of semantics is compositional by definition, at the cost of forgetting the internal computational behaviour of programs, and potentially excluding system level behaviour [6]. It uses categorical structure to describe its models and prove key results — in particular *soundness* with respect to an operational semantics.

By contrast, in *operational* game semantics [15,12], programs are interpreted as states in a labelled transition system based directly on their syntax and operational semantics. Internal computation is retained but can be factored out by restricting to observable behaviour. Soundness of these models "comes for free" — instead, the fundamental property requiring non-trivial proof is that they are *compositional* — that is, the equivalence induced on programs is a congruence. Basic structure which supports and systematizes these proofs would be useful (techniques such as Howe's method are not available in this intensional setting). We aim to show that defining operational game semantics in a Curry style gives the opportunity to formulate and apply such structure. This is complementary to characterization of the structure of operational game semantics at a categorical level [18], into which we believe our semantics can fit well. Our motivation and general methodology bears similarities to the programme of Berger, Honda and Yoshida [3] —- in which Curry-style types are used to characterize the π-calculus processes corresponding to functional and polymorphic programs — and to typing systems for process calculi such as those described in [10].

Hiding using types We will interpret (extrinsic) types as hiding operators: windows through which terms of a given type may interact with the world, while their internal behaviour is hidden from external observation — both passive and active. Our goal is to show that this interpretation can be used to model information hiding in two key areas of higher-order computation. The first, "parallel composition with hiding" is the fundamental operation on which game semantics is based. We axiomatize the notion of a typing system for an LTS with such an operation, in which a type is a state which characterizes precisely the possible interaction between a function and its argument at that type.

The second form of information hiding for which we give a Curry-style interpretation is hiding of implementation details using polymorphic (existential) types as abstract data types. Our key example of a typed labelled transition systems is a new model of the second-order $\lambda\mu$-calculus: we shall now discuss the background and significance of this contribution.

1.1 Program Equivalence and Polymorphism

Our starting point is the lazy λ-calculus — the pure, untyped λ-calculus, evaluated by weak head reduction — and its extension with first-class continuations, the corresponding version of Parigot's $\lambda\mu$-calculus [21]. As argued in [1], the lazy λ-calculus approximates well to the behaviour of lazy functional programming languages such as Haskell, and is thus an appropriate setting in which to explore properties such as program equivalence, for which there is now a rich and well-studied theory. For instance, *open* or *normal form* bisimilarity [25] is a coinductively defined equivalence which extends β-equivalence to infinitary behaviours. It gives a purely intensional characterization of program equivalence (by contrast to e.g. applicative bisimilarity, which involves quantifying over all possible arguments) and has a variety of alternative characterizations — for instance two terms are open bisimilar if and only if they have the same *Levy-Longo* trees [19], or their (call-by-name) translations in the π-calculus are weakly bisimilar [25,5]. (Or, indeed, if they are normal-form bisimilar as $\lambda\mu$-terms.)

Normal form bisimilarity of simply-typed λ-terms is just β-equivalence. However, extending to polymorphic types, such as those of the second-order λ-calculus (System F) [7,24] poses deeper questions. A primary motivation for introducing polymorphic types is that they can express abstract data types which hide implementation details [20] (cf. the module systems of Haskell and ML). A useful notion of program equivalence should therefore reflect this. As a simple example, the untyped λ-terms $\lambda f.f\,\lambda x.\lambda y.x$ and $\lambda f.f\,\lambda x.\lambda y.y$ are clearly not normal form bisimilar. But at the second-order type $\exists X.X \triangleq \forall Y.(\forall X.X \rightarrow Y) \rightarrow Y$ (which they both inhabit in a Curry-style presentation), they should be behaviourally equivalent — since any function of type : $\forall X.(X \rightarrow Y)$ will never call its argument. In other words, the existential type $\exists X.X$ "hides" the difference between $\lambda f.f\,\lambda x.\lambda y.x$ and $\lambda f.f\,\lambda x.\lambda y.y$. This is an observational equivalence, but of a particularly fundamental kind, since it (and other equivalences involving abstract data types) is robust in the presence or absence of side-effects. It can be captured by extensional methods such as applicative bisimilarity, which was extended to a polymorphic setting in [26], but this requires *quantification* over instantiating terms and types, whereas our semantics is based on *unification* of instantiating types.

The problem is that comparing the evaluation trees of terms (e.g. by normal form bisimulation) does not capture the capacity of their types to restrict interaction with the environment. Game semantics does reflect this interaction (in various manifestations), and therefore offers a potential solution. Although several games models for polymorphism do not capture data abstraction by existential types (including Hughes' semantics of System F [8], which is faithful with respect to $\beta\eta$-equivalence, and Curry-style models [16]) a series of related approaches does so. These include translation into the (polymorphically typed) π-calculus [4], and an operational form [17,27] and a traditional compositional presentation [14,13] of game semantics.

In these semantics, values of polymorphic variable type are interpreted as *pointers* to data of undisclosed type — e.g. a location where it is stored, or a channel on which it may be received. Instantiation of universally quantified type variables replaces this pointer-passing with copycat behaviour. This gives a natural interpretation of polymorphism in settings such as the π-calculus, or languages with general references, where pointers are first-class objects. However, it is closely associated with a Church-style presentation of second-order type systems — e.g. by the interpretation of type abstraction as an explicit creation of a pointer; in the case of "typed normal form bisimulation" [17] the translation of a term is explicitly determined by its type. This is significant because it is in the presence of polymorphism that key differences between Church-style and Curry-style emerge — for example, in allowing intersection types. The pointer-passing models also exhibit behaviours which go beyond untyped functional interaction, making their relationship to it unclear — in the game semantics [14], instantiation violates the fundamental innocence and visibility conditions on strategies; the π-calculus interpretation uses free name as well as bound name passing.

Curry-style semantics give a natural interpretation of second-order Curry-style typing, with a simple relationship to the semantics of the untyped $\lambda\mu$-calculus, by overlaying a more refined LTS of second order types on the same underlying LTS of computations.

2 Typed Labelled Transition Systems

In this section we describe a notion of typed labelled transition system and an associated equivalence: typed bisimulation. Based on this we axiomatize a simple typing system for parallel composition with hiding and show that it preserves typed bisimulation. Examples of typed LTS (in the form of models of the lazy $\lambda\mu$-calculus and lazy $\lambda\mu 2$-calculus) follow in the rest of the paper.

We work in the setting of *nominal sets* [23], which allows the introduction of fresh names (for store locations, communication channels, types etc). Assume a fixed, infinite set of *atoms* and a group G of permutations on them. A nominal set X is an action of G on a set $|X|$ such that each $x \in |X|$ has a finite supporting set of atoms such that if $\pi(a) = a$ for all atoms in this set then $\pi \cdot x = x$. We write $\sup(x)$ for the $\subseteq$-least of these sets (which is the intersection of all supporting sets for x).

Definition 1. *A nominal LTS is a labelled transition system $(\mathcal{S}, Act, \rightarrow)$ such that $\mathcal{S}$ (states) and Act (actions) are nominal sets and the transition relation $\rightarrow$ is* equivariant *— i.e. for any $\pi \in G$, $C \xrightarrow{a} C'$ if and only if $\pi \cdot C \xrightarrow{\pi \cdot a} \pi \cdot C'$.*

Similarly motivated notions of nominal LTS are developed in e.g. [22]. Our key example — an abstract machine for direct-style CPS evaluation — is given in the next section.

The directly observable part of a labelled transition system may be characterized by defining a *typing system* for it. (Similar notions of typing system for a process calculus are defined in [10], for example.)

Definition 2. *A* typing system *for a nominal LTS $(\mathcal{S}; Act; \rightarrow)$ is a nominal LTS $(\mathcal{T}; Obs; \hookrightarrow)$ such that $Obs \subseteq Act$, with a relation, $⨾$ (typing), from $\mathcal{S}$ to $\mathcal{T}$ which satisfies the following* subject reduction *properties for each $C ⨾ T$:*

- *If $C \xrightarrow{a} C'$ and $T \overset{a}{\hookrightarrow} T'$ then $C' ⨾ T'$ (we write $C ⨾ T \xrightarrow{a} C' ⨾ T'$).*
- *If $C \xrightarrow{a} C'$, where $a \notin Obs$ and $\sup(C') \cap \sup(T) \subseteq \sup(C) \cap \sup(T)$, then $C' ⨾ T$ (we write $C ⨾ T \longrightarrow C' ⨾ T$).*

Subject reduction requires that actions which are *observable* (i.e. in Obs) change a computation and its type in a way that respects the typing relation, and that those which are *internal* to a computation (i.e. in $Act \backslash Obs$) maintain its type (provided that any names fresh for the state are also fresh for its type).

Let $\Longrightarrow$ be the reflexive, transitive closure of the internal reduction $\longrightarrow$, and define $C \mathbin{⨾} T \stackrel{a}{\Longrightarrow} C' \mathbin{⨾} T'$ if $C \mathbin{⨾} T \Longrightarrow D \mathbin{⨾} T \stackrel{a}{\longrightarrow} D' \mathbin{⨾} T' \Longrightarrow C' \mathbin{⨾} T'$. To define weak bisimulation between typed states based on these relations, we need to take account of the fact that a name may be fresh for one, but already occur internally in the other (cf. [22]). So bisimulation is defined up to the equivalence on the states of type T which allows permutation of internal names: $C \simeq_T C'$ if there exists a permutation $\pi \in \mathrm{stab}(T)$ (i.e. $\pi \cdot T = T$) such that $C' = \pi \cdot C$.

Definition 3. *A* typed bisimulation *is a binary, symmetric, equivariant relation $\mathcal{R}$ between typed states ($C \mathbin{⨾} S$), such that if $(C \mathbin{⨾} S)\mathcal{R}(D \mathbin{⨾} T)$ then $S = T$ and:*

1. *If $C \mathbin{⨾} T \stackrel{a}{\longrightarrow} C' \mathbin{⨾} T'$ then there exists $D' \simeq_T D$ such that $(D' : T) \stackrel{a}{\Longrightarrow} (D'' : T')$, where $(C' \mathbin{⨾} T')\mathcal{R}(D'' : T')$.*
2. *If $C \mathbin{⨾} T \longrightarrow C' \mathbin{⨾} T$ then there exists $D' \simeq_T D$ such that $(D' : T) \Longrightarrow (D'' : T)$, where $(C' \mathbin{⨾} T)\mathcal{R}(D'' : T)$.*

Typed bisimilarity is the largest typed bisimulation: states C and D are bisimilar at type T ($C \sim_T D$) if ($C \mathbin{⨾} T$) and ($D \mathbin{⨾} T$) are typed bisimilar.

2.1 Parallel Composition with Hiding

Having proposed an interpretation of types as operators which hide internal communication, we now characterize the properties of a typing system for *parallel composition with hiding* which entail that it preserves typed bisimulation (i.e. the latter is a congruence).

Definition 4. *An* interaction structure *is a nominal LTS $(\mathcal{S}; Act; \rightarrow)$ such that $Act = \mathcal{L} \cup (\{+, -\} \times \mathcal{L})$ for some set of $\mathcal{L}$ of (unpolarized) labels, with an equivariant partial binary operation $|$ on $\mathcal{S}$ (*parallel composition*) such that if $C = C_1|C_2$ then $C \stackrel{a}{\longrightarrow} C'$ if and only if $C' = C_1'|C_2'$ for some C_1' and C_2' such that either:*

- *$C_1 \stackrel{a}{\longrightarrow} C_1'$ and $C_2' = C_2$, where $(\mathrm{sup}(C_1') \cup \mathrm{sup}(a)) \cap \mathrm{sup}(C_2) \subseteq \mathrm{sup}(C_1)$ or,*
- *$C_1' = C_1$ and $C_2 \stackrel{a}{\longrightarrow} C_2'$, where $(\mathrm{sup}(C_2') \cup \mathrm{sup}(a)) \cap \mathrm{sup}(C_1) \subseteq \mathrm{sup}(C_2)$ or,*
- *$C_1 \stackrel{pa}{\longrightarrow} C_1'$ and $C_2 \stackrel{\overline{p}a}{\longrightarrow} C_2'$, where $p \in \{+, -\}$.*

The nominal side-conditions require that any names which are fresh for the component to which they are introduced are fresh for the whole state.

Parallel composition is typed using a *ternary relation* between types: $T_1 \stackrel{T_2}{\multimap} T_3$ means "T_2 is an arrow type from T_1 to T_3" — there may be several arrow types between two types (or none).

Definition 5. *A typing system for an interaction structure* $(\mathsf{Comp}, \mathcal{L}, |)$ *is a typing system $(\mathcal{T}; (\{+, -\} \times \mathcal{L}); \hookrightarrow)$ for* Comp *with an equivariant ternary relation, $\multimap$, on $\mathcal{T}$ such that if $T_1 \stackrel{T_2}{\multimap} T_3$ then for any $C_1 \mathbin{⨾} T_1$ and $C_2 \mathbin{⨾} T_2$ such that*

$\mathsf{sup}(C_1) \cap \mathsf{sup}(C_2) \subseteq \mathsf{sup}(T_1)$, *the state* $C_1|C_2$ *is well-defined, has type* T_3 *and satisfies the following* interaction conditions*:*

1. *If* $C_1 \xrightarrow{pl} C'$ *and* $C_2 \xrightarrow{\overline{p}l} C_2'$ *then* $T_1 \overset{pl}{\hookrightarrow} T_1'$ *and* $T_2 \overset{\overline{p}l}{\hookrightarrow} T_2'$ *such that* $T_1' \overset{T_2'}{\multimap} T_3$.
2. *If* $C_2 \xrightarrow{a} C_2'$ *and* $T_3 \overset{a}{\hookrightarrow} T_3'$ *(with* $\mathsf{sup}(T_3') \cap \mathsf{sup}(T_2) \subseteq \mathsf{sup}(T_3)$*) then* $T_2 \overset{a}{\hookrightarrow} T_2'$ *such that* $T_1 \overset{T_2'}{\multimap} T_3'$.
3. *If* $C_1 \xrightarrow{a} C_1'$ *and* $T_3 \overset{a'}{\hookrightarrow} T_3'$ *then* $a \neq a'$.

Informally (1) requires that if C_1 and C_2 may communicate, then this is permitted by T_1 and T_2, and (2) and (3) require that the observable actions of $C_1|C_2$ permitted by T_3 correspond to actions of C_2 permitted by T_3. Note that for any $C_1 \mathbin{⨾} T_1$ and $C_2 \mathbin{⨾} T_2$ there exists $C_1' \simeq_{T_1} C_1$ such that $\mathsf{sup}(C_1') \cap \mathsf{sup}(C_2) \subseteq \mathsf{sup}(T_1)$ — i.e. there are no sidechannels of communication between C_1' and C_2 — and thus $C_1'|C_2$ is well-defined, has type T_3 and satisfies the interaction conditions. Moreover, these are sufficient to establish that typed bisimulation is a congruence with respect to parallel composition with hiding: a result that we will apply to our examples in the rest of the paper.

Proposition 1. *If* $C_1 \sim_{T_1} D_1$ *and* $C_2 \sim_{T_2} D_2$ *(and* $\mathsf{sup}(C_1) \cap \mathsf{sup}(C_2), \mathsf{sup}(D_1) \cap \mathsf{sup}(D_2) \subseteq \mathsf{sup}(T_1)$*) where* $T_1 \overset{T_2}{\multimap} T_3$ *then* $C_1|C_2 \sim_{T_3} D_1|D_2$.

Proof. We first establish the following renaming property: if $C_1 \mathbin{⨾} T_1 \longrightarrow C_1' \mathbin{⨾} T_1$ then there exists $\pi \in \mathsf{stab}(T_1) \cap \mathsf{stab}(T_2) \cap \mathsf{stab}(T_3)$ such that $C_1|C_2 \mathbin{⨾} T_3 \longrightarrow \pi(C_1')|C_2 \mathbin{⨾} T_3$ — by renaming any fresh names introduced by internal transition so that they are also fresh for C_2. Similarly, any internal reduction of C_2 corresponds to a reduction of $C_1|C_2$, up to such a renaming.

So suppose $C_1|C_2 \mathbin{⨾} T_3 \xrightarrow{pl} C' \mathbin{⨾} T_3$ (an observable transition). By definition of an interaction structure, and conditions (2) and (3), $C_2 \mathbin{⨾} T_2 \xrightarrow{pl} C_2' \mathbin{⨾} T_3'$ such that $T_1 \overset{T_2'}{\multimap} T_3'$. By assumption, there exists $D_2' \simeq_{T_2} D_2$ such that $D_2' \mathbin{⨾} T_2 \Longrightarrow D_2'' \xrightarrow{a} D_2''' \mathbin{⨾} T_2' \Longrightarrow D_2'''' \mathbin{⨾} T_2'$ and $D_2'''' \sim_{T_2'} C_2'$ and by the renaming property we may rename any fresh names in this reduction sequence to avoid clashes with D_1 — i.e. there exists $\pi \in \mathsf{stab}(T_1) \cap \mathsf{stab}(T_2) \cap \mathsf{stab}(T_3)$ such that:
$D_1|D_2' \mathbin{⨾} T_3 \Longrightarrow D_1|\pi(D_2'') \xrightarrow{\pi(a)} D_1|\pi(D_2''') \mathbin{⨾} T_3' \Longrightarrow D_1|\pi(D_2'''') \mathbin{⨾} T_3'$, and hence $\pi^{-1}(D_1)|\pi^{-1}(D_2') \mathbin{⨾} T_3 \overset{pl}{\Longrightarrow} \pi^{-1}(D_1)|D_2''''$ as required (since bisimilarity is closed under permutation of internal names).

If $C_1|C_2 \mathbin{⨾} T_3$ performs an *internal* action then this is either an internal action of $C_1 \mathbin{⨾} T_1$ or $C_2 \mathbin{⨾} T_2$, which is similar to the observable case, or else $C_1 \xrightarrow{pl} C_1'$ and $C_2 \xrightarrow{\overline{p}l} C_2'$ — so that $C_1|C_2$ performs the internal action l. Then by interaction condition (1), $T_1 \overset{pl}{\hookrightarrow} T_1'$ and $T_2 \overset{\overline{p}l}{\hookrightarrow} T_2'$ such that $T_1' \overset{T_2'}{\multimap} T_3$. So since $C_1 \sim_{T_1} D_1$ and $C_2 \sim_{T_2} D_2$, there exist $D_1' \sim_{T_1} D_1$ and $D_2' \sim_{T_2} D_2$ such that $D_1' \mathbin{⨾} T_1 \Longrightarrow D_1'' \mathbin{⨾} T_1 \xrightarrow{pl} D_1''' \mathbin{⨾} T_1' \Longrightarrow D_1'''' \mathbin{⨾} T_1'$ and $D_2' \mathbin{⨾} T_2 \Longrightarrow D_2'' \mathbin{⨾} T_2 \xrightarrow{\overline{p}l} D_2''' \mathbin{⨾} T_2' \Longrightarrow D_2'''' \mathbin{⨾} T_1'$ where $C_1' \sim_{T_1'} D_1''''$ and $C_2' \sim_{T_2'} D_2''''$. So using the renaming property we may obtain $\pi \in \mathrm{stab}(T_1) \cap \mathrm{stab}(T_2) \cap \mathrm{stab}(T_3)$ such that $D_1'|D_2' \fatsemi T_3 \Longrightarrow \pi(D_1'')|\pi(D_2'') \fatsemi T_3 \longrightarrow \pi(D_1''')|\pi(D_2''') \fatsemi T_3 \Longrightarrow \pi(D_1'''')|\pi(D_2''') \fatsemi T_3$ as required.

3 The Lazy $\lambda\mu$-calculus

We now define a typed interaction system giving an interpretation of the (untyped) lazy $\lambda\mu$-calculus — i.e. a direct-style CPS interpretation of lazy functional computation — yielding a novel, direct characterization of normal form bisimulation as typed bisimulation. This acts as a non-trivial example of a typed interaction system (as defined in the previous section) and a stepping stone to the polymorphic typing system for the same underlying language in the next section. First, we define an abstract machine for lazy CPS evaluation, in the form of a nominal LTS in which actions make explicit the calls made by a program to its environment. (Cf the analysis of $\lambda\mu$-calculus by π-calculus translation in [5].)

Definition 6. *The* unnamed *and* named *terms of the untyped $\lambda\mu$-calculus [21] are given (respectively) by the following grammars:*
$t ::= \; x \mid \lambda x.t \mid t\,t \mid \mu\alpha.M$
$M ::= [\alpha]t$

We equip the set of $\lambda\mu$-terms with a group action by assuming a set $\mathcal{N}$ of distinguished identifiers, partitioned into sorts (infinite subsets) of λ-variables $(x, y, z, \ldots)$ and μ-variables $(\alpha, \beta, \gamma \ldots)$ and (for later use) type variables $(X, Y, Z, \ldots$ The group of sort-preserving permutations on $\mathcal{N}$ acts pointwise on expressions (i.e. permuting elements of $\mathcal{N}$ and fixing symbols not in $\mathcal{N}$). We form a nominal set of $\lambda\mu$-terms consisting of the terms in which the free variables are all in $\mathcal{N}$ and those which occur bound (by λ or μ) are not, so that the support of a term is its set of free variables.

Based on this syntax, we define the sets of expressions (control terms) which determine the next transition of our abstract machine.

Definition 7. Control terms *are given by the grammar:* $\mathcal{A} ::= M \mid V \mid K \mid \bullet$

- *M ranges over the set of $\lambda\mu$* programs *(named terms) — i.e. $M ::= [\alpha]t$.*
- *V ranges over the set of $\lambda\mu$* values *(λ-abstractions) — i.e. $V ::= \lambda x.t$.*
- *K ranges over the set of $\lambda\mu$* continuations *(named contexts with a single hole at head position) — i.e. $K[\bullet] ::= [\alpha]\bullet \mid K[\bullet t]$.*
- *$\bullet$ is the empty context.*

As above we form a nominal set of control terms in which the support of each element is its set of free variables.

Definition 8. *An* environment *is a sort-respecting finite partial function $\mathcal{E}$ from $\mathcal{N}$ into the nominal sets of unnamed $\lambda\mu$-terms and continuations. The nominal set of environments has the G-action: $(\pi \cdot \mathcal{E})(a) = \pi \cdot (\mathcal{E}(\pi^{-1} \cdot a))$.*

Direct-style CPS evaluation of a program in an environment proceeds as follows:

- A variable inside a continuation $(\mathcal{E}; K[x])$ fetches the term bound to x and names it with a fresh μ-variable which is bound to K.
- A β-redex inside a continuation $(\mathcal{E}; K[\lambda x.t\,s])$ binds s to a fresh λ-variable y and K to a fresh μ-variable α and evaluates $[\alpha]t[y/x]$.
- A μ-abstraction inside a continuation $(\mathcal{E}; K[\mu\alpha.M)$ binds K to β and evaluates $M[\beta/\alpha]$.
- A named value $(\mathcal{E}; [\alpha]V)$ calls the continuation bound to α with V.

These transitions are labelled with actions of the form $a\langle \overrightarrow{b} \rangle$, where a is the variable called (if any) and $\overrightarrow{b}$ are the fresh variables created (if any). Except for μ-abstraction reduction, each of these evaluation rules decomposes into a complementary pair of input and output rules corresponding to the behaviour of the active (or "positive") part of the program and, a passive (or "negative" part). This decomposition is made precise in Definition 10 (parallel composition for configurations).

Definition 9. *The nominal labelled transition system* $\mathsf{Comp}_{\lambda\mu}$ *is defined:*

- States *are pairs* $(\mathcal{E}; \mathcal{A})$, *where* $\mathcal{E}$ *is an environment and* $\mathcal{A}$ *is a control term.*
- *The set of* actions *is* $\mathcal{L} \cup (\{+,-\} \times \mathcal{L})$, *where* $\mathcal{L}$ *is the nominal set of* labels

$$\bigcup_{x,\alpha \in \mathcal{N}_\lambda \times \mathcal{N}_\mu} \{\alpha, x\langle\alpha\rangle, \langle\alpha, x\rangle, \langle\alpha\rangle\}$$

- *The* transitions *are given in Table 1. By convention, a variable name mentioned on the right of a rule but not the left is assumed not to occur there.*

The *polarity* of a state is positive if the control term is a program or continuation, and negative if it is a value or the empty context (we write $V_\bullet$ for a passive term of either kind). Unpolarized transitions send positive states to positive states. Except for μ-abstraction reduction, each corresponds to complementary, positive and negative transitions, which send positive states to negative states and vice-versa.

$$(\mathcal{E}[\alpha \mapsto K]; [\alpha]V_\bullet) \xrightarrow{\alpha} (\mathcal{E}; K[V_\bullet])$$
$$(\mathcal{E}; K[(\lambda x.s)\,t]) \xrightarrow{\langle y,\alpha\rangle} (\mathcal{E}, (y \mapsto t), (\alpha \mapsto K); [\alpha]s[y/x])$$
$$(\mathcal{E}[x \mapsto t]; K[x]) \xrightarrow{x\langle\alpha\rangle} (\mathcal{E}, (\alpha \mapsto K); [\alpha]t)$$
$$(\mathcal{E}; K[\mu\alpha.M]) \xrightarrow{\langle\beta\rangle} (\mathcal{E}, (\beta \mapsto K); M[\beta/\alpha])$$

$(\mathcal{E}; [\alpha]V_\bullet) \xrightarrow{+\alpha} (\mathcal{E}; V_\bullet)$	$(\mathcal{E}[\alpha \mapsto K]; V_\bullet) \xrightarrow{-\alpha} (\mathcal{E}; K[V_\bullet])$
$(\mathcal{E}; K[\bullet\, t]) \xrightarrow{+\langle y,\alpha\rangle} (\mathcal{E}, (y \mapsto t), (\alpha \mapsto K); \bullet)$	$(\mathcal{E}; \lambda x.t) \xrightarrow{-\langle y,\alpha\rangle} (\mathcal{E}; [\alpha]t[y/x])$
$(\mathcal{E}; K[x]) \xrightarrow{+x\langle\alpha\rangle} (\mathcal{E}, (\alpha \mapsto K); \bullet)$	$(\mathcal{E}[x \mapsto t]; \bullet) \xrightarrow{-x\langle\alpha\rangle} (\mathcal{E}; [\alpha]t)$

Table 1: Abstract machine for CPS evaluation of lazy $\lambda\mu$-calculus

$$
\begin{array}{cc}
(\_; \lambda f.f\, \lambda x.x) & (\_; [\alpha] \bullet \lambda y.y) \\
\downarrow {\scriptstyle -\langle g,\beta\rangle} & \downarrow {\scriptstyle +\langle g,\beta\rangle} \\
(\_; [\beta]g\, \lambda x.x) & ((\beta \mapsto [\alpha]\bullet), (g \mapsto \lambda y.y); \bullet) \\
\downarrow {\scriptstyle +g\langle\gamma\rangle} & \downarrow {\scriptstyle -g\langle\gamma\rangle} \\
((\gamma \mapsto [\beta]\bullet\ \lambda x.x); \bullet) & ((\beta \mapsto [\alpha]\bullet), (g \mapsto \lambda y.y); [\gamma]\lambda y.y) \\
\downarrow {\scriptstyle -\gamma} & \downarrow {\scriptstyle +\gamma} \\
((\gamma \mapsto [\beta] \bullet \lambda x.x); [\beta]\bullet\ \lambda x.x) & ((\beta \mapsto [\alpha]\bullet), (g \mapsto \lambda y.y); \lambda y.y) \\
\downarrow {\scriptstyle +\langle z,\delta\rangle} & \downarrow {\scriptstyle -\langle z,\delta\rangle} \\
(\gamma \mapsto [\beta]\bullet\ \lambda x.x), (z \mapsto \lambda x.x), (\delta \mapsto [\beta]\bullet); \bullet) & ((\beta \mapsto [\alpha]\bullet), (g \mapsto \lambda y.y); [\delta]z) \\
\downarrow {\scriptstyle -z\langle\epsilon\rangle} & \downarrow {\scriptstyle +z\langle\epsilon\rangle} \\
(\gamma \mapsto [\beta]\bullet\ \lambda x.x), (z \mapsto \lambda x.x), (\delta \mapsto [\beta]\bullet); [\epsilon]\lambda x.x) & ((\beta \mapsto [\alpha]\bullet), (g \mapsto \lambda y.y), (\epsilon \mapsto [\delta]\bullet); \bullet) \\
\downarrow {\scriptstyle +\epsilon} & \downarrow {\scriptstyle -\epsilon} \\
(\gamma \mapsto [\beta]\bullet\ \lambda x.x), (z \mapsto \lambda x.x), (\delta \mapsto [\beta]\bullet); \lambda x.x) & ((\beta \mapsto [\alpha]\bullet), (g \mapsto \lambda y.y), (\epsilon \mapsto [\delta]\bullet); [\delta]\bullet) \\
\downarrow {\scriptstyle -\delta} & \downarrow {\scriptstyle +\delta} \\
(\gamma \mapsto [\beta]\bullet\ \lambda x.x), (z \mapsto \lambda xy.x), (\delta \mapsto [\beta]\bullet); [\beta]\lambda x.x) & ((\beta \mapsto [\alpha]\bullet), (g \mapsto \lambda y.y), (\epsilon \mapsto [\delta]\bullet); \bullet) \\
\downarrow {\scriptstyle +\beta} & \downarrow {\scriptstyle -\beta} \\
(\gamma \mapsto [\beta]\bullet\ \lambda x.x), (z \mapsto \lambda x.x), (\delta \mapsto [\beta]\bullet); \lambda x.x) & ((\beta \mapsto [\alpha]\bullet), (g \mapsto \lambda y.y), (\epsilon \mapsto [\delta]\bullet); [\alpha]\bullet)
\end{array}
$$

Fig. 1: Example traces evaluating $[\alpha](\lambda f.f\, \lambda x.x)\, \lambda y.y$

To define an *interaction structure* on $\mathsf{Comp}_{\lambda\mu}$ (Definition 4) we require a parallel composition operation on configurations.

Definition 10. *[Parallel Composition] On control terms, let* $|$ *be the (least) partial operation such that* $\mathcal{A}|\bullet = \bullet|\mathcal{A} = \mathcal{A}$ *and* $K|V = V|K = K[V]$.
Given configurations $C_1 = (\mathcal{E}_1; \mathcal{A}_1)$ *and* $C_2 = (\mathcal{E}_2; \mathcal{A}_2)$ *let* $C_1|C_2 \triangleq (\mathcal{E}_1 \cup \mathcal{E}_2; \mathcal{A}_1|\mathcal{A}_2)$, *provided* $\mathsf{dom}(\mathcal{E}) \cap \mathsf{dom}(\mathcal{E}) = \varnothing$ *and* $\mathcal{A}_1|\mathcal{A}_2$ *is well-defined. (*$C_1|C_2$ *is undefined, otherwise.)*

By inspection of the transitions in Table 1, we may see that $C_1|C_2$ has precisely the transitions of C_1 or C_2 (provided any fresh names are fresh for $C_1|C_2$), together with internal transitions arising from communication between C_1 and C_2. Therefore we have an interaction structure according to Definition 4. Figure 1 gives an illustrative example: the evaluation of $[\alpha](\lambda f.f\, \lambda x.x)\, \lambda y.y$ — which is the parallel composition $(\lambda f.f\, \lambda x.x)|([\alpha] \bullet \lambda y.y)$ — to $[\alpha]\lambda x.x$.

3.1 A Typing System

We now define a basic typing system for configurations which records minimal information about the control term (whether it is a program, value, continuation or empty context) but captures a more significant property of environments — acyclicity. This has practical relevance for memory management, but its immediate significance is that the second order typing in the next section relies on the fact that an acyclic environment may be contracted into a *valuation* by iteratively replacing variables bound in the environment until none occur as free variables.

Definition 11. *Given a nominal environment $\mathcal{E}$, define the binary relation on $\mathcal{N}$: $a \ll_{\mathcal{E}} b$ if $a \in \sup(\mathcal{E}(b))$ and let $\ll_{\mathcal{E}}^*$ be its transitive closure. Say that $\mathcal{E}$ is a* pre-valuation *(i.e. acyclic) if this is a strict partial order — i.e. $a \not\ll^* a$ for all $a \in \mathcal{N}$. $\mathcal{E}$ is a* valuation *if $\ll_{\mathcal{E}} = \ll_{\mathcal{E}}^*$ — i.e. $\sup(\mathcal{E}(a)) \cap \mathsf{dom}(\mathcal{E}) = \varnothing$ for all $a \in \mathsf{dom}(\mathcal{E})$.*

We assume a closure operation which takes an expression e and pre-valuation $\mathcal{E}$ to an expression $\mathcal{E}(e)$ obtained by replacing each atom $a \in \mathsf{dom}(\mathcal{E})$ with $\mathcal{E}(a)$ in e, having the property that $\sup(\mathcal{E}(e)) \cap \mathsf{dom}(\mathcal{E}) = \bigcup\{\sup(\mathcal{E}(a)) \mid a \in \sup(e) \cap \mathsf{dom}(\mathcal{E})\}$.

Lemma 1. *For any pre-valuation $\mathcal{E}$ there is a unique valuation $\mathcal{E}^*$ such that $\mathcal{E}^*(\mathcal{E}(e)) = \mathcal{E}^*(e)$ for all expressions e.*

Proof. Defining $\mathcal{E}^i$ by $\mathcal{E}^{i+1}(a) = \mathcal{E}^i(\mathcal{E}(a))$, the $\mathcal{E}^i$ form a chain of pre-evaluations such that the $\ll_{\mathcal{E}}$ downward closure of $\bigcup\{\sup(\mathcal{E}^i(a)) \cap \mathsf{dom}(\mathcal{E}) \mid a \in \mathsf{dom}(\mathcal{E})\}$ is empty or strictly decreasing, and thus is empty for some k — i.e. $\mathcal{E}^k$ is a pre-valuation and thus $\mathcal{E}^k(\mathcal{E}(a)) = \mathcal{E}(\mathcal{E}^k(a)) = \mathcal{E}^k(a)$ for all $a \in \mathsf{dom}(\mathcal{E})$, and so $\mathcal{E}^*(\mathcal{E}(e)) = \mathcal{E}^*(e)$ for all expressions e. If $\mathcal{E}^*(e) = \mathcal{E}^*(\mathcal{E}(e))$ for all expressions e, then $\mathcal{E}^*(e) = \mathcal{E}^*(\mathcal{E}^k(e)) = \mathcal{E}^k(e)$ for all e.

Definition 12. *The* basic *types for control terms are tuples $\Gamma \vdash \tau; \Delta$ where $\tau \in \{\top, \bot\}$ and Γ, Δ are non-repeating sequences — i.e. totally ordered finite sets — of λ and μ variables in $\mathcal{N}$, respectively.*

A control term $\mathcal{A}$ is well-typed with $\Gamma \vdash \tau; \Delta$ if $FV(\mathcal{A}) \subseteq \Gamma \cup \Delta$ and $\tau = \top$ if and only if $\mathcal{A}$ is a value or continuation. Basic types form a nominal set with the evident pointwise G-action.

Configurations are typed with polarized versions of these types. Given a polarized context (non-repeating sequence of polarized variables) $\Gamma = p_1x_1, \ldots, p_nx_n$ we write $|\Gamma|$ for the unpolarized context $x_1, \ldots, x_n$, $\overline{\Gamma}$ for the polarized context $\overline{p_1}x_1, \ldots, \overline{p_n}x_n$, and Γ^p for the (unpolarized) restriction of Γ to p-polarized elements.

Definition 13. *The nominal LTS $\mathsf{Ty}_{\lambda\mu}$ of basic $\lambda\mu$ configuration types:*

- States *are polarized configuration types — triples $\Gamma \vdash p\tau; \Delta$, where $p\tau \in \{+,-\} \times \{\top, \bot\}$ and Γ and Δ are polarized contexts of λ and μ variables in $\mathcal{N}$.*
- Actions *are the polarized actions of $\mathsf{Comp}_{\lambda\mu}$ — $Obs = \{+,-\} \times \mathcal{L}$*
- Transitions *are given by the rules in Table 2.*

We now define a typing relation from configurations to types. Let Γ be a polarized context. A pre-valuation for Γ is a pre-valuation $\mathcal{E}$ such that $\Gamma^+ \subseteq \mathsf{dom}(\mathcal{E})$, $\sup(\mathcal{E}(a)) \subseteq \mathsf{dom}(\mathcal{E}) \cup \Gamma^-$ for every $a \in \mathsf{dom}(\mathcal{E})$, and if $a, b \in \Gamma$ and $a \ll_{\mathcal{E}}^* b$ then $a <_\Gamma b$. Observe that if $\mathcal{E}$ is a pre-valuation for Γ, then $\mathcal{E}^*$ is a valuation for Γ such that for all $a \in \Gamma^+$, $FV(\mathcal{E}^*(a)) \subseteq \Gamma^-$.

$$\begin{array}{lcl}\Gamma \vdash p\top; \Delta & \overset{p\langle x,\alpha\rangle}{\hookrightarrow} & \Gamma, px \vdash \overline{p}\bot; \Delta, p\alpha \\ \Gamma[\overline{p}x] \vdash p\bot; \Delta & \overset{px\langle\alpha\rangle}{\hookrightarrow} & \Gamma \vdash \overline{p}\bot; \Delta, p\alpha \\ \Gamma \vdash p\top; \Delta[\overline{p}\alpha] & \overset{p\alpha}{\hookrightarrow} & \Gamma \vdash \overline{p}\bot; \Delta \\ \Gamma \vdash p\bot; \Delta[\overline{p}\alpha] & \overset{p\alpha}{\hookrightarrow} & \Gamma \vdash \overline{p}\top; \Delta\end{array}$$

Table 2: Transitions of basic configuration types

Definition 14 ($\lambda\mu$ Typing Relation). $(\mathcal{E}; \mathcal{A}) \mathbin{⨾} (\Gamma \vdash p\tau; \Delta)$ *if* $\mathsf{pol}(\mathcal{E}; \mathcal{A}) = p$ *and* $\mathcal{E}$ *is a pre-valuation for* $\Gamma \cup \Delta$ *such that* $\Gamma^- \vdash \mathcal{E}^*(\mathcal{A}) : \tau; \Delta^-$*, and for each* $x \in \Gamma^+$, $\Gamma^- \vdash \mathcal{E}^*(x) : \top; \Delta^-$ *and each* $\alpha \in \Delta^+$, $\Gamma^- \vdash \mathcal{E}^*(\alpha) : \top; \Delta^-$.

It is straightforward to check that this satisfies the subject reduction properties and thus defines a type system for $\mathsf{Comp}_{\lambda\mu}$.

Remark 1. We may apply a second constraint via our type system: *local control flow* — that continuations are called according to a LIFO discipline and thus may be stored on a stack (in game semantic terms, the *well-bracketing condition*). Evaluation of λ-terms by internal (and positive) transitions naturally satisfies this property — we can use types to ensure that the environment also does so.

Definition 15. *A configuration type* $\Gamma \vdash p\tau; \Delta$ *satisfies the local control condition if the polarities of* μ*-variables in* Δ *are alternating, and the polarity of the last element of* Δ *(if any) is* $\overline{p}$.

Transitions for local control types are given by refining the rules for calling a continuation to enforce stack discipline:

$$\begin{array}{l}\Gamma \vdash p\top; \Delta, \overline{p}\alpha \xrightarrow{p\alpha} \Gamma \vdash \overline{p}\bot; \Delta \\ \Gamma \vdash p\bot; \Delta, \overline{p}\alpha \xrightarrow{p\alpha} \Gamma \vdash \overline{p}\top; \Delta\end{array}$$

Subject reduction holds with respect to λ-configurations (in which the control term, and all terms and continuations in the environment, contain no μ-abstractions).

3.2 A Typed Interaction Structure

We now define an arrow relation, allowing a characterization of parallel composition with hiding for acyclic configurations. (Acyclicity is not preserved by union of environments in general, so the typing rules give a useful way of identifying pairs of configurations for which it does hold.)

Definition 16. *The* arrow relation *on configurations* $T_i = \Gamma_i \vdash p\tau_i; \Delta_i$ *is defined pointwise* — $T_1 \overset{T_2}{\multimap} T_3$ *if* $\Gamma_1 \overset{\Gamma_2}{\multimap} \Gamma_3$, $\Delta_1 \overset{\Delta_2}{\multimap} \Delta_3$*, and* $p\tau_1 \overset{p\tau_2}{\multimap} p\tau_3$ — *where*

- *For any polarized contexts,* $\Sigma_1 \overset{\Sigma_2}{\multimap} \Sigma_3$ *if* Σ_1 *and* Σ_3 *have disjoint underlying sets of elements and* Σ_2 *is an interleaving of* $\overline{\Sigma_1}$ *and* Σ_3.

– $p\tau_1 \overset{p\tau_2}{\multimap} p\tau_3$ *iff* $p\tau_1 = -\bot$ *and* $p\tau_2 = p\tau_3$ *or* $p\tau_3 = +\bot$ *and* $p\tau_2 = \overline{p}\tau_1$.

It remains to show that this satisfies Definition 5.

Proposition 2. $(\mathsf{Ty}_{\lambda\mu}, \multimap)$ *is a well-defined typing system for* $(\mathsf{Comp}_{\lambda\mu}, |)$.

Proof. Given $C_1 = (\mathcal{E}_1; \mathcal{A}_1)$ and $C_2 = (\mathcal{E}_2; \mathcal{A}_2)$, suppose $C_1 : T_1$, $C_2 : T_2$ and $\sup(C_1) \cap \sup(C_2) \subseteq \sup(T_1) = |\Gamma_1| \cup |\Delta_1|$:

- $\mathcal{A}_1|\mathcal{A}_2$ is well-defined, and has type τ_3, since either $\mathcal{A}_1 : -\bot$ (i.e. $\mathcal{A}_1 = \bullet$) and so $\mathcal{A}_1|\mathcal{A}_2 : \tau_2$, or $\mathcal{A}_1$ and $\mathcal{A}_2$ have complementary types, and so $\mathcal{A}_1|\mathcal{A}_2 : +\bot$ (i.e. they are a term and context which fit together to give a program).
- $\mathcal{E}_1 \cup \mathcal{E}_2$ is a pre-valuation, since the directed graph $(\ll_{\mathcal{E}_1} \cup \ll_{\mathcal{E}_2})$ is acyclic. (Any cycle in this graph would have to contain vertices from both $\ll_{\mathcal{E}_1}$ and $\ll_{\mathcal{E}_2}$, since both fragments are acyclic. Any path which enters and leaves one fragment must begin and end on points which are ordered by $\Gamma \cup \Delta$ and so composing such paths cannot lead to a cycle.)

Moreover, it is straightforward to verify that the interaction conditions are satisfied and that we therefore have a typed interaction structure.

Thus, by Proposition 1, typed bisimilarity is preserved by parallel composition plus hiding.

Proposition 3. *If* $C_1 \sim_{T_1} D_1$, $C_2 \sim_{T_2} D_2$ *and* $T_1 \overset{T_2}{\multimap} T_3$ *then* $C_1|C_2 \sim_{T_3} D_1|D_2$.

It immediately follows that (for example) bisimilarity of values is preserved by placing them inside the same continuation — i.e. if $(\_; v)$ and $(\_; v')$ are bisimilar at type $\Gamma \vdash -\top; \Delta$ then $(\_; K[v])$ and $(\_; K[v'])$ are bisimilar at type $\Gamma \vdash +\bot; \Delta$. Moreover, if typed bisimilarity is extended to an equivalence on all $\lambda\mu$-terms — $s \sim_{\Gamma;\Delta} t$ if $(\_; [\alpha]s) \sim_{-\Gamma \vdash +\bot; -\Delta, -\alpha} (\_; [\alpha]t)$, for $\alpha \notin \Delta$ — we may use Proposition 3 to show that if $s \sim_{\Gamma;\Delta} t$ then for any compatible context, $C[t] \sim_{\Gamma;\Delta} C[t']$.

4 A Polymorphic Type System

In this section we describe a more restrictive and informative typing system for the interaction structure of $\lambda\mu$ configurations. This yields a model of the lazy $\lambda\mu 2$-calculus — i.e. lazy $\lambda\mu$-calculus with polymorphic (second-order) Curry-style typing, which we now describe.

In order to fit such a type system to a semantics of lazy evaluation to weak head-normal form, we combine λ-abstraction and application with abstraction and instantiation of finite sequences of type variables — i.e. function types take the form $\forall(X_1 \ldots X_n).\sigma \to \tau$, where $X_1 \ldots X_n$ is a finite, non-repeating sequence of type variables. The judgments $\Theta \vdash \tau$ (τ is a well-formed type over the context of type-variables Θ) are derived according to the rules:

$$\frac{}{\Theta, X, \Theta' \vdash X} \qquad \frac{\Theta, X_1, \ldots, X_n \vdash \sigma \quad \Theta, X_1, \ldots, X_n \vdash \tau}{\Theta \vdash \forall(X_1 \ldots X_n).\sigma \to \tau}$$

Typing judgments are given with respect to an *equational context* (finite sequence of equations between types). These contexts play a key role in defining states in our LTS of types — they record constraints that type-instantiations must satisfy. For example, if a continuation K (with a hole) of type σ is called with an argument v of type τ then the type variables in σ and τ must have been instantiated so as to make these types equal. Formally, we define the judgment $\Theta \vdash \Xi$ (Ξ is a well-formed equational context over Θ) as follows:

$$\frac{}{\Theta \vdash \_} \qquad \frac{\Theta \vdash \Xi \quad \Theta \vdash \sigma \quad \Theta \vdash \tau}{\Theta \vdash \Xi, \sigma = \tau}$$

Type equality judgments with respect to an equational context, of the form $\Theta; \Xi \vdash \sigma = \tau$ (where $\Theta \vdash \Xi, \sigma, \tau$) are derived according to the rules:

$$\frac{}{\Theta;\Xi[\sigma=\tau]\vdash\sigma=\tau} \qquad \frac{}{\Theta;\Xi\vdash\tau=\tau} \qquad \frac{\Theta;\Xi\vdash\rho=\tau \quad \Theta;\Xi\vdash\sigma=\tau}{\Theta;\Xi\vdash\rho=\sigma}$$

$$\frac{\Theta;\Xi\vdash\forall\vec{X}.\sigma\to\tau=\forall\vec{X}.\sigma'\to\tau'}{\Theta,\vec{X};\Xi\vdash\sigma=\sigma'} \quad \frac{\Theta;\Xi\vdash\forall\vec{X}.\sigma\to\tau=\forall\vec{X}.\sigma'\to\tau'}{\Theta,\vec{X}\vdash\tau=\tau'} \quad \frac{\Theta,\vec{X};\Xi\vdash\sigma=\sigma' \quad \Theta,\vec{X}\vdash\tau=\tau'}{\Theta;\Xi\vdash\forall\vec{X}.\sigma\to\tau=\forall\vec{X}.\sigma'\to\tau'}$$

A valuation $\mathcal{V}$ for Θ *satisfies* an equational context $\Theta \vdash \sigma_1 = \tau_1, \ldots, \sigma_n = \tau_n$ if $\mathcal{V}(\sigma_i) \equiv \mathcal{V}(\tau_i)$ for each $i \leq n$.

Lemma 2. *$\Theta; \Xi \vdash \sigma = \tau$ if and only if for all valuations $\mathcal{V}$ which satisfy Ξ, $\mathcal{V}(\sigma) \equiv \mathcal{V}(\tau)$.*

A $\lambda\mu2$ type-in-context is a tuple $\Theta; \Xi; \Gamma \vdash \tau; \Delta$, where Θ is a context of type variables and Ξ is an equational context, τ is a $\lambda\mu2$-type (or $\bot$) and Γ and Δ are (respectively) sequences of λ-variables and μ-variables and their types (all over Θ). Assigning this type to a term may be understood as asserting that "for any valuation $\mathcal{V}$ of the type-variables in Θ which satisfies Ξ, the judgement $\mathcal{V}(\Gamma) \vdash t : \mathcal{V}(\tau); \mathcal{V}(\Delta)$ is valid". So, for example, $X, Y; Y = X \to X; \_ \vdash \lambda x.x : Y;$ is derivable according to the rules in Table 3. Note that there are no rules for introducing or discharging equational assumptions — they will be generated by the transitions of the LTS — so the terms of type $\Theta; \_; \Gamma \vdash t : \tau; \Delta$ are precisely those derivable in second-order $\lambda\mu$-calulus without type equality judgments.

$$\frac{}{\Theta;\Xi;\Gamma[x{:}\tau]\vdash x{:}\tau;\Delta} \qquad \frac{\Theta;\Xi;\Gamma\vdash t{:}\sigma;\Delta \quad \Theta;\Xi\vdash\sigma=\tau}{\Theta;\Xi;\Gamma\vdash t{:}\tau;\Delta} \qquad \frac{\Theta,X_1{:}\kappa,\ldots,X_n{:}\kappa_n;\Xi;\Gamma,x{:}\sigma\vdash t{:}\tau;\Delta}{\Theta;\Xi;\Gamma\vdash\lambda x.t{:}\forall X_1\ldots X_n.(\sigma\to\tau);\Delta}$$

$$\frac{\Theta;\Xi;\Gamma\vdash t{:}\forall X_1\ldots X_n.\sigma\to\tau;\Delta \quad \Theta\vdash\rho_1,\ldots,\rho_n \quad \Theta;\Xi;\Gamma\vdash s{:}\sigma[\rho_1/X_1\ldots\rho_n/X_n];\Delta}{\Theta;\Xi;\Gamma\vdash t\,s{:}\tau[\rho_1/X_1\ldots\rho_n/X_n];\Delta}$$

$$\frac{\Theta;\Xi;\Gamma\vdash t{:}\tau;\Delta[\alpha{:}\tau]}{\Theta;\Xi;\Gamma\vdash[\alpha]t{:}\bot;\Delta} \qquad \frac{\Theta;\Xi;\Gamma\vdash M{:}\bot;\Delta,\alpha{:}\tau}{\Theta;\Xi;\Gamma\vdash\mu x.M{:}\tau;\Delta}$$

Table 3: Typing Judgments for the lazy $\lambda\mu2$-Calculus

4.1 Second-Order Configuration Types

We now define a second-order typing system for the interaction structure $\mathsf{Comp}_{\lambda\mu}$ of $\lambda\mu$ configurations. Its states (second-order configuration types) capture the totality of information about the types of the control term and environment, and the instantiations for type variables by both a program and its environment, which may be inferred by an external observer of their interaction.

Definition 17. *A* second-order configuration type *is a polarized* $\lambda\mu 2$ *type-in-context — a tuple* $\Theta; \Xi; \Gamma; \vdash p\tau; \Delta$, *where* Θ *is a polarized context of type-variables, and* Ξ *is a polarized equational context,* Γ *and* Δ *are polarized contexts of typed* λ *and* μ *variables and* $p\tau$ *is a polarized* $\lambda\mu 2$*-type (or* $\perp$*), all over* Θ *.*

We place a further constraint — "polarized satisfiability" — on the configuration types which are permitted as states. This requires that their equational contexts can actually be satisfied by a program and environment successively instantiating type variables quantified positively and negatively (respectively), without knowing the types instantiated by the counterparty.

Definition 18. *A pre-valuation* $\mathcal{V}$ *for a polarized context of type variables* Θ positively satisfies *the polarized equational context* $\Theta \vdash \Xi$ *(written* $\mathcal{V} \vDash_\Theta \Xi$*) if for any pre-valuation* $\mathcal{W}$ *for* $\overline{\Theta}$*, the first formula in* Ξ not *satisfied by the valuation* $(\mathcal{V} \cup \mathcal{W})^*$ *for* $|\Theta|$ *(if any) is negative.* $\Theta \vdash \Xi$ *is (polarized)* satisfiable *if* $\Xi \vdash \Theta$ *and* $\overline{\Theta} \vdash \overline{\Xi}$ *are both positively satisfiable. Note that this implies that the underlying context* $|\Theta| \vdash |\Xi|$ *is satisfiable.*

Determining whether a polarized context is satisfiable is equivalent to a series of *conditional (first-order) unification* problems: these can be solved using the algorithm for first-order unification [11]. We place an equivalence relation on configuration types (cf. structural congruence of processes), allowing the principal type to be replaced by any of the (finitely many) types to which it is equivalent under Ξ.

Definition 19. $(\Theta; \Xi; \Gamma \vdash p\tau; \Delta) \eqsim (\Theta; \Xi; \Gamma \vdash p\tau'; \Delta)$ *if* $\Theta; \Xi \vdash \tau = \tau'$.

The (bipartite, nominal) LTS $\mathsf{Ty}_{\lambda\mu 2}$ of $\lambda\mu 2$ is defined:

- States are $\eqsim$-classes of satisfiable configuration types $\Theta; \Xi; \Gamma \vdash p\tau; \Delta$.
- *Actions* are polarized actions of $\mathsf{Comp}_{\lambda\mu}$: $Obs = \{+, -\} \times \mathcal{L}$.
- *Transitions* are given by the rules in Table 4.

To define a typing relation between configurations and $\lambda\mu 2$-configuration types, we first define typing judgements $\Theta; \Xi; \Gamma \vdash \mathcal{A} : \tau; \Delta$ for control terms. In the case of programs and values, these are as derived according to the rules in Table 3. For continuations, the rules

$$\frac{}{\Theta;\Xi;\Gamma\vdash[\alpha]\bullet:\tau;\Delta[\alpha:\tau]} \qquad \frac{\Theta;\Xi;\Gamma\vdash K:\tau[\rho_1/X_1\ldots\rho_n/X_n];\Delta \quad \Theta;\Xi;\Gamma\vdash s:\sigma[\rho_1/X_1\ldots\rho_n/X_n]}{\Theta;\Xi;\Gamma\vdash K[\bullet\, s]:\forall X_1\ldots X_n.\sigma\rightarrow\tau;\Delta}$$

are equivalent to typing $\Theta; \Xi; \Gamma \vdash K : \tau; \Delta$ if $\Theta; \Xi; \Gamma, \bullet : \tau \vdash K[\bullet] : \perp; \Delta$. The empty context has type $\perp$ in any well-formed context.

$\Theta; \Xi; \Gamma \vdash p\forall X_1 \ldots X_n.\sigma \to \tau; \Delta$	$\overset{p\langle x,\alpha\rangle}{\hookrightarrow}$	$\Theta, pX_1, \ldots, pX_n; \Xi; \Gamma, px : \sigma \vdash \overline{p}\bot; \Delta, p\alpha : \tau$
$\Theta; \Xi; \Gamma[\overline{p}x : \tau]; p\bot$	$\overset{px\langle\alpha\rangle}{\hookrightarrow}$	$\Theta; \Xi; \Gamma \vdash \overline{p}\bot; \Delta, p\alpha : \tau$
$\Theta; Xi; \Gamma \vdash p\bot; \Delta[\overline{p}\alpha : \tau]$	$\overset{p\alpha}{\hookrightarrow}$	$\Theta; \Xi; \Gamma \vdash \overline{p}\tau; \Delta$
$\Theta; \Xi; \Gamma \vdash p\sigma; \Delta[\overline{p}\alpha : \tau]$	$\overset{p\alpha}{\hookrightarrow}$	$\Theta; \Xi, p(\sigma = \tau); \Gamma \vdash \overline{p}\bot; \Delta$

Table 4: Transitions of second-order configuration types

Definition 20 (Typing Relation). *Let $\mathcal{V}$ be a valuation for Θ which positively satisfies Ξ, and define $\mathcal{V} \models (\mathcal{E}; \mathcal{A}) \,⨾\, \Theta; \Xi; \Gamma \vdash p\tau; \Delta$ if $\mathcal{E}$ is a pre-valuation for Γ, Δ, such that $\Theta^-; \mathcal{V}(\Xi^-); \mathcal{V}(\Gamma^-) \vdash \mathcal{E}^*(\mathcal{A})) : \mathcal{V}(\tau); \mathcal{V}(\Delta^-)$ and for each $x : \sigma \in \Gamma^+$, $\Theta^-; \mathcal{V}(\Xi^-); \mathcal{V}(\Gamma^-) \vdash \mathcal{E}^*(x) : \mathcal{V}(\sigma); \mathcal{V}(\Delta^-)$ and each $\alpha : \sigma \in \Delta^+$, $\Theta^-; \mathcal{V}(\Xi^-); \mathcal{V}(\Gamma^-) \vdash \mathcal{E}^*(\alpha) : \mathcal{V}(\sigma); \mathcal{V}(\Delta^-)$.*

Let $C \,⨾\, T$ if there exists a valuation $\mathcal{V}$ for Θ such that $\mathcal{V} \models C \,⨾\, T$.

Note that if $C \,⨾\, T$ and $T \approx T'$ then $C \,⨾\, T'$, so typing is a well-defined relation from configurations to equivalence classes of configuration types.

Proposition 4. $(\mathsf{Comp}_{\lambda\mu} \,⨾\, \mathsf{Ty}_{\lambda\mu 2})$ *satisfies the subject reduction property.*

Proof. For the observable transitions, this is a straightforward observation that the typing relation is preserved. For internal transitions (specifically, β reductions), we use the corresponding subject reduction property for $\lambda\mu 2$ substitutions — i.e. if $\Theta; \Xi; \Gamma \vdash K[\lambda x.t\, s] : \bot; \Delta$ then $\Theta; \Xi; \Gamma \vdash K[t[s/x]] : \bot; \Delta$ and if $\Theta; \Xi; \Gamma \vdash K[\mu\alpha.t] : \bot; \Delta$ then $\Theta; \Xi; \Gamma \vdash t[K/\alpha] : \bot; \Delta$.

Figure 2 gives an example illustrating the role of types in constraining behaviour: a trace of the value $\lambda f.f\, v \,⨾\, \exists X.X$, where v is an arbitrary typable value (recall that $\exists X.X \triangleq \forall Y.(\forall X.X \to Y) \to Y)$. Observe that there are no transitions from the the final state — a call to γ is not possible because $-Y, +X \vdash -(Y' = X')$ is not negatively satisfiable. In fact, the tree of transitions of $\exists X.X$ branches only on negative transitions (i.e. Opponent moves). It follows that any configuration of this type will have the same set of transitions, and that therefore $\lambda f.f\, \lambda xy.x \sim_{\exists X.X} \lambda f.f\, \lambda xy.y$ as proposed in the introduction.

4.2 A Second-Order Typed Interaction Structure

It remains to prove that $\mathsf{Ty}_{\lambda\mu 2}$ is a well-defined typing system for the interaction structure on $\mathsf{Comp}_{\lambda\mu}$, and that typed bisimulation is therefore a congruence. We need to establish that the pointwise extension of the arrow relation (Definition 16) to second-order configuration types (i.e. $T_1 \overset{T_2}{\multimap} T_3$ if $\Theta_1 \overset{\Theta_2}{\multimap} \Theta_3$, $\Xi_1 \overset{\Xi_2}{\multimap} \Xi_3$, $\Gamma_1 \overset{\Gamma_2}{\multimap} \Gamma_3$, $\Delta_1 \overset{\Delta_2}{\multimap} \Delta_3$, and $p\tau_1 \overset{p\tau_2}{\multimap} p\tau_3$) satisfies the conditions of Definition 5 — that if $C_1 = (\mathcal{E}_1; \mathcal{A}_1) \,⨾\, T_1$ and $C_2 = (\mathcal{E}_2; \mathcal{A}_2) \,⨾\, T_2$, where $T_1 \overset{T_2}{\multimap} T_3$ and

$$
\begin{array}{c}
(\_; \lambda f.f\,v) \fatsemi (\_;\_;\_ \vdash -(\forall X.X \to Y) \to Y; \_) \\
{\scriptstyle -\langle g,\alpha\rangle} \\
\downarrow \\
(\_; [\alpha] g\, v) \fatsemi (-Y'; \_; -g : \forall X.X \to Y' \vdash +\bot; -\alpha : Y') \\
{\scriptstyle +g\langle\beta\rangle} \\
\downarrow \\
((\beta \mapsto [\alpha] \bullet v); \bullet) \fatsemi (-Y'; \_; -g : \forall X.X \to Y' \vdash -\bot; -\alpha : Y', +\beta : \forall X.X \to Y' \\
{\scriptstyle -\beta} \\
\downarrow \\
((\beta \mapsto [\alpha] \bullet v); [\alpha] \bullet v) \fatsemi (-Y'; \_; -g : \forall X.X \to Y' \vdash +\forall X.X \to Y'; -\alpha : Y') \\
{\scriptstyle +\langle z,\gamma\rangle} \\
\downarrow \\
((\beta \mapsto [\alpha] \bullet v), (z \mapsto v), (\gamma \mapsto [\alpha]\bullet); \bullet) \fatsemi (-Y', +X'; \_; -g : \forall X.X \to Y', +z : X' \vdash -\bot; -\alpha : Y', +\gamma : Y') \\
{\scriptstyle -z\langle\delta\rangle} \\
\downarrow \\
((\beta \mapsto [\alpha] \bullet v), (z \mapsto v), (\gamma \mapsto [\alpha]\bullet); [\delta] v) \fatsemi (-Y', +X'; \_; -g : \forall X.X \to Y', +z : X' \vdash +\bot; -\alpha : Y', +\gamma : Y', -\delta : X') \\
{\scriptstyle +\delta} \\
\downarrow \\
((\beta \mapsto [\alpha] \bullet v), (z \mapsto v), (\gamma \mapsto [\alpha]\bullet); v) \fatsemi (-Y', +X'; \_; -g : \forall X.X \to Y', +z : X' \vdash -X'; -\alpha : Y', +\gamma : Y')
\end{array}
$$

Fig. 2: Trace of $\lambda f.f\,v : \exists X.X$

$\sup(C_1) \cap \sup(C_2) \subseteq \sup(T_1)$, then $C_1|C_2$ is well-defined, has type T_3 and satisfies the interaction conditions.

By Proposition 2, $C_1|C_2 = (\mathcal{E}_1 \cup \mathcal{E}_2; \mathcal{A}_1|\mathcal{A}_2)$ is a well-defined configuration, and $\mathcal{E} \triangleq \mathcal{E}_1 \cup \mathcal{E}_2$ is a pre-valuation for $\Gamma_3 \cup \Delta_3$. By the assumption that $C_1 \fatsemi T_1$ and $C_2 \fatsemi T_2$, there are valuations $\mathcal{V}_1 \vDash C_1 \fatsemi T_1$ and $\mathcal{V}_2 \vDash C_2 \fatsemi T_2$. Then $\mathcal{V} \triangleq \mathcal{V}_1 \cup \mathcal{V}_2$ is a pre-valuation for Θ_3. To show that $\mathcal{V}^* \vDash C_1|C_2 \fatsemi T_3$, we need to verify that:

Lemma 3. *$\mathcal{V}$ positively satisfies Ξ_3.*

Proof. Let $\mathcal{W}$ be a pre-valuation for $\overline{\Theta_3}$. The first formula in Ξ_2 (if any) which is not satisfied by $\mathcal{V} \cup \mathcal{W} = \mathcal{V}_1 \cup \mathcal{V}_2 \cup \mathcal{W}$ cannnot be positive in Ξ_1 (positively satisfied by $\mathcal{V}_1$) nor in Ξ_2 (positively satisfied by $\mathcal{V}_2$), and so must be a negative formula in Ξ_3.

Lemma 4. $\Theta_3^-; \mathcal{V}^*(\Xi_3); \mathcal{V}^*(\Gamma_3^-) \vdash \mathcal{E}^*(\mathcal{A}_1|\mathcal{A}_2) : \mathcal{V}(\tau); \mathcal{V}^*(\Delta_3^-)$

Proof. Observe that $\mathcal{E}^* = (\mathcal{E}_1^* \cdot \mathcal{E}_1^*)^i$ and $\mathcal{V} = (\mathcal{V}_2 \cdot \mathcal{V}_1)^i$ for some $i \leq n$. Hence, it suffices to prove by induction on i that $\Theta_2; (\mathcal{V}_2 \cdot \mathcal{V}_1)^i(\Xi_2); (\mathcal{V}_2 \cdot \mathcal{V}_1)^i(\Gamma_2^-) \vdash (\mathcal{E}_2^* \cdot \mathcal{E}_1^*)^i(\mathcal{A}_1|\mathcal{A}_2); (\mathcal{V}_2 \cdot \mathcal{V}_1)^i(\Delta_2^-)$.

Similarly, each term and continuation assigned to an output variable is well-typed under closure by $\mathcal{V}^*$ and $\mathcal{E}^*$ and thus:

Proposition 5. $C_1|C_2 \fatsemi T_3$.

It remains to show that the interaction conditions of Definition 5 are satisfied. The key is establishing condition 1 — that if $C_1 \xrightarrow{pl} C_1'$ and $C_2 \xrightarrow{\overline{pl}} C_2'$ then $T_1 \overset{pl}{\hookrightarrow} T_1'$ and $T_2 \overset{\overline{pl}}{\hookrightarrow} T_2'$ such that $T_1' \overset{T_2'}{\multimap} T_3$. This requires some further investigation of configuration types.

The interesting cases are those where $\mathcal{A}_1 \equiv \lambda x.t$ and $\mathcal{A}_2 \equiv K[\bullet s]$ (or vice-versa) and so they can perform the complementary actions $-\langle y, \alpha\rangle$ and $+\langle y, \alpha\rangle$. We need to show that $|\Theta_1|; |\Xi_1| \vdash \tau$ is *non-atomic* — that is, $|\Theta_1|; |\Xi_1| \vdash \tau = \forall X_1 \ldots X_m.\rho \rightarrow \sigma$ — for some ρ, σ. Observe that this implies that $|\Theta_2|; |\Xi_2| \vdash \tau$ is also non-atomic (since Ξ_2 contains the equations in Ξ_1) so that T_1 and T_2 can perform the complementary actions $-\langle y, \alpha\rangle$ and $+\langle y, \alpha\rangle$.

Since any derivation of a typing judgement for $\lambda x.t$ or $K[\bullet s]$ must conclude with $\rightarrow$-introduction followed by applications of the type-equality rule we have:

Lemma 5. *If $\Theta; \Xi; \Gamma \vdash \lambda x.t : \tau; \Delta$ or $\Theta; \Xi; \Gamma \vdash K[\bullet s] : \tau; \Delta$ then $\Theta; \Xi \vdash \tau$ is non-atomic.*

Hence, by the assumption that $(\mathcal{E}_1; \lambda x.t) \between (\Theta_1; \Xi_2; \Gamma_1 \vdash -\tau; \Delta_1)$ and $(\mathcal{E}_2; K[\bullet t]) \between (\Theta_2; \Xi_2; \Gamma_2 \vdash +\tau; \Delta)$ we know that $\Theta_1; \mathcal{V}_1(\Xi_1) \vdash \mathcal{V}_1(\tau)$ and $\Theta_2; \mathcal{V}_2^*(\Xi_2) \vdash \mathcal{V}_2^*(\tau)$ are non-atomic. From the latter we may infer that $\overline{\Theta_1}; \mathcal{V}_2^*(\overline{\Xi_1}) \vdash \mathcal{V}_2^*(\tau)$ is non-atomic, since Θ_2 and Ξ_2 are interleavings of $\overline{\Theta_1}$ and $\overline{\Xi_1}$ with the disjoint contexts Θ_3 and Ξ_3.

So to show that $|\Theta_1|; |\Xi_1| \vdash \tau$ is non-atomic is it is sufficient to prove the contrapositive.

Lemma 6. *Suppose $\mathcal{V}_+ \vDash_\Theta \Xi$ and $\mathcal{V}_- \vDash_{\overline{\Theta}} \overline{\Xi}$, where $|\Theta|; |\Xi| \vdash \tau$ is atomic. Then either $\Theta^-; \mathcal{V}_+(\Xi) \vdash \mathcal{V}_+(\tau)$ or $\Theta^+; \mathcal{V}_-(\Xi) \vdash \mathcal{V}_-(\tau)$ is atomic.*

Proof. We extend the grammar of types with an unbounded set of "neutral atoms" $A, B, C, \ldots$, which are equal only if syntactically identical, and prove the lemma for this extended set of types by an outer induction on the size of Θ, and an inner induction on the sum of the lengths of the types in Ξ.

At least one of $\mathcal{V}_+(\tau)$ and $\mathcal{V}_-(\tau)$ must be atomic and so if Ξ is empty then the hypothesis holds. Otherwise, $\Xi \equiv p(\sigma = \sigma'), \Xi'$ for some types σ, σ' and equational context Ξ' over Θ, and polarity $p \in \{+, -\}$.

If σ and σ' are both non-atomic, then by satisfiability $\sigma \equiv \forall X_1 \ldots X_n.\rho_1 \rightarrow \rho_2$ and $\sigma \equiv \forall X_1 \ldots X_n.\rho_1' \rightarrow \rho_2'$ for some $\rho_1, \rho_2, \rho_1', \rho_2'$. Letting $A_1, \ldots, A_n$ be fresh, distinct atomic types, define $\widehat{\rho} = \rho[A_1/X_1, \ldots, A_n/X_n]$. The equational context $\Xi'' = p(\widehat{\rho_1} = \widehat{\rho_1}'), p(\widehat{\rho_2} = \widehat{\rho_2}'), \Xi'$ is equivalent to (satisfied by the same valuations as) Ξ, and so $\Theta; \Xi'' \vdash \tau$ is atomic, and positively and negatively satisfied by $\mathcal{V}_+$ and $\mathcal{V}_-$. Hence, by inner induction hypothesis, one of $\Theta^-; \mathcal{V}_+(\Xi'') \vdash \mathcal{V}_+(\tau)$ or $\Theta^+; \mathcal{V}_-(\Xi'') \vdash \mathcal{V}_-(\tau)$ is atomic.

Otherwise at least one of σ and σ' is atomic. If $\sigma \equiv \sigma'$, then we may discard the tautology $\sigma = \sigma'$ and apply the (inner) inductive hypothesis to $\Theta; \Xi' \vdash \tau$. Otherwise at least one of σ, σ' must be a type-variable with polarity p in Θ (none of the other cases are p-satisfiable). So assume without loss of generality that $\Theta \equiv \Theta', pX, \Theta''$ and $\Xi \equiv p(\sigma = X), \Xi'$. We may show that:

- $\Theta', \Theta''; \Xi'[\sigma/X] \vdash \tau[\sigma/X]$ is atomic.
- $\Theta, \Theta'' \vdash \Xi'[\sigma/X]$ is positively satisfied by $\mathcal{V}_+$ and negatively satisfied by $\mathcal{V}_-$.

So by the outer inductive hypothesis, either $(\Theta', \Theta'')^-; \mathcal{V}_+(\Xi[\sigma/X]) \vdash \mathcal{V}_+(\tau)$ or $(\Theta, \Theta'')^+; \mathcal{V}_-(\Xi[\sigma/X]) \vdash \mathcal{V}_-(\tau)$ is atomic, and hence either $\Theta^-; \mathcal{V}_+(\Xi) \vdash \mathcal{V}_+(\tau)$ or $\Theta^+; \mathcal{V}_-(\Xi) \vdash \mathcal{V}_-(\tau)$ is atomic.

We have shown that the arrow relation satisfies the first interaction condition. 2 and 3 are straightforward to verify, establishing that $(\mathsf{Comp}_{\lambda\mu 2} \fatsemi \mathsf{Ty}_{\lambda\mu 2})$ is a well-defined typed interaction structure. Therefore, by Proposition 1, typed bisimulation is preserved by parallel composition plus hiding, and thus:

Theorem 1. *Typed bisimulation is a congruence for the $\lambda\mu 2$-calculus.*

5 Conclusions and Further Directions

We have described a "Curry-style" approach to game semantics, and used it to give new models of polymorphism. Various existing models may also be framed as typed interaction systems, such as the semantics of call-by-value in [12]. Nor are instances restricted to operational game semantics: for example we can present linear combinatory algebras of games and strategies in this way, and potentially other models of concurrent interaction. Unlike basic Church-style game semantics, these models give the opportunity to make finer distinctions between programs based on internal behaviour, which we have not explored here.

The notion of typed interaction structure reflects only limited structure of our models, but may be developed further. Having characterized parallel composition plus hiding within this setting, a natural next step would be a notion of copycat strategy, leading to structure for sharing and discarding information. One goal for such a development would be to put the generalization of congruence from configurations to terms on a systematic footing.

In another direction, our models of polymorphism may be developed further. In particular combining and fully exploiting generic and abstract data types often requires *higher-order* polymorphism, in which quantifiers range over *type operators* (functions which take types as arguments and return them as values). Whereas this is difficult to represent in game semantics, our model readily extends to a typing system based on System F_ω, which allows quantification over type-operators: the price to pay is that satisfiability of configuration types (and thus effective presentation of the states of our LTS) requires the solution of higher-order unification problems, which are undecidable, in general.

References

1. S. Abramsky. The lazy λ-calculus. In D. Turner, editor, *Research Topics in Functional Programming*, pages 65–117. Addison Wesley, 1990.
2. S. Abramsky, R. Jagadeesan, and P. Malacaria. Full abstraction for PCF. *Information and Computation*, 163:409–470, 2000.
3. M. Berger, K. Honda, and N. Yoshida. Sequentiality and the π-calculus. In *Proceedings of TLCA 2001*, volume 2044 of *Lecture Notes in Computer Science*. Springer-Verlag, 2001.
4. M. Berger, K. Honda, and N. Yoshida. Genericity and the π-calculus. *Acta Informatica*, 42, 2005.

5. M. Berger, K. Honda, and N. Yoshida. Process types as a descriptive tool for interaction: Control and the π-calculus. In *Proceedings of the Rewriting and Typed Lambda-calculi - joint international conference*, 2014.
6. D. R. Ghica and N. Tzevelekos. System level game semantics. *Proceedings of MFPS XXVIII*, ENTCS volume 286, pages 191 –211. 2012.
7. J.-Y. Girard. Linear logic. *Theoretical Computer Science*, 50, 1987.
8. D. Hughes. Games and definability for System F. In *Proceedings of the Twelfth International syposium on Logic in Computer Science, LICS '97.* IEEE Computer Society Press, 1997.
9. J. M. E. Hyland and C.-H. L. Ong. On full abstraction for PCF: I, II and III. *Information and Computation*, 163:285–408, 2000.
10. A. Igurashi and N. Kobayashi. A generic type system for the π-calculus. *Theoretical Computer Science*, 311:121–163, 2004.
11. Vladimir N. Krupski. The single-conclusion proof logic and inference rules specification. *Annals of Pure and Applied Logic*, 113:181 – 206, 2002.
12. J. Laird. A fully abstract trace semantics for general references. In *34th ICALP*, volume 4596 of *LNCS*, pages 667–679. Springer, 2007.
13. J. Laird. Game semantics of call-by-value polymorphism. In *Proceedings of ICALP '10*, number 6198 in LNCS. Springer-Verlag, 2010.
14. J. Laird. Game semantics for a polymorphic programming language. *Journal of the ACM*, 60(4), 2013.
15. S. B. Lassen and P. B. Levy. Typed normal form bisimulation. In *Proceedings 16th EACSL Conference on Computer Science and Logic*, number 4646 in LNCS, pages 283–297, 2007.
16. Joachim De Lataillade. Curry-style type isomorphisms and game semantics. *MSCS*, 18:647–692, 2008.
17. P. B. Levy and S. Lassen. Typed normal form bisimulation for parametric polymorphism. In *Proceedings of LICS 2008*, pages 341–552. IEEE press, 2008.
18. Paul Levy and Sam Staton. Transition systems over games. In *CSL-LICS '14*. ACM Press, 2014.
19. G. Longo. Set-theoretical models of lambda calculus: Theories, expansions and isomorphisms. *Annals of Pure and Applied Logic*, 24:153–188, 1983.
20. J. Mitchell and G. Plotkin. Abstract types have existential type. *ACM transactions on Programming Languages and Systems*, 10(3):470–502, 1988.
21. M. Parigot. $\lambda\mu$ calculus: an algorithmic interpretation of classical natural deduction. In *Proc. International Conference on Logic Programming and Automated Reasoning*, pages 190–201. Springer, 1992.
22. Joachim Parrow, Johannes Borgström, Lars-Henrik Eriksson, Ramunas Gutkovas, and Tjark Weber. Modal Logics for Nominal Transition Systems. In *26th International Conference on Concurrency Theory (CONCUR 2015)*, volume 42, pages 198–211, 2015.
23. A. M. Pitts. *Nominal Sets: Names and Symmetry in Computer Science.* Cambridge University Press, 2013.
24. J. C. Reynolds. Towards a theory of type structure. In *Proceedings of the Programming Symposium, Paris 1974*, number 19 in LNCS. Springer, 1974.
25. D. Sangiorgi. The lazy λ-calculus in a concurrency scenario. *Information and Computation*, 111:120 –153, 1994.
26. M. Smyth and G. Plotkin. The category-theoretic solution of recursive domain equations. *SIAM Journal on Computing*, 11(4):761–783, 1982.
27. N. Tzevelekos and G. Jaber. Trace semantics for polymorphic references. In *Proc. LICS'16*, pages 585–594. ACM, 2016.

An Axiomatic Approach to Reversible Computation*

Ivan Lanese[1] (✉), Iain Phillips[2], and Irek Ulidowski[3]

[1] Focus Team, University of Bologna/INRIA, Italy ivan.lanese@gmail.com
[2] Imperial College London, England i.phillips@imperial.ac.uk
[3] University of Leicester, England i.ulidowski@leicester.ac.uk

Abstract. Undoing computations of a concurrent system is beneficial in many situations, e.g., in reversible debugging of multi-threaded programs and in recovery from errors due to optimistic execution in parallel discrete event simulation. A number of approaches have been proposed for how to reverse formal models of concurrent computation including process calculi such as CCS, languages like Erlang, prime event structures and occurrence nets. However it has not been settled what properties a reversible system should enjoy, nor how the various properties that have been suggested, such as the parabolic lemma and the causal-consistency property, are related. We contribute to a solution to these issues by using a generic labelled transition system equipped with a relation capturing whether transitions are independent to explore the implications between these properties. In particular, we show how they are derivable from a set of axioms. Our intention is that when establishing properties of some formalism it will be easier to verify the axioms rather than proving properties such as the parabolic lemma directly. We also introduce two new notions related to causal consistent reversibility, namely causal safety and causal liveness, and show that they are derivable from our axioms.

Keywords: Reversible Computation, Labelled Transition System with Independence, Causal Safety, Causal Liveness

1 Introduction

Reversible computing studies computations which can proceed both in the standard, forward direction, and backward, going back to past states. Reversible computation has attracted interest due to its applications in areas as different as low-power computing [15], simulation [4], robotics [21], biological modelling [31] and debugging [23].

* This work has been partially supported by COST Action IC1405 on Reversible Computation - Extending Horizons of Computing. The first author has also been partially supported by French ANR project DCore ANR-18-CE25-0007 and by INdAM as a member of GNCS (Gruppo Nazionale per il Calcolo Scientifico).

J. Goubault-Larrecq and B. König (Eds.): FOSSACS 2020, LNCS 12077, pp. 442–461, 2020.
https://doi.org/10.1007/978-3-030-45231-5_23

There is widespread agreement in the literature about what properties characterise reversible computation in the sequential setting. Thus in reversible finite state automata [32], reversible cellular automata [13], reversible Turing machines [2] and reversible programming languages such as Janus [35] the main point is that the mapping from inputs to outputs is injective, and the reverse computation is deterministic.

Matters are less clear when it comes to reversible computation in the concurrent setting. Indeed, various reversible concurrent models have been studied, most notably in the areas of process calculi [6,29,18], event structures [34], Petri nets [1,25] and programming languages such as Erlang [20].

A main result of this line of research is that the notion of reversibility most suited for concurrent systems is *causal-consistent reversibility* (other notions are also used, e.g., to model biological systems [31]). According to an informal account of causal-consistent reversibility, any action can be undone provided that its consequences, if any, are undone beforehand. Following [6] this account is formalised using the notion of causal equivalent traces: two traces are causal equivalent if and only if they only differ for swapping independent actions, and inserting or removing pairs of an action and its reverse. According to [6, Section 3]

> Backtracking an event is possible when and only when a causally equivalent trace would have brought this event as the last one

which is then formalised as the so called causal consistency (CC) [6, Theorem 1], stating that coinitial computations are causal equivalent if and only if they are cofinal. Our new proof of CC (Proposition 3.6) shows that it holds in essentially any reversible formalism satisfying the Loop Lemma and the Parabolic Lemma, and we believe that CC is insufficient on its own to capture the informal notion.

A formalisation closer to the informal statement above is provided in [20, Corollary 22], stating that a forward transition t can be undone after a derivation iff all its consequences, if any, are undone beforehand. We are not aware of other discussions trying to formalise such a notion, except for [30], in the setting of reversible event structures. In [30], a reversible event structure is *cause-respecting* if an event cannot be reversed until all events it has caused have also been reversed; it is *causal* if it is cause-respecting and a reversible event can be reversed if all events it has caused have been reversed [30, Definition 3.34].

We provide (Section 4) a novel definition of the idea above, composed by:

Causal Safety (CS): an action cannot be reversed until any actions caused by it have been reversed;

Causal Liveness (CL): we should allow actions to reverse in any order compatible with CS, not necessarily the exact inverse of the forward order.

We shall see that CC does not capture the same property as CS+CL (Examples 4.15, 4.37), and that there are slightly different versions of CS and CL, which can all be proved under a small set of reasonable assumptions.

The main aim of this paper is to take an abstract model, namely labelled transition systems with independence equipped with reverse transitions (Section 2), and to show that the properties above (as well as others) can be derived

Acronym	Name	Defined in	Proved in	using
SP	Square Property	Def. 3.1	Axiom	-
BTI	Backward Transitions are Independent	Def. 3.1	Axiom	-
WF	Well-Founded	Def. 3.1	Axiom	-
CPI	Coinitial Propagation of Independence	Def. 4.2	Axiom	-
IRE	Independence Respects Events	Def. 4.12	Axiom	-
CIRE	Coinitial Independence Respects Events	Def. 4.29	Axiom	implied by IRE
IEC	Independence of Events is Coinitial	Def. 4.16	Axiom	-
PL	Parabolic Lemma	Def. 3.3	Prop. 3.4	BTI, SP
CC	Causal Consistency	Def. 3.5	Prop. 3.6	WF, PL
UT	Unique Transition	Def. 3.7	Cor. 3.8	CC
ID	Independence of Diamonds	Def. 4.6	Prop. 4.7	BTI, CPI
RPI	Reversing Preserves Independence	Def. 4.17	Prop. 4.18	SP, CPI, IRE, IEC
CS	Causal Safety	Def. 4.11	Thm. 4.13	SP, BTI, WF, CPI, IRE
CL	Causal Liveness	Def. 4.11	Thm. 4.14	SP, BTI, WF, CPI, IRE
$CS_<$	ordered Causal Safety	Def. 4.24	Prop. 4.39	SP, BTI, WF, CPI, NRE
$CL_<$	ordered Causal Liveness	Def. 4.24	Prop. 4.39	SP, BTI, WF, CPI, CIRE
CS_{ci}	coinitial Causal Safety	Def. 4.27	Thm. 4.28	SP, BTI, WF, CPI
CL_{ci}	coinitial Causal Liveness	Def. 4.27	Thm. 4.30	SP, BTI, WF, CPI, CIRE
NRE	No Repeated Events	Def. 4.35	Prop. 4.42	SP, BTI, WF, CPI, CIRE
RED	Reverse Event Determinism	Def. 4.40	Prop. 4.41	SP, BTI, WF, CPI, NRE

Table 1. Axioms and properties for causal reversibility.

from a small set of simple axioms (Sections 3, 4, 5). This is in sharp contrast with the large part of works in the literature, which consider specific frameworks such as CCS [6], CCS with broadcast [26], CCB [14], π-calculus [5], higher-order π [18], Klaim [11], Petri nets [25], μOz [22] and Erlang [20], and all give similar but formally unrelated proofs of the same main results. Such proofs will become instances of our general results. More precisely, our axioms will:

- exclude behaviours which are not compatible with causal-consistent reversibility (as we will discuss shortly);
- allow us to derive the main properties of reversible calculi which have been studied in the literature, such as CC (Proposition 3.6);
- hold for a number of reversible calculi which have been proposed, such as RCCS [6] and reversible Erlang [20] (Section 6).

Thus, when defining a new reversible formalism, one just has to check whether the axioms hold, and get for free the proofs of the most relevant properties. Notably, the axioms are normally easier to prove than the properties, hence the assessment of a reversible calculus gets much simpler.

As a reference, Table 1 lists the axioms and properties used in this paper.

In order to understand which kinds of behaviours are incompatible with a causal-consistent reversible setting, consider the following LTSs in CCS:

$a.\mathbf{0} \xrightarrow{a} \mathbf{0}$, $b.\mathbf{0} \xrightarrow{b} \mathbf{0}$: from state $\mathbf{0}$ one does not know whether to go back to $a.\mathbf{0}$ or to $b.\mathbf{0}$;

$a.\mathbf{0} + b.\mathbf{0} \xrightarrow{a} \mathbf{0}$, $a.\mathbf{0} + b.\mathbf{0} \xrightarrow{b} \mathbf{0}$: as above, but starting from the same process, hence showing that it is not enough to remember the initial configuration;

$P \xrightarrow{a} P$ **where** $P = a.P$: one can go back forever, against the idea that a state models a process reachable after a finite computation.

We remark that all such behaviours are perfectly reasonable in CCS, and they are dealt with in the reversible setting by adding history information about past actions. For example, in the first case one could remember the initial state, in the second case both the initial state and the action taken, and in the last case the number of iterations that have been performed.

Due to space constraints, some proofs and additional results can only be found in the companion technical report [16].

2 Labelled Transition Systems with Independence

We want to study reversibility in a setting as general as possible. Thus, we base on the core of the notion of *labelled transition system with independence* (LTSI) [33, Definition 3.7]. However, while [33] requires a number of axioms on LTSI, we take the basic definition and explore what can be done by adding or not adding various axioms. Also, we extend LTSI with reverse transitions, since we study reversible systems. We define first labelled transition systems (LTSs).

We consider the LTS of the entire set of processes in a calculus, rather than the transition graph of a particular process and its derivatives, hence we do not fix an initial state.

Definition 2.1. *A* labelled transition system (LTS) *is a structure* $(\mathsf{Proc}, \mathsf{Lab}, \rightharpoonup)$, *where* Proc *is the set of states (or processes),* Lab *is the set of action labels and* $\rightharpoonup \subseteq \mathsf{Proc} \times \mathsf{Lab} \times \mathsf{Proc}$ *is a* transition relation.

We let $P, Q, \ldots$ range over processes, $a, b, c, \ldots$ range over labels, and $t, u, v, \ldots$ range over transitions. We can write $t : P \xrightarrow{a} Q$ to denote that $t = (P, a, Q)$. We call a-transition a transition with label a.

Definition 2.2 (LTS with independence). *We say that* $(\mathsf{Proc}, \mathsf{Lab}, \rightarrow, \iota)$ *is an* LTS with independence *(LTSI) if* $(\mathsf{Proc}, \mathsf{Lab}, \rightarrow)$ *is an LTS and* ι *is an irreflexive symmetric binary relation on transitions.*

In many cases (see Section 6), the notion of independence coincides with the notion of concurrency. However, this is not always the case. Indeed, concurrency implies that transitions are independent since they happen in different processses, but transitions taken by the same process can be independent as well. Think, for instance, of a reactive process that may react in any order to two events arriving at the same time, and the final result does not depend on the order of reactions.

We shall assume that all transitions are reversible, so that the Loop Lemma [6, Lemma 6] holds. This does not hold in models of reversibility with control mechanisms such as irreversible actions [6,7] or a rollback operator [17]. Nevertheless, when showing properties of models with controlled reversibility it has proved sensible to first consider the underlying models where all transitions are reversible, and then study how control mechanisms change the picture [11,20]. The present work helps with the first step.

Definition 2.3. *Given* $(\mathsf{Proc}, \mathsf{Lab}, \rightharpoonup)$*, let the* reverse LTS *be* $(\mathsf{Proc}, \mathsf{Lab}, \rightsquigarrow)$*, where* $P \stackrel{a}{\rightsquigarrow} Q$ *iff* $Q \stackrel{a}{\rightharpoonup} P$*. It is convenient to combine the two LTSs (forward and reverse): let the reverse labels be* $\underline{\mathsf{Lab}} = \{\underline{a} : a \in \mathsf{Lab}\}$*, and define the combined LTS to be* $\rightarrow \subseteq \mathsf{Proc} \times (\mathsf{Lab} \cup \underline{\mathsf{Lab}}) \times \mathsf{Proc}$ *by* $P \stackrel{a}{\rightarrow} Q$ *iff* $P \stackrel{a}{\rightharpoonup} Q$ *and* $P \stackrel{\underline{a}}{\rightarrow} Q$ *iff* $P \stackrel{a}{\rightsquigarrow} Q$*.*

We stipulate that the union $\mathsf{Lab} \cup \underline{\mathsf{Lab}}$ is disjoint. We let $\alpha, \ldots$ range over $\mathsf{Lab} \cup \underline{\mathsf{Lab}}$. For $\alpha \in \mathsf{Lab} \cup \underline{\mathsf{Lab}}$, the *underlying* action label $\mathsf{und}(\alpha)$ is defined as $\mathsf{und}(a) = a$ and $\mathsf{und}(\underline{a}) = a$. Let $\underline{\underline{a}} = a$ for $a \in \mathsf{Lab}$. Given $t : P \stackrel{\alpha}{\rightarrow} Q$, let $\underline{t} : Q \stackrel{\underline{\alpha}}{\rightarrow} P$ be the transition which reverses t.

We let $\rho, \sigma, \ldots$ range over finite sequences $\alpha_1 \ldots \alpha_n$, with ε_P representing the empty sequence starting and ending at P. We shall write ε when P is understood. Given an LTS, a *path* is a sequence of forward or reverse transitions of the form $P_0 \stackrel{\alpha_1}{\rightarrow} P_1 \cdots \stackrel{\alpha_n}{\rightarrow} P_n$. We let $r, s, \ldots$ range over paths. We may write $r : P \stackrel{\rho}{\rightarrow}_* Q$ where the intermediate states are understood. On occasion we may refer to a path simply by its sequence of labels ρ. Given a path $r : P \stackrel{\rho}{\rightarrow}_* Q$, the inverse path is $\underline{r} : Q \stackrel{\underline{\rho}}{\rightarrow}_* P$ where $\underline{\varepsilon} = \varepsilon$ and $\underline{\alpha\rho} = \underline{\rho}\,\underline{\alpha}$. The length of a path r (notated $|r|$) is the number of transitions in the path. Paths $r : P \stackrel{\rho}{\rightarrow}_* Q$ and $R \stackrel{\sigma}{\rightarrow}_* S$ are *coinitial* if $P = R$ and *cofinal* if $Q = S$. We say that a path is *forward-only* if it contains no reverse transitions.

Let $(\mathsf{Proc}, \mathsf{Lab}, \rightharpoonup)$ be an LTS. The irreversible processes in $(\mathsf{Proc}, \mathsf{Lab}, \rightharpoonup)$ are $\mathsf{Irr} = \{P \in \mathsf{Proc} : P \not\rightsquigarrow\}$. A *rooted path* is a path $r : P \stackrel{\rho}{\rightarrow}_* Q$ such that $P \in \mathsf{Irr}$.

In the following we will consider LTSIs obtained by adding a notion of independence to combined LTSs as above. We will call the result a *combined LTSI.*

3 Basic Properties

In this section we show that most of the properties in the reversibility literature (see, e.g., [6,29,18,20]), in particular the parabolic lemma and causal consistency, can be proved under minimal assumptions on the combined LTSI under analysis.

We formalise the minimal assumptions using three axioms, described below.

Definition 3.1 (Basic axioms). *Let* $\mathcal{L} = (\mathsf{Proc}, \mathsf{Lab}, \rightarrow, \iota)$ *be a combined LTSI. We say* $\mathcal{L}$ *satisfies:*

Square Property (SP) *if whenever* $t : P \stackrel{\alpha}{\rightarrow} Q$*,* $u : P \stackrel{\beta}{\rightarrow} R$ *with* $t \;\iota\; u$ *then there are cofinal transitions* $u' : Q \stackrel{\beta}{\rightarrow} S$ *and* $t' : R \stackrel{\alpha}{\rightarrow} S$*;*

Backward Transitions are Independent (BTI) *if whenever* $t : P \stackrel{a}{\rightsquigarrow} Q$ *and* $t' : P \stackrel{b}{\rightsquigarrow} Q'$ *and* $t \neq t'$ *then* $t \;\iota\; t'$*;*

Well-Foundedness (WF) *if there is no infinite reverse computation, i.e. we do not have P_i (not necessarily distinct) such that $P_{i+1} \stackrel{a_i}{\rightarrow} P_i$ for all $i = 0, 1, \ldots$.*

WF can alternatively be formulated using backward transitions, but the current formulation makes sense also in non-reversible calculi (e.g., CCS), which can be used as a comparison. Let us discuss the intuition behind these axioms. SP takes its name from the Square Lemma, where it is proved for concrete calculi and languages in [6,18,20], and captures the idea that independent transitions can be executed in any order, that is they form commuting diamonds. SP can be seen as a sanity check on the chosen notion of independence. BTI generalises the key notion of backward determinism used in sequential reversibility (see, e.g., [32] for finite state automata and [35] for the imperative language Janus) to a concurrent setting. Backward determinism can be spelled as "two coinitial backward transitions do coincide". This can be generalised to "two coinitial backward transitions are independent". Finally, WF means that we consider systems which have a finite past. That is, we consider systems starting from some initial state and then moving forward and back.

Axioms SP and BTI are related to properties which are part of the definition of (occurrence) transition systems with independence in [33, Definitions 3.7, 4.1]. WF was used as an axiom in [28].

Using the minimal assumptions above we can prove relevant results from the literature. We first define causal equivalence, equating computations differing only for swaps of independent transitions and simplification of a transition with its reverse.

Definition 3.2 (cf. [6]). *Let $(\mathsf{Proc}, \mathsf{Lab}, \rightarrow, \iota)$ be an LTSI satisfying SP. Let $\approx$ be the smallest equivalence relation on paths closed under composition and satisfying:*

1. *if $t : P \stackrel{\alpha}{\rightarrow} Q$, $u : P \stackrel{\beta}{\rightarrow} R$ are independent, and $u' : Q \stackrel{\beta}{\rightarrow} S$, $t' : R \stackrel{\alpha}{\rightarrow} S$ (which exist by SP) then $tu' \approx ut'$;*
2. *$t\underline{t} \approx \varepsilon$ and $\underline{t}t \approx \varepsilon$.*

We first consider the Parabolic Lemma ([6, Lemma 10]), which states that each path is causal equivalent to a backward path followed by a forward path.

Definition 3.3. Parabolic Lemma (PL): *for any path r there are forward-only paths s, s' such that $r \approx \underline{s}s'$ and $|s| + |s'| \leq |r|$.*

Proposition 3.4. *Suppose an LTSI satisfies BTI and SP. Then PL holds.*

The proof of Proposition 3.4 (available in [16]) is very similar to that of [6, Lemma 10] except that in the latter BTI is shown directly as part of the proof.

A corollary of PL is that if a process is reachable from an irreversible process, then it is also forwards reachable from it. In other words, making a system reversible does not introduce new reachable states but only allows one to explore differently forwards reachable states. This is relevant in reversible debugging of concurrent systems [10,20], where one wants to find bugs that actually occur in forward-only computations. See the companion technical report [16, Corollary A.1]. We now move to causal consistency [6, Theorem 1].

Definition 3.5. Causal Consistency (CC): *if r and s are coinitial and cofinal then $r \approx s$.*

Essentially, causal consistency states that history information allows one to distinguish computations which are not causal equivalent, indeed, if two computations are cofinal, that is they reach the same final state (which includes the stored history information) then they need to be causal equivalent.

Causal consistency frequently includes the other direction, namely that coinitial causal equivalent computations are cofinal, meaning that there is no way to distinguish causal equivalent computations. This second direction follows easily from the definition of causal equivalence.

Notably, our proof of CC below is very much shorter than existing proofs.

Proposition 3.6. *Suppose an LTSI satisfies WF and PL. Then CC holds.*

Proof. Let $r : P \xrightarrow{\rho}_* Q$ and $r' : P \xrightarrow{\rho'}_* Q$. Using WF, let I, s be such that $s : I \xrightarrow{\sigma}_* P$, $I \in \mathsf{Irr}$. Now $sr\underline{sr'}$ is a path from I to I, and so by PL there are r_1, r_2 forward-only such that $r_1\underline{r_2} \approx sr\underline{sr'}$. But $I \in \mathsf{Irr}$ and so $r_1 = \varepsilon$ and $r_2 = \varepsilon$. Thus $\varepsilon \approx sr\underline{sr'}$, so that $sr \approx sr'$ and $r \approx r'$ as required. □

Causal consistency implies the unique transition property.

Definition 3.7. *An LTSI $(\mathsf{Proc}, \mathsf{Lab}, \rightarrow, \iota)$ satisfies* ***Unique Transition (UT)*** *if $P \xrightarrow{a} Q$ and $P \xrightarrow{b} Q$ imply $a = b$.*

Corollary 3.8. *If an LTSI satisfies CC then it satisfies UT.*

UT was shown in the forward-only setting of occurrence TSIs in [33, Corollary 4.4]; it was taken as an axiom in [28].

Example 3.9 (PL alone does not imply WF or CC). Consider the LTSI with states P_i for $i = 0, 1, \ldots$ and transitions $t_i : P_{i+1} \xrightarrow{a} P_i$, $u_i : P_{i+1} \xrightarrow{b} P_i$ with $a \neq b$ and $t_i \ \iota \ u_i$. BTI and SP hold. Hence PL holds by Proposition 3.4. However clearly WF fails. Also t_i and u_i are coinitial and cofinal, and $a \neq b$, so that UT fails, and hence CC fails using Corollary 3.8. Note that the ab diamonds here have the same side states so are degenerate (cf. Lemma 4.4).

4 Causal Safety and Causal Liveness

In the literature, causal consistent reversibility is frequently informally described by saying that "a transition can be undone if and only if each of its consequences, if any, has been undone". In this section we study this property, where the two implications will be referred to as causal safety and causal liveness. We provide three different versions of such properties, based on independence of transitions (Section 4.2), ordering of events (Section 4.3), and independence of events (Section 4.4), and study their relationships. In order to define such properties we need the concept of event.

4.1 Events

Definition 4.1 (Event, general definition). *Let* (Proc, Lab, $\rightarrow$, ι) *be an LTSI. Let* $\sim$ *be the smallest equivalence relation satisfying: if* $t : P \xrightarrow{\alpha} Q$, $u : P \xrightarrow{\beta} R$, $u' : Q \xrightarrow{\beta} S$, $t' : R \xrightarrow{\alpha} S$, *and* $t \;\iota\; u$, $\underline{u} \;\iota\; t'$, $\underline{t'} \;\iota\; \underline{u'}$, $u' \;\iota\; \underline{t}$, *and*

- $Q \neq R$ *if* α *and* β *are both forwards or both backwards;*
- $P \neq S$ *otherwise;*

then $t \sim t'$. *The equivalence classes of forward transitions, written* $[P, a, Q]$, *are the* events. *The equivalence classes of reverse transitions, written* $[P, \underline{a}, Q]$, *are the* reverse events. *Define a labelling function* ℓ *from* $\rightarrow / \sim$ *to* Lab *by setting* $\ell([P, \alpha, Q]) = \alpha$.

Events are introduced as a derived notion in an LTS with independence in [33], in the context of forward-only computation. We have changed their definition by using coinitial independence at all corners of the diamond, yielding rotational symmetry. This reflects our view that forward and backward transitions have equal status.

Our definition can be simplified if the LTSI, and independence in particular, are well-behaved. Thus, we now add a further axiom related to independence.

Definition 4.2 (Coinitial Propagation of Independence (CPI)). *If* $t : P \xrightarrow{\alpha} Q$, $u : P \xrightarrow{\beta} R$, $u' : Q \xrightarrow{\beta} S$ *and* $t' : R \xrightarrow{\alpha} S$ *with* $t \;\iota\; u$, *then* $u' \;\iota\; \underline{t}$.

CPI states that independence is a property of commuting diamonds more than of their specific pairs of edges. Indeed, it allows independence to propagate around a commuting diamond.

Definition 4.3. *If a combined LTSI satisfies axioms SP, BTI, WF and CPI, we say that it is* pre-reversible.

The name 'pre-reversible' indicates that we expect to require further axioms, but the present four are enough to ensure that LTSIs are well-behaved, with events compatible with causal equivalence. Pre-reversible axioms are separated from further axioms by a dashed line in Table 1.

The following non-degeneracy property was shown for occurrence transition systems with independence in [33, page 312], which have forward transitions only. We have to cope with backwards as well as forward transitions.

Lemma 4.4. *Suppose that an LTSI is pre-reversible. If we have a diamond* $t : P \xrightarrow{\alpha} Q$, $u : P \xrightarrow{\beta} R$ *with* $t \;\iota\; u$ *together with cofinal transitions* $u' : Q \xrightarrow{\beta} S$ *and* $t' : R \xrightarrow{\alpha} S$, *then the diamond is* non-degenerate, *meaning that* P, Q, R, S *are distinct states.*

If an LTSI is pre-reversible then by Lemma 4.4 and the use of CPI we can simplify the statement of Definition 4.1 to:

Definition 4.5 (Event, simplified definition). *Let* $(\mathsf{Proc}, \mathsf{Lab}, \rightarrow, \iota)$ *be a pre-reversible LTSI. Let* $\sim$ *be the smallest equivalence relation satisfying: if* $t : P \xrightarrow{\alpha} Q$, $u : P \xrightarrow{\beta} R$, $u' : Q \xrightarrow{\beta} S$, $t' : R \xrightarrow{\alpha} S$, *and* $t \,\iota\, u$, *then* $t \sim t'$.

We are now able to show independence of diamonds (ID), which can be seen as dual of SP.

Definition 4.6 (Independence of Diamonds (ID)). *An LTSI satisfies the Independence of Diamonds property (ID) if whenever we have a diamond* $t : P \xrightarrow{\alpha} Q$, $u : P \xrightarrow{\beta} R$, $u' : Q \xrightarrow{\beta} S$ *and* $t' : R \xrightarrow{\alpha} S$, *with*

- $Q \neq R$ *if* α *and* β *are both forwards or both backwards;*
- $P \neq S$ *otherwise;*

then $t \,\iota\, u$.

Proposition 4.7. *If an LTSI satisfies BTI and CPI then it satisfies ID.*

We now consider the interaction between events and causal equivalence. We need some notation first.

Definition 4.8. *Let* r *be a path in an LTSI* $\mathcal{L}$ *and let* e *be an event of* $\mathcal{L}$. *Let* $\sharp(r, e)$ *be the number of occurrences of transitions* t *in* r *such that* $t \in e$, *minus the number of occurrences of transitions* t *in* r *such that* $t \in \underline{e}$.

We now show that $\sharp(r, e)$ is invariant under causal equivalent traces.

Lemma 4.9. *Let* $\mathcal{L}$ *be a pre-reversible LTSI. Let* $r \approx s$. *Then for each event* e *we have that* $\sharp(r, e) = \sharp(s, e)$.

Lemma 4.9 generalises what was shown for the forward-only setting in [33, Corollary 4.3].

Proposition 4.10. *If an LTSI is pre-reversible, then for any rooted path* r *and any forward event* e *we have* $\sharp(r, e) \geq 0$.

4.2 CS and CL via Independence of Transitions

We first define causal safety and liveness using the independence relation.

Definition 4.11. *Let* $\mathcal{L} = (\mathsf{Proc}, \mathsf{Lab}, \rightarrow, \iota)$ *be a pre-reversible LTSI.*

1. *We say that* $\mathcal{L}$ *is* causally safe (CS) *if whenever* $P \xrightarrow{a} Q$, $r : Q \xrightarrow{\rho}_* R$, $\sharp(r, [P, a, Q]) = 0$ *and* $S \xrightarrow{a} R$ *with* $(P, a, Q) \sim (S, a, R)$, *then* $(P, a, Q) \,\iota\, t$ *for all* t *in* r *such that* $\sharp(r, [t]) > 0$.
2. *We say that* $\mathcal{L}$ *is* causally live (CL) *if whenever* $P \xrightarrow{a} Q$, $r : Q \xrightarrow{\rho}_* R$ *and* $\sharp(r, [P, a, Q]) = 0$ *and* $(P, a, Q) \,\iota\, t$, *for all* t *in* r *such that* $\sharp(r, [t]) > 0$, *then we have* $S \xrightarrow{a} R$ *with* $(P, a, Q) \sim (S, a, R)$.

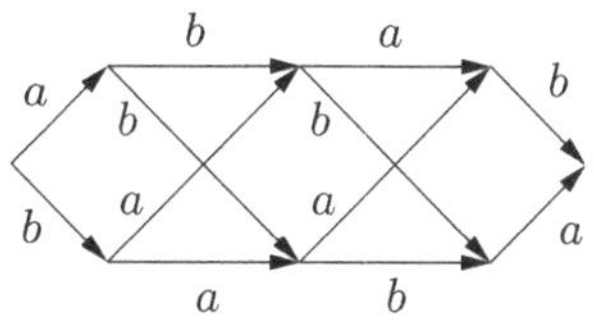

Fig. 1.

We may wish to close the independence relation over this axiom:

Definition 4.12 (Independence Respects Events (IRE)). *Whenever* $t \sim t' \iota u$ *we have* $t \iota u$.

IRE is one of the conditions in the definition of transition systems with independence [33, Definition 3.7]. Together with the axioms for pre-reversibility, it is enough to show both causal safety and causal liveness.

Theorem 4.13. *Let a pre-reversible LTSI satisfy IRE. Then it satisfies CS.*

Theorem 4.14. *Let a pre-reversible LTSI satisfy IRE. Then it satisfies CL.*

CS and CL are not derivable from CC; we give an example LTSI which satisfies CC but not CS and not CL.

Example 4.15. Consider the LTS in Figure 1. Independence is mostly coinitial and given by closing under BTI and CPI. Additionally we make the leftmost a-transition independent with all b-transitions. Note that all a-transitions belong to the same event, and all b-transitions belong to the same event. Also SP and WF hold, so that the LTSI is pre-reversible, and CC holds. However IRE does not hold. Furthermore CS fails using Definition 4.11. Indeed, consider any path $\overset{ba\underline{b}}{\rightarrow}_*$ from the start. CS would imply that the first b is independent with the a but this is not the case (we do have $\underline{b} \iota a$).

Also CL fails using Definition 4.11. Indeed, consider any path $\overset{abb}{\rightarrow}_*$ from the start. Since the leftmost a-transition is independent with all b-transitions, we should be able to reverse a at the end of the path, but this is not possible.

The next axiom states that independence is fully determined by its restriction to coinitial transitions. This is related to axiom (E) of [33, page 325], but here we allow reverse as well as forward transitions.

Definition 4.16 (Independence of Events is Coinitial (IEC)). *If* $t_1 \iota t_2$ *then there are* $t'_1 \sim t_1$, $t'_2 \sim t_2$ *such that* t'_1 *and* t'_2 *are coinitial and* $t'_1 \iota t'_2$.

Thanks to previous axioms, independence behaves well w.r.t. reversing.

Definition 4.17 (Reversing Preserves Independence (RPI)). *If* $t \iota t'$ *then* $\underline{t} \iota t'$.

Proposition 4.18. *If an LTSI satisfies SP, CPI, IRE, IEC then it also satisfies RPI.*

All the axioms that we have introduced are independent, i.e. none is derivable from the remaining axioms.

Proposition 4.19. *SP, BTI, WF, CPI, IRE, IEC are independent of each other.*

4.3 CS and CL via Ordering of Events

To define CS and CL via ordering of events, we define the causality relation $\leq$ on events.

Definition 4.20. *Let $\mathcal{L} = (\mathsf{Proc}, \mathsf{Lab}, \rightarrow, \iota)$ be an LTSI. Let e, e' be events of $\mathcal{L}$. Let $e \leq e'$ iff for all rooted paths r, if $\sharp(r, e') > 0$ then $\sharp(r, e) > 0$. As usual $e < e'$ means $e \leq e'$ and $e \neq e'$. If $e < e'$ we say that e is a* cause *of e'.*

Lemma 4.21. *If an LTSI satisfies SP, BTI, WF and CPI then $\leq$ is a partial ordering on events.*

Previously, orderings on events have been defined using forward-only rooted paths; in fact, the definitions coincide for pre-reversible LTSIs.

Definition 4.22 ([12,28]). *Let $\mathcal{L} = (\mathsf{Proc}, \mathsf{Lab}, \rightarrow, \iota)$ be an LTSI. Let e, e' be events of $\mathcal{L}$. Let $e \leq_{\mathsf{f}} e'$ iff for all rooted forward-only paths r, if r contains a representative of e' then r also contains a representative of e.*

Lemma 4.23. *For any LTSI, $e \leq e'$ implies $e \leq_{\mathsf{f}} e'$. If an LTSI satisfies SP, BTI, WF and CPI then $e \leq_{\mathsf{f}} e'$ implies $e \leq e'$.*

Proof. Straightforward using PL and Lemma 4.9. □

We now give definitions of causal safety and causal liveness using ordering on events.

Definition 4.24. *Let $\mathcal{L} = (\mathsf{Proc}, \mathsf{Lab}, \rightarrow, \iota)$ be an LTSI.*

1. *We say that $\mathcal{L}$ is* ordered causally safe ($\mathrm{CS}_<$) *if whenever $P \xrightarrow{a} Q$, $r : Q \xrightarrow{\rho}_* R$, $\sharp(r, [P, a, Q]) = 0$ and $S \xrightarrow{a} R$ with $(P, a, Q) \sim (S, a, R)$, then $[P, a, Q] \not< e'$ for all e' such that $\sharp(r, e') > 0$.*
2. *We say that $\mathcal{L}$ is* ordered causally live ($\mathrm{CL}_<$) *if whenever $P \xrightarrow{a} Q$, $r : Q \xrightarrow{\rho}_* R$ and $\sharp(r, [P, a, Q]) = 0$ and $[P, a, Q] \not< e'$ for all e' such that $\sharp(r, e') > 0$ then we have $S \xrightarrow{a} R$ with $(P, a, Q) \sim (S, a, R)$.*

We postpone giving proofs of $\mathrm{CS}_<$ and $\mathrm{CL}_<$ until we have introduced a further definition of causal safety and liveness using independence of events.

4.4 CS and CL via Independent Events

We now introduce a third version of causal safety and liveness, which uses independence like CS and CL, but on events rather than on transitions. First we lift independence from transitions to events.

Definition 4.25 (Coinitially independent events). *Let events e, e' be* (coinitially) independent, *written* $e \mathsf{~ci~} e'$, *iff there are coinitial transitions t, t' such that* $[t] = e$, $[t'] = e'$ *and* $t \iota t'$.

Lemma 4.26. *If an LTSI is pre-reversible, then if $e \mathsf{~ci~} e'$ we have also $\underline{e} \mathsf{~ci~} e'$.*

Thus in pre-reversible LTSIs, ci is fully determined just considering forward events. By Lemma 4.26, if we know $e \mathsf{~ci~} e'$ then we know $\mathsf{und}(e) \mathsf{~ci~} \mathsf{und}(e')$.

We can give a third formulation of causal safety and liveness using ci:

Definition 4.27. *Let $\mathcal{L} = (\mathsf{Proc}, \mathsf{Lab}, \rightarrow, \iota)$ be a pre-reversible LTSI.*

1. *We say that $\mathcal{L}$ is* coinitially causally safe *(CS_{ci}) if whenever $P \xrightarrow{a} Q$, $r : Q \xrightarrow{\rho}_* R$, $\sharp(r, [P, a, Q]) = 0$ and $S \xrightarrow{a} R$ with $(P, a, Q) \sim (S, a, R)$, then $[P, a, Q] \mathsf{~ci~} e$ for all forward events e such that $\sharp(r, e) > 0$.*
2. *We say that $\mathcal{L}$ is* coinitially causally live *(CL_{ci}) if whenever $P \xrightarrow{a} Q$, $r : Q \xrightarrow{\rho}_* R$ and $\sharp(r, [P, a, Q]) = 0$ and $[P, a, Q] \mathsf{~ci~} e$, for all forward events e such that $\sharp(r, e) > 0$, then we have $S \xrightarrow{a} R$ with $(P, a, Q) \sim (S, a, R)$.*

Note that in Definition 4.27 we operate at the level of events, rather than at the level of transitions as in Definition 4.11.

Theorem 4.28. *If an LTSI is pre-reversible then it satisfies CS_{ci}.*

We now introduce a weaker version of axiom IRE (Definition 4.12).

Definition 4.29 (Coinitial IRE (CIRE)). *If $[t] \mathsf{~ci~} [u]$ and t, u are coinitial then $t \iota u$.*

Theorem 4.30. *If a pre-reversible LTSI satisfies CIRE then it satisfies CL_{ci}.*

We next give an example where CC holds but not $\mathrm{CS}_{\mathsf{ci}}$ (and not CPI).

Example 4.31. Consider the cube with transitions a, b, c on the left in Figure 2, where the forward direction is from left to right. We add independence as given by BTI. So SP, BTI, WF hold, but not CPI. From the start we have an a-transition followed by a path $r = bc$ followed by $\underline{a}$. For $\mathrm{CS}_{\mathsf{ci}}$ to hold, we want $\underline{a}$ to be the reverse of the same event as the first a. They are connected by a ladder with sides cb. We add independence for all corners on the two faces of the ladder (ab and ac). Then we get $bc \approx cb$ (independence at a single corner is enough). However the bs are not the same event since the bc face does not have independence at each corner. Therefore we do not get $[a] \mathsf{~ci~} [b]$, and $\mathrm{CS}_{\mathsf{ci}}$ fails.

We next give an example where $\mathrm{CS}_{\mathsf{ci}}$ and $\mathrm{CL}_{\mathsf{ci}}$ hold but not CC.

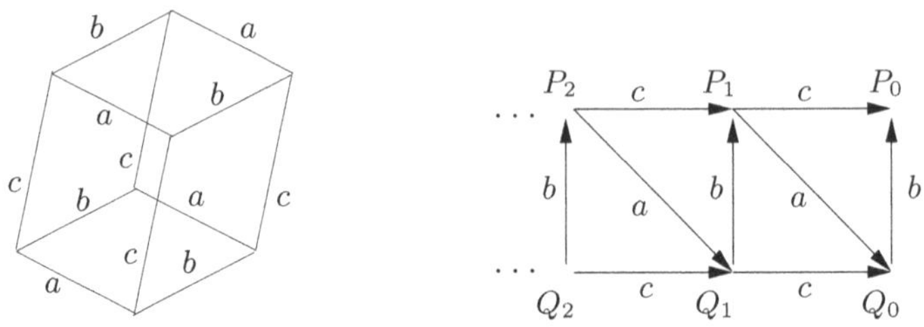

Fig. 2.

Example 4.32. Consider the LTSI with $Q_i \xrightarrow{b} P_i$, $P_{i+1} \xrightarrow{c} P_i$, $Q_{i+1} \xrightarrow{c} Q_i$, $P_{i+1} \xrightarrow{a} Q_i$ for $i = 0, 1, \ldots$. This is shown on the right in Figure 2. Clearly WF does not hold. We add coinitial independence to make BTI and CPI hold. Then also SP and CIRE hold. However, CC fails since, for example $P_1 \xrightarrow{a} Q_0 \xrightarrow{b} P_0$ and $P_1 \xrightarrow{c} P_0$ are coinitial and cofinal but not causally equivalent. Note that there are just three events a, b, c with $a \ \mathsf{ci} \ c$, $b \ \mathsf{ci} \ c$ but not $a \ \mathsf{ci} \ b$. $\mathrm{CS}_{\mathsf{ci}}$ and $\mathrm{CL}_{\mathsf{ci}}$ hold. Indeed, c is independent from every other action, and it can always be undone, while a and b are independent from c only and they can be undone after any path composed by c and no others.

4.5 Polychotomy

In this section we relate our three versions of causal safety and liveness, with the help of what we call *polychotomy*, which states that if events do not cause each other and are not in conflict, then they must be independent. We start by defining a conflict relation on events.

Definition 4.33. *Two forward events e, e' are in* conflict, *written $e \# e'$, if there is no rooted path r such that $\sharp(r, e) > 0$ and $\sharp(r, e') > 0$.*

Much as for orderings, conflict on events has been defined previously using forward-only rooted paths [12,28]; in fact, the definitions coincide for pre-reversible LTSIs. We omit the details.

Definition 4.34 (Polychotomy). *Let $\mathcal{L}$ be a pre-reversible LTSI. We say that $\mathcal{L}$ satisfies* polychotomy *if whenever e, e' are* forward *events, then exactly one of the following holds: 1. $e = e'$; 2. $e < e'$; 3. $e' < e$; 4. $e \# e'$; or 5. $e \ \mathsf{ci} \ e'$.*

Property NRE below is related to polychotomy.

Definition 4.35 (No Repeated Events (NRE)). *In any rooted path r, for any forward event e we have $\sharp(r, e) \leq 1$.*

Lemma 4.36 (Polychotomy). *Suppose that a pre-reversible LTSI satisfies NRE. Then polychotomy holds.*

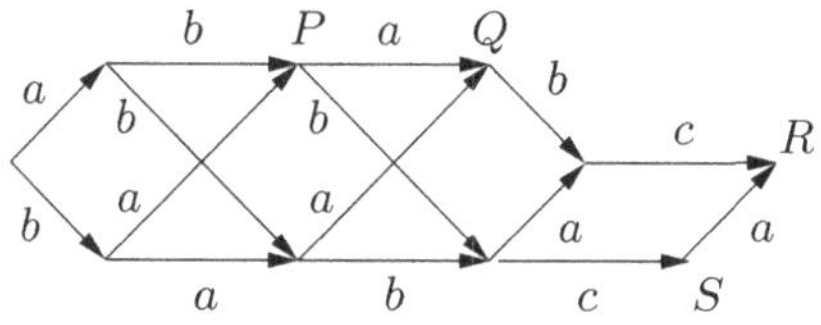

Fig. 3.

Example 4.37. Consider the LTSI in Figure 3. We add independence to make BTI and CPI hold. Both SP and WF hold. Hence, CC holds as well. There are three events, labelled with a, b, c. Clearly NRE fails for both a and b. We see that $a < c$ but also $a \text{ ci } c$, so that polychotomy fails. CS_{ci} holds by Theorem 4.28. However $\text{CS}_{<}$ fails: consider the transition $P \xrightarrow{a} Q$ together with the path $r : Q \xrightarrow{bc}_* R$ and $S \xrightarrow{a} R$, and note that $a < c$.

The next lemma allows us to connect ordered safety and liveness with coinitial safety and liveness.

Lemma 4.38. *Suppose that a pre-reversible LTSI satisfies NRE. Suppose* $P \xrightarrow{a} Q$, $e = [P, a, Q]$, $r : Q \xrightarrow{\rho}_* R$ *and* $\sharp(r, e') > 0$ *where* e' *is a forward event. Then exactly one of* $e \text{ ci } e'$ *and* $e < e'$ *holds.*

Proposition 4.39. *Suppose that a pre-reversible LTSI* $\mathcal{L}$ *satisfies NRE. Then*

1. $\mathcal{L}$ *satisfies* $CS_{<}$.
2. $\mathcal{L}$ *satisfies* CL_{ci} *iff* $\mathcal{L}$ *satisfies* $CL_{<}$.

Property RED below is also related to NRE and polychotomy.

Definition 4.40. *An LTSI satisfies* ***Reverse Event Determinism (RED)*** *if whenever* t, t' *are backward coinitial transitions and* $t \sim t'$ *then* $t = t'$.

Proposition 4.41. *If a LTSI* $\mathcal{L}$ *is pre-reversible then the following are equivalent: 1.* $\mathcal{L}$ *satisfies NRE; 2.* $\mathcal{L}$ *satisfies RED; 3. independence* ci *is irreflexive on events; and 4. polychotomy holds.*

Proposition 4.42. *Suppose that a pre-reversible LTSI satisfies CIRE. Then it also satisfies NRE.*

NRE was shown in the forward-only setting of occurrence transition systems with independence in [33, Corollary 4.6]. It was also shown in the reversible setting without independence in [28, Proposition 2.10].

Example 4.43. Consider the LTSI in Figure 4. Independence is given by closing under BTI and CPI. There are three events, labelled a, b, c, which are all independent of each other. We see that NRE holds but not CIRE. Also CL_{ci} and $\text{CL}_{<}$ fail: consider $P \xrightarrow{a} Q \xrightarrow{b} R$, where a cannot be reversed at R.

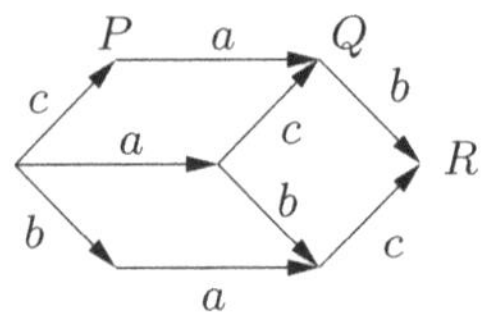

Fig. 4.

Proposition 4.44. *Let $\mathcal{L}$ be a pre-reversible LTSI.*

1. *If IEC holds then CL_{ci} implies CL.*
2. *If IEC and NRE hold then $CL_{<}$ implies CL.*

5 Coinitial Independence

In this section we consider coinitial LTSIs, defined as follows, and their relationship with LTSIs in general.

Definition 5.1. *Let $\mathcal{L} = (\mathsf{Proc}, \mathsf{Lab}, \rightarrow, \iota)$ be a combined LTSI. Then ι is* coinitial *if for all transitions t, u, if $t \; \iota \; u$ then t and u are coinitial. We say that $\mathcal{L}$ is coinitial if ι is coinitial.*

We define a mapping c restricting general independence to coinitial transitions and a mapping g extending independence along events.

Definition 5.2. *Given an LTSI $(\mathsf{Proc}, \mathsf{Lab}, \rightarrow, \iota)$, define $t \; g(\iota) \; u$ iff $t \sim t' \; \iota \; u' \sim u$ for some t', u'. Furthermore, define $t \; c(\iota) \; u$ iff $t \; \iota \; u$ and t, u are coinitial.*

Proposition 5.3. *Let $\mathcal{L} = (\mathsf{Proc}, \mathsf{Lab}, \rightarrow, \iota)$ be a pre-reversible LTSI.*

1. *If $\mathcal{L}$ is coinitial and satisfies CIRE then $\mathcal{L}' = (\mathsf{Proc}, \mathsf{Lab}, \rightarrow, g(\iota))$ is a pre-reversible LTSI and satisfies IRE and IEC.*
2. *if $\mathcal{L}$ satisfies IRE then $\mathcal{L}' = (\mathsf{Proc}, \mathsf{Lab}, \rightarrow, c(\iota))$ is a pre-reversible coinitial LTSI and satisfies CIRE.*

Thanks to Proposition 5.3, we can extend a coinitial pre-reversible LTSI satisfying CIRE in a canonical way to a pre-reversible LTSI satisfying IRE and IEC.

In some reversible calculi (such as RCCS) independence of coinitial transitions is defined purely by reference to the labels. If this is the case it is a simple matter to verify the axioms CPI and CIRE.

Proposition 5.4. *Let $\mathcal{L} = (\mathsf{Proc}, \mathsf{Lab}, \rightarrow, \iota)$ be a coinitial combined LTSI. Suppose that I is a binary relation on* Lab *such that for any coinitial transitions $t : P \xrightarrow{\alpha} Q$ and $u : P \xrightarrow{\beta} R$ we have $t \; \iota \; u$ iff $I(a, b)$, where a and b are the underlying labels $a = \mathsf{und}(\alpha)$, $b = \mathsf{und}(\beta)$. Then $\mathcal{L}$ satisfies CPI and CIRE.*

Proof. Straightforward, noting that labels on opposite sides of a diamond of transitions must be equal. □

Note that I must be irreflexive, since ι is irreflexive.

If we have a coinitial pre-reversible LTSI satisfying CIRE then $CS_<$ and $CL_<$ hold (using Proposition 4.42 and Proposition 4.39). Applying mapping g we get a general pre-reversible LTSI satisfying IRE and IEC by Proposition 5.3. This will satisfy CS and CL as a result of applying Theorem 4.13 and Theorem 4.14 respectively. It will also satisfy $CS_<$ and $CL_<$. Conversely, if we have a general pre-reversible LTSI satisfying IRE then CS and CL hold by Theorem 4.13 and Theorem 4.14 respectively. Applying mapping c we get a coinitial pre-reversible LTSI satisfying CIRE. This will satisfy $CS_<$ and $CL_<$.

6 Case Studies

We look at whether our axioms hold in various reversible formalisms. Remarkably, all the works below provide proofs of the Loop Lemma.

RCCS We consider here the semantics of RCCS in [6], and restrict the attention to coherent processes [6, Definition 2]. In RCCS, transitions $P \overset{\mu:\zeta}{\rightarrow} Q$ and $P \overset{\mu':\zeta'}{\rightarrow} Q'$ are concurrent if $\mu \cap \mu' = \emptyset$ [6, Definition 7]. This allows us to define coinitial independence as $t \ \iota \ u$ iff t and u are concurrent. We now argue that the resulting coinitial LTSI is pre-reversible and also satisfies CIRE. SP was shown in [6, Lemma 8]. BTI was shown in the proof of [6, Lemma 10]. WF is straightforward, noting that backward transitions decrease memory size. Hence, we obtain a very much simplified proof of CC. For CPI and CIRE we note that independence is defined on the underlying labels and thus Proposition 5.4 applies. Therefore $CS_<$ and $CL_<$ hold. Using Proposition 5.3, we can get an LTSI with general independence satisfying IRE and IEC, and therefore CS and CL. This is the first time these causal properties have been proved for RCCS.

HOπ We consider here the uncontrolled reversible semantics for HOπ [18]. We restrict our attention to reachable processes, called there consistent. The semantics is a reduction semantics; hence there are no labels (or, equivalently, all the labels coincide). To have more informative labels we can consider the transitions defined in [18, Section 3.1], where labels are composed of memory information and a flag denoting whether the transition is forward or backward. The notion of independence would be given by the concurrency relation on coinitial transitions [18, Definition 9]. All pre-reversible LTSI axioms hold, as well as CIRE which is needed for causal safety and liveness. Specifically, SP is proved in [18, Lemma 9]. BTI holds since distinct memories have disjoint sets of keys [18, Definition 3 and Lemma 3] and by the definition of concurrency [18, Definition 9]. WF holds as each backward step consumes a memory, which is finite to start with. Finally, CPI and CIRE are valid since the notion of concurrency is defined on the annotated labels and using our Proposition 5.4.

As a result we obtain a very much simplified proof of CC. Moreover, using CPI and CIRE, we get the $CS_<$ and $CL_<$ safety and liveness properties and, applying mapping g from Section 5, we get a general pre-reversible LTSI satisfying IRE and IEC, hence CS and CL are satisfied. This is the first time that causal properties have been shown for HOπ.

Rπ We consider the (uncontrolled) reversible semantics for π-calculus defined in [5]. We restrict the attention to reachable processes. The semantics is an LTS semantics. Independence is given as concurrency which is defined for consecutive transitions [5, Definition 4.1]. CC holds [5, Theorem 4.5].

Our results are not directly applicable to Rπ, since SP holds up to label equivalence of transitions on opposite sides of the diamond, rather than equality of labels as in our approach. We would need to extend axiom SP and the definition of causal equivalence to allow for label equivalence in order to handle Rπ using our axiomatic method.

Erlang We consider the uncontrolled reversible (reduction) semantics for Erlang in [20]. We restrict our attention to reachable processes. In order to have more informative labels we can consider the annotations defined in [20, Section 4.1]. We then can define coinitial transitions to be independent if they are concurrent [20, Definition 12].

We next discuss the validity of our axioms in reversible Erlang. SP is proved in [20, Lemma 13] and BTI is trivial from the definition of concurrency [20, Definition 12]. WF holds since the pairs of integers (total number of elements in memories, total number of messages queued) ordered under lexicographic order are always positive and decrease at each backward step. Intuitively, each step but the ones derived using the rule for reverse sched (see [20, Figure 11]) consumes an item of memory, and each step derived using rule reverse sched removes a message from a process queue. Finally, CPI and CIRE hold since the notion of concurrency is defined on the annotated labels, and by Proposition 5.4.

Since this the setting is very similar to the one of HOπ (both calculi have a reduction semantics and a coinitial notion of independence defined on enriched labels), we get the same results as for HOπ, including CC, and CS and CL.

Reversible occurrence nets Reversible occurrence nets [25,24] are traditional occurrence nets (safe and with no backward conflicts) extended with a reverse transition for each forward transition. They give rise to an LTS where states are pairis (N, m) with N a net and m a marking. A computation that represents firing a transition t in (N, m) and resulting in (N, m') is given by a firing relation $(N, m) \xrightarrow{t} (N, m')$. The notion of independence is the concurrency relation [25, Section 3] which is defined between arbitrary firings (transitions). Hence, we get a general LTSI. The CC property is shown by following the traditional approach in [6]. SP and PL are shown as well. PL and CC require several pages of proofs [24]. The causal safety and causal liveness properties are not considered in [25,24].

We can obtain CC, and additionally CS and CL, as follows. SP and BTI are proved for reversible occurrence nets in [24] as Lemma 4.3 and Lemma 3.3

respectively. WF holds because there are no forward cycles of firings in occurrence nets, hence no infinite reverse paths. In order to have CS and CL, we need to show CPI and IRE. Lemma 3.4 in [24] gives CPI. Events can be defined on firings as in our Definition 4.5, and then IRE holds as the concurrency relation preserves such events.

7 Conclusion, Related and Future Work

The literature on causal-consistent reversibility (see, for example the early survey [19]) has a number of proofs of results such as the parabolic lemma (PL) and the causal consistency property (CC), all of which are instantiated to a specific calculus, language or formalism. We have taken here a complementary approach, analysing the properties of interest in an abstract and language-independent setting. In particular, we have shown how to prove the most relevant of these properties from a small number of axioms.

Our approach builds upon [28], where a set of axioms for reverse LTSs was given and several interesting properties were shown. While the idea is similar, the development is rather different since we consider more basic axioms (we only share WF, while many of the axioms in [28], such as UT, follow from ours), and since the two papers focus on different properties. We focus on CC and various forms of CS and CL, while [28] considers correspondence with prime event structures and reversible bisimulations. Moreover, LTSs in [28] do not have a notion of independence.

In other related work, we may particularly mention [8], which like ours takes an abstract view, though based on category theory. However, its results concern irreversible actions, and do not provide insights in our setting, where all actions are reversible. The only other work which takes a general perspective is [3], which concentrates on how to derive a reversible extension of a given formalism. However, proofs concern a limited number of properties (essentially our CC), and hold only for extensions built using the technique proposed there. Also [27,29] are general, since they propose how to reverse a calculus that can be defined in a general format of SOS rules. However, the format has its syntactic constraints while our approach abstracts from them. Finally, [9] presents a number of properties such as, for example, backward confluence, which arise in the context of reversing of steps of executed transitions in Place/Transition nets.

The approach proposed in this paper opens a number of new possibilities. Firstly, when devising a new reversible formalism, our results provide a rich toolbox to prove (or disprove) relevant properties in a simple way. This is particularly relevant since causal-consistent reversibility is getting applied to more and more complex languages, such as Erlang [20], where direct proofs become cumbersome and error-prone. Secondly, our abstract proofs are relatively easy to formalise in a proof-assistant, which is even more relevant given that this will certify the correctness of the results for many possible instances. Another possible extension of our work concerns integrating into our framework irreversible actions [7]. In order to do that we could take inspiration from the above-mentioned [8].

References

1. Barylska, K., Koutny, M., Mikulski, Ł., Piątkowski, M.: Reversible computation vs. reversibility in Petri nets. Science of Computer Programming **151**, 48–60 (2018)
2. Bennett, C.H.: Logical reversibility of computation. IBM Journal of Research and Development **17**(6), 525–532 (1973)
3. Bernadet, A., Lanese, I.: A modular formalization of reversibility for concurrent models and languages. In: Bartoletti, M., Henrio, L., Knight, S., Vieira, H.T. (eds.) ICE. EPTCS, vol. 223, pp. 98–112 (2016)
4. Carothers, C.D., Perumalla, K.S., Fujimoto, R.: Efficient optimistic parallel simulations using reverse computation. ACM Transactions on Modeling and Computer Simulation **9**(3), 224–253 (1999)
5. Cristescu, I., Krivine, J., Varacca, D.: A compositional semantics for the reversible pi-calculus. In: LICS. pp. 388–397. IEEE Computer Society (2013)
6. Danos, V., Krivine, J.: Reversible communicating systems. In: Gardner, P., Yoshida, N. (eds.) CONCUR. LNCS, vol. 3170, pp. 292–307. Springer (2004)
7. Danos, V., Krivine, J.: Transactions in RCCS. In: Abadi, M., de Alfaro, L. (eds.) CONCUR. LNCS, vol. 3653, pp. 398–412. Springer (2005)
8. Danos, V., Krivine, J., Sobociński, P.: General reversibility. In: Amadio, R.M., Phillips, I. (eds.) EXPRESS. ENTCS, vol. 175(3), pp. 75–86. Elsevier (2006)
9. de Frutos Escrig, D., Koutny, M., Mikulski, Ł.: Reversing steps in Petri nets. In: Donelli, S., Haar, S. (eds.) Petri Nets. LNCS, vol. 11522. Springer (2019)
10. Giachino, E., Lanese, I., Mezzina, C.A.: Causal-consistent reversible debugging. In: Gnesi, S., Rensink, A. (eds.) FASE. LNCS, vol. 8411, pp. 370–384. Springer (2014)
11. Giachino, E., Lanese, I., Mezzina, C.A., Tiezzi, F.: Causal-consistent rollback in a tuple-based language. Journal of Logical and Algebraic Methods in Programming **88**, 99–120 (2017)
12. van Glabbeek, R., Vaandrager, F.: The difference between splitting in n and $n+1$. Information and Computation **136**(2), 109–142 (1997)
13. Kari, J.: Reversible cellular automata: From fundamental classical results to recent developments. New Generation Computing **36**(3), 145–172 (2018)
14. Kuhn, S., Ulidowski, I.: Local reversibility in a Calculus of Covalent Bonding. Science of Computer Programming **151**, 18–47 (2018)
15. Landauer, R.: Irreversibility and heat generated in the computing process. IBM Journal of Research and Development **5**, 183 –191 (1961)
16. Lanese, I., Phillips, I., Ulidowski, I.: An axiomatic approach to reversible computation (TR) (2020), http://www.cs.unibo.it/~lanese/work/axrev-TR.pdf
17. Lanese, I., Mezzina, C.A., Schmitt, A., Stefani, J.: Controlling reversibility in higher-order pi. In: Katoen, J., König, B. (eds.) CONCUR. LNCS, vol. 6901, pp. 297–311. Springer (2011)
18. Lanese, I., Mezzina, C.A., Stefani, J.: Reversibility in the higher-order π-calculus. Theoretical Computer Science **625**, 25–84 (2016)
19. Lanese, I., Mezzina, C.A., Tiezzi, F.: Causal-consistent reversibility. Bulletin of the EATCS **114** (2014)
20. Lanese, I., Nishida, N., Palacios, A., Vidal, G.: A theory of reversibility for Erlang. Journal of Logical and Algebraic Methods in Programming **100**, 71–97 (2018)
21. Laursen, J.S., Schultz, U.P., Ellekilde, L.: Automatic error recovery in robot assembly operations using reverse execution. In: IROS. pp. 1785–1792. IEEE (2015)
22. Lienhardt, M., Lanese, I., Mezzina, C.A., Stefani, J.: A reversible abstract machine and its space overhead. In: Giese, H., Rosu, G. (eds.) FMOODS/FORTE. LNCS, vol. 7273, pp. 1–17. Springer (2012)

23. McNellis, J., Mola, J., Sykes, K.: Time travel debugging: Root causing bugs in commercial scale software. CppCon talk, https://www.youtube.com/watch?v=l1YJTg_A914 (2017)
24. Melgratti, H.C., Mezzina, C.A., Ulidowski, I.: Reversing Place Transition nets. arXiv **1910.04266** (2019)
25. Melgratti, H.C., Mezzina, C.A., Ulidowski, I.: Reversing P/T nets. In: Nielson, H.R., Tuosto, E. (eds.) COORDINATION. LNCS, vol. 11533, pp. 19–36. Springer (2019)
26. Mezzina, C.A.: On reversibility and broadcast. In: Kari, J., Ulidowski, I. (eds.) RC 2018. LNCS, vol. 11106, pp. 67–83. Springer (2018)
27. Phillips, I., Ulidowski, I.: Reversing algebraic process calculi. In: Aceto, L., Ingólfsdóttir, A. (eds.) FoSSaCS. LNCS, vol. 3921, pp. 246–260. Springer (2006)
28. Phillips, I., Ulidowski, I.: Reversibility and models for concurrency. In: Hennessy, M., van Glabbeek, R. (eds.) SOS. ENTCS, vol. 192(1), pp. 93–108. Elsevier (2007)
29. Phillips, I., Ulidowski, I.: Reversing algebraic process calculi. Journal of Logic and Algebraic Programming **73**(1-2), 70–96 (2007)
30. Phillips, I., Ulidowski, I.: Reversibility and asymmetric conflict in event structures. Journal of Logical and Algebraic Methods in Programming **84**, 781–805 (2015)
31. Phillips, I., Ulidowski, I., Yuen, S.: A reversible process calculus and the modelling of the ERK signalling pathway. In: Glück, R., Yokoyama, T. (eds.) RC. LNCS, vol. 7581, pp. 218–232. Springer (2012)
32. Pin, J.: On the language accepted by finite reversible automata. In: Ottmann, T. (ed.) ICALP. LNCS, vol. 267, pp. 237–249. Springer (1987)
33. Sassone, V., Nielsen, M., Winskel, G.: Models of concurrency: Towards a classification. Theoretical Computer Science **170**(1-2), 297–348 (1996)
34. Ulidowski, I., Phillips, I., Yuen, S.: Reversing event structures. New Generation Computing **36**(3), 281–306 (2018)
35. Yokoyama, T., Glück, R.: A reversible programming language and its invertible self-interpreter. In: Ramalingam, G., Visser, E. (eds.) ACM SIGPLAN PEMP. pp. 144–153. ACM (2007)

An Auxiliary Logic on Trees: on the Tower-hardness of logics featuring reachability and submodel reasoning

Alessio Mansutti(✉)

LSV, CNRS, ENS Paris-Saclay, Université Paris-Saclay, mansutti@lsv.fr

Abstract. We describe a set of simple features that are sufficient in order to make the satisfiability problem of logics interpreted on trees TOWER-hard. We exhibit these features through an *Auxiliary Logic on Trees* (ALT), a modal logic that essentially deals with reachability of a fixed node inside a forest and features modalities from sabotage modal logic to reason on submodels. After showing that ALT admits a TOWER-complete satisfiability problem, we prove that this logic is captured by four other logics that were independently found to be TOWER-complete: two-variables separation logic, quantified computation tree logic, modal logic of heaps and modal separation logic. As a by-product of establishing these connections, we discover strict fragments of these logics that are still non-elementary.

1 Introduction

In mathematical logic there is a well-known trade-off between expressive power and complexity, where weaker languages cannot capture interesting properties of complex systems, whereas finding solutions of a given problem is infeasible for richer languages. For instance, many verification tasks, such as reachability and homomorphisms queries, happen to be expressible in monadic second-order logic (MSO) [15]. This logic is however not usable in practice, as its satisfiability problem SAT(MSO) is undecidable in general and was famously proved by Rabin [36] to be decidable but non-elementary when the logic is interpreted on trees or on one unary function. A more recent analysis that uses the hierarchy of non-elementary ranking functions [38] classifies SAT(MSO) on these two structures as TOWER-complete, i.e. complete for the class of problems of time complexity bounded by a tower of exponentials, whose height is an elementary function of the input.

In order to bypass these problems, a general approach is to design restrictions of MSO that can solve complex reasoning tasks while being more appealing complexity-wise. An example of this is given by the framework of temporal logics, formalisms that describe the evolution of reactive systems [24]. Among the various temporal logics, from the classical linear temporal logic (LTL) [39] and computation tree logic (CTL) [13], as well as their fragments [2,33], to the more recently developed interval temporal logics [7,8], the main common feature of this framework is perhaps the ability to check whether the system can evolve to a certain configuration, i.e. a *reachability* query. In this context, we recall the landmark result on the satisfiability of CTL, shown EXPTIME-complete by Fisher and Ladner [23]. Another possibility to deal with the complexity of MSO is to restrict the second-order quantifications to specific *submodels*. This is the idea behind ambient logic [16], separation logic [37] and more generally bunched logics [35]

J. Goubault-Larrecq and B. König (Eds.): FOSSACS 2020, LNCS 12077, pp. 462–481, 2020.
https://doi.org/10.1007/978-3-030-45231-5_24

and graphs logics [1]. These logics provide primitives for reasoning about resource composition, mainly by adding a *spatial conjunction* $\varphi * \psi$ which requires to split a model into two disjoint pieces, one satisfying φ and the other satisfying ψ. Similar ideas are developed in sabotage modal logics, where the formula $\blacklozenge \varphi$, headed by the *sabotage* modality $\blacklozenge$, states that φ must hold in a graph obtained by removing one edge from the current model [4,21]. Within these logics, we highlight the quantifier-free fragment of separation logic restricted to the $*$ operator, denoted here with SL($*$) and whose satisfiability problem is proved to be PSPACE-complete in [12].

Once a framework provides a solid foundation for reasoning tasks, a natural step is to extend its expressiveness while keeping its complexity in check. Sometimes the additional capabilities do not change the complexity of the logic, as for example SL($*$) extended with reachability predicates, whose satisfiability problem is still PSPACE-complete [20]. However, it often happens that the new features make the problem jump to higher complexity classes and, sometimes, reach MSO. We pinpoint two instances of this:

- SL($*$) enriched with first-order quantifiers, albeit less expressive than MSO interpreted on one unary function, has a TOWER-complete satisfiability problem [9].
- CTL enriched with propositional quantifiers has an undecidable satisfiability problem on general models. On trees (i.e. QCTL^T), the problem is TOWER-complete [28].

Consequently, it is natural to ask ourselves why the additional features made the problem harder. Answering this question requires to study the interplays between the various operators of the logic, searching for a sufficient set of conditions explaining its complexity.

Our motivation. Second-order features often lead to logics with TOWER-hard satisfiability problems, as illustrated above for first-order SL($*$) and QCTL^T. A good amount of research has been done independently on these logics [5,9,17,28], culminating with the TOWER-hardness of SL($*$) with two quantified variables [17] and the TOWER-hardness of QCTL^T with just one temporal operator between *exists-finally* EF and *exists-next* EX [5] (see Section 4 for the definitions). Connections between these two formalisms have not been explicitly developed so far, perhaps because of the quite different logics: QCTL^T is built on top of propositional calculus and it is interpreted on infinite trees, whereas SL($*$) does not feature propositional symbols and it is essentially interpreted on finite structures. Nevertheless, we argue that these and other logics are related not only as they are fragments of MSO, but also as they share a form of reachability and an ability of reasoning on submodels which is sufficient to obtain TOWER-hard logics.

Our contribution. We explicit these common features that lead to TOWER-hard logics by relying on an *Auxiliary Logic on Trees* (ALT), introduced in Section 2. ALT reasons about reachability of a fixed *target node* inside a finite forest and features modalities from sabotage logics to reason on submodels. Here, reachability should be understood as the ability to reach the target node in at least one step, starting from a "current" node which can be updated thanks to the existential modality *somewhere* $\langle U \rangle$ [26]. In Section 3, we take a look at the expressive power of ALT and show that SAT(ALT) is TOWER-hard. In Section 4, we then display how ALT is captured by first-order SL($*$) and QCTL^T, as well as modal logic of heaps (MLH) and modal separation logics (MSL), two other logics introduced in [17] and [18], respectively. In this context, beside exposing that all these logics are TOWER-hard because of the way they reason about reachability and submodels, we discover interesting sublogics that are still TOWER-complete:

- QCTL^T restricted to $\mathsf{E}(\varphi \mathbin{\mathsf{U}} \psi)$ modalities, where φ, ψ are Boolean combinations of atomic propositions, or to the EF modality, which can be nested at most once.
- the common fragment of MLH and MSL having Boolean connectives and the modalities $\Diamond$, $\langle \mathrm{U} \rangle$ and $*$. Notice that this logic does not have propositional symbols.

2 The definition of an Auxiliary Logic on Trees

We introduce an Auxiliary Logic on Trees (ALT). Its formulae are from the grammar:

$$\varphi := \top \mid \varphi \wedge \varphi \mid \neg\varphi \mid \mathtt{T} \mid \mathtt{G} \mid \langle \mathrm{U} \rangle\, \varphi \mid \blacklozenge \varphi \mid \blacklozenge^{*} \varphi$$

As we will soon clarify, the symbol $\langle \mathrm{U} \rangle$ is borrowed from Goranko and Passy paper on modal logic with universal modality [26]. Similarly, readers who are familiar with sabotage modal logics will recognise in $\blacklozenge$ the sabotage modality [4], and in $\blacklozenge^*$ its Kleene closure (i.e. $\blacklozenge$ applied an arbitrary number of times). These two operators modify the model during the evaluation of a formula, making ALT a *relation-changing* modal logic (following the terminology used in [3]). However, contrary to most modal logics, ALT does not feature classical propositional symbols. Instead, this logic only features two interpreted atomic propositions T and G. Roughly speaking, T stands for "the target node is reachable" whereas G stands for "the target node is not reachable". The formal definitions will be given soon in order to clarify these two sentences.

Let $\mathcal{N}$ be countably infinite set of *nodes*. A *(finite) forest* $\mathcal{F} : \mathcal{N} \rightarrow_{\mathtt{fin}} \mathcal{N}$ is a partial function (encoding the standard parent relation) that

- has a finite domain of definition, i.e. $\mathrm{dom}(\mathcal{F}) \stackrel{\text{def}}{=} \{\mathsf{n} \in \mathcal{N} \mid \mathcal{F}(\mathsf{n}) \text{ is defined}\}$ is finite;
- is acyclic, i.e. for every $\mathsf{n} \in \mathrm{dom}(\mathcal{F})$ and $\delta \geq 1$, $\mathcal{F}^{\delta}(\mathsf{n}) \neq \mathsf{n}$.

Here, $\mathcal{F}^{\delta}$ denotes $\delta \geq 0$ *functional composition(s)* of $\mathcal{F}$. Albeit non-standard, our definition of finite forests over an infinite set of nodes simplifies the forthcoming definitions. Besides, in Section 3.2 we show how restricting $\mathcal{N}$ to a finite set does not change the expressive power nor the complexity of ALT.

We denote the image of $\mathcal{F}$ as $\mathrm{ran}(\mathcal{F}) \stackrel{\text{def}}{=} \{\mathsf{n}' \mid \mathcal{F}(\mathsf{n}) = \mathsf{n}' \text{ for some } \mathsf{n} \in \mathrm{dom}(\mathcal{F})\}$. Given a finite set X, we denote with $|X|$ its cardinality. Let n, n' be two nodes. As usual, n is a *$\mathcal{F}$-descendant* of n' (alternatively, n' is an *$\mathcal{F}$-ancestor* of n) whenever $\mathcal{F}^{\delta}(\mathsf{n}) = \mathsf{n}'$ for some $\delta \geq 1$. In this case, if $\delta = 1$ then n is a *$\mathcal{F}$-child* of n' (alternatively, n' is the *$\mathcal{F}$-parent* of n). We drop the prefix $\mathcal{F}$- from these terms when it is clear from the context. Given two forests $\mathcal{F}, \mathcal{F}'$, we say that $\mathcal{F}'$ is a *subforest* of $\mathcal{F}$, written $\mathcal{F}' \sqsubseteq \mathcal{F}$, whenever $\mathcal{F}(\mathsf{n}) = \mathcal{F}'(\mathsf{n})$ for every $\mathsf{n} \in \mathrm{dom}(\mathcal{F}')$. Figure 1 intuitively represents two forests (every "▫" represents a node), the one on the left being a subforest of the one on the right.

ALT is interpreted on *pointed forests* $(\mathcal{F}, \mathsf{t}, \mathsf{n})$, where $\mathcal{F}$ is a forest and $\mathsf{t}, \mathsf{n} \in \mathcal{N}$ are respectively called the *target node* and the *current evaluation node*. The satisfaction relation $\models$ is defined (throughout the paper, we omit standard clauses for $\top, \wedge, \neg$) as:

$(\mathcal{F}, \mathsf{t}, \mathsf{n}) \models \mathtt{T} \quad \stackrel{\text{def}}{\Leftrightarrow} \quad \mathsf{n}$ is a $\mathcal{F}$-descendant of t.

$(\mathcal{F}, \mathsf{t}, \mathsf{n}) \models \mathtt{G} \quad \stackrel{\text{def}}{\Leftrightarrow} \quad \mathsf{n} \in \mathrm{dom}(\mathcal{F})$ and $(\mathcal{F}, \mathsf{t}, \mathsf{n}) \not\models \mathtt{T}$.

$(\mathcal{F}, \mathsf{t}, \mathsf{n}) \models \langle \mathrm{U} \rangle\, \varphi \quad \stackrel{\text{def}}{\Leftrightarrow} \quad$ there is $\mathsf{n}' \in \mathcal{N}$ s.t. $(\mathcal{F}, \mathsf{t}, \mathsf{n}') \models \varphi$.

$(\mathcal{F}, \mathsf{t}, \mathsf{n}) \models \blacklozenge \varphi \quad \stackrel{\text{def}}{\Leftrightarrow} \quad$ there is $\mathcal{F}'$ s.t. $\mathcal{F}' \sqsubseteq \mathcal{F}$, $|\mathrm{dom}(\mathcal{F}')|+1 = |\mathrm{dom}(\mathcal{F})|$, $(\mathcal{F}', \mathsf{t}, \mathsf{n}) \models \varphi$.

$(\mathcal{F}, \mathsf{t}, \mathsf{n}) \models \blacklozenge^{*} \varphi \quad \stackrel{\text{def}}{\Leftrightarrow} \quad$ there is $\mathcal{F}'$ s.t. $\mathcal{F}' \sqsubseteq \mathcal{F}$ and $(\mathcal{F}', \mathsf{t}, \mathsf{n}) \models \varphi$.

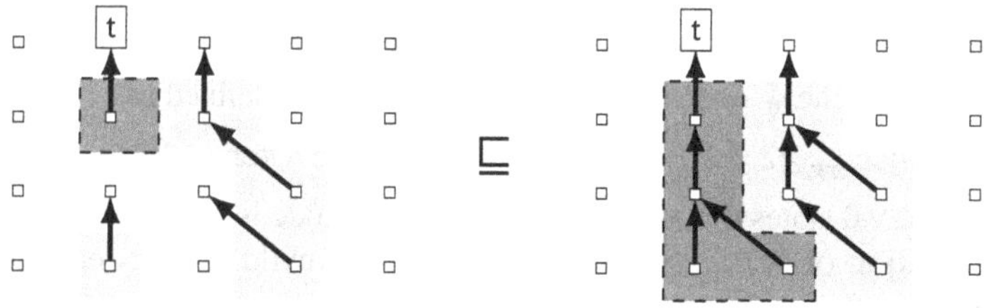

Fig. 1. Subforest relation

We denote with $\bot$ the contradiction $\neg\mathtt{T}$. The standard connectives $\vee$ and $\Rightarrow$ are defined as usual. The semantics of $\mathtt{T}$ and $\mathtt{G}$ is pretty straightforward. As a visual aid, the nodes in Figure 1 satisfying $\mathtt{T}$ are the ones in the dark grey area, whereas the ones in the light grey area satisfy $\mathtt{G}$. As stated before, the semantics given to $\langle\mathrm{U}\rangle\,\varphi$ is the one of the existential modality *somewhere* [26], stating that there is a way to change the current evaluation node so that φ becomes true. Its dual operator $[\mathrm{U}]\,\varphi \stackrel{\text{def}}{=} \neg\,\langle\mathrm{U}\rangle\,\neg\varphi$ is the universal modality *everywhere*. The semantics given to $\blacklozenge\,\varphi$ is the one of the *sabotage* modality from [4], which requires to find one edge of the forest that, when removed, makes the model satisfy φ. Lastly, the $\blacklozenge^*$ modality, here called *repeated sabotage* operator, can be seen as the operator obtained by applying $\blacklozenge$ an arbitrary number of times. Indeed, by inductively defining $\blacklozenge^k\,\varphi$ ($k \in \mathbb{N}$) as the formula φ for $k = 0$ and otherwise ($k \geq 1$) as $\blacklozenge\blacklozenge^{k-1}\,\varphi$, it is easy to see that $(\mathcal{F}, \mathsf{t}, \mathsf{n}) \models \blacklozenge^*\varphi$ is equivalent to $\exists k \in \mathbb{N}.\ (\mathcal{F}, \mathsf{t}, \mathsf{n}) \models \blacklozenge^k\,\varphi$.

Given a pointed forest $(\mathcal{F}, \mathsf{t}, \mathsf{n})$, we denote with $\mathcal{F}(\mathtt{G})_{\mathsf{t}}$ the set of its *garbage nodes*: the set of elements in $\mathrm{dom}(\mathcal{F})$ that are not descendants of t, i.e. $\mathcal{F}(\mathtt{G})_{\mathsf{t}} \stackrel{\text{def}}{=} \{\mathsf{n} \in \mathrm{dom}(\mathcal{F}) \mid \forall\delta \geq 1,\ \mathcal{F}^{\delta}(\mathsf{n}) \neq \mathsf{t}\}$. Then, $\mathcal{F}(\mathtt{G})_{\mathsf{t}}$ is equivalent to $\{\mathsf{n} \in \mathcal{N} \mid (\mathcal{F}, \mathsf{t}, \mathsf{n}) \models \mathtt{G}\}$. We omit the subscript t from $\mathcal{F}(\mathtt{G})_{\mathsf{t}}$ when it is clear from the context. We augment the standard precedence rules of propositional logic so that the modalities $\langle\mathrm{U}\rangle$, $\blacklozenge$ and $\blacklozenge^*$ have the same precedence as $\neg$. For instance, the formula $\langle\mathrm{U}\rangle\,\mathtt{T} \wedge \mathtt{G}$ should be read as $(\langle\mathrm{U}\rangle\,\mathtt{T}) \wedge \mathtt{G}$.

Satisfiability problem. As usual, given a logic $\mathfrak{L}$ and one of its interpretations $\models$ on a class of structures $\mathfrak{C}$, the satisfiability problem of $\mathfrak{L}$, denoted with $\mathsf{SAT}(\mathfrak{L})$ when the interpretation is clear from the context, takes as input a formula φ of $\mathfrak{L}$ and asks whether there is a structure $\mathfrak{M} \in \mathfrak{C}$ such that $\mathfrak{M} \models \varphi$. If the answer is positive, then φ is *satisfiable*.

Where does ALT come from? A preliminary definition of ALT was introduced in [31] to reason on the complexity of separation logic. As such, in [31] ALT features the separating conjunction $\varphi * \psi$ from separation logic, stating that the forest can be partitioned into two disjoint subforests, one satisfying φ and one satisfying ψ. This binary operator generalises both $\blacklozenge$ and $\blacklozenge^*$ operators (we show how in Section 4). Hence, the TOWER-hardness of the satisfiability problem for the logic defined here cannot be inherited from [31] and must be proved (Section 3). Unfortunately, the proof does not give any indication on whether or not the two versions of ALT have the same expressive power. What is clear is that the two logics analyse the model in a different way: the $*$ operator is able to reason on the model in a "concurrent" way, whereas $\blacklozenge$ and $\blacklozenge^*$ do it in a "sequential" one. Let us draw an example of this. Let $(\mathcal{F}, \mathsf{t}, \mathsf{n})$ be a pointed forest. We aim at defining a formula $\#\mathtt{ch}_{\mathtt{trg}} \geq 2$ stating that the target node t has at least two children. First, we define $\#\mathtt{ch}_{\mathtt{trg}} \geq 1$ (the formula for just one child) as $\langle\mathrm{U}\rangle(\mathtt{T} \wedge \neg\blacklozenge\,\mathtt{G})$. Intuitively, $\#\mathtt{ch}_{\mathtt{trg}} \geq 2$

can then be defined with the $*$ operator simply as the formula $\#\mathtt{ch}_{\mathtt{trg}} \geq 1 * \#\mathtt{ch}_{\mathtt{trg}} \geq 1$, stating that it is possible to partition the forest into two subforests having both at least one child of t. With the $\blacklozenge$ operator, this property is instead defined as

$$\#\mathtt{ch}_{\mathtt{trg}} \geq 2 \stackrel{\text{def}}{=} \langle \mathsf{U} \rangle \big(\mathtt{T} \land \neg \blacklozenge \mathtt{G} \land \blacklozenge (\neg\, \mathtt{inDom} \land \#\mathtt{ch}_{\mathtt{trg}} \geq 1) \big).$$

where $\mathtt{inDom} \stackrel{\text{def}}{=} \mathtt{T} \lor \mathtt{G}$ states that the current evaluation node is in the domain of the forest. This definition of $\#\mathtt{ch}_{\mathtt{trg}} \geq 2$ requires to find one child of t (as encoded by the "$\langle \mathsf{U} \rangle (\mathtt{T} \land \neg \blacklozenge \mathtt{G} \land \cdots$" part of the formula) and remove it from the model (as expressed by the "$\blacklozenge(\neg\, \mathtt{inDom} \land \cdots$" part). Only afterwards we check for the existence of a second child of t. This form of "sequential reasoning" (that can be often avoided when using the $*$ operator), is used in almost all the formulae of the next sections: we first find a node satisfying a certain property, we remove it from the structure, and only afterwards we check if the model satisfy a second property. This principle only works well for monotonic properties: with respect to the definition of $\#\mathtt{ch}_{\mathtt{trg}} \geq 2$, the set of children of t monotonically decreases when considering subforests. Thus, finding a child of t in the subforest, implies finding a child of t in the original forest.

3 On the complexity and expressive power of ALT

In this section, we show that SAT(ALT) is TOWER-hard by reduction from the satisfiability problem of Propositional Interval Temporal Logic on finite words (Section 3.3). The proof adapts the arguments used in [31] for the version of ALT featuring the separating conjunction $*$. The reduction is somewhat non-intuitive and in [31] it is given without explaining why more direct ways fail. Here, we clarify this issue which is related to the fact that ALT cannot deduce any property of the portion of a pointed forest $(\mathcal{F}, \mathsf{t}, \mathsf{n})$ corresponding to the nodes in $\mathcal{F}(\mathtt{G})$, except for the size of $\mathcal{F}(\mathtt{G})$ and the query $\mathsf{n} \in \mathcal{F}(\mathtt{G})$. This is done in Section 3.2, by relying on a notion of Ehrenfeucht-Fraïssé games for ALT.

3.1 Towards the TOWER-hardness of SAT(ALT): how to encode finite words.

As a first step, we define a correspondence between finite words and specific pointed forests. As usual, we define the set of finite words on a finite alphabet Σ as the closure under Kleene star Σ^*. To ease our modelling, we suppose $\Sigma \stackrel{\text{def}}{=} [1, n]$ to be the alphabet of natural numbers between 1 and n. Let $\mathfrak{w} = \mathsf{a}_1 \cdots \mathsf{a}_k$ be a k-symbols word in Σ^* and $\mathsf{M} = \{\mathsf{n}_1, \cdots, \mathsf{n}_k\}$ be a set of k nodes. Let N_i ($i \in [1, k]$) be a set of $\mathsf{a}_i + 1$ nodes different from $\mathsf{n}_1, \cdots, \mathsf{n}_k$ and so that for each distinct $i, j \in [1, k]$, $\mathsf{N}_i \cap \mathsf{N}_j = \emptyset$. Lastly, let t be a node not in $\mathsf{M} \cup \bigcup_{i \in [1,k]} \mathsf{N}_i$. A pointed forest $(\mathcal{F}, \mathsf{t}, \mathsf{n})$ encodes $\mathfrak{w}$ w.r.t. the sets $\mathsf{M}, \mathsf{N}_1, \cdots, \mathsf{N}_k$ iff **(I)** $\mathcal{F}(\mathsf{n}_k) = \mathsf{t}$, **(II)** for all $i \in [1, k-1]$ $\mathcal{F}(\mathsf{n}_i) = \mathsf{n}_{i+1}$, **(III)** for all $i \in [1, k]$ and $\mathsf{n}' \in \mathsf{N}_i$, $\mathcal{F}(\mathsf{n}') = \mathsf{n}_i$ and **(IV)** every $\mathcal{F}$-descendant of t belongs to a set among $\mathsf{M}, \mathsf{N}_1, \cdots, \mathsf{N}_k$.

We call the path from n_1 to n_k, the *main path* of $\mathcal{F}$. The nodes of this path are the ones in M, and can be characterised as being the only descendants of t with at least one child. Moreover, n_1 is the only node of the main path having the same number of descendants and children. We say that a node $\mathsf{n} \in \mathrm{dom}(\mathcal{F})$ *encodes* the symbol $\mathsf{a} \in \Sigma$ if it is a descendant of t and it has exactly $\mathsf{a} + 1$ children that are not in M. Then, the nodes in M are the only ones encoding symbols, where n_i encodes a_i for any $i \in [1, k]$. For instance, Figure 2 shows an encoding of the word 1121.

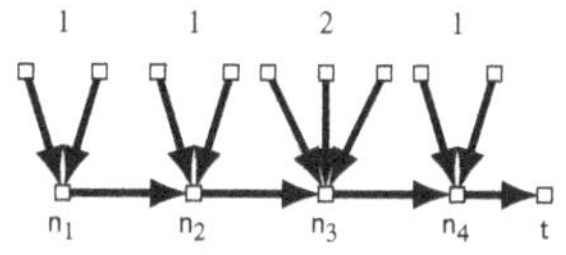

Fig. 2. Encoding of 1121.

Formula	Intended meaning
$\texttt{size(G)} \geq \beta$	$\lvert\mathcal{F}(\texttt{G})_{\mathsf{t}}\rvert \geq \beta$.
$\texttt{\#desc} \geq \beta$	$(\mathcal{F}, \mathsf{t}, \mathsf{n}) \models \mathrm{T}$ and n has at least β descendants.
$\texttt{\#child} \geq \beta$	$(\mathcal{F}, \mathsf{t}, \mathsf{n}) \models \mathrm{T}$ and n has at least β children.

Table 1. Formulae and their meaning on $(\mathcal{F}, \mathsf{t}, \mathsf{n})$.

In order to characterise the class of pointed forests encoding finite words, we adapt the formulae of [31] shown in Table 1 (where their semantics is described). Let $(\mathcal{F}, \mathsf{t}, \mathsf{n})$ be a pointed forest and let $\beta \in \mathbb{N}$. The formula $\texttt{size(G)} \geq \beta$ is inductively defined as:

$$\texttt{size(G)} \geq 0 \stackrel{\text{def}}{=} \mathrm{T}, \qquad \texttt{size(G)} \geq \beta{+}1 \stackrel{\text{def}}{=} \langle \mathrm{U} \rangle \big(\texttt{G} \land \blacklozenge(\neg\, \texttt{inDom} \land \texttt{size(G)} \geq \beta)\big).$$

Notice how, in the definition of $\texttt{size(G)} \geq \beta{+}1$, we use the same principle used to encode $\texttt{\#ch}_{\texttt{trg}} \geq 2$ at the end of Section 2: we first find a node in $\mathcal{F}(\texttt{G})$, remove it from the model, and then find other β elements of $\mathcal{F}(\texttt{G})$. The formulae $\texttt{\#desc} \geq \beta$ and $\texttt{\#child} \geq \beta$ (again, we refer to Table 1 for their semantics) are instead defined as:

$$\texttt{\#desc} \geq \beta \stackrel{\text{def}}{=} \blacklozenge^{*}\Big(\underbrace{[\mathrm{U}]\,\neg\texttt{G}}_{\mathcal{F}(\texttt{G}) \text{ is empty.}} \land \mathrm{T} \land \underbrace{\blacklozenge(\neg\, \texttt{inDom} \land \texttt{size(G)} \geq \beta)}_{\text{Removing n lead to a set of garbage nodes of size at least } \beta.} \Big)$$

$$\texttt{\#child} \geq 0 \stackrel{\text{def}}{=} \mathrm{T}, \qquad \texttt{\#child} \geq \beta{+}1 \stackrel{\text{def}}{=} \texttt{\#desc} \geq \beta{+}1 \land \underbrace{\neg \blacklozenge^{\beta}(\mathrm{T} \land \neg\,\texttt{\#desc} \geq 1)}_{\text{Whenever } \beta \text{ nodes of dom}(\mathcal{F}) \text{ are removed, if n still reaches t then it has at least one descendant.}}.$$

Given $\mathtt{s} \in \{\texttt{size(G)}, \texttt{\#ch}_{\texttt{trg}}, \texttt{\#desc}, \texttt{\#child}\}$, we write $\mathtt{s} = \beta$ for $\mathtt{s} \geq \beta \land \neg \mathtt{s} \geq \beta{+}1$. For instance, $\texttt{\#child} = \beta$ is the formula that checks whether n has *exactly* β children and it is a descendant of t. We can now conclude the encoding of finite words.

Let $(\mathcal{F}, \mathsf{t}, \mathsf{n})$ be a pointed forest encoding $\mathfrak{w} \in \Sigma^*$ and let M be the set of nodes in its main path. Let us recall two properties of our encoding: (I) a node n' encodes a symbol of $\mathfrak{w}$ iff $\mathsf{n}' \in \mathsf{M}$, and (II) the node encoding the first symbol of $\mathfrak{w}$ is the only node in M with the same number of descendants and children. To reflect (I) we denote with $\texttt{symb}$ the formula $\texttt{\#desc} \geq 1$. For (II), given $\mathsf{S} \subseteq \Sigma$, we introduce the formula $\texttt{1st}_{\mathsf{S}}$ that checks if the current evaluation node corresponds to the first node of the main path and encodes a symbol in S. It is defined as $\bigvee_{\beta \in \mathsf{S}}(\texttt{\#desc} = \beta + 1 \land \texttt{\#child} = \beta + 1)$. The following statement formalises the connection between this formula and property (II) stated above.

Lemma 1. *Let* $\mathfrak{w} \in \Sigma^+$. *Let* $(\mathcal{F}, \mathsf{t}, \mathsf{n})$ *be a pointed forest encoding* $\mathfrak{w}$. *Let* n_1 *be the first node in the main path of* $\mathcal{F}$. *For every* $\mathsf{S} \subseteq \Sigma$, $(\mathcal{F}, \mathsf{t}, \mathsf{n}) \models \langle \mathrm{U} \rangle\, \texttt{1st}_{\mathsf{S}}$ *iff* $(\mathcal{F}, \mathsf{t}, \mathsf{n}_1) \models \texttt{1st}_{\mathsf{S}}$.

We are finally ready to define the formula $\texttt{word}_{\Sigma}$, characterising the class of forests that encodes words in Σ^*. It is proved correct by Lemma 2, and it is defined as follows

$$\begin{aligned} \texttt{word}_{\Sigma} \stackrel{\text{def}}{=} & \overbrace{(\langle \mathrm{U} \rangle\, \mathrm{T} \Rightarrow \langle \mathrm{U} \rangle\, \texttt{symb})}^{\text{The target node has no descendants, or has a descendant that encodes a symbol.}} \land \neg\, \texttt{\#ch}_{\texttt{trg}} \geq 2 \land \\ & [\mathrm{U}](\texttt{symb} \Rightarrow \texttt{1st}_{\Sigma} \lor \underbrace{(\neg\, \texttt{1st}_{\{n+1\}} \land \blacklozenge\, \texttt{1st}_{\Sigma})}_{\text{The current node encodes a symbol in } [1,n] \text{ and exactly one of its children encodes a symbol.}}). \end{aligned}$$

Lemma 2. *A pointed forest* $(\mathcal{F}, \mathsf{t}, \mathsf{n})$ *is an encoding of a word in* Σ^* *iff* $(\mathcal{F}, \mathsf{t}, \mathsf{n}) \models \texttt{word}_{\Sigma}$.

game played on $((\mathcal{F}_1, \mathsf{t}_1, \mathsf{n}_1), (\mathcal{F}_2, \mathsf{t}_2, \mathsf{n}_2), (\mathsf{m}, \mathsf{s}, \mathsf{k}))$
if there is $p \in \{\mathtt{G}, \mathtt{T}\}$ s.t. not $((\mathcal{F}_1, \mathsf{t}_1, \mathsf{n}_1) \models p$ iff $(\mathcal{F}_2, \mathsf{t}_2, \mathsf{n}_2) \models p)$ **then** the spoiler wins, **otherwise** the spoiler chooses $i \in \{1, 2\}$ and plays on $(\mathcal{F}_i, \mathsf{t}_i, \mathsf{n}_i)$. The duplicator replies on $(\mathcal{F}_j, \mathsf{t}_j, \mathsf{n}_j)$ where $j \in \{1, 2\} \setminus \{i\}$. The spoiler **must** choose one of the following moves (else the duplicator wins).

$\langle \mathrm{U} \rangle$ **move**: if $\mathsf{m} \geq 1$ then the spoiler **can** choose to play a $\langle \mathrm{U} \rangle$ move. It selects a node $\mathsf{n}'_i \in \mathcal{N}$.
- Then, the duplicator **must** reply with some node $\mathsf{n}'_j \in \mathcal{N}$ (otherwise the spoiler wins).
- The game continues on $((\mathcal{F}_1, \mathsf{t}_1, \mathsf{n}'_1), (\mathcal{F}_2, \mathsf{t}_2, \mathsf{n}'_2), (\mathsf{m}-1, \mathsf{s}, \mathsf{k}))$.

$\blacklozenge$ **move**: if $\mathsf{s} \geq 1$ and $\mathrm{dom}(\mathcal{F}_i) \neq \emptyset$ then the spoiler **can** choose to play a $\blacklozenge$ move. It selects a finite forest $\mathcal{F}'_i$ such that $\mathcal{F}'_i \sqsubseteq \mathcal{F}_i$ and $|\mathrm{dom}(\mathcal{F}'_i)| = |\mathrm{dom}(\mathcal{F}_i)| - 1$.
- The duplicator **must** reply with some $\mathcal{F}'_j \sqsubseteq \mathcal{F}_j$ s.t. $|\mathrm{dom}(\mathcal{F}'_j)| = |\mathrm{dom}(\mathcal{F}_j)| - 1$.
- The game continues on $((\mathcal{F}'_1, \mathsf{t}_1, \mathsf{n}_1), (\mathcal{F}'_2, \mathsf{t}_2, \mathsf{n}_2), (\mathsf{m}, \mathsf{s}-1, \mathsf{k}))$.

$\blacklozenge^*$ **move**: if $\mathsf{k} \geq 1$ then the spoiler **can** choose to play a $\blacklozenge^*$ move. It selects a forest $\mathcal{F}'_i \sqsubseteq \mathcal{F}_i$.
- The duplicator **must** reply with some $\mathcal{F}'_j$ s.t. $\mathcal{F}'_j \sqsubseteq \mathcal{F}_j$.
- The game continues on $((\mathcal{F}'_1, \mathsf{t}_1, \mathsf{n}_1), (\mathcal{F}'_2, \mathsf{t}_2, \mathsf{n}_2), (\mathsf{m}, \mathsf{s}, \mathsf{k}-1))$.

Fig. 3. Ehrenfeucht-Fraïssé games for ALT

3.2 Inexpressibility results via the Ehrenfeucht-Fraïssé games for ALT

Now that we are more familiar with the logic, before completing the TOWER-hardness proof of SAT(ALT) we show some properties that ALT cannot express. Notably, these properties explain why the TOWER-hardness proof of the next section cannot be easily simplified. Moreover, inexpressibility results effectively reduce the set of forests that must be considered in order to solve SAT(ALT). This in turn makes reductions from SAT(ALT) to other logics more immediate, as we show throughout Section 4.

A standard way of proving inexpressibility results for logics interpreted on finite models is by adaptation of the Ehrenfeucht-Fraïssé games [29], as done for other relation-changing logics such as context logic for trees [10] and ambient logic [16].

We define the *rank* of a formula φ as the triple $(\mathsf{m}, \mathsf{s}, \mathsf{k}) \in \mathbb{N}^3$ where the *modal rank* m is the greatest nesting depth of the modal operator $\langle \mathrm{U} \rangle$ in φ, whereas the *sabotage rank* s (resp. *repeated sabotage rank* k) is the greatest nesting depth of the $\blacklozenge$ (resp. $\blacklozenge^*$) operator in φ. We denote with $\mathsf{ALT}(\mathsf{rk})$ the set of formulae with rank $\mathsf{rk} \in \mathbb{N}^3$.

The Ehrenfeucht-Fraïssé games (EF-games) for ALT are formally defined in Figure 3. A game is played by two players: the *spoiler* and the *duplicator*. A game state $((\mathcal{F}_1, \mathsf{t}_1, \mathsf{n}_1), (\mathcal{F}_2, \mathsf{t}_2, \mathsf{n}_2), \mathsf{rk})$ is a triple made of a rank rk and two pointed forests $(\mathcal{F}_1, \mathsf{t}_1, \mathsf{n}_1)$ and $(\mathcal{F}_2, \mathsf{t}_2, \mathsf{n}_2)$. The goal of the spoiler is to show that the two structures are different. The goal of the duplicator is to counter the spoiler and show that the two structures are similar. Let us make clear what we mean by two models being different: both players can only play following the rules of the logical formalism (in our case, ALT). Then, two models are different if and only if there is a formula $\varphi \in \mathsf{ALT}(\mathsf{rk})$ that it is satisfied by only one of the two models. This correspondence between the game and the logic is expressed by an adequacy result, formalised below in Lemma 3.

A player has a *winning strategy* if it can play in a way that guarantees it the victory, regardless what the other player does. We write $(\mathcal{F}_1, \mathsf{t}_1, \mathsf{n}_1) \approx_{\mathsf{rk}} (\mathcal{F}_2, \mathsf{t}_2, \mathsf{n}_2)$ whenever the duplicator has a winning strategy for the game $((\mathcal{F}_1, \mathsf{t}_1, \mathsf{n}_1), (\mathcal{F}_2, \mathsf{t}_2, \mathsf{n}_2), \mathsf{rk})$. By Martin's Theorem [32] our games are determined: if the duplicator does not have a winning strategy then spoiler has one, and vice-versa. Hence, $(\mathcal{F}_1, t_1, n_1) \not\approx_{rk} (\mathcal{F}_2, t_2, n_2)$ refers to the fact that the spoiler has a winning strategy.

Lemma 3. $(\mathcal{F}_1, t_1, n_1) \not\approx_{rk} (\mathcal{F}_2, t_2, n_2)$ *iff* $\exists \varphi \in \mathsf{ALT}(rk)$, $(\mathcal{F}_1, t_1, n_1) \models \varphi$ *and* $(\mathcal{F}_2, t_2, n_2) \not\models \varphi$.

The left-to-right direction is proved by induction on the rank and by cases on the first move that the spoiler makes in his winning strategy. The other direction is proved by structural induction on φ. We start to use the EF-games to derive three easy results.

Lemma 4. *Let φ be a formula.*

1. *φ is satisfiable iff it is satisfiable by a pointed forest $(\mathcal{F}, t, n)$ where $t \notin \mathrm{dom}(\mathcal{F})$.*
2. *Given a forest $\mathcal{F}$ and nodes $t \in \mathcal{N}$ and $n, n' \notin \mathrm{dom}(\mathcal{F})$, $(\mathcal{F}, t, n) \models \varphi$ iff $(\mathcal{F}, t, n') \models \varphi$.*
3. *If duplicator has a winning strategy for a game $((\mathcal{F}_1, t_1, n_1), (\mathcal{F}_2, t_2, n_2), rk)$ then it has a winning strategy where it always replies to $\langle \mathrm{U} \rangle$ moves by selecting nodes in $\mathrm{dom}(\mathcal{F}_i) \cup \mathrm{ran}(\mathcal{F}_i)$, for some $i \in \{1, 2\}$.*

Proof (sketch). We sketch the proof of (1) to show how EF-games are used. Let us consider a pointed forest $(\mathcal{F}, t, n)$ such that $(\mathcal{F}, t, n) \models \varphi$. We take a node $t' \notin \mathrm{dom}(\mathcal{F}) \cup \mathrm{ran}(\mathcal{F})$ and define the forest $\mathcal{F}'(n') \stackrel{\text{def}}{=}$ **if** $\mathcal{F}(n') = t$ **then** t' **else** $\mathcal{F}(n')$. Notice that $t' \notin \mathrm{dom}(\mathcal{F}')$. We then prove $\forall rk \in \mathbb{N}^3$ $(\mathcal{F}, t, n) \approx_{rk} (\mathcal{F}', t', n)$ by induction on rk, leading to (1) directly by Lemma 3. The proof of (3) essentially follows from (2). □

Interestingly enough, the third statement of Lemma 4 fundamentally implies that enforcing $\mathcal{N}$ to be finite, instead of infinite as we do throughout this work, does not change the expressive power nor the complexity of ALT.

Let $(\mathcal{F}, t, n)$ be a pointed forest. We now show that ALT has a very limited expressive power with respect to the garbage nodes. In particular, it can only check for the membership of n in $\mathcal{F}(\mathtt{G})$ (with the formula $\mathtt{G}$) and for the size of $\mathcal{F}(\mathtt{G})$ (with the formula $\mathtt{size}(\mathtt{G}) \geq \beta$). We formalise this inexpressibility result as follows.

Lemma 5. *Let $rk = (m, s, k)$. Let $\mathcal{F}, \mathcal{F}_1$ and $\mathcal{F}_2$ be three forests and let $n, t \in \mathcal{N}$, such that for every $i \in \{1, 2\}$, $\mathcal{F} \sqsubseteq \mathcal{F}_i$ and $\mathcal{F}_i(\mathtt{G})_t = \mathrm{dom}(\mathcal{F}_i) \setminus \mathrm{dom}(\mathcal{F})$. If we have*

$$n \in \mathcal{F}_1(\mathtt{G})_t \text{ iff } n \in \mathcal{F}_2(\mathtt{G})_t \quad \text{and} \quad \min(|\mathcal{F}_1(\mathtt{G})_t|, m + s + k) = \min(|\mathcal{F}_2(\mathtt{G})_t|, m + s + k)$$

then $(\mathcal{F}_1, t, n) \approx_{rk} (\mathcal{F}_2, t, n)$.

Let us informally explain Lemma 5, whose proof is by induction on rk and by cases on the moves of the spoiler. Let $(\mathcal{F}_1, t, n)$ be a pointed forest and suppose (ad absurdum) that it satisfies a formula φ of rank rk that express a property of the garbage nodes that is different form the ones cited above. For example, let us assume that φ characterise the set of pointed forests having a garbage node with at least two children. Consider the subforest $\mathcal{F} \sqsubseteq \mathcal{F}_1$ whose domain corresponds to the set of $\mathcal{F}_1$-descendants of t. In particular, $\mathcal{F}_1(\mathtt{G})_t = \mathrm{dom}(\mathcal{F}_1) \setminus \mathrm{dom}(\mathcal{F})$. We extend $\mathcal{F}$ to a forest $\mathcal{F}_2$ by (re)defining it on the nodes in $\mathcal{F}_1(\mathtt{G})_t$ so that $\mathcal{F}_2(\mathtt{G})_t = \mathcal{F}_1(\mathtt{G})_t$ and none of these nodes has more than one $\mathcal{F}_2$-child (this construction can always be done). This last equality implies that $n \in \mathcal{F}_1(\mathtt{G})_t \Leftrightarrow n \in \mathcal{F}_2(\mathtt{G})_t$ and $\min(|\mathcal{F}_1(\mathtt{G})_t|, m + s + k) = \min(|\mathcal{F}_2(\mathtt{G})_t|, m + s + k)$. By Lemma 5 $(\mathcal{F}_1, t, n) \approx_{rk} (\mathcal{F}_2, t, n)$, which implies $(\mathcal{F}_2, t, n) \models \varphi$ by Lemma 3. However, $(\mathcal{F}_2, t, n)$ is defined so that every node in $\mathcal{F}_2(\mathtt{G})_t$ has at most one child. Thus, φ cannot characterise the set of models having a garbage node with at least two children.

As shown in the next section, the inexpressibility result in Lemma 5 plays a central role in the development of the reduction that leads to the TOWER-hardness of SAT(ALT).

3.3 PITL on marked words and the TOWER-hardness of SAT(ALT)

We are now ready to show the non-elementarity of SAT(ALT). The proof is by reduction from the satisfiability problem of Propositional Interval Temporal Logic (PITL) under locality principle [34,25], which in turn is shown TOWER-hard by reduction from the non-emptiness problem of star-free regular languages (see [38] for the TOWER characterisation of this problem). PITL is a well-known logic that was introduced by Moszkowski in [34] for the verification of hardware components. It is interpreted on non-empty finite words over a finite alphabet of unary symbols Σ. Its formulae are from the grammar:

$$\varphi := \varphi \wedge \varphi \mid \neg\varphi \mid \mathsf{a} \mid 1 \mid \varphi \,|\, \varphi$$

where $\mathsf{a} \in \Sigma$. Under the *locality principle* interpretation, a word $\mathfrak{w} = \mathsf{a}_1 \cdots \mathsf{a}_k \in \Sigma^+$ satisfies a whenever $\mathsf{a}_1 = \mathsf{a}$. Moreover, $\mathfrak{w}$ satisfies 1 if it is a word of length one (i.e. $\mathfrak{w} \in \Sigma$). The main feature of this logic is its *chop* operator "$|$". Intuitively, $\varphi | \psi$ is satisfied by words that can be "chopped" into a prefix and a suffix sharing one symbol, so that the prefix satisfies φ and the suffix satisfies ψ. Formally,

$$\mathsf{a}_1 \cdots \mathsf{a}_k \models \varphi | \psi \quad \stackrel{\text{def}}{\Leftrightarrow} \quad \text{there is } i \in [1,k] \text{ such that } \mathsf{a}_1 \cdots \mathsf{a}_i \models \varphi \text{ and } \mathsf{a}_i \cdots \mathsf{a}_k \models \psi.$$

Translating $|$ in ALT is not easy. Indeed, given the encoding of words proposed in Section 3.1, chopping $\mathfrak{w}$ in two pieces means splitting in some way the main path $\mathsf{n}_1, \cdots, \mathsf{n}_k$ of a forest $(\mathcal{F}, \mathsf{t}, \mathsf{n})$ encoding $\mathfrak{w}$ to then check that the word encoded by $\mathsf{n}_1, \cdots, \mathsf{n}_i$ satisfies φ and the one encoded by $\mathsf{n}_i, \cdots, \mathsf{n}_k$ satisfies ψ. However, by doing this the elements $\mathsf{n}_1, \cdots, \mathsf{n}_i$ become garbage nodes. Thus, as a consequence of Lemma 5, ALT cannot check in any way what is the word encoded by these nodes. Easy translations from PITL to ALT seem therefore impossible and, as done in [31], we are required to go through an alternative interpretation of PITL based on *marking symbols* instead of chopping words.

A *marking* of an alphabet Σ is a bijection $\overline{(.)} : \Sigma \to \overline{\Sigma}$, relating a symbol $\mathsf{a} \in \Sigma$ to its *marked variant* $\overline{\mathsf{a}} \in \overline{\Sigma}$. We denote with $\mathbb{\Sigma}$ the extended alphabet $\Sigma \uplus \overline{\Sigma}$. A word is *marked* if it has some symbols from $\overline{\Sigma}$. We introduce the satisfaction relation $\models_{\bullet}$ on a marked word $\mathfrak{w} \in \mathbb{\Sigma}^+$. It is defined as usual for Boolean connectives. Moreover,

$$\mathfrak{w} \models_{\bullet} \mathsf{a} \stackrel{\text{def}}{\Leftrightarrow} \mathfrak{w} \text{ is headed by } \mathsf{a} \text{ or } \overline{\mathsf{a}}; \qquad \mathfrak{w} \models_{\bullet} 1 \stackrel{\text{def}}{\Leftrightarrow} \mathfrak{w} \text{ is headed by a marked symbol.}$$

The definition of $\varphi | \psi$ is more involved. Let $\mathfrak{w}' \in \Sigma^*$, $\overline{\mathsf{a}} \in \overline{\Sigma}$ and $\mathfrak{w}'' \in \mathbb{\Sigma}^*$ be such that $\mathfrak{w} = \mathfrak{w}'\overline{\mathsf{a}}\mathfrak{w}''$, so that $\overline{\mathsf{a}}$ is the first marked symbol occurring in $\mathfrak{w}$ (this decomposition is uniquely defined). Then, $\mathfrak{w}'\overline{\mathsf{a}}\mathfrak{w}'' \models_{\bullet} \varphi | \psi$ holds if and only if there is there is $\mathsf{b} \in \Sigma$ s.t.

(a) $\mathfrak{w}'$ is the empty word, $\mathsf{b} = \mathsf{a}$ and $\overline{\mathsf{a}}\mathfrak{w}'' \models_{\bullet} \varphi \wedge \psi$, or
(b) there is $\mathfrak{w}_2 \in \Sigma^*$ s.t. $\mathfrak{w}' = \mathsf{b}\mathfrak{w}_2$, $\overline{\mathsf{b}}\mathfrak{w}_2\overline{\mathsf{a}}\mathfrak{w}'' \models \varphi$ and $\mathsf{b}\mathfrak{w}_2\overline{\mathsf{a}}\mathfrak{w}'' \models \psi$, or
(c) $\mathfrak{w}'$ is not the empty word, $\mathsf{b} = \mathsf{a}$, $\mathfrak{w}'\overline{\mathsf{a}}\mathfrak{w}'' \models \varphi$ and $\overline{\mathsf{a}}\mathfrak{w}'' \models \psi$, or
(d) $\exists \mathfrak{w}_1 \in \Sigma^+, \exists \mathfrak{w}_2 \in \Sigma^*$ s.t. $\mathfrak{w}' = \mathfrak{w}_1\mathsf{b}\mathfrak{w}_2$, $\mathfrak{w}_1\overline{\mathsf{b}}\mathfrak{w}_2\overline{\mathsf{a}}\mathfrak{w}'' \models \varphi$ and $\mathsf{b}\mathfrak{w}_2\overline{\mathsf{a}}\mathfrak{w}'' \models \psi$.

On this semantics, the satisfaction of a formula only depends on the prefix $\mathsf{a}_1 \cdots \mathsf{a}_{i-1}\overline{\mathsf{a}}_i$ of $\mathfrak{w}$ that ends with the first marked symbol. To check whether $\mathfrak{w} \models_{\bullet} \varphi | \psi$ we search for a position $j \in [1,i]$ inside this prefix so that φ is satisfied by the word obtained from $\mathfrak{w}$ by marking the j-th symbol, whereas ψ is satisfied by the suffix of $\mathfrak{w}$ starting in j. In the definition above, this idea is split into four cases (a)–(d), depending on truthiness of $j = 1$ and $j = i$. This is done as it better reflects the encoding of PITL in ALT. The semantics on marked words is related to the standard semantics of PITL as follows.

Proposition 1 (from [31]). *Let $\mathfrak{w} \in \Sigma^*$, $\mathtt{a} \in \Sigma$ and $\mathfrak{w}' \in \mathbb{\Sigma}^*$. Let φ be a formula in PITL. $\mathfrak{w}\mathtt{a}$ satisfies φ under the standard interpretation of PITL if and only if $\mathfrak{w}\bar{\mathtt{a}}\mathfrak{w}' \models_\bullet \varphi$.*

The alternative interpretation of PITL allows us to reduce SAT(PITL) to SAT(ALT) in a neat way. Let $\Sigma = [1, n]$, $\mathbb{\Sigma} = \Sigma \cup \overline{\Sigma}$ and let $\mathfrak{f} : \mathbb{\Sigma} \to [1, 2n]$ be the bijection $\mathfrak{f}(\mathtt{a}) \stackrel{\text{def}}{=} 2\mathtt{a}$ for $\mathtt{a} \in \Sigma$ and $\mathfrak{f}(\bar{\mathtt{a}}) \stackrel{\text{def}}{=} 2\mathtt{a} - 1$ for $\bar{\mathtt{a}} \in \overline{\Sigma}$. $\mathfrak{f}(\mathtt{a}_1 \cdots \mathtt{a}_k)$ denotes the word $\mathfrak{f}(\mathtt{a}_1) \cdots \mathfrak{f}(\mathtt{a}_k)$. $\mathfrak{f}$ maps $\mathbb{\Sigma}$ into the alphabet $[1, 2n]$, whose words can be encoded into trees (as in Section 3.1). In these trees each symbol $\mathtt{a} \in \Sigma$ (resp. $\bar{\mathtt{a}} \in \overline{\Sigma}$) corresponds to a node in the main path having $2\mathtt{a} + 1$ (resp. $2\mathtt{a}$) children not in this path. Therefore, given a node n encoding a symbol in Σ, removing exactly one children of n that is not in the main path is equivalent to marking the symbol n encodes. Based on this description, we can check if the current evaluation node encodes a marked symbol from $\overline{\Sigma}$ with the following formula:

$$\mathtt{mark}_\Sigma \stackrel{\text{def}}{=} \bigvee_{\mathtt{a} \in \Sigma} \big((\#\mathtt{child} = 2\mathtt{a} \land \mathtt{1st}_{[1,2n]}) \lor (\#\mathtt{child} = 2\mathtt{a} + 1 \land \neg\, \mathtt{1st}_{[1,2n]}) \big)$$

As already stated, $\mathfrak{w} \models_\bullet \varphi$ examines the prefix of $\mathfrak{w}$ that ends with the first marked symbol. In pointed forests $(\mathcal{F}, \mathsf{t}, \mathsf{n})$ encoding $\mathfrak{w}$, this prefix corresponds to the subtree whose root encodes a marked symbol and is a $\mathcal{F}$-descendant of every other node encoding marked symbols. Therefore, to characterise this tree we need to track the number of nodes encoding marked symbols. We first define a formula $\mathtt{marks}_\Sigma \geq \beta$ stating that the forest has at least $\beta \in \mathbb{N}$ nodes encoding marked symbols. It is defined as $\top$ for $\beta = 0$, and otherwise ($\beta \geq 1$) as $\langle \mathrm{U} \rangle \big(\mathtt{mark}_\Sigma \land \blacklozenge(\neg\, \mathtt{inDom} \land \mathtt{marks}_\Sigma \geq \beta{-}1) \big)$. Again, this formula uses the same principle introduced in Section 2 for $\#\mathtt{ch}_{\mathtt{trg}} \geq 2$: we search for a node encoding a marked symbol, remove it from the structure and then search for $\beta{-}1$ other such nodes. Similarly, we introduce $\#\mathtt{markAnc}_\Sigma \geq \beta \stackrel{\text{def}}{=} \mathtt{symb} \land \blacklozenge(\neg\, \mathtt{inDom} \land \mathtt{marks}_\Sigma \geq \beta)$, the formula stating that the current evaluation node encodes a symbol and has at least β ancestors that encode marked symbols.

At last, for a formula φ in PITL having symbols from $\Sigma = [1, n]$, we introduce its translation $\nabla_\beta(\varphi)$ in ALT, where $\beta \geq 1$ tracks the number of nodes encoding marked symbols. It is homomorphic for Boolean connectives: $\nabla_\beta(\neg\varphi) \stackrel{\text{def}}{=} \neg\nabla_\beta(\varphi)$ and $\nabla_\beta(\varphi \land \psi) \stackrel{\text{def}}{=} \nabla_\beta(\varphi) \land \nabla_\beta(\psi)$. For $\mathtt{a} \in \Sigma$ and $\mathtt{1}$, it faithfully represent the $\models_\bullet$ relation: $\nabla_\beta(\mathtt{a}) \stackrel{\text{def}}{=} \langle \mathrm{U} \rangle\, \mathtt{1st}_{[2\mathtt{a}-1, 2\mathtt{a}]}$ and $\nabla_\beta(\mathtt{1}) \stackrel{\text{def}}{=} \langle \mathrm{U} \rangle (\mathtt{1st}_{[1,2n]} \land \mathtt{mark}_\Sigma)$. Lastly, the formula $\nabla_\beta(\varphi \,|\, \psi)$ is defined as

$$\begin{aligned}
&\langle \mathrm{U} \rangle \big(\mathtt{symb} \land \big((\mathtt{1st}_{[1,2n]} \land \mathtt{mark}_\Sigma \land \nabla_\beta(\varphi) \land \nabla_\beta(\psi)) \lor \\
&\quad (\mathtt{1st}_{[1,2n]} \land \neg\, \mathtt{mark}_\Sigma \land \blacklozenge(\mathtt{mark}_\Sigma \land \nabla_{\beta+1}(\varphi)) \land \nabla_\beta(\psi)) \lor \\
&\quad (\neg\, \mathtt{1st}_{[1,2n]} \land \mathtt{mark}_\Sigma \land \#\mathtt{markAnc}_\Sigma \geq \beta - 1 \land \nabla_\beta(\varphi) \land \blacklozenge(\mathtt{1st}_{[1,2n]} \land \nabla_\beta(\psi))) \lor \\
&\quad (\neg\, \mathtt{1st}_{[1,2n]} \land \neg\, \mathtt{mark}_\Sigma \land \#\mathtt{markAnc}_\Sigma \geq \beta \land \blacklozenge(\mathtt{mark}_\Sigma \land \nabla_{\beta+1}(\varphi)) \land \blacklozenge(\mathtt{1st}_{[1,2n]} \land \nabla_\beta(\psi))) \big) \big).
\end{aligned}$$

Notice how $\nabla_\beta(\varphi \,|\, \psi)$ follows closely the $\models_\bullet$ relation: it is split into four disjuncts, one for each case in the definition of $\varphi \,|\, \psi$. For example, the second disjunct of $\nabla_\beta(\varphi \,|\, \psi)$ encodes the case (b) in the definition of $\mathfrak{w}'\bar{\mathtt{a}}\mathfrak{w}'' \models_\bullet \varphi \,|\, \psi$, as schematised below:

PITL	$\exists \mathtt{b} \in \Sigma \ldots$	$\exists \mathfrak{w}_2 \in \Sigma^*$ s.t. $\mathfrak{w}' = \mathtt{b}\mathfrak{w}_2$	and $\bar{\mathtt{b}}\mathfrak{w}_2\bar{\mathtt{a}}\mathfrak{w}'' \models \varphi$	and $\mathtt{b}\mathfrak{w}_2\bar{\mathtt{a}}\mathfrak{w}'' \models \psi$
ALT	$\langle \mathrm{U} \rangle(\mathtt{symb} \ldots$	$\mathtt{1st}_{[1,2n]} \land \neg\, \mathtt{mark}_\Sigma$	$\land \blacklozenge(\mathtt{mark}_\Sigma \land \nabla_{\beta+1}(\varphi))$	$\land \nabla_\beta(\psi)$

The translation is proved correct (by induction on the structure of φ) in the next lemma.

Lemma 6. *Let $\Sigma = [1, n]$ and $\mathbb{\Sigma} = \Sigma \cup \overline{\Sigma}$. Let $\mathfrak{w} \in \mathbb{\Sigma}^+$ with $\beta \geq 1$ marked symbols. Let $(\mathcal{F}, \mathsf{t}, \mathsf{n})$ be an encoding of $\mathfrak{f}(\mathfrak{w})$. For every φ in PITL, $\mathfrak{w} \models_\bullet \varphi$ iff $(\mathcal{F}, \mathsf{t}, \mathsf{n}) \models \nabla_\beta(\varphi)$.*

Then, the reduction from SAT(PITL) on standard semantics follows as we are able to characterise the set of pointed forests encoding words in $\Sigma^*\overline{\Sigma}$ (first three conjuncts in the formula of Lemma 7). To conclude, we simply apply Lemma 6 and Proposition 1.

Lemma 7. *Every φ in PITL written with symbols from $\Sigma = [1, n]$ is satisfiable under the standard interpretation of PITL if and only if the following formula in* ALT *is satisfiable*

$$\underbrace{\mathtt{word}_{[1,2n]} \wedge \langle \mathrm{U} \rangle \top \wedge [\mathrm{U}](\mathtt{mark}_{\Sigma} \Leftrightarrow \top \wedge \neg \blacklozenge(\mathrm{G}))}_{\text{The forest encodes a non-empty word. The only node encoding a marked symbol is the child of the target node.}} \wedge \nabla_1(\varphi).$$

Because of the case distinction in the formula $\nabla_\beta(\varphi \,|\, \psi)$, the formula obtained via ∇_β is exponential (hence elementary) in the number of symbols used to write φ. Therefore, from the TOWER-hardness of SAT(PITL) we conclude that SAT(ALT) is TOWER-hard.

4 Revisiting TOWER-hard logics with ALT

We now display the usefulness of ALT as a tool for proving the TOWER-hardness of logics interpreted on tree-like structures. In particular, we provide semantically faithful reductions from SAT(ALT) to the satisfiability problem of four logics that were independently found to be TOWER-complete: first-order separation logic [9], quantified CTL on trees [28], modal logic of heaps [17] and modal separation logic [18]. Our reduction only use strict fragments of these formalisms, allowing us to draw some new results on these logics. Most notably, this section shows that all these logics are TOWER-hard because they fundamentally provide the reachability and submodel reasoning given by ALT.

4.1 From ALT to First-Order Separation Logic

Separation logic (SL) [37] is an assertion logic used in state-of-the-art tools [6,11] for Hoare-style verification of heap-manipulating programs. As already stated, a preliminary definition of ALT was defined in [31] to reason on the complexity of separation logic. Hence, here we briefly revisit the relation between ALT and SL.

Let VAR and LOC be two countably infinite sets of program variables and locations, respectively. Separation logic is interpreted on *memory states*: pairs (s, h) consisting of a function (the *store*) $s\colon \mathsf{VAR} \to \mathsf{LOC}$ and a partial function with finite domain (the *heap*) $h\colon \mathsf{LOC} \to_{\mathtt{fin}} \mathsf{LOC}$. Since $\mathcal{N}$ and LOC are both countably infinite sets, w.l.o.g. we assume $\mathsf{LOC} = \mathcal{N}$. We extend the notation of domain, image and function composition to stores and heaps. Two heaps h_1 and h_2 are said to be disjoint, written $h_1 \bot h_2$, whenever $\mathrm{dom}(h_1) \cap \mathrm{dom}(h_2) = \emptyset$, and when this holds the union $h_1 + h_2$ of h_1 and h_2 is defined as the standard sum of functions $(h_1 + h_2)(\ell) \stackrel{\text{def}}{=} \textbf{if } \ell \in \mathrm{dom}(h_1) \textbf{ then } h_1(\ell) \textbf{ else } h_2(\ell)$. Let $\mathtt{u} \in \mathsf{VAR}$ be a *fixed variable* that is reserved for quantification (quantification over other variables is not possible). We consider the separation logic $1\mathsf{SL}(*, \mathtt{alloc}, \hookrightarrow^+)$, whose formulae are built from the following grammar (as in [31]):

$$\varphi := \top \mid \varphi \wedge \varphi \mid \neg\varphi \mid \mathtt{emp} \mid \mathtt{x} = \mathtt{y} \mid \mathtt{x} \hookrightarrow \mathtt{y} \mid \mathtt{alloc}(\mathtt{x}) \mid \mathtt{x} \hookrightarrow^+ \mathtt{y} \mid \varphi * \varphi \mid \exists \mathtt{u}\, \varphi$$

where $\mathtt{x}, \mathtt{y} \in \mathsf{VAR}$. As shown below, the *reachability predicate* $\hookrightarrow^+$ can be seen as the transitive closure of the standard *points-to* predicate $\hookrightarrow$ of separation logic. For a memory state (s, h), the satisfaction relation $\models$ is defined as follows:

$$(s,h) \models \mathtt{emp} \stackrel{\text{def}}{\Leftrightarrow} \mathrm{dom}(h) = \emptyset. \qquad (s,h) \models \mathtt{x} \hookrightarrow \mathtt{y} \stackrel{\text{def}}{\Leftrightarrow} h(s(\mathtt{x})) = s(\mathtt{y}).$$
$$(s,h) \models \mathtt{x} = \mathtt{y} \stackrel{\text{def}}{\Leftrightarrow} s(\mathtt{x}) = s(\mathtt{y}). \qquad (s,h) \models \mathtt{alloc(x)} \stackrel{\text{def}}{\Leftrightarrow} s(\mathtt{x}) \in \mathrm{dom}(h).$$
$$(s,h) \models \mathtt{x} \hookrightarrow^+ \mathtt{y} \stackrel{\text{def}}{\Leftrightarrow} \text{there is } \exists \delta \geq 1 \text{ such that } h^{\delta}(s(\mathtt{x})) = s(\mathtt{y}).$$
$$(s,h) \models \varphi * \psi \stackrel{\text{def}}{\Leftrightarrow} \exists h_1, h_2 \text{ s.t. } h_1 \perp h_2,\ h_1 + h_2 = h,\ (s,h_1) \models \varphi \text{ and } (s,h_2) \models \psi.$$
$$(s,h) \models \exists \mathtt{u}\ \varphi \stackrel{\text{def}}{\Leftrightarrow} \text{there is a location } \ell' \in \mathsf{LOC} \text{ such that } (s[\mathtt{u} \leftarrow \ell'], h) \models \varphi,$$

where $s[\mathtt{u} \leftarrow \ell']$ is the store updated from s by only changing the evaluation of u from $s(\mathtt{u})$ to ℓ', i.e. for every $\mathtt{x} \in \mathsf{VAR}$, $s[\mathtt{u} \leftarrow \ell'](\mathtt{x}) \stackrel{\text{def}}{=}$ **if** $\mathtt{x} = \mathtt{u}$ (syntactically) **then** ℓ' **else** $s(\mathtt{x})$. The main ingredient of separation logic is the *separating conjunction* $\varphi * \psi$, that is satisfied whether h can be partitioned into h_1 and h_2 so that $(s, h_1) \models \varphi$ whereas $(s, h_2) \models \psi$. The $*$ operator captures the $\blacklozenge$ and $\blacklozenge^*$ operators as follows. Consider the formula $\mathtt{size} = 1 \stackrel{\text{def}}{=} \neg\mathtt{emp} \wedge \neg(\neg\mathtt{emp} * \neg\mathtt{emp})$, which is satisfied whenever $|\mathrm{dom}(h)| = 1$. We define $\blacklozenge_{\mathrm{SL}} \varphi \stackrel{\text{def}}{=} (\mathtt{size} = 1) * \varphi$ and $\blacklozenge^*_{\mathrm{SL}} \varphi \stackrel{\text{def}}{=} \top * \varphi$. The semantics of these formulae is related to the analogous operators of ALT as follows:

$$(s,h) \models \blacklozenge_{\mathrm{SL}} \varphi \iff \exists h_1, h_2 \text{ s.t. } h_1 \perp h_2,\ h_1 + h_2 = h,\ |\mathrm{dom}(h_1)| = 1 \text{ and } (s,h_2) \models \varphi.$$
$$(s,h) \models \blacklozenge^*_{\mathrm{SL}} \varphi \iff \exists h_1, h_2 \text{ s.t. } h_1 \perp h_2,\ h_1 + h_2 = h \text{ and } (s,h_2) \models \varphi.$$

In order to perform the reduction from SAT(ALT) to SAT(1SL($*$, alloc, $\hookrightarrow^+$)), we fix a variable $\mathtt{x} \in \mathsf{VAR}$ that is syntactically different from u and that plays the role of the target node. Then, the translation $\tau_{\mathtt{x}}(\varphi)$ of a formula φ in ALT is straightforward:

$$\tau_{\mathtt{x}}(\mathrm{T}) \stackrel{\text{def}}{=} \mathtt{u} \hookrightarrow^+ \mathtt{x}. \qquad \tau_{\mathtt{x}}(\blacklozenge \varphi) \stackrel{\text{def}}{=} \blacklozenge_{\mathrm{SL}} \tau_{\mathtt{x}}(\varphi). \qquad \tau_{\mathtt{x}}(\top) \stackrel{\text{def}}{=} \top.$$
$$\tau_{\mathtt{x}}(\mathrm{G}) \stackrel{\text{def}}{=} \mathtt{alloc(u)} \wedge \neg \tau_{\mathtt{x}}(\mathrm{T}). \qquad \tau_{\mathtt{x}}(\blacklozenge^* \varphi) \stackrel{\text{def}}{=} \blacklozenge^*_{\mathrm{SL}} \tau_{\mathtt{x}}(\varphi). \qquad \tau_{\mathtt{x}}(\neg\varphi) \stackrel{\text{def}}{=} \neg \tau_{\mathtt{x}}(\varphi).$$
$$\tau_{\mathtt{x}}(\langle \mathrm{U} \rangle \varphi) \stackrel{\text{def}}{=} \exists \mathtt{u}\ \tau_{\mathtt{x}}(\varphi). \qquad \tau_{\mathtt{x}}(\varphi \wedge \psi) \stackrel{\text{def}}{=} \tau_{\mathtt{x}}(\varphi) \wedge \tau_{\mathtt{x}}(\psi).$$

Given a pointed forest $(\mathcal{F}, \mathsf{t}, \mathsf{n})$ and a store s such that $s(\mathtt{x}) = \mathsf{t}$ and $s(\mathtt{u}) = \mathsf{n}$, by structural induction on φ we can easily show that $(\mathcal{F}, \mathsf{t}, \mathsf{n}) \models \varphi \Leftrightarrow (s, \mathcal{F}) \models \tau_{\mathtt{x}}(\varphi)$. This, together with the fact that $\forall \mathtt{u}\ \neg(\mathtt{u} \hookrightarrow^+ \mathtt{u})$ characterises the class of acyclic heaps (which correspond to the forests of ALT), directly implies the following result.

Lemma 8. *Let* $\mathtt{x} \in \mathsf{VAR} \setminus \{\mathtt{u}\}$. φ *in* ALT *and* $\tau_{\mathtt{x}}(\varphi) \wedge \forall \mathtt{u}\ \neg(\mathtt{u} \hookrightarrow^+ \mathtt{u})$ *are equisatisfiable.*

This lemma reproves that both 1SL($*$, alloc, $\hookrightarrow^+$) and first order separation logic with two quantified variables (denoted as 2SL($*$)) admit a TOWER-hard satisfiability problem. 2SL($*$), as introduced in [17], can be defined from 1SL($*$, alloc, $\hookrightarrow^+$) by removing alloc and $\hookrightarrow^+$ from the syntax and allowing a second variable, different from u, to be quantified. However, in [17] the authors show that both alloc and $\hookrightarrow^+$ are expressible in 2SL($*$), and with some very minor modification to their formulae we can show that both predicates are definable using $\blacklozenge$ and $\blacklozenge^*$ instead of $*$ and emp. Moreover, these logics are in TOWER by Rabin's Theorem [36], leading to the TOWER-completeness of SAT(ALT).

Theorem 1. SAT(2SL($*$)) *and* SAT(1SL($*$, alloc, $\hookrightarrow^+$)) *are* TOWER*-complete even when* emp *and* $*$ *are replaced with* $\blacklozenge_{\mathrm{SL}}$ *and* $\blacklozenge^*_{\mathrm{SL}}$. SAT(ALT) *is* TOWER*-complete.*

4.2 From ALT to Quantified Computation Tree Logic

We now consider Computation Tree Logic (CTL), a well-known logic for branching time model checking [14,13]. Among its extensions, in [5,22,28] the addition of propositional quantification is considered. The satisfiability problem of the resulting logic is undecidable on Kripke structures, and TOWER-complete on trees [28]. In [5], the authors show that the problem is TOWER-hard even when considering just one operator among *exists-next* EX or *exists-finally* EF (the definitions are below). Here, we reprove the result for EF by first tackling the TOWER-hardness of the logic with the *exists-until* $\mathsf{E}(\varphi \mathbin{\mathsf{U}} \psi)$, and then show that this operator can be defined using EF. Differently from [5] and thanks to the properties of ALT, our reduction does not imbricate until operators, showing that this extension of CTL remains TOWER-hard even when $\mathsf{E}(\varphi \mathbin{\mathsf{U}} \psi)$ is restricted so that φ and ψ are Boolean combinations of propositional symbols.

Let us first recall the standard definition of Kripke structure [27]. Let $\mathsf{AP} \stackrel{\text{def}}{=} \{p, q, \cdots\}$ be a countable set of *propositional symbols*. A *Kripke structure* is a triple $(\mathcal{W}, \mathcal{R}, \mathcal{V})$ where $\mathcal{W}$ is a countable set of *worlds*, $\mathcal{R} \subseteq \mathcal{W} \times \mathcal{W}$ is a left-total *accessibility relation* (left-total means that for each world $\mathsf{w} \in \mathcal{W}$ there is $\mathsf{w}' \in \mathcal{W}$ s.t. $(\mathsf{w}, \mathsf{w}') \in \mathcal{R}$) and $\mathcal{V} : \mathsf{AP} \to 2^{\mathcal{W}}$ is a *labelling function*. We define $\mathcal{R}(\mathsf{w}) \stackrel{\text{def}}{=} \{\mathsf{w}' \in \mathcal{W} \mid (\mathsf{w}, \mathsf{w}') \in \mathcal{R}\}$ as the set of worlds accessible from $\mathsf{w} \in \mathcal{W}$. Let $R \subseteq \mathcal{W} \times \mathcal{W}$ be an arbitrary relation on worlds (not necessarily left-total). A *path* π starting in w is a sequence of worlds $(\mathsf{w}_0, \mathsf{w}_1, \cdots)$ such that $\mathsf{w}_0 = \mathsf{w}$ and $(\mathsf{w}_i, \mathsf{w}_{i+1}) \in R$ for every two successive elements $\mathsf{w}_i, \mathsf{w}_{i+1}$ of the sequence. The path π is said to be *maximal* whenever it is not a strict prefix of any other path. We denote with $\Pi_R(\mathsf{w})$ the set of *maximal paths* starting in w. If R is left-total then $\Pi_R(\mathsf{w})$ is the set of all infinite paths starting in w. Lastly, $R^*(\mathsf{w})$ denotes the set of worlds reachable from w, i.e. those worlds belonging to a path in $\Pi_R(\mathsf{w})$.

We consider Quantified Computational Tree Logic interpreted under tree semantics (QCTL^T) and refer the reader to [28] for a complete description of the logic. The formulae of QCTL^T are built from the following grammar:

$$\varphi := \top \mid \varphi \land \varphi \mid \neg\varphi \mid p \mid \mathsf{EX}\,\varphi \mid \mathsf{E}(\varphi \mathbin{\mathsf{U}} \varphi) \mid \mathsf{A}(\varphi \mathbin{\mathsf{U}} \varphi) \mid \exists p\, \varphi$$

where $p \in \mathsf{AP}$. All temporal modalities of QCTL^T are from CTL: EX is the *exists-next* modality, $\mathsf{E}(\varphi \mathbin{\mathsf{U}} \psi)$ is the *exists-until* modality and $\mathsf{A}(\varphi \mathbin{\mathsf{U}} \psi)$ is the *all-until* modality.

QCTL^T is interpreted on Kripke trees. Formally, a Kripke structure $(\mathcal{W}, \mathcal{R}, \mathcal{V})$ is a *(finitely-branching) Kripke tree* if (I) $\mathcal{R}^{-1}$ is functional and acyclic, (II) for every world $\mathsf{w} \in \mathcal{W}$, $\mathcal{R}(\mathsf{w})$ is finite and (III) it has a *root*, i.e. $\mathcal{R}^*(\mathsf{r}) = \mathcal{W}$ for some $\mathsf{r} \in \mathcal{W}$. Given $\mathsf{w} \in \mathcal{W}$, the worlds in $\mathcal{R}^*(\mathsf{w}) \setminus \{\mathsf{w}\}$ are said to be *descendants* of w. As Kripke structures are left-total, Kripke trees can be seen as finitely-branching infinite trees. This leads to $\mathsf{SAT}(\mathsf{QCTL}^T)$ being in TOWER by reduction to MSO on trees [28]. Let $\mathcal{K} = (\mathcal{W}, \mathcal{R}, \mathcal{V})$ be a Kripke tree and $\mathsf{w} \in \mathcal{W}$. The satisfaction relation $\models$ of QCTL^T is defined as:

$(\mathcal{K}, \mathsf{w}) \models p \quad \stackrel{\text{def}}{\Leftrightarrow} \quad \mathsf{w} \in \mathcal{V}(p)$.

$(\mathcal{K}, \mathsf{w}) \models \mathsf{EX}\,\varphi \quad \stackrel{\text{def}}{\Leftrightarrow} \quad \exists \mathsf{w}' \in \mathcal{R}(\mathsf{w})$ s.t. $(\mathcal{K}, \mathsf{w}') \models \varphi$.

$(\mathcal{K}, \mathsf{w}) \models \mathsf{E}(\varphi \mathbin{\mathsf{U}} \psi) \quad \stackrel{\text{def}}{\Leftrightarrow} \quad$ there are $(\mathsf{w}_0, \mathsf{w}_1, \cdots) \in \Pi_{\mathcal{R}}(\mathsf{w})$ and $j \in \mathbb{N}$ such that $(\mathcal{K}, \mathsf{w}_j) \models \psi$ and for every $i < j$, $(\mathcal{K}, \mathsf{w}_i) \models \varphi$.

$(\mathcal{K}, \mathsf{w}) \models \mathsf{A}(\varphi \mathbin{\mathsf{U}} \psi) \quad \stackrel{\text{def}}{\Leftrightarrow} \quad$ for all $(\mathsf{w}_0, \mathsf{w}_1, \cdots) \in \Pi_{\mathcal{R}}(\mathsf{w})$, $\exists j \in \mathbb{N}$ such that $(\mathcal{K}, \mathsf{w}_j) \models \psi$ and for every $i < j$, $(\mathcal{K}, \mathsf{w}_i) \models \varphi$.

$(\mathcal{K}, \mathsf{w}) \models \exists p\, \varphi \quad \stackrel{\text{def}}{\Leftrightarrow} \quad$ there is $\mathcal{W}' \subseteq \mathcal{W}$ such that $(\mathcal{W}, \mathcal{R}, \mathcal{V}[p \leftarrow \mathcal{W}']) \models \varphi$,

where, similarly to the store update $s[\mathtt{u} \leftarrow \ell']$ of the previous section, $\mathcal{V}[p \leftarrow \mathcal{W}']$ stands for the function obtained from $\mathcal{V}$ by updating the evaluation of p from $\mathcal{V}(p)$ to $\mathcal{W}'$.

The formula $\exists p\ \varphi$ requires to update the satisfaction of p in a way such that φ is satisfied. This should already give a good clue on how to reduce ALT to QCTL^T: we represent the nodes of a forest as the set of worlds satisfying a propositional symbol $\mathtt{D}$. Then, for instance, the repeated sabotage operator $\blacklozenge^*$ is encoded by using an existential $\exists\mathtt{E}$ that changes the evaluation of a propositional symbol $\mathtt{E}$ so that it only holds in worlds where $\mathtt{D}$ holds. In this way, the set of worlds satisfying $\mathtt{E}$ represents a subforest of the original one. The universal quantification $\forall$ and the connectives $\Rightarrow$ and $\vee$ are defined as usual. So are the classical temporal operators from [14], *exists-finally* $\mathsf{EF}\ \varphi \stackrel{\text{def}}{=} \mathsf{E}(\top\ \mathsf{U}\ \varphi)$, *all-generally* $\mathsf{AG}\ \varphi \stackrel{\text{def}}{=} \neg\mathsf{EF}\ \neg\varphi$, *all-finally* $\mathsf{AF}\ \varphi \stackrel{\text{def}}{=} \mathsf{A}(\top\ \mathsf{U}\ \varphi)$, *exists-generally* $\mathsf{EG}\ \varphi \stackrel{\text{def}}{=} \neg\mathsf{AF}\ \neg\varphi$, and *exists-strong-release* $\mathsf{E}(\varphi\ \mathsf{M}\ \psi) \stackrel{\text{def}}{=} \mathsf{E}(\varphi\ \mathsf{U}\ \varphi \wedge \psi)$.

We now work towards a formal encoding of a pointed forest $(\mathcal{F}, \mathsf{t}, \mathsf{n})$ into a *pointed model* $(\mathcal{K}, \mathsf{w})$, where $\mathcal{K} = (\mathcal{W}, \mathcal{R}, \mathcal{V})$ is a Kripke tree and w is one of its worlds. We use w to play the role of the target node t. To encode the forest $\mathcal{F}$ and the current evaluation node n we use the worlds appearing in $\mathcal{R}^*(\mathsf{w})$ and three propositional symbols: $\mathtt{D}$, $\mathtt{end}$ and $\mathtt{n}$. The intended use of $\mathtt{D}$ is to state which elements of $\mathcal{R}^*(\mathsf{w})$ encode nodes in $\mathrm{dom}(\mathcal{F})$. We need to be careful here, as $\mathcal{R}^*(\mathsf{w})$ is an infinite set whereas $\mathrm{dom}(\mathcal{F})$ is finite. We use the propositional symbol $\mathtt{end}$ to solve this inconsistency: we constraint $\mathcal{K}$ to satisfy the formula $\mathsf{AF}(\mathtt{end})$ stating that every maximal path $(\mathsf{w}_0, \mathsf{w}_1, \cdots) \in \Pi_{\mathcal{R}}(\mathsf{w})$ has a finite prefix $(\mathsf{w}_0, \cdots, \mathsf{w}_{j-1})$ $(j \in \mathbb{N})$ of worlds not satisfying $\mathtt{end}$, whereas $\mathsf{w}_j \in \mathcal{V}(\mathtt{end})$. Then, a world in $\mathcal{W}$ encodes an element in $\mathrm{dom}(\mathcal{F})$ whenever it satisfies $\mathtt{D}$ and it belongs to one of these prefixes. We use the propositional symbol $\mathtt{n}$ to encode the current evaluation node. During the translation we require $\mathtt{n}$ to be satisfied by exactly one descendant of w, so that the modality $\langle \mathrm{U} \rangle$ roughly becomes a quantification over $\mathtt{n}$. From [28], checking whether a formula φ holds in exactly one descendant of w can be done with the formula $\mathtt{uniq}(\varphi) \stackrel{\text{def}}{=} \mathsf{EF}(\varphi) \wedge \forall p\ (\mathsf{EF}(\varphi \wedge p) \Rightarrow \mathsf{AG}(\varphi \Rightarrow p))$ where $p \in \mathsf{AP}$ does not appear in φ. For technical reasons, we treat in a similar way the world w, which encodes the target node, and require it to be the only world (among the ones in $\mathcal{R}^*(\mathsf{w})$) satisfying the auxiliary propositional symbol $\mathtt{t}$. Lastly, we use an additional propositional symbol $\mathtt{E}$ in order to encode subforests and deal with the encoding of $\blacklozenge$ and $\blacklozenge^*$ (as stated above).

We now formalise the encoding. For the remaining of this section, we fix a tuple $X \stackrel{\text{def}}{=} (\mathtt{end}, \mathtt{n}, \mathtt{t})$ of three different propositional symbols. Let $\mathtt{D}$ be an additional symbol not in X, and let $(\mathcal{F}, \mathsf{t}, \mathsf{n})$ be a pointed forest s.t. $\mathsf{t} \notin \mathrm{dom}(\mathcal{F})$ (by Lemma 4(1) it is sufficient to consider this class of structures in order to decide satisfiability of a formula in ALT). A pointed model $(\mathcal{K} = (\mathcal{W}, \mathcal{R}, \mathcal{V}), \mathsf{w})$, is an $(X, \mathtt{D})$*-encoding* of $(\mathcal{F}, \mathsf{t}, \mathsf{n})$, or simply *encoding* when $(X, \mathtt{D})$ is clear from the context, if there is an injection $\mathfrak{f} : \mathcal{N} \to \mathcal{R}^*(\mathsf{w})$ s.t.

1. $\mathfrak{f}(\mathsf{t}) \stackrel{\text{def}}{=} \mathsf{w}$ is the only world in $\mathrm{ran}(\mathfrak{f}) \cap \mathcal{V}(\mathtt{t})$, and $\mathfrak{f}(\mathsf{n})$ is the only world in $\mathrm{ran}(\mathfrak{f}) \cap \mathcal{V}(\mathtt{n})$;
2. for every $\mathsf{n}' \in \mathrm{dom}(\mathcal{F})$ it holds that $(\mathfrak{f}(\mathcal{F}(\mathsf{n}')), \mathfrak{f}(\mathsf{n}')) \in \mathcal{R}$;
3. for every infinite path $(\mathsf{w}_0, \mathsf{w}_1 \cdots) \in \Pi_{\mathcal{R}}(\mathsf{w})$ there is $i \geq 0$ s.t. $\mathsf{w}_i \in \mathcal{V}(\mathtt{end})$ and
   - $\forall j \in [0, i-1]$, $\mathsf{w}_j \notin \mathcal{V}(\mathtt{end})$ and $(\mathsf{w}_j \in \mathcal{V}(\mathtt{D}) \Leftrightarrow \exists \mathsf{n}' \in \mathrm{dom}(\mathcal{F})\ \mathfrak{f}(\mathsf{n}') = \mathsf{w}_j)$;
   - for every $j \geq i$ and every node $\mathsf{n}' \in \mathrm{dom}(\mathcal{F})$, $\mathfrak{f}(\mathsf{n}') \neq \mathsf{w}_j$.

It is easy to show that such an encoding always exists. Informally, the first property states that w encodes t and is the only world in $\mathcal{R}^*(\mathsf{w})$ satisfying $\mathtt{t}$. Similarly, the world $\mathfrak{f}(\mathsf{n})$ encoding n is the only world in $\mathcal{R}^*(\mathsf{w})$ that satisfies $\mathtt{n}$. The second property states that the forest must be correctly encoded in the Kripke structure. In particular, notice that the parent relation of the finite forest is inverted so that it becomes the child relation in the

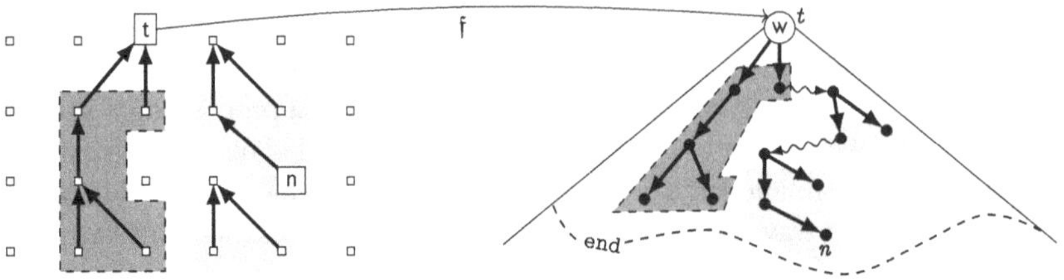

Fig. 4. A pointed forest (left) and one of its encoding as a finitely-branching Kripke tree (right).

Kripke structure (as shown in Figure 4). As $\mathfrak{f}$ is an injection, the encoding does not merge together trees that are disconnected in the forest. Lastly, the third property of $\mathfrak{f}$ states that the elements in $\mathrm{dom}(\mathcal{F})$ must be encoded by nodes in $\mathcal{R}^*(\mathsf{w})$ that precede every world satisfying $\mathtt{end}$. Moreover, among all the descendants of w preceding $\mathtt{end}$, the worlds encoding $\mathrm{dom}(\mathcal{F})$ are the only ones satisfying $\mathtt{D}$. This implies that w does not satisfy $\mathtt{D}$ (as $\mathsf{t} \notin \mathrm{dom}(\mathcal{F})$). Figure 4 shows a pointed forest and one of its possible encodings.

We now formalise the translation. Fix two different symbols $\mathtt{D}, \mathtt{E}$ not in X. In order to alternate between $\mathtt{D}$ and $\mathtt{E}$, we define $\overline{\mathtt{D}} \stackrel{\text{def}}{=} \mathtt{E}$ and $\overline{\mathtt{E}} \stackrel{\text{def}}{=} \mathtt{D}$. The translation $\tau_u(\varphi)$ of a formula φ in ALT, implicitly parametrised by X and where $u \in \{\mathtt{D}, \mathtt{E}\}$, is homomorphic for $\top$ and Boolean connectives (as in τ_{X}, see Section 4.1), and otherwise it is defined as

$$\tau_u(\mathrm{T}) \stackrel{\text{def}}{=} \mathsf{E}(((u \vee t) \wedge \neg \mathit{end})\, \mathsf{M}\, (u \wedge n)). \qquad \tau_u(\langle \mathrm{U} \rangle\, \varphi) \stackrel{\text{def}}{=} \exists n\, (\mathtt{uniq}(n) \wedge \tau_u(\varphi)).$$

$$\tau_u(\mathrm{G}) \stackrel{\text{def}}{=} \mathsf{E}(\neg \mathit{end}\, \mathsf{M}\, (u \wedge n)) \wedge \neg \tau_u(\mathrm{T}). \qquad \tau_u(\blacklozenge^* \varphi) \stackrel{\text{def}}{=} \exists \overline{u}\, (\mathsf{AG}\, (\overline{u} \Rightarrow u) \wedge \tau_{\overline{u}}(\varphi)).$$

$$\tau_u(\blacklozenge\, \varphi) \stackrel{\text{def}}{=} \exists \overline{u}\, (\mathsf{AG}\, (\overline{u} \Rightarrow u) \wedge \mathtt{uniq}(u \wedge \neg \overline{u}) \wedge \mathsf{E}(\neg \mathit{end}\, \mathsf{M}\, (u \wedge \neg \overline{u})) \wedge \tau_{\overline{u}}(\varphi)).$$

Let $(\mathcal{F}, \mathsf{t}, \mathsf{n})$ be a pointed forest s.t. $\mathsf{t} \notin \mathrm{dom}(\mathcal{F})$ and let $((\mathcal{W}, \mathcal{R}, \mathcal{V}), \mathsf{w})$ be one of its (X, u)-encodings w.r.t. the injection $\mathfrak{f}$. For instance, $\tau_u(\mathrm{T})$ requires that there is a path $(\mathsf{w}, \mathsf{w}_1, \cdots, \mathsf{w}_j)$ starting in $\mathfrak{f}(\mathsf{t}) = \mathsf{w}$ and whose worlds do not satisfy end and must satisfy u or t. Moreover, the last world w_j must satisfy u and n. From property (1) of the definition of $\mathfrak{f}$, the only element satisfying t is w, which does not satisfy u (as $\mathsf{t} \notin \mathrm{dom}(\mathcal{F})$). Then, this path of worlds encodes a path in the pointed forest, from the current evaluation node n (which is encoded by the only world satisfying n) to the target node t. The translation is shown correct (by structural induction on φ) for pointed forests that admit an encoding.

Lemma 9. *Let* $(\mathcal{F}, \mathsf{t}, \mathsf{n})$ *be a pointed forest s.t.* $\mathsf{t} \notin \mathrm{dom}(\mathcal{F})$*, and let* $(\mathcal{K}, \mathsf{w})$ *be a* (X, u)*-encoding of* $(\mathcal{F}, \mathsf{t}, \mathsf{n})$*. Given a formula* φ *in* ALT*,* $(\mathcal{F}, \mathsf{t}, \mathsf{n}) \models \varphi$ *if and only if* $(\mathcal{K}, \mathsf{w}) \models \tau_u(\varphi)$*.*

Then, to conclude the reduction we just need to characterise the set of models encoding a pointed forest. The formula $\mathtt{enc} \stackrel{\text{def}}{=} \neg \mathtt{D} \wedge t \wedge \mathtt{uniq}(t) \wedge \mathtt{uniq}(n) \wedge \mathsf{AF}(\mathit{end})$ does the job.

Lemma 10. φ *in* ALT *and* $\mathtt{enc} \wedge \tau_{\mathtt{D}}(\varphi)$ *in* QCTL^T *are equisatisfiable.*

We now take a closer look to the translation. Given a temporal modality $\mathcal{T}$ and $k \in \mathbb{N} \cup \{\omega\}$, $\mathsf{QCTL}^T(\mathcal{T}^k)$ denotes the fragment of QCTL^T restricted to formulae where the only temporal modality allowed is $\mathcal{T}$, which can be nested at most k times (ω stands for an arbitrary number of imbrications). For instance, $\mathsf{QCTL}^T(\mathsf{EF}^k)$ denotes the set of formulae restricted to the operator EF, which can be nested at most k times. This fragment of QCTL^T is shown to be k-NEXPTIME-hard in [5], which directly leads to the TOWER-hardness of $\mathsf{QCTL}^T(\mathsf{EF}^\omega)$ and $\mathsf{QCTL}^T(\mathsf{EU}^\omega)$. By analysing our translation it is easy to show that

$\mathsf{QCTL}^T(\mathsf{EU}^0)$, i.e. QCTL^T restricted to the only modality $\mathsf{E}(\varphi \mathbin{\mathsf{U}} \psi)$ where φ and ψ are Boolean combination of propositional symbols, and $\mathsf{QCTL}^T(\mathsf{EF}^1)$ are already TOWER-hard. First of all, the formula $\mathsf{E}(\varphi \mathbin{\mathsf{U}} \psi)$ in $\mathsf{QCTL}^T(\mathsf{EU}^0)$ is equivalent to the following formula in $\mathsf{QCTL}^T(\mathsf{EF}^1)$: $\exists p\big(\mathsf{AG}\,(\neg\varphi \wedge \neg\psi \Rightarrow p) \wedge \mathsf{AG}\,(p \Rightarrow \mathsf{AG}\,p) \wedge \mathsf{EF}\,(\psi \wedge \neg p)\big)$, where p does not appear in φ or ψ. Then, we just need to prove the result for $\mathsf{QCTL}^T(\mathsf{EU}^0)$.

Clearly, the translation τ_u is defined so that the resulting formula is in $\mathsf{QCTL}^T(\mathsf{EU}^0)$. However, we need to deal with the occurrence of $\mathsf{AF}\,(\boldsymbol{end})$ used inside the formula $\mathtt{enc}$. Let us first consider the formula $\mathsf{AG}\,(\varphi \Rightarrow \mathsf{AG}\,\psi)$ which is satisfied by models where once φ is found to hold in a certain world w, then ψ is satisfied in every world of $\mathcal{R}^*(\mathsf{w})$. Despite not being in $\mathsf{QCTL}^T(\mathsf{EU}^0)$, the formula $\mathsf{AG}\,(\varphi \Rightarrow \mathsf{AG}\,\psi)$ is equivalent to the following formula: $\forall p \forall q\,\big(\mathtt{uniq}(p) \wedge \mathtt{uniq}(q) \wedge \mathsf{EF}\,(p \wedge \varphi) \wedge \mathsf{EF}\,(q \wedge \neg\psi) \Rightarrow \mathsf{E}(\neg p \mathbin{\mathsf{M}} q)\big)$, where p and q do not appear in φ or ψ. We then define a formula $\chi_{\mathsf{EG}}\,(\varphi)$ that only uses EF modalities and is equivalent to $\mathsf{EG}\,\varphi$, so that then $\neg\chi_{\mathsf{EG}}\,(\neg\varphi)$ is equivalent to $\mathsf{AF}\,\varphi$:

$$\begin{aligned}\chi_{\mathsf{EG}}\,(\varphi) \stackrel{\text{def}}{=} \exists p\big(&\neg p \wedge \mathsf{AG}\,(\neg\varphi \Rightarrow p) \wedge \mathsf{AG}\,(p \Rightarrow \mathsf{AG}\,p) \wedge \\ &\forall q\big(\mathtt{uniq}(q) \wedge \mathsf{EF}\,(q \wedge \neg p) \Rightarrow \mathsf{EF}\,(q \wedge \mathsf{EF}\,(\neg q \wedge \neg p))\big)\big)\end{aligned}$$

where p does not appear in φ. This formula is expressible in $\mathsf{QCTL}^T(\mathsf{EU}^0)$, as every subformula that is not in this fragment is an instance of $\mathsf{AG}\,(\varphi \Rightarrow \mathsf{AG}\,\psi)$. Then, we conclude that $\mathsf{AF}\,(\boldsymbol{end})$ is expressible in $\mathsf{QCTL}^T(\mathsf{EU}^0)$, leading to the following result.

Theorem 2. *The satisfiability problems of* $\mathsf{QCTL}^T(\mathsf{EU}^0)$ *and* $\mathsf{QCTL}^T(\mathsf{EF}^1)$ *are* TOWER*-c.*

4.3 From ALT to Modal Logic of Heaps and Modal Separation Logic

In [17] and later in [18] two families of logics are presented, respectively called *modal logic of heaps* (MLH) and *modal separation logic* (MSL). At their core, both logics can be seen as modal logics extended with separating connectives, hence mixing separation logic (Section 4.1) with temporal aspects as in quantified CTL (Section 4.2). As we already shown how ALT is captured by these two latter logics, it is natural to ask ourselves if the same holds for MLH and MSL. In this section, we show that this is indeed the case and, as for the previous two sections, ALT allows us to refine the analysis on these logics. Both MLH and MSL are interpreted on finite Kripke functions. A *finite Kripke function* is a Kripke structure $(\mathcal{W}, \mathcal{R}, \mathcal{V})$ (see Section 4.2 for its definition) where $\mathcal{W}$ is infinite and $\mathcal{R}$, instead of being left-total, is finite and weakly functional, i.e. $|\mathcal{R}| \in \mathbb{N}$ and for every $\mathsf{w}, \mathsf{w}', \mathsf{w}'' \in \mathcal{W}$, if $(\mathsf{w}, \mathsf{w}') \in \mathcal{R}$ and $(\mathsf{w}, \mathsf{w}'') \in \mathcal{R}$ then $\mathsf{w}' = \mathsf{w}''$. As $\mathcal{N}$ and $\mathcal{W}$ are both countably infinite sets, without loss of generality we assume $\mathcal{W} = \mathcal{N}$. Two Kripke structures $\mathcal{K}_1 = (\mathcal{W}, \mathcal{R}_1, \mathcal{V})$ and $\mathcal{K}_2 = (\mathcal{W}, \mathcal{R}_2, \mathcal{V})$ are disjoint if $\mathcal{R}_1 \cap \mathcal{R}_2 = \emptyset$. When this holds, $\mathcal{K}_1 + \mathcal{K}_2$ denotes the model $(\mathcal{W}, \mathcal{R}_1 \cup \mathcal{R}_2, \mathcal{V})$. To shorten the presentation, in the following diagram we introduce a language having the operators from MSL and MLH, and summarise known and new results on these logics (where $p \in \mathsf{AP}$):

MSL: TOWER-complete from [18]. MLH: TOWER-complete from [17].

$$\varphi := p \mid \langle\neq\rangle\,\varphi \mid \top \mid \varphi \wedge \varphi \mid \neg\varphi \mid \Diamond\varphi \mid \varphi * \psi \mid \langle\mathrm{U}\rangle\,\varphi \mid \Diamond^{-1}\varphi$$

TOWER-hard by reduction from SAT(ALT), shown here.

As defined below, $\Diamond$ is the standard alethic modality from modal logic, $\Diamond^{-1}$ is its converse modality, and $\langle\neq\rangle$ is the *elsewhere* modality that generalises the somewhere modality $\langle U \rangle$ as $\langle U \rangle\, \varphi = \varphi \vee \langle\neq\rangle\, \varphi$. For a *pointed model* $(\mathcal{K}, \mathsf{w})$, where $\mathcal{K} = (\mathcal{W}, \mathcal{R}, \mathcal{V})$ is a finite Kripke function and $\mathsf{w} \in \mathcal{W}$, the satisfaction relation $\models$ is defined as follows:

$$
\begin{array}{lll}
(\mathcal{K}, \mathsf{w}) \models p & \overset{\text{def}}{\Leftrightarrow} & \mathsf{w} \in \mathcal{V}(p). \\
(\mathcal{K}, \mathsf{w}) \models \Diamond\varphi & \overset{\text{def}}{\Leftrightarrow} & \text{there is } \mathsf{w}' \in \mathcal{R}(\mathsf{w}) \text{ such that } (\mathcal{K}, \mathsf{w}) \models \varphi. \\
(\mathcal{K}, \mathsf{w}) \models \Diamond^{-1}\varphi & \overset{\text{def}}{\Leftrightarrow} & \text{there is } \mathsf{w}' \in \mathcal{W} \text{ such that } \mathsf{w} \in \mathcal{R}(\mathsf{w}') \text{ and } (\mathcal{K}, \mathsf{w}') \models \varphi. \\
(\mathcal{K}, \mathsf{w}) \models \langle\neq\rangle\, \varphi & \overset{\text{def}}{\Leftrightarrow} & \text{there is } \mathsf{w}' \in \mathcal{W} \text{ such that } \mathsf{w}' \neq \mathsf{w} \text{ and } (\mathcal{K}, \mathsf{w}') \models \varphi. \\
(\mathcal{K}, \mathsf{w}) \models \varphi * \psi & \overset{\text{def}}{\Leftrightarrow} & (\mathcal{K}_1, \mathsf{w}) \models \varphi \text{ and } (\mathcal{K}_2, \mathsf{w}) \models \psi \text{ for some } \mathcal{K}_1, \mathcal{K}_2 \text{ s.t. } \mathcal{K}_1 + \mathcal{K}_2 = \mathcal{K}.
\end{array}
$$

By looking at the diagram above, compared to the work in [18], ALT allows us to show that propositional symbols and the elsewhere modality can be removed from MSL without changing the complexity status of its satisfiability problem. Similarly, ALT allows us to refine the analysis on the complexity of SAT(MLH) by showing that the $\Diamond^{-1}$ modality is not needed in order to achieve non-elementary complexities.

Let $(\mathcal{F}, \mathsf{t}, \mathsf{n})$ be a pointed forest and let $(\mathcal{K}, \mathsf{w})$ be a pointed model where $\mathcal{K} = (\mathcal{W}, \mathcal{R}, \mathcal{V})$. For the reduction, we use w to encode the current node n. Encoding t is not so immediate, as MLH does not have propositional symbols. A possible solution is to encode it as a self-loop, so that the formula T is translated to a query stating that w reaches the self-loop. As done in Section 4.1 we define the formula $\texttt{size=1} \overset{\text{def}}{=} \langle U \rangle \Diamond \top \wedge \neg(\langle U \rangle \Diamond \top * \langle U \rangle \Diamond \top)$, that is satisfied whenever $|\mathcal{R}|{=}1$. We also define the modalities $\blacklozenge$ and $\blacklozenge^*$ in MLH: $\blacklozenge_{\text{ML}}\, \varphi \overset{\text{def}}{=} (\texttt{size=1}) * \varphi$ and $\blacklozenge^*_{\text{ML}} \varphi \overset{\text{def}}{=} \top * \varphi$. Lastly, we introduce the formula $\texttt{selfloop} \overset{\text{def}}{=} \blacklozenge^*_{\text{ML}}(\Diamond\Diamond\top \wedge \neg\blacklozenge_{\text{ML}}\blacklozenge_{\text{ML}}\top)$ that is satisfied by $(\mathcal{K}, \mathsf{w})$ if $(\mathsf{w}, \mathsf{w}) \in \mathcal{R}$. Suppose for a moment that we are able to use this formula to characterise the class of of every finite Kripke function $(\mathcal{W}, \mathcal{R}, \mathcal{V})$ where there is exactly one cycle, and this cycle is a self-loop on a world w_t. Then, we use w_t to encode the target node t of a finite forest $(\mathcal{F}, \mathsf{t}, \mathsf{n})$ while being careful that the $\blacklozenge$ and $\blacklozenge^*$ operators of ALT are translated in such a way that the self-loop on w_t is preserved. Because of the specific treatment of w_t, it is convenient to assume that the current evaluation node n is encoded by a world different from w_t, which reflects on the translation of $\langle U \rangle$. The admissibility of this assumption follows by Lemma 4.

We encode pointed forests as finite Kripke functions. Let $(\mathcal{F}, \mathsf{t}, \mathsf{n})$ be a pointed forest s.t. $\mathsf{t} \notin \text{dom}(\mathcal{F})$ and $\mathsf{n} \neq \mathsf{t}$. A finite Kripke function $((\mathcal{N}, \mathcal{R}, \mathcal{V}), \mathsf{n})$ (recall, $\mathcal{W} = \mathcal{N}$) is an *encoding* of $(\mathcal{F}, \mathsf{t}, \mathsf{n})$ iff for every $\mathsf{n}', \mathsf{n}'' \in \mathcal{N}$ we have $(\mathsf{n}', \mathsf{n}'') \in \mathcal{R} \Leftrightarrow (\mathcal{F}(\mathsf{n}') = \mathsf{n}''$ or $\mathsf{n}' = \mathsf{n}'' = \mathsf{t})$. Notice how $\mathcal{R}$ is essentially defined from $\mathcal{F}$ by adding the self-loop (t, t). The translation $\tau(\varphi)$ in MLH of a formula φ in ALT is homomorphic for $\top$ and Boolean connectives (as is the case for τ_x in Section 4.1), and otherwise it is defined as

$$
\begin{array}{llll}
\tau(\mathrm{T}) & \overset{\text{def}}{=} \blacklozenge^*_{\text{ML}}\big(\Diamond\top \wedge [U](\Diamond\top \Rightarrow \Diamond\Diamond\top)\big). & \tau(\blacklozenge\, \varphi) & \overset{\text{def}}{=} \blacklozenge_{\text{ML}}(\tau(\varphi) \wedge \langle U \rangle\, \texttt{selfloop}). \\
\tau(\mathrm{G}) & \overset{\text{def}}{=} \Diamond\top \wedge \neg\, \tau(\mathrm{T}). & \tau(\blacklozenge^*\varphi) & \overset{\text{def}}{=} \blacklozenge^*_{\text{ML}}(\tau(\varphi) \wedge \langle U \rangle\, \texttt{selfloop}). \\
\tau(\langle U \rangle\, \varphi) & \overset{\text{def}}{=} \langle U \rangle(\neg\texttt{selfloop} \wedge \tau(\varphi)). & &
\end{array}
$$

We highlight two points of this translation. First, $\tau(\mathrm{T})$ essentially asks to find a submodel where every path reaches the self-loop and the current evaluation node is in one of these paths. Second, notice how the translation of $\blacklozenge$ and $\blacklozenge^*$ checks that the model is updated so that the self-loop is not lost, as required by our encoding. It should be noted that this requirement cannot be met if we were translating the definition of ALT from [31], featuring the $*$ operator. Indeed, by partitioning the model into two pieces, this operator removes the self-loop from one of the two parts, breaking our encoding. The following lemma (proved by structural induction on φ) shows the correctness of our translation.

Lemma 11. *Let* $(\mathcal{F}, \mathsf{t}, \mathsf{n})$ *be a pointed model s.t.* $\mathsf{n} \neq \mathsf{t}$ *and* $\mathsf{t} \notin \mathrm{dom}(\mathcal{F})$. *Let* $(\mathcal{K}, \mathsf{n})$ *be an encoding of* $(\mathcal{F}, \mathsf{t}, \mathsf{n})$. *Given a formula* φ *in* ALT, $(\mathcal{F}, \mathsf{t}, \mathsf{n}) \models \varphi$ *iff* $(\mathcal{K}, \mathsf{n}) \models \tau(\varphi)$.

To conclude the reduction we show that we can characterise the class of models encoding pointed forests, i.e. the finite Kripke functions with exactly one cycle, which is a self-loop. We first define the formula $\mathtt{hascycl} \stackrel{\text{def}}{=} \blacklozenge^*_{\mathrm{ML}}\big(\langle \mathrm{U} \rangle \Diamond \top \wedge [\mathrm{U}](\Diamond \top \Rightarrow \Diamond\Diamond \top)\big)$ that checks if a finite Kripke function has at least one cycle. Then, the desired property can be simply defined by stating that there is a self-loop which, whenever removed, leads to an acyclic submodel: $\mathtt{1selfloop} \stackrel{\text{def}}{=} \langle \mathrm{U} \rangle \big(\mathtt{selfloop} \wedge \neg \blacklozenge_{\mathrm{ML}}(\Box \bot \wedge \mathtt{hascycl})\big)$.

Lemma 12. *Every formula* φ *in* ALT *is equisatisfiable with* $\tau(\varphi) \wedge \mathtt{1selfloop}$.

For the proof of Lemma 12, both Lemma 4(1) and (2) are used in order to restrict ourselves to pointed forest $(\mathcal{F}, \mathsf{t}, \mathsf{n})$ s.t. $\mathsf{n} \neq \mathsf{t}$ and $\mathsf{t} \notin \mathrm{dom}(\mathcal{F})$. Then, we apply Lemma 11.

Theorem 3. *The fragment of MLH and MSL with Boolean operators,* $\Diamond$ *and* $\langle \mathrm{U} \rangle$ *modalities, and* $*$ *(alternatively,* $\blacklozenge_{\mathrm{ML}}$ *and* $\blacklozenge^*_{\mathrm{ML}}$*) has a* TOWER*-complete satisfiability problem.*

5 Conclusions

We studied an *Auxiliary Logic on Trees* (ALT), a quite simple formalism that admits a TOWER-complete satisfiability problem. ALT is shown to be easily captured by various non-elementary logics: first-order separation logic, quantified CTL, modal logic of heaps and modal separation logic. Through ALT, we were not only able to connect these logics, but also to refine their analysis and find strict fragments that are still TOWER-hard. Most importantly, with ALT we hope to have shown a set of simple and concrete properties, centred around reachability and submodel reasoning, that when put together lead to logics having a non-elementary satisfiability problem.

This work leaves a few questions open. First, the fragments of ALT where $\blacklozenge$ or $\blacklozenge^*$ are removed from the logic have not being studied yet. The logic without $\blacklozenge^*$ is of particular interests, as it is connected with the sabotage logics from [4]. Second, the analysis done on first-order separation logic and on modal logic of heaps (Sections 4.1 and 4.3) reveals that the complexity of these logics does not change when the $*$ operator and the $\mathtt{emp}$ predicate are replaced with the less general operators $\blacklozenge$ and $\blacklozenge^*$. We find this point interesting, as from an overview of the literature, it seems that this result also holds for the separation logics considered in [9,17,19,30,31]. Moreover, for the logics whose expressiveness is known, i.e. the ones in [19,30], it seems that also the expressive power remains unchanged. However, we struggle to see how to uniformly express the operator $*$ with $\blacklozenge$ and $\blacklozenge^*$, as the resulting logics reason on the model in a different way (as as shown in Section 2). Lastly, this work illustrates the potential of ALT as a tool for proving the TOWER-hardness of logics interpreted on tree-like structures. As the operators of our logic are simple, we hope ALT to be useful to study logics with unknown complexities.

Acknowlegements. I would like to thank S. Demri and E. Lozes for their feedback.

References

1. T. Antonopoulos and A. Dawar. Separating graph logic from MSO. In *Foundations of Software Science and Computational Structures*, volume 5504 of *LNCS*, pages 63–77. Springer, 2009.
2. A. Artale, R. Kontchakov, V. Ryzhikov, and M. Zakharyaschev. The complexity of clausal fragments of LTL. In *Logic for Programming, Artificial Intelligence, and Reasoning*, volume 8312 of *LNCS*, pages 35–52. Springer, 2013.
3. G. Aucher, P. Balbiani, L. Fariñas del Cerro, and A. Herzig. Global and local graph modifiers. *Electronic Notes in Theoretical Computer Science*, 231:293–307, 2009.
4. G. Aucher, J. van Benthem, and D. Grossi. Sabotage modal logic: Some model and proof theoretic aspects. In *Logic, Rationality, and Interaction*, volume 9394 of *LNCS*, pages 1–13. Springer, 2015.
5. B. Bednarczyk and S. Demri. Why propositional quantification makes modal logics on trees robustly hard? In *Logic in Computer Science*, pages 1–13. IEEE, 2019.
6. J. Berdine, B. Cook, and S. Ishtiaq. Slayer: Memory safety for systems-level code. In *Computer-Aided Verification*, volume 6806 of *LNCS*, pages 178–183. Springer, 2011.
7. L. Bozzelli, A. Molinari, A. Montanari, and A. Peron. On the complexity of model checking for syntactically maximal fragments of the interval temporal logic HS with regular expressions. In *Games, Automata, Logics, and Formal Verification*, volume 256 of *EPTCS*, pages 31–45, 2017.
8. L. Bozzelli, A. Molinari, A. Montanari, A. Peron, and P. Sala. Interval vs. point temporal logic model checking: an expressiveness comparison. In *Foundations of Software Technology and Theoretical Computer Science*, volume 65 of *LIPIcs*, pages 26:1–26:14. Schloss Dagstuhl–Leibniz-Zentrum fuer Informatik, 2016.
9. R. Brochenin, S. Demri, and E. Lozes. On the almighty wand. *Information and Computation*, 211:106–137, 2012.
10. C. Calcagno, T. Dinsdale-Young, and P. Gardner. Adjunct elimination in context logic for trees. *Information and Computation*, 208:474–499, 2010.
11. C. Calcagno, D. Distefano, J. Dubreil, D. Gabi, P. Hooimeijer, M. Luca, P. W. O'Hearn, I. Papakonstantinou, J. Purbrick, and D. Rodriguez. Moving fast with software verification. In *Nasa Formal Methods*, volume 9058 of *LNCS*, pages 3–11. Springer, 2015.
12. C. Calcagno, H. Yang, and P. W. O'Hearn. Computability and complexity results for a spatial assertion language for data structures. In *Foundations of Software Technology and Theoretical Computer Science*, volume 2245 of *LNCS*, pages 108–119. Springer, 2001.
13. E. M. Clarke. *The Birth of Model Checking*, pages 1–26. Springer, 2008.
14. E. M. Clarke and E. A. Emerson. Design and synthesis of synchronization skeletons using branching time temporal logic. In *Logics of Programs*, volume 131 of *LNCS*, pages 52–71. Springer, 1982.
15. B. Courcelle. Graph structure and monadic second-order logic: Language theoretical aspects. In *Automata, Languages and Programming*, volume 5125 of *LNCS*, pages 1–13. Springer, 2008.
16. A. Dawar, P. Gardner, and G. Ghelli. Adjunct elimination through games in static ambient logic. In *Foundations of Software Technology and Theoretical Computer Science*, volume 3328 of *LNCS*, pages 211–223. Springer, 2004.
17. S. Demri and M. Deters. Two-variable separation logic and its inner circle. *Transactions on Computational Logic*, 16:15:1–15:36, 2015.
18. S. Demri and R. Fervari. On the complexity of modal separation logics. In *Advances in Modal Logic*, pages 179–198. College Publications, 2018.
19. S. Demri, D. Galmiche, D. Larchey-Wendling, and D. Méry. Separation logic with one quantified variable. *Theoretical Computer Science*, 61:371–461, 2017.

20. S. Demri, E. Lozes, and A. Mansutti. The effects of adding reachability predicates in propositional separation logic. In *Foundations of Software Science and Computational Structures*, volume 10803 of *LNCS*, pages 476–493. Springer, 2018.
21. R. Fervari. *Relation-Changing Modal Logics*. PhD thesis, 2014.
22. K. Fine. Propositional quantifiers in modal logic. *Theoria*, 36:336–346, 1970.
23. M. J. Fischer and R. E. Ladner. Propositional dynamic logic of regular programs. *Journal of Computer and System Sciences*, 18:194 – 211, 1979.
24. V. Goranko. Temporal logics of computations. Lecture Notes from ESSLLI'00, 2000.
25. V. Goranko, A. Montanari, and G. Sciavicco. A road map of interval temporal logics and duration calculi. *Journal of Applied Non-Classical Logics*, 14:9–54, 2004.
26. V. Goranko and S. Passy. Using the universal modality: Gains and questions. *Journal of Logic and Computation*, 2:5–30, 1992.
27. S. A. Kripke. Semantical considerations on modal logic. *Acta Philosophica Fennica*, 16:83–94, 1963.
28. F. Laroussinie and N. Markey. Quantified CTL: expressiveness and complexity. *Logical Methods in Computer Science*, 10, 2014.
29. L. Libkin. *Elements of Finite Model Theory*. Texts in Theoretical Computer Science. An EATCS Series. Springer, 2004.
30. E. Lozes. Adjuncts elimination in the static ambient logic. *Electronic Notes in Theoretical Computer Science*, 96:51–72, 2004.
31. A. Mansutti. Extending propositional separation logic for robustness properties. In *Foundations of Software Technology and Theoretical Computer Science*, pages 42:1–42:23. Schloss Dagstuhl–Leibniz-Zentrum fuer Informatik, 2018.
32. D. A. Martin. Borel determinacy. *Annals of Mathematics*, 102:363–371, 1975.
33. A. Meier, M. Mundhenk, M. Thomas, and H. Vollmer. The complexity of satisfiability for fragments of CTL and CTL*. *Electronic Notes in Theoretical Computer Science*, 223:201–213, 2008.
34. B. C. Moszkowski. *Reasoning About Digital Circuits*. PhD thesis, 1983.
35. P. W. O'Hearn and D. J. Pym. The logic of bunched implications. *Bulletin of Symbolic Logic*, 5:215–244, 1999.
36. M. O. Rabin. Decidability of second-order theories and automata on infinite trees. *Transactions of the American Mathematical Society*, 41:1–35, 1969.
37. J. C. Reynolds. Separation logic: A logic for shared mutable data structures. In *Logic in Computer Science*, pages 55–74. IEEE, 2002.
38. S. Schmitz. Complexity hierarchies beyond elementary. *ransactions on Computation Theory*, 8:3:1–3:36, 2016.
39. A. P. Sistla and E. M. Clarke. The complexity of propositional linear temporal logics. *Journal of the Association for Computing Machinery*, 32:733–749, 1985.

The Inconsistent Labelling Problem of Stutter-Preserving Partial-Order Reduction

Thomas Neele[1] (✉), Antti Valmari[2], and Tim A.C. Willemse[1]

[1] Eindhoven University of Technology, Eindhoven, The Netherlands
{t.s.neele, t.a.c.willemse}@tue.nl
[2] University of Jyväskylä, Jyväskylä, Finland
antti.valmari@jyu.fi

Abstract. In model checking, partial-order reduction (POR) is an effective technique to reduce the size of the state space. Stubborn sets are an established variant of POR and have seen many applications over the past 31 years. One of the early works on stubborn sets shows that a combination of several conditions on the reduction is sufficient to preserve stutter-trace equivalence, making stubborn sets suitable for model checking of linear-time properties. In this paper, we identify a flaw in the reasoning and show with a counter-example that stutter-trace equivalence is not necessarily preserved. We propose a solution together with an updated correctness proof. Furthermore, we analyse in which formalisms this problem may occur. The impact on practical implementations is limited, since they all compute a correct approximation of the theory.

1 Introduction

In formal methods, model checking is a technique to automatically decide the correctness of a system's design. The many interleavings of concurrent processes can cause the state space to grow exponentially with the number of components, known as the *state-space explosion* problem. *Partial-order reduction* (POR) is one technique that can alleviate this problem. Several variants of POR exist, such as *ample sets* [11], *persistent set* [7] and *stubborn sets* [16,21]. For each of those variants, sufficient conditions for preservation of stutter-trace equivalence have been identified. Since LTL without the next operator (LTL_{-X}) is invariant under finite stuttering, this allows one to check most LTL properties under POR.

However, the correctness proofs for these methods are intricate and not reproduced often. For stubborn sets, LTL_{-X}-preserving conditions and an accompanying correctness result were first presented in [15], and discussed in more detail in [17]. While trying to reproduce the proof for [17, Theorem 2] (see also Theorem 1 in the current work), we ran into an issue while trying to prove a certain property of the construction used in the original proof [17, Construction 1]. This led us to discover that stutter-trace equivalence is not necessarily preserved. We will refer to this as the *inconsistent labelling problem.* The essence of the problem is that POR in general, and the proofs in [17] in particular, reason mostly about actions, which label the transitions. The only relevance of

J. Goubault-Larrecq and B. König (Eds.): FOSSACS 2020, LNCS 12077, pp. 482–501, 2020.
https://doi.org/10.1007/978-3-030-45231-5_25

the state labelling is that it determines which actions are *visible*. On the other hand, stutter-trace equivalence and the LTL semantics are purely based on state labels. The correctness proof in [17] does not deal properly with this disparity. Further investigation shows that the same problem also occurs in two works of Beneš *et al.* [2,3], who apply ample sets to state/event LTL model checking.

Consequently, any application of stubborn sets in LTL_{-X} model checking is possibly unsound, both for safety and liveness properties. In literature, the correctness of several theories [9,10,18] relies on the incorrect theorem.

Our contributions are as follows:

- We prove the existence of the inconsistent labelling problem with a counter-example. This counter-example is valid for weak stubborn sets and, with a small modification, in a non-deterministic setting for strong stubborn sets.
- We propose to strengthen one of the stubborn set conditions and show that this modification resolves the issue (Theorem 2).
- We analyse in which circumstances the inconsistent labelling problem occurs and, based on the conclusions, discuss its impact on existing literature. This includes a thorough analysis of Petri nets and several different notions of invisible transitions and atomic propositions.

Our investigation shows that probably all practical implementations of stubborn sets compute an approximation which resolves the inconsistent labelling problem. Furthermore, POR methods based on the standard independence relation, such as ample sets and persistent sets, are not affected.

The rest of the paper is structured as follows. In Section 2, we introduce the basic concepts of stubborn sets and stutter-trace equivalence, which is not preserved in the counter-example of Section 3. A solution to the inconsistent labelling problem is discussed in Section 4, together with an updated correctness proof. Sections 5 and 6 discuss several settings in which correctness is not affected. Finally, Section 7 presents related work and Section 8 presents a conclusion.

2 Preliminaries

Since LTL relies on state labels and POR relies on edge labels, we assume the existence of some fixed set of atomic propositions *AP* to label the states and a fixed set of edge labels *Act*, which we will call *actions*. Actions are typically denoted with the letter a.

Definition 1. *A* labelled state transition system, *short* LSTS, *is a directed graph* $TS = (S, \rightarrow, \hat{s}, L)$, *where:*

- S *is the state space;*
- $\rightarrow \subseteq S \times Act \times S$ *is the transition relation;*
- $\hat{s} \in S$ *is the initial state; and*
- $L : S \rightarrow 2^{AP}$ *is a function that labels states with atomic propositions.*

We write $s \xrightarrow{a} t$ whenever $(s, a, t) \in \rightarrow$. A *path* is a (finite or infinite) alternating sequence of states and actions: $s_0 \xrightarrow{a_1} s_1 \xrightarrow{a_2} s_2 \ldots$. We sometimes omit the intermediate and/or final states if they are clear from the context or not relevant, and write $s \xrightarrow{a_1 \ldots a_n} t$ or $s \xrightarrow{a_1 \ldots a_n}$ for finite paths and $s \xrightarrow{a_1 a_2 \ldots}$ for infinite paths. Paths that start in the initial state $\hat{s}$ are called *initial paths*. Given a path $\pi = s_0 \xrightarrow{a_1} s_1 \xrightarrow{a_2} s_2 \ldots$, the *trace* of π is the sequence of state labels observed along π, *viz.* $L(s_0)L(s_1)L(s_2)\ldots$. An action a is *enabled* in a state s, notation $s \xrightarrow{a}$, if and only if there is a transition $s \xrightarrow{a} t$ for some t. In a given LSTS TS, $enabled_{TS}(s)$ is the set of all enabled actions in a state s. A set $\mathcal{I}$ of *invisible* actions is chosen such that if (but not necessarily only if) $a \in \mathcal{I}$, then for all states s and t, $s \xrightarrow{a} t$ implies $L(s) = L(t)$. Note that this definition allows the set $\mathcal{I}$ to be under-approximated. An action that is not invisible is called *visible*. We say TS is *deterministic* if and only if $s \xrightarrow{a} t$ and $s \xrightarrow{a} t'$ imply $t = t'$, for all states s, t and t' and actions a. To indicate that TS is not necessarily deterministic, we say TS is *non-deterministic*.

2.1 Stubborn sets

In POR, *reduction functions* play a central role. A reduction function $r : S \rightarrow 2^{Act}$ indicates which transitions to explore in each state. When starting at the initial state $\hat{s}$, a reduction function induces a *reduced LSTS* as follows.

Definition 2. *Let $TS = (S, \rightarrow, \hat{s}, L)$ be an LSTS and $r : S \rightarrow 2^{Act}$ a reduction function. Then the* reduced LSTS *induced by r is defined as $TS_r = (S_r, \rightarrow_r, \hat{s}, L_r)$, where L_r is the restriction of L on S_r, and S_r and $\rightarrow_r$ are the smallest sets such that the following holds:*

- *$\hat{s} \in S_r$; and*
- *If $s \in S_r$, $s \xrightarrow{a} t$ and $a \in r(s)$, then $t \in S_r$ and $s \xrightarrow{a}_r t$.*

Note that we have $\rightarrow_r \subseteq \rightarrow$. In the remainder of this paper, we will assume the reduced LSTS is finite. This is essential for the correctness of the approach detailed below. In general, a reduction function is not guaranteed to preserve almost any property of an LSTS. Below, we list a number of conditions that have been proposed in literature; they aim to preserve LTL$_{-X}$. Here, we call an action a a *key action* in s iff for all paths $s \xrightarrow{a_1 \ldots a_n} s'$ such that $a_1 \notin r(s), \ldots, a_n \notin r(s)$, it holds that $s' \xrightarrow{a}$. We typically denote key actions by a_{key}.

D0 If $enabled(s) \neq \emptyset$, then $r(s) \cap enabled(s) \neq \emptyset$.
D1 For all $a \in r(s)$ and $a_1 \notin r(s), \ldots, a_n \notin r(s)$, if $s \xrightarrow{a_1} \cdots \xrightarrow{a_n} s_n \xrightarrow{a} s'_n$, then there are states $s', s'_1, \ldots, s'_{n-1}$ such that $s \xrightarrow{a} s' \xrightarrow{a_1} s'_1 \xrightarrow{a_2} \cdots \xrightarrow{a_n} s'_n$.
D2 Every enabled action in $r(s)$ is a key action in s.
D2w If $enabled(s) \neq \emptyset$, then $r(s)$ contains a key action in s.
V If $r(s)$ contains an enabled visible action, then it contains all visible actions.
I If an invisible action is enabled, then $r(s)$ contains an invisible key action.
L For every visible action a, every cycle in the reduced LSTS contains a state s such that $a \in r(s)$.

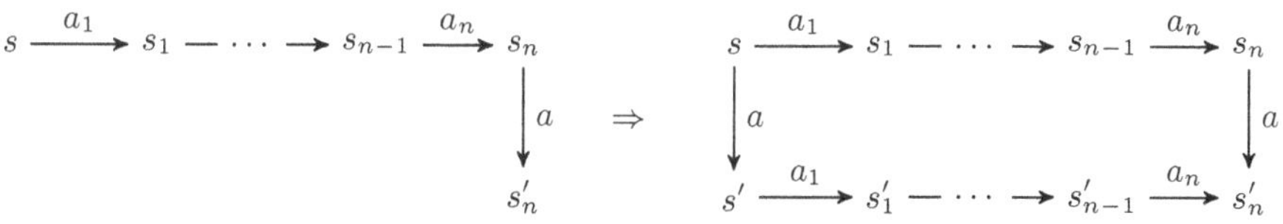

Fig. 1: Visual representation of condition **D1**.

These conditions are used to define *strong* and *weak* stubborn sets in the following way.

Definition 3. *A reduction function* $r : S \rightarrow 2^{Act}$ *is a* strong stubborn set *iff for all states* $s \in S$*, the conditions* ***D0, D1, D2, V, I, L*** *all hold.*

Definition 4. *A reduction function* $r : S \rightarrow 2^{Act}$ *is a* weak stubborn set *iff for all states* $s \in S$*, the conditions* ***D1, D2w, V, I, L*** *all hold.*

Below, we also use 'weak/strong stubborn set' to refer to the set of actions $r(s)$ in some state s. First, note that key actions are always enabled, by setting $n = 0$. Furthermore, a stubborn set can never introduce new deadlocks, either by **D0** or **D2w**. Condition **D1** enforces that a key action $a_{\mathsf{key}} \in r(s)$ does not disable other paths that are not selected for the stubborn set. A visual representation of condition **D1** can be found in Figure 1. When combined, **D1** and **D2w** are sufficient conditions for preservation of deadlocks. Condition **V** enforces that the paths $s \xrightarrow{a_1 \dots a_n a} s'_n$ and $s \xrightarrow{a a_1 \dots a_n} s'_n$ in **D1** contain the same sequence of visible actions. The purpose of condition **I** is to preserve the possibility to perform an invisible action, if one is enabled. Finally, we have condition **L** to deal with the *action-ignoring problem*, which occurs when an action is never selected for the stubborn set and always ignored. Since we assume that the reduced LSTS is finite, it suffices to reason in **L** about every cycle instead of every infinite path. The combination of **I** and **L** helps to preserve divergences (infinite paths containing only invisible actions).

Conditions **D0** and **D2** together imply **D2w**, and thus every strong stubborn set is also a weak stubborn set. Since the reverse does not necessarily hold, weak stubborn sets might offer more reduction.

2.2 Weak and Stutter Equivalence

To reason about the similarity of an LSTS TS and its reduced LSTS TS_r, we introduce the notions of *weak equivalence*, which operates on actions, and *stutter equivalence*, which operates on states. The definitions are generic, so that they can also be used in Section 6.

Definition 5. *Two paths* π *and* π' *are weakly equivalent with respect to a set of actions* A*, notation* $\pi \sim_A \pi'$*, if and only if they are both finite or both infinite and their respective projections on* $Act \setminus A$ *are equal.*

Definition 6. *The* no-stutter trace *under labelling* L *of a path* $s_0 \xrightarrow{a_1} s_1 \xrightarrow{a_2} \dots$ *is the sequence of those* $L(s_i)$ *such that* $i = 0$ *or* $L(s_i) \neq L(s_{i-1})$. *Paths* π *and* π' *are stutter equivalent under* L, *notation* $\pi \triangleq_L \pi'$, *iff they are both finite or both infinite, and they yield the same no-stutter trace under* L.

We typically consider weak equivalence with respect to the set of invisible actions $\mathcal{I}$. In that case, we write $\pi \sim \pi'$. We also omit the subscript for stutter equivalence when reasoning about the standard labelling function and write $\pi \triangleq \pi'$. Remark that stutter equivalence is invariant under finite repetitions of state labels, hence its name. We lift both equivalences to LSTSs, and say that TS and TS' are *weak-trace equivalent* iff for every initial path π in TS, there is a weakly equivalent initial path π' in TS' and vice versa. Likewise, TS and TS' are *stutter-trace equivalent* iff for every initial path π in TS, there is a stutter equivalent initial path π' in TS' and vice versa.

In general, weak equivalence and stutter equivalence are incomparable, even for initial paths. However, for some LSTSs, these notions can be related in a certain way. We formalise this in the following definition.

Definition 7. *Let* TS *be an LSTS and* π *and* π' *two paths in* TS *that both start in some state* s. *Then,* TS *is* labelled consistently *iff* $\pi \sim \pi'$ *implies* $\pi \triangleq \pi'$.

Note that if an LSTS is labelled consistently, then in particular all weakly equivalent initial paths are also stutter equivalent. Hence, if an LSTS TS is labelled consistently and weak-trace equivalent to a subgraph TS', then TS and TS' are also stutter-trace equivalent.

Stubborn sets as defined in the previous section aim to preserve stutter-trace equivalence between the original and the reduced LSTS. The motivation behind this is that two stutter-trace equivalent LSTSs satisfy exactly the same formulae [1] in LTL$_{-X}$. The following theorem, which is frequently cited in literature [9,10,18], aims to show that stubborn sets indeed preserve stutter-trace equivalence. Its original formulation reasons about the validity of an arbitrary LTL$_{-X}$ formula. Here, we give the alternative formulation based on stutter-trace equivalence.

Theorem 1. *[17, Theorem 2] Given an LSTS* TS *and a weak/strong stubborn set* r, *then the reduced LSTS* TS_r *is stutter-trace equivalent to* TS.

The original proof correctly concludes that the stubborn set method preserves the order of visible actions in the reduced LSTS, *i.e.*, $TS \sim TS_r$. However, this only implies preservation of stutter-trace equivalence ($TS \triangleq TS_r$) if the full LSTS is labelled consistently, so Theorem 1 is invalid in the general case. In the next section, we will see a counter-example which exploits this fact.

3 Counter-Example

Consider the LSTS in Figure 2, which we will refer to as TS^C. There is only one atomic proposition q, which holds in the grey states and is false in the other states. The initial state $\hat{s}$ is marked with an incoming arrow. First, note that this LSTS is deterministic. The actions a_1, a_2 and a_3 are visible and a and a_{key} are invisible. By setting $r(\hat{s}) = \{a, a_{\mathsf{key}}\}$, which is a weak stubborn set, we obtain a reduced LSTS TS_r^C that does not contain the dashed states and transitions. The original LSTS contains the trace $\emptyset\{q\}\emptyset\emptyset\{q\}^\omega$, obtained by following the path with actions $a_1 a_2 a a_3^\omega$. However, the reduced LSTS does not contain a stutter equivalent trace. This is also witnessed by the LTL$_{-X}$ formula $\Box(q \Rightarrow \Box(q \vee \Box\neg q))$, which holds for TS_r^C, but not for TS^C.

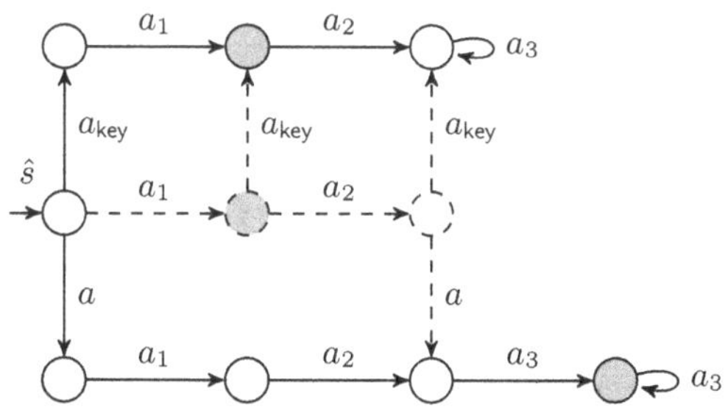

Fig. 2: Counter-example showing that stubborn sets do not preserve stutter-trace equivalence. Grey states are labelled with $\{q\}$. The dashed transitions and states are not present in the reduced LSTS.

A very similar example can be used to show that strong stubborn sets suffer from the same problem. Consider again the LSTS in Figure 2, but assume that $a = a_{\mathsf{key}}$, making the LSTS non-deterministic. Now, $r(\hat{s}) = \{a\}$ is a strong stubborn set and again the trace $\emptyset\{q\}\emptyset\emptyset\{q\}^\omega$ is not preserved in the reduced LSTS. In Section 4.3, we will see why the inconsistent labelling problem does not occur for deterministic systems under strong stubborn sets.

The core of the problem lies in the fact that condition **D1**, even when combined with **V**, does not enforce that the two paths it considers are stutter equivalent. Consider the paths $s \xrightarrow{a}$ and $s \xrightarrow{a_1 a_2 a}$ and assume that $a \in r(s)$ and $a_1 \notin r(s), a_2 \notin r(s)$. Condition **V** ensures that at least one of the following two holds: (i) a is invisible, or (ii) a_1 and a_2 are invisible. Half of the possible scenarios are depicted in Figure 3; the other half are symmetric. Again, the grey states (and only those states) are labelled with $\{q\}$.

The two cases delimited with a solid line are problematic. In both LSTSs, the paths $s \xrightarrow{a_1 a_2 a} s'$ and $s \xrightarrow{a a_1 a_2} s'$ are weakly equivalent, since a is invisible. However, they are not stutter equivalent, and therefore these LSTSs are not labelled consistently. The topmost of these two LSTSs forms the core of the counter-example TS^C, with the rest of TS^C serving to satisfy condition **D2**/**D2w**.

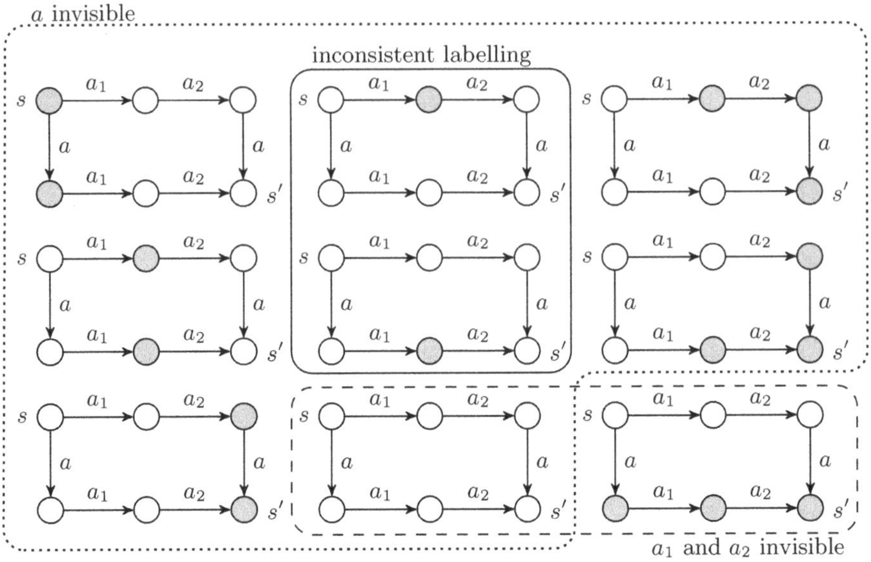

Fig. 3: Nine possible scenarios when $a \in r(s)$ and $a_1 \notin r(s), a_2 \notin r(s)$, according to conditions **D1** and **V**. The dotted and dashed lines indicate when a or a_1, a_2 are invisible, respectively.

4 Strengthening Condition D1

To fix the issue with inconsistent labelling, we propose to strengthen condition **D1** as follows.

D1' For all $a \in r(s)$ and $a_1 \notin r(s), \ldots, a_n \notin r(s)$, if $s \xrightarrow{a_1} s_1 \xrightarrow{a_2} \cdots \xrightarrow{a_n} s_n \xrightarrow{a} s'_n$, then there are states $s', s'_1, \ldots, s'_{n-1}$ such that $s \xrightarrow{a} s' \xrightarrow{a_1} s'_1 \xrightarrow{a_2} \cdots \xrightarrow{a_n} s'_n$. Furthermore, if a is invisible, then $s_i \xrightarrow{a} s'_i$ for every $1 \leq i < n$.

This new condition **D1'** provides a form of *local* consistent labelling when one of $a_1, \ldots, a_n$ is visible. In this case, **V** implies that a is invisible and, consequently, the presence of transitions $s_i \xrightarrow{a} s'_i$ implies $L(s_i) = L(s'_i)$. Hence, the problematic cases of Figure 3 are resolved; a correctness proof is given below.

Condition **D1'** is very similar to condition **C1** [5], which is common in the context of ample sets. However, **C1** requires that action a is *globally* independent of each of the actions $a_1, \ldots, a_n$, while **D1'** merely requires a kind of *local* independence. Persistent sets [7] also rely on a condition similar to **D1'**, and require local independence.

4.1 Implementation

In practice, most, if not all, implementations of stubborn sets approximate **D1** based on a binary relation $\rightsquigarrow_s$ on actions. This relation may (partly) depend on the current state s and it is defined such that **D1** can be satisfied by ensuring that if $a \in r(s)$ and $a \leadsto_s a'$, then also $a' \in r(s)$. A set satisfying **D0**, **D1**, **D2**, **D2w**, **V** and/or **I** can be found by searching for a suitable *strongly connected component* in the graph $(Act, \leadsto_s)$. Condition **L** is dealt with by other techniques.

Practical implementations construct $\leadsto_s$ by analysing how any two actions a and a' interact. If a is enabled, the simplest (but not necessarily the best possible) strategy is to make $a \leadsto_s a'$ if and only if a and a' access at least one variable in common. This can be relaxed, for instance, by not considering commutative accesses, such as writing to and reading from a FIFO buffer. As a result, $\leadsto_s$ can only detect reduction opportunities in (sub)graphs of the shape

$$\begin{array}{ccccccc}
s & \xrightarrow{a_1} & s_1 & - \cdots \rightarrow & s_{n-1} & \xrightarrow{a_n} & s_n \\
\downarrow a & & \downarrow a & & \downarrow a & & \downarrow a \\
s' & \xrightarrow{a_1} & s'_1 & - \cdots \rightarrow & s'_{n-1} & \xrightarrow{a_n} & s'_n
\end{array}$$

where $a \in r(s)$ and $a_1 \notin r(s), \dots, a_n \notin r(s)$. The presence of the vertical a transitions in $s_1, \dots, s_{n-1}$ implies that **D1'** is also satisfied by such implementations.

4.2 Correctness

To show that **D1'** indeed resolves the inconsistent labelling problem, we reproduce the construction in the original proof [17, Construction 1] in two lemmata and show that it preserves stutter equivalence. Below, recall that $\rightarrow_r$ indicates which transitions occur in the reduced state space.

Lemma 1. *Let r be a weak stubborn set, where condition **D1** is replaced by **D1'**, and $\pi = s_0 \xrightarrow{a_1} \cdots \xrightarrow{a_n} s_n \xrightarrow{a} s'_n$ a path such that $a_1 \notin r(s_0), \dots, a_n \notin r(s_0)$ and $a \in r(s_0)$. Then, there is a path $\pi' = s_0 \xrightarrow{a}_r s'_0 \xrightarrow{a_1} \cdots \xrightarrow{a_n} s'_n$ such that $\pi \triangleq \pi'$.*

Proof. The existence of π' follows directly from condition **D1'**. Due to condition **V** and our assumption that $a_1 \notin r(s_0), \dots, a_n \notin r(s_0)$, it cannot be the case that a is visible and at least one of $a_1, \dots, a_n$ is visible. If a is invisible, then the traces of $s_0 \xrightarrow{a_1} \cdots \xrightarrow{a_n} s_n$ and $s'_0 \xrightarrow{a_1} \cdots \xrightarrow{a_n} s'_n$ are equivalent, since **D1'** implies that $s_i \xrightarrow{a} s'_i$ for every $0 \leq i \leq n$, so $L(s'_i) = L(s_i)$. Otherwise, if all of $a_1, \dots, a_n$ are invisible, then the sequences of labels observed along π and π' have the shape $L(s_0)^{n+1} L(s'_0)$ and $L(s_0) L(s'_0)^{n+1}$, respectively. We conclude that $\pi \triangleq \pi'$. □

Lemma 2. *Let r be a weak stubborn set, where condition **D1** is replaced by **D1'**, and $\pi = s_0 \xrightarrow{a_1} s_1 \xrightarrow{a_2} \dots$ a path such that $a_i \notin r(s_0)$ for any a_i that occurs in π. Then, the following holds:*

- *If π is of finite length $n > 0$, there exist an action a_{key}, a state s'_n such that $s_n \xrightarrow{a_{\mathsf{key}}} s'_n$ and a path $\pi' = s_0 \xrightarrow{a_{\mathsf{key}}}_r s'_0 \xrightarrow{a_1} \cdots \xrightarrow{a_n} s'_n$.*
- *If π is infinite, there exists a path $\pi' = s_0 \xrightarrow{a_{\mathsf{key}}}_r s'_0 \xrightarrow{a_1} s'_1 \xrightarrow{a_2} \dots$ for some action a_{key}.*

In either case, $\pi \triangleq \pi'$.

Proof. Let K be the set of key actions in s. If a_1 is invisible, K contains at least one invisible action, due to **I**. Otherwise, if a_1 is visible, we reason that K is not empty (condition **D2w**) and all actions in $r(s_0)$, and thus also all actions in K, are invisible, due to **V**. In the remainder, let a_{key} be an invisible key action.

In case π has finite length n, the existence of $s_n \xrightarrow{a_{\mathsf{key}}} s'_n$ and $s_0 \xrightarrow{a_{\mathsf{key}}}_r s'_0 \xrightarrow{a_1} \cdots \xrightarrow{a_n} s'_n$ follows from the definition of key actions and **D1'**, respectively.

If π is infinite, we can apply the definition of key actions and **D1'** successively to obtain a path $\pi_i = s_0 \xrightarrow{a_{\mathsf{key}}} s'_0 \xrightarrow{a_1} \cdots \xrightarrow{a_i} s'_i$ for every $i \geq 0$, with $s_j \xrightarrow{a_{\mathsf{key}}} s'_j$ for every $1 \leq j < i$. Since the reduced state space is finite, infinitely many of these paths must use the same state as s'_0. At most one of them ends at s'_0 (the one with $i = 0$), so infinitely many continue from s'_0. Of them, infinitely many must use the same s'_1, again because the reduced state space is finite. Again, at most one of them is lost because of ending at s'_1. This reasoning can continue without limit, proving the existence of $\pi' = s_0 \xrightarrow{a_{\mathsf{key}}}_r s'_0 \xrightarrow{a_1} s'_1 \xrightarrow{a_2} \ldots$, with $s_j \xrightarrow{a_{\mathsf{key}}} s'_j$ for every $j \geq 0$.

Since a_{key} is invisible, we have $L(s_j) = L(s'_j)$ for every $j \geq 0$. This implies $\pi \triangleq \pi'$. □

Lemmata 1 and 2 coincide with branches 1 and 2 of [17, Construction 1], respectively, but contain the stronger result that $\pi \triangleq \pi'$. Thus, when applied in the proof of [17, Theorem 2] (see also Theorem 1), this yields the result that stubborn sets with condition **D1'** preserve stutter-trace equivalence.

Theorem 2. *Given an LSTS TS and weak/strong stubborn set r, where condition **D1** is replaced by **D1'**, then the reduced LSTS TS_r is stutter-trace equivalent to TS.*

We do not reproduce the complete proof, but provide insight into the application of the lemmata with the following example.

Example 1. Consider the path obtained by following $a_1a_2a_3$ in Figure 4. Lemmata 1 and 2 show that $a_1a_2a_3$ can always be mimicked in the reduced LSTS, while preserving stutter equivalence. In this case, the path is mimicked by the path corresponding to $a_{\mathsf{key}}a_2a_1a'_{\mathsf{key}}a_3$, drawn with dashes. The new path reorders the actions a_1, a_2 and a_3 according to the construction of Lemma 1 and introduces the key actions a_{key} and a'_{key} according to Lemma 2. □

We remark that Lemma 2 also holds if the reduced LSTS is infinite, but finitely branching.

4.3 Deterministic LSTSs

As already noted in Section 3, strong stubborn sets for deterministic systems do not suffer from the inconsistent labelling problem. The following lemma, which also appeared as [20, Lemma 4.2], shows why.

Lemma 3. *For deterministic LSTSs, conditions **D1** and **D2** together imply **D1'**.*

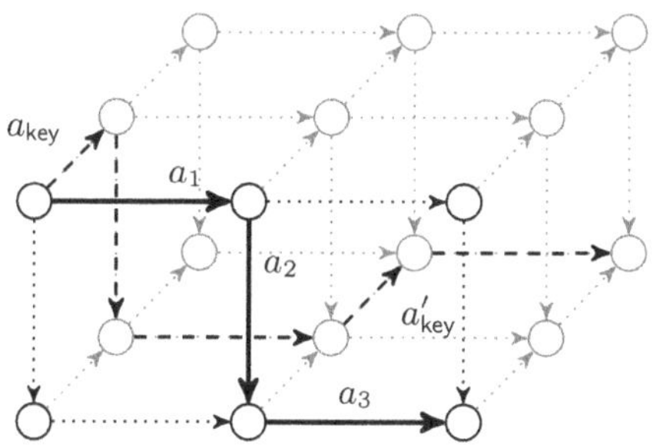

Fig. 4: Example of how the trace a_1, a_2, a_3 can be mimicked by introducing additional actions and moving a_2 to the front (dashed trace). Transitions that are drawn in parallel have the same label.

5 Safe Logics

In this section, we will identify two logics, *viz.* reachability and CTL_{-X}, which are not affected by the inconsistent labelling problem. This is either due to their limited expressivity or the extra POR conditions that are required.

5.1 Reachability properties

Although the counter-example of Section 3 shows that stutter-trace equivalence is in general not preserved by stubborn sets, some fragments of LTL_{-X} are preserved. One such class of properties is reachability properties, which are of the shape $\Box f$ or $\Diamond f$, where f is a formula not containing temporal operators.

Theorem 3. *Let TS be an LSTS, r a reduction function that satisfies either **D0**, **D1**, **D2**, **V** and **L** or **D1**, **D2w**, **V** and **L** and TS_r the reduced LSTS. For all possible labellings $l \subseteq AP$, TS contains a path to a state s such that $L(s) = l$ iff TS_r contains a path to a state s' such that $L(s') = l$.*

Proof. The 'if' case is trivial, since TS_r is a subgraph of TS. For the 'only if' case, we reason as follows. Let $TS = (S, \rightarrow, \hat{s}, L)$ be an LSTS and $\pi = s_0 \xrightarrow{a_1} \cdots \xrightarrow{a_n} s_n$ a path such that $s_0 = \hat{s}$. We mimic this path by repeatedly taking some enabled action a that is in the stubborn set, according to the following schema. Below, we assume the path to be mimicked contains at least one visible action. Otherwise, its first state would have the same labelling as s_n.

1. If there is an i such that $a_i \in r(s_0)$, we consider the smallest such i, *i.e.*, $a_1 \notin r(s_0), \ldots, a_{i-1} \notin r(s_0)$. Then, we can shift a_i forward by **D1**, move towards s_n along $s_0 \xrightarrow{a_i} s'_0$ and continue by mimicking $s'_0 \xrightarrow{a_1} \cdots \xrightarrow{a_{i-1}} s_i \xrightarrow{a_{i+1}} \cdots \xrightarrow{a_n} s_n$.
2. If all of $a_1 \notin r(s_0), \ldots, a_n \notin r(s_0)$, then, by **D0** and **D2** or by **D2w**, there is a key action a_{key} in s_0. By the definition of key actions and **D1**, a_{key} leads to a state s'_0 from which we can continue mimicking the path $s'_0 \xrightarrow{a_1} s'_1 \xrightarrow{a_2} \cdots \xrightarrow{a_n} s'_n$. Note that $L(s_n) = L(s'_n)$, since a_{key} is invisible by condition **V**.

The second case cannot be repeated infinitely often, due to condition **L**. Hence, after a finite number of steps, we reach a state s'_n with $L(s'_n) = L(s_n)$. □

We remark that more efficient mechanisms for reachability checking under POR have been proposed, such as condition **S** [21], which can replace **L**, or conditions based on *up-sets* [13]. Another observation is that model checking of LTL$_{-X}$ properties can be reduced to reachability checking by computing the cross-product of a Büchi automaton and an LSTS [1], in the process resolving the inconsistent labelling problem. Peled [12] shows how this approach can be combined with POR, but please see [14].

5.2 Deterministic LSTSs and CTL$_{-X}$ Model Checking

In this section, we will consider the inconsistent labelling problem in the setting of CTL$_{-X}$ model checking. When applying stubborn sets in that context, stronger conditions are required to preserve the branching structure that CTL$_{-X}$ reasons about. Namely, the original LSTS must be deterministic and one more condition needs to be added [5]:

C4 Either $r(s) = Act$ or $r(s) \cap enabled(s) = \{a\}$ for some $a \in Act$.

We slightly changed its original formulation to match the setting of stubborn sets. A weaker condition, called **Ä8**, which does not require determinism of the whole LSTS is proposed in [19]. With **C4**, strong and weak stubborn sets collapse, as shown by the following lemma.

Lemma 4. *Conditions **D2w** and **C4** together imply **D0** and **D2**.*

Proof. Let TS be an LSTS, s a state and r a reduction function that satisfies **D2w** and **C4**. Condition **D0** is trivially implied by **C4**. Using **C4**, we distinguish two cases: either $r(s)$ contains precisely one enabled action a, or $r(s) = Act$. In the former case, this single action a must be a key action, according to **D2w**. Hence, **D2**, which requires that all enabled actions in $r(s)$ are key actions, is satisfied. Otherwise, if $r(s) = Act$, we consider an arbitrary action a that satisfies **D2**'s precondition that $s \xrightarrow{a}$. Given a path $s \xrightarrow{a_1 \dots a_n}$, the condition that $a_1 \notin r(s), \dots, a_n \notin r(s)$ only holds if $n = 0$. We conclude that **D2**'s condition $s \xrightarrow{a_1 \dots a_n a}$ is satisfied by the assumption $s \xrightarrow{a}$. □

It follows from Lemmata 3 and 4 and Theorem 2 that CTL$_{-X}$ model checking of deterministic systems with stubborn sets does not suffer from the inconsistent labelling problem. The same holds for condition **Ä8**, as already shown in [19].

6 Petri Nets

Petri nets are a widely-known formalism for modelling concurrent processes and have seen frequent use in the application of stubborn-set theory [4,10,21,22]. A Petri net contains a set of *places* P and a set of *structural transitions* T.

Arcs between places and structural transitions are weighted according to a total function $W : (P \times T) \cup (T \times P) \to \mathbb{N}$. The state space of the underlying LSTS is the set $\mathcal{M}$ of all *markings*; a marking m is a function $P \to \mathbb{N}$, which assigns a number of *tokens* to each place. The LSTS contains a transition $m \xrightarrow{t} m'$ iff $m(p) \geq W(p,t)$ and $m'(p) = m(p) - W(p,t) + W(t,p)$ for all places $p \in P$. As before, we assume the LSTS contains some labelling function $L : \mathcal{M} \to 2^{AP}$. More details on the labels are given below. Note that markings and structural transitions take over the role of states and actions respectively. The set of markings reachable under $\to$ from some *initial marking* $\hat{m}$ is denoted $\mathcal{M}_{reach}$.

Example 2. Consider the Petri net with initial marking $\hat{m}$ below on the left. Here, all arcs are weighted 1, except for the arc from p_5 to t_2, which is weighted 2. Its LSTS is infinite, but the reachable substructure is depicted on the right. The number of tokens in each of the places $p_1, \dots, p_6$ is inscribed in the nodes, the state labels (if any) are written beside the nodes.

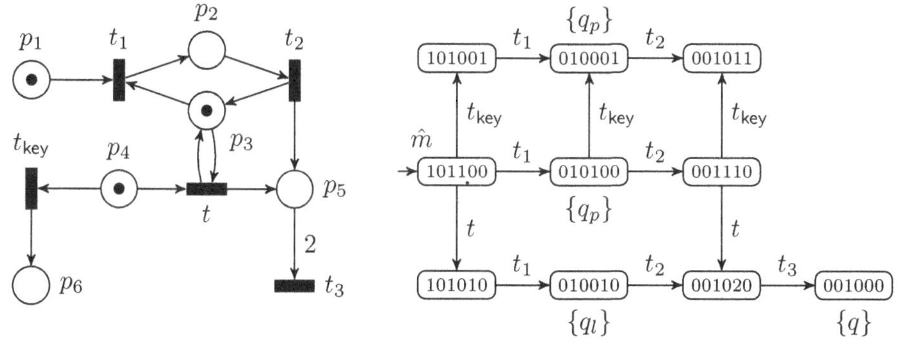

The LSTS practically coincides with the counter-example of Section 3. Only the self-loops are missing and the state labelling, with atomic propositions q, q_p and q_l, differs slightly; the latter will be explained later. For now, note that t and t_{key} are invisible and that the trace $\emptyset\{q_p\}\emptyset\emptyset\{q\}$, which occurs when firing transitions $t_1 t_2 t t_3$ from $\hat{m}$, can be lost when reducing with weak stubborn sets. □

In the remainder of this section, we fix a Petri net $(P, T, W, \hat{m})$ and its LSTS $(\mathcal{M}, \to, \hat{m}, L)$. Below, we consider three different types of atomic propositions. Firstly, polynomial propositions [4] are of the shape $f(p_1, \dots, p_n) \bowtie k$ where f is a polynomial over $p_1, \dots, p_n$, $\bowtie \in \{<, \leq, >, \geq, =, \neq\}$ and $k \in \mathbb{Z}$. Such a proposition holds in a marking m iff $f(m(p_1), \dots, m(p_n)) \bowtie k$. A linear proposition [10] is similar, but the function f over places must be linear and $f(0, \dots, 0) = 0$, *i.e.*, linear propositions are of the shape $k_1 p_1 + \dots + k_n p_n \bowtie k$, where $k_1, \dots, k_n, k \in \mathbb{Z}$. Finally, we have arbitrary propositions [22], whose shape is not restricted and which can hold in any given set of markings.

Several other types of atomic propositions can be encoded as polynomial propositions. For example, *fireable*(t) [4,10], which holds in a marking m iff t is enabled in m, can be encoded as $\prod_{p \in P} \prod_{i=0}^{W(p,t)-1} (p - i) \geq 1$. The proposition *deadlock*, which holds in markings where no structural transition is enabled, does not require special treatment in the context of POR, since it is already preserved by **D1** and **D2w**. The sets containing all linear and polynomial propositions are henceforward called AP_l and AP_p, respectively. The corresponding labelling functions are defined as $L_l(m) = L(m) \cap AP_l$ and $L_p(m) = L(m) \cap AP_p$ for all markings m. Below, the two stutter equivalences $\triangleq_{L_l}$ and $\triangleq_{L_p}$ that follow from the new labelling functions are abbreviated $\triangleq_l$ and $\triangleq_p$, respectively. Note that $AP \supseteq AP_p \supseteq AP_l$ and $\triangleq \subseteq \triangleq_p \subseteq \triangleq_l$.

For the purpose of introducing several variants of invisibility, we reformulate and generalise the definition of invisibility from Section 2. Given an atomic proposition $q \in AP$, a relation $\mathcal{R} \subseteq \mathcal{M} \times \mathcal{M}$ is *q-invisible* if and only if $(m, m') \in \mathcal{R}$ implies $q \in L(m) \Leftrightarrow q \in L(m')$. We consider a structural transition t q-invisible iff its corresponding relation $\{(m, m') \mid m \xrightarrow{t} m'\}$ is q-invisible. Invisibility is also lifted to sets of atomic propositions: given a set $AP' \subseteq AP$, relation $\mathcal{R}$ is *AP′-invisible* iff it is q-invisible for all $q \in AP'$. If $\mathcal{R}$ is AP-invisible, we plainly say that $\mathcal{R}$ is *invisible*. AP'-invisibility and invisibility carry over to structural transitions. We sometimes refer to invisibility as *ordinary invisibility* for emphasis. Note that the set of invisible structural transitions $\mathcal{I}$ is no longer an under-approximation, but contains exactly those structural transitions t for which $m \xrightarrow{t} m'$ implies $L(m) = L(m')$ (cf. Section 2).

We are now ready to introduce three orthogonal variations on invisibility. Firstly, relation $\mathcal{R} \subseteq \mathcal{M} \times \mathcal{M}$ is *reach q-invisible* [21] iff $\mathcal{R} \cap (\mathcal{M}_{reach} \times \mathcal{M}_{reach})$ is q-invisible, *i.e.*, all the pairs of reachable markings $(m, m') \in \mathcal{R}$ agree on the labelling of q. Secondly, $\mathcal{R}$ is *value q-invisible* if (i) q is polynomial and for all $(m, m') \in \mathcal{R}$, $f(m(p_1), \ldots, m(p_n)) = f(m'(p_1), \ldots, m'(p_n))$; or if (ii) q is not polynomial and $\mathcal{R}$ is q-invisible. Intuitively, this means that the value of polynomial f never changes between two markings $(m, m') \in \mathcal{R}$. Reach and value invisibility are lifted to structural transitions and sets of atomic propositions as before, *i.e.*, by taking $\mathcal{R} = \{(m, m') \mid m \xrightarrow{t} m'\}$ when considering invisibility of t. Finally, we introduce another way to lift invisibility to structural transitions: t is *strongly q-invisible* iff the set $\{(m, m') \mid \forall p \in P : m'(p) = m(p) + W(t, p) - W(p, t)\}$ is q-invisible. Strong invisibility does not take the presence of a transition $m \xrightarrow{t} m'$ into account, and purely reasons about the effects of t. Value invisibility and strong invisibility are new in the current work, although strong invisibility was inspired by the notion of invisibility that is proposed by Varpaaniemi in [22].

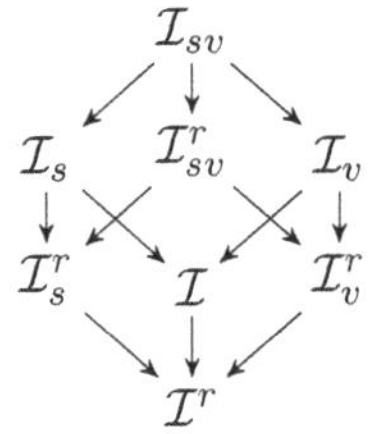

Fig. 5: Lattice of sets of invisible actions. Arrows represent a subset relation.

We indicate the sets of all value, reach and strongly invisible structural transitions with $\mathcal{I}_v$, $\mathcal{I}^r$ and $\mathcal{I}_s$ respectively. Since $\mathcal{I}_v \subseteq \mathcal{I}$, $\mathcal{I}_s \subseteq \mathcal{I}$ and $\mathcal{I} \subseteq \mathcal{I}^r$, the set of all their possible combinations forms the lattice shown in Figure 5. In the remainder, the weak equivalence relations that follow from each of the eight invisibility notions are abbreviated, *e.g.*, $\sim_{\mathcal{I}^r_{sv}}$ becomes $\sim^r_{sv}$.

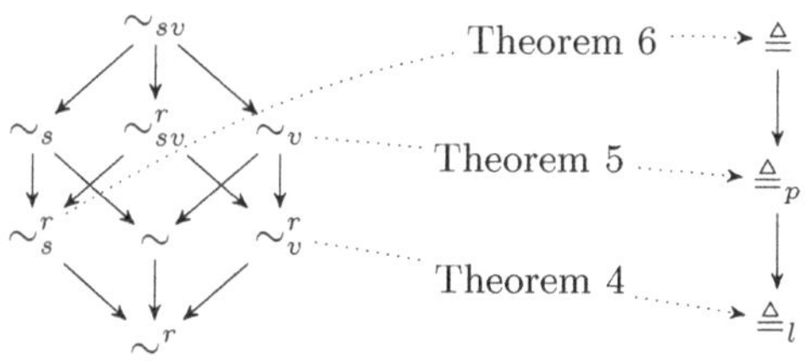

Fig. 6: Two lattices containing variations of weak equivalence and stutter equivalence, respectively. Solid arrows indicate a subset relation inside the lattice; dotted arrows follow from the indicated theorems and show when the LSTS of a Petri net is labelled consistently.

Example 3. Consider again the Petri net and LSTS from Example 2. We can define q_l and q_p as linear and polynomial propositions, respectively:

- $q_l := p_3 + p_4 + p_6 = 0$ is a linear proposition, which holds when neither p_3, p_4 nor p_6 contains a token. Structural transition t is q_l-invisible, because $m \xrightarrow{t} m'$ implies that $m(p_3) = m'(p_3) \geq 1$, and thus neither m nor m is labelled with q_l. On the other hand, t is not value q_l-invisible (by the transition $101100 \xrightarrow{t} 101010$) or strongly reach q_l-invisible (by 010100 and 010010). However, t_{key} is strongly value q_l-invisible: it moves a token from p_4 to p_6 and hence never changes the value of $p_3 + p_4 + p_6$.
- $q_p := (1 - p_3)(1 - p_5) = 1$ is a polynomial proposition, which holds in all reachable markings m where $m(p_3) = 0$ and $m(p_5) = 0$. Structural transition t is reach value q_p-invisible, but not q_p-invisible (by $002120 \xrightarrow{t} 002030$) or strongly reach q_p invisible. Strong value q_p-invisibility of t_{key} follows immediately from the fact that the adjacent places of t_{key}, *viz.* p_4 and p_6, do not occur in the definition of q_p.

This yields the state labelling which is shown in Example 2. □

Given a weak equivalence relation $R_\sim$ and a stutter equivalence relation $R_{\triangleq}$, we write $R_\sim \preceq R_{\triangleq}$ to indicate that $R_\sim$ and $R_{\triangleq}$ yield consistent labelling. We spend the rest of this section investigating under which notions of invisibility and propositions from the literature, the LSTS of a Petri net is labelled consistently. More formally, we check for each weak equivalence relation $R_\sim$ and each stutter equivalence relation $R_{\triangleq}$ whether $R_\sim \preceq R_{\triangleq}$. This tells us when existing stubborn set theory can be applied without problems. The two lattices containing all weak and stuttering equivalence relations are depicted in Figure 6; each dotted arrow represents a consistent labelling result. Before we continue, we first introduce an auxiliary lemma.

Lemma 5. *Let I be a set of invisible structural transitions and L some labelling function. If for all $t \in I$ and paths $\pi = m_0 \xrightarrow{t_1} m_1 \xrightarrow{t_2} \dots$ and $\pi' = m_0 \xrightarrow{t} m_0' \xrightarrow{t_1} m_1' \xrightarrow{t_2} \dots$, it holds that $\pi \triangleq_L \pi'$, then $\sim_I \preceq \triangleq_L$.*

Proof. We assume that the following holds for all paths and $t \in I$:

$$m_0 \xrightarrow{t_1} m_1 \xrightarrow{t_2} \cdots \triangleq_L m_0 \xrightarrow{t} m_0' \xrightarrow{t_1} m_1' \xrightarrow{t_2} \ldots \tag{\dagger}$$

We consider two initial paths π and π' such that $\pi \sim_I \pi'$ and prove that $\pi \triangleq_L \pi'$. The proof proceeds by induction on the combined number of invisible structural transitions (taken from I) in π and π'. In the base case, π and π' contain only visible structural transitions, and $\pi \sim_I \pi$' implies $\pi = \pi'$ since Petri nets are deterministic. Hence, $\pi \triangleq_L \pi'$.

For the induction step, we take as hypothesis that, for all initial paths π and π' that together contain at most k invisible structural transitions, $\pi \sim_I \pi'$ implies $\pi \triangleq_L \pi'$. Let π and π' be two arbitrary initial paths such that $\pi \sim_I \pi'$ and the total number of invisible structural transitions contained in π and π' is k. We consider the case where an invisible structural transition is introduced in π', the other case is symmetric. Let $\pi' = \sigma_1\sigma_2$ for some σ_1 and σ_2. Let $t \in I$ be some invisible structural transition and $\pi'' = \sigma_1 t \sigma_2'$ such that σ_2 and σ_2' contain the same sequence of structural transitions. Clearly, we have $\pi' \sim_I \pi''$. Here, we can apply our original assumption (†), to conclude that $\sigma_2 \triangleq t\sigma_2'$, *i.e.*, the extra stuttering step t thus does not affect the labelling of the remainder of π''. Hence, we have $\pi' \triangleq_L \pi''$ and, with the induction hypothesis, $\pi \triangleq_L \pi''$. Note that π and π'' together contain $k + 1$ invisible structural transitions.

In case π and π' together contain an infinite number of invisible structural transitions, $\pi \sim_I \pi'$ implies $\pi \triangleq_L \pi'$ follows from the fact that the same holds for all finite prefixes of π and π' that are related by $\sim_I$. □

The following theorems each focus on a class of atomic propositions and show which notion of invisibility is required for the LSTS of a Petri net to be labelled consistently. In the proofs, we use a function d_t, defined as $d_t(p) = W(t, p) - W(p, t)$ for all places p, which indicates how structural transition t changes the state. Furthermore, we also consider functions of type $P \to \mathbb{N}$ as vectors of type $\mathbb{N}^{|P|}$. This allows us to compute the pairwise addition of a marking m with d_t $(m + d_t)$ and to indicate that t does not change the marking $(d_t = 0)$.

Theorem 4. *Under reach value invisibility, the LSTS underlying a Petri net is labelled consistently for linear propositions,* i.e., $\sim_v^r \preceq \triangleq_l$.

Proof. Let $t \in \mathcal{I}_v^r$ be a reach value invisible structural transition such that there exist reachable markings m and m' with $m \xrightarrow{t} m'$. If such a t does not exist, then $\sim_v^r$ is the reflexive relation and $\sim_v^r \preceq \triangleq_l$ is trivially satisfied. Otherwise, let $q := f(p_1, \ldots, p_n) \bowtie k$ be a linear proposition. Since t is reach value invisible and f is linear, we have $f(m) = f(m') = f(m + d_t) = f(m) + f(d_t)$ and thus $f(d_t) = 0$. It follows that, given two paths $\pi = m_0 \xrightarrow{t_1} m_1 \xrightarrow{t_2} \ldots$ and $\pi' = m_0 \xrightarrow{t} m_0' \xrightarrow{t_1} m_1' \xrightarrow{t_2} \ldots$, the addition of t does not influence f, since $f(m_i) = f(m_i) + f(d_t) = f(m_i + d_t) = f(m_i')$ for all i. As a consequence, t also does not influence q. With Lemma 5, we deduce that $\sim_v^r \preceq \triangleq_l$. □

Whereas in the linear case one can easily conclude that π and π' are stutter equivalent under f, in the polynomial case, we need to show that f is constant under all value invisible structural transitions t, even in markings where t is not enabled. This follows from the following proposition.

Proposition 1. *Let $f : \mathbb{N}^n \to \mathbb{Z}$ be a polynomial function, $a, b \in \mathbb{N}^n$ two constant vectors and $c = a - b$ the difference between a and b. Assume that for all $x \in \mathbb{N}^n$ such that $x \geq b$, where $\geq$ denotes pointwise comparison, it holds that $f(x) = f(x + c)$. Then, f is constant in the vector c,* i.e., *$f(x) = f(x + c)$ for all $x \in \mathbb{N}^n$.*

Proof. Let f, a, b and c be as above and let $\mathbf{1} \in \mathbb{N}^n$ be the vector containing only ones. Given some arbitrary $x \in \mathbb{N}^n$, consider the function $g_x(t) = f(x + t \cdot \mathbf{1} + c) - f(x + t \cdot \mathbf{1})$. For sufficiently large t, it holds that $x + t \cdot \mathbf{1} \geq b$, and it follows that $g_x(t) = 0$ for all sufficiently large t. This can only be the case if g_x is the zero polynomial, *i.e.*, $g_x(t) = 0$ for all t. As a special case, we conclude that $g_x(0) = f(x + c) - f(x) = 0$. □

The intuition behind this is that $f(x + c) - f(x)$ behaves like the directional derivative of f with respect to c. If the derivative is equal to zero in infinitely many x, f must be constant in the direction of c. We will apply this result in the following theorem.

Theorem 5. *Under value invisibility, the LSTS underlying a Petri net is labelled consistently for polynomial propositions,* i.e., $\sim_v \preceq \triangleq_p$.

Proof. Let $t \in \mathcal{I}_v$ be a value invisible structural transition, m and m' two markings with $m \xrightarrow{t} m'$, and $q := f(p_1, \ldots, p_n) \bowtie k$ a polynomial proposition. Note that infinitely many such (not necessarily reachable) markings exist in $\mathcal{M}$, so we can apply Proposition 1 to obtain $f(m) = f(m + d_t)$ for all markings m. It follows that, given two paths $\pi = m_0 \xrightarrow{t_1} m_1 \xrightarrow{t_2} \ldots$ and $\pi' = m_0 \xrightarrow{t} m'_0 \xrightarrow{t_1} m'_1 \xrightarrow{t_2} \ldots$, the addition of t does not alter the value of f, since $f(m_i) = f(m_i + d_t) = f(m'_i)$ for all i. As a consequence, t also does not change the labelling of q. Application of Lemma 5 yields $\sim_v \preceq \triangleq_p$. □

Varpaaniemi shows that the LSTS of a Petri net is labelled consistently for arbitrary propositions under his notion of invisibility [22, Lemma 9]. Our notion of strong visibility, and especially strong reach invisibility, is weaker than Varpaaniemi's invisibility, so we generalise the result to $\sim^r_s \preceq \triangleq$.

Theorem 6. *Under strong reach visibility, the LSTS underlying a Petri net is labelled consistently for arbitrary propositions,* i.e., $\sim^r_s \preceq \triangleq$.

Proof. Let $t \in \mathcal{I}^r_s$ be a strongly reach invisible structural transition and $\pi = m_0 \xrightarrow{t_1} m_1 \xrightarrow{t_2} \ldots$ and $\pi' = m_0 \xrightarrow{t} m'_0 \xrightarrow{t_1} m'_1 \xrightarrow{t_2} \ldots$ two paths. Since, $m'_i = m_i + d_t$ for all i, it holds that either (i) $d_t = 0$ and $m_i = m'_i$ for all i; or (ii) each pair (m_i, m'_i) is contained in $\{(m, m') \mid \forall p \in P : m'(p) = m(p) + W(t, p) - W(p, t)\}$, which is the set that underlies strong reach invisibility of t. In both cases, $L(m_i) = L(m'_i)$ for all i. It follows from Lemma 5 that $\sim^r_s \preceq \triangleq$. □

To show that the results of the above theorems cannot be strengthened, we provide two negative results.

Theorem 7. *Under ordinary invisibility, the LSTS underlying a Petri net is not necessarily labelled consistently for arbitrary propositions,* i.e., $\sim \not\subseteq \triangleq$.

Proof. Consider the Petri net from Example 2 with the arbitrary proposition q_l. Disregard q_p for the moment. Structural transition t is q_l-invisible, hence the paths corresponding to $t_1t_2tt_3$ and $tt_1t_2t_3$ are weakly equivalent under ordinary invisibility. However, they are not stutter equivalent. □

Theorem 8. *Under reach value invisibility, the LSTS underlying a Petri net is not necessarily labelled consistently for polynomial propositions,* i.e., $\sim_v^r \not\subseteq \triangleq_p$.

Proof. Consider the Petri net from Example 2 with the polynomial proposition $q_p := (1-p_3)(1-p_5) = 1$ from Example 3. Disregard q_l in this reasoning. Structural transition t is reach value q_p-invisible, hence the paths corresponding to $t_1t_2tt_3$ and $tt_1t_2t_3$ are weakly equivalent under reach value invisibility. However, they are not stutter equivalent for polynomial propositions. □

It follows from Theorems 7 and 8 and transitivity of $\subseteq$ that Theorems 4, 5 and 6 cannot be strengthened further. In terms of Figure 6, this means that the dotted arrows cannot be moved downward in the lattice of weak equivalences and cannot be moved upward in the lattice of stutter equivalences. The implications of these findings on related work will be discussed in the next section.

7 Related Work

There are many works in literature that apply stubborn sets. We will consider several works that aim to preserve LTL$_{-X}$ and discuss whether they are correct when it comes to the problem presented in the current work.

Liebke and Wolf [10] present an approach for efficient CTL model checking on Petri nets. For some formulas, they can reduce CTL model checking to LTL model checking, which allows greater reductions under POR. They rely on the incorrect LTL preservation theorem, and since they apply the techniques on Petri nets with ordinary invisibility, their theory is incorrect (Theorem 7). Similarly, the overview of stubborn set theory presented by Valmari and Hansen in [21] applies reach invisibility and does not necessarily preserve LTL$_{-X}$. Varpaaniemi [22] also applies stubborn sets to Petri nets, but relies on a visibility notion that is stronger than strong invisibility. The correctness of these results is thus not affected (Theorem 6). The approach of Bønneland *et al.* [4] operates on two-player Petri nets, but only aims to preserve reachability and consequently does not suffer from the inconsistent labelling problem.

A generic implementation of weak stubborn sets is proposed by Laarman *et al.* [9]. They use abstract concepts such as guards and transition groups to implement POR in a way that is agnostic of the input language. The theory they present includes condition **D1**, which is too weak, but the accompanying implementation follows the framework of Section 4.1, and thus it is correct by Theorem 2 The implementations proposed in [21,23] are similar, albeit specific for Petri nets.

Others [6,8] perform action-based model checking and thus strive to preserve weak trace equivalence or inclusion. As such, they do not suffer from the problems discussed here, which applies only to state labels.

Although Beneš *et al.* [2,3] rely on ample sets, and not on stubborn sets, they also discuss weak trace equivalence and stutter-trace equivalence. In fact, they present an equivalence relation for traces that is a combination of weak and stutter equivalence. The paper includes a theorem that weak equivalence implies their new state/event equivalence [2, Theorem 6.5]. However, the counter-example on the right shows that this consistent labelling theorem does not hold. Here, the action τ is invisible, and the two paths in this transition system are thus weakly equivalent. However, they are not stutter equivalent, which is a special case of state/event equivalence. Although the main POR correctness result [2, Corollary 6.6] builds on the incorrect consistent labelling theorem, its correctness does not appear to be affected. An alternative proof can be constructed based on Lemmas 1 and 2.

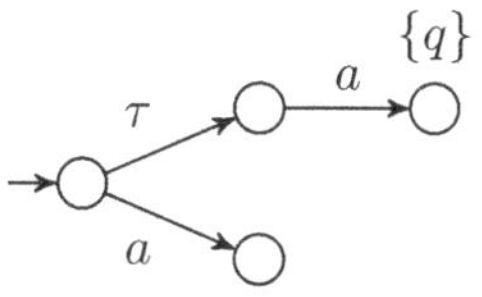

The current work is not the first to point out mistakes in POR theory. In [14], Siegel presents a flaw in an algorithm that combines POR and on-the-fly model checking [12]. In that setting, POR is applied on the product of an LSTS and a Büchi automaton. Let q be a state of the LSTS and s a state of the Büchi automaton. While investigating a transition $(q, s) \xrightarrow{a} (q', s')$, condition **C3**, which—like condition **L**—aims to solve the action ignoring problem, incorrectly sets $r(q, s') = \mathit{enabled}(q)$ instead of $r(q, s) = \mathit{enabled}(q)$.

8 Conclusion

We discussed the inconsistent labelling problem for preservation of stutter-trace equivalence with stubborn sets. The issue is relatively easy to repair by strengthening condition **D1**. For Petri nets, altering the definition of invisibility can also resolve inconsistent labelling depending on the type of atomic propositions. The impact on applications presented in related works seems to be limited: the problem is typically mitigated in the implementation, since it is very hard to compute **D1** exactly. This is also a possible explanation for why the inconsistent labelling problem has not been noticed for so many years.

Since this is not the first error found in POR theory [14], a more rigorous approach to proving its correctness, *e.g.* using proof assistants, would provide more confidence.

References

1. Baier, C., Katoen, J.P.: Principles of model checking. MIT Press (2008)

2. Beneš, N., Brim, L., Buhnova, B., Ern, I., Sochor, J., Vařeková, P.: Partial order reduction for state/event LTL with application to component-interaction automata. Science of Computer Programming **76**(10), 877–890 (2011). https://doi.org/10.1016/j.scico.2010.02.008
3. Beneš, N., Brim, L., Černá, I., Sochor, J., Vařeková, P., Zimmerova, B.: Partial Order Reduction for State/Event LTL. In: IFM 2009. LNCS, vol. 5423, pp. 307–321 (2009). https://doi.org/10.1007/978-3-642-00255-7_21
4. Bønneland, F.M., Jensen, P.G., Larsen, K.G., Muñiz, M.: Partial Order Reduction for Reachability Games. In: CONCUR 2019. vol. 140, pp. 23:1–23:15 (2019). https://doi.org/10.4230/LIPIcs.CONCUR.2019.23
5. Gerth, R., Kuiper, R., Peled, D., Penczek, W.: A Partial Order Approach to Branching Time Logic Model Checking. Information and Computation **150**(2), 132–152 (1999). https://doi.org/10.1006/inco.1998.2778
6. Gibson-Robinson, T., Hansen, H., Roscoe, A.W., Wang, X.: Practical Partial Order Reduction for CSP. In: NFM 2015. LNCS, vol. 9058, pp. 188–203 (2015). https://doi.org/10.1007/978-3-319-17524-9_14
7. Godefroid, P.: Partial-Order Methods for the Verification of Concurrent Systems, LNCS, vol. 1032. Springer (1996). https://doi.org/10.1007/3-540-60761-7
8. Hansen, H., Lin, S., Liu, Y., Nguyen, T.K., Sun, J.: Diamonds Are a Girl's Best Friend: Partial Order Reduction for Timed Automata with Abstractions. In: CAV 2014. LNCS, vol. 8559, pp. 391–406 (2014). https://doi.org/10.1007/978-3-319-08867-9_26
9. Laarman, A., Pater, E., van de Pol, J., Hansen, H.: Guard-based partial-order reduction. STTT **18**(4), 427–448 (2016). https://doi.org/10.1007/s10009-014-0363-9
10. Liebke, T., Wolf, K.: Taking Some Burden Off an Explicit CTL Model Checker. In: Petri Nets 2019. LNCS, vol. 11522, pp. 321–341 (2019). https://doi.org/10.1007/978-3-030-21571-2_18
11. Peled, D.: All from One, One for All: on Model Checking Using Representatives. In: CAV 1993. LNCS, vol. 697, pp. 409–423 (1993). https://doi.org/10.1007/3-540-56922-7_34
12. Peled, D.: Combining partial order reductions with on-the-fly model-checking. FMSD **8**(1), 39–64 (1996). https://doi.org/10.1007/BF00121262
13. Schmidt, K.: Stubborn sets for model checking the EF/AG fragment of CTL. Fundamenta Informaticae **43**(1-4), 331–341 (2000)
14. Siegel, S.F.: What's Wrong with On-the-Fly Partial Order Reduction. In: CAV 2019. LNCS, vol. 11562, pp. 478–495 (2019). https://doi.org/10.1007/978-3-030-25543-5_27
15. Valmari, A.: A Stubborn Attack on State Explosion. In: CAV 1990. LNCS, vol. 531, pp. 156–165 (1991). https://doi.org/10.1007/BFb0023729
16. Valmari, A.: Stubborn sets for reduced state space generation. In: Advances in Petri Nets. vol. 483, pp. 491–515 (1991). https://doi.org/10.1007/3-540-53863-1_36
17. Valmari, A.: A Stubborn Attack on State Explosion. Formal Methods in System Design **1**(4), 297–322 (1992). https://doi.org/10.1007/BF00709154
18. Valmari, A.: The state explosion problem. In: ACPN 1996. LNCS, vol. 1491, pp. 429–528 (1996). https://doi.org/10.1007/3-540-65306-6_21
19. Valmari, A.: Stubborn Set Methods for Process Algebras. In: POMIV 1996. DIMACS, vol. 29, pp. 213–231 (1997). https://doi.org/10.1090/dimacs/029/12
20. Valmari, A.: Stop It, and Be Stubborn! TECS **16**(2), 46:1–46:26 (2017). https://doi.org/10.1145/3012279

21. Valmari, A., Hansen, H.: Stubborn Set Intuition Explained. In: ToPNoC XII. LNCS, vol. 10470, pp. 140–165 (2017). https://doi.org/10.1007/978-3-662-55862-1_7
22. Varpaaniemi, K.: On Stubborn Sets in the Verification of Linear Time Temporal Properties. FMSD **26**(1), 45–67 (2005). https://doi.org/10.1007/s10703-005-4594-y
23. Wolf, K.: Petri Net Model Checking with LoLA 2. In: Petri Nets 2018. LNCS, vol. 10877, pp. 351–362 (2018). https://doi.org/10.1007/978-3-319-91268-4_18

Semantical Analysis of Contextual Types

Brigitte Pientka[1] and Ulrich Schöpp[2](✉)

[1] McGill University, Montreal, Canada, bpientka@cs.mcgill.ca
[2] fortiss GmbH, Munich, Germany, schoepp@fortiss.org

Abstract. We describe a category-theoretic semantics for a simply typed variant of Cocon, a contextual modal type theory where the box modality mediates between the weak function space that is used to represent higher-order abstract syntax (HOAS) trees and the strong function space that describes (recursive) computations about them. What makes Cocon different from standard type theories is the presence of first-class contexts and contextual objects to describe syntax trees that are closed with respect to a given context of assumptions. Following M. Hofmann's work, we use a presheaf model to characterise HOAS trees. Surprisingly, this model already provides the necessary structure to also model Cocon. In particular, we can capture the contextual objects of Cocon using a comonad $\flat$ that restricts presheaves to their closed elements. This gives a simple semantic characterisation of the invariants of contextual types (e.g. substitution invariance) and identifies Cocon as a type-theoretic syntax of presheaf models. We express our category-theoretic constructions by using a modal internal type theory that is implemented in Agda-Flat.

1 Introduction

A fundamental question when defining, implementing, and working with languages and logics is: How do we represent and analyse syntactic structures? Higher-order abstract syntax [19] (or lambda-tree syntax [17]) provides a deceptively simple answer to this question. The basic idea to represent syntactic structures is to map uniformly binding structures in our object language (OL) to the function space in a meta-language thereby inheriting α-renaming and capture-avoiding substitution. In the logical framework LF [10], for example, we can define a small functional programming language consisting of functions, function application, and let-expressions using a type `tm` as follows:

```
lam : (tm → tm) → tm.          letv: tm → (tm → tm) → tm.
app : tm → tm → tm.
```

The object-language term $(\mathsf{lam}\ x.\, \mathsf{lam}\ y.\, \mathsf{let}\ w = x\ y\ \mathsf{in}\ w\ y)$ is then encoded as `lam λx.lam λy.letv (app x y) λw.app w y` using the LF abstractions to model binding. Object-level substitution is modelled through LF application; for instance, the fact that $((\mathsf{lam}\ x.M)\ N)$ reduces to $[N/x]M$ in our object language is expressed as `(app (lam M) N)` reducing to `(M N)`.

This approach is elegant and can offer substantial benefits: we can treat objects equivalent modulo renaming and do not need to define object-level substitution.

J. Goubault-Larrecq and B. König (Eds.): FOSSACS 2020, LNCS 12077, pp. 502–521, 2020.
https://doi.org/10.1007/978-3-030-45231-5_26

However, we not only want to just construct HOAS trees, but also to analyse them and to select sub-trees. This is challenging, as sub-trees are context sensitive. For example, the term `letv (app x y) λw.app w y` only makes sense in a context `x:tm,y:tm`. Moreover, one cannot simply extend LF to allow syntax analysis. If one simply added a recursion combinator to LF, then it could be used to define many functions `M: tm → tm` for which `lam M` would not represent an object-level syntax term [12].

Contextual types [18,20] offer a type-theoretic solution to these problems by reifying the typing judgement, i.e. that `letv (app x y) λw.app w y` has type `tm` in the context `x:tm,y:tm`, as a *contextual type* $\lceil x{:}\mathtt{tm}, y{:}\mathtt{tm} \vdash \mathtt{tm}\rceil$. The contextual type $\lceil x{:}\mathtt{tm}, y{:}\mathtt{tm} \vdash \mathtt{tm}\rceil$ describes a set of terms of type `tm` that may contain variables x and y. In particular, the contextual object $\lceil x, y \vdash \mathtt{letv}\ (\mathtt{app}\ x\ y)\ \lambda w.\mathtt{app}\ w\ y\rceil$ has the given contextual type. By abstracting over contexts and treating contexts as first-class, we can now recursively analyse HOAS trees [20,25,21]. Recently, [23] further generalised these ideas and presented a contextual modal type theory, Cocon, where we can mix HOAS trees and computations, i.e. we can use (recursive) computations to analyse and traverse (contextual) HOAS trees and we can embed computations within HOAS trees. This line of work provides a syntactic perspective to the question of how to represent and analyse syntactic structures with binders, as it focuses on decidability of type checking and normalisation. However, its semantics remains not well-understood. What is the semantic meaning of a contextual type? Can we semantically justify the given induction principles? What is the semantics of a first-class context?

While a number of closely related categorical models of abstract syntax with bindings [12,8,9] were proposed around 2000, the relationship of these models to concrete type-theoretic languages for computing with HOAS structures was teneous. In this paper, we give a category-theoretic semantics for Cocon (for simply-typed HOAS). This provides semantic perspective of contextual types and first-class contexts. Maybe surprisingly, the presheaf model introduced by Hofmann [12] already provides the necessary structure to also model contextual modal type theory. Besides the standard structure of this model, we only need two additional concepts: a $\flat$-modality and a cartesian closed universe of representables. For simplicity and lack of space, we focus on the special case of Cocon where the HOAS trees are simply-typed. Concentrating on the simply-typed setting allows us to introduce the main idea without the additional complexity that type dependencies bring with them. We outline the dependently-typed case in Sec. 6.

Our work provides a semantic foundation to Cocon and can serve as a starting point to investigate connections to similar work. First, our work connects Cocon to other work on internal languages for presheaf categories with a $\flat$-modality, such as spatial type theory [27] or crisp type theory [16]. Second, it may help to understand the relations of Cocon to type theories that use a modality for metaprogramming and intensional recursion, such as [15]. While Cocon is built on the same general ideas, a main difference seems to be that Cocon distinguishes between HOAS trees and computations, even though it allows mixed use of them. We hope to clarify the relation by providing a semantical perspective.

2 Presheaves for Higher-Order Abstract Syntax

Our work begins with the presheaf models for HOAS of [12,8]. The key idea of those approaches is to integrate substitution-invariance in the computational universe in a controlled way. For the representation of abstract syntax, one wants to allow only substitution-invariant constructions. For example, `lam M` represents an object-level abstraction if and only if `M` is a function that uses its argument in a substitution-invariant way. For computation with abstract syntax, on the other hand, one wants to allow non-substitution-invariant constructions too. Presheaf categories allow one to choose the desired amount of substitution-invariance.

Let $\mathbb{D}$ be a small category. The presheaf category $\widehat{\mathbb{D}}$ is defined to be the category $\mathrm{Set}^{\mathbb{D}^{\mathrm{op}}}$. Its objects are functors $F\colon \mathbb{D}^{\mathrm{op}} \to \mathrm{Set}$, which are also called *presheaves*. Such a functor F is given by a set $F(\Psi)$ for each object Ψ of $\mathbb{D}$ together with a function $F(\sigma)\colon F(\Phi) \to F(\Psi)$ for any object Φ and $\sigma\colon \Psi \to \Phi$ in $\mathbb{D}$, subject to the functor laws. The intuition is that F defines sets of elements in various $\mathbb{D}$-contexts, together with a $\mathbb{D}$-substitution action. A morphism $f\colon F \to G$ is a natural transformation, which is a family of functions $f_\Psi\colon F(\Psi) \to G(\Psi)$ for any Ψ. This family of functions must be natural, i.e. commute with substitution $f_\Psi \circ F(\sigma) = F(\sigma) \circ f_\Phi$.

For the purposes of modelling higher-order abstract syntax, $\mathbb{D}$ will typically be the term model of some domain-level lambda-calculus. By domain-level, we mean the calculus that serves as the meta-level for object-language encodings. It is the calculus that contains constants like `lam` and `app` from the Introduction. We call it domain-level to avoid possible confusion between different meta-levels later. For simplicity, let us for now use a simply-typed lambda-calculus with functions and products as the domain language. It is sufficient to encode the example from the Introduction and allows us to explain the main idea underlying our approach.

The term model of the simply-typed domain-level lambda-calculus forms a cartesian closed category $\mathbb{D}$. The objects of $\mathbb{D}$ are contexts $x_1\colon A_1, \dots, x_n\colon A_n$ of simple types. We use Φ and Ψ to range over such contexts. A morphism from $x_1\colon A_1, \dots, x_n\colon A_n$ to $x_1\colon B_1, \dots, x_m\colon B_m$ is a tuple $(t_1, \dots, t_m)$ of terms $x_1\colon A_1, \dots, x_n\colon A_n \vdash t_i\colon B_i$ for $i = 1, \dots, m$. A morphism of type $\Psi \to \Phi$ in $\mathbb{D}$ thus amounts to a (domain-level) substitution that provides a (domain-level) term in context Ψ for each of the variables in Φ. Terms are identified up to $\alpha\beta\eta$-equality. One may achieve this by using a de Bruijn encoding, for example, but the specific encoding is not important for this paper. The terminal object is the empty context, which we denote by 1, and the product $\Phi \times \Psi$ is defined by context concatenation. It is not hard to see that any object $x_1\colon A_1, \dots, x_n\colon A_n$ is isomorphic to an object that is given by a context with a single variable, namely $x_1\colon (A_1 \times \cdots \times A_n)$. This is to say that contexts can be identified with product types. In view of this isomorphism, we shall allow ourselves to consider the objects of $\mathbb{D}$ also as types and vice versa. The category $\mathbb{D}$ is cartesian closed, the exponential of Φ and Ψ being given by the function type $\Phi \to \Psi$ (where the objects are considered as types).

The presheaf category $\widehat{\mathbb{D}}$ is a computational universe that both embeds the term model $\mathbb{D}$ and that can represent computations about it. Note that we cannot just enrich $\mathbb{D}$ with terms for computations if we want to use HOAS. In a simply-typed lambda-calculus with just the constant terms `app: tm → tm → tm` and `lam: (tm → tm) → tm`, each term of type `tm` represents an object-level term. This would not be the true anymore, if we were to allow computations in the domain language, since one could define `M` to be something like `(λx. if x represents an object-level application then M1 else M2)` for distinct `M1` and `M2`. In this case, `lam M` would not represent an object-level term anymore. If we want to preserve a bijection between the object-level terms and their representations in the domain-language, we cannot allow case-distinction over whether a term represents an object-level an application.

The category $\widehat{\mathbb{D}}$ unites syntax with computations by allowing one to enforce various degrees of substitution-invariance. By choosing objects with different substitution actions, one can control the required amount of substitution-invariance.

In one extreme, a set S can be represented by the constant presheaf ΔS with $\Delta S(\Psi) = S$ and $\Delta S(\sigma) = \mathsf{id}$ for all Ψ and σ. The substitution action is trivial. As a consequence, a morphism $\Delta S \to \Delta T$ amounts to a function from set S to set T, since the trivial choice of the substitution action makes the naturality condition vacuous.

The Yoneda embedding represents the other extreme. For any object Φ of $\mathbb{D}$, the presheaf $\mathrm{y}(\Phi)\colon \mathbb{D}^{\mathrm{op}} \to \mathrm{Set}$ is defined by $\mathrm{y}(\Phi)(\Psi) = \mathbb{D}(\Psi, \Phi)$, which is the set of morphisms from Ψ to Φ in $\mathbb{D}$. The functor action is pre-composition. The presheaf $\mathrm{y}(\Phi)$ should be understood as the type of all domain-level substitutions with codomain Φ. An important example is $\mathtt{Tm} := y(\mathtt{tm})$. In this case, $\mathtt{Tm}(\Psi)$ is the set of all morphisms of type $\Psi \to \mathtt{tm}$ in $\mathbb{D}$. By the definition of $\mathbb{D}$, these correspond to domain-level terms of type `tm` in context Ψ. In this way, the presheaf `Tm` represents the domain-level terms of type `tm`.

The Yoneda embedding does in fact embed $\mathbb{D}$ into $\widehat{\mathbb{D}}$ fully and faithfully. The Yoneda embedding becomes a functor $\mathrm{y}\colon \mathbb{D} \to \widehat{\mathbb{D}}$ if one defines the morphism action to be post-composition. This means that y maps a morphism $\sigma\colon \Psi \to \Phi$ in $\mathbb{D}$ to the natural transformation $\mathrm{y}(\sigma)\colon \mathrm{y}(\Psi) \to \mathrm{y}(\Phi)$ that is defined by post-composing with σ. This definition makes y into a functor $\mathrm{y}\colon \mathbb{D} \to \widehat{\mathbb{D}}$ that is moreover full and faithful: its action on morphisms is a bijection from $\mathbb{D}(\Psi, \Phi)$ to $\widehat{\mathbb{D}}(\mathrm{y}(\Psi), \mathrm{y}(\Phi))$ for any Ψ and Φ. This is because a natural transformation $f\colon \mathrm{y}(\Psi) \to \mathrm{y}(\Phi)$ is, by naturality, uniquely determined by $f_\Psi(\mathsf{id})$, where $\mathsf{id} \in \mathbb{D}(\Psi, \Psi) = \mathrm{y}(\Psi)(\Psi)$, and $f_\Psi(\mathsf{id})$ is an element of $y(\Phi)(\Psi) = \mathbb{D}(\Psi, \Phi)$.

Since $\mathbb{D}$ embeds into $\widehat{\mathbb{D}}$ fully and faithfully, the term model of the domain language is available in $\widehat{\mathbb{D}}$. Consider for example $\mathtt{Tm} = \mathrm{y}(\mathtt{tm})$. Since y is full and faithful, the morphisms from `Tm` to `Tm` in $\widehat{\mathbb{D}}$ are in one-to-one correspondence with the morphisms from `tm` to `tm` in $\mathbb{D}$. These, in turn, are defined to be substitutions and correspond to simply-typed (domain-level) lambda terms with one free variable. This shows that substitution invariance cuts down the morphisms from `Tm` to `Tm` in $\widehat{\mathbb{D}}$ just as much as one would like for HOAS encodings.

But $\widehat{\mathbb{D}}$ contains not just a term model of the domain language. It can also represent computations about the domain-level syntax and computations that are not substitution-invariant. For example, arbitrary functions on terms can be represented as morphisms from the constant presheaf $\Delta(\mathtt{Tm}(1))$ to $\mathtt{Tm}$. Recall that 1 is the empty context, so that $\mathtt{Tm}(1)$ is the set $\mathbb{D}(1, \mathtt{tm})$, by definition, which is isomorphic to the set of closed domain-level terms of type $\mathtt{tm}$. The morphisms from $\Delta(\mathtt{Tm}(1))$ to $\mathtt{Tm}$ in $\widehat{\mathbb{D}}$ correspond to arbitrary functions from closed terms to closed terms, without any restriction of substitution invariance.

The restriction to the constant presheaf of closed terms can be generalised to arbitrary presheaves. Define a functor $\flat \colon \widehat{\mathbb{D}} \to \widehat{\mathbb{D}}$ by letting $\flat F$ be the constant presheaf $\Delta(F(1))$, i.e. $\flat F(\Psi) = F(1)$ and $\flat F(\sigma) = \mathsf{id}$. Thus, $\flat$ restricts any presheaf to the set of its closed elements. The functor $\flat$ defines a comonad where the counit $\varepsilon_F \colon \flat F \to F$ is the obvious inclusion and the comultiplication $\nu_F \colon \flat F \to \flat\flat F$ is the identity. The latter means that the comonad $\flat$ is idempotent.

3 Internal Language

To explain how $\widehat{\mathbb{D}}$ models higher-order abstract syntax and contextual types, we need to expose more of its structure. Most of this structure is standard. Defining it directly in terms of functors and natural transformations is somewhat laborious and the technical details may obscure the basic idea of our approach.

We therefore use the internal type theory of $\widehat{\mathbb{D}}$ as a meta-language for working with its structure. The structure of $\widehat{\mathbb{D}}$ furnishes a model of a dependent type theory that supports dependent products, dependent sums and extensional identity types, among others, in a standard way [11]. We use Agda notation for the types and terms of this internal type theory. We write $(x \colon S) \to T$ for a dependent function type and write $\lambda x \colon S.m$ and $m\ n$ for the associated lambda-abstractions and applications. As usual, we will sometimes also write $S \to T$ for $(x \colon S) \to T$ if x does not appear in T. However, to make it easier to distinguish the function spaces at various levels, we will write $(x \colon S) \to T$ by default even when x does not appear in T. We use $\mathtt{let}\ x = m\ \mathtt{in}\ n$ as an abbreviation for $(\lambda x \colon T.n)\ m$, as usual. For two terms $m \colon T$ and $n \colon T$, we write $m =_T n$ or just $m = n$ for the associated identity type. Our notation is similar to Agda's, since the internal type theory can be seen as a fragment of Agda's type theory. Agda has been useful as a tool for type-checking our constructions in the internal type theory [1].

In the spirit of Martin-Löf type theory, we will define basic types and terms successively as they are needed. In the Agda development this corresponds to postulating constants that are justified by the interpretation in $\widehat{\mathbb{D}}$. In the following sections, we will expose the structure of $\widehat{\mathbb{D}}$ step by step until we have enough to interpret contextual types.

While much of the structure of $\widehat{\mathbb{D}}$ can be captured by adding rules and constants to standard Martin-Löf type theory, for the comonad $\flat$ such a formulation would not be very satisfactory. The issues are discussed by Shulman [27, p.7], for example. To obtain a more satisfactory syntax for the comonad, we refine the internal type theory into a modal type theory in which $\flat$ appears as a necessity modality. This approach goes back to [3,4,6] and is also used by recent work of Shulman [27], Licata et al. [16] and others on working with the $\flat$-modality in type theory. Agda has recently gained support for such a $\flat$-modality [29].

We summarise here the typing rules for the $\flat$-modality which we will rely on. To control the modality, one uses two kinds of variables. In addition to standard variables $x{:}T$, one has a second kind of so-called *crisp* variables $x{::}T$. Typing judgements have the form $\Delta \mid \Theta \vdash m{:}T$, where Δ collects the crisp variables and Θ collects the ordinary variables. In essence, a crisp variable $x{::}T$ represents an assumption of the form $x{:}\flat T$. The syntactic distinction is useful, since it leads to a type theory that is well-behaved with respect to substitution, see [6,27].

The typing rules are closely related to those in modal type systems [6,18], where Δ is the typing context for modal (global) assumptions and Θ for (local) assumptions, and type systems for linear logic [4], where Δ is the typing context for non-linear assumptions and Θ for linear assumptions.

$$\frac{}{\Delta, u{::}T, \Delta' \mid \Theta \vdash u{:}T} \qquad \frac{}{\Delta \mid \Theta, x{:}T, \Theta' \vdash x{:}T}$$

$$\frac{\Delta \mid \cdot \vdash m : T}{\Delta \mid \Theta \vdash \mathsf{box}\ m : \flat T} \qquad \frac{\Delta \mid \Theta \vdash m : \flat T \quad \Delta, x{::}T \mid \Theta \vdash n : S}{\Delta \mid \Theta \vdash \mathtt{let\ box}\ x = m\ \mathtt{in}\ n : S}$$

Given any term $m{:}\ T$ which only depends on modal variable context Δ, we can form the term $\mathsf{box}\ m{:}\ \flat T$. We have a let-term $\mathtt{let\ box}\ x = m\ \mathtt{in}\ n$ that takes a term $m{:}\ \flat T$ and binds it to a variable $x{::}T$. The rules maintain the invariant that the free variables in a type $\flat T$ or a term $\mathsf{box}\ m$ are all crisp variables from the crisp context Δ.

The other typing rules do not modify the crisp context. For examples, the rules for dependent products are:

$$\frac{\Delta \mid \Theta, x{:}T \vdash m{:}S}{\Delta \mid \Theta \vdash \lambda x{:}T.m : (x{:}T) \to S} \qquad \frac{\Delta \mid \Theta \vdash m{:}(y{:}T) \to S \quad \Delta \mid \Theta \vdash n{:}T}{\Delta \mid \Theta \vdash m\ n{:}[n/y]S}$$

When Δ is empty, we shall write just $\Theta \vdash m{:}T$ for $\Delta \mid \Theta \vdash m{:}T$.

4 From Presheaves to Contextual Types

Armed with the internal type theory, we can now explore the structure of $\widehat{\mathbb{D}}$.

4.1 A Universe of Representables

For our purposes, the main feature of $\widehat{\mathbb{D}}$ is that it embeds $\mathbb{D}$ fully and faithfully via the Yoneda embedding. In the type theory for $\widehat{\mathbb{D}}$, we may capture this embedding by means of a Tarski-style universe. Such a universe is defined by a type of codes for types together with a decoding function that maps codes to actual types.

The type of codes `Obj` represents the set of objects of $\mathbb{D}$ in the internal type theory of $\widehat{\mathbb{D}}$. We have seen above that any set can be represented as a presheaf with trivial substitution action, and `Obj` is one such example. Particular objects of $\mathbb{D}$ then appear as terms of type `Obj`. The cartesian closed structure of $\mathbb{D}$ gives us terms `unit`, `times`, `arrow` for the terminal object 1, finite products $\times$ and the exponential (function type). We also have a term for the domain-level type `tm`.

$$\vdash \mathtt{Obj}\ \text{type} \qquad \vdash \mathtt{tm}{:}\ \mathtt{Obj} \qquad \vdash \mathtt{times}{:}\ (a{:}\,\mathtt{Obj}) \to (b{:}\,\mathtt{Obj}) \to \mathtt{Obj}$$

$$\vdash \mathtt{unit}{:}\ \mathtt{Obj} \qquad \vdash \mathtt{arrow}{:}\ (a{:}\,\mathtt{Obj}) \to (b{:}\,\mathtt{Obj}) \to \mathtt{Obj}$$

Subsequently, we sometimes talk about objects of $\mathbb{D}$ when we intend to describe terms of type $\texttt{Obj}$ (and vice versa).

The morphisms of $\mathbb{D}$ could similarly be encoded as a constant presheaf with many term constants, but this is in fact not necessary. Instead, we can use the Yoneda embedding as a function that decodes elements of $\texttt{Obj}$ into actual types.

$$x{:}\,\texttt{Obj} \vdash \texttt{El}\, x \text{ type}$$

The function $\texttt{El}$ is almost direct syntax for the Yoneda embedding. The interpretation in $\widehat{\mathbb{D}}$ is such that, for any object A of $\mathbb{D}$, the type $\texttt{El}\, A$ is interpreted by the presheaf $\mathrm{y}(A)$. Such a presheaf is called *representable*. One can think of $\texttt{El}\, A$ as the type of all morphisms of type $\Psi \to A$ in $\mathbb{D}$ for arbitrary Ψ. Recall from above that a morphism of type $\Psi \to A$ in $\mathbb{D}$ amounts to a domain-level term of type A that may refer to variables in Ψ. In this sense, one should think of $\texttt{El}\, A$ as a type of domain-level terms of type A, both closed and open ones.

We get all morphisms of $\mathbb{D}$, and no more, in this way, since the Yoneda embedding is full and faithful, recall Sec. 2. In our case, this means that the type $(x{:}\,\texttt{El}\, A) \to \texttt{El}\, B$ represents the morphisms of type $A \to B$ in $\mathbb{D}$. Any closed term of type $(x : \texttt{El}\, A) \to \texttt{El}\, B$ corresponds to such a morphism and vice versa. This is because the naturality requirements in $\widehat{\mathbb{D}}$ enforce substitution-invariance, as outlined in Sec. 2. The type $(x : \texttt{El}\, A) \to \texttt{El}\, B$ thus does not represent arbitrary functions from terms of type A to terms of type B, but only substitution-invariant ones. If a function of this type maps a domain-level variable $x{:}\,A$ (encoded as an element of $\texttt{El}\, A$) to some term $M{:}\,B$ (encoded as an element of $\texttt{El}\, B$), then it must map any other $N{:}\,A$ to $[N/x]M$.

We note that the type dependency in $\texttt{El}$ is easy to work with. A term of type $(a{:}\texttt{Obj}) \to (b{:}\,\texttt{Obj}) \to (x{:}\texttt{El}\, a) \to \texttt{El}\, b$ corresponds to a family of terms $(x{:}\texttt{El}\, A) \to \texttt{El}\, B$ indexed by objects A and B in $\mathbb{D}$. This is because $\texttt{Obj}$ is just a set, so that the naturality constraints of $\widehat{\mathbb{D}}$ are vacuous for functions out of $\texttt{Obj}$.

To summarise, we get access to $\mathbb{D}$ in the internal type theory of $\widehat{\mathbb{D}}$ simply by considering the Yoneda embedding as the decoding function $\texttt{El}$ of a universe á la Tarski. Since is consists of the representable presheaves, we call it the *universe of representables.* The following lemmas state that the embedding preserves terminal object, binary products and the exponential.

Lemma 1. *The internal type theory of* $\widehat{\mathbb{D}}$ *has a term* $\vdash \textit{terminal}{:}\ \textit{El}\, \textit{unit}$, *such that* $x = \textit{terminal}$ *holds for any* $x{:}\ \textit{El}\, \textit{unit}$.

Lemma 2. *The internal type theory of* $\widehat{\mathbb{D}}$ *justifies the terms below, such that* $\textit{fst}\ (\textit{pair}\ x\ y) = x$, $\textit{snd}\ (\textit{pair}\ x\ y) = y$, $z = \textit{pair}\ (\textit{fst}\ z)\ (\textit{snd}\ z)$ *for all* x, y, z.

$c{:}\ \textit{Obj},\ d{:}\ \textit{Obj} \vdash \textit{fst}{:}\ (z : \textit{El}\,(\textit{times}\ c\ d)) \to \textit{El}\, c$
$c{:}\ \textit{Obj},\ d{:}\ \textit{Obj} \vdash \textit{snd}{:}\ (z : \textit{El}\,(\textit{times}\ c\ d)) \to \textit{El}\, d$
$c{:}\ \textit{Obj},\ d{:}\ \textit{Obj} \vdash \textit{pair}{:}\ (x : \textit{El}\, c) \to (y : \textit{El}\, d) \to \textit{El}\,(\textit{times}\ c\ d)$

Lemma 3. *The internal type theory of* $\widehat{\mathbb{D}}$ *justifies the terms below such that* $\textit{arrow-i}\ (\textit{arrow-e}\ f) = f$ *and* $\textit{arrow-e}\ (\textit{arrow-i}\ g) = g$ *for all* f, g.

$c{:}\ \textit{Obj},\ d{:}\ \textit{Obj} \vdash \textit{arrow-e}{:}\ (x{:}\,\textit{El}\,(\textit{arrow}\ c\ d)) \to (y{:}\,\textit{El}\, c) \to \textit{El}\, d$
$c{:}\ \textit{Obj},\ d{:}\ \textit{Obj} \vdash \textit{arrow-i}{:}\ (y{:}\,(\textit{El}\, c \to \textit{El}\, d)) \to \textit{El}\,(\textit{arrow}\ c\ d)$

4.2 Higher-Order Abstract Syntax

The last lemma in the previous section states that $\mathtt{El}\,A \to \mathtt{El}\,B$ is isomorphic to $\mathtt{El}\,(\mathtt{arrow}\ A\ B)$. This is particularly useful to lift HOAS-encodings from $\mathbb{D}$ to $\widehat{\mathbb{D}}$. For instance, the domain-level term constant $\mathtt{lam}\colon (\mathtt{tm} \to \mathtt{tm}) \to \mathtt{tm}$ gives rise to an element of $\mathtt{El}\,(\mathtt{arrow}\,(\mathtt{arrow}\ \mathtt{tm}\ \mathtt{tm})\ \mathtt{tm})$. But this type is isomorphic to $(\mathtt{El}\,\mathtt{tm} \to \mathtt{El}\,\mathtt{tm}) \to \mathtt{El}\,\mathtt{tm}$, by the lemma.

This means that the higher-order abstract syntax constants lift to $\widehat{\mathbb{D}}$:

$$\mathtt{app}\colon (m{:}\,\mathtt{El}\,\mathtt{tm}) \to (n{:}\,\mathtt{El}\,\mathtt{tm}) \to \mathtt{El}\,\mathtt{tm} \qquad \mathtt{lam}\colon (m{:}\,(\mathtt{El}\,\mathtt{tm} \to \mathtt{El}\,\mathtt{tm})) \to \mathtt{El}\,\mathtt{tm}$$

Once one recognises $\mathtt{El}\,A$ as $\mathrm{y}(A)$, the adequacy of this higher-order abstract syntax encoding lifts from $\mathbb{D}$ to $\widehat{\mathbb{D}}$ as in [12]. For example, an argument M to $\mathtt{lam}$ has type $\mathtt{El}\,\mathtt{tm} \to \mathtt{El}\,\mathtt{tm}$, which is isomorphic to $\mathtt{El}\,(\mathtt{arrow}\ \mathtt{tm}\ \mathtt{tm})$. But this type represents (open) domain-level terms $t\colon \mathtt{tm} \to \mathtt{tm}$. The term $\mathtt{lam}\ M\colon \mathtt{El}\,\mathtt{tm}$ then represents the domain-level term $\mathtt{lam}\ t\colon \mathtt{tm}$, so it just lifts the domain-level.

4.3 Closed Objects

One should think of $\flat T$ as the type of 'closed' elements of T. In particular, $\flat(\mathtt{El}\,A)$ represents morphisms of type $1 \to A$ in $\mathbb{D}$, recall the definition of $\flat$ from Sec. 2 and that $\mathtt{El}\,A$ corresponds to $\mathrm{y}A$. In the term model $\mathbb{D}$, the morphisms $1 \to A$ correspond to closed domain-language terms of type A. Thus, while $\mathtt{El}\,A$ represents both open and closed domain-level terms, $\flat(\mathtt{El}\,A)$ represents only the closed ones.

This applies also to the type $\mathtt{El}\,A \to \mathtt{El}\,B$. We have seen above that $\mathtt{El}\,A \to \mathtt{El}\,B$ is isomorphic to $\mathtt{El}\,(\mathtt{arrow}\ A\ B)$ and may therefore be thought of as containing the terms of type B with a distinguished variable of type A. But, these terms may contain other free domain language variables. The type $\flat(\mathtt{El}\,A \to \mathtt{El}\,B)$, on the other hand, contains only terms of type B that may contain (at most) one variable of type A.

Restricting to closed object with the modality is useful because it disables substitution-invariance. For example, the internal type theory for $\widehat{\mathbb{D}}$ justifies a function $\mathtt{is\text{-}lam}\colon (x{:}\flat(\mathtt{El}\,\mathtt{tm})) \to \mathtt{bool}$ that returns $\mathtt{true}$ if and only if the argument represents a domain language lambda abstraction. We shall define it in the next section. Such a function cannot be defined with type $\mathtt{El}\,\mathtt{tm} \to \mathtt{bool}$, since it would not be invariant under substitution. Its argument ranges over terms that may be open; which particularly includes domain-level variables. The function would have to return $\mathtt{false}$ for them, since a domain-level variable is not a lambda-abstraction. But after substituting a lambda-abstraction for the variable, it would have to return $\mathtt{true}$, so it could not be substitution-invariant.

We note that the type $\mathtt{Obj}$ consists only of closed elements and that $\mathtt{Obj}$ and $\flat\mathtt{Obj}$ happen to be definitionally equal types (an isomorphism would suffice, but equality is more convenient).

4.4 Contextual Objects

Using function types and the modality, it is now possible to work with contextual objects that represent domain level terms in a certain context, much like in [20,21]. A contextual type $\lceil \Psi \vdash A \rceil$ is a boxed function type of the form $\flat(\mathtt{El}\,\Psi \to \mathtt{El}\,A)$. It represents domain-level terms of type A with variables from Ψ. Here, we consider the domain-level context Ψ as a term that encodes it. The interpretation will make this precise.

For example, domain-level terms with up to two free variables now appear as terms of type $\flat(\mathtt{El}\,((\mathtt{times}\ (\mathtt{times}\ \mathtt{unit}\ \mathtt{tm})\ \mathtt{tm}) \to \mathtt{El}\,\mathtt{tm})$, as the following example illustrates.

$$\begin{aligned}\mathsf{box}\ (\lambda u{:}\,\mathtt{El}\,((\mathtt{times}\ (\mathtt{times}\ \mathtt{unit}\ \mathtt{tm})\ \mathtt{tm}).\ &\mathtt{let}\ x_1 = \mathtt{snd}\ (\mathtt{fst}\ u)\ \mathtt{in}\\ &\mathtt{let}\ x_2 = \mathtt{snd}\ u\ \mathtt{in}\\ &\quad \mathtt{app}\ (\mathtt{lam}\ (\lambda x{:}\,\mathtt{El}\,\mathtt{tm}.\,\mathtt{app}\ x_1\ x))\ x_2\ )\end{aligned}$$

The context variables x_1 and x_2 are bound at the meta level.

This representation integrates substitution as usual. For example, given crisp variables $m{::}\mathtt{El}\,(\mathtt{times}\ c\ \mathtt{tm}) \to \mathtt{tm}$ and $n{::}\mathtt{El}\,c \to \mathtt{tm}$ for contextual terms, the term $\mathsf{box}\ (\lambda u{:}\,\mathtt{El}\,c.\,m\ (\mathtt{pair}\ u\ (n\ u)))$ represents substitution of n for the last variable in the context of m.

For working with contextual objects, it is convenient to lift the constants app and lam to contextual types.

$$\begin{aligned}c{:}\,\mathtt{Obj} &\vdash \mathtt{app}'{:}\ \flat(\mathtt{El}\,c \to \mathtt{El}\,\mathtt{tm}) \to \flat(\mathtt{El}\,c \to \mathtt{El}\,\mathtt{tm}) \to \flat(\mathtt{El}\,c \to \mathtt{tm})\\ c{:}\mathtt{Obj} &\vdash \mathtt{lam}'{:}\ \flat(\mathtt{El}\,(\mathtt{times}\ c\ \mathtt{tm}) \to \mathtt{El}\,\mathtt{tm}) \to \flat(\mathtt{El}\,c \to \mathtt{El}\,\mathtt{tm})\end{aligned}$$

These terms are defined by:

$$\begin{aligned}\mathtt{app}' &:= \lambda m, n.\,\mathtt{let\ box}\ m' = m\ \mathtt{in\ let\ box}\ n' = n\ \mathtt{in}\\ &\qquad\qquad \mathsf{box}\ (\lambda u{:}\,\mathtt{El}\,c.\,\mathtt{app}\ (m'\ u)\ (n'\ u))\\ \mathtt{lam}' &:= \lambda m.\,\mathtt{let\ box}\ m' = m\ \mathtt{in}\ \mathsf{box}\ (\lambda u{:}\,\mathtt{El}\,c.\,\mathtt{lam}\ (\lambda x{:}\,\mathtt{El}\,\mathtt{tm}.\ m'\ (\mathtt{pair}\ u\ x)))\end{aligned}$$

A contextual type for domain-level variables (as opposed to arbitrary terms) can be defined by restricting the function space in $\flat(\mathtt{El}\,\Psi \to \mathtt{El}\,A)$ to consist only of projections. Projections are functions of the form $\mathtt{snd} \circ \mathtt{fst}_k$, where we write $\mathtt{fst}_k$ for the k-fold iteration $\mathtt{fst} \circ \cdots \circ \mathtt{fst}$. Let us write $S \to_v T$ for the subtype of $S \to T$ consisting only of projections. The contextual type $\flat(\mathtt{El}\,\Psi \to_v \mathtt{El}\,A)$ is then a subtype of $\flat(\mathtt{El}\,\Psi \to \mathtt{El}\,A)$.

With these definitions, we can express a primitive recursion scheme for contextual types. We write it in its general form where the result type A can possibly depend on x. This is only relevant for the dependently typed case; in the simply typed case, the only dependency is on c.

Lemma 4. *Let $c{:}\,\mathtt{Obj}$, $x{:}\,\flat(\mathtt{El}\,c \to \mathtt{El}\,\mathtt{tm}) \vdash A\ c\ x$ type and define:*

$$\begin{aligned}X_{\mathtt{var}} &:= (c{:}\,\mathtt{Obj}) \to (x{:}\,\flat(\mathtt{El}\,c \to_v \mathtt{El}\,\mathtt{tm})) \to A\ c\ x\\ X_{\mathtt{app}} &:= (c{:}\,\mathtt{Obj}) \to (x, y{:}\,\flat(\mathtt{El}\,c \to \mathtt{El}\,\mathtt{tm})) \to\ A\ c\ x \to A\ c\ y \to A\ c\ (\mathtt{app}'\ x\ \ y)\\ X_{\mathtt{lam}} &:= (c{:}\,\mathtt{Obj}) \to (x{:}\,\flat(\mathtt{El}\,(\mathtt{times}\ c\ \mathtt{tm}) \to \mathtt{El}\,\mathtt{tm})) \to A\ (\mathtt{times}\ c\ \mathtt{tm})\ x \to A\ c\ (\mathtt{lam}'\ x)\end{aligned}$$

Then, $\widehat{\mathbb{D}}$ justifies a term

$$\vdash \mathit{rec}\colon X_{var} \to X_{app} \to X_{lam} \to (c\colon \mathit{Obj}) \to (x\colon \flat(\mathit{El}\, c \to \mathit{El}\, \mathit{tm})) \to A\ c\ x$$

such that the following equations are valid.

$$\begin{aligned} &\mathit{rec}\ t_{var}\ t_{app}\ t_{lam}\ c\ x &&= t_{var}\ c\ x && \text{if } x\colon \flat(\mathit{El}\, c \to_v \mathit{El}\, \mathit{tm}) \\ &\mathit{rec}\ t_{var}\ t_{app}\ t_{lam}\ c\ (\mathit{app}'\ s\ t) &&= t_{app}\ c\ s\ t \\ &\mathit{rec}\ t_{var}\ t_{app}\ t_{lam}\ c\ (\mathit{lam}'\ s) &&= t_{lam}\ c\ s \end{aligned}$$

Proof (outline). To outline the proof idea, note first that a function of type $(c\colon \mathtt{Obj}) \to (x\colon \flat(\mathtt{El}\, c \to \mathtt{El}\, \mathtt{tm})) \to A\ c\ x$ in $\widehat{\mathbb{D}}$, corresponds to an inhabitant of $A\ \Phi\ t$ for each concrete object Φ of $\mathbb{D}$ and each inhabitant $t\colon \flat(\mathtt{El}\,\Phi \to \mathtt{El}\,\mathtt{tm})$. This is because naturality constraints for boxed types are vacuous (and $\mathtt{Obj} = \flat\mathtt{Obj}$). Next, note that inhabitants of $\flat(\mathtt{El}\,\Phi \to \mathtt{El}\,\mathtt{tm})$ correspond to domain-level terms of type $\mathtt{tm}$ in context Φ up to $\alpha\beta\eta$-equality. We can perform a case-distinction on whether it is a variable, abstraction or application and depending on the result use $t_{\mathtt{var}}$, $t_{\mathtt{app}}$ or $t_{\mathtt{lam}}$ to define the required inhabitant of $A\ \Phi\ t$.

As a simple example for `rec`, we can define the function `is-lam` discussed above by $\mathtt{rec}\ (\lambda c, x.\mathtt{false})\ (\lambda c, x, y, r_x, r_y.\mathtt{false})\ (\lambda c, x, r_x.\mathtt{true})$.

5 Simple Contextual Modal Type Theory

We have outlined informally how the internal dependent type theory of $\widehat{\mathbb{D}}$ can model contextual types. In this section, we make this precise by giving the interpretation of Cocon [23], a contextual modal type theory where we can work with contextual HOAS trees and computations about them, into $\widehat{\mathbb{D}}$. We will focus here on a simply-typed version of Cocon where we use a simply-typed domain-language with constants `app` and `lam` and also only allow computations about HOAS trees, but do not consider, for example, universes. Concentrating on a stripped down, simply-typed version of Cocon allows us to focus on the essential aspects, namely how to interpret domain-level contexts and domain-level contextual objects and types semantically. The generalisation to a dependently typed domain-level such as LF in Sec. 6 will be conceptually straightforward, although more technical. Handling universes is an orthogonal issue (see also [16]).

We first define our simply-typed domain-level with the type `tm` the term constants lam and app (see Fig. 1). Following Cocon, we allow computations to be embedded into domain-level terms via unboxing. The intuition is that if a program t promises to compute a value of type $\lceil x{:}\mathtt{tm}, y{:}\mathtt{tm} \vdash \mathtt{tm}\rceil$, then we can embed t directly into a domain-level object writing $\mathtt{lam}\ \lambda x.\mathtt{lam}\ \lambda y.\mathtt{app}\ \lfloor t\rfloor\ x$, unboxing t. Domain-level objects (resp. types) can be packaged together with their domain-level context to form a contextual object (resp. type). Domain-level contexts are formed as usual, but may contain context variables to describe a yet unknown prefix. Last, we include domain-level substitutions that allow us to move between domain-level contexts. The compound substitution σ, M extends the substitution σ with domain $\widehat{\Psi}$ to a substitution with domain $\widehat{\Psi}, x$, where M replaces x. Following [18,23], we do not store the domain (like $\widehat{\Psi}$) in the

Domain-level types	A, B	$::=$	$\mathtt{tm} \mid A \rightarrow B$
Domain-level terms	M, N	$::=$	$\lambda x.M \mid M\,N \mid x \mid \mathsf{lam} \mid \mathsf{app} \mid \lfloor t \rfloor_\sigma$
Domain-level contexts	Ψ, Φ	$::=$	$\cdot \mid \psi \mid \Psi, x{:}A$
Domain-level context (erased)	$\widehat{\Psi}, \widehat{\Phi}$	$::=$	$\cdot \mid \psi \mid \widehat{\Psi}, x$
Domain-level substitutions	σ	$::=$	$\cdot \mid \mathsf{wk}_{\widehat{\Psi}} \mid \sigma, M$
Contextual types	T	$::=$	$\Psi \vdash A \mid \Psi \vdash_{\overline{v}} A$
Contextual objects	C	$::=$	$\widehat{\Psi} \vdash M$
Domain of discourse	$\check{\tau}$	$::=$	$\tau \mid \mathtt{ctx}$
Types and Terms	$\tau, \mathcal{I}$	$::=$	$\lceil T \rceil \mid (y : \check{\tau}_1) \Rightarrow \tau_2$
	t, s	$::=$	$y \mid \lceil C \rceil \mid \mathsf{rec}^{\mathcal{I}}\ \boldsymbol{\mathcal{B}}\ \Psi\ t \mid \mathsf{fn}\ y \Rightarrow t \mid t_1\ t_2$
Branches	$\mathcal{B}$	$::=$	$\Gamma \mapsto t$
Contexts	Γ	$::=$	$\cdot \mid \Gamma, y : \check{\tau}$

Fig. 1. Syntax of COCON with a fixed simply-typed domain tm

substitution, it can always be recovered before applying the substitution. We also include *weakening substitution*, written as $\mathsf{wk}_{\widehat{\Psi}}$, to describe the weakening of the domain Ψ to $\Psi, \overrightarrow{x{:}A}$. Weakening substitutions are necessary, as they allow us to express the weakening of a context variable ψ. Identity is a special form of the $\mathsf{wk}_{\widehat{\Psi}}$ substitution, which follows immediately from the typing rule of $\mathsf{wk}_{\widehat{\Psi}}$. Composition is admissible.

We summarise the typing rules for domain-level terms and types in Fig. 2. We also include typing rules for domain-level contexts. Note that since we restrict ourselves to a simply-typed domain-level, we simply check that A is a well-formed type. We defer the reduction and expansion rules to the appendix and only remark here that equality for domain-level terms and substitution is modulo $\beta\eta$. In particular, $\lfloor \lceil \widehat{\Phi} \vdash N \rceil \rfloor_\sigma$ reduces to $[\sigma]N$.

In our grammar, we distinguish between the contextual type $\Psi \vdash A$ and the more restricted contextual type $\Phi \vdash_{\overline{v}} A$ which characterises only variables of type A from the domain-level context Φ. We give here two sample typing rules for $\Phi \vdash_{\overline{v}} A$ which are the ones used most in practice to illustrate the main idea. We embed contextual objects into computations via the modality. Computation-level types include boxed contextual types, $\lceil \Phi \vdash A \rceil$, and function types, written as $(y : \check{\tau}_1) \Rightarrow \tau_2$. We overload the function space and allow as domain of discourse both computation-level types and the schema ctx of domain-level context, although only in the latter case y can occur in τ_2. We use $\mathsf{fn}\ y \Rightarrow t$ to introduce functions of both kinds. We also overload function application $t\ s$ to eliminate function types $(y : \tau_1) \Rightarrow \tau_2$ and $(y : \mathsf{ctx}) \Rightarrow \tau_2$, although in the latter case s stands for a domain-level context. We separate domain-level contexts from contextual objects, as we do not allow functions that return a domain-level context.

The recursor is written as $\mathsf{rec}^{\mathcal{I}}\ \boldsymbol{\mathcal{B}}\ \Psi\ t$. Here, t describes a term of type $\lceil \Psi \vdash \mathtt{tm} \rceil$ that we recurse over and $\boldsymbol{\mathcal{B}}$ describes the different branches that we can take

$\boxed{\Gamma; \Psi \vdash M : A}$ Term M has type A in domain-level context Ψ and context Γ

$$\frac{\Gamma \vdash \Psi : \mathtt{ctx} \quad x{:}A \in \Psi}{\Gamma; \Psi \vdash x : A} \quad \frac{\Gamma \vdash \Psi : \mathtt{ctx}}{\Gamma; \Psi \vdash \mathsf{lam} : (\mathtt{tm} \to \mathtt{tm}) \to \mathtt{tm}} \quad \frac{\Gamma \vdash \Psi : \mathtt{ctx}}{\Gamma; \Psi \vdash \mathsf{app} : \mathtt{tm} \to \mathtt{tm} \to \mathtt{tm}}$$

$$\frac{\Gamma; \Psi \vdash M : A \to B \quad \Gamma; \Psi \vdash N : A}{\Gamma; \Psi \vdash M\ N : B} \quad \frac{\Gamma; \Psi, x{:}A \vdash M : B}{\Gamma; \Psi \vdash \lambda x.M : A \to B}$$

$$\frac{\Gamma \vdash t : \lceil \Phi \vdash A \rceil \quad \Gamma; \Psi \vdash \sigma : \Phi}{\Gamma; \Psi \vdash \lfloor t \rfloor_\sigma : A}$$

$\boxed{\Gamma; \Phi \vdash \sigma : \Psi}$ Substitution σ provides a mapping from the (domain) context Ψ to Φ

$$\frac{\Gamma \vdash \Psi, \overrightarrow{x{:}A} : \mathtt{ctx}}{\Gamma; \Psi, \overrightarrow{x{:}A} \vdash \mathsf{wk}_{\widehat{\Psi}} : \Psi} \quad \frac{\Gamma \vdash \Phi : \mathtt{ctx}}{\Gamma; \Phi \vdash \cdot : \cdot} \quad \frac{\Gamma; \Phi \vdash \sigma : \Psi \quad \Gamma; \Phi \vdash M : A}{\Gamma; \Phi \vdash \sigma, M : \Psi, x{:}A}$$

$\boxed{\Gamma \vdash \Psi : \mathtt{ctx}}$ Domain-level context Ψ is a well-formed

$$\frac{}{\Gamma \vdash \cdot : \mathtt{ctx}} \quad \frac{\Gamma(y) = \mathsf{ctx}}{\Gamma \vdash y : \mathtt{ctx}} \quad \frac{\Gamma \vdash \Psi : \mathtt{ctx}}{\Gamma \vdash \Psi, x{:}A : \mathtt{ctx}}$$

Fig. 2. Typing Rules for Domain-level Terms, Substitutions, Contexts

depending on the value computed by t. As is common when we have dependencies, we annotate the recursor with the typing invariant $\mathcal{I}$. Here, we consider only the recursor over domain-level terms of type $\mathtt{tm}$. Hence, we annotate it with $\mathcal{I} = (\psi : \mathsf{ctx}) \Rightarrow (y : \lceil \psi \vdash \mathtt{tm} \rceil) \Rightarrow \tau$. To check that the recursor $\mathsf{rec}^{\mathcal{I}}\ \mathcal{B}\ \Psi\ t$ has type $[\Psi/\psi]\tau$, we check that each of the three branches has the specified type $\mathcal{I}$. In the base case, we may assume in addition to $\psi : \mathsf{ctx}$ that we have a variable $p : \lceil \psi \vdash_v \mathtt{tm} \rceil$ and check that the body has the appropriate type. If we encounter a contextual object built with the domain-level constant app, then we choose the branch b_{app}. We assume $\psi{:}\,\mathsf{ctx}$, $m{:}\,\lceil \psi \vdash \mathtt{tm} \rceil$, $n{:}\,\lceil \psi \vdash \mathtt{tm} \rceil$, as well as f_n and f_m which stand for the recursive calls on m and n respectively. We then check that the body t_{app} is well-typed. If we encounter a domain object built with the domain-level constant lam, then we choose the branch b_{lam}. We assume $\psi{:}\,\mathsf{ctx}$ and $m{:}\,\lceil \psi, x{:}\mathtt{tm} \vdash \mathtt{tm} \rceil$ together with the recursive call f_m on m in the extended LF context $\psi, x{:}\mathtt{tm}$. We then check that the body t_{lam} is well-typed. The typing rules for computations are given in Fig. 3. We omit the reduction rules here and refer the interested reader to the appendix.

5.1 Interpretation

We now give an interpretation of simply-typed Cocon in a presheaf model with a cartesian closed universe of representables. Let us first extend the internal dependent type theory with the constant $\mathtt{tm}$ for modelling the domain-level type constant $\mathtt{tm}$ and with the constants $\mathtt{app}\colon \mathtt{El}\,\mathtt{tm} \to \mathtt{El}\,\mathtt{tm} \to \mathtt{El}\,\mathtt{tm}$ and

$\boxed{\Gamma \vdash C : T}$ Contextual object C has contextual type T

$$\frac{\Gamma; \Psi \vdash M : A}{\Gamma \vdash (\widehat{\Psi} \vdash M) : (\Psi \vdash A)} \qquad \frac{\Gamma \vdash \Psi : \mathtt{ctx} \quad x{:}A \in \Psi}{\Gamma \vdash (\widehat{\Psi} \vdash x) : (\Psi \vdash_{v} A)} \qquad \frac{x{:}\lceil \Phi \vdash_{v} A \rceil \in \Gamma \quad \Gamma; \Psi \vdash \mathsf{wk}_{\widehat{\Psi}} : \Phi}{\Gamma \vdash (\widehat{\Psi} \vdash \lfloor x \rfloor_{\mathsf{wk}_{\widehat{\Psi}}}) : (\Psi \vdash_{v} A)}$$

$\boxed{\Gamma \vdash t : \tau}$ Term t has computation type τ

$$\frac{y : \check{\tau} \in \Gamma}{\Gamma \vdash y : \check{\tau}} \qquad \frac{\Gamma \vdash C : T}{\Gamma \vdash \lceil C \rceil : \lceil T \rceil}$$

$$\frac{\Gamma \vdash t : (y : \check{\tau}_1) \Rightarrow \tau_2 \quad \Gamma \vdash s : \check{\tau}_1}{\Gamma \vdash t\ s : [s/y]\tau_2} \qquad \frac{\Gamma, y : \check{\tau}_1 \vdash t : \tau_2 \quad \Gamma \vdash (y : \check{\tau}_1) \Rightarrow \tau_2 : \text{type}}{\Gamma \vdash \mathsf{fn}\ y \Rightarrow t : (y : \check{\tau}_1) \Rightarrow \tau_2}$$

Recursor over domain-level terms $\mathcal{I} = (\psi : \mathsf{ctx}) \Rightarrow (y : \lceil \psi \vdash \mathtt{tm} \rceil) \Rightarrow \tau$

$$\frac{\Gamma \vdash t : \lceil \Psi \vdash \mathtt{tm} \rceil \quad \Gamma \vdash \mathcal{I} : u \quad \Gamma \vdash b_v : \mathcal{I} \quad \Gamma \vdash b_{\mathsf{app}} : \mathcal{I} \quad \Gamma \vdash b_{\mathsf{lam}} : \mathcal{I}}{\Gamma \vdash \mathsf{rec}^{\mathcal{I}}(b_v \mid b_{\mathsf{app}} \mid b_{\mathsf{lam}})\ \Psi\ t : [\Psi/\psi]\tau}$$

Branch for Variable $\dfrac{\Gamma, \psi : \mathsf{ctx}, p : \lceil \psi \vdash_{v} \mathtt{tm} \rceil \vdash t_v : \tau}{\Gamma \vdash (\psi, p \mapsto t_v) : \mathcal{I}}$

Branch for Application app $\dfrac{\Gamma, \psi : \mathsf{ctx}, m{:}\lceil \psi \vdash \mathtt{tm} \rceil, n{:}\lceil \psi \vdash \mathtt{tm} \rceil, f_m{:}\tau, f_n{:}\tau \vdash t_{\mathsf{app}} : \tau}{\Gamma \vdash (\psi, m, n, f_n, f_m \mapsto t_{\mathsf{app}}) : \mathcal{I}}$

Branch for Function lam $\dfrac{\Gamma, \phi : \mathsf{ctx}, m{:}\lceil \phi, x{:}\mathtt{tm} \vdash \mathtt{tm} \rceil, f_m{:}[(\phi, x{:}\mathtt{tm})/\psi]\tau \vdash t_{\mathsf{lam}} : [\phi/\psi]\tau}{\Gamma \vdash \psi, m, f_m \mapsto t_{\mathsf{lam}} : \mathcal{I}}$

Fig. 3. Typing Rules for Contextual Objects and Computations

$\mathtt{lam}$: $(\mathtt{El}\ \mathtt{tm} \to \mathtt{El}\ \mathtt{tm}) \to \mathtt{El}\ \mathtt{tm}$ to model the corresponding domain-level constants app and lam.

We can now translate domain-level and computation-level types of Cocon into the internal dependent type theory for $\widehat{\mathbb{D}}$. We do so by interpreting the domain-level terms, types, substitutions, and contexts (see Fig. 4). All translations are on well-typed terms and types. Domain-level types are interpreted as the terms of type $\mathtt{Obj}$ in the internal dependent type theory that represent them. Domain-level contexts are also interpreted as terms of type $\mathtt{Obj}$ by $[\![\Gamma \vdash \Psi : \mathtt{ctx}]\!]$. For example, a domain-level context $x{:}\mathtt{tm}, y{:}\mathtt{tm}$ is interpreted as $\mathtt{times}\ (\mathtt{times}\ \mathtt{unit}\ \mathtt{tm})\ \mathtt{tm} : \mathtt{Obj}$. A domain-level substitution with domain Ψ and codomain Φ becomes a term of type $\mathtt{El}\, e'$ that is parameterised by an element $u{:}\,\mathtt{El}\, e$, where $e = [\![\Gamma \vdash \Phi : \mathtt{ctx}]\!]$ and $e' = [\![\Gamma \vdash \Psi : \mathtt{ctx}]\!]$. As e' is some product, for example $\mathtt{times}\ (\mathtt{times}\ \mathtt{unit}\ \mathtt{tm})\ \mathtt{tm}$, the domain-level substitution is translated into an n-ary tuple. A weakening substitution $\Gamma; \Psi, x{:}\mathtt{tm} \vdash \mathsf{wk}_{\Psi} : \Psi$ is interpreted as $\mathtt{fst}\ u$ where $u{:}\,\mathtt{El}\,(\mathtt{times}\ e\ \mathtt{tm})$ and $e = [\![\Gamma \vdash \Psi : \mathtt{ctx}]\!]$. More generally, when we weaken a context Ψ by n declarations, i.e. $\overrightarrow{x{:}A}$, we interpret wk_{Ψ} as $\mathtt{fst}_n\ u$.

A well-typed domain-level term, $\Gamma; \Psi \vdash M : A$, is mapped to an object of type $\mathtt{El}\,[\![A]\!]$ that depends on $u{:}\mathtt{El}\,[\![\Gamma \vdash \Psi : \mathtt{ctx}]\!]$.

Hence the translation of a well-typed domain-level term is indexed by u that stands for the term-level interpretation of a domain-level context Φ. Initially, u

Interpretation of domain-level types

$$\llbracket \texttt{tm} \rrbracket = \texttt{tm}$$
$$\llbracket A \to B \rrbracket = \texttt{arrow}\ \llbracket A \rrbracket\ \llbracket B \rrbracket$$

Interpretation of domain-level contexts

$$\llbracket \Gamma \vdash \psi : \texttt{ctx} \rrbracket = \psi$$
$$\llbracket \Gamma \vdash \cdot : \texttt{ctx} \rrbracket = \texttt{unit}$$
$$\llbracket \Gamma \vdash (\Psi, x{:}A) : \texttt{ctx} \rrbracket = \texttt{times}\ e\ \llbracket A \rrbracket \qquad \text{where } \llbracket \Gamma \vdash \Psi : \texttt{ctx} \rrbracket = e$$

Interpretation of domain-level terms where $u{:}\,\texttt{El}\, e$ and $\llbracket \Gamma \vdash \Psi : \texttt{ctx} \rrbracket = e$

$$\llbracket \Gamma; \Psi \vdash x{:}\ A \rrbracket_u = \texttt{snd}\ (\texttt{fst}_k\ u) \qquad \text{where } \Psi = \Psi_0, x{:}A, y_k{:}A_k, \ldots, y_1{:}A_1$$
$$\llbracket \Gamma; \Psi \vdash \lambda x.\, M : A \to B \rrbracket_u = \texttt{arrow-i}\ (\lambda x{:}\texttt{El}\ \llbracket A \rrbracket.\ e) \qquad \text{where } \llbracket \Gamma; \Psi, x{:}A \vdash M : B \rrbracket_{(\texttt{pair}\ u\ x)} = e$$
$$\llbracket \Gamma; \Psi \vdash M\ N : B \rrbracket_u = \texttt{arrow-e}\ e_1\ e_2 \qquad \text{where } \llbracket \Gamma; \Psi \vdash M : A \to B \rrbracket_u = e_1 \text{ and } \llbracket \Gamma; \Psi \vdash N : A \rrbracket_u = e_2$$
$$\llbracket \Gamma; \Psi \vdash \lfloor t \rfloor_\sigma : A \rrbracket_u = \texttt{let box}\ x = e_1\ \texttt{in}\ x\ e_2 \qquad \text{where } \llbracket \Gamma \vdash t : \lceil \Phi \vdash A \rceil \rrbracket = e_1 \text{ and } \llbracket \Gamma; \Psi \vdash \sigma : \Phi \rrbracket_u = e_2$$
$$\llbracket \Gamma; \Psi \vdash \mathsf{app} : \texttt{tm} \to \texttt{tm} \to \texttt{tm} \rrbracket_u = \texttt{arrow-i}(\lambda x{:}\texttt{El}\,\texttt{tm}.\ \texttt{arrow-i}\ (\lambda y{:}\texttt{El}\,\texttt{tm}.\ \texttt{app}\ x\ y))$$
$$\llbracket \Gamma; \Psi \vdash \mathsf{lam} : (\texttt{tm} \to \texttt{tm}) \to \texttt{tm} \rrbracket_u = \texttt{arrow-i}(\lambda f{:}\texttt{El}\,(\texttt{arrow tm tm}).\ \texttt{lam}\ (\lambda x{:}\texttt{El}\,\texttt{tm}.\ \texttt{arrow-e}\ f\ x))$$

Interpretation of domain-level substitutions where $u{:}\,\texttt{El}\ e$ and $\llbracket \Gamma \vdash \Phi : \texttt{ctx} \rrbracket = e$

$$\llbracket \Gamma; \Psi \vdash \cdot : \cdot \rrbracket_u = \texttt{terminal}$$
$$\llbracket \Gamma; \Psi \vdash (\sigma, M) : \Phi, x{:}A \rrbracket_u = \texttt{pair}\ e_1\ e_2 \qquad \text{where } \llbracket \Gamma; \Psi \vdash \sigma : \Phi \rrbracket_u = e_1 \text{ and } \llbracket \Gamma; \Psi \vdash M : A \rrbracket_u = e_2$$
$$\llbracket \Gamma; \Psi, \overrightarrow{x{:}A} \vdash \mathsf{wk}_{\widehat{\Phi}} : \Phi \rrbracket_u = \texttt{fst}_n\ u \qquad \text{where } n = |\overrightarrow{x{:}A}|$$

Fig. 4. Interpretation of Domain-level Types and Terms

is simply a variable. However, when we translate $\Gamma; \Phi \vdash \lambda x.M : A \to B$ given $u{:}\,\texttt{El}\, e$ where $\llbracket \Gamma \vdash \Psi : \texttt{ctx} \rrbracket = e$, we need to recursively translate M in the extended domain-level context $\Psi, x{:}A$ and hence we also need to build a term $\texttt{pair}\ u\ x$ that inhabits $\texttt{El}\,(\texttt{times}\ e\ \llbracket A \rrbracket)$. The translation of $\Gamma; \Phi, x{:}A \vdash M : A$ will return a term e that may contain x. However, note that x will eventually be bound in $\texttt{arrow-i}\ (\lambda x{:}\texttt{El}\ \llbracket A \rrbracket.\ e)$ When we translate a variable x where $\Phi = \Phi_0, x{:}A, y_k{:}A_k, \ldots, y_1{:}A_1$, we return $\texttt{fst}_k\ (\texttt{snd}\ u)$. We translate $\Gamma; \Phi \vdash \lfloor t \rfloor_\sigma : A$ directly using let box-construct where the domain-level substitution σ is simply translated into a pair. As the computation t has the contextual type $\lceil \Psi \vdash \texttt{tm} \rceil$ its translation will be of type $\flat(\texttt{El}\, e \to \texttt{El}\,\texttt{tm})$ where $e = \llbracket \Gamma \vdash \Psi : \texttt{ctx} \rrbracket$. Hence we simply can extract a function $x{:}(\texttt{El}\, e \to \texttt{El}\,\texttt{tm})$ using let box construct and pass to it the interpretation of σ. The translation of domain-level applications and domain-level constants app and lam is straightforward.

The interpretation of a contextual types $\lceil \Psi \vdash A \rceil$ makes explicit the fact that they correspond to functions $\texttt{El}\, e \to \texttt{El}\ \llbracket A \rrbracket$ where $e = \llbracket \Gamma \vdash \Psi : \texttt{ctx} \rrbracket$ (see Fig. 5). Consequently, the corresponding contextual object $(\widehat{\Phi} \vdash M)$ is interpreted as a

Interpretation of contextual objects (C)

$$[\![\Gamma \vdash (\widehat{\Phi} \vdash M) : (\Phi \vdash A)]\!] = \lambda u{:}\,\mathtt{El}\,e.\,e' \quad \text{where } [\![\Gamma \vdash \Phi : \mathtt{ctx}]\!] = e \text{ and } [\![\Gamma; \Phi \vdash M : A]\!]_u = e'$$

$$[\![\Gamma \vdash (\widehat{\Phi} \vdash M) : (\Phi \vdash_v A)]\!] = \lambda u{:}\,\mathtt{El}\,e.\,e' \quad \text{where } [\![\Gamma \vdash \Phi : \mathtt{ctx}]\!] = e \text{ and } [\![\Gamma; \Phi \vdash M : A]\!]_u = e'$$

Interpretation of contextual types (T)

$$[\![\Gamma \vdash (\Phi \vdash A)]\!] = (u{:}\mathtt{El}\,e) \to \mathtt{El}\,[\![A]\!] \quad \text{where } [\![\Gamma \vdash \Phi : \mathtt{ctx}]\!] = e$$

$$[\![\Gamma \vdash (\Phi \vdash_v A)]\!] = (u{:}\mathtt{El}\,e) \to_v \mathtt{El}\,[\![A]\!] \quad \text{where } [\![\Gamma \vdash \Phi : \mathtt{ctx}]\!] = e$$

Fig. 5. Interpretation of Contextual Objects and Types

Interpretation of computation-level types ($\check{\tau}$)

$$[\![\lceil T \rceil]\!] = \flat[\![T]\!]$$

$$[\![(x{:}\check{\tau}_1) \Rightarrow \tau_2]\!] = (x{:}[\![\check{\tau}_1]\!]) \to [\![\tau_2]\!]$$

$$[\![\mathtt{ctx}]\!] = \mathtt{Obj}$$

Computation-level typing contexts (Γ)

$$[\![\cdot]\!] = \cdot$$

$$[\![\Gamma,\, x{:}\,\check{\tau}]\!] = [\![\Gamma]\!],\, x{:}\,[\![\check{\tau}]\!]$$

Interpretation of computations ($\Gamma \vdash t : \tau$; without recursor)

$$[\![\Gamma \vdash \lceil C \rceil : \lceil T \rceil]\!] = \mathtt{box}\ e \quad \text{where } [\![\Gamma \vdash C : T]\!] = e$$

$$[\![\Gamma \vdash t_1\ t_2 : \tau]\!] = e_1\ e_2 \quad \text{where } [\![\Gamma \vdash t_1 : (x{:}\check{\tau}_2) \Rightarrow \tau]\!] = e_1 \text{ and } [\![\Gamma \vdash t_2 : \check{\tau}_2]\!] = e_2$$

$$[\![\Gamma \vdash \mathsf{fn}\ x \Rightarrow t : (x{:}\check{\tau}_1) \Rightarrow \tau_2]\!] = \lambda x{:}\,[\![\check{\tau}_1]\!].\,e \quad \text{where } [\![\Gamma, x{:}\check{\tau}_1 \vdash t : \tau_2]\!] = e$$

$$[\![\Gamma \vdash x : \tau]\!] = x$$

Fig. 6. Interpretation of Computation-level Types and Terms – without recursor

function. Similarly, $\lceil \Psi \vdash_v A \rceil$ is mapped to the restricted function space denoted by $\to_v$, which describes functions with bodies that only contain projections.

Last, we give the interpretation of computation-level types, contexts and terms (see Fig. 6). It is mostly straightforward, as we simply map $\lceil T \rceil$ to $\flat[\![T]\!]$ and $\lceil C \rceil$ is simply interpreted as boxed term.

The interpretation of the recursor is also straightforward now (see Fig. 7). In Lemma 4, we expressed a primitive recursion scheme in our internal type theory and defined a term `rec` together with its type. We now interpret every branch of our recursor in the computation-level as a function of the required type in our internal type theory. While this is somewhat tedious, it is straightforward.

We can now show that all well-typed domain-level and computation-level objects are translated into well-typed constructions in our internal type theory. As a consequence, we can show that equality in Cocon is equivalent to the corresponding equivalence in our internal type theoretic interpretation.

Interpretation of recursor for $\mathcal{I} = (\psi : \mathsf{ctx}) \Rightarrow (y : \lceil \psi \vdash \mathtt{tm} \rceil) \Rightarrow \tau$:

$[\![\Gamma \vdash \mathsf{rec}^{\mathcal{I}}(b_v \mid b_{\mathsf{app}} \mid b_{\mathsf{lam}})\ \ \Psi\ t : [\Psi/\psi,\ t/y]\tau]\!] = \mathtt{rec}\ e_v\ e_{\mathtt{app}}\ e_{\mathtt{lam}}\ e_c\ e$
where $[\![\Gamma \vdash b_v : \mathcal{I}]\!] = e_v$, $[\![\Gamma \vdash b_{\mathsf{app}} : \mathcal{I}]\!] = e_{\mathsf{app}}$, $[\![\Gamma \vdash b_{\mathsf{lam}} : \mathcal{I}]\!] = e_{\mathtt{lam}}$,
$[\![\Gamma \vdash \Psi : \mathtt{ctx}]\!] = e_c$ and $[\![\Gamma \vdash t : \lceil \Psi \vdash \mathtt{tm} \rceil]\!] = e$

Interpretation of Variable Branch

$[\![\Gamma \vdash (\psi, x \mapsto t_v) : \mathcal{I}]\!] \qquad = \curlywedge \psi{:}\,\mathtt{Obj}.\ \curlywedge x{:}\,\flat(\mathtt{El}\,\psi \to_v \mathtt{El}\,\mathtt{tm}).\,e$
where $[\![\Gamma, \psi : \mathsf{ctx}, x : \lceil \psi \vdash_{\overline{v}} \mathtt{tm} \rceil \vdash t_v : [x/y]\tau]\!] = e$

Interpretation of Application Branch

$[\![\Gamma \vdash (\psi, m, n, f_n, f_m \mapsto t_{\mathsf{app}}) : \mathcal{I}]\!] = \curlywedge \psi{:}\,\mathtt{Obj}.\ \curlywedge m, n{:}\,\flat(\mathtt{El}\,\psi \to \mathtt{El}\,\mathtt{tm}).$
$\curlywedge f_m{:}\,[\![[m/y]\tau]\!].\ \curlywedge f_n{:}\,[\![[n/y]\tau]\!].\,e$
where $[\![\Gamma, \psi{:}\mathtt{ctx}, m{:}\lceil \psi \vdash \mathtt{tm} \rceil, n{:}\lceil \psi \vdash \mathtt{tm} \rceil \vdash t_{\mathsf{app}} : [\lceil \psi \vdash \mathsf{app}\ \lfloor m \rfloor\ \lfloor n \rfloor \rceil/y]\tau]\!] = e$

Interpretation of Lambda-Abstraction Branch

$[\![\Gamma \vdash (\psi, m, f_m \mapsto t_{\mathsf{lam}}) : \mathcal{I}]\!] \qquad = \curlywedge \psi{:}\,\mathtt{Obj}.\ \curlywedge m{:}\,\flat(\mathtt{El}\,(\mathtt{times}\ \psi\ \mathtt{tm}) \to \mathtt{El}\,\mathtt{tm}).$
$\curlywedge f_m{:}\tau_m.e$
where $[\![[(\psi, x{:}\mathtt{tm})/\psi,\ m/y]\tau]\!] = \tau_m$,
$[\![\Gamma, \psi{:}\mathtt{ctx}, m{:}\lceil \psi, x{:}\mathtt{tm} \vdash \mathtt{tm} \rceil \vdash t_{\mathsf{app}} : [\lceil \psi \vdash \mathsf{lam}\ \lambda x.\lfloor m \rfloor \rceil/y]\tau]\!] = e$

Fig. 7. Interpretation of Recursor

Lemma 5. *The interpretation maintains the following typing invariants:*

– *If* $\Gamma \vdash \Psi\colon \mathit{ctx}$ *then* $[\![\Gamma \vdash \Psi\colon \mathit{ctx}]\!]\colon \mathtt{Obj}$.
– *If* $\Gamma;\ \Psi \vdash M\colon A$ *then* $[\![\Gamma]\!],\ u\colon \mathit{El}\,[\![\Gamma \vdash \Psi : \mathit{ctx}]\!] \vdash [\![\Gamma; \Psi \vdash M\colon A]\!]_u\colon \mathit{El}\,[\![A]\!]$.
– *If* $\Gamma;\ \Psi \vdash \sigma\colon \Psi$ *then* $[\![\Gamma]\!],\ u\colon \mathit{El}\,[\![\Gamma \vdash \Psi : \mathit{ctx}]\!] \vdash [\![\Gamma; \Psi \vdash \sigma\colon \Psi]\!]_u\colon \mathit{El}\,[\![\Psi]\!]$.
– *If* $\Gamma \vdash C\colon T$ *then* $[\![\Gamma]\!] \vdash [\![\Gamma \vdash C : T]\!]\colon [\![T]\!]$.
– *If* $\Gamma \vdash t\colon \tau$ *then* $[\![\Gamma]\!] \vdash [\![\Gamma \vdash t\colon \tau]\!]\colon [\![\tau]\!]$.

The proof goes by induction on derivations.

Proposition 1 (Soundness). *The following are true.*

– *If* $\Gamma;\ \Psi \vdash M \equiv N\colon A$ *then*
$[\![\Gamma]\!],\ u\colon \mathit{El}\,[\![\Psi]\!] \vdash [\![\Gamma;\ \Psi \vdash M : A]\!]_u = [\![\Gamma;\ \Psi \vdash N : A]\!]_u\colon \mathit{El}\,[\![A]\!]$.
– *If* $\Gamma; \Psi \vdash \sigma \equiv \sigma'\colon \Phi$ *then*
$[\![\Gamma]\!], u\colon \mathit{El}\,[\![\Psi]\!] \vdash [\![\Gamma; \Psi \vdash \sigma : \Phi]\!]_u = [\![\Gamma; \Psi \vdash \sigma' : \Phi]\!]_u\colon \mathit{El}\,[\![\Phi]\!]$.
– *If* $\Gamma \vdash t_1 \equiv t_2\colon \tau$ *then* $[\![\Gamma]\!] \vdash [\![\Gamma \vdash t_1 : \tau]\!] = [\![\Gamma \vdash t_2 : \tau]\!]\colon [\![\tau]\!]$.

6 Presheaves on a Small Category with Attributes

To explain the core of our approach as simply as possible, we have concentrated on a simply-typed domain language. In the remaining space, we outline how our approach generalises to dependent domain languages like LF.

We follow the same approach as above. We start from a term model $\mathbb{D}$ of the domain language and then interpret contextual types in the presheaf category $\widehat{\mathbb{D}}$. In the simply-typed case above, $\mathbb{D}$ was a small cartesian closed category. In the dependent case, $\mathbb{D}$ is a small *Category with Attributes*. Categories with attributes (CwAs) [11] are a general notion of model for dependent type theories that is suitable for modelling dependent domain languages like LF.

With this change, we follow essentially the same approach as above. The main difference is that the universe of representables now makes available the CwA-structure of $\mathbb{D}$ instead of the cartesian closed structure. The following section outlines this in analogy to Sec. 4.1.

6.1 Yoneda CwA

In a Yoneda CwA we again have a type for the objects of $\mathbb{D}$, which we now denote `Ctx`. In the term model for LF, these would be the LF contexts. The type $\mathtt{Ty}\ c$ represents (possibly dependent) LF types in context c. Contexts can be built with the constants `nil` and `cons`.

$$\begin{array}{rl} \vdash \mathtt{Ctx}\ \text{type} & \qquad \vdash \mathtt{nil}\colon \mathtt{Ctx} \\ c\colon \mathtt{Ctx} \vdash \mathtt{Ty}\ c\ \text{type} & \qquad \vdash \mathtt{cons}\colon (c\colon \mathtt{Ctx}) \to (a\colon \mathtt{Ty}\ c) \to \mathtt{Ctx} \end{array}$$

Both `Ctx` and $\mathtt{Ty}\ c$ are constant presheaves, i.e. $\flat\mathtt{Ctx} = \mathtt{Ctx}$ and $\flat(\mathtt{Ty}\ c) = \mathtt{Ty}\ c$.

As in Sec. 4.1, we consider the contexts as codes of a universe.

$$c\colon \mathtt{Ctx} \vdash \mathtt{El}\, c\ \text{type}$$

The type $\mathtt{El}\, c$ has the same interpretation as before and is essentially just the Yoneda embedding. The morphisms $c \to d$ of the CwA $\mathbb{D}$ thus appear as functions of type $\mathtt{El}\, c \to \mathtt{El}\, d$.

The axioms of a CwA can be stated using terms and equations in the internal language of $\widehat{\mathbb{D}}$. For example, substitution on types and context projection morphisms are given by the following constants.

$$\begin{array}{c} c, d\colon \mathtt{Ctx} \vdash \mathtt{sub}\colon (a\colon \mathtt{Ty}\ d) \to (f\colon \mathtt{El}\, c \to \mathtt{El}\, d) \to \mathtt{Ty}\ c \\ c\colon \mathtt{Ctx},\ a\colon \mathtt{Ty}\ c \vdash p\colon \mathtt{El}\ \ (\mathtt{cons}\ c\ a) \to \mathtt{El}\, c \end{array}$$

The other components of a CwA are added similarly and the CwA-axioms [11] are expressed in terms of equations for these constants.

The inhabitants of a type can then be captured by the dependent type

$$c\colon \mathtt{Ctx},\ a\colon \mathtt{Ty}\ c,\ u\colon \mathtt{El}\, c \vdash \mathtt{I}\ a\ u\ \text{type}$$

defined by $\mathtt{I}\ a\ u\ :=\ \Sigma v\colon \mathtt{El}\,(\mathtt{cons}\ c\ a).\,(p\ v) = u$. This type contains all values in $\mathtt{El}\,(\mathtt{cons}\ c\ a)$ whose first projection is u. If one considers $u\colon \mathtt{El}\, c$ as a dependent tuple of LF terms (one term for each variable in the context represented by c), then $\mathtt{I}\ a\ u$ represents all the terms that can be appended to this tuple to make it into one of type $\mathtt{El}\,(\mathtt{cons}\ c\ a)$. Indeed, one can define a pairing operation by $\mathtt{pair} := \lambda u.\, \lambda\langle v, p\rangle.\, v$.

$$c\colon \mathtt{Ctx},\ a\colon (\mathtt{Ty}\ c) \vdash \mathtt{pair}\colon (u\colon \mathtt{El}\, c) \to \mathtt{I}\ a\ u \to \mathtt{El}\,(\mathtt{cons}\ c\ a)$$

With these definitions, we can represent dependent contextual types much like the simply-typed ones. Recall that we had interpreted $\Phi \vdash A$ by $\mathtt{El}\, [\![\Phi]\!] \to \mathtt{El}\, [\![A]\!]$ where both $[\![\Phi]\!]$ and $[\![A]\!]$ were terms of type $\mathtt{Obj}$. In the dependent case, A may depend on Φ. The interpretation of Φ is a term $[\![\Phi]\!] \colon \mathtt{Ctx}$, much as before. The interpretation of A takes the dependency into account: $u \colon \mathtt{El}\, [\![\Phi]\!] \vdash [\![A]\!]_u \colon \mathtt{Ty}\ u$. The interpretation of the contextual type $\Phi \vdash A$ will then be:

$$(u \colon \mathtt{El}\, [\![\Phi]\!]) \to \mathtt{I}\ [\![A]\!]_u\ u$$

It may be interesting to note that $(u \colon \mathtt{El}\, c) \to \mathtt{I}\ a\ u$ is isomorphic to the type of sections of $p \colon \mathtt{El}\,(\mathtt{cons}\ c\ a) \to \mathtt{El}\, c$.

Object-level term constants in LF can be lifted using $\mathtt{I}$. Consider, for example, an encoding of the simply-typed lambda-calculus in LF. It represents only well-typed terms by means of the constants $\mathtt{app} \colon \Pi a, b \colon \mathtt{ty}.\ \mathtt{tm}\,(\mathtt{arr}\ a\ b) \to \mathtt{tm}\ a \to \mathtt{tm}\ b$ and $\mathtt{lam} \colon \Pi a, b \colon \mathtt{ty}.\ (\mathtt{tm}\ a \to \mathtt{tm}\ b) \to \mathtt{tm}\ (\mathtt{arr}\ a\ b)$. Therein, the type $\mathtt{tm}$ of object-level terms is dependent on an object-level type $\mathtt{ty}$, which may be built using a constant $\mathtt{o} \colon \mathtt{ty}$ for a base type and a constant $\mathtt{arr} \colon \mathtt{ty} \to \mathtt{ty} \to \mathtt{ty}$ for function types. This encoding lifts to the Yoneda CwA as in simply-typed case:

$$c \colon \mathtt{Ctx} \vdash \mathtt{ty} \colon \mathtt{Ty}\ c \qquad \Gamma \vdash \mathtt{o} \colon \mathtt{I}\ \mathtt{ty}\ u$$
$$c \colon \mathtt{Ctx} \vdash \mathtt{tm} \colon \mathtt{Ty}\ (\mathtt{cons}\ c\ \mathtt{ty}) \qquad \Gamma \vdash \mathtt{arr} \colon \mathtt{I}\ \mathtt{ty}\ u \to \mathtt{I}\ \mathtt{ty}\ u \to \mathtt{I}\ \mathtt{ty}\ u$$
$$\Delta \vdash \mathtt{app} \colon \mathtt{I}\ \mathtt{tm}\ (\mathtt{pair}\ u\ (\mathtt{arr}\ a\ b)) \to \mathtt{I}\ \mathtt{tm}\ (\mathtt{pair}\ u\ a) \to \mathtt{I}\ \mathtt{tm}\ (\mathtt{pair}\ u\ b)$$
$$\vdash \mathtt{lam} \colon (\mathtt{I}\ \mathtt{tm}\ (\mathtt{pair}\ u\ a) \to \mathtt{I}\ \mathtt{tm}\ (\mathtt{pair}\ u\ b)) \to \mathtt{I}\ \mathtt{tm}\ (\mathtt{pair}\ u\ (\mathtt{arr}\ a\ b))$$

Here, Γ abbreviates $c \colon \mathtt{Ctx}$, $u \colon (\mathtt{El}\, c)$ and Δ abbreviates Γ, $a, b \colon (\mathtt{I}\ \mathtt{ty}\ u)$. Notice how $\mathtt{lam}$ uses higher-order abstract syntax at the meta level.

With these definitions, the interpretation of Cocon is essentially just as before. For working with the dependencies in a Yoneda CwA, we found it very useful to type-check our definitions in Agda, see our sources [1].

7 Conclusion

We have given a rational reconstruction of contextual type theory in presheaf models of higher-order abstract syntax. This provides a semantical way of understanding the invariants of contextual types independently of the algorithmic details of type checking. At the same time, we identify the contextual modal type theory, Cocon, which is known to be normalising, as a syntax for presheaf models of HOAS. By accounting for the Yoneda embedding with a universe á la Tarski, we obtain a manageable way of constructing contextual types in the model, especially in the dependent case. While various forms of universes are being studied in the context of functor categories, e.g. [2,16], we are not aware of previous uses of presheaves over CwAs or similar.

In future work, one may consider using the model as a way of compiling contextual types, by implementing the semantics. In another direction, it may be interesting to apply the syntax of contextual types to other presheaf categories. We also hope that the model will help to guide the further development of Cocon.

Acknowledgements. We thank the anonymous reviewers for helpful feedback.

References

1. The Agda sources for this paper are available from: http://github.com/uelis/contextual.
2. Guillaume Allais, Robert Atkey, James Chapman, Conor McBride, and James McKinna. A type and scope safe universe of syntaxes with binding: Their semantics and proofs. *Proc. ACM Program. Lang.*, 2(ICFP):90:1–90:30, July 2018.
3. Benton, P.N., Bierman, G.M., de Paiva, V., Hyland, M.: A term calculus for intuitionistic linear logic. In: Bezem, M., Groote, J.F. (eds.) Typed Lambda Calculi and Applications, International Conference on Typed Lambda Calculi and Applications, TLCA '93, Utrecht, The Netherlands, March 16-18, 1993, Proceedings. vol. 664, pp. 75–90. Springer (1993)
4. Andrew Barber and Gordon Plotkin. Dual intuitionistic linear logic. Technical Report, LFCS, University of Edinburgh, 1997.
5. John Cartmell. Generalised algebraic theories and contextual categories. *Annals of Pure and Applied Logic*, 32:209 – 243, 1986.
6. Rowan Davies and Frank Pfenning. A modal analysis of staged computation. *Journal of the ACM*, 48(3):555–604, 2001.
7. Peter Dybjer. Internal type theory. In *Types for Proofs and Programs (TYPES'95)*, pages 120–134, 1995.
8. M. Fiore, G. D. Plotkin, and D. Turi. Abstract syntax and variable binding. In *Logic in Computer Science (LICS'99)*, pages 193–202. IEEE Press, 1999.
9. Murdoch Gabbay and Andrew Pitts. A new approach to abstract syntax involving binders. In *Logic in Computer Science (LICS'99)*, pages 214–224. IEEE Press, 1999.
10. Robert Harper, Furio Honsell, and Gordon Plotkin. A framework for defining logics. *Journal of the ACM*, 40(1):143–184, January 1993.
11. Martin Hofmann. *Syntax and Semantics of Dependent Types*, page 79–130. Publications of the Newton Institute. Cambridge University Press, 1997.
12. Martin Hofmann. Semantical analysis of higher-order abstract syntax. In *Logic in Computer Science (LICS'99)*, pages 204–213. IEEE Press, 1999.
13. Furio Honsell, Marino Miculan, and Ivan Scagnetto. An axiomatic approach to metareasoning on nominal algebras in HOAS. In *International Colloquium on Automata, Languages and Programming (ICALP'01)*, LNCS 2076, pages 963–978. Springer, 2001.
14. Bart Jacobs. Comprehension categories and the semantics of type dependency. *Theor. Comput. Sci.*, 107(2):169–207, 1993.
15. Kavvos, G.A.: Intensionality, intensional recursion, and the Gödel-Löb axiom. CoRR **abs/1703.01288** (2017), http://arxiv.org/abs/1703.01288
16. Daniel R. Licata, Ian Orton, Andrew M. Pitts, and Bas Spitters. Internal universes in models of homotopy type theory. In *Formal Structures for Computation and Deduction (FSCD'18)*, pages 22:1–22:17, 2018.
17. Dale Miller and Catuscia Palamidessi. Foundational aspects of syntax. *ACM Comput. Surv.*, 31(3es), 1999.
18. Aleksandar Nanevski, Frank Pfenning, and Brigitte Pientka. Contextual modal type theory. *ACM Transactions on Computational Logic*, 9(3):1–49, 2008.
19. Frank Pfenning and Conal Elliott. Higher-order abstract syntax. In *Symposium on Language Design and Implementation (PLDI'88)*, pages 199–208, June 1988.
20. Brigitte Pientka. A type-theoretic foundation for programming with higher-order abstract syntax and first-class substitutions. In *Principles of Programming Languages (POPL'08)*, pages 371–382. ACM Press, 2008.

21. Brigitte Pientka and Andreas Abel. Well-founded recursion over contextual objects. In *Typed Lambda Calculi and Applications (TLCA'15)*, pages 273–287, 2015.
22. Brigitte Pientka, Andreas Abel, Francisco Ferreira, David Thibodeau, and Rébecca Zucchini. Cocon: Computation in contextual type theory. *CoRR*, abs/1901.03378, 2019.
23. Brigitte Pientka, Andreas Abel, Francisco Ferreira, David Thibodeau, and Rebecca Zucchini. A type theory for defining logics and proofs. In *34th IEEE/ ACM Symposium on Logic in Computer Science (LICS'19)*, pages 1–13, IEEE Computer Society, 2019.
24. Brigitte Pientka and Andrew Cave. Inductive Beluga: Programming Proofs (System Description). In *Conference on Automated Deduction (CADE-25)*, LNCS 9195, pages 272–281. Springer, 2015.
25. Brigitte Pientka and Joshua Dunfield. Programming with proofs and explicit contexts. In *Principles and Practice of Declarative Programming (PPDP'08)*, pages 163–173, 2008.
26. Brigitte Pientka and Joshua Dunfield. Beluga: a framework for programming and reasoning with deductive systems (System Description). In *International Joint Conference on Automated Reasoning (IJCAR'10)*, LNAI 6173, pages 15–21. Springer, 2010.
27. Shulman, M.: Brouwer's fixed-point theorem in real-cohesive homotopy type theory. Mathematical Structures in Computer Science **28**(6), 856–941 (2018)
28. Thomas Streicher. *Semantics of Type Theory.* Birkhäuser, 1991.
29. Andrea Vezzosi. Agda with a flat modality. Available from https://github.com/agda/agda/tree/flat, 2018.

Ambiguity, Weakness, and Regularity in Probabilistic Büchi Automata

Christof Löding and Anton Pirogov(✉) *

RWTH Aachen University, Templergraben 55, 52062 Aachen, Germany
{loeding,pirogov}@cs.rwth-aachen.de

Abstract. Probabilistic Büchi automata are a natural generalization of PFA to infinite words, but have been studied in-depth only rather recently and many interesting questions are still open. PBA are known to accept, in general, a class of languages that goes beyond the regular languages. In this work we extend the known classes of restricted PBA which are still regular, strongly relying on notions concerning ambiguity in classical ω-automata. Furthermore, we investigate the expressivity of the not yet considered but natural class of weak PBA, and we also show that the regularity problem for weak PBA is undecidable.

Keywords: probabilistic · Büchi · automata · ambiguity · weak

1 Introduction

Probabilistic finite automata (PFA) are defined similarly to nondeterministic finite automata (NFA) with the difference that each transition is equipped with a probability (a value between 0 and 1), such that for each pair of state and letter, the probabilities of the corresponding outgoing transitions sum up to 1. PFA have been investigated already in the 1960ies in the seminal paper of Rabin [18]. But while the development of the theory of automata on infinite words also started around the same time [7], the model of probabilistic automata on infinite words has first been studied systematically in [3]. The central model in this theory is the one of probabilistic Büchi automata (PBA), which are syntactically the same as PFA. The acceptance condition for runs is defined as for standard nondeterministic Büchi automata (NBA): a run on an infinite word is accepting if it visits an accepting state infinitely often (see [23,24] for an introduction to the theory of automata on infinite words). In general, for probabilistic automata one distinguishes different criteria of when a word is accepted. In the positive semantics, it is required that the probability of the set of accepting runs is greater than 0, in the almost-sure semantics it has to be 1, and in the threshold semantics it has to be greater than a given value λ between 0 and 1. It is easy to see that PFA with positive or almost-sure semantics can only accept regular languages, because these conditions correspond to the fact that there is an accepting run or

* This work is supported by the German research council (DFG) Research Training Group 2236 UnRAVeL

J. Goubault-Larrecq and B. König (Eds.): FOSSACS 2020, LNCS 12077, pp. 522–541, 2020.
https://doi.org/10.1007/978-3-030-45231-5_27

that all runs are accepting. For infinite words the situation is different, because single runs on infinite words can have probability 0. Therefore, the existence of an accepting run is not the same as the set of accepting runs having probability greater than 0 (similarly, almost-sure semantics is not equivalent to all runs being accepting). And in fact, it turns out that PBA with positive (or almost-sure) semantics can accept non-regular languages [3]. This naturally raises the question under which conditions a PBA accepts a regular language.

In [3] a subclass of PBA that accept only regular languages (under positive semantics) is introduced, called uniform PBA. The definition uses a semantic condition on the acceptance probabilities in end components of the PBA. A syntactic class of PBA that accepts only regular languages (under positive and almost-sure semantics) are the hierarchical PBA (HPBA) introduced in [8]. The state space of HPBA is partitioned into a sequence of layers such that for each pair of state and letter there is at most one transition that does not increase the layer. Decidability and expressiveness questions for HPBA have been studied in more detail in [11,10]. While HPBA accept only regular languages for positive and almost-sure semantics, it is not very hard to come up with HPBA that accept non-regular languages under the threshold semantics [8,11] (see also the example in Figure 2(a) on page 10). Restricting HPBA further such that there are only two layers and all accepting states are on the first layer leads to a class of PBA (called simple PBA, SPBA) that accept only regular languages even under threshold semantics [9].

In this paper, we are also interested in the question under which conditions PBA accept only regular languages. We identify syntactical patterns in the transition structure of PBA whose absence guarantees regularity of the accepted language. These patterns have been used before for the classification of the degree of ambiguity of NFA and NBA [25,19,16]. The degree of ambiguity of a nondeterministic automaton corresponds to the maximal number of accepting runs that a single input word can have. For NBA, the ambiguity can (roughly) be uncountable, countable, or finite. For positive semantics, we show that PBA whose transition structure corresponds to at most countably ambiguous NBA, accept only regular languages. For almost-sure semantics, we need a slightly stronger condition for ensuring regularity. But both classes that we identify are easily seen to strictly subsume the class of HPBA. For the emptiness and universality problems for these classes we obtain the same complexities as the ones for HPBA. In the case of threshold semantics, we show that finite ambiguity is a sufficient condition for regularity of the accepted language, generalizing a corresponding result for PFA from [12]. The class of finitely ambiguous PBA strictly subsumes the class of SPBA.

Besides the relation between regularity and ambiguity in PBA, we also investigate the class of weak PBA (abbreviated PWA). In weak Büchi automata, the set of accepting states is a union of strongly connected components of the automaton. We show that PWA with almost-sure semantics define the same class of languages as PBA with almost-sure semantics (which implies that with positive semantics PWA define the same class as probabilistic co-Büchi automata).

This is in correspondence to results for non-probabilistic automata: weak automata with universal semantics (a word is accepted if all runs are accepting) define the same class as Büchi automata with universal semantics, and nondeterministic weak automata correspond to nondeterministic co-Büchi automata (see, e.g., [17], where weak automata are called weak parity automata). Furthermore, it is known that universal Büchi automata, respectively nondeterministic co-Büchi automata, can be transformed into equivalent deterministic automata (with the same acceptance condition). An analogue of deterministic automata in the probabilistic setting are the so-called 0/1 automata, in which each word is either accepted with probability 0 or with probability 1. It is known that almost-sure PBA can be transformed into equivalent 0/1 PBA (see the proof of Theorem 4.13 in [4]). Concerning weak automata, a language can be accepted by a deterministic weak automaton (DWA) if, and only if, it can be accepted by a deterministic Büchi and by a deterministic co-Büchi automaton (this follows from results in [14], see [6] for a more direct construction). We show an analogous result in the probabilistic setting: The class of languages defined by 0/1 PWA corresponds to the intersection of the two classes defined by PWA with almost-sure semantics and with positive semantics, respectively. It turns out that this class contains only regular languages, that is, 0/1 PWA define the same class as DWA.

We also show that the regularity problem for PBA is undecidable (the problem of deciding for a given PBA whether its language is regular). For PBA with positive semantics this is not surprising, as for those already the emptiness problem is undecidable [4]. However, for PBA with almost-sure semantics the emptiness and universality problems are decidable [1,2,8]. We show that regularity is undecidable already for PWA with almost-sure or with positive semantics. The proof also yields that it is undecidable for a fixed regular language whether a given PWA accepts this language.

This work is organized as follows. After introducing basic notations in Section 2 we first characterize various regular subclasses of PBA that we derive from ambiguity patterns in Section 3 and then we derive some related complexity results in Section 4. In Section 5 we present our results concerning weak probabilistic automata and in Section 6 we conclude.

2 Preliminaries

First we briefly review some basic definitions.

If Σ is a finite alphabet, then Σ^* is the set of all finite and Σ^ω is the set of all infinite *words* $w = w_0 w_1 \dots$ with $w_i \in \Sigma$. For a word w we denote by $w(i)$ the i-th symbol w_i.

Classical automata used in this work have usually the shape $(Q, \Sigma, \Delta, Q_0, F)$, where Q is a finite set of states, Σ a finite alphabet, $\Delta \subseteq Q \times \Sigma \times Q$ is the transition relation and $Q_0, F \subseteq Q$ are the sets of initial and final states, respectively.

We write $\Delta(p, a) := \{q \in Q \mid (p, a, q) \in \Delta\}$ to denote the set of *successors* of $p \in Q$ on symbol $a \in \Sigma$, and $\Delta(P, w)$ for $P \subseteq Q, w \in \Sigma^*$ with the usual meaning, i.e., states reachable on word w from any state in P.

A *run* of an automaton on a word $w \in \Sigma^\omega$ is an infinite sequence of states $q_0, q_1, \ldots$ starting in some $q_0 \in Q_0$ such that $(q_i, w(i), q_{i+1}) \in \Delta$ for all $i \geq 0$. We say that a set of runs is *separated (at time i)* when the prefixes of length i of those runs are pairwise different.

As usual, an automaton is *deterministic* if $|Q_0| = 1$ and $|\Delta(p,a)| \leq 1$ for all $p \in Q, a \in \Sigma$, and *nondeterministic* otherwise. For deterministic automata we may use a transition function $\delta : Q \times \Sigma \to Q$ instead of a relation.

Probabilistic automata we consider have the shape $(Q, \Sigma, \delta, \mu_0, F)$, i.e., the transition relation is replaced by a function $\delta : Q \times \Sigma \times Q \to [0,1]$ which for each state and symbol assigns a probability distribution on successor states (i.e. $\sum_{q \in Q} \delta(p,a,q) = 1$ for all $p \in Q, a \in \Sigma$), and $\mu_0 : Q \to [0,1]$ with $\sum_{q \in Q} \mu_0(q) = 1$ is the initial probability distribution on states. The *support* of a distribution μ is the set $\mathsf{supp}(\mu) := \{x \mid \mu(x) > 0\}$. Similarly as above, we may write $\delta(\mu, w)$ and mean the resulting probability distribution after reading $w \in \Sigma^*$, when starting with probability distribution μ.

For a probabilistic automaton $\mathcal{A}$ the *underlying automaton* $\mathcal{A}^{\triangleleft}$ is given by recovering the transition relation $\Delta := \{(p,x,q) \mid \delta(p,x,q) > 0\}$ of positively reachable states and the initial state set $Q_0 := \mathsf{supp}(\mu_0)$.

As usual, a run of an automaton for finite words is *accepting* if it ends in a final state. For automata on infinite words, run acceptance is determined by the Büchi (run visits infinitely many final states) or Co-Büchi (run visits finitely many final states) conditions.

We write $p \xrightarrow{x} q$ if there exists a path from p to q labelled by $x \in \Sigma^+$ and $p \to q$ if there exists some x such that $p \xrightarrow{x} q$. The *strongly connected component (SCC)* of $p \in Q$ is $\mathsf{scc}(p) := \{q \in Q \mid p = q \text{ or } p \to q \text{ and } q \to p\}$. The set $\mathsf{SCCs}(\mathcal{A}) := \{\mathsf{scc}(q) \mid q \in Q\}$ is the set of all SCCs and partitions Q. An SCC is *accepting* (*rejecting*) if all (no) runs that stay there forever are accepting. An SCC is *useless* if no accepting run can continue from there. An automaton is *weak*, if the set of final states is a union of its SCCs. In this case, Büchi and Co-Büchi acceptance are equivalent and we treat weak automata as Büchi automata.

A classical automaton is *trim* if it has no useless SCCs, whereas a probabilistic automaton is trim if it has at most one useless SCC, which is a rejecting sink that we canonically call q_{rej}. We assume w.l.o.g. that all considered automata are trim, which also means that in an underlying automaton the sink q_{rej} is removed.

We call transitions of probabilistic automata that have probability 1 *deterministic* and otherwise *branching*. If there are transitions $p \xrightarrow{a} q$ and $p \xrightarrow{a} q'$ with $q \neq q'$, we call this pattern a *fork*. Every branching transition clearly has at least one fork. We call a (p, q, q') fork *intra-SCC*, if p, q, q' are all in the same SCC, otherwise it is an *inter-SCC* fork. A run of an automaton is *deterministic* if it never goes through forks, and *limit-deterministic* if it goes only through finitely many forks. We say that two deterministic runs *merge* when they reach the same state simultaneously. For a finite run prefix ρ, we call all valid runs with this prefix *continuations* of ρ.

A classical automaton $\mathcal{A}$ *accepts* $w \in \Sigma^\omega$ if there exists an accepting run on w, and the language $L(\mathcal{A})$ *recognized* by $\mathcal{A}$ is the set of all accepted words. If P is a set of states of an automaton, we write $L(P)$ for the language accepted by this automaton with initial state set P. For sets consisting of one state q, we write $L(q)$ instead of $L(\{q\})$.

For a probabilistic automaton $\mathcal{A}$ and an input word w (finite or infinite), the transition structure of $\mathcal{A}$ induces a probability space on the set of runs of $\mathcal{A}$ on w in the usual way. We do not provide the details here but rather refer the reader not familiar with these concepts to [4]. In general, we write $\mathsf{Pr}(E)$ for the probability of a measurable event E in a probability space. For probabilistic automata, we consider *positive*, *almost-sure* and *threshold* semantics, i.e., an automaton accepts w if the probability of the set of accepting runs on w is > 0, $=1$ or $>\lambda$ (for some fixed $\lambda \in]0,1[$), respectively. For an automaton $\mathcal{A}$ these languages are denoted by $L^{>0}(\mathcal{A}), L^{=1}(\mathcal{A})$ and $L^{>\lambda}(\mathcal{A})$, respectively, whereas $L(\mathcal{A}) := L(\mathcal{A}^{\triangleleft})$ is the language of the underlying automaton. A probabilistic automaton is $0/1$ if all words are accepted with either probability 0 or 1 (in this case the languages with the different probabilistic semantics coincide).

To denote the type of an automaton, we use abbreviations of the form XYA$^{(\gamma)}$ where the type of transition structure is denoted by X $\in$ { D (det.), N (nondet.), P (prob.) }, the acceptance condition is specified by Y $\in$ { F (finite word), B (Büchi), C (Co-Büchi), W (Weak) }, and for probabilistic transitions the semantics for acceptance is given by $\gamma \in \{>0, =1, >\lambda, 0/1\}$.

By $\mathbb{L}^{(\gamma)}(\mathsf{XYA})$ we denote the whole class of languages accepted by the corresponding type of automaton. If $\mathbb{L}$ is a set of languages, then $\overline{\mathbb{L}}$ denotes the set of all complement languages (similarly, for a language L, we denote by $\overline{L}$ its complement), and $\mathsf{BCl}(\mathbb{L})$ the set of all finite boolean combinations of languages in $\mathbb{L}$. We use the notion of *regular language* for finite words and for infinite words (the type of words is always clear from the context).

3 Ambiguity of PBA

Ambiguity of automata refers to the number of different accepting runs on a word or on all words. An automaton is *finitely ambiguous* (on w) if there are at most k different accepting runs (on w) for some fixed $k \in \mathbb{N}$, and in case of at most one accepting run it is called *unambiguous*. If on each word there are only finitely many accepting runs, but no constant upper bound over all words, then it is *polynomially ambiguous* if the number of different run prefixes that are possible for any word prefix of length n can be bounded by a polynomial in n, and otherwise *exponentially ambiguous*. Finally, if if there exist words that have infinitely many runs, but no word on which there are uncountably many accepting runs, then it is *countably ambiguous*, and otherwise it is *uncountably ambiguous*.

In [16] (see also [19]), a syntactic characterization of those classes is presented for NBA by simple patterns of states and transitions. We define those patterns here and refer to [16] for further details. An automaton $\mathcal{A}$ has an *IDA pattern* if there exist two states $p \neq q$ and a word $v \in \Sigma^*$ such that $p \xrightarrow{v} p$, $p \xrightarrow{v} q$ and $q \xrightarrow{v} q$. If additionally $q \in F$, then this is also an IDA_F pattern. Finally, $\mathcal{A}$ has an *EDA pattern* if there exists a state p and $v \in \Sigma^*$ such that there are two different paths $p \xrightarrow{v} p$, and if additionally $p \in F$, this is also an EDA_F pattern. If a PBA has no EDA pattern, we call it *flat*, reflecting the naming of a similar concept in other kinds of transition systems (e.g. [15]). The names IDA and EDA abbreviate "infinite/exponential degree of ambiguity", which they indicated in the original NFA setting, and we keep those names for consistency.

By k-NBA, n^k-NBA, 2^n-NBA, $\aleph_0$-NBA we denote the subsets of at most finitely, polynomially, exponentially and countably ambiguous NBA (and similarly for other types of automata). When speaking about ambiguity of some PBA $\mathcal{A}$, we mean the ambiguity of the trimmed underlying NBA $\mathcal{A}^{\triangleleft}$.

In [8], hierarchical PBA (HPBA) were identified as a syntactic restriction on PBA which ensures regularity under positive and almost-sure semantics. A PBA with a unique initial state is hierarchical, if it admits a ranking on the states such that at most one successor on a symbol has the same rank, and no successor has a smaller rank. A HPBA has k levels if it can be ranked with only k different values. Simple PBA (SPBA) were introduced in [9] and are restricted HPBA with two levels such that all accepting states are on level 0.

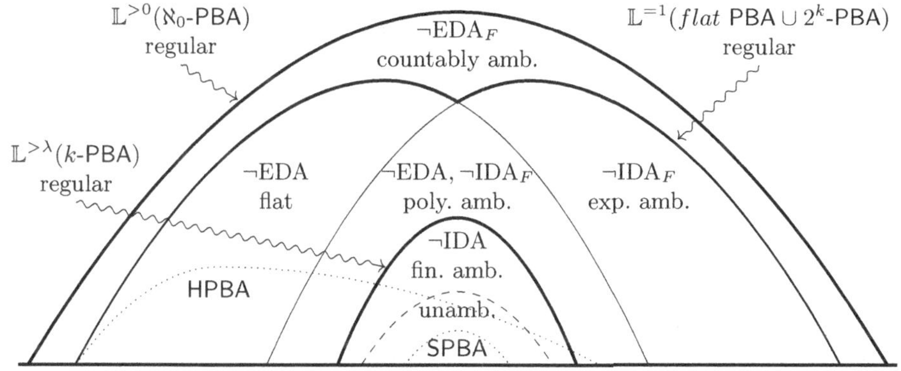

Fig. 1: Illustration of the automata classes with restricted ambiguity as presented for NBA in [16], which are characterized by the absence of the state patterns $\mathrm{IDA}, \mathrm{IDA}_F, \mathrm{EDA}$, and EDA_F and their relation to the restricted classes called "Hierarchical PBA" (HPBA) [8] and "Simple PBA" (SPBA) [9]. We identify classes in this hierarchy which can be seen as extensions "in spirit" of respectively SPBA and HPBA, subsuming them while also preserving their good properties, as e.g. definition by syntactic means, regularity under different semantics and several complexity results.

First, we show how HPBA relate to the ambiguity hierarchy, which can easily be derived by inspection of the definitions. A visual illustration is given in Figure 1.

Proposition 1 (Relation of HPBA and the ambiguity hierarchy).

1. HPBA $\subset$ *flat PBA* $\subset \aleph_0$-PBA.
2. k-PBA $\not\subseteq$ HPBA *and* HPBA $\not\subseteq$ k-PBA.
3. SPBA $\subset$ *unambiguous PBA* $\subset$ k-PBA.

Starting from these observations, this work was motivated by the question whether the ambiguity restrictions, which were only implicit in HPBA and SPBA, can be used explicitly to get larger classes with good properties. In the following we will positively answer this question.

3.1 From classical to probabilistic automata

First, we observe that probabilistic automata can recognize regular languages even under severe ambiguity restrictions.

Proposition 2. *Let $\mathcal{A}$ be a DBA. Then there exists an unambiguous PBA $\mathcal{B}$ such that $L^{>0}(\mathcal{B}) = L^{=1}(\mathcal{B}) = L(\mathcal{A})$.*

Proof. As $\mathcal{A}$ is a (w.l.o.g. complete) DBA, there exists exactly one run on each word and all transitions when seen as PBA must have probability 1. Clearly this unique natural 0/1 PBA obtained from $\mathcal{A}$ accepts the same language under both probable and almost-sure semantics and it is trivially unambiguous. □

Limit-deterministic NBA (LDBA) are NBA which are deterministic in all non-rejecting SCCs. The natural mapping of LDBA into PBA [4, Lemma 4.2] already trivially yields countably ambiguous automata (because the deterministic part of the LDBA cannot contain an EDA_F pattern, which implies uncountable ambiguity [16]). The following result shows that already unambiguous PBA under positive semantics suffice for all regular languages.

Theorem 1. *Let $L \subseteq \Sigma^\omega$ be a regular language.*
Then there exists an unambiguous PBA $\mathcal{B}$ such that $L^{>0}(\mathcal{B}) = L$.

Proof (sketch). Let $\mathcal{A} = (Q, \Sigma, \delta, q_0, c)$ be a deterministic parity automaton accepting L, i.e., a finite automaton with priority function $c : Q \to \{1, \ldots, m\}$ such that $w \in L(\mathcal{A})$ iff the smallest priority assigned to a state on the unique run of $\mathcal{A}$ on w which is seen infinitely often is even.

We construct an unambiguous LDBA for L, which then easily yields a $\text{PBA}^{>0}$ by assigning arbitrary probabilities ([4, Lemma 4.2]) without influencing the ambiguity. If the parity automaton $\mathcal{A}$ has m priorities, the LDBA $\mathcal{B}$ can be obtained by taking $m+1$ copies, where m of them are responsible for one priority each, and one is modified to guess which priority i on the input word is the most important one appearing infinitely often along the run of $\mathcal{A}$, and correspondingly switch into the correct copy. This switching is done unambiguously for the first position after which no priority more important than i appears. □

3.2 From probabilistic to classical automata

First we establish a result for flat PBA, i.e. PBA that have no EDA pattern. In automata without EDA pattern there are no states which are part of two different cycles labeled by the same finite word. Even though we defined flat PBA by using an ambiguity pattern, the set of flat PBA does not correspond to an ambiguity class, but it is useful for our purposes due to the following property:

Lemma 1. *If $\mathcal{A}$ is a flat PBA and $w \in \Sigma^\omega$, then the probability of a run of $\mathcal{A}$ on w to be limit-deterministic is 1.*

Proof. Let $\mathsf{Runs}(\mathcal{A}, w)$ denote the set of all runs of $\mathcal{A}$ on w and $\mathsf{nldRuns}(\mathcal{A}, w)$ denote the subset containing all such runs that are not limit-deterministic. As $\mathcal{A}$ is flat, it has no EDA and thus also no EDA_F pattern, hence $\mathcal{A}$ is at most countably ambiguous (by [16]). Moreover, there are not only at most countably many accepting runs on any word, but also countably many rejecting runs (which can be seen by a simple generalization of [16, Lemma 4]). But as all runs are disjoint events, each run ρ that uses infinitely many forks has probability 0, and the total number of runs is countable, we can see that

$$\mathsf{Pr}(\mathsf{Runs}(\mathcal{A}, w) \setminus \mathsf{nldRuns}(\mathcal{A}, w)) = \sum_{\rho \in \mathsf{Runs}(\mathcal{A}, w)} \mathsf{Pr}(\rho) \quad - \sum_{\rho \in \mathsf{nldRuns}(\mathcal{A}, w)} \mathsf{Pr}(\rho) = 1 - 0 = 1. \qquad \square$$

The following lemma characterizes acceptance of PBA under extremal semantics with restricted ambiguity and is crucial for the constructions in the following sections:

Lemma 2 (Characterizations for extremal semantics).
Let $\mathcal{A}$ be a PBA.

1. *If $\mathcal{A}$ is at most countably ambiguous, then* $w \in L^{>0}(\mathcal{A}) \Leftrightarrow$ *there exists an accepting run on w that is limit-deterministic.*
2. *If there are finitely many accepting runs of $\mathcal{A}$ on w, then* $w \in L^{=1}(\mathcal{A}) \Leftrightarrow$ *all runs on w are accepting and limit-deterministic.*
3. *If $\mathcal{A}$ is flat, then* $w \in L^{=1}(\mathcal{A}) \Leftrightarrow$ *there is no limit-deterministic rejecting run on w.*

Proof. (1.) : For contradiction, assume that every accepting run on w goes through forks infinitely often. But then the probability of every individual accepting run on w is 0. Each run is a measurable event (it is a countable intersection of finite prefixes) and clearly disjoint from other runs, as two different runs must eventually differ after a finite prefix. But as the number of accepting runs is countable by assumption, by σ-additivity it follows that the probability of all accepting runs is also 0, contradicting the fact that $w \in L^{>0}(\mathcal{A})$.

For the other direction, pick a limit-deterministic accepting run ρ of $\mathcal{A}$ on w and let $uv = w$ and $q \in Q$ such that the state of ρ after reading u is q and there are no forks visited on v. Clearly, the probability to be in q after u in a run of $\mathcal{A}$ is positive (because u is finite), and the probability that $\mathcal{A}$ continues like ρ from q on v is 1. Hence, the probability of ρ is positive.

(2.) : The ($\Leftarrow$) direction is obvious. We now proceed to show ($\Rightarrow$). Take some time t after which all accepting runs on w separated. Assume that some accepting run ρ is not limit-deterministic. But then ρ goes through infinitely many forks after t which with positive probability lead to a successor from which the probability to accept is 0, and the probability of following ρ is also 0. As the probability to follow ρ until time t is positive, but after that the probability to accept is 0, this implies that there is a positive probability that $\mathcal{A}$ rejects w. Therefore, all accepting runs on w must be limit-deterministic. Now assume that some run ρ on w is rejecting. Following this run until the time at which ρ is separated from all accepting runs has positive probability and all continuations must be also rejecting, so $\mathcal{A}$ must reject w.

(3.) : Clearly ($\Rightarrow$) holds, because a limit-deterministic rejecting run has positive probability, i.e., if such a run exists on w, then $\mathcal{A}$ cannot accept almost surely. For ($\Leftarrow$), observe that because $\mathcal{A}$ is flat, we know by Lemma 1 that with probability 1 runs are limit-deterministic. Hence, if there exists no limit-deterministic rejecting run on w (which would have positive probability), then with probability 1 runs are limit-deterministic and accepting. □

Using these characterizations, we can provide simple constructions from probabilistic to classical automata.

Theorem 2. *Let $\mathcal{A}$ be a PBA that is at most countably ambiguous. Then $L^{>0}(\mathcal{A})$ is a regular language.*

Proof (sketch). An NBA construction taking two copies of the PBA, where in the first copy no state is accepting and the second copy has no forks, with the purpose of guessing a limit-deterministic accepting run. □

Corollary 1. *If $L^{>0}(\mathcal{A})$ is not regular, then it contains an* EDA_F *pattern.*

Theorem 3. *Let $\mathcal{A}$ be a PBA that is at most exponentially ambiguous or flat. Then $L^{=1}(\mathcal{A})$ is regular and recognizable by DBA.*

Proof (sketch). Both cases (exp. ambiguous or flat) shown using a deterministic breakpoint construction resulting in a DBA. In one case it checks whether all runs are accepting, in the other it checks that there are no limit-deterministic rejecting runs. □

Corollary 2. *If $L^{=1}(\mathcal{A})$ is not regular, then $\mathcal{A}$ contains both an* EDA *and an* IDA_F *pattern.*

The corollaries above follow directly from the theorems and the syntactic characterization of ambiguity classes [16]. The following proposition states that these characterizations of regularity in terms of the ambiguity patterns are tight.

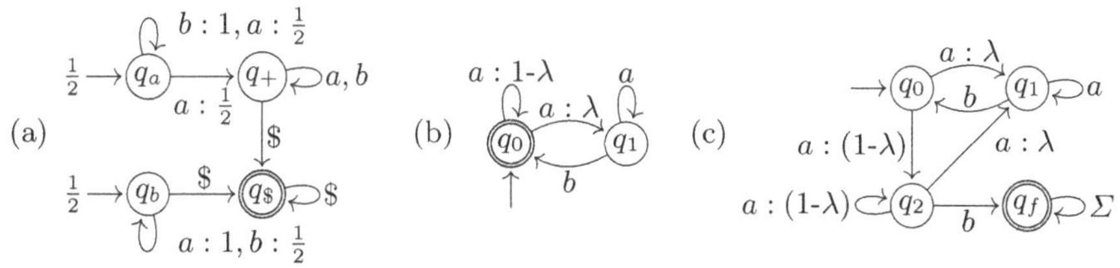

Fig. 2: (a) Some PWA which accepts the non-regular language $\{ w = (a+b)^*\$^\omega \mid \#_a(w) > \#_b(w) \}$ with a threshold of $\frac{1}{2}$, where $\#_x(w)$ denotes the number of occurrences of $x \in \Sigma$ in $w \in \Sigma^\omega$. (b) A family of PBA $\mathcal{P}_\lambda$ from [4] such that $\mathbb{L}^{>0}(\mathcal{P}_\lambda)$ is not regular for any $\lambda \in \mathbb{R}$. (c) A family of PWA $\tilde{\mathcal{P}}_\lambda$ (closely related to [4, Fig. 6]) such that $\mathbb{L}^{=1}(\tilde{\mathcal{P}}_\lambda)$ is not regular for any $\lambda \in \mathbb{R}$.

Proposition 3. *There exist PBA...*

1. *...with* EDA_F *pattern (i.e. uncountably ambiguous) that accept non-regular languages under positive semantics.*
2. *...with no* EDA_F *pattern (i.e. countably ambiguous) that accept non-regular languages under almost-sure semantics.*

Proof. (1.) Note that this statement just means that there are PBA accepting non-regular languages, which is well known. For example, the automata family from [4, Fig. 3], depicted in Figure 2(b), accepts non-regular languages under positive semantics and clearly contains an EDA_F pattern, e.g. there are two different paths from p_0 to p_0 on the word *aab*.

(2.) The automata family depicted in Figure 2(c) is a simple modification of the PBA family depicted in [4, Fig. 6] and recognizes the same non-regular languages under almost-sure semantics. It does not contain an EDA_F pattern, because the accepting state is a sink, but it does contain an IDA_F and an EDA pattern (both e.g. on *aab*), so it is countably ambiguous and not flat. □

This completes our classification of regular subclasses of PBA under extremal semantics that are defined by ambiguity patterns, showing that going beyond the restricted classes presented above (by allowing more patterns) in general leads to a loss of regularity.

Notice that the presented constructions do not track exact probabilities, just whether transitions have a probability > 0 or $= 1$. This is a noteworthy observation, as in general, the probabilities do matter for PBA, as shown in [4, Thm. 4.7, Thm. 4.11].

Proposition 4. *Let* $\mathcal{A}$ *be a PBA. The exact probabilities in* $\mathcal{A}$ *do not influence* $L^{>0}(\mathcal{A})$ *if* $\mathcal{A}$ *is at most countably ambiguous, and* $L^{=1}(\mathcal{A})$ *if* $\mathcal{A}$ *is at most exponentially ambiguous or flat.*

3.3 Threshold Semantics

In this section we consider PBA under threshold semantics and we will see that in this setting, we lose regularity much earlier than in the case of extremal semantics, but there is still the large and natural subclass of finitely ambiguous PBA that retains regularity. Before we can show this, we need to derive a suitable characterization of such languages.

We derive it from the following simple observation, which was also used more implicitly in the proof that Simple HPBA with threshold semantics are equivalent to DBA in [9].

Lemma 3. *Let $\mathcal{A}$ be a PBA. Then for every threshold $\lambda \in]0,1]$, there exists a finite set of probability values $V_{\geq\lambda} \subset [\lambda, 1]$ such that for every finite run prefix with probability v in $\mathcal{A}$ we have $v \geq \lambda \Rightarrow v \in V_{\geq\lambda}$.*

Proof. Observe that given a finite set of real numbers $R \subset [0,1]$, the set $R_{\geq\lambda} := \{r \mid r = \prod_i r_i \geq \lambda,\ r_i \in R\}$ must be finite, as in any sequence $p_1p_2\dots$ of $p_i \in R$, only at most $m = \lceil \log_\lambda(\max R) \rceil$ values can be < 1 and such that the product of the sequence remains $\geq \lambda$. In our case, let R be the set of distinct probabilities assigned to edges (including the initial edges) in $\mathcal{A}$. As every finite run prefix by definition has the probability given by the product of the edge probabilities, this implies the statement. □

If there is just one accepting run (i.e., the automaton is unambiguous), one can easily construct a nondeterministic automaton that guesses an accepting run and tracks it along with its probability value, of which there are only finitely many above the threshold. In the case that there are multiple accepting runs, for acceptance only the sum of their probabilities matters. As individual runs can in principle have arbitrarily small probability values, it is not obvious that the same approach (tracking a set of runs) can work. Determining a suitable cut-off point is not as simple, because it is not apparent when a single run becomes so improbable that it does not matter among the others. However, we will now show that such a cut-off point must exist:

Lemma 4. *Let $\mathcal{A}$ be a PBA, $\lambda \in]0,1]$ a threshold and $k \in \mathbb{N}$. There exists $\varepsilon_k \in]0,\lambda]$ such that for all sets $R^t = \{\rho_i^t\}_{i=1}^j$ of at most $j \leq k$ different run prefixes in $\mathcal{A}$ of the same length $t \in \mathbb{N}$, $\mathsf{Pr}(R^t) = \sum_{i=1}^j \mathsf{Pr}(\rho_i^t) < \lambda$ implies that $\mathsf{Pr}(R^t) < \lambda - \varepsilon_k$.*

Proof. We prove this by induction on the number of runs k. For $k = 1$, i.e. a single run prefix, let $V_{\geq\lambda}$ be the finite (by Lemma 3) set of different probability values $\geq \lambda$ and let E be the set of distinct probabilities in the automaton $\mathcal{A}$. Then clearly $v_{\max,<\lambda} := \max\{a \cdot b \mid a \cdot b < \lambda, a \in V_{\geq\lambda}, b \in E\}$ is the largest probability value $< \lambda$ that can correspond to a finite run prefix in $\mathcal{A}$. Hence, we can just pick an $\varepsilon_1 < \lambda - v_{\max,<\lambda}$ and immediately get that for any run prefix with probability $v < \lambda$, we have that $v \leq v_{\max,<\lambda} < \lambda - \varepsilon_1$.

Now assume the statement holds for all sets with at most k run prefixes. Let R^t be a set of $k+1$ of different run prefixes of the same length such that

$\mathsf{Pr}(R^t) < \lambda$ and let $\varepsilon := \varepsilon_k$. Then we know that for every subset S of at most k runs of R^t we have $\mathsf{Pr}(S) < \lambda - \varepsilon$. Also, every single run prefix can by Lemma 3 have one of only finitely many probability values in $V_{\geq \varepsilon}$ that are $\geq \varepsilon$ and there exists a value $v_{\max,<\varepsilon}$ denoting the largest possible probability value $< \varepsilon$ that a single run prefix can have.

If there exists a run prefix $\rho \in R^t$ with probability value $v < \varepsilon$, then we know that $\mathsf{Pr}(R^t) = \mathsf{Pr}(R^t \setminus \{\rho\}) + v < (\lambda - \varepsilon) + v_{\max,<\varepsilon} < \lambda$. If every run in R^t has a probability value $\geq \varepsilon$, then every run prefix in R^t has as probability one of the values in $V_{\geq \varepsilon}$. Consider all sums of k values from $V_{\geq \varepsilon}$, which are finitely many, and pick the largest sum s which is $< \lambda$. Choose ε_{k+1} such that $\varepsilon_{k+1} < \min(\varepsilon - v_{\max,<\varepsilon}, \lambda - s)$ to account for both cases. □

From this we can derive the following characterization of languages accepted by finitely ambiguous PBA under threshold semantics:

Lemma 5. *Let $\mathcal{A}$ be a k-ambiguous PBA and $\lambda \in]0,1]$ a threshold. There exists an $\varepsilon \in]0, \lambda]$ such that for all $w \in \Sigma^\omega$: $w \in L^{>\lambda}(\mathcal{A})$ iff there exists a set R of limit-deterministic accepting runs of $\mathcal{A}$ on w with $\mathsf{Pr}(R) > \lambda$, $\mathsf{Pr}(S) \leq \lambda$ for all $S \subset R$ and at most one run $\rho \in R$ with $\mathsf{Pr}(\rho) < \varepsilon$.*

Proof. Clearly ($\Leftarrow$) holds, as then w is accepted with probability $\geq \mathsf{Pr}(R) > \lambda$. We now show ($\Rightarrow$). In a finitely ambiguous PBA there are only finitely many different accepting runs on each word. Furthermore, as after finite time all accepting runs have separated and each accepting run that visits forks infinitely often has probability 0, accepting runs that visit forks infinitely often do not contribute positively to the acceptance probability and thus can be ignored. Hence, if $w \in L^{>\lambda}(\mathcal{A})$, there is a number of accepting runs that eventually all become deterministic and each such run has a positive probability, which must in total be $> \lambda$.

Let R be a set of different limit-deterministic accepting runs of $\mathcal{A}$ on w such that $\mathsf{Pr}(R) > \lambda$ and $\mathsf{Pr}(S) \leq \lambda$ for all $S \subset R$. As there are only finitely many accepting runs, such a set R must exist. Furthermore, notice that each limit-deterministic run has a finite prefix which has the same probability as the whole run, so there exists a time t such that the probability of the set of all different prefixes of runs in R of length t is exactly $\mathsf{Pr}(R)$, so that Lemma 4 applies.

Now pick an $\varepsilon := \varepsilon_k$ given by Lemma 4. We claim that at most one run $\rho \in R$ can have a probability less than ε. If there is no such run in R, we are done. Otherwise let ρ be a run with $\mathsf{Pr}(\rho) =: p < \varepsilon$ and notice that by choice of R, we have that $\mathsf{Pr}(R \setminus \{\rho\}) =: s \leq \lambda$. It cannot be the case that $s < \lambda$, as then by Lemma 4 we have $s < \lambda - \varepsilon$, which implies that $\mathsf{Pr}(R) = s + p < \lambda$, which is a contradiction. Hence, now assume that $s = \lambda$. But then, if there is any $\rho' \neq \rho \in R$ such that $\mathsf{Pr}(\rho') =: p' < \varepsilon$, by the same argument we get the contradiction that $s - p' < \lambda - \varepsilon$ and hence $s < \lambda$. Therefore, no other run in R can have a probability $< \varepsilon$. □

Now we can perform the intended automaton construction to show:

Theorem 4. *$L^{>\lambda}(\mathcal{A})$ is regular for each k-ambiguous PBA $\mathcal{A}$ and $\lambda \in]0, 1[$.*

Proof (sketch). We use the characterization of Lemma 5 to construct a generalized Büchi automaton accepting $L^{>\lambda}(\mathcal{A})$. Intuitively, the new automaton just guesses at most k different runs of $\mathcal{A}$ and verifies that the guessed runs are limit-deterministic and accepting. The automaton additionally tracks the probability of the runs over time, to determine whether the individual runs and their sum have enough "weight". The automaton rejects when the total probability of the guessed runs is $\leq \lambda$, one of the runs goes into the rejecting sink q_{rej} or a run does not see accepting states infinitely often.

By Lemma 5 we only need to consider sets of runs with at most one run that has a probability $< \varepsilon$, where $\varepsilon := \varepsilon_k$ is given by Lemma 4. For this single run we also do not need to track the exact probability value, as its only purpose is to witness that the acceptance probability is strictly greater than λ, whereas all other runs must have one of the finitely many different probabilities which are $\geq \varepsilon$ and must sum to λ. □

This generalizes the corresponding result for PFA [12, Theorem 3]. The proof in [12] uses similar concepts, though a rather different presentation. In the setting of infinite words we additionally have to deal with a single run that has arbitrarily low probability, and we have to ensure that this probability remains positive.

After seeing that finitely ambiguous PBA retain regularity, we show that this is the best we can do under threshold semantics:

Corollary 3. *There are polynomially ambiguous PBA $\mathcal{A}$, that is, with an* IDA *pattern and no* EDA, IDA$_F$ *patterns, such that $L^{>\lambda}(\mathcal{A})$ is not regular even for rational thresholds $\lambda \in]0, 1[$.*

Proof. Follows from the fact that the PWA $\mathcal{A}$ from Figure 2(a), which recognizes a non-regular language (and is used to show Proposition 6), has just an IDA pattern in the underlying NBA, but no EDA or IDA$_F$ patterns. □

This completes our characterization of languages which are recognized by PBA that are restricted by forbidden ambiguity patterns, so that we can state our main result of this section (see Figure 1 for a visualization):

Theorem 5. *The following results hold about PBA with restricted ambiguity:*

- $\mathbb{L}^{>0}(k\text{-PBA}) = \mathbb{L}^{>0}(\aleph_0\text{-PBA}) = \mathbb{L}(\text{NBA})$
- $\mathbb{L}^{=1}(k\text{-PBA}) = \mathbb{L}^{=1}(2^k\text{-PBA}) = \mathbb{L}^{=1}(\textit{flat}\ \text{PBA}) = \mathbb{L}(\text{DBA}) \subset \mathbb{L}^{=1}(\aleph_0\text{-PBA})$
- $\mathbb{L}^{>\lambda}(k\text{-PBA}) = \mathbb{L}(\text{NBA}) \subset \mathbb{L}^{>\lambda}(n^k\text{-PBA})$

Proof. The statements follow from the following inclusion chains:

$$\mathbb{L}(\text{NBA}) \overset{(1.)}{\subseteq} \mathbb{L}^{>0}(k\text{-PBA}) \overset{def.}{\subseteq} \mathbb{L}^{>0}(\aleph_0\text{-PBA}) \overset{(2.)}{\subseteq} \mathbb{L}(\text{NBA})$$

$$\mathbb{L}(\text{DBA}) \overset{(3.)}{\subseteq} \mathbb{L}^{=1}(k\text{-PBA}) \overset{def.}{\subseteq} \mathbb{L}^{=1}(2^k\text{-PBA} \cup \text{flat PBA}) \overset{(4.)}{\subseteq} \mathbb{L}(\text{DBA}) \overset{(5.)}{\subset} \mathbb{L}^{=1}(\aleph_0\text{-PBA})$$

$$\mathbb{L}(\text{NBA}) \overset{(1.)}{\subseteq} \mathbb{L}^{>0}(k\text{-PBA}) \overset{(6.)}{\subseteq} \mathbb{L}^{>\lambda}(k\text{-PBA}) \overset{(7.)}{\subseteq} \mathbb{L}(\text{NBA}) \overset{(8.)}{\subset} \mathbb{L}^{>\lambda}(n^k\text{-PBA})$$

Where the marked relationships hold due to: (1.) Theorem 1, (2.) Theorem 2, (3.) Proposition 2, (4.) Theorem 3, (5.) Proposition 3, (6.) Simple transformation by adding a new accepting sink q_{acc} and modifying the initial distribution μ_0 [4, Lemma 4.16], (7.) Theorem 4, (8.) Corollary 3, and (def.) by definition of the ambiguity-restricted automata classes. □

4 Complexity results

In this section, we state some upper and lower bounds on the complexity for deciding emptiness and universality for PBA with restricted ambiguity, derived from the characterizations and constructions presented above.

Theorem 6.

1. *the emptiness problem for* $\aleph_0$*-PBA*$^{>0}$ *is in* NL
2. *the universality problem for* $\aleph_0$*-PBA*$^{>0}$ *is in* PSPACE
3. *the universality problem for at most exp. ambiguous or flat PBA*$^{=1}$ *is in* NL

Proof. (1. + 2.) : By Theorem 2 the languages of $\aleph_0$-PBA$^{>0}$ are regular. The construction of an NBA just uses two copies of the given PBA. For emptiness, it thus suffices to guess an accepted ultimately periodic word and verify that it is accepted by the NBA, which can be done in NL. Since universality for NBA in in PSPACE [21], we also obtain (2.).

(3.): If the automaton is at most exponentially ambiguous, there are only finitely many accepting runs on each word and as we know by Lemma 2 that $w \in L^{=1}(\mathcal{A})$ iff all runs are accepting, it suffices to guess a rejecting run in $\mathcal{A}^{\triangleleft}$, which implies that the ultimately periodic word w labelling that run can not be in $L^{=1}(\mathcal{A})$. If the automaton is flat, then we know that for each rejected word there must exist a limit-deterministic rejecting run in the underlying NBA, which we also can guess. □

Type	regular? >0	$=1$	$>\lambda$	Emptiness >0	Emptiness $=1$	Universality >0	Universality $=1$
k-PBA							
n^k-PBA				∈ NL	∈ PSPACE	∈ PSPACE	∈ NL
2^n-PBA	✓		✗				
flat PBA				∈ NL c.	∈ PSPACE c.	∈ PSPACE c.	∈ NL c.
$\aleph_0$-PBA							∈ PSPACE

Table 1: Summary of main results from Theorems 5 and 6 concerning PBA with ambiguity restrictions. The completeness results follow from the hardness results for HPBA (which are subsumed by flat PBA) from [8, Section 5], the PSPACE inclusion of universality for almost-sure $\aleph_0$-PBA follows from [8, Theorem 4.4].

Observe that $\aleph_0$-PBA$^{>0}$ subsume HPBA$^{>0}$ and the union of flat PBA$^{=1}$ and exp. ambiguous PBA$^{=1}$ subsumes HPBA$^{=1}$, while preserving the same complexity of the emptiness and universality problems. A summary of the main results from Theorem 5 and Theorem 6 is presented in Table 1.

We conclude with an observation relevant to the question about feasibility of PBA with restricted ambiguity for the purpose of application in e.g. model-checking or synthesis.

Proposition 5 (Relationship to classical formalisms).

- *There is a doubly-exponential lower bound for translation from LTL formula to countably ambiguous PBA with positive semantics.*
- *There is an exponential lower bound for conversion from NBA to countably ambiguous PBA with positive semantics.*

Proof. It is known [20, Theorem 2] that there is a doubly-exponential lower bound from LTL to LDBA. It is also known that LTL to NBA has an exponential lower bound (e.g. [5, Theorem 5.42]), which implies an exponential lower bound from NBA to LDBA.

By Theorem 2 there is a polynomial transformation from countably ambiguous PBA with positive semantics into LDBA, which together with the aforementioned bounds implies the claimed lower bounds. □

5 Weakness in Probabilistic Büchi Automata

In this section we investigate the class of probabilistic weak automata (PWA), establishing the relation between different classes defined by PWA as shown in Figure 3 (see also the description of our contribution in the introduction).

As a first remark, notice that PWA can be "complemented" by inverting accepting and rejecting states and switching between dual semantics, e.g., for a PWA $\mathcal{A}$ we have $\overline{L^{>0}(\mathcal{A})} = L^{=1}(\overline{\mathcal{A}})$, where $\overline{\mathcal{A}}$ is just $\mathcal{A}$ with inverted accepting state set $F' = Q \setminus F$.

Since the overarching theme of this paper is trying to find regular subclasses of PBA, we will next establish the following result, showing that there is no hope to find a complete syntactical characterization of regularity in PBA:

Theorem 7. *The regularity of PWA (and therefore of PBA) under positive, almost-sure and threshold semantics is an undecidable problem.*

Proof (sketch). Since $\mathbb{L}^{>\lambda}(\text{PWA}) \supseteq \mathbb{L}^{>0}(\text{PWA})$ (see Theorem 10), $\mathbb{L}^{>0}(\text{PWA}) = \overline{\mathbb{L}^{=1}(\text{PWA})}$, and the class of regular ω-languages is closed under complement, it suffices to show the statement for $\text{PWA}^{=1}$. We do this by reduction from the value 1 problem for PFA, which is the question whether for each $\varepsilon > 0$ there exists a word accepted by the PFA with probability $> 1-\varepsilon$. This problem is known to be undecidable [13]. We consider a slightly modified version of the problem by assuming that no word is accepted with probability 1 by the given PFA. The problem remains undecidable under this assumption, because one can check if a PFA accepts a finite word with probability 1 by a simple subset construction.

Given some PFA $\mathcal{A}$, we construct a $\text{PWA}^{=1}$ $\mathcal{B}$ by taking a copy of $\mathcal{A}$ and extending it with a new symbol # such that from accepting states of $\mathcal{A}$ the automaton is "restarted" on #, while from non-accepting states # leads into a

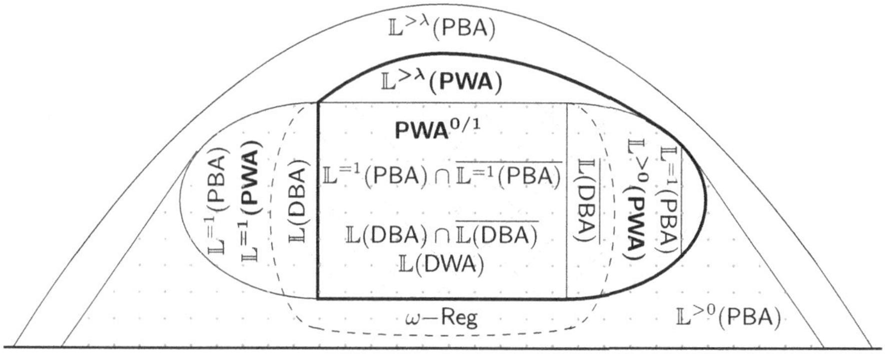

Fig. 3: Illustration of relationships between the class of languages accepted by weak probabilistic automata under various semantics with other already known classes. The overlapping patterns indicate intersection of classes, where dots mark $\mathbb{L}^{>0}(\mathsf{PBA})$, and different diagonal lines respectively $\mathbb{L}^{=1}(\mathsf{PBA})$ and $\overline{\mathbb{L}^{=1}(\mathsf{PBA})}$. The dashed line indicates intersections with different subclasses of regular languages. The class $\mathbb{L}^{>\lambda}(\mathsf{PBA})$ contains all the other depicted classes, $\mathbb{L}^{>\lambda}(\mathsf{PWA})$ contains the area inside the thick line. The depicted fact that $\mathbb{L}^{>0}(\mathsf{PWA}) = \mathbb{L}^{>\lambda}(\mathsf{PWA}) \cap \mathbb{L}^{>0}(\mathsf{PBA})$ is a conjecture, one direction is shown in Theorem 10.

new part which ensures that infinitely many # are seen and contains the only accepting state of $\mathcal{B}$. We show that $L^{=1}(\mathcal{B}) = (\Sigma^*\#)^\omega \setminus R$, where $R = \emptyset$ if $\mathcal{A}$ does not have value 1, and R is non-empty but does not contain an ultimately periodic word, otherwise. This implies that $L^{=1}(\mathcal{B})$ is regular iff $\mathcal{A}$ does not have value 1. □

We will now show that PWA with almost-sure semantics are as expressive as PBA, and with positive semantics as expressive as PCA.

Theorem 8. $\mathbb{L}^{>0}(\mathsf{PWA}) = \mathbb{L}^{>0}(\mathsf{PCA})$ *and* $\mathbb{L}^{=1}(\mathsf{PWA}) = \mathbb{L}^{=1}(\mathsf{PBA})$.

Proof (sketch). It suffices to show the first statement. The second then follows by duality, i.e., we can interpret a PBA$^{=1}$ $\mathcal{A}$ recognizing L as a PCA$^{>0}$ recognizing $\overline{L}$ and just apply the construction to get a PWA$^{>0}$ $\mathcal{B}$ for $\overline{L}$, such that $\overline{\mathcal{B}}$ (with inverted accepting and rejecting states) is a PWA$^{=1}$ for L. In the first statement the $\subseteq$ inclusion is trivial, hence we only need to show that $\mathbb{L}^{>0}(\mathsf{PCA}) \subseteq \mathbb{L}^{>0}(\mathsf{PWA})$.

We construct a PWA$^{>0}$ consisting of two copies of the original PCA$^{>0}$, a *guess* copy and a *verify* copy. In the first copy, the automaton can guess that no final states will be visited anymore and switch to the verify copy, which is accepting, but where all transitions into final states are redirected to a rejecting sink. □

Next, we show that languages that can be accepted by both, a PWA with almost-sure semantics, and by a PWA with positive semantics, are regular and can be accepted by a DWA. For the proof, we rely on a characterization of DWA languages in terms of the Myhill-Nerode equivalence relation from [22]. So we first define this equivalence, and show that languages defined by PBA with positive semantics have only finitely many equivalence classes. Then we come back to the result for PWA.

For $L \subseteq \Sigma^\omega$, define the Myhill-Nerode equivalence relation $\sim_L \subseteq \Sigma^* \times \Sigma^*$ by $u \sim_L v$ iff $uw \in L \Leftrightarrow vw \in L$ for all $w \in \Sigma^\omega$. Then the following holds:

Lemma 6 (Finitely many Myhill-Nerode classes).
Languages in $\mathbb{L}^{>0}(\mathsf{PBA})$ have finitely many Myhill-Nerode equivalence classes.

Proof. Let $\mathcal{A} = (Q, \Sigma, \delta, \mu_0, F)$ be some PBA$^{>0}$ and $u \in \Sigma^*$ some word and let $\mu_u := \delta^*(\mu_0, u)$ be the probability distribution on states of $\mathcal{A}$ after reading u. Pick any $w \in \Sigma^\omega$ and notice that $uw \in L = L^{>0}(\mathcal{A})$ iff there exists some state q such that $\mu_u(q) > 0$ and the probability to accept w from q is also > 0, as the product of two positive numbers clearly still is positive. But then, for any two $u, v \in \Sigma^*$ we have that whenever $\mu_u(q) > 0 \Leftrightarrow \mu_v(q) > 0$ for all q, then we have $uw \in L \Leftrightarrow vw \in L$ for all $w \in \Sigma^\omega$ by the reasoning above, as the exact value does not matter for acceptance, and therefore $u \sim_L v$. But as there are only at most $2^{|Q|}$ different possibilities how values in a distribution μ over Q are either equal to or greater than 0, this is an upper bound on the number of different equivalence classes. □

Theorem 9. $\mathbb{L}^{>0}(\mathsf{PWA}) \cap \mathbb{L}^{=1}(\mathsf{PWA}) = \mathbb{L}(\mathsf{DWA}) = \mathbb{L}(\mathsf{PWA}^{0/1})$

Proof. The inclusions $\mathbb{L}(\mathsf{DWA}) \subseteq \mathbb{L}(\mathsf{PWA}^{0/1}) \subseteq \mathbb{L}^{>0}(\mathsf{PWA}) \cap \mathbb{L}^{=1}(\mathsf{PWA})$ are trivial, hence it remains to show $\mathbb{L}^{>0}(\mathsf{PWA}) \cap \mathbb{L}^{=1}(\mathsf{PWA}) \subseteq \mathbb{L}(\mathsf{DWA})$.

So let L be a language from $\mathbb{L}^{>0}(\mathsf{PWA}) \cap \mathbb{L}^{=1}(\mathsf{PWA})$. We want to show that L can be accepted by a DWA. We use the following characterization of DWA languages [22, Theorem 21]: The DWA languages are precisely the languages with finitely many Myhill-Nerode classes in the class $G_\delta \cap F_\sigma$ in the Borel hierarchy. The classes G_δ and F_σ of the Borel hierarchy are often also referred to as Π_2 and Σ_2. We do not introduce the details of this hierarchy here, but rather refer the reader not familiar with these concepts to [22] and [8].

We already know that L has finitely many Myhill-Nerode classes by Lemma 6 (as PWA are special cases of PBA). It remains to show that L is in the class $G_\delta \cap F_\sigma$. It is known that PBA with almost-sure semantics define languages in G_δ [8, Lemma 3.2]. Hence L is in G_δ. Since L is accepted by a PWA with positive semantics, the complement of L is accepted by a PWA with almost-sure semantics (as noted at the beginning of this section). We obtain that the complement of L is also in G_δ again by [8, Lemma 3.2]. This means that L is in F_σ, which by definition consists of the complements of languages from G_δ. □

Concluding this section, we show a result about weak automata with threshold semantics, which (not surprisingly) turn out to be even more expressive. A careful analysis of the PWA $\mathcal{A}$ in Fig. 2(a) shows the following result:

Proposition 6. *For all thresholds $\lambda \in]0, 1[$ there exists a PWA $\mathcal{A}$ such that $L^{>\lambda}(\mathcal{A})$ is not regular and not PBA$^{>0}$ recognizable.*

Putting things together, we can say the following about threshold PWA, establishing the relation of $\mathbb{L}^{>\lambda}(\mathsf{PWA})$ to the other classes in Figure 3:

Theorem 10 (Expressive power of threshold PWA).

1. $\mathbb{L}^{>0}(\mathsf{PWA}) \subseteq \mathbb{L}^{>\lambda}(\mathsf{PWA}) \cap \mathbb{L}^{>0}(\mathsf{PBA})$.
2. $\mathbb{L}^{>\lambda}(\mathsf{PWA})$ *and* $\mathbb{L}^{>0}(\mathsf{PBA})$ *are incomparable (wrt. set inclusion).*
3. $\mathbb{L}^{>0}(\mathsf{PWA}) \subset \mathbb{L}^{>\lambda}(\mathsf{PWA}) \subset \mathbb{L}^{>\lambda}(\mathsf{PBA})$.

Proof. (1.) $\mathbb{L}^{>0}(\mathsf{PWA}) \subseteq \mathbb{L}^{>0}(\mathsf{PBA})$ by definition and $\mathbb{L}^{>0}(\mathsf{PWA}) \subseteq \mathbb{L}^{>\lambda}(\mathsf{PWA})$, as any $\text{PWA}^{>0}$ can be modified to a $\text{PWA}^{>\lambda}$ recognizing the same language by just adding an additional accepting sink and modifying the initial distribution, just as described in [4, Lemma 4.16] for general PBA.

(2.) By Proposition 6, there are languages recognized by $\text{PWA}^{>\lambda}$ that cannot be recognized with $\text{PBA}^{>0}$. To show that there are languages accepted by $\text{PBA}^{>0}$ that cannot be accepted by $\text{PWA}^{>\lambda}$ we can give a topological characterization of languages accepted by PWA by a simple adaptation of [8, Lemma 3.2] and combine it with other results shown in [8] to show that there are $\text{PBA}^{>0}$ that accept languages that cannot be accepted by $\text{PWA}^{>\lambda}$.

(3.) The first inclusion was discussed in (1.), the strictness follows from Proposition 6 and the fact that $\mathbb{L}^{>0}(\mathsf{PWA}) = \mathbb{L}^{=1}(\mathsf{PBA}) \subset \mathsf{BCl}(\mathbb{L}^{=1}(\mathsf{PBA})) = \mathbb{L}^{>0}(\mathsf{PBA})$, where the first equality is Theorem 8 and the second is shown in [8]. The second inclusion of the statement follows from (2.) and the fact from [4] that $\mathbb{L}^{>0}(\mathsf{PBA}) \subset \mathbb{L}^{>\lambda}(\mathsf{PBA})$. □

For the dual class $\mathbb{L}^{\geq\lambda}(\mathsf{PWA})$ one can show symmetric results that correspond to statements (1.) and (2.) above, for statement (3.) however there is no proof yet for the strictness of the inclusions (especially the second one), whereas the statement $\mathbb{L}^{=1}(\mathsf{PWA}) \subseteq \mathbb{L}^{\geq\lambda}(\mathsf{PWA}) \subseteq \mathbb{L}^{\geq\lambda}(\mathsf{PBA})$ is obvious. We leave this issue as an open question. Another interesting question is whether $> \lambda$ is equivalent to $< \lambda$ (or dually for $\geq / \leq$).

6 Conclusion

By using notions from ambiguity in classical Büchi automata, we were able to extend the set of easily (syntactically) checkable PBA which are regular under some or all of the usual semantics. As a consequence, ambiguity appears to be an even more interesting notion in the probabilistic setting, as here it in fact has consequences for the expressive power of automata, whereas in the classical setting there is no such effect. Our results also indicate that to get non-regularity, one requires the use of certain structural patterns which at least imply the existence of the ambiguity patterns that we used. It is an open question whether it is possible to identify more fine-grained syntactic characterizations, patterns or easily checkable properties which are just over-approximated by the ambiguity patterns and are required for non-regularity.

References

1. Baier, C., Bertrand, N., Größer, M.: On decision problems for probabilistic büchi automata. In: Foundations of Software Science and Computational Structures, 11th International Conference, FOSSACS 2008. Lecture Notes in Computer Science, vol. 4962, pp. 287–301. Springer (2008), https://doi.org/10.1007/978-3-540-78499-9
2. Baier, C., Bertrand, N., Größer, M.: Probabilistic automata over infinite words: Expressiveness, efficiency, and decidability. In: Proceedings Eleventh International Workshop on Descriptional Complexity of Formal Systems, DCFS 2009. EPTCS, vol. 3, pp. 3–16 (2009), https://doi.org/10.4204/EPTCS.3
3. Baier, C., Größer, M.: Recognizing omega-regular languages with probabilistic automata. In: 20th IEEE Symposium on Logic in Computer Science (LICS 2005), 26-29 June 2005, Chicago, IL, USA, Proceedings. pp. 137–146 (2005)
4. Baier, C., Größer, M., Bertrand, N.: Probabilistic ω-automata. Journal of the ACM (JACM) **59**(1), 1 (2012)
5. Baier, C., Katoen, J.: Principles of model checking. MIT Press (2008)
6. Boigelot, B., Jodogne, S., Wolper, P.: An effective decision procedure for linear arithmetic over the integers and reals. ACM Trans. Comput. Log. **6**(3), 614–633 (2005), https://doi.org/10.1145/1071596.1071601
7. Büchi, J.R.: On a decision method in restricted second order arithmetic. In: Studies in Logic and the Foundations of Mathematics, vol. 44, pp. 1–11. Elsevier (1966)
8. Chadha, R., Sistla, A.P., Viswanathan, M.: Power of randomization in automata on infinite strings. Logical Methods in Computer Science **7** (2011)
9. Chadha, R., Sistla, A.P., Viswanathan, M.: Probabilistic Büchi automata with non-extremal acceptance thresholds. In: International Workshop on Verification, Model Checking, and Abstract Interpretation. pp. 103–117. Springer (2011)
10. Chadha, R., Sistla, A.P., Viswanathan, M.: Emptiness under isolation and the value problem for hierarchical probabilistic automata. In: FOSSACS 2017. LNCS, vol. 10203, pp. 231–247 (2017), https://doi.org/10.1007/978-3-662-54458-7
11. Chadha, R., Sistla, A.P., Viswanathan, M., Ben, Y.: Decidable and expressive classes of probabilistic automata. In: FoSSaCS 2015. LNCS, vol. 9034, pp. 200–214. Springer (2015), https://doi.org/10.1007/978-3-662-46678-0
12. Fijalkow, N., Riveros, C., Worrell, J.: Probabilistic automata of bounded ambiguity. In: 28th International Conference on Concurrency Theory (CONCUR 2017). Schloss Dagstuhl-Leibniz-Zentrum fuer Informatik (2017)
13. Gimbert, H., Oualhadj, Y.: Probabilistic automata on finite words: Decidable and undecidable problems. In: International Colloquium on Automata, Languages, and Programming. pp. 527–538. Springer (2010)
14. Landweber, L.H.: Decision problems for ω-automata. Mathematical Systems Theory **3**, 376–384 (1969)
15. Leroux, J., Sutre, G.: On flatness for 2-dimensional vector addition systems with states. In: International Conference on Concurrency Theory. pp. 402–416. Springer (2004)
16. Löding, C., Pirogov, A.: On finitely ambiguous Büchi automata. In: Developments in Language Theory - 22nd International Conference, DLT 2018, Tokyo, Japan, September 10-14, 2018, Proceedings. pp. 503–515 (2018)
17. Löding, C., Thomas, W.: Alternating automata and logics over infinite words. In: Proceedings of the IFIP International Conference on Theoretical Computer Science, IFIP TCS2000. LNCS, vol. 1872, pp. 521–535. Springer (2000)

18. Rabin, M.O.: Probabilistic automata. Information and control **6**(3), 230–245 (1963)
19. Rabinovich, A.: Complementation of finitely ambiguous Büchi automata. In: Developments in Language Theory - 22nd International Conference, DLT 2018, Tokyo, Japan, September 10-14, 2018, Proceedings. pp. 541–552 (2018)
20. Sickert, S., Esparza, J., Jaax, S., Křetínský, J.: Limit-deterministic Büchi automata for linear temporal logic. In: Chaudhuri, S., Farzan, A. (eds.) Computer Aided Verification. pp. 312–332. Springer International Publishing, Cham (2016)
21. Sistla, A.P., Vardi, M.Y., Wolper, P.: The complementation problem for Büchi automata with applications to temporal logic (extended abstract). In: ICALP 1985. LNCS, vol. 194, pp. 465–474. Springer (1985), https://doi.org/10.1007/BFb0015725
22. Staiger, L.: Finite-state ω-languages. Journal of Computer and System Sciences **27**(3), 434–448 (1983)
23. Thomas, W.: Automata on infinite objects. In: Handbook of Theoretical Computer Science, vol. B: Formal Models and Semantics, pp. 133–192. Elsevier Science Publishers, Amsterdam (1990)
24. Thomas, W.: Languages, automata, and logic. In: Rozenberg, G., Salomaa, A. (eds.) Handbook of Formal Language Theory, vol. III, pp. 389–455. Springer (1997)
25. Weber, A., Seidl, H.: On the degree of ambiguity of finite automata. Theoretical Computer Science **88**(2), 325–349 (1991)

Local Local Reasoning: A BI-Hyperdoctrine for Full Ground Store*

Miriam Polzer and Sergey Goncharov(✉)

FAU Erlangen-Nürnberg, Erlangen, Germany
{miriam.polzer,sergey.goncharov}@fau.de

Abstract. Modelling and reasoning about dynamic memory allocation is one of the well-established strands of theoretical computer science, which is particularly well-known as a source of notorious challenges in semantics, reasoning, and proof theory. We capitalize on recent progress on categorical semantics of *full ground store*, in terms of a *full ground store monad*, to build a corresponding semantics of a higher order logic over the corresponding programs. Our main result is a construction of an *(intuitionistic) BI-hyperdoctrine*, which is arguably the semantic core of higher order logic over local store. Although we have made an extensive use of the existing generic tools, certain principled changes had to be made to enable the desired construction: while the original monad works over total heaps (to disable dangling pointers), our version involves partial heaps (*heaplets*) to enable compositional reasoning using separating conjunction. Another remarkable feature of our construction is that, in contrast to the existing generic approaches, our BI-algebra does not directly stem from an internal categorical partial commutative monoid.

1 Introduction

Modelling and reasoning about dynamic memory allocation is a sophisticated subject in denotational semantics with a long history (e.g. [19,15,14,16]). Denotational models for dynamic references vary over a large spectrum, and in fact, in two dimensions: depending on the expressivity of the features being modelled (*ground store* – *full ground store* – *higher order store*) and depending on the amount of *intensional* information included in the model (*intensional* – *extensional*), using the terminology of Abramsky [1].

Recently, Kammar et al [9] constructed an extensional monad-based denotational model of the *full ground store*, i.e. permitting not only memory allocation for discrete values, but also storing mutually linked data. The key idea of the latter work is an explicit delineation between the target presheaf category $[\mathbf{W}, \mathbf{Set}]$ on which the full ground store monad acts, and an auxiliary presheaf category $[\mathbf{E}, \mathbf{Set}]$ of *initializations*, naturally hosting a *heap functor* H. The latter category also hosts a *hiding monad* P, which can be loosely understood as a semantic

* Sergey Goncharov acknowledges support by German Research Foundation (DFG) under project GO 2161/1-2.

J. Goubault-Larrecq and B. König (Eds.): FOSSACS 2020, LNCS 12077, pp. 542–561, 2020.
https://doi.org/10.1007/978-3-030-45231-5_28

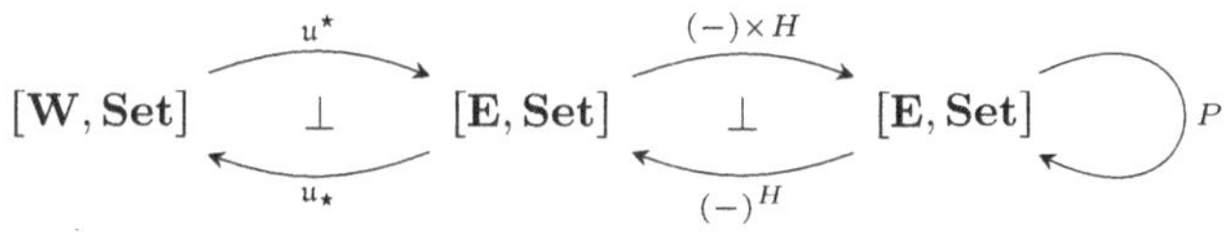

Fig. 1: Construction of the full ground store monad.

mechanism for idealized garbage collection. The full ground store monad is then assembled according to the scheme given in Fig. 1. As a slogan: the *local* store monad is a *global* store monad transform of the hiding monad sandwiched within a geometric morphism.

The fundamental reason, why extensional models of local store involve intricate constructions, such as presheaf categories is that the desirable program equalities include

$$\mathsf{let}\ \ell := \mathsf{new}\, v;\ \ell' := \mathsf{new}\, w\ \mathsf{in}\, p \ =\ \mathsf{let}\ \ell' := \mathsf{new}\, w;\ \ell := \mathsf{new}\, v\ \mathsf{in}\, p \qquad (\ell \not\equiv \ell')$$

$$\mathsf{let}\ \ell := \mathsf{new}\, v\ \mathsf{in}\ \mathsf{ret}\, \star \ =\ \mathsf{ret}\, \star$$

$$\mathsf{let}\ \ell := \mathsf{new}\, v\ \mathsf{in}\ \ (\mathsf{if}\ \ell = \ell'\ \mathsf{then\ true\ else\ false}) \ =\ \mathsf{false} \qquad (\ell \not\equiv \ell')$$

and these jointly do not have set-based models over countably infinite sets of locations [23, Proposition 6]. The first equation expresses irrelevance of the memory allocation order, the second expresses the fact that an unused cell is always garbage collected and the third guarantees that allocation of a fresh cell does indeed produce a cell different from any other. The aforementioned construction validates these equations and enjoys further pleasant properties, e.g. soundness and adequacy of a higher order language with user defined storable data structures.

The goal of our present work is to complement the semantics of programs over local store with a corresponding principled semantics of *higher order logic*. In order to be able to specify and reason modularly about local store, more specifically, we seek a model of higher order *separation logic* [21]. It has been convincingly argued in previous work on categorical models of separation logic [2,3] that a core abstraction device unifying such models is a notion of *BI-hyperdoctrine*, extending Lawvere's hyperdoctrines [10], which provide a corresponding abstraction for the first order logic. BI-hyperdoctrines are standardly built on *BI-algebras*, which are also standardly constructed from *partial commutative monoids (pcm)*, or more generally from *resource algebras* as in the Iris state of the art advanced framework for higher order separation logic [8]. One subtlety our construction reveals is that it does not seem to be possible to obtain a *BI-algebra* following general recipes from a pcm (or a resource algebra), due to the inherent local nature of the storage model, which does not allow one to canonically map store contents into a global address space. Another subtlety is that the devised logic is necessarily non-classical, which is intuitively explained by the fact that the semantics of programs must be suitably irrelevant to garbage collection, and in our case this follows from entirely formal considerations (Yoneda lemma). It is also worth mentioning that for this reason the logical theory that we obtain is incompatible with the standard (classical or intuitionistic) predicate logic. E.g. the formula $\exists \ell.\, \ell \hookrightarrow 5$ is always valid in our setup, which expresses the fact that a heap *potentially* contains a cell equal to 5 (which need not be reachable) – this is in accord with the second equation above – and correspondingly, the formula $\forall \ell.\, \neg(\ell \hookrightarrow 5)$ is unsatisfiable. This and other similar phenomena are explained by the fact that our semantics essentially behaves as a Kripke semantics along two orthogonal axes: (proof relevant) *cell allocation* and (proof irrelevant) *cell accessibility*. While the latter captures a *programming* view of locality, the latter captures a *reasoning* view of locality, and as we argue (e.g. Example 26), they are generally mutually irreducible.

Related previous work As we already pointed out, we take inspiration from the recent categorical approaches to modelling program semantics for dynamic references [9], as well as from higher order separation logic semantic frameworks [2]. Conceptually, the problem of combining separation logic with garbage collection mechanisms goes back to Reynolds [20], who indicated that standard semantics of separation logic in not compatible with garbage collection, which we also reinforce with our construction. Calcagno et al [4] addressed this issue by providing two models. The first model is based on total heaps, featuring the aforementioned effect of "potential" allocations. To cope with heap separation the authors introduced another model based on partial heaps, in which this effect again disappears, and has to be compensated by syntactic restrictions on the assertion language.

Plan of the paper After preliminaries (Section 2), we give a modified presentation of a call-by-value language with full ground references and the full ground store monad (Sections 3 and 4) following the lines of [9]. In Section 5 we provide some general results for constructing semantics of higher order separation logics. The main development starts in Section 6 where we provide a construction of a BI-hyperdoctrine. We show some example illustrating our semantics in Section 7 and draw conclusions in Section 8.

2 Preliminaries

We assume basic familiarity with the elementary concepts of category theory [12,6], all the way up to monads, toposes, (co)ends and Kan extensions. We denote by $|\mathbf{C}|$ the class of objects of a category $\mathbf{C}$; we often suppress subscripts of natural transformation components if no confusion arises.

In this paper, we work with special kinds of *covariant presheaf toposes*, i.e. functor categories of the form $[\mathbf{C}, \mathbf{Set}]$, where $\mathbf{C}$ is small and satisfies the following *amalgamation condition*: for any $f\colon a \to b$ and $g\colon a \to c$ there exist $g'\colon b \to d$ and $f'\colon c \to d$ such that $f' \circ g = g' \circ f$. Such toposes are particularly well-behaved, and fall into the more general class of *De Morgan* toposes [7]. As presheaf toposes, De Morgan toposes are precisely characterized by the condition

$$\textbf{(put)}\ \frac{\Gamma \vdash_{\mathsf{v}} \ell\colon \mathsf{Ref}_S \qquad \Gamma \vdash_{\mathsf{v}} v\colon \mathsf{CType}(S)}{\Gamma \vdash_{\mathsf{c}} \ell := v\colon 1} \qquad \textbf{(get)}\ \frac{\Gamma \vdash_{\mathsf{v}} \ell\colon \mathsf{Ref}_S}{\Gamma \vdash_{\mathsf{c}} !\ell\colon \mathsf{CType}(S)}$$

$$\textbf{(new)}\ \frac{\begin{array}{c}\Gamma, \ell_1\colon \mathsf{Ref}_{S_1}, \dots, \ell_n\colon \mathsf{Ref}_{S_n} \vdash_{\mathsf{v}} v_1\colon \mathsf{CType}(S_1)\\ \vdots \\ \Gamma, \ell_1\colon \mathsf{Ref}_{S_1}, \dots, \ell_n\colon \mathsf{Ref}_{S_n} \vdash_{\mathsf{v}} v_n\colon \mathsf{CType}(S_n)\\ \Gamma, \ell_1\colon \mathsf{Ref}_{S_1}, \dots, \ell_n\colon \mathsf{Ref}_{S_n} \vdash_{\mathsf{c}} p\colon A\end{array}}{\Gamma \vdash_{\mathsf{c}} \mathsf{letref}\, \ell_1 := v_1, \dots, \ell_n := v_n\, \mathsf{in}\, p\colon A}$$

Fig. 2: Term formation rules for memory management constructs.

that $2 = 1 + 1$ is a retract of the subobject classifier Ω. More specifically, our $\mathbf{C}$ support further useful structure, in particular, a strict monoidal tensor $\oplus$ with jointly epic injections in_1, in_2, forming an *independent coproduct* structure, as recently identified by Simpson [22]. Moreover, if the coslices $c \downarrow \mathbf{C}$ support independent products, we obtain *local independent coproducts* in $\mathbf{C}$, which are essentially cospans $c_1 \to c_1 \oplus_c c_2 \leftarrow c_2$ in $c \downarrow \mathbf{C}$. Given $\rho_1 : c \to c_1$ and $\rho_2 : c \to c_2$, we thus always have $\rho_1 \bullet \rho_2 : c_1 \to c_1 \oplus_c c_2$ and $\rho_2 \bullet \rho_1 : S_2 \to c_1 \oplus_c c_2$, such that $(\rho_1 \bullet \rho_2) \circ \rho_1 = (\rho_2 \bullet \rho_1) \circ \rho_2$, and as a consequence, $[\mathbf{C}, \mathbf{Set}]$ is a De Morgan topos. Intuitively, the category $\mathbf{C}$ represents worlds in the sense of *possible world semantics* [15,19]. A morphism $\rho\colon a \to b$ witnesses the fact that b is a *future* world w.r.t. a. Existence of local independent products intuitively ensures that diverse futures of a given world can eventually be unified in a canonical way.

Every functor $\mathfrak{f}\colon \mathbf{C} \to \mathbf{D}$ induces a functor $\mathfrak{f}^\star\colon [\mathbf{D}, \mathbf{Set}] \to [\mathbf{C}, \mathbf{Set}]$ by precomposition with $\mathfrak{f}$. By general considerations, there is a right adjoint $\mathfrak{f}_\star\colon [\mathbf{C}, \mathbf{Set}] \to [\mathbf{D}, \mathbf{Set}]$, computed as $\mathsf{Ran}_{\mathfrak{f}}$, the right Kan extension along $\mathfrak{f}$. This renders the adjunction $\mathfrak{f}^\star \dashv \mathfrak{f}_\star$, as a *geometric morphism*, in particular, $\mathfrak{f}^\star$ preserves all finite limits.

3 A Call-by-Value Language with Local References

To set the context, we consider the following higher order language of programs with local references by slightly adapting the language of Kammar et al [9] to match with the *fine-grain call-by-value* perspective [11]. This allows us to formally distinguish *pure* and *effectful* judgements. First, we postulate a collection of *cell sorts* $\mathcal{S}$ and then introduce further types with the grammar:

$$A, B \ldots ::= 0 \mid 1 \mid A \times B \mid A + B \mid A \to B \mid \mathsf{Ref}_S \qquad (S \in \mathcal{S}) \tag{1}$$

A type is *first order* if it does not involve the function type constructors $A \to B$. We then fix a map CType, assigning a first order type to every given sort from $\mathcal{S}$. We show three term formation rules over these data in Fig. 2 specific to local store.

Here the v-indices at the turnstiles indicate *values* and the c-indices indicate *computations*. In **(put)** the cell referenced by ℓ is updated with a value v, **(get)** returns a value under the reference ℓ and **(new)** simultaneously allocates new cells filled with the values $v_1, \ldots, v_n$ and makes them accessible in p under the corresponding references $\ell_1, \ldots, \ell_n$. A fine-grain call-by-value language is interpreted standardly in a category with a monad, which in our case must additionally provide a semantics to the rules **(put)**, **(get)** and **(new)**. We present this monad in detail in the next section.

Example 1 (Doubly Linked Lists). Let $\mathcal{S} = \{DLList\}$ and let $\mathsf{CType}(DLList) = 2 \times (\mathsf{Ref}_{DLList} + 1) \times (\mathsf{Ref}_{DLList} + 1)$, which indicates that a list element is a Boolean (i.e. an element of $2 = 1 + 1$) and two pointers (forwards and backwards) to list elements, each of which may be missing. Note that we thus avoid empty lists and null-pointers: every list contains at least one element, and the elements added by $+1$ cannot be dereferenced. This example provides a suitable illustration for the letref construct. E.g. the program

$$\mathsf{letref}\,\ell_1 := (0, \mathsf{inr}\,\star, \mathsf{inl}\,\ell_2); \ell_2 := (1, \mathsf{inl}\,\ell_1, \mathsf{inr}\,\star)\;\mathsf{in}\;\mathsf{ret}\,\ell_1$$

simultaneously creates two list elements pointing to each other and returns a reference to the first one.

4 Full Ground Store in the Abstract

We proceed to present the full ground store monad by slightly tweaking the original construction [9] towards higher generality. The main distinction is that we do not recur to any specific program syntax and proceed in a completely axiomatic manner in terms of functors and natural transformations. This mainly serves the purpose of developing our logic in Section 6, which will require a coherent upgrade of the present model. Besides this, in this section we demonstrate flexibility of our formulation by showing that it also instantiates to the model previously developed by Plotkin and Power [16] (Theorem 8).

Our present formalization is parametric in three aspects: the set of *sorts* $\mathcal{S}$, the set of *locations* $\mathcal{L}$ and a map range, introduced below for interpreting $\mathcal{S}$. We assume that $\mathcal{L}$ is canonically isomorphic to the set of natural numbers $\mathbb{N}$ under $\#\colon \mathcal{L} \cong \mathbb{N}$. Using this isomorphism, we commonly use the "shift of $\ell \in \mathcal{L}$ by $n \in \mathbb{N}$", defined as follows: $\ell + n = \#^{-1}(\#\ell + n)$.

Heap layouts and abstract heap(let)s Let $\mathbf{W}$ be a category of *(heap) layouts* and injections defined as follows: an object $w \in |\mathbf{W}|$ is a finitely supported partial function $w\colon \mathcal{L} \rightharpoonup_{fin} \mathcal{S}$ and a morphism $\rho\colon w \to w'$ is a type preserving injection $\rho\colon \mathsf{dom}\, w \to \mathsf{dom}\, w'$, i.e. for all $l \in \mathsf{img}\, w$, $w(\ell) = w'(\rho(\ell))$. We will equivalently view w as a left-unique subset of $\mathcal{L} \times \mathcal{S}$ and hence use the notation $(\ell\colon S) \in w$ as an equivalent of $w(\ell) = S$. Injections $\rho\colon w \to w'$ with the property that $w(\ell\colon S) = \ell\colon S$ for all $(\ell\colon S) \in w$ we also call *inclusions* and write $w \subseteq w'$ instead of $\rho\colon w \to w'$, for obviously there is at most one inclusion from w to w'. If $w \subseteq w'$

then we call w a *sublayout* of w'. We next postulate

$$\mathsf{range}\colon \mathcal{S} \to [\mathbf{W}, \mathbf{Set}].$$

The idea is, given a sort $S \in \mathcal{S}$ and a heap layout $w \in |\mathbf{W}|$, $\mathsf{range}(S)(w)$ yields the set of possible values for cells of type S over w.

Example 2. Assuming the grammar (1) and a corresponding map CType, a generic type A is interpreted as a presheaf $\underline{A}\colon \mathbf{W} \to \mathbf{Set}$, by obvious structural induction, e.g. $\underline{A \times B} = \underline{A} \times \underline{B}$, except for the clause for Ref, for which $(\underline{\mathsf{Ref}_S})w = w^{-1}(S)$. This yields the following definition for range: $\mathsf{range}(S) = \underline{\mathsf{CType}(S)}$ [9].

Example 3 (Simple Store). By taking $\mathcal{S} = \{\star\}$, $\mathcal{L} = \mathbb{N}$ (natural numbers) and $\mathsf{range}(\star)(w) = \mathcal{V}$ where $\mathcal{V}$ is a fixed set of *values*, we essentially obtain the model previously explored by Plotkin and Power [16]. We reserve the term *simple store* for this instance. Simple store is a ground store (since range is a constant functor), moreover this store is untyped (since $\mathcal{S} = \{\star\}$) and the locations $\mathcal{L}$ are precisely the natural numbers.

A *heap* over a layout w assigns to each $(\ell\colon S) \in w$ an element from $\mathsf{range}(S)(w)$. More generally, a *heaplet* over w assigns an element from $\mathsf{range}(S)(w)$ to *some*, possibly not all, $(\ell\colon S) \in w$. We thus define the following *heaplet bi-functor* $\mathcal{H}\colon \mathbf{W}^{\mathsf{op}} \times \mathbf{W} \to \mathbf{Set}$:

$$\mathcal{H}(w^-, w^+) = \prod\nolimits_{(\ell\colon S)\in w^-} \mathsf{range}(S)(w^+)$$

and identify the elements of $\mathcal{H}(w^-, w^+)$ with heaplets and the elements of $\mathcal{H}(w, w)$ with heaps. Of course, we intend to use $\mathcal{H}(w^-, w^+)$ for such w^- and w^+ that the former is a sublayout of the latter. The contravariant action of H is given by projection and the covariant action is induced by functoriality of $\mathsf{range}(S)$.

$$\begin{aligned}
\mathsf{pr}_{(\ell\colon S)}(\mathcal{H}(w^-, \rho_1\colon w_1^+ \to w_2^+)(\eta \in \mathcal{H}(w^-, w_1^+))) &= \mathsf{range}(S)(\rho_1)(\mathsf{pr}_{(\ell\colon S)}\,\eta)\\
\mathsf{pr}_{(\ell\colon S)}(\mathcal{H}(\rho_2\colon w_2^- \to w_1^-, w^+)(\eta \in \mathcal{H}(w_1^-, w^+))) &= \mathsf{pr}_{\rho_2(\ell\colon S)}\,\eta
\end{aligned}$$

The heaplet functor preserves independent coproduct, we overload the $\oplus$ operation with the isomorphism $\oplus\colon \mathcal{H}(w_1, w) \times \mathcal{H}(w_2, w) \cong \mathcal{H}(w_1 \oplus w_2, w)$.

Example 4. For illustration, consider the following simplistic example. Let $\mathcal{S} = \{\mathit{Int}, \mathsf{Ref}_{\mathit{Int}}, \mathsf{Ref}_{\mathsf{Ref}_{\mathit{Int}}}, \dots\}$ where Int is meant to capture the ground type of integers and recursively, Ref_A is the type of pointers to A. Then, we put

$$\mathsf{range}(\mathit{Int})(w) = \mathbb{Z}, \qquad \mathsf{range}(\mathsf{Ref}_S)(w) = w^{-1}(S) = \{\ell \in \mathsf{dom}\, w \mid w(\ell) = S\}.$$

For a heaplet example, consider $w^- = \{\ell_1\colon \mathit{Int}, \ell_2\colon \mathsf{Ref}_{\mathit{Int}}\}$ and $w^+ = \{\ell_1\colon \mathit{Int}, \ell_2\colon \mathsf{Ref}_{\mathit{Int}}, \ell_3\colon \mathit{Int}\}$. Hence, w^- is a sublayout of w^+. By viewing the elements of $\mathcal{H}(w^-, w^+)$ as lists of assignments on w^-, we can define $s_1, s_2 \in \mathcal{H}(w^-, w^+)$ as follows: $s_1 = [\ell_1\colon \mathit{Int} \mapsto 5, \ell_2\colon \mathsf{Ref}_{\mathsf{int}} \mapsto \ell_1]$, $s_2 = [\ell_1\colon \mathit{Int} \mapsto 3, \ell_2\colon \mathsf{Ref}_{\mathsf{int}} \mapsto \ell_3]$. The heaplets s_1 and s_2 can be graphically presented as follows:

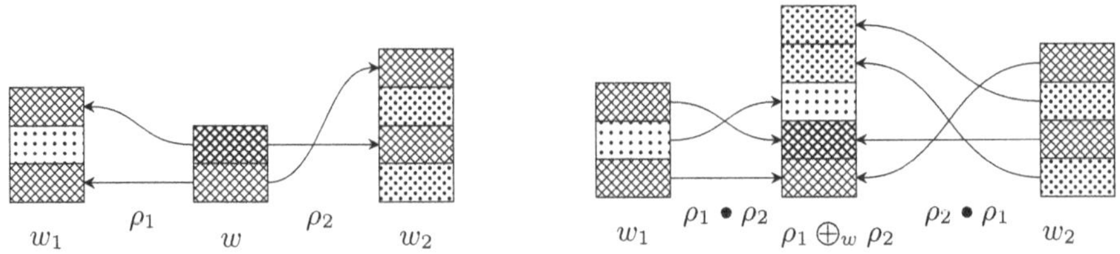

Fig. 3: Local independent coproduct

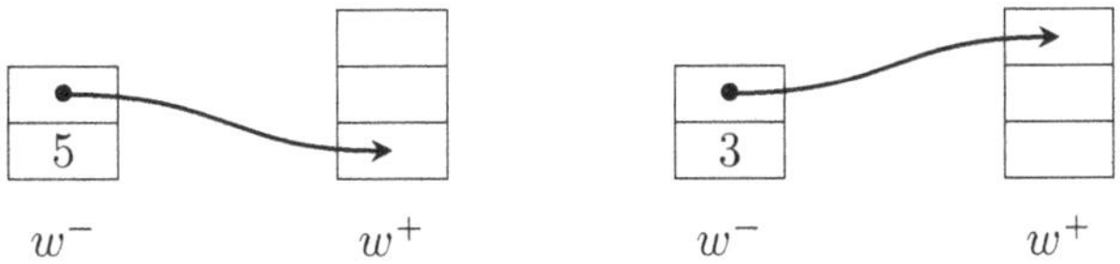

The category $\mathbf{W}$ supports (local) independent coproducts described in Section 2. These are constructed as follows. For $w, w' \in |\mathbf{C}|$, $w \oplus w' = w \cup \{\ell + n + 1 \colon S \mid (\ell, c) \in w'\}$ with n being the largest index for which w is defined on $\#^{-1}(n)$. This yields a strict monoidal structure $\oplus \colon \mathbf{W} \times \mathbf{W} \to \mathbf{W}$. Intuitively, $w_1 \oplus w_2$ is a canonical disjoint sum of w_1 and w_2, but note that $\oplus$ is not a coproduct in $\mathbf{W}$ (e.g. there is no $\nabla \colon 1 \oplus 1 \to 1$, for $\mathbf{W}$ only contains injections). For every $\rho \colon w_1 \to w_2$, there is a canonical complement $\rho^{\complement} \colon w_2 \ominus \rho \to w_2$ whose domain $w_2 \ominus \rho = w_2 \setminus \operatorname{img} \rho$ consists of all such cells $(\ell \colon S) \in w_2$ that ρ misses. Given two morphisms $\rho_1 \colon w \to w_1$ and $\rho_2 \colon w \to w_2$, we define the local independent coproduct $w_1 \oplus_w w_2$ as the layout consisting of the locations from w, and the ones from w_1 and w_2 which are neither in the image of ρ_1 nor in the image of ρ_2:

$$\rho_1 \oplus_w \rho_2 = w \oplus (w_1 \ominus \rho_1) \oplus (w_2 \ominus \rho_2).$$

There are morphisms $w_1 \xrightarrow{\rho_1 \bullet \rho_2} \rho_1 \oplus_w \rho_2$ and $w_2 \xrightarrow{\rho_2 \bullet \rho_1} \rho_1 \oplus_w \rho_2$ such that

$$\begin{array}{ccc} w & \xrightarrow{\rho_2} & w_2 \\ {\scriptstyle \rho_1}\downarrow & & \downarrow{\scriptstyle \rho_2 \bullet \rho_1} \\ w_1 & \xrightarrow{\rho_1 \bullet \rho_2} & \rho_1 \oplus_w \rho_2 \end{array}$$

Fig. 3 illustrates this definition with a concrete example.

Initialization and hiding Note that in the simple store model (Definition 3), $\mathcal{H}$ is equivalently a contravariant functor $H \colon \mathbf{W}^{\mathsf{op}} \to \mathbf{Set}$ with $Hw = \mathcal{V}^w$, hence $\mathcal{H}$ can be placed e.g. in $[\mathbf{W}^{\mathsf{op}}, \mathbf{Set}]$. In general, $\mathcal{H}$ is mix-variant, which calls for a more ingenious category where H could be placed. Designing such category is indeed the key insight of [9]. Closely following this work, we introduce a category $\mathbf{E}$, whose objects are the same as those of $\mathbf{W}$, and the morphisms $\epsilon \in \mathbf{E}(w, w')$, called *initializations*, consist of an injection $\rho \colon w \to w'$ and a heaplet $\eta \in \mathcal{H}(w' \ominus \rho, w')$:

$$\mathbf{E}(w, w') = \sum\nolimits_{\rho\colon w \to w'} \mathcal{H}(w' \ominus \rho, w').$$

Recall that the morphism $\rho\colon w \to w'$ represents a move from a world with w allocated memory cells a world with w' allocated memory cells. A morphism of $\mathbf{E}$ is a morphism of $\mathbf{W}$ augmented with a heaplet part η, which provides the information how the newly allocated cells in $w' \ominus \rho$ are filled. The heap functor now can be viewed as a representable presheaf $H\colon \mathbf{E} \to \mathbf{Set}$ essentially because by definition, $Hw = \mathcal{H}(w, w) \cong \mathbf{E}(\emptyset, w)$. Let us agree to use the notation $\epsilon\colon w \rightsquigarrow w'$ for morphisms in $\mathbf{E}$ to avoid confusion with the morphisms in $\mathbf{W}$.

Like $\mathbf{W}$, $\mathbf{E}$ supports local independent coproducts, but remarkably $\mathbf{E}$ does not have vanilla independent coproducts, due to the fact that $\mathbf{E}$ does not have an initial object. That is, in turn, because defining an inital morphism would amount to defining canonical fresh values for newly allocated cells, but those need not exist. The local independent coproducts of $\mathbf{W}$ and $\mathbf{E}$ agree in the sense that we can *promote* an initialization $(\rho_2, \eta)\colon w \rightsquigarrow w_2$ along an injection $\rho_1\colon w \to w_1$ to obtain an initialization $\rho_1 \bullet (\rho_2, \eta)\colon w_1 \rightsquigarrow \rho_1 \oplus_{w_1} \rho_2$. This is accomplished by mapping the heaplet structure η forward along $\rho_2 \bullet \rho_1\colon w_2 \to \rho_1 \oplus_w \rho_2$.

Hiding monad Recall that the local store is supposed to be insensitive to garbage collection. This is captured by identifying the stores that agree on their observable parts using the *hiding monad* P defined on $[\mathbf{E}, \mathbf{Set}]$ as follows:

$$(PX)w = \int^{\rho\colon w \to w' \in w \downarrow \mathrm{u}} Xw'. \tag{2}$$

Here, $\mathrm{u}\colon \mathbf{E} \to \mathbf{W}$ is the obvious heaplet discarding functor $\mathrm{u}(\rho, \eta) = \rho$. Intuitively, in (2), we view the locations of w as public and the ones of $w' \ominus \rho$ as private. The integral sign denotes a *coend*, which in this case is just an ordinary colimit on $\mathbf{Set}$ and is computed as a quotient of $\sum_{\rho\colon w \to w' \in w \downarrow \mathrm{u}} Xw'$ under the equivalence relation $\sim$ obtained as a symmetric-transitive closure of the relation

$$(\rho\colon w \to w_1, x \in Xw_1) \preceq (\mathrm{u}\epsilon \circ \rho\colon w \to w_2, (X\epsilon)(x) \in Xw_2) \qquad (\epsilon\colon w_1 \rightsquigarrow w_2)$$

Note that $\preceq$ is a preorder. Moreover, it enjoys the following *diamond property.*

Proposition 5. *If* $(\rho, x) \preceq (\rho_1, x_1)$ *and* $(\rho, x) \preceq (\rho_2, x_2)$ *then* $(\rho_1, x_1) \preceq (\rho', x')$ *and* $(\rho_2, x_2) \preceq (\rho', x')$ *for a suitable* (ρ', x').

Hence $(\rho_1, x_1) \sim (\rho_2, x_2)$ *iff* $(\rho_1, x_1) \preceq (\rho, x)$, $(\rho_2, x_2) \preceq (\rho, x)$ *for some* (ρ, x).

Example 6. To illustrate the equivalence relation $\sim$ behind P, we revisit the setting of Example 4. Consider the following situations:

Here, the solid lines indicate public locations and the dotted lines indicate private locations. The left equivalence holds because the private locations are not reachable from the public ones by references (depicted as arrows). On the right, although the public parts are equal, the reachable cells of the private parts reveal the distinction, preventing the equivalence under $\sim$. Intuitively, hiding identifies those heaps that agree both on their public and reachable private part.

The covariant action of PX (on $\mathbf{E}$) is defined via promotion of initializations:

$$\begin{aligned}(PX)(\epsilon\colon w_1 \rightsquigarrow w_2)(\rho\colon w_1 \to w_1', x \in Xw_1')_{\sim} \\ = (\mathfrak{u}\epsilon \bullet \rho\colon w_2 \to \rho \oplus_{w_1} \mathfrak{u}\epsilon, X(\rho \bullet \epsilon)(x))_{\sim}.\end{aligned}$$

Furthermore, there is a contravariant hiding operation (on $\mathbf{W}$) given by the canonical action of the coend: for $\rho\colon w \to w'$, we define $\mathsf{hide}_\rho\colon PXw' \to PXw$:

$$\mathsf{hide}_\rho(\rho'\colon w' \to w'', x \in Xw'')_{\sim} = (\rho' \circ \rho, x)_{\sim} \tag{3}$$

This allows us to regard P both as a functor $[\mathbf{E}, \mathbf{Set}] \to [\mathbf{E}, \mathbf{Set}]$ and as a functor $[\mathbf{E}, \mathbf{Set}] \to [\mathbf{W}^{\mathrm{op}}, \mathbf{Set}]$.

Full ground store monad We now have all the necessary ingredients to obtain the full ground store monad T on $[\mathbf{W}, \mathbf{Set}]$. This monad is assembled by composing the functors in Fig. 1 in the following way. First, observe that $(P(- \times H))^H$ is a standard (global) store monad transform of P on $[\mathbf{E}, \mathbf{Set}]$. This monad is sandwiched between the adjunction $\mathfrak{u}_\star \dashv \mathfrak{u}^\star$ induced by $\mathfrak{u}$ (see Section 2). Since any monad itself resolves into an adjunction, sandwiching in it between an adjunction again yields a monad. In summary,

$$T = \Big([\mathbf{W}, \mathbf{Set}] \xrightarrow{\mathfrak{u}^\star} [\mathbf{E}, \mathbf{Set}] \xrightarrow{P(- \times H)^H} [\mathbf{E}, \mathbf{Set}] \xrightarrow{\mathfrak{u}_\star} [\mathbf{W}, \mathbf{Set}]\Big). \tag{4}$$

Theorem 7. *The monad T, defined by* (4) *is strong.*

Proof. The proof is a straightforward generalization of the proof in [9]. □

We can recover the monad previously developed by Plotkin and Power [16] by resorting to the simple store (Example 3).

Theorem 8. *Under the simple store model T is isomorphic to the local store monad from [16]:*

$$(TX)w \cong \left(\int^{\rho\colon w \to w' \in w \downarrow \mathbf{W}} Xw' \times \mathcal{V}^{w'}\right)^{\mathcal{V}^w}.$$

Using (4), one obtains the requisite semantics to the language in Fig. 2 using the standard clauses of fine-grain call-by-value [11], except for the special clauses for **(put)**, **(get)** and **(new)**, which require special operations of the monad:

$$\begin{aligned}\mathsf{get}\colon & \mathfrak{u}^\star\underline{\mathsf{Ref}_S} \times H \to \mathfrak{u}^\star\underline{\mathsf{CType}(S)} \times H\\ \mathsf{put}\colon & (\mathfrak{u}^\star\underline{\mathsf{Ref}_S} \times \mathfrak{u}^\star\underline{\mathsf{CType}(S)}) \times H \to 1 \times H\\ \mathsf{new}\colon & \mathfrak{u}^\star(\underline{\mathsf{CType}(S)}^{\underline{\mathsf{Ref}_S}}) \times H \to P(\mathfrak{u}^\star\underline{\mathsf{Ref}_S} \times H)\end{aligned}$$

5 Intermezzo: BI-Hyperdoctrines and BI-Algebras

To be able to give a categorical notion of higher order logic over local store, following Biering et al [2], we aim to construct a *BI-hyperdoctrine.*

Note that algebraic structures, such as monoids and Heyting algebras can be straightforwardly internalized in any category with finite products, which gives rise to *internal monoids*, *internal Heyting algebras*, etc. The situation changes when considering non-algebraic properties. In particular, recall that a Heyting algebra A is *complete* iff it has arbitrary joins, which are preserved by binary meets. The corresponding categorical notion is essentially obtained from spelling out generic definitions from internal category theory [6, B2] and is as follows.

Definition 9 (Internally Complete Heyting Algebras). An internal Heyting (Boolean) algebra A in a finitely complete category $\mathbf{C}$ is *internally complete* if for every $f \in \mathbf{C}(I, J)$, there exist *indexed joins* $\bigvee_f : \mathbf{C}(I, A) \to \mathbf{C}(J, A)$, left order-adjoint to $(-) \circ f : \mathbf{C}(J, A) \to \mathbf{C}(I, A)$ such that for any pullback square on the left, the corresponding diagram on the right commutes (*Beck-Chevalley condition*):

$$\begin{array}{ccc} I & \xrightarrow{f} & J \\ {\scriptstyle g}\downarrow & \lrcorner & \downarrow{\scriptstyle h} \\ I' & \xrightarrow[f']{} & J' \end{array} \qquad\qquad \begin{array}{ccc} \mathbf{C}(J, A) & \xrightarrow{(-)\circ f} & \mathbf{C}(I, A) \\ {\scriptstyle \bigvee_h}\downarrow & & \downarrow{\scriptstyle \bigvee_g} \\ \mathbf{C}(J', A) & \xrightarrow[(-)\circ f']{} & \mathbf{C}(I', A) \end{array}$$

It follows generally that existence of indexed joins $\bigvee$ implies existence of indexed meets $\bigwedge$, which then satisfy dual conditions ([6, Corollary 2.4.8]).

Remark 10 (Binary Joins/Meets). The adjointness condition for indexed joins means precisely that $\bigvee_f \phi \leqslant \psi$ iff $\phi \leqslant \psi \circ f$ for every $\phi : I \to A$ and every $\psi : J \to A$. If $\mathbf{C}$ has binary coproducts, by taking $f = \nabla : X + X \to X$ we obtain that $\bigvee_\nabla \phi \leqslant \psi$ iff $\phi \leqslant [\psi, \psi]$ iff $\phi \circ \mathsf{inl} \leqslant \psi$ and $\phi \circ \mathsf{inr} \leqslant \psi$. This characterizes $\bigvee_\nabla [\phi_1, \phi_2] : X \to A$ as the binary join of $\phi_1, \phi_2 : X \to A$. Binary meets are characterized analogously.

Definition 11 ((First Order) (BI-)Hyperdoctrine). Let $\mathbf{C}$ be a category with finite products. A *first order hyperdoctrine over* $\mathbf{C}$ is a functor $S : \mathbf{C}^{\mathsf{op}} \to$ **Poset** with the following properties:

1. given $X \in |\mathbf{C}|$, SX is a Heyting algebra;
2. given $f \in \mathbf{C}(X, Y)$, $Sf : SY \to SX$ is a Heyting algebra morphism;
3. for any product projection $\mathsf{fst} : X \times Y \to X$, there are $(\exists Y)_X : S(X \times Y) \to SX$ and $(\forall Y)_X : S(X \times Y) \to SX$, which are respective left and right order-adjoints of $S\,\mathsf{fst} : S(X \times Y) \to SX$, naturally in X;
4. for every $X \in |\mathbf{C}|$, there is $=_X \in S(X \times X)$ such that for all $\phi \in S(X \times X)$, $\top \leqslant (S\langle \mathsf{id}_X, \mathsf{id}_X \rangle)(\phi)$ iff $=_X \leqslant \phi$.

If additionally

$$\frac{\Gamma \vdash_{\mathsf{v}} v\colon A \quad \Gamma \vdash \phi\colon \mathsf{P}A}{\Gamma \vdash \phi(v)\colon \mathrm{prop}} \qquad \frac{\Gamma, x\colon A \vdash \phi\colon \mathrm{prop}}{\Gamma \vdash x.\phi\colon \mathsf{P}A} \qquad \frac{\Gamma \vdash_{\mathsf{v}} \ell\colon \mathsf{Ref}_S \quad \Gamma \vdash_{\mathsf{v}} v\colon \mathsf{CType}(S)}{\Gamma \vdash \ell \hookrightarrow v\colon \mathrm{prop}}$$

$$\frac{\Gamma \vdash \phi\colon \mathsf{P}A}{\Gamma \vdash Q\,\phi\colon \mathrm{prop}}\ (Q \in \{\forall, \exists\}) \qquad \frac{\Gamma \vdash_{\mathsf{v}} v\colon A \qquad \Gamma \vdash_{\mathsf{v}} w\colon A}{\Gamma \vdash v = w\colon \mathrm{prop}}$$

$$\frac{}{\Gamma \vdash c\colon \mathrm{prop}}\ (c \in \{\top, \bot\}) \qquad \frac{\Gamma \vdash \phi\colon \mathrm{prop} \qquad \Gamma \vdash \psi\colon \mathrm{prop}}{\Gamma \vdash \phi \,\$\, \psi\colon \mathrm{prop}}\ (\$ \in \{\wedge, \vee, \Rightarrow, \star, -\!\!\star\})$$

Fig. 4: Term formation rules for the higher order separation logic.

5. given $X \in |\mathbf{C}|$, SX is a *BI-algebra*, i.e. a commutative monoid equipped with a right order-adjoint to multiplication;
6. given $f \in \mathbf{C}(X, Y)$, $Sf\colon SY \to SX$ is a BI-algebra morphism,

then S is called a *first order BI-hyperdoctrine.*

In a *(higher order) hyperdoctrine*, $\mathbf{C}$ is additionally required to be Cartesian closed and every SX is required to be poset-isomorphic to $\mathbf{C}(X, A)$ for a suitable internal Heyting algebra $A \in |\mathbf{C}|$ naturally in X. Such a hyperdoctrine is a *BI-hyperdoctrine* if moreover A is an internal BI-algebra.

Proposition 12. *Every internally complete Heyting algebra A in a Cartesian closed category $\mathbf{C}$ with finite limits gives rise to a canonical hyperdoctrine $\mathbf{C}(-, A)$: for every X, $\mathbf{C}(X, A)$ is a poset under $f \leqslant g$ iff $f \wedge g = f$.*

Proof. Clearly, every $\mathbf{C}(X, A)$ is a Heyting algebra and every $\mathbf{C}(f, A)$ is a Heyting algebra morphism. The quantifies are defined mutually dually as follows:

$$(\exists Y)_X(\phi\colon X \times Y \to A) = \bigvee\nolimits_{\mathsf{fst}\colon X \times Y \to X} \phi,$$
$$(\forall Y)_X(\phi\colon X \times Y \to A) = \bigwedge\nolimits_{\mathsf{fst}\colon X \times Y \to X} \phi.$$

Naturality in X follows from the corresponding Beck-Chevalley conditions.

Finally, internal equality $=_X : X \times X \to A$ is defined as $\bigvee_{\langle \mathsf{id}_X, \mathsf{id}_X \rangle} \top$. $\square$

A standard way to obtain an (internally) complete BI-algebra is to resort to ordered partial commutative monoids [18].

Definition 13 (Ordered PCM [18]). An *ordered partial commutative monoid (pcm)* is a tuple $(\mathcal{M}, \mathcal{E}, \cdot, \leqslant)$ where $\mathcal{M}$ is a set, $\mathcal{E} \subseteq \mathcal{M}$ is a set of *units*, *multiplication* $\cdot$ is a partial binary operation on $\mathcal{M}$, and $\leqslant$ is a preorder on $\mathcal{M}$, satisfying an number of axioms (see [18] for details).

We note that using general recipes [3], for every internal ordered pcm M in a topos $\mathbf{C}$ with subobject classifier Ω, $\mathbf{C}(- \times M, \Omega)$ forms a BI-hyperdoctrine, on particular, if $\mathbf{C} = \mathbf{Set}$ then $\mathbf{Set}(- \times M, 2)$ is a BI-hyperdoctrine.

6 A Higher Order Logic for Full Ground Store

We proceed to develop a local version of separation logic using semantic principles explored in the previous sections. That is, we seek an interpretation for the language in Fig. 4 in the category $[\mathbf{W}, \mathbf{Set}]$ over the type system (1), extended with *predicate types* $\mathsf{P}A$. The judgements $\Gamma \vdash \phi$: prop type formulas depending on a variable context Γ. Additionally, we have judgements of the form $\Gamma \vdash \phi\colon \mathsf{P}A$ for *predicates in context*. Both kinds of judgements are mutually convertible using the standard application-abstraction routine. Note that expressions for quantifiers $\exists x.\,\phi$ are thus obtained in two steps: by forming a predicate $x.\,\phi$, and subsequently applying $\exists$. Apart from the standard logical connectives, we postulate *separating conjunction* $\star$ and *separating implication* $\mathrel{-\!\!\star}$.

Our goal is to build a BI-hyperdoctrine, using the recipes, summarized in the previous section. That is, we construct a certain internal BI-algebra Θ in $[\mathbf{W}, \mathbf{Set}]$, and subsequently conclude that $[-, \Theta]$ is a BI-hyperdoctrine in question. In what follows, most of the effort is invested into constructing an internally complete Boolean algebra $\check{\mathcal{P}} \circ (\hat{P}\hat{H})$ (hence $[-, \check{\mathcal{P}} \circ (\hat{P}\hat{H})]$ is a hyperdoctrine), from which Θ is carved out as a subfunctor, identified by an upward closure condition. Here, $\check{\mathcal{P}}$ is a contravariant powerset functor, and $\hat{P}$ and $\hat{H}$ are certain modifications of the hiding and the heap functors from Section 4. As we shall see, the move from $\check{\mathcal{P}} \circ (\hat{P}\hat{H})$ to Θ remedies the problem of the former that the natural separation conjunction operator $\star$ on it does not have unit (Remark 19).

In order to model resource separation, we must identify a domain of logical assertions over partial heaps, i.e. heaplets, instead of total heaps. We thus need to derive a unary (covariant) heaplet functor from the binary, mix-variant one $\mathcal{H}$ used before. We must still cope not only with heaplets, but with partially hidden heaplets, to model information hiding. A seemingly natural candidate functor for hidden heaplets is the composition

$$P\big(\mathbf{E} \xrightarrow{\sum_{w \subseteq -} \mathcal{H}(w,-)} \mathbf{Set}\big) : \mathbf{W}^{\mathsf{op}} \to \mathbf{Set}.$$

One problem of this definition is that the equivalence relation $\sim$ underlying the construction of P (2) is too fine. Consider, for example, $e_w = (\emptyset \subseteq w, \star) \in \sum_{w' \subseteq w} \mathcal{H}(w', w)$. Then $(\mathsf{id}\colon w \to w, e_w) \not\sim (\mathsf{inl}\colon w \to w \oplus \{\star\colon 1\}, e_{w \oplus \{\star\colon 1\}})$, i.e. two hidden heaplets would not be equivalent if one extends the other by an inaccessible hidden cell. In order to arrive at a more reasonable model of logical assertions, we modify the previous model by replacing the category of initializations $\mathbf{E}$ is a category $\hat{\mathbf{E}}$ of *partial initializations*. This will induce a hiding monad $\hat{P}$ over $[\hat{\mathbf{E}}, \mathbf{Set}]$ using exactly the same formula (2) as for P.

A partial initialization is a pair (ρ, η) with $\rho \in \mathbf{W}(w_1^-, w_2^+)$ and $\eta \in \sum_{w^- \subseteq w_2^+ \ominus \rho} \mathcal{H}(w^-, w_2^+)$. Let $\hat{\mathbf{E}}$ be the category of heap layouts and partial initializations. Analogously to $\mathfrak{u}$, there is an obvious partial-heap-forgetting functor $\hat{\mathfrak{u}}\colon \hat{\mathbf{E}} \to \mathbf{W}$. Let $\hat{H}\colon \hat{\mathbf{E}} \to \mathbf{Set}$ be the following *heaplet functor*:

$$\hat{H}w = \sum\nolimits_{w' \subseteq w} \mathcal{H}(w', w).$$

Given a partial initialization $\epsilon = (\rho\colon w \to w', (w'' \subseteq w' \ominus \rho, \eta \in \mathcal{H}(w'', w')))\colon w \rightsquigarrow w'$, $\hat{H}\epsilon\colon \hat{H}w \to \hat{H}w'$ extends a given heaplet over w to a heaplet over w' via η:

$$(\hat{H}\epsilon)(w_1 \subseteq w, \eta' \in \mathcal{H}(w_1, w)) = (\rho[w_1] \cup w'' \subseteq w', \eta'')$$

where $\eta'' \in \mathcal{H}(\rho[w_1] \cup w'' \subseteq w', w')$ is as follows

$$\begin{aligned} \mathsf{pr}_{\rho(\ell\colon S)}\,\eta'' &= \mathsf{range}(S)(\rho)(\mathsf{pr}_{(\ell\colon S)}\,\eta') && ((\ell\colon S) \in w_1) \\ \mathsf{pr}_{(\ell\colon S)}\,\eta'' &= \mathsf{pr}_{(\ell\colon S)}\,\eta && ((\ell\colon S) \in w'') \end{aligned}$$

With $\hat{\mathbf{E}}$ and $\hat{H}$ as above instead of $\mathbf{E}$ and H, the framework described in Section 4 transforms coherently.

Remark 14. Let us fix a fresh symbol $\boxtimes$, and note that

$$\hat{H}w = \sum\nolimits_{w' \subseteq w} \prod\nolimits_{(\ell\colon S) \in w'} \mathsf{range}(S)(w) \cong \prod\nolimits_{(\ell\colon S) \in w} (\mathsf{range}(S)(w) \uplus \{\boxtimes\}),$$

meaning that the passage from $\mathbf{E}$, H and P to $\hat{\mathbf{E}}$, $\hat{H}$ and $\hat{P}$ is equivalent to extending the range function with designated values $\boxtimes$ for *inaccessible locations*. We prefer to think of $\boxtimes$ this way and not as a content of *dangling pointers*, to emphasize that we deal with a *reasoning phenomenon* and not with a *programming phenomenon*, for our programs neither create nor process dangling pointers.

For the next proposition we need the following concrete description of the set $\hat{\mathrm{u}}_\star(2^X)w$ as the end $\int_{\rho\colon w \to w' \in w \downarrow \hat{\mathrm{u}}} \mathbf{Set}(Xw', 2)$: this set is a space of dependent functions ϕ sending every injection $\rho\colon w \to w'$ to a corresponding subset of Xw', and satisfying the constraint: $x \in \phi(\rho)$ iff $(X\,\epsilon)(x) \in \phi(\hat{\mathrm{u}}\,\epsilon \circ \rho)$ for every $\epsilon\colon w' \rightsquigarrow w''$.

Proposition 15. *The following diagram commutes up to isomorphism:*

$$\begin{array}{ccc} [\hat{\mathbf{E}}, \mathbf{Set}] & \xrightarrow{\;2^{(-)}\;} & [\hat{\mathbf{E}}, \mathbf{Set}]^{\mathsf{op}} \\ {\scriptstyle \hat{P}}\downarrow & & \downarrow{\scriptstyle \hat{\mathrm{u}}_\star} \\ [\mathbf{W}, \mathbf{Set}^{\mathsf{op}}]^{\mathsf{op}} & \xrightarrow{\;\check{\mathcal{P}} \circ (-)\;} & [\mathbf{W}, \mathbf{Set}]^{\mathsf{op}} \end{array}$$

(using the fact that $[\mathbf{W}, \mathbf{Set}^{\mathsf{op}}]^{\mathsf{op}} \cong [\mathbf{W}^{\mathsf{op}}, \mathbf{Set}]$*) where* $\check{\mathcal{P}}$ *is the contravariant powerset functor* $\check{\mathcal{P}}\colon \mathbf{Set}^{\mathsf{op}} \to \mathbf{Set}$ *and for every* $X\colon \hat{\mathbf{E}} \to \mathbf{Set}$ *the relevant isomorphism* $\Phi_w\colon \hat{\mathrm{u}}_\star(2^X)w \cong \check{\mathcal{P}}(\hat{P}Xw)$ *is as follows:*

$$(\rho\colon w \to w', x \in Xw')_\sim \in \Phi_w(\phi \in \hat{\mathrm{u}}_\star(2^X)w) \iff x \in \phi(\rho). \tag{5}$$

Let us clarify the significance of Proposition 15. The exponential $2^{\hat{H}}$ in $[\hat{\mathbf{E}}, \mathbf{Set}]$ can be thought of as a carrier of Boolean predicates over $\hat{H}$, and as we see next those form an internally complete Boolean algebra, which is carried from $[\hat{\mathbf{E}}, \mathbf{Set}]$ to $[\mathbf{W}, \mathbf{Set}]$ by $\hat{\mathrm{u}}_\star$. The alternative route via $\hat{P}$ and $\check{\mathcal{P}}$ induces a Boolean algebra of predicates over hidden heaplets $\hat{P}\hat{H}$ directly in $[\mathbf{W}, \mathbf{Set}]$. The equivalence established in Proposition 15 witnesses agreement of these two structures.

Theorem 16. *For every* $X\colon \hat{\mathbf{E}} \to \mathbf{Set}$, $\check{\mathcal{P}} \circ (\hat{P}X)$ *is an internally complete Boolean algebra in* $[\mathbf{W}, \mathbf{Set}]$ *under*

$$\begin{aligned}
\Big(\bigvee\nolimits_f \phi\colon I \to \check{\mathcal{P}} \circ (\hat{P}X)\Big)_w (j \in Jw) \\
&= \{(\rho\colon w \to w', x \in Xw')_\sim \mid \exists\, \epsilon\colon w' \rightsquigarrow w'', \exists i \in Iw''. \\
&\qquad f_{w''}(i) = J(\hat{\mathrm{u}}\,\epsilon \circ \rho)(j) \wedge (\mathsf{id}_{w''}, (X\,\epsilon)(x))_\sim \in \phi_{w''}(i)\}, \\
\Big(\bigwedge\nolimits_f \phi\colon I \to \check{\mathcal{P}} \circ (\hat{P}X)\Big)_w (j \in Jw) \\
&= \{(\rho\colon w \to w', x \in Xw')_\sim \mid \forall\, \epsilon\colon w' \rightsquigarrow w'', \forall i \in Iw''. \\
&\qquad f_{w''}(i) = J(\hat{\mathrm{u}}\,\epsilon \circ \rho)(j) \Rightarrow (\mathsf{id}_{w''}, (X\,\epsilon)(x))_\sim \in \phi_{w''}(i)\}.
\end{aligned}$$

for every $f\colon I \to J$, *and the corresponding Boolean algebra operations are computed as set-theoretic unions, intersections and complements.*

By Theorem 16, we obtain a hyperdoctrine $[-, \check{\mathcal{P}} \circ (\hat{P}\hat{H})]$, which provides us with a model of (classical) higher order logic in $[\mathbf{W}, \mathbf{Set}]$. In particular, this allows us to interpret the language from Fig. 4 over $[\mathbf{W}, \mathbf{Set}]$ excluding the separation logic constructs, in such a way that

$$[\![\varGamma \vdash \phi\colon \mathrm{prop}]\!] \colon \underline{\varGamma} \to \check{\mathcal{P}} \circ (\hat{P}\hat{H}), \qquad [\![\varGamma \vdash \phi\colon \mathsf{P}A]\!] \colon \underline{\varGamma} \times \underline{A} \to \check{\mathcal{P}} \circ (\hat{P}\hat{H})$$

where $\underline{\varGamma} = \underline{A_1} \times \ldots \times \underline{A_n}$ for $\varGamma = (x_1 \colon A_1, \ldots, x_n \colon A_n)$ where, additionally to the standard clauses, $\underline{\mathsf{P}A} = \check{\mathcal{P}} \circ \hat{P}(\mathrm{u}^\star \underline{A} \times \hat{H})$. The latter interpretation of predicate types $\mathsf{P}A$ is justified by the natural isomorphism:

$$(\check{\mathcal{P}} \circ (\hat{P}\hat{H}))^X \cong (\hat{\mathrm{u}}_\star(2^{\hat{H}}))^X \cong \hat{\mathrm{u}}_\star((2^{\hat{H}})^{\hat{\mathrm{u}}^\star X}) \cong \check{\mathcal{P}} \circ (\hat{P}(\hat{\mathrm{u}}^\star X \times \hat{H})).$$

Here, the first and the last transitions are by $\varPhi$ from Proposition 15 and the middle one is due to the fact that clearly both $(\hat{\mathrm{u}}_\star(-))^X \vdash \hat{\mathrm{u}}^\star(X \times (-))$ and $\hat{\mathrm{u}}_\star((-)^{\hat{\mathrm{u}}^\star X}) \vdash \hat{\mathrm{u}}^\star(X \times (-))$.

Since every set $\hat{H}w$ models a heaplet in the standard sense [18], we can equip $\hat{H}w$ with a standard pointer model structure.

Proposition 17. *For every* $w \in |\mathbf{W}|$, $(\hat{H}w, \{(\emptyset \subseteq w, \star)\}, \cdot, \leqslant)$ *is an ordered pcm where for every* $w \in |\mathbf{W}|$, $\hat{H}w$ *is partially ordered as follows:*

$$(w_1 \subseteq w, \mathcal{H}(w_1 \subseteq w_2, w)\eta \in \mathcal{H}(w_1, w)) \leqslant (w_2 \subseteq w, \eta \in \mathcal{H}(w_2, w)) \quad (w_1 \subseteq w_2)$$

and for $w_1 \subseteq w$, $w_2 \subseteq w$ *and* $\eta_1 \in \mathcal{H}(w_1, w)$, $\eta_2 \in \mathcal{H}(w_2, w)$, $(w_1 \subseteq w, \eta_1) \cdot (w_2 \subseteq w, \eta_2)$ *equals* $(w_1 \cup w_2, \eta_1 \cup \eta_2)$ *if* $w_1 \cap w_2 = \emptyset$, *and otherwise undefined.*

As indicated in Section 5, we automatically obtain a BI-algebra structure over the set of all subsets of $\hat{H}w$. The same strategy does not apply to $\hat{P}\hat{H}w$, roughly because we cannot predict mutual arrangement of hidden partitions of two heaplets wrt to each other, for we do not have a global reference space for pointers as contrasted to the standard separation logic setting. We thus define a separating conjunction operator directly on every $\check{\mathcal{P}}(\hat{P}\hat{H}w)$ as follows:

$$\begin{aligned}\phi \star_w \psi = \{(\rho\colon w \to w', (w_1 \uplus w_2 \subseteq w', \eta \in \mathcal{H}(w_1 \uplus w_2, w')))_\sim \mid \\ (\rho, (w_1 \subseteq w', \mathcal{H}(w_1 \subseteq w_1 \uplus w_2, w')\eta))_\sim \in \phi, \\ (\rho, (w_2 \subseteq w', \mathcal{H}(w_2 \subseteq w_1 \uplus w_2, w')\eta))_\sim \in \psi\}.\end{aligned}$$

Lemma 18. *The operator $\star_w$ on $\check{\mathcal{P}}(\hat{P}\hat{H}w)$ satisfies the following properties.*

1. *$\star_w$ is natural in w.*
2. *$\star_w$ is associative and commutative.*
3. *$(\rho\colon w \to w', (w'' \subseteq w', \eta \in \mathcal{H}(w'', w')))_\sim \in \phi \star_w \psi$ if and only if there exist w_1, w_2 such that $w_1 \uplus w_2 = w''$, $(\rho, (w_1 \subseteq w', \mathcal{H}(w_1 \subseteq w'', w')\eta))_\sim \in \phi$ and $(\rho, (w_2 \subseteq w', \mathcal{H}(w_2 \subseteq w'', w')\eta))_\sim \in \psi$.*

Property (3) specifically tells us that any representative of an equivalence class contained in a separating conjunction can be split in such a way that the respective pieces belong to the arguments of the separating conjunction.

Remark 19. The only candidate for the unit of the separating conjunction $\star_w$ would be the emptiness predicate $\mathsf{empty}_w\colon 1 \to \check{\mathcal{P}}(\hat{P}\hat{H}w)$, identifying precisely the empty heaplets. However, empty_w is not natural in w. In fact, it follows by Yoneda lemma that there are exactly two natural transformations $1 \to \check{\mathcal{P}} \circ (\hat{P}\hat{H})$, which are the total truth and the total false, none of which is a unit for $\star_w$.

Remark 19 provides a formal argument why we cannot interpret classical separation logic over $\check{\mathcal{P}} \circ (\hat{P}\hat{H})$. We thus proceed to identify for every w a subset of $\check{\mathcal{P}}(\hat{P}\hat{H}w)$, for which the total truth predicate becomes the unit of the separating conjunction. Concretely, let Θ be the subfunctor of $\check{\mathcal{P}} \circ (\hat{P}\hat{H})$ identified by the following *upward closure condition:* $\phi \in \Theta w$ if

$$(\rho, \eta)_\sim \in \phi,\ \eta \leqslant \eta' \qquad \text{imply} \qquad (\rho, \eta')_\sim \in \phi.$$

Lemma 20. *Θ is an internal complete sublattice of $\check{\mathcal{P}} \circ (\hat{P}\hat{H})$, i.e. the inclusion $\iota\colon \Theta \hookrightarrow \check{\mathcal{P}} \circ (\hat{P}\hat{H})$ preserves all meets and all joins. This canonically equips Θ with an internally complete Heyting algebra structure.*

Proof (Sketch). The key idea is to establish a retraction (ι, cl) with $\mathsf{cl} \circ \iota = \mathsf{id}$. The requisite structure is then transferred from $\check{\mathcal{P}} \circ (\hat{P}\hat{H})$ to Θ along it. The Heyting implication for Θ is obtained using the standard formula $(\phi \Rightarrow \psi) = \bigvee\{\xi \mid \phi \wedge \xi \leqslant \psi\}$ interpreted in the internal language. □

Lemma 21. *Separating conjunction preserves upward closure: for $\phi, \psi \in \Theta w$, $\phi \star_w \psi = \mathsf{cl}_w(\phi \star_w \psi)$.*

Lemma 22. *Θ is a BI-algebra: $\star_w$ is obtained by restriction from $\check{\mathcal{P}}(\hat{P}\hat{H}w)$ by Lemma 21, $\hat{P}\hat{H}w$ is the unit for it and*

$$\begin{aligned}\phi \mathrel{-\!\!\star}_w \psi = \{(\rho, \eta)_\sim \in \Theta w \mid \forall \rho'\colon w \to w', \eta_1, \eta_2 \in \hat{H}w', \eta_1 \cdot \eta_2 \text{ defined} \wedge \\ (\rho, \eta) \sim (\rho', \eta_1) \wedge (\rho', \eta_2)_\sim \in \phi \Rightarrow (\rho', \eta_1 \cdot \eta_2)_\sim \in \psi\}.\end{aligned}$$

- $s, \rho, \eta \models \top$
- $s, \rho, \eta \models \phi \wedge \psi$ if $s, \rho, \eta \models \phi$ and $s, \rho, \eta \models \psi$
- $s, \rho, \eta \models \phi \vee \psi$ if $s, \rho, \eta \models \phi$ or $s, \rho, \eta \models \psi$
- $s, \rho, \eta \models \phi \Rightarrow \psi$ if for all $(\rho, \eta) \sim (\rho', \eta')$ and $\eta' \leqslant \eta''$,
  $s, \rho', \eta'' \models \phi$ implies $s, \rho', \eta'' \models \psi$
- $s, \rho, \eta \models \phi(v)$ if $s, \rho, ((\llbracket \Gamma \vdash_{\mathsf{v}} v \colon A \rrbracket_{w'} \circ \underline{\Gamma}\rho)s, \eta) \models \phi$
- $s, \rho, (a, \eta) \models x. \phi$ if $a = (X\rho)b$ and $(s, b), \rho, \eta \models \phi$
- $s, \rho, \eta \models \ell \hookrightarrow v$ if $\eta = (w'' \subseteq w', \delta \in \mathcal{H}(w'', w'))$ and
  $\delta(r \colon S) = (\llbracket \Gamma \vdash_{\mathsf{v}} v \colon \mathsf{CType}(S) \rrbracket_{w'} \circ \underline{\Gamma}\rho)s$
  where $(\llbracket \Gamma \vdash_{\mathsf{v}} \ell \colon \mathsf{Ref}_S \rrbracket_{w'} \circ \underline{\Gamma}\rho)s = (r \colon S) \in w''$
- $s, \rho, \eta \models v = u$ if $(\llbracket \Gamma \vdash_{\mathsf{v}} v \colon A \rrbracket_{w''} \circ \underline{\Gamma}\rho' \circ \underline{\Gamma}\rho)(s) = (\llbracket \Gamma \vdash_{\mathsf{v}} u \colon A \rrbracket_{w''} \circ \underline{\Gamma}\rho' \circ \underline{\Gamma}\rho)(s)$
  for some $\rho' \colon w' \to w''$
- $s, \rho, \eta \models \phi \star \psi$ if for suitable w_1, w_2, $\eta \in \mathcal{H}(w_1 \uplus w_2, w')$,
  $s, \rho, (w_1 \subseteq w', \mathcal{H}(w_1 \subseteq w_1 \uplus w_2, w')\eta) \models \phi$ and
  $s, \rho, (w_2 \subseteq w', \mathcal{H}(w_2 \subseteq w_1 \uplus w_2, w')\eta) \models \psi$
- $s, \rho, \eta \models \phi \mathrel{-\!\!\star} \psi$ if for all $(\rho', \eta_1) \sim (\rho, \eta)$ and for all η_2 such that $\eta_1 \cdot \eta_2$ is defined,
  $s, \rho', \eta_2 \models \phi$ implies $s, \rho', \eta_1 \cdot \eta_2 \models \psi$
- $s, \rho, \eta \models \exists \phi$ if $\underline{\Gamma}(\hat{\mathrm{u}}\epsilon \circ \rho)s, \mathsf{id}_{w''}, (a, \hat{H}\epsilon \circ \eta) \models \phi$ for some $\epsilon \colon w' \rightsquigarrow w''$, $a \in \underline{A}w''$
- $s, \rho, \eta \models \forall \phi$ if $\underline{\Gamma}(\hat{\mathrm{u}}\epsilon \circ \rho)s, \mathsf{id}_{w''}, (a, \hat{H}\epsilon \circ \eta) \models \phi$ for all $\epsilon \colon w' \rightsquigarrow w''$, $a \in \underline{A}w''$

Fig. 5: Semantics of the logic.

Proof. In view of Lemma 20, we are left to show that the given operations are natural and that Θ is an internal BI-algebra w.r.t. them. Since BI-algebras form a variety [5], it suffices to show that each Θw is a BI-algebra. By Lemma 18 (ii), it suffices to show that every $(-) \star_w \phi$ preserves arbitrary joins, for then we can use the standard formula to calculate $\phi \mathrel{-\!\!\star}_w \psi$, which happens to be natural in w:

$$\phi \mathrel{-\!\!\star}_w \psi = \bigcup \{\xi \mid \phi \star_w \xi \leqslant \psi\}.$$

By unfolding the right-hand side, we obtain the expression for $\mathrel{-\!\!\star}_w$ figuring in the statement of the lemma. □

Theorem 23. *Θ is an internally complete Heyting BI-algebra, hence $[-, \Theta]$ is a BI-hyperdoctrine.*

Proof. Follows from Lemmas 20 and 22. □

This now provides us with a complete semantics of the language in Fig. 4 with $\llbracket \Gamma \vdash \phi \colon \mathrm{prop} \rrbracket \colon \underline{\Gamma} \to \Theta$ and $\llbracket \Gamma \vdash \phi \colon \mathsf{P}A \rrbracket \colon \underline{\Gamma} \to \underline{\mathsf{P}A}$ where $\underline{\mathsf{P}A}$ is the upward closed subfunctor of $\check{\mathcal{P}} \circ (\hat{P}(\hat{\mathrm{u}}\underline{A} \times \hat{H}))$, with upward closure only on the $\hat{H}$-part, which is isomorphic to $\Theta^{\underline{A}}$. The resulting semantics is defined in Fig. 5 where we write $s, \rho, \eta \models \phi$ for $(\rho, \eta)_\sim \in [\![\Gamma \vdash \phi\colon \mathrm{prop}]\!](s)$ and $s, \rho, (a, \eta) \models \phi$ for $(\rho, (a, \eta))_\sim \in [\![\Gamma \vdash \phi\colon \mathsf{P}A]\!](s)$. The following properties [4] are then automatic.

Proposition 24. – (Monotonicity) *If $s, \rho, \eta \models \phi$ and $\eta \leqslant \eta'$ then $s, \rho, \eta' \models \phi$.*
– (Shrinkage) *If $s, \rho, \eta \models \phi$, $\eta' \leqslant \eta$ and η' contains all cells reachable from s and w then $s, \rho, \eta' \models \phi$.*

7 Examples

Let us illustrate subtle features of our semantics by some examples.

Example 25. Consider the formula $\exists \ell\colon \mathsf{Ref}_{\mathsf{Int}}\,.\, \ell \hookrightarrow 5$ from the introduction in the empty context –. Then $-, \rho, \eta \models \exists \ell.\, \ell \hookrightarrow 5$ iff for some $\epsilon\colon w' \rightsquigarrow w''$, and some $x \in \underline{\mathsf{Ref}_{\mathsf{Int}}} w''$, $x, \mathsf{id}_{w''}, (\hat{H}\epsilon)\eta \models \ell' \hookrightarrow 5$. The latter is true iff $\mathsf{pr}_x((\hat{H}\epsilon)\eta) = 5$. Note that w' may not contain ℓ and it is always possible to choose ϵ so that w'' contains ℓ and $\mathsf{pr}_x((\hat{H}\epsilon)\eta) = 5$. Hence, the original formula is always valid.

Example 26. The clauses in Fig. 5 are very similar to the standard Kripke semantics of intuitionistic logic. Note however, that the clause for implication strikingly differs from the expected one

– $s, \rho, \eta \models \phi \Rightarrow \psi$ if for all $\eta \leqslant \eta'$, $s, \rho, \eta' \models \phi$ implies $s, \rho, \eta' \models \psi$,

though. The latter is indeed not validated by our semantics, as witnessed by the following example. Consider the following formulas ϕ and ψ respectively:

$$\ell\colon \mathsf{Ref}_{\mathsf{Ref}_{Int}} \vdash \exists \ell'.\, \exists x.\, \ell \hookrightarrow \ell' \wedge \ell' \hookrightarrow x\colon \mathrm{prop} \tag{6}$$
$$\ell\colon \mathsf{Ref}_{\mathsf{Ref}_{Int}} \vdash \exists \ell'.\, \ell \hookrightarrow \ell' \wedge \ell' \hookrightarrow 6\colon \mathrm{prop} \tag{7}$$

The first formula is valid over heaplets, in which ℓ refers to a reference to some integer, while the second one is only valid over heaplets, in which ℓ refers to a reference to 6. Any $\eta' \geqslant \eta = (\mathsf{id}_w, (\{\ell''\} \subseteq \{\ell, \ell''\}, [\ell'' \mapsto 6]))$ satisfies both (6) and (7) or none of them. However, the implication $\phi \Rightarrow \psi$ still is not valid over η in our semantics, for

$$\begin{aligned}\eta &\sim (w \hookrightarrow w \oplus (\ell'\colon Int), (\{\ell', \ell''\} \subseteq \{\ell, \ell', \ell''\}, [\ell' \mapsto 5, \ell'' \mapsto 6]))\\ &\leqslant (w \hookrightarrow w \oplus (\ell'\colon Int), (\{\ell, \ell', \ell''\} \subseteq \{\ell, \ell', \ell''\}, [\ell \mapsto \ell', \ell' \mapsto 5, \ell'' \mapsto 6]))\end{aligned}$$

and the latter heaplet validates ϕ but not ψ.

Example 27. Least μ and greatest ν fixpoints can be encoded in higher order logic [2]. As an example, consider

$$\mathit{isList} = \mu\gamma.\, \ell.\; \ell \hookrightarrow \mathit{null} \vee \exists \ell', x.\, \ell \hookrightarrow (x, \ell') \star \gamma(\ell'),$$

which specifies the fact that ℓ is a pointer to a head of a list (eliding coproduct injections in $\mathsf{inl}\,\mathit{null}$ and $\mathsf{inr}(x, \ell')$). By definition, *isList* satisfies the following recursive equation:

$$\mathit{isList}(\ell) = \ell \hookrightarrow \mathit{null} \vee \exists \ell', x.\, \ell \hookrightarrow (x, \ell') \star \mathit{isList}(\ell')$$

Let us expand the semantics of the right hand side. We have

$$\begin{aligned}
&[\![\ell\colon \mathsf{Ref}_{\mathsf{list}}, \mathit{isList}\colon \mathsf{P}(\mathsf{Ref}_{\mathsf{list}}) \vdash l \hookrightarrow \mathit{null} \vee \exists \ell', x.\, \ell \hookrightarrow (x, \ell') \star \mathit{isList}(\ell')]\!]_w(\mathit{isList})\\
&\quad = \{(\rho\colon w \to w', (\underline{\mathsf{Ref}_{\mathsf{list}}}\rho)(\ell), \delta \in \hat{H}w')_\sim \mid \mathsf{pr}_{\rho(\ell)}(\delta) = \mathit{null}\} \cup\\
&\qquad\quad [\![\ell\colon \mathsf{Ref}_{\mathsf{list}}, \mathit{isList}\colon \mathsf{P}(\mathsf{Ref}_{\mathsf{list}}) \vdash \exists \ell', x.\, \ell \hookrightarrow (x, \ell') \star \mathit{isList}(\ell')]\!]_w(\mathit{isList})\\
&\quad = \{(\rho\colon w \to w', (\underline{\mathsf{Ref}_{\mathsf{list}}}\rho)(\ell), \delta \in \hat{H}w')_\sim \mid\\
&\qquad\quad \mathsf{pr}_{\rho(\ell)}(\delta) = \mathit{null} \;\vee\; \exists \ell',\; x.\; \mathsf{pr}_{\rho(\ell)}\,\delta = (x, \ell') \wedge (\rho, \ell', \delta \setminus \rho(\ell))_\sim \in \mathit{isList}\}
\end{aligned}$$

where $\delta \setminus \rho(\ell)$ denotes the δ with the cell $\rho(\ell)$ removed. In summary, $(\rho\colon w \to w', (\underline{\mathsf{Ref}_{\mathsf{list}}}\rho)(\ell), \delta \in \hat{H}w')_\sim$ is in $[\![\ell\colon \mathsf{Ref}_{\mathsf{list}}, \mathit{isList}\colon \mathsf{P}(\mathsf{Ref}_{\mathsf{list}}) \vdash \mathit{isList}(\ell)]\!]_w(\mathit{isList})$ if and only if either $\mathsf{pr}_{\rho(\ell)}\,\delta = \mathit{null}$ or there exists an $l' \in w'$ such that $\mathsf{pr}_{\rho(\ell)}\,\delta = (x, \ell')$ and $(\rho, \ell', \delta \setminus \rho(\ell))_\sim \in \mathit{isList}$.

8 Conclusions and Further Work

Compositionality is an uncontroversial desirable property in semantics and reasoning, which admits strikingly different, but equally valid interpretations, as becomes particularly instructive when modelling dynamic memory allocation. From the programming perspective it is desirable to provide compositional means for keeping track of integrity of the underlying data, in particular, for preventing *dangling pointers*. Reasoning however inherently requires introduction of partially defined data, such as *heaplets*, which due to the compositionality principle must be regarded as first class semantic units.

Here we have made a step towards reconciling recent extensional monad-based denotational semantic for full-ground store [9] with higher order categorical reasoning frameworks [2] by constructing a suitable intuitionistic BI-hyperdoctrine. Much remains to be done. A highly desirable ingredient, which is currently missing in our logic in Fig. 4 is a construct relating programs and logical assertions, such as the following dynamic logic style modality

$$\frac{\Gamma \vdash_{\mathsf{c}} p\colon A \qquad \Gamma \vdash \phi\colon \mathsf{P}A}{\Gamma \vdash [p]\phi\colon \mathrm{prop}}$$

which would allow us e.g. in a standard way to encode *Hoare triples* $\{\phi\}p\{\psi\}$ as implications $\phi \Rightarrow [p]\psi$. This is difficult due to the outlined discrepancy in the semantics for construction and reasoning. The categories of initializations for p and ϕ and the corresponding hiding monads are technically incompatible. In future work we aim to deeply analyse this phenomenon and develop a semantics for such modalities in a principled fashion.

Orthogonally to these plans we are interested in further study of the full ground store monad and its variants. One interesting research direction is developing algebraic presentations of these monads in terms of operations and equations [17]. Certain generic methods [13] were proposed for the simple store case (Example 3), and it remains to be seen if these can be generalized to the full ground store case.

References

1. Samson Abramsky. Intensionality, definability and computation. In Alexandru Baltag and Sonja Smets, editors, *Johan van Benthem on Logic and Information Dynamics*, pages 121–142. Springer, 2014.
2. Bodil Biering, Lars Birkedal, and Noah Torp-Smith. BI-hyperdoctrines, higher-order separation logic, and abstraction. *ACM Trans. Program. Lang. Syst.*, 29(5), 2007.
3. Ales Bizjak and Lars Birkedal. On models of higher-order separation logic. *Electr. Notes Theor. Comput. Sci.*, 336:57–78, 2018.
4. Cristiano Calcagno, Peter O'Hearn, and Richard Bornat. Program logic and equivalence in the presence of garbage collection. *Theoretical Computer Science*, 298(3):557 – 581, 2003. Foundations of Software Science and Computation Structures.
5. Nikolaos Galatos, Peter Jipsen, Tomasz Kowalski, and Hiroakira Ono. *Residuated Lattices: An Algebraic Glimpse at Substructural Logics, Volume 151*. Elsevier Science, San Diego, CA, USA, 1st edition, 2007.
6. Peter Johnstone. *Sketches of an elephant: A topos theory compendium*. Oxford logic guides. Oxford Univ. Press, New York, 2002.
7. Peter T Johnstone. Conditions related to De Morgan's law. In *Applications of sheaves*, pages 479–491. Springer, 1979.
8. Ralf Jung, Robbert Krebbers, Jacques-Henri Jourdan, Aleš Bizjak, Lars Birkedal, and Derek Dreyer. Iris from the ground up: A modular foundation for higher-order concurrent separation logic. *Journal of Functional Programming*, 28:e20, 2018.
9. Ohad Kammar, Paul Blain Levy, Sean K. Moss, and Sam Staton. A monad for full ground reference cells. In *32nd Annual ACM/IEEE Symposium on Logic in Computer Science, LICS 2017*, pages 1–12, 2017.
10. William Lawvere. Adjointness in foundations. *Dialectica*, 23(3-4):281–296, 1969.
11. Paul Blain Levy, John Power, and Hayo Thielecke. Modelling environments in call-by-value programming languages. *Inf. & Comp*, 185:2003, 2002.
12. Saunders Mac Lane. *Categories for the Working Mathematician*. Springer, 1971.
13. Kenji Maillard and Paul-André Melliès. A fibrational account of local states. In *30th Annual ACM/IEEE Symposium on Logic in Computer Science, LICS 2015*, pages 402–413. IEEE Computer Society, 2015.
14. Peter O'Hearn and Robert D. Tennent. Semantics of local variables. *Applications of categories in computer science*, 177:217–238, 1992.
15. Frank Joseph Oles. *A Category-theoretic Approach to the Semantics of Programming Languages*. PhD thesis, Syracuse University, Syracuse, NY, USA, 1982.
16. Gordon Plotkin and John Power. Notions of computation determine monads. In *FoSSaCS'02*, volume 2303 of *LNCS*, pages 342–356. Springer, 2002.
17. Gordon Plotkin and John Power. Algebraic operations and generic effects. *Appl. Cat. Struct.*, 11(1):69–94, 2003.
18. David J. Pym, Peter W. O'Hearn, and Hongseok Yang. Possible worlds and resources: the semantics of BI. *Theor. Comput. Sci.*, 315:257–305, May 2004.
19. John Reynolds. The essence of ALGOL. In Peter W. O'Hearn and Robert D. Tennent, editors, *ALGOL-like Languages, Volume 1*, pages 67–88. Birkhauser Boston Inc., Cambridge, MA, USA, 1997.
20. John Reynolds. Intuitionistic reasoning about shared mutable data structure. In *Millennial Perspectives in Computer Science*, pages 303–321. Palgrave, 2000.
21. John Reynolds. Separation logic: A logic for shared mutable data structures. In *17th Annual IEEE Symposium on Logic in Computer Science, LICS 2002*, pages 55–74. IEEE Computer Society, 2002.

22. Alex Simpson. Category-theoretic structure for independence and conditional independence. *Electr. Notes Theor. Comput. Sci.*, 336:281–297, 2018.
23. Sam Staton. Instances of computational effects: An algebraic perspective. In *Proc. 28th Annual ACM/IEEE Symposium on Logic in Computer Science (LICS 2013)*, pages 519–519, June 2013.

Quantum Programming with Inductive Datatypes: Causality and Affine Type Theory

Romain Péchoux[1], Simon Perdrix[1], Mathys Rennela[2], and Vladimir Zamdzhiev[1(✉)]

[1] Université de Lorraine, CNRS, Inria, LORIA, F 54000 Nancy, France
{romain.pechoux|simon.perdrix|vladimir.zamdzhiev}@loria.fr
[2] Leiden University, Leiden, The Netherlands
m.p.a.rennela@liacs.leidenuniv.nl

Abstract. Inductive datatypes in programming languages allow users to define useful data structures such as natural numbers, lists, trees, and others. In this paper we show how inductive datatypes may be added to the quantum programming language QPL. We construct a sound categorical model for the language and by doing so we provide the first detailed semantic treatment of user-defined inductive datatypes in quantum programming. We also show our denotational interpretation is invariant with respect to big-step reduction, thereby establishing another novel result for quantum programming. Compared to classical programming, this property is considerably more difficult to prove and we demonstrate its usefulness by showing how it immediately implies computational adequacy at all types. To further cement our results, our semantics is entirely based on a physically natural model of von Neumann algebras, which are mathematical structures used by physicists to study quantum mechanics.

Keywords: Quantum programming · Inductive types · Adequacy

1 Introduction

Quantum computing is a computational paradigm which takes advantage of quantum mechanical phenomena to perform computation. A quantum computer can solve problems which are out of reach for classical computers (e.g. factorisation of large numbers [24], solving large linear systems [8]). The recent developments of quantum technologies points out the necessity of filling the gap between theoretical quantum algorithms and the actual (prototypes of) quantum computers. As a consequence, quantum software and in particular quantum programming languages play a key role in the future development of quantum computing. The present paper makes several theoretical contributions towards the design and denotational semantics of quantum programming languages.

Our development is based around the quantum programming language QPL [23] which we extend with inductive datatypes. Our paper is the first to construct a denotational semantics for user-defined inductive datatypes in quantum programming. In the spirit of the original QPL, our type system is *affine* (discarding

J. Goubault-Larrecq and B. König (Eds.): FOSSACS 2020, LNCS 12077, pp. 562–581, 2020.
https://doi.org/10.1007/978-3-030-45231-5_29

of arbitrary variables is allowed, but copying is restricted). We also extend QPL with a copy operation for *classical data*, because this is an admissible operation in quantum mechanics which improves programming convenience. The addition of inductive datatypes requires a departure from the original denotational semantics of QPL, which are based on finite-dimensional quantum structures, and we consider instead (possibly infinite-dimensional) quantum structures based on *W*-algebras* (also known as *von Neumann algebras*), which have been used by physicists in the study of quantum foundations [25]. As such, our semantic treatment is physically natural and our model is more accessible to physicists and experts in quantum computing compared to most other denotational models.

QPL is a first-order programming language which has *procedures*, but it does not have lambda abstractions. Thus, there is no use for a !-modality and we show how to model the copy operation by describing the canonical comonoid structure of all classical types (including the inductive ones).

An important notion in quantum mechanics is the idea of *causality* which has been formulated in a variety of different ways. In this paper, we consider a simple operational interpretation of causality: if the output of a physical process is discarded, then it does not matter which process occurred [10]. In a symmetric monoidal category $\mathbf{C}$ with tensor unit I, this can be understood as requiring that for any morphism (process) $f : A_1 \to A_2$, it must be the case that $\diamond_{A_2} \circ f = \diamond_{A_1}$, where $\diamond_{A_i} : A_i \to I$ is the discarding map (process) at the given objects. This notion ties in very nicely with our affine language, because we have to show that the interpretation of values is causal, i.e., values are always discardable.

A major contribution of this paper is that we prove the denotational semantics is invariant with respect to both small-step reduction and big-step reduction. The latter is more difficult in quantum programming and our paper is the first to demonstrate such a result. As a corollary, we obtain computational adequacy.

2 Syntax of QPL

The syntax of QPL (including our extensions) is summarised in Figure 1. A well-formed type context, denoted $\vdash \Theta$, is simply a list of distinct type variables. A type A is well-formed in type context Θ, denoted $\Theta \vdash A$, if the judgement can be derived according to the following rules (see [1,6] for a more detailed exposition):

$$\frac{\vdash \Theta}{\Theta \vdash \Theta_i} \qquad \frac{\vdash \Theta}{\Theta \vdash I} \qquad \frac{\vdash \Theta}{\Theta \vdash \mathbf{qbit}} \qquad \frac{\Theta \vdash A \qquad \Theta \vdash B}{\Theta \vdash A \star B} \star \in \{+, \otimes\} \qquad \frac{\Theta, X \vdash A}{\Theta \vdash \mu X.A}$$

A type A is *closed* if $\cdot \vdash A$. Note that nested type induction is allowed. Henceforth, we implicitly assume that all types we are dealing with are well-formed.

Example 1. The type of natural numbers is defined as $\mathbf{Nat} \equiv \mu X.I + X$. Lists of a closed type $\cdot \vdash A$ are defined as $\mathbf{List}(A) \equiv \mu Y.I + A \otimes Y$.

Notice that our type system is not equipped with a !-modality. Indeed, in the absence of function types, there is no reason to introduce it. Instead, we specify

$$
\begin{array}{lll}
\text{Types} & A, B & ::= X \mid I \mid \mathbf{qbit} \mid A + B \mid A \otimes B \mid \mu X.A \\
\text{Classical Types} & P, R & ::= X \mid I \mid P + R \mid P \otimes R \mid \mu X.P \\
\text{Terms} & M, N & ::= \mathbf{new\ unit}\ u \mid \mathbf{discard}\ x \mid y = \mathbf{copy}\ x \mid \mathbf{new\ qbit}\ q \mid \\
& & \quad b = \mathbf{measure}\ q \mid q_1, \ldots, q_n \mathrel{*=} S \mid M; N \mid \mathbf{skip} \mid \\
& & \quad \mathbf{while}\ b\ \mathbf{do}\ M \mid x = \mathbf{left}_{A,B} M \mid x = \mathbf{right}_{A,B} M \mid \\
& & \quad \mathbf{case}\ y\ \mathbf{of}\ \{\mathbf{left}\ x_1 \to M \mid \mathbf{right}\ x_2 \to N\} \mid \\
& & \quad x = (x_1, x_2) \mid (x_1, x_2) = x \mid y = \mathbf{fold}\ x \mid y = \mathbf{unfold}\ x \mid \\
& & \quad \mathbf{proc}\ f :: \ x : A \to y : B\ \{M\} \ \mid y = f(x) \\
\text{Variable contexts} & \Gamma, \Sigma & ::= x_1 : A_1, \ldots, x_n : A_n \\
\text{Procedure contexts} & \Pi & ::= f_1 : A_1 \to B_1, \ldots, f_n : A_n \to B_n
\end{array}
$$

$$
\frac{}{\Pi \vdash \langle \Gamma \rangle\ \mathbf{new\ unit}\ u\ \langle \Gamma, u : I \rangle}
\qquad
\frac{}{\Pi \vdash \langle \Gamma, x : A \rangle\ \mathbf{discard}\ x\ \langle \Gamma \rangle}
$$

$$
\frac{P \text{ is a classical type}}{\Pi \vdash \langle \Gamma, x : P \rangle\ y = \mathbf{copy}\ x\ \langle \Gamma, x : P, y : P \rangle}
\qquad
\frac{}{\Pi \vdash \langle \Gamma \rangle\ \mathbf{skip}\ \langle \Gamma \rangle}
$$

$$
\frac{\Pi \vdash \langle \Gamma \rangle\ M\ \langle \Gamma' \rangle \qquad \Pi \vdash \langle \Gamma' \rangle\ N\ \langle \Sigma \rangle}{\Pi \vdash \langle \Gamma \rangle\ M; N\ \langle \Sigma \rangle}
$$

$$
\frac{\Pi \vdash \langle \Gamma, b : \mathbf{bit} \rangle\ M\ \langle \Gamma, b : \mathbf{bit} \rangle}{\Pi \vdash \langle \Gamma, b : \mathbf{bit} \rangle\ \mathbf{while}\ b\ \mathbf{do}\ M\ \langle \Gamma, b : \mathbf{bit} \rangle}
$$

$$
\frac{}{\Pi \vdash \langle \Gamma \rangle\ \mathbf{new\ qbit}\ q\ \langle \Gamma, q : \mathbf{qbit} \rangle}
\qquad
\frac{}{\Pi \vdash \langle \Gamma, q : \mathbf{qbit} \rangle\ b = \mathbf{measure}\ q\ \langle \Gamma, b : \mathbf{bit} \rangle}
$$

$$
\frac{S \text{ is a unitary of arity } n}{\Pi \vdash \langle \Gamma, q_1 : \mathbf{qbit}, \ldots, q_n : \mathbf{qbit} \rangle\ q_1, \ldots, q_n \mathrel{*=} S\ \langle \Gamma, q_1 : \mathbf{qbit}, \ldots, q_n : \mathbf{qbit} \rangle}
$$

$$
\frac{}{\Pi \vdash \langle \Gamma, x : A \rangle\ y = \mathbf{left}_{A,B}\ x\ \langle \Gamma, y : A + B \rangle}
$$

$$
\frac{}{\Pi \vdash \langle \Gamma, x : B \rangle\ y = \mathbf{right}_{A,B}\ x\ \langle \Gamma, y : A + B \rangle}
$$

$$
\frac{\Pi \vdash \langle \Gamma, x_1 : A \rangle\ M_1\ \langle \Sigma \rangle \qquad \Pi \vdash \langle \Gamma, x_2 : B \rangle\ M_2\ \langle \Sigma \rangle}{\Pi \vdash \langle \Gamma, y : A + B \rangle\ \mathbf{case}\ y\ \mathbf{of}\ \{\mathbf{left}_{A,B}\ x_1 \to M_1 \mid \mathbf{right}_{A,B}\ x_2 \to M_2\ \}\ \langle \Sigma \rangle}
$$

$$
\frac{}{\Pi \vdash \langle \Gamma, x_1 : A, x_2 : B \rangle\ x = (x_1, x_2)\ \langle \Gamma, x : A \otimes B \rangle}
$$

$$
\frac{}{\Pi \vdash \langle \Gamma, x : A \otimes B \rangle\ (x_1, x_2) = x\ \langle \Gamma, x_1 : A, x_2 : B \rangle}
$$

$$
\frac{}{\Pi \vdash \langle \Gamma, x : A[\mu X.A/X] \rangle\ y = \mathbf{fold}_{\mu X.A}\ x\ \langle \Gamma, y : \mu X.A \rangle}
$$

$$
\frac{}{\Pi \vdash \langle \Gamma, x : \mu X.A \rangle\ y = \mathbf{unfold}\ x\ \langle \Gamma, y : A[\mu X.A/X] \rangle}
$$

$$
\frac{\Pi, f : A \to B \vdash \langle x : A \rangle\ M\ \langle y : B \rangle}{\Pi \vdash \langle \Gamma \rangle\ \mathbf{proc}\ f :: \ x : A \to y : B\ \{M\}\ \langle \Gamma \rangle}
$$

$$
\frac{}{\Pi, f : A \to B \vdash \langle \Gamma, x : A \rangle\ y = f(x)\ \langle \Gamma, y : B \rangle}
$$

Fig. 1: Syntax and formation rules for QPL terms.

the subset of types where copying is an admissible operation. The *classical types* are a subset of our types defined in Figure 1. They are characterised by the property that variables of classical types may be copied, whereas variables of non-classical types may not be copied (see the rule for copying in Figure 1).

We use small Latin letters (e.g. x, y, u, q, b) to range over *term variables*. More specifically, q ranges over variables of type **qbit**, u over variables of unit type I, b over variables of type $\mathbf{bit} := I + I$ and x, y range over variables of arbitrary type. We use Γ and Σ to range over *variable contexts*. A variable context is a function from term variables to *closed types*, which we write as $\Gamma = x_1 : A_1, \ldots, x_n : A_n$.

We use f, g to range over *procedure names*. Every procedure name f has an *input type* A and an *output type* B, denoted $f : A \to B$, where A and B are closed types. We use Π to range over *procedure contexts*. A procedure context is a function from procedure names to pairs of procedure input-output types, denoted $\Pi = f_1 : A_1 \to B_1, \ldots, f_n : A_n \to B_n$.

Remark 2. Unlike lambda abstractions, procedures cannot be passed to other procedures as input arguments, nor can they be returned as output.

A *term judgement* has the form $\Pi \vdash \langle \Gamma \rangle\ M\ \langle \Sigma \rangle$ (see Figure 1) and indicates that term M is well-formed in procedure context Π with input variable context Γ and output variable context Σ. All types occurring within it are closed.

The intended interpretation of the quantum rules are as follows. The term **new qbit** q prepares a new qubit q in state $|0\rangle\langle 0|$. The term $q_1, \ldots, q_n \mathrel{*}= S$ applies a unitary operator S to a sequence of qubits in the standard way. The term $b = \mathbf{measure}\ q$ performs a quantum measurement on qubit q and stores the measurement outcome in bit b. The measured qubit is destroyed in the process.

The no-cloning theorem of quantum mechanics [28] shows that arbitrary qubits cannot be copied. Because of this, copying is restricted only to classical types, as indicated in Figure 1, and this allows us to avoid runtime errors. Like the original QPL [23], our type system is also *affine* and so any variable can be discarded (see the formation rule for the term **discard** x in Figure 1).

3 Operational Semantics of QPL

In this section we describe the operational semantics of QPL. The central notion is that of a *program configuration* which provides a complete description of the current state of program execution. It consists of four components that must satisfy some coherence properties: (1) the term which remains to be executed; (2) a *value assignment*, which is a function that assigns formal expressions to variables as a result of execution; (3) a *procedure store* which keeps track of what procedures have been defined so far and (4) the *quantum state* computed so far.

Value Assignments. A *value* is an expression defined by the following grammar:

$$v, w ::= * \mid n \mid \mathbf{left}_{A,B} v \mid \mathbf{right}_{A,B} v \mid (v, w) \mid \mathbf{fold}_{\mu X.A} v$$

where n ranges over the natural numbers. Think of $*$ as representing the unique value of unit type I and of n as representing a pointer to the n-th qubit of a quantum state ρ. Specific values of interest are $\mathtt{ff} := \mathbf{left}_{I,I}*$ and $\mathtt{tt} := \mathbf{right}_{I,I}*$ which correspond to **false** and **true** respectively.

A *qubit pointer context* is a set Q of natural numbers. A value v of type A is well-formed in qubit pointer context Q, denoted $Q \vdash v : A$, if the judgement is derivable from the following rules:

$$\frac{}{\cdot \vdash * : I} \qquad \frac{}{\{n\} \vdash n : \mathbf{qbit}} \qquad \frac{Q \vdash v : A}{Q \vdash \mathbf{left}_{A,B} v : A + B} \qquad \frac{Q \vdash v : B}{Q \vdash \mathbf{right}_{A,B} v : A + B}$$

$$\frac{Q_1 \vdash v : A \qquad Q_2 \vdash w : B \qquad Q_1 \cap Q_2 = \varnothing}{Q_1, Q_2 \vdash (v, w) : A \otimes B} \qquad \frac{Q \vdash v : A[\mu X.A/X]}{Q \vdash \mathbf{fold}_{\mu X.A} v : \mu X.A}$$

If v is well-formed, then its type and qubit pointer context are uniquely determined. If $Q \vdash v : P$ with P classical, then we say v is a *classical value.*

Lemma 3. *If $Q \vdash v : P$ is a well-formed classical value, then $Q = \cdot$.*

A *value assignment* is a function from term variables to values, which we write as $V = \{x_1 = v_1, \ldots, x_n = v_n\}$, where x_i are variables and v_i are values. A value assignment is *well-formed* in qubit pointer context Q and variable context Γ, denoted $Q; \Gamma \vdash V$, if V has exactly the same variables as Γ, so that $\Gamma = \{x_1 : A_1, \ldots, x_n : A_n\}$, and $Q = Q_1, \ldots, Q_n$, s.t. $Q_i \vdash v_i : A_i$. Such a splitting of Q is necessarily unique, if it exists, and some of the Q_i may be empty.

Procedure Stores. A *procedure store* is a set of procedure definitions, written as:

$$\Omega = \{f_1 :: x_1 : A_1 \to y_1 : B_1 \; \{M_1\}, \ldots, f_n :: x_n : A_n \to y_n : B_n \; \{M_n\}\}.$$

A procedure store is *well-formed* in procedure context Π, written $\Pi \vdash \Omega$, if the judgement is derivable via the following rules:

$$\frac{}{\cdot \vdash \cdot} \qquad \frac{\Pi \vdash \Omega \qquad \Pi, f : A \to B \vdash \langle x : A \rangle \; M \; \langle y : B \rangle}{\Pi, f : A \to B \vdash \Omega, f :: x : A \to y : B \; \{M\}}$$

Program Configurations. A *program configuration* is a quadruple $(M \mid V \mid \Omega \mid \rho)$, where M is a term, V is a value assignment, Ω is a procedure store and $\rho \in \mathbb{C}^{2^n \times 2^n}$ is a finite-dimensional density matrix with $0 \leq \mathrm{tr}(\rho) \leq 1$. The density matrix ρ represents a (mixed) quantum state and its trace may be smaller than one because we also use it to encode probability information (see Remark 4). We write $\mathit{dim}(\rho) = n$ to indicate that the dimension of ρ is n.

A *well-formed* program configuration is a configuration $(M \mid V \mid \Omega \mid \rho)$, where there exist (necessarily unique) Π, Γ, Σ, Q, such that: (1) $\Pi \vdash \langle \Gamma \rangle \; M \; \langle \Sigma \rangle$ is a well-formed term; (2) $Q; \Gamma \vdash V$ is a well-formed value assignment; (3) $\Pi \vdash \Omega$ is a well-formed procedure store; and (4) $Q = \{1, 2, \ldots, \mathit{dim}(\rho)\}$. We write $\Pi; \Gamma; \Sigma; Q \vdash (M \mid V \mid \Omega \mid \rho)$ to indicate this situation. The formation rules enforce that the qubits of ρ and the qubit pointers from V are in a 1-1 correspondence.

$$\frac{}{(\textbf{new unit } u \mid V \mid \Omega \mid \rho) \rightsquigarrow (\textbf{skip} \mid V, u = * \mid \Omega \mid \rho)}$$

$$\frac{}{(\textbf{discard } x \mid V, x = v \mid \Omega \mid \rho) \rightsquigarrow (\textbf{skip} \mid r_v(V) \mid \Omega \mid tr_v(\rho))}$$

$$\frac{}{(y = \textbf{copy } x \mid V, x = v \mid \Omega \mid \rho) \rightsquigarrow (\textbf{skip} \mid V, x = v, y = v \mid \Omega \mid \rho)}$$

$$\frac{}{(\textbf{new qbit } q \mid V \mid \Omega \mid \rho) \rightsquigarrow (\textbf{skip} \mid V, q = dim(\rho) + 1 \mid \Omega \mid \rho \otimes |0\rangle\langle 0|)}$$

$$\frac{}{(\vec{q} \mathrel{*}= S \mid V, \vec{q} = \vec{m} \mid \Omega \mid \rho) \rightsquigarrow (\textbf{skip} \mid V, \vec{q} = \vec{m} \mid \Omega \mid S_{\vec{m}}(\rho))}$$

$$\frac{}{(b = \textbf{measure } q \mid V, q = m \mid \Omega \mid \rho) \rightsquigarrow (\textbf{skip} \mid r_m(V), b = \texttt{ff} \mid \Omega \mid {}_m\langle 0|\rho|0\rangle_m)}$$

$$\frac{}{(b = \textbf{measure } q \mid V, q = m \mid \Omega \mid \rho) \rightsquigarrow (\textbf{skip} \mid r_m(V), b = \texttt{tt} \mid \Omega \mid {}_m\langle 1|\rho|1\rangle_m)}$$

$$\frac{}{(\textbf{skip}; P \mid V \mid \Omega \mid \rho) \rightsquigarrow (P \mid V \mid \Omega \mid \rho)} \qquad \frac{(P \mid V \mid \Omega \mid \rho) \rightsquigarrow (P' \mid V' \mid \Omega' \mid \rho')}{(P; Q \mid V \mid \Omega \mid \rho) \rightsquigarrow (P'; Q \mid V' \mid \Omega' \mid \rho')}$$

$$\frac{}{(\textbf{while } b \textbf{ do } M \mid V, b = \texttt{ff} \mid \Omega \mid \rho) \rightsquigarrow (\textbf{skip} \mid V, b = \texttt{ff} \mid \Omega \mid \rho)}$$

$$\frac{}{(\textbf{while } b \textbf{ do } M \mid V, b = \texttt{tt} \mid \Omega \mid \rho) \rightsquigarrow (M; \textbf{while } b \textbf{ do } M \mid V, b = \texttt{tt} \mid \Omega \mid \rho)}$$

$$\frac{}{(y = \textbf{left } x \mid V, x = v \mid \Omega \mid \rho) \rightsquigarrow (\textbf{skip} \mid V, y = \textbf{left } v \mid \Omega \mid \rho)}$$

$$\frac{}{(y = \textbf{right } x \mid V, x = v \mid \Omega \mid \rho) \rightsquigarrow (\textbf{skip} \mid V, y = \textbf{right } v \mid \Omega \mid \rho)}$$

$$\frac{}{(\textbf{case } y \textbf{ of } \{\textbf{left } x_1 \to M_1 \mid \textbf{right } x_2 \to M_2\ \} \mid V, y = \textbf{left } v \mid \Omega \mid \rho) \rightsquigarrow (M_1 \mid V, x_1 = v \mid \Omega \mid \rho)}$$

$$\frac{}{(\textbf{case } y \textbf{ of } \{\textbf{left } x_1 \to M_1 \mid \textbf{right } x_2 \to M_2\ \} \mid V, y = \textbf{right } v \mid \Omega \mid \rho) \rightsquigarrow (M_2 \mid V, x_2 = v \mid \Omega \mid \rho)}$$

$$\frac{}{(x = (x_1, x_2) \mid V, x_1 = v_1, x_2 = v_2 \mid \Omega \mid \rho) \rightsquigarrow (\textbf{skip} \mid V, x = (v_1, v_2) \mid \Omega \mid \rho)}$$

$$\frac{}{((x_1, x_2) = x \mid V, x = (v_1, v_2) \mid \Omega \mid \rho) \rightsquigarrow (\textbf{skip} \mid V, x_1 = v_1, x_2 = v_2 \mid \Omega \mid \rho)}$$

$$\frac{}{(y = \textbf{fold } x \mid V, x = v \mid \Omega \mid \rho) \rightsquigarrow (\textbf{skip} \mid V, y = \textbf{fold } v \mid \Omega \mid \rho)}$$

$$\frac{}{(y = \textbf{unfold } x \mid V, x = \textbf{fold } v \mid \Omega \mid \rho) \rightsquigarrow (\textbf{skip} \mid V, y = v \mid \Omega \mid \rho)}$$

$$\frac{}{(\textbf{proc } f :: \ x : A \to y : B\ \{M\} \mid V \mid \Omega \mid \rho) \rightsquigarrow (\textbf{skip} \mid V \mid \Omega, f :: \ x : A \to y : B\ \{M\} \mid \rho)}$$

$$\frac{}{\begin{array}{c}(y_1 = f(x_1) \mid V, x_1 = v \mid \Omega, f :: \ x_2 : A \to y_2 : B\ \{M\} \mid \rho) \rightsquigarrow \\ (M_\alpha \mid V, x_1 = v \mid \Omega, f :: \ x_2 : A \to y_2 : B\ \{M\} \mid \rho)\end{array}}$$

Fig. 2: Small Step Operational semantics of QPL.

The small step semantics is defined for configurations $(M \mid V \mid \Omega \mid \rho)$ by induction on M in Figure 2 and we now explain the notations used therein.

In the rule for discarding, we use two functions that depend on a value v. They are tr_v, which modifies the quantum state ρ by tracing out all of its qubits which are used in v, and r_v which simply reindexes the value assignment, so that the pointers within $r_v(V)$ correctly point to the corresponding qubits of $tr_v(\rho)$, which is potentially of smaller dimension than ρ. Formally, for a well-formed value v, let Q and A be the unique qubit pointer context and type, such that $Q \vdash v : A$. Then $tr_v(\rho)$ is the quantum state obtained from ρ by tracing out all qubits specified by Q. Given a value assignment $V = \{x_1 = v_1, \dots x_n = v_n\}$, then $r_v(V) = \{x_1 = r'_v(v_1), \dots, x_n = r'_v(v_n)\}$, where:

$$r'_v(w) = \begin{cases} *, & \text{if } w = * \\ k - |\{i \in Q \mid i < k\}|, & \text{if } w = k \in \mathbb{N} \\ \textbf{left } r'_v(w'), & \text{if } w = \textbf{left } w' \\ \textbf{right } r'_v(w'), & \text{if } w = \textbf{right } w' \\ (r'_v(w_1), r'_v(w_2)) & \text{if } w = (w_1, w_2) \\ \textbf{fold } r'_v(w'), & \text{if } w = \textbf{fold } w' \end{cases}$$

In the rule for unitaries, the superoperator $S_{\vec{m}}$ applies the unitary S to the vector of qubits specified by $\vec{m}$. In the rules for measurement, the m-th qubit of ρ is measured in the computational basis, the measured qubit is destroyed in the process and the measurement outcome is stored in the bit b. More specifically, $|i\rangle_m = I_{2^{m-1}} \otimes |i\rangle \otimes I_{2^{n-m}}$ and ${}_m\langle i|$ is its adjoint, for $i \in \{0, 1\}$, and where I_n is the identity matrix in $\mathbb{C}^{n \times n}$.

Remark 4. Because of the way we decided to handle measurements, reduction $(- \rightsquigarrow -)$ is a *nondeterministic* operation, where we encode the probabilities of reduction within the trace of our density matrices in a similar way to [9]. Equivalently, we may see the reduction relation as *probabilistic* provided that we normalise all density matrices and decorate the reductions with the appropriate probability information as specified by the Born rule of quantum mechanics. The nondeterministic view leads to a more concise and clear presentation and because of this we have chosen it over the probabilistic view.

The introduction rule for procedures simply defines a procedure which is added to the procedure store. In the rule for calling procedures, the term M_α is α-equivalent to M and is obtained from it by renaming the input x_2 to x_1, renaming the output y_2 to y_1 and renaming all other variables within M to some fresh names, so as to avoid conflicts with the input, output and the rest of the variables within V.

Theorem 5 (Subject reduction). *If $\Pi; \Gamma; \Sigma; Q \vdash (M \mid V \mid \Omega \mid \rho)$ and $(M \mid V \mid \Omega \mid \rho) \rightsquigarrow (M' \mid V' \mid \Omega' \mid \rho')$, then $\Pi'; \Gamma'; \Sigma; Q' \vdash (M' \mid V' \mid \Omega' \mid \rho')$, for some (necessarily unique) contexts Π', Γ', Q' and where Σ is invariant.*

Assumption 6. *From now on we assume all configurations are well-formed.*

```
while b do {
  new qbit q;
  q *= H;
  discard b;
  b = measure q
}
```

(a) A term M

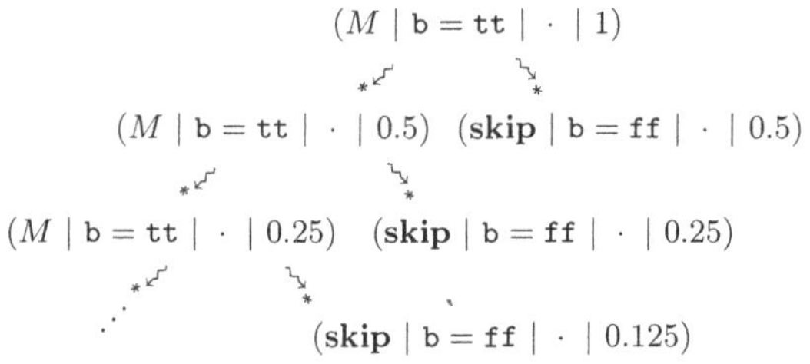

(b) A reduction graph involving M

Fig. 3: Example of a term and of a reduction graph.

A configuration $(M \mid V \mid \Omega \mid \rho)$ is said to be *terminal* if $M = \mathbf{skip}$. Program execution finishes at terminal configurations, which are characterised by the property that they do not reduce any further. We will use calligraphic letters $(\mathcal{C}, \mathcal{D}, \ldots)$ to range over configurations and we will use $\mathcal{T}$ to range over terminal configurations. For a configuration $\mathcal{C} = (M \mid V \mid \Omega \mid \rho)$, we write for brevity $\mathrm{tr}(\mathcal{C}) := \mathrm{tr}(\rho)$ and we shall say $\mathcal{C}$ is *normalised* whenever $\mathrm{tr}(\mathcal{C}) = 1$. We say that a configuration $\mathcal{C}$ is *impossible* if $\mathrm{tr}(\mathcal{C}) = 0$ and we say it is *possible* otherwise.

Theorem 7 (Progress). *If $\mathcal{C}$ is a configuration, then either $\mathcal{C}$ is terminal or there exists a configuration $\mathcal{D}$, such that $\mathcal{C} \rightsquigarrow \mathcal{D}$. Moreover, if $\mathcal{C}$ is not terminal, then $\mathrm{tr}(\mathcal{C}) = \sum_{\mathcal{C} \rightsquigarrow \mathcal{D}} \mathrm{tr}(\mathcal{D})$ and there are at most two such configurations $\mathcal{D}$.*

In the situation of the above theorem, the probability of reduction is given by $\Pr(\mathcal{C} \rightsquigarrow \mathcal{D}) := \mathrm{tr}(\mathcal{D})/\mathrm{tr}(\mathcal{C})$, for any possible $\mathcal{C}$ (see Remark 4) and Theorem 7 shows the total probability of all single-step reductions is 1. If $\mathcal{C}$ is impossible, then $\mathcal{C}$ occurs with probability 0 and subsequent reductions are also impossible.

Probability of Termination. Given configurations $\mathcal{C}$ and $\mathcal{D}$ let $\mathrm{Seq}_n(\mathcal{C}, \mathcal{D}) := \{\mathcal{C}_0 \rightsquigarrow \cdots \rightsquigarrow \mathcal{C}_n \mid \mathcal{C}_0 = \mathcal{C} \text{ and } \mathcal{C}_n = \mathcal{D}\}$, and let $\mathrm{Seq}_{\leq n}(\mathcal{C}, \mathcal{D}) = \bigcup_{i=0}^{n} \mathrm{Seq}_n(\mathcal{C}, \mathcal{D})$. Finally, let $\mathrm{TerSeq}_{\leq n}(\mathcal{C}) := \bigcup_{\mathcal{T} \text{ terminal}} \mathrm{Seq}_{\leq n}(\mathcal{C}, \mathcal{T})$. In other words, $\mathrm{TerSeq}_{\leq n}(\mathcal{C})$ is the set of all reduction sequences from $\mathcal{C}$ which terminate in at most n steps (including 0 if $\mathcal{C}$ is terminal). For every terminating reduction sequence $r = (\mathcal{C} \rightsquigarrow \cdots \rightsquigarrow \mathcal{T})$, let $\mathrm{End}(r) := \mathcal{T}$, i.e. $\mathrm{End}(r)$ is simply the (terminal) end-point of the sequence.

For any configuration $\mathcal{C}$, the sequence $\left(\sum_{r \in \mathrm{TerSeq}_{\leq n}(\mathcal{C})} \mathrm{tr}(\mathrm{End}(r))\right)_{n \in \mathbb{N}}$ is increasing with upper bound $\mathrm{tr}(\mathcal{C})$ (follows from Theorem 7). For any possible $\mathcal{C}$, we define:

$$\mathrm{Halt}(\mathcal{C}) := \bigvee_{n=0}^{\infty} \sum_{r \in \mathrm{TerSeq}_{\leq n}(\mathcal{C})} \mathrm{tr}(\mathrm{End}(r))/\mathrm{tr}(\mathcal{C})$$

which is exactly the *probability of termination* of $\mathcal{C}$. This is justified, because $\mathrm{Halt}(\mathcal{T}) = 1$, for any terminal (and possible) configuration $\mathcal{T}$ and $\mathrm{Halt}(\mathcal{C}) = \sum_{\substack{\mathcal{C} \rightsquigarrow \mathcal{D} \\ \mathcal{D} \text{ possible}}} \Pr(\mathcal{C} \rightsquigarrow \mathcal{D}) \mathrm{Halt}(\mathcal{D})$. We write $\rightsquigarrow_*$ for the transitive closure of $\rightsquigarrow$.

```
proc GHZnext :: l : ListQ -> l : ListQ {
  new qbit q;
  case l of
      nil -> q*=H;
            l = q :: nil
    | q' :: l' -> q',q *= CNOT;
                  l = q :: q' :: l'
}

proc GHZ :: n : Nat -> l : ListQ {
  case n of
      zero -> l = nil
    | s(n') -> l = GHZnext(GHZ(n'))
}
```

(a) Procedures for generating GHZ_n.

$$(\texttt{l = GHZ(n)} \mid \texttt{n} = \texttt{s(s(s(zero)))} \mid \Omega \mid 1)$$
$$\rightsquigarrow^*$$
$$(\texttt{l = GHZnext(l)} \mid \texttt{l} = 2 :: 1 :: \texttt{nil} \mid \Omega \mid \gamma_2)$$
$$\rightsquigarrow$$
$$(\texttt{new qbit q;}\cdots \mid \texttt{l} = 2 :: 1 :: \texttt{nil} \mid \Omega \mid \gamma_2)$$
$$\rightsquigarrow$$
$$(\texttt{case l of } \cdots \mid \texttt{l} = 2 :: 1 :: \texttt{nil}, \texttt{q} = 3 \mid \Omega \mid \gamma_2 \otimes |0\rangle\langle 0|)$$
$$\rightsquigarrow^*$$
$$(\texttt{q',q *=CNOT;}\cdots \mid \texttt{l'} = 1 :: \texttt{nil}, \texttt{q} = 3, \texttt{q'} = 2 \mid \Omega \mid \gamma_2 \otimes |0\rangle\langle 0|)$$
$$\rightsquigarrow$$
$$(\texttt{l = q :: q' :: l'} \mid \texttt{l'} = 1 :: \texttt{nil}, \texttt{q} = 3, \texttt{q'} = 2 \mid \Omega \mid \gamma_3)$$
$$\rightsquigarrow^*$$
$$(\mathbf{skip} \mid \texttt{l} = 3 :: 2 :: 1 :: \texttt{nil} \mid \Omega \mid \gamma_3)$$

(b) A reduction sequence producing GHZ_3.

Fig. 4: Example with lists of qubits and a recursive procedure.

Example 8. Consider the term M in Figure 3. The body of the **while** loop (3a) has the effect of performing a fair coin toss (realised through quantum measurement in the standard way) and storing the outcome in variable `b`. Therefore, starting from configuration $\mathcal{C} = (M \mid \texttt{b} = \texttt{tt} \mid \cdot \mid 1)$, as in Subfigure 3b, the program has the effect of tossing a fair coin until `ff` shows up. The set of terminal configurations reachable from $\mathcal{C}$ is $\{(\mathbf{skip} \mid \texttt{b} = \texttt{ff} \mid \cdot \mid 2^{-i}) \mid i \in \mathbb{N}_{\geq 1}\}$ and the last component of each configuration is a 1×1 density matrix which is exactly the probability of reducing to the configuration. Therefore $\mathrm{Halt}(\mathcal{C}) = \sum_{i=1}^{\infty} 2^{-i} = 1$.

Example 9. The GHZ_n state is defined as $\gamma_n := (|0\rangle^{\otimes n} + |1\rangle^{\otimes n})(\langle 0|^{\otimes n} + \langle 1|^{\otimes n})/2$. In Figure 4, we define a procedure `GHZ`, which given a natural number n, generates the state γ_n, which is represented as a list of qubits of length n. The procedure (4a) uses an auxiliary procedure `GHZnext`, which given a list of qubits representing the state γ_n, returns the state γ_{n+1} again represented as a list of qubits. The two procedures make use of some (hopefully obvious) syntactic sugar. In 4b, we also present the last few steps of a reduction sequence which produces γ_3 starting from configuration $(\texttt{l = GHZ(n)} \mid \texttt{n} = \texttt{s(s(s(zero)))} \mid \Omega \mid 1)$, where Ω contains the above mentioned procedures. In the reduction sequence we only show the term in evaluating position and we omit some intermediate steps. The type `ListQ` is a shorthand for $\mathbf{List}(\mathbf{qbit})$ from Example 1.

4 W*-algebras

In this section we describe our denotational model. It is based on W*-algebras, which are algebras of observables (i.e. physical entities), with interesting domain-theoretic properties. We recall some background on W*-algebras and their categorical structure. We refer the reader to [25] for an encyclopaedic account on W*-algebras.

Domain-theoretic Preliminaries. Recall that a directed subset of a poset P is a non-empty subset $X \subseteq P$ in which every pair of elements of X has an upper bound in X. A poset P is a *directed-complete partial order* (*dcpo*) if each directed subset has a supremum. A poset P is *pointed* if it has a least element, usually denoted by $\bot$. A monotone map $f : P \to Q$ between posets is *Scott-continuous* if it preserves suprema of directed subsets. If P and Q are pointed and f preserves the least element, then we say f is *strict*. We write **DCPO** ($\mathbf{DCPO}_{\bot!}$) for the category of (pointed) dcpo's and (strict) Scott-continuous maps between them.

Definition of W-algebras.* A *complex algebra* is a complex vector space V equipped with a bilinear multiplication $(- \cdot -) : V \times V \to V$, which we write as juxtaposition. A *Banach algebra* A is a complex algebra A equipped with a submultiplicative norm $\| - \| : A \to \mathbb{R}_{\geq 0}$, i.e. $\forall x, y \in A : \|xy\| \leq \|x\|\|y\|$. A $*$-*algebra* A is a complex algebra A with an involution $(-)^* : A \to A$ such that $(x^*)^* = x$, $(x+y)^* = (x^* + y^*)$, $(xy)^* = y^*x^*$ and $(\lambda x)^* = \overline{\lambda} x^*$, for $x, y \in A$ and $\lambda \in \mathbb{C}$. A *C*-algebra* is a Banach $*$-algebra A which satisfies the C*-identity, i.e. $\|x^*x\| = \|x\|^2$ for all $x \in A$. A C*-algebra A is *unital* if it has an element $1 \in A$, such that for every $x \in A : x1 = 1x = x$. All C*-algebras in this paper are unital and for brevity we regard unitality as part of their definition.

Example 10. The algebra $M_n(\mathbb{C})$ of $n \times n$ complex matrices is a C*-algebra. In particular, the set of complex numbers $\mathbb{C}$ has a C*-algebra structure since $M_1(\mathbb{C}) \cong \mathbb{C}$. More generally, the $n \times n$ matrices valued in a C*-algebra A also form a C*-algebra $M_n(A)$. The C*-algebra of qubits is $\mathbf{qbit} := M_2(\mathbb{C})$.

An element $x \in A$ of a C*-algebra A is called *positive* if $\exists y \in A : x = y^*y$. The *poset of positive elements* of A is denoted A^+ and its order is given by $x \leq y$ iff $(y - x) \in A^+$. The *unit interval* of A is the subposet $[0,1]_A \subseteq A^+$ of all positive elements x such that $0 \leq x \leq 1$.

Let $f : A \to B$ be a linear map between C*-algebras A and B. We say that f is *positive* if it preserves positive elements. We say that f is *completely positive* if it is n-positive for every $n \in \mathbb{N}$, i.e. the map $M_n(f) : M_n(A) \to M_n(B)$ defined for every matrix $[x_{i,j}]_{1 \leq i,j \leq n} \in M_n(A)$ by $M_n(f)([x_{i,j}]_{1 \leq i,j \leq n}) = [f(x_{i,j})]_{1 \leq i,j \leq n}$ is positive. The map f is called *multiplicative*, *involutive*, *unital* if it preserves multiplication, involution, and the unit, respectively. The map f is called *subunital* whenever the inequalities $0 \leq f(1) \leq 1$ hold. A *state* on a C*-algebra A is a completely positive unital map $s : A \to \mathbb{C}$.

Although W*-algebras are commonly defined in topological terms (as C*-algebras closed under several operator topologies) or equivalently in algebraic terms (as C*-algebras which are their own bicommutant), one can also equivalently define them in domain-theoretic terms [19], as we do next.

A completely positive map between C*-algebras is *normal* if its restriction to the unit interval is Scott-continuous [19, Proposition A.3]. A *W*-algebra* is a

C*-algebra A such that the unit interval $[0,1]_A$ is a dcpo, and A has a separating set of normal states: for every $x \in A^+$, if $x \neq 0$, then there is a normal state $s : A \to \mathbb{C}$ such that $s(x) \neq 0$ [25, Theorem III.3.16].

A linear map $f : A \to B$ between W*-algebras A and B is called an *NCPSU-map* if f is normal, completely positive and subunital. The map f is called an *NMIU-map* if f is normal, multiplicative, involutive and unital. We note that every NMIU-map is necessarily an NCPSU-map and that W*-algebras are closed under formation of matrix algebras as in Example 10.

Categorical Structure. Let $\mathbf{W}^*_{\text{NCPSU}}$ be the category of W*-algebras and NCPSU-maps and let $\mathbf{W}^*_{\text{NMIU}}$ be its full-on-objects subcategory of NMIU-maps. Throughout the rest of the paper let $\mathbf{C} := (\mathbf{W}^*_{\text{NCPSU}})^{\text{op}}$ and let $\mathbf{V} := (\mathbf{W}^*_{\text{NMIU}})^{\text{op}}$. QPL types are interpreted as functors $[\![\Theta \vdash A]\!] : \mathbf{V}^{|\Theta|} \to \mathbf{V}$ and closed QPL types as objects $[\![A]\!] \in \text{Ob}(\mathbf{V}) = \text{Ob}(\mathbf{C})$. One should think of $\mathbf{V}$ as the category of *values*, because the interpretation of our values from §3 are indeed $\mathbf{V}$-morphisms. General QPL terms are interpreted as morphisms of $\mathbf{C}$, so one should think of $\mathbf{C}$ as the category of *computations*. We now describe the categorical structure of $\mathbf{V}$ and $\mathbf{C}$ and later we justify our choice for working in the opposite categories.

Both $\mathbf{C}$ and $\mathbf{V}$ have a symmetric monoidal structure when equipped with the spatial tensor product, denoted here by $(- \otimes -)$, and tensor unit $I := \mathbb{C}$ [11, Section 10]. Moreover, $\mathbf{V}$ is symmetric monoidal closed and also complete and cocomplete [11]. $\mathbf{C}$ and $\mathbf{V}$ have finite coproducts, given by direct sums of W*-algebras [2, Proposition 4.7.3]. The coproduct of objects A and B is denoted by $A + B$ and the coproduct injections are denoted $\text{left}_{A,B} : A \to A + B$ and $\text{right}_{A,B} : B \to A + B$. Given morphisms $f : A \to C$ and $g : B \to C$, we write $[f, g] : A + B \to C$ for the unique cocone morphism induced by the coproduct. Moreover, coproducts distribute over tensor products [2, §4.6]. More specifically, there exists a natural isomorphism $d_{A,B,C} : A \otimes (B + C) \to (A \otimes B) + (A \otimes C)$ which satisfies the usual coherence conditions. The initial object in $\mathbf{C}$ is moreover a zero object and is denoted 0. The W*-algebra of bits is $\mathbf{bit} := I + I = \mathbb{C} \oplus \mathbb{C}$.

The categories $\mathbf{V}, \mathbf{C}$ and $\mathbf{Set}$ are related by symmetric monoidal adjunctions:

$$\mathbf{Set} \underset{G}{\overset{F}{\rightleftarrows}} \mathbf{V} \underset{R}{\overset{J}{\rightleftarrows}} \mathbf{C} \qquad (F \dashv G,\ J \dashv R) \qquad [26, \text{pp. } 11]$$

and the subcategory inclusion J preserves coproducts and tensors up to equality.

Interpreting QPL within $\mathbf{C}$ and $\mathbf{V}$ is not an ad hoc trick. In physical terms, this corresponds to adopting the *Heisenberg picture* of quantum mechanics and this is usually done when working with infinite-dimensional W*-algebras (like we do). Semantically, this is necessary, because (1) our type system has conditional branching and we need to interpret QPL terms within a category with finite coproducts; (2) we have to be able to compute parameterised initial algebras to interpret inductive datatypes. The category $\mathbf{W}^*_{\text{NCPSU}}$ has finite products, but it does *not* have coproducts, so by interpreting QPL terms within $\mathbf{C} = (\mathbf{W}^*_{\text{NCPSU}})^{\text{op}}$ we solve problem (1). For (2), the monoidal closure of $\mathbf{V} = (\mathbf{W}^*_{\text{NMIU}})^{\text{op}}$ is crucial, because it implies the tensor product preserves ω-colimits.

$\mathrm{tr} : M_n(\mathbb{C}) \to \mathbb{C}$	$\mathrm{new}_\rho : \mathbb{C} \to M_{2^n}(\mathbb{C})$	$\mathrm{meas} : M_2(\mathbb{C}) \to \mathbb{C} \oplus \mathbb{C}$	$\mathrm{unitary}_S : M_{2^n}(\mathbb{C}) \to M_{2^n}(\mathbb{C})$
$\mathrm{tr} :: A \mapsto \sum_i A_{i,i}$	$\mathrm{new}_\rho :: a \mapsto a\rho$	$\mathrm{meas} :: \begin{pmatrix} a & b \\ c & d \end{pmatrix} \mapsto (a\ d)$	$\mathrm{unitary}_S :: A \mapsto SAS^\dagger$
$\mathrm{tr}^\dagger : \mathbb{C} \to M_n(\mathbb{C})$	$\mathrm{new}_\rho^\dagger : M_{2^n}(\mathbb{C}) \to \mathbb{C}$	$\mathrm{meas}^\dagger : \mathbb{C} \oplus \mathbb{C} \to M_2(\mathbb{C})$	$\mathrm{unitary}_S^\dagger : M_{2^n}(\mathbb{C}) \to M_{2^n}(\mathbb{C})$
$\mathrm{tr}^\dagger :: a \mapsto aI_n$	$\mathrm{new}_\rho^\dagger :: A \mapsto \mathrm{tr}(A\rho)$	$\mathrm{meas}^\dagger :: (a\ d) \mapsto \begin{pmatrix} a & 0 \\ 0 & d \end{pmatrix}$	$\mathrm{unitary}_S^\dagger :: A \mapsto S^\dagger AS$

Fig. 5: A selection of maps in the Schrödinger picture ($f : A \to B$) and their Hermitian adjoints ($f^\dagger : B \to A$) used in the Heisenberg picture.

Convex Sums. In both $\mathbf{C}$ and $\mathbf{W}^*_{\mathrm{NCPSU}}$, morphisms are closed under *convex sums*, which are defined pointwise, as usual. More specifically, given NCPSU-maps $f_1, \ldots, f_n : A \to B$ and real numbers $p_i \in [0,1]$ with $\sum_i p_i \leq 1$, then the map $\sum_i p_i f_i : A \to B$ is also an NCPSU-map.

Order-enrichment. For W*-algebras A and B, we define a partial order on $\mathbf{C}(A,B)$ by : $f \leq g$ iff $g - f$ is a completely positive map. Equipped with this order, our category $\mathbf{C}$ is $\mathbf{DCPO}_{\perp!}$-enriched [3, Theorem 4.3]. The least element in $\mathbf{C}(A,B)$ is also a zero morphism and is given by the map $\mathbf{0} : A \to B$, defined by $\mathbf{0}(x) = 0$. Also, the coproduct structure and the symmetric monoidal structure are both $\mathbf{DCPO}_{\perp!}$-enriched [2, Corollary 4.9.15] [3, Theorem 4.5].

Quantum Operations. For convenience, our operational semantics adopts the *Schrödinger picture* of quantum mechanics, which is the picture most experts in quantum computing are familiar with. However, as we have just explained, our denotational semantics has to adopt the Heisenberg picture. The two pictures are equivalent in finite dimensions and we will now show how to translate from one to the other. By doing so, we provide an explicit description (in both pictures) of the required quantum maps that we need to interpret QPL.

Consider the maps in Figure 5. The map *tr* is used to trace out (or discard) parts of quantum states. Density matrices ρ are in 1-1 correspondence with the maps new_ρ, which we use in our semantics to describe (mixed) quantum states. The *meas* map simply measures a qubit in the computational basis and returns a bit as measurement outcome. The $unitary_S$ map is used for application of a unitary S. These maps work as described in the Schrödinger picture of quantum mechanics, i.e., the category $\mathbf{W}^*_{\mathrm{NCPSU}}$. For every map $f : A \to B$ among those mentioned, $f^\dagger : B \to A$ indicates its Hermitian adjoint [3]. In the Heisenberg picture, composition of maps is done in the opposite way, so we simply write $f^\ddagger := (f^\dagger)^{\mathrm{op}} \in \mathbf{C}(A,B)$ for the Hermitian adjoint of f when seen as a morphism in $(\mathbf{W}^*_{\mathrm{NCPSU}})^{\mathrm{op}} = \mathbf{C}$. Thus, the mapping $(-)^\ddagger$ translates the above operations from the Schrödinger picture (the category $\mathbf{W}^*_{\mathrm{NCPSU}}$) to the Heisenberg picture (the category $\mathbf{C}$) of quantum mechanics.

[3] This adjoint exists, because A and B are *finite-dimensional* W*-algebras which therefore have the structure of a Hilbert space when equipped with the Hilbert-Schmidt inner product [27, pp. 145].

Parameterised Initial Algebras. In order to interpret inductive datatypes, we need to be able to compute parameterised initial algebras for the functors induced by our type expressions. $\mathbf{V}$ is ideal for this, because it is cocomplete and monoidal closed and so all type expressions induce functors on $\mathbf{V}$ which preserve ω-colimits.

Definition 11 (cf. [6, §6.1]). *Given a category $\mathbf{A}$ and a functor $T : \mathbf{A}^n \to \mathbf{A}$, with $n \geq 1$, a* parameterised initial algebra *for T is a pair $(T^\sharp, \phi^T)$, such that:*

- *$T^\sharp : \mathbf{A}^{n-1} \to \mathbf{A}$ is a functor;*
- *$\phi^T : T \circ \langle Id, T^\sharp \rangle \Rightarrow T^\sharp : \mathbf{A}^{n-1} \to \mathbf{A}$ is a natural isomorphism;*
- *For every $A \in \mathrm{Ob}(\mathbf{A}^{n-1})$, the pair $(T^\sharp A, \phi^T_A)$ is an initial $T(A, -)$-algebra.*

Proposition 12. *Every ω-cocontinuous functor $T : \mathbf{V}^n \to \mathbf{V}$ has a parameterised initial algebra $(T^\sharp, \phi^T)$ with $T^\sharp : \mathbf{V}^{n-1} \to \mathbf{V}$ being ω-cocontinuous.*

Proof. $\mathbf{V}$ is cocomplete, so this follows from [13, §4.3]. □

5 Denotational Semantics of QPL

In this section we describe the denotational semantics of QPL.

5.1 Interpretation of Types

The interpretation of a type $\Theta \vdash A$ is a functor $[\![\Theta \vdash A]\!] : \mathbf{V}^{|\Theta|} \to \mathbf{V}$, defined by induction on the derivation of $\Theta \vdash A$ in Figure 6. As usual, one has to prove this assignment is well-defined by showing the required initial algebras exist.

Proposition 13. *The assignment in Figure 6 is well-defined.*

Proof. By induction, every $[\![\Theta \vdash A]\!]$ is an ω-cocontinuous functor and thus it has a parameterised initial algebra by Proposition 12. □

Lemma 14 (Type Substitution). *Given types $\Theta, X \vdash A$ and $\Theta \vdash B$, then:*

$$[\![\Theta \vdash A[B/X]]\!] = [\![\Theta, X \vdash A]\!] \circ \langle Id, [\![\Theta \vdash B]\!] \rangle.$$

Proof. Straightforward induction. □

For simplicity, the interpretation of terms is only defined on closed types and so we introduce more concise notation for them. For any closed type $\cdot \vdash A$ we write for convenience $[\![A]\!] := [\![\cdot \vdash A]\!](*) \in \mathrm{Ob}(\mathbf{V})$, where $*$ is the unique object of the terminal category $\mathbf{1}$. Notice also that $[\![A]\!] \in \mathrm{Ob}(\mathbf{C}) = \mathrm{Ob}(\mathbf{V})$.

Definition 15. *Given a closed type $\cdot \vdash \mu X.A$, we define an isomorphism (in $\mathbf{V}$):*

$$\mathrm{fold}_{\mu X.A} : [\![A[\mu X.A/X]]\!] = [\![X \vdash A]\!][\![\mu X.A]\!] \cong [\![\mu X.A]\!] : \mathrm{unfold}_{\mu X.A}$$

where the equality is Lemma 14 and the iso is the initial algebra structure.

Example 16. The interpretation of the types from Example 1 are $[\![\mathbf{Nat}]\!] = \bigoplus_{i=0}^{\omega} \mathbb{C}$ and $[\![\mathbf{List}(A)]\!] = \bigoplus_{i=0}^{\omega} [\![A]\!]^{\otimes i}$. Specifically, $[\![\mathbf{List}(\mathbf{qbit})]\!] = \bigoplus_{i=0}^{\omega} \mathbb{C}^{2^i \times 2^i}$.

$$
\begin{aligned}
[\![\Theta \vdash A]\!] &: \mathbf{V}^{|\Theta|} \to \mathbf{V} \\
[\![\Theta \vdash \Theta_i]\!] &= \Pi_i \\
[\![\Theta \vdash I]\!] &= K_I \\
[\![\Theta \vdash \mathbf{qbit}]\!] &= K_{\mathbf{qbit}} \\
[\![\Theta \vdash A + B]\!] &= + \circ \langle [\![\Theta \vdash A]\!], [\![\Theta \vdash B]\!] \rangle \\
[\![\Theta \vdash A \otimes B]\!] &= \otimes \circ \langle [\![\Theta \vdash A]\!], [\![\Theta \vdash B]\!] \rangle \\
[\![\Theta \vdash \mu X.A]\!] &= [\![\Theta, X \vdash A]\!]^\sharp
\end{aligned}
$$

Fig. 6: Interpretations of types. K_A is the constant-A-functor.

$$
\begin{aligned}
[\![\cdot \vdash * : I]\!] &:= \mathrm{id}_I \\
[\![\{n\} \vdash n : \mathbf{qbit}]\!] &:= \mathrm{id}_{\mathbf{qbit}} \\
[\![Q \vdash \mathbf{left}_{A,B} v : A + B]\!] &:= \mathrm{left} \circ [\![v]\!] \\
[\![Q \vdash \mathbf{right}_{A,B} v : A + B]\!] &:= \mathrm{right} \circ [\![v]\!] \\
[\![Q_1, Q_2 \vdash (v, w) : A \otimes B]\!] &:= [\![v]\!] \otimes [\![w]\!] \\
[\![Q \vdash \mathbf{fold}_{\mu X.A} v : \mu X.A]\!] &:= \mathrm{fold} \circ [\![v]\!]
\end{aligned}
$$

Fig. 7: Interpretation of values.

$[\![\Pi \vdash \langle \Gamma \rangle \ \mathbf{new\ unit}\ u \ \langle \Gamma, u : I \rangle]\!] := \pi \mapsto r^{-1}$
$[\![\Pi \vdash \langle \Gamma, x : A \rangle \ \mathbf{discard}\ x \ \langle \Gamma \rangle]\!] := \pi \mapsto (r \circ (\mathrm{id} \otimes \diamond))$
$[\![\Pi \vdash \langle \Gamma, x : P \rangle \ y = \mathbf{copy}\ x \ \langle \Gamma, x : P, y : P \rangle]\!] := \pi \mapsto (\mathrm{id} \otimes \triangle)$
$[\![\Pi \vdash \langle \Gamma \rangle \ \mathbf{new\ qbit}\ q \ \langle \Gamma, q : \mathbf{qbit} \rangle]\!] := \pi \mapsto \left((\mathrm{id} \otimes \mathrm{new}^\dagger_{|0\rangle\langle 0|}) \circ r^{-1}\right)$
$[\![\Pi \vdash \langle \Gamma, q : \mathbf{qbit} \rangle \ b = \mathbf{measure}\ q \ \langle \Gamma, b : \mathbf{bit} \rangle]\!] := \pi \mapsto \left(\mathrm{id} \otimes \mathrm{meas}^\dagger\right)$
$[\![\Pi \vdash \langle \Gamma, \vec{q} : \overrightarrow{\mathbf{qbit}} \rangle \ \vec{q} \mathrel{*}= S \ \langle \Gamma, \vec{q} : \overrightarrow{\mathbf{qbit}} \rangle]\!] := \pi \mapsto \left(\mathrm{id} \otimes \mathrm{unitary}^\dagger_S\right)$
$[\![\Pi \vdash \langle \Gamma \rangle \ M; N \ \langle \Sigma \rangle]\!] := \pi \mapsto ([\![N]\!](\pi) \circ [\![M]\!](\pi))$
$[\![\Pi \vdash \langle \Gamma \rangle \ \mathbf{skip} \ \langle \Gamma \rangle]\!] := \pi \mapsto \mathrm{id}$
$[\![\Pi \vdash \langle \Gamma, b : \mathbf{bit} \rangle \ \mathbf{while}\ b\ \mathbf{do}\ M \ \langle \Gamma, b : \mathbf{bit} \rangle]\!] := \pi \mapsto \mathrm{lfp}(W_{[\![M]\!](\pi)})$
$[\![\Pi \vdash \langle \Gamma, x : A \rangle \ y = \mathbf{left}_{A,B}\ x \ \langle \Gamma, y : A + B \rangle]\!] := \pi \mapsto (\mathrm{id} \otimes \mathrm{left}_{A,B})$
$[\![\Pi \vdash \langle \Gamma, x : B \rangle \ y = \mathbf{right}_{A,B}\ x \ \langle \Gamma, y : A + B \rangle]\!] := \pi \mapsto \left(\mathrm{id} \otimes \mathrm{right}_{A,B}\right)$
$[\![\Pi \vdash \langle \Gamma, y : A + B \rangle \ \mathbf{case}\ y\ \mathbf{of}\ \{\mathbf{left}\ x_1 \to M_1 \mid \mathbf{right}\ x_2 \to M_2\} \ \langle \Sigma \rangle]\!] :=$
$\quad \pi \mapsto ([[\![M_1]\!](\pi), [\![M_2]\!](\pi)] \circ d)$
$[\![\Pi \vdash \langle \Gamma, x_1 : A, x_2 : B \rangle \ x = (x_1, x_2) \ \langle \Gamma, x : A \otimes B \rangle]\!] := \pi \mapsto \mathrm{id}$
$[\![\Pi \vdash \langle \Gamma, x : A \otimes B \rangle \ (x_1, x_2) = x \ \langle \Gamma, x_1 : A, x_2 : B \rangle]\!] := \pi \mapsto \mathrm{id}$
$[\![\Pi \vdash \langle \Gamma, x : A[\mu X.A/X] \rangle \ y = \mathbf{fold}\ x \ \langle \Gamma, y : \mu X.A \rangle]\!] := \pi \mapsto (\mathrm{id} \otimes \mathrm{fold})$
$[\![\Pi \vdash \langle \Gamma, x : \mu X.A \rangle \ y = \mathbf{unfold}\ x \ \langle \Gamma, y : A[\mu X.A/X] \rangle]\!] := \pi \mapsto (\mathrm{id} \otimes \mathrm{unfold})$
$[\![\Pi \vdash \langle \Gamma \rangle \ \mathbf{proc}\ f :: \ x : A \to y : B \ \{M\} \ \langle \Gamma \rangle]\!] := \pi \mapsto \mathrm{id}$
$[\![\Pi, f : A \to B \vdash \langle \Gamma, x : A \rangle \ y = f(x) \ \langle \Gamma, y : B \rangle]\!] := (\pi, f) \mapsto (\mathrm{id} \otimes f)$,
where r is the right monoidal unit. For simplicity, we omit the monoidal associator.

Fig. 8: Interpretation of QPL terms.

5.2 Copying and Discarding

Our type system is affine, so we have to construct discarding maps at all types. The tensor unit I is a terminal object in $\mathbf{V}$ (but not in $\mathbf{C}$) which leads us to the next definition.

Definition 17 (Discarding map). *For any W*-algebra A, let $\diamond_A : A \to I$ be the unique morphism of $\mathbf{V}$ with the indicated domain and codomain.*

We will see that all values admit an interpretation as $\mathbf{V}$-morphisms and are therefore discardable. In physical terms, this means values are causal (in the sense mentioned in the introduction). Of course, this is not true for the interpretation of general terms (which correspond to $\mathbf{C}$-morphisms).

Our language is equipped with a copy operation on classical data, so we have to explain how to copy classical values. We do this by constructing a copy map defined at all *classical* types using results from [13,14].

Proposition 18. *Using the categorical data of* $\mathbf{Set} \underset{G}{\overset{F}{\rightleftarrows}} \mathbf{V}$ (with $F \dashv G$)*, one can define a copy map $\triangle_{[\![P]\!]} : [\![P]\!] \to [\![P]\!] \otimes [\![P]\!]$ for every classical type $\cdot \vdash P$, such that the triple $\left([\![P]\!], \triangle_{[\![P]\!]}, \diamond_{[\![P]\!]}\right)$ forms a cocommutative comonoid in $\mathbf{V}$.*

We shall later see that the interpretations of our *classical* values are comonoid homomorphisms (w.r.t. Proposition 18) and therefore they may be copied.

5.3 Interpretation of Terms

Given a variable context $\Gamma = x_1 : A_1, \ldots, x_n : A_n$, we interpet it as the object $[\![\Gamma]\!] := [\![A_1]\!] \otimes \cdots \otimes [\![A_n]\!] \in \mathrm{Ob}(\mathbf{C})$. The interpretation of a procedure context $\Pi = f_1 : A_1 \to B_1, \ldots, f_n : A_n \to B_n$ is defined to be the pointed dcpo $[\![\Pi]\!] := \mathbf{C}(A_1, B_1) \times \cdots \times \mathbf{C}(A_n, B_n)$. A term $\Pi \vdash \langle\Gamma\rangle\ M\ \langle\Sigma\rangle$ is interpreted as a Scott-continuous function $[\![\Pi \vdash \langle\Gamma\rangle\ M\ \langle\Sigma\rangle]\!] : [\![\Pi]\!] \to \mathbf{C}([\![\Gamma]\!], [\![\Sigma]\!])$ defined by induction on the derivation of $\Pi \vdash \langle\Gamma\rangle\ M\ \langle\Sigma\rangle$ in Figure 8. For brevity, we often write $[\![M]\!] := [\![\Pi \vdash \langle\Gamma\rangle\ M\ \langle\Sigma\rangle]\!]$, when the contexts are clear or unimportant.

We now explain some of the notation used in Figure 8. The rules for manipulating qubits use the morphisms $\mathrm{new}^{\ddagger}_{|0\rangle\langle 0|}$, $\mathrm{meas}^{\ddagger}$ and $\mathrm{unitary}^{\ddagger}_S$ which are defined in §4. For the interpretation of **while** loops, given an arbitrary morphism $f : A \otimes \mathbf{bit} \to A \otimes \mathbf{bit}$ of $\mathbf{C}$, we define a Scott-continuous endofunction

$$W_f : \mathbf{C}\left(A \otimes \mathbf{bit}, A \otimes \mathbf{bit}\right) \to \mathbf{C}(A \otimes \mathbf{bit}, A \otimes \mathbf{bit})$$
$$W_f(g) = \left[\mathrm{id} \otimes \mathrm{left}_{I,I},\ g \circ f \circ (\mathrm{id} \otimes \mathrm{right}_{I,I})\right] \circ d_{A,I,I},$$

where the isomorphism $d_{A,I,I} : A \otimes (I + I) \to (A \otimes I) + (A \otimes I)$ is explained in §4. For any pointed dcpo D and Scott-continuous function $h : D \to D$, its *least fixpoint* is $\mathrm{lfp}(h) := \bigvee_{i=0}^{\infty} h^i(\bot)$, where $\bot$ is the least element of D.

Remark 19. The term semantics for defining and calling procedures does not involve any fixpoint computations. The required fixpoint computations are done when interpreting procedure stores, as we shall see next.

5.4 Interpretation of Configurations

Before we may interpret program configurations, we first have to describe how to interpret values and procedure stores.

Interpretation of Values. A qubit pointer context Q is interpreted as the object $[\![Q]\!] = \mathbf{qbit}^{\otimes |Q|}$. A value $Q \vdash v : A$ is interpreted as a morphism in $\mathbf{V}$ $[\![Q \vdash v : A]\!] : [\![Q]\!] \to [\![A]\!]$, which we abbreviate as $[\![v]\!]$ if Q and A are clear from context. It is defined by induction on the derivation of $Q \vdash v : A$ in Figure 7.

For the next theorem, recall that if $Q \vdash v : A$ is a classical value, then $Q = \cdot$.

Theorem 20. *Let $Q \vdash v : A$ be a value. Then:*

1. *$[\![v]\!]$ is discardable (i.e. causal). More specifically, $\diamond_{[\![A]\!]} \circ [\![v]\!] = \diamond_{[\![Q]\!]} = \mathrm{tr}^{\ddagger}$.*
2. *If A is classical, then $[\![v]\!]$ is copyable, i.e., $\triangle_{[\![A]\!]} \circ [\![v]\!] = ([\![v]\!] \otimes [\![v]\!]) \circ \triangle_I$.*

We see that, as promised, interpretations of values may always be discarded and interpretations of classical values may also be copied. Next, we explain how to interpret value contexts. For a value context $Q; \Gamma \vdash V$, its interpretation is the morphism:

$$[\![Q; \Gamma \vdash V]\!] = \left([\![Q]\!] \xrightarrow{\cong} [\![Q_1]\!] \otimes \cdots \otimes [\![Q_n]\!] \xrightarrow{[\![v_1]\!] \otimes \cdots \otimes [\![v_n]\!]} [\![\Gamma]\!] \right),$$

where $Q_i \vdash v_i : A_i$ is the splitting of Q (see §3) and $[\![\Gamma]\!] = [\![A_1]\!] \otimes \cdots \otimes [\![A_n]\!]$. Some of the Q_i can be empty and this is the reason why the definition depends on a coherent natural isomorphism. We write $[\![V]\!]$ as a shorthand for $[\![Q; \Gamma \vdash V]\!]$. Obviously, $[\![V]\!]$ is also causal thanks to Theorem 20.

Interpretation of Procedure Stores. The interpretation of a well-formed procedure store $\Pi \vdash \Omega$ is an element of $[\![\Pi]\!]$, i.e. a $|\Pi|$-tuple of morphisms from $\mathbf{C}$. It is defined by induction on $\Pi \vdash \Omega$:

$$[\![\cdot \vdash \cdot]\!] = ()$$
$$[\![\Pi, f : A \to B \vdash \Omega, f :: x : A \to y : B \ \{M\}]\!] = ([\![\Omega]\!], \mathrm{lfp}([\![M]\!]([\![\Omega]\!], -))).$$

Interpretation of Configurations. Density matrices $\rho \in M_{2^n}(\mathbb{C})$ are in 1-1 correspondence with $\mathbf{W}^*_{\mathrm{NCPSU}}$-morphisms $\mathrm{new}_\rho : \mathbb{C} \to M_{2^n}(\mathbb{C})$ which are in turn in 1-1 correspondence with $\mathbf{C}$-morphisms $\mathrm{new}_\rho^{\ddagger} : I \to \mathbf{qbit}^{\otimes n}$. Using this observation, we can now define the interpretation of a configuration $\mathcal{C} = (M \mid V \mid \Omega \mid \rho)$ with $\Pi; \Gamma; \Sigma; Q \vdash (M \mid V \mid \Omega \mid \rho)$ to be the morphism

$$[\![\Pi; \Gamma; \Sigma; Q \vdash (M \mid V \mid \Omega \mid \rho)]\!] :=$$
$$\left(I \xrightarrow{\mathrm{new}_\rho^{\ddagger}} \mathbf{qbit}^{\otimes \dim(\rho)} \xrightarrow{[\![Q;\Gamma \vdash V]\!]} [\![\Gamma]\!] \xrightarrow{[\![\Pi \vdash \langle \Gamma \rangle \ M \ \langle \Sigma \rangle]\!]([\![\Pi \vdash \Omega]\!])} [\![\Sigma]\!] \right).$$

For brevity, we simply write $[\![(M \mid V \mid \Omega \mid \rho)]\!]$ or even just $[\![\mathcal{C}]\!]$ to refer to the above morphism.

5.5 Soundness, Adequacy and Big-step Invariance

Since our operational semantics allows for branching, *soundness* is showing that the interpretation of configurations is equal to the sum of small-step reducts.

Theorem 21 (Soundness). *For any non-terminal configuration $\mathcal{C}$:*

$$[\![\mathcal{C}]\!] = \sum_{\mathcal{C} \rightsquigarrow \mathcal{D}} [\![\mathcal{D}]\!].$$

Proof. By induction on the shape of the term component of $\mathcal{C}$. □

Remark 22. The above sum and all sums that follow are well-defined convex sums of NCPSU-maps where the probability weights p_i have been encoded in the density matrices.

A natural question to ask is whether $[\![\mathcal{C}]\!]$ is also equal to the (potentially infinite) sum of all terminal configurations that $\mathcal{C}$ reduces to. In other words, is the interpretation of configurations also invariant with respect to big-step reduction. This is indeed the case and proving this requires considerable effort.

Theorem 23 (Big-step Invariance). *For any configuration $\mathcal{C}$, we have:*

$$[\![\mathcal{C}]\!] = \bigvee_{n=0}^{\infty} \sum_{r \in \mathrm{TerSeq}_{\leq n}(\mathcal{C})} [\![\mathrm{End}(r)]\!]$$

The above theorem is the main result of our paper. This is a powerful result, because with big-step invariance in place, computational adequacy[4] at all types is now a simple consequence of the causal properties of our interpretation. Observe that for any configuration $\mathcal{C}$, we have a subunital map $\diamond \circ [\![\mathcal{C}]\!] : \mathbb{C} \to \mathbb{C}$ and evaluating it at 1 yields a real number $(\diamond \circ [\![\mathcal{C}]\!])(1) \in [0, 1]$.

Theorem 24 (Adequacy). *For any normalised $\mathcal{C}$: $(\diamond \circ [\![\mathcal{C}]\!])(1) = \mathrm{Halt}(\mathcal{C})$.*

If $\mathcal{C}$ is not normalised, then adequacy can be recovered simply by normalising: $(\diamond \circ [\![\mathcal{C}]\!])(1) = \mathrm{tr}(\mathcal{C})\mathrm{Halt}(\mathcal{C})$, for any possible configuration $\mathcal{C}$. The adequacy formulation of [17] and [5] is now a special case of our more general formulation.

Corollary 25. *Let M be a closed program of unit type, i.e. $\cdot \vdash \langle\cdot\rangle\ M\ \langle\cdot\rangle$. Then:*

$$[\![(M \mid \cdot \mid \cdot \mid 1)]\!](1) = \mathrm{Halt}(M \mid \cdot \mid \cdot \mid 1).$$

Proof. By Theorem 24 and because $\diamond_I = \mathrm{id}$. □

[4] Recall that a computational adequacy result has to establish an equivalent *purely denotational* characterisation of the operational notion of non-termination.

6 Conclusion and Related Work

There are many quantum programming languages described in the literature. For a survey see [7] and [16, pp. 129]. Some circuit programming languages (e.g. Proto-Quipper [21,22,15]), generate quantum circuits, but do not necessarily support executing quantum measurements. Here we focus on quantum languages which support measurement and which have either inductive datatypes or some computational adequacy result.

Our work is the first to present a detailed semantic treatment of user-defined inductive datatypes for quantum programming. In [17] and [5], the authors show how to interpret a quantum lambda calculus extended with a datatype for lists, but their syntax does not support any other inductive datatypes. These languages are equipped with lambda abstractions, whereas our language has only support for procedures. Lambda abstractions are modelled using constructions from quantitative semantics of linear logic in [17] and techniques from game semantics in [5]. We believe our model is simpler and certainly more physically natural, because we work only with mathematical structures used by physicists in their study of quantum mechanics. Both [17] and [5] prove an adequacy result for programs of unit type. In [20], the authors discuss potential categorical models for inductive datatypes in quantum programming, but there is no detailed semantic treatment provided and there is no adequacy result, because the language lacks recursion.

Other quantum programming languages without inductive datatypes, but which prove computational adequacy results include [9,12]. A model based on W*-algebras for a quantum lambda calculus without recursion or inductive datatypes was described in a recent manuscript [4]. In that model, it appears that currying is *not* a Scott-continuous operation, and if so, the addition of recursion renders the model neither sound, nor adequate. For this reason, we use procedures and not lambda abstractions in our language.

To conclude, we presented two novel results in quantum programming: (1) we provided a denotational semantics for a quantum programming language with inductive datatypes; (2) we proved that our denotational semantics is invariant with respect to big-step reduction. We also showed that the latter result is quite powerful by demonstrating how it immediately implies computational adequacy.

Our denotational model is based on W*-algebras, which are used by physicists to study quantum foundations. We hope this would make it useful for developing static analysis methods (based on abstract interpretation) that can be used for entanglement detection [18] and we plan on investigating this in future work.

Acknowledgements. We thank Andre Kornell, Bert Lindenhovius and Michael Mislove for discussions regarding this paper. We also thank the anonymous referees for their feedback. MR acknowledges financial support from the Quantum Software Consortium, under the Gravitation programme of the Dutch Research Council NWO. The remaining authors were supported by the French projects ANR-17-CE25-0009 SoftQPro, ANR-17-CE24-0035 VanQuTe and PIA-GDN/Quantex.

References

1. Abadi, M., Fiore, M.P.: Syntactic Considerations on Recursive Types. In: Proceedings, 11th Annual IEEE Symposium on Logic in Computer Science, New Brunswick, New Jersey, USA, July 27-30, 1996. pp. 242–252. IEEE Computer Society (1996). https://doi.org/10.1109/LICS.1996.561324
2. Cho, K.: Semantics for a Quantum Programming Language by Operator Algebras (2014), Master Thesis, University of Tokyo.
3. Cho, K.: Semantics for a Quantum Programming Language by Operator Algebras. New Generation Comput. **34**(1-2), 25–68 (2016). https://doi.org/10.1007/s00354-016-0204-3
4. Cho, K., Westerbaan, A.: Von Neumann Algebras form a Model for the Quantum Lambda Calculus (2016), `http://arxiv.org/abs/1603.02133`, manuscript.
5. Clairambault, P., de Visme, M., Winskel, G.: Game semantics for quantum programming. PACMPL **3**(POPL), 32:1–32:29 (2019). https://doi.org/10.1145/3290345
6. Fiore, M.P.: Axiomatic Domain Theory in Categories of Partial Maps. Ph.D. thesis, University of Edinburgh, UK (1994)
7. Gay, S.J.: Quantum programming languages: survey and bibliography. Mathematical Structures in Computer Science **16**(4), 581–600 (2006). https://doi.org/10.1017/S0960129506005378
8. Harrow, A.W., Hassidim, A., Lloyd, S.: Quantum Algorithm for Linear Systems of Equations. Phys. Rev. Lett. **103**, 150502 (Oct 2009). https://doi.org/10.1103/PhysRevLett.103.150502
9. Hasuo, I., Hoshino, N.: Semantics of higher-order quantum computation via geometry of interaction. Ann. Pure Appl. Logic **168**(2), 404–469 (2017). https://doi.org/10.1016/j.apal.2016.10.010
10. Kissinger, A., Uijlen, S.: A categorical semantics for causal structure. In: 32nd Annual ACM/IEEE Symposium on Logic in Computer Science, LICS 2017, Reykjavik, Iceland, June 20-23, 2017. pp. 1–12. IEEE Computer Society (2017). https://doi.org/10.1109/LICS.2017.8005095
11. Kornell, A.: Quantum collections. International Journal of Mathematics **28**(12), 1750085 (2017). https://doi.org/10.1142/S0129167X17500859
12. Lago, U.D., Faggian, C., Valiron, B., Yoshimizu, A.: The geometry of parallelism: classical, probabilistic, and quantum effects. In: Castagna, G., Gordon, A.D. (eds.) Proceedings of the 44th ACM SIGPLAN Symposium on Principles of Programming Languages, POPL 2017, Paris, France, January 18-20, 2017. pp. 833–845. ACM (2017), `http://dl.acm.org/citation.cfm?id=3009859`
13. Lindenhovius, B., Mislove, M., Zamdzhiev, V.: LNL-FPC: The Linear/Non-linear Fixpoint Calculus `https://arxiv.org/abs/1906.09503`, submitted.
14. Lindenhovius, B., Mislove, M., Zamdzhiev, V.: Mixed linear and non-linear recursive types. Proc. ACM Program. Lang. **3**(ICFP), 111:1–111:29 (Jul 2019). https://doi.org/10.1145/3341715
15. Lindenhovius, B., Mislove, M.W., Zamdzhiev, V.: Enriching a Linear/Non-linear Lambda Calculus: A Programming Language for String Diagrams. In: Dawar, A., Grädel, E. (eds.) Proceedings of the 33rd Annual ACM/IEEE Symposium on Logic in Computer Science, LICS 2018, Oxford, UK, July 09-12, 2018. pp. 659–668. ACM (2018). https://doi.org/10.1145/3209108.3209196
16. Mosca, M., Roetteler, M., Selinger, P.: Quantum Programming Languages (Dagstuhl Seminar 18381). Dagstuhl Reports **8**(9), 112–132 (2019). https://doi.org/10.4230/DagRep.8.9.112

17. Pagani, M., Selinger, P., Valiron, B.: Applying quantitative semantics to higher-order quantum computing. In: Jagannathan, S., Sewell, P. (eds.) The 41st Annual ACM SIGPLAN-SIGACT Symposium on Principles of Programming Languages, POPL '14, San Diego, CA, USA, January 20-21, 2014. pp. 647–658. ACM (2014). https://doi.org/10.1145/2535838.2535879
18. Perdrix, S.: Quantum Entanglement Analysis Based on Abstract Interpretation. In: Alpuente, M., Vidal, G. (eds.) Static Analysis, 15th International Symposium, SAS 2008, Valencia, Spain, July 16-18, 2008. Proceedings. Lecture Notes in Computer Science, vol. 5079, pp. 270–282. Springer (2008). https://doi.org/10.1007/978-3-540-69166-2_18
19. Rennela, M.: Operator Algebras in Quantum Computation (2013), Master Thesis, Université Paris 7 Denis Diderot.
20. Rennela, M., Staton, S.: Classical Control and Quantum Circuits in Enriched Category Theory. Electr. Notes Theor. Comput. Sci. **336**, 257–279 (2018). https://doi.org/10.1016/j.entcs.2018.03.027
21. Rios, F., Selinger, P.: A Categorical Model for a Quantum Circuit Description Language. In: QPL (2017). https://doi.org/10.4204/EPTCS.266.11
22. Ross, N.J.: Algebraic and Logical Methods in Quantum Computation (2015), Ph.D. thesis, Dalhousie University.
23. Selinger, P.: Towards a quantum programming language. Mathematical Structures in Computer Science **14**(4), 527–586 (2004). https://doi.org/10.1017/S0960129504004256
24. Shor, P.W.: Polynomial-Time Algorithms for Prime Factorization and Discrete Logarithms on a Quantum Computer. SIAM Review **41**(2), 303–332 (1999). https://doi.org/10.1137/S0036144598347011
25. Takesaki, M.: Theory of Operator Algebras. Vol. I, II and III. Springer-Verlag, Berlin (2002)
26. Westerbaan, A.: Quantum Programs as Kleisli Maps. In: Duncan, R., Heunen, C. (eds.) Proceedings 13th International Conference on Quantum Physics and Logic, QPL 2016, Glasgow, Scotland, 6-10 June 2016. EPTCS, vol. 236, pp. 215–228 (2016). https://doi.org/10.4204/EPTCS.236.14
27. Westerbaan, B.: Dagger and Dilation in the Category of Von Neumann algebras. Ph.D. thesis, Radboud University (2018), `http://arxiv.org/abs/1803.01911`
28. Wootters, W.K., Zurek, W.H.: A single quantum cannot be cloned. Nature **299**(5886), 802–803 (1982)

Spinal Atomic Lambda-Calculus

David Sherratt[1] (✉), Willem Heijltjes[2], Tom Gundersen[3], and Michel Parigot[4]

[1] Friedrich-Schiller-Universität Jena, Germany.
david.rhys.sherratt@uni-jena.de
[2] University of Bath, United Kingdom.
w.b.heijltjes@bath.ac.uk
[3] Red Hat, Inc. Norway.
teg@jklm.no
[4] Institut de Recherche en Informatique Fondamentale, CNRS, Université de Paris. France.
parigot@irif.fr

Abstract. We present the spinal atomic λ-calculus, a typed λ-calculus with explicit sharing and atomic duplication that achieves spinal full laziness: duplicating only the direct paths between a binder and bound variables is enough for beta reduction to proceed. We show this calculus is the result of a Curry–Howard style interpretation of a deep-inference proof system, and prove that it has natural properties with respect to the λ-calculus: confluence and preservation of strong normalisation.

Keywords: Lambda-Calculus · Full laziness · Deep inference · Curry–Howard

1 Introduction

In the λ-calculus, a main source of efficiency is *sharing*: multiple use of a single subterm, commonly expressed through graph reduction [27] or explicit substitution [1]. This work, and the *atomic λ-calculus* [16] on which it builds, is an investigation into sharing as it occurs naturally in intuitionistic *deep-inference* proof theory [26]. The atomic λ-calculus arose as a Curry–Howard interpretation of a deep-inference proof system, in particular of the *distribution* rule given below left, a variant of the characteristic *medial* rule [10, 26]. In the term calculus, the corresponding *distributor* enables duplication to proceed *atomically*, on individual constructors, in the style of sharing graphs [21]. As a consequence, the natural reduction strategy in the atomic λ-calculus is *fully lazy* [27, 4]: it duplicates only the minimal part of a term, the *skeleton*, that can be obtained by lifting out subterms as explicit substitutions. (While duplication is atomic *locally*, a duplicated abstraction does not form a redex until also its bound variables have been duplicated; hence duplication becomes fully lazy *globally*.)

This work was supported by EPSRC Project EP/R029121/1 *Typed Lambda-Calculi with Sharing and Unsharing* and ANR project 15-CE25-0014 *The Fine Structure of Formal Proof Systems and their Computational Interpretations (FISP)*

J. Goubault-Larrecq and B. König (Eds.): FOSSACS 2020, LNCS 12077, pp. 582–601, 2020.
https://doi.org/10.1007/978-3-030-45231-5_30

Distribution: $\dfrac{A \to (B \wedge C)}{(A \to B) \wedge (A \to C)}\,d$ Switch: $\dfrac{(A \to B) \wedge C}{A \to (B \wedge C)}\,s$

We investigate the computational interpretation of another characteristic deep-inference proof rule: the *switch* rule above right [26].[5] Our result is the *spinal atomic λ-calculus*, a λ-calculus with a refined form of full laziness, *spine duplication*. In the terminology of [4], this strategy duplicates only the *spine* of an abstraction: the paths to its bound variables in the syntax tree of the term.[6]

We illustrate these notions in Figure 1, for the example $\lambda x.\lambda y.((\lambda z.z)y)x$. The *scope* of the abstraction λx is the entire subterm, $\lambda y.((\lambda z.z)y)x$ (which may or may not be taken to include λx itself). Note that with explicit substitution, the scope may grow or shrink by lifting explicit substitutions in or out. The *skeleton* is the term $\lambda x.\lambda y.(wy)x$ where the subterm $\lambda z.z$ is lifted out as an (explicit) substitution $[\lambda z.z/w]$. The *spine* of a term, indicated in the second image, cannot naturally be expressed with explicit substitution, though one can get an impression with *capturing* substitutions: it would be $\lambda x.\lambda y.wx$, with the subterm $(\lambda z.z)y$ extracted by a capturing substitution $[(\lambda z.z)y/w]$. Observe that the skeleton can be described as the *iterated spine*: it is the smallest subgraph of the syntax tree closed under taking the spine of each abstraction, i.e. that contains the spine of every abstraction it contains.

These notions give rise to four natural duplication regimes. For a shared abstraction to become available as the function in a β-redex: *laziness* duplicates its *scope* [22]; *Full laziness* duplicates its *skeleton* [27]; *Spinal full laziness* duplicates its *spine* [8]; *optimal reduction* duplicates only the abstraction λx and its bound variables x [21, 3].[7]

While each of these duplication strategies has been expressed in graphs and labelled calculi, the atomic λ-calculus is the first term calculus with Curry–Howard corresponding proof system to naturally describe full laziness. Likewise, the spinal atomic λ-calculus presented here is the first term calculus with Curry–Howard corresponding proof system to naturally describe spinal full laziness.

Switch and Spine. One way to describe the skeleton or the spine of an abstraction within a λ-term is through explicit end-of-scope markers, as explored by Berkling and Fehr [7], and more recently by Hendriks and Van Oostrom [18]. We use their *adbmal* (⅄) to illustrate the idea: the constructor ⅄$x.N$ indicates that the subterm N does not contain occurrences of x (or that any that do occur are

[5] The switch rule is an intuitionistic variant of *weak* or *linear distributivity* [12] for multiplicative linear logic.

[6] There is a clash of (existing) terminology: the *spine of an abstraction*, as we use here, is a different notion from the *spine of a λ-term*, which is the path from the root to the leftmost variable, as used e.g. in head reduction and abstract machines.

[7] Interestingly, Balabonski [5] shows that for *weak* reduction (where one does not reduce under an abstraction) full laziness and spinal full laziness are both optimal (in the number of beta-steps required to reach a normal form).

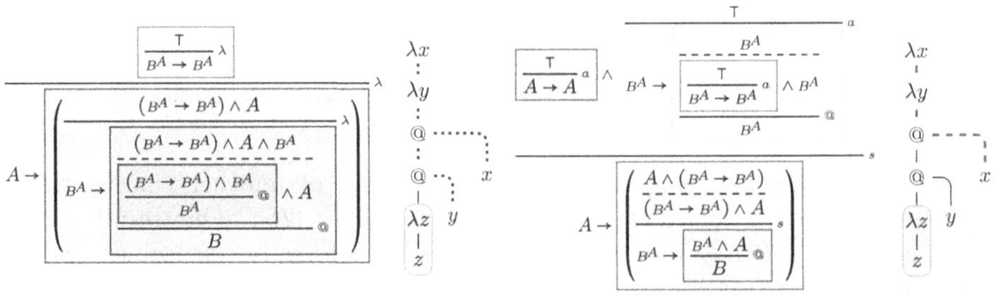

Fig. 1: Balanced and unbalanced typing derivations for $\lambda x.\lambda y.((\lambda z.z)y)x$, with corresponding graphical representations of the term. The variable x has type A and y, z type $A \to B$, shortened to B^A. The left derivation isolates the skeleton of λx, and the right derivation its spine, both by the subderivations in braces.

not available to a binder λx outside $\reflectbox{$\lambda$}x.N$). The scope of an abstraction thus becomes explicitly indicated in the term. This opens up a distinction between *balanced* and *unbalanced* scopes: whether scopes must be properly nested, or not; for example, in $\lambda x.\lambda y.N$, a subterm $\reflectbox{$\lambda$}y.\reflectbox{$\lambda$}x.M$ is balanced, but $\reflectbox{$\lambda$}x.\reflectbox{$\lambda$}y.M$ is not. With balanced scope, one can indicate the skeleton of an abstraction; with unbalanced scope (which Hendriks and Van Oostrom dismiss) one can indicate the spine. We do so for our example term $\lambda x.\lambda y.((\lambda z.z)y)x$ below.

Balanced scope/skeleton:	$\underline{\lambda x.\lambda y.(\reflectbox{λ}y.}(\reflectbox{λ}x.\lambda z.z)\underline{y)(\reflectbox{λ}y.x)}$
Unbalanced scope/spine:	$\underline{\lambda x.\lambda y.}(\reflectbox{λ}x.(\reflectbox{λ}y.\lambda z.z)y)\underline{(\reflectbox{λ}y.x)}$

A closely related approach is *director strings*, introduced by Kennaway and Sleep [19] for combinator reduction and generalized to any reduction strategy by Fernández, Mackie, and Sinot in [13]. The idea is to use nameless abstractions identified by their nesting (as with De Bruijn indices), and make the paths to bound variables explicit by annotating each constructor with a string of *directors*, that outline the paths. The primary aim of these approaches is to eliminate α-conversion and to streamline substitution. Consequently, while they can *identify* the spine, they do not readily isolate it for duplication.

The present work starts from our observation that the *switch* rule of open deduction functions as a proof-theoretic end-of-scope construction (see [25] for details). However, it does so in a *structural* way: it forces a deconstruction of a proof into readily duplicable parts, which together may form the spine of an abstraction. The derivations in Figure 1 demonstrate this, as we will now explain—see the next section for how they are formally constructed.

The abstraction λx corresponds in the proof system to the implication $A\to$, explicitly scoping over its right-hand side. On the left, with the *abstraction* rule (λ), scopes must be balanced, and the proof system may identify the *skeleton*; here, that of λx as the largest blue box. Decomposing the abstraction (λ) into *axiom* (a) and *switch* (s), on the right the proof system may express unbalanced scope. It does so by separating the scope of an abstraction into multiple parts; here, that of λx is captured as the two top-level red boxes. Each box is ready to be duplicated; in this way, one may duplicate the spine of an abstraction only.

These two derivations correspond to terms in our calculus. The subterms not part of the skeleton (i.e. $\lambda z.z$) remain shared and we are able to duplicate the skeleton alone. This is also possible in [16]. In our calculus we are also able to duplicate just the spine by using a *distributor*. We require this construct as otherwise we break the binding of the y-abstraction. The distributor manages and maintains these bindings. The y-abstraction in the spine ($y\langle a\rangle$) is a *phantom-abstraction*, because it is not real and we cannot perform β-reduction on it. However, it may become real during reduction. It can be seen as a placeholder for the abstraction. The variables in the *cover* (a) represent subterms that both remain shared and are found in the distributor.

$$\text{Skeleton:} \quad \underline{\lambda x.\lambda y.(a\, y)\, x}\, [a \leftarrow \lambda z.z]$$

$$\text{Spine:} \quad \underline{\lambda x.y\langle a\rangle.(a)\, x}\, [y\langle a\rangle \,|\, \lambda y.\, [a \leftarrow (\lambda z.z)y]]$$

Our investigation is then focused on the interaction of switch and distribution (later observed in the rewrite rule l_5). The use of the distribution rule allows us to perform duplication atomically, and thus provides a natural strategy for spinal full laziness. In Figure 1 on the right, this means duplicating the two top-level red boxes can be done independently from duplicating the yellow box.

2 Typing a λ-calculus in open deduction

We work in *open deduction* [15], a formalism of deep-inference proof theory, using the following proof system for (conjunction–implication) intuitionistic logic. A *derivation* from a *premise* formula X to a *conclusion* formula Z is constructed inductively as in Figure 2a, with from left to right: a propositional atom a, where $X = Z = a$; *horizontal composition* with a connective $\rightarrow$, where $X = Y \rightarrow X_2$ and $Z = Y \rightarrow Z_2$; *horizontal composition* with a connective $\wedge$, where $X = X_1 \wedge X_2$ and $Z = Z_1 \wedge Z_2$; and *rule composition*, where r is an inference rule (Figure 2b) from Y_1 to Y_2. The boxes serve as parentheses (since derivations extend in two dimensions) and may be omitted. Derivations are considered up to associativity of rule composition. One may consider formulas as derivations that omit rule composition. We work modulo associativity, symmetry, and unitality of conjunction, justifying the n-ary contraction, and may omit $\top$ from the axiom rule. A 0-ary contraction, with conclusion $\top$, is a *weakening*. Figure 2b: the abstraction rule (λ) is derived from axiom and switch. *Vertical composition* of a derivation from X to Y and one from Y to Z, depicted by a dashed line, is a defined operation, given in Figure 2c, where $* \in \{\wedge, \rightarrow\}$.

2.1 The Sharing Calculus

Our starting point is the *sharing calculus* (Λ^S), a calculus with an explicit sharing construct, similar to explicit substitution.

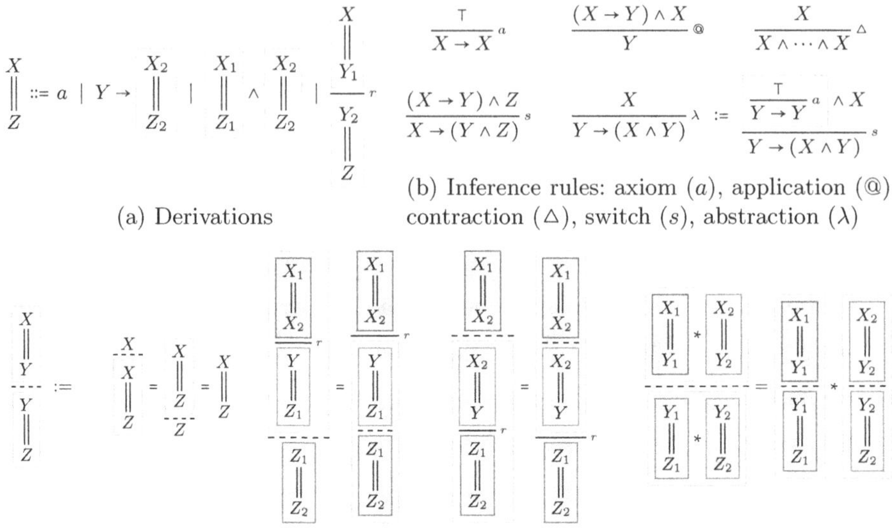

(a) Derivations

(b) Inference rules: axiom (a), application $(@)$, contraction $(\triangle)$, switch (s), abstraction (λ)

$$\frac{\top}{X \to X}\, a \qquad \frac{(X \to Y) \wedge X}{Y}\, @ \qquad \frac{X}{X \wedge \cdots \wedge X}\, \triangle$$

$$\frac{(X \to Y) \wedge Z}{X \to (Y \wedge Z)}\, s \qquad \frac{X}{Y \to (X \wedge Y)}\, \lambda \;:=\; \frac{\dfrac{\top}{Y \to Y}\, a \;\wedge\; X}{Y \to (X \wedge Y)}\, s$$

(c) Vertical composition

Fig. 2: Intuitionistic proof system in open deduction

Definition 1. *The **pre-terms** r, s, t, u and **sharings** $[\Gamma]$ of the Λ^S are defined by:*

$$s, t ::= x \mid \lambda x.t \mid s\,t \mid t[\Gamma] \qquad [\Gamma] ::= [x_1, \ldots, x_n \leftarrow s]$$

*with from left to right: a **variable**; an **abstraction**, where x occurs free in t and becomes bound; an **application**, where s and t use distinct variable names; and a **closure**; in $t[\vec{x} \leftarrow s]$ the variables in the vector $\vec{x} = x_1, \ldots, x_n$ all occur in t and become bound, and s and t use distinct variable names. **Terms** are pre-terms modulo **permutation** equivalence $(\sim)$:*

$$t[\vec{x} \leftarrow s][\vec{y} \leftarrow r] \sim t[\vec{y} \leftarrow r][\vec{x} \leftarrow s] \qquad (\{\vec{y}\} \cap (s)_{fv} = \{\})$$

*A term is in **sharing normal form** if all sharings occur as $[\vec{x} \leftarrow x]$ either at the top level or directly under a binding abstraction, as $\lambda x.t[\vec{x} \leftarrow x]$.*

Note that variables are *linear*: variables occur at most once, and bound variables must occur. A vector $\vec{x}$ has length $|\vec{x}|$ and consist of the variables $x_1, \ldots, x_{|\vec{x}|}$. An **environment** is a sequence of sharings $\overline{[\Gamma]} = [\Gamma_1] \ldots [\Gamma_n]$. Substitution is written $\{t/x\}$, and $\{t_1/x_1\} \ldots \{t_n/x_n\}$ may be abbreviated to $\{t_i/x_i\}_{i \in [n]}$.

Definition 2. *The **interpretation** $\llbracket - \rrbracket : \Lambda \to \Lambda^S$ is defined below.*

$$\llbracket x \rrbracket = x \quad \llbracket \lambda x.t \rrbracket = \lambda x.\llbracket t \rrbracket \quad \llbracket s\,t \rrbracket = \llbracket s \rrbracket \llbracket t \rrbracket \quad \llbracket t[\vec{x} \leftarrow s] \rrbracket = \llbracket t \rrbracket \{\llbracket s \rrbracket / x_i\}_{i \in [n]}$$

The **translation** $(\!|N|\!)$ of a λ-term N is the unique sharing-normal term t such that $N = \llbracket t \rrbracket$. A term t will be typed by a derivation with restricted types,

Basic Types: $A, B, C := a \mid A \to B$ Context Types: $\Gamma, \Delta, \Omega := A \mid \top \mid \Gamma \wedge \Delta$

$$
x : A^x \qquad t\,s : \quad
\cfrac{
\begin{array}{cc}
\Gamma & \Delta \\
\Big\|t & \Big\|s \\
A \to B & A
\end{array}
}{B}@
\qquad
\lambda x.t : \quad
\cfrac{
\cfrac{\Gamma}{\Gamma \wedge A^x}\lambda}{
\begin{array}{c}
\Big\|t \\
B
\end{array}
}\,A \to
\qquad
t[\vec{x} \leftarrow s] : \quad
\Gamma \wedge \;
\cfrac{
\cfrac{
\begin{array}{c}
\Delta \\
\Big\|s \\
A
\end{array}
}{A \wedge \cdots \wedge A}\Delta
}{
\cfrac{
\Gamma \wedge (A \wedge \cdots \wedge A)^{\vec{x}}
}{
\begin{array}{c}
\Big\|t \\
B
\end{array}
}
}
$$

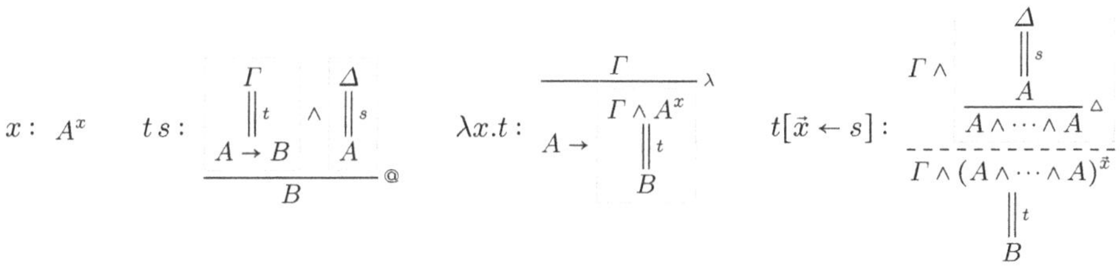

Fig. 3: Typing system for Λ^S

as shown below, where the *context type* $\Gamma = A_1 \wedge \cdots \wedge A_n$ will have an A_i for each free variable x_i of t. We connect free variables to their premises by writing A^x and $\Gamma^{\vec{x}}$. The Λ^S is then typed as in Figure 3.

3 The Spinal Atomic λ-Calculus

We now formally introduce the syntax of the spinal atomic λ-calculus (Λ_a^S), by extending the definition of the sharing calculus in Definition 1 with a *distributor* construct that allows for atomic duplication of terms.

Definition 3 (Pre-Terms). *The **pre-terms** r, s, t, **closures** $[\Gamma]$, and environments $\overline{[\Gamma]}$ of the Λ_a^S are defined by:*

$$
t ::= x \mid st \mid x\langle \vec{y} \rangle.t \mid t[\Gamma] \qquad \overline{[\Gamma]} ::= [\Gamma] \mid \overline{[\Gamma]}[\Gamma]
$$
$$
[\Gamma] ::= [\vec{x} \leftarrow t] \mid [\vec{x} \mid y\langle \vec{z} \rangle \overline{[\Gamma]}]
$$

Our generalized abstraction $x\langle \vec{y} \rangle.t$ is a **phantom-abstraction**, where x a **phantom-variable** and the **cover** $\vec{y}$ will be a subset of the free variables of t. It can be thought of as a "delayed" abstraction: x is a binder, but possibly not in t itself, and instead in the terms substituted for the variables $\vec{y}$; in other words, x is a *capturing* binder for substitution into $\vec{y}$. We define standard λ-abstraction as the special case $\lambda x.t \equiv x\langle x \rangle.t$, and generally, when we refer to $x\langle \vec{y} \rangle$ as a phantom-abstraction (rather than an abstraction) we assume $\vec{y} \neq x$. The **distributor** $u[\vec{x} \mid y\langle \vec{z} \rangle \overline{[\Gamma]}]$ binds the phantom-variables $\vec{x}$ in u, while its environment $\overline{[\Gamma]}$ will bind the variables in their covers; intuitively, it represents a set of explicit substitutions in which the variables $\vec{x}$ are expected to be captured.

The distributor is introduced when we wish to duplicate an abstraction, as depicted in Figure 4a. The sharing node (∘) duplicates the abstraction node, creating a distributor (depiced as the sharing and unsharing node (•), together with the bindings of the phantom-variables (depicted with a dashed line). The variables captured by the environment are the variables connected to sharing nodes linked with a dotted line. Notice one sharing node can be linked with multiple unsharing nodes, and vice versa. Duplication of applications also duplicates

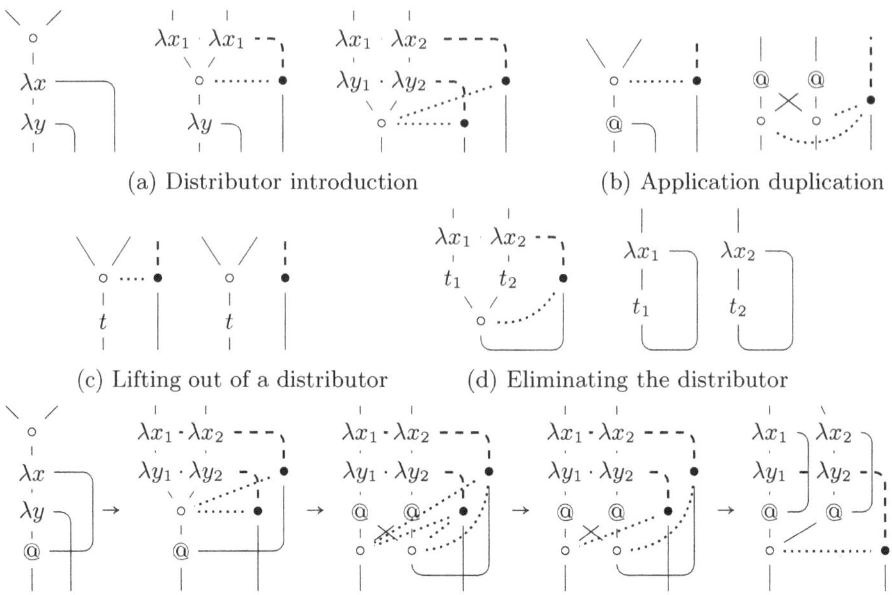

(a) Distributor introduction

(b) Application duplication

(c) Lifting out of a distributor

(d) Eliminating the distributor

(e) Duplicating the spine

Fig. 4: Graphical illustration of the distributor

the dotted line (Figure 4b), but these can be removed later if the term does not contain the variable bound to the unsharing (Figure 4c). These subterms are those which are not part of the spine. Eventually, we will reach a state where the only sharing node connected to the unsharing node is the one that shared the variable bound to the unsharing, allowing us to eliminate the distributor (Figure 4d). The purpose of the dotted line is similar to the brackets of optimal reduction graphs [21, 24], to supervise which sharing and unsharing match.

Terms are then pre-terms with sensible and correct bindings. To define terms, we first define *free* and *bound* variables and phantom variables; variables are bound by abstractions (not phantoms) and by sharings, while phantom-variables are bound by distributors.

Definition 4 (Free and Bound Variables). *The* ***free variables*** $(-)_{fv}$ *and* ***bound variables*** $(-)_{bv}$ *of a pre-term* t *are defined as follows*

$$
\begin{aligned}
(x)_{fv} &= \{x\} & (x)_{bv} &= \{\} \\
(s\,t)_{fv} &= (s)_{fv} \cup (t)_{fv} & (s\,t)_{bv} &= (s)_{bv} \cup (t)_{bv} \\
(x\langle\, x \,\rangle.t)_{fv} &= (t)_{fv} - \{x\} & (x\langle\, x \,\rangle.t)_{bv} &= (t)_{bv} \cup \{x\} \\
(x\langle\, \vec{y} \,\rangle.t)_{fv} &= (t)_{fv} & (x\langle\, \vec{y} \,\rangle.t)_{bv} &= (t)_{bv} \\
(u[\vec{x} \leftarrow t])_{fv} &= (u)_{fv} \cup (t)_{fv} - \{\vec{x}\} & (u[\vec{x} \leftarrow t])_{bv} &= (u)_{bv} \cup (t)_{bv} \cup \{\vec{x}\} \\
(u[\vec{x} \,|\, y\langle\, y \,\rangle\, \overline{[\Gamma]}])_{fv} &= (u\overline{[\Gamma]})_{fv} - \{y\} & (u[\vec{x} \,|\, y\langle\, y \,\rangle\, \overline{[\Gamma]}])_{bv} &= (u\overline{[\Gamma]})_{bv} \cup \{y\}
\end{aligned}
$$

$$(u[\vec{x}\,|\,y\langle\,\vec{z}\,\rangle\,\overline{[\Gamma]}])_{fv} = (u\overline{[\Gamma]})_{fv} \cup \{y\} \qquad (u[\vec{x}\,|\,y\langle\,\vec{z}\,\rangle\,\overline{[\Gamma]}])_{bv} = (u\overline{[\Gamma]})_{bv}$$

Definition 5 (Free and Bound Phantom-Variables). *The* ***free phantom-variables*** $(-)_{fp}$ *and* ***bound phantom-variables*** $(-)_{bp}$ *of the pre-term* t *are defined as follows*

$$\begin{aligned}
(x)_{fp} &= \{\} & (x)_{bp} &= \{\}\\
(s\,t)_{fp} &= (s)_{fp} \cup (t)_{fp} & (s\,t)_{bp} &= (s)_{bp} \cup (t)_{bp}\\
(x\langle\,x\,\rangle.t)_{fp} &= (t)_{fp} & &\\
(c\langle\,\vec{x}\,\rangle.t)_{fp} &= (t)_{fp} \cup \{c\} & (c\langle\,\vec{x}\,\rangle.t)_{bp} &= (t)_{bp}\\
(u[\vec{x} \leftarrow t])_{fp} &= (u)_{fp} \cup (t)_{fp} & (u[\vec{x} \leftarrow t])_{bp} &= (u)_{bp} \cup (t)_{bp}\\
(u[\vec{x}\,|\,c\langle\,c\,\rangle\,\overline{[\Gamma]}])_{fp} &= (u\overline{[\Gamma]})_{fp} - \{\vec{x}\} & &\\
(u[\vec{x}\,|\,c\langle\,\vec{y}\,\rangle\,\overline{[\Gamma]}])_{fp} &= (u\overline{[\Gamma]})_{fp} \cup \{c\} - \{\vec{x}\} & (u[\vec{x}\,|\,c\langle\,\vec{y}\,\rangle\,\overline{[\Gamma]}])_{bp} &= (u\overline{[\Gamma]})_{bp} \cup \{\vec{x}\}
\end{aligned}$$

The **free covers** $(u)_{fc}$ and **bound covers** $(u)_{bc}$ are the covers associated with the free phantom-variables $(u)_{fp}$ respectively the bound phantom-variables $(u)_{bp}$ of u; that is, if x occurs as $x\langle\,\vec{a}\,\rangle$ in u and $x \in (u)_{fp}$ then $\langle\vec{a}\rangle \in (u)_{fc}$. When bound, x and the variables in $\vec{a}$ may be alpha-converted independently. When a distributor $u[\vec{x}\,|\,y\langle\,\vec{z}\,\rangle\,\overline{[\Gamma]}]$ binds the phantom-variables $\vec{x} = x_1,\ldots,x_n$ where each x_i occurs as $x_i\langle\,\vec{a}_i\,\rangle$ in u, then for technical convenience we may make the covers explicit in the distributor itself, and write

$$u[x_1\langle\,\vec{a}_1\,\rangle \ldots x_n\langle\,\vec{a}_n\,\rangle\,|\,y\langle\,\vec{z}\,\rangle\,\overline{[\Gamma]}]\ .$$

The environment $\overline{[\Gamma]}$ is expected to bind *exactly* the variables in the covers $\langle\vec{a}_i\rangle$. We apply this and other restrictions to define the terms of the calculus.

Definition 6. ***Terms*** $t \in \Lambda_a^S$ *are pre-terms with the following constraints*

1. *Each variable may occur at most once.*
2. *In a phantom-abstraction* $x\langle\,\vec{y}\,\rangle.t$, $\{\vec{y}\} \subseteq (t)_{fv}$.
3. *In a sharing* $u[\vec{x} \leftarrow t]$, $\{\vec{x}\} \subseteq (u)_{fv}$.
4. *In a distributor* $u[x_1\langle\,\vec{a}_1\,\rangle \ldots x_n\langle\,\vec{a}_n\,\rangle\,|\,y\langle\,\vec{z}\,\rangle\,\overline{[\Gamma]}]$
   (a) $\{x_1,\ldots,x_n\} \subseteq (u)_{fp}$;
   (b) *the variables in* $\bigcup_{i\le n}\{\vec{a}_i\}$ *are free in* u *and bound by* $\overline{[\Gamma]}$.
   (c) *the variables in* $\{\vec{z}\}$ *occur freely in the environment* $\overline{[\Gamma]}$.

Example 1. Here we show some pre-terms that are not terms.

- $c\langle\,x\,\rangle.y$ (violates condition 2)
- $x\,y[x,z \leftarrow w]$ (violates condition 3)
- $e_2\langle\,w_2\,\rangle.w_2\,((e_1\langle\,w_1\,\rangle.w_1)\,z)[e_1\langle\,w_1\,\rangle,e_2\langle\,w_2\,\rangle\,|\,c\langle\,z\,\rangle\,[w_1,w_2 \leftarrow x\langle\,x\,\rangle.x\,y]]$ (violates condition 4*a*)

We also work modulo permutation with respect to the variables in the cover of phantom-abstractions. Let $\vec{x}$ be a list of variables and let $\vec{x_P}$ be a permutation of that list, then the following terms are considered equal.

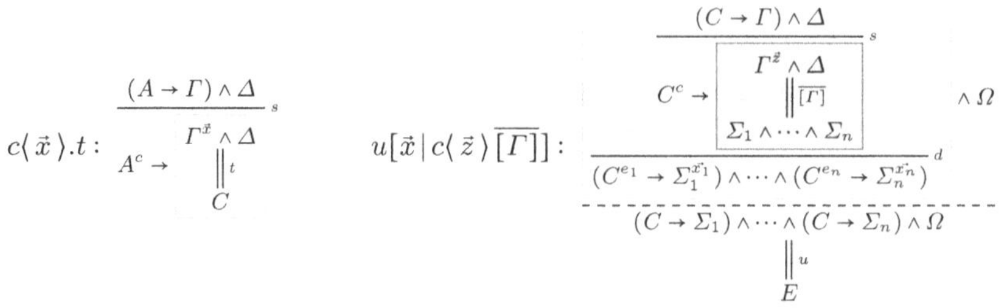

Fig. 5: Typing derivations for phantom-abstractions and distributors

$$u[\vec{x} \leftarrow t] \sim u[\vec{x_P} \leftarrow t] \qquad\qquad y\langle\, \vec{x}\,\rangle.t \sim y\langle\, \vec{x_P}\,\rangle.t$$

Terms are typed with the typing system for Λ^S extended with the *distribution* inference rule. This rule is the result of computationally interpreting the medial rule as done in [16]. We obtain this variant of the medial rule due to the restriction for implications and to avoid introducing disjunction to the typing system. The terms of Λ^S_a are then typed as in both Figure 3 and Figure 5. Note environments are typed by the derivations of all its closures composed horizontally with the conjunction connective. Also note that in the case for phantom-abstraction is similar for that of an abstraction, where we replace one occurrence of the simple type A by the conjunction Γ.

3.1 Compilation and Readback.

We now define the translations between Λ^S_a and the original λ-calculus. First we define the interpretation $\Lambda \to \Lambda^S_a$ (*compilation*). Intuitively, it replaces each abstraction $\lambda x.-$ with the term $x\langle\, x\,\rangle.-[x_1,\ldots,x_n \leftarrow x]$ where $x_1,\ldots,x_n$ replace the occurrences of x. Actual substitutions are denoted as $\{t/x\}$. Let $|\,M\,|_x$ denote the number of occurrences of x in M, and if $|\,M\,|_x = n$ let $M\frac{n}{x}$ denote M with the occurrences of x replaced by fresh, distinct variables $x_1,\ldots,x_n$. First, the translation of a *closed* term M is $(\!|\, M\,|\!)'$, defined below

Definition 7 (Compilation). *The interpretation of λ terms, $(\!|\,\Lambda\,|\!)' : \Lambda \to \Lambda^S_a$, is defined as*

$$(\!|\, M\frac{n_1}{x_1}\cdots\frac{n_k}{x_k}\,|\!)'[x_1^1,\ldots,x_1^{n_1} \leftarrow x_1]\ldots[x_k^1,\ldots,x_k^{n_k} \leftarrow x_k]$$

where $x_1,\ldots,x_k$ are the free variables of M such that $|\,M\,|_{x_i} = n_i > 1$ and $(\!|-|\!)'$ is defined on terms as (where $n \neq 1$ in the abstraction case):

$$\begin{aligned}(\!|\, x\,|\!)' &= x\\ (\!|\, M\, N\,|\!)' &= (\!|\, M\,|\!)'\,(\!|\, N\,|\!)'\end{aligned} \qquad (\!|\,\lambda x.M\,|\!)' = \begin{cases} x\langle\, x\,\rangle.(\!|\, M\,|\!)' & \text{if } |\,M\,|_x = 1\\ x\langle\, x\,\rangle.(\!|\, M\frac{n}{x}\,|\!)'[x_1,\ldots,x_n \leftarrow x] & \text{if } |\,M\,|_x = n\end{cases}$$

The readback into the λ-calculus is slightly more complicated, specifically due to the bindings induced by the distributor. Interpreting a distributor construct as a λ-term requires (1) converting the phantom-abstractions it binds in

u into abstractions (2) collapsing the environment (3) maintaining the bindings between the converted abstractions and the intended variables located in the environment.

Definition 8. *Given a total function σ with domain D and codomain C, we* ***overwrite*** *the function with case $x \mapsto v$ where $x \in D$ and $v \in C$ such that*

$$\sigma[x \mapsto v](z) \quad := \quad \textit{if } (x = z) \textit{ then } v \textit{ else } \sigma(z)$$

We use the map σ as part of the translation, the intuition is that for all bound variables x in the term we are translating, it should be that $\sigma(x) = x$. The purpose of the map γ is to keep track of the binding of phantom-variables.

Definition 9. *The interpretation $[\![-|-|-]\!] : \Lambda_a^S \times (V \to \Lambda) \times (V \to V) \to \Lambda$ is defined as*

$$[\![x\,|\,\sigma\,|\,\gamma]\!] = \sigma(x) \qquad [\![s\,t\,|\,\sigma\,|\,\gamma]\!] = [\![s\,|\,\sigma\,|\,\gamma]\!]\,[\![t\,|\,\sigma\,|\,\gamma]\!]$$

$$[\![c\langle\,c\,\rangle.t\,|\,\sigma\,|\,\gamma]\!] = \lambda c.[\![t\,|\,\sigma[c \mapsto c]\,|\,\gamma]\!]$$

$$[\![c\langle\,x_1,\ldots,x_n\,\rangle.t\,|\,\sigma\,|\,\gamma]\!] = \lambda c.[\![t\,|\,\sigma[x_i \mapsto \sigma(x_i)\{c/\gamma(c)\}]_{i\in[n]}\,|\,\gamma]\!]$$

$$[\![u[x_1,\ldots,x_n \leftarrow t]\,|\,\sigma\,|\,\gamma]\!] = [\![u\,|\,\sigma[x_i \mapsto [\![t\,|\,\sigma\,|\,\gamma]\!]]_{i\in[n]}\,|\,\gamma]\!]$$

$$[\![u[e_1\langle\,\vec{w}_1\,\rangle,\ldots,e_n\langle\,\vec{w}_n\,\rangle\,|\,c\langle\,c\,\rangle\,\overline{[\Gamma]}]\,|\,\sigma\,|\,\gamma]\!] = [\![u\overline{[\Gamma]}\,|\,\sigma\,|\,\gamma[e_i \mapsto c]_{i\in[n]}]\!]$$

$$[\![u[e_1\langle\,\vec{w}_1\,\rangle,\ldots,e_n\langle\,\vec{w}_n\,\rangle\,|\,c\langle\,x_1,\ldots,x_m\,\rangle\,\overline{[\Gamma]}]\,|\,\sigma\,|\,\gamma]\!] = [\![u\overline{[\Gamma]}\,|\,\sigma'\,|\,\gamma[e_i \mapsto c]_{i\in[n]}]\!]$$

where $\sigma' = \sigma[x_i \mapsto \sigma(x_i)\{c/\gamma(c)\}]_{i\in[n]}$

The following Proposition justifies working modulo permutation equivalence.

Proposition 1. *For $s,t \in \Lambda_a^S$, if $s \sim t$ then $[\![\,s\,]\!] = [\![\,t\,]\!]$.*

3.2 Rewrite Rules.

Both the spinal atomic λ-calculus and the atomic λ-calculus of [16] follow atomic reduction steps, i.e. they apply on individual constructors. The biggest difference is that our calculus is capable of duplicating not only the skeleton but also the spine. The rewrite rules in our calculus make use of 3 operations, *substitution, book-keeping*, and *exorcism*. The operation **substitution** $t\{s/x\}$ propagates through the term t, and replaces the free occurences of the variable x with the term s. Moreover, if x occurs in the cover of a phantom-variable $e\langle\,\vec{y}\cdot x\,\rangle$, then substitution replaces the x in the cover with $(s)_{fv}$, resulting in $e\langle\,\vec{y}\cdot(s)_{fv}\,\rangle$. Although substitution performs some book-keeping on phantom-abstractions, we define an explicit notion of **book-keeping** $\{\vec{y}/e\}_b$ that updates the variables stored in a free cover i.e. for a term t, $e\langle\,\vec{x}\,\rangle \in (t)_{fc}$ then $e\langle\,\vec{y}\,\rangle \in (t\{\vec{y}/e\}_b)_{fc}$. The last operation we introduce is called **exorcism** $\{c\langle\,\vec{x}\,\rangle\}_e$. We perform exorcisms on phantom-abstractions to convert them to abstractions. Intuitively, this will be performed on phantom-abstractions with phantom-variables bound to a distributor when said distributor is eliminated. It converts phantom-abstractions to abstractions by introducing a sharing of the phantom-variable that captures the variables in the cover, i.e. $(c\langle\,\vec{x}\,\rangle.t)\{c\langle\,\vec{x}\,\rangle\}_e = c\langle\,c\,\rangle.t[\vec{x} \leftarrow c]$.

Proposition 2. *The translation* $[\![u \,|\, \sigma \,|\, \gamma]\!]$ *commutes with substitutions, bookkeepings*[1], *and exorcisms*[2] *in the following way*

$$[\![u\{t/x\} \,|\, \sigma \,|\, \gamma]\!] = [\![u \,|\, \sigma[x \mapsto [\![t \,|\, \sigma \,|\, \gamma]\!]] \,|\, \gamma]\!]$$

$$[\![u\{\vec{x}/c\}_b \,|\, \sigma \,|\, \gamma]\!] = [\![u \,|\, \sigma \,|\, \gamma]\!]$$

$$[\![u\{c\langle x_1, \ldots, x_n \rangle\}_e \,|\, \sigma \,|\, \gamma]\!] = [\![u \,|\, \sigma[x_i \mapsto c]_{i \in [n]} \,|\, \gamma]\!]$$

(1) Given $c\langle \vec{y} \rangle \in (u)_{fc}$ *where* $\vec{x} \subseteq \vec{y}$ *and for* $z \in \vec{y}/\vec{x}$, $\gamma(c) \notin (\sigma(z))_{fv}$
(2) Given $c\langle \vec{x} \rangle \in (u)_{fc}$ *or* $\{\vec{x}\} \cap (u)_{fv} = \{\}$

Proof. See [25], proof of Proposition 18, 19, 20, 21.

Using these operations, we define the rewrite rules that allow for spinal duplication. Firstly we have beta reduction ($\rightsquigarrow_\beta$), which strictly requires an abstraction (not a phantom).

$$(x\langle x \rangle.t)\, s \rightsquigarrow_\beta t\{s/x\}$$

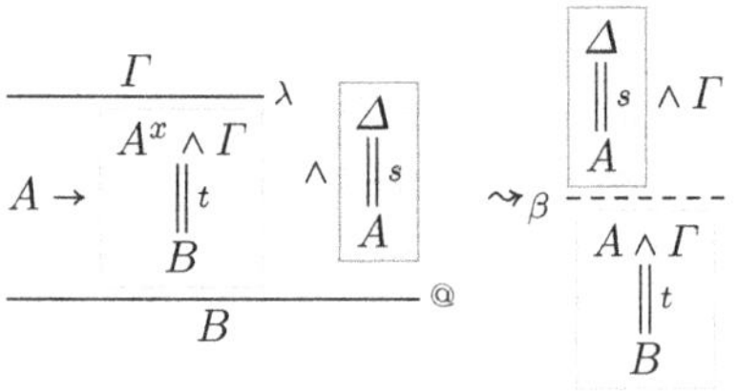

(β)

Here β-reduction is a linear operation, since the bound variable x occurs exactly once in the body t. Any duplication of the term t in the atomic λ-calculus proceeds via the sharing reductions.

The first set of sharing reduction rules move closures towards the outside of a term. Most of these rewrite rules only change the typing derivations in the way that subderivations are composed, with the exception of moving a closure out of scope of a distributor.

$$s[\Gamma]\, t \rightsquigarrow_L (s\, t)[\Gamma] \qquad (l_1)$$
$$s\, t[\Gamma] \rightsquigarrow_L (s\, t)[\Gamma] \qquad (l_2)$$
$$d\langle \vec{x} \rangle.t[\Gamma] \rightsquigarrow_L (d\langle \vec{x} \rangle.t)[\Gamma] \text{ if } \{\vec{x}\} \cap (t)_{fv} = \{\vec{x}\} \qquad (l_3)$$
$$u[\vec{x} \leftarrow t[\Gamma]] \rightsquigarrow_L u[\vec{x} \leftarrow t][\Gamma] \qquad (l_4)$$

For the case of lifting a closure outside a distributor, we use a notation $\parallel [\Gamma] \parallel$ to identify the variables captured by a closure, i.e.$\parallel [\vec{x} \leftarrow t] \parallel = \{\vec{x}\}$ and $\parallel [e_1\langle \vec{x_1} \rangle, \ldots, e_n\langle \vec{x_x} \rangle \,|\, c\langle c \rangle \overline{[\Gamma]}] \parallel = \{\vec{x_1}, \ldots, \vec{x_n}\}$. Then let $\{\vec{z}\} = \parallel [\Gamma] \parallel$ in the following rewrite rule, where we remove $\vec{z}$ from the covers, that can only occur if $\{\vec{x}\} \cap ([\Gamma])_{fv} = \{\}$.

$$\begin{aligned} & u[e_1\langle \vec{w_1} \rangle \ldots e_n\langle \vec{w_n} \rangle \,|\, c\langle \vec{x} \rangle \overline{[\Gamma]}[\Gamma]] \\ & \rightsquigarrow_L u\{(\vec{w_i} \setminus \vec{z})/e_i\}_{b_{i \in [n]}}[e_1\langle \vec{w_1} \setminus \vec{z} \rangle \ldots e_n\langle \vec{w_n} \setminus \vec{z} \rangle \,|\, c\langle \vec{x} \rangle \overline{[\Gamma]}][\Gamma] \end{aligned} \qquad (l_5)$$

The graphical version of this rule is shown in Figure 4c, where we remove the edge only if there is no edge between t and the unsharing node. The proof rewrite rule corresponding with the rewrite rule l_5 can be broken down into two parts. The first part is readjusting how the derivations compose as shown below.

$$\cfrac{\cfrac{\cfrac{(C \to \Gamma) \wedge \Delta \wedge \Omega}{\boxed{\Gamma \wedge \Delta \wedge \boxed{\begin{array}{c} \Omega \\ \| \\ A \wedge \cdots \wedge A \end{array}}}}\ s}{C \to \begin{array}{c} \| \\ \Sigma_1 \dots \Sigma_n \end{array}}}{(C \to \Sigma_1) \wedge \cdots \wedge (C \to \Sigma_n)}\ d \quad \rightsquigarrow_L \quad \cfrac{\cfrac{\cfrac{(C \to \Gamma) \wedge \Delta \wedge \boxed{\begin{array}{c} \Omega \\ \| \\ A \wedge \cdots \wedge A \end{array}}}{\Gamma \wedge \Delta \wedge A \dots A}\ s}{C \to \begin{array}{c} \| \\ \Sigma_1 \dots \Sigma_n \end{array}}}{(C \to \Sigma_1) \wedge \cdots \wedge (C \to \Sigma_n)}\ d$$

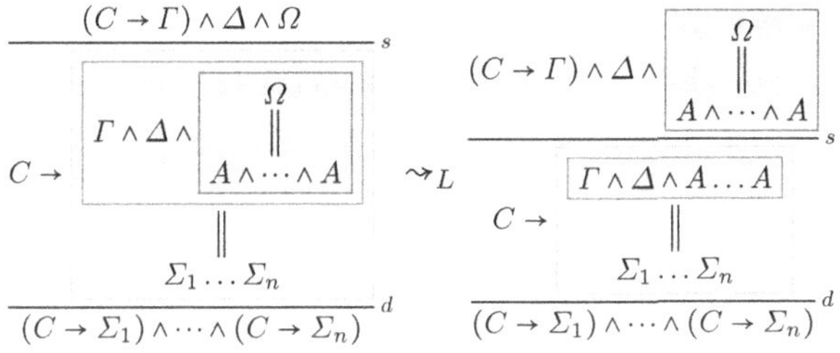

The second part of the rewrite rule justifies the need for the book-keeping operation. In the rewrite below, let A be the type of a variable z where $z \in \bar{z}$. After lifting, we want to remove the variable from the cover as to ensure correctness since the variables in the cover denote the variables captured by the environment. Book-keeping allows us to remove these variables simultaneously.

$$\cfrac{\cfrac{\cfrac{(C \to \Gamma) \wedge \Delta \wedge A}{\boxed{\begin{array}{c} \Gamma \wedge \Delta \\ \| \\ \Sigma_1 \wedge \cdots \wedge \Sigma_n \end{array}} \wedge A}\ s}{C \to \begin{array}{c} \\ \Sigma_1 \wedge \cdots \wedge \Sigma_i \wedge A \wedge \cdots \wedge \Sigma_n \end{array}}}{\cdots \wedge (C \to \Sigma_i \wedge A) \wedge \cdots}\ d \quad \rightsquigarrow \quad \cfrac{\cfrac{\cfrac{(C \to \Gamma) \wedge \Delta}{\boxed{\begin{array}{c} \Gamma \wedge \Delta \\ \| \\ \Sigma_1 \wedge \cdots \wedge \Sigma_n \end{array}}}\ s \wedge A}{C \to \begin{array}{c} \\ \Sigma_1 \wedge \cdots \wedge \Sigma_i \wedge \cdots \wedge \Sigma_n \end{array}}}{\cdots \wedge (C \to \Sigma_i) \wedge \cdots}{\cdots \wedge \boxed{\cfrac{(C \to \Sigma_i) \wedge A}{C \to \Sigma_i \wedge A}\ s} \wedge \cdots}\ d$$

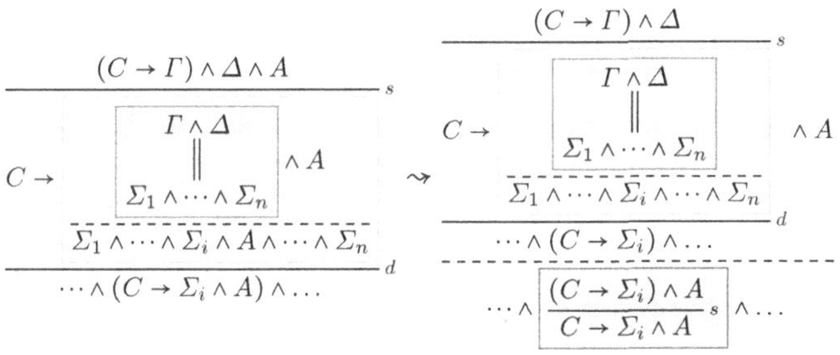

The lifting rules (l_i) are justified by the need to lift closures out of the distributor, as opposed to duplicating them. The second set of rewrite rules, consecutive sharings are compounded and unary sharings are applied as substitutions. For simplicity, in the equivalent proof rewrite step we only show the binary case.

$$u[\bar{w} \leftarrow y][y \cdot \bar{y} \leftarrow t] \rightsquigarrow_C u[\bar{w} \cdot \bar{y} \leftarrow t] \tag{c_1}$$

$$u[x \leftarrow t] \rightsquigarrow_C u\{t/x\} \tag{c_2}$$

$$\cfrac{\cfrac{A}{A \wedge \cfrac{A}{A \wedge A}\ \Delta}}{}\ \Delta \quad \rightsquigarrow_C \quad \cfrac{\cfrac{A}{A \wedge A \wedge A}}{}\ \Delta \qquad \cfrac{A}{A}\ \Delta \quad \rightsquigarrow_C \quad A$$

The atomic steps for duplicating are given in the third and final set of rewrite rules. The first being the atomic duplication step of an application, which is the same rule used in [16]. The binary case proof rewrite steps for each rule are also provided. There are also shown graphically in (respectively) Figure 4b (where we maintain links between sharings and unsharings), Figure 4a, and Figure 4d (where the unsharing node is linked to exactly one connecting sharing node).

$$u[x_1 \dots x_n \leftarrow s t] \rightsquigarrow_D u\{z_1 y_1/x_1\} \dots \{z_n y_n/x_n\}[z_1 \dots z_n \leftarrow s][y_1 \dots y_n \leftarrow t] \tag{d_1}$$

$$\cfrac{\cfrac{(A \to B) \wedge A}{B}\,@}{B \wedge B}\,\triangle \quad \rightsquigarrow_D \quad \begin{array}{c} \cfrac{(A \to B)}{(A \to B) \wedge (A \to B)}\,\triangle \wedge \cfrac{B}{B \wedge B}\,\triangle \\ \hdashline \cfrac{(A \to B) \wedge A}{B}\,@ \wedge \cfrac{(A \to B) \wedge A}{B}\,@ \end{array}$$

$$\begin{aligned} & u[x_1, \dots, x_n \leftarrow c\langle \vec{y} \rangle.t] \rightsquigarrow_D \\ & \quad u\{e_i\langle w_i \rangle.w_i/x_i\}_{i \in [n]}[e_1\langle w_1 \rangle \dots e_n\langle w_n \rangle \mid c\langle \vec{y} \rangle [w_1, \dots, w_n \leftarrow t]] \end{aligned} \tag{d_2}$$

$$\cfrac{\cfrac{(A \to B) \wedge \Gamma}{A \to \boxed{\begin{array}{c} B \wedge \Gamma \\ \| \\ C \end{array}}}\,s}{(A \to C) \wedge (A \to C)}\,\triangle \quad \rightsquigarrow_D \quad \cfrac{\cfrac{(A \to B) \wedge \Gamma}{A \to \boxed{\begin{array}{c} B \wedge \Gamma \\ \| \\ \cfrac{C}{C \wedge C}\,\triangle \end{array}}}\,s}{(A \to C) \wedge (A \to C)}\,d$$

$$u[e_1\langle \vec{w_1} \rangle \dots e_n\langle \vec{w_n} \rangle \mid c\langle c \rangle [\vec{w_1}, \dots, \vec{w_n} \leftarrow c]] \rightsquigarrow_D u\{e_1\langle \vec{w_1} \rangle\}_e \dots \{e_n\langle \vec{w_n} \rangle\}_e \tag{d_3}$$

$$\cfrac{\cfrac{}{A \to \cfrac{A}{A \wedge A}\,\triangle}\,a}{(A \to A) \wedge (A \to A)}\,d \quad \rightsquigarrow_D \quad \cfrac{}{A \to A}\,a \wedge \cfrac{}{A \to A}\,a$$

Example 2. The following example, illustrated in Figure 4e, is a reduction in the term calculus where we duplicate the spine of the term $[a_1, a_2 \leftarrow \underline{\lambda x.\lambda y.(}(\lambda z.z)y\underline{)x}]$.

$$\begin{aligned} & \rightsquigarrow_D \{\underline{x_1}\langle b_1\rangle.b_1/a_1\}\{\underline{x_2}\langle b_2\rangle.b_2/a_2\}[x_1\langle b_1\rangle, x_2\langle b_2\rangle \mid x\langle x\rangle[b_1, b_2 \leftarrow \underline{\lambda y.(}(\lambda z.z)y\underline{)x}]] \\ & \rightsquigarrow_D \{\underline{x_1}\langle c_1\rangle.\underline{y_1}\langle c_1\rangle c_1/a_1\}\{\underline{x_2}\langle c_2\rangle.\underline{y_2}\langle c_2\rangle.c_2/a_2\} \\ & \qquad [x_1\langle c_1\rangle, x_2\langle c_2\rangle \mid x\langle x\rangle[y_1\langle c_1\rangle, y_2\langle c_2\rangle \mid y\langle y\rangle[c_1, c_2 \leftarrow \underline{(}(\lambda z.z)y\underline{)x}]]] \\ & \rightsquigarrow_D \{\underline{x_1}\langle d_1, e_1\rangle.\underline{y_1}\langle d_1, e_1\rangle d_1 e_1/a_1\}\{\underline{x_2}\langle d_2, e_2\rangle.\underline{y_2}\langle d_2, e_2\rangle.d_2 e_2/a_2\} \\ & \qquad [x_1\langle d_1, e_1\rangle, x_2\langle d_2, e_2\rangle \mid x\langle x\rangle[y_1\langle d_1, e_1\rangle, y_2\langle d_2, e_2\rangle \mid y\langle y\rangle[d_1, d_2 \leftarrow (\lambda z.z)y][e_1, e_2 \leftarrow \underline{x}]]] \\ & \rightsquigarrow_L \{\underline{x_1}\langle d_1, e_1\rangle.\underline{y_1}\langle d_1\rangle d_1 e_1/a_1\}\{\underline{x_2}\langle d_2, e_2\rangle.\underline{y_2}\langle d_2\rangle.d_2 e_2/a_2\} \\ & \qquad [x_1\langle d_1, e_1\rangle, x_2\langle d_2, e_2\rangle \mid x\langle x\rangle[y_1\langle d_1\rangle, y_2\langle d_1\rangle \mid y\langle y\rangle[d_1, d_2 \leftarrow (\lambda z.z)y]]\,[e_1, e_2 \leftarrow \underline{x}]] \\ & \rightsquigarrow_L \{\underline{x_1}\langle e_1\rangle.\underline{y_1}\langle d_1\rangle d_1 e_1/a_1\}\{\underline{x_2}\langle e_2\rangle.\underline{y_2}\langle d_2\rangle.d_2 e_2/a_2\} \\ & \qquad [x_1\langle e_1\rangle, x_2\langle e_2\rangle \mid x\langle x\rangle[e_1, e_2 \leftarrow \underline{x}]]\ [y_1\langle d_1\rangle, y_2\langle d_2\rangle \mid y\langle y\rangle[d_1, d_2(\lambda z.z)y]] \\ & \rightsquigarrow_D \{\underline{\lambda x_1}.\underline{y_1}\langle d_1\rangle d_1 \underline{x_1}/a_1\}\{\underline{\lambda x_2}.\underline{y_2}\langle d_2\rangle.d_2 \underline{x_2}/a_2\}\ [y_1\langle d_1\rangle, y_2\langle d_2\rangle \mid y\langle y\rangle[d_1, d_2 \leftarrow (\lambda z.z)y]] \end{aligned}$$

Reduction ($\rightsquigarrow_{(L,C,D,\beta)}$) preserves the conclusion of the derivation, and thus the following proposition is easy to observe.

Proposition 3. *If $s \rightsquigarrow_{(L,C,D,\beta)} t$ and $s : A$, then $t : A$.*

Definition 10. *For a term $t \in \Lambda_a^S$, if there does not exists a term $s \in \Lambda_a^S$ such that $t \rightsquigarrow_{(L,C,D)} s$ then it is said that t is in* ***sharing normal form****.*

The following Lemma not only proves we have good translations in Section 3.1, and shows duplication preserves denotation.

Lemma 1. *For a* $t \in \Lambda_a^S$ *in sharing normal form and a* $N \in \Lambda$.

$$[\![(\!| N |\!)]\!] = N \qquad (\!| [\![t]\!] |\!) = t \qquad \exists_{M \in \Lambda}. t = (\!| M |\!)$$

Otherwise if $s \rightsquigarrow_{(L,D,C)} t$ *then* $[\![s \,|\, \sigma \,|\, \gamma]\!] = [\![t \,|\, \sigma \,|\, \gamma]\!]$.

Proof. See [25, Lemma 24, Lemma 25].

Lemma 2. *Given a term* $t \in \Lambda_a^S$, *then* $(\!| [\![t]\!] |\!)$ *is* t *in sharing normal form.*

Proof. We can prove this by induction on the longest sharing reduction path from t. Our base case is already covered by Lemma 1. We are then interested in the inductive case, where t is not in sharing normal form. By Lemma 1, $[\![t]\!] = [\![t']\!]$ where $t \rightsquigarrow_{(D,L,C)} t'$. By induction hypothesis, $(\!| [\![t']\!] |\!)$ is in sharing normal form. Hence $(\!| [\![t]\!] |\!)$ is in sharing normal form. □

4 Strong Normalisation of Sharing Reductions

In order to show our calculus is strongly normalising, we first show that the sharing reduction rules are strongly normalising. We indite a measure on terms and show that this measure strictly decreases as sharing reduction progresses. Similar ideas and results can be found elsewhere: with *memory* in [20], the λ-*I calculus* in [6], the λ-*void calculus* [2], and the weakening $\lambda\mu$-calculus [17]. Our measure will consist of three components. First, the **height** of a term is a multiset of integers, that measures the number of constructors from each sharing node to the root of the term in its graphical notation. The height is defined on terms as $\mathcal{H}^i(-)$, where i is an integer. We say $\mathcal{H}(t)$ for $\mathcal{H}^1(t)$. We use $\uplus$ to denote the disjoint union of two multisets. We denote $\mathcal{H}^i([\Gamma_1]) \uplus \cdots \uplus \mathcal{H}^i([\Gamma_n])$ as $\mathcal{H}^i(\overline{[\Gamma]})$ for the environment $\overline{[\Gamma]} = [\Gamma_1], \ldots, [\Gamma_n]$.

Definition 11 (Sharing Height). *The sharing height* $\mathcal{H}^i(t)$ *of a term* t *is given below, where* n *is the number of closures in* $\overline{[\Gamma]}$*:*

$$\mathcal{H}^i(x) = \{\} \qquad \mathcal{H}^i(s\,t) = \mathcal{H}^{i+1}(s) \uplus \mathcal{H}^{i+1}(t)$$
$$\mathcal{H}^i(c\langle \vec{x} \rangle.t) = \mathcal{H}^{i+1}(t) \qquad \mathcal{H}^i(t[\Gamma]) = \mathcal{H}^i(t) \uplus \mathcal{H}^i([\Gamma]) \uplus \{i^1\}$$
$$\mathcal{H}^i([x_1, \ldots, x_n \leftarrow t]) = \mathcal{H}^{i+1}(t) \quad \mathcal{H}^i([\vec{w} \,|\, c\langle \vec{x} \rangle \overline{[\Gamma]}]) = \mathcal{H}^{i+1}(\overline{[\Gamma]}) \uplus \{(i+1)^n\}$$

This measure then strictly decreases for the rewrite rules l_1, l_2, l_3, l_4 and l_5, i.e. if $t \rightsquigarrow_L u$ then $\mathcal{H}^i(t) > \mathcal{H}^i(u)$. The second measure we consider is the **weight** of a term. Intuitively this quantifies the remaining duplications, which are performed with $\rightsquigarrow_D$ reductions. If a term would be deleted, we assign it with a weight '1' to express that it is not duplicated. Calculating the weight requires an auxiliary function that assigns integer weights to the variables of a term. This function is defined on terms $\mathcal{V}^i(-)$, where i is an integer. To measure variables independently of binders is vital. It allows to measure distributors, which duplicate λ's but not the bound variable. Also, only bound variables for abstractions are measured since variables bound by sharings are substituted in the interpretation.

Definition 12 (Variable Weights). *The function $\mathcal{V}^i(t)$ returns a function that assigns integer weights to the free variables of t. It is defined by the below, where $f = \mathcal{V}^i(t)$ and $g = f(x_1) + \cdots + f(x_n)$ for each $x_i \in \vec{x}$.*

$$\begin{aligned}
\mathcal{V}^i(x) &= \{x \mapsto i\} & \mathcal{V}^i(s\,t) &= \mathcal{V}^i(s) \uplus \mathcal{V}^i(t)\\
\mathcal{V}^i(c\langle c\rangle.t) &= \mathcal{V}^i(t)/\{c\} & \mathcal{V}^i(c\langle \vec{x}\rangle.t) &= \mathcal{V}^i(t) \uplus \{c \mapsto i\}\\
\mathcal{V}^i(t[\vec{x} \leftarrow s]) &= \mathcal{V}^i(t)/\{\vec{x}\} \uplus \mathcal{V}^g(s) & \mathcal{V}^i(t[\leftarrow s]) &= \mathcal{V}^i(t) \uplus \mathcal{V}^1(s)
\end{aligned}$$

$$\begin{aligned}
\mathcal{V}^i(t[e_1\langle \vec{w_1}\rangle \ldots e_n\langle \vec{w_n}\rangle \,|\, c\langle c\rangle\,\overline{[\Gamma]}]) &= \mathcal{V}^i(t\overline{[\Gamma]})/\{c, e_1, \ldots, e_n\}\\
\mathcal{V}^i(t[e_1\langle \vec{w_1}\rangle \ldots e_n\langle \vec{w_n}\rangle \,|\, c\langle \vec{x}\rangle\,\overline{[\Gamma]}]) &= \mathcal{V}^i(t\overline{[\Gamma]})/\{e_1, \ldots, e_n\} \uplus \{c \mapsto i\}
\end{aligned}$$

The weight of a term can then be defined via the use of this auxiliary function. The auxiliary function is used when calculating the weight of a sharing, where the sharing weight of the variables bound by the sharing play a significant role in calculating the weight of the shared term. In the case of a weakening $[\leftarrow t]$, we assign an initial weight of 1. Again we say $\mathcal{W}(t) = \mathcal{W}^1(t)$.

Definition 13 (Sharing Weight). *The sharing weight $\mathcal{W}^i(t)$ of a term t is a multiset of integers computed by the function defined below, where $f = \mathcal{V}^i(t)$ and $g = f(x_1) + \cdots + f(x_n)$ for each $x_i \in \vec{x}$.*

$$\begin{aligned}
\mathcal{W}^i(x) &= \{\} & \mathcal{W}^i(s\,t) &= \mathcal{W}^i(s) \uplus \mathcal{W}^i(t) \uplus \{i\}\\
\mathcal{W}^i(c\langle c\rangle.t) &= \mathcal{W}^i(t) \uplus \{i\} \uplus \{\mathcal{V}^i(t)(c)\} & \mathcal{W}^i(c\langle \vec{x}\rangle.t) &= \mathcal{W}^i(t) \uplus \{i\}\\
\mathcal{W}^i(t[\vec{x} \leftarrow s]) &= \mathcal{W}^i(t) \uplus \mathcal{W}^g(s) & \mathcal{W}^i(t[\leftarrow s]) &= \mathcal{W}^i(t) \uplus \mathcal{W}^1(s)
\end{aligned}$$

$$\begin{aligned}
\mathcal{W}^i(t[e_1\langle \vec{w_1}\rangle \ldots e_n\langle \vec{w_n}\rangle \,|\, c\langle c\rangle\,\overline{[\Gamma]}]) &= \mathcal{W}^i(t\overline{[\Gamma]}) \uplus \{\mathcal{V}^i(t\overline{[\Gamma]})(c)\}\\
\mathcal{W}^i(t[e_1\langle \vec{w_1}\rangle \ldots e_n\langle \vec{w_n}\rangle \,|\, c\langle \vec{x}\rangle\,\overline{[\Gamma]}]) &= \mathcal{W}^i(t\overline{[\Gamma]})
\end{aligned}$$

This measure then strictly decreases on the rewrite rules d_1, d_2, d_3 and is unaffected by all the other sharing reduction rules, i.e. if $t \rightsquigarrow_D u$ then $\mathcal{W}^i(t) > \mathcal{W}^i(u)$. If $t \rightsquigarrow_{(L,C)} u$ then $\mathcal{W}^i(t) = \mathcal{W}^i(u)$. The third and last measure we consider is the **number of closures** in the term, where it can be easily observed that the rewrite rules c_1 and c_2 strictly decrease this measure, and that the $\rightsquigarrow_L$ rules do not alter the number of closures. We then use this along with height and weight to define a *sharing measure* on terms.

Definition 14. *The* **sharing measure** *of a Λ^S_a-term t is a triple $(\mathcal{W}(t), \mathcal{C}, \mathcal{H}(t))$, where $\mathcal{C}$ is the number of closures in the term t. We compare sharing measures by using the lexicographical preferences according to $\mathcal{W} > \mathcal{C} > \mathcal{H}$.*

Theorem 1. *Sharing reduction $\rightsquigarrow_{(D,L,C)}$ is strongly normalising.*

Now that we have proven the sharing reductions are strongly normalising, we can prove that they are confluent for closed terms.

Theorem 2. *The sharing reduction relation $\rightsquigarrow_{(D,L,C)}$ is confluent.*

Proof. Lemma 1 tells us that the preservation is preserved under reduction i.e. for $s \rightsquigarrow_{(D,L,C)} t$, $[\![s]\!] = [\![t]\!]$. Therefore given $t \rightsquigarrow^*_{(D,L,C)} s_1$ and $t \rightsquigarrow^*_{(D,L,C)} s_2$, $[\![t]\!] = [\![s_1]\!] = [\![s_2]\!]$. Since we know that sharing reductions are strongly normalising, we know there exists terms u_1 and u_2 in sharing normal form such that $s_1 \rightsquigarrow^*_{(D,L,C)} u_1$ and $s_2 \rightsquigarrow^*_{(D,L,C)} u_2$. Lemma 1 tells us that terms in sharing normal form are in correspondence with their denotations i.e. $(\!| [\![t]\!] |\!) = t$. Since by Lemma 1 we know $[\![u_1]\!] = [\![s_1]\!] = [\![s_2]\!] = [\![u_2]\!]$, and by Lemma 1 $(\!| [\![u_1]\!] |\!) = u_1$ and $(\!| [\![u_2]\!] |\!) = u_2$, we can conclude $u_1 = u_2$. Hence, we prove confluence. □

5 Preservation of Strong Normalisation and Confluence

A β-step in our calculus may occur within a weakening, and therefore is simulated by zero β-steps in the λ-calculus. Therefore if there is an infinite reduction path located inside a weakening in Λ_a^S, then the reduction path is not preserved in the corresponding λ-term as there are no weakenings. To deal with this, just as done in [2, 16, 17], we make use of the **weakening calculus**. A β-step is non-deleting precisely because of the weakening construct. If a β-step would be deleting, then the weakening calculus would instead keep the deleted term around as 'garbage', which can continue to reduce unless explicitly 'garbage-collected' by extra (non-β) reduction steps. PSN has already be shown for the weakening calculus through the use of a perpetual strategy in [16]. A part of proving PSN is then using the weakening calculus to prove that if $t \in \Lambda_a^S$ has a infinite reduction path, then its translation into the weakening calculus also has an infinite reduction path.

Definition 15. *The* w*-terms of the weakening calculus ($\Lambda_\mathcal{W}$) are*

$$T, U, V \quad ::= \quad x \quad | \quad \lambda x.T^* \quad | \quad U\,V \quad | \quad T[\leftarrow U] \quad | \quad \bullet \text{ (*) where } x \in (T)_{fv}$$

The terms are variable, abstraction, application, weakening, and a bullet. In the weakening $T[\leftarrow U]$, the subterm U is *weakened.* The interpretation of atomic terms to weakening terms $[\![- | - | -]\!]_\mathcal{W}$ can be seen as an extension of the translation into the λ-calculus (Definition 9).

Definition 16. *The interpretation* $[\![- | - | -]\!]_\mathcal{W} : \Lambda_a^S \times (V \to \Lambda_\mathcal{W}) \times (V \to V) \to \Lambda_\mathcal{W}$ *with maps* $\sigma : V \to \Lambda_\mathcal{W}$ *and* $\gamma : V \to V$ *is defined as an extension of the translation in (Definition 9) with the following additional special cases.*

$$\begin{aligned} [\![u[\leftarrow t] | \sigma | \gamma]\!]_\mathcal{W} &= [\![u | \sigma | \gamma]\!]_\mathcal{W}[\leftarrow [\![t | \sigma | \gamma]\!]_\mathcal{W}] \\ [\![u[\,|\,c\langle c \rangle \overline{[\Gamma]}] | \sigma | \gamma]\!]_\mathcal{W} &= [\![u\overline{[\Gamma]} | \sigma[c \mapsto \bullet] | \gamma]\!]_\mathcal{W} \\ [\![u[\,|\,c\langle x_1, \ldots, x_n \rangle \overline{[\Gamma]}] | \sigma | \gamma]\!]_\mathcal{W} &= [\![u\overline{[\Gamma]} | \sigma' | \gamma]\!]_\mathcal{W} \end{aligned}$$

where $\sigma'(z) :=$ *if* $z \in \{x_1, \ldots, x_n\}$ *then* $\sigma(z)\{\bullet/\gamma(c)\}$ *else* $\sigma(z)$

We say $[\![t]\!]^\mathcal{W} = [\![t | I | I]\!]_\mathcal{W}$ where I is the identity function. We also have translations of the weakening calculus to and from the λ-calculus. Both of these translations were provided in [16]. The interpretation $\lfloor - \rfloor$ from weakening terms to λ-terms discards all weakenings.

Definition 17. *The interpretation* $M \in \Lambda$, $(\!| - |\!)^{\mathcal{W}} : \Lambda \to \Lambda_{\mathcal{W}}$ *is defined below.*

$$(\!| x |\!)^{\mathcal{W}} = x \quad (\!| M\,N |\!)^{\mathcal{W}} = (\!| M |\!)^{\mathcal{W}}\,(\!| N |\!)^{\mathcal{W}} \quad (\!| \lambda x.N |\!)^{\mathcal{W}} = \begin{cases} \lambda x.(\!| N |\!)^{\mathcal{W}} & \text{if } x \in (N)_{fv} \\ \lambda x.(\!| N |\!)^{\mathcal{W}}[\leftarrow x] & \text{otherwise} \end{cases}$$

The following equalities can be observed, where $\sigma^{\Lambda}(z) = \lfloor \sigma^{\mathcal{W}}(z) \rfloor$.

Proposition 4. *For* $N \in \Lambda$ *and* $t \in \Lambda_a^S$ *the following properties hold*

$$\lfloor [\![t \mid \sigma^{\mathcal{W}} \mid \gamma]\!]_{\mathcal{W}} \rfloor = [\![t \mid \sigma^{\Lambda} \mid \gamma]\!] \qquad [\![(\!| N |\!)]\!]^{\mathcal{W}} = (\!| N |\!)^{\mathcal{W}} \qquad \lfloor (\!| N |\!)^{\mathcal{W}} \rfloor = N$$

where for each $\{x \mapsto M\} \in \sigma^{\mathcal{W}}$, $\{x \mapsto \lfloor M \rfloor\} \in \sigma^{\Lambda}$.

Definition 18. *In the weakening calculus,* β*-reduction is defined as follows, where* $\overline{[\Gamma]}$ *are weakening constructs.* $((\lambda x.T)\overline{[\Gamma]})\,U \to_{\beta} T\{U/x\}\overline{[\Gamma]}$

Proposition 5. *If* $N \in \Lambda$ *is strongly normalising, then so is* $(\!| N |\!)^{\mathcal{W}}$.

When translating from Λ_a^S to $\Lambda_{\mathcal{W}}$, weakenings are maintained whilst sharings are interpreted via substitution. Thus the reduction rules in the weakening calculus cover the spinal reductions for nullary distributors and weakenings.

Definition 19. *Weakening reduction* ($\to_{\mathcal{W}}$) *proceeds as follows.*

$$\begin{array}{rlrl} U[\leftarrow T]\,V & \to_{\mathcal{W}} (U\,V)[\leftarrow T] & U\,V[\leftarrow T] & \to_{\mathcal{W}} (U\,V)[\leftarrow T] \\ T[\leftarrow U[\leftarrow V]] & \to_{\mathcal{W}} T[\leftarrow U][\leftarrow V] & T[\leftarrow \lambda x.U] & \to_{\mathcal{W}} T[\leftarrow U\{\bullet/x\}] \\ T[\leftarrow U\,V] & \to_{\mathcal{W}} T[\leftarrow U][\leftarrow V] & T[\leftarrow \bullet] & \to_{\mathcal{W}} T \\ T[\leftarrow U] & \to_{\mathcal{W}} T^{(1)} & \lambda x.T[\leftarrow U] & \to_{\mathcal{W}} (\lambda x.T)[\leftarrow U]^{(2)} \end{array}$$

(1) if U *is a subterm of* T *and (2) if* $x \notin (U)_{fv}$

It is easy to see that these rules correspond to special cases of the sharing reduction rules for Λ_a^S. This resemblance is confirmed by the following Lemma, proven in [25, pp. 82-86]. We use this to show how Λ_a^S enjoys PSN.

Lemma 3. *If* $t \rightsquigarrow_{\beta} u$ *then* $[\![t]\!]^{\mathcal{W}} \to_{\beta}^{+} [\![u]\!]^{\mathcal{W}}$. *If* $t \rightsquigarrow_{(C,D,L)} u$ *and for any* $x \in (t)_{bv} \cup (t)_{fp}$ *such that for all* z, $x \notin (\sigma(z))_{fv}$.

$$[\![t \mid \sigma \mid \gamma]\!]_{\mathcal{W}} \to_{\mathcal{W}}^{*} [\![u \mid \sigma \mid \gamma]\!]_{\mathcal{W}}$$

Lemma 4. *For* $t \in \Lambda_a^S$ *has an infinite reduction path, then* $[\![t]\!]^{\mathcal{W}}$ *also has an infinite reduction path.*

Proof. Due to Theorem 2, we know that the infinite reduction path contains infinite β-steps. This means in the reduction sequence, between each β-step, there are finite many $\rightsquigarrow_{(D,L,C)}$ reduction steps. Lemma 3 says each $\rightsquigarrow_{(D,L,C)}$ step in Λ_a^S corresponds to zero or more weakening reductions ($\rightsquigarrow_{\mathcal{W}}^{*}$). Lemma 3 says that each beta step in Λ_a^S corresponds to one or more β-steps in $\Lambda_{\mathcal{W}}$. Therefore, it must be that $[\![t]\!]^{\mathcal{W}}$ also has an infinite reduction path. □

Theorem 3. *If $N \in \Lambda$ is strongly normalising, then so is $(\!| N |\!)$.*

Proof. For a given $N \in \Lambda$ that is strongly normalising, we know by Lemma 5 that $(\!| N |\!)^{\mathcal{W}}$ is strongly normalising. Then $[\![(\!| N |\!)]\!]^{\mathcal{W}}$ is strongly normalising, since Proposition 4 states that $(\!| N |\!)^{\mathcal{W}} = [\![(\!| N |\!)]\!]^{\mathcal{W}}$. Then by Lemma 4, which states that if $[\![t]\!]^{\mathcal{W}}$ is strongly normalising, then t is strongly normalising, proves that $(\!| N |\!)$ is strongly normalising. □

We also prove confluence, which is already known for the λ-calculus [11]. We first observe that a β-step in the λ-calculus is simulated in Λ_a^S by one β-step followed by zero or more sharing reductions.

Lemma 5. *Given $N, M \in \Lambda$. If $N \rightsquigarrow_\beta M$, then $(\!| N |\!) \rightsquigarrow_\beta \rightsquigarrow^*_{(D,L,C)} (\!| M |\!)$.*

Proof. This is proven by Sherratt in [25, Lemma 67].

Theorem 4. *Given $t, s_1, s_2 \in \Lambda_a^S$. If $t \rightsquigarrow^*_{(\beta,D,L,C)} s_1$ and $t \rightsquigarrow^*_{(\beta,D,L,C)} s_2$, there exists a $u \in \Lambda_a^S$ such that $s_1 \rightsquigarrow^*_{(\beta,D,L,C)} u$ and $s_2 \rightsquigarrow^*_{(\beta,D,L,C)} u$.*

Proof. Suppose $t \rightsquigarrow^*_{(\beta,D,L,C)} s_1$ and $t \rightsquigarrow^*_{(\beta,D,L,C)} s_2$. Then we have $[\![t]\!] \rightsquigarrow^*_\beta [\![s_1]\!]$ and $[\![t]\!] \rightsquigarrow^*_\beta [\![s_2]\!]$. By the Church-Rosser theorem, there exists a $M \in \Lambda$ such that $[\![s_1]\!] \rightsquigarrow^*_\beta M$ and $[\![s_2]\!] \rightsquigarrow^*_\beta M$. Due to Lemma 2, $(\!| [\![s_1]\!] |\!) = s_1'$ and $(\!| [\![s_2]\!] |\!) = s_2'$ where $s_1', s_2' \in \Lambda_a^S$ in sharing normal form. Then thanks to Lemma 5 we know $s_1' \rightsquigarrow^*_{(\beta,D,L,C)} (\!| M |\!)$ and $s_2' \rightsquigarrow^*_{(\beta,D,L,C)} (\!| M |\!)$. Combined, we get confluence. □

6 Conclusion, related work, and future directions

We have studied the interaction between the switch and the medial rule, the two characteristic inference rules of deep inference. We built a Curry–Howard interpretation based on this interaction, whose resulting calculus not only has the ability to duplicate terms atomically but can also duplicate solely the spine of an abstraction such that beta reduction can proceed on the duplicates. We show that this calculus has natural properties with respect to the λ-calculus.

This work, which started as an investigation into the Curry-Howard correspondence of the switch rule [25], fits into a broader effort to give a computational interpretation to intuitionistic deep-inference proof theory. Brünnler and McKinley [9] give a natural reduction mechanism without medial (or switch), and observe that preservation of strong normalization fails. Guenot and Straßburger [14] investigate a different switch rule, corresponding to the implication-left rule of sequent calculus. He [17] extends the atomic λ-calculus to the $\lambda\mu$-calculus.

Our future goal is to develop the intuitionistic open deduction formalism towards optimal reduction [23, 21, 3], via the remaining medial and switch rules [26].

Acknowledgements We thank the anonymous reviewers for their comments.

References

1. Abadi, M., Cardelli, L., Curien, P.L., Lévy, J.J.: Explicit substitutions. Journal of Functional Programming **1**(4), 375–416 (1991)
2. Accattoli, B., Kesner, D.: Preservation of strong normalisation modulo permutations for the structural lambda-calculus. Logical Methods in Computer Science **8**(1) (2012)
3. Asperti, A., Guerrini, S.: The Optimal Implementation of Functional Programming Languages. Cambridge University Press (1998)
4. Balabonski, T.: A unified approach to fully lazy sharing. ACM SIGPLAN Notices **47**(1), 469–480 (2012)
5. Balabonski, T.: Weak Optimality, and the Meaning of Sharing. In: International Conference on Functional Programming (ICFP). pp. 263–274. Boston, United States (Sep 2013). https://doi.org/10.1145/2500365.2500606, https://hal.archives-ouvertes.fr/hal-00907056
6. Barendregt, H.P.: The Lambda Calculus – Its Syntax and Semantics, Studies in Logic and the Foundations of Mathematics, vol. 103. North-Holland (1984)
7. Berkling, K.J., Fehr, E.: A consistent extension of the lambda-calculus as a base for functional programming languages. Information and Control **55**, 89–101 (1982)
8. Blanc, T., Lévy, J.J., Maranget, L.: Sharing in the weak lambda-calculus. Processes, Terms and Cycles: Steps on the Road to Infinity: Essays Dedicated to Jan Willem Klop on the Occasion of His 60th Birthday **3838**, 70 (2005)
9. Brünnler, K., McKinley, R.: An algorithmic interpretation of a deep inference system. In: International Conference on Logic for Programming Artificial Intelligence and Reasoning (LPAR). pp. 482–496 (2008)
10. Brünnler, K., Tiu, A.: A local system for classical logic. In: 8th International Conference on Logic for Programming Artificial Intelligence and Reasoning (LPAR). LNCS, vol. 2250, pp. 347–361 (2001)
11. Church, A., Rosser, J.B.: Some properties of conversion. Transactions of the American Mathematical Society **39**(3), 472–482 (1936), http://www.jstor.org/stable/1989762
12. Cockett, R., Seely, R.: Weakly distributive categories. Journal of Pure and Applied Algebra **114**(2), 133–173 (1997)
13. Fernández, M., Mackie, I., Sinot, F.R.: Lambda-calculus with director strings. Applicable Algebra in Engineering, Communication and Computing **15**(6), 393–437 (2005)
14. Guenot, N., Straßburger, L.: Symmetric normalisation for intuitionistic logic. In: Joint Meeting of the Twenty-Third EACSL Annual Conference on Computer Science Logic (CSL) and the Twenty-Ninth Annual ACM/IEEE Symposium on Logic in Computer Science (LICS) (2014)
15. Guglielmi, A., Gundersen, T., Parigot, M.: A proof calculus which reduces syntactic bureaucracy. In: 21st International Conference on Rewriting Techniques and Applications (RTA). pp. 135–150 (2010)
16. Gundersen, T., Heijltjes, W., Parigot, M.: Atomic lambda-calculus: a typed lambda-calculus with explicit sharing. In: 28th Annual ACM/IEEE Symposium on Logic in Computer Science (LICS). pp. 311–320 (2013)
17. He, F.: The Atomic Lambda-Mu Calculus. Ph.D. thesis, University of Bath (2018)
18. Hendriks, D., van Oostrom, V.: Adbmal. In: 19th International Conference on Automated Deduction (CADE). LNCS, vol. 2741, pp. 136–150 (2003)

19. Kennaway, R., Sleep, R.: Director strings as combinators. ACM Transactions on Programming Languages and Systems (1988)
20. Klop, J.W.: Combinatory Reduction Systems. Ph.D. thesis, Utrecht University (1980)
21. Lamping, J.: An algorithm for optimal lambda calculus reduction. In: Proceedings of the 17th ACM SIGPLAN-SIGACT symposium on Principles of programming languages. pp. 16–30 (1990)
22. Launchbury, J.: A natural semantics for lazy evaluation. In: 20th ACM SIGPLAN-SIGACT symposium on Principles of programming languages (POPL). pp. 144–154 (1993)
23. Lévy, J.J.: Optimal reductions in the lambda-calculus. In: To H.B. Curry: Essays in Combinatory Logic, Lambda Calculus and Formalism. Academic Press (1980)
24. van Oostrom, V., van de Looij, K.J., Zwitserlood, M.: Lambdascope: another optimal implementation of the lambda-calculus. In: Workshop on Algebra and Logic on Programming Systems (ALPS) (2004)
25. Sherratt, D.R.: A lambda-calculus that achieves full laziness with spine duplication. Ph.D. thesis, University of Bath (2019)
26. Tiu, A.: A local system for intuitionistic logic. In: International Conference on Logic for Programming Artificial Intelligence and Reasoning (LPAR). pp. 242–256 (2006)
27. Wadsworth, C.P.: Semantics and Pragmatics of the Lambda-Calculus. Ph.D. thesis, University of Oxford (1971)

Learning Weighted Automata over Principal Ideal Domains*

Gerco van Heerdt[1], Clemens Kupke[2](✉), Jurriaan Rot[1,3], and Alexandra Silva[1]

[1] University College London, United Kingdom
{gerco.heerdt,alexandra.silva}@ucl.ac.uk
[2] University of Strathclyde, United Kingdom
clemens.kupke@strath.ac.uk
[3] Radboud University, The Netherlands
jrot@cs.ru.nl

Abstract. In this paper, we study active learning algorithms for weighted automata over a semiring. We show that a variant of Angluin's seminal $\mathtt{L}^\star$ algorithm works when the semiring is a principal ideal domain, but not for general semirings such as the natural numbers.

1 Introduction

Angluin's seminal $\mathtt{L}^\star$ algorithm [4] for active learning of deterministic automata (DFAs) has been successfully used in many verification tasks, including in automatically building formal models of chips in bank cards or finding bugs in network protocols (see [27,14] for a broad overview of successful applications of active learning). While DFAs are expressive enough to capture interesting properties, certain verification tasks require more expressive models. This motivated several researchers to extend $\mathtt{L}^\star$ to other types of automata, notably Mealy machines [28,24], register automata [15,22,1], and nominal automata [20].

Weighted finite automata (WFAs) are an important model made popular due to their applicability in image processing and speech recognition tasks [11,21]. The model is prevalent in other areas, including bioinformatics [2] and formal verification [3]. Passive learning algorithms and associated complexity results have appeared in the literature (see e.g. [5] for an overview), whereas active learning has been less studied [6,7]. Furthermore, the existing learning algorithms, both passive and active, have been developed assuming the weights in the automaton are drawn from a field, such as the real numbers.[4] To the best of our knowledge, no learning algorithms, whether passive or active, have been developed for WFAs in which the weights are drawn from a general semiring.

* The research leading to this work was partially funded by the European Union's Horizon 2020 research and innovation programme under the ERC Starting Grant ProFoundNet (grant code 679127) and the Marie Skłodowska-Curie Grant Agreement No. 795119, by the EPSRC Standard Grant CLeVer (EP/S028641/1) and by GCHQ via the VeTSS grant "Automated black-box verification of networking systems" (4207703/RFA 15845).

[4] Balle and Mohri [6] define WFAs generically over a semiring but then restrict to fields from Section 3 onwards as they present an overview of existing learning algorithms.

J. Goubault-Larrecq and B. König (Eds.): FOSSACS 2020, LNCS 12077, pp. 602–621, 2020.
https://doi.org/10.1007/978-3-030-45231-5_31

In this paper, we explore *active learning* for WFAs over a general semiring. The main contributions of the paper are as follows:

1. We introduce a weighted variant of $\mathtt{L}^\star$ parametric on an arbitrary semiring, together with sufficient conditions for termination (Section 4).
2. We show that for general semirings our algorithm might not terminate. In particular, if the semiring is the natural numbers, one of the steps of the algorithm might not converge (Section 5).
3. We prove that the algorithm terminates if the semiring is a *principal ideal domain*, covering the known case of fields, but also the integers. This yields the first active learning algorithm for WFAs over the integers (Section 6).

We start in Section 2 by explaining the learning algorithm for WFAs over the reals and pointing out the challenges in extending it to arbitrary semirings.

2 Overview of the Approach

In this section, we give an overview of the work developed in the paper through examples. We start by informally explaining the general algorithm for learning weighted automata that we introduce in Section 4, for the case where the semiring is a field. More specifically, for simplicity we consider the field of real numbers throughout this section. Later in the section, we illustrate why this algorithm does not work for an arbitrary semiring.

Angluin's $\mathtt{L}^\star$ algorithm provides a procedure to learn the minimal DFA accepting a certain (unknown) regular language. In the weighted variant we will introduce in Section 4, for the specific case of the field of real numbers, the algorithm produces the minimal WFA accepting a weighted rational language (or formal power series) $\mathcal{L}\colon A^* \to \mathbb{R}$.

A WFA over $\mathbb{R}$ consists of a set of states, a linear combination of initial states, a transition function that for each state and input symbol produces a linear combination of successor states, and an output value in $\mathbb{R}$ for each state (Definition 5). As an example, consider the WFA over $A = \{a\}$ below.

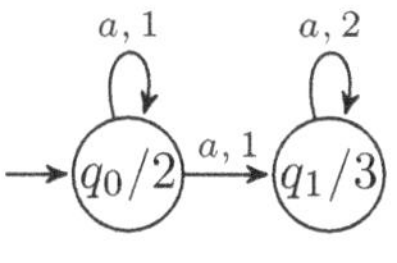

Here q_0 is the only initial state, with weight 1, as indicated by the arrow into it that has no origin. When reading a, q_0 transitions with weight 1 to itself and also with weight 1 to q_1; q_1 transitions with weight 2 just to itself. The output of q_0 is 2 and the output of q_1 is 3.

The language of a WFA is determined by letting it read a given word and determining the final output according to the weights and outputs assigned to individual states. More precisely, suppose we want to read the word aaa in the example WFA above. Initially, q_0 is assigned weight 1 and q_1 weight 0. Processing the first a then leads to q_0 retaining weight 1, as it has a self-loop with weight 1, and q_1 obtaining weight 1 as well. With the next a, the weight of q_0 still remains 1, but the weight of q_1 doubles due to its self-loop of weight 1 and is added to the weight 1 coming from q_0, leading to a total of 3. Similarly, after the last a the weights are 1 for q_0 and 7 for q_1. Since q_0 has output 2 and q_1 output 3, the final result is $2 \cdot 1 + 3 \cdot 7 = 23$.

The learning algorithm assumes access to a *teacher* (sometimes also called *oracle*), who answers two types of queries:

- *membership queries*, consisting of a single word $w \in A^*$, to which the teacher replies with a weight $\mathcal{L}(w) \in \mathbb{R}$;
- *equivalence queries*, consisting of a hypothesis WFA $\mathcal{A}$, to which the teacher replies **yes** if its language $\mathcal{L}_\mathcal{A}$ equals the target language $\mathcal{L}$ and **no** otherwise, providing a counterexample $w \in A^*$ such that $\mathcal{L}(w) \neq \mathcal{L}_\mathcal{A}(w)$.

In practice, membership queries are often easily implemented by interacting with the system one wants to model the behaviour of. However, equivalence queries are more complicated—as the perfect teacher does not exist and the target automaton is not known they are commonly approximated by testing. Such testing can however be done exhaustively if a bound on the number of states of the target automaton is known. Equivalence queries can also be implemented exactly when learning algorithms are being compared experimentally on generated automata whose languages form the targets. In this case, standard methods for language equivalence, such as the ones based on bisimulations [9], can be used.

The learning algorithm incrementally builds an *observation table*, which at each stage contains partial information about the language $\mathcal{L}$ determined by two finite sets $S, E \subseteq A^*$. The algorithm fills the table through membership queries. As an example, and to set notation, consider the following table (over $A = \{a\}$).

		E: ε	a	aa
S	ε	0	1	3
S, $S \cdot A$	a	1	3	7
$S \cdot A$	aa	3	7	15

$$\mathsf{row}\colon S \to \mathbb{R}^E$$
$$\mathsf{row}(u)(v) = \mathcal{L}(uv)$$
$$\mathsf{srow}\colon S \cdot A \to \mathbb{R}^E$$
$$\mathsf{srow}(ua)(v) = \mathcal{L}(uav)$$

This table indicates that $\mathcal{L}$ assigns 0 to ε, 1 to a, 3 to aa, 7 to aaa, and 15 to $aaaa$. For instance, we see that $\mathsf{row}(a)(aa) = \mathsf{srow}(aa)(a) = 7$. Since row and srow are fully determined by the language $\mathcal{L}$, we will refer to an observation table as a pair (S, E), leaving the language $\mathcal{L}$ implicit.

If the observation table (S, E) satisfies certain properties described below, then it represents a WFA (S, δ, i, o), called the *hypothesis*, as follows:

- $\delta\colon S \to (\mathbb{R}^S)^A$ is a linear map defined by choosing for $\delta(s)(a)$ a linear combination over S of which the rows evaluate to $\mathsf{srow}(sa)$;
- $i\colon S \to \mathbb{R}$ is the initial weight map defined as $i(\varepsilon) = 1$ and $i(s) = 0$ for $s \neq \varepsilon$;
- $o\colon S \to \mathbb{R}$ is the output weight map defined as $o(s) = \mathsf{row}(s)(\varepsilon)$.

For this to be well-defined, we need to have $\varepsilon \in S$ (for the initial weights) and $\varepsilon \in E$ (for the output weights), and for the transition function there is a crucial property of the table that needs to hold: closedness. In the weighted setting, a table is closed if for all $t \in S \cdot A$, there exist $r_s \in \mathbb{R}$ for all $s \in S$ such that

$$\mathsf{srow}(t) = \sum_{s \in S} r_s \cdot \mathsf{row}(s).$$

If this is not the case for a given $t \in S \cdot A$, the algorithm adds t to S. The table is repeatedly extended in this manner until it is closed. The algorithm then constructs a hypothesis, using the closedness witnesses to determine transitions, and poses an equivalence query to the teacher. It terminates when the answer is **yes**; otherwise it extends the table with the counterexample provided by adding all its suffixes to E, and the procedure continues by closing again the resulting table. In the next subsection we describe the algorithm through an example.

Remark 1. The original $\mathtt{L}^{\star}$ algorithm requires a second property to construct a hypothesis, called *consistency*. Consistency is difficult to check in extended settings, so the present paper is based on a variant of the algorithm inspired by Maler and Pnueli [19] where only closedness is checked and counterexamples are handled differently. See [13] for an overview of consistency in different settings.

2.1 Example: Learning a Weighted Language over the Reals

Throughout this section we consider the following weighted language:

$$\mathcal{L}\colon \{a\}^* \to \mathbb{R} \qquad\qquad \mathcal{L}(a^j) = 2^j - 1.$$

The minimal WFA recognising it has 2 states. We will illustrate how the weighted variant of Angluin's algorithm recovers this WFA.

We start from $S = E = \{\varepsilon\}$, and fill the entries of the table on the left below by asking membership queries for ε and a. The table is not closed and hence we build the table on its right, adding the membership result for aa. The resulting table is closed, as $\mathsf{srow}(aa) = 3 \cdot \mathsf{row}(a)$, so we construct the hypothesis $\mathcal{A}_1$.

	ε
ε	0
a	1

	ε
ε	0
a	1
aa	3

$\mathcal{A}_1 =$ 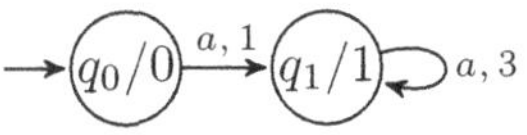

$q_0 = \varepsilon$

$q_1 = a$

The teacher replies **no** and gives the counterexample aaa, which is assigned 9 by the hypothesis automaton $\mathcal{A}_1$ but 7 in the language. Therefore, we extend $E \leftarrow E \cup \{a, aa, aaa\}$. The table becomes the one below. It is closed, as $\mathsf{srow}(aa) = 3 \cdot \mathsf{row}(a) - 2 \cdot \mathsf{row}(\varepsilon)$, so we construct a new hypothesis $\mathcal{A}_2$.

	ε	a	aa	aaa
ε	0	1	3	7
a	1	3	7	15
aa	3	7	15	31

$\mathcal{A}_2 =$ 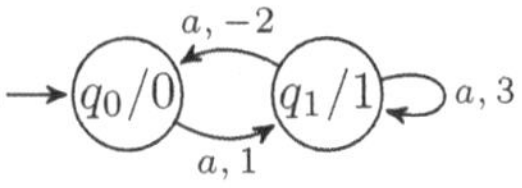

The teacher replies **yes** because $\mathcal{A}_2$ accepts the intended language assigning $2^j - 1 \in \mathbb{R}$ to the word a^j, and the algorithm terminates with the correct automaton.

2.2 Learning Weighted Languages over Arbitrary Semirings

Consider now the same language as above, but represented as a map over the semiring of natural numbers $\mathcal{L}\colon \{a\}^* \to \mathbb{N}$ instead of a map $\mathcal{L}\colon \{a\}^* \to \mathbb{R}$ over the reals. Accordingly, we consider a variant of the learning algorithm over the semiring $\mathbb{N}$ rather than the algorithm over $\mathbb{R}$ described above. For the first part, the run of the algorithm for $\mathbb{N}$ is the same as above, but after receiving the counterexample we can no longer observe that $\mathsf{srow}(aa) = 3 \cdot \mathsf{row}(a) - 2 \cdot \mathsf{row}(\varepsilon)$, since $-2 \notin \mathbb{N}$. In fact, there are no $m, n \in \mathbb{N}$ such that $\mathsf{srow}(aa) = m \cdot \mathsf{row}(\varepsilon) + n \cdot \mathsf{row}(a)$. To see this, consider the first two columns in the table and note that $\frac{3}{7}$ is bigger than $\frac{0}{1} = 0$ and $\frac{1}{3}$, so it cannot be obtained as a linear combination of the latter two using natural numbers. We thus have a closedness defect and update $S \leftarrow S \cup \{aa\}$, leading to the table below.

	ε	a	aa	aaa
ε	0	1	3	7
a	1	3	7	15
aa	3	7	15	31
aaa	7	15	31	63

Again, the table is not closed, since $\frac{7}{15} > \frac{3}{7}$. In fact, these closedness defects continue appearing indefinitely, leading to non-termination of the algorithm. This is shown formally in Section 5.

Note, however, that there does exist a WFA over $\mathbb{N}$ accepting this language:

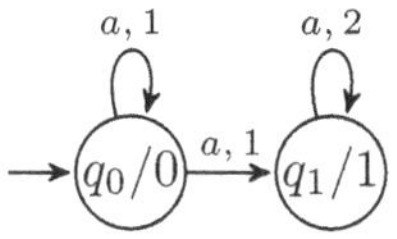

(1)

The reason that the algorithm cannot find the correct automaton is closely related to the algebraic structure induced by the semiring. In the case of the reals, the algebras are vector spaces and the closedness checks induce increases in the dimension of the hypothesis WFA, which in turn cannot exceed the dimension of the minimal one for the language. In the case of commutative monoids, the algebras for the natural numbers, the notion of dimension does not exist and unfortunately the algorithm does not terminate. In Section 6 we show that one can get around this problem for a class of semirings which includes the integers.

We mentioned earlier that during experimental evaluation the target WFA is known, and equivalence queries may be implemented via standard language equivalence methods. A further issue with arbitrary semirings is that language equivalence can be undecidable; that is the case, e.g., for the tropical semiring.

In Section 3 we recall basic definitions used throughout the paper, after which Section 4 introduces our general algorithm with its (parameterised) termination proof of Theorem 14. We then proceed to prove non-termination of the example discussed above over the natural numbers in Section 5 before instantiating our algorithm to PIDs in Section 6 and showing that it terminates in Theorem 28. We conclude with a discussion of related and future work in Section 7.

3 Preliminaries

Throughout this paper we fix a semiring[5] $\mathbb{S}$ and a finite alphabet A. We start with basic definitions related to semimodules and weighted languages.

Definition 2 (Semimodule). *A (left) semimodule M over $\mathbb{S}$ consists of a monoid structure on M, written using $+$ as the operation and 0 as the unit, together with a scalar multiplication map $\cdot\colon \mathbb{S} \times M \to M$ such that:*

$$s \cdot 0_M = 0_M \qquad 0_{\mathbb{S}} \cdot m = 0_M \qquad 1 \cdot m = m$$
$$s \cdot (m+n) = s \cdot m + s \cdot n \quad (s+r) \cdot m = s \cdot m + r \cdot m \quad (sr) \cdot m = s \cdot (r \cdot m).$$

When the semiring is in fact a ring, we speak of a module *rather than a semimodule. In the case of a field, the concept instantiates to a vector space.*

As an example, commutative monoids are the semimodules over the semiring of natural numbers. Any semiring forms a semimodule over itself by instantiating the scalar multiplication map to the internal multiplication. If X is any set and M is a semimodule, then M^X with pointwise operations also forms a semimodule. A similar semimodule is the *free semimodule* over X, which differs from M^X in that it fixes M to be $\mathbb{S}$ and requires its elements to have *finite support.* This enables an important operation called *linearisation.*

Definition 3 (Free semimodule). *The* free semimodule *over a set X is given by the set*

$$V(X) = \{f\colon X \to \mathbb{S} \mid \mathsf{supp}(f) \text{ is finite}\}$$

with pointwise operations. Here $\mathsf{supp}(f) = \{x \in X \mid f(x) \neq 0\}$. We sometimes identify the elements of $V(X)$ with formal sums over X. Any semimodule isomorphic to $V(X)$ for some set X is called free.

If X is a finite set, then $V(X) = \mathbb{S}^X$. We now define *linearisation* of a function into a semimodule, which uniquely extends it to a semimodule homomorphism, witnessing the fact that $V(X)$ is free.

[5] Rings and semirings considered in this paper are taken to be unital.

Definition 4 (Linearisation). *Given a set X, a semimodule M, and a function $f\colon X \to M$, we define the* linearisation *of f as the semimodule homomorphism $f^\sharp\colon V(X) \to M$ given by*

$$f^\sharp(\alpha) = \sum_{x \in X} \alpha(x) \cdot f(x).$$

The $(-)^\sharp$ operation has an inverse that maps a semimodule homomorphism $g\colon V(X) \to M$ to the function $g^\dagger\colon X \to M$ given by

$$g^\dagger(x) = g(\partial_x), \qquad \partial_x(y) = \begin{cases} 1 & \text{if } y = x \\ 0 & \text{if } y \neq x. \end{cases}$$

We proceed with the definition of WFAs and their languages.

Definition 5 (WFA). *A* weighted finite automaton (WFA) *over $\mathbb{S}$ is a tuple (Q, δ, i, o), where Q is a finite set, $\delta\colon Q \to (\mathbb{S}^Q)^A$, and $i, o\colon Q \to \mathbb{S}$.*

A *weighted language* (or just *language*) over $\mathbb{S}$ is a function $A^* \to \mathbb{S}$. To define the language accepted by a WFA $\mathcal{A} = (Q, \delta, i, o)$, we first introduce the notions of *observability map* $\mathsf{obs}_\mathcal{A}\colon V(Q) \to \mathbb{S}^{A^*}$ and *reachability map* $\mathsf{reach}_\mathcal{A}\colon V(A^*) \to V(Q)$ as the semimodule homomorphisms given by

$$\begin{aligned} \mathsf{reach}^\dagger_\mathcal{A}(\varepsilon) &= i & \qquad \mathsf{obs}_\mathcal{A}(m)(\varepsilon) &= o^\sharp(m) \\ \mathsf{reach}^\dagger_\mathcal{A}(ua) &= \delta^\sharp(\mathsf{reach}^\dagger_\mathcal{A}(u))(a) & \qquad \mathsf{obs}_\mathcal{A}(m)(au) &= \mathsf{obs}_\mathcal{A}(\delta^\sharp(m)(a))(u). \end{aligned}$$

The *language accepted by a WFA* $\mathcal{A} = (Q, \delta, i, o)$ is the function $\mathcal{L}_\mathcal{A}\colon A^* \to \mathbb{S}$ given by $\mathcal{L}_\mathcal{A} = \mathsf{obs}_\mathcal{A}(i)$. Equivalently, one can define this as $\mathcal{L}_\mathcal{A} = o^\sharp \circ \mathsf{reach}^\dagger_\mathcal{A}$.

4 General Algorithm for WFAs

In this section we define the general algorithm for WFAs over $\mathbb{S}$, as described informally in Section 2. Our algorithm assumes the existence of a *closedness strategy* (Definition 8), which allows one to check whether a table is closed, and in case it is, provide relevant witnesses. We then introduce sufficient conditions on $\mathbb{S}$ and on the language $\mathcal{L}$ to be learned under which the algorithm terminates.

Definition 6 (Observation table). *An* observation table *(or just* table*) (S, E) consists of two sets $S, E \subseteq A^*$. We write $\mathsf{Table}_{\mathsf{fin}} = \mathcal{P}_f(A^*) \times \mathcal{P}_f(A^*)$ for the set of finite tables (where $\mathcal{P}_f(X)$ denotes the collection of finite subsets of a set X). Given a language $\mathcal{L}\colon A^* \to \mathbb{S}$, an observation table (S, E) determines the* row function *$\mathsf{row}_{(S,E,\mathcal{L})}\colon S \to \mathbb{S}^E$ and the* successor row *function $\mathsf{srow}_{(S,E,\mathcal{L})}\colon S \cdot A \to \mathbb{S}^E$ as follows:*

$$\mathsf{row}_{(S,E,\mathcal{L})}(w)(v) = \mathcal{L}(wv) \qquad \mathsf{srow}_{(S,E,\mathcal{L})}(wa)(v) = \mathcal{L}(wav).$$

We often write $\mathsf{row}_\mathcal{L}$ and $\mathsf{srow}_\mathcal{L}$, or even row and srow, when the parameters are clear from the context.

A table is *closed* if the successor rows are linear combinations of the existing rows in S. To make this precise, we use the linearisation $\mathsf{row}^\sharp$ (Definition 4), which extends row to linear combinations of words in S.

Definition 7 (Closedness). *Given a language $\mathcal{L}$, a table (S, E) is* closed *if for all $w \in S$ and $a \in A$ there exists $\alpha \in V(S)$ such that $\mathsf{srow}(wa) = \mathsf{row}^\sharp(\alpha)$.*

This corresponds to the notion of closedness described in Section 2.

A further important ingredient of the algorithm is a method for checking whether a table is closed. This is captured by the notion of closedness strategy.

Definition 8 (Closedness strategy). *Given a language $\mathcal{L}$, a* closedness strategy *for $\mathcal{L}$ is a family of computable functions*

$$\left(\mathsf{cs}_{(S,E)} \colon S \cdot A \to \{\bot\} \cup V(S)\right)_{(S,E) \in \mathsf{Table}_{\mathsf{fin}}}$$

satisfying the following two properties:

- *if $\mathsf{cs}_{(S,E)}(t) = \bot$, then there is no $\alpha \in V(S)$ s.t. $\mathsf{row}^\sharp(\alpha) = \mathsf{srow}(t)$, and*
- *if $\mathsf{cs}_{(S,E)}(t) \neq \bot$, then $\mathsf{row}^\sharp(\mathsf{cs}_{(S,E)}(t)) = \mathsf{srow}(t)$.*

Thus, given a closedness strategy as above, a table (S, E) is closed iff $\mathsf{cs}_{(S,E)}(t) \neq \bot$ for all $t \in S \cdot A$. More specifically, for each $t \in S \cdot A$ we have that $\mathsf{cs}_{(S,E)}(t) \neq \bot$ iff the (successor) row corresponding to t already forms a linear combination of rows labelled by S. In that case, this linear combination is returned by $\mathsf{cs}_{(S,E)}(t)$. This is used to close tables in our learning algorithm, introduced below.

Examples of semirings and (classes of) languages that admit a closedness strategy are described at the end of this section. Important for our algorithm will be that closedness strategies are computable. This problem is equivalent to solving systems of equations $A\underline{x} = \underline{b}$, where A is the matrix whose columns are $\mathsf{row}(s)$ for $s \in S$, $\underline{x}$ is a vector of length $|S|$, and $\underline{b}$ is the vector consisting of the row entries in $\mathsf{srow}(t)$ for some $t \in S \cdot A$. These observations motivate the following definition.

Definition 9 (Solvability). *A semiring $\mathbb{S}$ is* solvable *if a solution to any finite system of linear equations of the form $A\underline{x} = \underline{b}$ is computable.*

We have the following correspondence.

Proposition 10. *For any language accepted by a WFA over any semiring there exists a closedness strategy if and only if the semiring is solvable.*

Proof. If the semiring is solvable, we obtain a closedness strategy by the remarks prior to Definition 9. Conversely, we can construct a language that is non-zero on finitely many words and encode in a table (S, E) a given linear equation. To be able to freely choose the value in each table cell, we can consider a sufficiently large alphabet to make sure S and E contain only single-letter words. This avoids dependencies within the table. □

Algorithm 1 Abstract learning algorithm for WFA over $\mathbb{S}$

1: $S, E \leftarrow \{\varepsilon\}$
2: **while** `true` **do**
3:     **while** $\mathsf{cs}_{(S,E)}(t) = \bot$ for some $t \in S \cdot A$ **do**
4:         $S \leftarrow S \cup \{t\}$
5:     **for** $s \in S$ **do**
6:         $o(s) \leftarrow \mathsf{row}_{\mathcal{L}}(s)(\varepsilon)$
7:         **for** $a \in A$ **do**
8:             $\delta(s)(a) \leftarrow \mathsf{cs}_{(S,E)}(sa)$
9:     **if** $\mathsf{EQ}(S, \delta, \varepsilon, o) = w \in A^*$ **then**
10:         $E \leftarrow E \cup \mathsf{suffixes}(w)$
11:     **else**
12:         **return** $(S, \delta, \varepsilon, o)$

We now have all the ingredients to formulate the algorithm to learn weighted languages over a general semiring. The pseudocode is displayed in Algorithm 1.

The algorithm keeps a table (S, E), and starts by initialising both S and E to contain just the empty word. The inner while loop (lines 3–4) uses the closedness strategy to repeatedly check whether the current table is closed and add new rows in case it is not. Once the table is closed, a hypothesis is constructed, again using the closedness strategy (lines 5–8). This hypothesis $(S, \delta, \varepsilon, o)$ is then given to the teacher for an equivalence check. The equivalence check is modelled by EQ (line 9) as follows: if the hypothesis is incorrect, the teacher non-deterministically returns a counterexample $w \in A^*$, the condition evaluates to `true`, and the suffixes of w are added to E; otherwise, if the hypothesis is correct, the condition on line 9 evaluates to `false`, and the algorithm returns the correct hypothesis on line 12.

4.1 Termination of the General Algorithm

The main question remaining is: under which conditions does this algorithm terminate and hence learns the unknown weighted language? We proceed to give abstract conditions under which it terminates. There are two main assumptions:

1. A way of measuring progress the algorithm makes with the observation table when it distinguishes linear combinations of rows that were previously equal, together with a bound on this progress (Definition 11).
2. An assumption on the *Hankel matrix* of the input language (Definition 12), which makes sure we encounter finitely many closedness defects throughout any run of the algorithm. More specifically, we assume that the Hankel matrix satisfies a finite approximation property (Definition 13).

The first assumption is captured by the definition of progress measure:

Definition 11 (Progress measure). *A* progress measure *for a language $\mathcal{L}$ is a function* $\mathsf{size}\colon \mathsf{Table}_{\mathsf{fin}} \to \mathbb{N}$ *such that*

(a) there exists $n \in \mathbb{N}$ *such for all* $(S, E) \in \mathsf{Table}_{\mathsf{fin}}$ *we have* $\mathsf{size}(S, E) \leq n$*;*
(b) given $(S, E), (S, E') \in \mathsf{Table}_{\mathsf{fin}}$ *and* $s_1, s_2 \in V(S)$ *such that* $E \subseteq E'$ *and* $\mathsf{row}^\sharp_{(S,E,\mathcal{L})}(s_1) = \mathsf{row}^\sharp_{(S,E,\mathcal{L})}(s_2)$ *but* $\mathsf{row}^\sharp_{(S,E',\mathcal{L})}(s_1) \neq \mathsf{row}^\sharp_{(S,E',\mathcal{L})}(s_2)$*, we have* $\mathsf{size}(S, E') > \mathsf{size}(S, E)$*.*

A progress measure assigns a 'size' to each table, in such a way that (a) there is a global bound on the size of tables, and (b) if we extend a table with some proper tests in E, i.e., such that some combinations of rows in $\mathsf{row}^\sharp$ that were equal before get distinguished by a newly added test, then the size of the extended table is properly above the size of the original table. This is used to ensure that, when adding certain counterexamples supplied by the teacher, the size of the table, measured according to the above size function, properly increases.

The second assumption that we use for termination is phrased in terms of the Hankel matrix associated to the input language $\mathcal{L}$, which represents $\mathcal{L}$ as the (semimodule generated by the) infinite table where both the rows and columns contain all words. The Hankel matrix is defined as follows.

Definition 12 (Hankel matrix). *Given a language* $\mathcal{L}\colon A^* \to \mathbb{S}$*, the* semimodule generated by a table (S, E) *is given by the image of* $\mathsf{row}^\sharp$*. We refer to the semimodule generated by the table* (A^*, A^*) *as the* Hankel matrix *of* $\mathcal{L}$*.*

The Hankel matrix is approximated by the tables that occur during the execution of the algorithm. For termination, we will therefore assume that this matrix satisfies the following finite approximation condition.

Definition 13 (Ascending chain condition). *We say that a semimodule* M *satisfies the* ascending chain condition *if for all inclusion chains of subsemimodules of* M*,*

$$S_1 \subseteq S_2 \subseteq S_3 \subseteq \cdots,$$

there exists $n \in \mathbb{N}$ *such that for all* $m \geq n$ *we have* $S_m = S_n$*.*

Given the notions of progress measure, Hankel matrix and ascending chain condition, we can formulate the general theorem for termination of Algorithm 1.

Theorem 14 (Termination of the abstract learning algorithm). *In the presence of a progress measure, Algorithm 1 terminates whenever the Hankel matrix of the target language satisfies the ascending chain condition (Definition 13).*

Proof. Suppose the algorithm does not terminate. Then there is a sequence $\{(S_n, E_n)\}_{n\in\mathbb{N}}$ of tables where (S_0, E_0) is the initial table and (S_{n+1}, E_{n+1}) is formed from (S_n, E_n) after resolving a closedness defect or adding columns due to a counterexample.

We write H_n for the semimodule generated by the table (S_n, A^*). We have $S_n \subseteq S_{n+1}$ and thus $H_n \subseteq H_{n+1}$. Note that a closedness defect for (S_n, E_n) is also a closedness defect for (S_n, A^*), so if we resolve the defect in the next step, the inclusion $H_n \subseteq H_{n+1}$ is strict. Since these are all included in the Hankel matrix, which satisfies the ascending chain condition, there must be an n such that for all $k \geq n$ we have that (S_k, E_k) is closed.

In [13, Section 6] it is shown that in a general table used for learning automata with side-effects given by a monad there exists a suffix of each counterexample for the corresponding hypothesis that when added as a column label leads to either a closedness defect or to distinguishing two combinations of rows in the table. Since WFAs are automata with side-effects given by the free semimodule monad[6] and we add all suffixes of the counterexample to the set of column labels, this also happens in our algorithm. Thus, for all $k \geq n$ where we process a counterexample, there must be two linear combinations of rows distinguished, as closedness is already guaranteed. Then the semimodule generated by (S_k, E_k) is a strict quotient of the semimodule generated by (S_{k+1}, E_{k+1}). By the progress measure we then find $\mathsf{size}(S_k, E_k) < \mathsf{size}(S_{k+1}, E_{k+1})$, which cannot happen infinitely often. We conclude that the algorithm must terminate. □

To illustrate the hypotheses needed for Algorithm 1 and its termination (Theorem 14), we consider two classes of semirings for which learning algorithms are already known in the literature [7,13].

Example 15 (Weighted languages over fields). Consider any field for which the basic operations are computable. Solvability is then satisfied via a procedure such as Gaussian elimination, so by Proposition 10 there exists a closedness strategy. Hence, we can instantiate Algorithm 1 with $\mathbb{S}$ being such a field.

For termination, we show that the hypotheses of Theorem 14 are satisfied whenever the input language is accepted by a WFA. First, a progress measure is given by the dimension of the vector space generated by the table. To see this, note that if we distinguish two linear combinations of rows, we can assume without loss of generality that one of these linear combinations in the extended table uses only basis elements. This in turn can be rewritten to distinguishing a single row from a linear combination of rows using field operations, with the property that the extended version of the single row is a basis element. Hence, the row was not a basis element in the original table, and therefore the dimension of the vector space generated by the table has increased. Adding rows and columns cannot decrease this dimension, so it is bounded by the dimension of the Hankel matrix. Since the language we want to learn is accepted by a WFA, the associated Hankel matrix has a finite dimension [10,12] (see also, e.g., [5]), providing a bound for our progress measure.

Finally, for any ascending chain of subspaces of the Hankel matrix, these subspaces are of finite dimension bounded by the dimension of the Hankel matrix. The dimension increases along a strict subspace relation, so the chain converges.

Example 16 (Weighted languages over finite semirings). Consider any finite semiring. Finiteness allows us to apply a brute force approach to solving systems of equations. This means the semiring is solvable, and hence a closedness strategy exists by Proposition 10.

For termination, we can define a progress measure by assigning to each table the size of the image of $\mathsf{row}^\sharp$. Distinguishing two linear combinations of rows

[6] We note that [13] assumes the monad to preserve finite sets. However, the relevant arguments do not depend on this.

increases this measure. If the language we want to learn is accepted by a WFA, then the Hankel matrix contains a subset of the linear combinations of the languages of its states. Since there are only finitely many such linear combinations, the Hankel matrix is finite, which bounds our measure. A finite semimodule such as the Hankel matrix in this case does not admit infinite chains of subspaces. We conclude by Theorem 14 that Algorithm 1 terminates for the instance that the semiring $\mathbb{S}$ is a finite, if the input language is accepted by a WFA over $\mathbb{S}$.

For the Boolean semiring, an instance of the above finite semiring example, WFAs are non-deterministic finite automata. The algorithm we recover by instantiating Algorithm 1 to this case is close to the algorithm first described by Bollig et al. [8]. The main differences are that in their case the hypothesis has a state space given by a minimally generating subset of the distinct rows in the table rather than all elements of S, and they do apply a notion of consistency.

In Section 6 we will show that Algorithm 1 can learn WFAs over principal ideal domains—notably including the integers—thus providing a strict generalisation of existing techniques.

5 Issues with Arbitrary Semirings

We concluded the previous section with examples of semirings for which Algorithm 1 terminates if the target language is accepted by a WFA. In this section, we prove a negative result for the algorithm over the semiring $\mathbb{N}$: we show that it does not terminate on a certain language over $\mathbb{N}$ accepted by a WFA over $\mathbb{N}$, as anticipated in Section 2.2. This means that Algorithm 1 does not work well for arbitrary semirings. The problem is that the Hankel matrix of a language recognised by WFA does not necessarily satisfy the ascending chain condition that is used to prove Theorem 14. In the example given in the proof below, the Hankel matrix is not even finitely generated.

Theorem 17. *There exists a WFA $\mathcal{A}_{\mathbb{N}}$ over $\mathbb{N}$ such that Algorithm 1 does not terminate when given $\mathcal{L}_{\mathcal{A}_{\mathbb{N}}}$ as input, regardless of the closedness strategy used.*

Proof. Let $\mathcal{A}_{\mathbb{N}}$ be the automaton over the alphabet $\{a\}$ given in (1) in Section 2.2. Formally, $\mathcal{A}_{\mathbb{N}} = (Q, \delta, i, o)$, where

$$Q = \{q_0, q_1\} \qquad i = q_0 \qquad o(q_0) = 0$$
$$\delta(q_0)(a) = q_0 + q_1 \qquad \delta(q_1)(a) = 2q_1 \qquad o(q_1) = 1.$$

As mentioned in Section 2.2, the language $\mathcal{L}\colon \{a\}^* \to \mathbb{N}$ accepted by $\mathcal{A}_{\mathbb{N}}$ is given by $\mathcal{L}(a^j) = 2^j - 1$. This can be shown more precisely as follows. First one shows by induction on j that $\mathsf{obs}_{\mathcal{A}_{\mathbb{N}}}(q_1)(a^j) = 2^j$ for all $j \in \mathbb{N}$—we leave the straightforward argument to the reader. Second, we show, again by induction on j, that $\mathsf{obs}_{\mathcal{A}_{\mathbb{N}}}(q_0)(a^j) = 2^j - 1$. This implies the claim, as $\mathcal{L} = \mathsf{obs}_{\mathcal{A}_{\mathbb{N}}}(q_0)$. For $j = 0$ we have $\mathsf{obs}_{\mathcal{A}_{\mathbb{N}}}(q_0)(a^j) = o(q_0) = 0 = 2^0 - 1$ as required. For the inductive step, let $j = k + 1$ and assume $\mathsf{obs}_{\mathcal{A}_{\mathbb{N}}}(q_0)(a^k) = 2^k - 1$. We calculate

$$\begin{aligned}\mathsf{obs}_{\mathcal{A}_{\mathbb{N}}}(q_0)(a^{k+1}) &= \mathsf{obs}_{\mathcal{A}_{\mathbb{N}}}(q_0 + q_1)(a^k)\\ &= \mathsf{obs}_{\mathcal{A}_{\mathbb{N}}}(q_0)(a^k) + \mathsf{obs}_{\mathcal{A}_{\mathbb{N}}}(q_1)(a^k)\\ &= (2^k - 1) + 2^k\\ &= 2^{k+1} - 1.\end{aligned}$$

Note that in particular the language $\mathcal{L}$ is injective.

Towards a contradiction, suppose the algorithm does terminate with table (S, E). Let $J = \{j \in \mathbb{N} \mid a^j \in S\}$ and define $n = \mathsf{max}(J)$. Since the algorithm terminates with table (S, E), the latter must be closed. In particular, there exist $k_j \in \mathbb{N}$ for all $j \in J$ such that $\sum_{j \in J} k_j \cdot \mathsf{row}_{\mathcal{L}}(a^j) = \mathsf{srow}_{\mathcal{L}}(a^n a)$. We consider two cases. First assume $E = \{\varepsilon\}$ and let $\mathcal{A} = (Q', \delta', i', o')$ be the hypothesis. For all $l \in \mathbb{N}$ we have $\mathsf{row}^{\sharp}_{\mathcal{L}}(\mathsf{reach}^{\dagger}_{\mathcal{A}}(a^l))(\varepsilon) = 2^l - 1$ because $\mathcal{A}$ must be correct. Thus, if $a^l \in S \cdot A$, then $\mathsf{row}^{\sharp}_{\mathcal{L}}(\mathsf{reach}^{\dagger}_{\mathcal{A}}(a^l)) = \mathsf{srow}_{\mathcal{L}}(a^l)$. In particular,

$$\mathsf{row}^{\sharp}_{\mathcal{L}}(\mathsf{reach}^{\dagger}_{\mathcal{A}}(a^n a)) = \mathsf{srow}_{\mathcal{L}}(a^n a) = \sum_{j \in J} k_j \cdot \mathsf{row}_{\mathcal{L}}(a^j).$$

Note that we can choose the k_j such that $\mathsf{reach}^{\dagger}_{\mathcal{A}}(a^n a) = \sum_{j \in J} k_j \cdot a^j$. Since

$$\begin{aligned}\mathsf{row}^{\sharp}_{\mathcal{L}}\left(\delta'^{\sharp}\left(\sum_{j \in J} k_j \cdot a^j\right)(a)\right) &= \mathsf{row}^{\sharp}_{\mathcal{L}}\left(\sum_{j \in J} k_j \cdot \delta'(a^j)(a)\right)\\ &= \sum_{j \in J} k_j \cdot \mathsf{row}_{\mathcal{L}}(\delta'(a^j)(a))\\ &= \sum_{j \in J} k_j \cdot \mathsf{srow}_{\mathcal{L}}(a^j a),\end{aligned}$$

we have $\mathsf{row}^{\sharp}_{\mathcal{L}}(\mathsf{reach}^{\dagger}_{\mathcal{A}}(a^n aa)) = \sum_{j \in J} k_j \cdot \mathsf{srow}_{\mathcal{L}}(a^j a)$ and therefore

$$\sum_{j \in J} k_j \cdot \mathsf{srow}_{\mathcal{L}}(a^j a)(\varepsilon) = \mathsf{row}^{\sharp}_{\mathcal{L}}(\mathsf{reach}^{\dagger}_{\mathcal{A}}(a^n aa))(\varepsilon) = 2^{n+2} - 1.$$

Then

$$\begin{aligned}2^{n+2} - 1 &= \sum_{j \in J} k_j \cdot \mathsf{srow}_{\mathcal{L}}(a^j a)(\varepsilon) = \sum_{j \in J} k_j (2^{j+1} - 1)\\ &= 2\left(\sum_{j \in J} k_j (2^j - 1)\right) + \sum_{j \in J} k_j = 2(2^{n+1} - 1) + \sum_{j \in J} k_j,\end{aligned}$$

so $\sum_{j \in J} k_j = 1$. This is only possible if there is $j_1 \in J$ s.t. $k_{j_1} = 1$ and $k_j = 0$ for all $j \in J \setminus \{j_1\}$. However, this implies that $\mathsf{row}_{\mathcal{L}}(a^{j_1}) = \mathsf{srow}_{\mathcal{L}}(a^n a)$, which contradicts injectivity of $\mathcal{L}$ as $n \geq j_1$. Thus, the algorithm did not terminate.

For the other case, assume there is $a^m \in E$ such that $m \geq 1$. We have

$$2^{n+1} - 1 = \mathsf{srow}_{\mathcal{L}}(a^n a)(\varepsilon) = \sum_{j \in J} k_j \cdot \mathsf{row}_{\mathcal{L}}(a^j)(\varepsilon) = \sum_{j \in J} k_j (2^j - 1),$$

so

$$\begin{aligned}
\sum_{j \in J} k_j (2^{j+m} - 1) &= \sum_{j \in J} k_j \cdot \mathsf{row}_{\mathcal{L}}(a^j)(a^m) \\
&= \mathsf{srow}_{\mathcal{L}}(a^n a)(a^m) \\
&= 2^{n+m+1} - 1 \\
&= 2^m (2^{n+1} - 1) + 2^m - 1 \\
&= 2^m \left(\sum_{j \in J} k_j (2^j - 1) \right) + 2^m - 1 \\
&= \left(\sum_{j \in J} k_j (2^{j+m} - 2^m) \right) + 2^m - 1 \\
&= \left(\sum_{j \in J} k_j (2^{j+m} - 1) \right) + \left(\sum_{j \in J} k_j (1 - 2^m) \right) + 2^m - 1.
\end{aligned}$$

Then

$$\left(\sum_{j \in J} k_j (1 - 2^m) \right) + 2^m - 1 = 0.$$

Since $m \geq 1$ this is only possible if there is $j_1 \in J$ s.t. $k_{j_1} = 1$ and $k_j = 0$ for all $j \in J \setminus \{j_1\}$. However, this implies $\mathsf{row}_{\mathcal{L}}(a^{j_1}) = \mathsf{srow}_{\mathcal{L}}(a^n a)$, which again contradicts injectivity of $\mathcal{L}$ as $n \geq j_1$. Thus, the algorithm did not terminate. □

Remark 18. Our proof shows non-termination for a bigger class of algorithms than Algorithm 1; it uses only the definition of the hypothesis, that closedness is satisfied before constructing the hypothesis, that S and E contain the empty word, and that termination implies correctness. For instance, adding the prefixes of a counterexample to S instead of its suffixes to E will not fix the issue.

We have thus shown that our algorithm does not instantiate to a terminating one for an arbitrary semiring. To contrast this negative result, in the next section we identify a class of semirings not previously explored in the learning literature where we can guarantee a terminating instantiation.

6 Learning WFAs over PIDs

We show that for a subclass of semirings, namely *principal ideal domains (PIDs)*, the abstract learning algorithm of Section 4 terminates. This subclass includes the integers, Gaussian integers, and rings of polynomials in one variable with coefficients in a field. We will prove that the Hankel matrix of a language over a PID accepted by a WFA has analogous properties to those of vector spaces—finite rank, a notion of progress measure, and the ascending chain condition. We also give a sufficient condition for PIDs to be solvable, which by Proposition 10 guarantees the existence of a closedness strategy for the learning algorithm.

To define PIDs, we first need to introduce ideals. Given a ring $\mathbb{S}$, a *(left) ideal* I of $\mathbb{S}$ is an additive subgroup of $\mathbb{S}$ s.t. for all $s \in \mathbb{S}$ and $i \in I$ we have $si \in I$. The ideal I is *(left) principal* if it is of the form $I = \mathbb{S}s$ for some $s \in \mathbb{S}$.

Definition 19 (PID). *A* principal ideal domain $\mathbb{P}$ *is a non-zero commutative ring in which every ideal is principal and where for all* $p_1, p_2 \in \mathbb{P}$ *such that* $p_1 p_2 = 0$ *we have* $p_1 = 0$ *or* $p_2 = 0$.

A module M over a PID $\mathbb{P}$ is called *torsion free* if for all $p \in \mathbb{P}$ and any $m \in M$ such that $p \cdot m = 0$ we have $p = 0$ or $m = 0$. It is a standard result that a module over a PID is torsion free if and only if it is free [17, Theorem 3.10].

The next definition of *rank* is analogous to that of the dimension of a vector space and will form the basis for the progress measure.

Definition 20 (Rank). *We define the* rank *of a finitely generated free module* $V(X)$ *over a PID as* $\mathsf{rank}(V(X)) = |X|$.

This definition extends to any finitely generated free module over a PID, as $V(X) \cong V(Y)$ for finite sets X and Y implies $|X| = |Y|$ [17, Theorem 3.4].

Now that we have a candidate for a progress measure function, we need to prove it has the required properties. The following lemmas will help with this.

Lemma 21. *Given finitely generated free modules* M, N *over a PID s.t.* $\mathsf{rank}(M) \geq \mathsf{rank}(N)$, *any surjective module homomorphism* $f\colon N \to M$ *is injective.*

Proof. Since $\mathsf{rank}(M) \geq \mathsf{rank}(N)$, there exists a surjective module homomorphism $g\colon M \to N$. Therefore $g \circ f\colon N \to N$ is surjective and by [23] an iso. In particular, f is injective. □

Lemma 22. *If* M *and* N *are finitely generated free modules over a PID such that there exists a surjective module homomorphism* $f\colon N \to M$, *then* $\mathsf{rank}(M) \leq \mathsf{rank}(N)$. *If* f *is not injective, then* $\mathsf{rank}(M) < \mathsf{rank}(N)$.

Proof. Let $f\colon N \to M$ be a surjective module homomorphism. Suppose towards a contradiction that $\mathsf{rank}(M) > \mathsf{rank}(N)$. By Lemma 21 f is injective, so M is isomorphic to a submodule of N and $\mathsf{rank}(M) \leq \mathsf{rank}(N)$ [17]; contradiction.

For the second part, suppose f is not injective and assume towards a contradiction that $\mathsf{rank}(M) \geq \mathsf{rank}(N)$. Again by Lemma 21 f is injective, which is a contradiction with our assumption. Thus, in this case $\mathsf{rank}(M) < \mathsf{rank}(N)$. □

The lemma below states that the Hankel matrix of a weighted language over a PID has finite rank which bounds the rank of any module generated by an observation table. This will be used to define a progress measure, used to prove termination of the learning algorithm for weighted languages over PIDs.

Lemma 23 (Hankel matrix rank for PIDs). *When targeting a language accepted by a WFA over a PID, any module generated by an observation table is free. Moreover, the Hankel matrix has finite rank that bounds the rank of any module generated by an observation table.*

Proof. Given a WFA $\mathcal{A} = (Q, \delta, i, o)$, let M be the free module generated by Q. Note that the Hankel matrix is the image of the composition $\mathsf{obs}_{\mathcal{A}} \circ \mathsf{reach}_{\mathcal{A}}$. Consider the image of the module homomorphism $\mathsf{reach}_{\mathcal{A}} \colon V(A^*) \to M$, which we write as R. Since R is a submodule of M, we know from [17] that R is free and finitely generated with $\mathsf{rank}(R) \leq \mathsf{rank}(M)$. The Hankel matrix can now be obtained as the image of the restriction of $\mathsf{obs}_{\mathcal{A}} \colon M \to \mathbb{S}^{A^*}$ to the domain R. Let H be this image, which we know is finitely generated because R is. Since H is a submodule of the torsion free module $\mathbb{S}^{A^*}$, it is also torsion free and therefore free. We also have a surjective module homomorphism $s \colon R \to H$, so by Lemma 22 we find $\mathsf{rank}(H) \leq \mathsf{rank}(R)$.

Let N be the module generated by an observation table (S, E). We have that N is a quotient of the module generated by (S, A^*), which in turn is a submodule of H. Using again [17] and Lemma 22 we conclude that N is free and finitely generated with $\mathsf{rank}(N) \leq \mathsf{rank}(H)$. □

The second part of Lemma 23 would follow from a PID variant of Fliess' theorem [12]. We are not aware of such a result, and leave this for future work.

Proposition 24 (Progress measure for PIDs). *There exists a progress measure for any language accepted by a WFA over a PID.*

Proof. Define $\mathsf{size}(S, E) = \mathsf{rank}(M)$, where M is the module generated by the table (S, E). By Lemma 23 this is bounded by the rank of the Hankel matrix. If M and N are modules generated by two tables such that N is a strict quotient of M, then by Lemma 22 we have $\mathsf{rank}(M) > \mathsf{rank}(N)$. □

Recall that, for termination of the algorithm, Theorem 14 requires a progress measure, which we defined above, and it requires the Hankel matrix of the language to satisfy the ascending chain condition (Definition 13). Proposition 25 shows that the latter is always the case for languages over PIDs.

Proposition 25 (Ascending chain condition PIDs). *The Hankel matrix of a language accepted by a WFA over a PID satisfies the ascending chain condition.*

Proof. Let H be the Hankel matrix, which has finite rank by Lemma 23. If

$$M_1 \subseteq M_2 \subseteq M_3 \subseteq \cdots$$

is any chain of submodules of H, then $M = \bigcup_{i \in \mathbb{N}} M_i$ is a submodule of H and therefore also of finite rank [17]. Let B be a finite basis of M. There exists $n \in \mathbb{N}$ such that $B \subseteq M_n$, so $M_n = M$. □

The last ingredient for the abstract algorithm is solvability of the semiring: the following fact provides a sufficient condition for a PID to be solvable.

Proposition 26 (PID solvability). *A PID $\mathbb{P}$ is solvable if all of its ring operations are computable and if each element of $\mathbb{P}$ can be* effectively *factorised into irreducible elements.*

Proof. It is well-known that a system of equations of the form $A\underline{x} = \underline{b}$ with integer coefficients can be efficiently solved via computing the Smith normal form [25] of A. The algorithm generalises to principal ideal domains, if we assume that the factorisation of any given element of the principal ideal domain[7] into irreducible elements is computable, cf. the algorithm in [16, p. 79-84]. To see that all steps in this algorithm can be computed, one has to keep in mind that the factorisation can be used to determine the greatest common divisor of any two elements of the principal ideal domain. □

Remark 27. In the case that we are dealing with an Euclidean domain $\mathbb{P}$, a sufficient condition for $\mathbb{P}$ to be solvable is that Euclidean division is computable (again this can be deduced from inspecting the algorithm in [16, p. 79-84]). Such a PID behaves essentially like the ring of integers.

Putting everything together, we obtain the main result of this section.

Theorem 28 (Termination for PIDs). *Algorithm 1 can be instantiated and terminates for any language accepted by a WFA over a PID of which all ring operations are computable and of which each element can be effectively factorised into irreducible elements.*

Proof. To instantiate the algorithm, we need a closedness strategy. According to Proposition 10 it is sufficient for the PID to be solvable, which is shown by Proposition 26. Proposition 24 provides a progress measure, and we know from Proposition 25 that the Hankel matrix satisfies the ascending chain condition, so by Theorem 14 the algorithm terminates. □

The example run given in Section 2.1 is the same when performed over the integers. We note that if the teacher holds an automaton model of the correct language, equivalence queries are decidable by lifting the embedding of the PID into its *quotient field* to the level of WFAs and checking equivalence there.

7 Discussion

We have introduced a general algorithm for learning WFAs over arbitrary semirings, together with sufficient conditions for termination. We have shown an inherent termination issue over the natural numbers and proved termination for PIDs. Our work extends the results by Bergadano and Varricchio [7], who showed that WFAs over fields could be learned from a teacher. Although we note that a PID can be embedded into its corresponding field of fractions, the WFAs produced when learning over the field potentially have weights outside the PID.

[7] Note that factorisations exist as each principal ideal domain is also a unique factorisation domain, cf. e.g. [17, Thm. 2.23].

Algorithmic issues with WFAs over arbitrary semirings have been identified before. For instance, Krob [18] showed that language equivalence is undecidable for WFAs over the tropical semiring.

On the technical level, a variation on WFAs is given by probabilistic automata, where transitions point to convex rather than linear combinations of states. One easily adapts the example from Section 5 to show that learning probabilistic automata has a similar termination issue. On the positive side, Tappler et al. [26] have shown that deterministic MDPs can be learned using an $L^\star$ based algorithm. The deterministic MDPs in *loc.cit.* are very different from the automata in our paper, as their states generate observable output that allows to identify the current state based on the generated input-output sequence.

One drawback of the ascending chain condition on the Hankel matrix is that this does not give any indication of the number of steps the algorithm requires. Indeed, the submodule chains traversed, although converging, may be arbitrarily long. We would like to measure and bound the progress made when fixing closedness defects, but this turns out to be challenging for PIDs. The rank of the module generated by the table may not increase. We leave an investigation of alternative measures to future work.

We would also like to adapt the algorithm so that for PIDs it always produces minimal automata. At the moment this is already the case for fields,[8] since adding a row due to a closedness defect preserves linear independence of the image of row. For PIDs things are more complicated—adding rows towards closedness may break linear independence and thus a basis needs to be found in $\mathsf{row}^\sharp$. This complicates the construction of the hypothesis.

Our results show that, on the one hand, WFAs can be learned over finite semirings and arbitrary PIDs (assuming computability of the relevant operations) and, on the other hand, that there exists an infinite commutative semiring for which they cannot be learned. However, there are many classes of semirings in between commutative semirings and PIDs, of which we would like to know whether their WFAs can be learned by our general algorithm.

Finally, we would like to generalise our results to extend the framework introduced in [13], which focusses on learning automata with side-effects over a monad. WFAs as considered in the present paper are an instance of those, where the monad is the free semimodule monad $V(-)$. At the moment, the results in [13] apply to a monad that preserves finite sets, but much of our general WFA learning algorithm and termination argument can be extended to that setting. It would be interesting to see if crucial properties of PIDs that lead to a progress measure and to satisfying the ascending chain condition could also be translated to the monad level.

Acknowledgments. We thank Joshua Moerman for comments and discussions.

[8] There is one exception: the language that assigns 0 to every word, which is accepted by a WFA with no states. The algorithm initialises the set of row labels, which constitute the state space of the hypothesis, with the empty word.

References

1. Fides Aarts, Paul Fiterau-Brostean, Harco Kuppens, and Frits W. Vaandrager. Learning register automata with fresh value generation. In Martin Leucker, Camilo Rueda, and Frank D. Valencia, editors, *ICTAC*, volume 9399 of *LNCS*, pages 165–183. Springer, 2015.
2. Cyril Allauzen, Mehryar Mohri, and Ameet Talwalkar. Sequence kernels for predicting protein essentiality. In William W. Cohen, Andrew McCallum, and Sam T. Roweis, editors, *ICML*, volume 307 of *ACM International Conference Proceeding Series*, pages 9–16. ACM, 2008.
3. Benjamin Aminof, Orna Kupferman, and Robby Lampert. Formal analysis of online algorithms. In Tevfik Bultan and Pao-Ann Hsiung, editors, *ATVA*, volume 6996 of *LNCS*, pages 213–227. Springer, 2011.
4. Dana Angluin. Learning regular sets from queries and counterexamples. *Information and computation*, 75(2):87–106, 1987.
5. Borja Balle and Mehryar Mohri. Spectral learning of general weighted automata via constrained matrix completion. In Peter L. Bartlett, Fernando C. N. Pereira, Christopher J. C. Burges, Léon Bottou, and Kilian Q. Weinberger, editors, *NIPS*, pages 2168–2176, 2012.
6. Borja Balle and Mehryar Mohri. Learning weighted automata. In Andreas Maletti, editor, *CAI*, volume 9270 of *LNCS*, pages 1–21. Springer, 2015.
7. Francesco Bergadano and Stefano Varricchio. Learning behaviors of automata from multiplicity and equivalence queries. *SIAM J. Comput.*, 25(6):1268–1280, December 1996.
8. Benedikt Bollig, Peter Habermehl, Carsten Kern, and Martin Leucker. Angluin-style learning of NFA. In Craig Boutilier, editor, *IJCAI*, pages 1004–1009, 2009.
9. Michele Boreale. Weighted bisimulation in linear algebraic form. In *CONCUR*, volume 5710 of *LNCS*, pages 163–177. Springer, 2009.
10. Jack W. Carlyle and Azaria Paz. Realizations by stochastic finite automata. *J. Comput. Syst. Sci.*, 5(1):26–40, 1971.
11. Karel Culik II and Jarkko Kari. Image compression using weighted finite automata. *Computers & Graphics*, 17(3):305–313, 1993.
12. Michel Fliess. Matrices de Hankel. *J. Math. Pures Appl*, 53(9):197–222, 1974.
13. Gerco van Heerdt, Matteo Sammartino, and Alexandra Silva. Optimizing automata learning via monads. *arXiv preprint arXiv:1704.08055*, 2017.
14. Falk Howar and Bernhard Steffen. Active automata learning in practice - an annotated bibliography of the years 2011 to 2016. In Amel Bennaceur, Reiner Hähnle, and Karl Meinke, editors, *Machine Learning for Dynamic Software Analysis: Potentials and Limits - International Dagstuhl Seminar 16172*, volume 11026 of *LNCS*, pages 123–148. Springer, 2018.
15. Malte Isberner, Falk Howar, and Bernhard Steffen. The open-source learnlib - A framework for active automata learning. In Daniel Kroening and Corina S. Pasareanu, editors, *CAV*, volume 9206 of *LNCS*, pages 487–495. Springer, 2015.
16. Nathan Jacobson. *Lectures in Abstract Algebra*, volume 31 of *GTM*. Springer, 1953.
17. Nathan Jacobson. *Basic algebra I*. Courier Corporation, 2012.
18. Daniel Krob. The equality problem for rational series with multiplicities in the tropical semiring is undecidable. *International Journal of Algebra and Computation*, 4(3):405–425, 1994.

19. Oded Maler and Amir Pnueli. On the learnability of infinitary regular sets. *Inform. and Comput.*, 118:316–326, 1995.
20. Joshua Moerman, Matteo Sammartino, Alexandra Silva, Bartek Klin, and Michal Szynwelski. Learning nominal automata. In Giuseppe Castagna and Andrew D. Gordon, editors, *POPL*, pages 613–625. ACM, 2017.
21. Mehryar Mohri, Fernando Pereira, and Michael Riley. Weighted automata in text and speech processing. *CoRR*, abs/cs/0503077, 2005.
22. Malte Mues, Falk Howar, Kasper Søe Luckow, Temesghen Kahsai, and Zvonimir Rakamaric. Releasing the PSYCO: using symbolic search in interface generation for java. *ACM SIGSOFT Software Engineering Notes*, 41(6):1–5, 2016.
23. Morris Orzech. Onto endomorphisms are isomorphisms. *The American Mathematical Monthly*, 78(4):357–362, 1971.
24. Muzammil Shahbaz and Roland Groz. Inferring Mealy machines. In *FM*, volume 5850 of *LNCS*, pages 207–222, Berlin, Heidelberg, 2009. Springer-Verlag.
25. Henry J. Stephen Smith. On systems of linear indeterminate equations and congruences. *Philosophical Transactions of the Royal Society of London*, 151:293–326, 1861.
26. Martin Tappler, Bernhard K. Aichernig, Giovanni Bacci, Maria Eichlseder, and Kim G. Larsen. L^*-Based Learning of Markov Decision Processes. In Maurice H. ter Beek, Annabelle McIver, and José N. Oliveira, editors, *FM*, volume 11800 of *LNCS*, pages 651–669. Springer, 2019.
27. Frits W. Vaandrager. Model learning. *Commun. ACM*, 60(2):86–95, 2017.
28. Juan Miguel Vilar. Query learning of subsequential transducers. In *ICGI*, volume 1147 of *LNCS*, pages 72–83. Springer, 1996.

The Polynomial Complexity of Vector Addition Systems with States

Florian Zuleger (✉)
zuleger@forsyte.tuwien.ac.at

TU Wien

Abstract. Vector addition systems are an important model in theoretical computer science and have been used in a variety of areas. In this paper, we consider vector addition systems with states over a parameterized initial configuration. For these systems, we are interested in the standard notion of computational time complexity, i.e., we want to understand the length of the longest trace for a fixed vector addition system with states depending on the size of the initial configuration. We show that the asymptotic complexity of a given vector addition system with states is either $\Theta(N^k)$ for some computable integer k, where N is the size of the initial configuration, or at least exponential. We further show that k can be computed in polynomial time in the size of the considered vector addition system. Finally, we show that $1 \leq k \leq 2^n$, where n is the dimension of the considered vector addition system.

1 Introduction

Vector addition systems (VASs) [13], which are equivalent to Petri nets, are a popular model for the analysis of parallel processes [7]. Vector addition systems with states (VASSs) [10] are an extension of VASs with a finite control and are a popular model for the analysis of concurrent systems, because the finite control can for example be used to model shared global memory [12]. In this paper, we consider VASSs over a parameterized initial configuration. For these systems, we are interested in the standard notion of computational time complexity, i.e., we want to understand the length of the longest execution for a fixed VASS depending on the size of the initial configuration. VASSs over a parameterized initial configuration naturally arise in two areas: 1) *The parameterized verification problem.* For concurrent systems the number of system processes is often not known in advance, and thus the system is designed such that a template process can be instantiated an arbitrary number of times. The problem of analyzing the concurrent system for all possible system sizes is a common theme in the literature [9, 8, 1, 11, 4, 2, 3]. 2) *Automated complexity analysis of programs.* VASSs (and generalizations) have been used as backend in program analysis tools for automated complexity analysis [18–20]. The VASS considered by these tools are naturally parameterized over the initial configuration, modelling the dependency of the program complexity on the program input. The cited papers have proposed practical techniques but did not give complete algorithms.

J. Goubault-Larrecq and B. König (Eds.): FOSSACS 2020, LNCS 12077, pp. 622–641, 2020.
https://doi.org/10.1007/978-3-030-45231-5_32

Two recent papers have considered the computational time complexity of VASSs over a parameterized initial configuration. [15] presents a PTIME procedure for deciding whether a VASS is polynomial or at least exponential, but does not give a precise analysis in case of polynomial complexity. [5] establishes the precise asymptotic complexity for the special case of VASSs whose configurations are linearly bounded in the size of the initial configuration. In this paper, we generalize both results and fully characterize the asymptotic behaviour of VASSs with polynomial complexity: We show that the asymptotic complexity of a given VASS is either $\Theta(N^k)$ for some computable integer k, where N is the size of the initial configuration, or at least exponential. We further show that k can be computed in PTIME in the size of the considered VASS. Finally, we show that $1 \leq k \leq 2^n$, where n is the dimension of the considered VASS.

1.1 Overview and Illustration of Results

We discuss our approach on the VASS $\mathcal{V}_{run}$, stated in Figure 1, which will serve as running example. The VASS has dimension 3 (i.e., the vectors annotating the transitions have dimension 3) and four states s_1, s_2, s_3, s_4. In this paper we will always represent vectors using a set of variables Var, whose cardinality equals the dimension of the VASS. For $\mathcal{V}_{run}$ we choose $Var = \{x, y, z\}$ and use x, y, z as indices for the first, second and third component of 3-dimensional vectors. The configurations of a VASS are pairs of states and valuations of the variables to non-negative integers. A step of a VASS moves along a transition from the current state to a successor state, and adds the vector labelling the transition to the current valuation; a step can only be taken if the resulting valuation is non-negative. For the computational time complexity analysis of VASSs, we consider traces (sequences of steps) whose initial configurations consist of a valuation whose maximal value is bounded by N (the parameter used for bounding the size of the initial configuration). The computational time complexity is then the length of the longest trace whose initial configuration is bounded by N. For ease of exposition, we will in this paper only consider VASSs whose control-flow graph is *connected*. (For the general case, we remark that one needs to decompose a VASS into its strongly-connected components (SCCs), which can then be analyzed in isolation, following the DAG-order of the SCC decomposition; for this, one slightly needs to generalize the analysis in this paper to initial configurations with values $\Theta(N^{k_x})$ for every variable $x \in Var$, where $k_x \in \mathbb{Z}$.) For ease of exposition, we further consider traces over arbitrary initial states (instead of some fixed initial state); this is justified because for a fixed initial state one can always restrict the control-flow graph to the reachable states, and then the two options result in the same notion of computational complexity (up to a constant offset, which is not relevant for our asymptotic analysis).

In order to analyze the computational time complexity of a considered VASS, our approach computes *variable bounds* and *transition bounds*. A variable bound is the maximal value of a variable reachable by any trace whose initial configuration is bounded by N. A transition bound is the maximal number of times a transition appears in any trace whose initial configuration is bounded by N. For

$\mathcal{V}_{run}$, our approach establishes the linear variable bound $\Theta(N)$ for x and y, and the quadratic bound $\Theta(N^2)$ for z. We note that because the variable bound of z is quadratic and not linear, $\mathcal{V}_{run}$ cannot be analyzed by the procedure of [5]. Our approach establishes the bound $\Theta(N)$ for the transitions $s_1 \to s_3$ and $s_4 \to s_2$, the bound $\Theta(N^2)$ for transitions $s_1 \to s_2$, $s_2 \to s_1$, $s_3 \to s_4$, $s_4 \to s_3$, and the bound $\Theta(N^3)$ for all self-loops. The computational complexity of $\mathcal{V}_{run}$ is then the maximum of all transition bounds, i.e., $\Theta(N^3)$. In general, our main algorithm (Algorithm 1 presented in Section 4) either establishes that the VASS under analysis has at least exponential complexity or computes asymptotically precise variable and transition bounds $\Theta(N^k)$, with k computable in PTIME and $1 \leq k \leq 2^n$, where n is the dimension of the considered VASS. We note that our upper bound 2^n also improves the analysis of [15], which reports an exponential dependence on the number of transitions (and not only on the dimension).

We further state a family $\mathcal{V}_n$ of VASSs, which illustrate that k can indeed be exponential in the dimension (the example can be skipped on first reading). $\mathcal{V}_n$ uses variables $x_{i,j}$ and consists of states $s_{i,j}$, for $1 \leq i \leq n$ and $j = 1, 2$. We note that $\mathcal{V}_n$ has dimension $2n$. $\mathcal{V}_n$ consists of the transitions

- $s_{i,1} \xrightarrow{d} s_{i,2}$, for $1 \leq i \leq n$, with $d(x_{i,1}) = -1$ and $d(x) = 0$ for all $x \neq x_{i,1}$,
- $s_{i,2} \xrightarrow{d} s_{i,1}$, for $1 \leq i \leq n$, with $d(x) = 0$ for all x,
- $s_{i,1} \xrightarrow{d} s_{i,1}$, for $1 \leq i \leq n$, with $d(x_{i,1}) = -1$, $d(x_{i,2}) = 1$, $d(x_{i+1,1}) = d(x_{i+1,2}) = 1$ in case $i < n$, and $d(x) = 0$ for all other x,
- $s_{i,2} \xrightarrow{d} s_{i,2}$, for $1 \leq i \leq n$, with $d(x_{i,1}) = 1$, $d(x_{i,2}) = -1$, and $d(x) = 0$ for all other x,
- $s_{i,1} \xrightarrow{d} s_{i+1,1}$, for $1 \leq i < n$, with $d(x_{i,1}) = -1$ and $d(x) = 0$ for all $x \neq x_{i,1}$,
- $s_{i+1,2} \xrightarrow{d} s_{i,2}$, for $1 \leq i < n$, with $d(x) = 0$ for all x.

$\mathcal{V}_{exp}$ in Figure 1 depicts $\mathcal{V}_n$ for $n = 3$, where the vector components are stated in the order $x_{1,1}, x_{1,2}, x_{2,1}, x_{2,2}, x_{3,1}, x_{3,2}$. It is not hard to verify for all $1 \leq i \leq n$ that $\Theta(N^{2^{i-1}})$ is the precise asymptotic variable bound for $x_{i,1}$ and $x_{i,2}$, that $s_{i,1} \to s_{i,2}$, $s_{i,2} \to s_{i,1}$ and $s_{i,1} \to s_{i+1,1}$, $s_{i+1,2} \to s_{i,2}$ in case $i < n$, and that $\Theta(N^{2^i})$ is the precise asymptotic transition bound for $s_{i,1} \to s_{i,1}$, $s_{i,2} \to s_{i,2}$ (Algorithm 1 can be used to find these bounds).

1.2 Related Work

A celebrated result on VASs is the EXPSPACE-completeness [16,17] of the boundedness problem. Deciding termination for a VAS with a *fixed* initial configuration can be reduced to the boundedness problem, and is therefore also EXPSPACE-complete; this also applies to VASSs, whose termination problem can be reduced to the VAS termination problem. In contrast, deciding the termination of VASSs for *all* initial configurations is in PTIME. It is not hard to see that non-termination over all initial configurations is equivalent to the existence of non-negative cycles (e.g., using Dickson's Lemma [6]). Kosaraju and Sullivan have given a PTIME procedure for the detection of zero-cycles [14], which can be easily be adapted to non-negative cycles. The existence of zero-cycles is decided

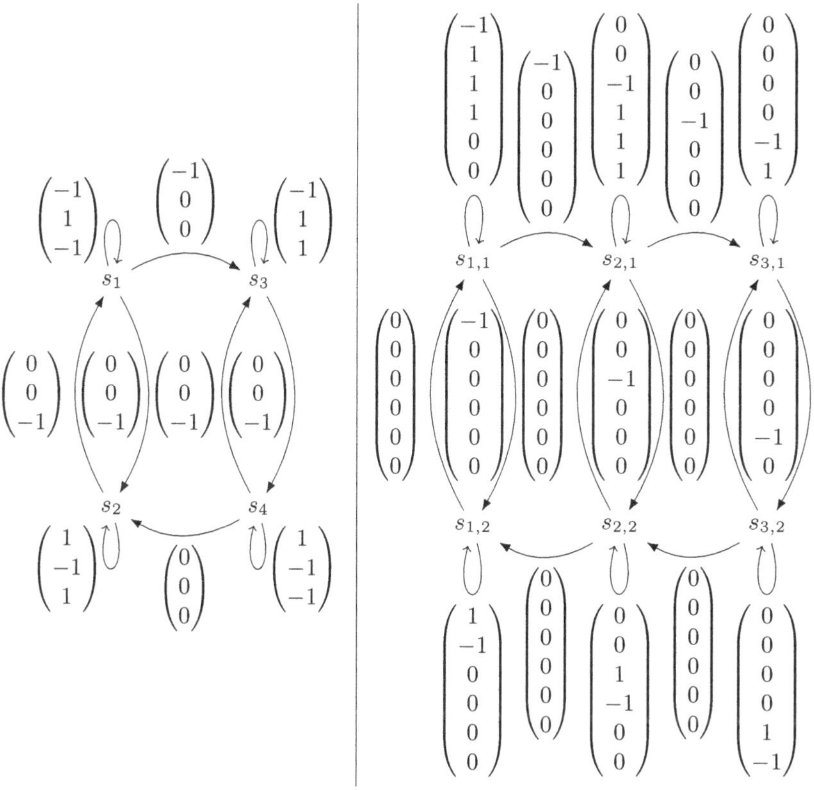

Fig. 1. VASS $\mathcal{V}_{run}$ (left) and VASS $\mathcal{V}_{exp}$ (right)

by the repeated use of a constraint system in order to remove transitions that can definitely not be part of a zero-cycle. The algorithm of Kosaraju and Sullivan forms the basis for both cited papers [15, 5], as well as the present paper.

A line of work [18–20] has used VASSs (and their generalizations) as backends for the automated complexity analysis of C programs. These algorithms have been designed for practical applicability, but are not complete and no theoretical analysis of their precision has been given. We point out, however, that these papers have inspired the Bound Proof Principle in Section 5.

2 Preliminaries

Basic Notation. For a set X we denote by $|X|$ the number of elements of X. Let $\mathbb{S}$ be either $\mathbb{N}$ or $\mathbb{Z}$. We write $\mathbb{S}^I$ for the set of vectors over $\mathbb{S}$ indexed by some set I. We write $\mathbb{S}^{I \times J}$ for the set of matrices over $\mathbb{S}$ indexed by I and J. We write $\mathbf{1}$ for the vector which has entry 1 in every component. Given $a \in \mathbb{S}^I$, we write $a(i) \in \mathbb{S}$ for the entry at line $i \in I$ of a, and $\|a\| = \max_{i \in I} |a(i)|$ for the maximum absolute value of a. Given $a \in \mathbb{S}^I$ and $J \subseteq I$, we denote by $a|_J \in \mathbb{S}^J$ the restriction of a to J, i.e., we set $a|_J(i) = a(i)$ for all $i \in J$. Given $A \in \mathbb{S}^{I \times J}$, we write $A(j)$ for the vector in column $j \in J$ of A and $A(i,j) \in \mathbb{S}$ for the entry in column $i \in I$ and row $j \in J$ of A. Given $A \in \mathbb{S}^{I \times J}$ and $K \subseteq J$, we denote by $A|_K \in \mathbb{S}^{I \times K}$ the restriction of A to K, i.e., we set $A|_K(i,j) = A(i,j)$ for all $(i,j) \in I \times K$. We write **Id** for the square matrix which has entries 1 on the diagonal and 0 otherwise. Given $a, b \in \mathbb{S}^I$ we write $a+b \in \mathbb{S}^I$ for component-wise addition, $c \cdot a \in \mathbb{S}^I$ for multiplying every component of a by some $c \in \mathbb{S}$ and $a \geq b$ for component-wise comparison. Given $A \in \mathbb{S}^{I \times J}$, $B \in \mathbb{S}^{J \times K}$ and $x \in \mathbb{S}^J$, we write $AB \in \mathbb{S}^{I \times K}$ for the standard matrix multiplication, $Ax \in \mathbb{S}^I$ for the standard matrix-vector multiplication, $A^T \in \mathbb{S}^{J \times I}$ for the transposed matrix of A and $x^T \in \mathbb{S}^{1 \times J}$ for the transposed vector of x.

Vector Addition System with States (VASS). Let *Var* be a finite set of variables. A vector addition system with states (VASS) $\mathcal{V} = (St(\mathcal{V}), Trns(\mathcal{V}))$ consists of a finite set of *states* $St(\mathcal{V})$ and a finite set of *transitions* $Trns(\mathcal{V})$, where $Trns(\mathcal{V}) \subseteq St(\mathcal{V}) \times \mathbb{Z}^{Var} \times St(\mathcal{V})$; we call $n = |Var|$ the *dimension* of $\mathcal{V}$. We write $s_1 \xrightarrow{d} s_2$ to denote a transition $(s_1, d, s_2) \in Trns(\mathcal{V})$; we call the vector d the *update* of transition $s_1 \xrightarrow{d} s_2$. A *path* π of $\mathcal{V}$ is a finite sequence $s_0 \xrightarrow{d_1} s_1 \xrightarrow{d_2} \cdots s_k$ with $s_i \xrightarrow{d_{i+1}} s_{i+1} \in Trns(\mathcal{V})$ for all $0 \leq i < k$. We define the *length* of π by $length(\pi) = k$ and the *value* of π by $val(\pi) = \sum_{i \in [1,k]} d_i$. Let $\texttt{instance}(\pi, t)$ be the number of times π contains the transition t, i.e., the number of indices i such that $t = s_i \xrightarrow{d_i} s_{i+1}$. We remark that $length(\pi) = \sum_{t \in Trns(\mathcal{V})} \texttt{instance}(\pi, t)$ for every path π of $\mathcal{V}$. Given a finite path π_1 and a path π_2 such that the last state of π_1 equals the first state of π_2, we write $\pi = \pi_1\pi_2$ for the path obtained by joining the last state of π_1 with the first state of π_2; we call π the *concatenation* of π_1 and π_2, and $\pi_1\pi_2$ a *decomposition* of π. We say π' is a *sub-path* of π, if there is a decomposition $\pi = \pi_1\pi'\pi_2$ for some π_1, π_2. A *cycle* is a path that has the same start- and end-state. A *multi-cycle* is a finite set of cycles. The value $val(M)$ of a multi-cycle M is the sum of the values of its cycles. $\mathcal{V}$ is *connected*, if for all $s, s' \in St(\mathcal{V})$ there is a path from s to s'. VASS $\mathcal{V}'$ is a *sub-VASS* of $\mathcal{V}$, if $St(\mathcal{V}') \subseteq St(\mathcal{V})$ and $Trns(\mathcal{V}') \subseteq Trns(\mathcal{V})$. Sub-VASSs $\mathcal{V}_1$ and $\mathcal{V}_2$ are *disjoint*, if $St(\mathcal{V}_1) \cap St(\mathcal{V}_2) = \emptyset$. A *strongly-connected component (SCC)* of a VASS $\mathcal{V}$ is a maximal sub-VASS S of $\mathcal{V}$ such that S is connected and $Trns(S) \neq \emptyset$.

Let $\mathcal{V}$ be a VASS. The set of *valuations* $Val(\mathcal{V}) = \mathbb{N}^{Var}$ consists of *Var*-vectors over the natural numbers (we assume $\mathbb{N}$ includes 0). The set of *configurations* $Cfg(\mathcal{V}) = St(\mathcal{V}) \times Val(\mathcal{V})$ consists of pairs of states and valuations. A *step* is a triple $((s_1, \nu_1), d, (s_2, \nu_2)) \in Cfg(\mathcal{V}) \times \mathbb{Z}^{dim(\mathcal{V})} \times Cfg(\mathcal{V})$ such that $\nu_2 = \nu_1 + d$ and $s_1 \xrightarrow{d} s_2 \in Trns(\mathcal{V})$. We write $(s_1, \nu_1) \xrightarrow{d} (s_2, \nu_2)$ to denote a step $((s_1, \nu_1), d, (s_2, \nu_2))$ of $\mathcal{V}$. A *trace* of $\mathcal{V}$ is a finite sequence $\zeta = (s_0, \nu_0) \xrightarrow{d_1} (s_1, \nu_1) \xrightarrow{d_2} \cdots (s_k, \nu_k)$ of steps. We lift the notions of length and instances from paths to traces in the obvious way: we consider the path $\pi = s_0 \xrightarrow{d_1} s_1 \xrightarrow{d_2} \cdots s_k$ that consists of the transitions used by ζ, and set $length(\zeta) := length(\pi)$ and $\texttt{instance}(\zeta, t) = \texttt{instance}(\pi, t)$, for all $t \in Trns(\mathcal{V})$. We denote by $\texttt{init}(\zeta) = \|\nu_0\|$ the maximum absolute value of the starting valuation ν_0 of ζ. We say that ζ *reaches* a valuation ν, if $\nu = \nu_k$. The *complexity* of $\mathcal{V}$ is the function $comp_{\mathcal{V}}(N) = \sup_{\text{trace } \zeta \text{ of } \mathcal{V}, \texttt{init}(\zeta) \leq N} length(\zeta)$, which returns for every $N \geq 0$ the supremum over the lengths of the traces ζ with $\texttt{init}(\zeta) \leq N$. The *variable bound* of a variable $x \in Var$ is the function $\texttt{vbound}_x(N) = \sup_{\text{trace } \zeta \text{ of } \mathcal{V}, \texttt{init}(\zeta) \leq N, \zeta \text{ reaches valuation } \nu} \nu(x)$, which returns for every $N \geq 0$ the supremum over the the values of x reachable by traces ζ with $\texttt{init}(\zeta) \leq N$. The *transition bound* of a transition $t \in Trns(\mathcal{V})$ is the function $\texttt{tbound}_t(N) = \sup_{\text{trace } \zeta \text{ of } \mathcal{V}, \texttt{init}(\zeta) \leq N} \texttt{instance}(\zeta, t)$, which returns for every $N \geq 0$ the supremum over the number of instances of t in traces ζ with $\texttt{init}(\zeta) \leq N$.

Rooted Tree. A *rooted tree* is a connected undirected acyclic graph in which one node has been designated as the root. We will usually denote the root by ι. We note that for every node η in a rooted tree there is a unique path of η to the root. The *parent* of a node $\eta \neq \iota$ is the node connected to η on the path to the root. Node η is a *child* of a node η', if η' is the parent of η. η' is a *descendent* of η, if η lies on the path from η' to the root; η' is a *strict* descendent, if furthermore $\eta \neq \eta'$. η is an *ancestor* of η', if η' a descendent of η; η is a *strict* ancestor, if furthermore $\eta \neq \eta'$. The *distance* of a node η to the root, is the number of nodes $\neq \eta$ on the path from η to the root. We denote by $\texttt{layer}(l)$ the set of all nodes with the same distance l to the root; we remark that $\texttt{layer}(0) = \{\iota\}$.

All proofs are presented in the extended version [21] for space reasons.

3 A Dichotomy Result

We will make use of the following matrices associated to a VASS throughout the paper: Let $\mathcal{V}$ be a VASS. We define the *update matrix* $D \in \mathbb{Z}^{Var \times Trns(\mathcal{V})}$ by setting $D(t) = d$ for all transitions $t = (s, d, s') \in Trns(\mathcal{V})$. We define the *flow matrix* $F \in \mathbb{Z}^{St(\mathcal{V}) \times Trns(\mathcal{V})}$ by setting $F(s, t) = -1$, $F(s', t) = 1$ for transitions $t = (s, d, s')$ with $s' \neq s$, and $F(s, t) = F(s', t) = 0$ for transitions $t = (s, d, s')$ with $s' = s$; in both cases we further set $F(s'', t) = 0$ for all states s'' with $s'' \neq s$ and $s'' \neq s'$. We note that every column t of F either contains exactly one -1 and 1 entry (in case the source and target of transition t are different) or only 0 entries (in case the source and target of transition t are the same).

Example 1. We state the update and flow matrix for $\mathcal{V}_{run}$ from Section 1:

$$D = \begin{pmatrix} -1 & 1 & -1 & 1 & 0 & 0 & 0 & 0 & -1 & 0 \\ 1 & -1 & 1 & -1 & 0 & 0 & 0 & 0 & 0 & 0 \\ -1 & 1 & 1 & -1 & -1 & -1 & -1 & -1 & 0 & 0 \end{pmatrix}, F = \begin{pmatrix} 0 & 0 & 0 & 0 & 1 & -1 & 0 & 0 & -1 & 0 \\ 0 & 0 & 0 & 0 & -1 & 1 & 0 & 0 & 0 & 1 \\ 0 & 0 & 0 & 0 & 0 & 0 & 1 & -1 & 1 & 0 \\ 0 & 0 & 0 & 0 & 0 & 0 & -1 & 1 & 0 & -1 \end{pmatrix},$$

with column order $s_1 \to s_1$, $s_2 \to s_2$, $s_3 \to s_3$, $s_4 \to s_4$, $s_2 \to s_1$, $s_1 \to s_2$, $s_4 \to s_3$, $s_3 \to s_4$, $s_1 \to s_3$, $s_4 \to s_2$ (from left to right) and row order x, y, z for D resp. s_1, s_2, s_3, s_4 for F (from top to bottom).

We now consider the constraint systems (P) and (Q), stated below, which have maximization objectives. The constraint systems will be used by our main algorithm in Section 4. We observe that both constraint systems are always satisfiable (set all coefficients to zero) and that the solutions of both constraint systems are closed under addition. Hence, the number of inequalities for which the maximization objective is satisfied is unique for optimal solutions of both constraint systems. The maximization objectives can be implemented by suitable linear objective functions. Hence, both constraint systems can be solved in PTIME over the integers, because we can use linear programming over the rationales and then scale rational solutions to the integers by multiplying with the least common multiple of the denominators.

constraint system (P):	constraint system (Q):
there exists $\mu \in \mathbb{Z}^{Trns(\mathcal{V})}$ with	there exist $r \in \mathbb{Z}^{Var}, z \in \mathbb{Z}^{St(\mathcal{V})}$ with
$D\mu \geq 0$ $\mu \geq 0$ $F\mu = 0$	$r \geq 0$ $z \geq 0$ $D^T r + F^T z \leq 0$
Maximization Objective: Maximize the number of inequalities with $(D\mu)(x) > 0$ and $\mu(t) > 0$	Maximization Objective: Maximize the number of inequalities with $r(x) > 0$ and $(D^T r + F^T z)(t) < 0$

The solutions of (P) and (Q) are characterized by the following two lemmata:

Lemma 2 (Cited from [14]). *$\mu \in \mathbb{Z}^{Trns(\mathcal{V})}$ is a solution to constraint system (P) iff there exists a multi-cycle M with $val(M) \geq 0$ and $\mu(t)$ instances of transition t for every $t \in Trns(\mathcal{V})$.*

Lemma 3 (Cited from [5][1]). *Let r, z be a solution to constraint system (Q). Let $rank(r, z) : Cfg(\mathcal{V}) \to \mathbb{N}$ be the function defined by $rank(r, z)(s, \nu) = r^T \nu + z(s)$. Then, $rank(r, z)$ is a* quasi-ranking function *for $\mathcal{V}$, i.e., we have*

1. *for all $(s, \nu) \in Cfg(\mathcal{V})$ that $rank(r, z)(s, \nu) \geq 0$;*
2. *for all transitions $t = s_1 \xrightarrow{d} s_2 \in Trns(\mathcal{V})$ and valuations $\nu_1, \nu_2 \in Val(\mathcal{V})$ with $\nu_2 = \nu_1 + d$ that $rank(r, z)(s_1, \nu_1) \geq rank(r, z)(s_2, \nu_2)$; moreover, the inequality is strict for every t with $(D^T r + F^T z)(t) < 0$.*

We now state a dichotomy between optimal solutions to constraint systems (P) and (Q), which is obtained by an application of Farkas' Lemma. This dichotomy is the main reason why we are able to compute the precise asymptotic complexity of VASSs with polynomial bounds.

[1] There is no explicit lemma with this statement in [5], however the lemma is implicit in the exposition of Section 4 in [5]. We further note that [5] does not include the constraint $z \geq 0$. However, this difference is minor and was added in order to ensure that ranking functions always return non-negative values, which is more standard than the choice of [5]. A proof of the lemma can be found in the extended version [21].

Lemma 4. *Let r and z be an optimal solution to constraint system (Q) and let μ be an optimal solution to constraint system (P). Then, for all variables $x \in Var$ we either have $r(x) > 0$ or $(D\mu)(x) \geq 1$, and for all transitions $t \in Trns(\mathcal{V})$ we either have $(D^T r + F^T z)(t) < 0$ or $\mu(t) \geq 1$.*

Example 5. Our main algorithm, Algorithm 1 presented in Section 4, will directly use constraint systems (P) and (Q) in its first loop iteration, and adjusted versions in later loop iterations. Here, we illustrate the first loop iteration. We consider the running example $\mathcal{V}_{run}$, whose update and flow matrices we have stated in Example 1. An optimal solution to constraint systems (P) and (Q) is given by $\mu = (1441111100)^T$ and $r = (220)^T$, $z = (0011)^T$. The quasi-ranking function $rank(r, z)$ immediately establishes that $\mathtt{tbound}_t(N) \in O(N)$ for $t = s_1 \to s_3$ and $t = s_4 \to s_2$, because 1) $rank(r, z)$ decreases for these two transitions and does not increase for other transitions (by Lemma 3), and because 2) the initial value of $rank(r, z)$ is bounded by $O(N)$, i.e., we have $rank(r, z)(s, \nu) \in O(N)$ for every state $s \in St(\mathcal{V}_{run})$ and every valuation ν with $\|\nu\| \leq N$. By a similar argument we get $\mathtt{vbound}_x(N) \in O(N)$ and $\mathtt{vbound}_y(N) \in O(N)$. The exact reasoning for deriving upper bounds is given in Section 5. From μ we can, by Lemma 2, obtain the cycles $C_1 = s_1 \to s_2 \to s_2 \to s_2 \to s_2 \to s_2 \to s_1 \to s_1$ and $C_2 = s_3 \to s_4 \to s_4 \to s_4 \to s_4 \to s_4 \to s_4 \to s_4$ with $\nu(C_1) + \nu(C_2) \geq (001)^T$ (*). We will later show that the cycles C_1 and C_2 give rise to a family of traces that establish $\mathtt{tbound}_t(N) \in \Omega(N^2)$ for all transitions $t \in Trns(\mathcal{V}_{run})$ with $t \neq s_1 \to s_3$ and $t \neq s_4 \to s_2$. Here we give an intuition on the construction: We consider a cycle C of $\mathcal{V}_{run}$ that visits all states at least once. By (*), the updates along the cycles C_1 and C_2 cancel each other out. However, the two cycles are not connected. Hence, we execute the cycle C_1 some $\Omega(N)$ times, then (a part of) the cycle C, then execute C_2 as often as C_1, and finally the remaining part of C; this we repeat $\Omega(N)$ times. This construction also establishes the bound $\mathtt{vbound}_z(N) \in \Omega(N^2)$ because, by (*), we increase z with every joint execution of C_1 and C_2. The precise lower bound construction is given in Section 6.

4 Main Algorithm

Our main algorithm – Algorithm 1 – computes the complexity as well as variable and transition bounds of an input VASS $\mathcal{V}$, either detecting that $\mathcal{V}$ has at least exponential complexity or reporting precise asymptotic bounds for the transitions and variables of $\mathcal{V}$ (up to a constant factor): Algorithm 1 will compute values $\mathtt{vExp}(x) \in \mathbb{N}$ such that $\mathtt{vbound}_N(x) \in \Theta(N^{\mathtt{vExp}(x)})$ for every $x \in Var$ and values $\mathtt{tExp}(t) \in \mathbb{N}$ such that $\mathtt{tbound}_N(t) \in \Theta(N^{\mathtt{tExp}(t)})$ for every $t \in Trns(\mathcal{V})$.

Data Structures. The algorithm maintains a rooted tree T. Every node η of T will always be labelled by a sub-VASSs $\mathtt{VASS}(\eta)$ of $\mathcal{V}$. The nodes in the same layer of T will always be labelled by disjoint sub-VASS of $\mathcal{V}$. The main loop of Algorithm 1 will extend T by one layer per loop iteration. The variable l always contains the next layer that is going to be added to T. For computing variable and transition bounds, Algorithm 1 maintains the functions $\mathtt{vExp} : Var \to \mathbb{N} \cup \{\infty\}$ and $\mathtt{tExp} : Trns(\mathcal{V}) \to \mathbb{N} \cup \{\infty\}$.

Initialization. We assume D to be the update matrix and F to be the flow matrix associated to $\mathcal{V}$ as discussed in Section 3. At initialization, T consists of the root node ι and we set $\texttt{VASS}(\iota) = \mathcal{V}$, i.e., the root is labelled by the input $\mathcal{V}$. We initialize $l = 1$ as Algorithm 1 is going to add layer 1 to T in the first loop iteration. We initialize $\texttt{vExp}(x) = \infty$ for all variables $x \in \mathit{Var}$ and $\texttt{tExp}(t) = \infty$ for all transitions $t \in \mathit{Trns}(\mathcal{V})$.

The constraint systems solved during each loop iteration. In loop iteration l, Algorithm 1 will set $\texttt{tExp}(t) := l$ for some transitions t and $\texttt{vExp}(x) := l$ for some variables x. In order to determine those transitions and variables, Algorithm 1 instantiates constraint systems (P) and (Q) from Section 3 over the set of transitions $U = \bigcup_{\eta \in \texttt{layer}(l-1)} \mathit{Trns}(\texttt{VASS}(\eta))$, which contains all transitions associated to nodes in layer $l-1$ of T. However, instead of a direct instantiation using $D|_U$ and $F|_U$ (i.e., the restriction of D and F to the transitions U), we need to work with an extended set of variables and an extended update matrix. We set $\mathit{Var}_{ext} := \{(x,\eta) \mid \eta \in \texttt{layer}(l - \texttt{vExp}(x))\}$, where we set $n - \infty = 0$ for all $n \in \mathbb{N}$. This means that we use a different copy of variable x for every node η in layer $l - \texttt{vExp}(x)$. We note that for a variable x with $\texttt{vExp}(x) = \infty$ there is only a single copy of x in Var_{ext} because $\iota \in \texttt{layer}(0)$ is the only node in layer 0. We define the extended update matrix $D_{ext} \in \mathbb{Z}^{\mathit{Var}_{ext} \times U}$ by setting

$$D_{ext}((x,\eta),t) := \begin{cases} D(x,t), \text{ if } t \in \mathit{Trns}(\texttt{VASS}(\eta)), \\ 0, \qquad\qquad \text{otherwise.} \end{cases}$$

Constraint systems (I) and (II) stated in Figure 2 can be recognized as instantiation of constraint systems (P) and (Q) with matrices D_{ext} and $F|_U$ and variables Var_{ext}, and hence the dichotomy stated in Lemma 4 holds.

We comment on the choice of Var_{ext}: Setting $\mathit{Var}_{ext} = \{(x,\eta) \mid \eta \in \texttt{layer}(i)\}$ for any $i \leq l - \texttt{vExp}(x)$ would result in correct upper bounds (while $i > l - \texttt{vExp}(x)$ would not). However, choosing $i < l - \texttt{vExp}(x)$ does in general result in sub-optimal bounds because fewer variables make constraint system (I) easier and constraint system (II) harder to satisfy (in terms of their maximization objectives). In fact, $i = l - \texttt{vExp}(x)$ is the optimal choice, because this choice allows us to prove corresponding lower bounds in Section 6. We will further comment on key properties of constraint systems (I) and (II) in Sections 5 and 6, when we outline the proofs of the upper resp. lower bound.

We note that Algorithm 1 does not use the optimal solution μ to constraint system (I) for the computation of the $\texttt{vExp}(x)$ and $\texttt{tExp}(t)$, and hence the computation of the optimal solution μ could be removed from the algorithm. The solution μ is however needed for the extraction of lower bounds in Sections 6 and 8, and this is the reason why it is stated here. The extraction of lower bounds is not explicitly added to the algorithm in order to not clutter the presentation.

Discovering transition bounds. After an optimal solution r, z to constraint system (II) has been found, Algorithm 1 collects all transitions t with $(D_{ext}^T r + F|_U^T z)(t) < 0$ in the set R (note that the optimization criterion in constraint system (II) tries to find as many such t as possible). Algorithm 1 then sets $\texttt{tExp}(t) := l$ for all $t \in R$. The transitions in R will not be part of layer l of T.

Input: a connected VASS $\mathcal{V}$ with update matrix D and flow matrix F
$T :=$ single root node ι with $\texttt{VASS}(\iota) = \mathcal{V}$;
$l := 1$;
$\texttt{vExp}(x) := \infty$ for all variables $x \in \mathit{Var}$;
$\texttt{tExp}(t) := \infty$ for all transitions $t \in \mathit{Trns}(\mathcal{V})$;
repeat
    let $U := \bigcup_{\eta \in \texttt{layer}(l-1)} \mathit{Trns}(\texttt{VASS}(\eta))$;
    let $\mathit{Var}_{ext} := \{(x, \eta) \mid \eta \in \texttt{layer}(l - \texttt{vExp}(x))\}$, where $n - \infty = 0$ for $n \in \mathbb{N}$;
    let $D_{ext} \in \mathbb{Z}^{\mathit{Var}_{ext} \times U}$ be the matrix defined by

$$D_{ext}((x,\eta), t) = \begin{cases} D(x,t), & \text{if } t \in \mathit{Trns}(\texttt{VASS}(\eta)) \\ 0, & \text{otherwise} \end{cases};$$

    find optimal solutions μ and r, z to constraint systems (I) and (II);
    let $R := \{t \in U \mid (D_{ext}^T r + F|_U^T z)(t) < 0\}$;
    set $\texttt{tExp}(t) := l$ for all $t \in R$;
    foreach $\eta \in \texttt{layer}(l-1)$ **do**
        let $\mathcal{V}' := \texttt{VASS}(\eta)$ be the VASS associated to η;
        decompose $(\mathit{St}(\mathcal{V}'), \mathit{Trns}(\mathcal{V}') \setminus R)$ into SCCs;
        foreach *SCC S of* $(\mathit{St}(\mathcal{V}'), \mathit{Trns}(\mathcal{V}') \setminus R)$ **do**
            create a child η' of η with $\texttt{VASS}(\eta') = S$;
    foreach $x \in \mathit{Var}$ *with* $\texttt{vExp}(x) = \infty$ **do**
        if $r(x, \iota) > 0$ **then** set $\texttt{vExp}(x) := l$;
    if *there are no* $x \in \mathit{Var}$, $t \in \mathit{Trns}(\mathcal{V})$ *with* $l < \texttt{vExp}(x) + \texttt{tExp}(t) < \infty$ **then**
        return "$\mathcal{V}$ has at least exponential complexity"
    $l := l + 1$;
until $\texttt{vExp}(x) \neq \infty$ *and* $\texttt{tExp}(t) \neq \infty$ *for all* $x \in \mathit{Var}$ *and* $t \in \mathit{Trns}(\mathcal{V})$;

Algorithm 1: Computes transition and variable bounds for a VASS $\mathcal{V}$

constraint system (I):	constraint system (II):
there exists $\mu \in \mathbb{Z}^U$ with	there exist $r \in \mathbb{Z}^{\mathit{Var}_{ext}}, z \in \mathbb{Z}^{\mathit{St}(\mathcal{V})}$ with
$D_{ext}\mu \geq 0$	$r \geq 0$
$\mu \geq 0$	$z \geq 0$
$F\|_U \mu = 0$	$D_{ext}^T r + F\|_U^T z \leq 0$
Maximization Objective: Maximize the number of inequalities with $(D_{ext}\mu)(x) > 0$ and $\mu(t) > 0$	Maximization Objective: Maximize the number of inequalities with $r(x,\eta) > 0$ and $(D_{ext}^T r + F\|_U^T z)(t) < 0$

Fig. 2. Constraint Systems (I) and (II) used by Algorithm 1

Construction of the next layer in T. For each node η in layer $l-1$, Algorithm 1 will create children by removing the transitions in R. This is done as follows: Given a node η in layer $l-1$, Algorithm 1 considers the VASS $\mathcal{V}' = \texttt{VASS}(\eta)$ associated to η. Then, $(\mathit{St}(\mathcal{V}'), \mathit{Trns}(\mathcal{V}') \setminus R)$ is decomposed into its SCCs. Finally, for each SCC S of $(St(\mathcal{V}'), Trns(\mathcal{V}') \setminus R)$ a child η' of η is created with $\texttt{VASS}(\eta') = S$. Clearly, the new nodes in layer l are labelled by disjoint sub-VASS of $\mathcal{V}$.

The transitions of the next layer. The following lemma states that the new layer l of T contains all transitions of layer $l-1$ except for the transitions R; the lemma is due to the fact that every transition in $U \setminus R$ belongs to a cycle and hence to some SCC that is part of the new layer l.

Lemma 6. *We consider the new layer constructed during loop iteration l of Algorithm 1: we have $U \setminus R = \bigcup_{\eta \in \texttt{layer}(l)} Trns(\texttt{VASS}(\eta))$.*

Discovering variable bounds. For each $x \in Var$ with $\texttt{vExp}(x) = \infty$, Algorithm 1 checks whether $r(x, \iota) > 0$ (we point out that the optimization criterion in constraint systems (II) tries to find as many such x with $r(x, \iota) > 0$ as possible). Algorithm 1 then sets $\texttt{vExp}(x) := l$ for all those variables.

The check for exponential complexity. In each loop iteration, Algorithm 1 checks whether there are $x \in Var$, $t \in Trns(\mathcal{V})$ with $l < \texttt{vExp}(x) + \texttt{tExp}(t) < \infty$. If this is not the case, then we can conclude that $\mathcal{V}$ is at least exponential (see Theorem 9 below). If the check fails, Algorithm 1 increments l and continues with the construction of the next layer in the next loop iteration.

Termination criterion. The algorithm proceeds until either exponential complexity has been detected or until $\texttt{vExp}(x) \neq \infty$ and $\texttt{tExp}(t) \neq \infty$ for all $x \in Var$ and $t \in Trns(\mathcal{V})$ (i.e., bounds have been computed for all variables and transitions).

Invariants. We now state some simple invariants maintained by Algorithm 1, which are easy to verify:

- For every node η that is a descendent of some node η' we have that $\texttt{VASS}(\eta)$ is a sub-VASS of $\texttt{VASS}(\eta')$.
- The value of vExp and tExp is changed at most once for each input; when the value is changed, it is changed from ∞ to some value $\neq \infty$.
- For every transition $t \in Trns(\mathcal{V})$ and layer l of T, we have that either $\texttt{tExp}(t) \leq l$ or there is a node $\eta \in \texttt{layer}(l)$ such that $t \in Trns(\texttt{VASS}(\eta))$.
- We have $\texttt{tExp}(t) = l$ for $t \in Trns(\mathcal{V})$ if and only if there is a $\eta \in \texttt{layer}(l-1)$ with $t \in Trns(\texttt{VASS}(\eta))$ and there is no $\eta \in \texttt{layer}(l)$ with $t \in Trns(\texttt{VASS}(\eta))$.

Example 7. We sketch the execution of Algorithm 1 on $\mathcal{V}_{run}$. In iteration $l = 1$, we have $Var_{ext} = \{(x, \iota), (y, \iota), (z, \iota)\}$, and thus matrix D_{ext} is identical to the matrix D. Hence, constraint systems (I) and (II) are identical to constraint systems (P) and (Q), whose optimal solutions $\mu = (1441111100)^T$ and $r = (220)^T$, $z = (0011)^T$ we have discussed in Example 5. Algorithm 1 then sets $\texttt{tExp}(s_1 \to s_3) = 1$ and $\texttt{tExp}(s_4 \to s_2) = 1$, creates two children η_A and η_B of ι labeled by $\mathcal{V}_A = (\{s_1, s_2\}, \{s_1 \to s_1, s_1 \to s_2, s_2 \to s_2, s_2 \to s_1\})$ and $\mathcal{V}_B = (\{s_3, s_4\}, \{s_3 \to s_3, s_3 \to s_4, s_4 \to s_4, s_4 \to s_3\})$, and sets $\texttt{vExp}(x) = 1$ and $\texttt{vExp}(y) = 1$. In iteration $l = 2$, we have $Var_{ext} = \{(x, \eta_A), (y, \eta_A), (x, \eta_B), (y, \eta_B), (z, \iota)\}$ and the matrix D_{ext} stated in Figure 3. Algorithm 1 obtains $\mu = (11110000)^T$ and $r = (12211)^T$, $z = (0000)^T$ as optimal solutions to (I) and (II). Algorithm 1 then

$$D_{ext} = \begin{pmatrix} -1 & 1 & 0 & 0 & 0 & 0 & 0 & 0 \\ 1 & -1 & 0 & 0 & 0 & 0 & 0 & 0 \\ 0 & 0 & -1 & 1 & 0 & 0 & 0 & 0 \\ 0 & 0 & 1 & -1 & 0 & 0 & 0 & 0 \\ -1 & 1 & 1 & -1 & -1 & -1 & -1 & -1 \end{pmatrix}$$

with column order $s_1 \to s_1$, $s_2 \to s_2$, $s_3 \to s_3$, $s_4 \to s_4$, $s_2 \to s_1$, $s_1 \to s_2$, $s_4 \to s_3$, $s_3 \to s_4$ (from left to right) and row order $(x, \eta_A), (y, \eta_A), (x, \eta_B), (y, \eta_B), (z, \iota)$ (from top to bottom)

$$D_{ext} = \begin{pmatrix} -1 & 0 & 0 & 0 \\ 1 & 0 & 0 & 0 \\ 0 & 1 & 0 & 0 \\ 0 & -1 & 0 & 0 \\ 0 & 0 & -1 & 0 \\ 0 & 0 & 1 & 0 \\ 0 & 0 & 0 & 1 \\ 0 & 0 & 0 & -1 \\ -1 & 1 & 0 & 0 \\ 0 & 0 & 1 & -1 \end{pmatrix}$$

with column order $s_1 \to s_1$, $s_2 \to s_2$, $s_3 \to s_3$, $s_4 \to s_4$, (from left to right) and row order $(x, \eta_1), (y, \eta_1), (x, \eta_2), (y, \eta_2), (x, \eta_3), (y, \eta_3), (x, \eta_4), (y, \eta_4), (z, \eta_A), (z, \eta_B)$ (from top to bottom)

Fig. 3. The extended update matrices during iteration $l = 2$ (left) and $l = 3$ (right) of Algorithm 1 on the running example $\mathcal{V}_{run}$ from Section 1.

sets $\texttt{tExp}(s_1 \to s_2) = \texttt{tExp}(s_2 \to s_1) = \texttt{tExp}(s_3 \to s_4) = \texttt{tExp}(s_4 \to s_3) = 2$, creates the children η_1, η_2 resp. η_3, η_4 of η_A resp. η_B with η_i labelled by $\mathcal{V}_i = (\{s_i\}, \{s_i \to s_i\})$, and sets $\texttt{vExp}(z) = 2$. In iteration $l = 3$, we have $Var_{ext} = \{(x, \eta_1), (y, \eta_1), (x, \eta_2), (y, \eta_2), (x, \eta_3), (y, \eta_3), (x, \eta_4), (y, \eta_4), (z, \eta_A), (z, \eta_B)\}$ and the matrix D_{ext} stated in Figure 3. Algorithm 1 obtains $\mu = (0000)^T$ and $r = (1113311111)^T$, $z = (0000)^T$ as optimal solutions to (I) and (II). Algorithm 1 then sets $\texttt{tExp}(s_i \to s_i) = 3$, for all i, and terminates.

We now state the main properties of Algorithm 1:

Lemma 8. *Algorithm 1 always terminates.*

Theorem 9. *If Algorithm 1 returns "$\mathcal{V}$ has at least exponential complexity", then $comp_{\mathcal{V}}(N) \in 2^{\Omega(N)}$, and we have $\texttt{tbound}_t(N) \in 2^{\Omega(N)}$ for all $t \in Trns(\mathcal{V})$ with $\texttt{tExp}(t) = \infty$ and $\texttt{vbound}_t(N) \in 2^{\Omega(N)}$ for all $x \in Var$ with $\texttt{vExp}(x) = \infty$.*

The proof of Theorem 9 is stated in Section 8. We now assume that Algorithm 1 does not return "$\mathcal{V}$ has at least exponential complexity". Then, Algorithm 1 must terminate with $\texttt{tExp}(t) \neq \infty$ and $\texttt{vExp}(x) \neq \infty$ for all $t \in Trns(\mathcal{V})$ and $x \in Var$. The following result states that tExp and vExp contain the precise exponents of the asymptotic transition and variable bounds of $\mathcal{V}$:

Theorem 10. $\texttt{vbound}_N(x) \in \Theta(N^{\texttt{vExp}(x)})$ *for all $x \in Var$ and* $\texttt{tbound}_N(t) \in \Theta(N^{\texttt{tExp}(t)})$ *for all $t \in Trns(\mathcal{V})$.*

The upper bounds of Theorem 10 will be proved in Section 5 (Theorem 16) and the lower bounds in Section 6 (Corollary 20).

We will prove in Section 7 that the exponents of the variable and transition bounds are bounded exponentially in the dimension of $\mathcal{V}$:

Theorem 11. *We have* $\texttt{vExp}(x) \leq 2^{|Var|}$ *for all $x \in Var$ and* $\texttt{tExp}(t) \leq 2^{|Var|}$ *for all $t \in Trns(\mathcal{V})$.*

Finally, we obtain the following corollary from Theorems 10 and 11:

Corollary 12. *Let $\mathcal{V}$ be a connected VASS. Then, either $comp_{\mathcal{V}}(N) \in 2^{\Omega(N)}$ or $comp_{\mathcal{V}}(N) \in \Theta(N^i)$ for some computable $1 \leq i \leq 2^{|Var|}$.*

4.1 Complexity of Algorithm 1

In the remainder of this section we will establish the following result:

Theorem 13. *Algorithm 1 (with the below stated optimization) can be implemented in polynomial time with regard to the size of the input VASS $\mathcal{V}$.*

We will argue that A) every loop iteration of Algorithm 1 only takes polynomial time, and B) that polynomially many loop iterations are sufficient (this only holds for the optimization of the algorithm discussed below).

Let $\mathcal{V}$ be a VASS, let $m = |\mathit{Trns}(\mathcal{V})|$ be the number of transitions of $\mathcal{V}$, and let $n = |\mathit{Var}|$ be the dimension of $\mathcal{V}$. We note that $|\mathtt{layer}(l)| \leq m$ for every layer l of T, because the VASSs of the nodes in the same layer are disjoint.

A) Clearly, removing the decreasing transitions and computing the strongly connected components can be done in polynomial time. It remains to argue about constraint systems (I) and (II). We observe that $|\mathit{Var}_{ext}| = |\{(x, \eta) \mid \eta \in \mathtt{layer}(l - \mathtt{vExp}(x))\}| \leq n \cdot m$ and $|U| \leq m$. Hence the size of constraint systems (I) and (II) is polynomial in the size of $\mathcal{V}$. Moreover, constraint systems (I) and (II) can be solved in PTIME as noted in Section 3.

B) We do not a-priori have a bound on the number of iterations of the main loop of Algorithm 1. (Theorem 11 implies that the number of iterations is at most exponential; however, we do not use this result here). We will shortly state an improvement of Algorithm 1 that ensures that polynomially many iterations are sufficient. The underlying insight is that certain layers of the tree do not need to be constructed explicitly. This insight is stated in the lemma below:

Lemma 14. *We consider the point in time when the execution of Algorithm 1 reaches line $l := l + 1$ during some loop iteration $l \geq 1$. Let $\mathit{RelevantLayers} = \{\mathtt{tExp}(t) + \mathtt{vExp}(x) \mid x \in \mathit{Var}, t \in \mathit{Trns}(\mathcal{V})\}$ and let $l' = \min\{l' \mid l' > l, l' \in \mathit{RelevantLayers}\}$. Then, $\mathtt{vExp}(x) \neq i$ and $\mathtt{tExp}(t) \neq i$ for all $x \in \mathit{Var}$, $t \in \mathit{Trns}(\mathcal{V})$ and $l < i < l'$.*

We now present the optimization that achieves polynomially many loop iterations. We replace the line $l := l+1$ by the two lines $\mathit{RelevantLayers} := \{\mathtt{tExp}(t) + \mathtt{vExp}(x) \mid x \in \mathit{Var}, t \in \mathit{Trns}(\mathcal{V})\}$ and $l := \min\{l' \mid l' > l, l' \in \mathit{RelevantLayers}\}$. The effect of these two lines is that Algorithm 1 directly skips to the next relevant layer. Lemma 14, stated above, justifies this optimization: First, no new variable or transition bound is discovered in the intermediate layers $l < i < l'$. Second, each intermediate layer $l < i < l'$ has the same number of nodes as layer l, which are labelled by the same sub-VASSs as the nodes in l (otherwise there would be a transition with transition bound $l < i < l'$); hence, whenever needed, Algorithm 1 can construct a missing layer $l < i < l'$ on-the-fly from layer l.

We now analyze the number of loop iterations of the optimized algorithm. We recall that the value of each $\mathtt{vExp}(x)$ and $\mathtt{tExp}(t)$ is changed at most once from ∞ to some value $\neq \infty$. Hence, Algorithm 1 encounters at most $n \cdot m$ different values in the set $\mathit{RelevantLayers} = \{\mathtt{tExp}(t) + \mathtt{vExp}(x) \mid x \in \mathit{Var}, t \in \mathit{Trns}(\mathcal{V})\}$ during execution. Thus, the number of loop iterations is bounded by $n \cdot m$.

5 Proof of the Upper Bound Theorem

We begin by stating a proof principle for obtaining upper bounds.

Proposition 15 (Bound Proof Principle). *Let $\mathcal{V}$ be a VASS. Let $U \subseteq Trns(\mathcal{V})$ be a subset of the transitions of $\mathcal{V}$. Let $w : Cfg(\mathcal{V}) \to \mathbb{N}$ and $\mathtt{inc}_t : \mathbb{N} \to \mathbb{N}$, for every $t \in Trns(\mathcal{V}) \setminus U$, be functions such that for every trace $\zeta = (s_0, \nu_0) \xrightarrow{d_1} (s_1, \nu_1) \xrightarrow{d_2} \cdots$ of $\mathcal{V}$ with $\mathtt{init}(\zeta) \leq N$ we have for every $i \geq 0$ that*

1) $s_i \xrightarrow{d_i} s_{i+1} \in U$ implies $w(s_i, \nu_i) \geq w(s_{i+1}, \nu_{i+1})$, and
2) $s_i \xrightarrow{d_i} s_{i+1} \in Trns(\mathcal{V}) \setminus U$ implies $w(s_i, \nu_i) + \mathtt{inc}_t(N) \geq w(s_{i+1}, \nu_{i+1})$.

We call such a function w a complexity witness *and the associated $\mathtt{inc}_t$ functions the* increase certificates.

Let $t \in U$ be a transition on which w decreases, *i.e., we have $w(s_1, \nu_1) \geq w(s_2, \nu_2) - 1$ for every step $(s_1, \nu_1) \xrightarrow{d} (s_2, \nu_2)$ of $\mathcal{V}$ with $t = s_1 \xrightarrow{d} s_2$. Then,*

$$\mathtt{tbound}_t(N) \leq \max_{(s,\nu) \in Cfg(\mathcal{V}), \|\nu\| \leq N} w(s, \nu) + \sum_{t' \in Trns(\mathcal{V}) \setminus U} \mathtt{tbound}_{t'}(N) \cdot \mathtt{inc}_{t'}(N).$$

Further, let $x \in Var$ be a variable such that $\nu(x) \leq w(s, \nu)$ for all $(s, \nu) \in Cfg(\mathcal{V})$. Then,

$$\mathtt{vbound}_x(N) \leq \max_{(s,\nu) \in Cfg(\mathcal{V}), \|\nu\| \leq N} w(s, \nu) + \sum_{t' \in Trns(\mathcal{V}) \setminus U} \mathtt{tbound}_{t'}(N) \cdot \mathtt{inc}_{t'}(N).$$

Proof Outline of the Upper Bound Theorem. Let $\mathcal{V}$ be a VASS for which Algorithm 1 does not report exponential complexity. We will prove by induction on loop iteration l that $\mathtt{vbound}_N(x) \in O(N^l)$ for every $x \in Var$ with $\mathtt{vExp}(x) = l$ and that $\mathtt{tbound}_N(t) \in O(N^l)$ for every $t \in Trns(\mathcal{V})$ with $\mathtt{tExp}(t) = l$.

We now consider some loop iteration $l \geq 1$. Let $U = \bigcup_{\eta \in \mathtt{layer}(l-1)} Trns(\mathtt{VASS}(\eta))$ be the transitions, Var_{ext} be the set of extended variables and $D_{ext} \in \mathbb{Z}^{Var_{ext} \times U}$ be the update matrix considered by Algorithm 1 during loop iteration l. Let r, z be some optimal solution to constraint system (II) computed by Algorithm 1 during loop iteration l. The main idea for the upper bound proof is to use the quasi-ranking function from Lemma 3 as witness function for the Bound Proof Principle. In order to apply Lemma 3 we need to consider the VASS associated to the matrices in constraint system (II): Let $\mathcal{V}_{ext}$ be the VASS over variables Var_{ext} associated to update matrix D_{ext} and flow matrix $F|_U$. From Lemma 3 we get that $rank(r, z) : Cfg(\mathcal{V}_{ext}) \to \mathbb{N}$ is a quasi-ranking function for $\mathcal{V}_{ext}$. We now need to relate $\mathcal{V}$ to the extended VASS $\mathcal{V}_{ext}$ in order to be able to use this quasi-ranking function. We do so by extending valuations over Var to valuations over Var_{ext}. For every state $s \in St(\mathcal{V})$ and valuation $\nu : Var \to \mathbb{N}$, we define the *extended valuation* $\mathtt{ext}_s(\nu) : Var_{ext} \to \mathbb{N}$ by setting

$$\mathtt{ext}_s(\nu)(x, \eta) = \begin{cases} \nu(x), \text{ if } s \in St(\mathtt{VASS}(\eta)), \\ 0, \quad\quad \text{otherwise.} \end{cases}$$

As a direct consequence from the definition of extended valuations, we have that $(s, \mathtt{ext}_s(\nu)) \in \mathit{Cfg}(\mathcal{V}_{ext})$ for all $(s, \nu) \in \mathit{Cfg}(\mathcal{V})$, and that $(s_1, \mathtt{ext}_{s_1}(\nu_1)) \xrightarrow{D_{ext}(t)} (s_2, \mathtt{ext}_{s_2}(\nu_2))$ is a step of $\mathcal{V}_{ext}$ for every step $(s_1, \nu_1) \xrightarrow{d} (s_2, \nu_2)$ of $\mathcal{V}$ with $s_1 \xrightarrow{d} s_2 \in U$. We now define the witness function w by setting

$$w(s, \nu) = \mathit{rank}(r, z)(s, \mathtt{ext}_s(\nu)) \qquad \text{for all } (s, \nu) \in \mathit{Cfg}(\mathcal{V}).$$

We immediately get from Lemma 3 that w maps configurations to the non-negative integers and that condition 1) of the Bound Proof Principle is satisfied. Indeed, we get from the first item of Lemma 3 that $w(s, \nu) \geq 0$ for all $(s, \nu) \in \mathit{Cfg}(\mathcal{V})$, and from the second item that $w(s_1, \nu_1) \geq w(s_2, \nu_2)$ for every step $(s_1, \nu_1) \xrightarrow{d} (s_2, \nu_2)$ of $\mathcal{V}$ with $t = s_1 \xrightarrow{d} s_2 \in U$; moreover, the inequality is strict if $(D_{ext}^T r + F|_U^T z)(t) < 0$, i.e., the witness function w decreases for transitions t with $\mathtt{tExp}(t) = l$. It remains to establish condition 2) of the Bound Proof Principle. We will argue that we can find increase certificates $\mathtt{inc}_t(N) \in O(N^{l - \mathtt{tExp}(t)})$ for all $t \in \mathit{Trns}(\mathcal{V}) \setminus U$. We note that $\mathtt{tExp}(t) < l$ for all $t \in \mathit{Trns}(\mathcal{V}) \setminus U$, and hence the induction assumption can be applied for such t. We can then derive the desired bounds from the Bound Proof Principle because of $\sum_{t \in \mathit{Trns}(\mathcal{V}) \setminus U} \mathtt{tbound}_t(N) \cdot \mathtt{inc}_t(N) = \sum_{t \in \mathit{Trns}(\mathcal{V}) \setminus U} O(N^{\mathtt{tExp}(t)}) \cdot O(N^{l - \mathtt{tExp}(t)}) = O(N^l)$.

Theorem 16. $\mathtt{vbound}_N(x) \in O(N^{\mathtt{vExp}(x)})$ *for all* $x \in \mathit{Var}$ *and* $\mathtt{tbound}_N(t) \in O(N^{\mathtt{tExp}(t)})$ *for all* $t \in \mathit{Trns}(\mathcal{V})$.

6 Proof of the Lower Bound Theorem

The following lemma will allow us to consider traces ζ_N with $\mathtt{init}(\zeta_N) \in O(N)$ instead of $\mathtt{init}(\zeta_N) \leq N$ when proving asymptotic lower bounds.

Lemma 17. *Let* $\mathcal{V}$ *be a VASS, let* $t \in \mathit{Trns}(\mathcal{V})$ *be a transition and let* $x \in \mathit{Var}$ *be a variable. If there are traces* ζ_N *with* $\mathtt{init}(\zeta_N) \in O(N)$ *and* $\mathtt{instance}(\zeta_N, t) \geq N^i$, *then* $\mathtt{tbound}_N(t) \in \Omega(N^i)$. *If there are traces* ζ_N *with* $\mathtt{init}(\zeta_N) \in O(N)$ *that reach a final valuation* ν *with* $\nu(x) \geq N^i$, *then* $\mathtt{vbound}_N(x) \in \Omega(N^i)$.

The lower bound proof uses the notion of a *pre-path*, which relaxes the notion of a path: A pre-path $\sigma = t_1 \cdots t_k$ is a finite sequence of transitions $t_i = s_i \xrightarrow{d_i} s'_i$. Note that we do not require for subsequent transitions that the end state of one transition is the start state of the next transition, i.e., we do not require $s'_i = s_{i+1}$. We generalize notions from paths to pre-paths in the obvious way, e.g., we set $\mathit{val}(\sigma) = \sum_{i \in [1,k]} d_i$ and denote by $\mathtt{instance}(\sigma, t)$, for $t \in \mathit{Trns}(\mathcal{V})$, the number of times σ contains the transition t. We say the pre-path σ *can be executed from valuation* ν, if there are valuations $\nu_i \geq 0$ with $\nu_{i+1} = \nu_i + d_{i+1}$ for all $0 \leq i < k$ and $\nu = \nu_0$; we further say that σ *reaches* valuation ν', if $\nu' = \nu_k$. We will need the following relationship between execution and traces: in case a pre-path σ is actually a path, σ can be executed from valuation ν, if and only if there is a trace with initial valuation ν that uses the same sequence of transitions as σ. Two pre-paths $\sigma = t_1 \cdots t_k$ and $\sigma' = t'_1 \cdots t'_l$ can be *shuffled* into a pre-path $\sigma'' = t''_1 \cdots t''_{k+l}$, if σ'' is an order-preserving interleaving of σ and σ'; formally, there are injective monotone functions $f : [1,k] \to [1,k+l]$ and $g : [1,l] \to [1,k+l]$ with $f([1,k]) \cap g([1,l]) = \emptyset$ such that $t''_{f(i)} = t_i$ for all $i \in [1,k]$ and $t''_{g(i)} = t'_i$ for all $i \in [1,l]$. Further, for $d \geq 1$ and pre-path σ, we denote by $\sigma^d = \underbrace{\sigma\sigma\cdots\sigma}_{d}$ the pre-path that consists of d subsequent copies of σ.

For the remainder of this section, we fix a VASS $\mathcal{V}$ for which Algorithm 1 does not report exponential complexity and we fix the computed tree T and bounds $\texttt{vExp}$, $\texttt{tExp}$. We further need to use the solutions to constraint system (I) computed during the run of Algorithm 1: For every layer $l \geq 1$ and node $\eta \in \texttt{layer}(l)$, we fix a cycle $C(\eta)$ that contains $\mu(t)$ instances of every $t \in \mathit{Trns}(\texttt{VASS}(\eta))$, where μ is an optimal solution to constraint system (I) during loop iteration l. The existence of such cycles is stated in Lemma 18 below. We note that this definition ensures $\mathit{val}(C(\eta)) = \sum_{t \in \mathit{Trns}(\texttt{VASS}(\eta))} D(t) \cdot \mu(t)$. Further, for the root node ι, we fix an arbitrary cycle $C(\iota)$ that uses all transitions of $\mathcal{V}$ at least once.

Lemma 18. *Let μ be an optimal solution to constraint system (I) during loop iteration l of Algorithm 1. Then there is a cycle $C(\eta)$ for every $\eta \in \texttt{layer}(l)$ that contains exactly $\mu(t)$ instances of every transition $t \in \mathit{Trns}(\texttt{VASS}(\eta))$.*

Proof Outline of the Lower Bound Theorem.
Step I) We define a pre-path τ_l, for every $l \geq 1$, with the following properties:

1) $\texttt{instance}(\tau_l, t) \geq N^{l+1}$ for all transitions $t \in \bigcup_{\eta \in \texttt{layer}(l)} \mathit{Trns}(\texttt{VASS}(\eta))$.
2) $\mathit{val}(\tau_l) = N^{l+1} \sum_{\eta \in \texttt{layer}(l)} \mathit{val}(C(\eta))$.
3) $\mathit{val}(\tau_l)(x) \geq 0$ for every $x \in \mathit{Var}$ with $\texttt{vExp}(x) \leq l$.
4) $\mathit{val}(\tau_l)(x) \geq N^{l+1}$ for every $x \in \mathit{Var}$ with $\texttt{vExp}(x) \geq l+1$.
5) τ_l is executable from some valuation ν with
   a) $\nu(x) \in O(N^{\texttt{vExp}(x)})$ for $x \in \mathit{Var}$ with $\texttt{vExp}(x) \leq l$, and
   b) $\nu(x) \in O(N^l)$ for $x \in \mathit{Var}$ with $\texttt{vExp}(x) \geq l+1$.

The difficulty in the construction of the pre-paths τ_l lies in ensuring Property 5). The construction of the τ_l proceeds along the tree T using that the cycles $C(\eta)$ have been obtained according to solutions of constraint system (I).

Step II) It is now a direct consequence of Properties 3)-5) stated above that we can choose a sufficiently large $k > 0$ such that for every $l \geq 0$ the pre-path $\rho_l = \tau_0^k \tau_1^k \cdots \tau_l^k$ (the concatenation of k copies of each τ_i, setting $\tau_0 = C(\iota)^N$), can be executed from some valuation ν and reaches a valuation ν' with

1) $\|\nu\| \in O(N)$,
2) $\nu'(x) \geq kN^{\texttt{vExp}(x)}$ for all $x \in \mathit{Var}$ with $\texttt{vExp}(x) \leq l$, and
3) $\nu'(x) \geq kN^{l+1}$ for all $x \in \mathit{Var}$ with $\texttt{vExp}(x) \geq l+1$.

The above stated properties for the pre-path $\rho_{l_{\max}}$, where $l_{\max}$ is the maximal layer of T, would be sufficient to conclude the lower bound proof except that we need to extend the proof from pre-paths to proper paths.

Step III) In order to extend the proof from pre-paths to paths we make use of the concept of shuffling. For all $l \geq 0$, we will define paths γ_l that can be obtained by shuffling the pre-paths $\rho_0, \rho_1, \ldots, \rho_l$. The path $\gamma_{l_{\max}}$, where $l_{\max}$ is the maximal layer of T, then has the desired properties and allows to conclude the lower bound proof with the following result:

Theorem 19. *There are traces ζ_N with $\texttt{init}(\zeta_N) \in O(N)$ such that ζ_N ends in configuration (s_N, ν_N) with $\nu_N(x) \geq N^{\texttt{vExp}(x)}$ for all variables $x \in Var$ and we have $\texttt{instance}(\zeta_N, t) \geq N^{\texttt{tExp}(t)}$ for all transitions $t \in Trns(\mathcal{V})$.*

With Lemma 17 we get the desired lower bounds from Theorem 19:

Corollary 20. $\texttt{vbound}_N(x) \in \Omega(N^{\texttt{vExp}(x)})$ *for all* $x \in Var$ *and* $\texttt{tbound}_N(t) \in \Omega(N^{\texttt{tExp}(t)})$ *for all* $t \in Trns(\mathcal{V})$.

7 The Size of the Exponents

For the remainder of this section, we fix a VASS $\mathcal{V}$ for which Algorithm 1 does not report exponential complexity and we fix the computed tree T and bounds $\texttt{vExp}$, $\texttt{tExp}$. Additionally, we fix a vector $z_l \in \mathbb{Z}^{St(\mathcal{V})}$ for every layer l of T and a vector $r_\eta \in \mathbb{Z}^{Var}$ for every node $\eta \in \texttt{layer}(l)$ as follows: Let r, z be an optimal solution to constraint system (II) in iteration $l+1$ of Algorithm 1. We then set $z_l = z$. For every $\eta \in \texttt{layer}(l)$ we define r_η by setting $r_\eta(x) = r(x, \eta')$, where $\eta' \in \texttt{layer}(l - \texttt{vExp}(x))$ is the unique ancestor of η in layer $l - \texttt{vExp}(x)$. The following properties are immediate from the definition:

Proposition 21. *For every layer l of T and node $\eta \in \texttt{layer}(l)$ we have:*

1) *$z_l \geq 0$ and $r_\eta \geq 0$.*
2) *$r_\eta^T d + z_l(s_2) - z_l(s_1) \leq 0$ for every transition $s_1 \xrightarrow{d} s_2 \in Trns(\texttt{VASS}(\eta))$; moreover, the inequality is strict for all transitions t with $\texttt{tExp}(t) = l + 1$.*
3) *Let $\eta' \in \texttt{layer}(i)$ be a strict ancestor of η. Then, $r_{\eta'}^T d + z_i(s_2) - z_i(s_1) = 0$ for every transition $s_1 \xrightarrow{d} s_2 \in Trns(\texttt{VASS}(\eta))$.*
4) *For every $x \in Var$ with $\texttt{vExp}(x) = l+1$ we have $r_\eta(x) > 0$ and $r_\eta(x) = r_{\eta'}(x)$ for all $\eta' \in \texttt{layer}(l)$.*
5) *For every $x \in Var$ with $\texttt{vExp}(x) > l + 1$ we have $r_\eta(x) = 0$.*
6) *For every $x \in Var$ with $\texttt{vExp}(x) \leq l$ there is an ancestor $\eta' \in \texttt{layer}(i)$ of η such that $r_{\eta'}(x) > 0$ and $r_{\eta'}(x') = 0$ for all x' with $\texttt{vExp}(x') > \texttt{vExp}(x)$.*

For a vector $r \in \mathbb{Z}^{Var}$, we define the *potential* of r by setting $\texttt{pot}(r) = \max\{\texttt{vExp}(x) \mid x \in Var, r(x) \neq 0\}$, where we set $\max \emptyset = 0$. The motivation for this definition is that we have $r^T \nu \in O(N^{\texttt{pot}(r)})$ for every valuation ν reachable by a trace ζ with $\texttt{init}(\zeta) \leq N$. We will now define the *potential* of a set of vectors $Z \subseteq \mathbb{Z}^{Var}$. Let M be a matrix whose columns are the vectors of Z and whose rows are ordered according to the variable bounds, i.e., if the row associated to variable x' is above the row associated to variable x, then we have

$\mathtt{vExp}(x') \geq \mathtt{vExp}(x)$. Let L be some lower triangular matrix obtained from M by elementary column operations. We now define $\mathtt{pot}(Z) = \sum_{\text{column } r \text{ of } L} \mathtt{pot}(r)$, where we set $\sum \emptyset = 0$. We note that $\mathtt{pot}(Z)$ is well-defined, because the value $\mathtt{pot}(Z)$ does not depend on the choice of M and L.

We next state an upper bound on potentials. Let $l \geq 0$ and let $B_l = \{\mathtt{vExp}(x) \mid x \in \mathit{Var}, \mathtt{vExp}(x) < l\}$ be the set of variable bounds below l. We set $\mathtt{varsum}(l) = 1$, for $B_l = \emptyset$, and $\mathtt{varsum}(l) = \sum B_l$, otherwise. The following statement is a direct consequence of the definitions:

Proposition 22. *Let $Z \subseteq \mathbb{Z}^{\mathit{Var}}$ be a set of vectors such that $r(x) = 0$ for all $r \in Z$ and $x \in \mathit{Var}$ with $\mathtt{vExp}(x) > l$. Then, we have $\mathtt{pot}(Z) \leq \mathtt{varsum}(l+1)$.*

We define $\mathtt{pot}(\eta) = \mathtt{pot}(\{r_{\eta'} \mid \eta' \text{ is a strict ancestor of } \eta\})$ as the *potential* of a node η. We note that $\mathtt{pot}(\eta) \leq \mathtt{varsum}(l+1)$ for every node $\eta \in \mathtt{layer}(l)$ by Proposition 22. Now, we are able to state the main results of this section:

Lemma 23. *Let η be a node in T. Then, every trace ζ with $\mathtt{init}(\zeta) \leq N$ enters $\mathtt{VASS}(\eta)$ at most $O(N^{\mathtt{pot}(\eta)})$ times, i.e., ζ contains at most $O(N^{\mathtt{pot}(\eta)})$ transitions $s \xrightarrow{d} s'$ with $s \notin St(\mathtt{VASS}(\eta))$ and $s' \in St(\mathtt{VASS}(\eta))$.*

Lemma 24. *For every layer l, we have that $\mathtt{vExp}(x) = l$ resp. $\mathtt{tExp}(t) = l$ implies $\mathtt{vExp}(x) \leq \mathtt{varsum}(l)$ resp. $\mathtt{tExp}(t) \leq \mathtt{varsum}(l)$.*

The next result follows from Lemma 24 only by arithmetic manipulations and induction on l:

Lemma 25. *Let l be some layer. Let k be the number of variables $x \in \mathit{Var}$ with $\mathtt{vExp}(x) < l$. Then, $\mathtt{varsum}(l) \leq 2^k$.*

Theorem 11 is then a direct consequence of Lemma 24 and 25 (using $k \leq |\mathit{Var}|$).

8 Exponential Witness

The following lemma from [15] states a condition that is sufficient for a VASS to have exponential complexity[2]. We will use this lemma to prove Theorem 9:

Lemma 26 (Lemma 10 of [15]). *Let $\mathcal{V}$ be a connected VASS, let U, W be a partitioning of Var and let $C_1, \ldots, C_m$ be cycles such that a) $val(C_i)(x) \geq 0$ for all $x \in U$ and $1 \leq i \leq m$, and b) $\sum_i val(C_i)(x) \geq 1$ for all $x \in W$. Then, there is a $c > 1$ and paths π_N such that 1) π_N can be executed from initial valuation $N \cdot \mathbf{1}$, 2) π_N reaches a valuation ν with $\nu(x) \geq c^N$ for all $x \in W$ and 3) $(C_i)^{c^N}$ is a sub-path of π_N for each $1 \leq i \leq m$.*

We now outline the proof of Theorem 9: We assume that Algorithm 1 returned "$\mathcal{V}$ has at least exponential complexity" in loop iteration l. According to Lemma 18, there are cycles $C(\eta)$, for every node $\eta \in \mathtt{layer}(l)$, that contain $\mu(t)$ instances of every transition $t \in \mathit{Trns}(\mathtt{VASS}(\eta))$. One can then show that the cycles $C(\eta)$ and the sets $U = \{x \in \mathit{Var} \mid \mathtt{vExp}(x) \leq l\}$, $W = \{x \in \mathit{Var} \mid \mathtt{vExp}(x) > l\}$ satisfy the requirements of Lemma 26, which establishes Theorem 9.

[2] Our formalization differs from[15], but it is easy to verify that our conditions a) and b) are equivalent to the conditions on the cycles in the 'iteration schemes' of [15].

References

1. Parosh Aziz Abdulla, Giorgio Delzanno, and Laurent van Begin. A language-based comparison of extensions of Petri nets with and without whole-place operations. In *LATA*, pages 71–82, 2009.
2. Benjamin Aminof, Sasha Rubin, and Florian Zuleger. On the expressive power of communication primitives in parameterised systems. In *LPAR*, pages 313–328, 2015.
3. Benjamin Aminof, Sasha Rubin, Florian Zuleger, and Francesco Spegni. Liveness of parameterized timed networks. In *ICALP*, pages 375–387, 2015.
4. Roderick Bloem, Swen Jacobs, Ayrat Khalimov, Igor Konnov, Sasha Rubin, Helmut Veith, and Josef Widder. Decidability in parameterized verification. *SIGACT News*, 47(2):53–64, 2016.
5. Tomás Brázdil, Krishnendu Chatterjee, Antonín Kucera, Petr Novotný, Dominik Velan, and Florian Zuleger. Efficient algorithms for asymptotic bounds on termination time in VASS. In *LICS*, pages 185–194, 2018.
6. Leonard Dickson. Finiteness of the odd perfect and primitive abundant numbers with n distinct prime factors. *Am. J. Math*, 35:413—-422, 1913.
7. Javier Esparza and Mogens Nielsen. Decidability issues for Petri nets - a survey. *Elektronische Informationsverarbeitung und Kybernetik*, 30(3):143–160, 1994.
8. Alain Finkel, Gilles Geeraerts, Jean-François Raskin, and Laurent van Begin. On the *omega*-language expressive power of extended Petri nets. *TCS*, 356(3):374–386, 2006.
9. Steven M. German and A. Prasad Sistla. Reasoning about systems with many processes. *J. ACM*, 39(3):675–735, 1992.
10. John E. Hopcroft and Jean-Jacques Pansiot. On the reachability problem for 5-dimensional vector addition systems. *TCS*, 8:135–159, 1979.
11. Annu John, Igor Konnov, Ulrich Schmid, Helmut Veith, and Josef Widder. Parameterized model checking of fault-tolerant distributed algorithms by abstraction. In *FMCAD*, pages 201–209, 2013.
12. Alexander Kaiser, Daniel Kroening, and Thomas Wahl. A widening approach to multithreaded program verification. *TOPLAS*, 36(4):14:1–14:29, 2014.
13. Richard M. Karp and Raymond E. Miller. Parallel program schemata. *J. Comput. Syst. Sci.*, 3(2):147–195, 1969.
14. S. Rao Kosaraju and Gregory F. Sullivan. Detecting cycles in dynamic graphs in polynomial time (preliminary version). In *STOC*, pages 398–406, 1988.
15. Jérôme Leroux. Polynomial vector addition systems with states. In *ICALP*, pages 134:1–134:13, 2018.
16. Richard J. Lipton. *The Reachability Problem Requires Exponential space.* Research report 62. Department of Computer Science, Yale University, 1976.
17. Charles Rackoff. The covering and boundedness problems for vector addition systems. *TCS*, 6:223–231, 1978.
18. Moritz Sinn, Florian Zuleger, and Helmut Veith. A simple and scalable static analysis for bound analysis and amortized complexity analysis. In *CAV*, pages 745–761, 2014.
19. Moritz Sinn, Florian Zuleger, and Helmut Veith. Difference constraints: An adequate abstraction for complexity analysis of imperative programs. In *FMCAD*, pages 144–151, 2015.
20. Moritz Sinn, Florian Zuleger, and Helmut Veith. Complexity and resource bound analysis of imperative programs using difference constraints. *JAR*, 59:3–45, 2017.

21. Florian Zuleger. The polynomial complexity of vector addition systems with states. *CoRR*, abs/1907.01076, 2019.

Author Index

Printed in the United States
By Bookmasters